Hydrocolloids

Part 1

Physical Chemistry and Industrial Application of Gels, Polysaccharides, and Proteins

Hydrocolloids

Part 1

Physical Chemistry and Industrial Application of Gels, Polysaccharides, and Proteins

Edited by

Katsuyoshi Nishinari

Osaka City University

Osaka, Japan

2000

ELSEVIER
Amsterdam – Lausanne – New York – Oxford – Shannon – Singapore – Tokyo

ELSEVIER SCIENCE B.V.
Sara Burgerhartstraat 25
P.O. Box 211, 1000 AE Amsterdam, The Netherlands

First edition 2000

Library of Congress Cataloging in Publication Data
A catalog record from the Library of Congress has been applied for.

ISBN 0 444 50178 9

∞ The paper used in this publication meets the requirements of ANSI/NISO Z39.48-1992 (Permanence of Paper).

Printed in The Netherlands.

Preface to Part 1

This volume is based on the presentations given at Osaka City University International Symposium 98 - Joint meeting with the 4th International Conference on Hydrocolloids - held on 4-10 October 1998 in Osaka.

The first article, by Professor Israelachvili, describes concisely the concept and the experimental aspects of molecular forces, which govern the structure and the physico-chemical properties of colloidal dispersed systems.

Section 2 covers the gel-sol transition, the structure of gels, the volume phase transition of gels, which are hot topics in the science of soft materials.

Section 3 consists of articles on the production, structure, gelation, conformation, functional properties of polysaccharides. New polysaccharides with emulsifying abilities, produced by plant cell culture and from soybean, attracted much attention because of its competing functionality with gum arabic. Physico-chemical studies clarifying the structure-property relation continue to be important, and describe the recent development. Production and functional properties of other new polysaccharides are also reported and expected to be useful hydrocolloids in the near future.

Sections 4 and 5 deal with cellulose and starch. Although both of these polysaccharides have been studied extensively, there are still many problems to be solved. Papers in Sections 4 and 5 challenge new problems which have not been studied so much.

Section 6 discloses the rich world of proteins. Although most proteins described here are not necessarily new faces, this section sheds new light on the fundamental and industrially important aspects such as emulsification, foaming and gelation.

I am sure that this volume provides valuable information and stimulating problems based on the enthusiastic discussions, questions, comments and answers during the conference. All the articles included in this volume have been reviewed and rewritten carefully according to comments and criticisms. I hope that the readers will share the pleasure to get the experience on many exciting aspects and infinite possibilities of hydrocolloids.

I would like to thank especially Drs. G.O. Phillips, M. Doi, N. Nemoto, F. Tanaka, H. Maeda, M. Tokita, T. Yano, M. Annaka, Y. Izumi, K. Kajiwara, A. Takada, V.J. Morris, K. Nakamura, E.R. Morris, M. Rinaudo, K.I. Draget, D. Klemm, H. Hatakeyama, P.A. Williams, T. Norisuye, M.A. Rao, H. Fuwa, A. Misaki, T. Matsumoto, K. Katsuta, E.A. Foegeding, T. van Vliet, S.B. Ross-Murphy, K. Kubota, S. Hayakawa, T. Nagano, and D. Oakenfull for their valuable comments.

Katsuyoshi Nishinari
Department of Food and Nutrition
Faculty of Human Life Science
Osaka City University
3-3-138 Sugimoto, Sumiyoshi-ku
Osaka 558-8585, Japan
Tel : +81-6-6605-2818
Fax : +81-6-6605-3086
e-mail : nisinari@life.osaka-cu.ac.jp

CONTENTS

3. POLYSACCHARIDES

4. CELLULOSE

5. STARCH

6. PROTEINS

1. INTRODUCTORY LECTURE

HYDROCOLLOIDS – PART 1
Edited by K. Nishinari
2000 Elsevier Science B.V.

3

Short-range and long-range forces between hydrophilic surfaces and biopolymers in aqueous solutions

J. N. Israelachvili

Department of Chemical Engineering, and Materials Department,
University of California, Santa Barbara, California 93106, USA.

The Surface Forces Apparatus (SFA) allows one to measure various interaction forces between surfaces as a function of their separation in aqueous solutions. In addition, the optical technique used allows one to directly visualize various interfacial phenomena (such as slow structural rearrangements) that may be occurring during an interaction. In this way, complex colloidal interactions – such as typically occur in gels, polyelectrolyte solutions and biocolloidal systems – may be studied at the molecular level both in space and time. Recent SFA (and other) results on a variety of hydrocolloidal and biopolymer systems show that these have much more complex and time-dependent interaction potentials than normally occur between simpler colloidal surfaces, i.e., their interactions are not simply described by van der Waals attraction and electrostatic repulsion which are the two principal forces of the Derjaguin-Landau-Verwey-Overbeek (DLVO) theory [1].

After briefly describing how such interactions can be measured and visualized, a review will be given of the various types of forces that can arise, sometimes simultaneously, between complex biomolecules and surfaces in aqueous solutions. These include van der Waals, ionic, electrostatic, structural, hydration, hydrophobic, polymer-mediated, thermal fluctuation and bio-specific interactions, and specific examples are given of how of these arises.

These recent results show that – even though all forces have a common origin – biocolloidal interactions can differ from normal, non-specific "colloidal" interactions in three important ways: (1) biological, especially bio-specific, interactions are qualitatively different in that many molecular groups are often involved "sequentially" (in different regions of space and time) in such a way that "the whole is greater than the sum of the parts", (2) interacting bio-colloidal surfaces are usually 'asymmetric' which, as will be shown, gives rise to very different interactions than those that arise between similar (symmetric) surfaces, and (3) non-equilibrium and time effects often play a crucial role in regulating biological interactions.

It is unlikely that a single, generic interaction potential can be written that covers all possible situations, but careful consideration of the surfaces, the molecules and solution conditions should allow reasonable predictions to be made in many cases.

1. INTRODUCTION

The intermolecular forces between hydrophilic surfaces and molecular groups in aqueous solutions ultimately determine the structure of 'complex fluid' structures such as gels and biological assemblies composed of 'soft matter': surfactants, lipids, polyelectrolytes, biopolymers and proteins. These forces also determine the *dynamic* properties of these structures such as their non-equilibrium and time-dependent interactions, their shape fluctuations, molecular exchange processes, and rheology. Because each type of interaction can have very different effects at large and small separation, both quantitatively and qualitatively, it is useful to distinguish between the long-range (colloidal) forces and short-range (adhesion) forces. For example, the electrostatic interaction between two surfaces in water can be repulsive at large separations (due to the 'double-layer' interaction) but attractive at small separations (due to ion-correlation and complementary Coulombic interactions). Here, the different interactions that occur in water and aqueous solutions will be critically reviewed in the light of recent experiments and theoretical modelling.

2. DIRECT MEASUREMENTS OF SURFACE AND INTERMOLECULAR FORCE FUNCTIONS

There are various techniques for directly measuring the forces between two surfaces or molecules. Most of these techniques use cantilever springs for measuring the forces and an optical or electric technique (strain gauges, capacitance plates) for measuring distances. The Surface Forces Apparatus (Fig. 1) has long been used to measure the forces between two extended surfaces in liquids or vapours. This method allows for the full force-law (force *vs* surface separation) to be directly measured between different types of surfaces of known surface geometry in different liquids [2]. A similar technique – the Atomic Force Microscope (Fig. 2) – allows measurement of the force between a very small tip and a surface or even between two molecules. Comparing the SFA with the AFM, the AFM has a much higher sensitivity to measuring forces (10 pN compared to 10 nN), and both techniques have a similar distance resolution (about 1Å). However, the AFM does not allow for a direct measurement of the absolute surface-surface separation or local surface geometry, but it does allow for "imaging" of surfaces, i.e., characterizing their morphology.

Many other force-measuring techniques have been developed that are suitable for specific systems. For example, the Osmotic Stress (OS) technique allows for the measurements of the repulsive forces between surfactant or lipid bilayers [4], and the Total Internal Reflectance (TIR) technique allows for the measurements of the forces between a large colloidal particle and a surface [5].

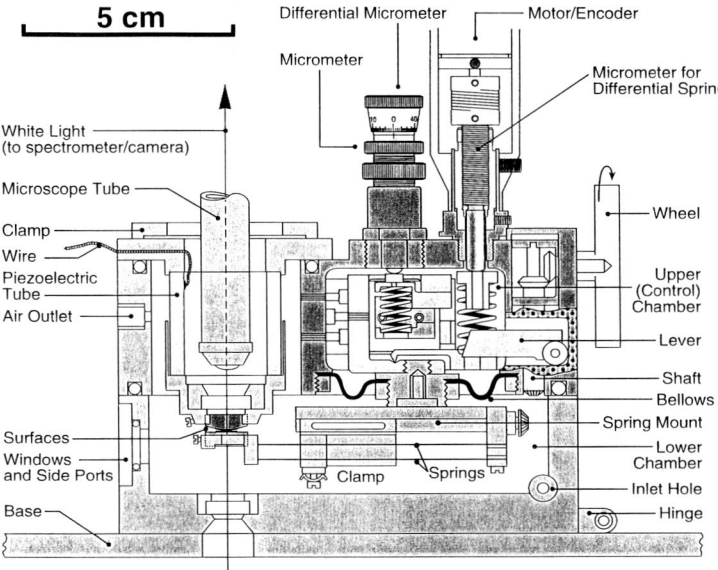

Figure 1. Schematic of a Surface Forces Apparatus (SFA) for measuring the interaction forces between and deformations of two surfaces as a function of their separation [2, 3].

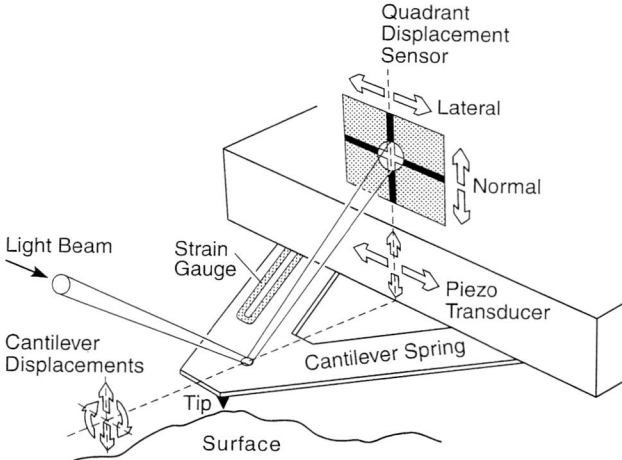

Figure 2. Schematic of an Atomic Force Microscope (AFM) for measuring the forces between specific molecular groups. The forces between individual molecules can be measured by this technique, but absolute tip-surface separations cannot be unambiguously determined.

6

3. LONG-RANGE FORCES (DLVO THEORY)

The two major long-range forces that almost always exist between two molecules or surfaces in a polar liquid such as water are the Van der Waals and Electrostatic "double-layer" forces (Fig. 3), which together make up the two forces of the DLVO theory of colloid stability [1]. These forces have been found to be well-accounted for by 'continuum' theories of long-range interactions (based on the assumption that the properties of matter in thin films are the same as the bulk properties) even for a complex, asymmetric biological surfaces, as illustrated in Fig. 4.

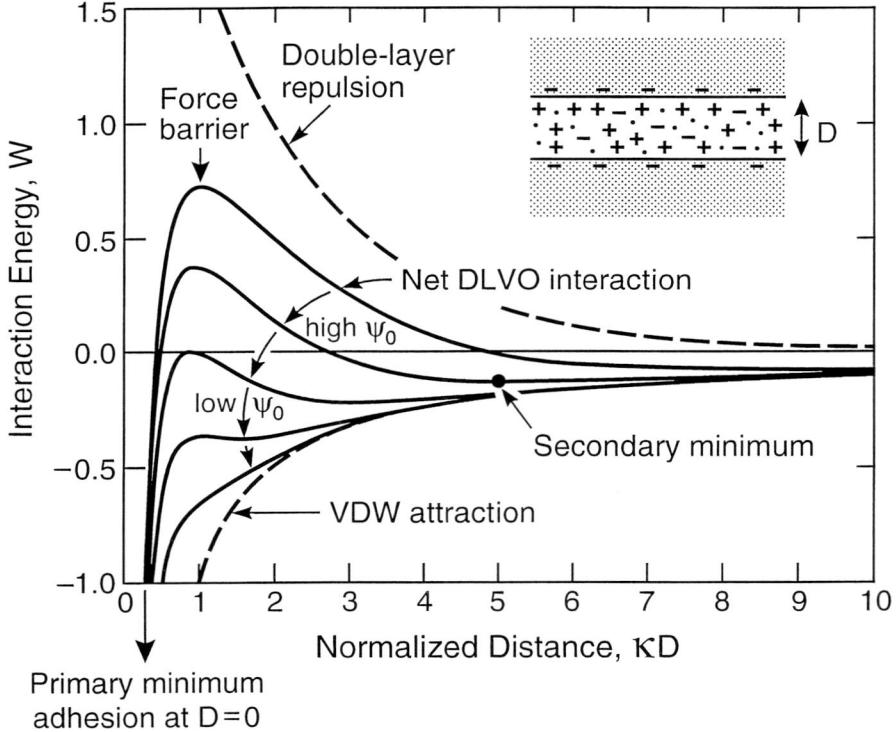

Figure 3. The DLVO theory [1] adequately describes the long-ranged van der Waals and electrostatic 'double-layer' forces between charged surfaces in water. However, at smaller separations, below 10–20Å or below 1 Debye length, other forces often take-over and determine the final adhesion.

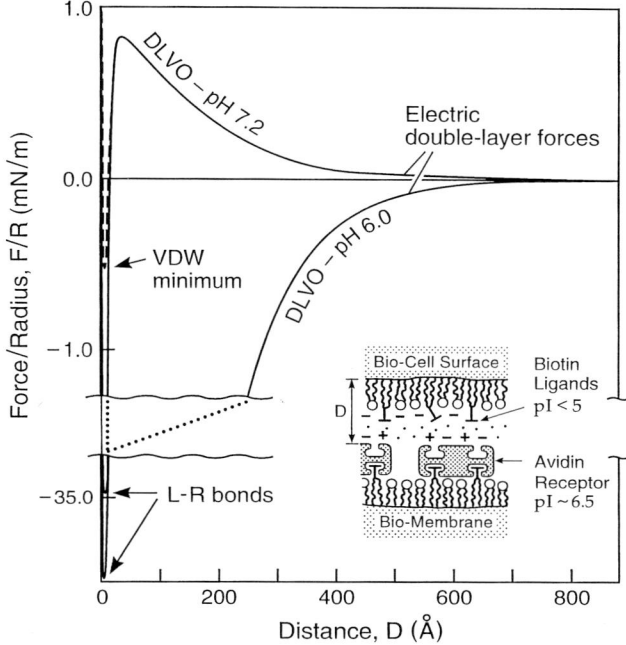

Figure 4. Measured DLVO forces between two dissimilar surfaces of Biotin (pK 3) and Streptavidin (pH 6.5) at different pH. At pH 7.2 both surfaces are negatively charged and the double-layer interaction is repulsive. At pH 6.0 the surfaces have opposite charges (Biotin negative, Streptavidin positive) and the double-layer force is now attractive. However, the van der Waals attraction and lock-and-key binding force at contact are unchanged. Adapted from Ref. [6].

4. ATTRACTIVE SHORT-RANGE ADHESION FORCES: COMPLEMENTARY INTERACTIONS

At small separations, however, less than about 20 Å, either the continuum approach breaks down or additional, short-range, forces come into play. This has the effect of modifying the interaction as two molecules or surfaces come into molecular or atomic contact, and the adhesion force or energy at contact cannot normally be calculated by simply applying continuum theories down to molecular contact separations ("adhesion" typically occurs at surface or molecular separations of $D = D_0 = 2$ to 4 Å). We first consider what recent experiments and theory have revealed about what happens when the continuum approximation breaks down, or in other words, what happens to the long-range DLVO forces at short-range – as D approaches D_0.

Recent work has shown that the electrostatic force at short-range can be very complex, even between symmetric surfaces, and that it can even change sign compared to the long-range interaction. How this occurs is illustrated in Fig. 5 which shows two amphoteric (polyelectrolyte) surfaces whose mean or surface-averaged charge density is negative, i.e., both surfaces have more negative charges than positive charges. This gives rise to a long-range *repulsive* double-layer force because this force is determined by the *mean* charge density of the surfaces. However, at smaller separations, depending on the distance between the charges on the surfaces and the Debye length, individual ions begin to see each other as discrete entities rather than a smeared out charge, and the individual Coulombic interactions of ionic pairs now determine the net interaction.

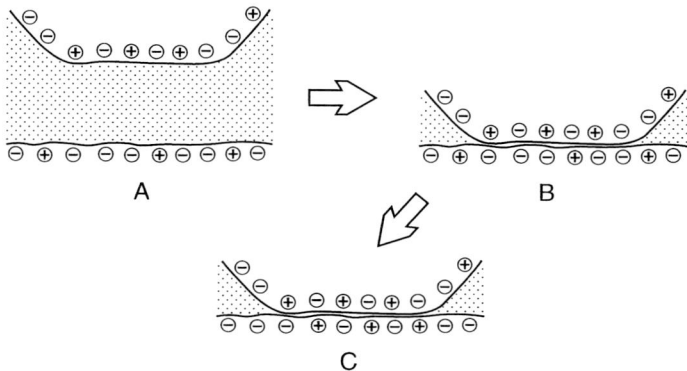

Figure 5. Schematic of two amphoteric surfaces, composed of both positive (basic) and negative (acidic) groups that repel at large separation (A) due to the double-layer interaction, but attract each other when in contact (B and C) due to favourable ion-pair interactions.

As can be seen in Fig. 5B, strong *attractive* ionic bonds can develop between most of the ions on the two surfaces, leaving only a few that repel each other. If the surface ionic groups can also rearrange on the surfaces so that each cation faces an anion on the opposite surface, then it is possible to end up with a two-dimensional ionic lattice at the interface and where *all* of the ionic bonds are now attractive (Fig. 5C) but at the expense of some entropy. In biology, this type of attractive short-range electrostatic interaction is known as "electrostatic complementarity" [7], and it can be very important in determining the adhesion and binding of amphoteric polyelectrolytes and proteins whose primary structure generally consist of a combination of both positive (basic) and negative (acidic) groups. This type of interaction is enhanced when the secondary and tertiary conformations of the molecules or surfaces are flexible, which allows for local changes in structure to occur so as to maximize the number and magnitude of ionic bonds. One should note that electrostatic complementarity cannot arise between polyelectrolytes or surfaces that contain only one kind of charged group, e.g., acidic or basic groups. Thus, the short-range adhesion forces between amphoteric polyelectrolytes such as proteins, gelatin, etc., are fundamentally very different from those between simple purely anionic or cationic polyelectrolytes.

A similar type of short-range "complementary adaptation" interaction also occurs involving the Van der Waals forces when the discrete nature (i.e., the size and shape) of atoms and molecular groups are taken into account. This is known as "geometric complementarity" [8] or "induced fit" binding and is illustrated in Fig. 6 B and C. Again, when macromolecules have some flexibility, a quasi-2D lattice can form at the interface and the strength of the adhesion can be much enhanced.

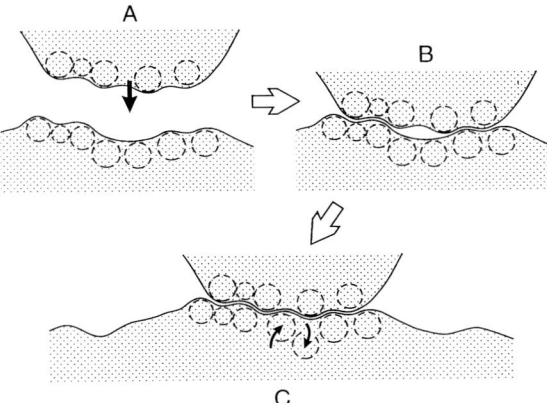

Figure 6. Geometric complementarity where the final adhesion of two surfaces (A) depends on the way they can fit together at the atomic scale, either naturally (B) or after some slight rearrangement of the atomic positions (C).

The way complementary and induced-fit interactions enhance adhesion is difficult to quantify or simulate: they are not described by a simple force-law or potential function. But they are very important in biological processes and in the interactions of charged surfaces composed of "soft matter".

5. FORCES DUE TO LIQUID OR SOLVENT STRUCTURE

Whereas complementary and induced-fit interactions are due to a particular ordering (or reordering) of *surface* molecular groups, similar effects arise due to reordering of the molecules of the *liquid* or solvent medium. Such interactions are sometimes called "solvation", "structural" or "hydration" forces, and it was previously believed that these forces depended only on the structure and properties of the liquid. But recent experimental and theoretical work has shown that their strength and range depend on a synergistic interaction involving both the liquid and the surface molecules [9].

Just as the surface ionic or molecular groups can reorder to enhance the attraction over what would be expected from continuum mean field theories, so can the counterions in the diffuse double-layer reorder to enhance the electrostatic attraction (or lower the repulsion) in aqueous solution. Such effects have been termed "ion-correlation" or "Manning Condensation" forces [10–12], in which the monotonically repulsive double-layer force can become suddenly attractive at small separations (D<2nm). If the surface co-ions are on a lattice, the counterions may also be induced to order into a lattice so that at contact, the two surfaces are separated by a thin crystalline ionic lattice film. As in the case of complementary interactions, ion-correlation forces can significantly enhance adhesion, but they are difficult to quantify or simulate and, so far, do not appear to be describable by a simple force-law or potential function.

Similar effects arise with van der Waals forces and excluded volume interactions, which can lead to a quasi-crystalline ordering of the liquid molecules in thin films. Historically, this type of structural force was the first to be studied in detail and is now fairly well understood. The most common type is the so-called oscillatory force which is entropic in origin and is due to the geometric packing constraints that modify the ordering of solvent or solute molecules trapped between two approaching surfaces. The resulting oscillations in the density and forces give rise to multiple adhesive minima at discrete surface separations corresponding to some dimension of the confined molecules, with a final deep "primary minimum" at contact (where no solvent molecules remain between the surfaces) whose strength can exceed the contribution from the continuum van der Waals force [15], as shown in Fig. 7 A. However, if there is a strong solute-solvent or surface-solvent bond (Fig. 7 B), a protective solvent shell or layer is formed, and the primary minimum is now much weaker and occurs farther out – at a separation corresponding to two solvent molecules (at $D=2\sigma$). Such effects are particularly common for hydrophilic groups and surfaces in aqueous solutions [9].

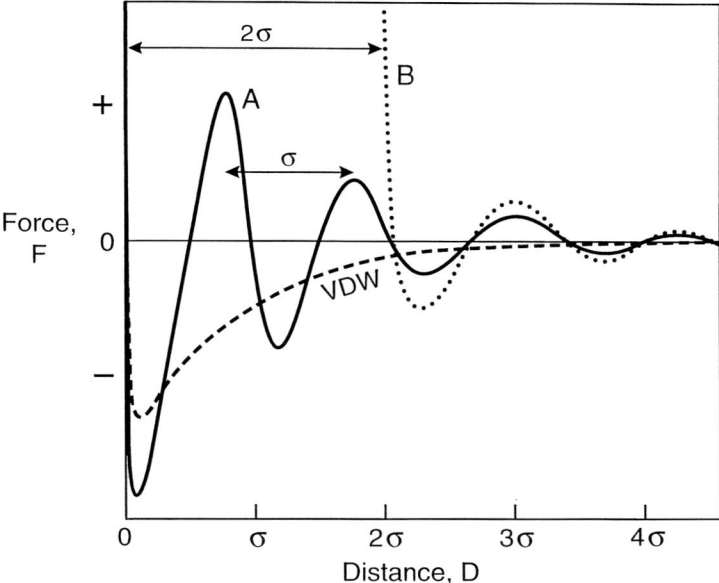

Figure 7. (A) Schematic of oscillatory "structural" or "solvation" force between two neutral, uncharged surfaces in a liquid such as water [13–15]. The double-layer force due to the presence of a surface charge appears to be simply additive with the oscillatory force [16]. (B) Force profile between "solvated' or "hydrated" hydrophilic (but uncharged) surfaces in water, where the first layer of strongly bound water molecules provide a steric barrier at $D \approx 2\sigma$. The strong binding of a single layer of water molecules to a surface or polyelectrolyte in aqueous solution (forming a primary "hydration shell") can come about from strong H-bonds and does not require that the surface groups be charged.

A primary solvation or hydration shell around a macromolecule, ionic group or polymer chain acts not only to reduce the adhesion strength, but also to increase the excluded volume of the molecule in solution, which enhances the repulsive entropic (excluded volume osmotic) interaction to the point where it can exceed the van der Waals attraction [9]. When such hydrated groups are loosely attached to a biopolymer or surface (Fig. 8) they experience an overall repulsion in water. This causes the polymers to swell or the surfaces to repel each other even when they are uncharged, as occurs with strongly hydrophilic lipid headgroups, PEO (PEG) and polysaccharides (Figs 8 and 9).

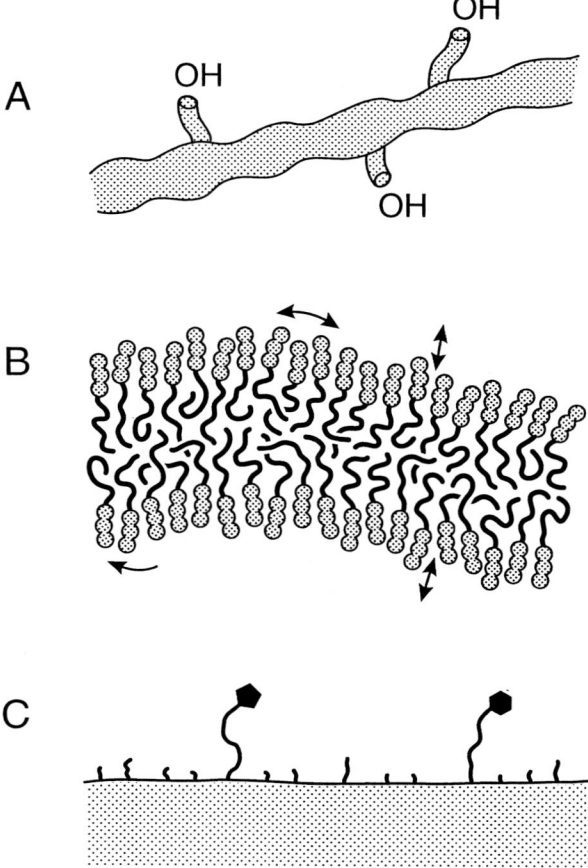

Figure 8. Examples of (A) hydrophilic groups extending from a polymer backbone, (B) lipid headgroups attached to amphiphilic molecules in a micelle or bilayer, and (C) tethered hydrated groups protruding from a protein or membrane surface. The solvated chains interact like end-grafted polymer chains ("mushroom" or "brush" layers, depending on the coverage) and give rise to a short-range entropic repulsion in water which is often erroneously attributed to "water structure" [9].

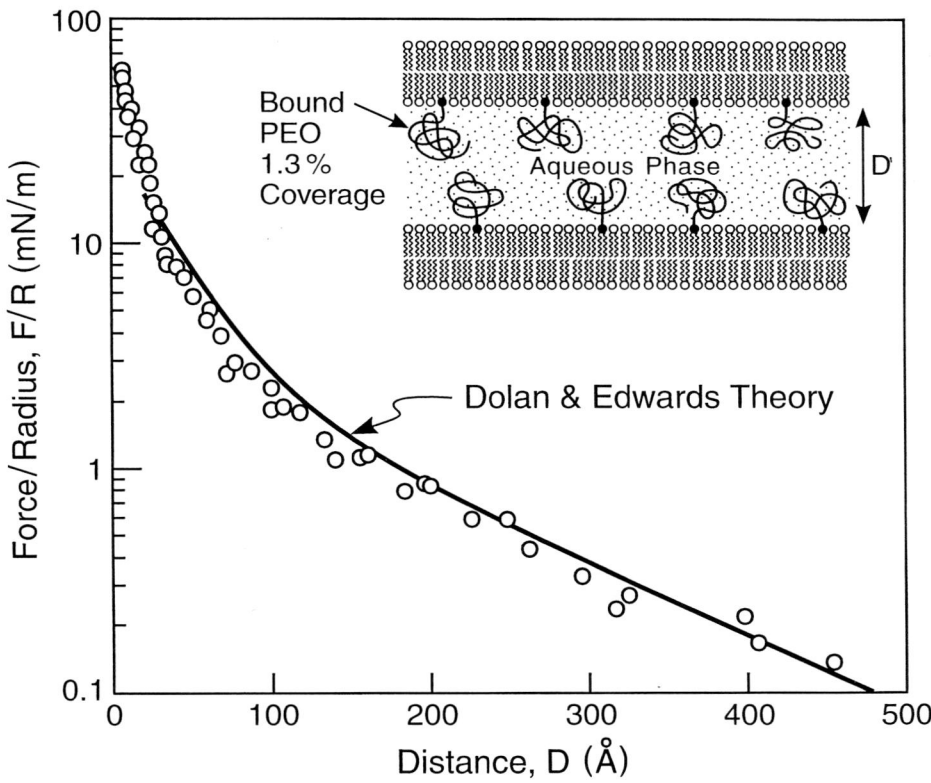

Figure 9. Forces between two DSPE bilayers exposing surface-grafted PEG chains with 44 segments per chain at a coverage of 1 chain per 80 lipid molecules which corresponds to the non-overlapping or "mushroom" configuration [17]. Similar forces arise in the high coverage "brush" regime [17] and also for much shorter PEG chains [9].

6. DEPLETION FORCES

Depletion forces have a similar entropic origin to the oscillatory forces. They were first proposed by Oosawa [18] to account for the effect of dissolved polymer and small colloidal particles on the forces between surfaces or large colloidal particles in solution. These forces may be thought of as a smeared out oscillatory force where only the primary minimum (and, sometimes, the first repulsive peak and second minimum) remain. Such forces can give rise to a weak adhesion between colloidal particles in concentrated solutions of non-adsorbing low MW polymer or small colloidal particles (cf. Fig. 10). Such forces are often used in biomedical experiments to induce cells to adhere [19].

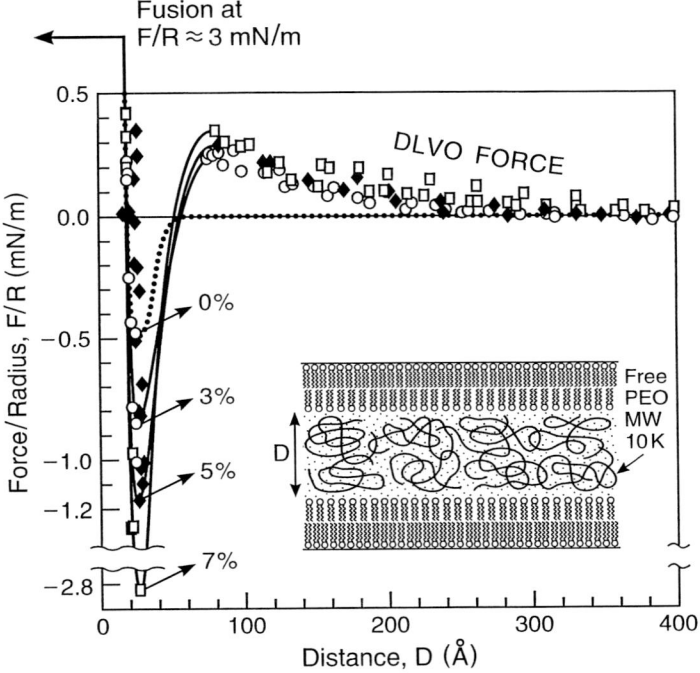

Figure 10. Depletion force between two lecithin bilayers mediated by free PEG (PEO) of MW=10 K in solution [18]. The attractive force in the absence of PEG is the van der Waals force, which is enhanced with increasing PEG concentration.

7. OTHER NON-DLVO FORCES IN WATER: HYDRATION AND HYDROPHOBIC FORCES

Hydration forces have always been something of a mystery, both experimentally and theoretically. They were first proposed many years ago by Langmuir [20] and others to "explain" the unexpected stability of uncharged particles, such as "coacervates", in solution, and subsequent experiments showed them to be monotonically repulsive, roughly exponential, and with a range of about 20Å. The idea that these forces are due to "water structuring" at surfaces gained steady popularity, and the exponential decay length was believed to be due to some characteristic property of water. However, as theoretical work and computer modelling failed to confirm the existence of any type of exponentially repulsive hydration force associated with water structuring other interpretations for their origin – based on the properties of the interacting *surfaces* rather than the *medium* – were offered and later supported by theoretical work [9, 21].

Monotonically repulsive "hydration" forces are probably a specific example of a more general type of "thermal fluctuation" force that can arise between certain types of solvated surfaces in any liquid medium. The main requirement is that the surfaces are not smooth and rigid (solid-like), but dynamically rough or fluid-like at the nanometer scale. Such situations occur with surfaces that expose mobile groups, for example, surfactant and lipid bilayers that expose hydrophilic headgroups or surfaces with attached low MW polymer layers (cf. Figs 8 and 9) and certain solid surfaces such as silica that have protruding "hairs" or a soft gel layer [22]. In most cases, the protruding molecular groups are "solvated" or "hydrated" by the solvent molecules, viz., there is a strongly bound first layer of liquid molecules to the "solvophilic" or "hydrophilic" surface groups, but beyond that the liquid behaves more or less with its bulk fluid properties. Whenever true hydration forces (i.e., forces due to liquid structure) have been unambiguously measured and theoretically accounted for, they have been oscillatory, as shown in Fig. 7.

In other cases, a monotonically *attractive* force has been measured between hydrophobic surfaces, i.e., surfaces or groups that do not strongly bind water molecules. Hydrophobic forces have been measured in many different systems, especially between biological molecules exposing hydrocarbon groups, but their origin has still not been properly explained [23]. Their long range (cf. Fig. 11) makes it unlikely that they are due to water structuring effects, and various models based on fluctuating electrostatic or modified van der Waals type interactions have been proposed. What is certain is that these forces are very important in many colloidal and biological processes, such as detergency, froth flotation, and the adhesion and fusion of biological membranes, as will now be described.

16

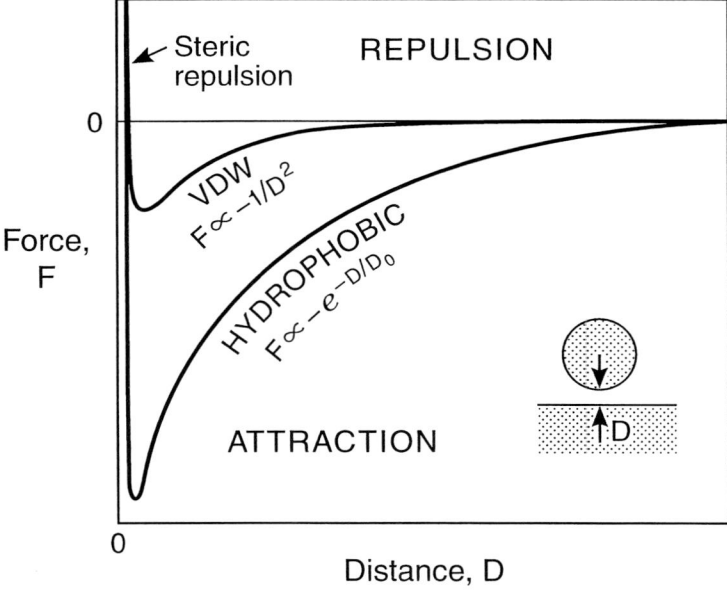

Figure 11. Hydrophobic interaction potential (schematic) between two fully hydrophobic surfaces (contact angle with water >90°) [26].

Hydrophobic interactions have been implicated in the adhesion and fusion of amphiphilic assemblies such as proteins, bilayers and biological membranes. Their mode of action is often subtle, and does not involve a simple coming together of two surfaces, but also a rearrangement of the molecular configuration around the adhesion or fusion "site" in such a way that the hydrophobic interaction can take place. For example, the moving apart of the hydrophilic headgroups on the surface of a lipid bilayer exposes a hydrophobic "pocket" to the outside aqueous medium. When this occurs on two apposing bilayers that face each other (Fig. 12), the exposed hydrophobic chains then jump across the water gap towards each other, meet at the centre, and thereby constitute the first (nucleation) stage of a fusion event. The initial opening up or exposure of hydrophobic groups can be triggered by a local clustering or declustering of molecules due to a transient mechanical stress induced by a curvature change, calcium, pH or flow [24, 25]. Similar mechanisms are believed to occur when proteins bind to surfaces or when "pores" or "channels" open up through a protein or membrane. As can be seen, these processes are much more complex than a simple coming together of two surfaces.

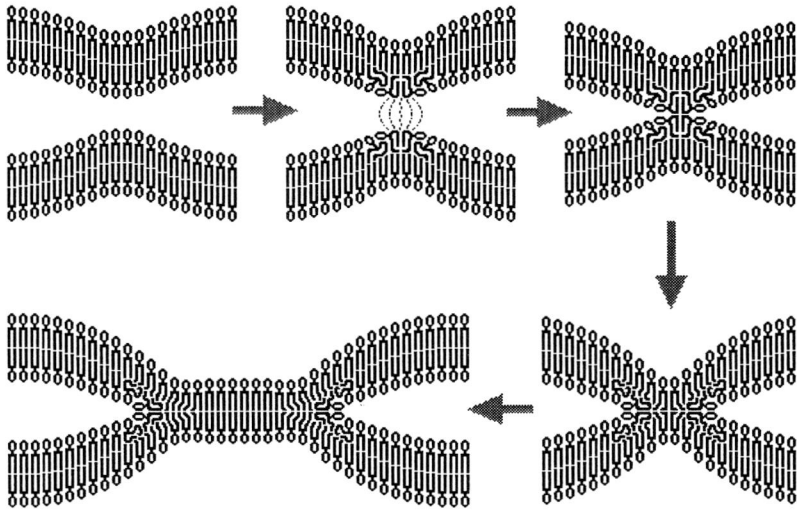

Figure 12. Adhesion and fusion of lipid bilayer involving local rearrangements of the lipid molecules [24, 25, 27].

8.   NON-EQUILIBRIUM INTERACTIONS

Many interactions and self-assembly processes involving complex multicomponent structures can involve different processes occurring simultaneously but sequentially at different locations in space and time. An example of this was shown in Fig. 12 above, and another is shown in Fig. 13 below, which shows the interaction potential of a tethered ligand with a receptor [28]. The force-distance curve depends on the rate of approach of the ligand to the receptor because a certain time is needed for the ligand to find and lock into the binding site on the receptor. Likewise, on separation, the adhesion force will depend on the rate of separation: if the surfaces are separated quickly, breakage will occur where the *force* is weakest, which involves lipid pull-out; but if the separation is carried out slowly, breakage will occur where the *energy* is weakest, which now involves the cleavage of the ligand-receptor bond [29].

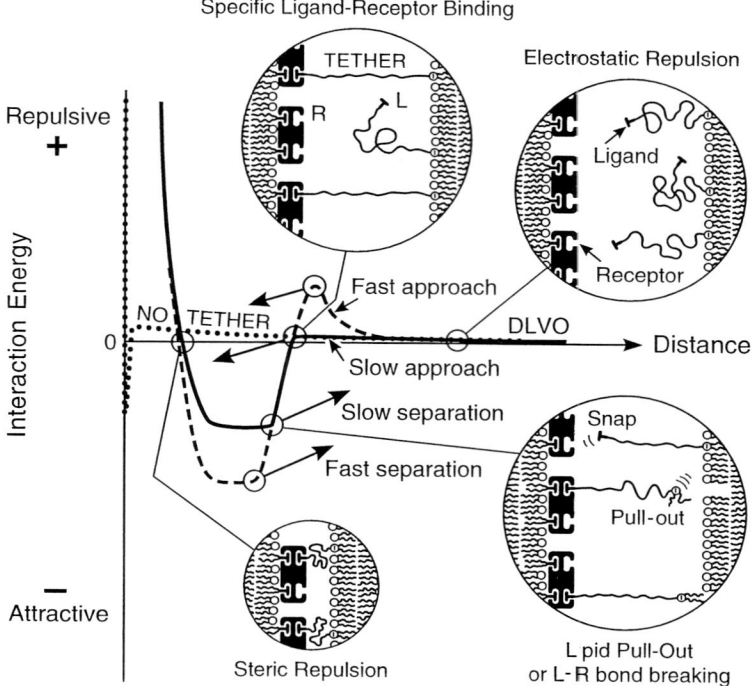

Figure 13. Interaction force between a tethered ligand (biotin) and a receptor (streptavidin). The long-range force is the repulsive electrostatic double-layer interaction. The shorter-ranged attractive force is due to the bridging ligand-receptor bond. On rapid approach, the repulsion is enhanced (dotted curve) because the ligand and receptor do not find each other until the surfaces have come closer together. On rapid separation, the lipid molecules become pulled out of the membranes (instead of the ligand-receptor bond breaking) and the force curve is now more attractive (dashed curve). Adapted from Refs 28 – 30.

9. DISCUSSION: SEQUENTIAL INTERACTIONS IN SPACE AND TIME

Recent research has shown that the interactions of complex molecular assemblies ("soft materials" or "complex fluids") can be far more complicated than can be described by a simple interaction potential or force function (Fig. 14). The whole process can involve both normal and lateral rearrangements of molecules (spatial effects) as well as non-equilibrium (time-dependent) effects. This means that different things can be going on simultaneously but sequentially in different regions of space, both in the z-direction (away from a surface) and x-y plane (at different locations away from the "centre of action"). Biological cells clearly have developed means for controlling interactions at different distance- and time- regimes. For example, a transient change in the local pH or calcium ion concentration at a membrane surface could modify the long-range electrostatic interaction but not the short-range hydrophobic interaction; or a change in local fluidity could alter the diffusion rates of a tethered ligand and thereby prevent or enhance its probability of capture. Living systems clearly make full use of all of these subtle effects in their control and modulation of complex processes. It should be possible to exert similar controls in practice by carefully changing the processing protocols of gels, hydrocolloids and 'smart' biomaterials.

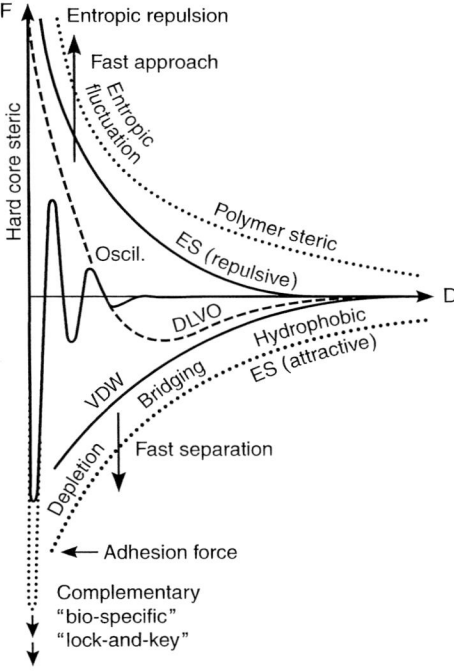

Figure 14. Generic interaction potential for a complex multicomponent system. No one system is likely to display all of these interactions at the same time.

REFERENCES

1. E. J. W. Verwey and J. Th. G. Overbeek (1948) *Theory of Stability of Lyophobic Colloids*, Elsevier, Amsterdam.
2. J. N. Israelachvili and G. E. Adams, *J. Chem. Soc., Faraday Trans. I* **74** (1978) 975; J. N. Israelachvili and P. M. McGuiggan, *J. Mater. Res.* **5**, No.10 (1990) 2223.
3. M. Heuberger, G. Luengo, J. Israelachvili, *Langmuir* **13** (1997) 3839.
4. D. M. LeNeveu, R. P. Rand and V. A. Parsegian, *Nature* **259** (1976) 601.
5. D. C. Prieve and N. A. Frej *Langmuir* **6** (1990) 396; D.C. Prieve, S. G. Bike and N. A. Frej *Faraday Discuss. Chem. Soc.* **90** (1990) 209.
6. D. E. Leckband, F-J Schmitt, J. N. Israelachvili and W. Knoll *Biochemistry* **33** (1994) 4611.
7. P. M. Dean, *Molecular foundations of drug-receptor interactions*, Cambridge University press, London & Cambridge (UK), 1987, pp. 254.
8. P. M. Dean, *Molecular foundations of drug-receptor interactions*, Cambridge University press, London & Cambridge (UK), 1987, pp. 121.
9. J. Israelachvili and H. Wennerström, *Langmuir* **6** (1990) 873; J. Israelachvili and H. Wennerström, *J. Phys. Chem.* **96** (1992) 520.
10. B. Jönsson, H. Wennerström and B. Halle *J. Phys. Chem.* **84** (1980) 2179; L. Guldbrand, B. Jönsson, H. Wennnerström and P. Linse. *J. Chem. Phys.* **80** (1984) 2221.
11. D. Stigter, *Biophys. J.* **69** (1995) 380.
12. I. Rouzina and V. A. Bloomfield, *J. Phys. Chem.* **100** (1996) 9977.
13. J. N. Israelachvili, Academic Press: London, First ed. (1985), Second ed. (1991).
14. D. Henderson and M. Lozada-Cassou *J. Colloid Interface Sci.* **162** (1994) 508; *ibid.* **114** (1986) 180; A. Trokhymchuk, D. Henderson and D. T. Wasan (in press).
15. H. K. Christenson and V. V. Yaminsky. *Langmuir* **9** (1993) 2448.
16. H. K. Christenson and R. G. Horn *Chem. Phys. Lett.* **28** (1983) 45.
17. T. L. Kuhl, D. E. Leckband, D. D. Lasic and J. N. Israelachvili *Biophys. J.* **66** (1994) 1479; T. L. Kuhl, D. E. Leckband, D. D. Lasic and J. N. Israelachvili In *CRC Handbook on Stealth Liposomes*, Dan Lasic & Frank Martin, Eds, CRC Press, Boca Raton, Florida, **Ch. 8**, pp. 73–91 (1995).
18. S. Asakura and F. Oosawa *J. Polymer Sci.* **33** (1958) 183.
19. T. L. Kuhl, Y. Guo, J. L. Aldferfer, A. Berman, D. Leckband, J. Israelachvili and S. W. Hui, *Langmuir* **12** (1996) 3003; T. L. Kuhl, A. D. Berman, S. W. Hui, J. N. Israelachvili, *Macromolecules* (in press); T. L. Kuhl, A. D. Berman, S. W. Hui and J. N. Israelachvili, *Macromolecules* (in press).
20. I. Langmuir *J. Chem. Phys.* **6** (1938) 873.
21. S. Marcelja, Nature **385** (1997) 689; Hydration in electric double layers — reply *Nature* **385** (1997) 690.

22. G. Vigil, Z. Xu, S. Steinberg and J. Israelachvili, *J. Colloid Interface Sci.* **165** (1994) 367.
23. H. K. Christenson, in *"Modern Approaches to Wettability: Theory and Applications"*, M. E. Schrader & G. Loeb, eds., Plenum, New York, 1992.
24. C. A. Helm, J. N. Israelachvili and P. M. McGuiggan, *Biochemistry* **31** (1992) 1794.
25. D. E. Leckband, C. A. Helm and J. Israelachvili, *Biochemistry* **32** (1993) 1127.
26. R-H. Yoon, D. H. Flinn and Y. I. Rabinovich, *J. Colloid Interface Sci.* **185** (1997) 363.
27. C. A. Helm, J. N. Israelachvili and P. M. McGuiggan, *Science* **246** (1989) 919; Horn, R. G. *Biochim Ciophys. Acta* **778** (1984) 224.
28. J. Y. Wong, T. L. Kuhl, J. N. Israelachvili, N. Mullah and S. Zalipsky, *Science* **275** (1997) 820.
29. G. I. Bell, *Science* **200** (1978) 618; J. N. Israelachvili and A. Berman. *Israel J. Chemistry* **35** (1995) 85; D. Leckband, W. Muller, F-J. Schmitt and H. Ringsdorf. *Biophys. J.* **69** (1995) N3:1162.
30. D.E. Leckband, F-J. Schmitt, W. Knoll and J. Israelachvili, *Science* **255** (1992) 1419.

2. STRUCTURE OF GELS AND GELATION

HYDROCOLLOIDS – PART 1
Edited by K. Nishinari
2000 Elsevier Science B.V.

Thermoreversible gelation with multiple junctions in associating polymers

F. Tanaka

Department of Polymer Chemistry, Graduate School of Engineering
Kyoto University, Kyoto 606-8501, Japan

This paper studies sol/gel transition of associating polymers with multiple cross-link junctions. Paying special attention on the multiplicity and sequence length of the network junctions, we derive phase diagrams of thermoreversible gels competing with phase separation. New methods to analize the molecular structure of network junctions, their lifetimes, elastically effective chains are proposed. The effect of added surfactants on the formation of reversible gels in hydrophobically modified polymers is also studied under the assumption of the existence of a minimum multiplicity required for stable cross-links. Transition from intramolecular closed association (flower micelles) to more open intermolecular association (bridges) with increase in the polymer concentration is also theoretically studied.

1. INTRODUCTION

Associating polymers are polymers carrying associative groups (or segment blocks) sparsely distributed along the backbone or on the chain side. These groups form aggregates, or micelles, through hydrogen bonds, ionic attraction, hydrophobic interaction, *etc.* Polymers with associative interactions exhibit a variety of condensed phases, typical examples of which are microscopically ordered phases, gels, and liquid crystals. All of these phases have their counterparts formed by covalently connected polymers, but, since association is thermally controllable, associating phases provide a new pathway to modelling statistical clusters, block copolymers, and reversible networks[1, 2, 3].

The strength of association is described by the *association constant* defined by

$$\lambda(T) \equiv \exp(-\beta \Delta f_0), \tag{1}$$

where $\beta \equiv 1/k_B T$ the reciprocal temperature and Δf_0 the standard free energy change on binding a single functional group into a junction. If the group (or block) consists of ζ statistical segments, as in micro-crystalization and hydrophobic aggregation of short chains, the free energy change can be written as $\Delta f_0 = \zeta(\Delta h - T\Delta s)$ by the use of the binding enthalpy Δh and entropy Δs per statistical unit. The number ζ is called *sequence length* of a junction.

Another important structural parameter of the junction is its *multiplicity*. The multiplicity k is defined by the number of groups combined together in a junction. It is often referred to as *aggregation number* in the case of hydrophobic groups.

Most thermoreversible gels have multiple cross-links, markedly different from pairwise bonding of the chemical cross-linking.

The reorganizability of the network junction is characterized by the *average duration time* $\tau_\times$ for an associative group to be in a bound state. It is governed by the free energy barrier ΔF_0 separating the bound state from the free one :

$$\tau_\times = \tau_0 \exp(\beta \Delta F_0). \tag{2}$$

(τ_0 being the typical microscopic time scale.) The timescale for reorganizing transient structures can be adjusted by this barrier height of the associative interaction.

2. MODELS OF NETWORK JUNCTIONS

We consider a model mixture of functional molecules (or primary polymer chains) in a solvent. The molecules are distinguished by the number f of the functional groups ("stickers") they bear, each functional group being capable of taking part in the junctions which may bind together any number k of such groups[4]. We include $k = 1$ to indicate unreacted groups. In what follows, we allow junctions with multiplicities lying in a certain range to coexist in proportions determined by the thermodynamic equilibrium conditions. Let n_f be the number of statistical segments on an f-functional molecule and let N_f be the number of f-functional primary molecules in the solution. The weight fraction w_f of the molecules with specified f relative to the total weight is then given by $w_f = f N_f / \sum f N_f$. In thermal equilibrium, the solution has a distribution of clusters with a population distribution fixed by the equilibrium conditions. Following the notation in reference [4], we define a cluster of the type $(\mathbf{j}; \mathbf{l})$ to consist of l_f molecules of functionality f ($f = 1, 2, 3, ...$) and j_k junctions of multiplicity k ($k = 1, 2, 3, ...$). The bold letters $\mathbf{j} \equiv \{j_1, j_2, j_3, ...\}$ and $\mathbf{l} \equiv \{l_1, l_2, l_3, ...\}$ denote the sets of indices. An isolated molecule of functionality f, for instance, is indicated by $\mathbf{j}_{0f} \equiv \{f, 0, 0, ...\}$, and $\mathbf{l}_{0f} \equiv \{0, ..., 1, 0, ...\}$.

We now introduce a specific model of the multiple junctions with lower bound s_{min} and the upper bound s_{max}:

$$k = 1(\text{unassociated}), \quad k = s_{min}, ..., s_{max}(\text{associated}). \tag{3}$$

In the case of micro-crystalline junctions, for instance, it is natural to assume that a minimum number s_{min} greater than 2 of the crystalline chains is required for a junction formation. This is because the surface energy terms will prevent small-k units from being stable, leading to the existence of the critical multiplicity for the nucleation of the crystallites. Similarly a minimum aggregation number is required for the stability of micelles formed by hydrophobes on water-soluble polymers. As we will see later, surfactants added to the solution cause complex interaction with hydrophobically modified polymers due to the existence of this minimum multiplicity. On the other hand, saturation of the junction multiplicity is caused by the dense packing of the chains near the junction zones which prevents access of excessive functional groups. When $s_{min} = s_{max} \equiv s$, only one multiplicity is allowed. We call this special case *fixed multiplicity model*.

3. SOL/GEL TRANSITION

Let $\nu(\mathbf{j};\mathbf{l})$ be the number density of the clusters of type $(\mathbf{j};\mathbf{l})$, and let $\phi(\mathbf{j};\mathbf{l}) \equiv (n\sum l_f)\nu(\mathbf{j};\mathbf{l})$ be their volume fraction. (n being the number of statistical units on a chain.) The free energy change on passing from the standard reference state (polymers and solvent molecules being separated in hypothetical crystalline states) to the final solution, at equilibrium with respect to cluster formation, is given by the expression[4]

$$\frac{\beta\Delta F}{\Omega} = \nu_0 \ln \phi_0 + \sum_{\mathbf{j},\mathbf{l}} \nu(\mathbf{j};\mathbf{l})[\Delta(\mathbf{j};\mathbf{l}) + \log \phi(\mathbf{j};\mathbf{l})] + \chi\phi_0\phi, \tag{4}$$

where ϕ is the volume fraction of the polymer, Ω the total number of lattice cells in the lattice-theoretical picture of polymer solutions, and χ conventional Flory's interaction parameter. The subscript zero denotes the solvent, with volume fraction $\phi_0 = 1 - \phi$. The quantity $\Delta(\mathbf{j};\mathbf{l})$ involves the free energy change accompanying the formation of a $(\mathbf{j};\mathbf{l})$-cluster in a hypothetical undiluted amorphous state from the separate primary molecules in their standard states.

By minimizing this free energy with respect to the volume fraction $\phi(\mathbf{j};\mathbf{l})$, the most probable distribution of clusters are found. Using the result of multiple tree statistics for the combinatorial entropy in the free energy $\Delta(\mathbf{j};\mathbf{l})$ of cluster formation, we find the volume fraction of clusters as a function of the temperature and concentration. In the pre-gel regime, the total sum over all volume fractions of clusters must give the volume fraction of polymers in the solution. For example, this normalization relation for the fixed multiplicity model of monodisperse polymers (f and n definite) is given by

$$\lambda(T)\phi/n = \alpha^{1/s'}/f(1-\alpha)^{s/s'}, \tag{5}$$

where $s' \equiv s - 1$ and $\lambda(T)$ is the association constant. This relation connects the extent α of association to the (scaled) polymer concentration. The extent α of association (or conversion) is defined by the probability for a randomly chosen associative group to be associated. It is the counterpart of the extent of reaction in conventional chemical gels.

We next calculate the weight-average molecular weight of the clusters by using the most probable distribution. From its divergence, we find the sol/gel transition point. The gel condition is most generally given by [4]

$$(f_w - 1)(\mu_w - 1) = 1, \tag{6}$$

where $f_w \equiv \sum_{f\geq 1} f w_f$ is the weight average functionality of the primary chains, and $\mu_w \equiv \sum_{k\geq 1} k p_k$ the average multiplicity of the junctions. (p_k being the probability for an associative group to be associated into the junction of multiplicity k.)

Specifically for the fixed multiplicity model of monodisperse primary chains discussed above, the gelation condition is given by $\alpha = \alpha^* \equiv 1/f's'$, leading to the critical concentration

$$\lambda(T)\phi^*/n = f's'/f(f's' - 1)^{s/s'}, \tag{7}$$

where ϕ^* is the volume fraction of the polymer at gelation, and $f' \equiv f - 1$ and $s' \equiv s - 1$. As the multiplicity is increased, with other parameters being fixed, gelation concentration changes and sol/gel line shifts on the temperature-concentration plane.

Taking the logarithm of the gelation concentration (7), we find an important relation

$$\ln \phi^* = \zeta \frac{\Delta h_0}{k_B T} + \ln[\frac{f's'n}{f(f's' - 1)^{s/s'}}] - \zeta \frac{\Delta s_0}{k_B}. \tag{8}$$

We can find s and ζ by comparing this relation with the experimental sol/gel transition concentration. For the micro-crystalline junction formed by homopolymers, each ζ sequence of repeat units along a chain may be regarded as a functional group. A polymer chain is then regarded as carrying $f = n/\zeta$ functional groups. Since we have large n, and hence large f, we can neglect 1 compared to n or f. We are thus led to an equation

$$\ln c^* = \zeta \frac{\Delta h_0}{k_B T} - \frac{1}{s - 1} \ln M + constant, \tag{9}$$

where weight concentration c^* has been substituted for the volume fraction.

Poly(vinyl alcohol)/water

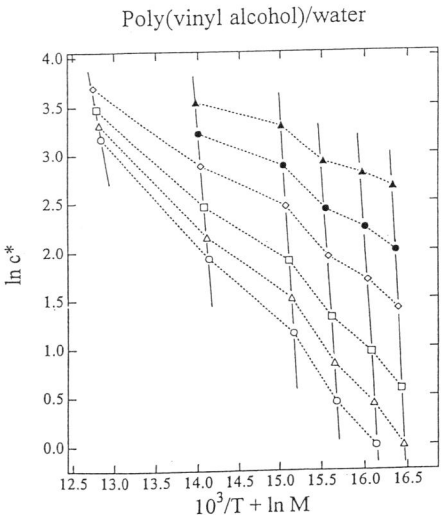

Figure 1: Modified Eldridge-Ferry plot for poly(vinyl alcohol) gel in water. Gelation concentration at constant molecular weight (solid lines) and at constant temperature (dotted lines) are plotted against a combined variable $10^3/T + \ln M$.

Thus, from the slope of the plot $\ln c^*$ against $10^3/T + \ln M$, we can find ζ and s independently. This is called *Modified Eldridge-Ferry* method[5]. Fig. 1 shows an example of our method for the gel melting concentration of poly(vinyl alcohol) in water. The slope of the solid lines with constant molecular weight gives $-A = 13.43$ almost independently of their molecular weights. Hence we find $\zeta = 26.7$ kcal/mol/$|(\Delta h_0)_{mol}|$. If we use the heat of fusion $(\Delta h_0)_{mol} = 1.64$ kcal/mol in the bulk crystal, we find $\zeta = 16.3$. On the other hand, the slope of the dotted lines with constant temperature depends on their temperature. At the highest temperature $T = 91°C$ in the measurement, it is -0.38, while it gives a larger value -0.9 at $T = 71°C$. The multiplicity is estimated to decrease from 3.6 for high-temperature melting to 2.1 for low-temperature melting. From thermodynamic stability of the junctions it is only natural that a gel which melts at lower temperature has smaller multiplicity. The real thickness of the junction zone may, however, be much larger than the dimensions of the chain diameter because of the chain folding.

4. ELASTIACLLY EFFECTIVE CHAINS IN THE GEL NETWORK

To study the dynamics of associating polymers, we introduce a model network made up of polymers of uniform molecular weight M (or the number n of statistical units) carrying associative functional groups at their both chain ends (named *telechelic polymer*).

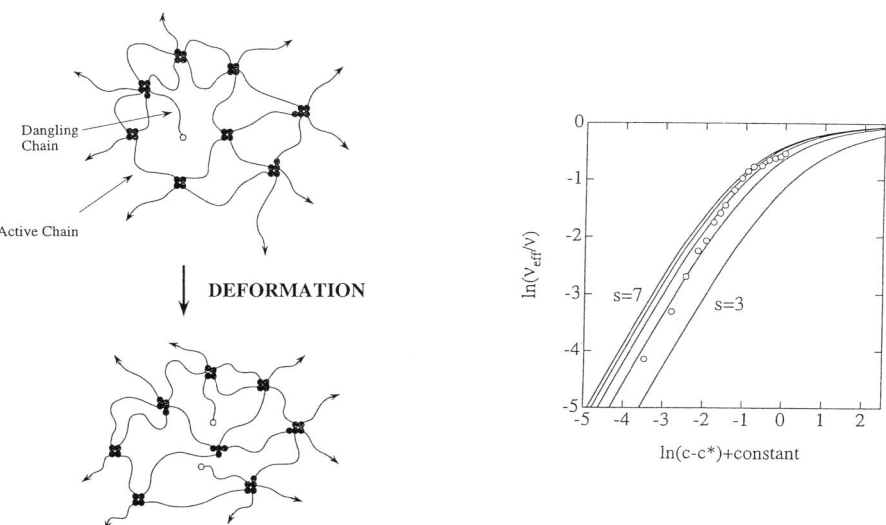

Figure 2: (a)Internal reorganization of the transient network induced by a macroscopic deformation. Associative groups on the chains with high tension disengage from the junctions and form dangling ends, while some dangling chains catch junctions in the neighborhood. (b)Comparison of the experimental data on the elastically effective chains in HEUR 16C/35K in water with theoretical calculation. To eliminate temperature prefactor $\lambda(T)$, the experimantal data are horizontally shifted.

We focus our attention specifically on the *unentangled regime* where M is smaller than the entanglement molecular weight M_e, so that each chain obeys Rouse dynamics modified by end-association. In the following, we assume the lifetime of a junction is sufficiently long so that it is well separated from the Rouse relaxation spectrum, the longest time of which is given by the Rouse relaxation time τ_R.

Consider a time interval dt smaller than $\tau_\times$ but still larger than τ_R. Under a macroscopic deformation $\hat{\lambda}(t)$ given to the network, either end of a chain, being stretched above a critical length, snaps from the junction, and the chain relaxes to a Gaussian conformation, whilst some of the free dangling ends recapture the junctions in their neighborhood (Fig. 2(a)). Since the stress is transmitted only through the chains whose both ends are connected to the network junctions, we call these chains *elastically effective* (or *active*) *chains*. On longer time scale, transition between elastically effective chains and dangling chains is frequently taking place in the network, so that the model is called *transient network model*[6].

On the basis of these assumptions, we can derive the complex frequency-dependent modulus. Specifically we find that a modulus-frequency curve at any temperature T can be superimposed onto a single curve at the reference temperature T_0, if it is horizontally and vertically shifted properly. Specifically we have

$$\frac{G(\omega)}{\nu_e(T_0)kT_0}b_T = g(\frac{\omega}{\beta(T_0)}a_T) \tag{10}$$

for both storage modulus $G = G'$ and loss modulus $G = G''$, where

$$a_T \equiv \beta(T_0)/\beta(T) = \exp[-\frac{W}{k}(\frac{1}{T_0} - \frac{1}{T})] \tag{11}$$

is the frequency (horizontal) shift factor, and

$$b_T \equiv \nu_e(T_0)kT_0/\nu_e(T)kT \tag{12}$$

is the mudulus (vertical) shift factor. Here, $\nu_e(T)$ is the number of elastically effective chains in the network at a given temperature T, $\beta(T)$ the chain dissociation rate. The fact that the frequency shift factor depends exponentially on the reciprocal of the temperature indicates that the linear viscoelasticity is dominated by the activation process with the free energy barrier height W for the chain dissociation. In fact, experiments on HEUR estimated the average lifetime $\tau_\times$ of the junction from the peak position of the loss moduli, and found the activation energy as a function of the length of end-chain hydrophobes.

On the other hand, at high frequencies $\omega >> \beta_0$, plateau value of the storage modulus gives the equilibrium number ν_e of elastically active chains in the network.

In order to derive the number of elastically effective chains in equilibrium, we next employ the criterion of Scanlan and Case[7] that only subchains bound at both ends to junctions connected by at least *three paths* to the network matrix are elastically effective. We thus have $i, i' \geq 3$ for an effective chain, where i indicates the number of paths leading to the network matrix. A junction with one path ($i = 1$) to the gel unites a group of subchains dangling from the network matrix whose conformations are not affected by an applied stress. A junction with two paths ($i = 2$) to the gel merely extends the length of an already counted effective subchain.

Fig. 2(b) compares our theoretical calculation[8] with the experimental data on the high frequency dynamic modulus for HEUR measured by Annable *et al.*[9](HEUR C16/35K, end-capped with $C_{16}H_{33}$, molecular weight 35,000). In fitting the data, we have horizontally shifted the experimental data because of the temperature prefactor $\lambda(T)$ and also of the difference in the unit of the polymer concentration. Multiplicity is changed from curve to curve. Although fitting by a single theoretical curve with a fixed multiplicity is impossible due to the existence of polydispersity in the multiplicity, our theory produces correct behavior over a wide range of the concentration.

5. EFFECT OF ADDED SURFACTANTS

Existence of a limited range in the multiplicity of the network junctions results in high sensitivity of their rheological properties to external perturbation. To see the effect, let us consider how aqueous thermoreversible gels of hydrophobically modified polymers interact with added surfactants. Hydrophobic tails of the surfactants may join in the junctions and form mixed micelles. To model the associating polymer-surfactant system, we consider a mixture of polymers and low molecular-weight surfactant molecules in a solvent. Each polymer is assumed to carry the number $f(\geq 2)$ of associative groups (for instance, hydrophobes in the case of HEUR) along its chain comprising of n_f statistical units, and each surfactant molecule is modeled as a molecule with n_1 statistical units carrying a single hydrophobe connected to the hydrophilic head. General theory developed in the above section to find the sol/gel transition point of polydisperse mixtures of associating polymers can directly be applied to this simple but important case.

When polymer concentration is low and the number of hydrophobes is not enough to form junctions, addition of surfactants combines the unassociated hydrophobes into a micelle until its aggregation number exceeds s_{min}. In this situation, the surfactant works as a cross-linking agency. On the contrary, when the polymer concentration is large and many junctions are already formed, some of the polymer hydrophobes in the junctions are replaced by surfactant hydrophobes, and lead to the dissociation of network junctions. Fig. 3(a) shows how junctions are formed and destroyed by added surfactants in the special case where the multiplicity is fixed at $s_{min} = s_{max} = 5$ for telechelic polymers ($f = 2$ at chain ends). From these considerations, we expect that there is no surfactant-mediated process if no minimum multiplicity exists, $i.e.$, if $s_{min} = 2$. In such a special case, hydrophobes form stable junctions no matter how small their aggregation number may be. The addition of surfactants therefore simply destroys the already existing junctions.

We show in Fig. 3(b) the result of theoretical calculation[10] of the gelation concentration for the telechelic ($f = 2$) polymers as a function of the concentration of the added surfactant. Both polymer and surfactant concentrations are expressed in terms of the reduced concentration, the number of hydrophobes (per lattice cell) times association constant. To see the effect of the minimum multiplicity, s_{min} is varied from curve to curve, while the maximum multiplicity is fixed at $s_{max} = 8$. It is clear that the sol/gel concentration c_f^* monotonically increases with the surfactant concentration for $s_{min} = 2$ (no lower bound), $i.e.$, gelation is blocked by the surfactant. But if there is a forbidden range between $k = 1$ (unassociated) and $k = s_{min}$, a minimum in c_f^* appears. At this surfactant concentration gelation is most promoted as can be seen for $s_{min} \geq 3$. The surfactant concentration at which c_f^* becomes minimum (referred to as *surfactant-mediated gelation* point SMG) increases as the gap becomes larger.

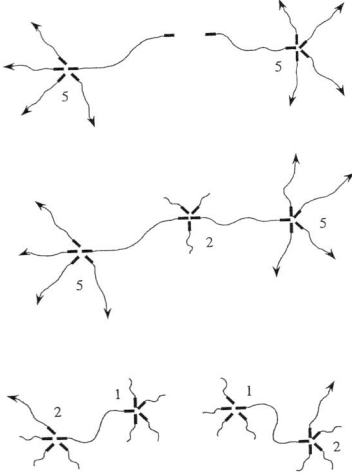

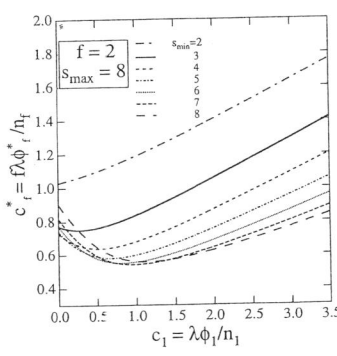

Figure 3: (a) Formation of a junction with the help of surfactant molecules (surfactant-mediated association) and destruction of a junction by excess surfactant molecules. The allowed multiplicity is fixed at $s = 5$. Figures near the junctions indicate their branching numbers. Average branching number is $(5+5)/2=5$ (top figure), $(5+5+2)/3=4$ (middle figure), $(2+2+1+1)/4=1.5$ (bottom figure). It monotonically decreases with surfactant concentration. (b) Polymer concentration at sol/gel transition as a function of the reduced concentration of added surfactant. Minimum multiplicity s_{min} is varied from curve to curve under a fixed maximum multiplicity $s_{max} = 8$. There appears a minimum at a certain surfactant concentration for $s_{min} \geq 3$. Gelation is promoted by the surfactant molecules, and referred to as *surfactant-mediated gelation*.

6. The LOOP/BRIDGE TRANSITION

To stress the importance of the junction multiplicity, we next consider the effect of loop formation on the gelation of associating polymers. Because of the high probability of encounter between the neighboring groups along the polymer chain, association commences to form isolated flower-like micelles at low concentration. With increase in the concentration, the number of chains bridging the flower micelles increases in order to release conformational entropy, eventually leading to the network formation. To see how each chain changes its conformation, let us study the simplest case of telechelic polymers. For the telechelic associating polymers, there are six categories of the chains in a solution: isolated chains, isolated loops, loops forming (pure) flower micelles, loops adsobed into the network junctions, bridge chains, dangling chains. The probability for a single chain to form a loop (petal) depends on its molecular weight and the binding free energy. It is given by

$$\theta = Be^{-\beta\Delta f_0}/n^{3\nu+\gamma-1}, \tag{13}$$

where n is the number of the statistical unit on the chain, $3\nu + \gamma - 1 = 1.96$ (ν Flory's exponent of the chain dimension, γ the critical exponent of the total number of self-avoiding random walks), B a numerical constant. We can regard the petal as a psudo molecule carrying a single effective associative groups made up of two original groups. Hence, the solution is seen as a mixture of telechelic polymers and psudomolecules carrying a single composite functional group. This system, therefore, can be mathematically mapped onto the polymer/surfactant mixture problem. The relevant difference is that the population of the loops relative to the open chains is thermodynamically controlled depending on the parameter θ.

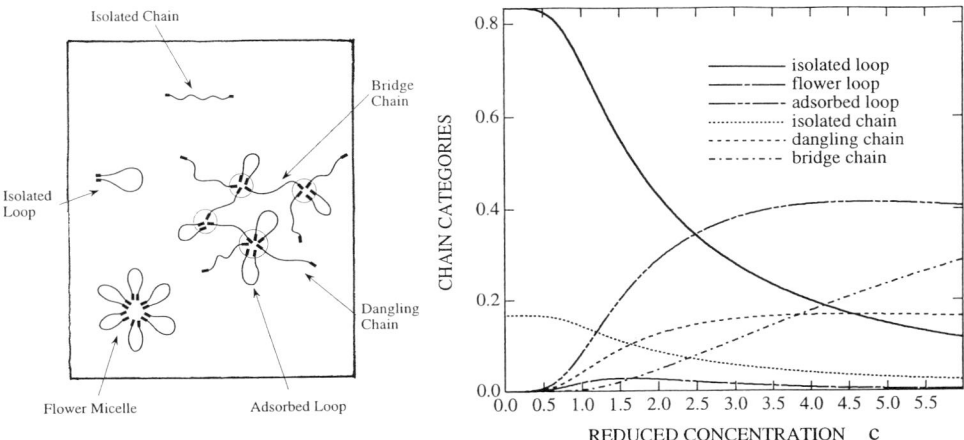

Figure 4: Relative population of loops and chains is plotted against the total polymer concentration. The loop parameter is fixed at $\theta = 5.0$ for $s_{min} = 5, s_{max} = 8$. Isolated loops and isolated chains start at the ratio of 5 to 1. The flower loops show a peak below the gel concentration ($c* = 2.2$ in this case).

In Fig. 4 we show the typical example of the numerical calculation[11] on how the relative population of the six categories listed above changes against the (scaled) polymer concentration $c \equiv 2\lambda(T)\phi/n$ for $\theta = 5$. The minimum and maximum of the multiplicity are 5 and 8. The isolated loop and isolated chain start with the ratio 5 to 1. In the dilute regime, isolated loops are dominant. As the concentration increases, the isolated loops rapidly decreases, while the absorbed loops increase. This is because of the high probability of mixed micelle formation. The pure flowers gradually appear and their number reaches a maximum at a certain concentration. The sol/gel transition concentration is given by 2.2 in this example. If we take the the position of initial rise of pure flowers as the measure of their cmc, then it lies below the sol/gel concentration in this case of the parameter θ. The flowers appear before the solution gels. For smaller values of θ, however, the cmc of the pure flowers can lie above the gel point because of the low probability of forming loops. In such a case the solution first gels and then flowers are formed in the network. All populations change continuously across the gel point. No singularity accompanies the physical quantities in this flower/bridge transition.

7. CONCLUSIONS

We have sdudied thermoreversible gelation of associating polymers with special emphasis on the multiple cross-link junctions. It is shown that the existence of the lower and upper limit in the multiplicity may lead to a variety of new phenomena such as the shift of the gel point, surfactant-mediated gelation, simultaneous occurence of gelation and cmc, loop/bridge transition. The structural complexity and reorganizability of the network junctions thus produce unique properties of thermoreversible gels.

References

[1] R.S. Russo, "Reversible Polymeric Gels and Related Systems ", American Chemical Society, ACS Symposium Series, **350** (1987).

[2] J.M. Guenet, "Thermoreversible Gelation of Polymers and Biopolymers", Academic Press, Harcourt Brace Jovanovich Publishers (1992).

[3] K. te Nijenhuis, Ad. Polym. Sci., **130**, (1997) 1.

[4] F.Tanaka and W.H.Stockmayer, Macromolecules **27** (1994) 3943.

[5] F.Tanaka and K.Nishinari, Macromolecules **29** (1996) 3625.

[6] F.Tanaka and S.F.Edwards, Macromolecules **25** (1992) 1516; F.Tanaka and S.F.Edwards, J. Non-Newtonian Fluid Mech. **43** (1992) 247; 273; 289.

[7] J. Scanlan, J. Polym. Sci., **43**, (1960) 501; L.C. Case, J. Polym. Sci., **45**, (1960) 397.

[8] F.Tanaka and M.Ishida, Macromolecules **29** (1996) 7571.

[9] T. Annable, R. Buscall, R. Ettelaie and D. Whittlestone, J. Rheol., **37** (1993) 695 .

[10] F. Tanaka, Macromolecules, **31**, (1997) 384.

[11] F. Tanaka and T. Koga, to be published in Computational and Theoretical Polymer Science (1999).

HYDROCOLLOIDS – PART 1
Edited by K. Nishinari
2000 Elsevier Science B.V.

35

Effect of electric charges on the volume phase transition of thermosensitive gels

H. Maeda, H. Kawasaki and S. Sasaki

Department of Chemistry, Faculty of Science, Kyushu University, Hakozaki, Higashi-ku, Fukuoka, 812-8581, Japan

Effects of the introduced electric charges on the swelling behavior of a thermosensitive gel, N-isopropylacrylamide (NIPA), were examined with the copolymer gels with sodium acrylate(NaA) or sodium styrene sulfonate (NaSS). The introduced charge altered the volume change behavior of NIPA gels from the discontinuous type (the volume phase transition) to the continuous one by a very small amount, less than 0.01 mole fraction. This charge effect was suppressed by the addition of a salt, NaCl, and at the NaCl concentration of about 0.3 M or higher, the transition behavior was resumed. These effects of the charge and the salt were both understood in a unified manner in terms of the Donnan osmotic pressure of the shrunken state of the gels. In parallel with the change of the swelling behavior, the endothermic peak observed at the transition temperature of NIPA gels in the DSC measurements became smaller and broader as the mole fraction of the charged monomer increased. The solution nature of the shrunken state of NIPA gels was suggested from the effects of sugars and/or salts on the volume phase transition. This solution nature was suggested to be kept in the shrunken state of ionized NIPA gels. This is the prerequisite to the above interpretation of the charge effect in terms of the Donnan osmotic pressure of the shrunken state. Temperature-induced volume changes of the copolymer gels of NIPA with sodium acrylate were examined at various pH. The volume change of the gel was discontinuous against temperature at pH below 6.3, although it was continuous at pH above 7.5. Potentiometric titrations of the linear copolymer of NIPA and acrylic acid (PNIPA-AA) revealed that the dissociation constant of carboxyl groups in solutions decreased with raising temperature above 34 ^{0}C. The similar decrease in the carboxyl ionization with raising temperature was also observed for NIPA-AA gel. The resumed discontinuous volume change at neutral or acidic pH was correlated with the two effects arising from the protonation of the carboxylate groups of NIPA-AA gel : the reduced Donnan osmotic pressure and the presence of a significant fraction of unionized carboxyl groups in the gel at the collapsed state. The nature of the shrunken state under these pH values was suggested to differ considerably from that of NIPA or ionized NIPA gels. It is not solution-like but a space of low dielectric constant similar to the interior of globular proteins.

1. INTRODUCTION

Polymer gels change their volume according to various perturbations in external conditions; temperature, solvent composition or pH. Some gels change their volume discontinuously between a swollen and a shrunken state. The behavior is known as the volume phase transition of gels [1, 2]. N-isopropylacrylamide (NIPA) gel in water is known to show a temperature - induced volume phase transition at about 34 ^{0}C [3]. The driving force of the volume phase transition of NIPA gel has been suggested to be due to the hydrophobic interaction among isopropyl groups in the side chains [3]. Hirotsu et al. have observed a temperature-induced volume phase transition of copolymer gels of N-isopropylacrylamide and sodium acrylate in water as a function of the copolymer composition [4, 5]. They observed both the transition temperature and the volume change at the transition increased, as the amount of charges increased. The increase in the discontinuity with ionization was explained by the increased Donnan osmotic pressure due to counter ions on the basis of the Flory-Huggins theory combined with the ideal Donnan osmotic pressure [4, 5]. Prausnitz et al., however, have reported experimentally that the volume change behavior of the same copolymer gels is continuous as the ionization increased by changing solution pH [6]. The cause for the inconsistent results on the volume change of ionized NIPA gel should be clarified. We report here the temperature-induced volume change behavior of the binary copolymer gels of N-isopropylacrylamide (NIPA) with sodium acrylate (NIPA-NaA) and with sodium styrene sulfonate (NIPA-NaSS) and also of the ternary copolymer gels of NIPA with sodium acrylate/acrylic acid (NIPA-NaA-AA).

2. EXPERIMENTAL

NIPA-NaA gels were prepared by radical copolymerization in aqueous solutions of N-isopropylacrylamide (700-693 mM), sodium acrylate (0-7 mM) and N, N'-methylenebis(acrylamide) (3.5 mM). Gels with a fed mole fraction of the acrylate x is denoted as NIPA-AA(x). The polymerization was initiated by ammonium persulfate, accelerated by N', N', N', N' - tetramethylethylendiamine(TEMED) and carried out at 5°C for 24hrs. Copolymer gels with sodium styrene sulfonate were prepared similarly. Gels synthesized in a capillary (0.30 mm diameter) were cut into rods (20mm length), rinsed thoroughly with distilled water and then dried . The dried gel fixed in a sample holder (silicone rubber) was immersed in the solution. Gel volume V was determined from the diameter d measured under an optical microscope. Swelling ratio was defined as $V/V_0=(d/d_0)^3$ where d_0 was a diameter of the capillary in which the gel was synthesized. The temperature in the gel swelling experiment was controlled within ± 0.1 °C.

DSC experiments were carried out with a DSC 120 calorimeter (Seiko Inc.) from 5 to 60 °C at a heating rate of 1.0 °C /min. In the present study, gels of small sizes ($d_0 = 0.14$ mm) were used to avoid the complexity arising from the slow volume change of large gels. The effects of the gel size have been examined [7].

3. THE VOLUME PHASE TRANSITION OF N-ISOPROPYL-ACRYLAMIDE (NIPA) GELS

It is pertinent to summarize some characteristic aspects of the volume phase transition of NIPA gel before examining the effects of the electric charge on it. NIPA is soluble in water at temperatures lower than about 34°C, probably due to the hydration of amide groups. When warmed up to a temperature above about 35 °C, the solutions undergo the phase separation thus exhibiting a lower critical solution temperature(LCST) [8]. The heat of phase separation was reported to be about 6.9 - 16 kJ per mole residue [8]. When the polymer chains are crosslinked, the resulted gels undergo dramatic volume changes reaching as large as 100 fold at LCST. This has been known as the volume phase transition. An endothermic peak has been observed at the transition temperature T_C (Fig. 1). This is associated with the destruction of so-called 'icebergs' or the hydrophobic hydration when the swollen gels are transformed into the collapsed phase [9-11]. When the swelling behavior is analyzed in terms of the Flory-Huggins theory, the χ parameter is obtained as a function of temperature. As shown in Fig. 2, the χ parameter increases gradually with the temperature T in the swollen state (at low temperatures), jumps at T_C and then approximately levels off at temperatures higher than T_C. It should be stated that we could not detemine

exactry the χ values because of large relative errors about the volume fraction estimates in the high temperature region. The jump of the χ parameter at T_C is a direct consequence of the assumed strong dependence on the network concentration [5]. Recently, we have developed a simple theory on the basis of the coupling between the hydration of the network and the excluded volume increase [12]. According to this theory, the two phases are characterized by different χ values and hence its jump at T_C is a natural consequence.

It is to be noted that the parameter χ takes almost the maximum value in the temperature range higher than T_C (Fig. 2). This strongly suggests that the transition disappears with any perturbation that favors the swollen phase and this disappearance cannot be recovered by elevating the temperature. If the transition resumes at a higher temperature in the presence of the perturbation, then it is highly likely that there should be involved some attractive contribution in favor of the shrunken phase.

The volume phase transition of NIPA gels is affected by the addition of inorganic salts [13, 14] or sugars [15]. These additives in most cases lower the T_C from that in pure water $T_C{}^0(34°C)$. It has been also observed that the transition is induced at a given temperature T_1 by a concentration change of the additive. A striking aspect of the effect of the additives on the volume phase transition of NIPA gels is that the transition is controlled by the chemical potential of water. We have found that the chemical potentials of water at the transition points are always characterized with the same value [16].

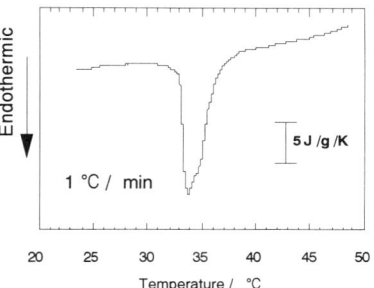

Figure 1. DSC thermogram of NIPA gel in water

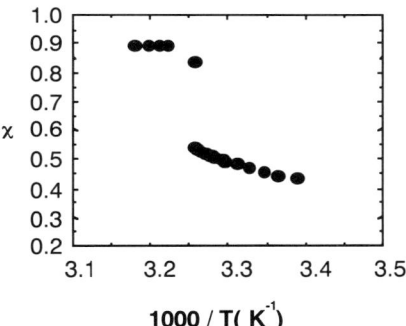

Figure 2. The χ parameter of NIPA gel in water as a function of temperature.

$$\mu_w{}^0(Tc^0) = \mu_w[Tc(C), C] = \mu_w(T_1, C_1) \qquad (1)$$

Here, Tc(C) denotes the transition temperature at a given additive concentration C. Also, C_1 denotes the transition concentration of additive under a given temperature T_1. This simple relation is understood if the effects of the additives are identical to both the swollen and the shrunken phases suggesting the solution nature of the shrunken state. It also suggests that the network does not play a significant role as far as the effect of the additives is concerned. It is natural under this situation that the transition is controlled by the chemical potential of solvent water.

4. SWELLING BEHAVIOR OF BINARY COPOLYMER GELS

4. 1. Swelling behavior of binary copolymer gels with sodium acrylate(NaA) at pH 9.5.

We have reported the swelling behavior of the gels of low mole fractions x of acrylate(NaA) (0.004 < x< 0.100) at pH 9.5 either in salt free or NaCl aqueous solutions [17]. The pH of the gel was kept to be 9.5± 0.2 by equilibrating the gel with the 5mM sodium polyacrylate (NaPAA) solution. The introduction of electric charges into the network switched the temperature-induced volume change of the gel from the discontinuous type to the continuous type in salt free aqueous solutions. (Fig. 3). The switching of the regime took place in a very small range of x around a very small value of 0.002. The continuous volume change

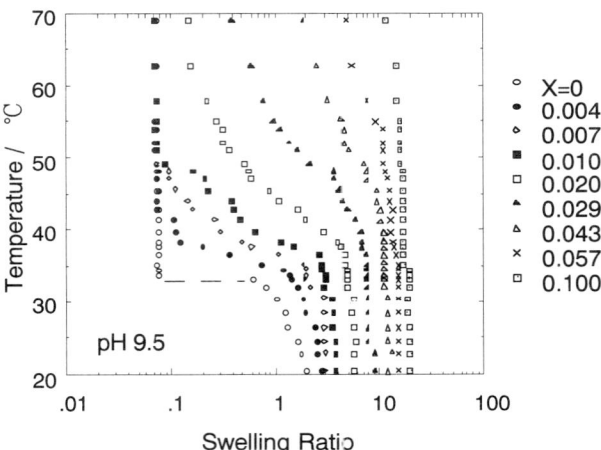

Figure 3.   Temperature dependence of the swelling ratios for NIPA-NaA gels with various compositions at pH9.5.
The mole fraction of acrylate is denoted as x.

of the gels with x greater than 0.002 in salt free aqueous solution was transformed into a discontinuous one in NaCl aqueous solutions of 0.1 and 0.3 M [17].

4. 2. Swelling behavior of binary copolymer gels of NIPA - sodium styrene sulfonate.

Similar results as found on the copolymer gel NIPA-NaA were also obtained on the copolymer gels NIPA - sodium styrene sulfonate (NIPA - NaSS).   The regime change by the addition of NaCl is shown in Fig. 4.

4. 3. The Donnan osmotic pressure.

Effects of the two factors, the charged monomer fraction x and the ionic strength, were explained in a unified manner in terms of the Donnan osmotic pressure. Donnan pressure Π/RT of NIPA-NaA gels is given as eq. (2), since the charge contents were very low and the gels were immersed in 5mM sodium polyacrylate solution.

$$\Pi/RT = Ce + 2Cs - (\varphi_p \, Ce' + 2Cs') \tag{2}$$

In eq. (2), the additivity of the   osmotic pressure is assumed [18-20]. Ce and Ce'(= 5 mM) refer to the carboxylate concentrations of the gel and the external polyacrylate buffer solution, respectively.   Cs and Cs' denote NaCl concentrations inside and outside the gel, respectively.

φ_p (= 0.14) refers to the osmotic coefficient of salt - free polyacrylate solution [21]. The only unknown quantity Cs, the salt concentration inside the gel, was given by the Donnan equilibrium according to eq. (3) to a good approximation in the present study.

$$(Ce + Cs)Cs = (\gamma'_{\pm})^2(Ce' + Cs')Cs' \qquad (3)$$

In eq. (3), $\gamma'_{\pm}$ was evaluated according to Manning's limiting law [20]. When salt is not present in the external solution(Cs'=0), eq. (3) reduces to $\Pi/RT = Ce - \varphi_p Ce'$. In the case of NIPA - NaSS gels, $\Pi/RT = Ce$, since no buffer was used and only the data in salt-free solutions were analyzed.   We introduce the maximum value Q of the derivative of the swelling curve as a measure of the sharpness of the volume change [17]. In Fig. 5, the Q values are plotted against the calculated Donnan osmotic pressure in the shrunken state. It is clearly seen that the continuous (B) and the discontinuous (A) regimes of NIPA-NaA gel volume change were well characterized, respectively, by high and low osmotic pressures.   The above calculation of the Donnan osmotic

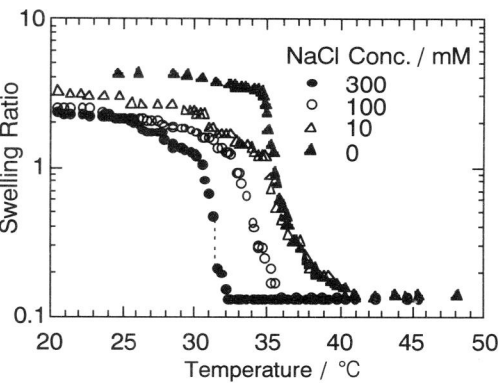

Figure 4. Swelling behavior of NIPA - NaSS gels in the presence of NaCl

pressure implicitly assumes the solution nature of the shrunken state.   The effects of sugars and salts on the transition discussed in relation to eq. (1) support this assumption.   On the other hand, we have found that the sugars or salts are excluded, more or less, from the shrunken state gels (unpublished result).   It is likely that the nature of the shrunken state at the transition point differs from that at higher temperatures (completely collapsed state).

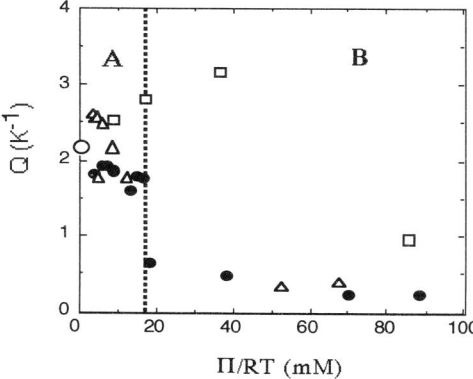

Figure 5.   Q values as a function of Π/RT for NIPA-NaA gels in salt free polyelectrolyte buffer or NaCl solutions and NIPA-NaSS gels in water. Closed circles and open triangles represent the results on NIPA-NaA gels in salt free polyelec- trolyte buffer and NaCl aqueous solutions, respectively. Open squears represent those on NIPA-NaSS gels in water. A broken line is tentatively drawn to indicate a border between the two regimes of the continuous (B) and the discontinuous (A) volume change. An open circle refers to NIPA gels.

4. 4. Calorimetric behavior of NIPA-NaA gel and NIPA-NaSS gel.

The effect of the introduced charge on the calorimetric behavior of N-isopropylacrylamide (NIPA) gels was investigated by the differential scanning calorimetry (DSC) on the copolymer gels of NIPA and sodium acrylate (NIPA-NaA) in water. Since the polyelectrolyte buffer was not used in the DSC measurements, the pH was about 6. As the NaA content increased, the enthalpy change per NIPA monomer (ΔH), which was related to the amount of the dehydration of NIPA chains, decreased and the sharp endothermic peak of NIPA gel changed into a broad one (Fig. 6). Similar results have been observed on NIPA - NaSS gels. According to the DSC thermograms and the swelling curves, it was found that a coupling between the shrinking of the network chains and the dehydration of NIPA chains. The derivative curve of the swelling curve was almost identical with the corresponding DSC curve. On the basis of

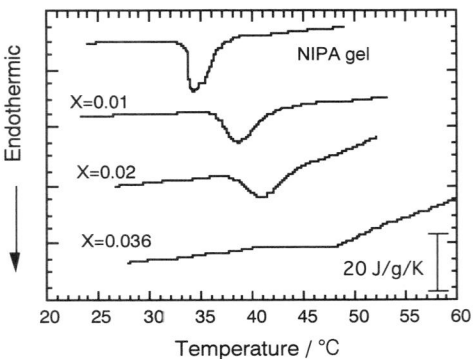

Figure 6. DSC thermogram of NIPA-NaA gels in water (pH 5).
The mole fraction of Na A residue is denoted as x.

this similarity, the effect of the introduced charge on the DSC thermogram can be explained in terms of the Donnan osmotic pressure of counter ions : the introduced charged groups produce the Donnan osmotic pressure, leading to the suppression of the dehydration of NIPA chains and hence the shrinkage of the gel-chains. Consequently, the NIPA-NaA gels in water exhibit the reduced ΔH value and a broad peak. For the NIPA-NaSS gel in NaCl solution, the Donnan pressure decreases with the increase of the NaCl concentration. We have found in NaCl solutions the increased ΔH value and the sharp peak again. Relevant data are summarized in Table 1.

Table 1. Transition Temperature (T_t) and Transition Heat for NIPA-AA gels at pH 6			
Sample	T_t from DSC (°C)	T_t from swelling curve (°C)	transition heat (kJ / mol)
NIPA gel	33.8	33.8	4.7
NIPA-AA(X=0.01) gel	35.8	38.0 (37.0)[2]	4.6
NIPA-AA(X=0.02) gel	38.1	40.0 (39.0)[2]	4.1
NIPA-AA(X=0.036) gel	41.5	44.0(41.2)[2]	(1)

(1) The transition heat for NIPA-AA(X=0.036) gel was not estimated because the base line was not clear due to the slow volume shrinkage.
(2) The bending temperature of the swelling curve is shown with a bracket.

Until now ther is no convincing mechanism to account for the decrease of ΔH with the charge amount . However, two different mechanisms have been proposed. The destruction of thehydrophobic hydration due to the electric field and/or ionic hydration caused by the introduced charges is one of them [22]. Judged from the very small amounts of the introduced charges, however, this mechanism meets some difficulty in the quantitative aspect. For example, suppose $\Delta H = 0$ at $x = 0.1$. Then, one charge destroyes the hydrohobic hydration layer of 50 - 60 water molecules. Also, this mechanism meets difficulty to account for the observed salt effect. According to another model, the reduction of ΔH is attributed to the reduced amount of the hydrophobic hydration at elevated transition temperatures caused by the presence of charges [23]. This is proposed based on their finding that the heat of phase separation of the solutions of nonpolar polymers decreases (in magnitude) linearly with LCST [23]. On the basis of the observed close correspondence between the swelling and calorimetric behaviors concerning the effects of charges and salts, we propose another mechanism. In the case of the transition like volume change, theformation/destruction of the icebergs takes place cooperatively and hence the associated heat is effectively detected by DSC, resulting in a large and sharp endothermic peak. When the volume change takes place continuously by the introduction of charges, on the other hand, the formation/destruction of the icebergs does not take place cooperatively any more and the associated heat escapes detection of DSC more or less, resulting in a small and broad peak.

4. 5.   A brief discussion on the observed charge effects.

The crossover of the swelling regime as a result of the introduction of charges will be interpreted as follows in a simple and qualitative way. In this simple approach, the swelling behavior of NIPA gels will be described by eq. (4) (valid for crosslinking procedure in pure melt), in terms of the network volume fraction φ_2, the number of chains v in the gel of initial volume V_0, and the molar volume of water V_1. The equilibrium swelling is given by the condition $\Delta\mu_w = 0$. Here, $\Delta\mu_w$ denotes the chemical potential of water in the gel measured from that of outside soution in equilibrium with the gel.

$$\Delta\mu_w / RT = \ln(1 - \varphi_2) + \varphi_2 + \chi\varphi_2^2 + [V_1(v/N_A)/V_0](\varphi_2^{1/3} - \varphi_2/2) \qquad (4)$$

Since NIPA gels exhibit discontinuous volume changes, $\Delta\mu_w$ should have two maxima and a minimum when plotted against φ_2 (Fig. 7 curve 1). In Fig. 7, the χ parameter is allowed to vary with the concentration as eq. (5).

$$\chi = \chi_1 + \chi_2 \varphi_2. \qquad (5)$$

When electric charges are introduced to NIPA gels, a negative contribution $(\Delta\mu_w)_{osmotic}$ should be added to $\Delta\mu_w$ as eq. (6).

$$(\Delta\mu_w)_{osmotic} = [-ig(v/N_A)(V_1/V)] \qquad (6)$$

Here, g and N_A represent, the osmotidc coefficient and the Avogadro number.

Due to the Donnan osmotic contribution, the maximum value of $\Delta\mu_w$ of curve 1 becomes negative and the discontinuous transition disappears eventually (Fig. 7, curve 3). An important prerequisite of this mechanism resides in the assumption that χ parameter cannot increase significantly by elevating the temperature above T_c^0. This implicit assumption is supported by the observed temperature dependence of χ on the temperature (Fig. 2). In Fig. 3, the gel with $x = 0.020$ exhibits no transition behavior but its volume decreases considerably with raising the temperature. When temperature T is incrreased, χ_1 is expected to increase for this gel ($x=0.020$) but χ_2 is independent of T. Consequently, the equilibrium volume fraction, represented as φ_A.in Fig. 7, increases with T but the maximum point C of curve 3 does not shift upward to the extent as reaching the horizontal line of $\Delta\mu_w = 0$.

It is to be noted that the effects of the introduction of charged residues include those other than the Donnan osmotic contribution. Changes in the hydration property are likely to make χ values different. The collapsed state of the ionic NIPA copolymer gel will differ from that of pure nonionic NIPA.

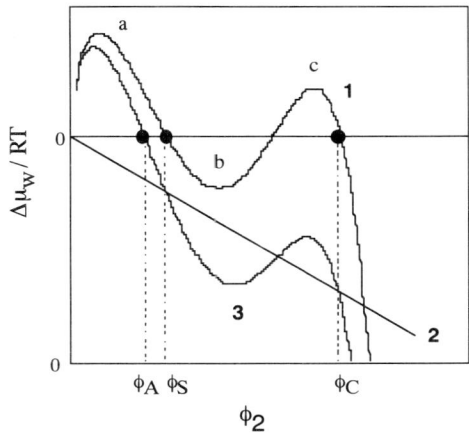

Figure 7. A model calculation indicating the disappearing of the volume phase transition due to the Donnan osmotic pressure.
$\chi = 0.334 + 0.855\,\phi_2$. $V_1\nu / N_A V_0 = 0.0058$.
Curve 1 : $i = 0$.   Curve 3 : $i = 0.005$, $g = 1$

5. Effects of pH on the Swelling Behavior of Copolymer Gels (NIPA - NaA - AA) in Salt - free Sodium Polyacrylate Solutions

As shown in section 4. 1, the temperature-induced volume change of NIPA - NaA gel was continuous at pH above pH 7.5. However, it was again discontinuous at pH below 6.3 [24]. The results obtained at pH 5.6 are similar to those found in the literature [4] (Fig.8). To elucidate this remarkable pH effect, hydrogen ion titrations were carried out mostly on the linear copolymer solutions. The results have revealed that the dissociation constant of carboxyl groups pK_0 in NIPA-AA gel decreased with raising temperature above 34°C. No such a temperature dependence of pK_0 was observed on the copolymers of NIPA with acrylamide [24]. A significant fraction of the carboxyl group was unionized in the gel at high temperatures i.e., at the deswollen state (Fig. 9). The discontinuous volume change was correlated with the decrease in the degree of ionization α of carboxyl groups of NIPA-AA gel, since the Donnan osmotic pressure was supposed to decrease as a result of the decrease of ionization. Thus, it is suggested that the decrease of α with raising temperature plays an essential role in the discontinuous volume change of NIPA-NaA/AA gel. It is to be noted that the reduction of the Donnan osmotic pressure is not so large as to explain the recovery of the volume phase transition, it is highly probable that another contribution in favor of the shrunken state will be introduced accompanying the protonation of carboxylate groups. Hydrogen bonds between $-CO_2^-$ and CO_2H, between $-CO_2H$ and amide group, and between two carboxyl groups .

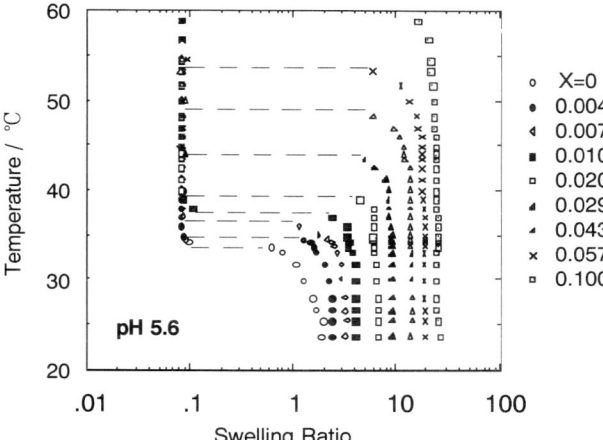

Figure 8. Temperature dependence of the swelling ratios for NIPA-AA gels with various compositions at pH 5.6. The mole fraction of the acrylate X is shown in the figure.

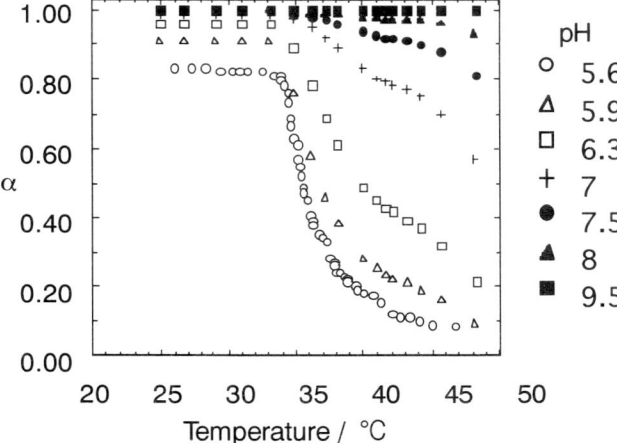

Figure 9. The degree of ionization α of the carboxyl group of the NIPA-AA(x=0.007) gels at various pH values is plotted as a function of temperature.

44

REFERENCES
1. K. Dusek and D. J. Patterson, Polym. Sci. Part A-2, 6 (1963) 1209.
2. T. Tanaka, Phys. Rev. Lett., 40 (1978) 820.
3. Y. Hirokawa and T. Tanaka, J. Chem. Phys., 81 (1984) 6379.
4. S. Hirotsu, Y. Hirokawa and T. Tanaka, J. Chem. Phys., 87 (1987) 1392.
5. S.Hirotsu, Phase Transitions, 47 (1994) 183.
6. S. Beltran, J. P. Baker, H. H. Hooper, H. W. Blanch and J. M. Prausnitz, Macromolecules, 24 (1991) 549.
7. H. Kawasaki, S. Sasaki, S and H. Maeda, Langumir, 14(1998)773.
8. M. Heskins and J. E. Guillet, J. Macromol. Sci. Chem., A2 (1968) 1441.
9. K. Otake, H. Inomata, M. Konno and S. Saito, Macromolecules, 23(1990) 283.
10. H. Inomata, S. Goto and S. Saito, Macromolecules, 23 (1990) 4887.
11. G.S. Howard and A.T. David, J. Phys. Chem., 94 (1990) 4352.
12. S. Sasaki and H. Maeda, Phys. Rev. E, 54 (1996) 2761.
13. H. Inomata, S. Goto and K. Otake and S. Saito, Langumir, 8 (1992) 687.
14. A. Suzuki, Adv. Polym. Sci., 110 (1993) 199.
15. H. Kawasaki, S. Sasaki, S, H. Maeda, S. Mihara, M. Tokita and T. Komai, J. Phys. Chem., 40 (1996) 16282.
16. S. Sasaki, H. Kawasaki and H. Maeda, Macromolecules, 30 (1997)1847.
17. H. Kawasaki, S. Sasaki and H. Maeda, J. Phys. Chem. B, 101(1997) 4184.
18. Z. Alexandrowicz and A. Katchalsky, J. Polym. Sci., A1 (1963) 3231.
19. F. Oosawa, "Polyelectrolytes", 1971, Marcel Dekker, NewYork.
20. G.S. Manning, J. Chem. Phys., 51(1969) 924.
21. Z. Alexandrowicz, J. Polym. Sci., 56(1962)115.
22. M. Shibayama, M. Morimoto and S. Nomura, Macromclecules, 27 (1994) 5060 ; M. Shibayama, S. Mizutani and S. Nomura, Macromolecules, 29 (1996) 2019.
23. H. Feil, Y. H. Bae, J. Feijen, S. W. Kim, Macromolecules, 26 (1993) 2496.
24. H. Kawasaki, S. Sasaki and H. Maeda, J. Phys. Chem. B, 101(1997) 5089.

HYDROCOLLOIDS – PART 1
Edited by K. Nishinari
2000 Elsevier Science B.V.

Structure and dynamics of ovalbumin gels

N. Nemoto

Department of Molecular and Material Sciences, IGSES, Kyushu University, Hakozaki, Fukuoka 812-8581, JAPAN

This paper briefly discusses solvent effects on global structure and dynamics of ovalbumin(OVA) gels induced by high-temperature heat treatment in glycerin, ethylene glycol and their mixtures with water used as solvent using data obtained from DSC, SEM, and rheological measurements. We also report preliminary results of small-angle neutron scattering measurements on the 5wt% aqueous solution and the 15wt% OVA gel in deutreated water, which gives a new information on kinetics of gel formation for this OVA gel.

1. INTRODUCTION

Ovalbumin(OVA) is a globular protein, a major component in egg white, with molecular weight $M = 46,000$ and the diameter $d = 5.6$nm in the native state[1]. When native OVA is heated in water, thermal denaturation occurs above temperature $T = 75°C$ as is evidenced by an endothermic peak from DSC measurements[2], and hydrophobic parts of the denatured proteins exposed on the surface are so unstable in water as to form aggregates by hydrophobic interaction. Thermal denaturation at 80°C without added salt or at very low ionic strength produces linear aggregates whose M reaches to several millions for prolonged heating time, and the second heating of the solution in the presence of added salt gives rise to a transparent elastic gel at a protein concentration C of 50mg/cm^3. It is to be noted that gels, though often turbid, can be obtained directly from one-step heating of 50mg/cm^3 OVA solution at 80°C in brine of high ionic strength[3].

Transparent OVA gels can be also prepared by high-temperature heat treatment such as heating for 5min at 160°C and by subsequent rapid quenching of the sol to room temperature as described in a previous report[4]. Circular dichroism measurements on the secondary structure of an individual protein revealed that thermal denaturation at 160°C brought about a decrease in α-helix content and an increase in β-sheet structure content for a short interval of heating time and also that prolonged heating gave rise to conformation change of the molecule to a random-coil form, which lost gel-forming ability[2]. Furthermore thermal behaviors of OVA molecules were found to be susceptible to the kind of solvent used in preparation of samples. These informations are in good agreement with the postulate that hydrophobic interaction between denatured proteins, while keeping their spherical shapes, plays an essential role for their aggregation process in water as well as gel formation at sufficiently high C[5].

In this paper, we mainly describe a recent study on solvent effects on global structure and dynamics of 15wt% OVA gels induced by high-temperature heat treatment in glycerin, ethylene glycol and their mixtures with water used as solvent using viscoelasticity, DSC, and SEM methods[6]. We also report results of complementary rheological measurements as well as preliminary small-angle neutron scattering experiments on the 5wt% solution and the 15wt% OVA gels in deuterated water, which gave a new information on kinetics of gel formation for this OVA gel.

2. EXPERIMENTAL

Highly purified OVA samples were obtained with the method described in detail in Ref. 4, which also gives the preparation procedure of gels by high-temperature heat treatment. Distilled water, glycerin and ethylene glycol of reagent grade, and also their mixtures with water were used as solvent, and the sample code, OVA/X-Y where X = W, G, and EG denotes that the gels were prepared in water(W), glycerin(G), and ethylene glycol(EG), respectively and that Y represents water fraction by percent .

Dynamic viscoelastic measurements and shear creep measurements were performed with a stress-controlled rheometer CSL-100(Carri-MED, ITS Japan). DSC measurements were performed with a heat-flux type of apparatus(DSC-8240B, Rigaku) with a TAS-100 controller at a heating rate of 2K/min. Micrographs of dried gel specimens obtained after complicated processing were taken with a Hitachi S-9000 scanning electron microscope. Small-angle neutron scattering(SANS) experiments were conducted at the research reactor located at the Japan Atomic Energy Research Institute, Tokai, Japan.

3. RESULTS AND DISCUSSION

3. 1. Viscoelastic Behaviors

A pronounced aging effect was observed for the dynamic viscoelastic behaviors of the

OVA/G and OVA/EG gels at room temperature where the OVA/W gel was stable once prepared. For example, as Fig. 1 shows, G' of the 15wt% OVA/G gel at $\omega =$ 1.0rad/s is larger than that of the OVA/W gel with the same protein concentration, and increased monotonously as time elapsed and leveled off to a constant value about one week later. G' of the OVA/EG gel increased with aging time to the same extent as the modulus of the OVA/G gel, but it did not reach an equilibrium value even 1 month later. These aging effects may be related to stability of native OVA in three solvents revealed by DSC measurements on OVA solutions that thermal denaturation temperature T_d decreased from 75℃ in pure water to room temperature with increasing EG content, while slight decrease in T_d occurred for the G content > 0.9 in G/W mixtures. The latter behavior is consistent with the fact that the aging effect was not observed for OVA/G-Y gels with Y > 10 as shown in Fig.1.

There were big differences in angular frequency (ω) dependencies of the storage and the loss shear moduli, $G'(\omega)$ and $G''(\omega)$, among the three OVA gels prepared in water, glycerin, and ethylene glycol. The time-temperature superposition principle was applicable for data of the 15wt% OVA/W gel measured at $T = 5$, 25, and 45℃ using the

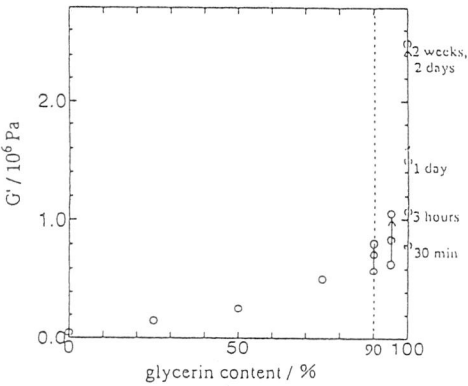

Fig. 1 The storage modulus G' of OVA gels prepared in mixtures of water and glycerin at ω = 1rad/s and the aging effect for the OVA/G gel.

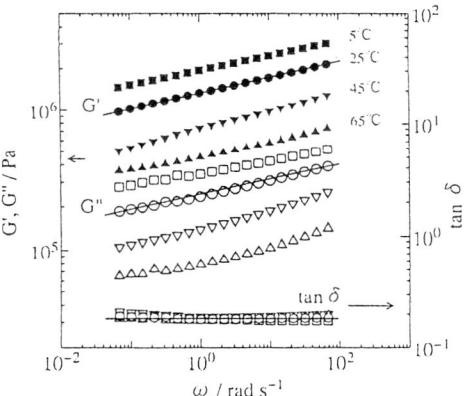

Fig 2 ω dependencies of $G'(\omega)$, $G''(\omega)$, and tan δ of the 15wt% OVA gel at $T = 5$, 25, 45, and 65℃. Straight lines are drawn for the data at $T =$ 25℃, for which Eqs 1 and 2 are applicable[6].

ratio of viscosity of water at measuring temperature and the reference temperature as the shift factor, and the G' became almost independent of ω at the low ω end in consistence with a low value of tan $\delta \sim 0.1$[4]. Thus, the 15wt% OVA/W gel may be regarded as an elastic gel in the temperature range measured. In contrast, superposition was found to be unsuccessful for the 15wt% OVA/G and OVA/EG gels. G' and G'' of the 15wt% OVA/G gel, on the other hand, followed the power law at 25℃ with the same value of the relaxation exponent n = 0.11 ± 0.01 as shown in Fig. 2 and given by Eq 1.

$$G'(\omega) \sim G''(\omega) \sim \omega^n \qquad (1)$$

Consequently, tan δ became independent of ω and was found to satisfy the equation:

$$\tan \delta = \tan(n \pi/2) \qquad (2)$$

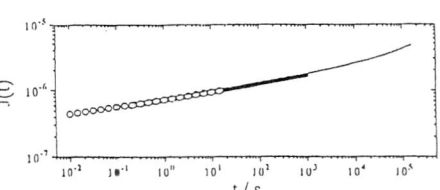

Fig. 3   Superposition of $J(t)$(○) from G' and G'' onto $J(t)$(●) from creep measurement.   The solid curve is $J(t)$ from the latest measurement.   In the relation $J(t) = J_g + \Phi(t) + t/\eta$, $J_g = 0$ and η is infinite for this gel

Applicability of Eqs 1 and 2 indicates that, on the basis of the Kramers-Kronig relationship, the power law may hold over a much wider ω range than the measured range of 0.07 ~ 68.2rad/s[7].   This conjecture was examined by creep measurements, and the creep compliance $J(t)$ curve obtained from the latest measurement, given by the solid line in Fig. 3, is nicely superposed on earlier data, which indicates that the power law holds over nearly 6 decades.   The upturn of $J(t)$ at the long time end may be due to non-linear response of the gel at large strain.

Winter and his associates[8] showed that, for the gelation process by chemical crosslinking, the gel point could be accurately determined from rheological measurements as the point at which G' and G'' simultaneously satisfied Eqs 1 and 2, and postulated that the gel at the gel point must take a self-similar structure, i.e., a fractal structure.   If their arguments would be assumed to apply for physical gels, the OVA/G gel prepared by thermal denaturation at high temperature might take a fractal structure characterized by the relaxation exponent $n = 0.11$.   It is to be noted that OVA/G gels with higher C also satisfied Eqs 1 and 2 simultaneously at higher temperature, for example, at 65℃ for the 45wt% OVA/G gel. However, G' and G'' of the 15wt% OVA/EG gel did not satisfy Eq 1 in such a manner that G' followed the power law with $n = 0.07$ and G'' was concave to the horizontal axis.

3. 2. Gel Morphology

Direct observation on the dried gels by SEM has revealed that (1)at low magnification the both OVA/G and OVA/W gels take structures similar to each other, as if plates with rough surfaces are piled up, and each plate is formed by adhesion of spherical particles with an average diameter of 30nm, which is much larger than the diameter d = 5.6nm of native OVA sphere; (2)at the largest magnification, as Fig. 4 shows, there is a distinct difference in gel morphology between the two gels in such a manner that some spheres melt away in glycerin so that network-like structure is formed microscopically, whereas most spheres adhere to one another in water while keeping their shapes.   In taking into account that this gel is a cold-set gel, we speculated that gelation took place by the two-step mechanisms; aggregation of thermally denatured OVA molecules at high temperature everywhere in the sample volume was followed by their adhesion to form gels mainly at room temperature. The aggregates are not considered as the stable state in glycerin irrespective of T in comparison with those in water as judged from the thermal behavior of native OVA in the two solvents, which explains melting of spheres to the more tightly connected network which is related to higher G' values in glycerin than G' in water and also to the aging effect at room temperature.

The above speculation is based on the SEM photographs of the dried gel samples obtained after rather complicated processing.   This motivated us to make SANS experiments on a fresh 15wt% OVA/W gel sample which was prepared using deuterated water instead of distilled water.   The SANS measurement was also made on the 5wt% D_2O solution of native OVA for comparison.   As Fig. 5 shows, the scattering profile of $I(q)$ of the solution, where q is the absolute value of the scattering vector $\mathbf{q}$, is characterized by $I(q)$ independent of q at low q and by monotonous decrease with increasing q at high q.   We found that a Guinier plot was applicable for the data, from which the diameter of native OVA at finite protein concentration of 5wt% was estimated as 5.6±0.2nm. This value is a little larger than the value of 5.04nm reported by Matsumoto and Chiba[9] but very close to 5.6nm obtained from dynamic light scattering measurements[1].   It is to be noted that the latter is the hydrodynamic diameter at infinite dilution. Thus we infer that native OVA is present as stable

——— 50nm

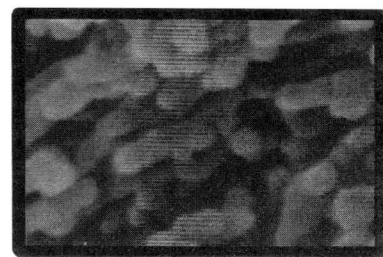

——— 50nm

Fig. 4 Electron micrographs of 15wt% OVA gels in water(top) and in glycerin (bottom).   Fig. 9 of Ref. 6 gives micrographs at low magnification.

monomer in 5wt% solution. This is not in agreement with small-angle X-ray scattering data of the 4.8% OVA solution reported by Matsumoto et al[10]. They observed a distinct formation of solution structure at this concentration and showed that $I(q)$ was proportional to q^{-4} for q above $8 \times 10^{-2} nm^{-1}$. Unfortunately, our SANS measurements did not cover such a high q range. On the other hand, a plot of scattering intensity $I(q)$ against q exhibits a distinct maximum at q_{max} for the gel. From the value of q_{max}, we obtain $2 \pi / q_{max} = 28.8nm$, being in harmony with formation of spherical particles of an average diameter of 30nm observed for the dried 15wt% OVA/W sample with SEM. The scattering profile in the low q region was fitted by the linearized Cahn-Hilliard theory[11,12] for spinodal decomposition(SD) at the early stage given by Eq 3.

$$I(q^*,t^*) = [I(0,0)/(1 + q^{*2})] \times$$
$$\exp[-2q^{*2}t^*(-1 + q^{*2})] \quad (3)$$

Here, $q^* = q \lambda / 2 \pi$ is a dimensionless wave vector involving the wavelength of concentration fluctuations λ in the gel and t^* the time in dimensionless units evolved since thermal denaturation of OVA took place at 160℃. The result is shown by a dotted curve in Fig. 6. It is likely that the scattering profile in the low q region may be explained with this treatment, but the behavior in the high q region obviously needs another interpretation.

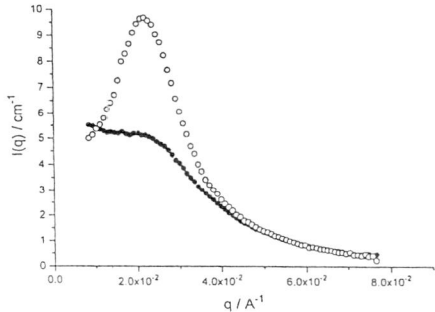

Fig. 5 SANS profiles of scattering intensity for the 15wt% OVA/W gel(unfilled circles) and for the 5wt% native OVA solution(filled ones).

This finding with previous SEM and DSC results may allow us to make the following conjecture. Denaturation of the protein instantaneously occurs at 160℃ and denatured ones are quite unstable in either water or glycerin still, so that the phase separation kinetics may take place based on the SD mechanism. Since concentration fluctuation at the early stage of the SD process is known to grow predominantly at a constant wave vector independent of heating time, we expect aggregation of OVA molecules to spherical particles with the diameter $d = 2 \pi / q_{max}$ everywhere in the sample volume, which form gels at sufficiently high protein concentration by further connection of the particles. At this point we should remark why we applied the linearized Cahn-Hilliard theory for interpretation of thermal denaturation process which is a irreversible process. The theory has been formulated in itself to explain a phase transition phenomenon which is thermorevesible. However, the theory essentially describes kinetics of a system suddenly brought to an unstable non-equilibrium state to a new equilibrium state. Therefore it does not seem unreasonable to assume that an aggregation process of denatured protein molecules follows a spinodal type of mechanism, though there remains a suspicion if the process were a linear one.

The scattering profile in the high q region is well described by the power law as shown in Fig. 6 and the slope is -3. It is reasonable to assume that scattering in this q region is mainly originated from the surface of the particles and q dependence of $I(q)$ is expressed using the surface fractal dimension d_s as

$$I(q) \sim q^{d_s - 6} \qquad (4)$$

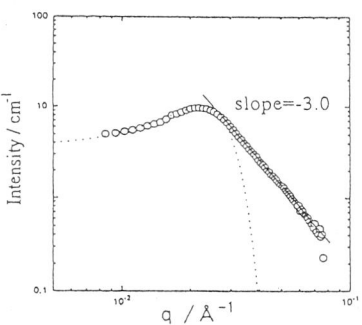

Fig. 6 Double logarithmic plot of $I(q)$ against q for the 15wt% OVA/ W gel. The dotted curve is a prediction of Cahn-Hilliard theory for the SD process.

When the surface is smooth, $d_s = 2$, which leads to $I(q) \sim q^{-4}$ known as the Porod law[13]. When the surface is rough, d_s lies between 2 and 3. The q^{-3} dependence of $I(q)$ observed for the OVA/W gel corresponds to $d_s = 3$, which strongly suggests that the surface of the particle composed of more than 100 OVA molecules is quite rough or irregular. SANS measurements are in progress to clarify how the scattering profile changes with heating time, which is expected to give a clear-cut answer if the spinodal decomposition theory is applicable for the aggregation process of globular proteins denatured at high temperature.

ACKNOWLEDGMENT

The author is grateful to Drs. A. Koike, T. Kanaya, A. Takada, Y. Urabe, and H. Maeda, and Mrs. K. Nakamura and M. Kiriyama for their helps in respective measurements.

Note added in proof. Quite recently we performed SANS experiments on 15wt% solutions of OVA thermally denatured for several heating time periods from 2 to 30min at 160 and 130°C. Intensity profiles clearly exhibited features characteristic for the SD mechanism for short heating periods, but a peak intensity observed at the same q position started to decrease as heating time elapsed, being probably due to that OVA molecules could not keep their

52

spherical shape and transformed to a random coil which lost gel-forming ability.   Details of
the work will be reported quite soon.

REFERENCES

1.   N. Nemoto, A. Koike, K. Osaki, T. Koseki, and E. Doi, Biopolymers, **33** (1993) 551.

3.   M. Kiriyama, A. Takada, Y. Urabe, and N. Nemoto, Netsu Sokutei, **24** (1997) 118 (in
     Japanese).

3.   A. Koike, A. Takada, and N. Nemoto, Polym. Gels Networks, **6** (1998) 257.

4.   A. Koike, N. Nemoto, and E. Doi, Polymer, **37** (1996) 587.

5.   T. Koseki, N. Kitabatake, and E. Doi, Food Hydrocoll., **3** (1989) 123.

6.   K. Nakamura, M. Kiriyama, A. Takada, H. Maeda, and N. Nemoto, Rheol Acta, **36**
     (1997) 252.

7.   J. D. Ferry, Viscoelastic Properties of Polymers, 3rd ed., Wiley, New York, 1980.

8.   H. H. Winter and F. Chambon, J. Rheol., **30** (1986) 367.

9.   T. Matsumoto and J. Chiba, J. Chem. Soc. Faraday Trans. **86**, (1990) 2877.

10.   T. Matsumoto, H. Inoue, and J. Chiba, J. Appl. Phys., **71** (1992) 1020.

11.   J. W. Cahn and J. E. Hilliard, J. Chem. Phys., **29** (1958) 258.

12.   J. W. Cahn, J. Chem. Phys., **42** (1965) 931.

13.   A. Guinier and G. Fournet, Small-Angle Scattering of X-Rays, translated by C. B
     Walker, John Wiley & Sons, New York, 1955.

HYDROCOLLOIDS – PART 1
Edited by K. Nishinari
2000 Elsevier Science B.V.

Thermoreversible gelation strongly coupled to polymer conformational transition

F. Tanaka and T. Koga

Department of Polymer Chemistry, Graduate School of Engineering
Kyoto University, Kyoto 606-8501, Japan

Effects of polymer conformational change on the thermoreversible gelation of natural and synthetic polymers are studied on the basis of the recent theory of associating polymers cross-linked with multiple junctions. The effective number of associative groups carried by each chain is not a fixed number but varies depending on the chain conformation. Transition from intra- to intermolecular bonding due to the denaturation of some proteins on heating leads to the high-temperature gelation, while coil-to-helix (or coil-to-globule) transition, followed by the aggregation of helices (globules), leads to the low-temperature gelation. Transition from intramolecular flower-like micelles to intermolecular bridge chains with increase in the concentration in hydrophobically modified associating polymers is another important example of the gelation coupled to the polymer conformational change.

1. INTRODUCTION

Most natural polymers undergo conformational transition preceding gelation, thereby activation of the particular functional groups on the polymer chain accompanied by a proper three dimensional conformation is a necessary prerequisite for the interchain cross-linking. For instance, water-soluble natural polymers such as agarose and κ-carrageenan first change their conformation from the random coil state to a partially helical state, and then the helical parts aggregate to form network junctions[1, 2]. Another important examples are globular proteins. Proteins such as ovalbumin, or human serum albumin, are believed to form gels after some of the intramolecular bonds in a native state are broken during denaturation, with their functional groups being exposed to the outer space, followed by the intermolecular recombination of the groups[1, 3]. Certain degree of unfolding is a neccessary condition for the gelation in these examples.

Gelation strongly coupled to polymer conformational change can also be found in synthetic polymers. An important example of current interest is coil-to-globule transition preceding gelation. Upon cooling in organic solvents, polymers are partially collapsed by strongly attractive force between the spatially neighboring monomers, and then the formed globular segments aggregate into the network junctions.

Another important synthetic polymers whose gelation is strongly coupled to the polymer conformational change is the associating polymers. Associating polymers are water-soluble polymers partially modified by hydrophobic groups. These groups form aggregates, or micelles by hydrophobic interaction. At low polymer concentrations, intramolecular association in the form of flower-like micelles is dominant, but with increase in the concentration, more open association (inter-micellar bridging) prevails, and such bridge chains eventually form networks with multiple cross-link junctions[4, 5]. Similar transition from flower-micelles to bridge chains is observed in triblock copolymers in selective solvents[6, 7].

To study these examples systematically, let us here classify the types of gelation in the following way:

- *Intra/Inter Transition* The functional groups hidden inside a polymer molecule are activated by the change of environmental conditions such as the temperature, polymer concentration, pH, concentration of another component,*etc.*, and form a gel by intermolecular bonds.

- *Coil/Helix or Coil/Globule Transition* Polymers in random coil conformation first partially form helices (or globules) as the temperature is lowered, and then helices (globules) aggregates into network junctions.

- *Two-State Transition* Each monomeric unit A along a polymer chain can take two states, *i.e.*, active state A^* and inactive state A. The active monomeric units form cross-links $(A^*)_k$ $(k = 2, 3, 4, ...)$ by multiple association.

At this stage, it should be remarked that the equilibrium polymerization of sulfur is a special case of the above intra/inter transition[8]. A ring polymer S_8 (called λ-sulfur), which is inactive at room temperature, first opens the ring into a linear chain carrying reactive unfilled covalent bond on its both ends (called μ-sulfur) as the temperature is raised, and then polymerized through interchain bonding at 160°C. Since the association takes place pairwisely and the functionality f (number of active sites on a molecule) of a μ-sulfur is two, molecules form linear chains instead of three dimensional networks. In analogy to sulfur polymerization, we may therefore generally call a molecule staying in the inactive state "λ-*molecule*", and one in the active state "μ-*molecule*" in our gel forming polymer solutions. The λ/μ-transition followed by intermolecular association is schematically summarized in Figure 1.

2. FREE ENERGY AND MODELS OF THE JUNCTION

We now consider the free energy of the system at temperature T and polymer volume fraction ϕ on the basis of the classical Flory-Huggins theory for multicomponent polymer solutions. The chosen standard reference states are pure solvent and separated pure unmixed amorphous primary solute polymers in the λ state. The free energy change on passing from the reference states to the final solution, at equilibrium with respect to cluster formation, consists of three parts: $\Delta F = \Delta F_{conf} + \Delta F_{reac} + \Delta F_{mix}$, where ΔF_{conf} is the free energy associated with the change in molecular conformation, ΔF_{reac}

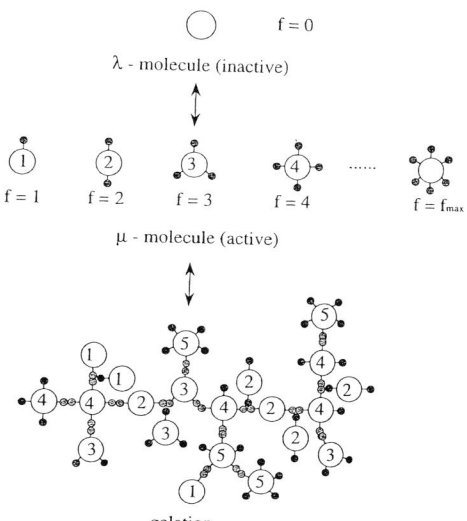

Figure 1: Gelation by the λ/μ transition of primary molecules

the free energy of reaction required to connect μ-molecules into clusters, and ΔF_{mix} the free energy produced on mixing all clusters with the solvent[9].

The first term is newly introduced here in this work to study cooperative effect between conformational change and gelation. It is written as

$$\Delta F_{conf} = A_\lambda N_\lambda + \sum_{\mathbf{j},\mathbf{l}} (\sum_{f\geq 1} A_f l_f) N(\mathbf{j};\mathbf{l}) + \sum_{f\geq 1} A_f N_f^G, \tag{1}$$

where N_λ is the number of λ-molecules, $N(\mathbf{j};\mathbf{l})$ the number of clusters consisting of $\mathbf{l} \equiv (l_1, l_2, ...)$ functional molecules connected by the junction $\mathbf{j} \equiv (j_1, j_2, ...)$, N_f^G the number of primary molecules in the gel network that carry f activated functional groups, A_λ the conformational free energy of a single λ-molecule, A_f the same of a μ-molecule with f active groups. The free energy required for the activation of a molecule is therefore given by $\Delta A_f \equiv A_f - A_\lambda$. In this context, the λ-state of a chain should be regarded as the "reference conformation". We minimize this free energy to find the most probable distribution of clusters.

The structure of the junctions is in principle decided by the thermodynamic requirement for a given associative interaction. In the present study, we introduce a model junction in which the multiplicities lying in a certain range covering from $k = s_{min}$ to s_{max} are allowed. When only a single value is allowed, i.e., $s_{min} = s_{max} \equiv s$, we call the model fixed multiplicity model.

The fundamental relation connecting the degree of association α with the total polymer concentration then takes the form

$$< f > \frac{\lambda}{n}\phi = S_1(\alpha), \tag{2}$$

where $< f > \equiv F_1(\alpha)/[1 + F_0(\alpha)]$ is the average functionality of the primary molecules including the λ-molecules, $S_1(\alpha) \equiv \alpha^{1/s'}/f_n(\alpha)(1 - \alpha)^{s/s'}$ is the first moment of the multiple tree distribution. The average functionalities are $f_n(\alpha) = F_1(\alpha)/F_0(\alpha)$, $f_w(\alpha) = F_2(\alpha)/F_1(\alpha)$, where the functions F_m are defined in terms of α as

$$F_m(\alpha) \equiv \sum_{f \geq 1} f^m e^{-\beta \Delta A_f}/(1 - \alpha)^f. \tag{3}$$

From the divergent condition of the weight average molecular weight of the clusters, we find the gel point is given by the solution of the equation $(s - 1)[f_w(\alpha) - 1]\alpha = 0$.

3. MODELS OF EXCITATION

We now discuss specific forms of the functions $F_m(z)$. By definition it depends on the excitation free energy A_f with possible number f of active groups measured relative to the reference conformation.

- *Independent Excitation Model* In this model a polymer chain is assumed to carry the number f of associative groups, each of which may independently take either an active or inactive state. The energy difference between the two states is assumed to be given by ΔA_1. Then the functions $F_m(z)$ are given by $F_m(z) = (xd/dx)^m(1+x)^f$, where $x \equiv \zeta u(z)$ with $\zeta \equiv e^{-\beta \Delta A_1}$.

- *All-or-None Model* This model assumes that all associative groups are either active or inactive simultaneously. We then have functionality f for the excited state and 0 for the ground state, so that $F_m(z) = f^m \zeta u(z)^f$, where $\zeta \equiv e^{-\beta \Delta A_f}$. A typical phase diagram for this model is shown in Figure 2.

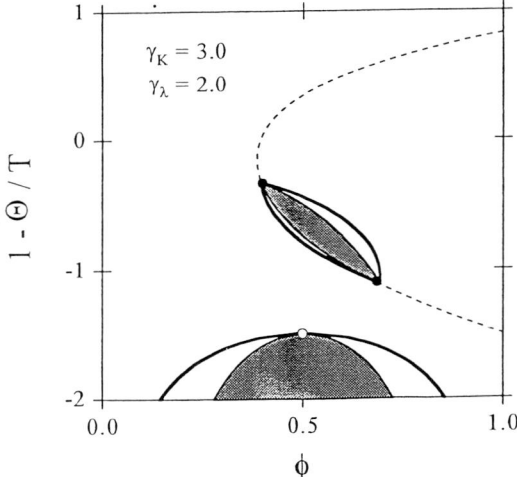

Figure 2: Gelation and phase separation of "all-or-none" model for $f = 3$ and $s = 3$.

4. SIMULATION ON MICELLE FORMATION AND CROSS-LINKING

To investigate the gelation coupled to the polymer conformational change numerically, we performed Monte Carlo (MC) simulations using an off-lattice bead-spring model in three dimensions. The potential energy for a polymer chain of N beads consists of three terms: $H = H_b + H_\theta + H_{nb}$. H_b is the following harmonic spring potential: $H_b = \sum_{i=1}^{N-1} k_b(l_i - l_0)^2/2$, where l_i is the length of bond i and l_0 is the equilibrium bond length. The bond angle θ_i between successive bonds is maintained close to the equilibrium value θ_0 by the potential H_θ: $H_\theta = \sum_{i=2}^{N-1} k_\theta(\cos\theta_i - \cos\theta_0)^2/2$. H_{nb} represents non-bonded interactions between monomers. The excluded volume interaction with the hard-sphere diameter σ acts between all monomers. In addition, we introduce another kind of monomers, called "*sticker*" in this paper, which interact via the following square-well potential each other: $H_{nb} = \sum_{i<j} \phi(r_{ij})$

$$\phi(r_{ij}) = \begin{cases} \infty & r_{ij} < \sigma \\ -\varepsilon & \sigma \le r_{ij} < d \\ 0 & d \le r_{ij} \end{cases} \tag{4}$$

where r_{ij} is the distance between beads i and j.

In the MC simulations, we employed the Metropolis algorithm and used the following parameters: $k_b = 50$, $k_\theta = 10$, $l_0 = 0.4$, $\sigma = 2l_0/3$, $d = 2\sigma$, and $k_B T = 1$. The strength of the square-well potential, ε, varies from 0 to 100. The details of the simulation method and the quantitative features of the system will be described elsewhere. Here we briefly describe the behavior of the system qualitatively.

5. RESULTS

First, we investigate effects of the attractive interaction on conformation of single polymer chains. In the system, there is a molecule with the degree of polymerization $N = 101$ and the number of stickers on the chain $N_s = 26$. The stickers are periodically arranged along the chain.

In the case of strong attractive interactions (typically $\varepsilon > 10$), the stickers are strongly bonded and form aggregates, which are surrounded by the corona of non-attractive monomers. These aggregates can be regarded as intramolecular micelles. Since the attractive interaction between stickers is screened by such surrounded monomers, there seems to be a fixed size of aggregates determined by the parameters and the architecture of the molecule. In the case that the strength of the attractive interaction is weak (typically $\varepsilon \simeq 2 \sim 5$), the stickers form weakly bonded large aggregates. If ε is small enough, the stickers do not form such aggregates.

Based on the behavior of single chains mentioned above, we consider that if stickers of a molecule are involved in an intramolecular aggregate formed in another molecule, the aggregate forms a cross-linking region between the two molecules. To confirm this consideration, we performed MC simulations with two molecules. The other conditions used in the simulation are the same as those in the single chain case. Typical snapshot obtained by the simulation is presented in Figure 3. There are three aggregates in this

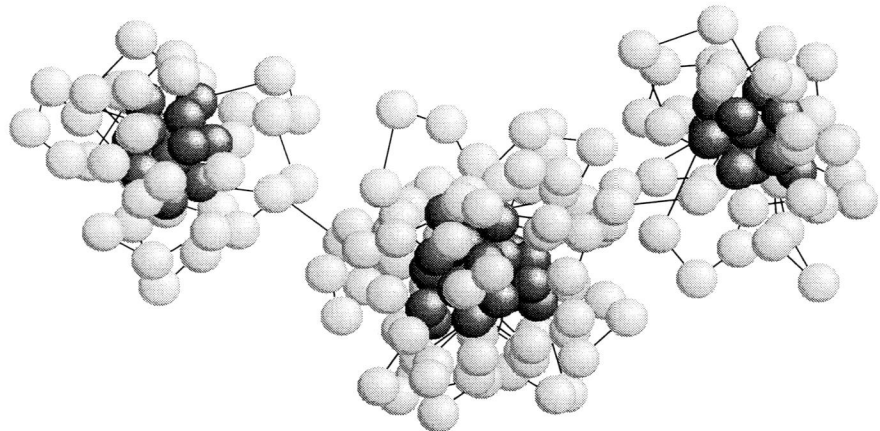

Figure 3: Typical snapshot obtained by the MC simulation in three dimensions at $\varepsilon = 2$. Dark monomers correspond to the stickers.

figure; the left and the right ones are the intramolecular aggregates, but the one in the center is formed by the stickers belonging to two molecules. Therefore, it provides an intermolecular cross-linking region.

In principle, since each aggregate is formed by many stickers, the aggregates can become multiple cross-linking regions in many chain systems. If the polymer concentration increases in such systems, the gelation should take place. This is an example of the gelation coupled to the polymer conformational change.

References

1. A. H. Clark, S. B. Ross-Murphy, *Ad. Polym. Sci.* **83**, (1987) 57.

2. W. Burchard, *British Polym. J.* **17**, (1985) 154; J. M. Guenet, *Thermoreversible Gelation of Polymers and Biopolymers*; Academic Press, Harcourt Brace Jovanovich Publishers (1992); K. Te Nijenhuis, *Ad. Polym. Sci.* **130**, (1997) 1.

3. A. Tobitani, S. B. Ross-Murphy, *Macromolecules*, **30**, (1997) 4845; 4855.

4. T. Annable, R. Buscall, R. Ettelaie, D. Whittlestone, *J. Rheol.* **37**, (1993) 695; T. Annable, R. Buscall, R. Ettelaie, P. Shepherd, D. Whittlestone, *Langmuir* **10**, (1994) 1060.

5. B. Xu, A. Yekta, M. A. Winnik, *Langmuir* **13**, (1997) 6903.

6. E. Raspaud, D. Lairez and M. Adam, Macromolecules, **27**, (1994) 2956

7. E. Alami, M. Almgren, W. Brown and J. Francois, *Macromolecules*, **29**, (1996) 2229.

8. R. L. Scott, *J. Phys. Chem.* **69**, (1965) 261.

9. F. Tanaka, W. H. Stockmayer, *Macromolecules* **27**, (1994) 3943.

HYDROCOLLOIDS – PART 1
Edited by K. Nishinari
2000 Elsevier Science B.V.

59

Hydrogels from N-isopropylacrylamide Oligomer

H. Hachisako, S. Miyagawa and R. Murakami

Department of Applied Chemistry, Kumamoto Institute of Technology, 4-22-1 Ikeda, Kumamoto 860-0082, Japan.

Oligo(N-isopropylacrylamide)s with single terminal groups such as stilbazole (**1**), azobenzene (**2**), biphenyl (**3**), n-hexadecyl (**4**), and propionic acid (**5**), were prepared to investigate their hydrogel formation ability. Aqueous solution of the oligomer **1** with a stilbazole terminal group gelled upon heating to the temperatures above its cloud point (CP; 32 °C) in water in the pH range below the pKa (pKa ≅ 5) of the pyridinium ring in the stilbazole moiety. On the other hand, shape-remembered, shrunken hydrogel was formed in the pH range above the pKa upon heating. Both the pH-dependent hydrogels from **1** were found to be composed of highly organized terminal stilbazolium (at the pH below the pKa) and stilbazole (at the pH above the pKa) as evidenced by their excimers formed at the expense of monomeric species with increasing the concentration of **1**. These results indicate that the gelation behaviors are related to the extent of excimer formation of the terminal stilbazole chromophore. No similar gelations were observed for other oligomers (**2**-**5**) under the same conditions. This indicates that the terminal stilbazole group of **1** under highly organized aggregation is essential for the hydrogel formation. Formation of the shape-remembered hydrogels was found to be dependent on the kind and the combination of external stimuli such as temperature, pH and UV light. Morphological difference of the xerogels cast from water at pH 2 and pH 11 was studied by scanning electron microscopy (SEM).

1. INTRODUCTION

It is known that poly(N-isopropyl-acrylamide) (PNIPAM) exhibits thermally reversible soluble-insoluble changes in response to temperature changes across its cloud point (CP) in aqueous solution [1]. Similarly, hydrogels from chemically cross-linked poly(N-ispropylacrylamide)s undergo thermally reversible swelling-deswelling behaviors. Therefore, PNIPAM and its chemically cross-linked hydrogels have been attracting much attention from both the theoretical and practical points of view. On the other hand, synthetic polymers which respond to various external stimuli such as temperature, UV light, pH, chemicals, etc. are referred to as stimuli-responsive polymers. It is known that PNIPAM is one of the most representative stimuli-responsive polymers. Primary structure of the stimuli-responsive polymers can be designed by the combination of several kinds of stimuli receptors. To construct more

Figure 1   NIPAM oligomers.

versatile system, it is considered effective to utilize self-assembling (aggregation) behavior instead of chemical cross-linking by which most of the conventional PNIPAM gels have been prepared. The highly organized structure would be more advantageous for molecular recognition. It is considered that the segments of the conventional PNIPAM and its cross-linked gels would be less ordered in water. Therefore, NIPAM oligomer would be more superior in segment orientation to the conventional PNIPAM and its cross-linked gels. However, it is difficult to prepare hydrogels from NIPAM oligomers without using chemical cross-linking. We prepared NIPAM oligomers **1**-**5** with various kinds of terminal groups as shown in Figure 1. This paper describes a hydrogel formation from a novel *N*-isopropylacrylamide oligomer **1** bearing single stilbazole terminal group which is capable of response to several kinds of stimuli such as pH, UV light, etc. Response to these external stimuli has also been investigated.

2. MATERIALS AND METHODS

N-isopropylacrylamide (NIPAM) was purchased from KOHJIN (Tokyo, Japan) and was purified by recrystallization from *n*-hexane and dried at room temperature in vacuo. 2, 2'-azobisisobutyronitrile (AIBN; Tokyo Chemical Industry, Tokyo, Japan) was used after recrystallization from methanol. Terephthalaldehydic acid, 4-methylpyridine (4-picoline), 4-phenylazobenzenesulfonyl chloride, biphenyl-4-carboxylic acid, 2-mercaptoethylamine, triethylamine, and 3-mercaptopropionic acid were purchased from Tokyo Chemical Industry (Tokyo, Japan) and used without further purification. Thionyl chloride, acetic anhydride, tetrahydrofuran (THF), and diethyl ether were obtained from Kanto Chemicals (Tokyo, Japan) and used as obtained.

A stilbazole **6** with a carboxylic acid was prepared by coupling of 4-methylpyridine and terephthalaldehydic acid in acetic anhydride under reflux as shown in Scheme 1. The precipitate was filtrated and dried in vacuo: yield 50%, m.p.>300 °C. Anal.Found: C, 74.23; H, 4.97; N, 6.13. Calcd for $C_{14}H_{11}NO_2$: C, 74.65; H, 7.92; N, 6.23 %. Then **6** (3.0 g, 13 mmol) was reacted with thionyl chloride (30 cm^3) under reflux to obtain the corresponding acid chloride. The acid chloride was suspended in THF, and the solution was added dropwise to the chloroform solution of 2-mercaptoethylamine (1.2 g, 16 mmol) in the presence of triethylamine (4.0 g, 40 mmol) with ice-cooling to obtain **7**: yield

Scheme 1

80%; m.p. 219-222 °C. Anal.Found: C, 65.59; H, 5.42; N, 9.51. Calcd for $C_{16}H_{16}N_2OS \cdot 0.5H_2O$: C, 65.59; H, 5.83; N, 9.56 %. A NIPAM oligomer **1** with single stilbazole terminal group was prepared by free radical telomerization of NIPAM (3.2 g, 28 mmol) in the presence of **7** (0.40 g, 1.4 mmol) in benzene under reflux for 2h, using AIBN (40 mg) as an initiator [2,3]. After the reaction, the solution was poured into diethyl ether to precipitate the polymer. The polymer **1** was further purified by repeated precipitation from THF into diethyl ether twice. Average degree of polymerization was estimated to be ca. 20 by ^{1}H-NMR measurement and confirmed by elemental analysis: **1**; white powder, yield 59 %, IR(KBr): 3438, 2976, 1657, 1545 cm^{-1}; ^{1}H-NMR(CDCl$_3$): δ 0.80-1.28 (120H, m, C$\underline{H}_3$), 1.40-1.91 (40H, m, C$\underline{H}_2$CH), 1.91-2.49 (20H, m, CH$_2$C$\underline{H}$), 2.95-3.09 (2H, m, C$\underline{H}_2$S), 3.76-3.86 (2H, m, NHC$\underline{H}_2$), 4.05 (20H, s, NHC$\underline{H}$), 6.90-7.48 (2H, m, olefinic H), 7.31-7.44 (2H, m, aromatic H), 7.49 (2H, d, aromatic H), 7.89 (2H, d, aromatic H), 8.60 (2H, d, aromatic H). Anal. Found: C, 60.875; H, 9.105; N, 11.485. Calcd for (x=20) $C_{135}H_{250}N_{22}O_{28}S \cdot 7.5H_2O$: C, 60.875; H, 9.446; N, 11.485 %. An oligomer **2** was prepared as described above using the corresponding thiol (prepared by coupling of 4-phenylazobenzenesulfonyl chloride and 2-mercaptoethylamine in the presence of triethylamine in THF, m.p. 108-115 °C) as a chain transfer agent: **2**; orange powder, yield 50 %, IR(KBr): 3440, 1649, 1547, 1369, 1174 cm^{-1}; ^{1}H-NMR(CDCl$_3$): δ 1.13-1.24 (240H, m, CH$_3$), 1.48-1.94 (80H, m, CH$_2$CH), 1.94-2.54 (40H, m, CH$_2$CH), 3.0-3.2 (2H, m, CH$_2$S), 3.6-3.8 (2H, m, NHCH$_2$), 3.99 (40H, s, NHCH), 7.54-8.17 (9H, m, aromatic H). An oligomer **3** was prepared as described above using the corresponding thiol (prepared by the coupling of biphenyl-4-carboxylic acid and 2-mercaptoethylamine in the presence of triethylamine in THF, m.p. 238-243 °C) as a chain transfer agent: **3**; white powder, yield 30%, IR(KBr): 3330, 2976, 1655, 1543 cm^{-1}; ^{1}H-NMR (CDCl$_3$): δ 1.14 (108H, s, CH$_3$), 1.47-1.95 (36H, m, CH$_2$CH), 1.95-2.51 (18H, m, CH$_2$CH), 2.98-3.08 (2H, m, CH$_2$S), 3.80-3.90 (2H, m, NHCH$_2$), 4.01 (18H, s, NHCH), 7.36-7.93 (9H, m, aromatic H). Oligomers **4** and **5** were prepared using commercially available 1-hexadecanethiol and 3-mercaptopropionic acid, respectively, as chain transfer agents; **4**: yield 70%; IR (KBr): 2974, 2870, 1651, 1547 cm^{-1}; ^{1}H-NMR(CDCl$_3$) δ 1.00-1.50 (m, 3H, CH$_3$, 150H, CH(CH$_3$)$_2$, 28H, (CH$_2$)$_{14}$), 1.50-2.40 (m, 50H, CH$_2$CH, 25H, CH$_2$CH), 2.40-2.60 (m, 2H, CH$_2$S), 4.00 (m, 25H, NHCH); Found: C, 62.08; H, 9.49; N, 10.89. Calcd for (x=24.7) $C_{164.2}H_{305.7}N_{24.7}O_{24.7}S \cdot 6.8H_2O$: C, 62.08; H, 10.13; N, 10.89%. **5**: yield 70%; IR (KBr): 3300, 2974, 1710, 1651, 1547 cm^{-1}; ^{1}H-NMR (CDCl$_3$) δ 1.01-1.25 (102H, m, CH$_3$), 1.41-2.31 (34H, m, CH$_2$, 17H, m, CH), 2.5-3.0 (m, 2H, CH$_2$S), 4.24 (17H, s, NHCH). Anal. Found: C, 59.49; H, 9.85; N, 11.34. Calcd for (x=16.5) $C_{123}H_{226}N_{20}O_{22}S \cdot 5.8H_2O$: C, 59.73; H, 9.68; N, 11.33%.

The NIPAM oligomer **1** was dissolved in water by sonication with ice-cooling to give translucent solution. The cloud points (CPs) of aqueous solutions of **1** (8.0 mM) were measured by differential scanning calorimetry (DSC) with a SEIKO I&E DSC 120 using a heating rate of 2 °C min^{-1}. Fluorescence spectra were recorded with a SHIMADZU RF-5300PC spectrofluorophotometer with a band width of 1.5 nm. UV irradiation for photodimerization of stilbazolium moiety in **1** (pH 2) was conducted for 10 min using a high-pressure mercury lamp in conjunction with a Toshiba UV-D35 color filter (330 < λ < 380 nm). UV-visible absorption spectral changes were measured with a JASCO V-530 UV/VIS spectrophotometer.

SEM investigations of the xerogels of **1** cast from water (pH2 and pH11) were conducted with a JEOL JSM 6301F scanning electron microscope. The acceleration voltage was 6 kV, and the working distance was 24, 25 or 39 mm. The xerogel cast on a cover glass was pasted on a brass support with a silver conducting glue. Before observation, the xerogel was sputtered with a 140 Å thickness gold layer.

3. RESULTS AND DISCUSSION

It was confirmed from both UV-visible absorption and fluorescence spectra that the pKa of pyridinium ring of the protonated form of stilbazole terminal group was ca. 5. Cloud points (CPs) of aqueous solutions of **1** (20 mM) were 32°C for pH 2 and 31 °C for pH 11. These values are almost identical with those of conventional PNIPAM [1,4]. It was revealed from the excimer fluorescence spectra that **1** self-assembled in water with an increase in its concentration at the temperatures below its CP (Figure 2). Aqueous solution of **1** (4.8 mM in Figure 2, pH 7) gelled upon heating to the temperatures above its CP. The gel was shrunken into a shape similar to that of a quartz cell for fluorescence measurement. However, at pH 2, the solution gelled without shrinking, probably because of electrostatic repulsion between protonated stilbazolium terminal groups. These results indicate that the shrinking behaviors of the gels are pH-responsive based on the pKa. It is also noted that both the gels (4.8 mM) at pH 3 and pH7, respectively, consisted of excimer of the terminal chromophoric groups. This indicates

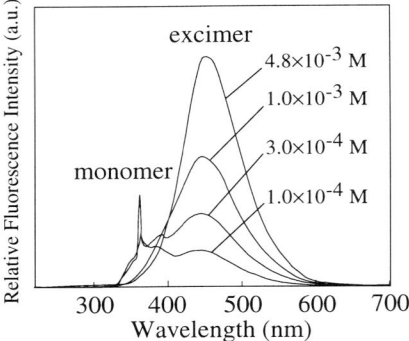

Figure 2 Concentration dependence of fluorescence spectra for aqueous solutions of **1**. λex = 360-370 nm; Temp., 20 °C; pH7.0.

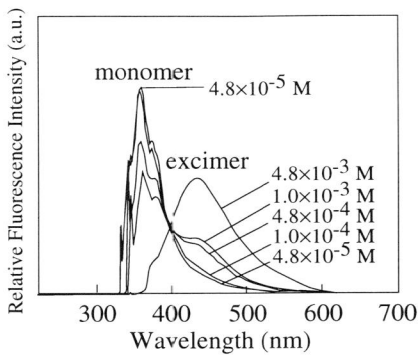

Figure 3 Concentration dependence of fluorescence spectra for ethanol solutions of **1**. λex = 350-360nm; Temp., 20 °C.

that the terminal groups of the NIPAM oligomers are highly organized under gelation. On the other hand, no gelation was observed for ethanol solution under the same conditions in spite of the excimer formation (Figure 3), indicating that the excimer formation of the terminal group does not necessarily lead to gelation. This suggests that hydrophobic interaction between dehydrated NIPAM segments in water is one of the necessary conditions for gelation upon heating. The hydrogels produced at the pH above 5 were hardly redissolved upon cooling to the temperature below the CP, whereas those produced at the pH below 5 was easily redissolved. NIPAM oligomers with other kinds of terminal groups, e.g., azobenzene group (**2**), biphenyl group (**3**), n-hexadecyl group [2,3] (**4**), and propionic acid [5] (**5**), were precipitated without gelation under the same conditions. These results indicate that appropriate terminal group enables the gelation of the aqueous solution of the NIPAM oligomer. In conclusion, hydrogel was formed from NIPAM oligomer **1** with a single stilbazole terminal group without chemical cross-linking.

Effect of UV-irradiation on the hydrogel formation of **1** was investigated. As mentioned above, no shrinking behavior was observed for hydrogel of **1** under protonated condition of the terminal stilbazole group. However, after UV-irradiation for 10 min at room temperature, shrunken hydrogel was formed from the aqueous solution of **1** (pH 2) upon heating. This may be ascribed to the photodimerization of the terminal group because protonated stilbazoles are known to undergo photodimerization when exposed to UV light [6,7]. Figure 4 shows UV-visible absorption spectral changes of aqueous solution of **1** (pH 2) before and after UV-irradiation, suggesting that the protonated **1** is likely to be photodimerized when exposed to UV light. It should be noted that UV-exposure to aqueous solution of **1** led to the formation of shrunken hydrogel upon heating even at pH 2 at which non-shrunken hydrogel was formed without UV-irradiation. Formation of the

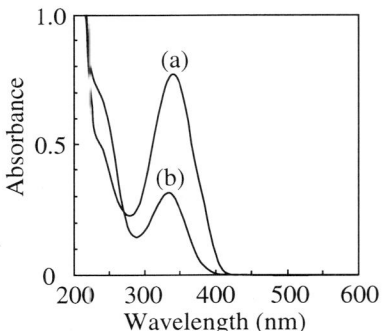

Figure 4 UV-visible absorption spectra for aqueous solution of **1**. [**1**] = 4.8 ×10^{-3} M, pH 2.0, 20 °C; path length of quartz cell, 0.1 mm. (a) before UV irradiation, (b) after UV-irradiation for 10 min using high pressure mercury lamp; filter, TOSHIBA UV-D35 (330 < λ < 380 nm).

shape-remembered, shrunken hydrogels was found to be dependent on the kind and the combination of external stimuli such as temperature, pH, and UV light, as summarized schematically in Figure 5.

Figures 6 and 7 show scanning electron micrographs of xerogels of **1** cast from water of pH 2 and pH 11, respectively. The pictures (c) and (d) in Figure 6 show the xerogels produced at 60 °C. It should be noted that the dendritic growths were observed for the heated sample prepared under hydrogel formation. On the other hand, as shown in Figure 7, no dendritic growths were observed for a xerogel prepared at pH 11. Such a remarkable difference in morphologies between xerogels prepared at pH 2 and pH 11 by

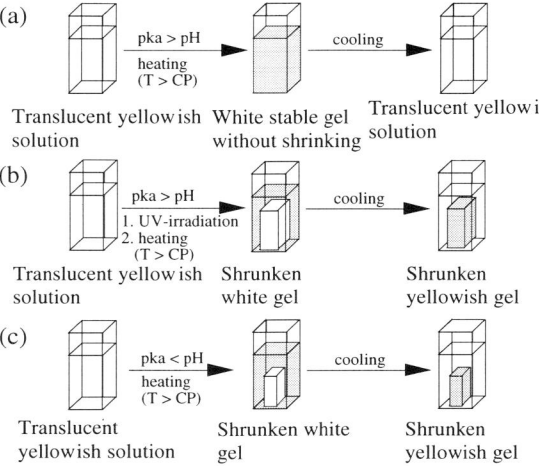

(a) Translucent yellowish solution — pka > pH, heating (T > CP) → White stable gel without shrinking — cooling → Translucent yellowish solution

(b) Translucent yellowish solution — pka > pH, 1. UV-irradiation, 2. heating (T > CP) → Shrunken white gel — cooling → Shrunken yellowish gel

(c) Translucent yellowish solution — pka < pH, heating (T > CP) → Shrunken white gel — cooling → Shrunken yellowish gel

Figure 5 Schematic representation of gelation behaviors of 5mM of **1** under various conditions. (a) and (b) pH < 5; (c) pH>5.

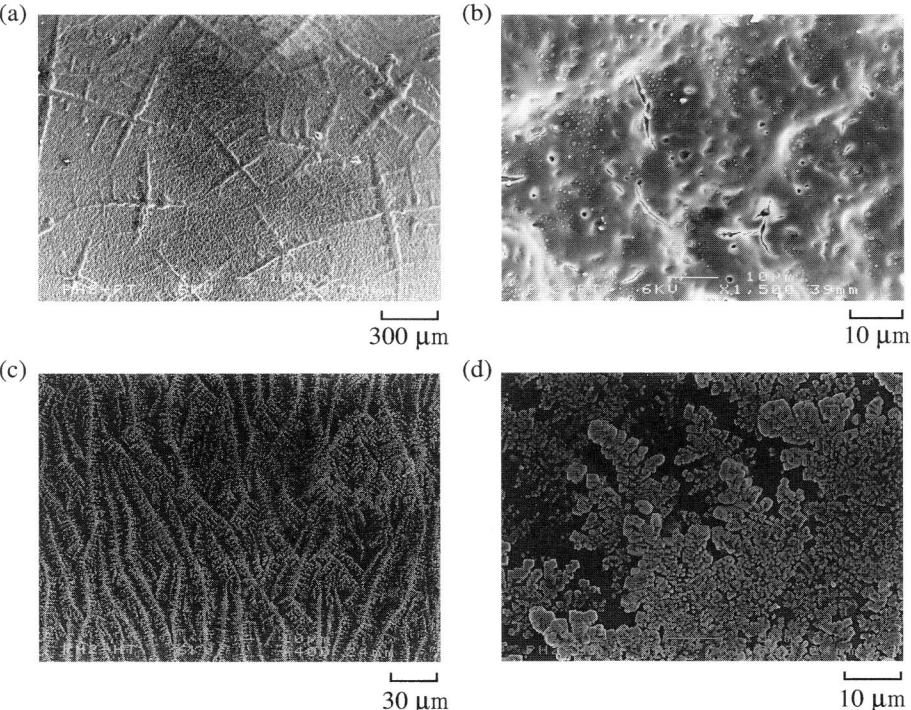

(a) 300 µm

(b) 10 µm

(c) 30 µm

(d) 10 µm

Figure 6 Scanning electron micrographs of xerogels of **1** cast from water at pH 2: (a) and (b), evaporated at room temperature; (c) and (d), evaporated at 60 °C.

64

(a)   (b)

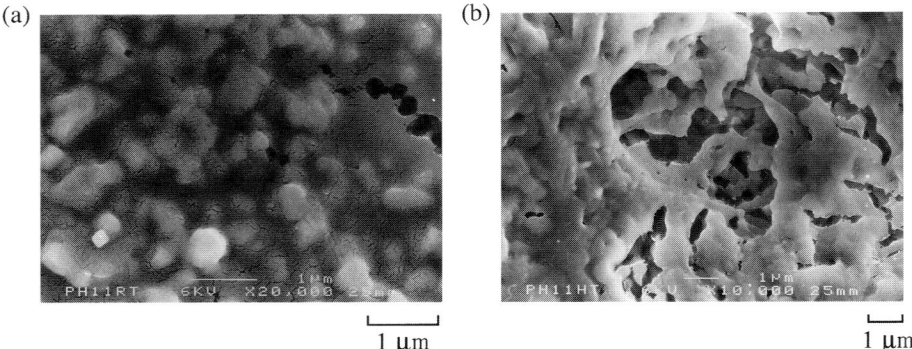

       1 µm   1 µm

Figure 7 Scanning electron micrographs of xerogels of **1** cast from water at pH 11:
(a), evaporated at room temperature; (b), evaporated at 60 °C.

heating is considered to be related to the difference in bulk density of both the xerogels, because
it has been reported that less compact calcite deposits with lower bulk densityexhibited
dendritic growths in spite of no development of dendrites for more compact calcite deposits
with higher bulk density [8].   The dendritic morphologies were also observed for the
Langmuir-Blodgett monolayer systems of fatty acids, demonstrating that the growth process
was unstable [9].   Our present result that the dendrites are observed for the xerogel prepared
from the aqueous solution of pH 2 by heating also suggests that the bulk density of the
corresponding hydrogel formed from the aqueous solution of pH 2 is lower than that of the the
hydrogel formed from the aqueous solution of pH 11. Visual observation of the shrunken
hydrogel at pH 11 and the non-shrunken hydrogel at pH 2 is consistent with the difference in
the density, because it is obvious that the bulk density of the shrunken hydrogel is higher
than that of the non-shrunken hydrogel.

ACKNOWLEDGEMENT

The authors would like to express their sincere thanks to Professor H. Shosenji of
Kumamoto University for elemental analyses, Dr. R. Tomoshige of Kumamoto Institute of
Technology for SEM observation, and M. Yoshimoto, T. Hirabara, and A. Morikawa of
Kumamoto Institute of Technology for their technical supports.

REFERENCES

1. M. Heskins and J. E. Guillet, J. Macromol. Sci.-Chem., A2 (1968) 1441.
2. H. Hachisako, R. Murakami, and K. Yamada, Proceedings of '97 Kyushu-Seibu/Pusan-
Kyeongnam Joint Symposium on High Polymers (8th) and Fibers (6th), (1997) 155.
3. H. Hachisako, M. Yoshimoto, R. Murakami, and K. Yamada, Polymer Preprints, Japan, 46
(1997) 567.
4. S. Ito, Kobunshi Ronbunshu, 46 (1989) 437.
5. H. Hachisako, R. Murakami, and K. Yamada, Polymer Preprints, Japan, 43 (1994) 675.
6. F. H. Quina and D. G. Whitten, J. Am. Chem. Soc., 99 (1977) 877.
7. K. Takagi, B. R. Suddaby, S. L. Vadas, C. A. Backer, and D. G. Whitten, J. Am. Chem.
Soc., 108 (1986) 7865.
8. N. Andritsos, A. J. Karabelas, and P. G. Koutsoukos, Langmuir, 13 (1997) 2873.
9. H. D. Sikes and D. K. Schwartz, Langmuir, 13 (1997) 4704.

HYDROCOLLOIDS – PART 1
Edited by K. Nishinari
2000 Elsevier Science B.V.

Kinetic effects of the gel size on the thermal behavior of poly (*N*-isopropylacrylamide) gels : a calorimetric study

H. Kawasaki, S. Sasaki, and H. Maeda

Department of Chemistry, Faculty of Science, Kyushu University 33, Hakozaki, Higashi-ku, Fukuoka, 812-8581, Japan

Effects of the gel size on the thermal behavior and the shrinking profile of chemically cross-linked poly(*N*-isopropylacrylamide)(NIPA)gel in water were mainly investigated by differential scanning calorimetry(DSC) and the volume measurement at the same heating rate as the DSC. The DSC thermogram of NIPA gel depended on the shrinking process at the transition. The DSC thermogram of NIPA gel of the smallest size (0.3 mm diameter at the preparation) gave double endothermic peaks and the gel underwent a discontinuous volume shrinkage (i.e. a volume transition) at about 34.5 °C within the time scale of DSC run of heating rate of 0.1 °C/min. This suggests that the two-step cooperative dehydration of NIPA chains takes places in the volume shrinkage. It was proposed that the peak at the low temperature (peak 1) originated from the dehydration of NIPA chains without the volume transition while the peak at the high temperature (peak 2) was associated with the discontinuous volume shrinkage. A single broad endothermic peak was observed for NIPA gel of the largest size (3 mm diameter at the preparation) that shrank only slightly during DSC run. For the largest gel, the volume transition cann0
ot occur within the time scale of DSC run due to the slow volume shrinkage while dehydration of the polymer chains takes place locally. The peak of the largest gel and the peak 1 were suggested to be due to the dehydration of NIPA chains coupled with the spinodal decomposition without the volume transition.

1. INTRODUCTION

It is well known that chemically cross-linked poly(*N*-isopropylacrylamide) (NIPA) gels in water undergo a discontinuous volume shrinkage at about 34 °C with increasing temperature[1]. This behavior is known as volume phase transition of gels [2,3]. Thermal analyses by differential scanning calorimetry (DSC) on the transition of NIPA gel and poly(*N*-isopropylacrylamide) (PNIPA) aqueous solution showed a single endothermic peak at the transition [4,5]. The endothermic transition heat has been suggested to originate from the dehydration of NIPA chains at the transition(decrease of the amount of the structured water around the hydrophobic residue).

In the case of PNIPA, the peak shape and the temperature of the endothermic peak have been confirmed to be independent of heating rate of less than 30 °C / h [6]. This reflects the fast collapse process of the isolated chains at the transition of PNIPA. On the application of DSC experiments to the volume phase transition of NIPA gel, due attention should be paid to the shrinking kinetics of the sample during DSC run. The volume shrinkage at the transition stage has been well known to be a slow process for large gels of several millimeters in size due partly to the collective diffusive process of gel-network [7-9] and partly to the formation of a denser outer region (so-called skin layer) on the surface of the gel [10]. The time required to attain the equilibrium gel volume has been known to be proportional to the square of a characteristic size of the gel [7]. The skin layer is believed to be impermeable to the inner fluid and it usually slows down the velocity of the volume shrinkage at the transition [10]. These kinetic factors on the volume shrinking may affect the thermal behavior of NIPA gel.

In this study, we mainly examined the relation between the thermal behavior of NIPA gel and the volume shrinkage by comparing the thermogram of the NIPA gel with the volume shrinkage profile at the same heating rate as the DSC. In order for NIPA gel to undergo the volume transition within the time scale of DSC run, low heating rates and small size of sample gel were both required[11]. The DSC experiments were mainly performed at low heating rate of 0.1 °C/min, using the gel of small size (0.3 mm diameter at the preparation)[11]. Similar experiments were also performed for NIPA gel with 3 mm diameter at the preparation and for copolymer gel of NIPA and acrylamide (NIPA-AAm) with 0.3 mm diameters at the preparation. The volume of the large NIPA gel is expected to be almost constant (i.e. isochore path) during the DSC run because of the slow volume change. The NIPA-AAm gel of high AAm content exhibits a continuous volume shrinkage.

2. EXPERIMENTAL SECTION

2.1. Sample. N-isopropylacrylamide(NIPA)gels were prepared by free radical copolymerization in the aqueous solutions of N-isopropylacrylamide (696.5mM) and N,N'-methylenebis(acrylamide) (3.5mM). Copolymer gel of NIPA and acrylamide (NIPA-AAm) gels were prepared by radical copolymerization in aqueous solutions of N-isopropylacrylamide, acrylamide and N,N'-methylenebis(acrylamide) (3.5mM). The total residue concentration of NIPA-AAm gel was fixed to be 700mM. NIPA-AAm gel with a mole fraction of acrylamide residue, Q is denoted as NIPA-AAm(X=Q) gel. The polymerization was initiated by ammonium persulfate(APS), accelerated by N',N',N',N'-tetramethylethylendiamine(TEMED) and carried out at 5 °C for 24h. Cylindrical gels of different sizes for NIPA gels were prepared with glass capillary. The inner diameters of the glass capillary were 0.3 and 3 mm for NIPA gel and 0.3 mm for NIPA-AAm gel. NIPA gel prepared in a capillary of a diameter, D (mm) is denoted as NIPA(D)gel.

2.2 Experiments. DSC experiments were carried out with a DSC 120 calorimeter (Seiko Inc.) from 25 to 60 °C (heating rate of 0.1, 0.5 and 1.0 °C / min) and water was used as a reference. The dried sample gels of about 2 mg were sealed in a aluminum pan with 70mL of water (about 3 wt %). The transmittance (= I_t/I_0 x 100) of the sample was monitored by a UV spectrometer operating at a wavelength of 550 nm. Where I_t and I_0 are the transmitted light intensity and the incident intensity, respectively. In the swelling measurement, the diameter of the gel was measured with an optical microscope coupled with a video monitor system. The temperature in swelling measurement was controlled within ± 0.05 °C.

3. RESULTS AND DISCUSSION

Figure 1 (a) shows a shrinking profile of NIPA (0.3)gel at a heating rate of 0.1 °C / min as a function of temperature. For comparison, equilibrium swelling ratios of he sample gel are also shown in Fig. 1 (a)

The swelling ratio is defined as (d/d_0) where d_0 and d are diameters of the gel in water at 25 °C and a given temperature, respectively. The two swelling curves coincide with each other in the temperature range lower than 33.8 °C, which is the transition temperature obtained from the equilibrium swelling measurement. In the temperature range of 33.8 - 34.5 °C, the gel shows a continuous volume shrinkage. At about 34.5 °C, the gel surface was wrinkled and underwent a discontinuous volume shrinkage. The shrinking profile indicates that the size of NIPA(0.3) gel is small enough for the discontinuous volume shrinkage to occur at about 34.5 °C at the heating rate of 0.1 °C / min. In Fig.

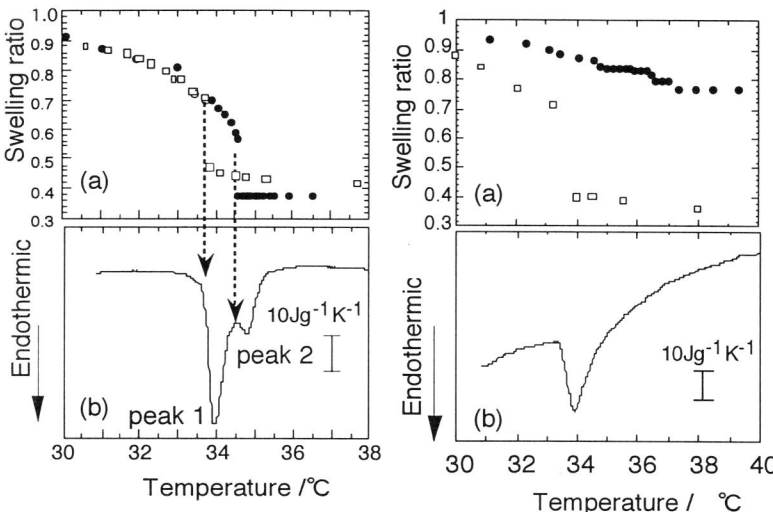

Figure 1   The shrinking profile (a) and the DSC thermogram (b) of NIPA (0.3) gel at a heating rate of 0.1 ℃/ min. (●): The shrinking profile of the gel at the heating rate.   For comparison, the equilibrium swelling ratios of NIPA (0.3) gel are also shown with open squares(□). Swelling ratio is   defined as (d /d_0) where d_0 is a diameter of the gel at 25 ℃.

Figure 2.   The shrinking profile (a) and the DSC thermogram (b) of NIPA (3) gel at the heating rate of 0.1 ℃/ min. (●): The   shrinking profile of the gel at the heating   rate.   For comparison,   the equilibrium   swelling ratios of NIPA (3) gel are also shown with open squares (□).

DSC thermogram of NIPA (0.3)gel   at the same heating rate of 0.1 ℃ / min is shown. The thermogram of NIPA(0.3) gel gives double endothermic peaks.   The shrinking profile of NIPA (3) gel   at the heating rate of   0.1 ℃/ min   (a) and the DSC thermogram at the same heating rate (b)   are   shown in Figure 2.   NIPA(3) gel   shrinks only slightly   in   the temperature   range   (up to 60 ℃) due to   the   slow   volume   change.   The thermogram   of NIPA(3) gel gives   a single   broad endothermic peak. The DSC thermograms   at a heating rate of 0.1 ℃/ min   and the equilibrium swelling ratios for NIPA-AAm gels   are shown in Fig 3(a) and 3(b),   respectively. The swelling ratio is defined as $(d /d_0)^3$ where d_0 (= 0.3 mm) and d   are the diameter at the preparation and a given temperature, respectively. The increase of   acrylamide content in the NIPA- AAm gels AAm gels leads to the decrease in the width of the discontinuous volume jump as well   as the raise of   the transition temperature. For   the NIPA-AAm gels of 0.3 mm size, it was confirmed that the shrinking profile of the NIPA-AAm gel   at the heating rate of   0.1 ℃/ min   almost coincided with the equilibrium swelling curve(not shown).   The shoulder peak of   NIPA-AAm gel   becomes obscure as the acrylamide content of NIPA-AAm gels increases as shown in Fig 3(b).

The transition temperature (T_t) and the transition heat per NIPA monomer mol (ΔH) for NIPA gels and NIPA-AAm gels   are shown in Table I. The transition temperature T_t was determined from the intersection of the baseline and   the leading edge of the peak.   The ΔH value was estimated from the peak area of the DSC peak. The temperature T_t of NIPA gel from DSC curve is   close to that from the equilibrium swelling measurement. As for NIPA-AAm gels which shows continuous volume shrinkage, the   temperature T_t is close   to

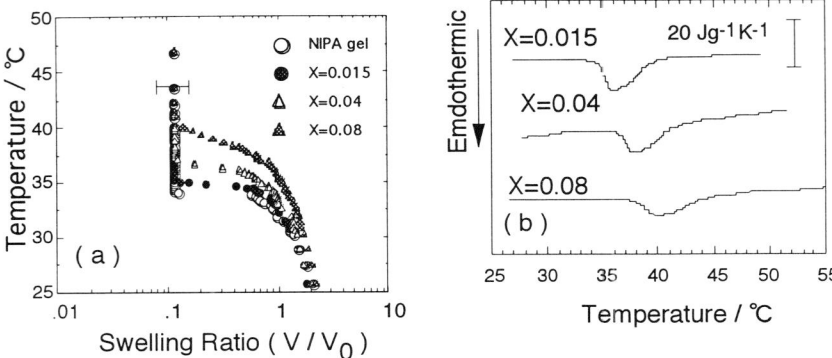

Fig. 3   (a)   Temperature dependence of the swelling ratios of NIPA-AAm gels   in water   when increasing temperature.   (b)   DSC thermograms of NIPA-AAm gels and NIPA gel in water at a heating rate of 0.1 ℃ / min .

the temperature determed by   the inflection ponit on   the equilibrium   swelling curve, where the gel starts to shrink abruptly   with   increasing   temperature. The transition heat of NIPA gel depends weakly on   the gel-size and the heating rate. The transition heat of NIPA(0.3) gel   was estimated to be   4.5 (kJ / mol) from   the area of the double endothermic peaks. This value   agrees with those by Otake et al. (3.3-4.5   kJ / mol)[4].

The ΔH value of NIPA-AAm gel decreases with increasing AAm content.

In this study, it was demonstrated that the DSC thermogram of NIPA gel   depended on the gel size.   The   gel size dependence of the thermal behavior is considered to originate from   different transition processes during DSC run : NIPA(0.3) gel underwent the volume transition during DSC run while   NIPA(3) gel did not.   The thermogram of NIPA(3) gel gave a single   endothermic peak as shown in Fig 2(b).   This suggests   that   the dehydration of NIPA chains of NIPA (3) gel   takes place under a nearly constant volume corresponding to the   swollen state.   Figure 4   shows   the   transmittance   of NIPA(3)

Table I.   Transition Temperature (T_t) and Transition Heat for
NIPA gels and NIPA-AAm gels

sample gel	T_t from DSC (℃)	T_t from swelling curves	ΔH (kJ /mol)	heating rate (℃ / min)
NIPA(0.3) gel	33.6 (34.8)[(1)]	33.8	4.5	0.1
NIPA(0.3) gel	33.7	33.8	4.2	0.5
NIPA(0.3) gel	33.7	33.8	4.5	1.0
NIPA(3) gel	33.4	34	—	0.1
NIPA(3) gel	33.4	34	4.0	0.5
NIPA(3) gel	33.6	34	4.0	1.0
NIPA-AAm gel (X=0.015)	34.4	34.7	4.6	0.1
NIPA-AAm gel (X=0.04)	35.9	36.2	3.9	0.1
NIPA-AAm gel (X=0.08)	37.6	38.2	3.6	0.1

(1) As for the peak at the low temperature of NIPA(0.3) gel, the temperature
      at the maxim of the endothermic peak is shown with a bracket.

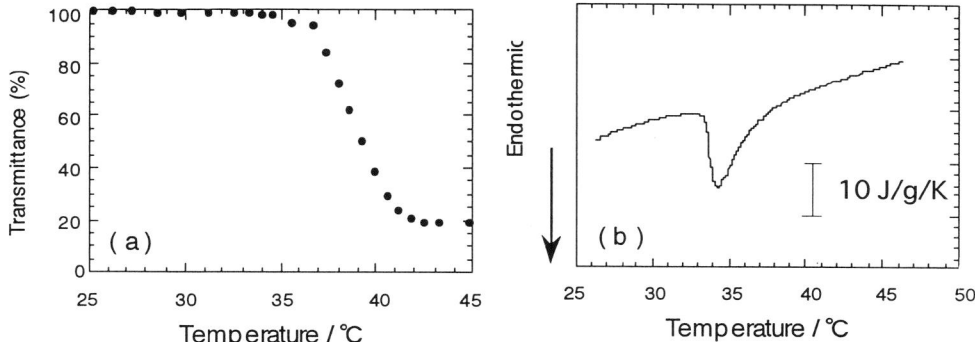

Figure 4 (a) the turbidity of NIPA(3) gel as a function of temperature when increasing temperature at a heating rate of 1 °C / min. (b) the DSC thermogram of NIPA(3) gel at a heating rate of 1 °C / min.

gel at a heating rate of 1 °C / min. The finding that NIPA(3) gel becomes opaque above the transition temperature of 34 °C suggests the occurrence of the network segregation within the gel. The endothermic peak of NIPA(3) gel is considered to correspond to the dehydration of NIPA chains coupled with a spinodal decomposition of NIPA gel which has been reported by Li et al. [12] and Bansil et al[13] with isochore NIPA gel. We believe that the peak 1 of NIPA (0.3)gel also corresponds to the local network chain association related to the spinodal decomposition as suggested for the NIPA(3) gel. To clarify the relation between the peak 1 and the local network chain association, further experiments are necessary.

The double endothermic peaks of NIPA (0.3) gel suggest the two step cooperative dehydration of NIPA chains takes place. The peak at the high temperature (peak 2) most likely originates from the cooperative dehydration of NIPA chains accompanying the volume transition (i. e. a discontinuous volume shrinkage) from the following reasons. (1) The peak 2 position corresponded to the temperature of the discontinuous volume shrinkage at the same heating rate as the DSC as shown in Figs. 1. (2) No counterpart was observed for NIPA(3)gel that underwent little volume shrinkage during the DSC measurement as shown in Fig. 2. (3) For another NIPA (0.3) gel sample with the small size and with a high cross linker mole fraction of 0.04 { NIPA (1152 mM) and BIS (48 mM) }, a continuous volume shrinkage was found in the equilibrium swelling measurement. This gel showed a single endothermic peak (not shown). Similar results were observed for NIPA-AAm(X=0.08) gel which underwent a continuous volume shrinkage as shown in Fig. 4. Recently, an endothermic peak with a shoulder was also found by L. M. Mikheeva et al. on another thermo-shrinking gel of chemically

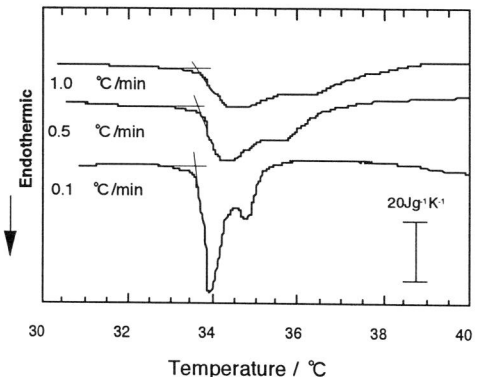

Figure 5 The dependence of the heating rate on the DSC thermogram of NIPA(0.3) gel.The heating rates are shown in Fig. 5.

cross-linked poly (*N*-vinylcaprolactam)(PVCa gel)[14]. They asserted that the peak with a shoulder was due to the multiple phase transitions. In the case of NIPA gel, however, it should be stressed that the double endothermic peaks are responsible for the kinetic process in the volume phase transition. We, therefore, consider that there would be only one peak for NIPA(0.3) gel if the temperature could be changed infinitely slowly.

We also examined a possibility that the double endothermic peak of NIPA(0.3) gel was a consequence of any difference in the packed condition from NIPA(3) gel because the difference may cause the inhomogeneity in the thermal conductivity within the sample pan leading to a large temperature distribution. However, this possibility can be ruled out by the following reasons. (1) To avoid the possible loose thermal contact, the DSC experiments were performed with the samples gels immersed in water. (2) The doubleendothermic peaks of NIPA(0.3) gel were repeatedly observed with reproducibility when the sample was replaced twice or more. (3) The NIPA-AAm(X=0.08) gel of the small size packed in the same way showed a single endothermic peak, as mentioned above. (4) If a large temperature distribution within the sample is responsible for the double endothermic peaks of NIPA(0.3) gel, it is expected that clearer double endothermic peaks should be observed at faster scan rates in DSC. However, faster scan rates actually made the double peak into shoulder peak as shown in Figure 5.

REFERENCES

1. Y. Hirokawa and T. Tanaka, J. Chem. Phys., 81(1984),6379.
2. K. Dusek and D. Patterson, J. Polym. Sci. Part A-2., 6(1963)1209.
3. T. Tanaka, and D. J. Fillmore, S.-T Sun, I. Nishio, G. Swislow and A. Shah, Phys. Rev. Lett., 45(1980) 1636.
4. K. Otake, H. Inomata, M. Konno and S. Saito, Macromolecules, 23(1990)283.
5. M. Shibayama, M. Morimoto and S. Nomura, Macromolecules, 27(1994) 5060.
6. G.S. Howard and A.T. David, J. Phys. Chem., 94(1990) 4352.
7. T. Tanaka, and D. J. Fillmore, J. Chem. Phys., 70(1979)1214.
8. Y. Li and T. Tanaka, J. Chem. Phys., 90(1989) 5161.
9. H. Hirose and M. Shibayama, Macromolecules, 31(1998) 5336.
10. R. Yoshida, K. Uchida, Y. Kaneko, K. Sakai, A. Kikuchi, Y. Sakurai and T. Okano, Nature 240(1995) 374.
11. H. Kawasaki, S. Sasaki and H. Maeda, Langumir, 14 (1998) 773.
12. Y. Li, G. Wang and Z. Hu, Macromolecules, 28(1995)4194.
13. R. Bansil, G. Liao and P. Falus, Physica A, 231 (1996)346.
14. L. M. Mikheeva, N. V. Grinberg, A. Y. Mashkevich, V. Y. Grinberg, L. T. M. Thanh, E. E. Makhaeva, A. R. Khokhlov. Macromolecules 30(1997) 2693.

HYDROCOLLOIDS – PART 1
Edited by K. Nishinari
2000 Elsevier Science B.V.

Characterizations of dehydrated polyacrylamide gel and its formation process

K. Hara, A. Nakamura[a], N. Hiramatsu[a] and A. Matsumoto[b]

Institute of Environmental Systems, Faculty of Engineering,
Kyushu University, Hakozaki, Higashi-ku, Fukuoka 812-8581, Japan

[a]Department of Applied Physics, Faculty of Science, Fukuoka University,
Nanakuma, Jonan-ku, Fukuoka 814-0180, Japan

[b]Osaka Municipal Technical Research Institute,
Morinomiya, Joto-ku, Osaka 536-8553, Japan

In the present study, we have made some characterizations on property evolution of the polyacrylamide (PAAm) gel during dehydration process (DP), and then, that of the dehydrated PAAm gel during temperature-increasing process (TIP). During DP, the amplitude of complex elastic stiffness grew around 10^3 times as large as the initial value and the loss tangent showed a peak around the time when the logarithm of weight showed a bend point (t_g'). Besides, around t_g', a pair of broad peaks emerged in the low frequency region of Raman spectrum, which looked like boson peak usually observed in the amorphous materials. These results indicated that some freezing transition analogous to the glass transition took place around t_g'. During TIP, in DSC measurement, a thermal anomaly was observed at some temperature (T_g), where the storage modulus decreased remarkably and the loss modulus showed a peak. Besides, intensity of the low-lying Raman peak decreased remarkably with increasing temperature. Because these features resembled those commonly observed around the glass transition in noncrystalline polymers, the similar transition in the dehydrated gel around T_g was confirmed. However, we have also noticed a difference between the spectral change of the dehydrated gel and that of usual glasses during TIP; the low-lying Raman peak in the dehydrated PAAm gel could be still distinguished even far above T_g.

1. INTRODUCTION

Gels are composed of a three-dimensional polymer network soaked in solvent. They show unique properties, such as the volume phase transition [1] due to the interaction between these two constituents, namely, the polymer network and solvent. Among the fascinating phenomena, we have been much interested in the property evolution when the solvent is being taken away. The first scientific study on this phenomenon was made by Takushi et al. [2]. They found that an opaque heat-treated egg-white gel becomes a transparent glasslike substance by dehydration. Since this first report on the *gel-to-glasslike transition* in heat-treated egg white gel, the investigations on the property evolution during the gel dehydration process have been carried out extensively.

During the dehydration process of the heat-treated egg white gel (abbreviated as HTEWG hereafter), the logarithm of weight decreases with time; the slope alters at a certain time (t_g') [2]. The linear decrease with a steep slope in the early period is explained by the loss of free water, while that after t_g' with a gentle slope, by the loss of bound water [2]. The features during the dehydration process are much influenced by the interaction between the polymer network and solvent. The weight decreases in different fashions according to the difference between the solvent surface tension and the network strength [3, 4].

Change in the interaction between the network and the solvent can reflect to the elastic anomaly, as well as the change of weight and volume with time. Koshoubu *et al.* [5] measured the complex elastic stiffness of HTEWG during the dehydration process, and found that the amplitude of the complex elastic stiffness (E) increased up to 10^3 times the initial value around t_g', and that the elastic loss tangent (tan δ) showed a peak around t_g'. Such features resembled the elastic anomalies in the glass transition of noncrystalline polymers with decreasing temperature. On this similarity, a comparison of the properties of the dehydrated HTEWG with those of common glass attracted much interest. One of the most interesting subjects was whether the *glass transition* would take place in the dehydrated HTEWG with increasing temperature. Therefore, Kanaya *et al.* performed DTA measurements of the dehydrated HTEWG with elevating temperature [6], and found an endothermic peak of which the intensity and position increased with elevation of the heating rate. Judging from these characteristics which are also observed in the glass transition [7], they concluded that the dehydrated egg white gel could be regarded as a *glass*, which supported the similarity in the formation feature, namely freezing of the micro-Brownian motion, between the dehydrated HTEWG and glass.

The above-mentioned studies highlighted the interesting aspects of the gel-to-glasslike transition. However, because egg white consists of various kinds of protein and other elements, physical interpretations of observations are often very difficult. Therefore, it should be also necessary to investigate the dehydration process of much simpler gels in order to understand these behaviors more. The polyacrylamide (PAAm) gel, as a candidate, is one of the most familiar chemical gels. In this way, we investigated property evolutions of the *wet* PAAm gel during the dehydration process, and then those of the *dehydrated* gel with increasing temperature.

2. EXPERIMENTALS AND RESULTS

2.1. Sample Preparation

By dissolving 2g of acrylamide, 106mg of NN'-methylenbisacrylamide and 40mg of ammonium persulfate in 40mL water, the precursor solution was prepared. PAAm gel specimen was obtained by maintaining the solution at 50°C for 2h. Then, the samples were soaked in water for one day in order to get rid of unreacted starting materials.

2.2. Weight, Volume and Temperature Change during Dehydration Process

In order to measure the time dependence of the temperature during the dehydration process of the wet gel, we prepared two identical disk-shaped PAAm gel samples. One of them was used for the weight and volume measurement, and the other, for its temperature measurement. The temperature difference between the specimen and room temperature was measured by diverting a differential thermal analysis equipment; attaching the specimen at one side of two oppositely connected thermocouples, and setting the other side in the atmosphere ($19.0 \pm 0.3°C$; $70 \pm 5\%$ in humidity). During the dehydration process, the weight and the volume decreased remarkably with time and the logarithm of which could be approximately fitted by two straight lines altered at a certain period. The weight and the volume finally decreased up to around 5% and 7% of the initial values, respectively. In contrast, the density change was not remarkable, namely, the specimen shrank in the same manner with the weight. This means that the gel networks were not strong enough to maintain their shape in this stage. The temperature of the specimen showed a step-like time dependence (in a range from $\Delta T=-7.8°C$ to $-0.12°C$) around t_g'.

Let us explain the time dependence of the specimen's temperature: In the early stage, the large heat loss due to the free water evaporation can maintain the lower temperature than the atmosphere. When the free water has been used up, the heat inflow from environmental atmosphere may compensate the small evaporation heat loss due to the bound water evaporation. Then, the temperature of the gel can approach room temperature around t_g'.

2.3. Elastic Properties

A sample for the measurement of elastic properties was prepared in a rectangular parallelepiped shape ($5.4mm \times 6.1$ mm $\times$ $38.1mm$ in size, $1267.1mg$ in weight). The elastic property was measured by a dynamical elastic stiffness measurement system at 10Hz. Temperature and humidity were kept within a range of 14.4 to 15.9°C and of 49 to 53%, respectively. At the beginning of dehydration, the amplitude E of the complex elastic stiffness increased gradually. The increasing rate became large with time by dehydration and the value of E hopped up with the magnitude of three order when the weight started to saturate. Then, E finally settled down to a value around 10^8N/ m^2. The loss tangent simultaneously showed an anomalous peak. Such features were quite similar to those of HTEWG during the dehydration process mentioned above.

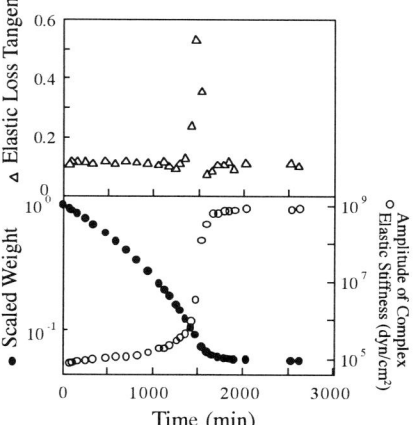

Fig.1 Evolution of weight, amplitute of complex elastic stiffness and elastic loss tangent during the dehydration process of PAAm gel.

Resemblance between PAAm and HTEWG were also revealed in the similar temporal changes of elastic properties during the dehydration process. Somewhat clearer feature in PAAm gel may comes from its simpler ingredient and structure compared with HTEWG. Besides, the absolute value of the elastic stiffness at a glass states was almost equal to that of HTEWG, which was about 100 times smaller than that in usual noncrystalline polymers. This difference can be attributed to the existence of the layer of the bound water cohered to the networks in the dehydrated gel. Such layers may obstruct the networks from making a hard structure by combining with each other.

2.4. Raman Scattering Measurements

2.4.1. Experimental Configurations

In the Raman scattering experiment, a light source of 5145 Å from a polarized Ar^+ ion laser was utilized with an output of 400mW, and the right angle scattering geometry was adopted with the VV-polarization configuration. The scattered light from the specimen was analyzed by a triple monochromator. The resolution of the monochromator was $0.5cm^{-1}$ in the observations of a low-frequency region from $-550cm^{-1}$ to $550cm^{-1}$ and $20cm^{-1}$ in those in the higher frequency region.

2.4.2. Evolution of Raman Scattering Spectrum during Dehydration Process

Two same-sized rectangular parallelepiped specimens ($6.0 \times 6.0 \times 10.0mm^3$, 407mg in weight) were prepared by cutting the purified gel block: one was used for Raman scattering experiment, and the other, for the weight measurement. The weight measurement was performed simultaneously with Raman scattering measurement under almost the same conditions in a temperature range from 22 to 24 °C and in a humidity range from 45 to 55%.

The weight of PAAm gel decreased to 4% of the initial value by the dehydration. On close examination, the time dependence of the logarithm of the weight could be roughly represented by three straight lines: the first stage until t_c' (65000s), the second one with a steeper slope until t_g' (93000s) and the last one with the gentlest slope. In previous studies, attention was paid to the bending point from the second to the last stage (t_g'). In order to denote water amount in the gel more quantitatively, we estimated the water content w by dividing the water weight in the gel by the minimum weight measured at 313440s. w decreased from 22 to 0.14, which was around 3.4 at

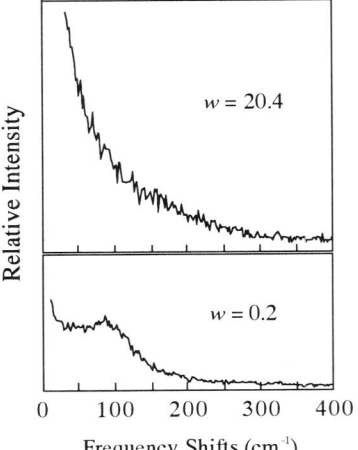

Fig. 2 Low Frequency Raman Spectra during the dehydration process of PAAm gel. (w : water content)

t_c' and 0.30 at t_g'. In an early period of the first stage, a diffusive central mode was observed in Raman spectrum (the upper part of Fig.1). After t_c', the diffusive central mode began to narrow, and a shoulder appeared on the skirt of the diffusive central mode. With the continuous narrowing of the diffusive central mode in the second stage, the shoulder became like a peak (the lower part of Fig.1). After t_g', the diffusive central mode became negligible and the peak could be clearly observed. Thereafter the spectrum hardly changed up to the end of the measurement. As for the high-frequency Raman spectrum, except for some modes of which intensity was increased by the density increase due to the shrinkage, a remarkable change in the spectrum was observed around 3200cm^{-1}: the intensity of the band decreased with time, and it disappeared after around t_g'.

As mentioned above, a rapid increase in ΔT took place around t_g' due to the complete exhaustion of the free water in the gel. Moreover, it is well known that the band at 3200cm^{-1} is the O-H stretching mode of pure water molecules [8]. Because the free water molecules in the gel are said to be in almost the same conditions with those in pure water [8, 9], the marked decrease in the intensity of the 3200cm^{-1} mode after t_g' can be the direct evidence of the complete free water exhaustion of the gel, which confirms the occurrence of the *gel-to-glasslike transition* at t_g' in PAAm gel.

Aging [10], drying [10] and the heat induced gel-to-glass transition [11] have been extensively investigated in the sol-gel science. Some of their results can be taken as keys to explain the spectral anomalies observed in the present investigation. The gel-to-glass transition is a heat-induced transition in which the porous dried gel changes into glass by the collapse of the pores. In a sense, it resembles the gel-to-glasslike transition during the dehydration process. There have been many reports on the low-frequency Raman measurements in the studies on gel-to-glass transition [12-16]. Among them, they reported the emergence of a low-lying Raman band in the heat-induced transition. The low-lying mode is called *boson peak*, which is commonly observed in glass or amorphous materials, although its nature remains a matter of controversy [17]. All results of this system such as elastic properties and Raman scattering spectra suggested that the gel-to-glasslike transition can be ascribed to *a* gel-to-glass transition. In that case, the low-lying Raman band appearing during the dehydration process of PAAm gel can be a boson peak.

However, we have also noticed that there was a difference in the feature of the low-lying mode between the dehydrated gels and usual glasses: the peak of the former lay around 80cm^{-1}, while the latter, usually around 30cm^{-1} [12, 13], as well as the remarkable difference in the value of the elastic stiffness mentioned above. The difference can come from the complicated structure of the dehydrated gel.

Besides, there were no experimental results demonstrating the existence of the glass transition in the dehydrated PAAm gel. In these situation, we have carried out DSC analysis and the observation of the low-lying Raman spectrum of the dehydrated PAAm gel with elevating temperature.

2.4.3. DSC measurement and Evolution of Raman Scattering Spectrum with Increasing Temperature

The purified gel was gently dried by leaving in the atmosphere for 6 days. After the dehydration, a rectangular parallelepiped specimen of $7 \times 8 \times 2mm^3$ was shaped out of the dehydrated gel block. The water content of the dehydrated gel was around 0.17.

The thermal property of the specimen with elevating temperature was measured with a DSC analysis system. The heating temperature was 10°C/min during the measurement. In the DSC curve, a clear bend point was perceived around 60°C.

At room temperature (20°C), a low-lying peak with a maximum around $80cm^{-1}$ was clearly observed. The spectral feature well corresponded to that in Sec.2.4.3. With elevating temperature, the intensity of the peak became much lower, however, the peak could be still distinguished even at 90°C.

The feature observed in the DSC measurement means that the glass transition also occurred in the dehydrated PAAm gel around 60°C at the heat rate of 10°C/min. The value of the glass transition temperature at the same heating rate coincides with that of the dehydrated egg white gel [6], which may suggest the same glass transition mechanism regardless of the difference in the materials. As observed in the previous DTA measurements of HTEWG [6], the glass transition temperature becomes lowered with decreasing the heating rate. Therefore, the glass transition temperature in the present Raman scattering measurement was considered to be below 60°C because the measurement of Raman spectrum took much time. This result indicates that Raman spectrum at 90°C was measured far above the glass transition temperature. Considering the circumstances mentioned above, the dehydrated PAAm gel is thought to be composed of weakly joined hard amorphous clusters of which the frozen structure can be maintained even far above 60°C, the glass transition temperature as a whole.

ACKNOWLEDGMENT

This work was partly supported by Grant-in-Aid for Scientific Research of The Ministry of Education, Science and Culture.

REFERENCES

1. T. Tanaka, Sci. Am. 244 (1981) 124.
2. E. Takushi, L. Asato and T. Nakada, Nature 345 (1990) 298.
3. H. Kanaya, K. Ishida, K. Hara, H. Okabe, S. Taki, K. Matsushige and E. Takushi, Jpn. J. Appl. Phys. 31 (1992) 3754.
4. H. Kanaya, K. Hara, E. Takushi and K. Matsushige, Jpn. J. Appl. Phys. 32 (1993) 2905.
5. N. Koshoubu, H. Kanaya, K. Hara, S. Taki, E. Takushi and K. Matsushige, Jpn. J. Appl. Phys. 32 (1993) 4038.
6. H. Kanaya, T. Nishida, M. Ohara, K. Hara, K. Matsushige, E. Takushi and Y. Matsumoto, Jpn. J. Appl. Phys. 33 (1994) 226.
7. T. Nishida, T. Ichii and Y. Takashima, J. Mater. Chem. 2 (1992) 733.
8. T. Terada, Y. Maeda and H. Kitano, J. Phys. Chem. 97 (1993) 3619.
9. K. Ogino, Y. Osada, T. Fushimi and A. Yamauchi, Geru --Sofutomateriaru no Kiso to Oyo- (Gels -Fundamentals and Applications of Soft Materials-) (Sangyo Tosho,

Tokyo, 1991) p.14 [in Japanese].

10. G. W. Scherer, J. Non-Cryst. Solids 100 (1988) 77.

11. P. F. James, J. Non-Cryst. Solids 100 (1988) 93.

12. G. Mariotto, M. Montagna, G. Viliant, R. Campostrini and G. Carturan, J. Non-Cryst. Solids 106 (1988) 384.

13. J. L. Rousset, E. Duval, A. Boukenter, B. Champagnon, A. Monteil, J. Serughetti and J. Dumas, J. Non-Cryst. Solids 107 (1988) 27.

14. J. Dumas, J. Serughetti, J. L. Rousset, A. Boukenter, B. Champagnon, E. Duval and J. F. Quinson, J. Non-Cryst. Solids 121 (1990) 128.

15. K. Dahmouche, C. Bovier, A. Boukenter, J. Dumas, E. Duval, C. Mai and J. Serughetti, J. Phys. IV 2 (1992) Colloq. C2-127.

16. A. Chmel, A. Krivda, E. Mazurina, V. Shashkin, and V. Zhizhenkov, J. Am. Ceram. Soc. 76 (1993) 1563.

17. R. A. Ramos, Phys. Rev. B49 (1994) 702.

HYDROCOLLOIDS – PART 1
Edited by K. Nishinari
2000 Elsevier Science B.V.

Viscoelastic behavior of tungstic acid gel during the gelation process

H. Okabe[a], K. Kuboyama[b], K. Hara[c] and S. Kai[a]

[a]Department of Applied Quantum Physics and Nuclear Engineering, Graduate School of Engineering, Kyushu University, Fukuoka, 812-8581, Japan

[b]Venture Business Laboratory, Kyoto University, 606-8317, Japan

[c]Institute of Environmental Systems, Faculty of Engineering, Kyushu University, Fukuoka 812-8581 Japan

The viscoelastic properties of tungstic acid during the gelation process was investigated by observing surface waves and rotational resistivity of cylindrical probe. Around gelation point, marked changes were observed in the surface wave velocity and amplitude. From the dispersion relations of the surface wave in the sol and gel states, we found that the surface tension wave drastically changed to the Rayleigh wave at gelation point. These experimental results agreed qualitatively with the theoretical results of viscoelastic model. Moreover, we tried to measure the change in the real elastic modulus near the gelation point for the comparison of the experimental results with theoretical results. From the preliminary experimental results, real elastic modulus probably emerge at the gelation point.

1. INTRODUCTION

Tungstic acid which is a kind of transition metal hydroxide exhibits electrochromism and photochromism.[1-3] Because of these features and its gelling ability, it is expected to be used as display devices or shading devices utilizing the sol-gel method, which is a processing technology of glass at relatively low temperatures and used to get a device having large area easily.[4,5] It is apparent that the formation process may influence the properties of products considerably.[5] However, although the properties of the products have been extensively investigated, there have been very few studies on the gel structure and formation process of tungstic acid.[6-9]

Moreover, although some viscoelastic measurements of other gels have been reported,[10,11] they were performed after completion of gelation at several temperatures, and they focused on the static features of elasticity at some stages of the gelation related to the percolation theory, which essentially holds in an equilibrium system.[10] Therefore, we have been studying the kinetic features during the gelation process in the nonequilibrium system.[12-14]

In the gelation process, a colloidal solution becomes a solid (gel) by the formation of a network in which solvents are captured; the composite structure determines the characteristics of the gels. Such a solidification process leads to change in the viscoelasticity of the network of the gel. However, conventional viscoelastic measurements (such as ultrasonic methods) could not detect such changes which occur during the sol-gel transition because of the majority of the solvent which may not suffer much change. Therefore, in the present study, we mainly paid attention to the changes of the surface wave properties, which could be expected to reflect the dynamical viscoelasticity at a high sensitivity.[12-14]

2. EXPERIMENTAL

Figure 1 shows the phase diagrams of tungstic acid gels.[15] As the usual oxidation state of tungsten is W^{6+}, the condition changes along the broken line in Fig. 1. When the pH value becomes lower than 6, the precursor arises through the olation and/or oxidation as follows,

$$2\,\text{T-OH} \rightarrow \text{T}\underset{\text{OH}}{\overset{\text{OH}}{<}}\text{T},\ \ 2\,\text{T-OH} \rightarrow \text{T-O-T} + H_2O.\ [4,15]$$

The sample of tungstic acid was prepared from sodium tungstate aqueous solution (0.75M) by $Na^+ \Rightarrow H^+$ ion exchange. In the procedure, the pH value was lowered from 9.9 to 1.4, and the gelation was initiated.

Figure 2(a) shows the surface wave measuring system constructed in our laboratory.[12] In the measurement, the sample prepared in the manner described above was poured into a cell. A glass tip attached to the edge of a PZT bimorph vibrator was set so as to touch the surface of the sample at the center of the vessel. To excite the surface wave, a pulse that consists of six cycles (50,100,150,200Hz) was applied to the bimorph vibrator. The waves propagating on the surface of the sol or gel were measured using the optical deflection method. To observe waves at two points on the sample surface, two He-Ne laser beams were directed to the two isolated points on the sample surface.[13] The distance between the two laser-beam spots was about 2mm (distance from the vibrating tip to the first detection

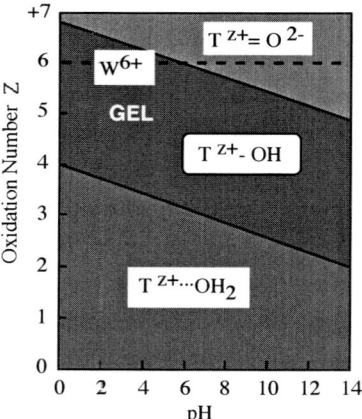

Fig. 1 Phase diagram of chemical equilibrium in aqueous solution of transition metal oxides. The ligand is present as H₂O, OH⁻(hydroxide), O²⁻(oxide).[15]

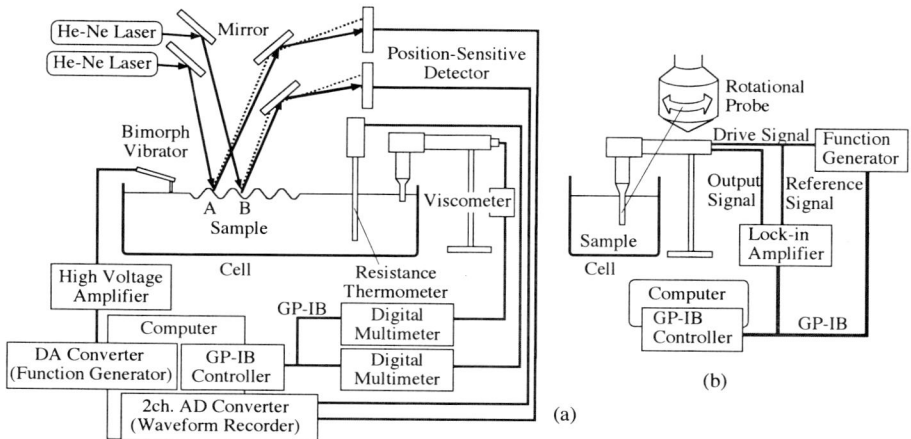

Fig.2 Schematic diagram of (a) surface wave, and (b) the rotational probe type viscoelastic measuring systems.

point was ~7.5mm, and was ~9.5mm to the second point). The reflection angles of the beams at the two points were detected by two position sensitive detectors (PSDs). The reflection angle changes with propagation of the surface waves. The signals detected by the PSDs were recorded by an analog-digital converter board in a computer. We also performed viscosity measurements (Yamaichi Denki : VM-1A-L) and tilting tests[12], simultaneously.

From the results of tilting tests, we confirmed that the gelation occurred homogeneously within a range of wavelength.

Moreover, we measured frequency dispersion of viscoelasticity independently. Figure 2(b) shows the schematic diagram of the measuring system (to be exact, the frequency dispersion measuring system of the rotational resistivity of the cylindrical probe). In this system, the sensor output V is formulated as a following equation.

$$V(f,t) = \frac{K}{r(f) + R(f,t)} v$$

where K is a constant, v and f are the drive voltage and frequency, r and R are the mechanical impedances of the probe and the sample, respectively.

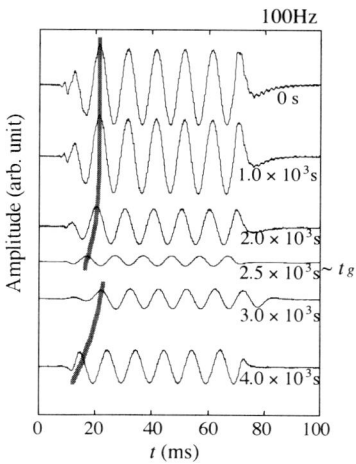

Fig. 3. Time variations of the surface wave profiles of tungstic acid (100Hz).The gelation occurred at around the time of 2.5×10^3s.

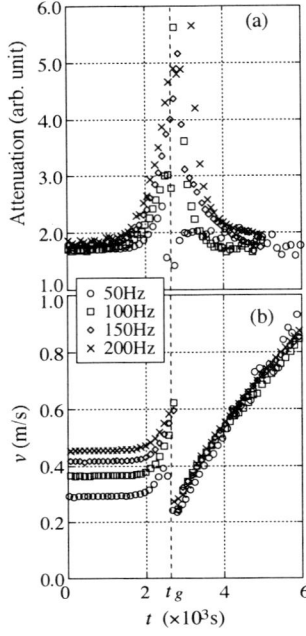

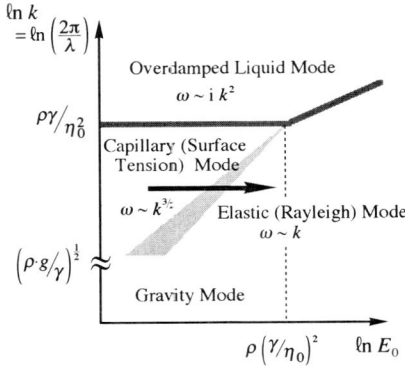

Fig. 5 Surface mode diagram for gels as a function of the wave number k and elastic modulus E_0, where ρ γ and η_0 are the density and the surface tension of the gel and the viscosity of the solvent, respectively. The arrow at the center of the figure shows the change of elasticity during the gelation process.

Fig.4 Time dependences of (a) attenuation coefficient and (b) wave velocity at several frequencies.

3. RESULTS AND DISCUSSION

Figure 3 shows the waveforms of the surface waves at first points (100Hz) at several times during the gelation process of tungstic acid. Tungstic acid was in the sol state before $\sim 2 \times 10^3$s and the gel state after $\sim 3 \times 10^3$s. The lines in Fig. 3 indicate a movement of the wave train in the time axis.

The time variations of the attenuation coefficient at several frequencies during the gelation process are shown in Fig. 4(a). The attenuation becomes the largest near the time of 2.5×10^3s and this point was the gelation time determined by the tilting test as described in previous paper.[12] In this region, the energy loss related to the construction of the gel network, namely, the growing cluster-cluster and cluster-network interactions, must be maximal. The data points of 50Hz around gelation point deviates from tendencies of other frequencies. This may be due to the nonlinearity and lower S/N ratio.

Figure 4(b) shows the evolution of the velocity of the surface waves. It was found that the surface wave velocity in the sol state is close to that of water, and after the gelation time, it increases considerably. This result reflects the fact that as gelation proceeds, the density of crosslink increases and elasticity emerges. Moreover, near the gelation point, we found interesting features in the time variation of the velocities. It has a maximum near the gelation point, and after the discontinuous decrease it again increases gradually as the gelation proceeds as shown in Fig. 4(b). The velocity increased before the gelation point is possibly caused by the increase of the effective surface tension coupled with the elasticity of the networks.

There are some theoretical studies of surface waves on gels.[16-18] Figure 5 shows surface mode diagram for gels, which shows the region of capillary (surface tension), Rayleigh (elastic), overdamped and gravity modes as a function of wave number k and elastic modulus E_0.[16,17] Since this diagram is based only on the simple model of viscoelasticity and not on the detailed structure, we may assume that the progress of gelation is the same as the increase of the elastic modulus indicated by the arrow in the center of Fig. 5.

At the sol-gel transition, the surface waves changed from the surface tension waves to the surface elastic (Rayleigh) waves, which is clearly shown in the time dependence of the wave number exponent x ($\omega \propto k^x$). As shown in Fig. 6(a), the exponent x was about 1.4 (the value of 1.5 and 1.0 indicate surface tension and elastic wave respectively) in the sol state and decreased rapidly at the gelation time. As the gelation progressed, it approached a constant value. We measured the absolute value of the complex viscosity (to be exact, the rotation resistivity of cylindrical probe) simultaneously as shown in Fig. 6(b). Figure 7 shows wave number exponent x (in the dispersion relation $\omega \propto k^x$) scaled by an absolute value of the complex viscosity. In the figure, the change of the exponent x agreed qualitatively with the diagram.

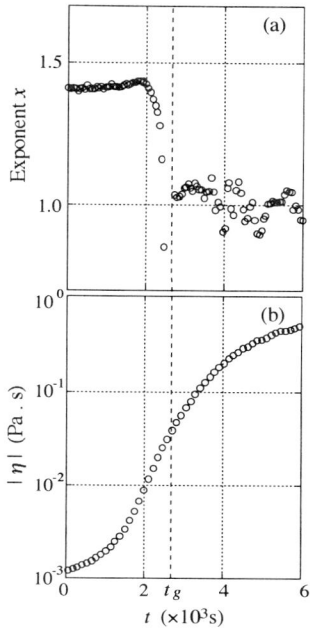

Fig. 6. Time dependence of (a) wave number exponent x in the dispersion relation $\omega \propto k^x$ and (b) absolute value of the complex viscosity (500Hz).

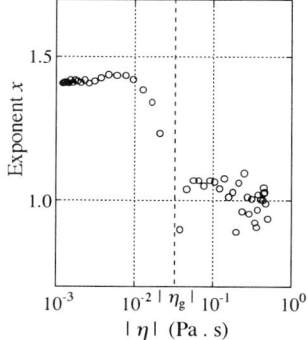

Fig.7. Absolute value of complex viscosity dependence of wave number exponent x in the dispersion relation $\omega \propto k^x$.

84

Figure 8 shows the frequency dependence of the sensor outputs V at first (sol states) and last (gel states) measurements. The sensor outputs at peak frequencies decrease as the gelation proceeds, and the peak frequency increases slightly (from 514Hz to 515Hz). Therefore, the elastic modulus of tungstic acid may emerge at gelation point. Since the probe made of steel is rigid, $r(f)$ is larger than $R(f,t)$ except the vicinity of mechanical resonant frequency ($r(f)$ becomes minimum at the frequency). For more precise measurements, the improvement of the probe that reduce $r(f)$ is needed.

As described above, the viscoelasticity of tungstic acid gel changes drastically at the gelation point. Further experiments and analysis are now in progress.

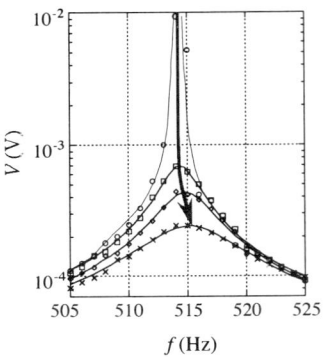

Fig. 8 Frequency dependences of sensor outputs. The arrow indicates the time change of sensor outputs.

ACKNOWLEDGMENT

This work was partly supported by the Grant-in-Aid from the Ministry of Education, Science, Culture and Sports.

REFERENCES

1. S. K. Deb, Philos. Mag., **22** (1973) 801.
2. K. Kuboyama, K. Hara and K. Matsushige, Jpn. J. Appl. Phys., **33** (1994) 4135.
3. K. Kuboyama, K. Hara, K. Matsushige and S. Kai, Jpn. J. Appl. Phys., **36** (1997) L443.
4. J. Livage, J. Solid State Chem., **64** (1986) 322.
5. J. Livage and J. Lemerle, Ann. Rev. Mater. Sci., **12** (1982) 103.
6. K. Hara, H. Kanaya, H. Okabe and K. Matsushige, J. Phys. Soc. Jpn., **60** (1991) 3568.
7. K. Hara, K. Kuboyama, H. Okabe, K. Matsushige and Y. Ishibashi, J. Phys. Soc. Jpn., **61** (1992) 2147.
8. K. Hara, K. Kuboyama, H. Okabe and K. Matsushige, Jpn. J. Appl. Phys., **32** (1993) 996.
9. K. Kuboyama, K. Hara, H. Okabe and K. Matsushige, J. Phys. Soc. Jpn., **62** (1993) 357.
10. P.-K. Choi, Jpn. J. Appl. Phys, **31** Suppl. 31-1 (1992) 54.
11. H. Kikuchi, K. Sakai and K. Takagi, Phys. Rev. B, **49** (1994) 3061.
12. K. Motonaga, H. Okabe, K. Hara and K. Matsushige, Jpn. J. Appl. Phys., **33** (1994) 3514.
13. K. Motonaga, H. Okabe, K. Hara and K. Matsushige, Jpn. J. Appl. Phys. **33** (1994) 2905.
14. H. Okabe, K. Kuboyama, K. Hara and S. Kai, Jpn. J. Appl. Phys., **37** (1998) 2815.
15. D. L. Kepert, The Early Transition Metals, Academic Press, New York, London, 1972.
16. J. L. Harden, H. Pleiner and P. A. Pincus, J. Chem. Phys., **94** (1991) 5208.
17. S. Kubota and H. Nakanishi, Prog. Theor. Phys., Suppl., **126** (1997) 359.
18. J. Jäckle and K. Kawasaki, J. Phys., Condens. Matter, 7 (1995) 4351.

HYDROCOLLOIDS – PART 1
Edited by K. Nishinari
2000 Elsevier Science B.V.

The volume phase transition of DNA and RNA gels

E. Takushi

Department of Physics, University of the Ryukyus, Okinawa 903-0129

The first order volume phase transitions of DNA gel, RNA/DNA gel and DNA/ magnetic fluid mixtures gel in acetone-water mixtures are described. The phase transition is measured as a function of solvent composition, temperature and pH, and pattern formations on the surface of DNA gels are observed to be shown as 5- and 6-sided honeycomb cells like quasi-crystals. A phenomenological theory which describes size effect on the transition is tentatively demonstrated by considering the surface tension acting on the ionic gel such as DNA, RNA/DNA mixtures and qualitatively compared with the experimental results.

1. INTRODUCTION

Under a certain condition in a disordered system such as the polymer network of chemically crosslinked gel, there occurs a discontinuous transition in equilibrium volume for a continuous change in solvent composition, pH, and temperature. This is mainly due to the ionization of the microscopic environment of the network distribution.

Recently, Amiya and Tanaka [1] clearly demonstrated the existence of a first order volume phase transition of natural polymer gels in acetone-water mixtures such as polypeptides (gelatin), polysaccharides (agarose) and polynucleotides (DNA). These three kinds of natural polymers cross-linked by ethylene-glycol-diglycidyl-ether (EGDE) undergo a discontinuous volume phase transition when their solvent composition is varied; and infinitesimal reversible volume change of the order of 10-100 times. In this paper, we wish to suggest the existence of the size effect of ionic gel such as DNA and $RNA_X DNA_{1-X}$ gels on the volume phase transition. Volume change of the larger gel occurs at higher concentration of solvent. This effect may have an interesting novel application for the biological elementary process of organs.

A size dependence on the phase transition for submicron or small gels ($\sim$mm) has already been suggested by

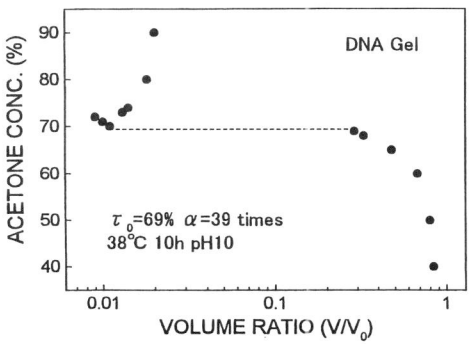

Fig. 1   Volume phase transition of DNA gel.

Li[2], and Hirose, *et al.* in Tanaka's group of MIT[3]. They kindly made their manuscript available to us in advance of publication. Their discussions seem to be consist of the following. (1) For a swollen and small gel the osmotic pressure of a gel network due to the surface tension should be considered and they investigated the number of dissociated counter ions per chain f, the surface tension γ and gel structure parameter ν. (2) Shape dependence in acrylamide gel system on the transition concentration of solvent are pointed out[3].

In 1989, Li[2] studied the linear swelling ratio of the surface tension induced the volume phase transition in polyacrylamide gels with different initial diameter in water and showed when the size of gel is the order of several hundred microns (sub-millimeter), the surface tension is not negligible. We have tried to adapt this idea to explain an origin of the size effect on the phase transition of DNA gel found by the chance in case of a lack of a series of the same inner radius of glass tubes[4]. Considering the surface tension of gel in the Flory-Huggins theory of osmotic pressure of gel a qualitative feature was obtained for the size dependence on the phase transition of DNA gel, RNA/DNA mixed gel, and DNA/magnetic fluids mixed gel.

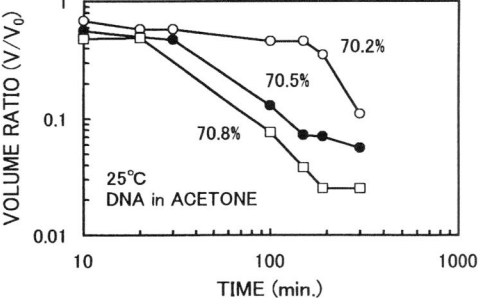

Fig. 2   The time course of collapse for DNA gels in acetone-water mixtures.

2.   EXPERIMENTAL RESULTS AND DISCUSSION

DNA gels were prepared as follows[1]. DNA fiber (Sigma, type III, Sodium salt from salmon) in water (20%) is crosslinked by using EGDE (50% of DNA) at pH 10~11 at 55 °C for 3 hours keeping in the furnace after 4 hours stirring. Figure 1 shows a typical volume phase transitions for a DNA gel as the acetone concentration $\tau(\%)$ is varied. The quantity V/V_0 represent the ratio of the final volume (normally the collapsed state) to initial volume in the swollen state. The ratio is given by $(d/d_0)^3$, where d_0 and d are the initial and final equilibrium diameters of the gel, respectively. The volume change at the first order discrete phase transition is $V/V_0 = 5 \sim 56$ times at 70%(acetone) $\sim$

Fig. 3   Pattern formation of DNA gel in ethanol-water solution (72%). Irregular (a) and regular (b) patterns appear on the surface.

80%(ethanol) concentration in water.

In Figure 2, the collapsing of the gel, the volume ratio V/V_0 is shown as a function of time for DNA gels shrinking in acetone-water mixtures very near the critical concentration. The delay before significant volume change at the concentration $\tau_0 = 70.2\%$ suggests that it is the closest concentration to the transition point among the three.

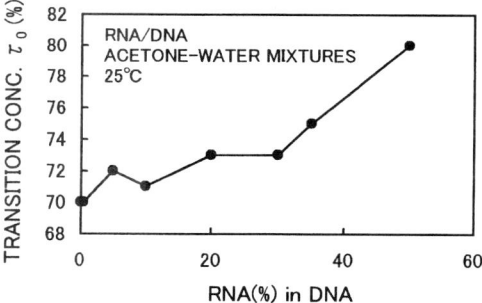

Fig. 4　Volume phase transition points (τ_0) as a function of the weight of RNA in DNA gel in acetone-water mixtures at 25 ℃.

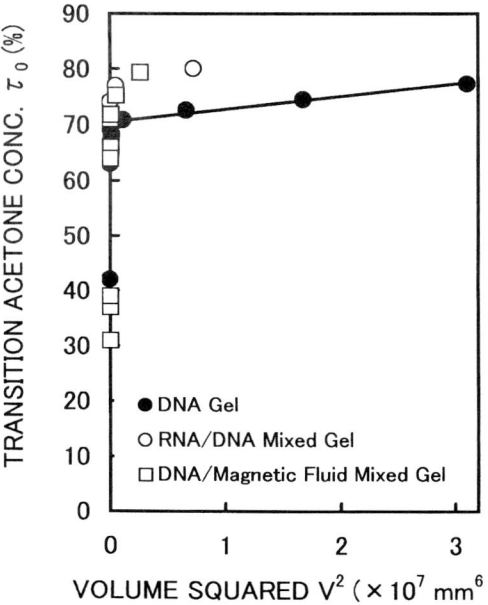

Fig. 5　Size effect of DNA, RNA/DNA and magnetic DNA gels in acetone-water mixtures at 25 ℃.

Next we measured the equilibrium volume ratio as a function of temperature, for a fixed acetone-water concentration of 65%. At the temperature higher than about 10 ℃, the gel network shrinks only a little, but below it, the degree of the collapse is about one-thirteenth.

Geometrical pattern formation during volume change are observed in DNA gels (Fig. 3). The patterns disappear before the volume reaches equilibrium, where the original shape and a homogenous density return. When DNA gel undergoes swelling in ethanol-water mixtures, or in pure water, patterns are similarly manifested. Though the initial and final gels, (2.5 mm and 4.5 mm in diameter, respectively), are smooth and homogeneous, on swelling dimples appear at first, then raised walls between neighboring dimples as their size increase. The walls form predominantly 5- and 6-sided honeycomb cells, which seem to be extended into the gel, similar to the structure of quasi-crystals like Al-Mn alloy. These cells coalesce, forming larger and larger cells, until the pattern finally disappears.

In our experiments, a preparation of RNA/DNA mixed gels, because pure RNA gel is still unsuccessful, are carried out, for the first time in the measurement of the volume phase transition, by chemically crosslinked EGDE stirring for 2 hours keeping at 38 ℃ for 10 hours in the furnace. The first order volume phase transition exists in the RNA/DNA mixed gels and the mass of the RNA dependence (wt%) in DNA on the transition τ_0 are obtained as a function of the RNA mass until 50% of mixtures (Fig. 4).

To demonstrate the afore mentioned size effect, the disk-shaped DNA gels of 5 mm, 8 mm and 18 mm in diameter were prepared, cut, and immersed in the acetone-water or ethanol-water mixtures. After gels were reached an equilibrium state, diameter of gels were measured by a travelling microscope. The transition concentration τ_0 increase with the diameter of gels ($\tau_0 = 69\%$ for $d_0 = 5$ mm, 71.0% for 8 mm and 77.5% for 15 mm, respectively) (Fig. 5). The size effect of DNA gel is 6 times stronger at lower temperatures (-13 ℃) than the room temperatures ($10{\sim}38$ ℃).

3. A PHENOMENOLOGICAL THEORY OF SIZE EFFECT

To understand a mechanism of this size effect on the degree of the swelling and the volume phase transition existed in the ionic gels, we assume that $\pi = 0$, for a flat surface gel, where π is the osmic pressure. If the surface is curved, the equilibrium condition should be $\pi = \gamma/\rho$, where γ is the surface tension and ρ is the radius of the curvature of the surface of the gel. As long as γ is small, this effect is negligible. However, if γ is unusually large, the dependence of the bulk properties on the surface tension will become detectable.

Here we try to explain the size effect of an ionic gel on the phase transition qualitatively by using the Flory-Huggins theory of the osmotic pressure of the gel π_F.

In general, when the gel is small the osmotic pressure due to the surface tension π_s should be considered to reach equilibrium ($\pi_{total} = 0$), that is

$$\pi_F + \pi_s = 0, \tag{1}$$

while

$$\pi_F = -\frac{NkT}{v}\left[\phi + \ln(1-\phi) + \left(\frac{1}{2}\frac{\Delta F}{kT}\phi^2\right)\right] + \nu kT\left[\frac{1}{2}\frac{\phi}{\phi_0} - \left(\frac{\phi}{\phi_0}\right)^{\frac{1}{3}}\right] + \nu fkT\left(\frac{\phi}{\phi_0}\right) \tag{2}$$

and

$$\pi_s = \frac{\gamma}{d_0}\left(\frac{\phi}{\phi_0}\right)^{\frac{1}{3}}, \tag{3}$$

where N is Avogadro's number, k is the Boltzman constant, T is the temperature, v is the molar volume of the solvent, ϕ is the volume fraction of the network, ΔF is the free-energy decrease associated with the formation of a contact between polymer segment, ϕ_0 is the volume fraction of the network at the condition of the constituent polymer chains have random-walk configuration and ν is the number of constituent chains per unit volume at $\phi = \phi_0$[6].

For a swollen gel, if we define α as the relative linear size of the gel, $\alpha = (V/V_0)^{1/3} = (\phi_0/\phi)^{1/3}$, then at equilibrium α is given by:

$$\alpha_e = \left[\left(f + \frac{1}{2}\right)\left(1 + \frac{\gamma}{\nu kTd_0}\right)\right]^{\frac{1}{2}}. \tag{4}$$

For a collapsed gel at equilibrium, the second term in Eq.(2) is small and may be neglected.

Then, the reduced temperature can be written in the form:

$$\frac{\gamma}{d_0} + \frac{NkT}{v}\left[\phi + \ln(1-\phi) + \frac{\Delta F}{kT}\phi^2\right] = 0. \tag{5}$$

Then the reduced temperature (transition point) can be written as

$$\tau_0 = 1 - \frac{\Delta F}{kT} = \left(\frac{2v}{NkT}\frac{\gamma}{d_0}\frac{1}{\phi^2}\right) - \frac{2}{3}\phi = \beta_1 V^2 - \beta_2 V^{-1}, \tag{6}$$

where, $\beta_1 = (2v/NkT)(\gamma/d_0)$ and $\beta_2 = 2/3$. Equation (6) when associated with surface tension seems to describe qualitatively the experimental results in Figure 5 for the case of ionic gels such as DNA gel. Though equation (6) is valid only for the condition of $\tau_0 = \beta_1 V^2 - \beta_2/V \geq 0$, that is, $\beta_2/\beta_1 = V^3$; or $\tau_0 > 0$ is valid at which V has the larger values, namely, β_2/V approach to zero asymptotically, the phase transition point τ_0 is approximately proportional to the volume squared V^2 of gel in a large scale. Similar size effect on the transition takes place in RNA/DNA mixed gels and the magnetic fluids doped-DNA gel shown in Figure 5 as well.

REFERENCES

1. T. Amiya and T. Tanaka, Macromol. 20 (1987) 1162.
2. Y. Li, Ph.D Thesis, p132 (MIT, 1989).
3. Y. Hirose, T. Amiya, Y. Hirokawa and T. Tanaka, Macromol. 20 (1987) 1342.
4. E. Takushi and Z-Q. Qi, Extended Abstract, Fall Mtg. Jpn. Phys. Soc. p533 (Gifu, 1990).
5. S. Hirotsu, Extended Abstract, Fall Mtg. Jpn. Phys. Soc. p535 (Gifu, 1990).
6. T. Tanaka, et al., Phys. Rev. Lett. 45 (1980) 1636.

HYDROCOLLOIDS – PART 1
Edited by K. Nishinari
2000 Elsevier Science B.V.

91

Rheological properties and microstructure of monodispersed O/W emulsion agar gel

S. Gohtani, K-H. Kim and Y. Yamano

Department of Biochenistry and Food Science, Kagawa University, Miki, Kagawa 761-0701, Japan

Effect of oil droplet on the rheological properties and microstructure of monodispersed O/W emulsion agar gel prepared by membrane emulsification technique was investigated by mechanical measurements and SEM observation. The compressive gel strength of emulsion gel decreased with an increase in oil droplet size, while that of emulsion gel determined by puncture test showed almost no change. Cryo-SEM observation revealed that oil droplets aggregated in the emulsion gel and that the gel had some void spaces between gel network and the oil droplet aggregate.

1. INTRODUCTION

An emulsion gel containing oil droplets is a good food model. Oil droplet size and oil volume fraction have important effects on the texture and rheological properties of the emulsion gel [1-5]. Most studies, however, have reported on the behavior of polydipersed emulsions. It is necessary to use an emulsion gel containing the monodispersed emulsion for rheological discussion of the emulsion gel.

The purposes of this study are to investigate the effects of oil droplet size and volume fraction on the rheological properties and to observe microstructure of emulsion agar gel.

2. MATERIALS AND METHODS

Commercial corn oil of food grade quality (Ajinomoto Co. Inc., Tokyo) and reagent grade agar powder (Daishin Co., Tokyo) were used for gel preparation. Emulsifier (polyglycerolesters of fatty acid) was purchased from Sakamoto Co. Ltd., Osaka.

2.1. Emulsion Preparation

The aqueous phase was prepared by dissolving 1% emulsifier in distilled water. Emulsions of different droplet sizes were produced by using different pore size porous glass membranes [6]. The emulsions with mean drop diameters of 1.5, 3.3, 6.5 and 12.2 μm were prepared. An Image-Analyzer was used to measure the distributions of oil droplets in o/w emulsions [7] and the mean number diameter was calculated. The photographed oil droplets of the emulsions

92

are shown in Figure 1.

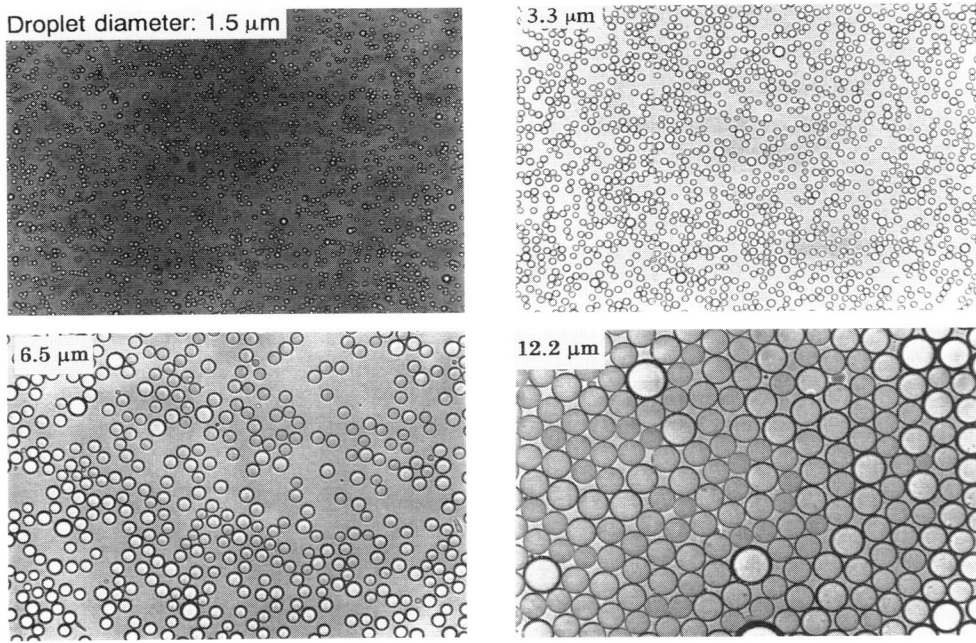

Figure 1 Microphotograph of emulsion prepared by membrane emulsification [8]

2.2. Gel Formation

Agar powder (1%) was dispersed in the emulsions containing 0.1, 0.2 and 0.3 volume fractions of oil, after which these agar emulsion dispersions were stirred gently for 60 min at 20°C , heated for 30 min at 70°C and then heated again for 30 min at 90°C during swelling. After being degassed, these sols were poured into cylindrical glass containers (inside diameter 24 mm, height 38 mm). These glass containers were held for 1 h at room temperature and then held in a water bath for 24 h at 25°C before measurements. In each gel, the uniformity of dispersibility of emulsion droplets was confirmed by determining droplet number and diameter at top, center and bottom of emulsion gel as described by Yamano et al [9].

2.3. Mechanical Analysis

Gel samples used for determining gel strength by Rheolometer (RE-33005, Yamaden Co., Tokyo) at 25°C were cut into sections 20 mm in height and 24 mm in diameter.   For determining compression gel strength, the gels were compressed to 80% deformation with a 2 kg compression load cell by a plunger with a diameter of 40 mm at a cross head speed of 0.5 mm/s. For puncture test, the gels were punctured at 1 mm/s to a depth of 16 mm by a flat plunger with a diameter of 5 mm. The results for compressive and puncture properties were expressed in terms of relative values: R = Se/So, where Se and So are the data of the filled gel and the droplet free gel, respectively [3].

2.4. Cryo-Scanning Electron Microscopy (Cryo-SEM)

The center of the gels were cut into cubes (approx. 3 x 3 x 3 mm) and were dehydrated in a 50% ethanol for 2 min. After being frozen by immersing in liquid nitrogen, the specimens were fractured in liquid nitrogen. After the ice on the specimens was sublimed, the specimens were coated with gold and observed in a Cryo-SEM (S-800, Hitachi Co., Ltd., Tokyo) at -120°C[10].

3. RESULTS AND DISCUSSION

3.1 Stability and Uniformity of Oil Droplets in Preparation of O/W Emulsion Agar Gel [8, 9]

It may be thought that it would be difficult to maintain the monodispersity of the O/W emulsion during this procedure, since the uniformity of oil droplet distribution may be lost by creaming and coalescence before the gel gelatinizes. We examined the stability of oil droplets in the O/W emulsion and uniformity of oil droplet distribution in the emulsion gels during emulsion sol and gel preparation.

At each sol preparation stage, almost no differences in the droplet size or the droplet number per unit area in excess of the range of measurement error were found (Table 1). From these results, it can be concluded that the stability of oil droplets was maintained throughout the preparation period of the emulsion sol. It is thought that the monodispersity of the emulsion droplets imparted high stability to the emulsion and that agar acts as a barrier against coalescence through collision of droplets.

Table 1 Size and number of oil droplets in emulsion containing agar [9]

Oil volume fraction		Emulsion in agar sol		
		20°C 1h	20°C 1h + 70°C 30min	20°C 1h +70°C 30min + 90°C 30min
0.05	Size (μm)	3.1±0.0	3.1±0.1	3.2±0.0
	Number*	1630.0±59.9	1613.6±33.6	1632.0±47.0
0.20	Size (μm)	3.2±0.1	3.3±0.1	3.3±0.0
	Number*	1493.8±25.0	1503.0±58.8	1491.6±57.2

*. Droplet number was estimated in an area of 0.1 mm^2 under a microscope.

Table 2 shows the mean droplet size and the number of droplets in a prepared emulsion gel with 0.1 oil volume fraction and a droplet mean diameter of 12 mm for which creaming is most rapid. There were no significant differences in the mean droplet size and the number of droplets among the three sections, top, middle and bottom of the emulsion gel. We believe that the viscosity of the emulsion sol is sufficient to prevent creaming of oil droplets before gelation of the emulsion agar sol.

Table 2: Distribution of oil droplets of 12 μm mean diameter in an emulsion gel
with 0.1 of oil volume fraction [8]

	Upper	Middle	Bottom
Mean diameter(μ m)	12.03±0.30	12.13±10.10	12.00±0.03
Number of oil droplets*	238.6±8.5	241.3±8.74	242.2±8.08

*. Droplet number was estimated in an area of 0.1 mm ^{2} under a microscope.

3.2 Effect of Oil Droplets on Compression of Emulsion Gel [8]

The compressive properties of emulsion gels containing different sizes of oil droplets are shown in terms of relative values in Figure 2. The change in compressive stress of emulsion gels decreased as the oil volume fraction increased (Figure 2A). The compressive stress showed a much greater difference between the gels of oil volume fractions 0.1 and 0.2 than between those of oil volume fractions 0.2 and 0.3. For each oil volume fraction gel, the change in compressive stress decreased as the oil droplet size increased. Little difference in compressive strain with the oil volume fraction was found in Figure 2B. For each oil volume fraction gel, the decrease in compressive strain of the gels was significant with increasing oil droplet size. As shown in Figure 2C, the compressive energy of the emulsion gels decreased with increasing oil volume fraction. The compressive energy of the emulsion gel with oil volume fraction 0.1 decreased remarkably up to 6.5 μm but only gradually thereafter, while those of 0.2 and 0.3 oil volume fractions decreased linearly.

All relative compressive values of emulsion gel were less than 100%. The emulsion gels were week in compression with the oil-free gel. It is suggested that emulsion droplets act as structure breakers for the agar network system.

3.3 Effect of Oil Droplets on Puncture of Emulsion Gel [8]

The puncture properties of emulsion gels containing different sizes of oil droplets are shown in terms of relative values in Figure 3. The changes in puncture values of emulsion gels with oil droplets were very small in comparison with the changes in compressive values (Figure 2) and puncture test is not available to compare mechanical properties for the emulsion gels with different size oil droplets. We think that it might be impossible to detect the changes in puncture properties of emulsion gels with different oil droplet sizes since the puncture plunger had only very small surface contact with the samples. The force measurements of the emulsion agar gels depend on the applied force and the diameter of the plunger. Thus, compressive measurement is more useful to investigate the effect of oil droplet size on the physical properties of the emulsion gel than puncture measurement.

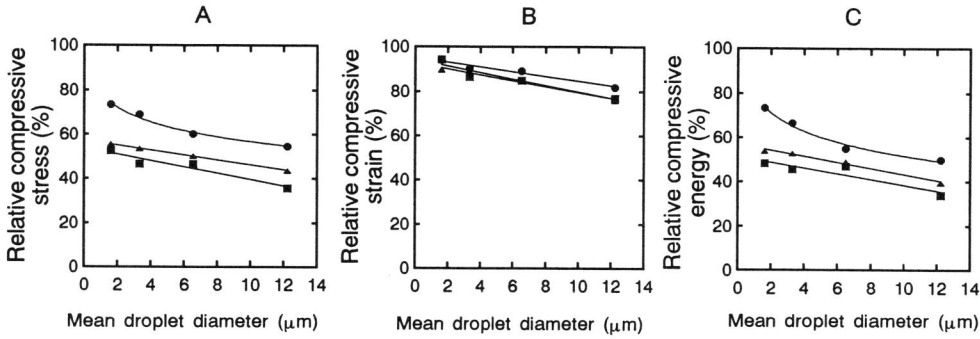

Figure 2 Effects of Droplet Sizes on Compressive Properties of Emulsion Gels[8]
          A, B and C show the relations of mean droplet diameter and the relative compressive stress, strain and energy, respectively. The symbols represent the gels of oil volume fractions 0.1(●), 0.2(▲) and 0.3(■), respectively.

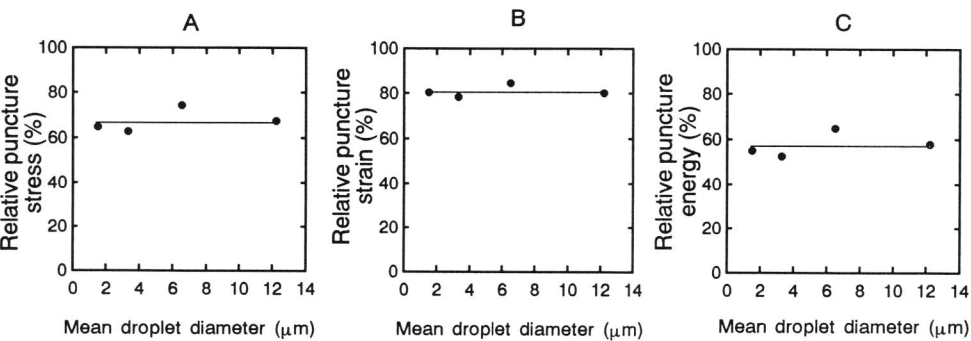

Figure 3 Effects of Droplet Sizes on Puncture Properties of Emulsion Gels of Oil Volume Fraction 0.1[8]
          A, B and C show the relations of mean droplet diameter and the relative puncture stress, strain and energy, respectively.

3.4. Microstructure of Emulsion Agar Gel [10]

The oil droplets in the gel network were not separately dispersed but tend to aggregate. And emulsion gel had some void spaces between the gel networks and the oil droplets aggregate and gel strands of the emulsion gel did not cover the oil droplets completely. We think the void spaces were not originated from dehydration and/or freezing treatments before Cryo-SEM observation since the void spaces were not observed for other emulsion gels (gelatin, whey protein, gellan gum etc.) by the treatments.

It is presumed that the strength of agar emulsion gels containing oil droplets is weakened compared with an oil-free gel, since there are many liquid spaces (the oil droplets and void spaces are all liquid) in emulsion agar gels. And we think that the effects of oil droplet sizes on the compressive properties of the emulsion agar gels may be due to the different distortion resistance of the liquid spaces in the gels [10].

REFERENCES

1. J. E. Kinsella, Crit. Rev. Food Sci. Nutr., 21 (1984) 197.
2. K. R. a. G. Langley, M. L., J. Texture Studies, 20 (1989) 191.
3. D. J. McClements, F. J. Monahan and J. E. Kinsella, J. Texture Studies, 24 (1993) 411.
4. Y. Matsumura, I. J. Kang, H. Sakamoto, M. Motoki and T. Mori, Food Hydrocolloids, 7 (1993) 227.
5. H. Rohm and A. Kovac, J. Texture Studies, 24 (1994) 411.
6. E. S. Chen, S. Gohtani, T. Nakashima and Y. Yamano, J. Jpn. Oil Chem. Soc. (Yukagaku), 42 (1993) 972.
7. Y. Yamano, S. Gohtani and S. Nakayama, Nippon Nogeikagaku kaishi, 58 (1984) 161.
8. K.-H. Kim, S. Gohtani and Y. Yamano, J. Texture Studies, 27 (1996) 655.
9. Y. Yamano, Y. Kagawa, K.-H. Kim and S. Gohtani, Food Sci. Technol., Int., 2 (1996) 16.
10. K.-H. Kim, S. Gohtani, R. Matsuno and Y. Yamano, J. Texture Studies, 30 (1999) in press.

3. POLYSACCHARIDES

HYDROCOLLOIDS – PART 1
Edited by K. Nishinari
2000 Elsevier Science B.V.

Viewing biopolymer networks, their formation and breakdown by AFM

V.J. Morris, A.P. Gunning, A.R. Kirby, A.R. Mackie and P.J. Wilde

Institute of Food Research, Norwich Research Park, Colney, Norwich NR4 7UA, UK.

The technique of Atomic Force Microscopy (AFM) generates images by 'feeling' surfaces with a sharp probe. This permits a magnification range spanning that of both the light and electron microscopes. The imaging process allows biological samples to be imaged in liquid or gaseous environments without elaborate sample preparation methods. Thus AFM provides a powerful technique for probing the structure of networks formed by biopolymers, their formation and/or breakdown. Methods have been developed for the routine and reliable imaging of polysaccharides and for proteins at interfaces. This allows AFM to be applied to study biopolymer functionality. Polysaccharide interactions responsible for bulk structure and rheology will be illustrated through studies on xanthan microgels, investigations of the mechanisms of gelation of gellan gum and images of gellan networks in hydrated films and gels, mixed carrageenan - cellulose networks in semi - refined carrageenan, microbial cellulose, and the cellulose based structures of hydrated plant cell walls. Biopolymers play an important role in the stabilisation of foams and emulsions. The structures formed by proteins at interfaces can be visualised through the use of AFM and Langmuir - Blodgett methods for sampling the interfacial structures. Such studies will be illustrated by investigations of the formation of gelatin networks at an air - water interface and through the discovery of new mechanisms for the surfactant destabilisation of protein networks at air - water interfaces.

1. INTRODUCTION.

The AFM is perhaps the most versatile member of a family of microscopes known as probe microscopes. These microscopes differ from conventional microscopes in generating images by 'feeling' surface structures with a sharp probe, rather than 'looking' at specimens by reflected or transmitted radiation. The resolution of an AFM is determined by the effective sharpness of the probe, and the accuracy with which the probe can be positioned relative to the sample surface. With modern AFMs it is possible to achieve submolecular resolution images of biopolymers. Because of the novel imaging process it is possible to obtain such images under gaseous or liquid environments. Thus, the AFM offers a magnification range spanning that of both the light and electron microscopes, but high magnification images can be obtained under the 'natural' conditions normally only associated with the light microscope, and with minimal sample preparation. The AFM allows the study of heterogeneous structures formed by association of biopolymers. At present the nature of such structures formed in the bulk, or at interfaces, are inferred from physical and physical chemical studies. Thus it becomes possible to image the network structures of films, gels and hydrated cell wall material, and the nature of biopolymer interactions at interfaces. AFM images are generated by monitoring changes in the forces acting between the surface and the probe, as the surface is scanned relative to the probe.

In the most common mode of operation (direct contact mode) the surface is positioned close to the probe, such that it experiences a preset repulsive interaction. As the surface is scanned beneath the probe the change in the character, or roughness of the surface, lead to variations in the force between the probe and the surface. At each image point the surface - probe separation is manipulated in order to restore the preset force value. The resultant changes in probe - surface separation are then amplified in order to generate images of the surface. Thus images are generated under constant force conditions. In the early days of the use of AFM biopolymers were deposited onto mica substrates and imaged in air. Under these conditions the imaging process was unreliable: residual water on the surface and tip coalesced resulting in large adhesive forces which damaged or displaced molecules when the sample was scanned [1]. Elimination of such adhesive forces by contact mode imaging under liquids [2], tapping [3], or non-contact imaging methods [4] has led to reproducible imaging of biopolymers. For delicate biopolymer structures, or weak networks, it is still necessary to monitor and control the imaging force, in order to prevent damaging the structures being imaged [2]. Stiffer networks can resist damage and can be imaged in air [5,6].

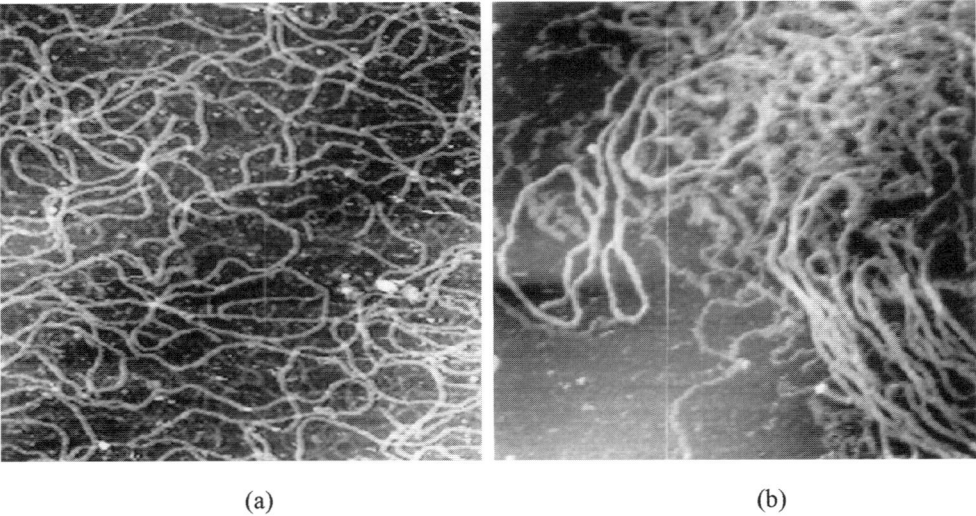

(a)   (b)

Figure 1. AFM images of xanthan preparations. (a) Entangled network of xanthan molecules. Image size 1 x 1 μm. (b) Xanthan microgel. Image size 1.4 x 1.4 μm.

The present studies describe the use of AFM to image heterogeneous network structures formed by biopolymers in the bulk, or at air - water interfaces.

2. MICROGELS AND GELS.

The solubility and rheology of bacterial polysaccharides such as xanthan gum is dependent on the methods of extraction, purification, drying and rehydration of the material. Precipitation of xanthan from the fermentation broth, in the presence of sufficient salt to maintain the helical conformation, leads to a freely soluble product. Precipitation from the broth in the absence of salt yields an insoluble product: in the absence of salt the helix is at least partially denatured, and precipitation and drying leads to concentration of the polymer under conditions of increasing ionic strength, resulting in aggregation due to intermolecular, rather than intramolecular helix formation. Aggregated xanthan molecules, or microgels, are considered to

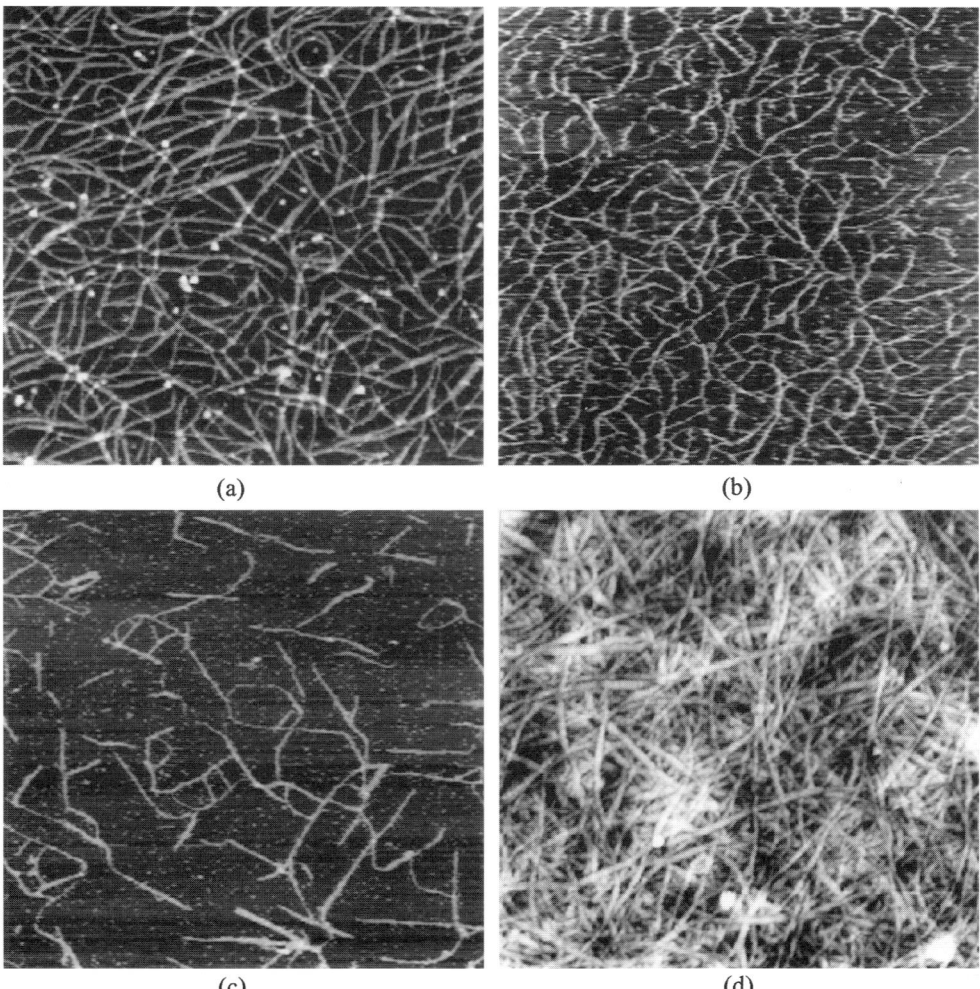

(a)

(b)

(c)

(d)

Figure 2. AFM images of gellan networks and gels. (a) Aqueous K gellan film; deposited from 10 μg ml^{-1} sample onto mica. Image size 800 x 800 nm. (b) Gellan aggregates formed by the TMA salt form, deposited from 10 μg ml^{-1} sample onto mica. Image size 1.5 x 1.5 μm. (c) Gellan gel precursors formed by the K salt form; deposited from 3 μg ml^{-1} sample onto mica. Image size 800 x 800 nm. (d) 1.2% acid set hydrated gellan gel. Image size 2 x 2 μm.

contribute to the thixotropy of xanthan dispersions responsible for its use as a thickening and suspending agent. Figure 1 shows AFM images of xanthan deposited from aqueous samples onto freshly cleaved mica, air dried for about 10 minutes and imaged under butanol. Figure 1a shows an entangled network of individual xanthan molecules whereas figure 1b shows a xanthan aggregate or microgel. Failure to swell such aggregates results in an insoluble product, whereas swelling but incomplete dissolution of the microgels results in an aqueous, thixotropic dispersion, rather than a true xanthan solution. The ability to image such structures allows one to monitor the effects of preparation conditions on the properties of the polysaccharide.

The image shown in figure 1a is the simplest type of polysaccharide network. For gelling polysaccharides concentration during air drying will lead to association and the formation of an aqueous film. Such films can be imaged under butanol and provide an image of the polysaccharide gel network. For the bacterial polysaccharide gellan AFM images of such films [6] reveal a continuous fibrous network structure (figure 2a). The thickness of the fibres varies throughout the network suggesting bundles of helical polymers associated to form the branched network. Some of the fibres show small stubs consisting of precursors to branches, small enough to be individual helices, which have not grown into proper branches. The gelation of gellan is known to occur in two separable stages. Cooling the sol in the absence of gel

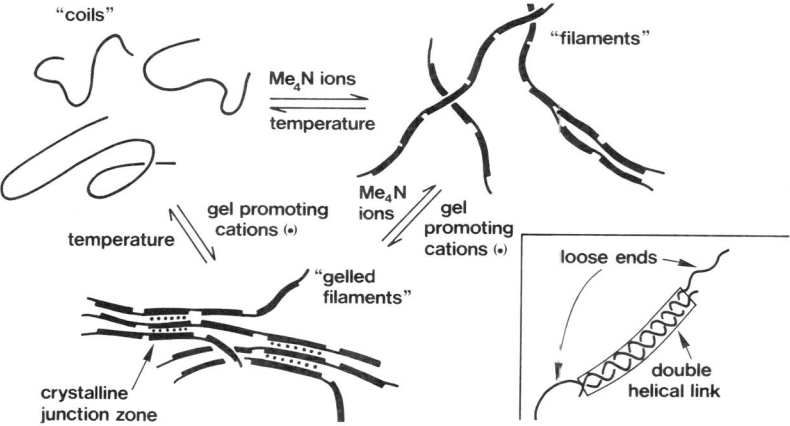

Figure 3. Schematic model for the gelation of gellan gum.

promoting cations leads to aggregation and even, at sufficiently high polysaccharide concentrations, weak gel formation. The association process involves helix formation and is thermoreversible with no hysteresis in the setting and melting behaviour. Furthermore, the weak gel structures break easily on shearing producing an aggregated dispersion. In the presence of gel promoting cations further association occurs. This process involves cation binding, leads to a permanent gel network and to thermal hysteresis in the setting and melting behaviour. Different models, based largely upon physical chemical data, have been proposed to explain gelation [7,8]. These models differ in the predicted large scale structure of the gels and the origins of the elasticity of the gel network. It is possible to use AFM to discriminate between these models [6]. Figure 2b shows images of gellan aggregates formed in the presence of non - gel promoting bulky tetra methyl ammonium (TMA) cations. The shape and size of the aggregates suggests that the aggregates are formed from helical molecules. The aggregates are all of the same size suggesting no side-by-side aggregation However, the molecules have clearly aggregated through helix formation alone into an heterogeneous population of elongated and branched structures. In the presence of the gel promoting potassium cation, at concentrations below the threshold concentration for gelation, similar branched aggregates are formed (figure 2c). However, in this case the fibrous structures are of variable thickness suggesting additional side-by-side aggregation of the gellan helices. These types of study permit a model to be built up of the gelation process (figure 3). Gellan is generally considered to be a model system for thermoreversible polysaccharide gels. The model shown in figure 3 differs from the present models for thermosetting polysaccharide gels in that it is a continuous fibrous structure, rather than ordered junction zones linked by disordered polymer chains [9]. In the present model (figure 3) ion binding occurs throughout the entire structure and the elasticity of the gel arises due to stretching or bending of the fibres, rather than the entropic 'rubber - like' elasticity assumed in conventional models for these polysaccharide gels. The question still remains as to whether such structures, based on model studies, actually exist in real bulk gels? In principle it should be possible to image hydrated gels by AFM. In practice it is only possible to image very stiff gels: as the probe traverses the gel surface it will compress the gel blurring any image of the network structure. For weak gels (shear modulus approx. 350 Pa for a 1.2 % potassium gellan gel) it is possible to obtain images of molecular structure by scanning the surface quickly, thus not allowing sufficient time for the network to deform. However these images are poor and the network appears aligned in the scan direction. Acid set gellan gels are very stiff (shear modulus approx. 10^4 Pa for a 1.2 % polymer gel) and can be imaged (figure 2d). The AFM images [9] of these gels show the fibrous structures seen previously in images of the flat films (figure 2a). It should be possible to extend the methodology applied here to study polysaccharide gels in order to investigate protein aggregation and gelation. Here the use of AFM would be more valuable because most of these gels are opaque, and not amenable to study by the spectroscopic methods, which have proved very useful for studying polysaccharide gels.

3. NATURAL NETWORKS.

   As well as imaging gel networks it is also possible to image naturally occurring structures within which polysaccharide association occurs. This can be illustrated by studies on cellulose based networks of different origins. Cellulose is secreted by the *Acetobacter xylinum* bacteria. Cellulose networks produced form the structural basis for the fermented food product Nata. Extracted bacterial cellulose preparations which have been modified to aid dispersion are now marketed as industrial hydrocolloids. The cellulose network is stiff and can be imaged in air without damage. Figures 4a, b show images of bacterial cellulose pellicles after they have been deposited onto mica and air dried [10]. Figure 4a illustrates a limitation in the use of AFM. Because the surface structure is rough then molecular structure is only discernible in certain parts of the image. The images can be improved by subtracting the low frequency surface undulations from the image thus projecting the molecular information onto a flat plane [11]. The processed image clearly reveals the flat ribbon - like structure of the bacterial cellulose. By contrast the cellulose based cell wall structures of land plants are very different in appearance [6,11]. Figure 4c shows a processed image of a hydrated fragment of Chinese water chestnut cell wall. In this case the image is viewed from the face adjacent to the plasma membrane looking down through the cell wall towards the middle lamella. The cell wall structure is seen to contain cylindrical cellulose based fibres organised into layered structures. It is also possible to use such images to monitor the effects of various processing conditions on the structure of the cell wall. A good example of this is shown in figure 4d. This is an image of an extract from an aqueous dispersion of a semi-refined iota carrageenan [11]. The milder extraction method used to prepare semi-refined carrageenans does not remove all the cellulose from the algal cell

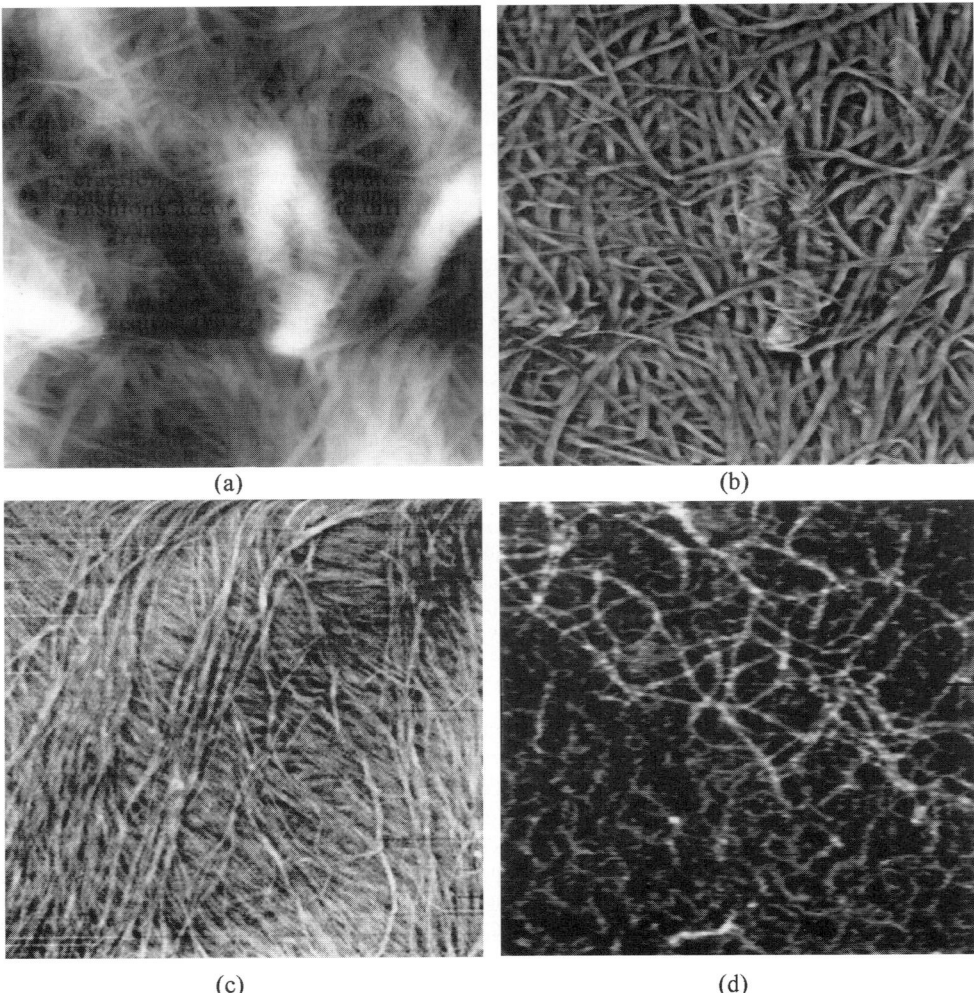

(a)   (b)

(c)   (d)

Figure 4. AFM images of natural polysaccharide networks. (a) Topograhical image of a bacterial cellulose network. Image size 8 x 8 μm. (b) Processed version of the topographical image of the bacterial cellulose network. The image has been processed by subtracting the background curvature (roughness) of the sample. Image size 8 x 8 μm. (c) Processed topographical image of an hydrated Chinese water chestnut plant cell wall. Image size 2 x 2 μm. (d) Mixed carrageenan and cellulose networks observed in semi - refined iota carrageenan sample. Image size 700 x 700 nm.

wall. Figure 4d shows an interpenetrating network of the more flexible carrageenan network and the stiffer, more extended, cellulose network. The two polymers are sufficiently different in structure to provide contrast in the image, without the need to label the two polymers. It can be seen that the extraction has disrupted the ordered assembly of the cellulose microfibrils resulting in a more open network structure

4. INTERFACIAL NETWORKS.

Biopolymers provide an important source of emulsifiers and foam stabilisers. At present there is no direct method for studying the structures formed by these polymers at oil - water or

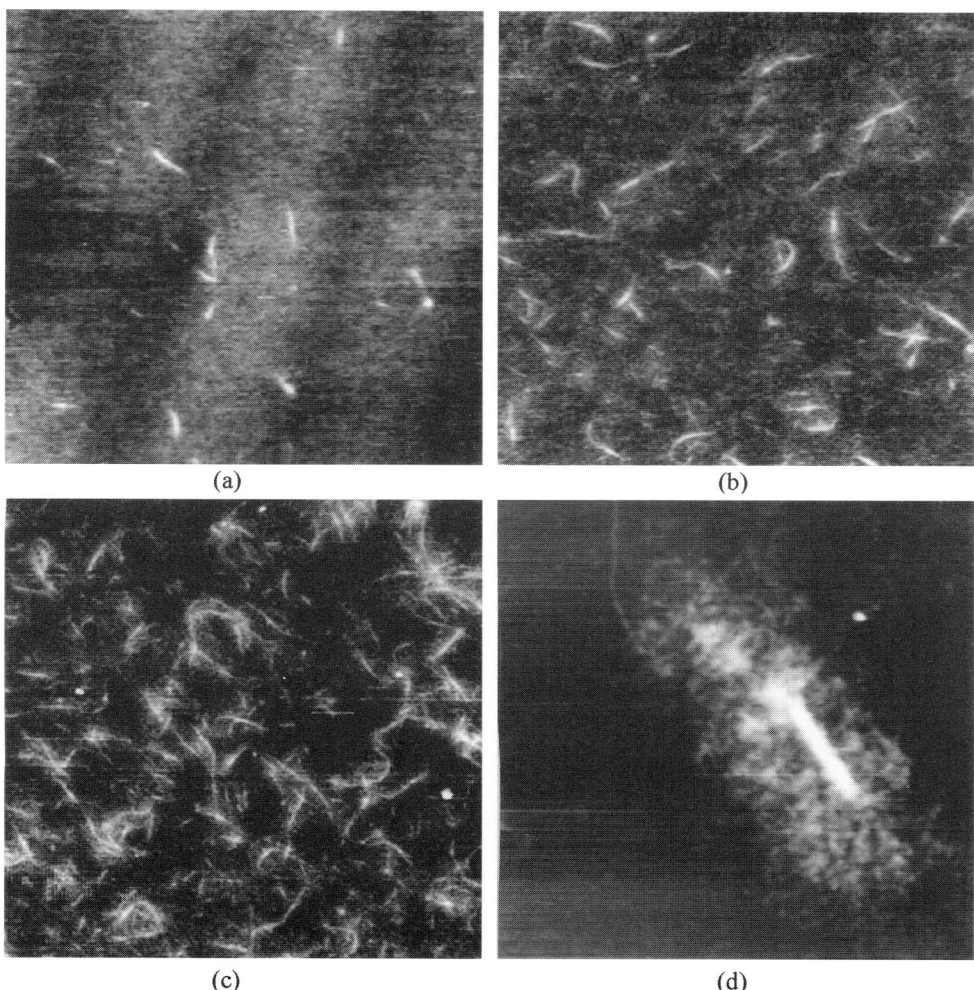

(a)   (b)

(c)   (d)

Figure 5. AFM images showing the development of a gelatin network at an air - water interface. (a) 1 hour, image size 4 x 4 μm, (b) 3 hours, image size 3.2 x 3.2 μm, and (c) 16 hours after addition of the gelatin solution, image size 4 x 4 μm. (d) Image of a fibrous gelatin aggregate separated from the bulk gelatin sample. Image size 800 x 800 nm.

air - water interfaces. It is possible [12] to build AFMs which can be used to look directly at polymer networks at the air - water or oil - water interface. However, the mobility of such structures will make it difficult to image the structure of these interfacial networks at the molecular level. Instead it is better to sample the structure at the interface by using Langmuir - Blodgett methods to deposit the films onto mica for imaging by AFM. The images obtained by AFM can be correlated with conventional studies (surface tension, surface rheology) of the interfacial films. As well as obtaining images of the interfacial films it is also possible to make quantitative measurements of film thickness, surface area and volume. The potential of this approach can be illustrated by studies of the formation and breakdown of protein films at air - water interfaces.

Gelatin is surface active and surface tension measurements can be used to monitor its adsorption at air - water interfaces. Rheological studies show the formation of elastic films at the interface, and have been taken to suggest that the gelatin gels forming a 2 - dimensional network structure analogous to that formed in bulk aqueous gels [13]. Use of AFM offers the chance to visualise such a network and to correlate network formation with changes in surface rheology [14]. For the studies reported here the gelatin (300 Bloom strength) was prepared as a stock solution (400 μg ml^{-1}), heated to 60^0C for 1 hour and then poured into the Langmuir trough. Measurements of surface tension and surface modulus were made as a function of time after addition of the gelatin. After each measurement of surface modulus a Langmuir - Blodgett (LB) film was drawn from the interface. LB films were obtained by lowering a freshly cleaved piece of mica (approx. 1x1 cm square) down through the interface and then back out of the interface at constant rate of 0.14 mm s^{-1}. Excess water was removed by blotting at the edges of the sample, which was then air dried for about 10 minutes, mounted into the AFM liquid cell and imaged under butanol (imaging forces 2 - 4 nN). The development of the surface modulus was found to be accompanied by the appearance, growth and aggregation of fibrous structures visualised by AFM (figure 5a, b, c) [14]. The AFM images taken at intervals of 1, 3 and 16 hours show the size and number density of the fibres increases with time eventually forming an interconnected network. By 16 hours the surface modulus had reached a plateau value of about 12 mN m^{-1}. These structures are similar to those seen by AFM for the bulk gelation of gelatin (figure 5d) [14]. The fibres appear to grow from a background of smaller filamentous structures typically about 1 nm in height. From dimensions and shapes of these filaments it seems likely that they are partially or completely reformed triple helical structures. The brighter fibrous structures are typically about 4 nm in height and between 300 - 450 nm in length suggesting that they are probably bundles of gelatin triple helices. These studies provide direct visual evidence for the formation of aggregated gelatin networks at the air - water interface and correlate network formation with the development of surface elasticity.

Both proteins and surfactants act as foam stabilisers. However, mixtures of proteins and surfactants do not show a synergistic effect: the presence of a small quantity of surfactant will breakdown a protein stabilised foam. This surprising effect occurs because proteins and surfactants stabilise air - water interfaces by very different mechanisms. For surfactants it is proposed that stabilisation occurs through the Gibbs - Marangoni effect [15], which is dependent on the high surface mobility of the surfactant molecules. By contrast proteins are believed to function through the formation of rigid elastic networks which confer mechanical strength to the foam film, slow drainage and resist stretching of the film [16]. It is the incompatibility of these two mechanisms which accounts for the deleterious effect of small quantities of surfactant on protein foams. The surfactant is believed to act by destroying the protein network, although the exact mechanism remains unclear. The AFM provides an unique facility for visualising the protein structures formed at the interface and their breakdown. Such studies are illustrated in figure 6 which shows the effect of added Tween 20 on the structure of

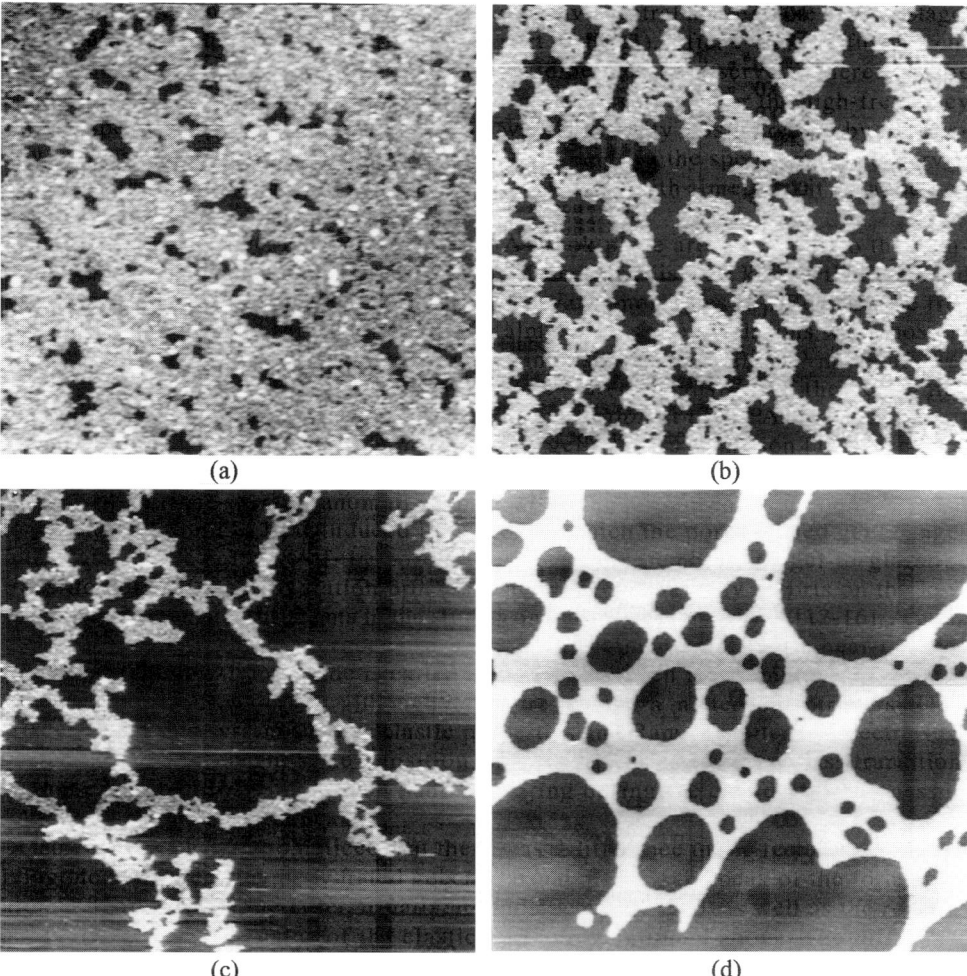

(a)   (b)

(c)   (d)

Figure 6. AFM images of the formation and growth of surfactant domains within protein networks formed at air - water interfaces. β - lactoglobulin - Tween 20 system (a) Π = 18.6 mN m^{-1}, image size 1 x 1 μm, (b) Π = 21.8 mN m^{-1}, image size 3.2 x 3.2 μm, and (c) Π = 24.6 mN m^{-1}, image size 3.2 x 3.2 μm. β - casein - Tween 20 system. (d) Π = 19.2 mN m^{-1}, image size 6.4 x 6.4 μm.

spread films of β lactoglobulin and β casein. A spread film of β lactoglobulin (surface concentration 2.5 mg m^{-2}) gave rise to a surface pressure of 10 mN m^{-1} after 30 minutes. Addition of surfactant to the subphase resulted in a rise in surface pressure (Π) as surfactant adsorbed at the interface. At each surface pressure LB methods were used to deposit monolayer surface films onto freshly cleaved mica. These samples were imaged by AFM under butanol. Figures 6a, b, c show the effect of increasing surfactant concentration on the structure of the protein film. At Π= 18.6 mN m^{-1} the protein layer is seen (figure 6a) to contain numerous small irregularly shaped holes corresponding to the nucleation of surfactant domains, believed to arise at 'defects' in the protein network. At higher surface pressures (eg Π= 21.8 mN m^{-1}) the domains have grown in size but still retain their irregular shape (figure 6b). Growth of the surfactant domains leads to compression of the protein network and the equilibrium at a given surface pressure arises when the elastic energy stored in the protein network is sufficient to inhibit further growth of the surfactant domains. At higher surface pressures (Π= 24.6 mN m^{-1}) the protein network has been reduced to interconnected strings of protein molecules (figure 6c). Finally, at sufficiently high surface pressures the protein network collapses, allowing complete desorption of proteins from the interface, and leaving a continuous surfactant phase containing islands of aggregated protein. The mechanism by which the growing surfactant domains compress and disrupt the protein network suggest that the process should be termed an 'orogenic mechanism' for protein displacement. The detailed behaviour is found to be sensitive to the type and structure of the protein. Figure 6d shows data for a spread film of β casein film displaced by Tween 20. A surface concentration of 1.4 mg m^{-2} also resulted in a surface pressure of about 10 mN m^{-1} after 30 minutes. As with β lactoglobulin, displacement of β casein with Tween 20 arose due to the nucleation and growth of surfactant domains but, in this case, the domain growth was isotropic (figure 6d). The breakdown of the β casein network is isotropic, whereas the irregular shapes of the domains formed in the β lactoglobulin system imply restricted propagation of stress through the network, stress concentration and progressive failure of the network at the weakest points. The use of AFM has provided direct evidence showing the structure of protein networks formed at the air - water interface, identified a new orogenic mechanism of protein displacement by surfactant, and demonstrated differences in behaviour relatable to protein structure and the effectiveness of different proteins to act as foam stabilisers. Given that the surfactant domains are heterogeneous in size and shape, and that at the early nucleation stage are comparable in size to individual proteins, AFM is at present the only experimental technique which could have been used to discover this new mechanism of protein displacement. In addition to imaging the surface structures it is also possible to obtain quantitative data on protein film thickness, area and volume allowing a detailed description of the displacement process. Such studies are in progress for both spread and adsorbed protein films.

6. CONCLUSIONS.

The development of routine and reproducible methods for AFM imaging of biopolymers has permitted use of AFM to investigate the functional behaviour of proteins and polysaccharides. The unique feature of the use of AFM is the study of heterogeneous structures at molecular, or submolecular resolution, under 'natural' conditions. Such studies have provided unique insights into the origins of thixotropy and gelation for polysaccharides and new data on the nature, formation and breakdown of protein networks at air - water interfaces.

ACKNOWLEDGEMENT.

The present research was supported by the BBSRC through core funding to the Institute.

REFERENCES.

1. H.G.Hansma and H.J.Hoh, Ann. Rev. Biophys. Biomol. Struct. 23 (1994) 115.

2. A.R.Kirby, A.P.Gunning and V.J.Morris, Biopolymers 38 (1996) 355.

3. A.P. Gunning, A.R.Kirby and V.J.Morris, Ultramicroscopy 63 (1996) 1.

4. T. McIntire, R.M.Penner and D.A.Brant, Macromolecules 28 (1995) 6375.

5. A.R. Kirby, A.P.Gunning, K.W.Waldron, V.J.Morris and A. Ng, Biophys. J.
70 (1996) 1138.

6. A.P.Gunning, A.R.Kirby, M.J.Ridout, G.J.Brownsey and V.J.Morris, Macromolecules 29
(1996) 6791; Macromolecules 30 (1997) 163.

7. V.J.Morris, In A.M.Stephen (ed.) Food Polysaccharides and their Applications, chapter 11
341, Marcel Dekker, New York,1995.

8. G.Robinson, C.E.Manning and E.R.Morris, In E.Dickinson (ed.) Food Polymers and
Colloids, 22, Royal Society of Chemistry, London,1991.

9. D.A.Rees, E.R.Morris, D.Thom and J.Madden, In G.O.Aspinall (ed.) The Polysaccharides,
volume 1,195, Academic Press, New York,1982.

10. A.P.Gunning, P.Cairns, A.N.Round, H.J.Bixler, A.R.Kirby and V.J.Morris, Carbohydr.
Polym. 36 (1998) 67.

11. A.N.Round, A.R.Kirby and V.J.Morris, Microscopy & Analysis 55 (1996) 33.

12. L.M.Eng, Ch.Seuret, H.Looser and P.J.Gunter, Vac. Sci. Technol. B 14 (1996) 1386.

13. E.Dickinson, B.S.Murray and G.Stainsby, In E.Dickinson and G.Stainsby (eds.), chap. 4,
123, Advances in Food Emulsions and Foams, Elsevier Applied Science, London & Oxford,
1988.

14. A.R.Mackie, A.P.Gunning, M.J.Ridout and V.J.Morris, Biopolymers 46 (1998) 245.

15. W.E.Ewers and K.L.Sutherland, Aust. J. Sci. Res. Ser. A5 (1952) 697.

16. P.Halling. Rev. Food Sci. Nutr. 15 (1981) 155.

HYDROCOLLOIDS – PART 1
Edited by K. Nishinari
2000 Elsevier Science B.V.

111

Thermally induced gels obtained with some amphiphilic polysaccharide derivatives: synthesis, mechanism and properties

M. RINAUDO* and J. DESBRIERES

Centre de Recherches sur les Macromolécules Végétales, affiliated with the Joseph Fourier University, BP 53, 38041 Grenoble cedex 9, France

*to whom correspondence should be addressed.

This paper concerns, in the first part, the preparation of homogeneously modified methylcelluloses such as to discuss the role of the substituent distribution along the chain on the physical properties of the polymers. The mechanism of gelation is described as a two step process in relation with the existence of trimethylated glucose zones necessary to get the first clear loose gel (35°C< T< 60°C). The second step is a phase separation occurring around 60°C with the formation of a turbid strong gel. Alkylchitosans prepared with alkyl chain larger than C8 give very large increase of the viscosity of aqueous dilute solution; the length of the alkyl chain and the degree of alkylation are very important parameters to control the solution properties. In addition, these amphiphilic polymers have some interfacial properties playing a role on stabilization of foams and emulsions.

1. INTRODUCTION

Associating polymers based on their amphiphilic character are now recognized as polymers having original behavior especially when the influence of temperature on solution properties is considered [1]. Synthetic polymers as well as modified natural polymers are more and more developed for many industrial applications. Depending on thermodynamic conditions (temperature, pH, ionic strength, …), these polymers give large increase in viscosity of fluids, leading to gelation. The hydrophilic-hydrophobic balance is essential to control these physical properties.

Our work concerns the synthesis and properties of polysaccharide derivatives based on cellulose and chitin. These two natural polysaccharides represent the most important sources of renewable polymers having interesting basic physical characteristics such as their relative stiffness.

In the first part of this paper, the role of the methyl substituent distribution along the cellulosic chain will be discussed for methylcellulose, a neutral polymer, in relation with the mechanism of gelation. In the second part, the synthesis of alkylated chitosans and their properties will be described. These polymers are also polyelectrolytes allowing us to demonstrate the role of the balance between electrostatic repulsions and hydrophobic interactions. Some applications in the food industry will be mentioned [2].

2. EXPERIMENTAL

2.1. Materials.

Commercial samples of methylcellulose were used; their references are Methocel A4C, A15C and A4 M from Dow Chemical Cy. The average degrees of substitution (DS) are around 1.7; their intrinsic viscosities were measured in water at 20°C to determine the viscometric average molar masses according to the relation given in literature [3]. Laboratory samples, with different DS, were prepared in homogeneous conditions as described later [4, 5]. The initial cellulose is a dissolving pulp with a DPv ~ 1300. Purification of methylcellulose was realized by dialysis against water with a membrane Spectra/Pore from Bioblock with a cutoff M = 12-14,000. The alkylchitosans (Ch-Cx) were also prepared in homogeneous conditions in our laboratory [6]. Cx is related to the number of carbons of the aldehyde used for the reaction of reductive amination. The original chitosan was from Protan (Norway), with an average degree of acetylation DA = 0.12 as determined by nmr [7], the viscometric average molar mass Mv around 170,000 g/mol obtained from intrinsic viscosity in the conditions given previously [8]. Different samples having different degrees of alkylation (τ) with different Cx were prepared and characterized.

2.2. Methods.

The rheological measurements were performed with a Couette type rheometer (Contraves Low Shear 40) or a stress-controlled rheometer (CarriMed CS 50) according to the polymer concentration and the temperature tested. Calorimetric measurements were performed on a Micro DSC III calorimeter from Setaram (France). The fluorescence spectra were obtained using a LS 50 B spectrometer from Perkin-Elmer using pyrene as a probe to test for hydrophobic domains. Pyrene concentration was 10^{-7} M and excitation wavelength was 334 nm. The ratio of the intensities of the first and third peaks (I_1/I_3) of fluorescence spectrum of pyrene in polymeric solution depends on the hydrophobicity of the environment and then proves the existence of hydrophobic domains. The chemical modifications are characterized by nmr using a Bruker AC 300 spectrometer at 353 K using DMSO-d6 for methylcellulose samples or D2O/HCl for chitosan derivatives.

Methylcellulose, used in a freeze dried-form, is dissolved in water at 5°C for 24 hours before experiments. Alkyl chitosans are dissolved directly in water after isolation of the corresponding chlorhydrate or in the presence of acetic acid (0.1 M-0.3 M).

3. RESULTS AND DISCUSSION

3.1. Synthesis and characterization of polysaccharide derivatives

3.1.1. Methylcellulose

The substitution of methylcellulose was characterized using ^{13}C nmr giving the distribution of methyl groups on the individual C2, C3 and C6 atoms [4]. From the total –CH3 substituents in reference to the C1 of glucose unit, one gets the average DS (Figure 1). Total hydrolysis of the polymer was carried out in 2N trifluoroacetic acid for 4 h at 120°C. After purification, the methylated glucoses were separated by HPLC with a Waters Millipore chromatograph equipped with a differential refractometer (Waters 410); a reversed phase column Nucleosil C18 5μm (250 x 4.6 mm) with water as eluent was used at ambient temperature. The unsubstituted, mono, di and trimethylated glucose units were separated to characterize the modified polymer. The main characteristics of the commercial samples used are given in Table 1.

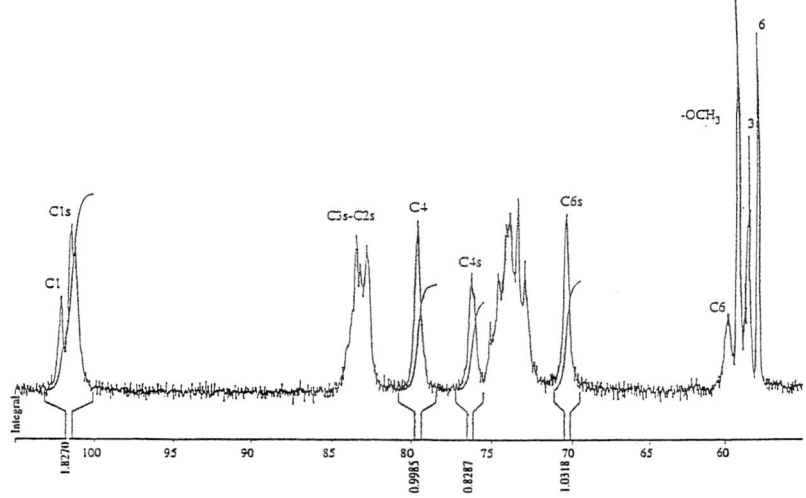

Figure 1. ^{13}C nmr spectrum for methylcellulose A4C in DMSO at 353K. Polymer concentration around 30g/L. (s) referred to the substituted carbon.

Table 1
Characteristics of some commercial methylcelluloses.

Samples	Mv g/mol.	DS	UnS %	MonoS (%)	Di S (%)	Tri S (%)	[η] ml/g in H$_2$O, 20°C
A4C	149,000	1.7	10	29	39	22	509
A15C	273,000	1.7	10	27	39	24	745
A4 M	409,000	1.8	11	27	40	22	961

It is known that industrial preparation of cellulose derivatives is usually performed in the heteregeneous phase; as the cellulose is a semi-crystalline polymer, it follows that the reactivity in amorphous or crystalline regions is not the same and controls the heterogeneity of the substituent distribution along the chains. These commercial samples will behave as a blockwise copolymer. In order to clarify the role of the substituent distribution on the mechanism of gelation, one prepared methylcelluloses under homogeneous conditions.

The method proposed was described previously [4]; to summarize, the procedure is the following: cellulose is activated in water and in DMAc before dissolving in DMAc containing 6 wt % LiCl. Dimsyl sodium solution was added to the cellulose solution and stirred at ambient temperature for one night. After cooling to 10°C, iodomethane was added and the mixture was stirred for different periods of time (from 8 h to 5 days). After, acetic acid was added to neutrality and the solution was dialyzed against water for purification before freeze drying. Different ratios cellulose/ DMAc and anion NaH/DMSO ratios were adopted [4, 5]. The time of reaction is the main parameter which controls the DS. The conditions of synthesis and characteristics of few of our polysaccharides are given in Table 2.

Table 2
Conditions of preparation (a) and characteristics of some laboratory samples (b).

(a)

Samples	Cellulose/DMAC g/g	NaH/DMSO g/mL	CH$_3$ I (ml)	Time (h)
M$_{10}$	6: 700	20:100	30	8
M$_{29}$	6: 740	20:100	30	36
M$_{18}$	6: 700	10:50	20	168

(b)

Samples	DS	UnS %	MonoS (%)	Di S (%)	Tri S (%)	[η] ml/g in H$_2$O, 30°C
M$_{10}$	0.9	-	-	-	-	-
M$_{29}$	1.7	4	45	32	19	544
M$_{18}$	2.2	9	12	36	43	339

From Table 2b, it is pointed out that a more regular substitution exists in these samples. In particular less unsubstituted units are left, whatever the average DS, for DS > 1. They will probably behave as statistical copolymers having average thermodynamic characteristics depending on the average DS.

3.1.2. Alkylchitosans

Alkylchitosans were prepared by reductive amination following the procedure described by Yalpani [9]. It involves the free –NH$_2$ on chitosan which reacts with an aldehyde having variable alkyl chain in homogeneous conditions. The reaction is the following :

$$Ch - NH_2 \; + \quad R - C \underset{H}{\overset{O}{\vphantom{|}}} \qquad \rightarrow \qquad Ch - NH - CH_2 - R$$

in which Ch-NH$_2$ means chitosan and R is the alkyl chain. Chitosan is dissolved in acidic media; then, pH is adjusted to 5.1, some ethanol is added to allow solubilization of aldehyde; then, the aldehyde in ethanol is added. Sodium cyanohydroborate NaCNBH$_4$ is introduced in excess for reduction. The mixture is stirred during 24 h at room temperature. Then the polymer is recovered after precipitation with ethanol and neutralization. Chitosan and alkylated chitosans were characterized by ^{1}H nmr when the samples are completely soluble in acidic conditions [6].

The solubility of the alkylchitosans decreases largely when the degree of alkylation and the length of the alkyl chain increases. To obtain a water soluble polymer one has to reduce the degree of alkylation, when R increases, down to very low τ.

These derivatives are amphiphilic charged polymers which behave as grafted copolymers. The originality of these polymers is that the modification is realized in mild and homogeneous conditions giving a regular distribution of the hydrophobic chains without change in the molecular weight of the parent polymer. Also the intrinsic stiffness of chitosan (persistence length in the range of 50 Å) seems to favour interchain interactions involving the hydrophobic grafted chains.

Table 3
Synthesis and characteristics of alkylated chitosans.

R group	Stoichiometric condition [NH$_2$] /aldehyde (monomol.)	τ	Concentration g/L	Solution at 20°C	Gel at 60°C
C8	1/0.1	0.12	10	Yes	Loose gel
C10	1/0.1	0.12	10	Yes	Yes
C12	1/0.05	0.04	5	Yes	Loose gel
			7	Yes	Yes

It was also demonstrated that these polymers have some surface active properties in relation with their amphiphilic character [10].They will be able to stabilize the interfacial film in foam and/or emulsion as known for polymeric surfactant [11].

3.2. Mechanism of gelation

3.2.1. Methylcellulose
First the mechanism of gelation of blockwise methylcellulose was investigated; then, the behaviour of homogeneous derivatives was compared such as to demonstrate the role of substituent distribution.

Methylcellulose with a high enough DS (0.9 with homogeneous distribution of methyl group or 1.3-1.5 for heterogeneous) perfectly solubilizes in water at low temperature (T < 30°C).

The phase diagram for the A4C sample was recently drawn by Axelos et al. [12];.this polymer is characterized by a LCST type phase diagram characterized by critical values Tc = 29°C and Φc= 0.045. The phase diagram was related to the rheological behavior obtained for this system when temperature increases [13]. For the heterogeneous methylcellulose a two step mechanism of gelation clearly occurs which was investigated especially by microcalorimetry and rheology. This behavior seems to be independent of molecular weight at least in the range covered in this work (Figure 2). When temperature increases the expansion decreases in very dilute solution in relation with intrachain coiling but in the semi-dilute, interchain interactions develop in a range of temperature 35°C < T < 60°C giving a first reversible sol-gel transition (Figure 3). This gel is a clear and loose gel stabilized by cooperative hydrophobic interactions involving blocks of trimethylated glucose units [13].The evidence of hydrophobic domains was demonstrated by fluorescence.

When the temperature continues to increase, a very cooperative transition occurs forming an elastic turbid gel when T > 60°C and when the polymer concentration is larger than the overlap concentration (here C > 2 g/L). This process is a phase separation corresponding to the binodal established separately. The location of this phase transition depends on the average DS as shown with laboratory samples. It must be pointed out that the two steps of gelation are controlled by kinetics as shown in Figure 4.

With homogeneous methylcelluloses, the situation is strongly modified.

For DS < 1, no hydrophobic interactions were demonstrated ; especially no enthalpic signal in DSC experiments was obtained. For DS > 1, a clear loose gel is formed for a temperature in the range of 70°C but no phase separation was ever observed for T < 120°C in relation with the good solubility of this polymer [14]. In the same time, an enthalpic signal is shown near 63°C on cooling in DSC experiments [15].

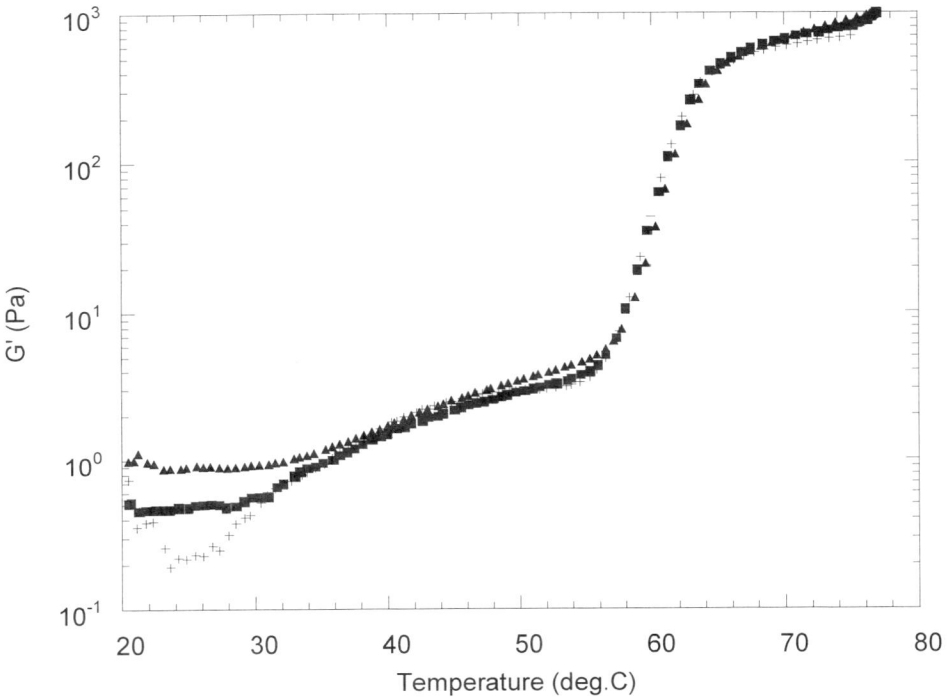

Figure 2. Elastic modulus G' at 1Hz as a function of the temperature for different molecular weight samples. Polymer concentration: 15g/L. Rate of temperature increase: 0.5 degree/min. + A4C ■ A15C ▲ A4M

For DS > 1.3, an aggregation mechanism was shown in very dilute solution by SEC experiment [4,5], intrinsic viscosity and Huggins constant analysis [4] when for the semi dilute regime a clear phase gel is formed. For DS = 1.7, the DSC measurement gives only a very small enthalpic peak (6.3 J/g) on heating compared with A4C with the same DS (22 J/g) and it is displaced to higher temperature (73°C compared to 60°C for A4C) (Figures 5-6).

In addition, only the clear gel is formed with a phase separation over 100°C [15].Only when DS is equal to 2.2, as shown previously, a two step mechanism is obtained in rheology; the behavior looks like that of A4C but the hysteresis is thinner due to the more regular distribution of substitution (Figures 6-7); only one peak appears on cooling around 38°C.

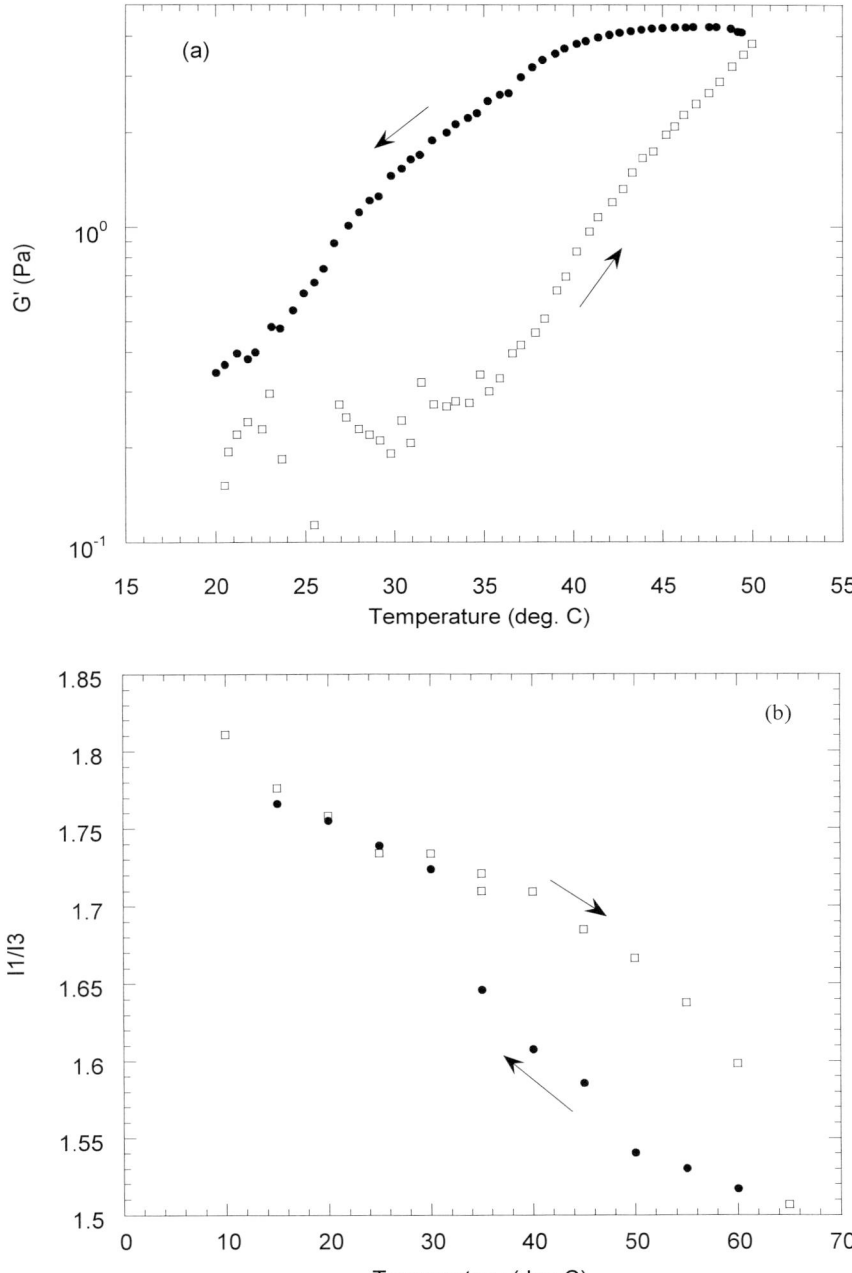

Figure 3.(a) Evolution of the elastic modulus (G') at 1 Hz as a function of the temperature in a limited range of temperature (20-50-20°C).Sample A4C; polymer concentration: 20g/L in H2O; rate of temperature change: 0.5 degree/min. □ heating   • cooling
(b) Evolution of the fluorescence ratio I1/I3 in the clear phase for A4C. Polymer concentration: 2g/L.

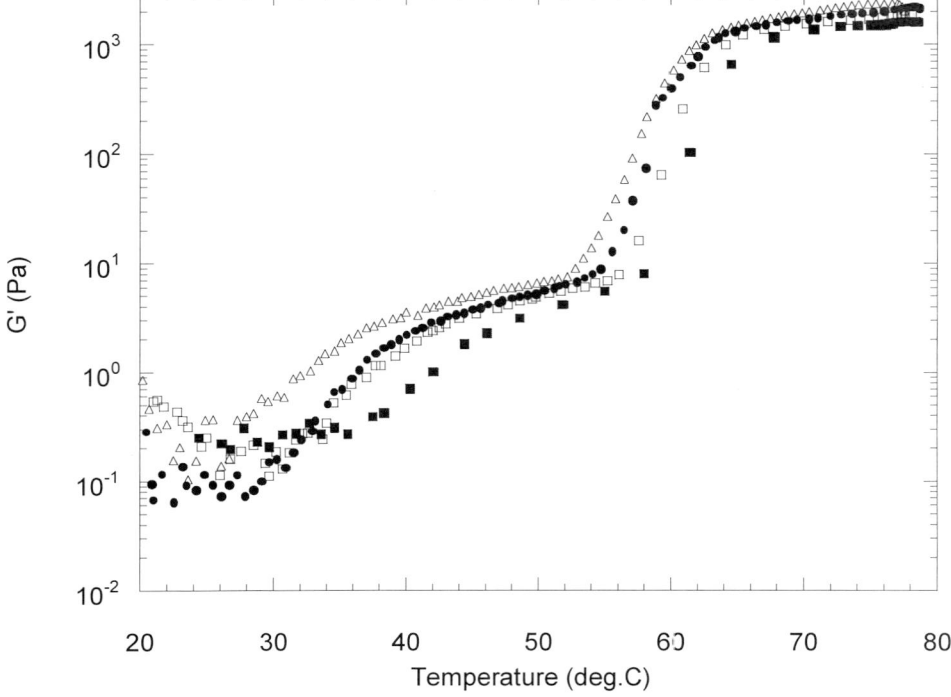

Figure 4. Role of the rate of temperature increase on the evolution of the elastic modulus at 1Hz with temperaure. Sample: A15C; polymer concentration: 18g/L.
△ 0.25 deg/min.   • 0.5 deg/min.   □ 1 deg/min.   + 2 deg/min.

On heating, the DSC peak is very sharp and related to the second step in rheology (~ 52°C). The relative amplitude of the first wave in rheology is larger than for A4C due to the larger probability to have blocks of trimethylated glucoses.

These results demonstrate clearly the essential role of the heterogeneity of substituent distribution along the cellulosic backbone. For the main application in food, for stabilization in a moderate range of temperature, the lower temperature of gelation of commercial methylcellulose is of considerable interest. The tensioactive character must also be related to this blockwise character.

3.2.2. Alkylchitosans

Chitosan is soluble in acidic conditions as soon as the net degree of ionization is larger than 0.5 [16]. Introduction of alkyl chain reduces considerably the solubility but in a limited domain, they present very interesting properties in dilute solution. First, when the number of C atoms of the alkyl chain R of the aldehyde used is larger than 6, for degree of alkylation in the range of 10% (on the basis of the monomolar ratio), a large increase of the viscosity with a non Newtonian character is observed compared to the initial chitosan. This increase in viscosity is directly related with the length of alkyl chain [6], polymer concentration [17], temperature and ionic composition in the solution [6].

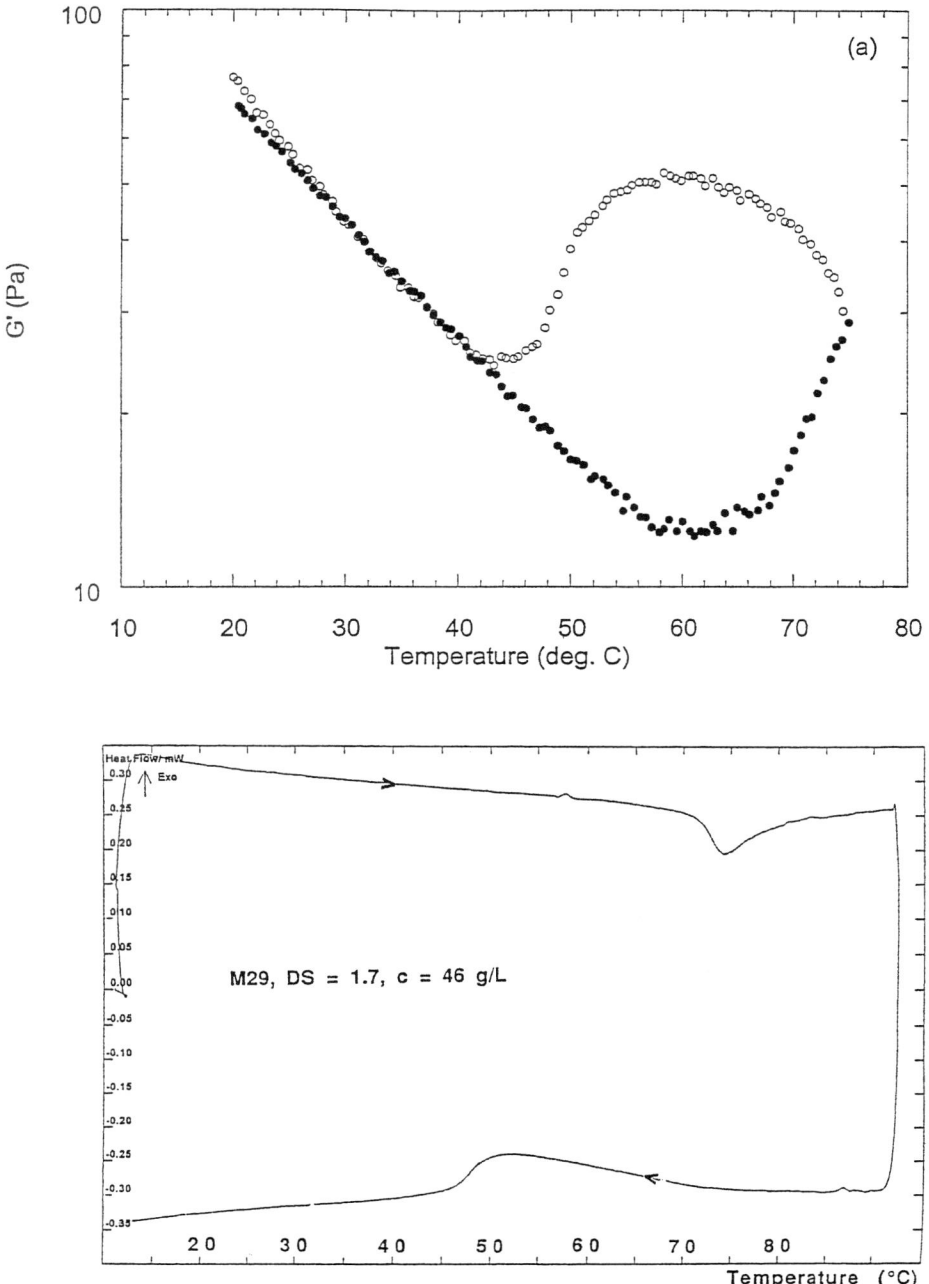

Figure 5. Rheological (a) and microcalorimetric (b) behaviours on heating and cooling runs for homogeneous methylcellulose (M29; DS=1.7). Polymer concentration: 46g/L. ω= 1Hz. Rate of temperature variation: 0.5 degree/min.   o heating   • cooling.

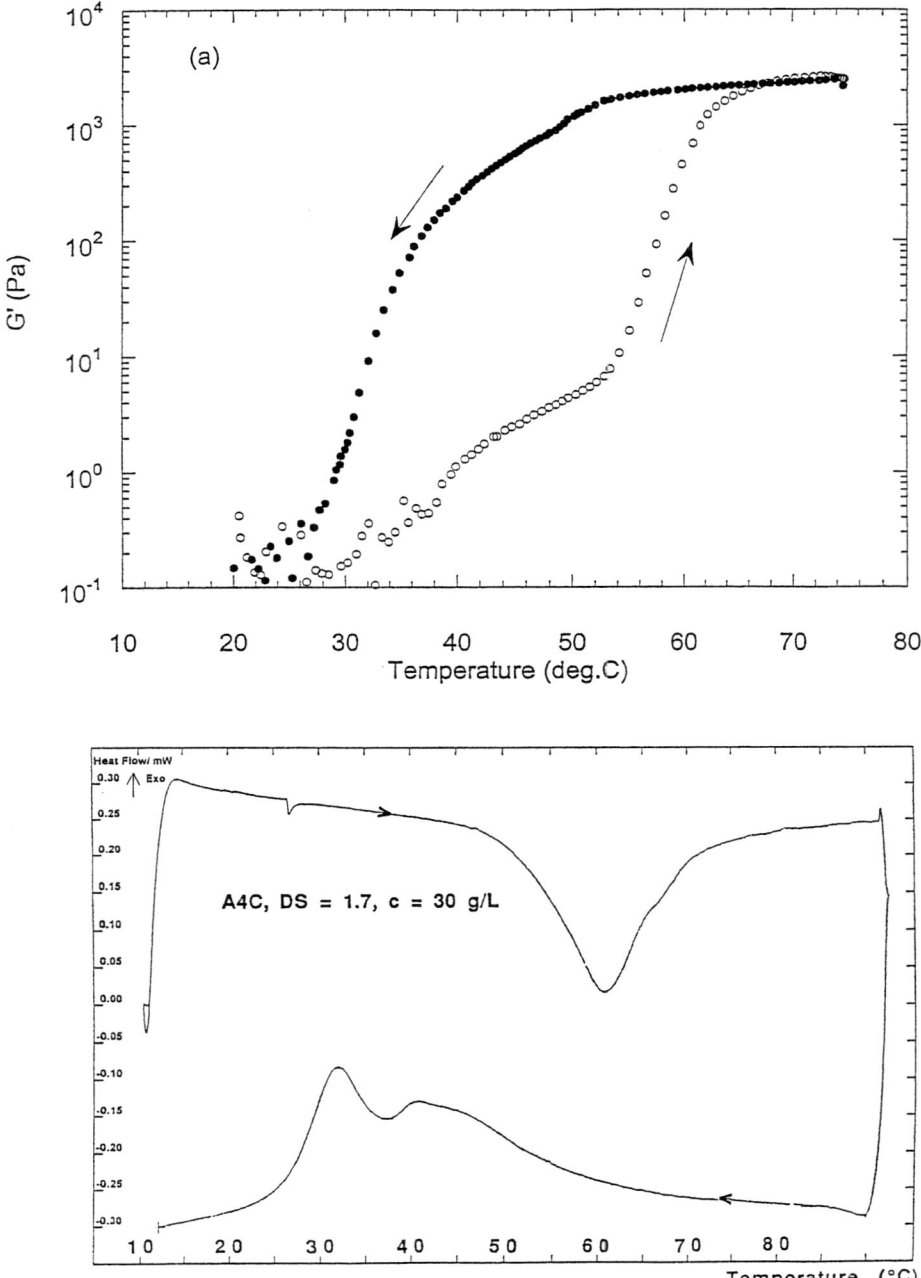

Figure 6. Rheological (a) and microcalorimetric (b) behaviours on heating and cooling runs for heterogeneous methylcellulose (A4C; DS=1.7). Polymer concentration: (a) 20 g/L; (b) 30g/L . ω= 1Hz. Rate of temperature variation: 0.5 degree/min. o heating   • cooling.

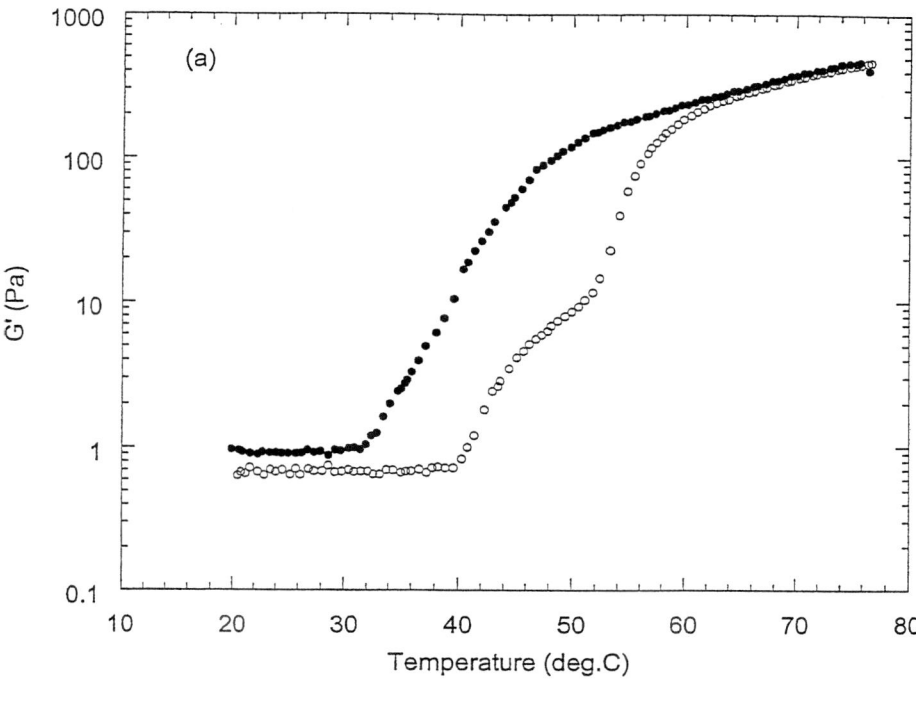

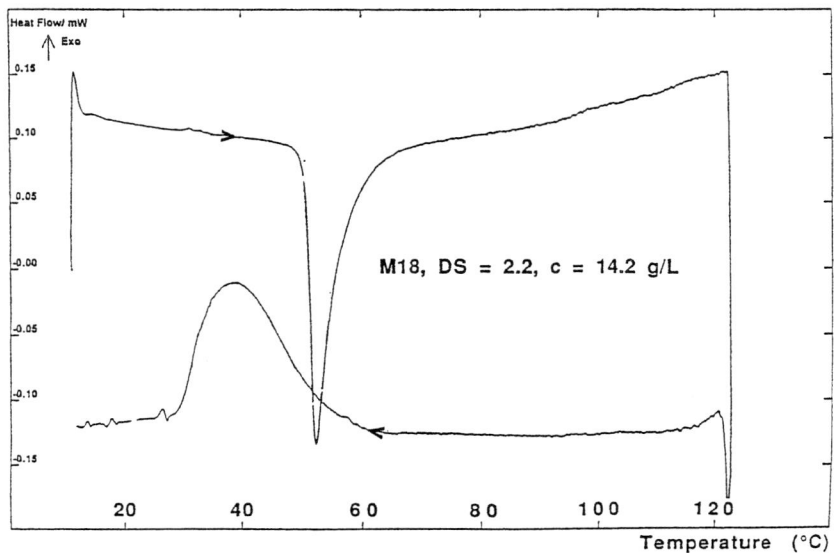

Figure 7. Rheological (a) and microcalorimetric (b) behaviours on heating and cooling runs for homogeneous methylcellulose (M18; DS=2.2). Polymer concentration: (a) 22.4g/L; (b) 14.2g/L. ω= 1Hz. Rate of temperature variation: 0.5 degree/min.   o heating   • cooling.

The increase in viscosity was shown to appear at a polymer concentration much lower than the overlap concentration (0.2 g/L compared with $C^* \sim 1$ g/L as determined from the intrinsic viscosity). In the same time, the existence of hydrophobic domains is demonstrated in presence of pyrene used as a probe.

The mechanism of gelation agrees with that usually proposed for amphiphilic polymers involving a micelle-like structure formed by aggregation of the alkylated hydrophobic entities [18].

It was also demonstrated that depending on pH, i.e. on the net charge of alkylchitosan, the performance of the alkylchitosan can be modulated based on the balance between hydrophobic and electrostatic interaction. Especially at low temperature (5 or 30°C), the viscosity increases very rapidly when the degree of ionisation decreases, passes through a maximum before phase separation [17].

The alkylchitosan behaves as a semi-rigid grafted polymer and gives especially interesting properties for very low degree of alkylation ($\tau < 0.05$) with C^{12}–C^{14} chains.

In addition it was shown some small interfacial properties compared with initial chitosan. This may be an advantage for stabilization of emulsion , chitosan by itself being proposed for whey proteins foaming but limited to acidic conditions (pH < 6) which is the condition for solubility. One has previously shown that the range of pH solubility of alkylchitin derivative can be extended after carboxylation of the chitin followed by alkylation [19].

Interesting interactions were also demonstrated for these alkylated ionic chitin derivatives and surfactants. Oppositely charged surfactants in a first step form cooperative electrostatic complexes and neutral or identically charged surfactants interact with the alkylchains [20].
All these interactions control the interfacial properties of the systems considered.

4. CONCLUSION

This paper summarizes our investigation on two different hydrophobically modified polysaccharides, cellulose and chitosan. The two-step mechanism of gelation for heterogeneously modified cellulose is especially described; for this purpose, homogeneous derivatives were prepared and compared. In the range of 35-55°C, a clear reversible gel is formed involving highly modified blocks (crystalline-like junction zones). Their length and distribution of the blocks controlled the physical properties of the gel. The absence of trimethylated glucose was shown to prevent gelation.

The second step is a cooperative phase separation around 60°C giving a reversible turbid gel. The gelation process is nearly independent on molecular weight in the range covered but imposed by the distribution of the substituents.

The physical properties of methylcelluloses are imposed by the conditions of derivatization and clearly, in this case, regarding applications, heterogeneity is an advantage giving gels in a moderate range of temperature. The risk is the irreproducibility of these properties when the cellulose source or conditions of modification are changed.

Alkylchitosans, on the other side, are charged hydrophobic polymers for which the ionic charge density plays a very large role. For very low polymer concentrations, a large increase of the viscosity of aqueous solution is induced by alkylation even at ambient temperature.

These polymers form a disordered network in which the crosslinking points consist of an aggregation of alkyl chains forming hydrophobic domains (micelle-like). In addition, the alkyl chitosan plays a role on interfacial film stability and may stabilize emulsion.

Aknowledgements. Muriel Hirrien is thanked for her technical help in the methylcellulose project.

REFERENCES

1. "Food Hydrocolloïds: structures, properties and functions",Edit: K.Nishinari, E.Doi, Plenum Press, New-York, 1994.
2. J.A.Grover , In "Food Hydrocolloïds", Edit.M.Glicksman, CRC Press, Boca Raton (USA) Vol 2 (1986) 122.
3. K. Uda, G. Meyerhoft, Makromol. Chem., 47 (1961) 168.
4. M. Hirrien, J. Desbrières, M. Rinaudo, Carbohydr. Polym., 31 (1996) 243.
5. J. Desbrières, M. Hirrien, M. Rinaudo, In "Cellulose derivatives. Modification, characterization and nanostructures", Edit. T.J. Heinze, W.G. Glasser, ACS Symposium Series, chap. 24, 332 (1997).
6. J. Desbrières, C. Martinez, M. Rinaudo, Int. J. Biol. Macromol., 19 (1996) 21.
7. P. Le Dung, M. Milas, M. Rinaudo, J. Desbrières, Carbohydr.Polym., 24 (1994) 209.
8. M. Rinaudo, M. Milas, P. Le Dung, Int. J.Biol.Macromol., 15 (1993) 281.
9. M. Yalpani, L.D. Hall, Macromolecules, 17 (1984) 272.
10. V. Babak, I. Lubina, G. Vibhoreva, J. Desbrières, M. Rinaudo, Colloïds and Surfaces. A: Physicochemical and engineering aspects (in press).
11. J. Desbrieres, M. Rinaudo, J.M. Klein, B. Malher, Patent FR 9408314, EU 95420177.8
12. C. Chevillard, M.A.V. Axelos, Colloïd. Polym. Sci., 275 (1997) 537.
13. M. Hirrien, C. Chevillard, J. Desbrières, M.A.V. Axelos, M. Rinaudo, Polymer, 39 (1998) 6251.
14. M. Vigouret, M. Rinaudo, J. Desbrières, J. Chim. Phys., 96 (1996) 858.
15. J. Desbrières, M. Hirrien, M. Rinaudo, Carbohydr. Polym., 37 (1998) 145.
16. M. Rinaudo, G. Pavlov, J. Desbrières Int. J. Polym. Anal. Charact. (in press)
17. J. Desbrières, M. Rinaudo, L. Chtcheglova, Macromol. Symp., 113 (1997) 135.
18. D. Hourdet, F. L'Alloret, R. Audebert, Polymer, 35 (1994) 2624.
19. J. Desbrières, M. Rinaudo, V. Babak, G. Vikloreva, Polymer Bull., 39 (1997) 209.
20. J. Desbrières, M. Rinaudo, ACS Symposium Series, Boston (1998) (submitted)

HYDROCOLLOIDS – PART 1
Edited by K. Nishinari
2000 Elsevier Science B.V.

125

Industrial production of new emulsifying polysaccharide by plant cell culture

Alan Lane, Cooperative Research Centre for Industrial Plant Biopolymers
Food Science Australia, PO Box 52, North Ryde, NSW 2113, Australia.
alan.lane@foodscience.afisc.csiro.au

The Cooperative Research Centre (CRC) for Industrial Plant Biopolymers was established in 1992 to develop the science and technology for producing plant polysaccharides ("gums") by large scale culture of plant cells. A screening program resulting in a library of cell lines producing a range of polysaccharides with potential for applications the food and other industries. One of these cell lines produces an emulsifying agent with functional properties and applications similar to Gum Arabic, but which is effective at much lower concentrations. The strategies used to obtain progressive improvement in productivity are described and the fermentation has been scaled up to a 10,000-litre fermenter. The technology is now ready for commercialisation and is being reviewed by prospective licensees.

1.   INTRODUCTION

The Cooperative Research Centre for Industrial Plant Biopolymers was established in 1992, as part of the Australian Government's Cooperative Research Centres Program. The goal of the CRC was to develop the science and technology for commercial production of functional plant polysaccharides by large-scale cultivation of plant cells in suspension. This paper presents an overview of the work of the CRC, culminating in a commercial process which is now being evaluated by prospective licensees.

The overall procedure for establishing plant cells in culture then scaling up to a commercial operation is illustrated in Fig. 1.

2.   SCREENING PROGRAM

A list of 130 target species was compiled on the basis of the likelihood of producing biopolymers having commercially useful functional properties. Suspension cultures were established from 28 species; from these, 24 polysaccharide products were obtained for chemical and physico-chemical characterisation and applications trials.

This resulted in a collection of cell lines producing a variety of biopolymers having a range of chemical and physico-chemical properties, and hence a range of potential applications in foods and industrial uses as gelling agents, thickeners, emulsifiers and encapsulating agents, film forming agents and emulsifiers.

126

3.   THE CRC EMULSIFYING BIOPOLYMER

One of these materials was identified as an emulsifying biopolymer with much greater emulsifying capacity than Gum Arabic. Over the past 2 years, work has focussed on developing the processes for producing and recovering the emulsifier, and consolidating knowledge of its chemical and functional characteristics.

A reliable supply of an alternative emulsifying biopolymer, with consistent performance, offering a good balance between cost and efficacy, and which could supplement supplies of Gum Arabic would assist in overcoming the problems inherent in the Gum Arabic industry.

FIGURE 1
Diagrammatic representation of the procedure

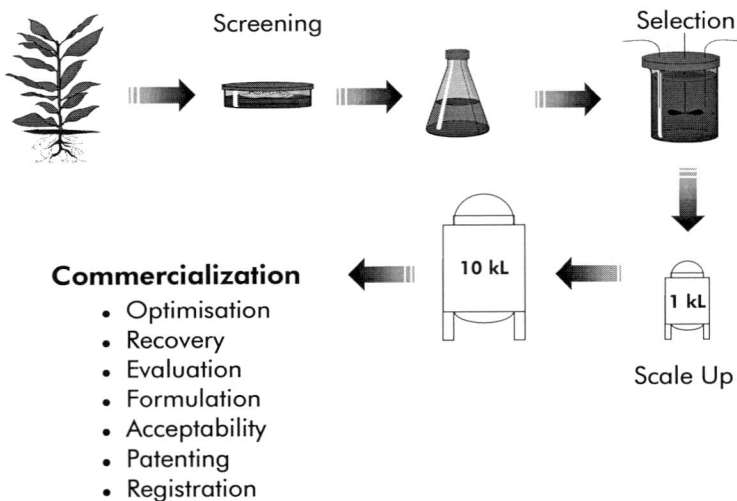

4.   PRODUCTIVITY ENHANCEMENT

The CRC has carried out extensive studies to maximise productivity of the emulsifying biopolymer, so as to reduce production costs and make its manufacture commercially attractive. These studies have followed a number of paths, aimed at increasing the growth rate of the cells, increasing their biopolymer secretion rate, and optimising the operational strategies for the fermentation.

4.1   Medium optimisation by multi-variant experimental design

Multi-variant experimental design, using Analysis of Variance (ANOVA) statistical analysis has proved to be a very powerful technique for optimisation of culture medium. An example of the improvements obtained in growth rate and biopolymer secretion rate using this approach are shown in Figs 2 and 3, respectively.

4.2   Feeding experiments

Data obtained from nutrient uptake studies indicated that the best growth rate and biopolymer yield would result if some of the nutrients were fed to the culture during growth, rather than being all added to the original medium at the start. An example of this phenomenon is shown in Fig. 4.

FIGURE 2
Example of improved growth rate obtained using ANOVA optimisation

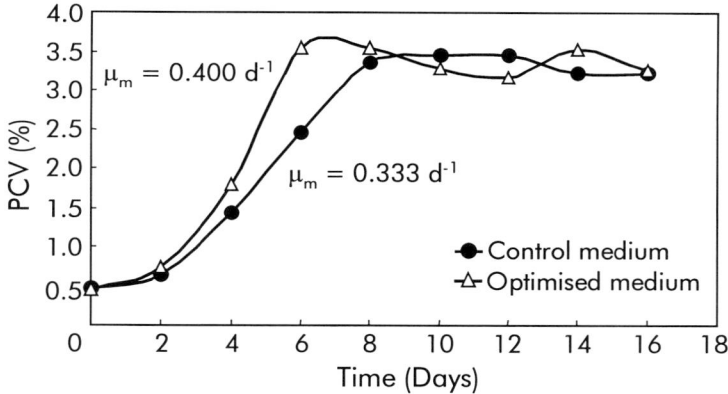

4.3   Fill and draw (semi-continuous) operation of 1000-l fermenter

From theoretical considerations, we would expect a fermenter operating in semi-continuous mode to have a higher productivity than in batch mode. We have explored this operating strategy in detail, at both laboratory scale and pilot scale. Our 1000-l fermenter was operated in semi-continuous mode for 28 successive cycles of 4 days, a total fermentation run of 112 days. No sign of deterioration was seen, either in the growth of the cells or in the emulsifying capacity of the product (Fig. 5).

Our calculations indicate that producing the emulsifier is commercially viable if fermenter productivity reaches 80 arbitrary units per litre per day. With a combination of strategies such as those outlined above, the productivity of the biopolymer fermentation has increased to the point where it is now commercially viable. As shown in Fig. 6, this commercially viable goal has been reached and exceeded.

4.4 Scaling up the fermentation to 10,000 litres

Semi-continuous fermentation has been carried out successfully in a 10,000-l stirred fermenter. Industrial grade chemicals and tap water were used for medium preparation, to simulate conditions of a full-scale industrial process. Cell growth and emulsifier production kinetics was identical to those obtained in smaller fermenters using high-grade ingredients, and no operational difficulties were encountered. This success has given us confidence that the process can now be scaled up further, to a full commercial scale fermentation of 50,000 or even 100,000 litres.

FIGURE 3
Example of improved biopolymer secretion rate obtained using ANOVA optimisation

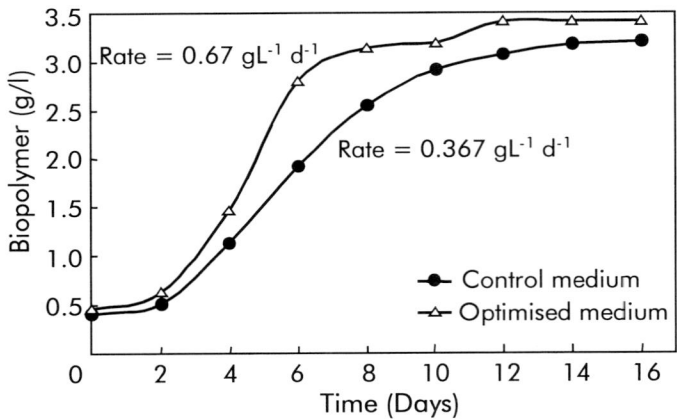

5 EFFICACY OF CRC GUM

There is one further requirement for the CRC process to be economically successful, once the target productivity has been reached. That is that the product should be 20 times as effective as an emulsifier as Gum Arabic on a weight for weight basis.

The unique property of Gum Arabic is its ability to produce oil droplets of fairly uniform 1-2 micron diameter and to stabilise the emulsion so that it neither aggregates nor coalescences in acid solutions, even over periods of several months. This makes Gum Arabic the ideal emulsifier for use in soft drinks, which is still one of the largest applications for it. A disadvantage of Gum Arabic, however, is that high concentrations (normally 20% w/v) are needed to stabilise a soft drink emulsion. The effectiveness of Gum Arabic declines as the concentration is reduced, with both initial droplet size and the rate of aggregation of the droplets increasing rapidly, until at a concentration of 1%, Gum Arabic becomes quite ineffective as an emulsifier (data not shown).

FIGURE 4

Beneficial effects of nutrient feeding (Flasks 1 and 2 – two different feeding regimens),
compared with the control in which all the nutrients were included in the initial medium

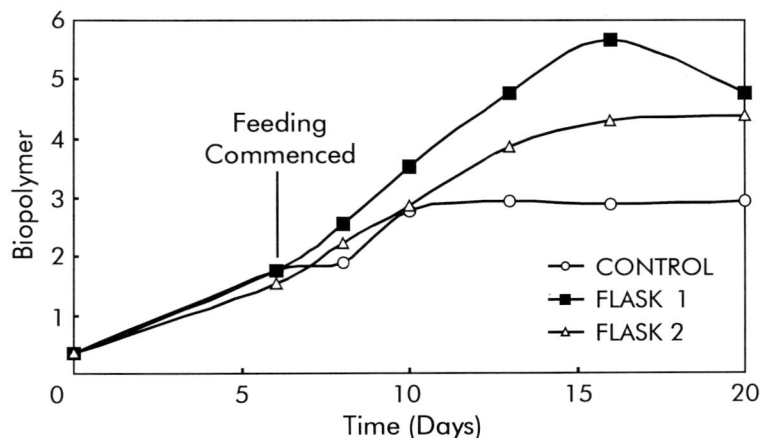

FIGURE 5

Semi-continuous operation of 1000-l fermenter:
Cell growth and product functionality

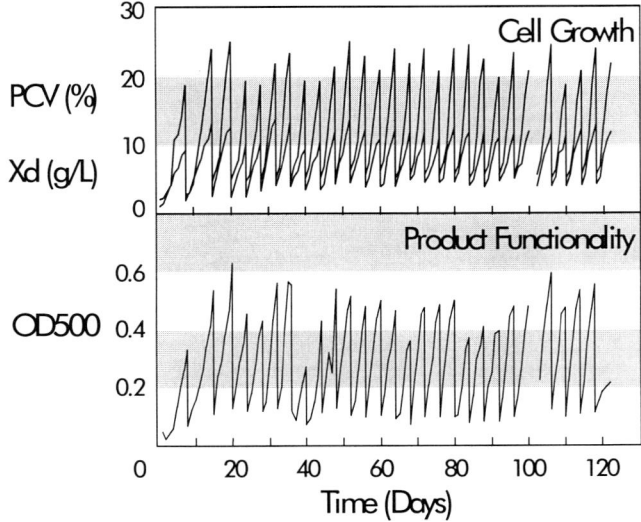

PCV – Packed Cell Volume
Xd – Cell Dry Weight
OD500 – Emulsifying Capacity

FIGURE 6

Milestones in achieving target productivity

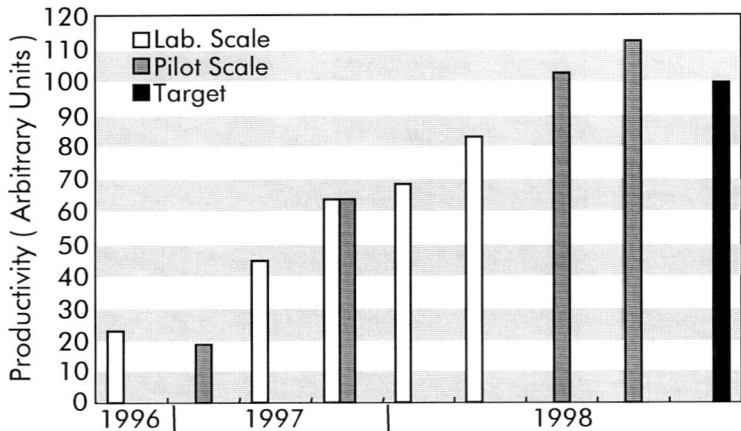

By contrast, the CRC Biopolymer is highly effective at producing stable emulsions with small droplet size at much lower concentrations, using either high shear mixing or a 2-stage pressure homogeniser. Fig. 7 shows the effect of gum concentration on droplet size, between 1.0 and 1.5 microns diameter, as measured by Coulter LS130 particle size analyser. Three different CRC gums are compared with Gum Arabic for their ability to produce droplets in the range 1.0 - 1.5 microns. It can be seen that between 1.2 - 2.7 % (w/v) CRC Biopolymer produced 1.0 micron droplets, while a concentration of 22.2 % was required for Gum Arabic. For droplets of 1.5 micron diameter, concentrations of CRC gums needed were in the range 0.3 - 0.6 % (w/v), while 14.8 % Gum Arabic was needed.

6.   STABILITY OF EMULSIONS

Soft drinks are prepared in two stages. The first is the production of the concentrated "cloud base" emulsion, using 20% (w/v) Gum Arabic and, typically 20 – 26% (v/v) citrus oil containing a weighting agent such as SAIB (sucrose acetyl iso-butyrate). This cloud base emulsion is diluted several hundred-fold in carbonated water to make the final soft drink beverage. Both the cloud base emulsion and the final product need to be stable on the shelf for 6 months. In the soft drinks industry, stability trials are therefore carried out over several months at ambient temperatures, or in an accelerated version of the test, for shorter periods of time at 37°C. We have carried out both long term and accelerated stability trials, comparing CRC Biopolymer with Gum Arabic.

Fig. 8 compares the median droplet size during storage for 6 months at 20°C, of both cloud base emulsion and soft drinks made from either 20 % (w/v) Gum Arabic or 1 % CRC Biopolymer. A slight increase is seen in droplet size in all cases but no appreciable difference was found between the performance of the different emulsifiers. Fig. 9 compares droplet size distributions in emulsions made from 20% Gum Arabic with those made from 1.0 and 1.3 % CRC Biopolymer, after 3 months storage at 37°C.

FIGURE 7
Efficacy of CRC Biopolymers compared with Gum Arabic:
Median droplet size vs. concentration

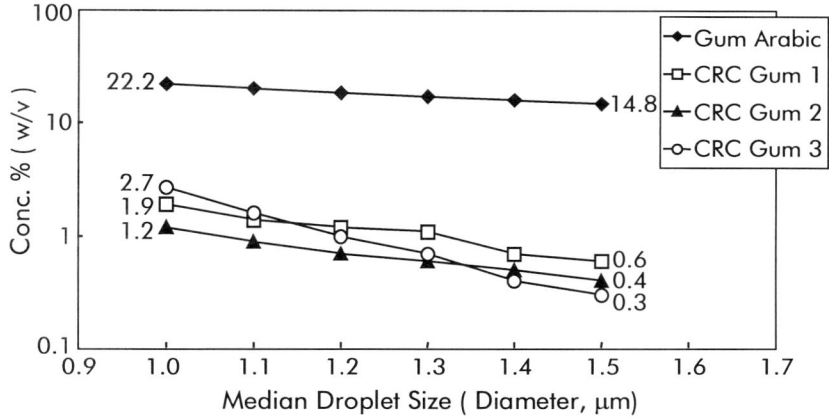

7.   MECHANISM OF STABILISATION

Calculation of zeta potential, obtained from electrophoretic mobility measurements, provides information about the mechanism of emulsion stabilisation. Measurements were carried out to determine the effect of pH on zeta potential of Gum Arabic (20%) and CRC Biopolymer (2%). As shown in Fig. 10, the isoelectric points (the pH at which the zeta potential is zero) were similar and very low (1.7 - 2.0) for the two materials. This indicates that the mechanism of stabilisation in both Gum Arabic and CRC Biopolymer is steric, rather than electrostatic, and argues for a strong structural similarity between the emulsifying components of the two materials. This could also explain one of the most prized characteristics of Gum Arabic as an emulsifier - its insensitivity to ionic environment, whether pH or salts. Our preliminary results show that the CRC Biopolymer is also very insensitive to pH and the presence of salts.

We have isolated what we believe to be the active emulsifying components of Gum Arabic and the CRC Biopolymer. Our preliminary data indicate that these components may be similar and that the CRC Biopolymer contains about 10 times as much of it as does Gum Arabic.

FIGURE 8
Comparison of CRC Biopolymer and Gum Arabic:
Storage trials of emulsions and soft drinks at 20°C

Sample	Day 1	Median Droplet Size (µm)			
		1 month	2 months	4 months	6 months
Emulsions					
Gum Arabic (20%)	0.99	1.3	1.5	1.8	2.4
CRC Gum (1%)	0.94	1.4	1.7	1.8	2.2
Drinks					
Gum Arabic (20%)	-	1.4	1.5	1.5	1.6
CRC Gum (1%)	-	1.5	1.4	1.5	1.7

8. MONOSACCHARIDE COMPOSITION OF CRC BIOPOLYMER

It is important in a commercial process that the composition and functional performance of the product coming from the factory should be consistent in both composition and functional performance from batch to batch over long periods. One way of characterising a biopolymer is to determine its detailed monosaccharide composition. We carried this out on samples taken from batches widely separated in time, and on samples produced in different types and sizes of fermenters and under differing conditions.

Very little differences have been found between Biopolymer produced in the 100-l and 1000-l airlift fermenters, or between those from airlift and stirred fermenter. Representative data in Fig. 11 show typical monosaccharide composition of CRC Biopolymer.

9. REGISTRATION

Procedures are under way to obtain approval for use of the CRC Biopolymer in soft drinks and other foods. Toxicological testing is being carried out as a pre-requisite for approval; the results will be forwarded to the regulatory authorities in Australia, the USA and other countries to enable approval for its use in foods to be granted.

FIGURE 9

Droplet size distribution in emulsions made with CRC Biopolymer (1.0 and 1.3%) and Gum Arabic (20%), after storage 3 months at 37°C

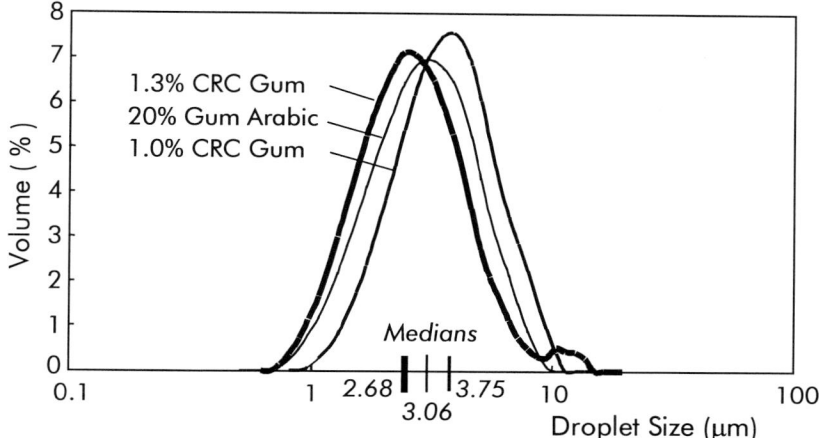

FIGURE 10

Zeta potential vs. pH: CRC Biopolymer and Gum Arabic

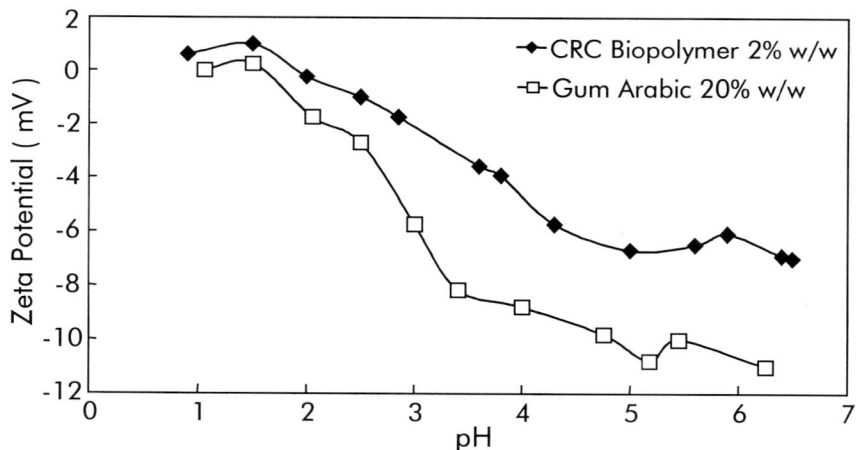

FIGURE 11

Typical Biopolymer Monosaccharide Composition

Ara	Rham	Fuc	Xyl	Man	Gal	Glc	GalA
10.2	2.6	1.2	9.7	2.1	32.2	15.8	23.2

10.   CONCLUSIONS

Experience has given us confidence that both the cell line and the process for producing the CRC Biopolymer have the stability, robustness and reliability needed for a viable commercial manufacturing process.

The CRC has comprehensive world-wide patent protection on all gums produced by plant cell culture, and also on their uses as emulsifying and texture modifying agents in a comprehensive range of food and other industrial applications.

The CRC process is now ready for commercialisation.   We have a potential licensee who has a first option on the technology and the intention of making a decision regarding commercialisation in 1999.

11.   ACKNOWLEDGMENTS

The CRC for Industrial Plant Biopolymers acknowledges the contributions made by all participants, both past and present, from the 4 nodes of the CRC and thanks them for their commitment and dedication in bringing this long and complex project to fruition.

Data shown here were drawn from the work of:

University of Melbourne, School of Botany; Judy Webster, Dr Carolyn Schultz, Dr Shiao-Lim Mau
University of Melbourne, Dept of Chemical Engineering; Dr David Dunstan, Eugene Chai
Tridan/Albright & Wilson (Australia) Ltd Partnership; Dr David McManus, Dr Kian Kwok, Tissa Habarakada, Andrew Redman
Food Science Australia; Nastaran Mahmoudifar, Dr Paul Gibbs, Michael Joss, Kay Middleton, Janet Sharp

The Cooperative Research Centre for Industrial Plant Biopolymers was established in 1992 and supported under the Australian Government's Cooperative Research Centres Program.

HYDROCOLLOIDS – PART 1
Edited by K. Nishinari
2000 Elsevier Science B.V.

135

Production and applications of novel plant cell culture polysaccharides

N. Mahmoudifar[a], E. Chai[b], D. Dunstan[b], and A. Lane[a]

Cooperative Research Centre for Industrial Plant Biopolymers

[a] Food Science Australia, Sydney Laboratory
P.O. Box 52, North Ryde NSW 1670, Australia

[b] University of Melbourne, Department of Chemical Engineering
Parkville VIC 3052, Australia

Plant cell cultures are candidates for production of polysaccharides (gums) with various functional properties. The Cooperative Research Centre (CRC) for Industrial Plant Biopolymers has been developing the science and technology for production of commercially viable gums from plant cell cultures.

More than 130 plants targeted for production of polysaccharide by plant cell culture technology were screened. Callus and suspension cultures were established from the selected cell lines. Suspension cultures were grown in shake flasks and bioreactors up to 100 litres capacity and medium composition and process conditions were optimised. Production was then scaled-up to 1000 litres by the CRC commercial partner, Tridan-Albright and Wilson (Australia) Ltd Partnership. Large batches of products were produced at this scale for product development and application testing.

Gums produced by plant cell cultures have a variety of functional properties depending on the plant cell line used. They have been tested in a number of applications and have been very effective emulsifying, thickening and gelling agents. This paper will describe the application of CRC gums in preparation of emulsions, cosmetic products, fruit toppings and desserts.

1. INTRODUCTION

Plant cell cultures are an alternative source of natural plant products usually harvested from the whole plant, including polysaccharides (gums) with various functional properties. The Cooperative Research Centre (CRC) for Industrial Plant Biopolymers was established in 1992 under the Australian Government Cooperative Research Centre program to develop the science and technology for production of commercially viable gums from plant cell cultures. The core participants in the CRC are Food Science Australia, the University of Melbourne (Department of Chemical Engineering and School of Botany) and Tridan-Albright and Wilson (Australia) Ltd Partnership. The CRC for Industrial Plant Biopolymers is interested in gums with emulsifying, thickening and gelling properties covering a range of food and non-food application.

2. PRODUCTION OF GUMS BY PLANT CELL CULTURE TECHNOLOGY

More than 130 plants, which were known to produce commercial gums or gums in culture, were screened for production of polysaccharides. Suspension cultures were initially established in typical plant cell culture media, containing salts, vitamins, sucrose and plant growth hormones in shake flasks. In suspension cultures polysaccharides are produced extracellularly and secreted into culture medium. These polysaccharides were tested for functional properties and chemical structure and cell lines were identified and selected for production of gums with desired properties.

Cultures were then scaled-up to 7,15 and 100 litres in laboratory scale bioreactors. Media formulations and process conditions such as temperature, pH and dissolved oxygen were optimised systematically to increase the gum productivity and to enhance functional capacity. Scale-up was successfully carried out to 1000 litres in the CRC pilot-scale facility. Large batches of gums were produced at this scale for product development and application testing. Downstream processing was developed by separating cells from gum solution by filtration, and recovering and purifying of gum by steps such as washing, ultrafiltration, dialysis, precipitation and drying.

The selected cell lines for commercial production of gums have been maintained in callus and suspension cultures for at least 3 years and have been subcultured more than 100 times, showing excellent reproducibility in growth, product formation and functional capacity at all scales. As a result of media and process optimisation, and use of novel cell culture strategies, cell growth and product formation rates as well as productivity and functional capacity of CRC gums have been increased, making this technology commercially attractive.

3. APPLICATION OF CRC GUMS

Gums produced by plant cell culture vary from cell line to cell line and have a variety of functional properties, making them suitable for a range of applications. Some have been found to be excellent emulsifiers, gelling and thickening agents and can be utilised in preparation of a number of products. Also, they can be mixed together to exhibit a combination of these properties. As emulsifiers they can be used in soft drinks and salad dressings. As thickening (viscosity building) agents they can be used in preparation of fruit toppings, lotions and creams. As gelling agents they can be used in preparation of desserts. The following data (Figures 1-3 and Table 1) demonstrate some of these applications for the CRC gums.

Figure 1. Application of CRC gum as emulsifying agent and its comparison with Gum Arabic

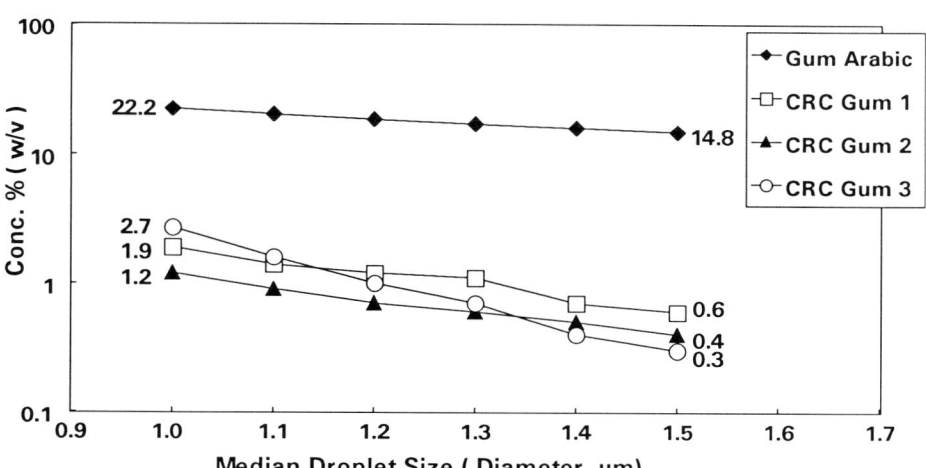

Note: Emulsions were prepared in a pressure homogeniser (with 26 %w/w D-limonene oil mix). The median droplet size was measured by a Coulter LS130 Particle Sizer.

Figure 2. Application of CRC gum as thickening agent and its comparison with commercial body lotion

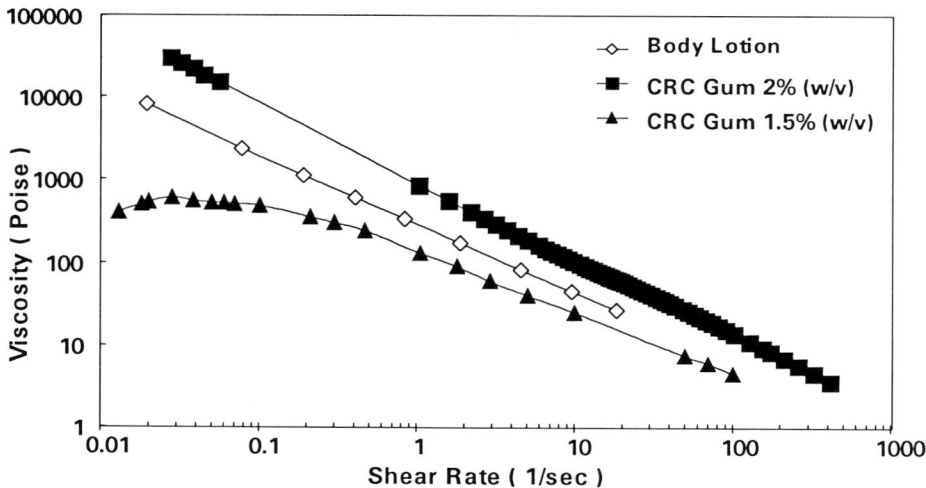

Note: CRC gum at 1.5 % and 2 % concentrations was mixed with mineral oil (10 %) and water to prepare the formulation. Viscosity was measured by a CSL2 Carri-med rheometer.

Figure 3. Application of CRC gum as thickening agent in preparation of strawberry topping and its comparison with commercial strawberry topping

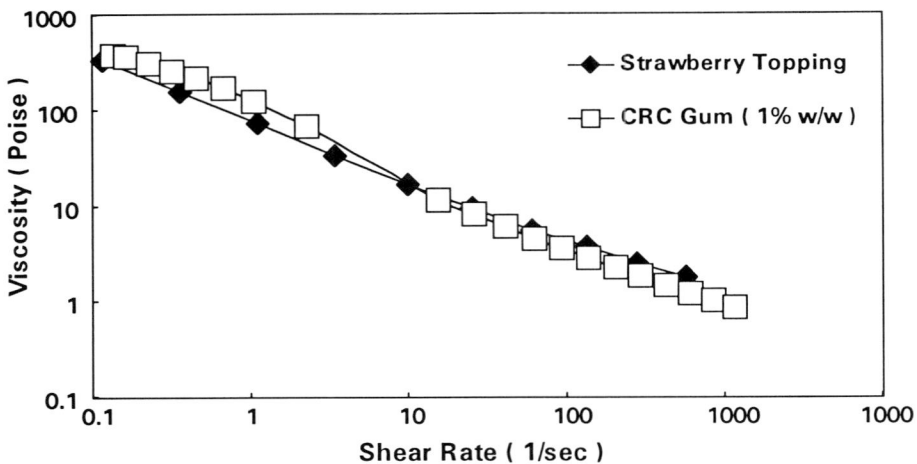

Note: The strawberry topping consisted of 10% fruit puree, 15% sugar and 1% CRC gum. Viscosity was measured by the CSL2 Carri-med rheometer.

Table 1. Application of CRC gum as gelling agent in preparation of chocolate dessert

CRC Gum (% w/w)	Initial Modulus (N/M^2)
0.5	3.5×10^3
1	1.5×10^4
2	3.0×10^4
Commercial Raspberry Flavour Jelly Dessert	2.0×10^4

Note: The dessert gel consist of 5% cocoa powder which was emulsified with a 2% CRC gum solution at 60^0C and cooled to room temperature. Gel strength was measured by a texture analyser, and the initial modulus was from the TPA analysis.

HYDROCOLLOIDS – PART 1
Edited by K. Nishinari
2000 Elsevier Science B.V.

Structural features of polysaccharide of *Hericium erinaceum*

T. Inakuma[a], K. Aizawa[a], R. Yamauchi[b] and K. Kato[b]

[a]Research Institute, Kagome Co., Ltd.,
17 Nishitomiyama, Nishinasuno-Machi, Nasu-Gun, Tochigi 329-2762, Japan

[b]Faculty of Agriculture, Gifu University,
1-1, Yanagido, Gifu 501-1112, Japan

A polysaccharide (PHE) was isolated from a hot-water extract of the fruiting bodies of *Hericium erinaceum*. The structure of PHE was investigated by the combination of chemical and spectroscopic methods. The results indicated that PHE had a main chain composed of β-(1→6)-D-glucan and β-(1→3)-linked glucosyl residues as side chains. The molecular weight of PHE was estimated to be 13,000~22,000.

1. INTRODUCTION

The fruiting bodies of *Hericium erinaceum* belonging to the *Aphllopheralles* have been used as a food and as a traditional Chinese medicine. It has been reported that hot-water extracts of the fruiting bodies of *H. erinaceum* included β-glucan and showed remarkable antitumor activities [1], however, detailed structure of that polysaccharide remained obscure. So we describe the isolation and structural analysis of polysaccharide of the fruiting bodies of *H. erinaceum*.

2. EXPERIMENTAL

2.1.Materials

The fruiting bodies of *H. erinaceum* growing in Yamagata, Japan were obtained in frozen form and submitted to the experiment.

2.2.Measurement

Specific rotations were determined at 20-25℃ with a Union PM-201 polarimeter. The i.r. spectra were recorded with a Perkin Elmer System 2000 FT-IR infrared spectrometer. G.C.-Mass was conducted with a Shimadzu GC-MS QP-1000 apparatus equipped with a Shimadzu capillary column CBP-1-M25-025 (ϕ0.2 mm $\times$ 25 mm) and programmed from 170 to 230 ℃ at 3 ℃/min. Mass spectra were recorded at an ionizing potential of 70 eV. The ^{1}H-, and ^{13}C-n.m.r. spectra were recorded at 499.895 MHz (^{1}H-n.m.r.) and 125.712 MHz (^{13}C-n.m.r.) with a Varian FT-NMR Unity inova 500 spectrometer for a solution in D_2O. 3-(trimethylsilyl)-propanesulfonate (DSS) was used as the internal standard. The temperature was 80.0 ℃ (^{1}H-n.m.r.) and 40.0 ℃ (^{13}C-n.m.r.).

3. RESULTS AND DISCUSSION

The defatted fruiting bodies of *Hericium erinaceum* were extracted with hot water and the extract was deproteined by fractional precipitation with ammonium sulfate. Then, the supernatant liquor was desalted by dialysis, concentrated and lyophilized. It was applied to a column (ϕ2.2$\times$20.5 cm) of DEAE-Sephadex A-50 to give a PHE in 0.5 % yield (Chart 1).

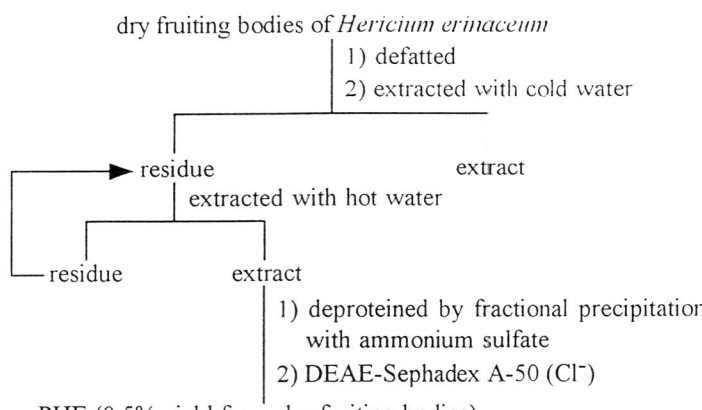

Chart 1. Isolation Procedure for PHE

The PHE had $[\alpha]_D$ -28.0 (c 0.2, H_2O) and was almost composed of D-glucose on hydrolysis (Table 1). The i.r. spectrum showed absorption at 890 cm^{-1} but not at 850 cm^{-1}. So PHE was presumed to be a β-glucan.

Table 1

Some properties of PHE

Yield[*1]	$[\alpha]_D$	Sugar	Protein	Sugar Composition (%)			
(%)	(c=0.2;H_2O)	(%)	(%)	Fuc	Man	Gal	Glc
0.5	-28.0	79.4	5	0.9	2.8	n.d.[*2]	96.3

*1: Yield (%) of the dry fruiting bodies of *Hericium erinaceum*

*2: not detected

To characterize linkages of PHE, G.C.-Mass analysis after methylation was attempted. A sample of PHE was methylated by the Hakomori procedure [2]. The reaction mixture was dialyzed against running water for 3 days, and nondialyzable fraction was dried. The methylation procedure was repeated until the product showed no absorption for free hydroxy groups in its i.r. spectrum. The methylated polysaccharide was then hydrolyzed with 90% formic acid for 3 h at 100 ℃, and then with 0.25 M sulfuric acid for 14 h at 100 ℃. The acid was neutralized with barium carbonate, and the neutral hydrolyzate was evaporated to dryness. The sugars thus obtained were converted into their alditol acetates [3] for G.C.-Mass analysis. The result was that four peak were detected, corresponding to alditol acetates from 2,3,4,6-tetra-O-methylglucose, 2,4,6-tri-O-methylglucose, 2,3,4-tri-O-methylglucose and 2,4-di-O-methylglucose (Table 2). This indicated that the polysaccharide had a main chain composed of β-(1→6)-D-glucan and β-(1→3)-linked glucosyl side chains.

It was confirmed by the ^1H-, and ^{13}C-n.m.r. spectral analysis. The results were summarized in Table 3, 4.

Table 2

Molar Ratio of Partially Methylated Alditol Acetates from PHE

Alditol acetate of	Molar Ratio
2,3,4,6-Me$_4$-Glc	1.00
2,4,6-Me$_3$-Glc	0.38
2,3,4-Me$_3$-Glc	3.75
2,4-Me$_2$-Glc	1.06

Table 3

^{1}H--n.m.r. Chemical Shifts* and Coupling Constants
of Anomeric Protons of PHE

	H-1→3 ppm ($J_{1,2}$)	H-1→6 ppm ($J_{1,2}$)
PHE	4.72 (5.86)	4.52 (7.81)

*Chemical shifts were expressed in ppm relative to
DDS as internal standard at 80.0 °C.

Table 4

^{13}C-n.m.r. Chemical Shifts* of PHE

	C-1	C-2	C-3	C-4	C-5	C-6
Sugar 1	105.66	75.76	78.32	72.30	76.18	70.88
Sugar 2	105.66	75.50	87.15	72.30	77.61	71.59
Sugar 3	105.66	75.76	78.32	72.30	76.18	70.88
Sugar 4	105.66	75.50	78.69	72.30	77.61	63.47

*Chemical shifts were expressed in ppm relative to DDS as internal standard at 40.0 °C.

To elucidate the detailed structures of side chains, controlled Smith-degradation was attempted [4].

PHE was oxidized with 0.01 M sodium metaperiodate for 10 days at 4.0 °C. After complete oxidation, 0.1 M sodium hydroxide was added to neutralize and the reaction mixture was dialyzed against running water for 1 day. The nondialyzable fraction was reduced with sodium borohydride, neutralized with 1 M acetic acid, and dialyzed to give the corresponding polyalchol. The polyalchol was submitted to controlled degradation with 10 M sulfuric acid for 12 h at room temperature and the hydrolyzate was applied to a column (2.5 × 900 cm) of Bio-Gel P-2, being monitored by the phenol sulfuric acid method. Three compounds were isolated as the main degradation products. These were identified by TLC, HPLC, Methylation analysis and i. r. spectra. The results indicated that three compounds were 1-O-β-D-glucopyranosyl-glycerol and 1-O-β-laminarabiosyl-glycerol. Isolation of three compounds shows that the side chains are composed of both β-glucopyranosyl and β-laminarabiosyl units.

PHE was dissolved in 20 mM phosphate buffer (pH 7.0). To this solution the enzyme (37.5 unit) of endo-β-D-glucanase from Achromobacter lunatus, commercial name "Amano YL-15" was added and then the solution was incubated for 24 h at 37 °C. After incubation with enzyme, the was analyzed by TLC. The products was not at all detected, suggesting that β-(1→3)-

linked glusosyl chains as side chain was not long, and that main chain of PHE was β-(1→6)-linked glucosyl chain (Figure 1) .

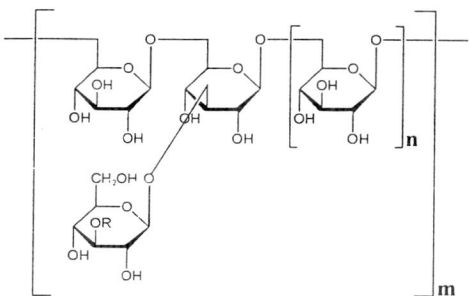

Figure 1. Structure of PHE

To estimate the molecular weight of PHE, it was dissolved in 50 mM sodium chloride solution (0.5 mL) and the solution was applied to a column (ϕ 10 mm ×30 cm) of Shodex OH pac KB-803 equilibrated with 50 mM sodium chloride solution. PHE on the column was eluted with the same solvent (1.0 ml/min). The molecular weight of the sample was determined according to the calibration curve prepared using dextrans T-40 (Mw=43,500), T-20 (Mw=20,400) and T-10 (Mw=10,500) from Pharmacia Fine Chemicals. The result was indicated that the molecular weight of PHE was estimated to be 13,000~22,000.

REFERENCES

1. T. Mizuno, T. Wasa, H. Ito, C. Suzuki and N. Ukai, Biosci. Biotech. Biochem., 56 (1992) 347.
2. S. Hakomori, J. Biochem., 55 (1964) 205.
3. J. H. Sloneker, Methods Carbohydr. Chem., 6 (1972) 20.
4. I. J. Goldstein, G. W. Hay. B. A. Lewis and F. Smith, Method of Carbohydr. Chem., V (1965) 361.

HYDROCOLLOIDS – PART 1
Edited by K. Nishinari
2000 Elsevier Science B.V.

Structural and physical features of polysaccharide of *Tremella aurantia*

K. Aizawa[a], T. Inakuma[a] and T. Kiho[b]

[a]Research Institute, Kagome Co., Ltd.,
17 Nishitomiyama, Nishinasuno-Machi, Nasu-Gun, Tochigi 329-2762, Japan

[b]Gifu Pharmaceutical University,
6-1, Mitahora-higashi 5-chome, Gifu 502, Japan

An acidic polysaccharide (TAP) was isolated from a hot-water extract of the fruiting bodies of *Tremella aurantia*, and composed of mannose, xylose, glucuronic acid, glucose and *O*-acetyl groups. TAP was found to have almost same physical properties as the stabilizers on the market, and was superior to them in the points of the stabilizers after thawing and in existence of salt.

1. INTRODUCTION

The fruiting bodies of *Tremella aurantia* [1] belonging to the *Tremellaceae* have been used as a food and as a traditional Chinese medicine. It has been reported that the fruiting bodies of *T. aurantia* contained polysaccharides in high-ratio (more than 25 % w/w of the dry fruiting bodies) and showed remarkable anti-diabetic activities [2], however, detailed structure and physical feature of that polysaccharide remained obscure. So we describe the structural and physical features of polysaccharide of the fruiting bodies of *T. aurantia*.

2. EXPEIMENTAL

2.1. Material

The fruiting bodies of *T. aurantia* growing in Yunnan, China were obtained in dried form and submitted to the experiment.

2.2.Measurement

Specific rotation was measured with JASCO DIP-4 automatic polarimeter. Paper chromatography (PPC) was performed on Advantec Toyo No. 51B filter paper with a solvent system (v/v) of 1-butanol-pyridine-water (6:4:3) by double ascending method. Gas chromatography (GC) was carried out on a Shimadzu GC-4CM apparatus equipped with a hydrogen-flame ionization detector. The viscosity was measured at 20 °C by viscometer (HAKKE VT 500).

3. RESULTS AND DISCUSSION

The dried fruiting bodies (52 g) of the fungus were extracted with hot water (1000 ml) for 4 h in a boiling water bath. The residue was extracted again with hot water (1000 ml) for 4 h in a boiling water bath. These extracts then were concentrated and 2 volumes of ethanol were added to the solution. After leaving the mixture for 1 day in a cool room, the precipitate obtained by centrifugation was lyophilized to give a crude polysaccharide. It was dissolved in water (200 ml) and the solution was mixed with 2 volumes of ethanol again and left for 1 day in a cool room. The precipitate obtained was dissolved in water (100 ml), and then dialyzed against water and lyophilized to give TAP in 42.0 % yield.

TAP had an $[\alpha]_D$ -7° (c=0.3, H_2O). For the estimation of the molecular weight of TAP, it was dissolved in 0.1 M sodium chloride solution and the solution was applied to a column of Toyopeal HW-75 equilibrated with 0.1 M sodium chloride solution. TAP on a column was eluted with the same solvent. The molecular weight of TAP was estimated to be about 1,500,000 according to the calibration curve made using marker (MW=70,000, 260,000, 580,000, 2,000,000).

TAP was hydolyzed with 2 N sulfuric acid for 8 h at 100 °C and the hydrolysates were analyzed by Paper chromatography (PPC) and by Gas chromatography (GC) as alditol acetates derivatives, as described previously [3]. Sugars were detected with alkaline silver nitrate reagent [4]. The results showed that TAP was composed of mannose, xylose, glucuronic acid and glucose (molar ratio, 4:2:1:0.5)   (Table 1).

Table 1

Some Properties of TAP

Yield[*1]	$[\alpha]_D$	Molecular Weight	Sugar Composition Ratio			
(%)	(c=0.3;H_2O)	(MW)	Mannose	Xylose	Glucuronic acid	Glucose
42.0	-7	1,500,000	4	2	1	0.5

*1: Yield (%) of the dry fruiting bodies of *Tremella aurantia*

TAP was a slightly bitter tasting water-soluble white powder. It was thought to be of value as if food stabilizer, because its solution had a high viscosity. For this reason various physical properties (viscosity, freeze thaw stability and stability to heat, salt and acid) were compared with xanthan gum, carrageenan and guar gum.

The viscosity of 1 % solutions of stabilizer was measured at 20 ℃ by viscometer (HAKKE VT 500). The result was summarized in Figure 1. To test their stabilities, 1 % solutions of TAP and above stabilizers were treated by heating at 80 ℃ in a boiling water (Figure 2), by freezing at -20 ℃ for 15 h and thawed at room temperature (Figure 3), by heating the 1 % solutions in 50 mM citrate buffer (pH 3.8) at 80 ℃ in a boiling water bath (Figure 4), and measured their viscosity at 20 ℃. Also, the viscosity of 1 % samples dissolved in 0.5~2.0 % sodium chloride solution were measured at 20 ℃ (Figure 5).

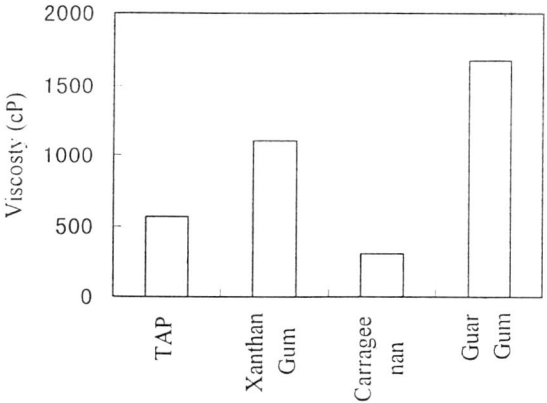

Figure 1. Viscosity of TAP and stabilizers. The viscosity were
measured 20 ℃ by viscometer (HAKKE VT 500).

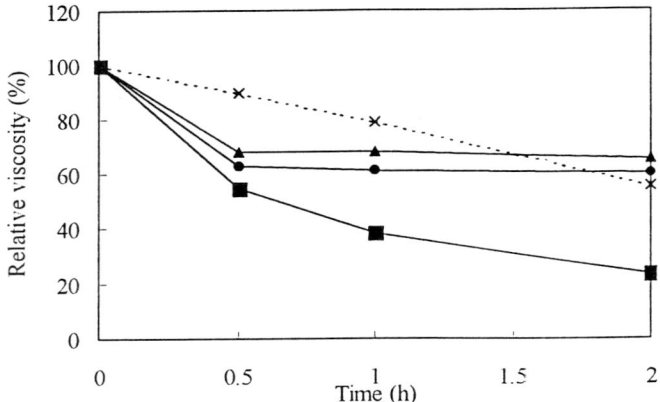

Figure 2. Time course of change in viscosity of the TAP and stabilizers by heating at 80 °C. The viscosity were measured 20 °C by viscometer (HAKKE VT 500). ■, TAP ; ▲, Xanthan Gum ; ●, Carrageenan ; ×, Guar Gum.

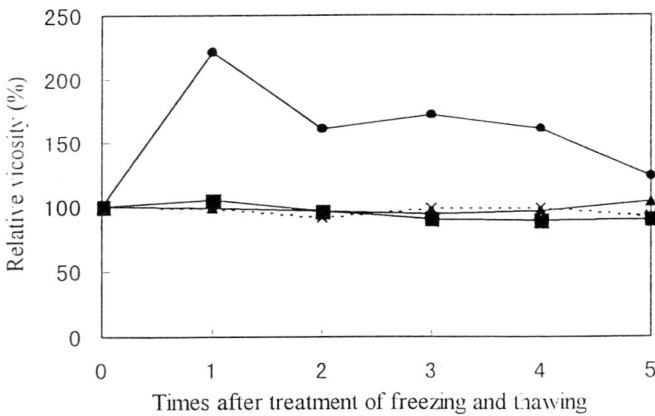

Figure 3. Change in viscosity of the TAP and stabilizers after treatment of freezing (-20 °C, for 15 h) and thawing at room temperature. The viscosity were measured 20 °C by viscometer (HAKKE VT 500). ■, TAP ; ▲, Xanthan Gum ; ●, Carrageenan ; ×, Guar Gum.

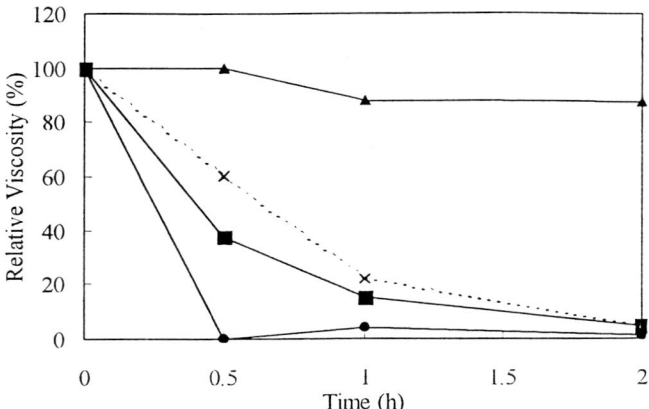

Figure 4. Change in viscosity of the TAP and stabilizers after heated in the acidic
condition (50 mM citrate buffer ; pH 3.8). The viscosity were measured
20 ℃ by viscometer (HAKKE VT 500). ■, TAP ; ▲, Xanthan Gum ;
●, Carrageenan ; ×, Guar Gum.

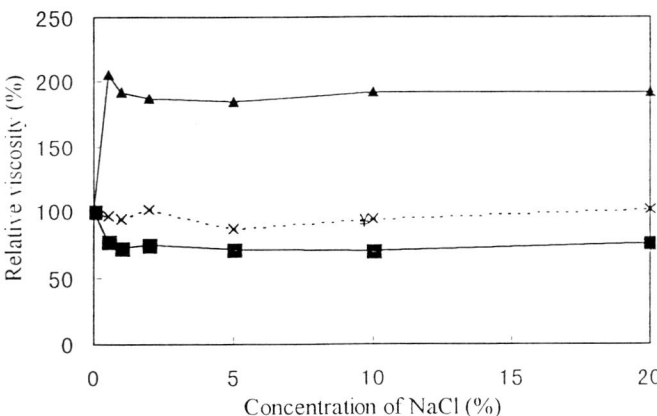

Figure 5. Change in viscosity of the TAP and stabilizers in the presence of salt
(0.5∼2.0 % sodium chloride). The viscosity were measured 20 ℃
by viscometer (HAKKE VT 500). ■, TAP ; ▲, Xanthan Gum ;
●, Carrageenan ; ×, Guar Gum. *Carrageenan was gelled in the
presence of salt.

The results indicated that TAP had almost same features in water-solubility and viscosity as other stabilizers. In the stabilities, TAP was not superior to another stabilizers in acidic and thermal conditions, but superior to them after thawing and in existence of salt (Table 2).

Table 2
Comparison of physical properties of TAP and stabilizers on the market.

	TAP	Xanthan Gum	Carrageenan	Guar Gum
Solubility	soluble in cold water	soluble in cold water	soluble in cold water	soluble in cold water
Gelation	—	—	+	—
Transparency	low	low	high	low
Viscosity (1 % sol. ; cP)	564.3	1101.6	303.2	1673.6
Stability after heating	not stable	stable	stable	stable
Stability after thawing	much stable	much stable	much stable	much stable
Stability in acidic condition	unstable	stable	unstable	unstable
Stability in presence of salt	stable	unstable	gelled	much stable

These works are carried out as the part of the undertaking of the Japanese Research and Development Association for New Functional Foods.

REFERENCES

1. Chinese name: 金耳; Liu B., Chinese Medical Fungi, Shansi People's Publisher, China, (1978) 51.
2. T. Kiho, H. Morimoto, M. Sakushima, S. Usui and S. Ukai, Biol., Pharm., Bull., 18 (1995) 1627.
3. T. Kiho, H. Morimoto, M. Sakushima, S. Usui and S. Ukai, Chem., Pharm., Bull., 35 (1987) 4286.
4. Trevelyan W. E., Procter D. P. and Harrison J. S., Nature (London), 166 (1950) 444.

HYDROCOLLOIDS – PART 1
Edited by K. Nishinari
2000 Elsevier Science B.V.

Dynamic light scattering of dilute and semi-dilute Xanthan solutions and comparison with Rheological Characteristics

[a]Andrew B Rodd, [b]Dave E. Dunstan, and [a]David V. Boger

[a]Department of Chemical Engineering, The University of Melbourne, Parkville 3052, Australia.

[b]Cooperative Research Centre For Industrial Plant Biopolymers, Department of Chemical Engineering, The University of Melbourne, Parkville 3052, Australia.

The 'weak gel' formation mechanism of xanthan gum solutions has been examined using rheological and light scattering methods. Significant differences are observed between c* measured by each technique. The effect of shear on c* for rod like molecules such as xanthan is demonstrated. Doi and Edwards' theory is used to explain the differences as observed between rheological measurements involving shear alignment effects and light scattering measurements conducted in a truly equilibrium environment. Xanthan gum solutions are further complicated by the widely reported existence of anisotropic-aggregates. The process of aggregation has resulted in the introduction of a second concentration regime parameter, the critical concentration for aggregation (c**). Measurements in various electrolytes using dynamic light scattering show both the transition from pure Brownian dilute behaviour to semi-dilute behaviour (c*) and the critical concentration for aggregation (c**).

1. INTRODUCTION

Xanthan gum is a high molecular weight extracellular polysaccharide produced by bacteria of the genus *Xanthomonas*. Xanthan gum chemically may be considered as an anionic polyelectrolyte, with a backbone chain identical to cellulose in structure with a trisaccharide sidechain on alternate residues [1]. The xanthan molecule undergoes a conformation transition from an ordered double helix to a random coil when heated between 40 and 80°C, depending on the ionic strength of solution [2]. The ordered conformation of the xanthan molecule is a semi-flexible rod with a hydrodynamic length varying between 600 and 2000nm [3] and a hydrodynamic diameter of the order of 2nm [4].

Xanthan gum is one of the most intensively studied polysaccharides both in terms of its rheological properties and its physical chemistry. The behaviour of xanthan gum in both dilute and semi-dilute solutions has been extensively examined [5-7]; however, a non-pertabative technique to accurately characterize the transition between dilute and semi-dilute has not been previously found. The concentration at which individual polymer molecules begin to physically interact with each other is defined as c*[8]. The concept of c* is based on the theory that the polymer coils rotate in an approximately spherical volume and at a certain concentration a critical close packing of these spheres occurs above which the molecules interact. Due to the complicated nature of xanthan gum solutions (i.e. rod-like secondary ordered conformation [9] and a tertiary structure of micro-aggregates [10]), c* obtained from shear measurements has not been accurate. The molecular alignment introduced by the application of shear forces, results in c* determined through shear rheology or capillary viscometry, not being indicative of molecular behaviour in a true Brownian state. Dynamic light scattering may be used to trace the diffusivity of molecules in solutions through a range of concentrations and thus gain insight into the solution behaviour at true equilibrium.

The complex nature of xanthan systems, where significant aggregation occurs, has led to the introduction of the parameter c** for the second concentration regime transition. c** is reported to the concentration at which anisotropic aggregaton occurs [11-14], and is observed to occur at a concentration much higher than that of which initial molecular interaction is observed (c*). Previous determinations of c* and c** have involved the combination of a number of techniques most of which involve the application of shear forces to solutions of xanthan gum.

Dynamic light scattering is a non purtabative technique commonly used to measure the diffusion of polymer molecules in solution and is proposed to be used here for the determination of both c* and c**. Experiments have been conducted to quantify c* and c** as defined by previous work [11,13,14] in the absence of imposed shear forces. The prior measurements have shown varying electrolyte concentrations have also been used to determine the effect of electrostatic screening on light scattering and rheological data. Results from the different methods are compared.

2. EXPERIMENTAL SECTION

The commercial food grade xanthan (Keltrol, lot no PX67769) was provided by Keltrol. It was determined as having a molecular weight of approximately $3.4*10^6$ g/mol and a polydisperisty of 1.12. Xanthan gum solutions were prepared in 1-liter batches at 0.55 wt % in triple distilled water. All xanthan concentrations were prepared by dilution of the original batch with the appropriate ionic strength electrolyte solutions.

The rheological experiments were performed on a Carrimed CSL^2100 controlled stress rheometer. The stress range of the Carrimed with this measurement system is 1.8×10^{-3} - 40 Pascal, which allowed accurate determination of the low shear Newtonian plateau by a shear stress ramp for all concentration ranges measured. Equilibrium flow measurements were performed for higher concentration samples to extend the low-end range.

Light Scattering measurements were conducted on a Malvern Series 4700 PCS100 spectrometer, using a 488nm argon ion laser operating at 1CmW. All measurements were conducted at 25°C and a solvent viscosity of 0.890 cP was assumed. An optimum scattering angle of 40° was determined, in agreement with previous work completed on xanthan gum [15]. A scattering angle of 40° was determined as the optimum based on the observation that the average time constant (Γ_i) was linearly related to the square of the scattering vector. The results were analysed using a Malvern CONTIN algorithm that fitted mono-modal and bi-modal distributions [11,16] as shown in equations 1 and 2.

$$C(\tau) = A^2 e^{-2\Gamma\tau} + B \qquad \text{(mono-modal)} \qquad (1)$$
$$C(\tau) = (A_1 e^{-\Gamma_1\tau} + A_2 e^{-\Gamma_2\tau})^2 \qquad \text{(bi-modal)} \qquad (2)$$

Optical clarification of the various electrolyte solutions was achieved by centrifugation at 15000g for 2 hours followed by filtration through a 0.22μm filter. A detailed description of the experimental technique is reported elsewhere [17].

3. RESULTS AND DISCUSSION

3.1 Rheological Measurements

Determination of c* using rheological methods has been conducted for a range of xanthan solutions with varying electrolyte conditions for the purposes of comparison with earlier work[18]. Both shear stress ramps and equilibrium flow measurements were conducted for a range of concentrations in each of the electrolyte conditions. Figure 1 illustrates the comparison of equilibrium flow and shear stress ramps for a number of concentrations. Extrapolation into a zero shear Newtonian viscosity was done using the Herschel Bulkley model and the high shear rate data was fitted to a power law model. A zero shear Newtonian viscosity was not extrapolated from higher concentration samples where hysteresis was observed. Figure 2 illustrates the zero shear Newtonian viscosities plotted versus concentration for each electrolyte condition. The plot shows c* as indicated by the discontinuity in the linear fit to range from 0.065 wt% to 0.13 wt% depending on electrolyte. The observed slopes are in agreement with literature which show the slope in the dilute region to vary between 1.07 and 1.3 and the semi-dilute region is reported to vary from 3.9 - 4.2[19,20].

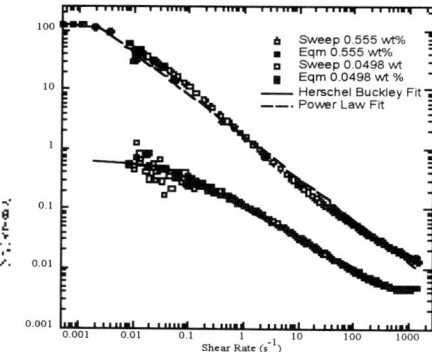

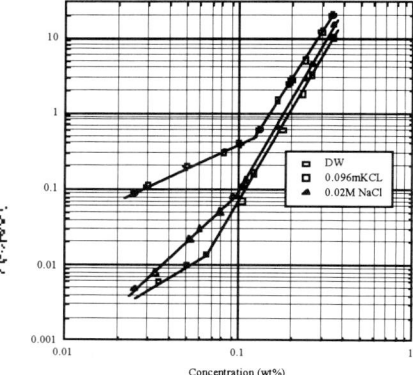

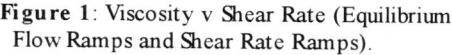

Figure 1: Viscosity v Shear Rate (Equilibrium Flow Ramps and Shear Rate Ramps).

Figure 2: Extrapolated Zero Shear Newtonian Viscosity v Concentration.

3.2 Theoretical Calculations

The rheological results obtained for c* agree with previous findings [19,21]. However, based on molecular properties the onset of interaction will occur at much lower concentrations. For the case of rigid rods and semi flexible rods in solutions, a deviation from dilute behaviour can be expected when the number of molecules in solution is such that the effective volumes occupied by rods rotating in true Brownian motion overlap. Berry and Russel [22] define dilute behaviour as:

$$nl^3 \ll 1 \tag{3}$$

where n is the number density of polymer molecules in solution and l is the hydrodynamic length of a single molecule in solution.

For rigid rods in solution the hydrodynamic length (l) can be obtained from the hydrodynamic radius by a relationship proposed by Broersma [23]. For the xanthan gum systems a hydrodynamic length of between 3474 nm and 4168 nm was calculated. These lengths are over-estimates, as the xanthan molecule flexibility is not considered. However the lengths are in agreement with the scale of xanthan in solution reported in the literature from electron microscopy [24]. Table 1 shows the resultant limit of diluteness from Berry and Russel [22] as calculated using $nl^3 = 1$.

Doi and Edwards [25] suggest that rods will be constrained to move longitudinally in a tube formed by their nearest neighbours. The discrepancy between calculated values and rheological results are thought to be a result of the alignment of rods due to imposed hydrodynamic forces during rheological measurements with increase c* resulting as is suggested by Doi and Edwards [25].

3.3 Dynamic Light Scattering

Dynamic Light Scattering presents a technique absent of imposed shear forces to investigate the diffusional behavior of xanthan over a large range of concentrations. A number of electrolyte conditions were investigated at polymer concentrations between 0.5 and 10^{-6} wt%. Figure 3 illustrates the measured diffusion coefficient as a function of concentration for each electrolyte condition studied. At infinite dilution all solutions showed diffusion coefficients that plateau to a constant value. It is expected that as the concentration of molecules in solution increases, the volume in solution mapped out due to rotation in true Brownian motion will be such that they would be expected to overlap and result in a reduced diffusion coefficient. Doi and Edwards [25] predicts the diffusion in the semi-dilute region to reduce as a power law function, this behaviour is observed in the data. The concentration at which the diffusion coefficient deviates markedly from the infinitely dilute case, is allocated as the onset of interaction between molecules. Measurements in sample cells of varying diameter showed multiple scattering is not contributing to the observed trend.

Electrolyte [ppm, NaCl]	$nl^3=1$ Calculated [wt%]	Diffusion Deviation c*. [wt%]	Min. in Diffusion c**. [wt%]
15000	$11.1*10^{-05}$	2×10^{-04}	0.07
10000	$4.72*10^{-05}$	2×10^{-04}	0.07
5000	$3.34*10^{-05}$	4×10^{-04}	0.07

Table 1: Diffusion critical points for all electrolyte conditions and comparison with calculated values. (c* and c**)

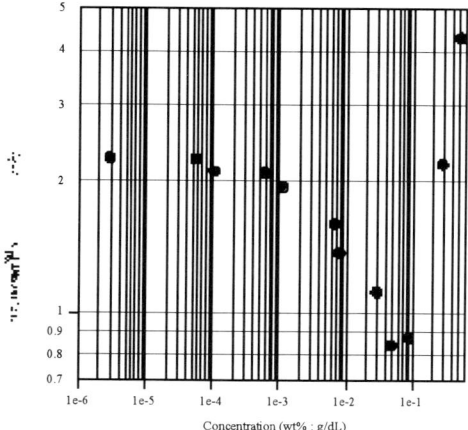

Figure 3: Diffusion v Concentration for Xanthan in 1000ppm NaCl.

Agreement between the observed diffusion deviation (measured c*) and calculated values are shown in Table 1, illustrating the importance of removing imposed hydrodynamics when measuring c*.

Theoretical predictions of the dependence of diffusivity on concentration, for rigid rods [25] are thought to provide a possible explanation for the observed increase in diffusion at higher concentrations. Doi and Edwards model the diffusional behavior for polymer molecules as constrained in a tube of radius r. The radius of the tube (r) corresponds to the distance the polymer may move perpendicular to its own axis without being hindered by other molecules. Doi and Edwards [25] propose that at a higher concentration, a dispersed anisotropic phase is predicted to appear where static correlation of the polymers occurs in the absence of any external field. Doi and Edwards [25] propose that as a local area of polymers orient in the one direction, the average diameter of the tube surrounding the test polymer will increase, resulting in tube dilation and a resultant increase in measured diffusivity. It is postulated that the minimum in diffusion represents a minimum concentration at which an anisotropic phase dominates solution behaviour and is referred to as c**. A value of c** = 0.07wt% is lower than that observed by Meyer *et al.* [13] and Milas *et al* [14]; however, it is in agreement with a second concentration regime observed using photon correlation spectroscopy (PCS) by Southwick *et al.* [11]. Both Meyer *et al.* [13] and Milas *et al.* [14] suggest c** represents the onset of anisotropic aggregation dominating the solution behaviour.

The increase in diffusion above a critical concentration herein denoted c** has been observed for polystyrenes of varying molecular weights in a θ solvent [26]. Munch *et al.* [26] observed a minimum in diffusion as a function of concentration and consequently applied the theoretical predictions of Brochard and de Gennes [27] to obtain an expression that for samples of a constant molecular weight, which predicts diffusion will increase linearly with concentration at $c > c^*$. The concept of cooperative chain diffusion may be applied qualitatively to the aggregation of rods in solution with diffusion predicted to increase as internal translation modes dominate rods restricted within the aggregate[27].

4. CONCLUSIONS

Dynamic light scattering has successfully been used to quantify two critical concentrations, the introduction of molecular interactions (c^*) and the onset of anisotropic aggregation (c^{**}) in the absence of imposed shear forces. Results obtained from shear rheology give a biased measure of the introduction of molecular interaction (c^*) due to alignment of the molecules to yield a higher value of c^*. Initial interactions of molecules are indicated by a decrease in diffusivity at a concentration of the order $c^* \approx 10^{-5}$ wt%. c^* agrees quantitatively with theoretical calculations based on a number density of molecules in pure Brownian motion. An observed minimum in diffusion at 0.07wt% has been described as the onset of an anisotropic phase of aligned rods in solution and referred to as c^{**}.

REFERENCES

1. E.R.Morris, V.J.Morris and S.B.Ross-Murphy, Journal of Polymer Science: Polymer Letters Edition, 20 (1982) 531.
2. T.Sato, T.Norisuye and H.Fujita, Macromolecules, 17 (1984) 2696.
3. S.B.Ross-Murphy, V.J.Morris and E.R.Morris, Faraday Symp Chem Soc,. 18 (1983) 115.
4. B.T.Stokke, A.Elgsaeter and O.Smidsrod, International Journal of Biological Macromolecules, 8 (1986) 217.
5. T.Lim, J.T.Uhl and R.K.Prud'Homme, Journal of rheology,. 28 (1984) 367.
6. M.V.Pastor, E.Costell, L.Izquierdo and L.Duran, Food Hydrocolloids, 8 (1994) 265.
7. C.J.Carriere, E.J.Amis, J.L.Schrag and J.D.Ferry,Journal of rheology,.37 (1993) 469.
8. J. D. Ferry. Viscoelastic Properties of Polymers, New York: Wiley, 1980.
9. G.Paradossi and D.A.Brant, Macromolecules,. 15 (1982) 874.
10. J.G.Southwick, H.Lee, A.M.Jamieson and J.Blackwell, Carbohydrate Research, 84 (1980) 287.
11. J.G.Southwick, A.M.Jamieson and J.Blackwell, Macromolecules, 14 (1981) 1728.
12. A.G.Jamieson, J.G.Southwick and J.Blackwell, J. Polymer Science, Polymer Physics Edition, 20 (1982) 1513.
13. E.L.Meyer, G.G.Fuller, R.C.Clark and W.-M.Kulicke, Macromolecules, 26 (1993) 504.
14. M.Milas, M.Rinaudo, M.Knipper and J.L.Schuppsier, Macromolecules, 23 (1990) 2506.
15. H.Brenner, Int Journal Multiphase Flow, 1 (1997) 195.
16. S.W.Provencher, Computer Physics Communicaitons, 27 (1982) 229.
17. A.B Rodd, D.E.Dunstan and D.V.Boger, Carbohydrate Polymers, Submitted, (1998).
18. A.Parker, P.A.Gunning, K.Ng and M.M.Robins, Food Hydrocolloids, 9 (1995) 333.
19. G.Gravanis, M.Milas, M Rinaudo and B.Tinland, Carbohydrate Research, 160 (1987) 259.
20. M.Milas and M.Rinaudo, Carbohydrate Research, 158 (1986) 191.

21. P.J.Whitcomb and C.W.Macosko, Journal of rheology, 22 (1978) 493.

22. D.H.Berry and W.B.Russel, Journal of Fulid Mechanics, 180 (1987) 475.

23. S.Broersma, The Journal of Chemical Physics, 32 (1960) 1626.

24. G.Holzwarth, Carbohydrate Research, 66 (1978) 173.

25. M. Doi and S. F. Edwards. Semidilute Solutions of Rigid Rodlike Polymers. In: The Theory of Polymer Dynamics, Anonymous Oxford: Clarendon Press, 1984.pp. 324-349.

26. J.Munch, G.Hild and S.Candau, Macromolecules, 16 (1983) 71.

27. F.Brochard and P.G.de Gennes, Macromolecules, 10 (1977) 1157.

HYDROCOLLOIDS – PART 1
Edited by K. Nishinari
2000 Elsevier Science B.V.

Relationships between structural features, molecular weight and rheological properties of cereal β-D- glucans

W. Cui and P. J. Wood

Food Research Program, Southern Crop Protection and Food Research Centre
Agriculture and Agri-Food Canada, 43 McGilvray, Guelph, Ontario, Canada N1G 2W1

Although cereal β-D-glucans exhibit similar C^{13}-NMR spectra, their structural features are significant different from each other as evident by the ratios of tri- and tetrasaccharides released from lichenase hydrolysis (4, 3 and 2 for wheat, barley and oat β-D-glucans, respectively). The present paper demonstrates the correlation between the structural features, molecular weight and some physical properties including solubility of freeze-dried samples and gel forming ability. Rheologically, cereal β-D-glucans behave similarly to random coil polysaccharides; however, some β-D-glucans can form gels under certain conditions, which is not observed for other random-coil polysaccharides including galactomannans. The gelling ability and insolubility of freeze-dried cereal β-D-glucans followed the order of wheat > barley > oat; this trend corresponds with the ratio of tri- and tetrasaccharides of cereal β-D-glucans. In addition to the structural features, molecular weight also played an important role in the conformation of cereal β-D-glucans in aqueous systems. It appeared that lower molecular weight β-D-glucan was easier to form a gel than larger molecules within the molecular weight range investigated.

1. INTRODUCTION

(1→3)(1→4)-β-D-Glucans (ß-glucans) occur in cereals in the endosperm and aleurone cell walls. The content of β-glucan in barley, oats, rye and wheat is generally in the range 3-11%, 3-7%, 1-2%, and <1% respectively. In allowing a health claim for an association between consumption of oatmeal, rolled oats and oat bran, and reduced risk of coronary heart disease, the Food and Drug Administration (FDA) of the USA has accepted that oat β-glucan is an active ingredient[1]. Barley β-D-glucan has been widely studied because of its significance in brewing and feed uses. High levels of unmodified barley β-D-glucan produce viscous solutions that can lead to problems in brewing and in performance as chicken feed. We have developed an interest in wheat β-glucan because of a wheat pre-processing technology which enriched β-glucan content from 0.5% in whole wheat to 2.6% in one of the bran fractions[2]. β-glucans from the above three cereals show differences in both structural features and molecular weights. The physical properties, such as ease of dissolution and rheological behavior of cereal β-D-glucans, are determined by linkage patterns, conformation and molecular weight distribution. The objectives of this paper were to review recent advances in determination of structure, molecular weight and solution and functional properties of cereal β-D-glucans and investigate the relationships between these.

Structure A: trisaccharide unit

Structure B: tetrasaccharide unit

2. STRUCTURE OF CEREAL β-D-GLUCANS

Cereal β-glucans are linear, unbranched polysaccharides composed mostly of 4-O-linked β-D-glucopyranosyl (70-72%) and 3-O-linked β-D-glucopyranosyl (28-30%) units[3]. Their structure is predominantly β-(1→3)-linked cellotriosyl (**Structure A**) (58-72%) and cellotetraosyl (**Structure B**) (20-34%) by weight. Up to 10 or more consecutive (1→4) linkages can also occur in a much lesser amounts. The structural similarity of β-glucans of different origin is evident from their ^{13}C-NMR spectra appeared almost identical (Figure 1). However, significant differences are revealed when these β-glucans are subjected to enzyme digestion by lichenase, a highly specific (1→3)(1→4)-β-D-glucan-4-glucanohydrolase (EC 3.2.1.73). The four arrows in **Structure A** and **B** indicate the point of hydrolysis by this enzyme. Because the enzyme is specific for β-(1→4)-linked 3-substituted glucopyranosyl units, all oligosaccharides containing β-(1→4)-linked glucopyranosyl units are terminated at the reducing end by a β-(1→3)-linked glucose (i.e., cellodextrins terminated at the reducing end by a 3-linked unit). The relative amounts of the oligosaccharides produced by lichenase constitute a fingerprint of structure as shown in Table1. The total of tri- and tetrasaccharide of cereal β-D-glucan is similar (92-93%) among cereal β- D-glucans. However, the amount of trisaccharide follows the order of wheat (72%), barley (64%) and oat (58%). The molar ratio of tri- to tetrasaccharides, another parameter defining the fingerprint of the structures of cereal β-glucans, also follows the order of wheat (4.2-4.5), barley (3.0) and oat (2.3), as shown in Table 1.

In addition to 3-O-β-cellobiosyl-, cellotriosyl-D-glucose and a number of soluble oligosaccharides up to a DP of 9, an insoluble precipitate was also produced by the action of lichenase. This was composed of longer (1→4)-linked cellodextrin-like chains (predominantly DP 9) terminated by a single 3-linked β-D-glucose at the reducing end [3, 4]. These higher DP oligosaccharides arise from cellulose-like portions of the intact polysaccharide which likely affect the solubility and conformation, and hence the rheological properties in solution [5,6]. Others have suggested that consecutive β-(1→3) linkages occur in cereal β-glucans[7, 8], but these have not been detected in our studies (see

Figure 1). If these exist, their overall presence would be too small to affect the essential conformational and rheological properties of the polymer.

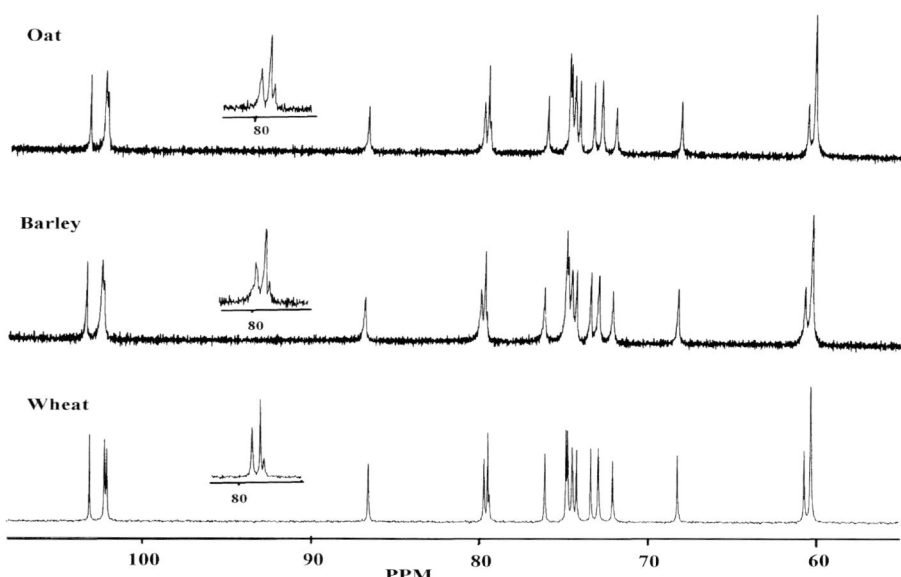

Figure 1. ^{13}C-NMR spectra of cereal β-D-glucans (4% w/v) in deuterated dimethylsulfoxide. The experiment was carried out on a Bruker WM500 NMR spectrometer at 90°C.

Table 1. Comparison of structural features of β-D-glucans from cereals after lichenase hydrolysis.

β-D-Glucan Source	Peak Area Percent %		Total%		Molar Ratio
	tri[a]	tetra[a]	tri+tetra[b]	penta~nona[b]	tri/tetra[c]
Wheat (Preprocessed)	72.3	21.0	93.3	6.7	4.5
Wheat Bran (AACC, Soft White)	70.6	22.9	93.5	6.5	4.2
Wheat Bran (AACC, Red Hard)	72.0	21.0	93.0	7.0	4.5
Barley	63.7	28.5	92.2	7.8	3.0
Oat	58.3	33.5	91.9	8.1	2.3

a: HPLC peak area percentage of tri and tetrasaccharides after lichenase hydrolysis
b: total percentage of tri and tetrasaccharides and penta to nonasaccharides, respectively
c: the molar ratio of trisaccharide over tetrasaccharide from lichenase hydrolysis

The possibility of the presence of charged groups, or protein/peptide attachment, has also been raised[9] - even at low levels these might profoundly affect conformation and rheological characteristics.

3. MOLECULAR WEIGHT OF CEREAL β-D-GLUCANS

The reported molecular weights (MW) of cereal β-glucans range from 2.7 x 10^4 to 40 x 10^6 as shown in Table 2. This large variation may reflect the diversity of origin or methodology. The extraction methods, solvents, conditions and sample history also have an impact on the MW determined [3, 9,10,11,12].

Size exclusion chromatography is a simple method for determining MW and MW distribution. The common commercially available MW standards are dextrins and pullulans. Because of the differences in molecular conformation (shape) between the α-linked glucan standards and β-glucans the MW values for β-glucans may be over-estimated. In our laboratory, high performance size exclusion chromatography (HPSEC) in combination with molecular weight sensitive detectors have been used to characterize the molecular weight and MW distributions of β-D-glucan standards[13]. A fluorescence detector exploits the specific dye-binding of Calcofluor by β-glucan to give a specific response to β-glucan in crude mixtures, excluding responses to any non-binding polymers also present. This system may also be used to quantitate the β-D-glucan[13,14]. To our knowledge, our laboratory was the first to report the use of β-D-glucan standards, the MW of which had been determined using HPLC fitted with an on-line low angle laser light scattering detector (LALLS)[13]. These standards were then used to estimate the MW of β-D-glucans from various sources. These are more realistic values than those obtained using dextrins and pullulans as standards without adjustment for conformational differences. Values found were in the order, oat (~3 x 10^6), barley (2-2.5 x 10^6), rye (1 x 10^6) and wheat (0.3-0.7 x 10^6). More recently, using HPLC with right angle laser light scattering (RALLS) and viscometric detection, we established somewhat lower MW in oat and barley extracts, but significant differences between oats and barley were still evident [15] (Table 2).

Autio et al. [16] reported cultivar variations in MW with a highest value of 1.5 x 10^6. Recent studies by Gómez and co-workers[17,18] reported a MW value of 616,000-641,000 for barley β-glucan using RI and MALLS detectors, which is much lower than the values (mostly 1.4 - 1.7x 10^6) found in our laboratory for crude extracts, but significantly higher than the value (160,000 to 290,000) reported by Woodward and co-workers for isolated and purified materials[10,19]. More recently, we have extracted a β-glucan from pre-processed wheat bran. The wheat β-glucan was extracted with 1M NaOH because it was not extractable in hot water possibly due to entrapment within arabinoxylans cross-linked with phenolic compounds in the cell wall[11]. This wheat β-glucan gave a MW value of 307,000. Prior to purification, when it co-exists with arabinoxylans, the peak molecular weight was 613,000[19]. This observation suggests that the xylanase (T. Viride, EC 3.2.1.8) used to remove the arabinoxylans from the initial extract may have some endo-β-D-glucanase activities. This was confirmed by the fact that significant degradation was observed when the same xylanase was added to oat and barley β-glucans[19].

Table 2. Comparison of molecular weight of β–D-glucan from cereals.

Reference	Source	PT[a]	Extraction Solvent and condition			Detection Method[b]	MW (10^{-6})
			Solvent	T°C	Duration (min)		
Cui and Wood [19]	Wheat Bran Preprocessed	2	1M NaOH	25	120	B, C	0.3 –0.6
Cui and Wood [19]	Wheat Bran AACC Soft White	2	1M NaOH	25	120	B, C	0.4- 0.7
Cui and Wood [19]	Wheat Bran AACC Red Hard	2	1M NaOH	25	120	B, C	0.27-0.6
Cui and Wood [19]	Barley Flour	2	Na_2CO_3	45	90	B, C	0.8
Gómez et al. [17]	Barley	2	H_2O	65	120	A,C	0.6
Forrest and Wainwright [9]	Barley cell wall (endosperm)	2	H_2O	40-65	---	D	~40
Autio et al[16]	Oat Bran	1	H_2O	70	60	A	1.5
Beer et al [15]	Oat Flour	2	H_2O	90	120	B, C	2.0 - 2.5
Beer et al [20]	Oat Bran	1	Na_2HPO_4 20mM	37	135	B, C	1.1 - 1.9
Malkki et al [21]	Oat Bran	1	Na_2CO_3	70	120	A	0.4 -1.5
Sundberg et al [22]	Oat Bran	2	Na_2CO_3	60	120	B, C	2.7
Wood et al [14]	Oat Flour	2	Na_2CO_3	60	120	B, C	3.0
	Oat Bran	2	Na_2CO_3	60	120	B, C	3.1

[a]Sample pre-treatment: 1 = no pre-treatment; 2 = Sample treated with hot aqueous or ethanol prior to extraction.

[b]Detection Method: A = Multi angle laser light scattering detection; B = Calcofluor post-column detection with β-glucan standards; C = Refractive index detection with pullulan standards; D=Sugar Concentration

4. RHEOLOGICAL PROPERTIES OF CEREAL β-D-GLUCANS

4.1. Flow Behaviour

Rheologically, solutions of cereal β-D-glucans fall into the category of viscoelastic fluids. They behave similarly to the well-characterized random coil type polysaccharides such as guar and locust bean gums. Typical flow curve of random coil type polysaccharides are shown in Figure 2. A Newtonian plateau is evident in the low shear rate range. As the shear rate increases beyond a certain value, the viscosity decreases. This shear thinning flow behavior is caused by the disruption of molecular entanglements of the polysaccharide by applied shear. It is believed that at low shear rate, there is sufficient time for new entanglements to form between different chain-partners leading to no net change in the degree of entanglement and therefore, the viscosity remains unchanged. At high shear rates, the rate of the re-entanglement is slower than the rate of the disruption of the existing entanglements, the overall content of three dimensional network decreases progressively and the resistance to flow correspondingly decreases[23].

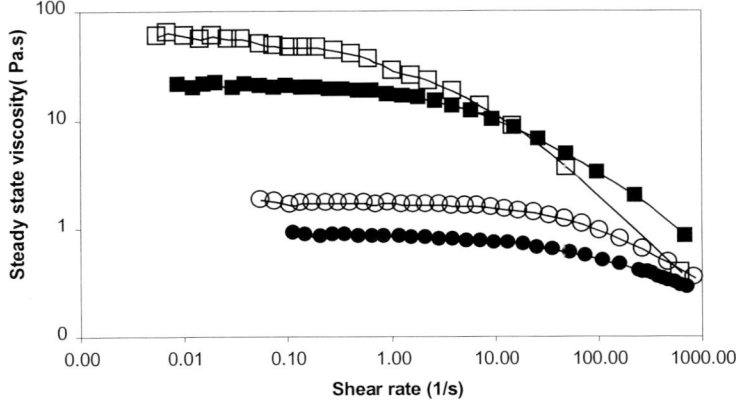

Figure 2. Flow curves of cereal β-D-glucans at 2.0% (w/w), 25°C. Oat β-D-glucan: - (Peak Molecular Weight, MW=1.2 Million); Barley β-D-glucan: ■-■(MW=0.8 Million); Wheat (AACC Soft White) β-D-glucan: O-O(MW=0.4 Million); Wheat (Preprocessed Bran) β-D-glucan: ●-●(MW=0.3 Million).

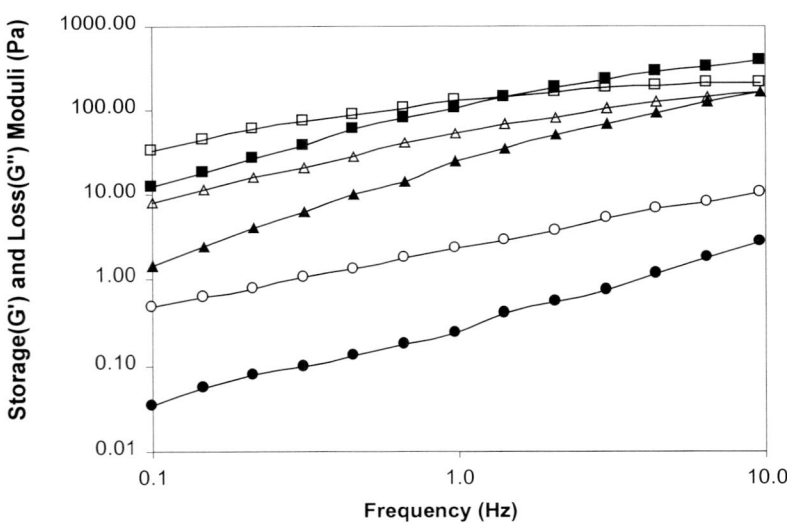

Figure 3. Mechanical spectra of cereal β-D-glucans of 2.0% (w/w), at 25°C. Solid symbols are the value of storage modulus G′ while the open symbols represent the loss modulus G″. Oat β-D-glucan: ■ & ; Barley β-D-glucan: ▲ & △; Wheat (Preprocessed Bran) β-D-glucan: ● & ○.

4.2. Viscoelastical Properties

The mechanical spectra of typical random-coil polysaccharide solutions have a characteristic in which the loss modulus G" is greater than the storage modulus G' at low frequencies. G' increases more rapidly with frequency and becomes greater than G" at high frequencies, as shown in Figure 3. At low frequencies, there is sufficient time for breaking and reforming of the entanglements with the period of oscillation, and the response is similar to that of non-entangled dilute solution, i.e. G" > G'. At higher frequencies, as the oscillation rate exceeds the time scale of molecular re-arrangement, the interchain entanglements do not have sufficient time to come apart within the period of one oscillation cycle. Thus the overall response approaches that of a gel, i.e. G' > G".

4.3. Gelling Properties

Although cereal β-glucan solutions show typical random-coil type behaviour similar to galactomannans as shown in Figures 2 and 3, unlike galactomannans they have a tendency to form gels. Doublier and Wood[24] reported a "weak-gel" property of partially hydrolyzed oat β-glucan (MW ~100,000) but not for unhydrolyzed samples (MW ~1,200,000). Recently, Gomez and co-workers[25] reported gel structures in two barley β-D-glucans: sample A (MW 573,000) at concentration 0.16% and 25°C and sample N (MW 9,2000) at concentration 1.5% and 70°C. In our laboratory, the gel forming ability of β-D-glucan samples of different molecular weight from oat, barley and wheat, was examined at 2.0% (w/w) and 4°C. After 24 hr, wheat β-glucan solution formed a firm gel with a small but visible amount of water separated from the gel. The mechanical spectrum of the gel is shown in Figure 4. This gel started to melt at 35°C and the melting process was completed at 60°C (Figure 5). Under the same experimental conditions, a clear solution of barley β-glucan standard (MW 216,000) from Megzyme (Bray, Co. Wicklow, Ireland) showed cloudiness after 24 hr at 4°C, and a firm gel was obtained after 72 hr. However, a high molecular weight barley β-glucan (Mw ~ 800,000) did not show any tendency to gel after seven days under the same storage conditions. None of the oat β-D-glucan solutions showed any tendency to form a gel under the same experiment conditions although weak gel properties have been reported for low MW oat-β-glucan[24].

4.4. Solubility of Freeze-Dried Samples

Drying methods have a significant effect on the ease with which cereal β-glucan may be dispersed into solutionsamples. For example, freeze-dried water-soluble barley β-glucan extracted at 65°C proved difficult to dissolve in aqueous buffer; it required several hours at 90°C with constant stirring before dissolution was completed[4]. In contrast, freeze-dried 40°C water-soluble barley β-glucan dissolved after 1-2 hr at 70°C[4]. Freeze-dried wheat β-glucan was essentially insoluble in water regardless of heating temperature and time[19]. The past experience in our laboratory also showed that freeze-dried oat β-glucan is very difficult to dissolve in water. However, all cereal β-glucan samples prepared by solvent exchange method in our laboratory exhibited good solubility in water, i.e. all samples are aqueous soluble with moderate heating.

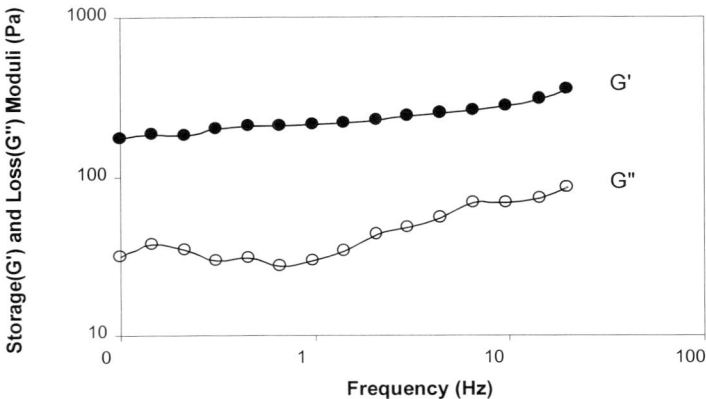

Figure 4. Mechanical spectrum of 2.0% (w/w) wheat β-glucan gel at 4°C (Bohlin CVO Rheometer, Cone/plate 4°).

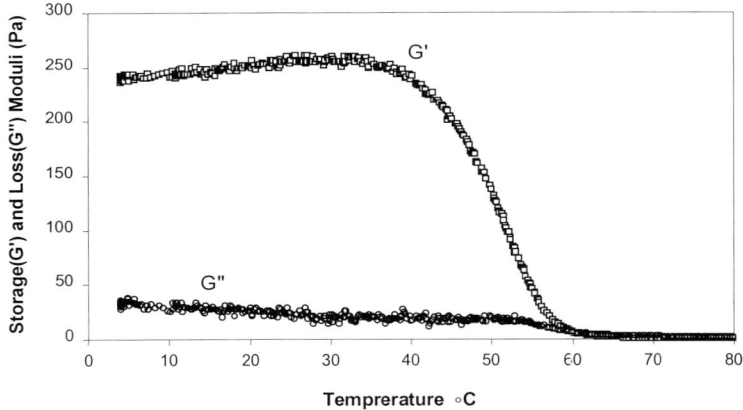

Figure 5. Melting curve of 2.0% wheat β-glucan gel (w/w) at 1 Hz of frequency on a Bohlin CVO Rheometer (cone/plate 4°).

5. CONCLUSIONS

In summary (Table 3), there is a correlation between the combination of both chemical and physical characteristics of the polysaccharides, the tri/tetrasaccharide ratio and molecular weight respectively of cereal β-glucans and the properties in aqueous systems, solubility of freeze-dried samples and the gel forming ability.

Table 3: Comparison of structural feature and physical properties of cereal β-glucans.

β-D Glucan Source	MW x 10³	Ratio of Tri/Tetra	Freeze-Dried Sample Solubility	Aggregation/ Gel Formation Ability
Oat (High MW)	976	2.2	Difficult	Low
Oat (Low MW)	263	2.2	Difficult	Medium
Barley (High W)	767	3.0	Difficult	Low
Barley (Medium)	290	3.0	70°C, 1-2 hr*	Medium High
Barley (Low MW)	72- 213	3.0	90°C, several hr*	Medium High
Wheat	300	4.5	Insoluble	High

*Reference [4], Woodward et al., 1988.

Summary

1   Cereal β-glucans can form gels under certain conditions in contrast to other random-coil polysaccharides, such as galactomannans.

2   The gelling ability and "insolubility" of freeze-dried cereal β-glucans follow the order of wheat > barley > oat; this trend corresponds to the tri/tetra saccharide ratio.

3   In addition to the structural features, molecular weight also plays an important role in the organization of cereal β-D-glucans in aqueous systems. It appeared that lower molecular weight β-D-glucan has a better chance to form a gel than larger molecules within the molecular weight range investigated.

Acknowledgement: The authors thank Ms Cathy Wang, Dr. Shane Wang and Mr. John Weisz for their technical assistant during the preparation of this paper. Dr. B. Blackwell and Mr. John Nikiforuk are also acknowledged for their expertise on NMR spectroscopy.

REFERENCES

1.   P.J. Wood, Canadian Chemical News, Nov./Dec. (1997) 17.
2.   W. Cui, P.J. Wood, J. Weisz, and J. Mullin, In P.A. Williams and G.O. Phillips (eds.) Gums and Stabilisers for the Food Industry 9, pp. 34-42, Royal Society of Chemistry, London, 1998.
3.   P.J. Wood, J. Weisz and B.A. Blackwell, Cereal Chem., 71(1994) 301.
4.   J.R. Woodward, D.R. Phillips and G.B. Fincher, Carbohydr. Polym., 8(1988)85.
5.   G. S. Buliga, D.A. Brant and G. B. Fincher, Carbohydr. Res., 157(1986)139.
6.   K.M. Värum and O. Smidsrød, Carbohydr. Polym., 9(1988)103.
7.   M. Fleming and D.J. Manners, Biochem. J., 100(1966) 4.
8.   M. Fleming and K. Kawakami, Carbohydr. Res., 57(1977)15.
9.   I.S. Forrest and T. Wainwright, J. Inst. Brew., 139(1977)535.
10. J.R. Woodward, D.R. Phillips and G.B. Fincher, Carbohydr. Polym. 3(1983)143.
11. W. Cui, P.J. Wood, J. Weisz and M.U. Beer, Cereal Chem., 76(1999)129.

12. K.M. Värum, A. Martinsen and O. Smidsrød, Food Hydrocolloids, 5(1991)363.
13. P.J. Wood, J. Weisz and W. Mahn, Cereal Chem., 68(1991)530.
14. T. Suortti, J. Chromatogr., 632(1993)105.
15. M.U. Beer, P.J. Wood and J. Weisz, Cereal Chem., 74(1997)476.
16. K. Autio, O. Myllymaki, T. Suortti, M. Saastamoinen and K. Poutanen, Food Hydrocolloids, 5(1992)513.
17. C. Gomez, A. Navarro, P. Manzanares, A. Horta and J.V. Carbonell, Carbohydr. Polym., 32(1997)7.
18. C. Gomez, A. Navarro, P. Manzanares, A. Horta and J.V. Carbonell, Carbohydr. Polym., 32(1997)17.
19. W. Cui and P.J. Wood, unpublished data, 1998.
20. M.U. Beer, P.J. Wood, J. Weisz and N. Fillion, Cereal Chem., 74(1997)705.
21. Y., Malkki, K. Autio, O. Hanninen, O. Myllimaki, K. Pelkonen, T. Suortti and R. Tarranen, Cereal Chem., 69(1992)647.
22. B. Sundberg, P.J. Wood, A. Lia, H. Andersson, A.S. Sandberg, G. Hallmans and P. Xman, Am. J. Clin. Nutr., 64(1996)878.
23. E.R. Morris, In R.P Millane, J.M. BeMiller and R. Chandrasekaran (eds.) Frontier in Carbohydrate Research I. Food Applications, pp132-163, Elsevier, New York, 1989.
24. J-L. Doublier and P.J. Wood, Cereal Chem., 72(1995)335.
25. C. Gomez, A. Navarro, P. Manzanares, A. Horta and J.V. Carbonell, Carbohydr. Polym., 32(1997)17.

HYDROCOLLOIDS – PART 1
Edited by K. Nishinari
2000 Elsevier Science B.V.

New biopolymers produced by nitrogen fixing microorganisms for use in foods

A. Scamparini[a]; C. Vendruscolo[b]; I. Maldonade[a]; J. Druzian[c] and D. Mariuzzo[a]

[a] Food Science Department, Food Engineering Faculty, State University
of Campinas, São Paulo, Brazil-POBox 6121, CEP 13081-970,
e-mail:mzz@obelix.unicamp.br

[b] Biotechnology Centre and CAVG - Federal University of Pelotas -Campus
Universitário. Pelotas, RS, Brazil - CEP 96010-900

[c] Food and Chemistry Science Department, West Santa Catarina University,
Videira, Santa Catarina, Brazil - POBox 867, CEP 89560-000

Two new biopolymers, PS-32 and NFB, obtained by aerobic fermentation from *Beijerinckia* sp and *Rhizobium* NFB bacteria, respectively, are being developed at our laboratory. HPLC, GC-MS, IR and NMR were used to study the chemical structure of PS-32 polysaccharide. It was found to contain glucose, galactose and fucose in a molar ratio of 3:1:3, and a possible chemical structure was proposed. Initial studies of NFB polysaccharide showed that it is composed of glucose, galactose, glucuronic and pyruvic acids in the proportion of 6:11:1:1. Rheological studies of both polysaccharide aqueous solutions showed pseudoplastic behaviour.

1. INTRODUCTION

Biopolymers are produced by most microorganisms, but for commercial production, fungi and bacteria show the easiest adaptation. Such microorganisms have the capabitily to grow in large scale batch fermentations and they produce biopolymers with potential applications in a variety of industries. In the food industry there is a large application of biopolymers as thickening agents, suspending or gelling polymers. Industrial interest is concentrated on the extracellular polysaccharides, due their ease of production and purification, with high yields. For a biopolymer to satisfy a large range of applications, it must have rheological properties, in which the viscosity decreases in the presence of a shear stress. In addition, these properties must be maintained during temperature, pH and ionic strength changes. These properties are closely related to the primary chemical structure. In this work the structural characteristics and rheological behaviour of PS-32 and NFB polysaccharides were studied. PS-32 was obtained from aerobic fermentation of *Beijerinckia* sp (strain 32) bacterium, isolated from soil used for the cultivation of sugar cane (*Saccharum officinarum*) in Ribeirão Preto, S.P. and NFB was obtained from aerobic fermentation of *Rhizobium* isolated from soil used for cultivation of cowpea bean (*Vigna unguiculata*) from Sergipe state.

2. MATERIALS AND METHODS

2.1. Polysaccharides production and purification

PS-32 production - a strain of *Beijerinckia* sp. isolated from sugar cane roots was used to produce PS-32 polysaccharide under aerobic fermentation. The fermentation medium contained sucrose (5%) as carbon source, $MgSO_4$ (0.05%), K_2HPO_4 (0.01%), KH_2PO_4 (0.05%) and tryptose (0.5%). It was sterilized at 121°C for 15 min. Fermentation conditions were 200 rpm agitation, during 72 h at 25°C in a G25 shaker (New Brunswick Scientific, Co.). After this period, the fermentation broth was centrifugated at 11.500 x *g*, 30 min and the polysaccharide was precipitated with ethanol (80%), yielding 7-9 g/l. PS-32 polysaccharide was dried at 55°C under vacuum and powdered. In order to purify PS-32 polysaccharide, its 1% aqueous solution was treated with a papain solution (0.05g/mL) at pH 6.5, 60°C for 3 h, then dialyzed against distilled water for 3 days, then precipitated with ethanol, dried under vacuum and powdered.

NFB production - one colony of *Rhizobium* NFB was inoculated in a 500 mL flask containing 200 mL YM broth (Difco) and incubated at 30°C and 200 rpm agitation for 6 days in a G25 shaker (New Brunswick Scientific, Co.). This culture was used as an inoculum at a ratio of 5.0% (v/v), for all subsequent experiments in a 1.6 L MULTIGEN fermenter (New Brunswick Scientific, Co.) with a working volume of 1 L of malt extract 0.3%, yeast extract 0.3%, peptone 0.5% and manitol 3%, for 10 days. After cultivation, the cells were harvested by centrifugation at 13,000 x *g* for 20 min. Dried cell weight was determined at 50°C to constant weight. After centrifugation, the biopolymer was precipitated with ethanol at a ratio of 1:3 (v/v), respectively, separated by centrifugation and dried at 50°C to constant weight. NFB was purified from protein by precipitation with trichloroacetic acid (TCA) according to an AOAC method [1]. NFB was solved in water and a TCA 50% solution was added. The mixture was heated at 50-55°C for 15 min. Protein was separated by centrifugation for 30 min., at 13.000 x *g*, at 5°C and the NFB in supernatant was precipitated with acetone 80%, redissoluted in water, reprecipitated in acetone, dried at 50-55°C under vacuum and powdered.

2.2. HPLC analysis

PS-32 hydrolysis - HPLC was used in order to determine and quantify the monosaccharides present. Two kind of hydrolysis were tested: (i) 1M HCl, 70°C, 16 h and (ii) 2N TFA, 100°C, 16 h. The hydrolysates were analysed in a 7.9 mm x 30 cm column (SCR-101-P, Shimadzu Co.), with sulfonated polystyrene bonded to Pb^{+2} cation (ligand exchange) and gel filtration separation. Column temperature was 80°C. Ultrapure water was used as mobile phase, with flow of 0.6 mL/minute. The injection volume was 40 μL and the automatic injection of samples was used. A refractive index detector was used to detect sugars. Sugars were identified by the retention time of corresponding standards (Sigma) and confirmed by mass spectroscopy. Quantification was by the external standard method, with three calibration levels for each standard and least squares method curve fit.

NFB hydrolysis - Samples (0.01g) of purified NFB were soaked in 0.5mL water overnight, and hydrolysed with 0.5mL of 0.2 M trifluoroacetic acid (to give 0.1 M TFA) at 100°C for 8 hours [2]. Identification of sugars and acids from NFB was by paper chromatography using descending elution in 10:3:3 1-butanol-pyridine-water. The sugars were detected using silver nitrate-sodium hydroxide, *p*-anisidine, and aniline phthalate spray reagent. Confirmation and quantitation were made using the same HPLC equipment and procedures as before. For acids a SCR-101H (polystyrene-divinylbenzene) column was used, with 0.6 ml/min perchloric acid solution (pH=1.9) at 50°C.

2.3. GLC-MS analysis

GLC-MS conditions - In order to confirm the sugar composition in PS-32 polysaccharide and determine glycosidic linkage positions, PS-32 polysaccharide was methylated according to Hakomori [3], hydrolysed, reduced and acetylated. The alditols were analysed in a Shimadzu Co. GC-MS, QP 5000 model, using a fused-silica bonded capillary column DB-5, 22 mm X 50 m, split of 1/100 and a temperature program: 130°C during the 5 initial minutes, rising to 300°C, 7°C/min. Helium was used as carrier at a flow rate of 30 mL/min. Injector and interface temperatures were 250 and 280°C, respectively. Automatic injection was used. MS analysis were made in a range of 40 to 600 a.m.u., with scans of 0.5 s, using electronic impact. PS-32 polysaccharide oxidation followed Fukagawa [4] and its products were identified by TLC and GLC-MS.

2.4. ^{13}C and ^{1}H N.M.R. analysis

PS-32 polysaccharide was hydrolysed for 16 h and silanised using Sigma reagent kit. Spectra were recorded at 300 MHz and 75.6 MHz at 70°C. A General Electric QE 300 spectrometer was used.

2.5. Infra-Red analysis

Infra-red analysis was made using a Perkin-Elmer FT-IR spectrophotometer model 16C with the PS-32 polysaccharide in KBr pellets.

2.6. Polysaccharides rheological measurements

PS-32 aqueous solution preparation - A Haake Rheometer, model CV20N, plus Rotovisco 303 software was used. It was a rotational rheometer, equipped with a coaxial cylinder sensor. The inner cylinder was held at rest and the outer was subjected to a defined shear stress. Fluids were placed in the annular gap. The rheometer was adjusted to produce a shear rate from 0 to 300 s^{-1}, following the return to 0 s^{-1}. Sample temperature was kept at 15, 25 or 45°C using a constant temperature thermal liquid bath and circulator. Purified polysaccharide 1% aqueous solution were prepared with ultrapure water. Changes of pH were made by adding 2N HCl or 50% (w/v) NaOH. All samples were sonnicated before measurements were made in order to eliminate bubbles. Storage and bulk modulus (G' and G'') were observed as a function of frequency.

NFB aqueous solution preparation - Concentrations of the *Rhizobium* polysaccharide varied from 0.25% to 2.0%. For solutions containing NFB in combination with sodium chloride and calcium chloride, the polysaccharide was dispersed in a small quantity of ultrapure water. The appropriate mass of additive was added and dissolved, and the solution made up to 20 mL. To determine the effect of pH on viscosity, 0.2 M acetate buffer were used at pH 3.0 and 5.0. In addition, 0.2M phosphate buffer were used at pH 7.0 and 9.0 pHs. Apparent viscosity was determined at 0.0463 s^{-1} in the same rheometer, at 25°C.

3. RESULTS AND DISCUSSION

HPLC was used for analysis of monosaccharides present in the hydrolysys products of PS-32 polysaccharide. It was composed of glucose:galactose:fucose in a ratio of 3:1:3, respectively. TFA hydrolysate showed best results, (Figure 1). Through the retention time of standard alditols and mass specrtrometry, it was possible to identify: 1,4,5-tri-*O*-acetyl-2,3-di-*O*-methyl-fucositol, 1,4,5-tri-*O*-acetyl-2,3,6-tri-*O*-methyl-glucitol, 1,5-di-*O*-acetyl-2,3,4,6-tetra-*O*-methyl-glucitol and 1,4,5,6-tetra-*O*-acetyl-2,3-di-*O*-methyl-galactitol, (in order to confirm the sugars composition in PS-32 polysaccharide and determine glycosidic linkages positions between these sugar residues). Together with above data, Smith degradation (periodate oxidation, borohydride reduction and hydrolysis with acid), periodate oxidation data that indicated the presence of glycoaldehyde, glycerol, threitol, erythritol and 1,2,3-butanetriol, lead us to conclude that glucose residues showed C1 linkages at the terminal ends, and (1→4) linkages; that galactose residues showed branched linkages at C4 and C6 linkages, and fucose residues showed (1→4) linkages. The periodate oxidation showed the formation of one mol of formic acid per basic unit. The ^{1}H NMR data (Figures 2), although showing interference due the high viscosity of PS-32 polysaccharide solutions, gave information about anomeric configuration of sugar residues with δ=5.02 ppm (α configuration to *L*-fucose), δ=4.88 ppm and 4.47 ppm (β configuration to *D*-glucose and *D*-galactose, respectively). The ^{13}C NMR chemical shifts in the δ=100 ppm region (Figure 3) indicated anomeric carbons with equatorial linkages. Rings sizes were confirmed by infra-red spectroscopy (Figure 4).

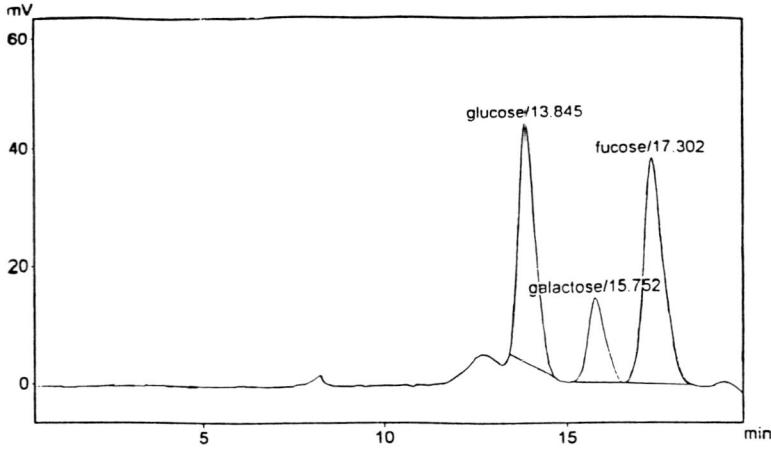

Figure 1 - PS-32 polysaccharide composition obtained by HPLC analysis. HPLC Conditions: SCR-101P column, T=80°C, mobile phase ultrapure water at 0.6 mL/min, refractive index detector.

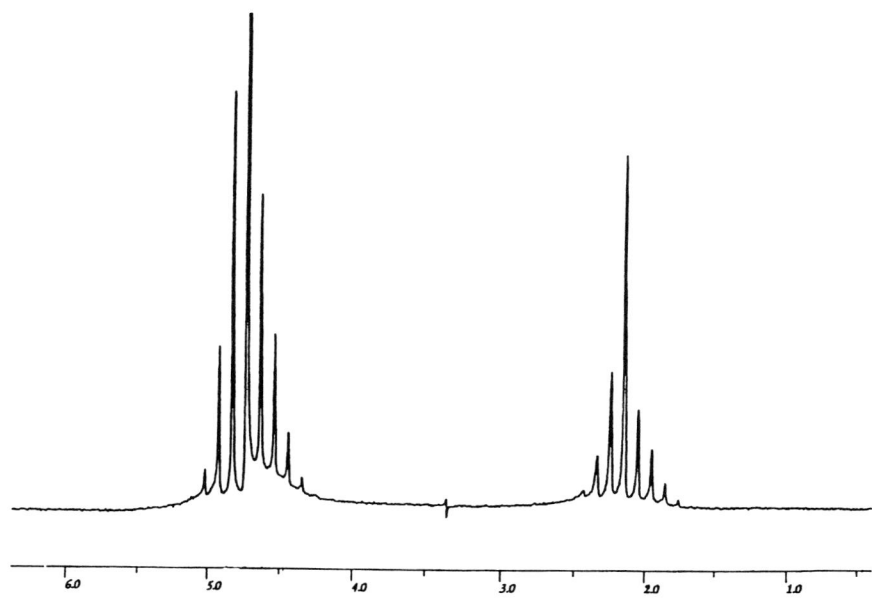

Figure 2 - PS-32 polysaccharide ^{1}H NMR spectrum recorded at 75.6 MHz.

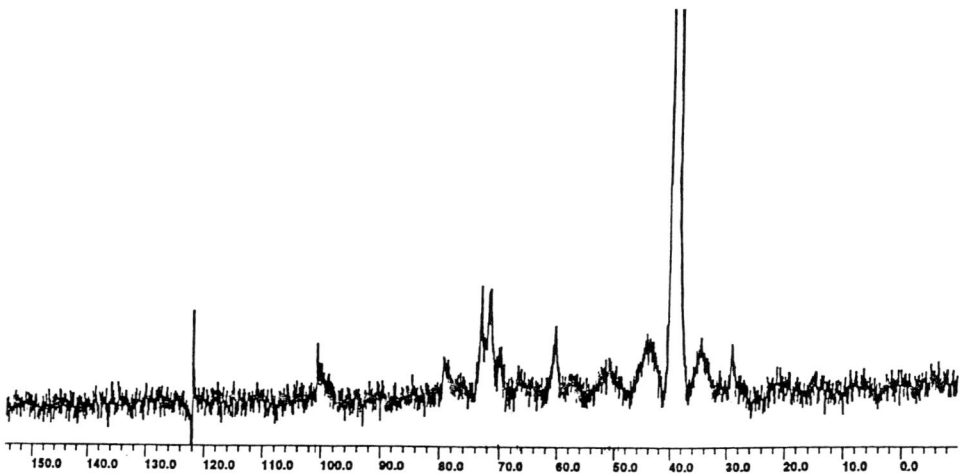

Figure 3 - PS-32 polysaccharide ^{13}C NMR spectrum recorded at 300 MHz.

T%

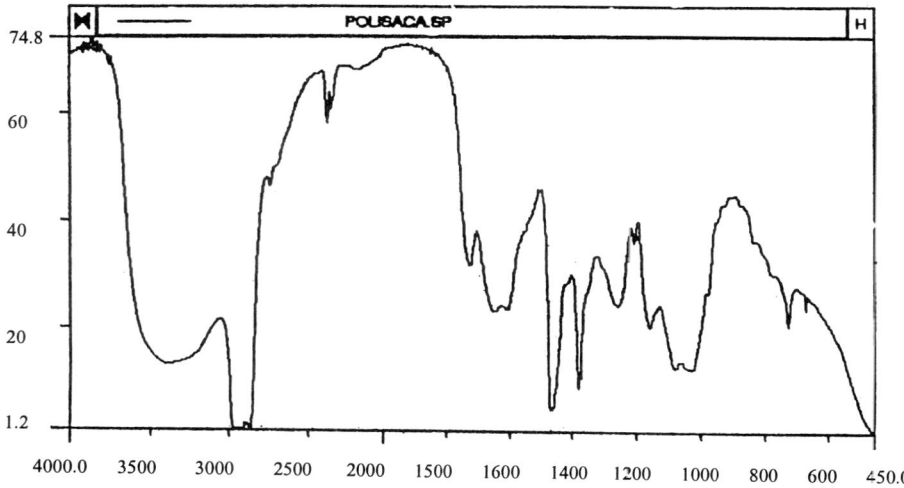

Figure 4 - PS-32 polysaccharide infra-red spectrum. CM^{-1}

Glucose and galactose residues were in the β configuration and fucose, in the α. Its main chain is formed by β-**D**-galactose and α-**L**-fucose units, linked through (1→4). A side chain is formed by β-**D**-glucose, which is branched through 1→4 linkages. The non-reducing terminal residues are formed by β-**D**-glucose units. α-**L**-fucose units may form the main chain through (1→4) linkages. A possible structure of the PS-32 polysaccharide is shown below:

β-**DG**lp(1→4)β-**DG**lp(1→4)β-**DG**lp
1
↓
6
→4)β-**DG**ap(1→4)α-**LF**cp(1→4)α-**LF**cp(1→4)α-**LF**cp(1→

HPLC analysis showed that the NFB polysaccharide contains glucuronic and pyruvic acids. Chromatograms are shown in Figures 5 and 6. Glucose, galactose, glucuronic acid and pyruvic acid were in a molar ratio of 6:11:1:1, respectively. Previous investigations [5-7] showed that the polymers produced by *Rhizobium* have 35-70% glucose, 7-25% galactose, 0-20% galacturonic acid, 0-16% glucuronic acid and 0-13% pyruvic acid depending on strains that generated the biopolymer. The polysaccharide produced from *Rhizobium* NFB isolated from cowpea bean showed higher proportions of galactose, and a very small proportion of glucuronic acids. Due to this proportion its hydrolysis was easier than other acidic polysaccharides obtained from *Rhizobium* strains.

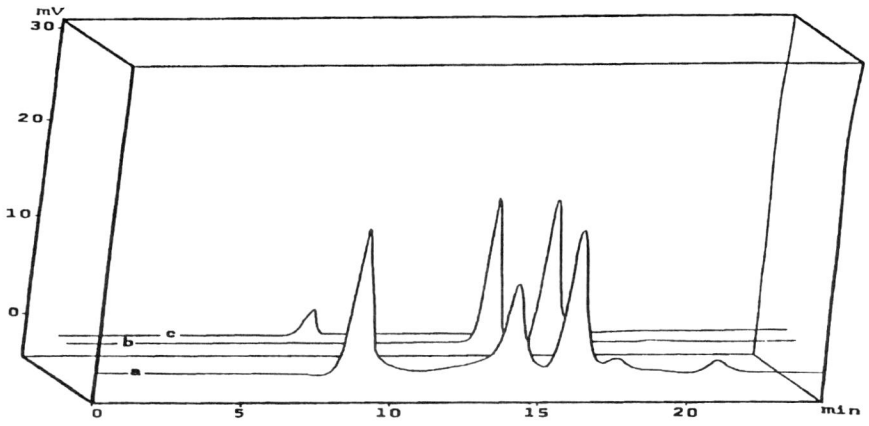

Figure 5 - Sugars composition of NFB polysaccharide obtained by HPLC analysis. HPLC Conditions: SCR-101P column, T=80°C, mobile phase ultrapure water at 0.6 mL/min, refractive index detector. (a) NFB, (b) glucose (Rt = 14.01) and galactose (Rt =16.00) standards, (c) TFA (Rt = 8.9).

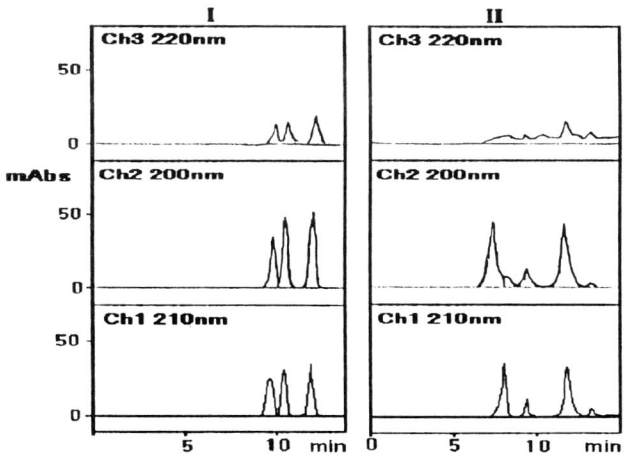

Figure 6 - Acid composition of NFB polysaccharide obtained by HPLC analysis. HPLC Conditions: SCR-101H column, T=50°C, mobile phase: perchloric acid solution (pH 1.9) at 0.6 mL/min, diode-array detector. (I) acids standards and (II) samples NFB hydrolysed with TFA at 200, 210 and 220nm. (a) glucoronic acid(Rt = 9.4), (b) galacturonic acid (Rt = 10.1), (c) pyruvic acid (Rt= 11.5) and (d) TFA residue (Rt = 7.2).

All PS-32 aqueous solutions showed typical properties of pseudoplastic fluids (Figures 7, 8, 9 and 10). Apparent viscosities strongly decreased with increasing of shear rate. It was possible to observe a small yield point in P solutions at 45°C (Figure 7), and at 25°C, pH 12 (Figure 8). With increasing temperature the apparent viscosity of the solutions decreased. pH changes of PS-32 apparent viscosity was highest at the natural pH (6.5). A smaller apparent viscosity was observed for PS-32 aqueous solutions in strongly basic (pH 12) solutions. Strongly acid solutions showed less change. Another important characteristic observed in the rheograms was the high yield stress and consequently high apparent viscosity for low shear rate. It was found a peak of yield stress followed by shear thinning region. According to Szczesniak [8] high yield shear is an important characteristic of substances utilized as binder and suspension agents. Manresa *et al.* [9] presented similar results with the extracellular biopolymer GSP-910 obtained from *Pseudomonas*. It can also be observed on the rheograms the flow dependence on time with hysteresis tixotropic. It is not known yet the extension in which the phenomena interphere on the functional properties of hydrocolloids. However, Szczesniak [8] stated that this behaviour is important in the manufacture of products thickened with gums.

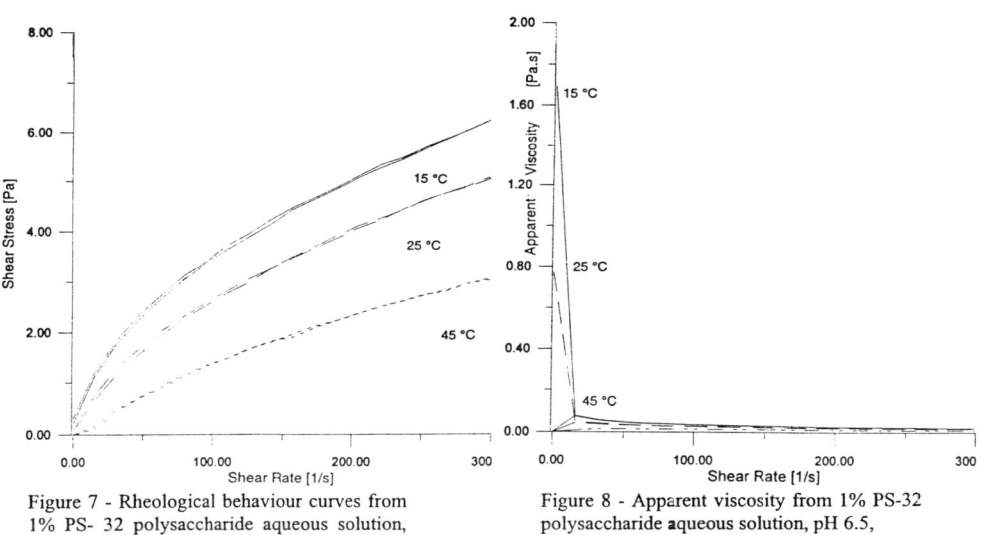

Figure 7 - Rheological behaviour curves from 1% PS- 32 polysaccharide aqueous solution, pH 6.5, at 15, 25 and 45°C.

Figure 8 - Apparent viscosity from 1% PS-32 polysaccharide aqueous solution, pH 6.5, at 15, 25 and 45°C.

The effect of shear rate on the apparent viscosity of polysaccharide aqueous solutions of *Rhizobium* NFB is shown in Figure 11. Solutions are non-Newtonian and pseudoplastic. Concentration of polysaccharide is important since it is directly related to cost. When the functional requirements of food additive allows a choice, the final decision always favours the cheaper alternative. Figure 11 shows that at constant shear rate, the apparent viscosity is directly proportional to polysaccharide concentration. The NFB polysaccharide aqueous solutions had apparent viscosities of 1.3 Pa.s at a concentration of 2.0% and 0.02 Pa.s at a concentration of 0.25% when measured at a shear rate of 8 s^{-1}.

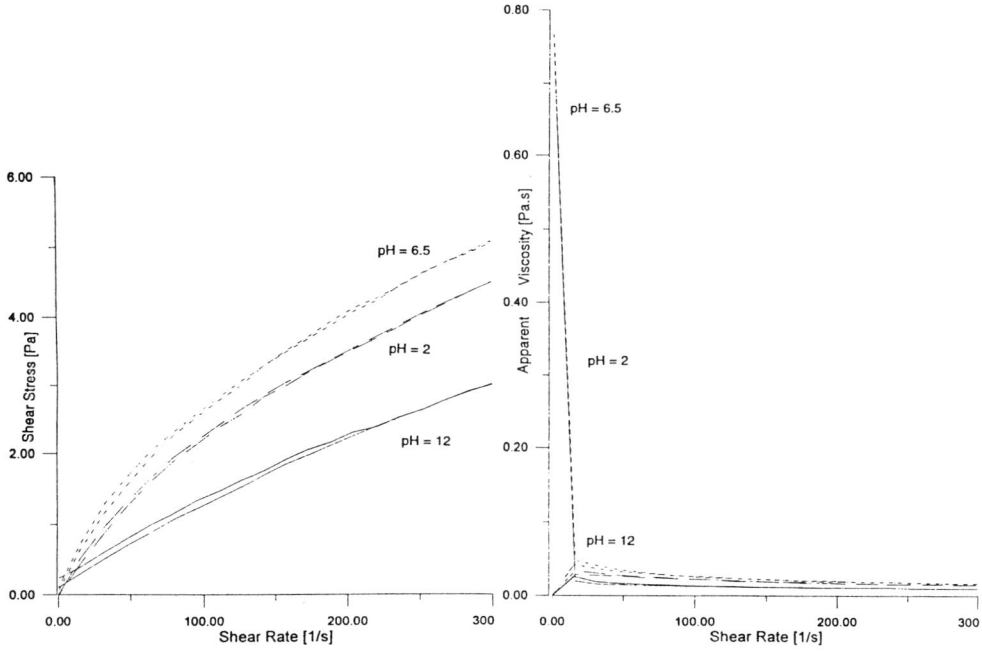

Figure 9 - Rheological behaviour curves from 1% PS-32 polysaccharide aqueous solution, pH 2.0, 6.5 and 12.5 at 25°C.

Figure 10 - Apparent viscosity from 1% PS-32 polysaccharide aqueous solution, pH 2.0, 6.5 and 12.5 at 25°C.

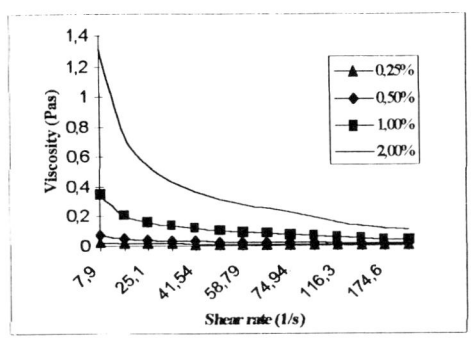

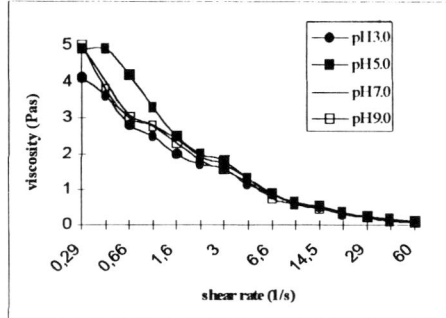

Figure 11 - Apparent viscosity from NFB polysaccharide aqueous solutions, different concentrations, 0.25, 0.5, 1.0 and 2.0%, at 25°C.

Figure 12 - Apparent viscosity from 1% NFB polysaccharide aqueous solutions, pH 3.0, 5.0, 7.0 and 9.0 at 25°C.

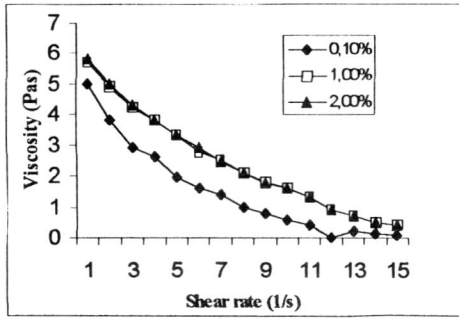

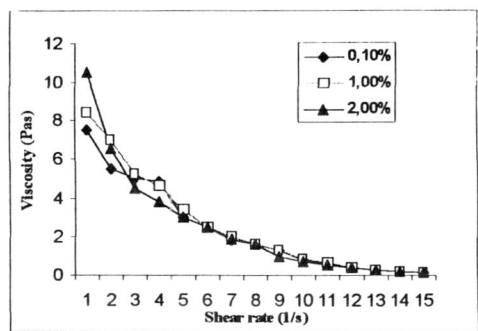

Figure 13 - Apparent viscosity from 1% NFB polysaccharide aqueous solutions, with differents CaCl₂ concentration at 25°C.

Figure 14 - Apparent viscosity from 1% NFB polysaccharide aqueous solutions, with differents NaCl concentration at 25°C

These values are comparable with those measured for the commonly used gums, including xanthan, guar, locust bean, sodium alginate and sodium carboxymethlcellulose [10]. Decreasing pH from 9.0 to 3.0 did not produce a significant effect on viscosity of the NFB aqueous solution, (Figure 12). In contrast, the apparent viscosity of some acidic polysaccharides may be modified at pH, and it appears that NFB has few acidic groups in its structure, in accordance with results obtained from HPLC analysis. Hydrocolloids are required to perform their functions in food products in the presence of other components. Aqueous solutions of other polysaccharide have been shown to be influenced by sodium chloride and calcium ions. In Figures 13 and 14 it is possible to observe that NFB polysaccharide had a small interaction only with NaCl. Addition of CaCl₂ showed no effect on the apparent viscosity of aqueous solutions of the polysaccharide, indicating that NFB polysaccharide has few acids and these are not available to make cross-linkages.

REFERENCES

1. M. A. Clarke, In "Sugars and Sugars Products", Sugar Processing Research Institute, Inc., A. O. A. C. Official Methods of Analysis, cap 44, 2-3, vol 2, part II, Edition 16th, 1995.
2. G. A.Ruiter, H. A. Schols; G, J. Voragen and F. M. Rombouts, Analytical Biochemistry, 207 (1992) 176.
3. S.I Hakomori, J. Biochem. (Tokio), 55 (1964).
4. K. Fukagawa, H. Yamaguchi, O. Uotani, T. Tsuimoto and D.Yonezawa, Agr. Biol. Chem 39 (1975) 1703.
5. W. F Dudan, Carbohydr. Res, 46 (1976) 97.
6. R. Somme, Carbohydr. Res, 33 (1974) 89.
7. B. K. Robertsen, P. Aman, A. G. Darvill, M. McNeil and P. Albersheim, Plant Physiol., 67 (1981) 389.
8. R.Somme, Carbohydr. Res., 43 (1975) 145.
9. A. S. Szczesniak, In: Phillips, G. O., D. J. Wedlock and P. A. Williams, ed. Gums and stabilisers for the food industry, 3. London, Elsevier, p 311-323, 1985.
10. A.Manresa; M. J. Espuny; J. Giunea and F. Commeles. Appl. Microbiol. Biotechnol., 26 (1987) 347.

HYDROCOLLOIDS – PART 1
Edited by K. Nishinari
2000 Elsevier Science B.V.

179

Studies on production and rheology of a polysaccharide synthesized by *Beijerinckia* sp strain 7070

F.F.Padilha[a], **J.L.Vendruscolo[b],** **O.A.Dellagostin[a],** **A.R.P.Scamparini[d],** **C.T. Vendruscolo[a,c]**

[a]Biotechnology Center, UFPEL, P.O.Box 354, 96010-900, Pelotas, RS, Brazil
[b]EMBRAPA - CPACT, P.O.Box 403, 96001-970, Pelotas, RS, Brazil
[c]Department of Food Science, F.C.D, UFPEL, P.O.Box 354, 96010-900, Pelotas, RS, Brazil
[d]Food Science Department, College of Food Engineering, UNICAMP, P.O.Box   13081, Campinas, SP, Brazil

The *Beijerinckia* sp strain 7070, which was isolated from soil of a sugar cane plantation in the state of São Paulo, Brazil, produces a water-soluble extracellular polysaccharide. We have studied different growth conditions and their effect on the rheological properties and yield. The production of the biopolymer was obtained using the inoculation of the strain 7070 in a solution of 150 mL (in Erlenmeyer of 250 mL) containing (gL^{-1}): glucose 50.0, tryptose 2.0, MgSO$_4$.7H$_2$O 2.7, K$_2$HPO$_4$ 2.0, KH$_2$PO$_4$ 3.6; pH 6.5, the incubation conditions were: 24 h or 36 h at 24°C and 180 rpm. The biopolymer obtained was retrieved, showing two types of fractions called long fiber and short fiber. The productivity was determined for the two periods of incubation (24 h and 36 h) and for each fraction was performed: analyses of apparent viscosity of the aqueous solutions at 6%, and viscoelasticity of the aqueous solutions at 3%. The analyses were obtained using a Haake rheometer at 25°C. The results showed that the apparent viscosity was influenced by the incubation time. For the biopolymer of 24 h incubation period, the result of viscosity was 62.900 cP for the short fibers and 30.000 cP for long fibers. For the biopolymer of 36 h it was 137.500 cP for short fibers and 65.000 cP for long fibers in a shear rate of 10 s^{-1}. The productivity was higher after 24 h (10.51 gL^{-1}) than after 36 h (7.89 gL^{-1}) of incubation. This biopolymer has highly viscous aqueous solutions and has a pseudoplastic behaviour with a large yield stress. The results show viscolelastic properties and also the possibility of producing biopolymer with similar characteristics to the ones obtained by the traditional method, however, reducing the production time by 60 h. This polysaccharide has potential applications in the food industry.

1. INTRODUCTION

Biopolymers are key ingredients in many processed foods. They are used essentially as texturing, stabilizing, gelling agents and, in some cases, as protective colloids. They are produced by plants, seaweed or microorganisms [1-3]. In the last decades water soluble polysaccharide research has increased, mostly the extracellulars produced by microorganisms [4].

The industrial importance of this research lies in the time and conditions required to obtain the biopolymer and the possibility of controlling the quality of the product. The goal is to obtain biopolymers that can economically compete with the starch and other conventional polysaccharides. The advantage of utilizing these biopolymers instead of conventional polysaccharide is that they are stable over a wide range of pH and temperature, and are active in low concentrations. In this research rheological properties are the parameters that define the potential applications of the biopolymers, pointing to their possible applications in the food industry [5-10].

Among the rheological properties is the apparent vicosity, which is the first parameter to be used to determine if the biopolymer has potential for the industry. This is also a key property for testing the organoleptic characteristics in food [11].

In the present work apparent viscosity and viscoelastic properties of a hydrocolloid obtained from *Beijerinckia* sp strain 7070 was investigated with the objective of identifying its potential for commercial application as a food ingredient; the productivity was also analysed.

2. MATERIAL AND METHODS

2.1. Microorganism and culture conditions

The bacteria *Beijerinckia* sp strain 7070 was isolated from soil of a sugar cane plantation in the region of Ribeirão Preto, State of São Paulo, Brazil [12]. The incubation was done with 150 mL of medium (in Erlenmeyer flasks of 250 mL) for 24 h and 36 h at 180 rpm, at 24°C. The incubator used was New Brunswick Scientific, model Innova 4230. The medium composition used for the biopolymer production was (gL^{-1}): glucose 50.0, tryptose 2.0, $MgSO_4.7H_2O$ 2.7, K_2HPO_4 2.0, KH_2PO_4 3.6, at pH 6.5.

2.2. Exopolysaccharide recovery and purification

The fermentation broth was centrifuged at 16.266 x g for 10 min. The polysaccharide in the cell-free supernatant was precipitated by shaking with three volumes of ethanol (99.5%) [13-15].

The recovery was in two steps. In the first step we used a sieve with 2 mm mesh to filtrate the medium and the fraction retained was classified as long fibers, the second step was made by centrifugation of the fibers that passed trough the sieve, the fibers obtained were classified as short fibers. The centrifugation (to recover the biopolymer) was at 2 000 x g for 30 min [15]. After centrifugation the biopolymer was dried in an oven at 55°C and stored in sealed glass containers. The dry biopolymer was then purified by dialysis against ultra pure water at 4°C for 48 h and then recovered with the same conditions as above. When it was used for rheological analyses it was dissolved again in dionized water.

2.3. Productivity

The dried biopolymer was weighed one more time to allow the calculation of the weight concentration of biopolymer in the incubation medium. The total weight (short fiber plus long fiber) was considered the productivity of the biopolymer in the two different incubation times.

2.4. Preparation of aqueous solutions

Aqueous solutions were prepared with 3% of biopolymer for oscillatory measurements and 6% for apparent viscosity, in dionized water, with pH 7.0, and heated at 70°C for 40 min. The biopolymer used was 24 h long and short fiber, and 36 h long and short fiber. They were analyzed separately. After cooling the solutions, the same were placed under vacuum to remove bubbles.

2.5. Rheological behaviour

Measurements were performed on a controlled strain Haake rheometer, model CV20, fitted with cone and a plate geometry PQ45, shear rate of 10 s^{-1}, at 25°C. Dynamic measurements were carried out to investigate the viscoelastic properties. From the oscillatory shear measurements, the rheological functions (G', G'', n* and φ) were obtained, over a frequency range of 0.01-100 rad/s.

3. RESULTS AND DISCUSSION

3.1. Productivity

The results obtained showed a productivity of 10.5 gL^{-1} and 7.8 gL^{-1} for biopolymers produced in incubation times of 24 h and 36 h, respectively. In the biopolymers literature on *Beijerinckia* sp, the best result reported with strain 7070 was 10.7 gL^{-1} [12]. However, this productivity required two media and 96 h incubation time. The biopolymer of 36 h showed similar characteristics to the 96 h biopolymer. Our results showed a very significant economy of time (60 h) and a reduction of one medium.

3.2. Rheological Behaviour
3.2.1. Viscosity

Biopolymers produced in 24 h have better yield than biopolymers obtained in 36 h, which have a higher apparent viscosity in aqueous solutions at 6% at the temperature tested.

The apparent viscosity was higher for the short fiber biopolymer of 36 h than the short fiber and long fiber biopolymer obtained in 24h. The viscosity reached 62,900 cP for short fiber 24 h and 137,000 cP for short fiber 36 h. For long fiber biopolymer 24 h and 36 h we obtained, 30,000 cP and 65,000 cP, respectively, at a shear rate of 10 s^{-1}.

The rheograms demonstrated that the biopolymer has pseudoplastic behaviour.

The only reference we could find in the literature to fibers does not mention size but acknowledges their presence and disappearance with the extended incubation time [16]. Our results do not agree with this information, as seen on Section 2.2.

The biopolymer obtained at 36 h is much superior to B-27. The results obtained by Vendruscolo [12], for the biopolymer B-27 produced by *Beijerinckia* sp strain 7070, with apparent viscosity of 6% was 175,000 cP using a shear rate of 0.3 s^{-1}, however, reproducing those conditions but using a shear rate of 10 s^{-1} the result decreased to 10,200 cP.

We conclud that the apparent viscosity is directly related to incubation time, since our results showed a definite increase of apparent viscosity between the biopolymer produced at 24 h and the biopolymer produced at 36 h.

3.2.2. Oscillatory measurements

Viscoelastic behaviour of the samples was confirmed by the oscillatory test. Gel-like properties were observed in the biopolymer obtained in 36 h short fiber as shown in the Figure 1. The storage modulus (G') was superior to the loss modulus (G'') over a large range of frequency.

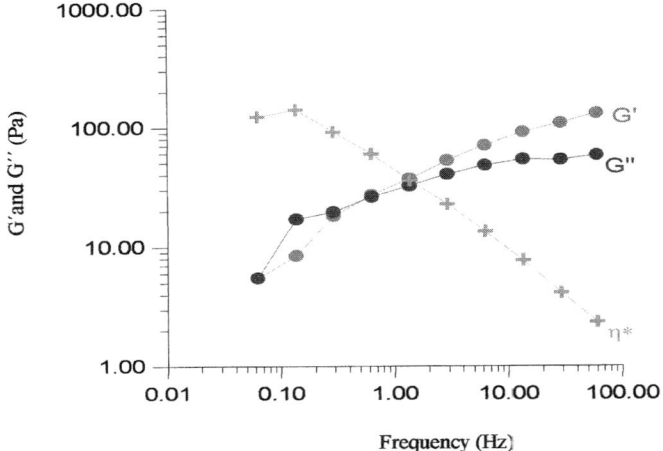

Figure 1- Dynamic test of aqueous solutions with 3% of the 36 h short
fiber biopolymer *Beijerinckia* sp strain 7070. The
measurements were performed in a Haake rheometer model
CV20, sensor system Q45, at 25°C

However, 24 h short fiber, long fiber, and 36 h long fiber biopolymers did not show gel-like properties because the loss modulus was always superior to storage modulus in all range of frequency used (Figures 2, 3, 4).

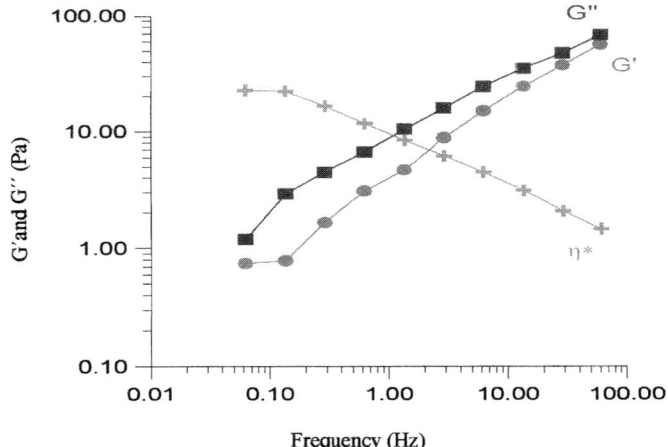

Figure 2- Dynamic test of aqueous solutions with 3% of the 24 h short
fiber biopolymer *Beijerinckia* sp strain 7070. The
measurements were performed in a Haake rheometer model
CV20, sensor system Q45, at 25°C

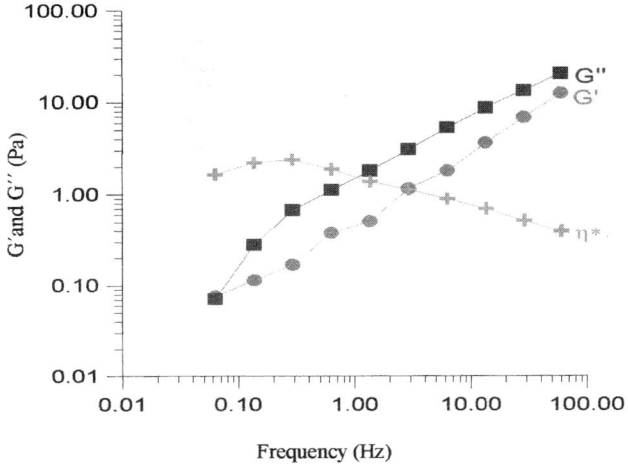

Figure 3- Dynamic test of aqueous solutions with 3% of the 24 h long fiber biopolymer *Beijerinckia* sp strain 7070. The measurements were performed in a Haake rheometer CV20, sensor system Q45, at 25°C

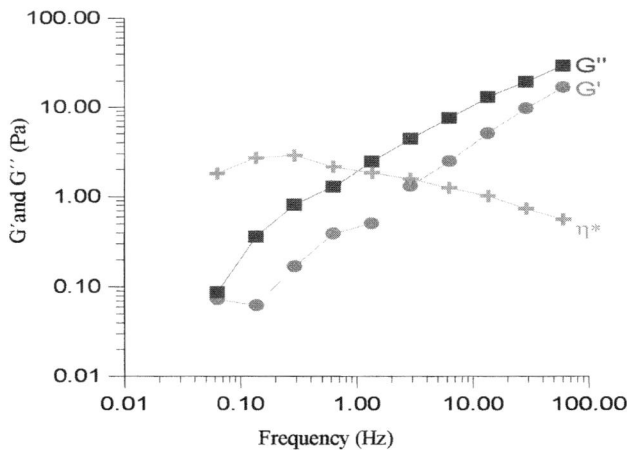

Figure 4- Dynamic test of aqueous solutions with 3% of the 36 h long fiber biopolymer *Beijerinckia* sp strain 7070. The measurements were performed in a Haake rheometer CV20, sensor system Q45, at 25°C

184

4. CONCLUSIONS

The results obtained in this study of the apparent viscosity of aqueous solutions of a biopolymer synthesized from *Beijerinckia* sp strain 7070 show that:

The incubation time influences the type of fiber produced.

#The best productivity obtained was the polymer of 24 h.

#The biopolymer of 36 h short fiber exhibited the best results for apparent viscosity.

The biopolymer produced in 36 h, showed a superior apparent viscosity and viscoelasticity compared with the biopolymer produced in 96 h Having a reduction of 60 h in the process of production, and besides, it was not necessary to use cell production medium. In conclusion this biopolymer has good characteristics of productivity and rheology which allow us to suggest its potential use in the food industry.

Acknowledgements: We thank FAPERGS and CAPES and T&T Techworks Inc. (FL-USA) for the financial support for our research.

4. REFERENCES

1. I.W. Cottrell, in: Sandford, P.A. & Matsuda, K. (eds.) Washington, D.C. American Chemical Society (1979) 251 (ACS Symposium Series, 126).
2. P.A. Sandford and J. Baird, The Polysaccharides, New York:Academic Press, 2 (1983) 412.
3. A.M. Marqués, I. Estãnol; J.M. Alsina; C. Fusté; D. Simon-Pujol; J. Guinea and F. Congregado, Appl. Environm. Microbiol., 52, n° 5, (1986), 1221.
4. K.S. Kang, G.T. Veeder and I.W. Cottrell, in: M.E. Bushell (ed.) Progress in industrial microbiology, 18 (1983) 231.
5. P.A. Sandford, Adv. Carbohydr. Chem. Biochem., 36 (1979), 292.
6. I.W. Sutherland, Adv. Microbiol. Physiol., 23 (1982) 79.
7. I.W. Sutherland, in: H. Dellwey, (ed.) Biotechnology. Weinhein: Verlag Chemie, 3 (1983) 553.
8. E.R. Morris, in: G.O. Phillips, D.J. Wedlock, and P.A. Williams (eds.) Gums and stabilisers for the food industry. Oxford: Pergamon Press. (1984) 57.
9. T. Harada, A. Misaki and H. Saito, Arch. Biochem. Biophys., 124 (1986) 292.
10. V. Morris, Food Biotechnol. 4, n° 1, (1990) 45.
11. M. Glicksman, in: M. Glicksman, (ed.), Food hydrocolloids. Boca Raton:CRC Press, 1 (1982) 3.
12. C.T. Vendruscolo, Produção e caracterização do biopolímero produzido por *Beijerinckia* sp isolada do solo da região de Ribeirão Preto - SP Brasil. Campinas: Curso de Pós-Graduação em Engenharia de Alimentos da UNICAMP, 1995. (Food Enginnering, PhD thesis).
13. F. Congregado, I. Estãnol, M.J. Espuny, M.C. Fusté, M.A. Manresa, A.M. Marqués, J. Guinea and M.D. Simon-Pujol, Biotechnol. Letters, 7, n° 12 (1985) 883.
14. A.C. Mochi and A.R.P. Scamparini, In: K. Nishinari & E. Doi (eds.) Food hydrocolloids. New York: Plenum Press, (1994) 147.

15. L. Lebrun, G.A Junter, T. Jouenne and L. Mignot, Enzyme and Microbial Technology, 16, n° 12 (1994) 1048.
16. R. Lopez and J.H. Backing, Microbiol. Espan., 21 (1968) 53.

HYDROCOLLOIDS – PART 1
Edited by K. Nishinari
2000 Elsevier Science B.V.

187

Heteropolysaccharides produced by *Xanthomonas campestris* pv *pruni* C24

C.T. Vendruscolo[a*] ; **A. da S. Moreira**[b]; **A. da S. Souza**[b]; **R. Zambiazi**[a]; **A. R. P. Scamparini**[c].

[a]Department of Food Science and [b]Center for Biotechnology ; UFPEL, P.O. Box 354, CEP 96010-900, Pelotas, RS, Brazil
[c]Food Science Department. College of Food Engineering, UNICAMP, P.O. Box 13081, Campinas, SP., Brazil

Several pathovars of the plant pathogenic bacteria *Xanthomonas campestris,* and few other *Xanthomonas* species, can produce a high-molecular weight polysaccharide called xanthan gum. This biopolymer is widely used in the modern food, cosmetics, and other industries. Therefore, the microorganisms that have the capability to produce this gum have been extensively studied. The *Xanthomonas campestris* pv *pruni* is the causal agent of *Prunus* bacterial spots (PBS) and naturally infects all cultivated *Prunus* species. In the south of Rio Grande do Sul, Brazil, a great number of strains of *Xanthomonas campestris* pv *pruni* were isolated from peach leaves (*Prunus persica)*. Many of them have been tested for their ability to synthesize polysaccharide in batch incubation. The biopolymers obtained have been analyzed for their productivity and chemical composition. The heteropolysaccharides produced by *Xanthomonas campestris* pv *pruni* C24 were obtained in two different fermentation media: PMII and PMI+II. Both media contained sucrose as the carbon source and salts. The type of monosaccharide and derivatives present in the biopolymers obtained at different incubation times were detected with thin layer chromatography. The chromatographic analysis revealed the presence of glucose, rhamnose, and glucuronic acid in polymers produced after 72 h of fermentation. In polymers obtained at 96 h mannose was also detected. The PM I+II medium showed higher productivity than PM II, but the viscosity was higher in polymers produced in the PM II medium. The incubation time also had an influence on productivity and viscosity.

1. INTRODUCTION

Bacteria belonging to the genus *Xanthomonas* are plant-associated bacteria, with the exception of *X. maltophilia.* They are not usually found in other environments, but they are not invariably plant pathogens[1]. It was shown that the infection caused by these bacteria occurs on at least 124 monocotyledonous plants, comprising 70 genera, and 268 dicotyledonous plants species, totaling more than 170 genera[2].

---

* Corresponding author

The bacteria of *Xanthomonas campestris* species are very significant with least 125 pathovars [3] which infect and cause diseases in various hosts. The pathovar *X. campestris* pv *pruni* is the causal agent of Prunus Bacterial Spot (PBS). This disease is known to occur in all continents, and it is more serious in areas where the growing season is under warm and wet conditions [4]. In the southern part of Rio Grande do Sul, Brazil, this bacterium naturally infects all cultivated *Prunus* species. Researchers from EMBRAPA-CPACT have collected and isolated a great number of strains of *X. campestris* pv *pruni* from peach leaves (*Prunus persica*) in order to control the PBS disease.

Several pathovars of *Xanthomonas campestris,* and a few other *Xanthomonas* species, produce xanthan gum in abundance as extracellular slime polysaccharide [5]. Because of the industrial applications and the very unusual properties of the polysaccharide in aqueous solution [6-8], the microorganisms that have the capability to produce this gum have been extensively studied. However, *Xanthomonas campestris* pv *pruni* have not being tested as a biopolymer producer.

Many *Xanthomonas campestris* pv *pruni* strains that were previously isolated have been tested for their capability of synthesizing polysaccharide in incubation batches. The polymers obtained have been analyzed for their productivity and chemical composition. This work describes the characterization of some biopolymers obtained from *X. campestris* pv *pruni* C24.

2. MATERIALS AND METHODS

2.1 Polysaccharide production

Strains of *Xanthomonas campestris* pv *pruni* C24 were isolated from peach leaves by Brazilian phytopathologists from EMBRAPA-CPACT. The strains were preserved by the replication technique. A strain of *Xanthomonas campestris* pv *pruni* C24 was used to produce the polysaccharide under aerobic fermentation. The Yeast Malt (YM) medium was used for propagation of cells during 24 h at 28°C and 150 rpm.

The media PM II [9] and PM I+II [9-10] were used for production of biopolymers. They contained: (gL^{-1}) $NH_4H_2PO_4$ 1.5, K_2HPO_4 2.5, $MgSO_4$ 0.1, sucrose 50.0 (medium II) and $NH_4H_2PO_4$ 1.5, K_2HPO_4 2.5, $MgSO_4$ 0.3, KH_2PO_4 5.0, H_3BO_3 0.006, $(NH_4)_2SO_4$ 2.0, $FeCl_3$ 0.0024, $CaCl_2.2H_2O$ 0.002, $ZnSO_4$ 0.002, sucrose 50.0 (medium I+II), respectively. The components were sterilized at 121°C for 15 min. The incubation conditions were 200 rpm for 72 h and 96 h at 28°C. After these periods the fermentation broth were centrifuged at 16266 g during 30 min to remove cells and other debris. The exopolysaccharides dissolved in the supernatant were precipitated with ethanol at 95% (v/v) and dried at 50°C to constant weight [11].

2.2 Yield and viscosity analysis

The yield of biopolymers (w/v) was calculated in terms of the volume of production medium.

In order to purify the polysaccharides, the 3% (v/v) aqueous solution was dialyzed against ultra pure water during 72 h, dried at 50°C, and it was finally powdered.

The rheological measurements were made with polymers not dialyzed. Samples were prepared with 3% of biopolymer in deionized water at pH 7,0, and bubbles removed under vacuum. The rheological measurements were made within 24 h, using a Haake rheometer model CV-20, at 25°C.

2.3 TLC analysis

The samples of polysaccharides were hydrolyzed with hydrochloric acid (2N) at 80°C for 16h. They were then evaporated at 45°C under vacuum, and the solvent-free samples were suspended in methanol.

The monosaccharides and their acid derivatives present in the polysaccharides were determined through comparative thin layer chromatography (TLC) and co-chromatography, by comparison with standards of monosaccharides and glucuronic acid. The plates were silica gel and eluted with trichloromethane-methanol-acetic acid-water (40:40:10:10; v/v/v/v) [12]. The spots were visualized through spraying of the plates with the detection agent anisaldehyde-sulfuric and heat at 100°C for 5 min [13].

3. RESULTS AND DISCUSSION

3.1. Polysaccharides yield and rheological analysis

The *Xanthomonas campestris* pv *pruni* C24 produced extracellular polysaccharide. This was obtained in two different media, PM II and PM I+II, using two incubation times: 72 and 96 h. The medium PM I+II with a higher variety of inorganic salts showed a higher productivity at 72 and 96 h of incubation times (12.4 gL^{-1} and 16.4 gL^{-1}, respectively). But the polymers produced in the PM II medium showed a higher viscosity (Table 1).

Table 1
Productivity ($g.L^{-1}$ fermentation medium) and viscosity (cP) at 25°C of biopolymers not dialyzed (3% w/v)

Time (h)	Productivity ($g.L^{-1}$)		Viscosity (cP)			
	MP II	MP I+II	MP II		MP I+II	
72	2.5	12.4	14000*	28**	15000*	30**
96	12.2	16.4	80000*	160**	26000*	52**

*Shear rate of 5 s^{-1}
**Shear rate of 500 s^{-1}

3.2. Chemical analysis

In order to identify the monosaccharides and their acid derivatives present in biopolymers produced by *Xanthomonas campestris* pv *pruni* C24, the hydrolyzed samples were submitted to comparative TLC and co-chromatography.

The chromatographic analysis showed that the *Xanthomonas campestris* pv *pruni* C24 produces heteropolysaccharides with chemical composition different (in sugar type) from that of commercial xanthan gum [14-15] (from *Xanthomonas campestris* pv *campestris*). Polymers produced with in 72 h of fermentation contained glucose, rhamnose, and glucuronic acid polymers, but obtained in 96 h, also contained mannose. These results showed that the *Xanthomonas campestris* pv *pruni* C24 produced a biopolymer with a chemical composition distinct from that of commercial xanthan gum (Table 2). The incubation time also had an influence on the type of sugars present in biopolymers, but there appeared to be no differences between biopolymers obtained on the different fermentation media.

Table 2
Monosaccharides and derivatives detected in commercial xanthan gum [14-15] and heteropolysaccharides produced by *Xanthomonas campestris* pv *pruni*

Monosaccharides and derivatives	Xanthan	Heteropolysaccharides	
		72h	96h
Glucose	+	+	+
Mannose	+	-	+
Rhamnose	-	+	+
Glucuronic acid	+	+	+

+Detected
-Not detected

4. CONCLUSION

The *Xanthomonas campestris* pv *pruni* C24 strain produced heteroexopolysaccharides with chemical composition different from commercial xanthan gum.

The production medium influenced the productivity and viscosity of biopolymers, but had no influence on the sugar type of polysaccharides. However, the incubation time influenced productivity, viscosity and chemical composition.

5. ACKNOWLEDGEMENTS

The authors are grateful to Olinda M. Martins (CPACT-EMBRAPA, Brazil) for supplying us with strains of *Xanthomonas campestris* pv pruni, and FAPERGS and CAPES for financial support.

6. REFERENCES

1. A. C. Hayward. In:, J. G. Swings and E. L. Civerolo (eds.), Xanthomonas, London, Chapman & Hall, 1993, 1.
2. F. Leyns, M. De Cleene, J Swings, De Ley, The Botanical Rev., No. 50 (1984) 308.
3. J.F. Bradbury. In: Bergey's Manual of Systematic Bacteriology, N. R. Krieg, J. G. Holt (eds), Baltimore, Williams and Wilkins, 1984, v. 1. 199.
4. E. L. Civerolo and M.J Hattingh. In: J. G. Swings and Civerolo, E. L. Xanthomonas, London, Chapman & Hall, 1993, 60.
5. I. W. Sutherland. In: J. G. Swings, E. L. Civerolo, Xanthomonas, London, Chapman & Hall, 1993, 363.
6. D. A. Betz, Food Technol. in Australia, 31 (1979) 11.

7. J. Baird, P. Sandford, I. Cottrell, Bio Technol., 1 (1983) 778.
8. K. S. Kang, G. T. Veeder, I. W. In: M. E. Bushell, (eds). Progress in Industrial Microbiology, 1993, v.18.
9. M. C. Cadmus, C. A. Knutson, A. A. Lagoda, J. E. Pittsley, K. A. Burton, Biotechnol. Bioeng., 20 (1978) 1003.
10. P. Souw, A. L. Demain, Appl. Env. Microbiol., 31 (1979), p.1186-1192, 1979.
11. C. T Vendruscolo,. Produção e caracterização do Biopolímero produzido por *Beijerinckia* sp isolada do solo cultivado com cana de açucar da região de Ribeirão Preto-São Paulo-Brasil. Faculdade de Engenharia de Alimentos, UNICAMP, Campinas, 1995.(Food Enginnering, PhD thesis).
12. A. da S. Moreira, A. da S. Souza and C. T. Vendruscolo, Agrociência, 3 (1998) 222.
13. S.C. Churms. In: G. Sherma and J. Zweig, (eds.), Handbook of chromatography, CRC Press, Boca Raton, 1982, v. 1.
14. P. E. Jansson, L. Kenne, B. Lindberg, Carbohydr. Res, No. 45 (1975) 275.
15. J. Stankowski, B. Mueller, S. Zeller, Carbohydr. Res, No. 241 (1993) 321.

HYDROCOLLOIDS – PART 1
Edited by K. Nishinari
2000 Elsevier Science B.V.

Rheological properties of guar galactomannan solutions filled with particulate inclusions

P. Rayment, P.R. Ellis and S.B. Ross-Murphy

Biopolymers Group, Division of Life Sciences, King's College London, Campden Hill Road, Kensington, London, W8 7AH.

The effect of particulate inclusions, or 'fillers', on the rheological properties of a typical polysaccharide entanglement solution (guar galactomannan/water) have been studied. Earlier steady shear experiments have shown an apparent upswing of the viscosity-shear rate profiles at low shear rates, with higher filler concentrations [1, 2]. The Cross model [3], typically used to describe the flow behaviour of semi-dilute polysaccharide solutions, has been modified to included a yield stress term.

In creep experiments, a low constant stress is applied to the sample. The creep experimental data has been compared to the steady shear flow behaviour fitted by the yield stress modified Cross equation [4]. The apparent zero-shear viscosity and yield stress parameters have been constrained in a number of ways in the model to establish the effect of these parameters on the modified Cross model. Although the creep data appears to alter the precision of the modified Cross equation, due to the low shear rates accessed, the apparent upswing of the steady shear data at low shear rates appears to be supported by the creep data. By fitting of data to mathematical models which incorporate volume fraction parameters, a new master curve has been produced that can be applied to a number of semi-dilute dispersions. Potential applications of this approach include functional foods such as soluble-fibre drinks and yoghurts, as well as fat spreads, thickened sauces and breakfast cereals.

1. INTRODUCTION

Guar galactomannan, or guar gum as it is commonly known, is the reserve polysaccharide of the endosperm of the seed of the Indian cluster bean, *Cyamopsis tetragonoloba* (L.) Taub. It is a galactomannan based on a β-D-(1-4) mannan (M) backbone solubilized by substitution with α-D-(1-6)-linked galactose (G) side groups [5].

Guar gum is a soluble non-starch polysaccharide (s-NSP). It is a non-α-glucan polysaccharide and therefore resists digestion by the enzymes secreted into the small intestine of man. For a number of years, soluble dietary NSPs have been considered to have nutritional benefits due to their modulation of carbohydrate and lipid metabolism [6, 7]. Certain polysaccharide gums, such as guar gum, reduce the hyperglycaemic response to a meal in both normal and diabetic patients [8, 9]. These effects appear to be mainly dependent on the ability of such gums to hydrate and form a viscous solution in the gastrointestinal tract.

One of the main drives of this work was to develop rheological methods to characterise such gums so that reliable predictions of biological activity could be made. The study of model systems may go some way to clarifying these therapeutic effects. Another important objective was to study the effect of filler particle size and shape on the flow properties of guar galactomannan.

For this study, a rice starch filler was chosen for its small size and homogeneity of the starch grain. The filler was added in increasing amounts to the guar galactomannan solution and the system was measured both in steady shear and creep. By carrying out these two types of rheological experiment, it was hoped that information on this guar galactomannan/rice starch mixture could be extended over a wide range of shear rates.

The Cross equation [3] is often used to describe the shear-thinning behaviour of guar gum dispersions [10]. This is reasonable for the pure polysaccharide system but, obviously, cannot be used to describe filled systems.

$$\eta = \eta_\infty + [\eta_{0X} - \eta_\infty]/[1 + (a\dot{\gamma})^p] \qquad (1)$$

where η_{0X} and η_∞ are limiting (Cross) viscosities, at zero and infinite shear rates, a and $\dot{\gamma}$ are a relaxation time and shear rate, respectively, whilst p is an exponent.

It has been known for some time that one of the major effects of increasing filler concentration is to modify the overall shear rate-viscosity profile. For filled Newtonian fluids, a more 'power-law' like $\eta(\dot{\gamma})$ profile is seen, with an apparent upswing at the lowest shear rates. This could be related to a yield stress developing at lower (zero) shear rates.

Such a system would require, for example, the Cross equation to be modified to include a yield stress term (τ_X) (1→2):

$$\eta = \eta_\infty + [\eta_{0X} - \eta_\infty]/[1 + (a\dot{\gamma})^p] + (\tau_X/\dot{\gamma}) \qquad (2)$$

In this, and our earlier work [1, 4], the yield stress modified Cross equation was used to describe the flow behaviour of our starch-filled systems.

2. EXPERIMENTAL

2.1 Materials

A guar gum sample (Meyprogat 90) was kindly provided by Meyhall Chemicals, A.G., Rhône Poulenc Group, Switzerland. The sample was purified using a modified isolation procedure devised by Girhammar and Nair [11]. The soluble galactose and mannose content of the purified sample was determined as 92.7±2.3% with a galactose to mannose ratio of 0.63±0.025. The intrinsic viscosity and weight averaged molecular weight of the sample were determined as 10.5 dlg^{-1} and 1.39 x 10^6, respectively, using dilute solution viscometry [12].

Guar gum solutions were prepared by dissolving a known weight of the dried material in distilled water to give a nominal polymer concentration of 1%. The solution was heated to 80°C for 5 minutes followed by stirring at room temperature overnight to ensure complete hydration of the polymer. An insoluble, cosmetic rice starch sample was kindly donated by Cairn Foods (Remy Industries, S.A., Belgium). The starch was added to the guar gum solution so that the galactomannan concentration in the aqueous medium was constant, and the filler concentration varied from 0 to 41% (w/w). Microscope examination of the rice starch sample showed the particle size to range between 6-10μm.

2.2 Measurements

Steady shear measurements were performed using a Rheometrics Fluids Spectrometer (RFSII, Rheometric Scientific Ltd., Epsom, Surrey, UK) at 25°C with a 25mm diameter parallel plate with a 1mm gap. For steady shear measurements it is preferable to employ cone and plate geometry since the shear rate approaches constancy throughout the sample. In previous work, a comparison of both geometries was made and no significant effect was demonstrated [12]. The parallel plate was used in order to compare with further work involving larger particulate material [1, 12]. The shear rate was ranged from $0.05\text{-}1000\text{s}^{-1}$, with a reduction in rate necessary at the higher filler concentration regimes. A high viscosity paraffin oil was applied to the exposed surface of the sample to prevent drying, along with the use of a solvent trap.

Creep experiments were performed at 25°C using a Rheometric Dynamic Stress Rheometer (DSR200) with a 40mm diameter and 0.04 radian cone and plate geometry. Again, problems encountered with drying of the samples were reduced by using a high viscosity paraffin oil and a solvent trap. Figure 1 shows the data collection set-up for the creep experiment.

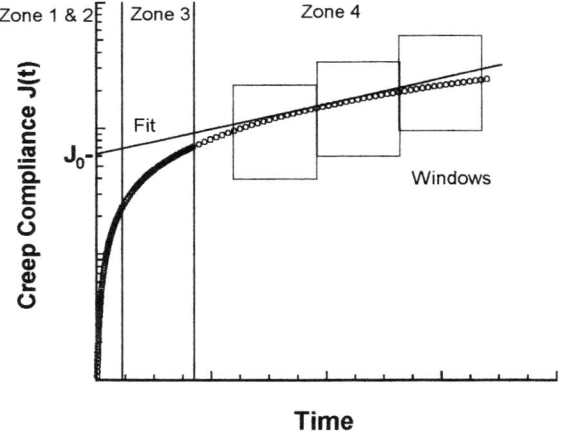

Figure 1. Graph showing data collection set-up in creep experiment. A least squares fit to time zero determines the instantaneous compliance, J_0.

The creep experiment is divided into creep and recovery phases. During the creep phase, a command stress is applied and the change in strain with time is observed. This can be investigated by monitoring the creep compliance parameter $J(t)$, which is the resulting strain divided by the applied stress. During the recovery phase the stress is 'commanded' to zero and the response of the sample to the removed load is observed. The creep analysis within the Rheometrics software (RHIOS version 4.3.2) allows determination of viscosity for the steady state region of the creep portion of the test. A least squares fit is performed from the end of the curve working back to time zero and the steady state is determined as the point where the difference between the slopes of adjacent windows is less than the slope tolerance.

In order to investigate the existence of a yield stress, lower shear rates should be accessed. This was made possible by using the cone and plate geometry where a lower command stress could be applied.

3. RESULTS AND DISCUSSION

Figure 2 shows the effect of rice starch concentration on the steady shear viscosity profile of a 1% guar galactomannan solution.

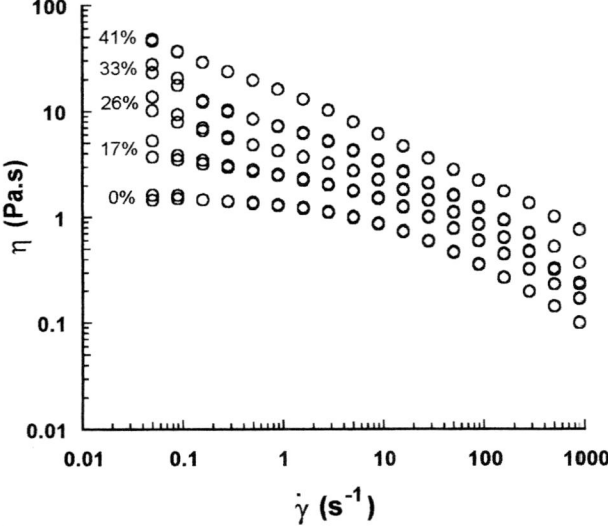

Figure 2. Effect of Rice Starch Concentration on the Steady Shear Viscosity Profile of Guar Galactomannan.

As the filler concentration is increased, the system response begins to develop some additional features. For the pure guar galactomannan system, a Newtonian plateau is displayed at low shear rates followed by shear-thinning at higher shear rates. However, as the rice starch concentration is increased from 0-41%, the initial Newtonian flow properties become more rate dependent at low shear rates and the so-called 'power law' behaviour is displayed.

From this initial data, it was determined that an extra yield stress term was required for the higher concentration range studied. The flow behaviour has been described, with reasonable accuracy, by a yield stress modified Cross equation [1]. In our earlier work, we have shown both zero-shear viscosity and yield stress parameters to increase as the filler concentration approaches the maximum packing fraction of the system. In this case, the reduced viscosity could be scaled independent of polymer concentration and therefore allowed creep experiments to be undertaken at one polymer concentration only.

Figure 3 shows the effect of increasing rice starch concentration on the creep compliance-time flow curve of a 1% guar galactomannan solution. The effect of particulate inclusions in the guar galactomannan/rice starch mixture is, primarily, to decrease the creep compliance parameter below that of the pure system.

The creep compliance profiles become more linear with time as the filler concentration is increased and on the log time scale appear to 'flatten off'. This is most pronounced between the 17% and 26% rice starch concentrations and this is displayed more clearly in figure 4, where the apparent viscosity, as calculated in the creep analysis software, is plotted against the reciprocal of time.

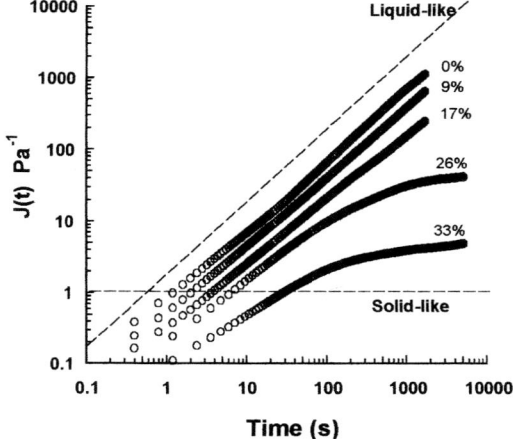

Figure 3. Effect of increasing rice starch concentration on the creep compliance-time flow curve of a 1% guar galactomannan solution, and the limiting liquid-like and solid-like behaviour.

The viscosity-shear rate data displayed in figure 2 are included in the same figure along with the creep data. Again, a power-law behaviour is displayed at low shear rates. It is important to appreciate that this superposition does not include the (geometric) shift to convert the $1/t$ scale to shear rate. The superposition presumably works because this (undetermined and not constant factor, since the stress (rate) is being controlled rather than the strain rate) must be of order unity. The creep experiment appears to have lowered the viscosity-shear rate information obtained in the steady shear experiment by two decades. The creep data shows Newtonian behaviour for the lower rice starch concentration mixtures but this behaviour becomes more shear rate dependent as the rice starch concentration increases. However, at very low shear rates the data does appear to flatten off although the scatter is high. Previously a 'pseudo-Newtonian plateau' has been described but comparison in this case is uncertain. However, the extension of the viscosity-shear rate data to such low shear rates has provided some interesting additional information.

By increasing filler concentration, the effect on the creep compliance-time profile was considerable (figure 3). The position and shape of the curves were strongly influenced. The overall effect of increasing filler concentration is to reduce the creep compliance parameter below that of the pure system and to make it more linear with respect to time, i.e. more solid-like profile. The data produced in the steady shear experiments has been expanded by approximately two decades of shear rate in the creep experiments. The presence of an apparent yield stress in the filled systems in steady shear appears to be supported by the creep data.

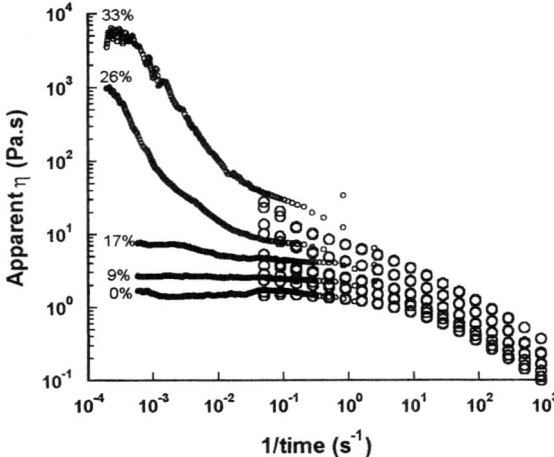

Figure 4. Effect of increasing rice starch concentration on a double log plot of apparent viscosity versus 1/t for a 1% guar galactomannan solution (small symbols) superposed on the viscosity-shear rate data of figure 2 (see Ref. 4 for details).

ACKNOWLEDGEMENTS

PR acknowledges the receipt of a Ministry of Agriculture, Fisheries and Food studentship (1993-1996) and Leatherhead Food Research Association for their collaboration and financial support. The authors acknowledge the support of the BBSRC for the purchase of the rheometers. We are grateful to Professor Iain Dea of Quest International, Naarden, The Netherlands for helpful discussions.

REFERENCES

1. P.Rayment, S.B.Ross-Murphy and P.R.Ellis, Carbohydr. Polym., 28 (1995) 121.
2. P.Rayment, S.B.Ross-Murphy and P.R.Ellis, Gums and Stabilisers for the Food Industry 8, G.O.Phillips, P.A.Williams and D.J.Wedlock (eds.) (1996) 237.
3. M.M.Cross, J.Colloid Science, 20 (1965) 417.
4. P.Rayment, S.B.Ross-Murphy and P.R.Ellis, Carbohydr. Polym., 35 (1998) 55.
5. I.C.M.Dea and A.Morrison, Adv. Carb. Chem. Biochem., 31 (1975) 241.
6. D.J.A.Jenkins, T.M.S.Wolever, A.R.Leeds, M.A.Gassull, P.Haisman, A.Dilawari, D.V.Goff, G.L.Metz and K.G.M.M.Alberti, British Medical Journal, 1 (1978) 1392.
7. P.R.Ellis, E.C.Apling, A.R.Leeds and D.B.Peterson, British Journal of Nutrition, 46 (1981) 267.
8. U.Smith and G.Holm, Atherosclerosis, 45 (1982) 1.
9. P.R.Ellis, F.M.Dawoud and E.R.Morris, British Journal of Nutrition, 66 (1991) 363.
10. W.R.Sharman, E.L.Richards and G.N.Malcolm, Biopolymers, 17 (1978) 2817.
11. U.Girhammar and B.N.Nair, Food Hydrocolloids, 6 (1992) 285.
12. P.Rayment, PhD thesis, University of London (1996).

HYDROCOLLOIDS – PART 1
Edited by K. Nishinari
2000 Elsevier Science B.V.

199

Characterization of chitosan film and structure in solution

S.Y. Park[a], H.J. Park[a], X.Q. Lin[b] and Y. Sano[b]

[a]Korea university, Graduate School of Biotechnology, Sungbuk-ku, Seoul 156-701, Korea
[b]National Food Research Institute, Tsukuba, Ibaraki 305, Japan

This study was conducted to determine the relation between molecular structure of chitosan in acidic aqueous solution and physical properties of chitosan films. The molecular weights (M_W) of chitosan dissolved in acidic solutions were varied and depend on the type of acid used. M_W of chitosan in acetic acid, citric acid, lactic acid and malic acid was 11.9×10^5, 5.64×10^5, 6.1×10^5 and 5.45×10^5, respectively. The Rg increased directly proportional to the increasing molecular weight. Tensile strength (TS), elongation (E), and water vapor permeability (WVP) of chitosan films depend on the type of acid used. The TS of acetic acid cast chitosan films is the highest (40.05MPa), and E of citric acid cast chitosan films is the highest (137%). In an acetic acid solution chitosan form dimer indicating that the intermolecular interaction is relatively strong. WVP of malic acid cast chitosan films was 1.02 ng.m/m^2.sPa and was lower than those of the films dissolved in three other acidic solutions.

1. INTRODUCTION

Chitosan [(1-4)-2-amino-2-deoxy-β–D-glucan)] is one of a linear polysaccharides obtained by deacetylating chitin, which is extracted from the shells of crab, shrimp and krill. After cellulose has long been used, chitin is noted as the second most abundant natural polymer.

Chitosan is insoluble in water but it becomes soluble and cationic when it is dissolved in organic acids such as acetic acid, malic acid, citric acid and formic acid[1,2]. Chitosan is an excellent polymer to make a functional film. Besides, chitosan is useful in various parts such as coagulant or adsorbent in waste water treatment, immobilizing matrix for microorganisms and cultured plant cell, matrix for the controlled release of pharmaceutical, tablet binder, membranes for reverse osmosis and biodegradable films and surgical sutures[3,4]. Obviously, many properties of chitosan are predictable from its polycationic macromolecular structure.

These characters of chitosan are primarily due to the polyelectrolytic nature of materials, strong intramolecular and intermolecular. It has a tendency for the polymer to self-aggregate in solution. In addition, solvent selection is very important to dissolve chitosan. The physicochemical behavior of polymer solutions and the numerous other physical properties of polymers are known to be functions of their averages molecular weight, molecular weight distributions, molecular size and diffusion coefficient. Molecular weight plays an important part in functional properties of chitosan. Muzzarelli[5] reported that formation of film and physical strength of chitosan increased with the increase of molecular weight. Molecular weight and structure of chitosan in an aqueous acidic solutions could affect physical properties of the chitosan film. In the present paper, the main purpose of research aims to make clear the relation between film properties and molecular structure of chitosan in four different solutions such as acetic acid, citric acid, lactic acid and malic acid.

2. MATERIALS AND METHOD

2.1. Material
Chitosan of 10cp (degree of deacetylate above 90%) obtained[6,7] from red crab were purchased from a local market in Korea.

2.2 Light-Scattering Instrument
Chitosan was dissolved in four kinds of acidic solutions (acetic acid, citric acid, lactic acid and malic acid). Measurements of the intensity of the light scattering were done with a modified Ellipsometer, an automatic light scattering analyzer AEP-700 (Shimadzu Co., Ltd.) at 20°C. The linearly polarized monochromatic incident light passed through the cylindrical scattering cell at wavelength of 488nm. The light scattered at scattering angle θ was detected through the linear analyzer by a photomultiplier in a telescope arm that can rotate from 15° to 135° by a stepping motor. The sample was made optically clean by filtering with a Millipore filter (pore size, 0.45 μm).

2.3. Preparation of Chitosan
2% of each chitosan was dissolved in four different acidic solutions (acetic acid, citric acid, malic acid and lactic acid) to make each film. The chitosan solutions were poured onto glass plate and dried at 25°C by casting method. All films were conditioned at 25°C, 50% RH for 48hr in a chamber (Model AT/I-150, Jeio Tech. Inc., Korea). The dried chitosan films were used for measuring tensile strength, elongation and water vapor permeability.

2.4. Measurement of Water Vapor Permeability (WVP)
WVP of films was measured by the cup method. Cups were filled to depth of 0.9 cm with distilled water and covered with a films to be tested. Then, cups and sealing rings were tightened with four screws and placed in a chamber conditioned at 25°C, 50% RH. Weight change of the cups versus time was measured and plotted. Linear regression was used to calculate the slope of a fitted straight line. WVP was then calculated as described by Park and Chinnan[8].

2.5. Mechanical Properties
Twelve samples, 2.54 wide and 10cm length were cut from film samples prepared on three glass plates (28×28cm). Samples were conditioned for 48hr at 25°C and 50% relative humidity(RH) in an environmental chamber before measuring tensile strength(TS) and elongation(E). A texture analyzer(TA-XT2i, MHK TRADING Co., England) was used to measure TS and E at breakage, according to the ASTM standard method D882-88.

3. RESULT AND DISCUSSION

3.1. Light-Scattering

The molecular weight of chitosan dissolved in four acidic solutions such as acetic acid, citric acid, lactic acid and malic acid were measured with a light scattering method[9]. Fig. 1 shows the light-scattering pattern of the chitosan dissolved in an acetic acid solution, in which the scattering function I(Q) as a function of scattering vector $Q=4\pi \sin(\theta /2) / \lambda$, where θ is the scattering angle, λ is the wavelength).

Scattered intensity profiles of citric acid, lactic acid and malic acid were similar results of acetic acid.

$$KC/R_\theta = 1/MP(\theta) + 2A_2c \qquad (1)$$

where K is the optical constant, C is the concentration, θ is the scattering angle, R_θ is the reduced scattering intensity observed at the scattering angle θ, Mw is the molecular weight, P is the particle scattering factor and A_2 is the second virial coefficient.

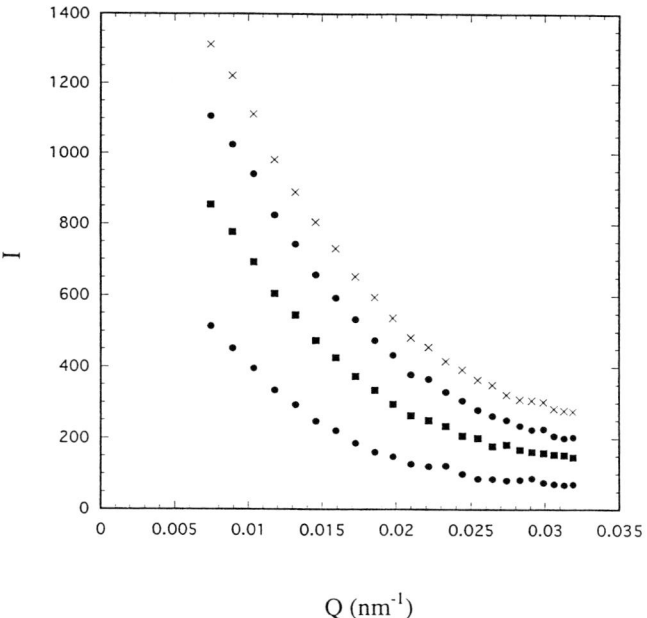

Fig. 1. Angular dependence of light scattering for the chitosan in acetic acid solution. Concentration of chitosan is ● : 1mg/ml, ■ : 2mg/ml, ● : 3mg/ml, and × : 4mg/ml.

A plotted of $KC/R_{\theta \to 0}$ values extrapolated to $\theta \to 0$ against each C as is shown in Fig. 2 gives molecular weight from the intercept of the ordinate and the second virial coefficient from the inclination of the straight line.

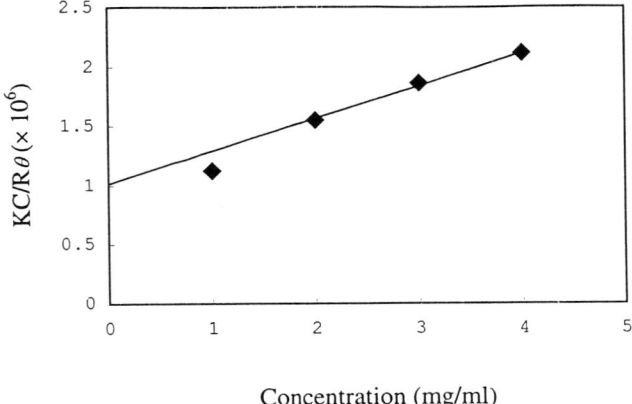

Fig. 2. Concentration dependence of inverse scattering intensity
extrapolated to θ=0 for the chitosan in acetic acid solution.

Fig. 2 shows concentration dependence of inverse scattering intensity extraplolated to θ = 0 for chitosan dissolved in an acetic acid solution. The similar straight line was also obtained for other acidic solutions such as citric acid, lactic acid and malic acid. The molecular weight of chitosan dissolved in acetic acid was higher than those of chitosan dissolved in three other acidic solutions. Molecular weight of chitosan in acetic acid, citric acid, lactic acid and malic acid was 11.9×10^5, 5.64×10^5, 6.1×10^5 and 5.45×10^5, respectively.

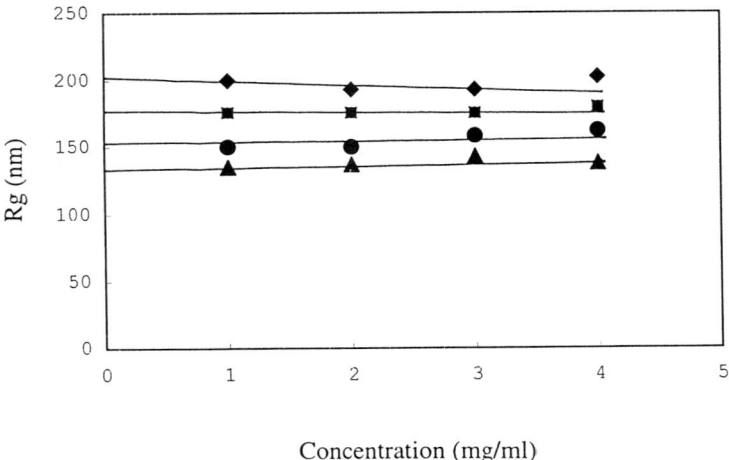

Fig. 3. Radius of gyration (Rg) obtained from light-scattering pattern
for the chitosan in various acidic solutions (◆ :acetic acid, ● :citric
acid, ■ :lactic acid and ▲:malic acid).

Lee et al. obtained the Zimm plots from light-scattering to measure M_W of a commercial chitosan, and extrapolated weight-averaged molecular of the chitosan was 1.2×10^5[10]. Radius of gyration (Rg) of chitosan dispersed in acetic acid was the highest among four acidic solutions. The value of Rg of chitosan in acetic acid, citric acid, lactic acid and malic acid was 200, 152, 176 and 138nm, respectively (Fig. 3).

3.2. Physical properties of films

Tensile Strength (TS), Elongation (E) and Water Vapor Permeability (WVP) of chitosan films show in Table 1. Mechanical properties (TS, E) and WVP of the four acidic solutions cast chitosan films were measured and then compared molecular weights with results by the light scattering method. Mechanical properties (TS, E) of the films were affected by the type of acid used. TS of acetic acid cast chitosan films is the highest (40.05MPa) among four acidic solutions cast chitosan films. E of citric acid cast chitosan films is the highest (137%) among four acidic solvents cast chitosan films. WVP of the films also were affected by the type of acid used. WVP of malic acid cast chitosan films were 1.02 ng.m/m^2.sPa and were generally lower than those of the films dissolved in three other acidic solutions. Chitosan films made with malic acid had the lowest WVP followed by the acetic acid, citric acid, malic acid and lactic acid.

The solution properties such as molecular weight of chitosan and radius of gyration were different from various acidic solutions and fairly related with the physical properties of films.

Muzzarelli et al. suggested that the TS of chitosan films increases as the increasing of molecular weight of chitosan[5]. The physical properties of chitosan in an aqueous solution differed slightly from that of chitosan films. In an acetic acid solution, chitosan molecules forms dimer from the light scattering results that the intermolecular interaction is relatively strong. Therefore, the tensile strength of chitosan films made in acetic acid is the highest among four acidic solutions. The tertiary structure of chitosan molecules in solutions is considerably related to the physical properties of film. The intermolecular arrangement of chitosan in an aqueous solution is influenced by the peculiarity (ion strength and the degree of dissociation) of each acid when films is formed.

Table 1. Mechanical properties and water vapor permeability of chitosan films prepared in various acidic solutions.

Acidic solution	Tensile strength (MPa)	Elongation (%)	Water vapor permeability (ng.m/m^2.sPa)
Acetic acid	40.1^a±8.18	11.4^c± 3.88	1.2^b±0.08
Citric acid	4.0^c±0.82	137.0^a±32.34	1.2^b±0.15
Lactic acid	2.9^c±0.82	88.2^b±23.62	1.9^a±0.33
Malic acid	20.0^b±6.13	20.7^c± 4.77	1.0^b±0.08

Each value is the mean of three replicates with the standard deviation. Means in the same column with diferent letter superscripts are significantly different by Duncan's multiple range test ($\alpha > 0.05$).

REFERENCES

1. P.A. Sandfordin, Skjak-Braek, Tanthonsen, and Sandford, (Ed.) Chitin / Chitosan: Sources, Chemistry, Biochemistry, Physical Properties, and Applications, Elsevier, Amsterdam, p.51 (1990).
2. O. Skaugrud, Manuf. Chem., Oct., p. 33 (1989).
3. E.L. Jonson and Q.P., Peniston, Utilization of shellfish waste for chitin and chitosan production. Flick, C.E. Hebard, and Ward, D.R. (Ed.) AVI Publishing Co., Westport, CT, p.415 (1982).
4. J.K. Robert, J. Polymer Science, 19 (1981) 1081.
5. R.A.A. Muzzarelli, Chitin. Pergamon press, Oxford, p. 69 (1977).
6. K. Toei, and T. Kohata, Anal. Chem. Acta, 83 (1976) 59.
7. R. H. Chen and H. D. Hwa, Carbohydrate Polymers, 29 (1996) 353.
8. H. J. Park, M.S. Chinnan, J. Food Engineering, 25 (1995) 497.
9. Y. Sano, J. Colloid Interface Sci., 124 (1988) 403.
10. V. Lee, Solution and shear properties of chitin and chitosan. Univ. Microfilms, Ann Arbor 74. 29 (1976) 446.

HYDROCOLLOIDS – PART 1
Edited by K. Nishinari
2000 Elsevier Science B.V.

205

Elsinan, a potential food hydrocolloid produced by *elsinoe* species; properties and enzymatic degradation

A. Misaki[a], Y. Tsumuraya[b] and M. Kakuta[c]

[a] Osaka City University, Sumiyoshi, Osaka 558, [b]Saitama University, Urawa, Saitama 338, and [c]Konan Women's University, Kobe 668, Japan

Elsinan, an extracellular α–glucan produced by *Elsinoe leuicospila* and other *Elsinoe* species, i.e., *E. fawcetti* and *E. ampelina*, is an essentially linear polymer (M_w, 2 - 6 x 10^5) consisting of maltotriose and maltotetraose units (ratio, 2:1) joined by α-1,3-linkage in the sequence : $\rightarrow$3)G(α1,4)G(α1,4)G(1$\rightarrow$ and $\rightarrow$3)G(α1,4)G(α1,4)G(α1,4)G(1$\rightarrow$. The X-ray diffraction analysis indicated that elsinan molecule has a five-hold helical conformation. The rheological properties of elsinan conferred by its unique structure, *e.g.*, high viscosity of its aqueous. solution, stable over wide range of pH (3 -11), temperature (30 - 70°), and salt concentrations, and the ability to form a strong, resilient and oxygen-impervious film, may provide this glucan useful applications for foodstuffs and pharmaceutical ingredients. It was also noted that elsinan exhibited interesting dietary-fiber effect,., *i.e*, significant reduction of the serum cholesterol level in hypercholesterolemic rats, probably due to its high hydrocolloidal property and the partial digestibility. The actions of α-amylases of. human salivary, pancreas and *B. subtilis* on elsinan afforded quantitative release of elsinotriose [G(α1,3)G(α1,4)G], whereas *Asp. oryzae* α-amylase (Taka amylase) cleaves specifically the 1,4-linkage in the maltotetraose units, but not maltotriose units, resulting in release of G(α,3)G(α1,4)G(α1,4)G and a series of higher oligosaccharides, *i.e.*,[G]7, [G]10, [G]13 and [G]16, containing 1,3-linkages in internal and at the non-reducing end.
In chemical derivatization of elsinan, partial introduction of 3,6-anhydro ring into the O-4 substituted glucose residues was achieved by sulfation and subsequent desulfation by alkali treatment. The 3.6-anhydro elsinan (DS,0.38) was resistant to amylolytic enzyme action, but susceptible to mild acid hydrolysis.

1. INTRODUCTION

Many microorganisms are known to produce considerable amounts of extracellular polysaccharides during growth in appropriate cultural conditions. Several of these microbial polysaccharides, either neutral or acidic polymers, such as pullulan, curdlan and other fungal β-1,3/1,6 glucans , and acidic heteroglycans, *e.g.* xanthan and gellan have been successively utilized in the food and pharmaceutical industries, because of their unique functional properties. Other interesting microbial polysaccharides, not in commercial development, may include succinyl glycan, microbial alginic acid, glucuronoxylomannan of *Tremella fuciformis* rhamsan and wellan (by Kelco Inc.). Beijeran, a unique acidic polysaccharide synthesized by *Azotobacter beijerinckii* has also recently been   discovered [1].

Elsinan, a unique fungal glucan, was first isolated from a slime when the conidia of *Elsinoe leucospila* , a fungus responsible for the white scab of tea leaves was cultured on sucrose-potato extract agar [2]. The chemical investigation suggested that it is neither starch nor dextran, but rather a new type of α–D-glucan containing 1,4- and 1,3-glucosidic linkages. It was named elsinan, and was elaborated by other *Elsinoe* species, e.g. *E. fwacetti* . These fungi appeared to produce considerable amount of elsinan when grown in an appropriate medium (28-30 g / L from 5% sucrose), and was shown to possess interesting properties as a food hydrocolloid.

In this symposium, we report production of elsinan, its structure and rheological properties, chemical derivatization and also distinct α-amylolytic degradation to produce novel oligosaccharides.

2. PRODUCTION

For production of elsinan, *E.leucospila* CS-1, originally isolated from the white scar spot on tea leaves, or other stocked strains, is cultured in a liquid medium containing 3-5% sucrose or glucose as a carbon source, 0.5% potato extract or corn steep liquor, 0.2% $NaNO_3$, 0.1% K_2HPO_4,0.05% $MgSO_4$, 0,05% KCl;pH adjusted to 6.8. The fermentation was carried out in a 500 ml flask (100-120 ml medium) or a 10L jar fermenter, at 24-26° for 5-7 days. After removal of mycelial cells elsinan was precipitaed from the clarified cultural fluid by addition of acetone or ethanol (2 volumes). Purification was effected by dialysis, followed by repeated ethanol precipitation, or by size exclusion chromatography and then lyophilized; yield 23-25 g, and 15g per L, from 5% sucrose and glucose medium, respectively. Frucose and maltose were also good carbon source, but galactose was not utilized for production of elsinan [3].We examined the productivity of other fungi, belonging to *Elisinoe* species, such as *E. fawcetti* produces elsinan, essentially same glucan of *E. leucospila* in a high yield (Table I). In addition to elsinan these fungi also elaborate an antitumor-active, cell wall *O*-6 branched β–1,3 glucan, and the cell surface galactomannan, which contains β–galactofuranosyl side chains [4].

Table 1 Polysaccharides elaborated by some *Elsinoe* fungi

	Host plant	Extracellular polysaccharide	Yield(g/dL)*	Cell wall
E. leucospila	tea leaves	elsinan	16	Galfman[a) , β-1,6/1,3 glucan
E. fawcetti	citrus	elsinan	15	Galfman, β-1,6/1,3 glucan
E. araliae	grape	β-1,6/1,3 glucan	0. 6	
		Gal, Glc, Man (1:1:1)	0.3	

*Shake flask culture at 26 °C, 5 days in a medium containing 3% glucose, 1% $NaNO_3$, 0.5% yeast extract .
a) Galactomannan containing galactofuranose residues.

3. STRUCTURE AND CHEMICAL PROPERTIES

The purified elsinan obtained as white powder is water soluble polysaccharide, composed solely of D-glucose. It showed a symmetrical sharp peak on ultracentrifugal analysis ($S20w$, 5.927 x 10^{-13}) and had M_w 2 - 6 x 10^5 (gel filtration on a sepharose 4B column). The native elsinan showed a high optical rotation $[\alpha]$ + 243 ° (c 0.8 in water) and an absorption at 840 cm^{-1} in the *IR* spectrum, indicating that it is an α−D-glucan [5].

Elsinan is soluble in warm water forming a viscous solution (intrinsic viscosity, $[\eta]$ 1.86 dl/g at 25°C), and at concentration above 5% it tends to form a gel. Some rheological properties will be described in later in this article.

The general structural features of elsinan was elucidated by precise chemical methods, involving methylation and fragmentation analysis, *i.e.*, Smith degradation, mild acid hydrolysis and also acetolysis[5]. The methylated elsinan yielded on acid hydrolysis 2,3,6-tri- and 2,4,6-tri-*O*-methyl glucose in a molar ratio of 2.6: 1.0. No appreciable tetramethyl sugar was detected, confirming that this glucan is an essentially unbranched chain consisting of α−1,4 and 1,3-glucosidic linkages (ratio, 5:2). The mild Smith degradation, involving periodate oxidation, borohydride reduction and mild hydrolysis with 0.1 M acid at 25°, 12 hr, followed by g.lc analysis of the hydrolysis products, gave 2-0-α-glucosyl erythritol and erythritol (ratio, 1.44:1), confirming that 1,3-linked glucose residue is flanked by 1,4-linked glucose residues [2]. This structure is supported by the ^{13}C-n.m.r. spectrum (Fig. 1).

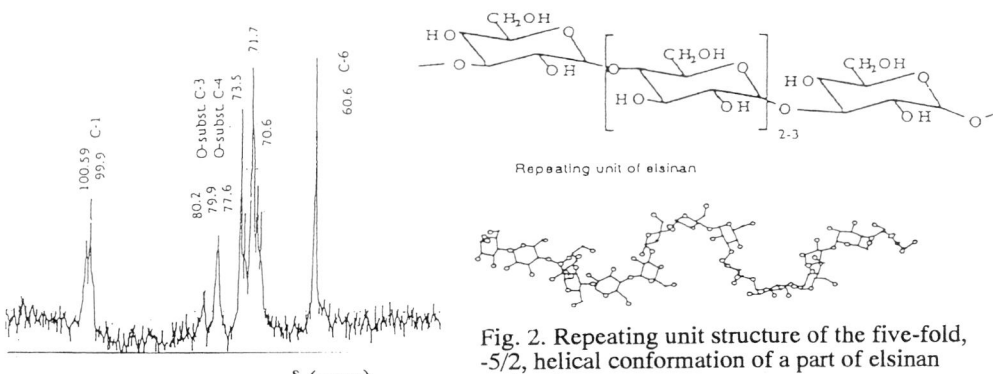

Repeating unit of elsinan

Fig. 1. ^{13}C-N.m.r spectrum of elsinan in Me₂SO-d6

δ (ppm)

Fig. 2. Repeating unit structure of the five-fold, -5/2, helical conformation of a part of elsinan molecule

Sequential analysis by partial acid hydrolysis (0.5M sulfuric acid, at 85°, for 4 hrs), and isolation of various oligosaccharides, including maltose, nigerose, 3-*0*-α−glucosylmaltose, 4-*0*-α−glucosyl nigerose, and maltotetraose, indicates that the elsinan molecule is composed of 3-*0*-α−glucosylmaltose and 3-*0*-α−glucosylmaltotriose units.

The above experimental data affords a possible repeating unit of the elsinan molecule, as depicted in Fig. 2. X-ray diffraction study suggested that the elsinan chain (poly[1,3-α−maltotriose] adopts a five-fold helical conformation with energetically possible, left handed, -5/1 or -5/2 helices [6]. More detailed sequential analysis of the elsinan molecule was accomplished by use of two different amylase actions, as discussed later.

4. Physical and rheological properties

Because elsinan is an essentially unbranched, water soluble α–glucan consisting of regular arrangements of α–1,4-, and 1,3-glucosidic linkages (Fig. 2), it exhibits interesting physical properties[3], such as high viscosity at a low concentration, gelation at a high concentration, and forming resilient, oxygen-impervious films, useful application to food processing.

4.1 Viscosity

Elsinan is soluble in water by warming to give a viscous solution at low concentrations. It showed a higher viscosity than the structurally related polysaccharide such as pullulan[a]. For instance, 2% aqueous solution of elsinan showed 100 cp at 30°C, a higher viscosity than that of pullulan. Under the same condition the 2% solution of pullulan sample (Mw, 2 x 10^5) gives approximately 10 cp (from technical data by Hayashibara Biochem. Inc). At concentrations above 5% the elsinan samples became a gel, as have been observed with some other neutral polysaccharides, *e.g.*, tamarind xyloglucan, guran and curdlan. The elsinan solutions showed interesting viscosity properties, as shown in Fig.3-1, and Fig. 3-2. The viscosity of the aqueous solution of elsinan (2%) was stable at a wide range of pH, 3 to 11 (Fig. 3-1), and the viscosity was not affected by added salt concentrations, even with 30% sodium chloride (Fig. 3-2). These properties are probably due to the non-ionic nature of the elsinan molecule.

The elsinan solutions showed highly pseudoplastic; the viscosity rapidly decreased with increasing the shear rate (Fig.3-3). As regards the effect of temperature on the viscosity of elsinan solution, when a 2% sample was dispersed at 30°C and gradually heated to 80°C, 5°C/min, the viscosity increased on heating up to 60°C and gave a peak at 65°C, then rapidly decreased with rising temperature. However, the pre-heated 2% elsinan, at 80° for 30 min, showed a constant viscosity without a peak., as shown in Fig. 3-4. These viscosity characteristics must be due, at least, to irreversible changes in the inter-molecular association of the helical elsinan chains.

4-2. Film formation

Because it is a linear molecule containing the regular arrangement of D-glucose residues elsinan is capable of forming a strong and resilient film by evaporation of the aqueous solutions. Interesting physical properties of elsinan films are listed in Table 2.

The film is resilient and impervious to oxygen, like pullulan films. For instance, when oleic acid or other long chain fatty acids are enclosed in the elsinan film, and kept over three months at room temperate, no coloration was observed. Also, when fresh sardines were coated with the elsinan film, or dipped in a dilute aqueous elsinan solution, and then air-dried, no coloration due to auto-oxidation was observed over a period of 4 months.

In addition, the elsinan film, unlike pullulan film, is resistant to dilute acids (pH 1-4), but it is partially digestible with human pancreatic α-amylase. These characteristics of the films suggest that they may be applicable for use in the food industry, as an edible packing film for fresh foods.

---

a)A. Misaki, *Food Hydrocolloids*, ed. K. Nishinari, p.17, 1991

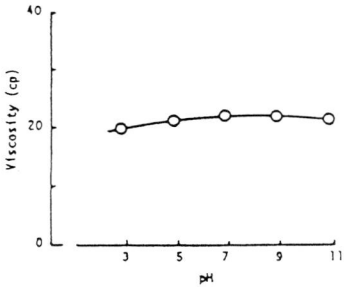

Fig. 3-1. Effect of pH on the viscosity of 2% aqueous solution.

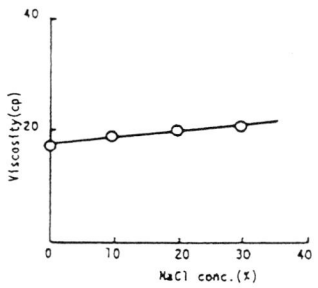

Fig. 3-2. Effect of Ionic strength (NaCl)

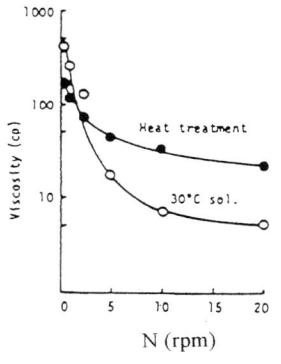

Fig. 3-3. Effect of pre-heat treatment on pseudo-plasticity of elsinan samples, at 2%. ●: dispersed at 100°C. ○: dispersed 30°C without pre-heat treatment. The viscosity was measured at 30°C.

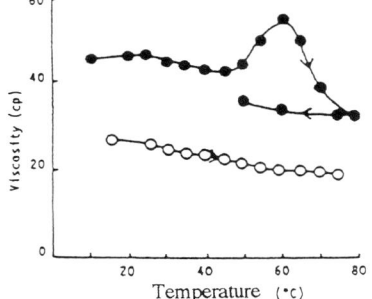

Fig, 3-4. Effect of temperature on the viscosity, measured at 10°C to 80°C (5°/min).
●:dispersed at 30°C , and ○: pre-heated at 80°C for 30 min.

Table 2. Properties of transparent film

Transparency	excellent	Film stability	stable at pH 1-4, 25 °C
Oxygen permeability	1.0 ml/m ^{2} atm	Hydroscopicity	11.1% (at RH 33%)
Bending strength	973 times		15.2% (at RH 65%)
Tensile strength	950 kg/cm ^{2}		19.5% (at RH 90%)
Heat-sealing property	excellent		

5. ALPHA-AMYLOLYTIC DEGRADATION OF ELSINAN AND ELUCIDATION OF FINE STRUCTURAL FEATURES

As already described elsinan exhibits physical properties which may be useful for food processing, such as he viscosity stability of its aqueous preparation over a wide range of pH (3-11) at low concentrations, temperature (30-70°), salt concentrations, and its acid -resistant and oxygen-impervious film. In addition, elsinan showed partial digestibility with a remarkable dietary-fiber effect, such as the capability to reduce the serum cholesterol level of hypercholesterolemic rats (A. Misaki, M. Kakuta and H. Okamatsu, *Abstr. paper Annual meeting of Jap. Food and Nutrition*, 1983. p.35).

Our preliminary study indicated that the susceptibility of elsinan to salivary-type amylases and fungal amylase, especially *Asp. oryzae* amylase (Taka-amylase) differed considerably. This prompted us to compare the action pattern of various amylolytic enzymes, and to isolate novel oligosaccharides.

Fig.4 shows comparison of the hydrolysis course of 0.5% elsinan and soluble starch by purified human salivary α-amylase, at 37°, pH 6.8 . The apparent hydrolysis of elsinan and starch was 29%, and 44% (as glucose). After complete degradation, the degradation product was isolated by connected columns of Bio-gel P-4 and P-2, which gave glucose and a novel trisaccharide, designated elsinotriose 1, O-α–glucosyl(1→3)-O-α-glucosyl(1→4)glucose, or 3-glucosylmaltose (Fig.5), as revealed by methylation analysis and the enzymatic hydrolysis [7,8]. In the hydrolysis process we also isolated an intermediate tetrasaccharide, 3-maltosyl-maltose 2, which is further hydrolyzed to glucose and 1 . Table 3 shows that salivary, pancreatic and bacterial saccharifying α–amylase produce essentially the same product, 1, almost quantitatively. Another experiment carried out in connection with the lowering of cholesterol in the rat serum, revealed that elsinotriose could slowly be hydrolyzed (55% as glucose) by the purified rat intestinal α–glucosidase [9].

Interestingly, Taka-amylase requires longer (1→4)-linked glucose segments, to cleave 1,4-linkages in the maltotetraose units, especially the site of the glucose joined to O-3 of the adjacent glucose unit; but not maltotriose units (cf. Fig.5). Thus, it produces a series of novel oligosaccharides, *i.e.*, DP4, 7, 10, 13, 16 (molar ratio, 2.78:1.0:0.61:0.26:0.11) etc (Table 3). These Taka amylase-released oligosaccharides were fractionated by gel-filtration chromatography, and on a thick paper chromatogram, and the structure of each oligosaccharide was elucidated [8].Thus, the tetrasaccharide (**TG4**), characterized as 0-α-glucosyl(1→3)-O-α-glucosyl(1→4)-O-α-glucosyl (1→4) glucose, or 3-α-glucosyl-maltotriose, should have arisen from single maltotriose residue franked by maltotetraose units.**TG7** must be from segments

Table 3. Comparison of the hydrolysis products by the actin of several α-amylases

α-Amylase	Hydrolysis (% as glucose)	Molar, proportion (%)*							
		G_1	G_2	G_3	G_4	G_7	G_{10}	G_{13}	G_{16}
Human salivary	29	29.7	0.7	62.2	7.4	—	—	—	—
Hog pancreatic	33	38.5	0.5	70.0	tr*	—	—	—	—
B. subtilis (saccharifying)	35	19.5	—	81.5	—	—	—	—	—
A. oryzae (Taka-amylase)	11	—	—	—	58.4	21.0	12.8	5.5	2.3

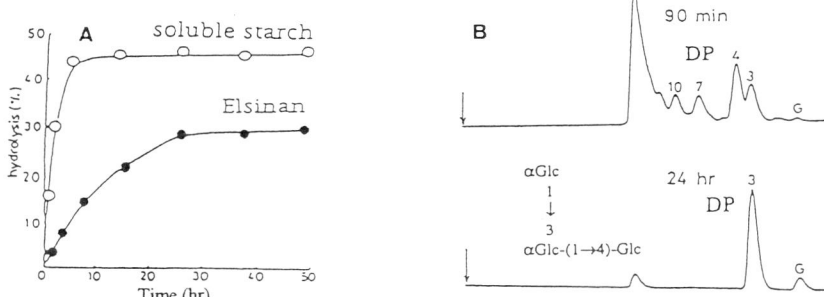

Fig. 4.-A. Hydrolysis of elsinan and soluble starch with human salivary α-amylase (pH 6.8, at 37°C, and B. The time course of degradation of elsinan, as analyzed by HPLC (column, LiChrosorb, acetonitrile: water, 70:30).

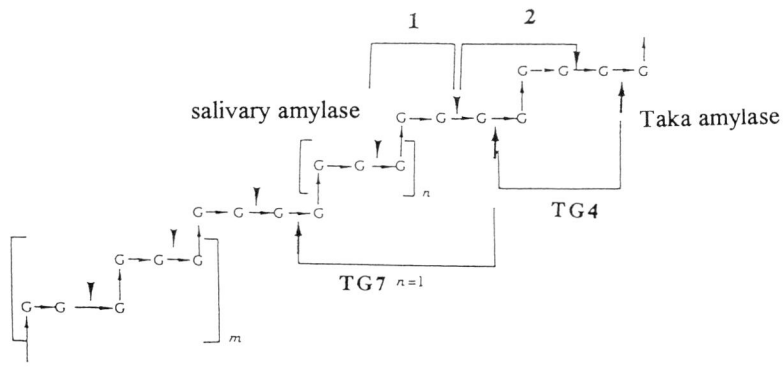

→ G-, (1→4)-linkage; G;(1→3)-linkage

1. 3-α-glucosyl maltose, 2. 3-maltosyl maltose
n=0 TG4, n=1 TG7, n=2 TG10, n=3 TG7 , n=4 TG16

Fig. 5. Distribution of maltotriose and -tetraose units in elsinan, and cleavage sites of the action with salivary amylase and Taka amylase.

from single maltotriose residue franked by maltotetraose units. **TG7** must be from segments containing two consecutive maltotriose units. Similarly **TG10**, **TG13**, and **TG16** should originate, respectively, from segments of two, three and four consecutive maltotriose units, flanked by maltotetraose units.

The aforementioned results strongly indicate that the sequence of maltotriose and maltotetraose units is not always in the regular arrangement, but rather as random distribution along the elsinan chain. It was interesting that a high molecular weight segment consists entirely of maltotriose residues can also be recovered in 33% yield. The heterogeneity of sequence in the elsinan chain may provide a clue for the mechanism of the chain elongation.

Thus, elsinan may provide a useful substrate examining the specificities of new amylolytic enzymes. This unique α−glucan may also be a useful source for the supply of a series of functional oligosaccharides

5. CHEMICAL DERIVATIZATION OF ELSINAN

Chemical modification and derivertization of food polysaccharides, *e.g.*, carboxymethyl(CM)-, hydroxyl-, diethylaminoethyl (DEAE) derivatives etc., have been developed to meet new applications. In chemical derivatization of elsinan we also prepared CM- and DEAE-derivatives for improvment of the rheological properties.

However, our interest was drawn to the partial introduction of 3,6-anhydro-linkages to the 1,4-linked D-glucose residues of elsinan by a new chemical strategy, since 3,6-ahydro-L-galctose rings in seaweed polysaccharides would be introduced by enzymatic sulfation of O-6-OH followed by a subsequent desulfation to result in simultaneous anhydro linkage formation.

The method adopted here for introduction of the 3,6-anhydro linkage into elsinan involved sulfation of the OH groups by the reaction with DMSO-SO$_3$ complex, and subsequent alkali treatment of the partial sulfated elsinan with 2M NaOH at 80°C under the reduced condition [10], forming the anhydro-linkage between O-3 and O-6 position of the 1,4-linked glucose residues. After neutralization and dialysis, the resulting anhydro-elsinan was lyophilized (yield 70%, degree of substitution, D.S, 0.38). For comparison, pullulan (M$_w$ 70,000) and a low molecular amylose (M$_w$ 5,000), both containing α-1,4-linked glucose residues, were also treated by the same procedure, as shown in Fig.6. The yields, D.S, and [η] of the anhydro derivatives of elsinan, pullulan and amylose are compared in Table 4. The resulting anhydro glucans gave low intrinsic viscosity, but since depolymerization during derivertization did not occur the viscosity change must be attributed to the conformation change of the 3,6-anhydro glucose residues, C1 to IC form. The anhydro elsinan showed, unlike the original elsinan, resistance toward the action of salivary α−amylase.

The glycosidic linkages of the 3,6-anhydro D-glucose residues (AG) with 1C conformation should be readily hydrolyzed with dilute acid. Therefore, it is possible to obtain various novel oligosaccharides, which contain 3,6-anhydro sugar residue at the reducing end, by mild acid treatment.

Table 5 summarizes the yields and structures of the oligosaccharides isolated from mild hydrolysis of 3,6-anhydro-derivatives of elsinan, pullulan and amylose. These oligo-saccharides were purified by chromatography on a Sephadex G-10 column (2.5 x 100 cm), which were identified as G(α1,4)G(α1,4)AG and G(α1,3)G(α1,4)AG etc.

The partial introduction of 3,6-anhydro rings into the O-4 substituted glucose residues in the glucans could be achieved by sulfation and subsequent desulfation by alkali treatment. Thus,

3,6-anhydro α–glucans, like 3,6-anhydro elsinan, may provide a useful source for obtaining novel oligosaccharides containing 3,6-anhydro glucosyl ends.

Fig.6. Procedure for chemical synthesis of 3,6-anhydro-elsinan

Table 4. Properties of sulfated and 3,6-anhydro-α-D-glucan derivatives of elsinan, pullulan and amylose

Derivative	Elsinan	Pullulan	Amylose
Sulfated glucan			
$[\alpha]^D$ (water, 25°)	+126.3	+108.6	+114.5
Sulfate content (%)	8.39	8.82	12.13
D.S.	0.61	0.62	1.0
3,6-Anhydro glucan			
$[\alpha]^D$ (water,25°)	+71.7	+70.7	+43.4
D.S.	0.38	0.35	0.67
M_w (parent glucan)	2×10^5	7×10^4	5×10^3
M_w (anhydro-glucan)	2×10^5	7×10^4	5×10^3
$[\eta]$ (dl/g)[a]	0.029	0.046	n.d[b].

a) measured with Ostwald type viscometer. b) not determined.

Table 5. Oligosaccharides isolated from anhydro-α-D-glucans
by mild acid hydrolysis

Oligosaccharide	Yield(%)	$[\alpha]^D_{25}$(c,1.0, water)
3,6-anhydro elsinan		
Gα(1→4)AG	14.4	+60.5
Gα(1→4) Gα(1→4)AG	10.0	+43.5
Gα(1→4) Gα(1→4) Gα(1→4)AG	13.1	+52.3
Gα(1→4) Gα(1→4) Gα(1→4) Gα(1→4)AG	44.4	+30.9
3,6-anhydro pullulan		
Gα(1→4)AG	17.1	+60.1
Gα(1→4) Gα(1→4)AG	6.5	+43.5
Gα(1→6) Gα(1→4) Gα(1→4)AG	44	+35.4
3,6-anhydro amylose		
Gα(1→4)AG	10.5	+60.5
Gα(1→4) Gα(1→4)AG	45.9	+43.5

AG: 3,6-anhydro-glucose, G: glucose

REFERENCES
1. A.Misaki, and Y. Ooiso, *Abstract. paper of Industrial Hydrocolloids Conference*, p.12, 1994, Ohio.
2. A. Misaki, Y. Tsumuraya and S. Takaya,*Agri.Biol..Chem..* **42** (1978) 491.
3. A. Misaki and Y.Tsumuraya, *Fungal polysaccharides* in ACS Symposium Ser. **126** (1980) 197.
4. N. Shirasugi and A. Misaki, *Biosci. Biotech. Biochem*, **56** (1992) 29.
5. Y. Tsumuraya, and A. Misaki, *Carbohydr. Res.*, **66** (1979) 53.
6. K.Ogawa, T.Yui, K. Okamura and A. Misaki, *Biosci. Biotech. Biochem..* **57**, (1993) 1338.
7.Y. Tsumuraya, and A. Misaki, *J. Appl. Biochem.*, **1** (1979) 235.
8. A. Misaki, H. Nishio, Y. Tsumuraya, *Carbohydr. Res.*, **109**(1982) 207.
9. Y. Sone, T. Sato, M.Yano, H.Kaku and A. Misaki, *Trace Nutrition Res*, **3**, (1986) 119.
10. Y. Ohe, K. Ohtani, Y.Sone and A. Misaki, *Biosci. Biotech. Biochem.*, **57** (1993) 227.

HYDROCOLLOIDS – PART 1
Edited by K. Nishinari
2000 Elsevier Science B.V.

215

Effects of alkali metal salts on the viscoelasticity of funoran and λ-carrageenan

M.Watase[a], T.Aihara[a], K.Nishinari[b]

[a] Faculty of Agriculture, Shizuoka University, Ohya, Shizuoka 422, Japan
[b] Faculty of Human Life Science, Osaka City University, Sumiyoshi, Osaka 558-8585, Japan

Dynamic viscoelastic measurements and differential scanning calorimetry (DSC) were carried out for the aqueous solutions of λ-carrageenan and funoran with and without alkali metal salts. DSC endothermic peak temperature of 2wt% λ-carrageenan solution remarkably shifted to higher temperatures by the addition of alkali metal salts with comparison to that in funoran solution. Both storage and loss shear moduli G' and G"of 2wt% solutions of λ-carrageenan and funoran were strongly frequency dependent at 25℃, typical of a behaviour of dilute polymer solutions. The addition of alkali metal salts increased G' and made G' less frequency dependent. This tendency was more conspicuous in λ-carrageenan than in funoran. The difference in the effect of alkali metal salts was attributed to the different modes of linkage of sulfate groups: a sulfate group is axially linked to C_6 in λ-carrageenan, whilst it is equatorially linked to C_6 in funoran.

1. INTRODUCTION

Both carrageenan and funoran are polysaccharides extracted from red seaweeds. Carrageenans are extracted from *Eucheuma cottonii*, and consist of D-galactose, 3,6-anhydro-D-galactose, and sulfte esters, and have been used widely in dairy products, sausages and hams because (1)they react with protein, (2)their gelling ability increases remarkably by the addition of monovalent or divalent cations, (3)they interact synergistically with locust bean gum [1]. The gelling ability of carrageenans is enhanced by removing sulfate groups and by increasing 3,6-anhydro-D-galactose.

Funoran is extracted from *Gloiopeltis furcata* and the structure is similar to that of λ-carrageenan, i.e. it consists of D-galactose and 3,6-anhydro-L-galactose, and some sulfate esters. A sulfate group is attached to C_6 axially in λ-carrageenan, whilst it is attached equatorially in funoran. Effects of alkali metal ions on the gel-sol transition of λ-carrageenan and funoran were studied by dynamic viscoelasticity and DSC measurements in the present work. Specimens of λ-carrageenan and funoran with approximately the same content of sulfate group and 3,6-anhydro-L-galactose were used in the present work.

2. EXPERIMENTAL

2.1. Materials

λ-carrageenan was extracted from *Eucheuma spinosum* by filtration using a Buchner. Funoran was extracted from *Gloiopeltis furcata* using autoclave at 110℃. The content of 3,6-anhydro-L-galactose was determined as 12.8% for λ-carrageenan and 13.2% for funoran by FTIR using polyvinylpirrolidon as a reference. Sulfur content was determined as 9.6% for λ-carrageenan and 9.2% for funoran by elemental analysis.

2.2. Measurements

IR measurements were carried out to determine the linkage mode of the sulfate group at C_6. The absorption band at 1250 cm^{-1} is proportional to the content of total sulfate groups. The main absorption band appeared at 850 cm^{-1} in λ-carrageenan, and at 820 cm^{-1} in funoran respectively. The sulfate group in λ-carrageenan is attached axially whilst that in funoran is attached equatorially [1] .

The apparatus used for dynamic viscoelastic measurements was MR-3 from Rheology Co.Ltd.(Kyoto). Frequency was changed from 0.01 to 2 Hz. The temperature was fixed at 25℃.

Differential scanning calorimetry (DSC) was carried out using a DSC 120 from Seiko Electronics Co.Ltd. (Tokyo). A silver pan of 70 μ l was used, and 45 mg sample solution was hermetically sealed in it. Distilled water of the same quantity was used as a reference. The temperature was raised at 2 ℃/min.

3. RESULTS AND DISCUSSION

3.1. Mechanical spectra

Storage shear modulus G', loss shear modulus G'', and loss tangent of aqueous solutions of 2wt% λ-carrageenan and 2wt% funoran without salt at 25℃ as a function of frequency are shown in Fig.1. Both moduli of a 2wt% λ-carrageenan solution were strongly frequency dependent, and increased with increasing frequency, and G'' is always larger than G': the behaviour of these solutions is typical of a dilute polymer solution (Fig.1(a)). In contrast, both moduli were less frequency dependent in a 2wt% funoran solution (Fig.1(b)). The difference should be attributed to the linkage mode of sulfate groups. Since the sulfate groups are attached equatorially in funoran, they donot inhibit the helix formation and their aggregation, and the rheological behaviour is not influenced so much by the addition of potassium ions.

The addition of LiCl and NaCl did not change this tendency so much, whilst the addition of KCl and CsCl increased both moduli and made them less frequency dependent especially in λ-carrageenan, which should be attributed to the screening of electrostatic repulsion of sulfate groups by potassium and cesium ions (Fig.2(a)). The addition of 0.02 M alkali metal salts KCl or CsCl increased only slightly both moduli of funoran solutions (Fig.2(b)), and the

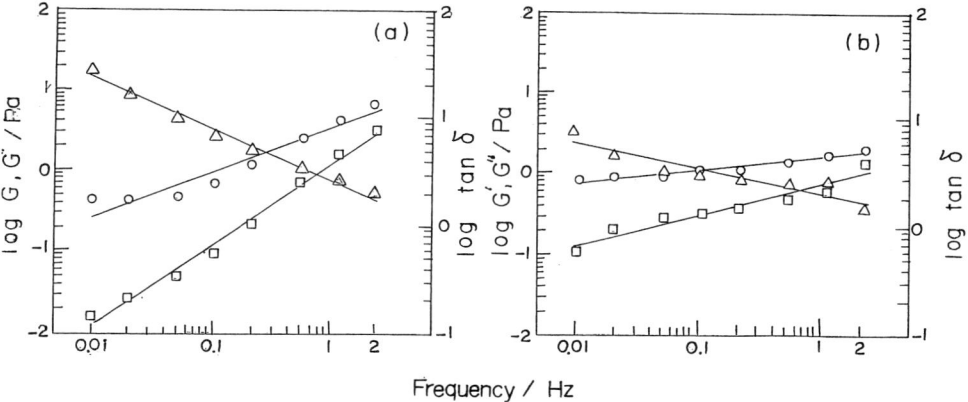

Fig.1   Frequency dependence of storage shear modulus G' (□),
loss shear modulus G" (○) and tan δ (△) of 2wt%
solution of λ-carrageenan (a) and funoran (b) at 25℃.

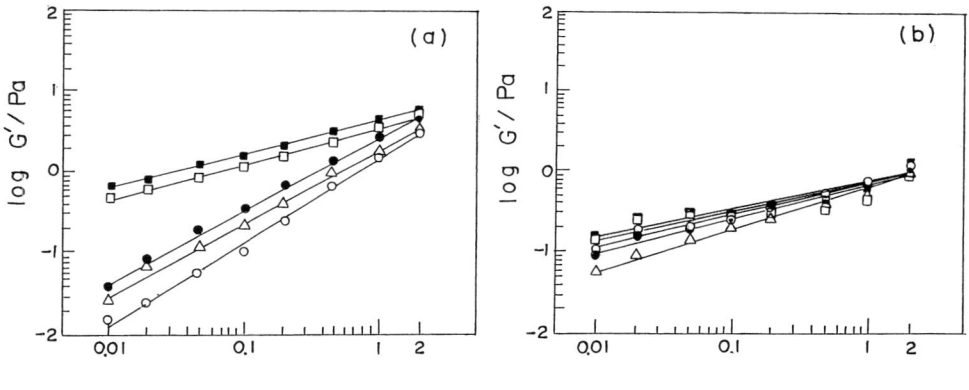

Fig.2   Frequency dependence of storage shear modulus G' of 2wt%
λ-carrageenan (a)and funoran (b) without salt (○) and
in the presence of 0.02M LiCl (△), . 0.02M NaCl (●),
.0.02M KCl (□), . 0.02M CsCl (■) at 25℃.

addition of LiCl or NaCl decreased slightly the storage modulus.   As is well known [2] ,
lithium and sodium ions are structure ordering ions, whilst potassium and cesium ions are

structure disordering ions.   Therefore, the former ions shield the electrostatic repulsion between sulfate groups more effectively than the latter ions, as has been observed before [3] . The reason why LiCl or NaCl decreased slightly the storage modulus of funoran gels, although both these salts shifted the endothermic peak temperature in DSC curves to higher temperatures as mentioned below, is not clear at present.

3.2. Differential scanning calorimetry

Fig.3 shows DSC heating curves of solutions of λ-carrageenan (a) and funoran (b) of various concentrations.   An endothermic peak became sharper and shifted remarkably to higher temperatures with increasing concentration of λ-carrageenan, whilst the shape of the endothermic peak for funoran solutions became broader and shifted slightly to higher temperatures with increasing concentration of funoran. The endothermic peak temperature T_m in heating DSC curves without salt was far higher for funoran than for λ-carrageenan because sulfate groups are attached equatorially in funoran whilst they are attached axially in λ-carrageenan.   Bulky sulfate groups attached axially inhibit the aggregation of helices than sulfate groups attached equatorially.

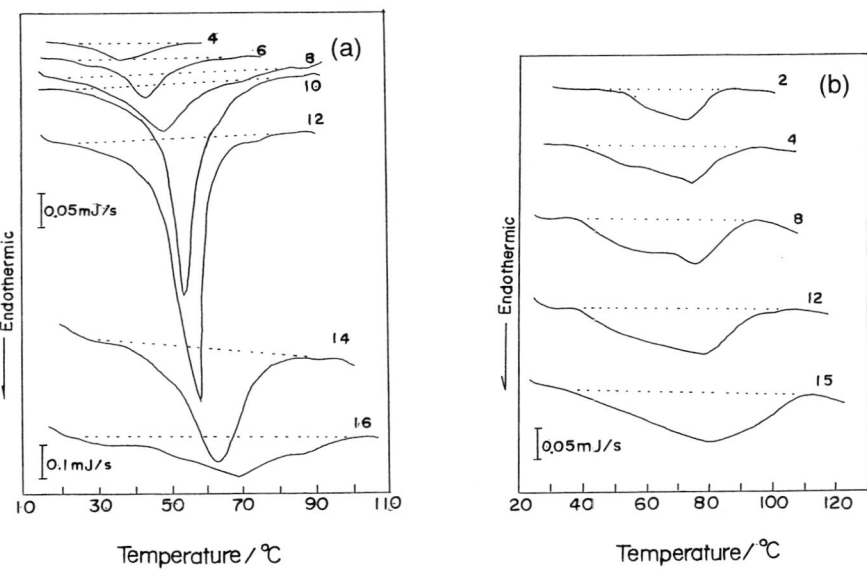

Fig.3   DSC heating curves of λ-carrageenan (a)and funoran (b)
of various concentrations.   Heating rate 2 ℃/min.
Numbers beside each curve represent the concentration
of the polysaccharides in wt%.

The endothermic peak temperature T_m in heating DSC curves shifted to higher temperatures remarkably by the addition of alkali metal salts in the order of LiCl < NaCl < KCl < CsCl for 2wt% λ-carrageenan and for 2wt% funoran gels as shown in Fig.4.

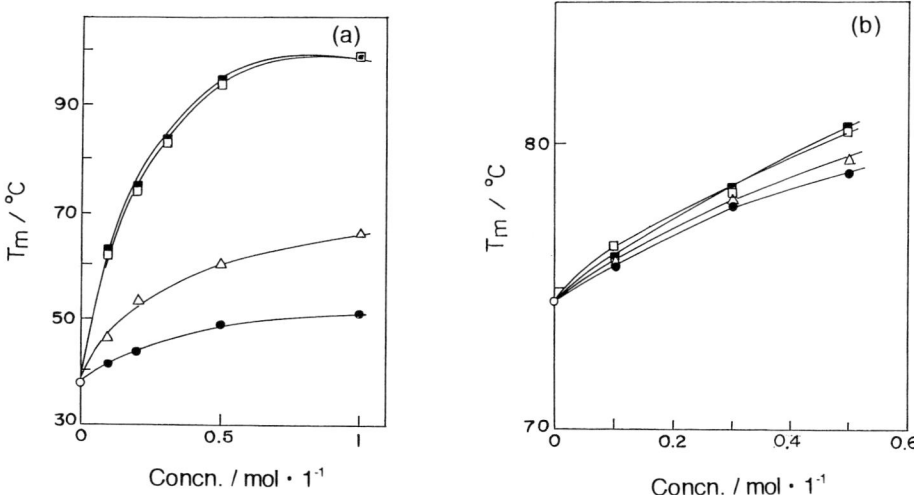

Fig.4   Endothermic peak temperature Tm of 2wt%
λ-carrageenan and 2wt% funoran solutions as a function
of the concentration of the added alkali metal salts.
LiCl ($\bullet$), NaCl ($\triangle$), KCl ($\square$), CsCl ($\blacksquare$).

The endothermic peak shifted remarkably to higher temperatures in λ-carrageenan with comparison to funoran.   This is again explained by the different linkage mode of sulfate groups in these two polysaccharides.   Although Tm shifted to higher temperatures by the addition of LiCl and NaCl as shown in Fig.4, the storage modulus decreased by the addition of these salts as shown in Fig.2.   It seems that the structure of junction zones is thermally stabilised by the addition of alkali metal salts, it does not necessarily mean the increase of the number of elastically active network chains.   This point should be examined in the near future.

Fig.5 (a) shows Eldridge-Ferry plot $\log C = \triangle H_m / 2.303 R T_m$ + const, where C is the polymer concentration , $\triangle H_m$ is the heat absorbed on forming a mole of junction zones that stabilise the network structure, R the gas constant, T_m the absolute temperature of the transition.   The value of $\triangle H_m$ determined from the slope of the straight line in the Eldridge-Ferry plot was 33.5kJ/mol and 135.2 kJ/mol for λ-carrageenan and funoran respectively.   This experimental finding that $\triangle H_m$ was far larger for funoran than for λ-carrageenan is in agreement with the above mentioned reason: since sulfate froups are attached equatorially in funoran and they are attached axially in λ-carrageenan, ordered structure made by the aggregation of helices are more stable in the former than in the latter.

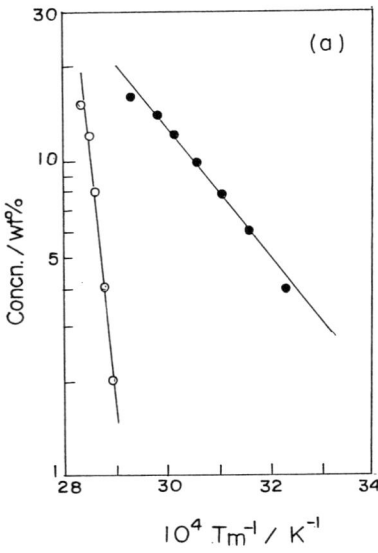

Fig.5   Eldridge - Ferry plot for λ-carrageenan ($\bullet$)and funoran ($\bigcirc$).

REFERENCES
1. D.A.Rees, in H.L.Kornberg and D.C.Phillips (eds.) Biochemistry of. Carbohydrates, Biochemistry Series One, Vol.5, pp.1-42, Butterworths, London, 1975.
2. H.Uedaira, in K.Nishinari and T.Yano (eds.) The Science of Food Hydrocolloids, pp.7-22, Asakura Shoten, Tokyo, 1990. (in Japanese)
3. M.Watase and K.Nishinari, Colloid & Polym. Sci., 260 (1982) 971.
4. M.Watase and K.Nishinari, Colloid & Polym. Sci., 263 (1985) 744.

HYDROCOLLOIDS – PART 1
Edited by K. Nishinari
2000 Elsevier Science B.V.

221

Texture and structure of high-pressure-frozen konjac

A. Teramoto and M. Fuchigami

Department of Nutritional Science, Faculty of Health and Welfare Science,
Okayama Prefectural University, 111 Kuboki, Soja, Okayama 719-1197, Japan

To determine the effect of high-pressure-freezing and thawing on the quality of konjac, konjac was frozen for 45 min at ca. -20°C at 100 MPa (ice I), 200 MPa (liquid phase), 340 MPa (ice III), 400, 500 or 600 MPa (ice V), 700 MPa (ice IV). Then they were thawed at atmospheric pressure (A: frozen 45 min; B: frozen 45 min then stored 1 day at -30°C; C: frozen 115 min) or thawed at the same pressure as high-pressure-freezing (D: frozen 45 min). Texture and structure were compared with frozen konjac (atmospheric pressure at 0.1 MPa, freezing at -20°C, -30°C or -80°C) and unfrozen konjac. Texture (stress-strain curves) and the structure of all frozen konjac (thawed at 0.1 MPa) differed greatly from the original gel; final rupture stress increased and strain decreased. Conversely, texture and structure of konjac frozen-then-thawed at 200 – 400 MPa were the same as the original gel. This suggested that phase transitions (ice VI → ice V → ice III → liquid → ice I) occurred either during reduction of pressure at -20°C, or during storage in a freezer. Thus, high-pressure-freezing-thawing at 200 – 400 MPa was effective in improving the quality of frozen konjac.

1. INTRODUCTION

With food gels of high-water content (e.g. konjac, tofu, agar), damage to structures through freezing is extensive and the texture after thawing becomes unacceptable. When water is frozen at atmospheric pressure (0.1 MPa), volume increases. This causes tissue damage during freezing at 0.1 MPa. However, under high pressure, several kinds of ices (ice II – IX) with different structures and physical properties are formed. These do not expand in volume during phase transition from water to ices [1][2]. If the high density property of high pressure ice is applied to frozen gels, damage of gels may be retained. In previous studies of tofu, results suggested that the damage of texture and structure were reduced by freezing at 200 MPa – 400 MPa [3][4]. The objective of the present work was to study the effect of high-pressure-freezing and high-pressure-thawing on the improvement in quality of frozen konjac (glucomannan gel made by the addition of calcium hydroxide). Although konjac was not frozen during pressurization at 200 MPa at -20°C (liquid phase), in this report the freezing method at 100 – 700 MPa at ca. -20°C is referred to as high-pressure-freezing.

2. MATERIALS & METHODS

2.1. Sample preparation and method of high-pressure-freezing

Konjac (sashimi-konjac, Soryo-konnyaku Co. Ltd., Hiroshima, Japan) was cut into disks 15 mm in diameter and 10 mm thick. The 8 pieces of konjac were vacuum packed in heat-sealed polyethylene bags, then placed into propylene glycol (pressure medium) in a pressure vessel (6 cm inside diameter and 20 cm high) previously kept at ca. -20°C by a circulation-type cooler. They were immediately pressurized at 100 – 700 MPa using a *Dr. Chef* high pressure food processor (Kobe Steel Ltd., Kobe)[3][4][5]. After pressurization for 45 min, samples were immersed in a pressure medium for 10 min to ensure freezing.

2.2. Methods of thawing

After high-pressure-freezing for 45 min, samples were thawed as follows (Figure 1):

A. pressure was reduced and konjac was thawed at 20 °C for 45 min in a low temperature incubator.

B. pressure was reduced and konjac was stored for 1 day in a freezer (-30°C) and then thawed at 20°C.

C. pressure was reduced and konjac was thawed at 20 °C. (High-pressure-frozen for 115 min; the same pressurization time as D).

D. to thaw, pressure vessel was heated by circulation-type heater (60°C) for about 70 min at the same pressure as high-pressure-freezing. When the temperature of the pressure medium reached 20°C, then pressure was reduced.

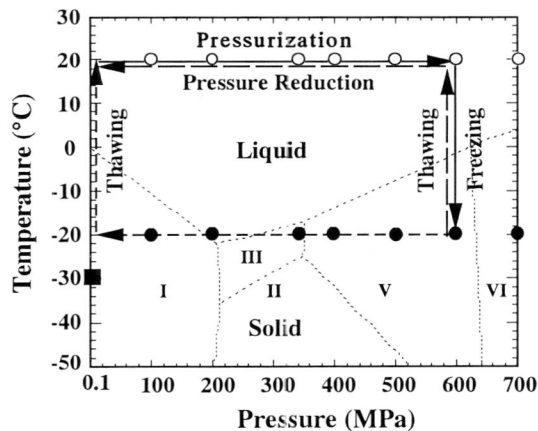

Figure 1. Phase diagram of freezing and thawing methods. ○ pressurization at 20°C, ● pressurization at -20°C, ■ storage at -30°C, — high-pressure-freezing, — — high-pressure-thawing, ---- atmospheric-pressure-thawing.

Texture and structure of high-pressure-frozen konjac (A – D) were compared with untreated konjac (control), boiled konjac (10 min or 20 min), unfrozen pressurized konjac

(20°C for 45 min), and frozen konjac (-20°C, -30°C or -80°C) frozen in freezers at atmospheric pressure for 1 day.

2.3. Texture measurement

Texture was measured by a creepmeter (Rheoner, RE-33005, Yamaden Co. Ltd., Tokyo). Thickness of samples was measured using a sample-height counter (HC-3305, Yamaden, Tokyo), then the konjac was punctured by using a plunger (cylindrical shape: 5 mm diameter, 22 mm long) at 0.5 mm / sec, and stopping at 99% of the thickness using a loadcell of 2 kg. Rupture stress and strain were indicated.

2.4. Structure measurement

The central part of konjac (untreated, boiled, pressurized at 20°C, frozen at atmospheric pressure and high-pressure-frozen) was cut into 6 mm × 1 mm × 1 mm, and dehydrated withfirst 20%, 40% and then 50% ethanol. Specimens were contained in metal holders and quickly frozen by immersing in LN_2. Frozen specimens were transferred to the cold stage of a cryo-system for scanning electron microscopy (S-4500, Hitachi Co. Ltd, Tokyo), and then cut with a knife (ca. -150°C). After etching at -85°C, the surface was observed at -105°C under low acceleration voltage (1 kV, magnification of ice pores: × 1000, gel network: × 30000)

3. RESULTS & DISCUSSION

3.1. High-pressure-frozen konjac thawed at atmospheric pressure

Pressurization at 20°C or boiling did not affect texture and structure of konjac (Figures 2, 4 and 6): a comparatively coarse gel network was observed in these konjac gels. However, texture (stress and strain) and structure of all frozen konjac thawed at atmospheric pressure differed greatly from the original gel due to syneresis and a volumetric shrinkage of gel (Figures 3, 5, 7 and 8). At 100 – 400 MPa and 700 MPa, rupture stress increased, and rupture strain decreased. In other words, the konjac lost elastic properties and became brittle. On the other hand, at 500 and 600 MPa, the rupture stress increased, but rupture strain did not change much (Figure 3).

Ice crystals were observed in all frozen konjac thawed at atmospheric pressure (Figures 5 and 7). Freezing temperature, pressure and time of storage did not affect the size of ice crystals greatly. A coarse gel network was not observed in all frozen-thawed konjac (Figures 5 and 8). It was supposed that the gel network was compressed by formation of ice crystals. These results suggest that ice I might be formed during reduction of pressure or during the storage of konjac in a freezer. When frozen konjac was thawed at atmospheric pressure, high-pressure-freezing was not effective at all compared to high-pressure-frozen tofu which was effective [3][4].

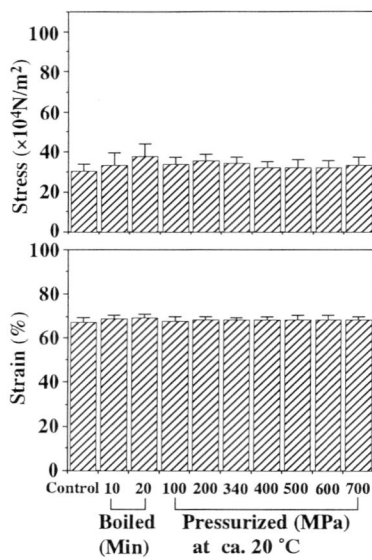

Figure 2. Effects of cooking and pressurization on rupture stress and strain of konjac.

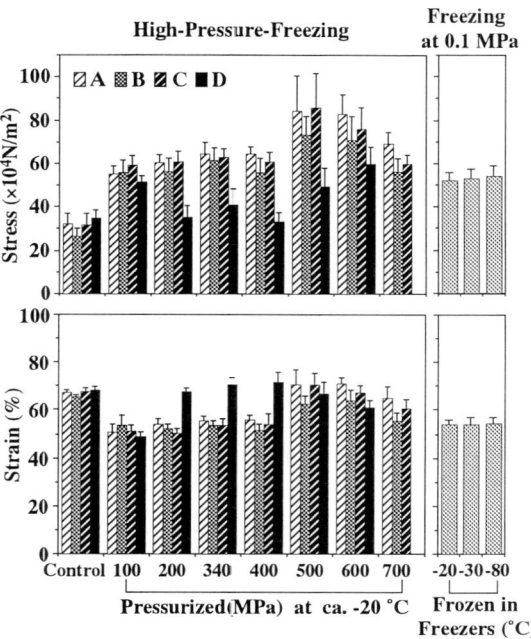

Figure 3. Effects of thawing methods on rupture stress and strain of high-pressure-frozen konjac. A, B, C and D: See "Methods of thawing".

3.2. High-pressure-frozen konjac thawed at high pressure

To determine phase transition of ices during reduction of pressure, high-pressure-frozen konjac was thawed at high pressure then pressure was reduced. Stress and strain of konjac frozen-then-thawed at 200 – 400 MPa were the same as the original gel (Figure 3). Ice crystals were not found (Figure 7, D) but a coarse gel network was observed (Figure 8, D) in frozen-thawed konjac at 200 – 500 MPa. These results indicated that phase transitions (ice VI → ice V → ice III → liquid → ice I) occurred either during reduction of pressure at -20°C or during storage in a pressure medium or in a freezer. This probably caused texture and structure damage. Thus high-pressure-freezing-thawing at 200 – 400 MPa was effective on the improvement in quality of frozen konjac. However, the liquid-solid equilibrium points might be different in pure water as compared to konjac liquid: freezing point for konjac might be lower than that of water. Thus, when konjac was frozen at 200 – 400 MPa and -20°C, it appeared as though the konjac was not frozen (supercooling). Therefore, a coarse gel network might be maintained after high-pressure-thawing. It may be necessary to define the liquid-solid equilibrium points for konjac under high pressure. This would then help determine the temperature changes.

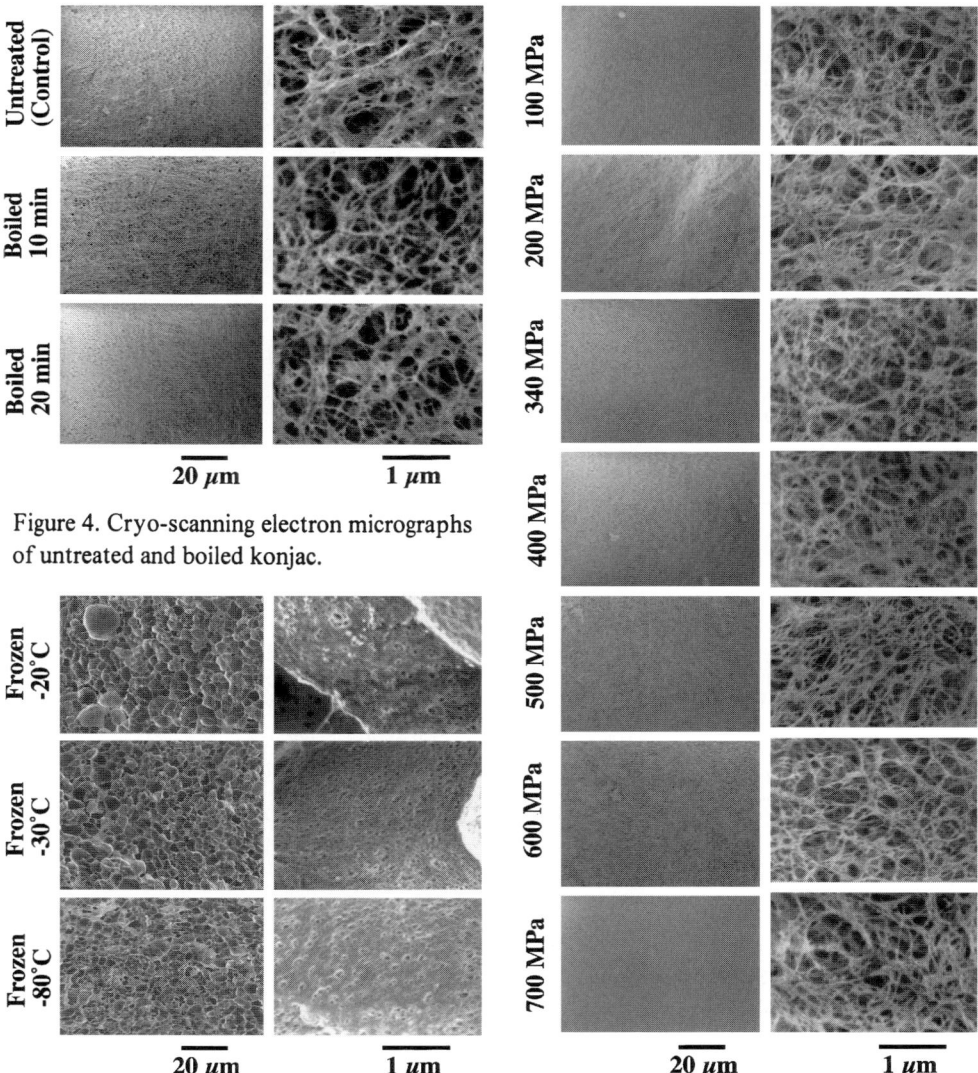

Figure 4. Cryo-scanning electron micrographs of untreated and boiled konjac.

Figure 5. Cryo-scanning electron micrographs of konjac frozen in freezers at atmospheric pressure then thawed at 20°C.

Figure 6. Cryo-scanning electron micrographs of konjac pressurized at 20°C (unfrozen).

226

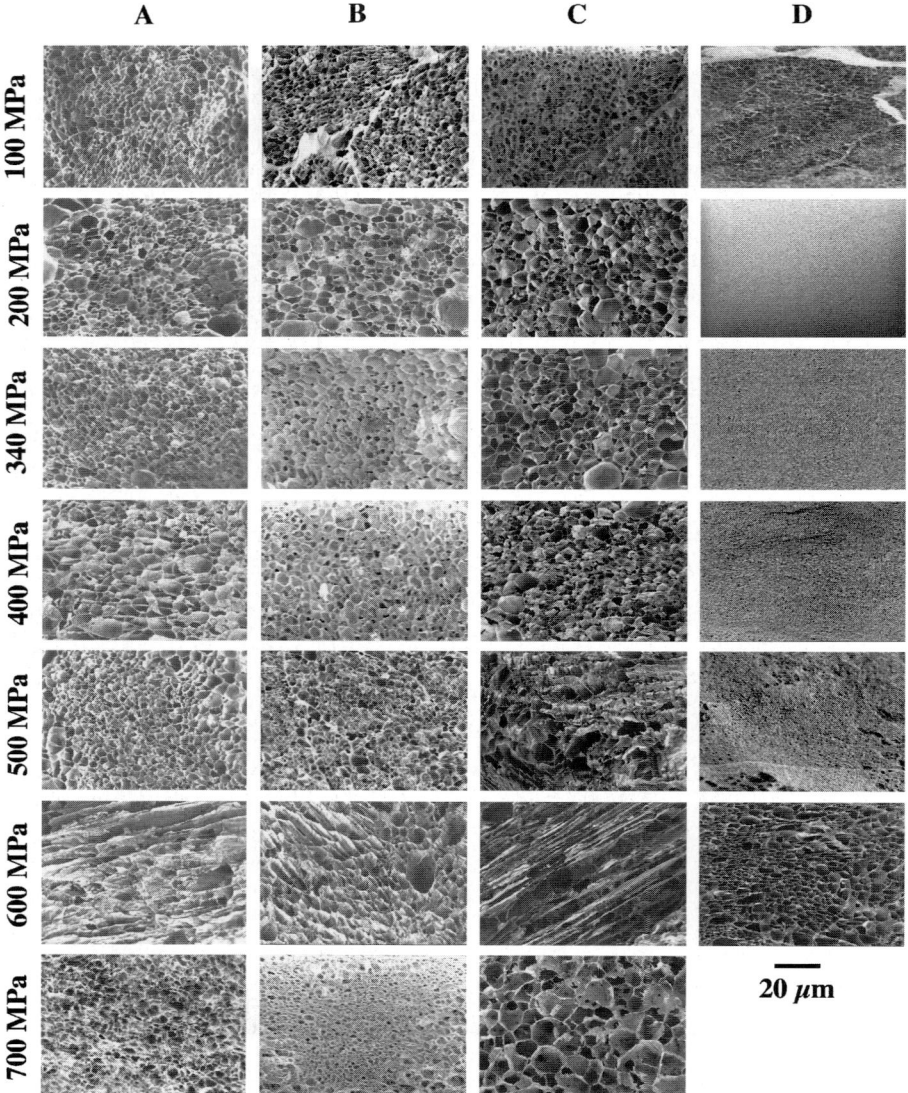

Figure 7. Cryo-scanning electron micrographs of high-pressure-frozen konjac (low magnification, ×1,000). A, B, C and D: See "Methods of thawing".

227

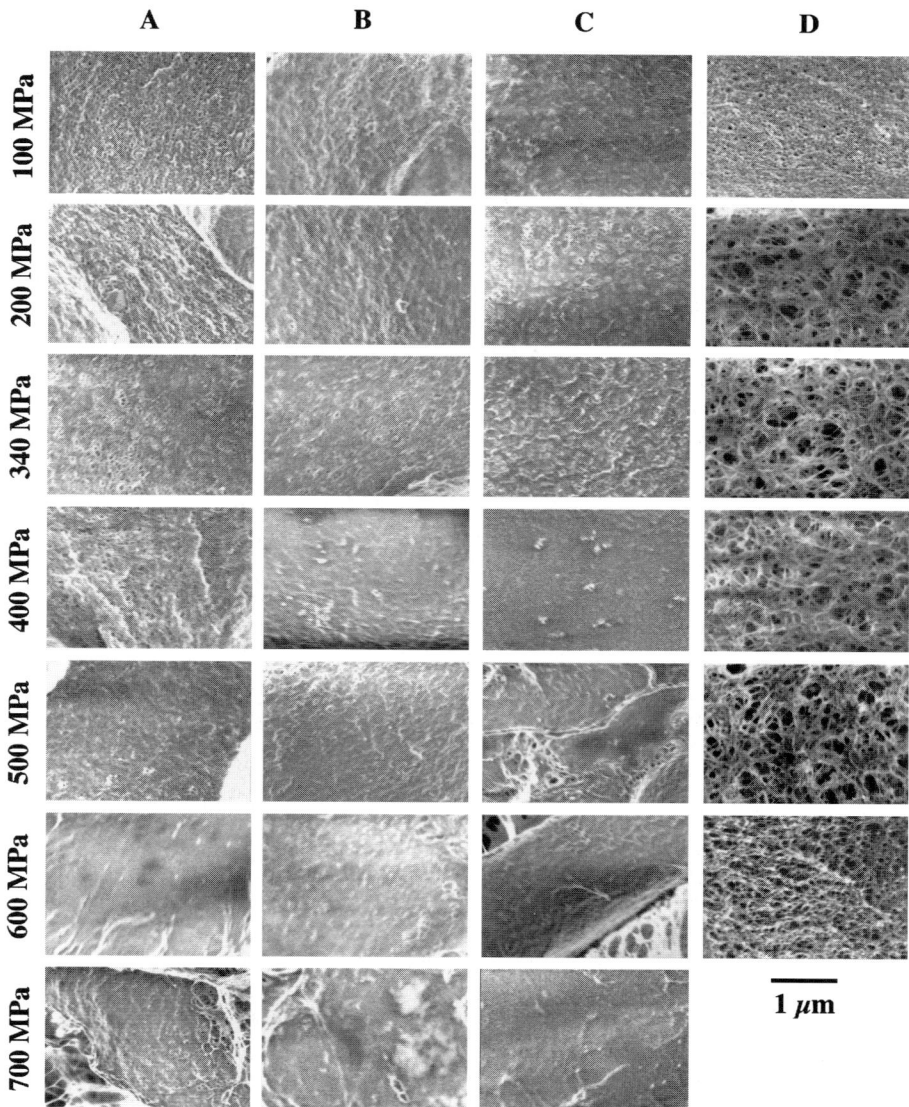

Figure 8. Cryo-scanning electron micrographs of high-pressure-frozen konjac (high magnification, ×30,000). A, B, C and D: See "Methods of thawing".

REFERENCES

1. N. H. Fletcher, The Chemical Physics of Ice, Cambridge University Press, London, 1970.
2. P. V. Hobbs, Ice Physics, Oxford University Press, London, 1974.
3. M. Fuchigami and A. Teramoto, J. Food Sci., 62 (1997) 828.
4. M. Fuchigami, A. Teramoto and N. Ogawa, J. Food Sci., 63 (1998) 1054.
5. N. Kato, A. Teramoto and M. Fuchigami, J. Food Sci., 62 (1997) 359.

HYDROCOLLOIDS – PART 1
Edited by K. Nishinari
2000 Elsevier Science B.V.

229

Extraction of highly gelling pectins from sugar beet pulp

T. Turquois[a], M. Rinaudo[b], F.R. Taravel[b] and A. Heyraud[b]

[a]NESTEC Ltd, Research Centre, Vers-chez-les-Blanc, P.O. Box 44, CH-1000 Lausanne 26, Switzerland

[b]Centre de Recherches sur les Macromolécules Végétales, BP 53, 38 041 Grenoble Cedex 09, France

This study concerns the recovery and characterisation of pectic extracts from sugar beet with a low degree of esterification and good gelling properties in the presence of calcium.

1. INTRODUCTION

Due to its high pectin content, about 20 % on a dry matter basis, and availability in large quantities, sugar beet pulp was considered as potential new source of pectins. Up to now, no pectins with gel forming properties comparable to the pectins extracted from apple pomace and citrus peels have ever been extracted from this by-product [1]. This poor gelling ability has been ascribed to the high acetyl content and low molecular weights of these pectins and to their high proportion of side-chains [1].
In the present work, an extraction process that maintains the structural integrity of the pectins as much as possible has been developed. This procedure yielded products which possessed both a high pectic substance content with a low degree of esterification and a high gelling ability in the presence of calcium. The composition of the extracted product was determined and the effect of extrinsic factors such as concentration of the extracted product, calcium content, sequestrant content and hydration temperature, on gel properties was evaluated by rheological measurements.

2. MATERIALS AND METHODS

2.1. Materials
Sugar beet was provided in the wet form (20-30% dry matter) by Saint Louis Sucre (former Générale Sucrière, France).

2.2. Methods
Product characterisation
Moisture and ash contents were determined according to AACC method 44.15A, 1983 [2] and AOAC method 14.006, 1984 [3] respectively. Fat was given by the Weibull-Stoldt method. Kjeldahl method was used to characterize the protein content [3]. The composition in neutral sugars was performed after hydrolysis of a 50 mg sample in 12 N H_2SO_4 for 60 min at 30 °C and then, in 0.4 N H_2SO_4 for 75 min at 130 °C.

Analysis of monosaccharides was realised using High Performance Anion Exchange Chromatography (HPAE) with Pulsed Amperometric Detection (PAD). The galacturonic acid content was determined by estimation of the carbon dioxide produced upon decarboxylation [4]. Finally, the degree of methoxylation was measured by gas chromatographic determination of methanol released by saponification [5].

Extraction procedure
The wet sugar beet pulp was dried in an oven at 45 °C and then ground (0.5 mm-60 mesh). 500 g of pulp was dispersed in 12.5 L tap water for 20 min at 85°C to wash the pulp and inactivate the enzymes. The material was dewatered with a spin-dryer, then suspended in 10 L 50 mM NaOH (pH 12.0) and stirred at room temperature for 2 h. After spin-drying, the liquid fraction was neutralised to pH 6.5-7.0 with 6 N HCl. The pectin was precipitated by adding 10 L 95 % ethanol to give a final ethanol concentration of 50 %. After stirring for 10 min, the system was stored overnight at 4 °C. The precipitate was recovered by centrifugation at 5000 rpm for 15 min and washed successively with 70, 80 and 90 % ethanol. Finally, the extracted product was dried in a vacuum oven at room temperature.

Gel preparation
The extracted product containing pectic substances was dispersed in deionized water containing 8 mM sodium citrate. The suspension was stirred at the chosen solubilization temperature for 2 h. Then, an appropriate volume of a 2.9 % w/v $CaSO_4.2H_2O$ suspension - preheated or not - was added to reach the desired amount of calcium in the system. Sodium citrate and the partially soluble calcium salt were used to lower the velocity of the calcium diffusion. The solution was stirred at the chosen temperature for 1 min, poured into a mould and stored at 4 °C for 24 h.

Rheological measurements
Elastic moduli (gel strength) were measured by compressing gel pieces (17 mm x 17 mm) between parallel plates at room temperature using a Micro System TA-XT2 Texture Analyser. The strain applied was 10 % and the speed was 0.8 mm/s.

3. RESULTS AND DISCUSSION

3.1. Recovery of pectic substances
This paper concerns the recovery and characterisation of pectic extracts from sugar beet with a low degree of esterification and good gelling properties in the presence of calcium. To our knowledge such polymeric materials have never been obtained from these agricultural residues.

Extracted product containing pectins
Taking into account the structural features of pectins contained in sugar beet pulp, especially their high degree of esterification (about 50 %), the extractions were carried in an alkaline medium without any sequestering agent. The alkaline procedure (pH 12.0 and 25 °C followed by neutralisation and alcohol precipitation) produced a product which formed gels in the presence of calcium. Furthermore, this process had the advantage to be carried out at room temperature. Table 1 displays the composition of the raw material and of the extracted product.

Table 1.

Composition of the raw material and of the extracted product containing pectins from sugar beet pulp using the alkaline procedure proposed.

a On the dried pulp.

Characteristics	Dried pulp	Extracted product
Moisture, %	2.5	14.4
Extraction yield, % [a]	-	10.0
Galacturonic acid, %	20.1	46.5
Degree of esterification, %	52.8	8.9
Rhamnose, %	1.4	0.7
Arabinose, %	18.4	9.6
Galactose, %	4.2	3.1
Glucose, %	19.2	1.3
Ashes, %	13.3	10.2
Fats, %	0.9	0.8
Proteins, %	8.2	6.0

Taking into account the yield of the alkaline process (10 % of the raw dried pulp, Table1), the pectin content of the extracted product and of the raw material (46.5 % and 20.1 % pectins based on the galacturonic acid content, respectively, Table 1), 23.1 % of the pectic substances were extracted from the by-product [6]. These pectins exhibited a low methoxyl content of 8.9 %. This structural characteristic could ensure the thermostability of the gels. Neutral sugars, including rhamnose, arabinose, galactose and glucose were found. Rhamnose is normally inserted into the main polygalacturonic chain. Arabinose and galactose are present as side-chains of the pectic polymer. Glucose is provided mainly by cellulose. Thus, with respect to the composition of the raw material, the extraction process reduces the glucose (cellulose) content from 19.2 to 1.3 % and the arabinose and galactose contents from 18.4 to 9.6 % and from 4.2 to 3.1 %, respectively. Finally, the ash content, which can affect the gelling properties, is rather high. This alkaline procedure can be applied to by-products containing low methoxyl pectins e.g. potato pulp, by adding a sequestering agent [7].

3.2. Characterization of gelling properties.
The ability of low methoxyl pectins to form gels is affected both by intrinsic (molecular weight, degree of esterification) and extrinsic factors (concentration of the extracted product, calcium content, sequestrant content and hydration temperature). The effect of extrinsic parameters on gel strength (elastic modulus) was, therefore, evaluated.

Extracted product from sugar beet pulp by the alkaline procedure
Figures 1 to 4 show the effect of extrinsic parameters on the elastic modulus of gels made with the product extracted from sugar beet pulp by the alkaline procedure. As expected, increasing the concentration of the extracted product increased the elastic modulus (Figure 1).

232

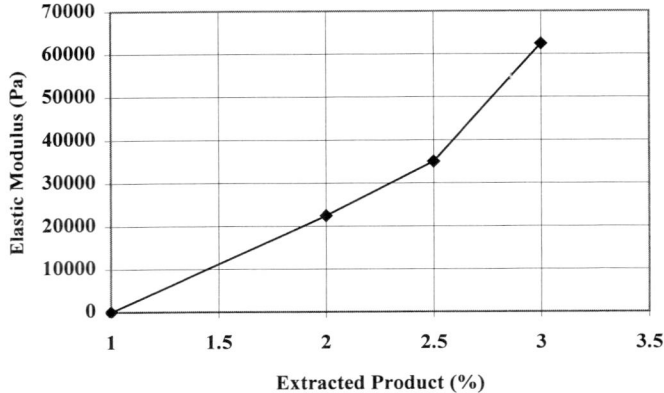

Figure 1. Plot of the elastic modulus as a function of concentration of the extracted product. Gels were made at 25 °C and at a calcium concentration of 40 mg/g extracted product (172 mg $CaSO_4.2H_2O$/g extracted product).

Moreover, it appears that the optimum calcium content (Ca^{2-}) varied from 60 to 100 mg/g extracted product (from 258 to 430 mg $CaSO_4.2H_2O$/g extracted product) (Figure 2).

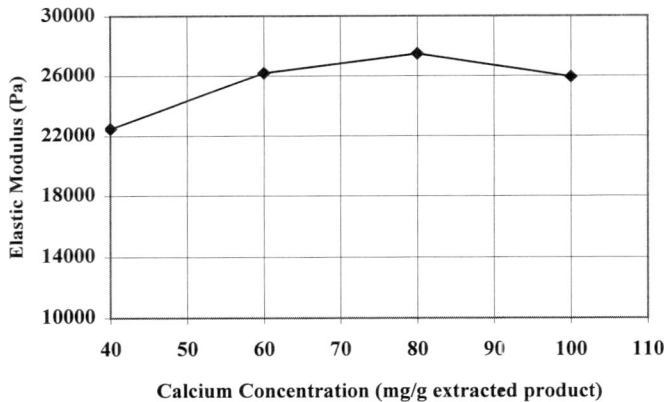

Figure 2. Plot of the elastic modulus as a function of calcium concentration. Gels were made at 25 °C and at 2 % w/v in extracted product.

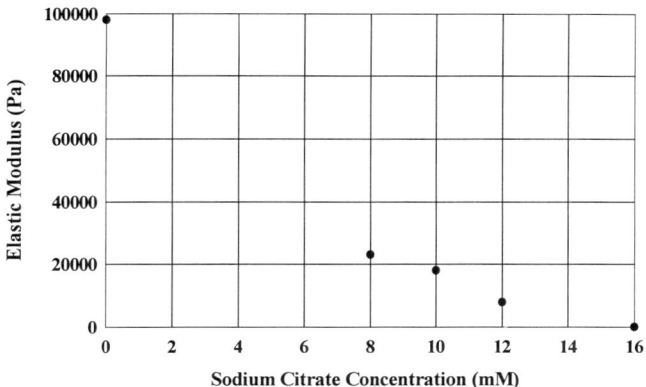

Figure 3. Plot of the elastic modulus as a function of sodium citrate concentration. Gels were made at 25 °C, at 2 % w/v in extracted product and a calcium concentration of 40 mg/g extracted product (172 mg $CaSO_4.2H_2O$/g extracted product).

The addition of sodium citrate, as calcium sequestrant to delay the gelation rate, dramatically decreased gel strength (Figure 3). In the absence of sodium citrate, a high elastic modulus was measured (98 000 Pa-Figure 3). However, since gelation was quite fast, the use of a sequestering agent would be advisable for an industrial application.

Finally, solubilization at room temperature was better than solubilization by heating (Figure 4).

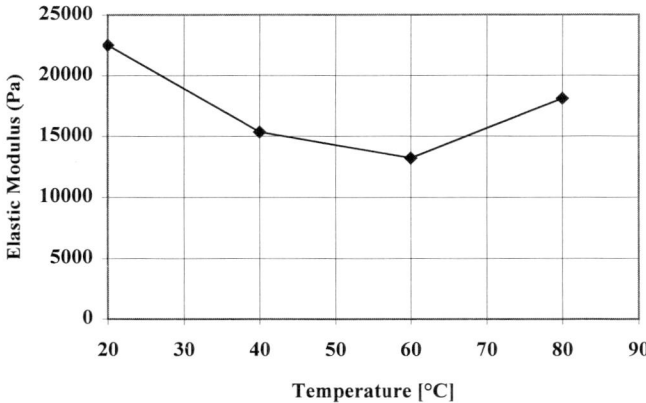

Figure 4. Plot of the elastic modulus as a function of temperature of dissolution. Gels were made at 2 % w/v in extracted product and at a calcium concentration of 40 mg/g extracted product (172 mg $CaSO_4.2H_2O$/g extracted product).

4. CONCLUSION

In this investigation, it has been shown that sugar beet could be used as new source of pectins. The extraction process developed (5, 6) gave products containing a high content in pectic substances with a high gelling ability in the presence of calcium.

REFERENCES

1. W. Pilnik and A.G.J. Voragen, Advances in Plant Cell Biochemistry and Biotechnology, JAI Press Ltd, Vol. 1, 1992.
2. AACC. Approved Methods of the AACC. American Association of Cereal Chimists, St. Paul, MN, 1983.
3. AOAC. Official Methods of Analysis, 14th ed. Asscciation of Official Analytical Chemists, Washington, DC, 1984.
4. G.J. Dutton-Dutton (eds), Glucuronic Acid, Academic Press, New York and London, 1986.
5. R.F. McFeeters and S.A. Armstrong, Anal. Biochem., 139 (1984) 212.
6. T. Turquois, E. Best, P. Vanacker, G.Saint Clair, M. Rinaudo, F.R. Taravel and A. Heyraud Patent No WO 97/49734 (1997a).
7. T. Turquois, E. Best, P. Vanacker, G.Saint Clair, M. Rinaudo, F.R. Taravel and A. Heyraud Patent No WO 97/49298 (1997b).

HYDROCOLLOIDS – PART 1
Edited by K. Nishinari
2000 Elsevier Science B.V.

235

The structure of soluble soybean polysaccharide

A. Nakamura[a], H. Furuta[a], H. Maeda[a], Y. Nagamatsu[b] and A. Yoshimoto[b]

[a]Hannan R&D Center, Fuji Oil Co., Ltd. 1 Sumiyoshi-cyo Izumisano-shi, Osaka, 598-8540, Japan

[b]Department of Applied Biochemistry, Faculty of Applied Biological Science, Hiroshima University, Higashi-Hiroshima, Hiroshima, 739, Japan

The chemical structure of soluble soybean polysaccharides (SSPS) was examined. SSPS was exhaustively degraded by three kinds of pectinase and four kinds of hemicellulase. The molecular structure of SSPS was examined using an enzymic-HPLC method. SSPS has three types of galacturonic main backbone, namely G-1, G-2 and G-3 unit, with longer homogeneous galactosyl and arabinosyl neutral sugar side chains than galacturonosyl main backbone. The G-1, G-2, and G-3 units consisted of rhamnogalacturonan (RG) subunits containing of rhamnose and galacturonic acid mainly as the sugar components, molecular ratio were nearly 1 : 1. Structural differences of main backbone between SSPS and citrus pectin were indicated by the analysis of the galacturonan region and galacturonate residues in the G-1, G-2, and G-3 units, using the polygalacturonase degradation. SSPS consisted of short-chain homogalacturonan (HG) and long-chain RG subunits. On the other hand, citrus pectin is made up of long-chain HG and short-chain RG subunits.

1. INTRODUCTION

Soybean *okara* is the residue after oil and protein extraction of soybean, and it is rich in soluble and non-soluble dietary fibers. For the soluble polysaccharides from soybeans, the sugar compositions and sequences have already been reported [1-3], but many aspects of their structure are poorly understood.

Generally, pectic polysaccharides are composed of two distinct regions, a smooth HG region and a hairy RG region [4-7]. The side chains in the RG region are composed mainly of arabino-4-galactans, which in turn are made up of β-1, 4-D-galactans which are partially substituted by arabinose and arabinosyl oligosaccharides [8]. The determination of the amount and distribution of the HG region in pectin molecules, and structure of arabino-4-galactan have been examined by enzymic-HPLC methods [9-12]. Recently, for soybean pectic polysaccharides, the structure of the HG region has been examined using three kinds of pectinase and two types of exo-hemicellulase [13]. The arabinan and galactan side chains were analysed previously with two exo-hemicellulases, and the HG region with three pectinases. Although SSPS is rich in neutral sugars, such as arabinose and galactose, it appears that pectinase cannot hydrolyse galacturonan regions completely. Therefore, the complete and accurate structure of SSPS has not been clarified.

In this study, we report on the amount and distribution of the HG and RG main chains and the structure of side chains, composed with neutral sugars, in soluble soybean polysaccharides using an enzymic-HPLC method.

2. EXPERIMENTAL

2.1. Materials
The soluble soybean polysaccharides were extracted from Soybean *okara*. Sepharose CL-4B and Sephacryl S-400HR were obtained from Pharmacia, and DEAE-toyopearl 650M from Toso.

2.2. Enzymes
Endo-α-(1→4)-polygalacturonase (endo-PG) from *Kluyveromyces fragilis* was purified using the procedure of Inoue *et al.* [14]. Exo-α-(1→4)-polygalacturonase (exo-PG) from carrot root using the method of Hatanaka *et al.* [15]. Exo-α-(1→4)-poly-galacturonate lyase (exo-PGL) was prepared from the culture fluid of *Erwinia carotovora* subsp. *carotovora* IFO 13921 according to Kegoya *et al.* [16]. Driselase (Kyowa Hakkou Kogyo Co., Tokyo, Japan) solution was prepared as described by Matsuhashi *et al.* [17]. Hemicellulase M (from *Aspergillus niger*) was a gift from Tanabe Pharmaceutical Co., Ltd. (Osaka, Japan). Cellulase A Amano 3 (from *Aspergillus niger*) was a gift from Amano Pharmaceutical Co., Ltd. (Nagoya, Japan).

2.3. Extraction and purification of soluble soybean polysaccharides
The acidic polysaccharide fraction (SSPS) was prepared from Soybean *okara* by hot-water extraction at 100℃ for 3.0 h ; the pH was adjusted to 4.0 - 5.0 during extraction. The extract was centrifuged at 8,000 rpm and supernatant pooled. The extract was dialysed against deionized water. Ethanol was added to the resulting solution to give a composition of 50% by volume. The precipitates were collected by centrifugation and washed successively with 80% ethanol, 99% ethanol and acetone, then air-dried.

The crude SSPS solution (4% solutions, 120ml) was added to a DEAE-toyopearl column (5.0 × 25 cm) equilibrated with 10 mM NaHCO3 buffer (pH 8.0). After washing the sample with the same buffer, the column was stepwise eluted with 1,500 ml of 0.05M NaHCO3, 1,500 ml of 0.1M NaHCO3, 1,500 ml of 0.1M Na2CO3, and 1,500 ml of 0.1M NaOH. The flow rate was 10 ml/min, and 30 ml fractions were collected. The total sugar content and the uronic acid content in each fraction were analysed by the phenol-sulfate method [18], and carbazole sulfate method [19], respectively. Fractions of each sugar peaks were collected, dialysed against deionized water, successively, and freeze-dried.

The main fractions of ion-exchange chromatography were dissolved in 50 mM acetate buffer (pH 5.0) and 10 ml of a 12% solution placed on a Sephacryl S-400 HR column (2.4 × 96 cm) and equilibrated with 50 mM acetate buffer (pH 5.0). The flow rate was 0.5 ml/min, and 5 ml fractions were collected. Fractions of main sugar peaks were collected, dialysed against deionized water, successively, freeze-dried, and used as purified SSPS.

2.4. Separation of enzyme-degraded purified SSPS
Hemicellulase degradation products from purified SSPS were separated by gel-filtration chromatography, placed on a Sepharose CL-4B column (2.4 × 96 cm) and equilibrated with

50 mM acetate buffer (pH 5.0). The flow rate was 2 ml/min, and 3 ml fractions were collected. Fractions of separated peaks were collected, successively dialysed against deionized water, successively, and freeze-dried.

2.5. Estimation of molecular weight distribution of SSPS

Molecular weight distributions were estimated by gel-filtration. Operating conditions were as follows: Column, TSK-gel G-5000PWXL (7.8 × 300 mm), TSK-gel G-3000PWXL (7.8 × 300 mm); guard column, TSKguardcolumn PWXL (6.0 × 40 mm); solvent, 0.02M acetate buffer (pH 5.0); flow rate, 0.6 ml/min; detector, RI (Jacso TRI ROTAR-IV); injection: 50 μ l. Pullulans (Showa denko, Tokyo, Japan) were used as standards for the estimation.

2.6. Measurement of neutral sugars and galacturonates

Both of neutral sugar and uronic acid contents of SSPS were analysed by enzymic-hydrolysis (A) and chemical-hydrolysis (B). The hydrolytic and operating conditions were as follows. (A) The reaction mixtures, containing 0.1% SSPS solution, 0.1% glycerol (internal standard), driselase solution (0.5 ml/ml) and 50 mM sodium acetate buffer (pH4.0), were incubated at 35℃ for 48 h, and then filtered with a Millipore Molcut II GC (1 × 10⁴ Da exclusion limit for globular proteins). The filtrates were analysed by HPLC using the following conditions: Column, Shodex SUGAR SH-1821 (8.0 × 300 mm); guard column, Shodex SUGAR SG1011P (6.0 × 50 mm); column temperature, 40℃; flow rate, 1.0 ml/min; mobile phase, 0.001M H₂SO₄ buffer. (B) The reaction mixtures containing 0.1% SSPS, 2N CF₃COOH (TFA) in tubes with Teflon seal were heated for 2 h at 121℃ [20]. After evaporation of TFA from the hydrolysates, these were concentrated to 0.5 ml with deionized water, then filtered with a Millipore Molcut II GC. The filtrates were analysed by HPLC using the same conditions already described.

Neutral sugar compositions were measured as follows: The hydrolysates were added to Amberlite CG-50 (0.5 × 20 mm) and Bio-Rad AG 1-X2 (0.5 × 20 mm) columns, equilibrated with 50% of ethanol, and eluted with 10 volumes of the same solvent. The effluents were concentrated and air-dried. The samples were then dissolved in deionized water, and analysed by HPLC using the following conditions: Column, TSKgel SUGAR AXI column (4.6 × 150 mm); column temperature, 65℃; flow rate, 0.25 ml/min; mobile phase, 0.1M Borate buffer containing 1.0% monoethanolamine, pH7.90.

2.7. Enzyme assays

Polysaccharide-degradation was determined by measuring the reducing sugars generated using the modified Somogyi method [21]. One unit of the enzyme was defined as the amount which liberated 1 μ mol of reduced sugars per minute under the above conditions.

2.8. HPLC measurement of the degradation limit of SSPS by various hemicellulases

The reaction mixtures (total volume 1.0 ml) containing 0.5% purified SSPS, 0.1% glycerol (internal standard), the enzyme preparations such as exo-α-arabinosidase, endo-α-L-arabanase, and endo-β-D-galactosidase from *Aspergillus niger* and exo-β-D-galactosidase from *Persimmon Saijyo*, and 50 mM acetate buffer, pH 4.0, and incubated at 35℃ for 48 h. The reaction mixtures were filtered with a Millipore Molcut II GC. The filtrates were used for HPLC analysis using a Shodex SUGAR SH-1821 column. Degradation limits were determined by the measuring neutral sugars with HPLC.

2.9. HPLC measurement of the degradation limits of SSPS by various pectinases

Degradation limits of SSPS by pectinases were measured by the method of Matsuhashi *et al.* [7]. The reaction mixtures containing purified SSPS (0.1%), glycerol (0.1%), sodium acetate buffer (50 mM, pH 5.0 at 35℃, for exo- and endo-PG) or tris-HCl buffer (50 mM, pH 9.0 at 35℃, for exo-PGL), and the enzymes (50 mU) were incubated at 35℃ for 48 h. For the combined action of endo-PG and exo-PG, 100 mU/ml of each enzymes was used. In the case of degradation by exo-PGL first and exo-PG secondly, the reaction mixtures contained 0.1% SSPS, 0.1% glycerol, 20 mM tris-HCl buffer (pH 9.0 at 35℃), and exo-PGL were incubated at 35℃ for 48 h. The reaction mixtures were added to an equal volume of soduim acetate buffer (100 mM, pH 5.0) containing exo-PG (25 mU). The reaction mixture was re-incubated at 35 ℃ for 48 h. In the case of degradation by exo-PG first and exo-PGL secondly, the reaction mixtures containing 0.1% SSPS, 0.1% glycerol, 20 mM acetate buffer (pH 5.0 at 35℃), and exo-PG were incubated at 35℃ for 48 h. The reaction mixtures were added to an equal volume of tris-HCl buffer (100mM, pH 9.0) containing exo-PGL (25 mU), and the reaction mixture was re-incubated at 35 ℃ for 48 h. HPLC analysis conditions were carried out as before.

3. RESULTS AND DISCUSSION

3.1. Purification of soluble soybean polysaccharides

Anion exchange chromatography using alkaline conditions is a proven method for separating acidic polysaccharides, but this method cannot be applied to pectins because of alkaline de-esterification and β-elimination of the galacturonide bonds [22]. DEAE-cellulose chromatography was used by Yamaguchi *et al.* [13] for the purification of pectic polysaccharides, but cellulose is hydrolysed under alkaline conditions. Therefore, DEAE-toyopearl chromatography with alkaline or acidic buffer was used in this study.

The DEAE-toyopearl chromatogram developed with carbonate buffer is shown in Fig. 1(A). The pectic polysaccharides separated into four fractior.s, D-1, D-2, D-3 and D-4. The yields of the four fractions were 14.8, 28.0, 17.2 and 6.2%, respectively. For DEAE-toyopearl chromatography with the carbonate buffer, most of galacturonate residues were eluted in the D-3 fractions (data not shown). Gel-filtration patterns of each fractions are shown in Fig. 1(B). High molecular weight components, which were the main components of crude SSPS, were found in D-3 the fraction. The gel-filtration chromatogram of the D-3 fraction is shown in Fig. 1(C). The D-3 fraction is eluted as a single peak. This result demonstrates that the purity of the main fraction of S-400 HR column chromatography is very high. The molecular weight of D-3 fraction was estimated to be about 550,000.

3.2. Analysis of the structure of side chains by exo and endo hemicellulase

The degradation limits of purified SSPS were determ:ned by exo-α-L-arabinosidase (AFase), exo-β-D-galactosidase (GPase). AFase can liberate up to 94% of all the arabinose. GPase can liberate about 89% of all the galactose. From these data, it can be concluded that side chains of soluble soybean polysaccharides consist of homogeneous arabinan and galactan residues. After almost all of the arabinan and galactan were removed from purified SSPS, endo-α-L-arabanase (ANase) and endo-β-D-galactosidase (GNase) were added to the mixtures of exo-hemicellulases limit degraded SSPS. The GNase can liberate some galactosyl oligosaccharides. In contrast, ANase liberate nothing. This result indicates that homogeneous arabinan is linked to RG regions through galactosyl oligosaccharides.

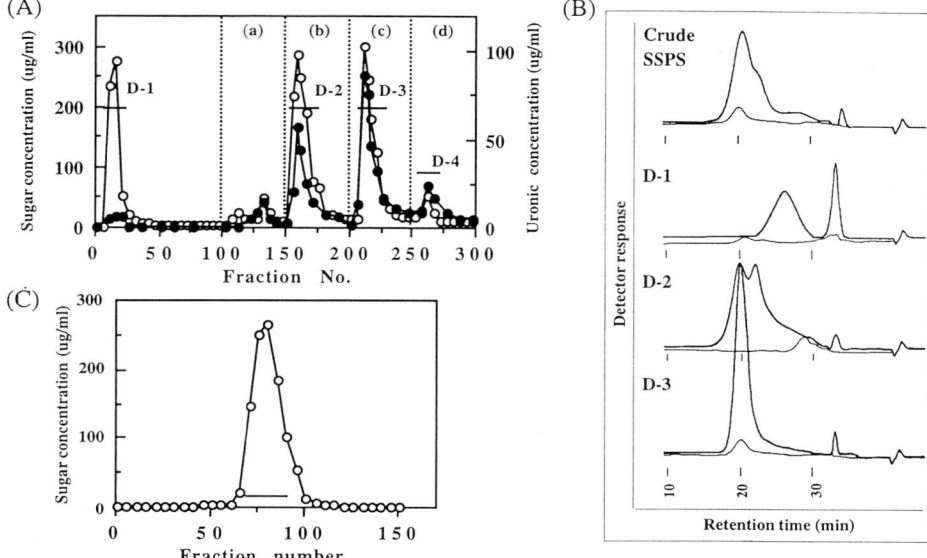

Figure 1. (A) DEAE-toyopearl chromatography of SSPS. The crude SSPS solution (120 ml of a 4% solution) was added to a DEAE-toyopearl column (5.0 × 25 cm) , and equilibrated with 10 mM NaHCO₃ buffer (pH 8.0). After washing the sample with the same buffer, stepwise elution with 0.05M NaHCO₃ (a), 0.1M NaHCO₃ (b), 0.1M Na₂CO₃ (c) and 0.1M NaOH (d). The flow rate was 10 ml/min, and 50 ml fractions were collected. (B) Gel-filtration patterns of SSPS fractionated by DEAE-toyopearl chromatography. Operating conditions: Column, TSK-gel G-5000PWxL (7.8 × 300 mm), TSK-gel G-3000PWxL (7.8 × 300 mm); guard column, TSKguardcolumn PWxL (6.0 × 40 mm); solvent, 0.02M acetate buffer (pH 5.0); flow rate, 0.6 ml/min; detector, RI (—) and UV (—, abs at 280 nm); injection: 50 μ l. (C) Gel-filtration patterns of D-3 fraction. D-3 fraction was dissolved in 50 mM acetate buffer (pH 5.0) and (10ml of a 12% solution) was put on a Sephacryl S-400 HR column (2.4 × 96 cm) equilibrated with 50 mM acetate buffer (pH 5.0). The flow rate was 0.5 ml/min, and 5 ml fractions were collected. ○, total sugar; ●, uronic acid .

3.3. Residual number and molecular weight of the rhamnogalacturonan backbone

Figure 2 shows the gel-filtration chromatogram of AFase and GPase treated purified SSPS. The purified SSPS is separated into three types of molecules which differ in molecular weight after removing the side chains (data not shown) using exo-hemicellulases. This result indicates that the purified SSPS consists of at least three types of galacturonate backbone, namely G-1, G-2, and G-3 units. In general, a rhamnogalacturonan with neutral side chains is liberated from pectic polysaccharide by the hydrolysis of homogalacturonan regions with endo-PG. Using this method, endo-PG is prevented from hydrolysing homogalacturonan because of solid obstacles made up of neutral sugars. Therefore we first hydrolysed neutral sugar side chains from purified SSPS, and then treated the remainder with endo-PG. The G-1, G-2, and G-3 units are also made up of three types of molecular weight different RG regions. Table 1 shows the residual number and molecular weight of RG backbone from purified SSPS. The G-1 unit consists of a G-1-a subunit, molecular weight is 63,500, and a G-1-b subunit, molecular weight 12,800. The G-2 unit consists of a G-2-a subunit, molecular weight 22,000, and a G-2-b subunit (similar to G-1-b), molecular weight 13,100. The G-3 unit consists of a G-3-a subunit (similar to G-1-b), molecular weight 11,200. These subunits contain mainly rhamnose and galacturonic acid as sugar components. The molecular ratio of rhamnose and galacturonic acid is nearly 1 : 1.

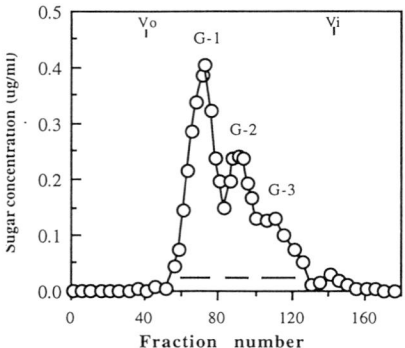

Figure 2. Gel-filtration pattern of exo-hemicellulases degrated SSPS on a Sepharose CL-4B column.

Table 1. Residual Number and Molecular Weight of RG backbone derived from Purified SSPS

Residure	Sugar composition (residue)				
	G-1-a	G-1-b	G-2-a	G-2-b	G-3-a
Rhamnose	84.2	23.4	38.6	21.3	16.1
Mannose	0.0	1.2	0.0	2.0	8.4
Fucose	6.6	2.8	6.1	2.8	3.4
Arabinose	26.1	10.1	18.9	13.4	10.8
Galactose	46.3	12.4	24.3	14.6	10.2
Xylose	18.5	6.1	15.4	8.5	4.6
Glucose	6.4	1.9	5.8	1.4	4.8
Galacturonic acid	100.2	24.6	47.1	22.1	18.2
Molecular weight	63500	12800	22000	13100	11200
MW of main chain	29958	7754	13939	7006	5559
Number of RG	1.0	3.0	1.0	1.3	

3.4. Distribution of the galacturonan backbone in purified SSPS

Fig. 3 shows the distribution of the galacturonan region and galacturonate residue in purified SSPS and citrus pectin. In general, pectic polysaccharides, such as citrus pectin, consist of HG regions and RG regions which are located between HG regions, and the degree of polymerizations (DPs) of HG regions is very high. On the other hand, in purified SSPS, the backbones of G-1, G-2, and G-3 units have three types of RG regions and are located between the HG regions. The HG regions are distributed at both the reducing and non-reducing ends of these molecules. The DPs of the HG regions is thought to be very low in comparision with pectin. Recently, Yamaguchi et al. [13] has reported that no HG regions is interposed between two RG regions of soybean pectic polysaccharides. This probably arises because pectinases cannot hydrolyse HG regions located between RG regions because of structural obstacles, caused by neutral sugar side chains. These workers analysed the distribution of galacturonan region and galacturonate residue without removing the neutral sugars.

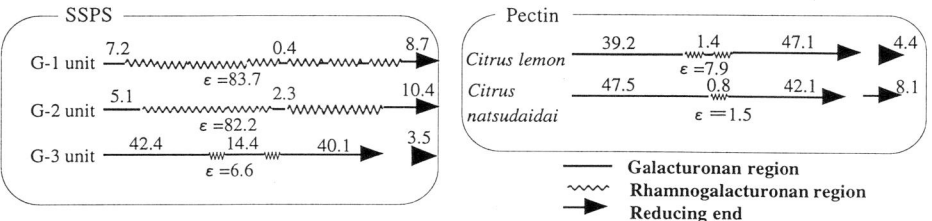

Figure 3. Distribution of galacturonan regions in the purified SSPS. Straight lines, notched lines and arrows indicate HG regions, RG regions and the reducing end, respectively. Figures express the percentages of galacturonate distribution.

REFERENCES

1. G. O. Aspinall, I. W. Cottrell, S. V. Egan, I. M. Morrison & J. N. C. Whyte, *J. Chem. Soc.* (C), **1967**, (1967) 1071.
2. S. Kawamura, *Nippon Shokuhin Kougyo Gakkaishi* (in Japanese), **14,** (1967) 514, 553.
3. T. Kikuchi & H. Sugimoto, *Agric. Biol. Chem.*, **40,** (1976) 87.
4. A. J. Barret & D. H. Notrhcote, *Biochem. J.*, **94**, (1965) 617.
5. J. A. De Vries, A. G. J. Voragan, F. M. Rombouts & W. Pilink, American Chemical Society, Washington. DC, (1986) 157.
6. B. V. McCleary & N. K. Matheson, *Adv. Carbohydr. Chem. Biochem.*, **44,** (1986) 147.
7. S. Matsuhashi, N. Nishikawa, T. Negishi & C. Hatanaka, *J. Liquid. Chromatogr.*, **16**, (1993) 3203.
8. R. F. H. Dekker & G. N. Richards, *Adv. Carbohydr. Chem. Biochem.*, **32**, (1976) 277.
9. S. Matsuhashi, K. Yokohiki & C. Hatanaka, *Agric. Biol. Chem.*, **53**, (1989) 1417.
10. S. Emi, J. Fukumoto & T. Yamamoto, *Agric. Biol. Chem.*, **35**, (1971) 1891.
11. J. A. De Vries, C. H. den Ujil, A. G. J. Voragen, F. M. Rombouts & W. Pilinik, *Carbohydr. Polym.*, **3**, (1983) 193.
12. F. Yamaguchi, S. Inoue & C. Hatanaka, *Biosci. Biotech. Biochem.*, **59**, (1995) 1742.
13. F. Yamaguchi, Y. Ota & C. Hatanaka, *Carbohydr. Polym.*, **30**, (1996) 265.
14. S. Inoue, Y. Nagamatsu & C. Hatanaka, *Agric. Biol. Chem.*, **48**, (1984) 633.
15. C. Hatanaka & J. Ozawa, *Agric. Biol. Chem.*, **28**, (1964) 627.
16. Y. Kegoya, M. Setoguchi, K. Yokohiki & C. Hatanaka, *Agric. Biol. Chem.*, **48**, (1984) 1055.
17. S. Matsuhashi, S. Inoue & C. Hatanaka, *Biosci. Biotech. Biochem.*, **56**, (1992) 1053.
18. M. Dubois, K. A. Gilles, J. K. Hamilton, P. A. Robers & F. Smith, *Anal. Chem.*, **25**, (1956) 350.
19. J. T. Galambos, *Anal. Biochem.*, **19**, (1967) 119.
20. P. Albersheim, D. J. Nevins, P. D. English & A. Karr, *Carbohydr. Res.*, **5**, (1967) 340.
21. C. Hatanaka & Y. Kobara, *Agric. Biol. Chem.*, **44**, (1980) 2943.
22. C. Hatanaka & J. Ozawa, *Nippon Nogeikagaku kaishi*, **40**, (1966) 421.

HYDROCOLLOIDS – PART 1
Edited by K. Nishinari
2000 Elsevier Science B.V.

Comparison of vegetable xyloglucans by NMR and immunochemical analyses

Y. Sone, A. Wakai and T. Maekawa,

Department of Food and Nutrition, Faculty of Human Life Science, Osaka City University
Sumiyoshi, Osaka 558-8585, Japan

Seven xyloglucans were extracted by treatment of common vegetables, for examples tomato and Chinese cabbage, with aqueous 24% KOH and purified through iodine-xyloglucan complex formations. ELISA experiments on seven xyloglucans revealed that xyloglucan of tomato showed similar titration curves to that of tamarind xyloglucan and were more reactive to the antibody than those of other vegetables. ^{1}H-NMR spectroscopic analyses, including 2D NMR techniques, of the purified xyloglucans, showed that structural features of xyloglucans of vegetables other than tomato were similar to each other, and were different from those of tomato and tamarind seed in containing L-fucosyl residue as a component sugar.

1. INTRODUCTION

Xyloglucan, a kind of hemicelluloses of plant cell wall, firmly associates with cellulose microfibrils through hydrogen bonds to maintain the cell-wall architecture [1]. In the course of our studies on the nutritional roles of plant xyloglucans, we have reported the inhibitory effects of oligosaccharides derived from tamarind seed xyloglucan on intestinal glucose absorption in rats [2] and estimation of vegetable xyloglucan by ELISA using anti-tamarind xyloglucan antibody [3,4]. Upon application of this ELISA system for the determination of the content of xyloglucan as a dietary fiber in commercially available vegetables, we recently became aware that xyloglucans extracted from green vegetables reacted with the antibody in a different manner to that of tamarind xyloglucan. This fact prompted us to compare the structural features of the purified xyloglucans of several vegetables using ^{1}H-NMR methods involving 2D NMR techniques with reference to the difference in their reactivities to the antibody.

In this paper, we report the results of the determination of vegetable xyloglucans by ELISA as well as the structural features revealed by 500 MHz ^{1}H-NMR spectroscopy.

2. EXPERIMENTAL

2.1. Materials and Methods

Seven commercially available vegetables, Sakaina, Komatuna, Chinese cabbage, Broccoli, and Tomato (Minikyaroru and Sopi,) were kindly donated by Dr. Hiromi Tsuji, Osaka Prefectural Agriculture and Forest Center, Syakudo, Habikino-shi, Osaka. Xyloglucan of tamarind seed (*Tamarindus indica*) was purified from tamarind gum provided by Dainippon Pharmaceutical Co., Ltd., Suita, Osaka.

Preparation of anti-xyloglucan antibody is described in our previous papers [3,5].

Vegetable xyloglucans were extracted from the homogenized vegetables by treatment with aqueous 24% KOH in the manner described previously [4]. Xyloglucans were further purified through the iodine-xyloglucan complex formation after pronase and glucoamylase treatments according to the procedure described by Kato et al [6].

Determination of xyloglucan by ELISA using anti-tamarind xyloglucan antibody[5] was performed in the manner reported previously[4]. The xyloglucan sample solution was prepared in 0.02% Tween 20 containing phosphate buffered saline at the concentration of 1mg/ml.

For the ^{1}H-NMR spectroscopy, xyloglucans were repeatedly exchanged in D_2O with intermediate lyophilizations. 500-MHz ^{1}H-NMR, 2D COSY (*Correlated Spectroscopy*) and TOCSY (*Total Correlation Spectroscopy*) spectra [7] (internal standard; 3-trimethylsilylpropanoate) were recorded using a Valian Unity-500 spectrometer with its standard programs (GMQCOSY for COSY, and TNTOCSY for TOCSY) for solution D_2O at 25℃.

3. RESULTS AND DISCUSSION

3.1. Determination of xyloglucans by ELISA.

Figure 1A shows the typical titration curves for determinations of xyloglucans by ELISA, where open circles show titration curves for tamarind xyloglucan and closed circles show titration curve for xyloglucan extracted from Chinese cabbage. It is apparent from this figure that in the calibration curve for Chinese cabbage xyloglucan, the maximum absorbance at 410 nm is lower than that for tamarind xyloglucan. The maximum absorbance was ca. 0.35 for the former xyloglucan, while that for latter one was ca. 0.72, which indicated that the immunochemical affinity of the xyloglucan of Chinese cabbage was lower than that of tamarind xyloglucan. In contrast, figure 1B shows the titration curves of xyloglucans of tomato and lettuce, showing almost the same curve as that for tamarind xyloglucan except "weak" reactivity of xyloglucan of lettuce to the antibody. In this experiment, anti-xyloglucan antibody used for ELISA was prepared by the immunization of rabbit with nona-saccharide obtained from the cellulase digestion of tamarind xyloglucan. It is, therefore, specific to the nona-saccharide unit of tamarind xyloglucan which contains two Gal1→2Xyl1→6Glc units [4]. Consequently, these results suggest that the green vegetable xyloglucans tested in this experiment have different structural features than the tamarind xyloglucan.

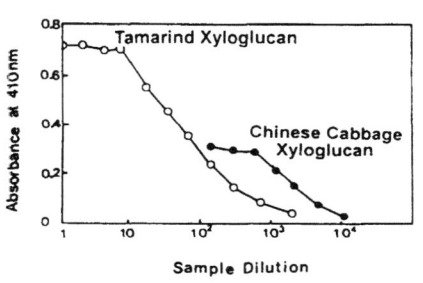

Figure 1A Reactivity of Chinese Cabbage Xylogucan on ELISA

Figure 1B. Reactivities of Tomato and Lettuce Xylogucan on ELISA

3.2. Structural features of vegetable xyloglucans revealed by ^{1}H-NMR.

Figure 2 shows 500 MHz ^{1}H-NMR spectrum of xyloglucan purified from the 24% KOH extracts of tomato. ^{1}H-signals were identified by COSY and TOCSY spectra referring to the published papers about the structure of tamarind xyloglucan [8~10]. As shown in Fig.2, α-D-xylosyl residues in which the H-1 signal occurs at δ 5.16 and 5.02 could be attributed to xylosyl residues substituted with Gal and those in which the H-1 signal occurs at δ 5.09 and 4.95 could be attributed to non-reducing xylose residues. In contrast, the spectrum of the xyloglucan of Chinese cabbage contained a methyl ^{1}H signal at δ 1.25 indicating the presence of 6-deoxy sugar, while no such a signal was present in the spectrum of xylolgucan of tomato.

Sugar analysis of the hydrolyzate of Chinese cabbage xyloglucan showed that it contained L-fucose as a deoxy sugar in addition to glucose, galactose and xylose. Concerning the H-1 resonances of β-anomer region, they were identified as β-D-galactopyranosyl and β-D-glucopyranosyl.residues.

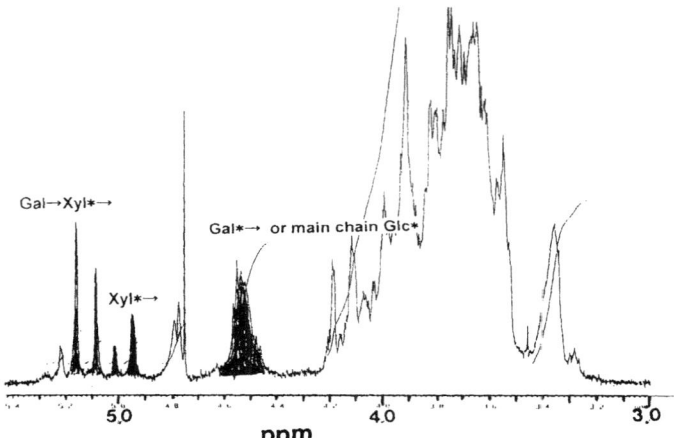

Figure 2. 500MHz ^{1}H-NMR spectroscopy of Purified Tomato Xyloglucan

By the same procedure as described above, assignments of ^{1}H resonances of ^{1}H-NMR spectra of the other vegetable xyloglucans were carried out and the results are summarized in Table 1. This table shows the chemical shifts of H-1 of constituent sugar residues of xyloglucans examined in this experiment, the estimated substituted side-chain of α-D-xylosyl residues, and the percentages of the intensities of each H-1 calculated from their integrals. In this Table, the intensities of H-1 signals of β-Glc and β-Gal are represented as the sum of their integral because their H-1 signals could not be separated enough to calculate each integral under these NMR conditions as shown in Fig. 2. It is apparent from this Table that xyloglucans of green vegetables have similar structural features, such as the degrees of substitution of xylosyl residues (10%~15%) and content of fucosyl residues (ca. 10%), while xyloglucans of lettuce and broccoli contained no xylosyl residues which are substituted by galactosyl residue.

Considering the plant taxonomy, it is interesting that Komatuna, Sakaina and Chinese cabbage belong to the same genus of *Brassica,* and have almost the same xyloglucan structural features. The small structural difference in the xyloglucan of broccoli, which belongs to the

Table 1. The chemical shifts (δ) of H-1 of constituent sugar residues of xyloglucans, substitution of α-D-xylosyl residues, and the percentages of the intensities of each H-1 of vegetable xyloglucans

Residue	H-1 (δ)	Substitution	Intensities of H-1 (%)						
			Sakaina	Komatuna	Chinese Cabbage	Lettuce	Broccoli	Tomato Minikyaroru	Sopi
α-Fuc	5.25-5.27		12.1	11.8	11.1	8.8	7.5	-	-
α-Ara	5.21-5.23		-	-	-	-	-	3.2	4.7
α-Xyl	5.15-5.16	Gal→	-	-	-	-	-	15.3	20.1
α-Xyl	5.15-5.17	Fuc→Gal→	7.3	7.3	7.6	4.4	5.0	-	-
α-Xyl	5.12-5.14	Fuc→Gal→	4.9	4.5	5.1	5.3	3.1	-	-
α-Xyl	5.00-5.02	Gal→	2.4	3.7	2.2	-	-	3.2	3.6
α-Xyl	5.08-5.09	Non-reducing	-	-	-	-	-	10.5	6.5
α-Xyl	4.94-4.95	Non-reducing	19.6	18.2	22.1	18.5	16.4	7.3	7.7
Glc+Gal	4.44-4.66		53.7	54.6	51.9	63.0	67.9	60.5	57.4

genus *Brassica*, is probably because due to the edible part of broccoli being the malformed flower and not the leaves. Xyloglucan of tomato also have a similar basic structure to those of green vegetables except for they have no fucosyl residues. Structural feature of tomato xyloglucan is very similar to that of tamarind xyloglucan reported by York et al. [9,10] except that the presence of arabinosyl residues in tomato xyloglucan suggested by this ^{1}H-NMR spectroscopy. Consequently, the lower reactivities of xyloglucans of green vegetables to anti-xyloglucan antibodies resulted from the presence of fucosyl residues at the non-reducing ends of the side chains as Fuc1→2Gal. As described in the previous paper [4], the antibody used in this ELISA is specific to Gal1→2Xyl1→6Glc unit. Therefore, fucosyl residues could prevent the green vegetable xyloglucan from binding to the antibody to some extent. The similar ELISA reactivities of xyloglucans of lettuce and broccoli to those of other green vegetables are probably explained by the hypotheses that the antibody mainly reacted with Fuc1→2Gal units by weak interaction, and the reactivities between the antibody and Gal1→2Xyl were weak due to the steric hindrance by Fuc1→2Gal unit.

From the results described above, we can summarize this study as the followings.

● Structural feature of tomato XG is very similar to that of tamarind xyloglucan except the presence of arabinose residues.

- XGs of green vegerables have similar structural features, such as the degree of substitution of xylosyl residue and content of fucosyl residue, while XGs of lettuce and broccoli contained no Gal1→Xyl side chain.
- The lower reactivities of XGs of green vegetables to anti-tamarind xyloglucan antibodies are resulted from the presence of fucosy residues at non-reducing ends of the side chains as Fuc1→Gal.

REFERENCES

1. M. McNeil, A.G. Darvill, S.C. Fry and P. Albersheim : *Annu. Rev. Biochem.*, **53** (1984) 625.
2. Y. Sone, C. Makino and A. Misaki : *J. Nutr. Sci. Vitaminol.*, **38** (1992) 391.
3. Y. Sone, K. Sato, and A. Misaki : *Agric. Biol. Chem.*, **55** (1991) 2155.
4. Y. Sone and Y. Fujikawa : *J. Nutr. Sci. Vitaminol.*, **39** (1993) 597.
5. Y. Sone, J. Kuramae, S. Shibata, and A. Misaki : *Agric. Biol. Chem.*, **53** (1989) 2821.
6. K. Kato and K. Matsuda: *Agric. Biol. Chem.*, **33** (1969) 1446.
7. J.K.M. Sanders and B.K. Hunter, Modern NMR Spectroscopy, 2nd edition, Oxford University Press, 1993
8. M. Hisamatsu, W.S. York, A.G. Darvill, and P. Albersheim: *Carbohydr. Res.*, **227** (1992) 45.
9. W. S. York, H. Halbeek, A.G. Darvill, and P. Albersheim: *Carbohydr. Res.*, **200** (1990) 9.
10. W.S. York, L.K. Harvey, R. Guillen, P. Albersheim, and A.G. Darvill: *Carbohydr. Res.*, **248** (1993) 285.

4. CELLULOSE

HYDROCOLLOIDS – PART 1
Edited by K. Nishinari
2000 Elsevier Science B.V.

251

Regiocontrolled cellulose functionalization: possibilities and problems

D. Klemm[a], W. Wagenknecht[b]

[a]Institut für Organische Chemie und Makromolekulare Chemie, Friedrich-Schiller-Universität Jena, Humboldtstraße 10, D-07743 Jena, Germany

[b]Fraunhofer Institut für Angewandte Polymerforschung, Kantstraße 55, D-14513 Teltow, Germany

Further progress in gel properties, surface interactions, and recognition processes of polysaccharides and their derivatives in an aqueous environment has to consider the site of functionalization, especially with hydrophilic groups, with regard to the properties aimed.

In this contribution the synthesis routes to regioselectively functionalized cellulose derivatives with sulfate, phosphate, carboxylate, and amino groups are compared with respect to advantages and shortcomings, centering on pathways via reactive intermediates.

Typical examples of the effect of regioselectivity on colloidal and surface/interaction properties are presented, e.g. on product solubility, anticlotting effects in connection with blood, analyt-sensitive cellulose reagent phases, and bacterial cellulose tubes for microsurgical uses.

1. INTRODUCTION

Cellulose as an hydrophilic polyhydroxy polymer is liable to gel formation in aqueous systems after rupture of inter-chain hydrogen bonds by partial derivatization or a suitable physical-mechanical treatment as demonstrated by the examples in Table 1.

Table 1
Aqueous gels from cellulose and its derivatives

Polymer	Remarks on gel formation and properties
Unsubstituted cellulose	Stabile gels from microcrystalline cellulose (degree of polymerization, DP 150 ... 200) after intense wet beating
Carboxymethylcellulose (CMC)	Adhesive gels, thickeners, and viscosity enhancers with CMC, degree of substitution, DS 0.5 ... 0.8
Cellulose xanthogenate	Gel formation with synersis on standing, porous cellulose sponges or cellulose beads after special routes of decomposition
Cellulose sulfate	Thermoreversible gels at DS 0.3 ... 0.5 with K^+
Cellulose acetate sulfate	Highly swollen insoluble gels (superabsorber)

These gel-forming cellulose derivatives were amply studied in the past decades with regard to the influence of DP and DS. But it is well-known that e.g. the rheological properties of aqueous CMC systems are not only depending on these macromolecular parameters, but also on the routes of syntheses and aftertreatment affecting the distribution of the functional groups. Thus, the problem of the role of the substitution pattern on properties arises not only with CMC but with all cellulosics forming supramolecular structures and effective inter-actions in an aqueous environment. It can be comprehensively solved by making samples available with a well-defined distribution of substituents within the repeating units as well as along the polymer chains of cellulose (Figure 1).

Figure 1. Molecular structure of cellulose.

The present contribution, largely based on a joint work of the groups in Jena and Teltow-Seehof, will therefore be centered on routes of syntheses of those types of gel-forming anionic cellulose derivatives. Moreover, the effect of regioselective functionalization on colloidal properties and intermolecular interactions of the samples in aqueous systems will be demonstrated by first examples.

2. ROUTES OF SYNTHESES

2.1. Some general remarks on regioselective cellulose derivatization
Among the reaction routes leading to regioselective cellulose derivatization [1-8] the following approaches are important:
(i) Primary regioselective partial functionalization of cellulose with transition from a hetero-geneous to a homogeneous reaction system and subsequent selective reaction of the free OH groups or the primary introduced groups. Example: Silylethers of cellulose.
(ii) Regioselective partial reaction of a trisubstituend (commercial) cellulose intermediate con-taining potential protecting or leaving groups. Examples: Selective deacetylation of cellulose acetate followed by sulfation; formation of cellulose trinitrite and its subsequent sulfation.
(iii) Regioselective competitive reaction of cellulosic OH group with two reagents.
Example: Acetosulfation of cellulose with acetic anhydride and chlorosulfonic acid followed by saponification of the acetate groups.
In the own work [6-8] the route via active intermediates containing protective groups for a subsequent reaction at the free OH groups (cellulose acetate, tritylcelluloses) or leaving groups acting in a consecutive substitution reaction (cellulose trinitrite, trimethylsilylethers, cellulosetosylate) was found to be convenient and suitable and thus had been preferred. The water content of the system can exert a decisive influence on the course of reaction and often

has to be kept down below 0.1 %. The properties of the products, especially the applicational once in connection with living matter, do by no means depend only on the proper route of synthesis itself, but also on the mode of product isolation especially from a homogeneous system, on product purification, and product aftertreatment.

But, if a block-like structure, i.e., a non-statistical substituent distribution along the chain being the goal of derivatization, the situation is somewhat different: The full availability of all the chains for the reaction is still a necessary prerequisite, but one of the reaction components has to be present in the form of fine solid particles in order to secure a point-like reaction at segments of contact with the polymer chains.

2.2. Syntheses of cellulose derivatives with a regioselective pattern of substitution within the anhydroglucose unit (AGU) via reactive intermediates

The acetyl groups in partially substituted cellulose acetates proved to be efficient protecting groups for regioselective synthesis of sulfuric acid half esters and phosphates of cellulose, as they resisted even the strongest acylation reagents in an acid medium, and as they could be split-off completely by a subsequent treatment with 4 % ethanolic NaOH without effecting the sulfuric acid half ester groups previously introduced (Figure 2).

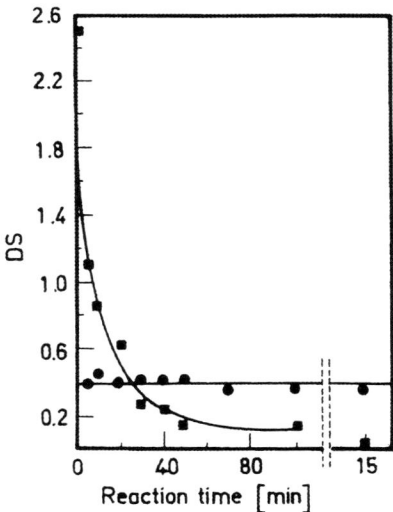

Figure 2.
Deacetylation of sodium cellulose acetate sulfate (degree of substitution, DS_{Acetyl} 2.5, DS_S 0.4) in ethanol containing 4 % NaOH at 20 °C (■ DS_{Acetyl}, ● DS_S) [9].

Figure 3 presents the course of deacetylation with reaction time in the 2,3 positions at one hand and the 6 position at the other in a mixture of dimethylsulfoxide (DMSO), water, and hexamethylenediamine. Also dimethylamine and hexylamine proved to be effective reagents for this purpose [7].

The effect of this regioselective deacetylation on a subsequent sulfation is shown

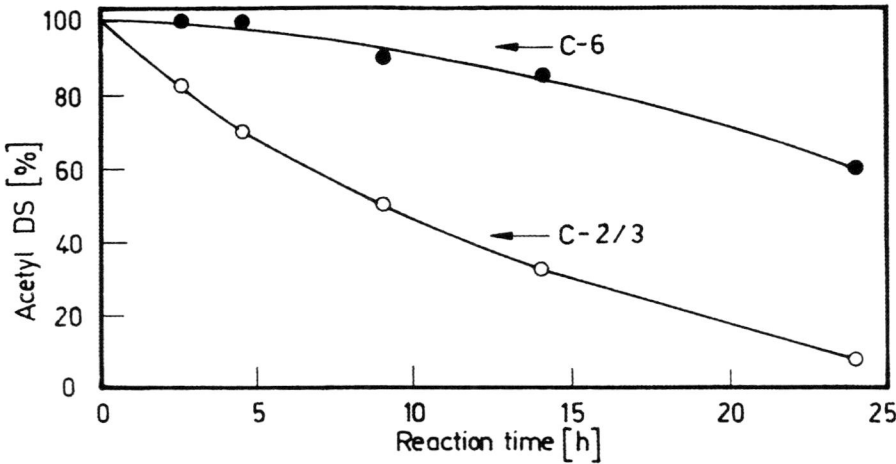

Figure 3. Distribution of acetyl groups after partial deacetylation of cellulose triacetate (DMSO, 80 °C, 2.3 mol hexamethylenediamine and 22 mol water per mol AGU).

in Table 2 with regard to substituent distribution and in comparison to results with a commercial acetate of the same total acetyl DS.

Table 2
Cellulose sulfates via partially 2,3 deacetylated cellulose triacetate in comparison to sulfates via commercial cellulose acetates (H_2NSO_3H, N,N-dimethylformamide (DMF), 80 °C, 2 h, subsequent complete deacetylation)

Cellulose acetate type	DS_{Acetyl}	Sulfating agent (mol/mol AGU)	DS_S	Sulfate group distribution		
				C-2	C-3	C-6
selective deac.	2.60	1	0.26	0.17	0.08	0.00
selective deac.	1.87	2	0.92	0.56	0.20	0.17
selective deac.	1.48	3	1.15	0.74	0.13	0.28
commercial	2.38	1	0.52	0.17	0.15	0.20
commercial	1.87	3	0.90	0.30	0.25	0.35

While sulfation of a regioselectively deacetylated sample with various sulfating agents leads to a pattern of substitution with the sulfate groups predominantly in the 2,3 positions, a cellulose 6-sulfate can be prepared up to a DS_S of about 0.8 with only minimal sulfation at C-2 and non substitution at all at C-3 by the competitive reaction of acetic anhydride and chlorosulfonic acid with a cellulose suspension in DMF/DMSO turning from a heterogeneous to a homogeneous system in the course of reaction (Table 3) [7].

As compared to sulfation, the synthesis of cellulose phosphates proceeds somewhat differently due to a lower reactivity of the oxides and chlorides of phosphorous in comparison to those of sulfur and presents additional problems due to formation of oligophosphates side chains and a strong tendency of crosslinking between the polymer chains by the trifunctional

Table 3
Cellulose sulfates via acetosulfation of cellulose (DMF, 50 °C, 15 h, subsequent complete de-acetylation)

Reagent (mol/mol AGU		DS_S	Sulfate group distribution		
$ClSO_3H$	$(CH_3CO)_2O$		C-2	C-3	C-6
0.7	8.0	0.50	< 0.10	0.00	0.40
2.0	8.0	0.75	0.05	0.00	0.70

reagents. While the sulfuric acid half esters in the form of the sodium salts generally show a complete water solubility above a DS of 0.2 ... 0.3, the anionic phosphates of cellulose are soluble in a rather small DS range only due to the counteracting effects of an increasing hydrophilicity at one hand, and an increasing crosslinking probability at the other with the DS of phosphate groups.

According to our experience, the formation of phosphates from partially substituted cellulose acetates can be performed with minor crosslinking only by the procedure of Whistler with polytetraphosphoric acid as the reagent in the presence of tributylamine. Also here the acetyl groups serve as reliable protecting groups. Commercial acetates with a statistic pattern of substitution exhibit some preference for the 6-phosphates resulting in products of rather good solubility at sufficient high phosphorous content. With a regioselectively deacetylated triacetate however, the anionic phosphate groups are forced predominantly into the C-2/C-3 positions resulting in products of rather poor solubility.

As an other choice along the route via reactive intermediates with a protective function trityl cellulose was employed in our studies. Especially with methoxy-substituted triphenylmethyl chloride a very high 6-regioselectivity inclusive a complete deprotection was observed permitting the synthesis of exclusively 2,3 functionalized products [10], e.g. of 2,3-carboxy- methylcelluloses [11]. The protective action of trityl or substituted trityl groups is very high even at strongly basic conditions, but a subsequent complete elimination of these protecting groups usually causes trouble and requires strong acidic conditions resulting in chain degradation.

Turning now to intermediates the functional groups of which serves as leaving groups in secondary reactions. The cellulose nitrite may be mentioned first [5]. The ONO-groups, which can easily be introduced up to a DS_N of 3, are very labile and thus can be replaced by other substituents. A sulfation of cellulose trinitrite in DMF solution can be achieved with various sulfating agents [12] with the final DS of sulfate groups increasing with the reagent to polymer ratio at first but then levelling-off at a DS_S usually below 1, as demonstrated in Figure 4 for sulfur trioxide, nitrosylsulfuric acid and also sulfur dioxide which forms nitrosylsulfuric acid *in situ* as the active agent.

The 6 position is the preferred site of reaction with all the sulfation agents studied at room temperature. A really good 6 regioselectivity was found with nitrosylsulfuric acid or SO_2, respectively, while with sulfuryl chloride a rather high 2 sulfation took place. Interesting to note is the strong effect of the reaction temperature on the pattern of substitution obtained with sulfur trioxide, a preferred 2-functionalization being found at low temperature in agreement with earlier observations of Schweiger.

A further suitable way of regioselective cellulose sulfation consists in the reaction of trimethylsilylcelluloses [13] with SO_3 in an aprotic medium the silyl ether groups can be replaced by sulfate groups via a cellulose silyl sulfate as intermediate and with the reaction

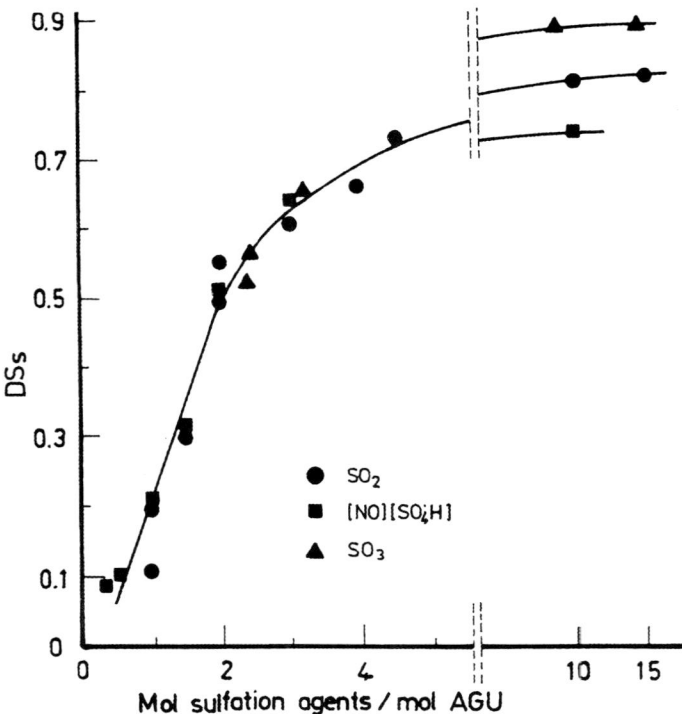

Figure 4. Sulfation of cellulose nitrite.

coming to a standstill at high reagent to polymer ratio at a DS level of sulfate groups equal to that of the silylether groups in the starting material (Figure 5) [5,7,14].

The free OH groups present in the AGU obviously do not participate in these reaction, with the cause of this phenomenon still being a matter of speculation. With a trimethylsilyl-cellulose of DS 1.5 exclusively 6 substituted cellulose sulfates could be obtained up to a DS_S of 0.9 by employing a limited amount of reagent only, while with a highly substituted silylether of DS 2.4 a complete sulfation of the positions at 6 and 2 and a partial sulfation of

Table 4
Cellulose sulfates via trimethylsilylcelluloses in tetrahydrofurane (THF)/DMF, 3 h at 20 °C

Trimethylsilylcellulose DS_{Si}	Reagent type	mol/mol AGU	DS_S	Sulfate group distribution C-2	C-3	C-6
1.55	SO_3	2	1.20	0.40	0.00	0.80
1.55	$ClSO_3H$	2	0.95	0.00	0.00	0.95
2.40	SO_3	3	1.80	0.60	0.20	1.00
2.40	SO_3	9	2.20	1.00	0.20	1.00

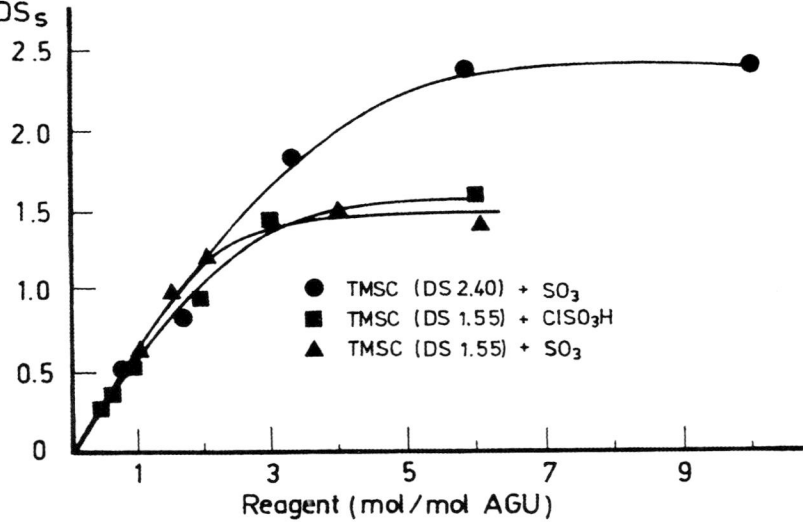

Figure 5. Sulfation of trimethylsilylcelluloses ($\bullet$DS$_{Si}$ 2.40 + SO$_3$, $\blacksquare$ DS$_{Si}$ 1.55 + ClSO$_3$H, $\blacklozenge$ DS$_{Si}$ 1.55 + SO$_3$).

the 3 position were found (compare Table 4).

Residual silyl groups can be eliminated also here by treatment with a suitable protic medium without effecting the sulfate groups, but the steps of isolation and purification the cellulose sulfates is somewhat more tedious and complicated due to a complete purification with respect to silicon-containing by-products.

A very interesting reactive intermediate with leaving group activity is a preferentially 6 substituted tosylcellulose obtained and subsequently functionalized according to Figure 6. The cellulose-deoxy-thiosulfates formed in a subsequent reaction are watersoluble and film-forming polymers suitable for different types of subsequent reactions e.g. crosslinking in the presence of oxidizing agents like hydrogen peroxide [8].

In DMSO at 100 °C the reaction of tosylcellulose (DS 2.3) with different types of diamines (e.g. p-phenylenediamine, PDA) leads by regioselective substitution of the 6-tosyl groups to the corresponding 6-deoxy-6-amino derivatives within 3 h. These PDA celluloses are suitable soluble and film-forming supports with controlled hydrophilic/hydrophobic properties for preparing biomedical sensors by coupling of enzymes like glucose oxidase and redox-chromogenic aromatic structural units [15].

Figure 6. Synthesis of cellulose-deoxy-thiosulfates (Bunte salts) via cellulose-p-toluenesul-
fonates]8].

2.3. Synthesis of block-like cellulose structures

A non-statistical substituent distribution along the cellulose chains should be possible by
reacting the dissolved polymer with two reaction components one of which is in the dissolved
state too, whilst the other one is present in the form of small solid particles. This principle was
recently realized by Heinze with a so-called induced phase separation technique [16-19]. As
described for carboxymethylation of cellulose to products with a combination of highly and
poorly substituted chain segments, cellulose is at first dissolved in a nonderivatizing system
like N,N-dimethylacetamide (DMA)/LiCl and subsequently reacted with monochloroacetic
acid also dissolved in the system and finely powdered sodium hydroxide present in the solid
state. A non-statistical substituent distribution along the chains could be proven by chromatographic analysis of the fragments obtained after hydrolytic cleavage of the glucosidic bonds.

Miyamoto et al. developed the synthesis of methyl celluloses with block-like structures [4]
and Mischnick et al. has been prepared and analyzed methyl starches of this type [20].

3. IMPACT OF REGIOSELECTIVITY OF FUNCTIONALIZATION ON PRODUCT PROPERTIES

A regioselective non-statistical substituent distribution within the AGU obviously affects
predominantly intermolecular non-covalent interactions in solution or at surfaces. We
compared the DS ranges of solubility of cellulose acetates with a rather statistical distribution

of functional groups and regioselectively in the 2,3 positions deacetylated samples [7]. A dominating influence of the 6 position in determining solubility can be concluded. In the case of many free OH groups in this position a good solubility in highly polar solvents like water results while, on the other hand, an acetyl group substitution at O-6 promotes solubility in less polar solvents. Quite in agreement with this observation is the complete water solubility of C-6 modified sodium cellulose sulfates already at a significantly lower DS than observed with 2,3 substituted samples.

A C-6 substituted cellulose sulfate is very effective in selective water transport through a polyelectrolyte complex membrane (cellulose sulfate as an anionic and polydimethyl-diallylammoniumchloride as the cationic component) in separating an ethanol/water mixture by pervaporation. A rather high C-2 substitution with sulfate groups, at the other hand, conveys to the cellulose molecule a strong anticlotting effect in connection with human blood (Table 5) [7, 21].

Table 5
Anticlotting effects of sodium cellulose sulfates in relation to the sulfation pattern

DS_S total	C-2	C-3	C-6	Thrombin time (s) 25 µg/ml blood	Partial thromboplastin time (s) 25 µg/ml blood
0.95	0.00	0.00	0.95	18.9	80.8
0.95	0.30	0.30	0.35	29.0	136.5
0.95	0.55	0.20	0.20	> 600	294.3
1.14	0.74	0.09	0.31	> 600	> 600

A regioselectively 2,3 substituted cellulose phosphate, but not a sample with the phosphate groups in the C-6 position, was found to decrease significantly the activation of detrimental proteins in hemodialysis, demonstrating again the sensitivity of cellulose interaction with living matter to the pattern of substitution within the AGU.

4. REGIO- AND STEREOSELECTIVE SYNTHESIS OF *BASYC* BIOMATERIALS

Starting from D-glucose *Acetobacter xylinum* produces regio- and stereoselectively cellulose microfibrils at the air/liquid interface of a culture medium. By using special technologies BActerial SYnthesized Cellulose (BASYC) hollow fibers (inside diameter of abaut 3 mm) have been prepared without application of additional physical methods. These tubes represent an unique supramolecular structure of a high crystalline and high molecular insoluble polymer with high water absorption capacity and mechanical stability in the wet state. In combination with inner surfaces properties, comparable with those of arteries and veins of the rat, the designed „nature like" biomaterial is useful in the microsurgery of vessels and nerves.

ACKNOWLEDGEMENT

The financial support of the author's work on regioselective cellulose functionalization by the Bundesministerium für Bildung, Forschung und Technologie, by the Deutsche Forschungsgemeinschaft, and by the Fonds der Chemischen Industrie of Germany is gratefully acknowledged. The authors are especially indebted to Prof. Dr. Burkart Philipp, Priv.-Doz. Dr. Thomas Heinze, Dr. Brigitte Heublein, Dr. Ute Heinze, Dr. Armin Stein, Dr. Juan Camacho Gomez, Dr. Lars Einfeldt, Ulrike Udhardt, and Jörg Tiller for their creative and helpful cooperation and experimental work in the field of the described cellulose derivatives.

REFERENCES
1. T. Kondo, Carbohydr. Res., 238 (1993) 231.
2. T. Kondo, J. Polym. Sci. B: Polym. Phys., 32 (1994) 1229.
3. M. Nojiri and T. Kondo, Macromolelcules, 29 (1996) 2392.
4. H-Q. Liu, L.-N. Zhang, A. Takaragi and T. Miyamoto, Macromol. Rapid Commun., 18 (1997) 921 .
5. D. Klemm, B. Philipp, T. Heinze, U. Heinze and W. Wagenknecht, Comprehensive Cellulose Chemistry (Vol. 1 and 2), Wiley-VCH, Weinheim, 1998.
6. D. Klemm, A. Stein, Th. Heinze, B. Philipp and W. Wagenknecht, In J.C. Salamone (ed.) Polymeric Materials Encyclopedia: Synthesis, Properties, and Applications, pp. 1043-1054, CRC Press, Inc., Boca Raton (USA), 1996.
7. D. Klemm, Th. Heinze, B. Philipp and W. Wagenknecht, Acta Polym., 48 (1997) 277.
8. D. Klemm, In Th. Heinze and G. Glasser (ed.) Cellulose Derivatives: Modification, Characterization and Nanostructures, ACS-Symp. Ser. No 688, pp. 19-37, Amer. Chem. Soc., Washington (USA), 1998.
9. W. Wagenknecht, I. Nehls, J. Kötz and J. Ludwig, Cellul. Chem. Technol., 25 (1991) 343; W. Wagenknecht, B. Philipp, I. Nehls, M. Schnabelrauch, D. Klemm and M. Hartmann, Acta Polym., 42 (1991) 213; 554.
10. J.A. Camacho Gomez, U. Erler and D. Klemm, Macromol. Chem. Phys., 197 (1996) 953.
11. Th. Heinze, K. Röttig and I. Nehls, Macromol. Rapid. Commun., 15 (1994) 311.
12. W. Wagenknecht, I. Nehls and B. Philipp, Carbohydr. Res., 240 (1993) 245.
13. D. Klemm and A. Stein, J. Macromol. Sci. A: Pure Appl. Chem., 32 (1995) 899; A. Stein, W. Wagenknecht, D. Klemm and B. Philipp, DD No. 298 644, (1989); Chem. Abstr. 117: 92 503.
14. W. Wagenknecht, I. Nehls, A. Stein, D. Klemm and B. Philipp, Acta Polym., 43 (1992) 266; I. Nehls, Habil. Thesis, University of Potsdam, 1994.
15 J. Tiller, P. Berlin and D. Klemm, Macromol. Chem. Phys., 200 (1999) 1.
16. Th. Heinze, Habil. Thesis, University of Jena, 1997; Th. Heinze, U. Erler, I. Nehls and D. Klemm, Angew. Macromol. Chem., 215 (1994) 93; T. Liebert, D. Klemm and Th. Heinze, J. Macromol. Sci. - Pure Appl. Chem., 33 (1996) 613.
17. Th. Heinze and K. Rahn, Macromol. Symp., 120 (1997) 103.
18. U. Heinze, Ph.D. Thesis, University of Jena, 1998.
19. U. Heinze, Th. Heinze and D. Klemm, Macromol. Chem. Phys., in press.
20. P. Mischnick and G. Kühn, Carbohydr. Res., 290 (1996) 199.
21. D. Klemm, Th. Heinze and W. Wagenknecht, Ber. Bunsenges., 100 (1996) 730.

HYDROCOLLOIDS – PART 1
Edited by K. Nishinari
2000 Elsevier Science B.V.

Interaction between cellulosic polysaccharides and water

H. Hatakeyama[a]   and T. Hatakeyama[b]

[a]Department of Applied Physics and Chemistry,   Fukui University of Technology,   3-6-1, Gakuen, Fukui 910-8505, Japan

[b]Department of Textile Science, Otsuma Women's University,   12, Sanban-cho, Chiyoda-ku, Tokyo 102-8357, Japan

Cellulosic polysaccharides with   hydrophilic groups, such as hydroxyl and carboxyl groups, have a strong interaction with water. Thermal properties of both polysaccharides and water are markedly influenced through this interaction.   The first-order phase transition of water fractions closely associated with the polysaccharide matrix is usually impossible to observe. Such fractions are called non-freezing water.   Less closely associated water fractions exhibit melting and crystallization showing considerable super cooling and significantly smaller enthalpy than that of   bulk water.   These water fractions   are referred to as freezing bound water. The sum of the freezing bound and non-freezing water fractions is the bound water content.   Water whose melting/crystallization temperature and enthalpy are not significantly different from those of normal (bulk)   water is designated as freezing water.

Bound water in water-insoluble hydrophilic polysaccharides,   such as cellulose breaks hydrogen bonding between the hydroxyl groups of the polysaccharide molecules. The bound water content   depends on the chemical and higher-order structure of each polysaccharide. It has also been observed that various kinds of   cellulosic polyelectrolytes with mono- and divalent cations, such as carboxymethylcellulose and cellulose sulfates,   form thermotropic /lyotropic liquid crystals in the water content ranging from 0. 5 to ca. 3.0   [(grams of water)/(gram of polymer)]. NMR studies have shown that water in the above liquid crystalline system is in a state ranging from non-rigid solid to viscous liquid and that the motion of counter ions in the system is profoundly influenced by the molecular motion of water.

1. INTRODUCTION

Various natural polymers with   hydrophilic groups, such as hydroxyl, carboxyl and carbonyl groups,   have a strong or weak interaction with water. Thermal properties of both polymers and water are markedly influenced through this interaction.
Cellulose is the most important hydrophilic but water-insoluble polysaccharide which consists in plant cell walls. However, cellulose derivatives with carboxyl or sulfate groups are water-soluble polymers and their interaction with water is different from that of the interaction between cellulose and water.
This paper reports the interaction between water and cellulosic polysaccharides such as cellulose,   sodium carboxymethylcellulose (NaCMC) and sodium cellulose sulfate (NaCS). Figure 1 shows the schematic chemical structures of cellulose,   NaCMC and NaCS.

2. INTERACTION BETWEEN HYDROPHILIC POLYMERS AND WATER

Figure 2 shows schematic   DSC cooling curves of water sorbed on hydrophilic polymers. As shown in Figure 2, when the amount of water sorbed on a hydrophilic polymer is very small, no first-order phase transition is detected until a critical amount of water is added to a polymer (curve A in Figure 1). The amount of water is defined as non-freezing water content (W_{nf}). The maximum

CH₂OH CH₂OH

Cellulose

CH₂OSO₃Na CH₂OH

Cellulcse Sulfate

CH₂OCH₂CO₂Na CH₂OCH₂CO₂Na

Carboxymethylcellulose

Figure 1. The schematic chemical structures of cellulose, NaCS and NaCMC.

W_{nf} depends on hydrophylicity of polymers. When W_c in the polymers exceeds a critical amount (the minimum amount of W_{nf}), a small peak (peak II) is observed at a temperature lower than the crystallization peak of bulk water (curve B). The amount of water is categorized as the freezing bound water content (W_{fb}). Free water (W_f) is shown as peak I in curves C and D. W_f is the unbound water (free or bulk water) content in polymers whose transition temperature and enthalpy are equal to those of pure water (curve E). Concerning the water content (W_c), the following definition will be used in this paper. Water content (W_c) = (grams of water)/(gram of dry polymer), (g/g)

$$W_c = W_{nf} + W_{fb} + W_f, (g/g)$$

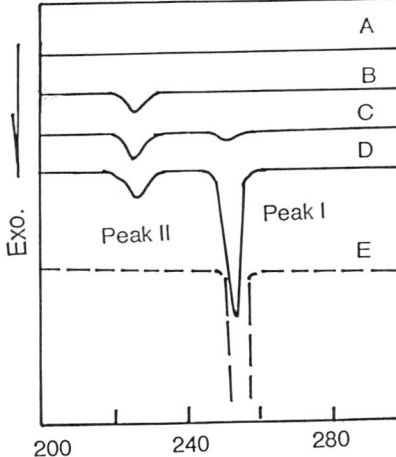

Figure 2. Schematic DSC cooling curves of water sorbed on hydrophilic polymers: A, non-freezing water (W_{nf}); B, W_{nf} + freezing bound water (W_{fb}); C and D, W_{nf} + W_{fb} + free water (W_f); E, free (bulk) water.

From the cooling cycle data, the proportion of the amount of free water (W_f) is calculated dividing the total area of the freezing water peak (Peak I) by the heat of crystallization of bulk water.

Ordinarily, the amount of W_{fb} is small compared with W_{nf} (g/g) and W_f (g/g). On this account, the total area of the $W_f + W_{fb}$ (g/g) (peak I + II) per gram of dry sample is plotted as a function of W_c (g/g). The intercept of the linear plot is adopted as the amount of W_{nf} (g/g).

Usually in the heating run, peak II merges to the peak I and in some cases, a shoulder peak is observed at the low temperature side of the melting peak. Since the melting peak appears at a lower temperature than 0 °C in the heating run of a hydrophilic polymer with water, the formation of irregular ice is indicated. In general, the enthalpy of melting (ΔH_m) is larger than that of crystallization (ΔH_c). This indicates the following facts: (1)super cooling of water and (2)a part of water recrystallizes during heating.

3. INTERACTION BETWEEN CELLULOSE AND WATER

Many authors have reported that water sorbed by cellulose has properties that are markedly different from free water [1-7]. It has been considered that this difference in water properties is caused by the restriction of the molecular motion of water.

Figure 3 shows the schematic DSC cooling curves of water sorbed on various cellulose samples. In the case of cotton and linen celluloses (curves A and B in Figure 3), the starting temperature of crystallization Peak II is observed at around 232 K. However, in the case of jute and wood cellulose (curves C and D), Peak II is observed as a larger peak than those of cotton and linen and the temperature range for Peak II is higher than that for cotton and linen. This suggests that the water which appears as Peak II is affected by the higher-order structure of cellulose samples, although Peak I does not show the significant dependency on the higher-order structure of cellulose.

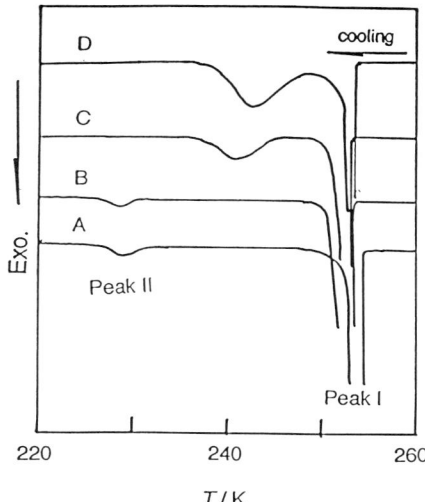

Figure 3. The schematic DSC cooling curves of water sorbed on various cellulose samples. A, cotton; B, linen; C, jute; D, wood cellulose

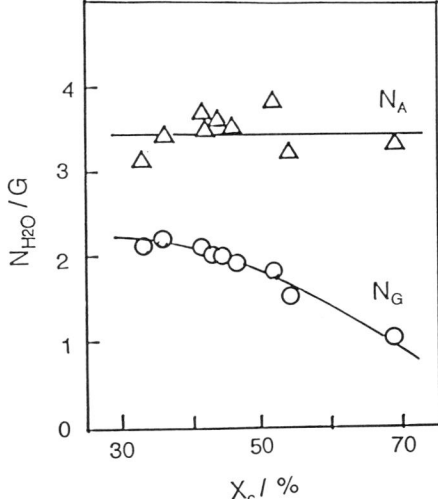

Figure 4. The relationship between degree of crystallinity of cellulose samples and the moles of bound water sorbed on the glucose unit of cellulose molecules and those in the amorphous region of cellulose. N_G: moles of bound water sorbed per glucose unit of cellulose samples. N_A: moles of bound water sorbed per glucose unit in the amorphous region of cellulose samples

In the case of the crystallization of water sorbed on cellulose samples, the sum of the mass of water calculated from the enthalpies of Peaks I and II is less than the total amount added to the dry sample. As mentioned previously, the amount of water corresponding to the difference between the added water and the amount of water calculated is known as non-freezing water. The non-freezing water has none of the first-order transition and does not have any kind of crystalline structure.

Accordingly, the mass of bound water (W_b) is shown as follows.

$$W_b = W_{nf} + W_{fb}$$

The percentage of bound water content, C_b, is calculated according to the following equation.

$$C_b = W_b/W_s \times 100 \ (\%)$$

where W_s is weight of sample.

The number of bound water molecules, N_G, sorbed on a glucose unit of cellulose can be calculated according to the following equation.

$$N_G = (162 \times W_b)/(18 \times W_s) = (9 \times W_b)/W_s = 0.09 \ C_b$$

where the numerals 162 and 18 are molecular weights of a glucose unit of cellulose and water. The number of bound water molecules sorbed on a glucose unit in the amorphous region of cellulose samples, N_A, is calculated according to the following equation,

$$N_A = (N_G \times 100)/(100\text{-degree of crystallinity})$$

Figure 4 shows the relationship between degree of crystallinity of cellulose and the number of bound water molecules (N_G) sorbed on a glucose unit of cellulose molecules and those (N_A) in the amorphous region of cellulose [6]. Crystallinity of cellulose samples was measured by x-ray diffractometry. As seen in Figure 5, N_G decreases with increasing degree of crystallinity of cellulose in a similar way to that of the bound water content, while N_A is nearly constant at about 3.4 moles per glucose unit. This fact suggests that bound water attaches to three hydroxyl groups in the amorphous region of cellulose.

4. INTERACTION BETWEEN CELLULOSIC POLYELCTROLYTES (NaCMC AND NaCS) AND WATER

The previous experimental results have revealed that polyelectrolytes such as NaCMC and NaCS in highly concentrated aqueous solutions have a unique property that is capable of forming a liquid crystalline phase. This characteristic originates from the fact that water molecules associated with polyelectrolytes have a conformation different from ordinary molecular packing. When water molecules are directly bonded to the ionic groups of polyelectrolytes, the mobility of the water decreases markedly [8-15].

Fig. 5 shows representative DSC curves of the water-sodium carboxymethylcellulose (NaCMC) system with $W_c = 1.26$ (g/g). Curve A shows two exotherms at 268 K and 245 K. When the sample was reheated, as shown in curve B, a step-wise change in the baseline at 190K, a broad exotherm at around 240 K and two endothermic peaks at 265 K and 330 K are observed. The large endothermic peak observed at 265 K is attributed to the melting of free water

(T_m). The small endothermic peak observed at 330 K is attributed to the transition (T^*) from the liquid crystalline state to the isotropic liquid state.

This was supported by observations of the presence of the liquid crystalline phase in the temperature ranging between T_m and T^* using a polarizing microscope [16-17]. The peak corresponding to T^* appeared when W_c was ca. 0.4 (g/g). The peak temperature (T^*) decreased with increasing W_c and was barely observable when W_c was 1.5 (g/g). When the sample was quenched, as shown in curve C, the step-wise change in the baseline at 190 K and the exotherm at around 240 K became prominent. Since the temperature corresponding to the step-wise change in the baseline showed a heating rate dependency and also the enthalpy relaxation by slow cooling or annealing at a temperature slightly lower than the endothermic deviation of the baseline, this transition was attributed to the glass transition (T_g). The exotherm at about 240 K was attributed to a cold crystallization (T_{cc}) of the system via which the glassy state was transformed to the crystalline state. Other water-CMC metal salt systems including LiCMC, KCMC and MgCMC showed similar phase transitions as shown in Figure 5 [18].

Figure 6 shows the DSC curves of the water-NaCS system with various W_c (curve A, pure water; curve B, W_c =2.03; curve C; W_c = 0.80; curve D, W_c =0.52). This figure shows that the water-NaCS system consists of glassy, crystalline, liquid

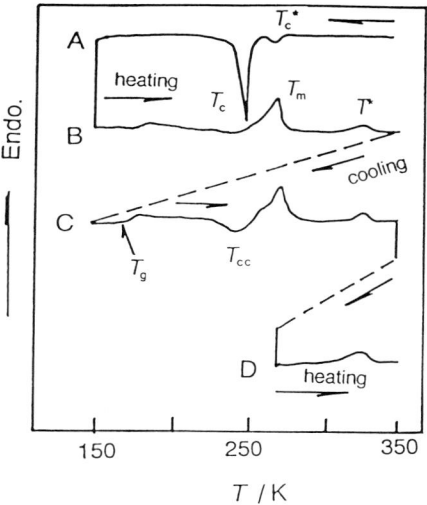

Figure 5. Schematic DSC curves of the water-sodium carboxymethylcellulose (NaCMC) system with W_c = 1.26 (g/g). Scanning rate: 10 K/min.

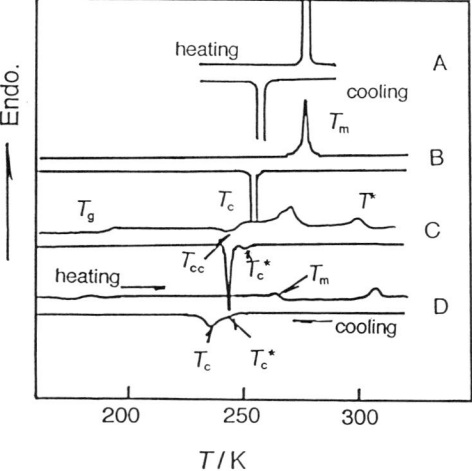

Figure 6. Schematic DSC curves of the water-sodium cellulose sulfate (NaCS) system with various W_c's (curve A, pure water; curve B, W_c =2.03; curve C; W_c = 0.80; curve D, W_c =0.52). Scanning rate: 10 K/min.

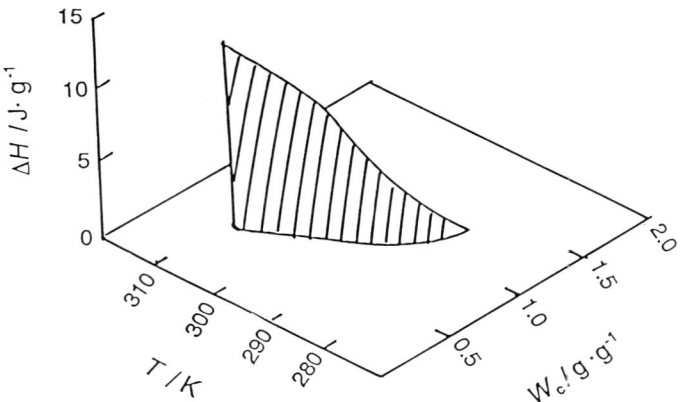

Figure 7. Three-dimensional relationship between enthalpy of liquid crystal transition (ΔH^*), temperature and W_c in the water-NaCS system.

crystalline and isotropic liquid states.

The transition temperature (T^*) from the liquid crystalline to the isotropic liquid phase decreased with increasing W_c in the W_c range between ca. 0.4 and 1.4 (g/g), although melting and cold crystallization temperatures (T_m and T_{cc}) increased with increasing W_c. However, T_g showed a minimum at around $W_c = 0.4$ (g/g). This phenomenon is explained as follows. Below W_c = 0.4 (g/g), T_g of the system decreases with increasing W_c because the molecular motion of the system increases with increasing W_{nf}, while above $W_c = 0.4$ (g/g) T_g of the system increases because the formation of ice begins to restrict the molecular motion of the system with increasing W_f.

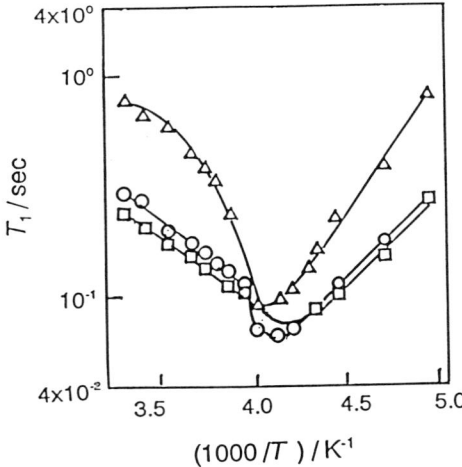

Figure 8. Temperature dependence of ^1H T_1 values of water in the NaCS-water system at various W_c's.

W_c : triangle, 1.58; circle, 1.24; square, 1.03

The enthalpy of liquid crystal transition (ΔH^*) in the water-NaCS system was calculated from the peak area of T^*. As shown in Figure 7 the enthalpy decreased with increasing W_C. The fact that both the heat of the transition (enthalpy, ΔH) from the liquid crystalline phase to the isotropic liquid phase and the transition temperature (T^*) decreases with increasing W_C suggests that the presence of mobile water molecules disturbs the regular alignment of NaCS molecules in the water-NaCS system.

Figures 8 and 9 show the changes in ^{1}H longitudinal relaxation time (T_1) and transverse relaxation time (T_2) of the NaCS-water system as a function of W_C at various temperatures, where temperature is shown as inverse absolute temperature ($1000/T$, K^{-1}) [16] . A broad minimum value for T_1 is observed at around $1000/T = 4.0$ K^{-1}). The T_2 value decreased with decreasing temperature. The T_2 value indicate the average molecular motion of water molecules in the NaCS-water system. Accordingly, the above facts suggest that the average motion of water molecules in the system decreases with decreasing temperature. In the lower temperature below 250 K ($1000/T = 4.0$ K^{-1}), the molecular motion of water is strongly restricted. As shown in Figures 6 and 7, the DSC measurements have indicated that the water molecules in the NaCS-water system are strongly associated with NaCS molecules, forming bound water which has an important role in the formation of the

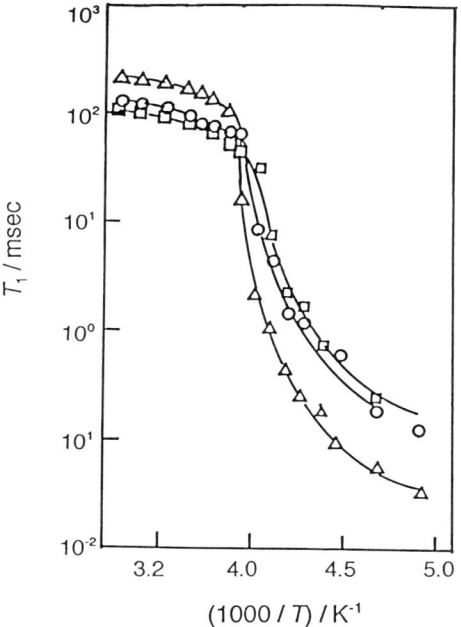

Figure 9. Temperature dependence of ^{1}H T_2 values of water in the NaCS-water system at various W_C 's.
W_C : triangle, 1.58; circle, 1.24; square, 1.03

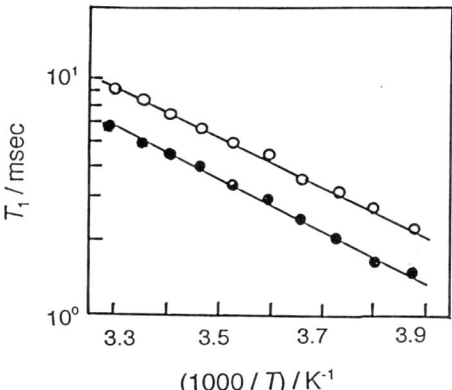

Figure 10. Temperature dependence of ^{23}Na T_1 values in the NaCS-water system at various W_C 's.
W_C : open circle, 1.84; closed circle, 0.96

268

liquid crystal phase of the system.

Figures 10 and 11 show the changes in ^{23}Na longitudinal relaxation time (T_1) and transverse relaxation time (T_2) of the NaCS-water system [16]. The T_1 values increase with increasing W_C and temperature. The logarithmic (ln) T_1 plots are linear in the temperature range where T_1 values were observed in this study.

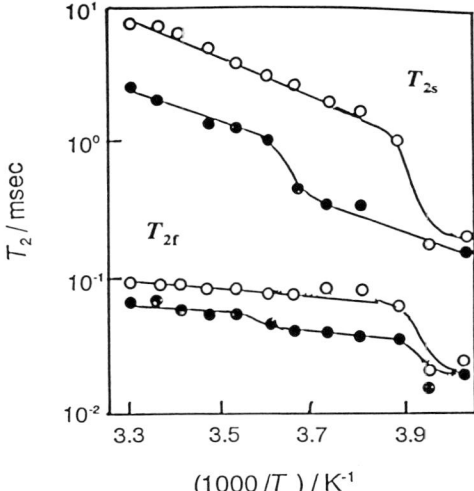

Figure 11. Temperature dependence of ^{23}Na T_2 values in the NaCS-water system at various W_C 's.

W_C : open circle, 1.84; closed circle, 0.96

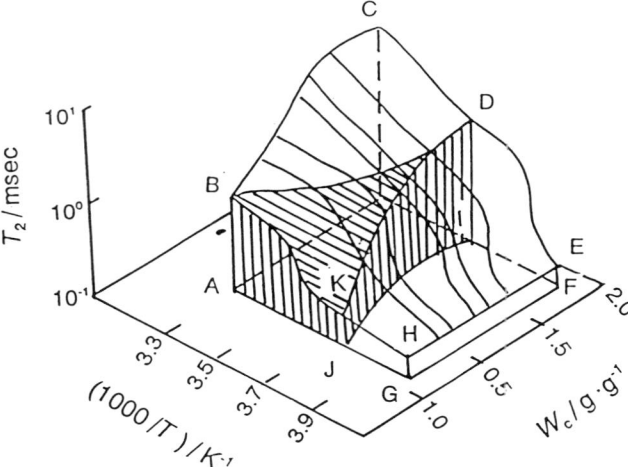

Figure 12. A three-dimensional diagram showing the relationship between ^{23}Na T_2 values, temperatures and W_C 's in the NaCS-water system. The shadowed region , ABDIJK, indicates the liquid crystalline region.

The apparent activation energy (E_a) of the relaxation process was calculated from the slope of the straight line of $\ln T_1$ vs. the inverse absolute temperature. The calculated E_a values were ca. 23 kJ/mol at $W_C = 0.8$ and 16 kJ/mol at $W_C = 1.8$, decreasing with increasing W_C. This fact suggests that the motion of ^{23}Na is strongly restricted in the low W_C region.

Figure 11 shows the presence of two types of transverse relaxation processes of ^{23}Na that are slow (s) and fast (f) relaxations. It is known that the transverse relaxation of ^{23}Na decays in a non-exponential manner and that the transverse relaxations which are produced by a quadrupole are the sum of two decaying exponentials [19]. The longer transverse relaxation time (T_{2s}) increases with increasing W_C. The T_{2s} values observed at around 270 K, when low W_C is lower than ca. 1.0, showed a sudden increase in the region where W_C is over 1.0. This fact probably reflects the change in the motion of ^{23}Na which is strongly influenced by the motion of water in the system. The shorter relaxation time (T_{2f}) does not change greatly with W_C.

Figure 12 is a three-dimensional diagram showing the relationship between ^{23}Na T_2 values, temperatures and W_C in the NaCS-water system. Usually the transition temperature observed by DSC does not agree well with the temperature observed by NMR, since the molecular motion corresponding to the frequencies measured by DSC and NMR is different. Accordingly, as shown in Figure 12, the shadowed region, ABDIJK, indicates the liquid crystalline region does not coincide well with the motion of ^{23}Na in the NaCS-water system. However, when W_C is low, the ^{23}Na T_2 relaxation process where the water molecules are strongly associated with NaCS seems to agree with the formation of the liquid crystal, although, when W_C is high, the motion of ^{23}Na does not agree well with the behaviour of the NaCS-water system, since the matrix of the NaCS-water system is strongly affected by the motion of water molecules. DSC and wide angle X-ray diffractometry (WAXS) data indicated that each sodium ion is sandwiched by two cellulose molecules coexisting with 4 - 20 water molecules in the liquid crystalline state. When W_c exceeds the critical value, free water molecules enhance the molecular motion of NaCS and molecular chains are arranged in a random form [20].

5. CONCLUSIONS

Cellulosic polysaccharides with hydrophilic groups, such as hydroxyl and carboxyl groups, have a strong interaction with water and the thermal properties of both polysaccharides and water are markedly influenced through this interaction.

The first-order phase transition of the above water fractions which are closely associated with the polysaccharide matrix is usually impossible to observe. Such fractions are called non-freezing water. Less closely associated water fractions exhibit melting and crystallization showing considerable super cooling and significantly smaller enthalpy than that of bulk water. These water fractions are referred to as freezing bound water.

Bound water in water-insoluble hydrophilic polysaccharides, such as cellulose breaks hydrogen bonding between the hydroxyl groups of the polysaccharide molecules. The bound water content depends on the chemical and high-order structure of each polysaccharide.

Various kinds of cellulosic polyelectrolytes with mono- and divalent cations, such as carboxymethylcellulose and cellulose sulfates, form thermotropic /lyotropic liquid crystals in the water content ranging from 0.5 to ca. 3.0 [(grams of water)/(gram of polymer)].

NMR studies of cellulosic polyelectrolytes with water have shown that water molecules in the above liquid crystalline system are in a state ranging from non-rigid solid to viscous liquid and that the motion of counter ions in the system is profoundly influenced by the molecular motion of water.

REFERENCES

1. F. C. Magne, H. J. Portas and H. A. Wakeham, J. Amer. Chem. Soc., 69 (1947) 1896.
2. F. C. Magne and E. L. Skau, Textile Res. J., 22 (1952) 748.
3. J. E. Ayer, J. Polym. Sci., 21 (1956) 455.
4. M. F. Froix and R. A. Nelson, Macromolecules, 8 (1975) 726.
5. R. A. Nelson, J. Appl. Polym. Sci., 21 (1977) 645.
6. K. Nakamura, T. Hatakeyama and H. Hatakeyama, Textile Res. J., 51 (1981) 607.
7. L. Salmen, (Ed. C. F. Baker), Transactions of the Tenth Fundamental Research Symposium, Oxford, 1993, p.369.
8. H. Hatakeyama, H. Yoshida and T. Hatakeyama, "Cellulose and Its Derivatives", (J. F. Kennedy, G. O. Phillips, D. J. Williams, eds.) Ellis Horwood, Chichester (1985) p. 255.
9. H. Hatakeyama, H. Iwata and T. Hatakeyama, "Wood and Cellulosics", J. F. Kennedy, G. O. Phillips, P. A. Williams, Eds., Ellis Horwood, Chichester (1987) p.39.
10. H. Hatakeyama, K. Nakamura and T. Hatakeyama, "Cellulose and Wood - Chemistry and Technology", (C. Schuerch ed.) John Wiley & Sons, N.Y. (1989) p. 419.
11. H. Hatakeyama and T. Hatakeyama, "Cellulose-Structural and Functional Aspects", (J. F. Kennedy, G. O. Phillips, P. A. Williams eds.) Ellis Horwood, Chichester (1989) p.131.
12. H. Hatakeyama, Kagaku to Kogyo (Chemistry and Chemical Industry, CSJ), 42 (1989) 878.
13. T. Hatakeyama and H. Hatakeyama, Poly. Adv. Tech., 1 (1990) 305.
14. H. Hatakeyama and T. Hatakeyama, "Properties of Ionic Polymer -Natural and Synthesis", L. Salmen, M. Htun eds., STFI Series A-989, Stockholm (1991) p. 123.
15. T. Hatakeyama and H. Hatakeyama, "Viscoelasticity of Biomaterials", (W. G. Glasser, H. Hatakeyama eds.) ACS Symp. Series 489 (1992) p. 329.
16. T. Hatakeyama, N. Bahar and H. Hatakeyama, Sen-i Gakkaishi, 47 (1991) 417.
17. H. Hatakeyama, J. National Inst. Material & Chemical Research, 1 (1993) 65.
18. T. Hatakeyama, H. Hatakeyama and K. Nakamura, Thermochimica Acta, 253 (1995) 1.
19. P. S. Hubbard, J. Chem. Phys., 53 (1970) 985.
20. T. Hatakeyama, H. Yoshida and H. Hatakeyama, Thermochimica Acta, 226 (1995) 343.

HYDROCOLLOIDS – PART 1
Edited by K. Nishinari
2000 Elsevier Science B.V.

271

Gelation of cellulose/calcium thiocyanate solution and it s application to spinning for fiber.

M.Hattori, Y.Shimaya and M.Saito

Central Research Laboratories, Asahi Chemical Industries Co. Ltd., 11-7 Hacchonawate, Takatsuki, Osaka 569-0096, Japan

For the purpose to develop regenerated cellulose fiber with high dry and wet tenacity, we tried gel spinning from cellulose/aqueous calcium thiocyanate system. The gelation conditions of this system were determined by temperature jump method. Under some conditions, this system forms gel in 1min enough for spinning. The system at 125°C was extruded into air at 10°C under various spinning draft, followed by extraction of the solvent using acetone or ethanol. As the result of spinning, we successfully obtained the new cellulose fiber with both strong tenacity in dry and wet state and good fibrillation resistance.

1. INTRODUCTION

Concentrated aqueous calcium thiocyanate ($Ca(SCN)_2$) solution dissolves natural cellulose at elevated temperature and its solution forms thermoreversible gel on cooling [1-4]. Authors have reported change of cellulose crystal form in aq.$Ca(SCN)_2$ and dissolved state in this system [5-7]. However, gelation behavior of this solution is not clear. In past, spinning from this system has never been succeeded [8]. In this study, we attempt to clarify the gelation behavior of the cellulose solution and try to spin from this system by gel spinning method.

2. EXPERIMENTAL

2.1. Sample

Conifer pulp (DP=970, Alaska Pulp Co.,U.S.A), conifer pulp hydrolyzed in 25%aq.H_2SO_4 (degree of polymerization (DP)=180, 290, 430, 550, 760), cuprammonium rayons (DP=870, Bemlise®, Asahi Chemical Co. Ltd., and Bemberg®, Asahi Chemical Co. Ltd.), viscose rayon (Silmax VI®, Asahi Chemical Co. Ltd.), lyocel (TENCEL®, Coutaulds Co.).

2.2.　Preparation of solution

A mixture of cellulose and 55wt%aq.Ca(SCN)$_2$ soln. was stocked for 1h-1day at room temperature, followed by heating for 20min-2h at 110-120℃.

2.3.　Phase diagram

8g of cellulose soln. was put into a test tube of 10ml with cap. The system was quenched by transferring the test tube from the bath controlled at T$_s$=110℃ to another at temperature T$_q$. Determination of gel state was made by the inversion method.

2.4.　Transmission electron microscope (TEM)

2wt% cellulose soln. was quenched at 0℃, aged for 1day, and then washed by water. In this time, the size of gel did not change. Thin cross section was prepared by slicing the gel solidified with acrylic resin (methyl methacrylate/n-butyl methacrylate=8/2) and was observed by a transmission electron microscope (TEM, HU-11B, Hitachi).

2.5.　Spinning

The 10wt% cellulose soln. kept at 125℃ was extruded into air at room temperature using a capirograph (CAPIROGRAPH1B,Toyoseiki) with monohole nozzle of 0.1mm diameter, and taken up to cylinder at draw ratio (Dr) =0.5-6. Those wet fibers were put into the extracting bath of acetone or EtOH, washed by water, and finally dried at 65℃.

2.6.　Analysis of mechanical property

Tensile strength and elongation in dry and wet state were measured using TENSILON UTM-III-100 (TOYO BALDWIN Co. Ltd.) under following conditions: Temp., 20℃; humidity, 65%; sample length, 30mm; tensile speed, 30mm/min. Fiber samples were conditioned at 20℃, 65% of humidity for 1day for dry state, and immersed water for 5min for wet state before measurement. Fibrillation resistance grade was measured as follows: Fiber was cut to 5mm length, and hydrolyzed in 3wt%aq.H$_2$SO$_4$ for 30min at 70℃. Fibrillation was occurred by mixing in 3wt%aq.H$_2$SO$_4$ for 5-30min using home mixer. Fibrillation resistance grade was determined according to the method reported by Sato et. al. [9].

2.7.　Birefringence

Birefringence ($\triangle$n) was calculated from n$_{//}$ and n$_{\perp}$ (n$_{//}$, reflection at the parallel direction to the fiber axis; n$_{\perp}$, reflection at the perpendicular direction to the fiber axis) measured using interference microscope (interphako®, CARLZEISS JENA).

3. RESULTS AND DISCUSSION

3.1. Gelation behavior

Figure 1 shows the phase diagrams for cellulose/55wt%aq. Ca(SCN)$_2$ system. Figure 2 shows the gelation time (t^G), defined as the time required for gelation to occur after quenching, plotted against cellulose concentration (wt%). Similar to other thermally induced gels, such as PVA, PE, Nylon66 [10-12], gelation occurs rapidly at higher degree of polymerization, higher concentration of cellulose, and lower temperature. Under some conditions, this system forms a gel in 1min enough for spinning.

Figure 3 shows the electron microphotograph of cross section of 2wt% cellulose gel which was prepared at 0°C and washed by water. Polymer network develops homogeneously.

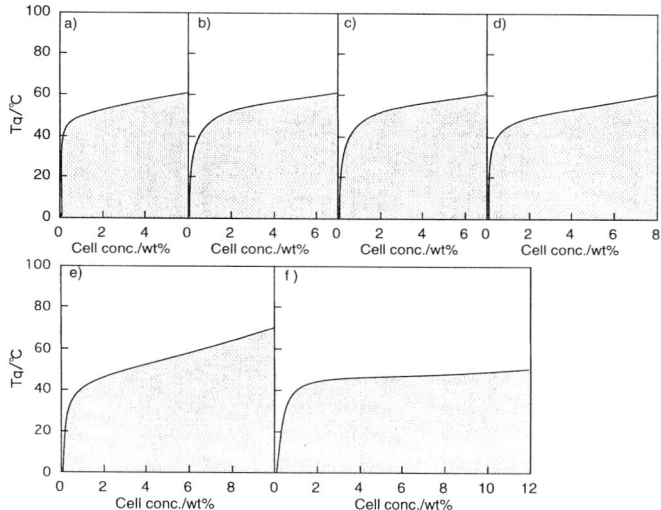

Fig.1 Phase diagram of cellulose/55wt%aq.Ca(SCN)$_2$ system with DP=970 (a), 760 (b), 550 (c), 430 (d), 290 (e), 180 (f), in 24h after quenching. Shadowed area, gel phase.

3.2. Spinning and fiber properties

Mechanical properties of new cellulose fiber were summarized in Table I. Figure 4 shows the relationship between mechanical properties of new fiber in the conditioned (dry) state and the draw ratio (Dr). Conditioned tenacity (TSd) and modulus (YMd) monotonically increased with an increase of Dr, but conditioned elongation (TEd) decreased. At Dr=6, we

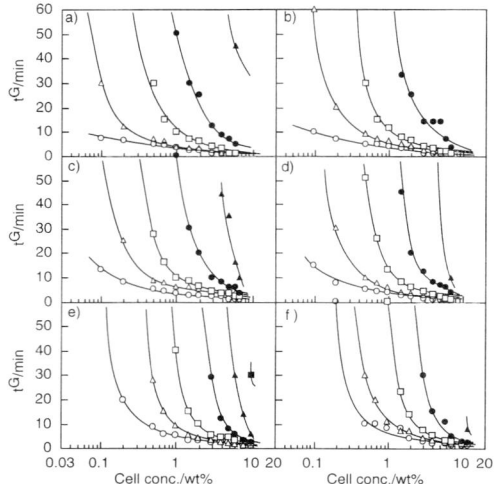

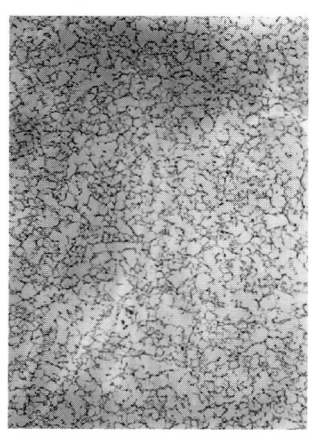

Fig.2 Relationship between gelation time t^G and cellulose concentration; DP=970 (a), 760 (b), 550 (c), 430 (d), 290 (e), 180 (f); T_q=10°C,O; 20°C,△; 30°C, □: 40°C, ●; 50°C, ▲; 60°C, ■.

━━ 1 μm

Fig.3 Electron microphotograph of cross section of 2wt% cellulose gel which was prepared at 0°C and washed by water

Table I. Mechanical properties of cellulose filament.

No.	denier[*1] d	Conditioned Tenacity /g · d⁻¹	Conditioned Elongation /%	Conditioned Modulus/ g · d⁻¹	Wet Tenacity /g · d⁻¹	Wet Elongation /%	Wet Modulus/ g · d⁻¹	Tenacity ratio of wet/conditioned
1	22.0	1.7	14	75	–	–	–	–
2	9.5	3.0	11	109	1.9	16	11	0.64
3	6.5	3.2	6	124	–	–	–	–
4	5.0	3.3	9	139	2.8	15	19	0.84
5	2.0	6.0	7	225	5.7	13	39	0.96
6	19.0	1.9	21	32	–	–	–	–
7	9.0	2.2	8	88	1.8	16	11	0.80
8	6.5	2.8	9	105	–	–	–	–
9	5.5	2.9	9	118	2.5	14	17	0.87
10	9.0	3.7	20	151	–	–	–	–
11	3.0	4.8	10	216	–	–	–	–
12	9.0	3.4	24	147	–	–	–	–
13	3.0	5.0	9	243	–	–	–	–
Viscose Rayon	2.2	1.8	17	85	0.9	23.1	–	0.49
Cupra	1.6	2.6	11	107	1.6	25.0	–	0.62
Lyocel[*2]	1.5	4.6	13	120	4.0	17.5	31	0.86

*1; denier of one filament

*2; Ref.13

successfully obtained the cellulose filaments with much higher TSd(=6.0g/d) and YMd(=225g/d) comparing to commercially available viscose, cupra and lyocel for cloth uses. Figure 5 shows the relationship between the wet/dry tenacity ratio(TSw/TSd) and Dr.

TSw/TSd also increased with Dr. Especially at Dr=6, the filament has the maximum TSw/TSd(=0.96) comparing to other commercially available regenerated cellulose fibers. Figure 6 shows the relationship between birefringence ($\triangle$n) and Dr. Increase of $\triangle$n means that the drawing of dope before formation of gel leads to increase of orientation of fiber. We reported that undissociated $Ca(SCN)_2$ – water complexes interact with ring oxygen and primary alcohol oxygen at C(6) position in glucopyranose ring of cellulose [6,7]. It seems that cellulose molecules are extended and rigid in the solution. As the result, cellulose molecules orient along the direction of shear by the drawing of dope, and the fiber revealed strong tenacity.

Figure 7 shows the dependence of fibrillation resistance grade on mixing time. New cellulose fiber reveals fibrillation resistance higher than cupra, and lyocel [9,13]. Schurz et. al. [14] reported for lyocel that increase of orientation brings about an increase of the strength of fiber and reduces the loss of tensile strength and modulus in water, but leads to high fibrillability. Fibrillation resistance is correlated with density of amorphous region [9], lateral crosslinks between elementary fibrils and aggregation of elementary fibrils

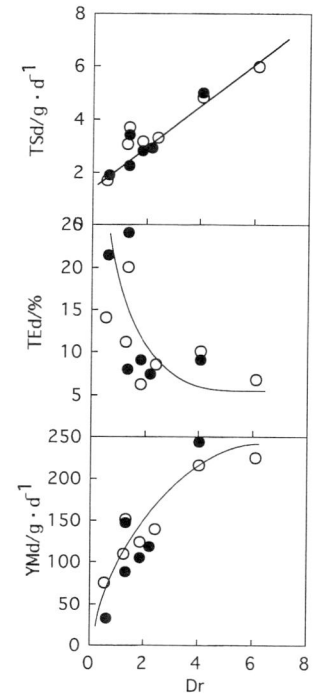

Fig.4   Relationship   between mechanical   properties   in   the conditioned state and Dr: a) TSd, b) TEd, c) YMd. Extracting agent; ○, acetone; ●,ethanol

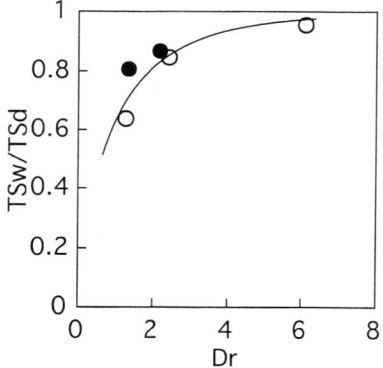

Fig. 5. Relationship between TSw/TSd and Dr.

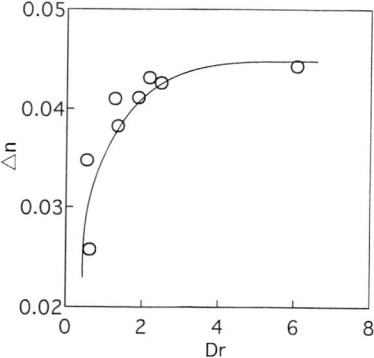

Fig. 6. Relationship between $\triangle$n and Dr.

[14]. In gel spinning process from cellulose/aq.Ca(SCN)$_2$ system, cellulose molecules orientated in the dope by drawing of dope, followed by formation of crosslinks by gelation homogeneously. These crosslinks seem to suppress the development of elementary fibril. And then organic solvent such as acetone and ethanol extracts calcium thiocyante solution from the system [5] forming tight amorphous without voids (defect in the fiber). As the result, we successfully obtained the new cellulose fiber with both strong tenacity in dry and wet state and good fibrillation resistance.

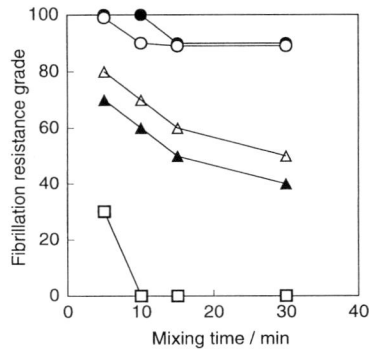

Fig. 7. Dependence of fibrillation resistance grade on the mixing time. Fiber No. from table I; ○, 10; △, 11; ●, 12; ▲, 13; □, cupra.

REFERENCES

1. J.O.Warwiker, in "Cellulose and Cellulose Derivatives", Part IV, N.M.Bikales and L.Segal. Ed., Wiley-Interscience, New York, 1971, Chapter XIII, Section H, p328.
2. H.E.Williams, J. Soc. Chem. Ind., **40**(1921) 221T.
3. H.Eebing and H.Geinitz, Kolloid-Z, **84**(1938) 25.
4. O.A.Wourinen and A.Visapää, Paperi puu, **41**(1959) 345.
5. M.Hattori, Y.Shimaya and M.Saito, Polym. J., **30**(1998) 37.
6. M.Hattori, T.Koga, Y.Shimaya and M.Saito, Polym. J., **30**(1998) 43.
7. M.Hattori, Y.Shimaya and M.Saito, Polym. J., **30**(1998) 49.
8. D.M.MacDONALD, ACS Symp. Ser., **58**(1977) 25.
9. J. Sato, N. O'Mara, ad M. Saito, Sen'i Gakkaishi, **54**(1998) 93.
10. See for example, P.S.Russo, Ed., "Reversible Polymeric Gels and Related Systems", ACS Symposium Series 350, American Chemical Society, Washington, D.C., 1987.
11. W.Burchard and S.B.Ross-Murphy, "Physical Networks", Elsevier Applied Science, New York, N.Y., 1990.
12. M.Hattori and M.Saito, Polym. J., **28**(1996) 139.
13. C. Yamane, M. Mori, M.Saito, and K.Okajima, Polym. J., **28**(1996) 1039.
14. J. Schurz, and J. Lenz, Macromol. Symp., **83**(1994) 273.

HYDROCOLLOIDS – PART 1
Edited by K. Nishinari
2000 Elsevier Science B.V.

277

SANS studies of aqueous suspension of microcrystalline cellulose

M. Sugiyama[a], K. Hara[b], N. Hiramatsu[c], A. Nakamura[c] and H. Iijima[d]

[a]Department of Chemistry and Physics of Condensed Matter, Kyushu University, Fukuoka 812-8581, Japan

[b]Institute of Environmental Systems, Kyushu University, Fukuoka 812-8581, Japan

[c]Department of Applied Physics, Fukuoka University, Fukuoka 814-0180, Japan

[d]Chemical Technology Department IV, Asahi Chemical Industry Co., Ltd., Nobeoka, Miyazaki 882-0847, Japan

H_2O and D_2O suspensions of microcrystalline cellulose with the mean diameter of around $3\mu m$ were investigated by the small-angle neutron scattering and ultra small-angle neutron scattering methods. It has been concluded that the scattered neutron observed in the small-angle region resulted from the internal structure of the cellulose particle which showed a fractal structure with a dimension of around 2.2. The fractal structure was stable in the temperature range between 25 and 80 °C. The microcrystalline cellulose particle maintained the fractal structure in the region from 6×10^{-3} up to 1.0×10^{1} μm at least.

1. INTRODUCTION

Cellulose has been widely used in various fields such as food, clothes, medicine and so on. Many kinds of the materials made of the cellulose are being developed in these days. A microcrystalline cellulose (MCC), which has been recently developed by Asahi Chemical Industry Co.,Ltd., is typical one of these materials. The characteristics of the MCC particle is the average size in around 3 μm; the minimum size of the cellulose particle was used to be around 10 μm.

It is well known that the cellulose usually consists of crystalline and amorphous regions because the degree of crystallinity cannot reach 100%. For example, the cotton composed of cellulose fibers has a fringed micelle structure[1]. Its elasticity and water absorptive power are due to this complex structure. It is estimated that the MCC particle also has a fringed micelle structure because it is made by the hydrolysis

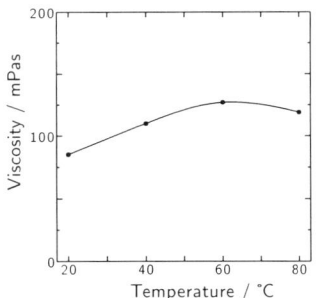

Figure 1. Temperature dependence of the viscosity of the aqueous suspension of MCC with the concentration of 2 wt%. The share rate is 16.4 Hz.

of woodpulp, however, the intra-particle structure is not yet fully understood. The first objective of our study is to make clear this point.

It is also interesting that the temperature dependence of the viscosity of the aqueous suspension of MCC exhibits the characteristic behavior from 20 to 80 °C, as is shown in Figure 1. In this temperature range, the viscosity of other food additives becomes lower with elevating temperature drastically[2]. On the other hand, we can say that the viscosity of the aqueous suspension of MCC is almost constant compared with the others; this feature is another objective to make clear in our study.

The small-angle neutron scattering (SANS) and ultra small-angle neutron scattering (USANS) methods are powerful tools to observe the structure in the scales from 0.001 to 0.1 μm and from 0.1 to 10 μm, respectively. The advantage of the neutron beam as a probe is its high transmission for materials which do not contain the special atoms, such as Li, B, Gd, Cd and so on. Therefore, we can easily observe the intra- and inter-particle structure of MCC in the aqueous suspension by the SANS and USANS methods. In addition, we can reduce the incoherent scattering from a hydrogen atom by replacing it to a deuterium atom. In the case of measuring a weak scattering intensity, such as that in the relatively higher scattering angle of MCC suspension, this method can be very powerful one.

In this paper, we will report the intra-particle structure of the MCC particle in the aqueous suspension observed by the SANS and USANS methods.

2.　EXPERIMENTAL

2.1.　Sample

The suspensions of MCC particles in H_2O and D_2O were specially supplied by Asahi Chemical Industry Co., Ltd, Japan. The exchange of H_2O for D_2O was carried out as fol-

Table 1. Sample characteristics.

Code	Concentration(wt%)	Solvent
Sample 1	10.0	H_2O
Sample 2	1.0	$H_2O+D_2O(1:9)$
Sample 3	15.2	D_2O
Sample 4	4.0	D_2O

lows. At first, the H_2O suspension was dissociated into the dilute and condensed MCC parts by centrifuge. Next, the condensed MCC part was dissolved in D_2O and the dissociation was repeated by centrifuge again. Then, the re-condensed MCC part was redissolved in D_2O. Table 1 shows the sample concentration and solvent. Sample 2 was prepared by being diluted Sample 1 with D_2O.

2.2.　SANS and USANS experiments

We used three SANS and USANS spectrometers: KUR-SANS spectrometer installed at Kyoto University Reactor, Kumatori, Osaka, Japan, and SANS-J and ULS (USANS) spectrometers installed at JRR-3M in the Japan Atomic Energy Research Institute, Tokai, Japan. The KUR-SANS and SANS-J spectrometers are the SANS cameras with optics of the pinhole-type. The ULS spectrometer is a Bonse-Hart type camera for observing ultra small-angle scattering region. The covered ranges of a magnitude of the scattering vector (q) by the KUR-SANS, SANS-J and ULS spectrometers were set to be from 1×10^{-2} to 1.2×10^{-1} Å^{-1}, from 5×10^{-3} to 5×10^{-2} Å^{-1}, from 6×10^{-5} to 3×10^{-3} Å^{-1}, respectively.

The SANS experiments with Samples 1, 2 and 3 at room temperature were carried out with the KUR-SANS and SANS-J spectrometers. The temperature dependence of the SANS intensity of Sample 1 at four temperature points, 25, 40, 60 and 80 °C, was observed with the KUR-SANS spectrometer. The USANS experiment of Sample 4 was performed with the ULS spectrometer.

3. RESULTS AND DISCUSSION

3.1. Dilution effect for SANS profiles

The scattering intensity of Samples 1 and 2 measured with the SANS-J are shown in Figures 2(a) and 2(b), respectively. The intense small-angle scattering observed in the both samples indicated the existence of fluctuation of the scattering length density with the contrast between crystalline and amorphous regions. It should be noted that there was almost no difference in the profile shapes of Samples 1 and 2, as shown in Figures 2(a) and 2(b). If we had observed the inter-particle structure, the shapes should have been different because the structure could be varied by dilution. Therefore, it was concluded that the observed profiles in the small q region were mainly due to the intra-particle structure of MCC.

Figures 3(a) and 3(b) show double-logarithmic plots for the SANS profiles of Samples 1 and 2, respectively. It is well known that the intensity $I(q)$ scattered from a fractal object is expressed with the following equation[3,4],

$$I(q) \sim q^{-D}, \qquad (1)$$

where q and D denote a magnitude of a scattering vector and a fractal dimension, respectively. Therefore, when a scattering object has a fractal structure, the q dependence of the scattered intensity is depicted by a straight line in the double logarithmic plot. As indicated by the straight lines in the figures, MCC particle has the fractal structure with $D = 2.2$.

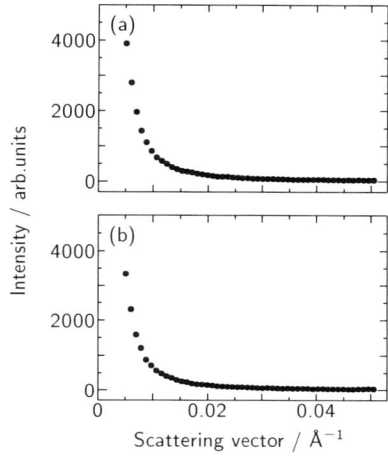

Figure 2. SANS profiles of (a) Sample 1 and (b) Sample 2 at room temperature.

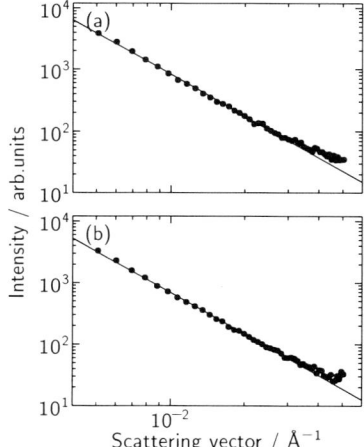

Figure 3. Double-logarithmic plots of SANS profiles of (a) Sample 1 and (b) Sample 2 at room temperature.

3.2. Temperature dependence of inner-structure

The temperature dependence of intra-particle structure of MCC in Sample 1 was measured with the KUR-SANS spectrometer. The double logarithmic plots of the scattered intensities at 25, 40, 60 and 80 °C are shown in Figure 4. It was found that the scattering profiles were almost same in the temperature range between 25 and 80 °C. In addition, as indicated by the straight lines in the figures, MCC particle has the fractal structure with $D = 2.2$ in this temperature range. This stability of the intra-particle structure is one of the reasons of the stability of the viscosity.

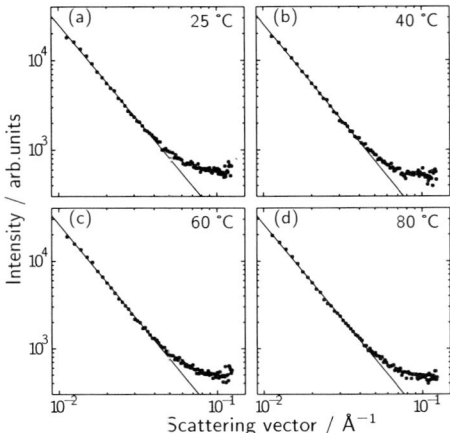

Figure 4. Double-logarithmic plots of Sample 1 at (a) 25 °C, (b) 40 °C, (c) 60 °C and (d) 80 °C.

3.3. Minimum fractal structure limit

The deviation from the Equation (1) in the higher q-range as shown in Figure 4 comes from an incoherent scattering of hydrogen and/or the limitations of the fractal structure. Respectively, Figures 5(a) and 5(b) show the scattering profiles of Samples 1 and 3 in the double-logarithmic scale. As indicated by the straight lines in the figures, both of the microcrystalline cellulose suspensions have the fractal structure with $D = 2.2$. In Figure 5(a), the scattering intensity deviates from the power-law line above 0.05 Å^{-1}, while, as shown in Figure 5(b), the data points of the suspension with D$_2$O fit the power-law line in a wider range, which means that the deviation of the data points from the power-law line in Figure 5(a) is due to the incoherent scattering of the hydrogen atom in H$_2$O. The q dependence of the scattered intensity from the deuterated suspension obeyed the Equation. (1) up to around 0.1 Å^{-1}. Therefore, the size of the unit cluster composing the fractal structure is smaller than 60 Å.

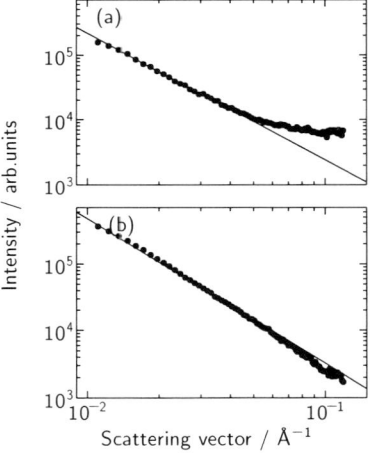

Figure 5. Double-logarithmic plots of (a) Sample 1 and (b) Sample 3 at room temperature.

3.4.   Maximum fractal structure limit

In order to make clear the maximum size of the internal fractal structure and/or the external complex matrix structure composed of MCC particles, we performed the USANS experiment of sample 4. Figure 6 shows the scattering profiles of Sample 4 in the double-logarithmic scale measured with the ULS spectrometer. The q dependence of the scattered intensity obeyed the power law with the index of -1.3. It has been known that an observed index is added one into a true index because of the smearing effect by the shape of the incident beam profile (length×width=2×1 cm^2) in the ULS spectrometer[5]. Therefore, the fractal dimension in this q-range is about 2.3 and almost same with the dimension observed in Figures 3, 4, and 5. In Figure 6, the smallest q value ($\sim 6 \times 10^{-5}$ Å^{-1}) corre-

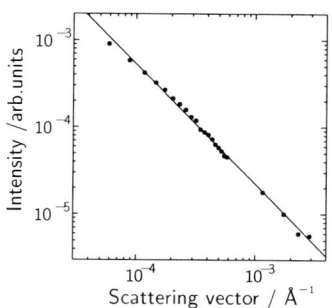

Figure 6.   Double-logarithmic plots of the USANS of Sample 4 at room temperature.

sponds to 10 μm in a real space. This is larger than the average size of the MCC particle of 3 μm. Therefore, the aggregation could occur in this concentration (4wt%). In addition, there is no peak indicating the ordering of the microcrystalline cellulose particles. This indicates that the aggregated units has no ordered structure.

ACKNOWLEDGEMENTS

The authors would like to thank Dr. S. Koizumi of JAERI for his kind assistance with the SANS-J experiments. The authors also would like to thank Dr. M. Hashimoto of ISSP for his kind assistance with the ULS experiments. This work was partly supported by the Grant-in-Aid from the Ministry of Education, Science, Culture and Sports, Japan.

REFFRENCES

1. A. K. Kulshreshtha and N. E. Dweltz: J. Polym. Sci. & Polym. Phys. **11** (1973) 487.
2. E. Kamata: New Food Industry, **36** (1994) 59 (in Japanese).
3. D. W. Schaefer and K. D. Keefer: Phys. Rev. Lett. **56** (1986) 2199.
4. J. Teixeira: J. Appl. Cryst. **21** (1988) 781.
5. Y. Izumi, A. Uchida, H. Nogami, K. Kajiwara, H. Urakawa, Y. Yuguchi, M. Hashimoto and T. Takahashi: Activity Rep. Neutron Scattering Res., **4** (1997) 205.

HYDROCOLLOIDS – PART 1
Edited by K. Nishinari
2000 Elsevier Science B.V.

Influence of surface charge on viscosity anomaly of microcrystalline cellulose suspensions

J. Araki, M. Wada, S. Kuga and T. Okano

Graduate School of Agricultural and Life Sciences, The University of Tokyo, Yayoi 1-1-1, Bunkyo-ku, Tokyo 113-8657, Japan

The effect of surface charge on the viscosity behavior of rodlike particle suspension was investigated by introducing different levels of surface charge to a charge-free cellulose microcrystal suspension, which was prepared by hydrolysis of kraft pulp with 4 N HCl. The HCl-hydrolyzed microcrystals were subsequently sulfated with 55% (w/w) sulfuric acid or phosphorylated with urea-phosphoric acid mixtures under various conditions. While the shape of microcrystal was similar for all samples, surface charge determined by conductometric titration and viscosity behavior of the suspensions were significantly different. Increase in surface charge of the microcrystals tended to reduce time- and concentration dependence of viscosity irrespective of the type of the ester group. These anomalous viscosity phenomena are ascribed to significant influences of surface charge on the state of interparticle association of the rodlike particles.

1. INTRODUCTION

We have shown that stable colloidal suspensions of cellulose microcrystals can be prepared by hydrolysis of native cellulose with hydrochloric acid, instead of conventionally used sulfuric acid [1]. While the HCl-hydrolyzed microcrystals have similar size and shape to those of the H_2SO_4-hydrolyzed microcrystals [2-8], the former have much lower levels of surface charge than the latter. Because of this difference, these two types of suspensions were significantly different in viscosity behavior [1]. We also attempted to establish the conditions of esterification treatments to control the surface charge and accompanying viscosity behavior [9].

In this study we tried to extend the range of surface charge control as well as to introduce phosphate groups by treating microcrystals with an urea-phosphoric acid mixture [10], a method originally developed for preparation of flame-proof textiles [11] or ion exchange materials [12, 13].

2. EXPERIMENTALS

2. 1. Preparation of cellulose microcrystal suspension

An industrial-grade bleached softwood kraft pulp (KP) was used as starting material without further purification. The preparation of suspension was described in Ref. 1 in detail; briefly, 10 g of air-dried KP was treated with 300 ml of 4N HCl at 80 °C for 225 min. The suspension was washed by repeated centrifugation (1600 g, for 5 min. at 10 °C). The turbid supernatant at pH over 4 was collected, dialyzed and concentrated by high-speed centrifugation, followed by sonication for 1 min. For comparison, 10 g of the KP was hydrolyzed with 65% (w/w) H_2SO_4 at 70 °C for 10 min and washed similarly [1, 3-5].

2. 2. Sulfation of cellulose microcrystals [9]

The HCl-treated microcrystals were sulfated with 55% (w/w) H_2SO_4 under various conditions; 2 h at 25 °C, 40 °C, and 60 °C, and 4 h at 60 °C. All samples were diluted with large amounts of water and washed as above.

2. 3. Phosphorylation of cellulose microcrystals [10]

Urea, phosphoric acid and the HCl-hydrolyzed suspension were mixed at a ratio of 50:32:18 by weight and heated at 70 °C for 30 min, then at 135 °C for 10 min or 25 min. The mixture was diluted with water immediately after the reaction, and the suspension was obtained by centrifugation as above. Since these suspensions had phosphate groups in ammonium salt form [14], they were treated twice with 0.1 N HCl for over 45 min at room temperature to convert the phosphate to the acid form [15].

2. 4. Characterization of cellulose microcrystals and suspensions

TEM observation of the microcrystals was performed with JEOL JEM-2000EX at 200 kV with defocus contrast technique. The surface charge of the microcrystals were determined by conductometric titration with 0.01 N NaOH [1, 9, 15]. The sulfur and phosphorus content was determined by X-ray fluorescence analysis with a HORIBA MESA-500. Viscosity of the suspensions was measured with a Brookfield-type viscometer (Brookfield LV-DV1+) thermostated at 20 ± 0.1 °C. For proper assessment of time-dependent viscosity (thixotropy or anti-thixotropy), all the samples were allowed to stand overnight at 20 °C before viscosity measurement.

3. RESULTS AND DISCUSSION

It is well known that viscosity of suspension depends on both of the size and shape of particles and their surface charge. Size and shape of microcrystals were not altered significantly (Figure 1) by the sulfation or phosphorylation treatments; therefore the changes in viscosity behavior described below can be ascribed to introduction of surface charge.

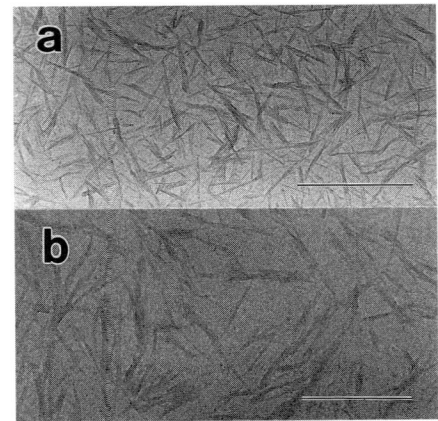

Figure 1. Micrographs of the cellulose microcrystals, sulfated for 2 hours at 60 °C (a) and phosphorylated for 25 min (b). Scale bars show 500 nm.

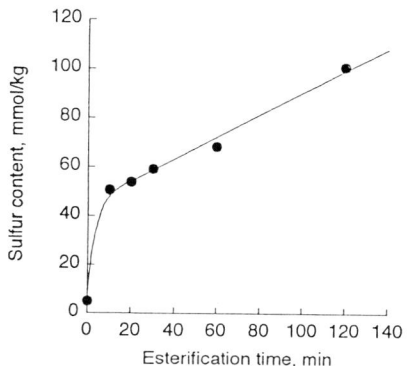

Figure 2. Sulfur contents of the micro-crystals at various reaction time, within a single batch sulfated at 40 °C.

Table 1. Amounts of strong and weak acid groups on various cellulose microcrystals.

samples	acid groups, mmol/kg	
	strong acid	weak acid
HCl hydrolyzed	0	16
H$_2$SO$_4$ hydrolyzed	84	26
sulfated 25 °C 2 h	0	N. D.*
sulfated 40 °C 2 h	53	29
sulfated 60 °C 2 h	60	28
sulfated 60 °C 4 h	107	21
phosphorylated 10min	0	N. D.*
phosphorylated 25min	30	6

* = Not detected because the second break in the titration curve was not clear.

Table 1 summarizes the amount of strong and weak acid groups of the various microcrystals determined by conducto-metric titration. With increasing extent of esterification, two clear breaks appeared in the titration curve of the suspensions, indicating introduction of strong acid groups. The strong and the weak acid groups in the sulfated samples correspond to sulfate and carboxylic acid groups, respectively.

The phosphate ester introduced by this reaction is believed to form single ester linkages with cellulose, resulting in divalent acid group. Two-step dissociation of this group is considered to correspond to strong and weak acid groups, consistent with the same amounts of strong and weak acid groups determined by titration [14]. In our results, however, the amount of strong acid was somewhat greater than that of weak acid for phosphorylated cellulose. The reason for this difference is not clear at this moment. As stated above, the surface charge of the cellulose microcrystals could be controlled by esterification conditions.

The amount of introduced ester groups, however, was not fully reproducible for different batches of same materials and conditions. For instance, several batches of sulfation at the same temperature gave a poor correlation between the reaction time and sulfur content. Within a single batch, however, the degree of sulfation increased monotonously with the reaction time. Figure 2 shows the relation between the reaction time and the sulfur content, showing the remarkable increase in the initial 5 min and subsequent slow increase.

The sulfur and phosphorus contents of the microcrystals determined by X-ray fluorescence analysis were considerably greater than those calculated from sulfate and phosphate contents (Table 1) [9]. Two possible causes for this discrepancy are i) presence of non-acidic sulfur or phosphorus groups, and ii) the inaccessibility of ester groups within the microcrystals.

Figure 3 shows the time dependence of viscosity of various microcrystalline cellulose suspensions. While the HCl-hydrolyzed suspension showed thixotropy at high solid contents

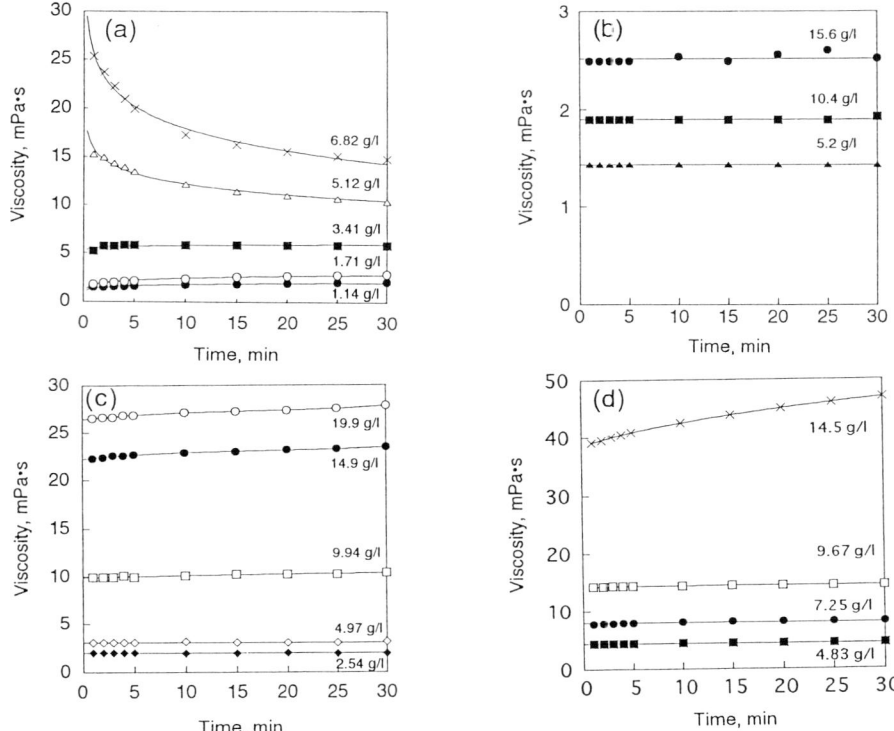

Figure 3. Time dependence of viscosity of the HCl-hydrolyzed suspension (a), the H₂SO₄-hydrolyzed suspension (b), the suspension sulfated at 40 °C (c), the suspension phosphorylated for 10 min (d).

(> 0.5%) and anti-thixotropy at low solid contents (< 0.3%) [1], such time dependence disappeared gradually with increase in sulfate content [9], and the fully charged H_2SO_4-hydrolyzed suspension showed time-independent viscosity.

The time-dependent viscosity indicates breakdown and formation of aggregated structure in the suspension [17, 18]. While remarkable time dependence was observed even at low solid content for the HCl-hydrolyzed suspension, the charged samples (the H_2SO_4-hydrolyzed and the esterified samples) showed no time dependence at low concentration, and only weak anti-thixotropy at very high concentrations. Though the introduction of surface charge causes dissociation of aggregates in the initial HCl-hydrolyzed suspension, certain aggregation still arises   in concentrated suspensions due to shorter distances between particles. Figure 3 also shows that thixotropy was observed only in the HCl-hydrolyzed (unesterified) suspension, while all the charged suspensions showed anti-thixotropy or time-independence.

Figure 4 shows the relation between solid content, phi, and relative viscosity for various suspensions. The value at 30 min after the start of measurement was taken for time-

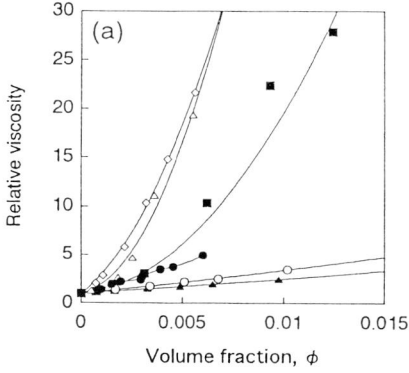 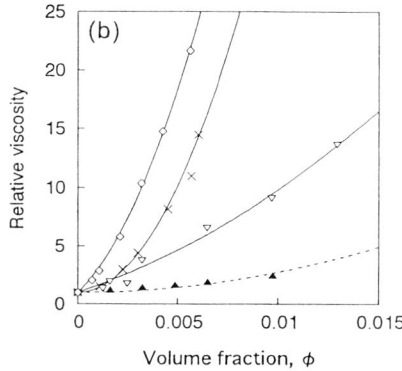

Figure 4. Relative viscosity of various suspensions plotted against their concentrations. (a) Sulfated suspensions, ◇: the HCl-hydrolyzed suspension, △: 25 ˚C 2 h, ■: 40 ˚C 2 h, ○: 60 ˚C 2 h, ●: 60 ˚C 4 h, ▲: the H₂SO₄-hydrolyzed suspension. (b) Phosphorylated suspension, ◇: the HCl-hydrolyzed suspension, ×: for 10 min, ▽: for 25 min, and the data of the H₂SO₄-hydrolyzed suspension (▲) were shown as comparison.

dependent viscosity. The aspect ratio of the particles in the H_2SO_4-hydrolyzed suspension calculated by Einstein-Simha equation [19, 20] agreed well with TEM observation [1]. This indicates that the particles are highly dispersed and isolated in the H_2SO_4-hydrolyzed suspension. In contrast, viscosity of the HCl-hydrolyzed suspension was several folds greater than that of the H_2SO_4-hydrolyzed suspension, apparently due to enhanced aggregation [1].

4. CONCLUSION

Our results show that the viscosity of the cellulose microcrystal suspension is significantly affected by the amount of surface charge. So far, the only reported case of rodlike colloidal particles with controlled surface charge is that of deacetylated chitin microcrystals [16]. Our method of introducing sulfate or phosphate groups to cellulose microcrystals provide another means to prepare such systems showing various unusual viscosity behavior. Such information should be useful in practical applications of colloidal cellulose suspensions and in quantitative interpretations of the impact of surface charge on anomalous phenomena characteristic to rodlike particles, such as chiral nematic phase separation [4-6, 16].

ACKNOWLEDGMENT

The authors are grateful to Dr. A. Isogai and Mr. M. Kato of the University of Tokyo for their help in X-ray fluorescence analysis. This work was supported in part by the Grant-in-Aid for Scientific Research from Ministry of Education, Science, Sports and Culture, Japan (Grant nos. 07406018, 09460073 and 10760105).

REFERENCES

1. J. Araki, M. Wada, S. Kuga and T. Okano, *Colloids Surfaces A*, **142** (1998) 75.
2. S. M. Mukherjee and H. J. Woods, *Biochim. Biophys. Acta*, **10** (1953) 499.
3. R. H. Marchessault, F. F. Morehead and N. M. Walter, *Nature*, **184** (1959) 632.
4. J.-F. Revol, H. Bradford, J. Giasson, R. H. Marchessault and D. G. Gray, *Int. J. Biol. Macromol.*, **14** (1992) 170.
5. J.-F. Revol, L. Godbout, X. M. Dong, D. G. Gray, H. Chanzy and G. Maret, *Liquid Crystals*, **16** (1994) 127.
6. X. M. Dong, T. Kimura, J.-F. Revol, and D. G. Gray, *Langmuir*, **12** (1996) 2076.
7. R. H. Marchessault, F. F. Morehead and M. Joan Koch, *J. Colloid Sci.*, **16** (1961) 327.
8. V. Favier, H. Chanzy and J. Y. Cavaillé, *Macromolecules*, **25** (1995) 6365.
9. J. Araki, M. Wada, S. Kuga and T. Okano, *J. Wood Sci.*, in press.
10. F. O. Lassen, U. S. Patent No. 3 997 647 (1976).
11. J. D. Guthrie, *Ind. Eng. Chem.*, **44** (1952) 2187.
12. J. D. Reid and L. W. Mazzeno Jr., *ibid.*, **41** (1949) 2828.
13. J. D. Reid, L. W. Mazzeno Jr. and E. M. Buras Jr., *ibid.*, **41** (1949) 2831.
14. K. Katsuura and S. Nonaka, *Sen-i Gakkaishi*, **13** (1957) 24. (in Japanese)
15. S. Kats, R. P. Beatson and A. M. Scallan, *Sven. Papperstidn.*, **6** (1984) 48.
16. J. Li, J.-F. Revol and R. H. Marchessault, *J. Appl. Polym. Sci.*, **65** (1997) 373.
17. J. Mewis and A. J. B. Spaull, *Adv. Colloid Interface Sci.*, **6** (1976) 173.
18. J. Mewis, *J. Non-Newtonian Fluid Mech.*, **6** (1979) 1.
19. A. Einstein, edited by R. Furth, Investigations on the Theory of the Brownian Movement, Dover Publications, Inc. (1956).
20. R. Simha, *J. Phys. Chem.*, **44** (1940) 25.

5. STARCH

HYDROCOLLOIDS – PART 1
Edited by K. Nishinari
2000 Elsevier Science B.V.

291

Rheological behavior of heated starch dispersions : role of starch granule*

M. A. Rao, E. K. Chamberlain, J. Tattiyakul, and W. H. Yang

Department of Food Science & Technology, Cornell University–Geneva, Geneva, New York 14456-0462, USA

The state of the starch granule influences whether a heated starch dispersion exhibits shear-thinning, shear-thickening, or antithixotropic rheological behavior. Increase in volume fraction of starch granules beyond a threshold imparts yield stress to the dispersions. Dynamic rheological data obtained during gelatinization reflect in part the changes in granule volume fraction as a result of corresponding changes in granule size.

1. INTRODUCTION

Starch is a major source of calories and raw material in food and process industries. In foods, one major use of starch is as a thickening/gelling agent. The swelling of starch granules is not hindered at low starch (excess moisture) concentrations [1]. One may also view gelatinized starch dispersions (STDs) as microgel systems whose flow and viscoelastic behavior are strongly influenced by the physical state of the granules (size and size distribution) [2, 3].

In corn and cowpea [4], and cross-linked waxy maize [5] STDs, increase in the average starch granule size resulted in increase in the notional volume fraction, cQ, where c is the concentration of starch (g dry starch / g dispersion) and Q is the swelling power of the starch granules (g swollen starch / g dry starch) [6], as well as in the power law (Equation 1) consistency index of the STDs.

$$\sigma = K\dot{\gamma}^n \tag{1}$$

where, σ is the shear stress, $\dot{\gamma}$ is the shear rate, n is the flow behavior index (dimensionless), and K is the consistency index (Pa s^n). In addition, the yield stress of gelatinized STDs (σ_{0c}), determined using the Casson model (Equation 2),

*Research work supported by USDA NRI Grant #97-35503-4493

increased with cQ [4, 7].

$$(\sigma)^{0.5} = (\sigma_{0c})^{0.5} + (\eta_\infty \dot\gamma)^{0.5} \qquad (2)$$

Magnitudes of σ_{0c} and the infinite-shear viscosity (η_∞) can be calculated from the intercept (K_{0c}) and slope(K_c) of a $\dot\gamma^{0.5}$ versus $\sigma^{0.5}$ plot, respectively.

Quemada et al. [8] proposed a rheological model (Equation 3) for dispersed systems based on the zero-shear, η_0, and infinite-shear, η_∞, viscosities, and a structural parameter, λ (Equation 4):

$$\eta = \frac{\eta_\infty}{\left\{ 1 - \left[1 - \left(\frac{\eta_\infty}{\eta_0} \right)^{0.5} \right] \lambda \right\}^2} \qquad (3)$$

$$\lambda = \frac{1}{\left[1 + (t_c \dot\gamma)^{0.5} \right]} \qquad (4)$$

The time constant t_c is related to the rate of aggregation of particles due to Brownian motion. For concentrated dispersed systems, η_∞ will be much lower than η_0, so that $(\eta_\infty/\eta_0) \ll 1$ and the dispersion may have a yield stress, and Equation (3) reduces to the Casson model (Equation 2) [9] with the Casson yield stress, $\sigma_{0c} = (\eta_\infty / t_c)$. The generalized Casson model, written in terms of either the shear stress (Equation 5) or the apparent viscosity (η_a) (Equation 6), can be used to study rheological behavior ranging from only shear-thinning to shear-thinning plus yield stress [10, 11].

$$(\sigma)^m = (\sigma_{0c})^m + (\eta_\infty \dot\gamma)^m \qquad (5)$$

$$(\eta_a)^m = (\sigma_{0c} / \dot\gamma)^m + (\eta_\infty)^m \qquad (6)$$

Cross-linked waxy maize (CWM) starch, composed primarily of amylopectin (ca. 98%), is commonly used in low-acid foods. Cross-linking adds chemical bond bridges to granules that originally only contained hydrogen bonds; it is attained by using a chemical with two or more moieties capable of reacting with hydroxyl groups with starch [12].

The results of recent studies on the role of granule size on the values of cQ and rheological behavior of gelatinized tapioca STDs [11] and the dynamic rheological behavior during gelatinization of a corn STD [13], and of starch granule morphology on the rheological behavior of heated CWM starch dispersions [5] are summarized.

2. EXPERIMENTAL

2.1. Starch dispersions

The tapioca starch (National Starch and Chemical Co.) used had 19.3% (dry wt. basis) amylose, determined using a rapid colorimetric method [14]. During the preparation of the 2.6%, w/w STDs, isothermal conditions were obtained rapidly using the procedure of [3]. The dispersion was mildly agitated using a magnetic spin bar (5.5cmx1.3cm) at a low fixed stirring rate that was just enough to keep the starch granules suspended. About 50mL of gelatinized starch were withdrawn into test tubes at various times that were immediately immersed into an ice bath to quickly cool the dispersion to room temperature in about one minute when aliquots were taken for granule size scan and rheometry.

CWM starch dispersions (2.6%, w/w) (National Starch and Chemical Co., Inc.) mixed in individual 211 x 400 cans with 0.5 cm headspace and allowed to hydrate approximately 2 hours were heated in a pilot scale agitated steam-heated retort (Steritort®, FMC) at 10 rpm and 120°C for 8, 15 and 30 min [5]. An unmodified corn starch (CERESTAR USA, Inc.) with the commercial specifications: 10.0% moisture, 5.5 pH, and 99.5% granulation through US 200 mesh was used in dynamic rheological studies [13].

2.2. Starch granule size

The granule size distribution of the gelatinized tapioca starch dispersions was determined using a laser diffraction particle size analyzer (LS130, Coulter Corporation) as described earlier [3].

2.3. Flow behavior

Steady shear data on the STDs were determined using a Carri-Med CSL 100 rheometer (TA Instruments) with a cone (acrylic, 2 degrees, 6 cm diameter) and plate geometry. Each sample was sheared at 20°C from 0.01 Pa shear stress to a maximum shear stress that depended on sample consistency and back to 0.01 Pa. The shearing time was 5 min each for the ascending and descending shear cycles. Only the descending shear data were analyzed for the flow behavior.

2.4. Dynamic rheological behavior

Dynamic rheological data on the 8% corn STDs were obtained using 4 cm diameter parallel plate (500 μm gap) geometry of a Carri-Med CSL-100 rheometer (TA Instruments). A plate-cone geometry was also tried, but the narrow gap (52 μm at the center) contributed to capillary suction of the few drops of paraffin oil that were placed on the exposed edge of the plate to minimize loss of water vapor during heating into the corn STD. In contrast, with the parallel plate geometry, the drops of paraffin oil placed soon after the storage modulus (G') reached values that were measurable did not penetrate into the corn STD [13]. Magnitudes of η^* were recorded during temperature sweeps: 60-95°C at 1% strain and several frequencies.

3. RESULTS AND DISCUSSION

3.1. Starch granule size distribution

The granule size distribution curve of raw hydrated tapioca starch (Figure 1) was left-skewed and platykurtic with the size range of 0.1-34 μm. A slight shoulder at the low end of the size distribution curve represents small granules or granule fragments that may have been generated during the starch production process. The volume average diameter of the raw hydrated starch granules was 12.9 μm with a standard deviation of 6.8 μm. Upon heating at a specific temperature, the size distribution curves shifted to a wider range of about 0.4-100 μm with increasing mean granule sizes that depended on heating times, and they were right skewed and platykurtic. Granule size distribution data of dispersions heated at 67°C are also shown in Figure 1. Similarly shaped distributions were observed for the 2.6% tapioca starch dispersion heated at the other temperatures. The maximum average granule size of around 48 μm (SD. 22 μm) was found in the starch dispersion heated at 70°C for 5 min.

Heating the tapioca STDs at 75°C and 80°C resulted in a decrease in the average granule diameters. For example, in tapioca STDs heated at 75°C for 5 min and 80°C for 5 min, they were 43.6 μm and 40.5 μm, respectively. Sustained heating at the lower temperatures also reduced the average granule diameter.

3.2. Volume fraction (cQ)-granule size relationship

Single linear relationships were found between cQ and $(D_t / D_0)^3$ over a wide range of the latter for heated STDs of corn and cowpea [4], and cross-linked waxy maize starches [5]. For the tapioca STDs two linear relationships provided a better fit of the data: one for values of $(D_t / D_0)^3$ from about 15 to 40 and the other from 40 to about 50 (Figure 2):

$$cQ = -0.07 + 0.012 \ (D_t / D_0)^3 \tag{7}$$

$$cQ = -1.4 + 0.044 \ (D_t / D_0)^3 \tag{8}$$

In the determination of Q, centrifuging and siphoning off the top layer of supernatant fluid are not precise procedures that contributed to the spread of the cQ data in Figure 2. For tapioca starch dispersions, a large increase in cQ at high values of $(D_t / D_0)^3$ is noteworthy. In both equations, the small physically meaningless negative intercepts were due to errors in extrapolation of the data. A single exponential equation also described the data in Figure 2 [11].

For the purpose of comparison, the values of intercept, slope, and R^2 for other heated STDs were: 0.16, 0.011, and 0.86 for 2.6% unmodified corn starch [3], 0.09, 0.013, and 0.84 for 2.6% cowpea starch [15], and 0.23, 0.0015, 0.98 for 2.6% cross-linked waxy maize starch [5], respectively. Because tapioca starch had a higher

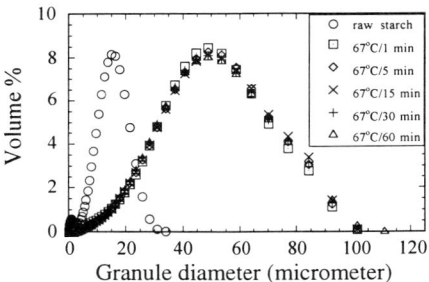

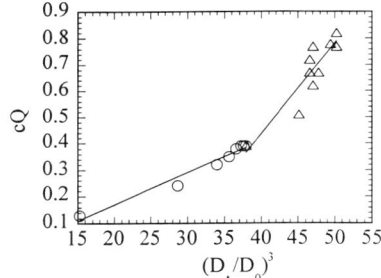

Figure 1. Tapioca starch granule size distribution.

Figure 2. Volume fraction versus cube of the starch granule diameter ratio

value of slope, especially at high $(D_t / D_0)^3$ values, the starch granule diameter ratio contributed to a higher increase of cQ of tapioca STDs.

3.3. Flow behavior of tapioca STDs

Shear-thickening behavior was observed in STDs gelatinized at 61°C up to 15 min (Figure 3). Starch/water systems exhibit shear-thickening behavior when the starch granules are rigid enough to resist shear and the concentrations are high enough for particle crowding [6]. Shear-thickening behavior was observed in heated STDs of wheat [16], native and cross-linked waxy maize [17], corn [3], cowpea [15], and cross-linked waxy maize [5]. In Figure 3, it can be seen that the shear-thickening behavior of the dispersions heated at 5 and 10 min was preceded by shear-thinning behavior. The critical shear rate $(\dot{\gamma}_c)$ at which the transition from shear-thinning to shear-thickening flow behavior occurred was about 300 s^{-1} for the dispersion heated for 5 min and about 500 s^{-1} for that heated for 10 min. Because of the low magnitude of viscosity of the dispersion heated for 1 min, the magnitude of $\dot{\gamma}_c$ could not be established. Nevertheless, the transition from shear-thinning to shear-thickening flow behavior occurred at high shear rates. Values of $\dot{\gamma}_c$ for a 2.6% cowpea starch dispersion heated at 67°C for 0.5-90 min were also high, in the range 337-372 s^{-1} [15]. With increase in heating temperature and time, the starch granules softened and became more deformable to shear forces resulting in shear-thinning behavior.

3.4. Casson-Quemada models for shear-thinning tapioca STDs

Increase in the volume fraction of solids in a starch dispersion can be achieved by either increasing the concentration of the granules or through volume expansion of the granules. η_a versus $\dot{\gamma}$ data of dispersions with starch concentrations 1.3%, 1.6%, 2.0%, and 2.3% that were heated at 74°C for 1 min are shown in Figure 4. While data of the 1.3% and the 1.67% dispersions exhibited no yield value, the other dispersions clearly exhibited yield stresses. The progression from only shear-thinning rheological behavior to shear-thinning plus yield stress behavior as the starch concentration, hence cQ of granules, was increased is

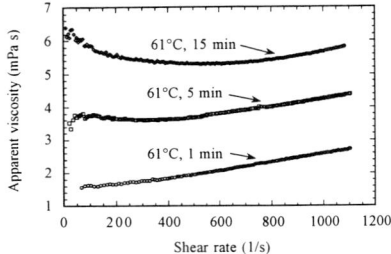
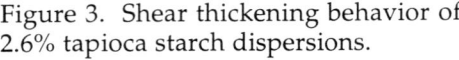
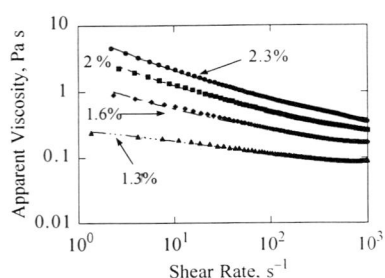

Figure 3. Shear thickening behavior of 2.6% tapioca starch dispersions.

Figure 4. Shear thinning and shear thinning plus yield stress behavior of 1.3% tc 2.3% tapioca dispersions.

noteworthy. In Figure 4, the Quemada model (Equation 3) described well the η_a versus $\dot{\gamma}$ data of the 1.30% dispersion (η_0=0.346 Pa s, η_∞=0.070 Pa s, and t_c =0.021 s; R^2=0.97) and of the 1.67% dispersion (η_0=2.161 Pa s, η_∞=0.115 Pa s, and t_c =0.014 s; R^2=1.00), while the generalized Casson model (Equation 6) with m=0.25 described well the data of the 2.07% dispersion (σ_{0c} =0.610 Pa, η_∞=0.100 Pa s; R^2=1.00) and of the 2.37% dispersion (σ_{0c}=1.370 Pa, η_∞=0.125 Pa s; R^2=1.00).

3.5. Power law model

Because of the relatively low magnitudes of yield stresses, the flow behavior of the dispersions was also described by the simple power law model (Equation 1); an additional reason for using the power law model was to compare the power law parameters of the tapioca starch STDs with those of corn, cowpea, and cross-linked waxy maize starches [4, 5]. The consistency index (K) and flow behavior index (n) were calculated using a power fit function in the shear rate range of 200-1100s^{-1}. With increase in the gelatinization temperature and time, K increased while n decreased indicating increase in shear-thinning behavior.

The dispersed phase characteristics (starch concentration, granule size and size distribution, granule shape, swelling pattern of the starch, and granule rigidity) affect the rheological behavior of a STD [18]. Figure 5 shows the relationship between the granule diameter ratio (D_t / D_0) and the consistency index (K) of gelatinized 2.6% tapioca STDs, that can be expressed by the equation:

$$K = K_0 \exp\left\{\varepsilon\left(\frac{D_t}{D_0}\right)\right\} \tag{9}$$

where, K_0 and ε are empirical constants. K increased gradually at the beginning and rapidly as the granule diameter ratio approached its maximum value. At (D_t / D_0) = 3.4, values of K started to increase rapidly. The estimated values of K_0, ε, and R^2 for the 2.6% tapioca STDs were 6.14x10^{-10} (mPa s^n), 7.40, and 0.90, respectively; the corresponding values for 2.6% corn starch were: $K_0 = 2.07 \times 10^{-5}$ (mPa s^n), $\varepsilon = 3.69$, and $R^2 = 0.99$ [3], for 2.6% cowpea starch were: $K_0 = 2.40 \times 10^{-5}$ (mPa s^n), $\varepsilon = 4.84$, and $R^2 = 0.96$ [15], and 2.6% cross-linked waxy maize starch were: $K_0 = 2.13 \times 10^{-7}$ (mPa s^n), $\varepsilon = 3.36$, and $R^2 = 0.92$ [5]. The higher value of ε for tapioca starch indicates that, compared to corn, cowpea and cross-linked waxy maize starches, the starch granule diameter ratio had the highest influence on the consistency index (K) of the tapioca starch dispersions.

3.6. Antithixotropic behavior of CWM starch dispersions

The antithixotropic behavior of the thermally processed (120°C, 8 min) CWM starch dispersions was clearly seen in three consecutive ascending and descending shear cycles: (1) a sample of the starch dispersion was pipetted onto the stage of the rheometer, and the first cycle of shearing was completed (5 min ascending, 5 min descending), (2) immediately, upon completion of the first shear cycle, a second shear cycle was performed, and (3) a third shear cycle was performed immediately following the second shear cycle. The response curves of the dispersion heated for 30 min at 120°C (Figure 6) indicate that the dispersion became more viscous with each shearing cycle, i.e., they exhibited time-dependent shear-thickening (anti-thixotropic) behavior. Similar behavior was seen with the dispersions heated for 15 min (not shown here). The anti-thixotropic behavior of the thermally processed dispersions seems to be due to granule cluster formation [5]. As the dispersions were sheared, starch granules rearranged to form clusters that, in turn, resulted in an increase in apparent viscosity.

3.7. Complex viscosity versus temperature profiles of corn STD

At low heating rates (e.g., 2.1 °C min⁻¹), a two stage increase in η^* with temperature can be seen (Figure 7): the first peak in viscosity, usually located between 55 and 75°C was due to a limited swelling of the starch granules [19, 20] that lead to an increase in the granule-granule interactions due to their larger dimensions. Because of continued swelling of the starch granules, the viscosity increased until about 90°C leading to the formation of a transient network of granules touching each other and sometimes resulting in granule disruption and amylose leaching [18, 21].

After the peak viscosity was reached, increasing the temperature further resulted in decrease in viscosity that has been attributed to rupture of the granules [18]. Two distinct types of curves were seen after the peak (Figure 7) viscosity was reached depending on whether: (1) the temperature was maintained at 92°C for about 2.5 minutes or at 95°C for about 5.5 minutes, or (2) the temperature was increased up to 95°C continuously after the peak viscosity temperature. The former was characterized by a slower rate of decrease in η^* after the temperature was held constant and the latter by a rapid rate of decrease in η^* up to the final temperature (95°C) suggesting that the network structure formed during gelatinization was drastically weakened by increase in the temperature [13].

298

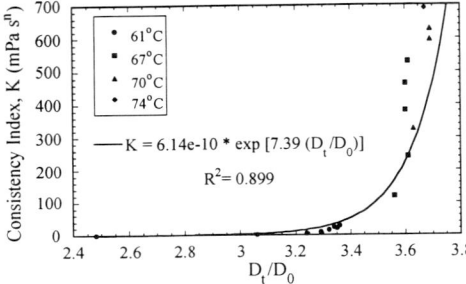

Figure 5. Effect of starch granule size (D_t / D_0) on the consistency index of gelatinized 2.6% tapioca STDs.

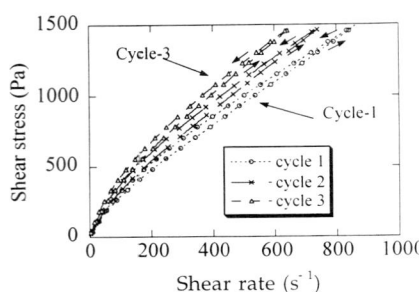

Figure 6. Antithixotropic behavior of 2.6% cross-linked waxy maize STDs heated at 120°C for 8 min

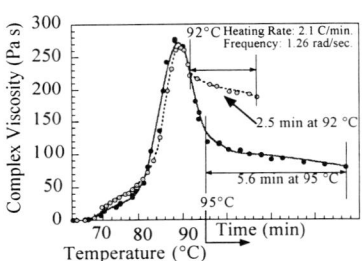

Figure 7. Complex viscosity data of 8% corn STD held at 92°C and 95°C.

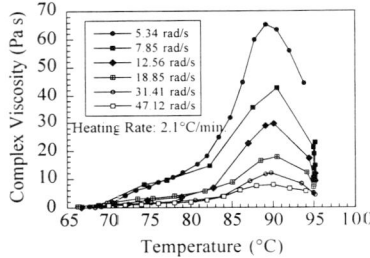

Figure 8. Viscosity-temperature profiles at different frequencies were similar.

3.8. Functional viscosity model

At a heating rate of 2.1°C min^{-1} and over the range of frequencies employed from 1.26 to 47.12 rad s^{-1}, the profiles of η^* at a specific frequency versus temperature were similar (Figure 8), so that by choosing an arbitrary reference frequency (ω_r), all the η^* versus temperature curves at the different frequencies could be reduced to a single curve. Several different frequencies were suitable for use as ω_r. The resulting master curve of reduced complex viscosity $\eta_R^* = \eta^* \left(\dfrac{\omega}{\omega_r} \right)$

versus temperature obtained using ω_r=6.28 rad s^{-1} (1.0 Hz) at several heating rates is shown in Figure 9. When data from a limited number of experiments conducted at high ω values: 62.83 rad s^{-1} (10 Hz) and 78.54 rad s^{-1} (12.5 Hz) at the heating rate of 2.1°C min^{-1} were considered, scaling of frequency could be

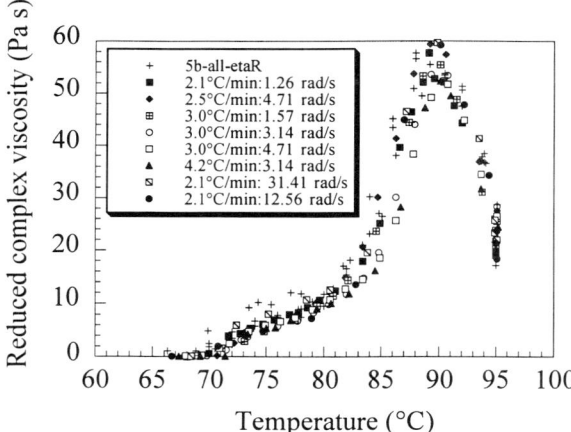

Figure 9. Reduced complex viscosity versus temperature master curve of 8% corn starch dispersion.

achieved by a more general relationship:

$$\eta_R^* = \eta^* \left(\frac{\omega}{\omega_r} \right)^\beta \tag{10}$$

where, $\left(\dfrac{\omega}{\omega_r} \right)^\beta$ is the frequency shift factor and the magnitude of the exponent, β, needs to be determined from experimental data at ω and ω_r. For complex viscosity data at 62.83 rad s^{-1} (10 Hz) and 78.54 rad s^{-1} (12.5 Hz), with ω_r = 6.28 rad s^{-1}, values of β were 0.913 and 0.922, respectively. It is noteworthy that both increasing and decreasing segments of the viscosity-temperature data at all the frequencies were reduced to a single curve. Figure 9 is the master curve of reduced complex viscosity η_R^* data obtained on the 8% corn STD at different heating rates and frequencies.

The reduced complex viscosity-temperature master curves could not be described satisfactorily by any one equation. For accurate description of similar data on 3.5% corn STDs for computer simulation of heat transfer to a corn STD, [22] found that three different equations were necessary over different ranges of temperature.

REFERENCES

1. H. Liu and J. Lelievre, Carbohydr. Polym., 20 (1993) 1.
2. I. D. Evans and A. Lips, J. Texture Studies, 23 (1992) 69.
3. P. E. Okechukwu and M. A. Rao, J. Texture Studies, 26 (1995) 501.
4. M. A. Rao, P. E. Okechukwu, P. M. S. Da Silva and J. C. Oliveira, Carbohydr.

Polym., 33 (1997) 273.

5. E. K. Chamberlain, M. A. Rao and C. Cohen, Int. J. Food Props., 2 (1999) 63.
6. D. D. Christianson and E. B. Bagley, Cereal Chem., 60 (1983) 116.
7. D. D. Christianson and E. B. Bagley, Cereal Chem., 61 (1984) 500.
8. D. Quemada, P. Fland and P. H. Jezequel, Chem. Eng. Comm., 32 (1985) 61.
9. C. Tiu, T. N. Fang, C. W. Chin, J. B. Watkins, N. Felton and H. Greaves, Chemical Eng. J., 45 (1990) B13.
10. T. N. Fang, C. Tiu, X. Wu and S. Dong, J. Texture Studies, 26 (1996) 203.
11. M. A. Rao and J. Tattiyakul, Carbohydr. Polym., 38 (1999) 123.
12. A. Rapaille, A. and J. Vanhemelrijck, in Modified Starches , A. Imeson (ed.) Blackie Academic and Professional, New York, 1992.
13. W. H. Yang and M. A. Rao, J. Food Proc. Eng., 21 (1998) 191.
14. P. C. Williams, F. D. Kuzina and I. Hlynka, Cereal Chem., 47 (1970) 411.
15. P. E. Okechukwu and M. A. Rao, J. Texture Studies, 27 (1996) 159.
16. E. B. Bagley and D. D. Christianson, J. Texture Studies, 13 (1982) 115.
17. R. V. Dail and J. F. Steffe, J. Food Sci., 55 (1990) 1764.
18. A. C. Eliasson, J. Texture Studies., 17 (1986) 253.
19. J. L. Doublier, Starch/Stärke, 33 (1981) 415.
20. Y. F. Champenois, M. A. Rao and L. P. Walker, J. Sci. Food Agric., 78 (1998) 119.
21. J. L. Doublier, Cereal Sci., 5 (1987) 247.
22. W. H. Yang, Rheological behavior and heat transfer to a canned starch dispersion: computer simulation and experiment, Ph.D. thesis, Cornell University, Ithaca, New York, 1997.

HYDROCOLLOIDS – PART 1
Edited by K. Nishinari
2000 Elsevier Science B.V.

301

The genetic effects on some physico-chemical properties of starch granules

N.Inouchi[a], D. V.Glover[b], H. Fuwa[a]
[a]Dept. of Food Science and Technology, Fukuyama University, Fukuyama 729-0251, Japan
[b]Department of Agronomy, Purdue University, West Lafayette, IN 47907 U.S.A.

We have demonstrated the effect of gene and plant species on the properties of commercial starches of several plant species, and wild type and two kinds of mutant maize starches.

1. INTRODUCTION

The properties of starch from plants predominantly depend on genetic factors and plant species. For example, the shape, size, crystallinity, gelatinization temperatures, enzymatic susceptibilities, amylose contents, chain-length distributions of amylopectins of starch granules differ for various genotypes and plant species. To identify genetic effects and the effect of plant species on the properties of starch, we have investigated some properties of commercial starches (with defined sourses) of several plant species, and normal type (wild type) and two kinds of mutant maize starches.

2. MATERIALS AND METHODS

MATERIALS Commercial starch granules of several plant species (normal maize, normal rice, wheat, sweet potato, and potato) and starch granules obtained from mature maize kernels with two kinds of genotypes (Table 1) were used as experimental samples.

Two endosperm mutant genotypes of maize (*Zea mays* L.), amylose-extender(*ae*) and sugary-2(*su2*) in the Oh43 inbred line, were used in these experiments. The maize species and mutants were grown at the Purdue University Agronomy farm. Starch granules were prepared according to Schoch's method [1]. A preparation of glucoamylase of *Aspergillus sp.*K-27 was kindly supplied by Nagase Biochemicals Co. Ltd., Fukuchiyama Japan. Crystalline *Pseudomonas* isoamylase (EC 3.2.1.68, debranching enzyme) was obtained from Hayashibara Biochemical Laboratories, Inc., Okayama, Japan.

Table 1 Starch sample

Starch	Source
Maize starch	
normal	Sanwa Denpun Kogyo Co., Ltd.
amylose-extender (*ae*)	Purdue University
sugary-2　　　(*su2*)	Purdue University
Rice starch	Shimada Kagaku Kogyo Co., Ltd.
Wheat starch	Sanwa Denpun Kogyo Co., Ltd.
Sweet potato starch	Kagoshima Keizairengokai
Potato starch	Tokachi Farmers' Cooperative

METHODS
2.1. Debranching of starch with isoamylase, and fractionation of
debranched materials by gel permeation chromatography(GPC)

Starches were debranched with crystalline *Pseudomonas* isoamylase by the method of Ikawa *et al.* [2]. Debranched materials were fractionated by gel filtration on a column (300 x 20mm) of Toyopearl HW55S connected in series to three column (300 x 20mm) HW50S[3]. Each fraction was divided according to a range of the wavelength at maximum absorbance (λ max) of absorption spectra of glucan-iodine complexes of each tube, Fraction I (Fr.I), λ max $\geqq$ 620nm, intermediate Fr. (Int.Fr.), 620nm $>$ λ max $\geqq$ 600nm, Fr.II, 600nm $>$ λ max $\geqq$ 540nm, and Fr.III, 540nm $>$ λ max. (In the only case of rice starch, the 525nm instead of 540nm was adopted.) Carbohydrate contents in each tube were determined by the phenol-sulfuric acid method [4].

2.2. Debranching of starch with isoamylase, and fractionation of
debranched materials by high performance anion exchange
chromatography with pulsed amperometric detection (HPAED-PAD)

Gelatinized starch (28mg) in 3.5ml of pure water at 100°C for 6min was added to 100 μ l of 1M acetate buffer (pH3.5) and 10 μ l of *Pseudomonas* isoamylase (10 μ g/10 μ l, 590 units/ μ g protein), and incubated at 45°C for 2.5h. The reaction mixture was added to 200 μ l of 0.1N sodium hydroxide solution, 1.0ml of 0.5M phosphate buffer(pH8.5)-0.1% sodium azide solution and 190 μ l of pure water, and filtered through a 0.22 μ m filter (Millipore). HPAEC-PAD was performed by a Dionex model DX-300 system (Dionex Corp., Sunnyvale, CA, USA) and a Model SC-PAD II pulsed amperometric detector consisting of an amperometric flow-through cell with a gold working electrode, a silver-silver reference electrode, and potentiostat, according to the method described by Koizumi *et al.*[5] with a minor modification. Briefly, the system was equipped with a Dionex CarboPac PA1 column (250 x 4 mm) in combination with a CarboPac PA1 Guard column (15 x 4 mm). Repeating sequences of potentials (volts) and durations(ms) on the PAD were as follows: E_1 0.10(t_1 300), E_2 0.60(t_2 120), E_3 $-$0.80(t_3 300). The sample injection loop size was 50 μ l. Results were recorded on a SC 8020 integrator (Tosoh, Japan). Eluent A was 0.15M NaOH, and eluent B was 0.15M NaOH containing 0.1M sodium nitrate. The stepwise gradient elution was programed as follows; 40% of eluent B at 0min, 54.1% at 5min, 63.5% at 10min, 70.7% at 15min, 77.0% at 20min, 82.9% at 25min, 88.6% at 30min, 94.3% at 35min, and 100% at 40min, respectively. All separations were carried out at ambient temperature with a flow rate of 1ml / min. The degree of polymerization(DP) of oligomers was assigned by spiking samples with maltohexaose, and standard response curves were obtained by using a mixture of malto-oligosaccharides (Fuji-oligo G67, Nihon Shokuhin Kako Co. Ltd., Fuji, Japan).

2.3. Differential Scanning Calorimetry (DSC)

Starch (4-5mg) was weighed into a DSC aluminum pan. Distilled water was added into the pan by a microsyringe in a ratio of 1:2(w/w) = dry starch : water, and the pan was hermetically sealed immediately to prevent moisture loss. DSC heating curves were recorded on a Rigaku 8240D-type DSC with distilled water as reference [6]. The heat of gelatinization (ΔH) was obtained from the area of the endothermic peak after fitting a baseline automatically by using the instrument software. The calibration coefficient for ΔH calculations was derived from the known heat of fusion of indium.

2.4. Rapid Visco-Analyzer (RVA) viscograms

The 25mls of distilled water and the culculated weight of starch samples required to make 8% starch suspension (dry weight basis) were added to the RVA aluminum cup. The viscogram of the starch suspension was measured by holding at 30℃ for 1min, heating from 30℃ to 95℃ at a rate of 5℃/min, holding at 95℃ for 6min, cooling from 95℃ to 50℃ at a rate of 5℃/min, and holding at 50℃ for 10min by using the RVA (model RVA-3D, Newport Scientific Pty, Ltd., Australia) with RVA data analysis software.

2.5. Swelling power

Swelling power were determined according to Kainuma *et al.* [7] with a minor modification. The 0.10 grams of starch (dry weight basis) was suspended in 10ml of distilled water at 85℃ in the centrifuge tube, placed in a waterbath and heated under gentle stirring. After 1h the suspension was centrifuged (13,000×g, 20℃, 20min). The weight of precipitate was measured. The swelling power was calculated from the weight of swelling starch which absorbed water per the weight of dry starch.

2.6. Hydrolysis of starch granules by glucoamylase of *Aspergillus sp.* K-27

Starch granules (20mg on dry weight basis) were suspended in 2.0ml of 0.1M acetate buffer (pH 5.5) containing 4mM potassium chloride, and added 0.5ml of 0.25% aqueous solution of a crude enzyme preparation of glucoamylase of *Aspergillus sp.* K-27. This enzyme which is capable of hydrolyzing native starch granules [8] was added to the suspension. The mixture was incubated at 40℃. Aliquots of the reaction mixture was removed at intervals and kept for 10 min in boiling water bath in order to inactivate the enzyme, and to gelatinize the starch granules. Total carbohydrate contents were measured by the phenol-sulfuric acid method, and liberated glucose contents by the glucose-oxidase method [9], using a kit from Toyobo Inc., Japan. Degree of degradation (hyrolysis) of starch granules was calculated from the total carbohydrate content and glucose content of the reaction mixture and expressed as a percentage value.

2.7. X-ray Diffractometry

X-ray diffractograms of fully moistured starch granules (exposed to 100% relative humidity for 3 days) were recorded by a spectrodiffractometer (ADG-502, Toshiba Co., Ltd., Japan) by the Hizukuri' s method [10]. The experimental conditions were wavelength $\lambda = 1.54\text{Å}$, Cu K α (Ni filtered), voltage 35kV, electric current; 15mA, and time constant ; 0.5s.

3. RESULTS AND DISCUSSION

Table 2 shows the characteristics of isoamylase-debranched starches of some plant species investigated by GPC. Fr.I, Fr.II and Fr.III correspond to amylose component, long unit-chains (long B chains) of amylopectin, and short unit-chains (short B chains and A chains) of amylopectin in starch, respectively. The Int.Fr. seems to be composed of short amylose chains and long unit-chains of amylopectin. The *ae* and *su2* maize starches are high amylose types, because the percentages of Fr.I of these starches are higher than that of normal maize starch. The *ae* maize starch has unusual chain-length distributions of amylopectin, namely the increased amount of long chains of amylopectin, and a large amount of intermediate fraction. Potato and sweet potato starches also contain the increased amount of long chains of amylopectin compared with cereal starches, such as rice and wheat starches. The percentages of Int.Fr. of potato and sweet potato starches were not as high as that of the *ae* maize starch.

Fig 1 shows that the differences in the unit-chain length distributions of debranched amylopectins in starches of several plant species with comparison to normal maize amylopectin by HPAEC-PAD. Elution profiles of unit-chains of all starches started at DP6. We have already found that elution profiles of unit-chains of glycogen and phytoglycogen started at DP 3. Differences of properties in unit-chain lengths of starch, glycogen, and phytoglycogen may play an important role in differences in properties among these α -glucans.

Table 3 summarizes some physico-chemical properties of starches of several plant species. The peak temperature of gelatinization of the *ae* maize starch determined by DSC was higher and that of *su2* maize starch was lower than that of the other starches. The peak viscosity measured by RVA and swelling power of potato and sweet potato starches were higher than those of cereal starches. The results indicate that starch granules of potato and sweet potato absorb larger amounts of hot water than cereal starch granules, and the paste of potato and sweet potato starches tend to have a higher viscosity than those of cereal starches. The peak viscosity (RVA) and swelling power of the *ae* starch granules were lower than those of other starches.The *ae* starch granules are not likely to gelatinize and absorb hot water at 85℃-95℃.

The susceptibility of starch granules to the crude glucoamylase was high in the descending order of *su2* maize, wheat, normal maize, sweet potato, and potato starches. This order is similar to the descending order of the increased amount of short unit-chain in the amylopectin. In general, the susceptibility of starch granules to either gluco- or α -amylase mainly depends on the size and crystallinity of the granules. The potato and *ae* maize starches had B-type X-ray diffractograms, sweet potato starch C-type, and cereal starches A-type. The susceptibility of starch granules with B-type crystallinity to amylase is, in general, low, and that with A-type crystallinity is high.

Thus our results indicate that the unit-chain lengths of amylopectins in starch granules are one of the important determinant for certain physico-chemical properties of the starch, such as pasting characteristics, susceptibility to amylases, and crystalline structure.

Table 2 Characteristics of isoamylase-debranched materials of starches of several plant species

Sample	Fr.I (%)	Int.Fr. (%)	Fr.II (%)	Fr.III (%)	Fr.III / Fr.II
Maize					
normal (+)	26.9	5.8	19.5	47.8	2.5
amylose-extender (*ae*)	37.4	11.9	24.5	26.1	1.1
sugary-2 (*su2*)	45.5	4.4	14.0	36.1	2.6
Rice	20.7	3.6	16.5	59.1	3.6
Wheat	33.8	3.7	15.1	48.1	3.2
Sweet potato	23.2	1.7	28.1	46.8	1.7
Potato	21.0	4.1	32.5	42.4	1.3

Fr.I : Amylose,   Int.Fr. : Short amylose and long B chains of amylopectin,

Fr.II : Long B chains of amylopectin,   Fr.III : Short B chains and A chains of amylopectin

Table 3 Some physico-chemical characteristics of starches of several plant species

Sample	Gelatinization temperature* (℃) (DSC)	Peak viscosity (RVU) (RVA)	Swelling power** (at 85℃)	Hydrolysis (1h) (%) (GA***)	Pattern of X-ray diffractogram
Maize					
normal (+)	66	148	8.2	20.4	A
amylose-extender (*ae*)	83	10	3.2	5.6	B
sugary-2 (*su2*)	53	N.D.	N.D.	61.2	A
Rice	68	158	8.5	37.2	A
Wheat	60	107	8.4	17.5	A
Sweet potato	75	267	21.1	13.9	C
Potato	65	550	52.0	4.3	B

*Peak temperature of gelatinization, **In hot water at 85℃,

***A preparation of crude glucoamylase

REFERENCES

1. T.J. Schoch : In "Methods in Enzymol.", III., ed.by Colowick, S.P., and N.O.Kaplan, Academic Press, Inc., (1954) 5
2. Y.Ikawa, D.V.Glover, Y.Sugimoto and H.Fuwa, Starch/Staerke, **33** (1981) 9
3. N.Inouchi, D.V.Glover, T.Takaya and H.Fuwa, Starch/Staerke, **35** (1983) 371
4. M.Dubois, K.A.Gilles, J.K.Hamilton, P.A. Rebers and F.Smith, Anal. Chem., **28** (1956) 35
5. K.Koizumi and M.Fukuda, J. Chromatogr., 585 (1991) 233
6. N.Inouchi, D.V.Glover, Y.Sugimoto and H.Fuwa, Starch/Staerke, **36** (1984) 8
7. K.Kainuma, T.Oda, S.Suzuki, Denpun Kogyo **14**, (1967) 24

8. J. Ave, F.W. bergmann, K.Obata, and S.Hizukuri, Appl. Microbiol. Biotechnol. **27** (1988) 447

9. J.B.Lloyd, and W.J.Whelan, Anal. Bioch., **30** (1973) 467

10. S.Hizukuri : In " Starch Science Handbook",ed.by M.Nakamura and S.Suzuki, Asakura Shoten, Ltd., Tokyo, Japan (1978) 209

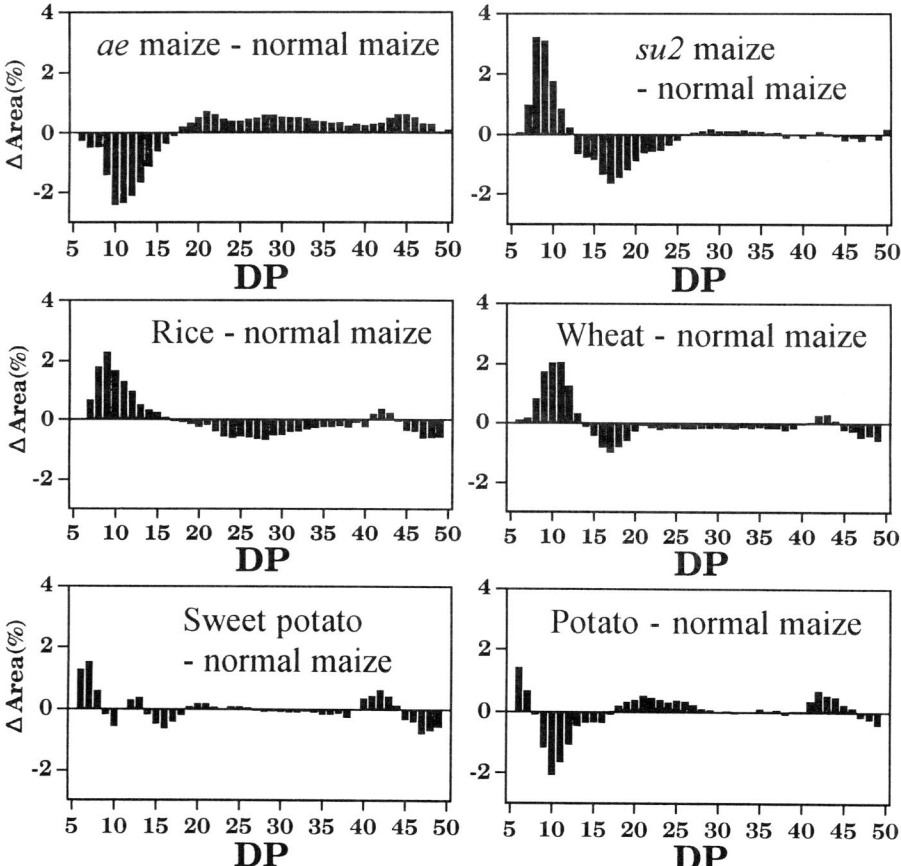

Fig.1 Differences in chain-length distributions of debranched amylopectins of several plant species with comparison to normal maize amylopectin

HYDROCOLLOIDS – PART 1
Edited by K. Nishinari
2000 Elsevier Science B.V.

A simultaneous measurement of frequency dependencies of viscoelastic properties during heating for starch disperse systems using Fourier transform technique

K. Katsuta[a], K. Tanaka[a], E. Maruyama[a], M. Kubo[b] and T. Ueda[b]

[a]Department of Food Science & Nutrition, Nara Women's University,
Kitauoyanishi-Machi, Nara 630-8506, Japan
[b]Material Property Research Center, Nippon Paint Co. Ltd.,
Neyagawa, Osaka 572-8501, Japan

Changes in the dynamic moduli, G' and G", for starch-water systems during heating could be described by a gelation and fusion profile. Since the temperature dependencies of moduli had been measured at a fixed frequency when a conventional oscillatory dynamic method was used, a limited information about the gelation and fusion process could be obtained.

A new method for measuring the frequency dependence of G' and G" using a Fourier transform multi-frequency wave, named as Fourier Transform Rheometry (FT-RM), was applied to the disperse systems of starch during heating. The transition state in gelation and fusion process of starch-water systems could be determined successfully by this FT-RM method.

1. INTRODUCTION

Katsuta had reported previously that the changes in storage shear modulus (G') with temperature, i.e., temperature dependence, for rice starch disperse systems could be described by a gelation and fusion profile which correspond to five stages for 10% starch [1] and four stages for 30% starch [2].

The frequency dependencies of dynamic parameters for disperse systems were needed to clarify the gelation and fusion behavior. The rheological behavior of chemical and physical gels [3-10] at near the sol-gel transition has been extensively measured. These reports demonstrate that the frequency dependencies of both moduli, G' (ω) and G" (ω), obey the power law and G' and G" are given by a (coincident) straight line in the log-log plot for stoichiometrically balanced gels but parallel lines for imbalanced stoichiometry. It take a long time, however, to detect G' and G" against some decades of frequency, ω. The gelling reaction needs to be stopped by some techniques before measurements.

Ueda [11] has developed a new method for measuring the frequency dependence of G' and

G" using a Fourier transform multi-frequency wave, named as Fourier Transform Rheometry (FT-RM). We, Katsuta, Ueda and other co-workers, will try to apply this FT-RM method to the disperse systems of starch during heating.

2. EXPERIMENTAL

2.1. Materials
Non-glutinous rice starch was kindly supplied by Shimada Kagaku Kogyo Co. Ltd. (Niigata, Japan).

2.2. Preparation of starch paste (sol)
Rice starch was weighed into a flask and degassed under vacuum to remove the air for 15 min. Then doubly distilled degassed water was added to the flask and degassed again whilst stirring for 90 min to obtain a homogeneous sol state. The sol (paste) was served for the dynamic viscoelastic measurements, FT-RM.

2.3. Measurement of dynamic viscoelasticity
A rheometer (Rheosol G-3000, UBM Co., Kyoto, Japan) was used for the dynamic viscoelastic measurements of starch paste.

An aliquot of starch paste was placed in the gap (50 μm and a revised gap for increasing temperature) between the cone (3.964 deg, 3.995 cm diameter) and plate. The exposed sample surface between the cone and plate was covered with silicone oil, and then test fixture, cone-and-plate, was wholly covered by a bob-comb type case filled with silicone oil to prevent the evaporation of water from the sample during testing.

Sinusoidal strain with multi-frequency by Fourier transform and with the angle of oscillation of 0.198 deg (γ =0.05) was applied.

2.3. Principle of FT-RM (Fourier Transform Rheometry)
In oscillatory shear, let

$$\gamma (t) = \gamma_0 \exp (i\omega t), \tag{1}$$

where $i = \sqrt{-1}$, ω is the frequency, t the time and γ_0 a strain amplitude which is small enough for the linearity to be satisfied. The corresponding stress can be expressed as follows,

$$\sigma (t) = \sigma_0 \exp \{i\omega (t +\delta)\}. \tag{2}$$

If the viscoelastic behavior is linear, this is substituted into the general integral equation by a Bolzmann's superposition principle.

$$\sigma (t) = \int_{-\infty}^{t} \phi (t-t') [\{d\gamma (t')/ dt'\} \cdot dt'] \tag{3}$$

When $\xi = t - t'$, we have

$$\sigma(t) = i\omega\,\gamma_0 \exp(i\omega t)\int_0^\infty \phi(\xi)\exp(-i\omega\xi)\,d\xi.$$ (4)

Denoting

$$\Phi(i\omega) = \int_0^\infty \phi(\xi)\exp(-i\omega\xi)\,d\xi,$$ (5)

substituting in equation (4), complex shear modulus G^* is given by

$$G^*(i\omega) = \sigma(t)/\gamma(t) = i\omega\,\Phi(i\omega).$$ (6)

The equation of Bolzmann's principle can be rewritten as follows through a Fourier transform,

$$\Sigma(i\omega) = \Phi(i\omega)\,i\omega\,\Gamma(i\omega),$$ (7)

where $\Sigma(i\omega)$ is Fourier transform for σ and $\Gamma(i\omega)$ for γ, respectively. Hence,

$$G^*(i\omega) = \Sigma(i\omega)/\Gamma(i\omega).$$ (8)

This shows that, in linear region, complex shear modulus G^* is given by the ratio of Fourier transform for stress to that for strain.

In this study, FT-RM, the strain with a Fourier transform multi-frequency wave, which is shown as follows, is applied.

$$\gamma(t) = \sum_{n=0}^m \gamma_n \sin(2\pi 2^n \omega t)$$ (9)

This means that the strains with higher degree of frequency are superposed on to a fundamental strain when $n=0$ and which is demonstrated in following equation.

$$\gamma(t) = \gamma_0 \sin(2\pi\omega t)$$ (10)

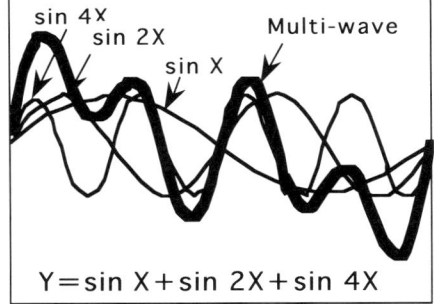

$$Y = \sin X + \sin 2X + \sin 4X$$

Figure 1. An example multi-wave composed by three sine curves at different frequencies.

An example for multi-sine wave is shown in Figure 1. The multi-wave corresponds to the bellow equation.

$$\gamma(t) = \sum_{n=0}^2 \gamma_n \sin(2\pi 2^n \omega t)$$ (11)

Thus, simultaneous measurement of frequency dependence for dynamic parameter can be made, if the composed strain with alternations at oscillatory amplitudes is applied.

It has been reported that if the viscoelastic behavior was non-linear, the response can be represented by a fundamental component and by the odd number of higher degree of frequent components, especially by 3rd and 5th order harmonics [12,13]. As seen in Equation 9 and 11, the degree of polynomials is 2^n. This means that the multi-frequency wave in the present FT-RM method is composed by only even harmonics and the fundamental frequency to ensure the linear viscoelastic response.

3. RESULTS AND DISCUSSION

3.1. Temperature dependence of G' at multi-frequency composed by FT-RM

Figure 2 show the temperature dependence of G' for 20% rice starch at multi-frequency composed by Fourier transform, when the fundamental frequency was 0.16Hz, i.e., $\omega =1$ (rad/sec). The results support our previous discussion which reported that the changes in modulus for starch disperse system with heating can be described by a gelation and fusion profile [1,2].

We also stated that the temperature dependence of G' for 30% rice starch could be divided into four stages; I , expanding the volume fraction of starch granules; II , filling-in between starch

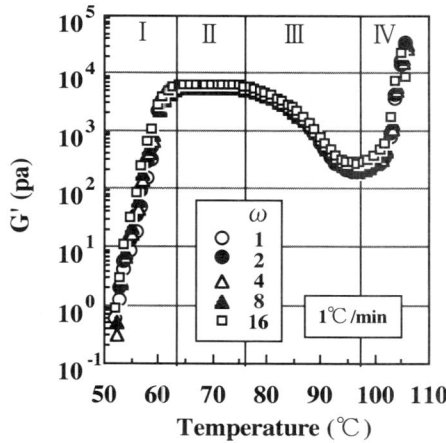

Figure 2. Temperature dependence of G' for 20% rice starch at multi-frequency composed by FT-RM method.

The fundamental frequency is 0.16Hz (ω =1) and the higher frequencies up to n=4 in Eq.(9), i.e., ω =16, are superposed.

particles (granules) and diffusion of water into crystalline region; III , fusion process of the crystalline region; IV, development of mechanical strength [2].

In stage I , i.e., at 50 °C through 60°C in Figure 2, there were large differences between the magnitudes of G' at various frequency (ω) 1 though 16, in which the frequency dependence of G' was great and the starch-water system still maintained the sol state. After passing 60°C, the frequency dependencies of G' were scarcely observed, corresponding to be gel state. At around 90°C, the frequency dependencies of G' became larger again. This is the evidence that the stage III can be defined as a fusion process.

3.2. Frequency dependence of G' in individual stages

The frequency dependencies of G' at various temperature, which are converted from the data shown in Figure 2, were illustrated in Figure 3 through Figure 5. Figure 3 corresponds to the stage I in Figure 2, Figure 4 the stage II and III , and Figure 5 the stage IV, respectively.

In Figure 3, the frequency dependence of G' at 50.7 °C scarcely appeared. This means that a plateau region exists in the starch disperse system when it is not yet gelatinized. This plateau region at longer time had been reported by Onogi and co-workers [12] for synthetic polymer disperse system and called "second plateau region". The appearance of the second plateau region demonstrates that the dispersion of starch granules is defined as the solid disperse system. At 52.7 ℃, it can be defined as the onset temperature for gelatinization of rice starch, the frequency dependence of G' for the system became larger and the indices of frequency dependence gradually increased with elevating the temperature up to 58.8 ℃. The frequency dependence of G' at 59.8 ℃ change drastically to be small. The gelling process might be concluded at 66.8℃.

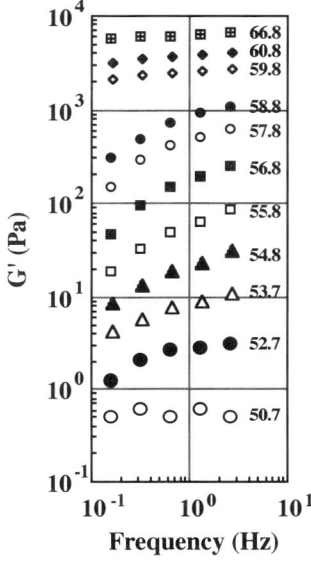

Figure 3. Frequency dependencies of G' for 20% starch in stage Ⅰ.

The numbers in figure show the temperature observed.

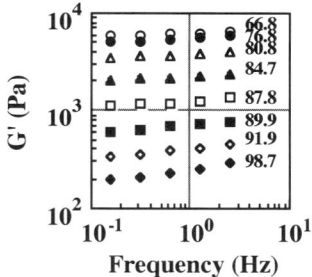

Figure 4. Frequency dependencies of G' for 20% starch in stage Ⅱ and Ⅲ.

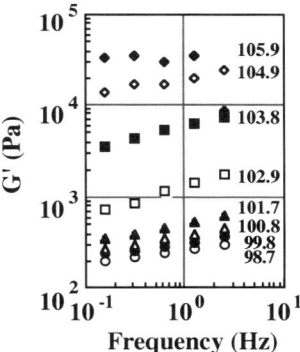

Figure 5. Frequency dependencies of G' for 20% starch in stage Ⅳ.

As seen in Figure 4, the viscoelastic behavior of the system at 66.8℃ through 76.8℃ showed the almost rubbery plateau regions. When the crystalline region in the starch system began to melt (at around 76.8℃), the magnitude of G' started to decrease and the indices of their frequency dependence became slightly larger. After the end of fusion process (at 98.7℃), the magnitude and of G' increased and the slope of G' (ω) curves also became more steeper (at 102.9 and 103.8℃), as seen in Figure 5. This shows that another gel-sol transition might exist in this stage, and the sol state finally transforms to solid, may be a glass, state at 105℃.

3.3. Sol-gel transition

Figure 6 shows the G' (ω) and G" (ω) for 20% starch containing 200mM NaCl obtained by FT-RM. The log-log plots of both moduli for frequency showed an almost straight line when the temperature reached 64.7℃, where the system might be a near gel point. The magnitude of the exponents of frequency dependence were 0.61 for G' (ω) and 0.49 for G" (ω), respectively. Both exponents were not perfectly coincident, but they were quite near to 0.5. In the case of 20% starch alone, the sol-gel transition point was observed at around 58 ～ 59℃, although the data are not shown.

4. CONCLUSION

A new method for measuring the frequency dependence of G' and G" using a Fourier transform multi-frequency wave was applied to the disperse systems of starch with and without salts during heating. The transition state in gelation and fusion process of starch-water systems could be determined successfully by this FT-RM method.

The argument for analysis of non-linear viscoelasticity is omitted here, but it will be discussed by another Fourier transform technique in the near future.

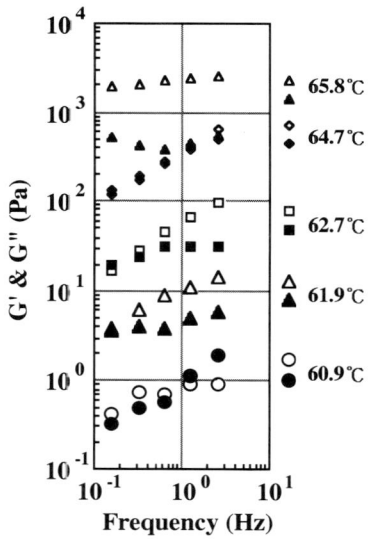

Figure 6. Frequency dependenices of G and G" for 20% rice starch containing 200mM NaCl.
The open symbols in figure show G' and solid ones G".

REFERENCES

1. K. Katsuta, *J. Appl. Glycosci.*, **43** (1996) 541.
2. K. Katsuta, In P.A. Williams and G.O. Phillips (eds.),*Gums and Stabilisers for the Food Industry 9*, pp.59-68, The Royal Soc. Chem., Cambridge, 1998.
3. F. Chambon and H.H. Winter, *Polym. Bull.*, **13** (1985) 499.
4. F. Chambon and H.H. Winter, *J. Rheology*, **31** (1987) 683.
5. H.H. Winter, *Polym. Eng. Sci.*, **27** (1987) 1698.
6. J.E. Martin, D. Adolf and J.P. Wilcoxon, *Phys. Rev. Letters*, **61** (1988) 2620.
7. C. Friedrich, L. Heymann and H.-R. Berger, *Rheol. Acta*, **28** (1989) 535.
8. G. Cuvelier, C. Peigney-Noury and B. Launay, *Gums and Stabilisers for the Food Industry 5*, G.O.Phillips and P.A. Williams (eds.) IRL Press, Oxford, pp.549-552 (1990).
9. A. Koike, N. Nemoto and E. Doi, *Polymer,* **37** (1996) 587.
10. K.S. Hossain, N. Nemoto and K. Nishinari, *J. Soc. Rheol. JPN*, **25** (1997) 135.
11. T. Ueda, *J. Soc. Rheol. JPN*, **23** (1995) 109.
12. S. Onogi, T. Masuda and T. Matsumoto, *Trans. Soc. Rheol.*, **14** (1970) 275.
13. T. Matsumoto, Y. Segawa, Y. Warashina and S. Onogi, *Trans. Soc. Rheol.*, **17** (1973) 47.

HYDROCOLLOIDS – PART 1
Edited by K. Nishinari
2000 Elsevier Science B.V.

Rheological study on physical modification starches

S. Akuzawa, S. Sawayama, A. Kawabata

Department of Nutritional Science, Faculty of Applied Bioscience
Tokyo University of Agriculture, 1-1-1 Sakuragaoka
Setagaya-ku, Tokyo 156-8502, Japan

The physical modifications, which were to incorporate several free fatty acids into starch and heat-moisture treated starch, were some useful methods. Especially, heat-moisture treated starches were changed in the property which completely differed from native each starch.

1. INTRODUCTION

Starch is commonly used in paste form on cooking and food processing as a viscosity enhancer and viscosity stabilizer. In this application, it is required to be viscostable, thermostable, colorless and transparent. Native starch paste that the control of the viscosity at the every temperature is difficult, and viscosity stability at each temperature is low, so that its use is restricted when these properties are required. Although starch chemically combined with many different materials has been developed to improve its physical and chemical properties, food applications are limited by law.

We examine some physical modifications to starch to obtain basic data for controlling the rheological properties of starch and starch paste. The physical modifications were to incorporate several free fatty acids into starch and heat-moisture treated starch. The thermal properties and viscoelastic behavior were investigated for the effects of added free fatty acids on their properties of starch granules. The most obvious effect caused by the heat-moisture treatment was found to be an alteration of the structure of the elementary cell, resulting in the gelatinization properties and rheological properties of these pastes being dynamically changed. The concentration dependence of the mechanical properties of starch pastes near the sol-gel transition point was analyzed by the scaling law [1, 4] derived from percolation theory. The three-dimensional network of the gel from native and heat-moisture treated starch pastes was also compared.

2. MATERIALS AND METHODS

2.1. Materials

The samples used in this study were cassava, corn and potato starches. Free fatty acids manufactured by Tokyo Kasei were used, including four saturated free fatty acids, i.e., lauric acid (C12:0), myristic acid (C14:0), palmitic acid (C16:0), and stearic acid (C18:0), and two unsaturated free fatty acids, i.e., oleic acid (C18:1) and linoleic acid (C18:2).

Corn and cassava starches were heat-moisture treated at 125°C for 20min (the saturated vapor), and potato starch was heat and moisture-treated at 110°C for 30min (the saturated vapor) (Sanwa Cornstarch Co., Ltd.).

2.2. Preparation of complex sample incorporated free fatty acid into starch granules [5]

Each free fatty acid was incorporated into granules of each starch sample (200g) by dissolving 5g of the free fatty acid in methanol (200ml) and then stirred for 5 hours. After filtration and washing away with methanol, the granules were air-dried to give a starch complex sample. Then, measurements of the incorporated free acid value were made by the gas chromatography and the thermal properties of each complex sample were investigated by DSC.

2.3. General characteristics and dynamic viscoelasticity of heat-moisture treated starch and starch pastes

The structure of heat-moisture treated starches were observed by microscopy. The gelatinization characteristics of the heat-moisture treated starch samples were investigated in respect of swelling-power and solubility, photopastegraphy, Brabender viscography and DSC [6].

Heat-moisture treated starch pastes, 0.5~4.0% concentration (w/w), were prepared by the method which had been reported previously [6]. Then dynamic viscoelasticity was measured in the frequency range of 0.06 ~ 60 (rad/s) by using a Rheo-Stress RS 100 rheometer (Haake Co.). The concentration dependence of the mechanical properties near the sol-gel transition point of native and heat-moisture treated corn, cassava and potato starches was analyzed by the scaling law derived from the percolation theory.

2.4. Application to the food

We examined application of heat-moisture treated starches to food materials for blancmange with native and heat-moisture treated corn starches. The textural properties were measured by using a Texture Analyzer TA-XT2i (Haake Co.) and a sensory evaluation was conducted by the semantic differential (SD) method.

3. RESULTS AND DISCUSSION

3.1. Effects of free fatty acids on the gelatinization characteristics of the starches

Figure 1 shows the analytical data of the free fatty acid values in the incorporated starches. It shows that myristic acid was the highest in the case of cassava and potato starch. The free fatty acid values for corn starch were significantly higher than those for the other starches. Photopastegrams indicated that the gelatinization characteristics were not changed when free fatty acids were incorporated into the starch samples.

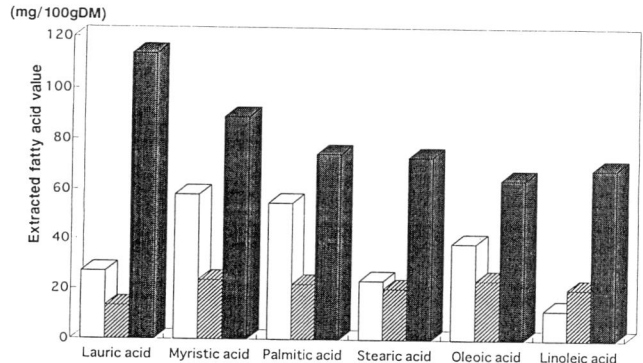

Figure 1. Each Fatty Acid Value from Starch with 75%-PrOH after Soxhlet-extractor

☐, Cassava ; ▨, Potato ; ■, Corn

However, the Brabender viscograms showed the different properties for each starches. The viscograms showed that these starch samples had a gelatinization temperature higher than that of defatted starch and stable viscosity. Defatted starch did not show any breakdown.

Table 1 shows the DSC characteristics and the incorporated free fatty acid values for corn starch. The corn starch alone showed a first peak and another peak at a higher temperature, which seemed to indicate an amylose/free fatty acid complex. The cassava starch and potato starch showed no second peak, probably due to the small content, although not zero, of the complex in the starch granules, making realistic observation of any DSC curves impossible.

Individual starch sample had specific characteristics resulting from the incorporation of a free fatty acid, the effect of stabilizing viscosity by mixing the starch paste with a free fatty acid being demonstrated. It is thus considered that the state of the complex between each free fatty acid and amylose might be understood by observing the thermal properties, which depended on the state of the complex of each free acid with amylose rather than on the value introduced into amylose [7].

Table 1   Characteristics of Corn Starch in High Temperature Region by DSC and the Incorporated Fatty Acid Values.

Fatty acid	To	Tp ($^{\circ}$C)	Tc	ΔH (mJ/mg)	Tc-To ($^{\circ}$C)	Fatty acid (mg/100g)
Lauric acid	------	----	----	------	-----	113.4
Myristic acid	93.0	101.0	108.0	0.35	15.0	89.0
Palmitic acid	99.0	104.0	108.0	0.24	9.0	74.2
Stearic acid	98.0	104.2	109.4	0.38	9.0	73.0
Oleic acid	95.8	101.4	108.0	0.44	12.2	64.2
Linoleic acid	86.8	95.8	103.8	0.25	17.0	68.4
Native	86.2	96.5	107.0	1.89	20.8	463.6

------ ; no peak

3.2. Effect of heat - moisture treatment on the rheological properties
1)   General characteristics of heat-moisture treated starches

A scanning electron microscope (SEM) photograph of the cross section of the heat-moisture treated potato and corn starch granules showed a large hollow area about 1/3 of the diameter at the center of the starch granules and a layered structure resembling growth rings [8].

The gelatinization characteristics of the heat-moisture treated starch samples were investigated with respect to swelling-power and solubility, photopastegraphy, Brabender viscography and DSC. Compared with the equivalent native starch samples, the swelling-power and solubility of each heat-moisture treated starch was considerably suppressed, except for cassava starch. The native starches showed an increasing transmittance in the photopastegrams with an increase in temperature from 4 to 9°C, whereas the decrease in transmittance was observed for the heat-moisture treated samples. The viscograms of heat-moisture treated potato and corn starches indicated considerably suppressed maximum viscosity and no breakdown. DSC curves showed two endothermic peaks in the low-and high-temperature regions for the heat-moisture treated potato and corn starches, but cassava did not show this. From these results, the effect by heat-moisture treatment was different by the type of the starch. Then the molecular association changed, not only in the crystalline region but also in the amorphous region. Viscosity stability in the gelatinization process considered to be practically effective characteristics as a new food material in food industry.

2)   Analysis of sol-gel transition of heat-moisture treated starches

The concentration dependence of the mechanical properties near the sol-gel transition point of native and heat-moisture treated corn, cassava and potato starches was analyzed by the scaling law derived from the percolation theory [9].   The critical concentration of the heat-moisture treated corn starch increased more than that of the native type, with the starch gel being observed to decrease in firmness with heat-moisture treated sample.   The critical concentration of the heat-moisture treated cassava starch also increased more than that of the native type, while the storage modulus increased more than that of the 4wt% native cassava sample, being soft at low concentration, and firm at higher concentrations. The critical concentration of the heat-moisture treated potato starch also increased more than that of the native type, but was softer than the native type.   The scaling law could not be applied to the native and heat-moisture treated potato starches, although the critical exponents were determined as 4.2 and 3.6 for the native and heat-moisture treated corn starches, and as 1.8 and 2.4 for the native and heat-moisture treated cassava starches, respectively (Figure 2).

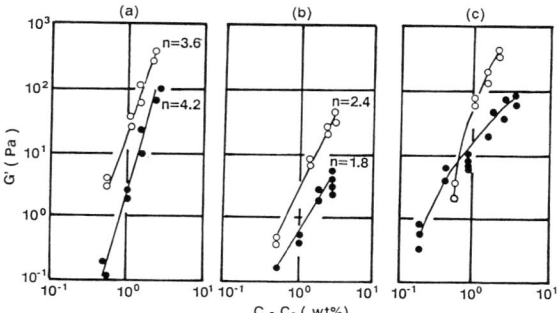

Figure 2. Critical Behavior of Storage Modulus G' near
the Percolation Threshold for Starch Pastes

●, native ; ○, heat-moisture treated ;
(a), corn ; (b) , cassava ; (c) , potato ;

Scaling low , $G' = k(C-C_c)^n$
G', elastic modulus ; C , concentration ; C_c, critical concentration ;
n , critical exponent

3)   Application of heat-moisture treated starches to food material

Results of sensory evaluation of blancmange are shown in Figure 3.   The texture of blancmange made with the heat-moisture treated corn starch had the best combination of softness and toughness cohesion than that with native one, this being confirmed by the results of sensory evaluation.   Heat-moisture treated corn starch is thus anticipated to be a useful material for food.

318

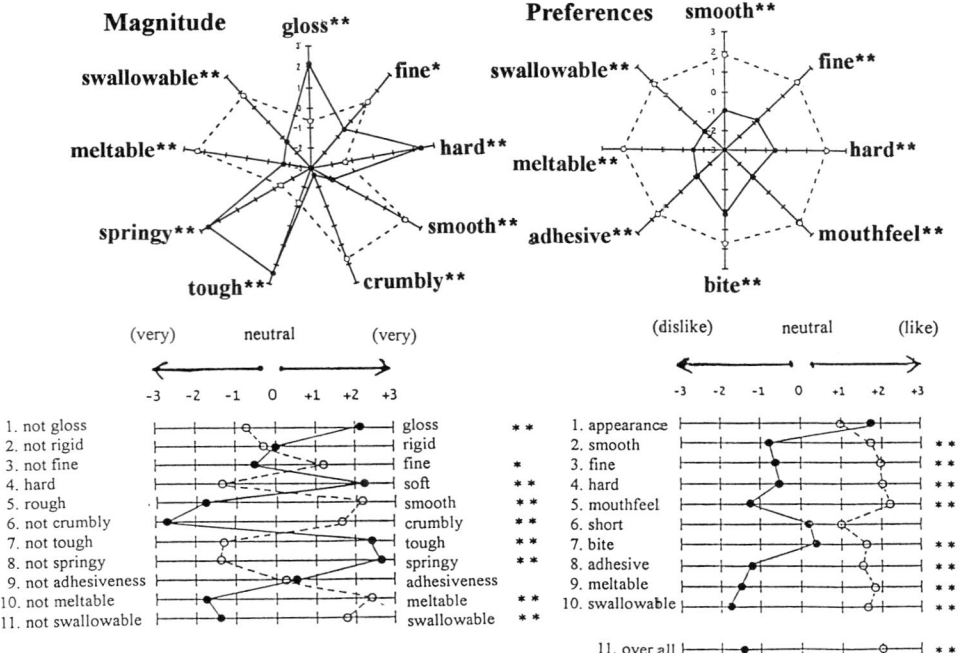

Figure 3. Average Score of Blancmange Made from Corn Starch with SD Method

●, native ; ○, heat-moisture treated ; *, 5% significant ; **, 1% significant

REFERENCES

1. A. Zosel, Rheol. Acta, 21 (1982) 72.
2. M. Tokita and K. Hikichi, Phys. Rev. A, 35 (1987) 4329.
3. Y. Otubo, Nippon Reoroji Gakkaishi, 19 (1991) 111.
4. T. Yano, H. Kumagai, T. Fujii and T. Inukai, Biosci. Biotech. Biochem., 57 (1993) 528.
5. S. Akuzawa, S. Sawayama and A. Kawabata, Biosci. Biotech. Biochem., 59 (1995) 1605.
6. A. Kawabata, N. Takase, S. Akuzawa and S. Sawayama, Oyo Toshitu Kagaku, 43 (1996) 471.
7. S. Akuzawa, S. Sawayama and A. Kawabata, Biosci. Biotech. Biochem., 61 (1997) 487.
8. A. Kawabata, N. Takase, E. Miyoshi, S. Sawayama, T. Kimura and K. Kudo, Starch/starke, 46 (1994) 463.
9. A. Kawabata, S. Akuzawa, T. Yazaki and Y. Otsubo, Oyo Toshitu Kagaku, 43 (1996) 479.

HYDROCOLLOIDS – PART 1
Edited by K. Nishinari
2000 Elsevier Science B.V.

Rheological and DSC studies of chemically modified starch

K. Morikawa[a]   and   K. Nishinari[b]

[a]Research Laboratory, Taito Co., Ltd., Nagata, Kobe 653-0023, Japan

[b]Faculty of Human Life Science, Osaka City University, Sumiyoshi, Osaka 558-8585, Japan

The effects of degree of modification on the physico-chemical properties of chemically modified starches (hydroxypropylated potato di-starch phosphates) were studied by dynamic viscoelasticity measurements and differential scanning calorimetry (DSC).   Dynamic viscoelastic measurements were performed for 10% pastes of native potato starch and chemically modified potato starches heated at different temperatures.   The storage shear modulus (G') and the loss shear modulus (G'') of pastes of modified starches heated at 50 to 100°C for 30 minutes were independent of heating temperature. G' and G'' were strongly dependent on the degree of modification.   Even a slight modification retarded the retrogradation remarkably.   Syneresis of pastes became more conspicuous with increasing level of modification.   Transparency of modified starch-pastes was constant at 5°C for 1 week and lower than that of native starch-paste.

The endothermic peak of chemically modified starches appeared at 48 to 50°C in the heating DSC curves.   The retrogradation phenomena of modified starch were scarcely observed, whilst a considerable retrogradation phenomenon was observed in a native one.

1. INTRODUCTION

Starch has been used widely in the food industry, however, it encounters difficulty sometimes because of its physical and/or chemical properties.   Chemical modification of starch is used to improve the gelatinization and cooking characteristics, to prevent retrogradation and gelling tendencies.   In recent years, there are many types of chemically modified starch, i.e. starches modified by acid hydrolysis, oxidation, etherification, esterification, and cross-linking.   Chemical modification has improved the functional properties of starches, such as resistance to severe processing conditions or storage at low-temperature.   Etherified starch has low-temperature stability, high clarity and good solubility. Cross-linking improves texture and temperature resistance of native starches.   In addition to these effects, it produces considerable change in the gelatinization and swelling properties of the starch.   The aim of this investigation, therefore, is to examine the effects of degree of modification on the physico-chemical properties of hydroxypropylated potato di-starch phosphates.

2. MATERIALS AND METHODS

Native potato starch and chemically modified potato starches were provided by Lyckeby Stärkelsen (Kristianstad, Sweden), and further purified by washing three times with 80%(v/v) methanol. The ratio of hydroxypropyl groups and phosphate cross linkages in native starch (NS) and chemically modified potato starches (P-HP1,2,3; P-:potato, H:hydroxypropylation, P:phosphate cross-linking) are shown in Table 1.

Table1.   The ratio(%) of hydroxypropyl groups and phosphate cross linkages

	hydroxypropyl groups [1]		phosphate cross linkages [2]	
	$(C_3H_7O)\%$	DS [3]	$(PO_4H)\%$	PCL [4]
N S	0.00	0.000	0.00	0.000
P-HP1	0.61	0.018	0.96	0.016
P-HP2	1.28	0.047	0.88	0.015
P-HP3	2.51	0.097	C.86	0.015

[1]; measured by modified Zeisel method
[2]; measured by spectrophotometric method using molybdenum blue
[3]; degree of substitution
[4]; phosphate cross linkages = number of cross linkages / number of glucose residues

2.1. Preparation of pastes for measurement

Dispersions of 3% NS and P-HP1,2 and 3 were prepared in distilled water using a motorized stirrer. Stirring was continued at 200 rpm for 30 minutes at 25°C. Then the dispersion was heated in an oil bath to 95°C for 15 minutes and then kept at 95 to 98°C for 30 minutes. The boiling distilled water was added to the hot dispersion to adjust the concentration.

2.2. Transparency

Transparency of starches was determined as described by Wu et al [1]. Gelatinized 3% pastes prepared by the above mentioned method were cooled at room temperature and stored at 5°C. The transparency of starch pastes was evaluated using percent transmittance at 650 nm against a distilled water blank in a UV-visible recording spectrophotometer UV-265FW (Shimadzu Co., Ltd., Kyoto, Japan).

2.3. Syneresis

Gelatinized 3% pastes of the starch were poured into glass tubes (30mm diameter and 300mm length) and were held vertically at 5°C for 7days. The extent of syneresis was estimated by the length of liquid phase Δh separated above the sedimented phase. The degree of syneresis was represented by $\Delta h/h_0$, where h_0 stands for the initial height of sample dispersion [2].

2.4. Dynamic viscoelasticity

Dynamic viscoelastic measurements were performed using a cone and plate geometry (25mm diameter, 1° angle) on a fluid spectrometer RFS II (Rheometrics Far East Co., Ltd., Tokyo, Japan). The hot sample dispersion was poured directly onto the plate of the instrument, which had been kept at 5°C. Frequency dependence of the storage shear modulus (G') and the loss shear modulus (G'') for starch-pastes was observed at 5°C after heating the dispersions at each temperature for 30 min. Immediately G' and G'' were measured as a function of time at 1 rad/s at 5°C for 15 hours.

2.5. DSC analysis

DSC measurements were performed with a Micro DSC III (Setaram, Caluire, France). Each 200mg (d.b.) of starch and 400mg of distilled water were directly weighed into a DSC cell. After sealing, a cell was left for one hour to equilibrate. A cell containing an equal amount of distilled water was used as a reference. The temperature was raised from 20°C to 90°C at 1.0°C/min (the first run) and decreased from 90°C to 5°C at 2.0°C/min. After the cell was stored for 5 days at 5°C, DSC measurement was performed again (the second run).

3. RESULT AND DISCUSSION

3.1. Transparency

Time dependence of transparency for NS and P-HP is shown in Figure 1. Transparency decreased with increasing degree of modification. Transparency of NS was high just after the gelatinization, and then was decreasing with time because of the retrogradation . On the contrary, transparencies of P-HP1,2 and 3 did not decrease with time indicating that chemically modified starch does not retrograde so fast.

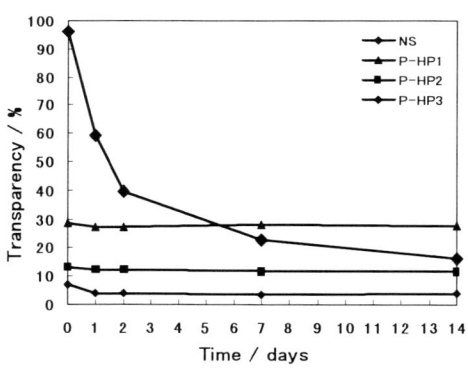

Figure1.   Time dependence of transparency %
(path length 1mm at the wave length of 650nm)
for 3% pastes of NS and P-HP1,2,3 stored at 5°C
after heated at 95°C for 30 min.

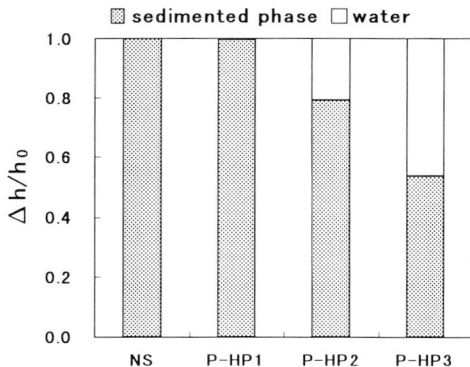

Figure2.   Syneresis for 3% pastes of NS and P-HP1,2,3
stored at 5°C for 1 week after heated at 95°C for 30
min.

3.2. Syneresis

Syneresis of 3% pastes for NS and P-HP is shown in Figure2.   The degree of syneresis increased with increasing degree of chemical modification. This is attributed to a decrease in the hydrophilicity with increasing substitution of hydroxy groups of the glucose residues with hydroxypropyl groups.

3.3. Dynamic viscoelasticity

Frequency dependence of G' and G'' for NS and P-HP1,2 and 3 pastes is shown in Figure3.   Both moduli of pastes decreased and became more frequency dependent with increasing degree of modification, i.e. tended to a more liquid like rheological behavior.   The fact that both moduli for P-HP1 are larger and less frequency dependent than those for NS is explained by the difference in the degree of disintegration of starch granules in P-HP1 and NS observed by light microscopy (data not shown); starch granules were completely disintegrated in NS after heated at 95°C for 30 minutes whilst some granules remained intact even after such a severe heat-treatment for P-HP1.

To understand the gelatinization properties, the frequency dependence of G' and G'' for NS and P-HP1 heated at various temperatures for 30 min was examined (Figure4).   In general, if starch is gelatinized at a high temperature, both moduli of the starch pastes will be smaller because of disintegration of starch granules.   The value of G' of NS was larger for the lower heating temperature, and became smaller when heated at 100°C because almost all starch granules were disintegrated.   On the other hand, P-HP1 was completely gelatinized at 50°C for 30 minutes, and it showed no change at higher temperatures over 50°C.

To understand the retrogradation properties, the time dependence of G' for dispersions of NS and P-HP1 heated at various concentrations was examined (Figure5).   The value of G' of NS increased with time and this tendency would be stronger if the concentration was higher. On the other hand, G' of P-HP1 of any concentration did not change with time.

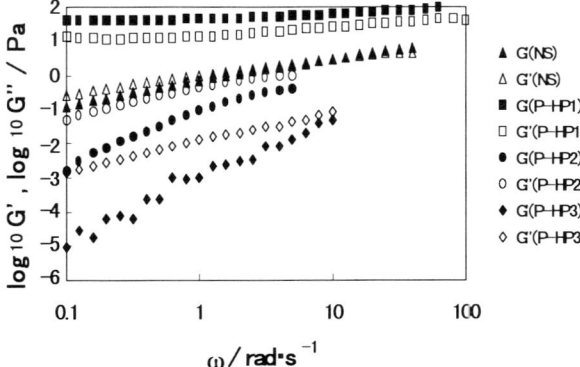

Figure3.   Frequency dependence of G' and G'' for 3% pastes of NS and P-HP1,2,3 stored at 5°C for 24 hours after heated at 95°C for 30 min.

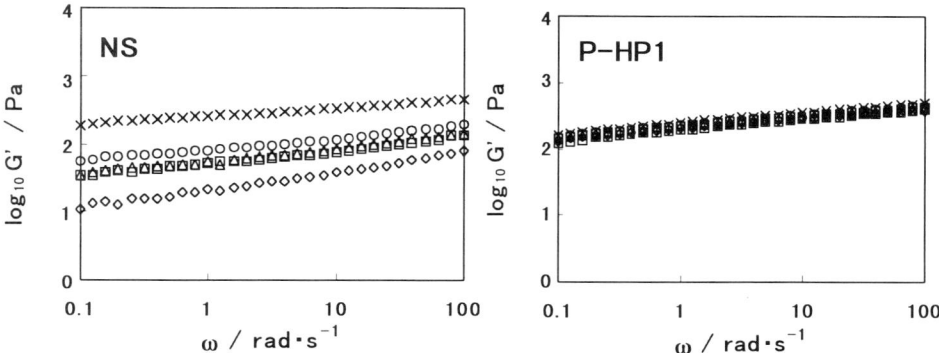

Figure4.　Frequency dependence of G' for 10% pastes of NS and P-HP1 heated at various temperatures for 30 min.　Temperature: +,50°C; ×,60°C; ○,70°C; △,80°C; □,90°C; ◇,100°C.

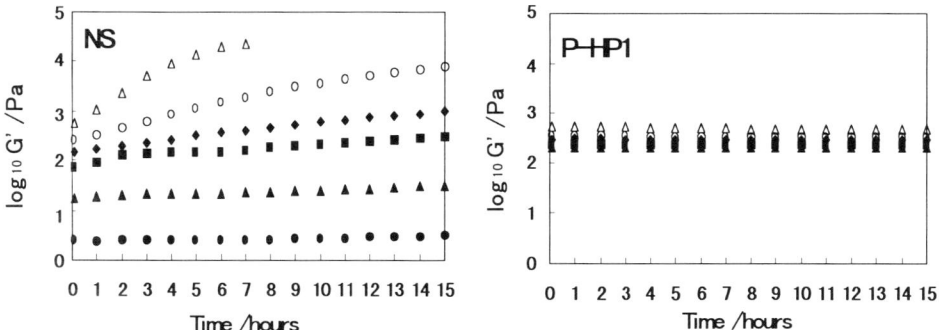

Figure5.　Time dependence of G' for various concentrations of NS and P-HP1 heated at 95°C for 30 min.　Measurement temperature: 5°C
Concentration: ●,5%; ▲,10%; ■,15%; ◆,20%; ○,25%; △,30%.

3.4. DSC analysis

Heating DSC curves of native potato starch and chemically modified starches are shown in Figure6.　In Table 2, onset temperature (To), peak temperature (Tp) and conclusion temperature (Tc) are shown. Tp of all P-HP ranged from 48°C to 50°C, and was about 12°C lower than that of NS.　The enthalpy of P-HP was also lower than that of NS.　In the second run, any endothermic peak was scarcely observed for chemically modified starch, whereas a broad endothermic peak was observed for native starch at a lower temperature than in the first run heating DSC curve.

324

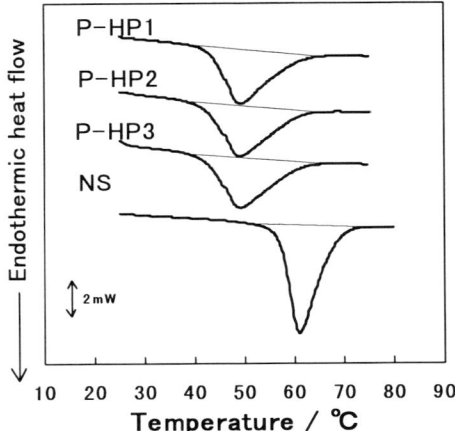

Figure6.   The heating DSC curves of NS and P-HP1,2,3. Heating rate: 1°C/min.
The second run was observed after storage for 5 days at 5°C.

Table 2.   Thermal transition characteristics of native and chemically modified starches

	To(°C)[*1]	Tp(°C)[*1]	Tc(°C)[*1]	ΔH_1(J/g)[*2]	ΔH_2(J/g)[*3]	Tp(°C)[*3]
Native starch	57.16	60.98	74.69	13.72	10.31	54.04
P-HP1	42.90	48.45	66.76	9.97	0.00	—
P-HP2	44.31	49.99	65.57	9.26	0.00	—
P-HP3	44.08	49.34	68.00	10.45	0.00	—

*1;To,Tp and Tc are the onset, peak and conclusion temperatures, respectivery.
* 2; ΔH_1 are the enthalpy determined from the 1st run DSC heating curves.
* 3; ΔH_2 are the enthalpy of 2nd runs after storage for 5 days at 5°C.

4. CONCLUSION

The results of this study indicate that G' and G'' were strongly dependent on the level of modification.   Values of G' and G'' for the pastes of modified starches heated at 50 to 100°C were almost independent of heating temperature.   Syneresis of pastes became more conspicuous with increasing level of modification.   Transparency of modified starch was constant and lower than that of native starch.

Judging from the gelatinization peak temperature in heating DSC curves, chemically modified starches are gelatinized at much lower temperatures than the native starch.

We are grateful to Prof. H. Fuwa (Fukuyama Univ.) for his valuable comments.

REFERENCES

1. Y.Wu and P.A.Seib, Cereal Chem., 67 (1990), 202.
2. M.Yoshimura, T.Takaya and K.Nishinari, Carbohydrate Polym., 35 (1998), 71.

HYDROCOLLOIDS – PART 1
Edited by K. Nishinari
2000 Elsevier Science B.V.

Effect of defatting of rice on the gelatinization

S. Miwa[ab] , H. Oda[a] , T. Takaya[b] , K. Nishinari[b]

[a]   Ishikawa Agriculture Research Center,

[b]   Department of Food and Nutrition, Faculty of Human Life Science, Osaka City University

Koshihikari and Matsumae (rice grains) were defatted by methanol-chloroform for five periods. Total fat and free content in rice decreased with increasing defatting time while the fat-by-hydrolysis content was not much changed. Heating DSC curves for the rice grains-water system showed two endothermic peaks around $60°C \sim 90°C$ (peak1 and peak2) and third endothermic peak around $90°C \sim 120°C$ (peak 3). The decrease in total fat content lowered the temperature of peak 2, increased the enthalpy of peak 1, and decreased the enthalpy of peak 2.

It was found that fat content has a drastic effect on the gelatinization of rice grains.

1.INTRODUCTION

Cooked rice is a staple Japanese food with a plain taste. Recently, Japanese consumers, concerned with quality of cooked rice, want to buy high quality rice. It is well known that the texture (stickiness and hardness) is an important factor for quality of cooked rice. Usually the texture varies according to the variety of rice and the growing district, mainly because of its starch composition (amylose, amylopectin) and other components such as lipids and protein.

The fat content of rice is about 1% to 2% in polished Japonica rice, while the fat-by-hydrolysis content of polished rice ranges from 0.4% to 1.1%[1]. There have been few reports on fat component in relation to quality of cooked rice because fat has been considered only a minor component compared with starch and protein [1-9].

Yoshizawa et. al. observed that with rice used for making Sake wine, the free fat component is distributed in the surface while the fat-by-hydrolysis component is distributed uniformly[10].

Tamaki et. al. reported that the content of amylose and fat-by-hydrolysis in rice was high in a year of high ripening temperatures[4]. They also reported that amylose content did not change with the storage time, but fat-by-hydrolysis content increased[5].

Yasumatsu compared the amylograph rheological properties of rice flour after storage as a function of storage time. The gelatinization temperature of stored rice flour shifted to higher temperatures with increasing storage time because free fatty acid and starch were combined[7]. It was reported that fat-by-hydrolysis content in rice grains is different for each variety. Rice varieties with good texture, for example Koshihikari, contained only a small amount of fat-by-hydrolysis[1].

The effects of fat content on the gelatinization of defatted rice are analyzed in the present work.

2.MATERIALS AND METHOD

2.1.Materials

Samples used were 10%polished Koshihikari and Matsumae rice grains that were cultivated in Ishikawa Agriculture Research Center. Matsumae was estimated to have low quality, while Koshihikari was estimated to have high quality. Koshihikari and Matsumae rice grains were defatted by methanol-chloroform for five periods: 0, 2, 8, 12, and 24hour.

Methanol-chloroform used to defat was removed by soaking defatted rice in methanol for 30 min., and then in water for 30 min..

Rice grains of Matsumae and Koshihikari were soaked in water for the same period for an additional measurement.

2.2.Fat content

Total fat content was determined as that which was extracted by hydrochloric acid. The free fat content was determined as that which was extracted by ethyl ether. Fat-by-hydrolysis content was calculated as the difference between the total fat and the free fat.

2.3.Differntial scanning calorimetry(DSC)

Gelatinization of rice grains were examined by heating DSC (Setaram Micro DSC Ⅲ, Caluire, France). Ten rice grains and 1.4 times the weight of rice grain in water were hermetically sealed in the DSC cell of 1.5 ml. The temperature was kept at 25°C for 10 min and then raised at 0.5°C/min to 120°C.

3.RESULTS

3.1.Fat content

Table 1 shows the change in fat content in rice grains as a function of defatting time. Total fat and free fat content decreased with increasing defatting time while the fat-by-hydrolysis content was much less changed even after 24 hours. This is consistent with the previous experimental findings of Yoshizawa et. al. which showed that the free fat content decreased with increasing rate of polishing, but fat-by-hydrolysis content did not decrease[2]. When rice grains were soaked in methanol-chloroform, fat-by-hydrolysis localized at the inner part of the grain could not be extracted.

Table 1 Change in fat content in rice grains by defatting (Matsumae)

Defatting time (hour)	Total fat (mg/g)	Free fat (mg/g)	Fat-by-hydrolysis (mg/g)
0	16.10	7.66	8.44
2	11.04	3.03	8.01
8	9.83	1.07	8.76
12	9.42	1.02	8.40
24	7.90	0.90	7.00

3.2. Gelatinization of defatted rice grains

Heating DSC curves for rice grains (Matsumae,Koshihikari)-water system are shown in Figure 1. Heating DSC curves showed two endothermic peaks around 60°C∼90°C, (Peak 1 and 2), and a third endothermic peak around 90°C∼120°C(Peak 3). The endothermic Peak 3 was attributed to the disintegration of the amylose-lipid complex.

In heating DSC curves of both varieties, the temperatures of Peaks 1 and 2 shifted slightly to lower temperatures with the defatting time (Figure 1). The temperatures of Peaks 1 and 2 of Koshihikari variety were lower than those of Matsumae for each defatting time (Figure 1). Figure 2 showed a correlation between the total fat content and peak temperature. We can confirm that decreasing total fat content induces the shift of Peak 2 to lower temperatures. Kiribuchi and Kubota also reported that gelatinization temperature of defatted rice powder was lower than that of non-treated rice powder on amylogram[9].

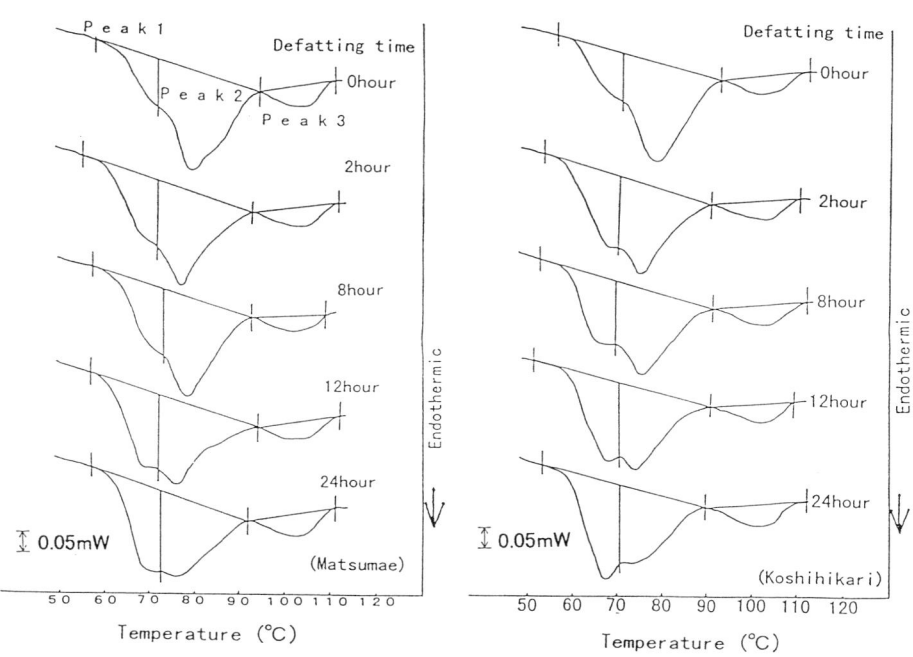

Figure 1   DSC heating curves of rice grains with different defatting times
Heating rate: 0.5°C/min

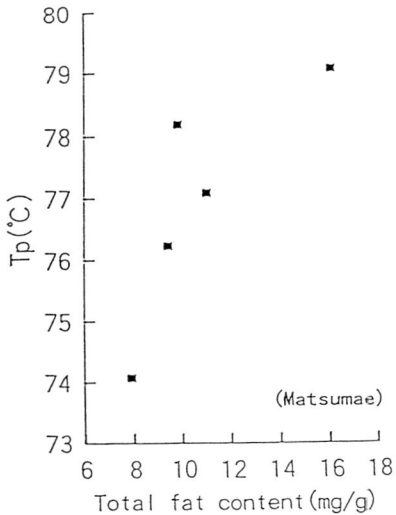

Figure 2 Correlation between total fat content

and peak 2 temperature

The enthalpy estimated from the area of Peak 1 increased while that of Peak 2 decreased with defatting time (Figure 2). Figure 3(a) shows the correlation between total fat content and the enthalpy of Peak 1. Figure 3(b) shows the correlation between total fat content and the enthalpy of Peak 2. As is shown clearly in these figures, the decrease in total fat content leads to an increase in the enthalpy of peak 1, and to a decrease in the enthalpy of peak 2.

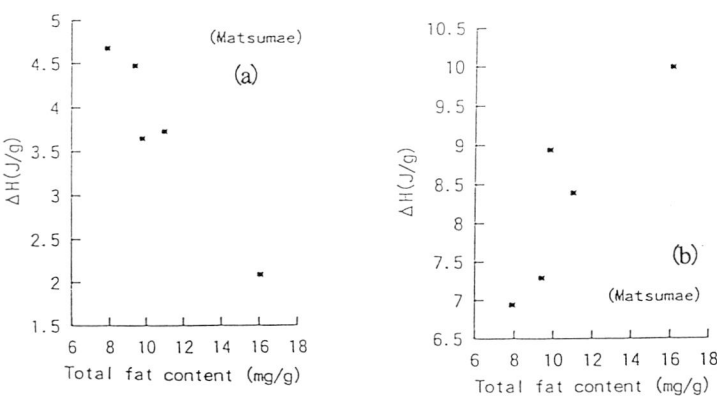

Figure 3 Correlation betweentotal fat content

and peak enthalpy

(a) : peak 1 enthalpy, (b) : peak2 enthalpy

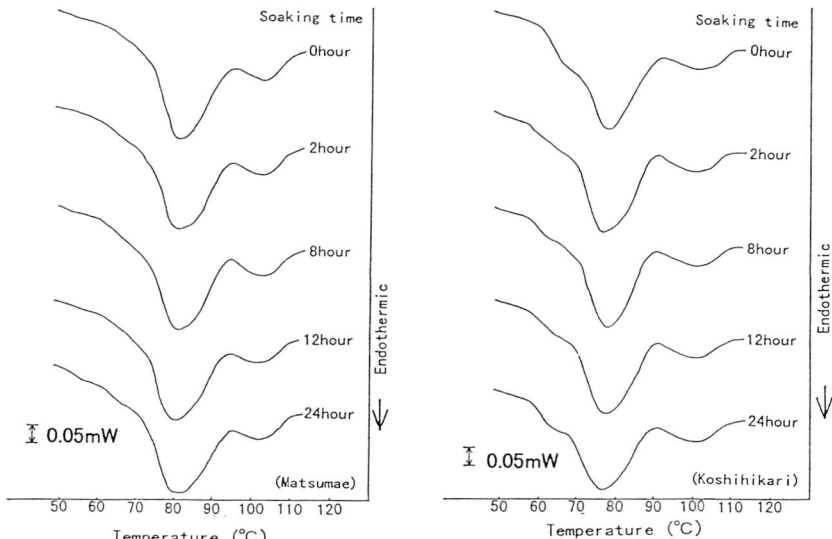

Figure 4   DSC heating curves of rice grains with different defatting times
Heating rate: 0.5°C/min

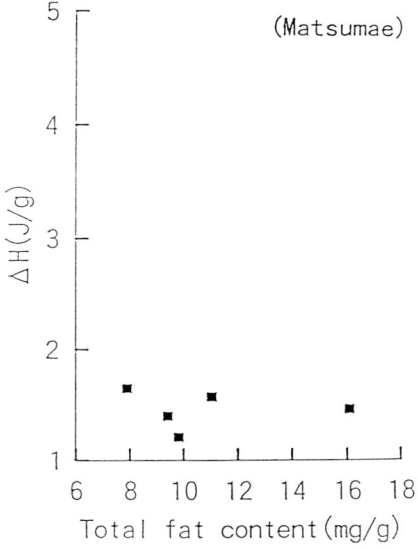

Figure 5   Correlation between total fat content
and peak 3 enthalpy

Heating DSC curves for rice grains (Matsumae,Koshihikari) which were soaked in water for various times were observed to see whether the DSC change observed for rice grains is induced only by defatting. The temperature and the enthalpy were not changed for peaks 1 and 2 when the soaking time in water was changed (Figure 4).

Both the temperature and the enthalpy of Peak 3 did not change when the defatting time and the soaking times were changed (Figure5).

4.CONCLUSION

Kugimiya et. al. reported the gelatinization of amylose-lipid complex is affected by fat-by-hydrolysis content[11-12]. In the present study, however, the results do not show that amylose-lipid complex in rice grains was influenced by defatting, because fat-by-hydrolysis in rice grains was not removed by this defatting method. While, gelatinization peak at a low temperature around 70℃ sifted slightly to lower temperatures and the enthalpy for Peak 1 increased while that for Peak 2 decreased by decreasing free fat content.

Marshall et. al. and Champagne et. al. also studied the effect of defatting whole grain milled rice using hexane and methanol-chloroform, and analyzed the starch gelatinization parameters by DSC. They reported that the shift of gelatinization on peak to lower temperatures by defatting is caused and not by decrease in fat content. They removed solvents (hexane,methanol-chloroform) by air drying, and observed the fissures on the surface of rice grain by SEM.[13-14].

However, the desolventing method from defatted rice in the present study is different from their method. Therefore, in the present study, we have not observed any fissures in the rice kernels.

In the present study, we found that free fat content has a drastic effect on the gelatinization of rice grains.

REFERENCES

1.S.Miwa, Shokuhin no Shiken to Kenkyu, No.30 (1996) 59.
2.K.Ohtsubo, H. Yanase and T. Ishima, Rep. Natl. Food Res. Inst., No.51 (1987) 59.
3.I.Shoji and H. Kurasawa, Kasegaku Zasshi, 32 (1981) 168.
4.M. Tamaki, M. Ebata and T. Tashiro, Japan. J. Crop Sci., 58 (1989) 659.
5.M. Tamaki, T. Tashiro, M. Ishikawa and M. Ebata, Japan. J. Crop Sci., 62 (1993) 540.
6.S. Kitamura, Denpun, No. 37 (1992) 174.
7.K. Yasumatsu, S. Moritaka and T. Kakinuma, Agric. Biol. Chem., 28 (1964) 265.
8.H. Fukuba and T. Kiribuchi, Denpun Kogyougaku Kaishi, 16 (1938) 116.
9.T. Kiribuchi and K. Kubota, Nippon Nogeikagaku Kaishi, 51 (1977) 621.
10.K. Yoshizawa, T. Ishikawa and K. Noshiro, Nippon Nogeikagaku Kaishi, 47 (1973) 713.
11.M. Kugimiya, J. W. Donovan and R. Y. Wong, Starch, 32 (1980) 265.
12.M. Kugimiya and J. W. Donovan, J. Food Sci., 46 (1981) 765.
13.E. T. Champagne, W. E. Marshall and W. R. Goynes, Cereal Chem., 67 (1989) 570.
14.W. E. Marshall, F. L. Normand and W. R. Goynes, Cereal Chem., 67 (1989) 458.

HYDROCOLLOIDS – PART 1
Edited by K. Nishinari
2000 Elsevier Science B.V.

Effects of maltose and fructose on the gelatinization and retrogradation of wheat starch

K. Y. Kim

Seoil college, Munmok-Dong, Jungrang-Ku, Seoul, Korea

The thermal properties of wheat starch in aqueous solution were observed by DSC analysis. The effect of saccharides on the gelatinization and retrogradation process of wheat starch was investigated. The retrogradation ratio was determined from the values of the heating enthalpy($\triangle H_1$) and reheating enthalpy($\triangle H_2$). The results showed that adding maltose than fructose reduced retrogradation. The viscosity of wheat starch with or without fructose or maltose was observed as a function of shear rate and temperature using a cone and plate viscometer. The values of viscosity decreased with increasing shear rate and temperature.

1. INTRODUCTION

Wheat starch has been used in most of the traditional applications for starch. These include, paper making, foods, and confectionery. Freshly gelatinized oxidized wheat starch has been recommended for stabilizing soured dairy products by increasing their consistency and viscosity. Wheat starch granules are biconvex discs with a fairly regular circular outline. When starch granules are heated in excess water, a phase transition from order to disorder called gelatinization occurs[1]. Several changes occur simultaneously during gelatinization, including uptake of heat by the starch granules and loss of crystallinity as measured by X-ray diffraction[2]. The previous papers about wheat starch have been concerned with the enthalpy determination as a function of hydration time[3], rigidity as a function of storage time[4] and the kinetics of recrystallization[5]. The effect[6] of additives on the gelatinization was investigated by DSC and the prevention[7] of retrogradation by adding sucrose has been reported. In this work, the rheological and thermal properties of wheat starch has been studied to determine the effect of fructose and maltose on the gelatinization and retrogradation process.

2. MATERIALS AND METHODS

2.1. Materials

Wheat starch - SIGMA CHEMICAL CO., Fructose & maltose - MERCK CHEMICAL CO.,

Pullulan (standard for GPC) – SHODEX CHEMICAL CO., Lithium Bromide, DMSO, DMF – SIGMA CHEMICAL CO.

2.2. Differential Scanning Calorimetry

Gelatinization enthalpy($\triangle H_1$) and gelatinization temperature (T_p) of 30%(w/w) wheat starch with varying ratios of fructose and maltose added {5%, 7.5%, 10%, 20%, 30%(w/w)} were measured using DSC (DSC 7-1022 Series, Perkin Elmer Co.) at a rate of 10℃/min. Regelatinization enthalpy($\triangle H_2$) of 30%(w/w) wheat starch added 10%(w/w) fructose and maltose were obtained as a function of storage time with(1, 3, 7, 14 days). Retrogradation ratio was calculated as $\triangle H_2/\triangle H_1$[8] .

2.3. Cone and plate viscometer

1% wheat starch solution in DMSO was prepared and after standing one day, the viscosity was measured as a function of shear rate at 15℃, 25℃, 35℃,45℃, 55℃, 65℃ temperatures by Brookfield Viscometer (Digital viscometer, Model DV-II, Brookfield Co. Ltd., German). The viscometer was a cone and plate type with cone radius (r)=1.2 cm, and cone angle (θ)=3.0° . The viscosity of wheat starch with fructose and maltose was observed using the same conditions.

3. RESULTS AND DISCUSSION

3.1. Gelatinization and Retrogradation

The thermal data of 30%(w/w) wheat starch with and without 10%(w/w) fructose and maltose are shown in Figure 1 and Table 1. With saccharides added, the enthalpy change($\triangle$H) increased, and was more pronounced for maltose than fructose.

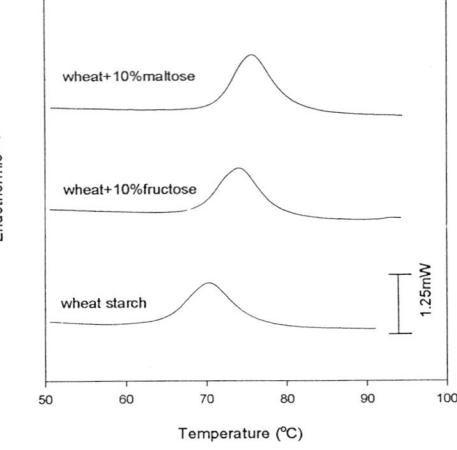

Fig.1. Heating DSC curves of 30%(w/w) wheat
starch with saccharides.

Table 1. Thermal characteristics of 30% wheat starch
with saccharides.

sample	T_0	T_p(℃)	$\triangle H_1$ (J/g)
wheat starch	65.94	70.20	9.81
wheat starch +10%fructose	69.94	74.16	10.46
wheat starch +10%maltose	70.14	75.08	10.78

The thermal data of 30%(w/w) wheat starch with varying ratios of saccharide is displayed in Figures 2, 3 and Tables 2, 3. The endothermic peak accompanying gelatinization shifted to higher temperature and increased in intensity with increasing saccharides ratios. This may be caused by the increasing concentration of gel (from 5% to 30%) by the loss of water mobility as free water is bound with fructose and maltose.

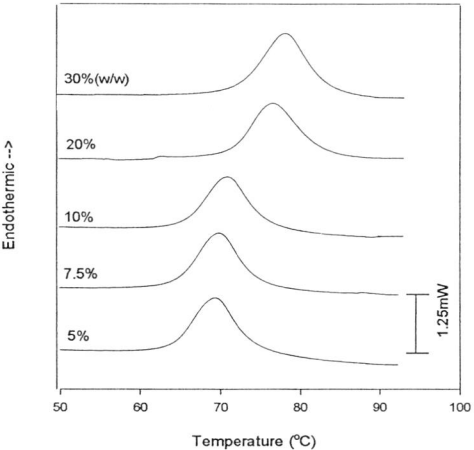

Fig.2. Heating DSC curves of 30%(w/w) wheat starch with fructose concentrations.

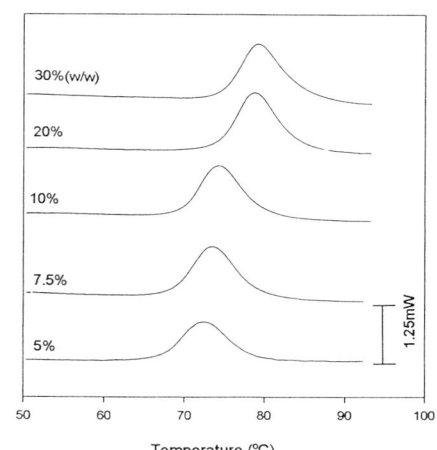

Fig.3. Heating DSC curves of 30%(w/w) wheat starch with maltose concentrations.

Table 2. Thermal characteristics of 30% wheat starch with fructose concentrations.

fructose conc.	T_0	$T_p(\text{℃})$	$\triangle H_1$ (J/g)
5%	69.34	73.43	10.45
7.5%	71.22	75.43	10.57
10%	73.31	77.50	11.00
20%	76.11	81.16	11.14
30%(w/w)	76.72	81.56	11.37

Table 3. Thermal characteristics of 30% wheat starch with maltose concentrations.

maltose conc.	T_0	$T_p(\text{℃})$	$\triangle H_1$ (J/g)
5%	66.58	70.76	10.85
7.5%	67.12	71.36	10.96
10%	67.90	72.33	11.08
20%	74.53	79.50	11.24
30%(w/w)	75.10	80.03	11.47

Figure 4 and Table 4 show the gelatinization enthalpy($\triangle H_1$) and regelatinization enthalpy($\triangle H_2$) of wheat starch with and without fructose and maltose as a function of storage time.

Regelatinization enthalpy decreased with increasing saccharides ratio with storage time. Retrogradation ratio is illustrated in Table 5. Maltose is more effective than fructose to prevent the retrogradation. This may be explained by the fact that there are more equatorial hydroxyl groups in maltose[9], and is in good agreement with a previous study on acorn starch[10]. We can see that saccharides are used as antistaling reagents and that the role of saccharide is to retard the retrogradation because they inhibit recrystallization.

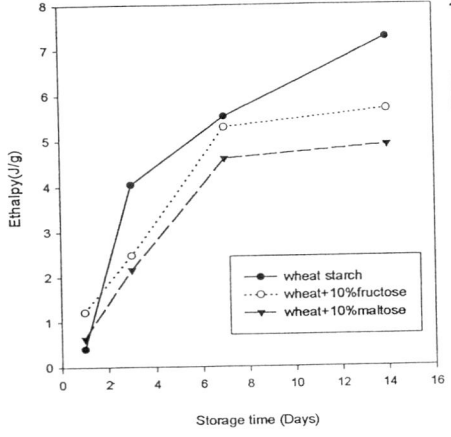

Fig.4. Regelatinization enthalpy of 30%(w/w) wheat starch with saccharides as a funtion of storage time.

Table 4. Thermal characteristics of 30% wheat starch with saccharides storage times.

sample	storage time (days)	$T_o(\mathtt{C})$	$T_p(\mathtt{C})$	$\triangle H_2$ (J/g) (Reheating enthalpy)
wheat starch	1	43.05	50.43	0.39
	3	43.51	53.09	4.04
	7	43.87	54.93	5.53
	14	43.23	56.76	7.25
wheat +10%fructose	1	47.79	56.43	1.22
	3	47.23	58.76	2.48
	7	47.72	60.10	5.30
	14	47.48	61.23	5.68
wheat +10%maltose	1	48.74	58.23	0.63
	3	49.77	60.76	2.16
	7	49.15	61.70	4.60
	14	49.30	62.76	4.89

Table 5. The retrogradation ratio of 30% wheat starch with saccharides.

sample	retrogradation ratio ($\triangle H_2 /\triangle H_1$)
wheat start	0.6885
wheat+fructose	0.5274
wheat+maltose	0.4503

3.2. Viscosity as shear rate

Figure 5 shows a plot of viscosity of wheat starch in DMSO against shear rate in the temperature range from 15℃ to 65℃. The viscosity decreased at all temperature studied with increasing shear rate as is typical for polymer solution and the temperature exponentially. The viscosity of wheat starch with added saccharides is plotted in Figure 6. The viscosity enhancement is greater for maltose than fructose. We can see that disaccharide promotes the rigidity of the hydrogen bond. It does this as a result of the interaction between starch and the hydroxyl group.

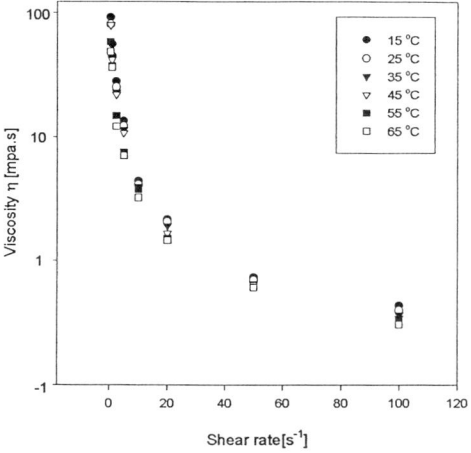

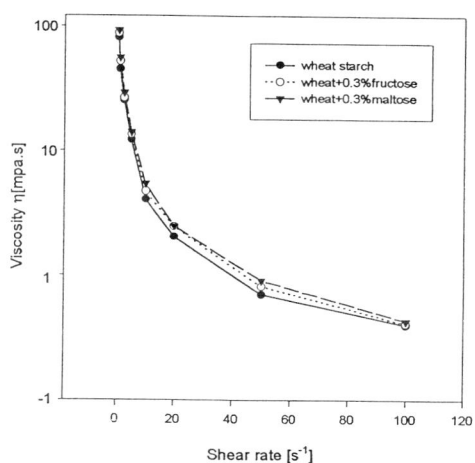

Fig.5. Viscosity of 1%(w/w) wheat starch
in DMSO at various temperature5.

Fig.6. Viscosity 1%(w/w) wheat starch solution
in DMSO with saccharides   at 25°C

REFERENCES

1. K.J. Zeleznak and R.C. Hoseney, Cereal Chem., 63 (1986) 407.
2. Ph. Roulet, W.M. Macinnes, P. Wursch, R.M. Sanchez and A. Raemy,
   Food Hydrocolloids. 2 (1988) 381.
3. D.A, Yost and R.C. Hoseney, Starch, 38 (1986) 289.
4. R.B. Won and K. Lelievre, J., Starch, 34 (1982) 231.
5. J.J. Swinkels, M., Starch, 37 (1985) 1.
6. H.L. Savage and E M. Osman, Cereal Chem., 57 (1980) 49.
7. R.D. Spies and R. C. Hoseney, Cereal Chem., 59 (1982) 128.
8. K. Kohyama and K. Nishinari, J. Agrie. Food Chem., 39 (1991) 1406.
9. H. Uedaira, in: K. Nishinari, T. Yano (Eds.), Science of Food Hydrocolloids,
   Asakura Shoten, 1990, p. 42.
10. Y. H. Lee, N. H. Kim and K. Nishinari, Thermochim. Acta, 322 (1998) 39.

HYDROCOLLOIDS – PART 1
Edited by K. Nishinari
2000 Elsevier Science B.V.

337

The effect of water interactions on the rheological behaviour of amylose, amylopectin, and mixtures from corn

S.J. McGrane, C.J. Rix, D.E. Mainwaring and H.J. Cornell

Department of Applied Chemistry, RMIT University, GPO Box 2476V, Melbourne, Victoria, 3001, Australia

The effect of water interactions on the concentrated rheology of amylose, amylopectin and their mixtures from corn starch was systematically examined using a binary dimethyl sulfoxide/ water solvent system. The effect of water content on natural corn starch has also been studied and the role of the starch granule in imparting mechanical properties to starch gels discussed.

1. INTRODUCTION

Starch is composed of two major polysaccharides, amylose (20 – 30% (w/ w) in normal starches) and amylopectin (70 – 80% (w/ w) in normal starches) [1, 2]. It is well known that starch behaviour is highly influenced by its interaction with water. High-, intermediate-, and low-moisture systems have unique behavioural characteristics and the results obtained at one moisture level may not necessarily be extrapolated to another [3]. However, despite the importance of water content on the behaviour of starch interactions, no systematic attempt has been made to quantify or correlate water interactions with the *concentrated* rheological properties of starch or its components. Since the manufacture of dough-based foods involves concentrated starch systems, the aim of this study was to use the non-aqueous co-solvent, dimethyl sulfoxide (DMSO) to systematically examine the influence of water content on the *concentrated* rheological behaviour of amylose, amylopectin and their mixtures. Hayashi et al. [4] and Cheetham and Tao [5] have studied the effect of water interactions on the behaviour of amylose solutions using a binary DMSO/ water solvent mix, but only using *dilute* systems.

2. EXPERIMENTAL

Corn amylose, corn amylopectin, native corn starch (Sigma Chemical Co.) and mixtures of amylose and amylopectin from corn were dissolved by heating at 95°C (amylose) or 85°C (amylopectin) in the non-aqueous, water-miscible solvent dimethyl sulfoxide. On cooling to room temperature, water was quantitatively added and mixed to yield homogeneous samples. The samples prepared in pure water were heated to either 160°C in an autoclave for amylose or 85°C for amylopectin. Samples were then allowed to stand overnight prior to analysis.

Rheological parameters were measured at 25.0 ± 0.1°C with a Rheometrics Fluids Spectrometer RFS-II (Rheometrics, Piscataway NJ) equipped with a parallel plate geometry having a radius of 25.0mm. Once the sample was loaded on the rheometer, a humidifying chamber was positioned around the sample to minimise water loss from the sample and a period of standing was allowed to ensure sample equilibration prior to the commencement of analysis. Steady shear and dynamic mechanical measurements were made in order to probe the viscous and elastic moduli of the samples.

3. RESULTS AND DISCUSSION

The effect of water content at varying amylose/ amylopectin concentrations on the apparent viscosity of each polysaccharide was studied initially. For the rate sweep measurements made on corn amylopectin at varying water contents, the solutions were pseudoplastic and the apparent viscosities remained fairly constant. Rate sweep measurements made on amylose indicated that at water contents below 30% (w/ w), the amylose samples were fairly Newtonian and there was little change in the apparent viscosity. At 30% (w/ w) water and above, the apparent viscosities all increased significantly and the samples exhibited pronounced pseudoplastic behaviour. The apparent viscosity of 2% (w/ w) amylose samples changed very little at all water contents, however, the apparent viscosity of 25% (w/ w) amylose in 20% (w/ w) water increased significantly and this was a result of the high amylose content of the sample.

The rate sweep data were then plotted on a solvent composition, polysaccharide concentration, low shear apparent viscosity orthogonal co-ordinate system that provides *surfaces*, which indicate the inter-relationships between these parameters. The low shear apparent viscosity is the point of maximum apparent viscosity for each solution examined under steady shear conditions. The use of such plots enables a large amount of data to be effectively compared and summarised.

Figure 1.a. shows the surface for amylose and is characterised by a large increase in low shear apparent viscosity as the water content in the solvent increases. There is little change in low shear viscosity at amylose concentrations of 2-5% (w/ w) and water contents 0-20% (w/ w). However, at amylose and water concentrations above these values, the low shear viscosities increase significantly up to about 3700 Pa.s for 10% (w/ w) amylose in 60% (w/ w) water. These increases in low shear apparent viscosity begin to occur at 30% (w/ w) water, indicating the onset of gelation of corn amylose requires a minimum water content of about 30% (w/ w). This result is in close agreement with studies by Wootton and Bamunuarachchi [6] and Eliasson [7] who found the minimum water content necessary for gelation of wheat starch to be 32% (w/ w) and 33% (w/ w) respectively using differential scanning calorimetry. Concentrations greater than 10% (w/ w) amylose and 60% (w/ w) water could not be measured using the fluids spectrometer as the gels produced were too rigid to be loaded between the parallel plates. The dotted lines on the graph represent interpolated data and have been added to show the overall trend in low shear apparent viscosity.

The surface produced for amylopectin is presented in Figure 1.b. and is very different to that for amylose. Since the low shear apparent viscosities obtained for the amylopectin samples are much lower than those obtained for the amylose, the plot has a reduced scale – one tenth of that used for the amylose. At 0% water content (i.e., in DMSO only), the low shear apparent viscosities of the amylopectin are greater than that of the amylose, which is expected due to amylopectin having a higher molecular weight. Increasing the water content, however, had little effect on the low shear viscosities obtained. After an initial decrease in low shear apparent viscosity from about 90 Pa.s, the viscosities stabilise to about 60 Pa.s for 25% (w/ w) amylopectin at the remaining water concentrations, consistent with the lower solvency of water causing a contraction in molecular volume [8, 9]. The results clearly illustrate that water content has little effect on the low shear apparent viscosities of the corn amylopectin samples.

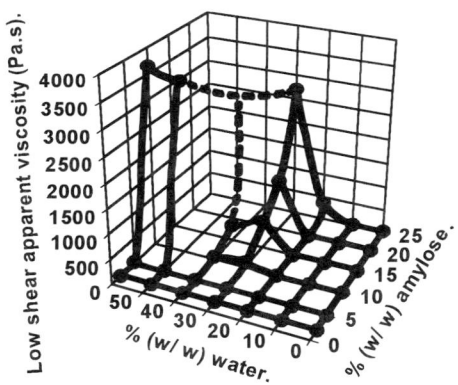 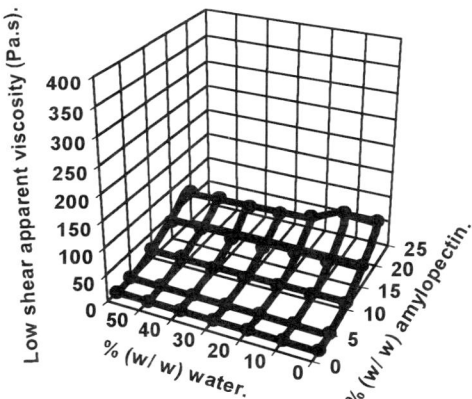

Figure 1.a. Three-dimensional surface plot for corn amylose.

Figure 1.b. Three-dimensional surface plot for corn amylopectin.

The strain and frequency sweep measurements made on 10% (w/ w) amylopectin samples at varying water contents are shown in Figure 2. The strain sweep measurements show that amylopectin has a characteristically large linear viscoelastic region; that is, the viscous and elastic moduli are independent of the applied strain up to about 100% strain. Both the strain and frequency sweeps show that in the absence of water (i.e., in pure DMSO), the viscous modulus (G'') dominates the elastic modulus (G'); indicating that the amylopectin solution is behaving as a viscous liquid. Here, the amylopectin chains are fully solvated and are not interacting. As the water content in the solvent is increased to 60% (w/ w), the solvent-chain interaction has decreased and the chain-chain interaction increased, giving comparable viscous and elastic moduli, that is G'' ≈ G'. Finally, in 100% water (i.e., no DMSO), the solvent-chain interactions have degraded to such an extent that chain-chain interactions now cause the elastic modulus to dominate the viscous modulus. The transition from a more viscous to a more elastic solution is therefore readily observed as the water content in the solvent is increased.

340

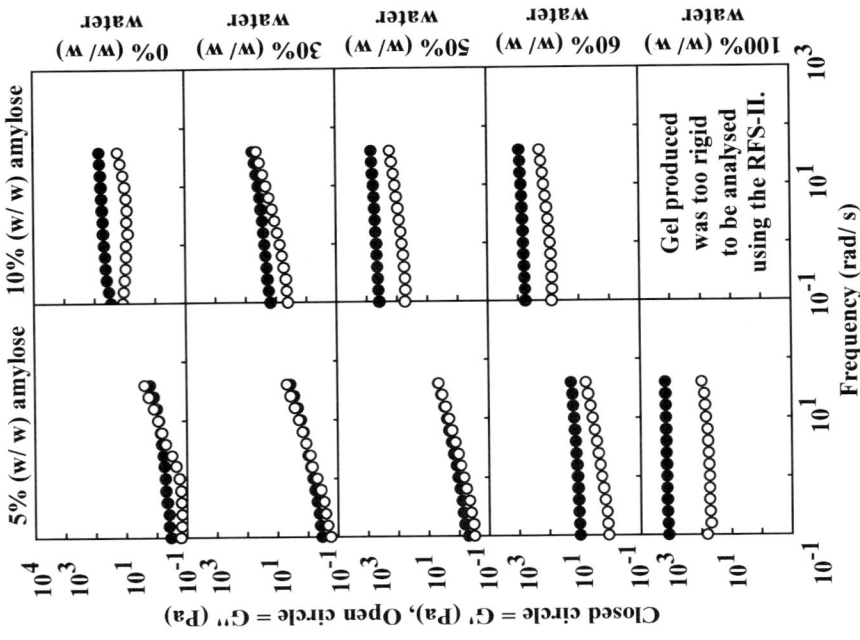

Figure 3. Frequency sweep measurements (at 5.0% strain) made on 5% and 10% (w/ w) corn amylose at varying water contents (25.0C).

Figure 2. Strain (at 1.0rad/ s frequency) and frequency (at 5.0% strain) sweep measurements made on 10% (w/ w) corn amylopectin at varying water contents (25.0C).

The frequency sweep measurements made on amylose at 5% and 10% (w/ w) concentration at varying water contents are presented in Figure 3. As expected, the gels formed by 10% (w/ w) amylose are stronger than those of 5% (w/ w) as the viscous and elastic moduli are larger in magnitude and more frequency independent. As the water content in the solvent increases, the magnitude of the moduli also increases and they move further apart. In pure water, 5% (w/ w) amylose produces a strong gel - the elastic modulus is frequency independent and G' is ten times G'' - typical behaviour of a true gel. Indeed, at 10% (w/ w) amylose, the gel produced is so strong that it is too rigid to be loaded on the rheometer. The frequency sweep measurements made on amylose and amylopectin in pure water agree with those made by Doublier and Llamas [10] on 1% (w/ w) potato amylose and 4% (w/ w) maize amylopectin who found amylose to behave as a gel and amylopectin as a viscoelastic solution.

A complete, systematic study similar to those made on the individual amylose and amylopectin components was also made on synthetic mixtures of varying ratios of amylose to amylopectin. No significant synergistic or antagonistic interactions were observed and the results obtained were essentially a summation of the individual behaviours.

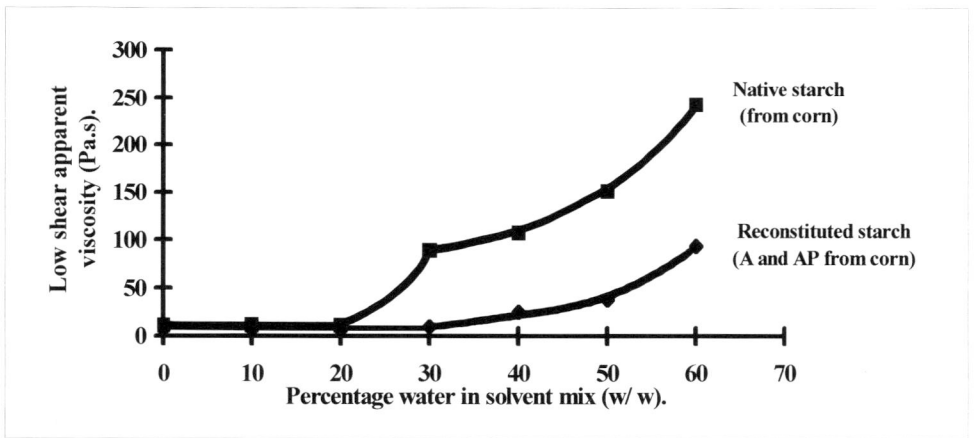

Figure 4. Plot of low shear apparent viscosity (Pa.s) versus percentage water (w/ w) in the solvent mix for native corn starch and reconstituted corn starch made from amylose (A) and amylopectin (AP). Sample concentrations were 10% (w/ w) and analysis temperature 25.0°C.

Finally, the effect of water content on natural corn starch was examined. Figure 4 compares the low shear apparent viscosities obtained at water contents up to 60% (w/ w) water in the solvent mix. Here, overall concentrations of the samples were held at 10% (w/ w). It can be readily seen that the low shear apparent viscosities of the synthetic mixtures of corn amylose and amylopectin (which was found to be 29% (w/ w) amylose: 71% (w/ w) amylopectin [11]) are significantly reduced compared to the natural corn starch, especially when the water content is ≥ 30% (w/ w). This suggests that the microstructure of corn starch, consisting of partially insolubilised microgranules of amylopectin acting as a filler within the interconnected amylose/ amylopectin network plays a significant role in gel rigidity and so the rheological behaviour, consistent with the observations of Ring and Stainsby [12]. We are currently further investigating the role of filler materials within starch gels.

4. CONCLUSION

The experimental results clearly illustrate the importance of water interactions on the concentrated rheological behaviour of corn amylose, amylopectin, and their mixtures. Water content has a significant effect on the low shear apparent viscosities of amylose, with large increases in apparent viscosity occurring at 30% (w/ w) water. The low shear apparent viscosity of amylopectin, however, is not significantly influenced by the water content, while the results for the mixtures are essentially a summation of the behaviours of the individual components. Amylopectin behaves as a viscoelastic solution at all water contents, however, amylose shows gel-forming properties; confirming that amylose is the dominant gel-forming component of starch. Amylose and amylopectin mixtures at the same ratio as in native corn starch do not enhance gel rigidity to the same extent as that exhibited by the natural starch gel, indicating the microstructure of the starch gel network plays a significant role in gel rigidity.

5. ACKNOWLEDGMENTS

The authors would like to thank the Australian Food Industry Science Centre for providing S.J. McGrane with the postgraduate research scholarship which funded this work, and also the Cooperative Research Centre for Industrial Plant Biopolymers for awarding S.J. McGrane the traveling scholarship to attend the Fourth International Hydrocolloids Conference.

REFERENCES

1. D. J. Manners, Carbohydr. Polym., **11** (1989) 87.
2. O.R. Fennema, Food Chemistry, 3rd Ed., Marcel Dekker Inc., New York, 1996.
3. R. B. Friedman, In Ingredient Interactions - Effects on Food Quality, Marcel Dekker Inc., New York, 1995.
4. A. Hayashi, K. Kinoshita, Y. Miyake and C.-H. Cho, Agric. Biol. Chem., **47** (1983) 1699.
5. N.W.H. Cheetham and L. Tao, Starch/ Stärke, **49** (1997) 407.
6. M. Wootton and A. Bamunuarachchi, Starch/ Stärke, **31** (1979) 262.
7. A.-C. Eliasson, Starch/ Stärke, **32** (1980) 270.
8. M.T. Kalichevsky and S.G. Ring, Carbohydr. Res., **162** (1987) 323.
9. V.M. Leloup, P. Colonna and A. Buleon, J. Cereal Sci., **13** (1991) 1.
10. J.-L. Doublier and G. Llamas, In Food Colloids and Polymers: Stability and Mechanical Properties, The Royal Society of Chemistry, Cambridge, 1993.
11. S.J. McGrane, H.J. Cornell and C.J. Rix, Starch/ Stärke, **50** (1998) 158.
12. S.G. Ring and G. Stainsby, Prog. Food Nutr. Sci., **6** (1982) 323.

HYDROCOLLOIDS – PART 1
Edited by K. Nishinari
2000 Elsevier Science B.V.

343

Rheological studies on the effects of the temperature of heat-moisture treatment on the retrogradation of corn starch

T. Takaya, C. Sano and K. Nishinari

Department of Food and Nutrition, Faculty of Human Life Science,
Osaka City University, Sumiyoshi, Osaka, Japan

Effects of the heating temperature of the heat-moisture treatment on the retrogradation of corn starch gels were studied by stress-strain curves and creep measurements. Gels were stored at 5℃ for one, three, five and seven days before the rheological measurements. Both breaking stress, Young's modulus increased and the breaking strain decreased with increasing storage time and the temperature of the heat-moisture treatment. Analysis of creep curves of gels stored at 5℃ for three days by four element model also showed that elastic coefficients increased remarkably by the heat-moisture treatment.

1. INTRODUCTION

Starch has been used in food industry to modify the texture because it is the cheapest among many hydrocolloids. Native starch, however, sometimes is not suitable because the rheological properties change drastically by the change in pH and temperature. Heat-moisture treated starch was prepared by heating starch in the presence of low level humidity so that starch does not gelatinise on heating. As a result of this treatment, structure of starch became inhomogeneous, which was shown by the broadening of the endothermic peak accompanying gelatinisation in the differential scanning calorimetry (DSC) [1]. It was also shown that the swelling power and the viscograph hot plate consistencies decreased whilst the water-binding capacities and enzyme susceptibilities increased by the heat-moisture treatment for starches from barley, arrow root and cassava [2].

The rheological change in gels, prepared from native and heat-moisture treated corn starch, during storage was studied in the present work.

2.EXPERIMENTAL

2.1. Material

Heat-moisture treated corn starch HMCS-120 (Lot No. 930619) and HMCS-130 (Lot No. 930611) and corn starch without treatment (NCS, Lot No. 930125) are gifts from Sanwa Starch Co. Ltd. HMCS-120 and HMCS-130 were prepared by heat-moisture treatment under the condition of a saturated humidity for 20 min at 120℃ and 130℃, respectively after evacuating to 10 kPa [3].

Forty five grams of starch powder were dispersed in 255 g distilled water using a motorised stirrer. It was stirred at 200rpm at room temperature for 30min, and then heated in an oil bath to 95℃ for about 10min. and then kept stirring at 450rpm at 95℃ for 30min. The distilled water was added to the hot dispersion to adjust the concentration. Then the dispersion was poured into teflon moulds of 20mm diameter and 20 mm height. It was kept at 25±1℃ for 60min, and then stored at 5℃ for one day, three days, and 7days.

2.2. Measurements

Stress-strain curves of cylindrical starch gels were observed as a function of storage time by a Rheoner RE-3305 (Yamaden Co. Ltd, Tokyo). The measurement temperature was 25±1℃, and the compression speed was 0.5mm/s. Breaking stress and the breaking strain were determined as the stress and strain at the breaking point, and Young's modulus was determined from the initial slope of the stress-strain curve. Creep measurements were also carried out for the gels at 25±1℃.

3.RESULTS AND DISCUSSION

Stress-strain curves of 15% NCS and HMCS gels stored at 5℃ for 3 days are shown in Figure 1. The breaking stress increased and the breaking strain decreased with increasing temperature of heat-moisture treatment. This rheological change could be recognised by prodding a gel with a finger, i.e. gels became harder and more brittle.

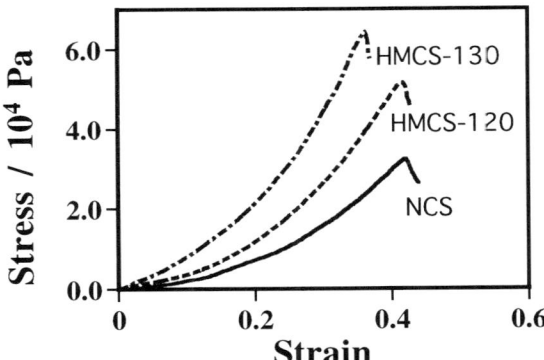

Figure 1   Stress-strain curves for 15% gels of native corn starch and heat-moisture treated corn starch stored at 5℃ for three days. Measurement temperature 25℃. Compression speed:0.5mm/s.

Figure 2 shows stress-strain curves of 15% gels of native corn starch and   heat-moisture treated corn starch stored at 5℃ for one, three, five and seven days.   Both breaking stress

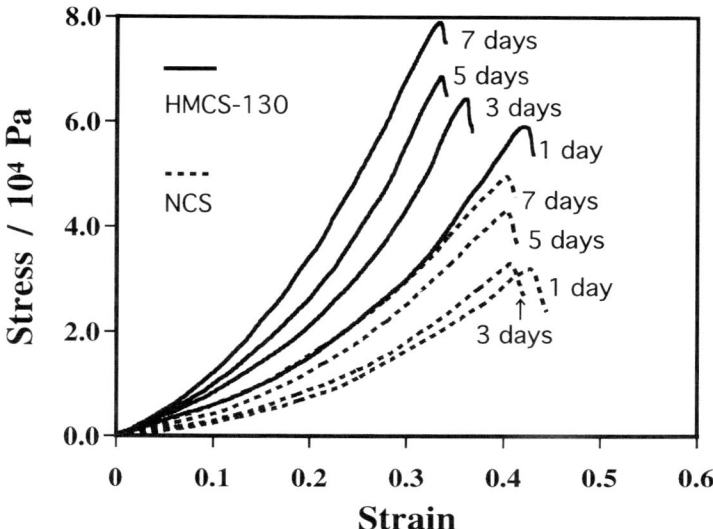

Figure 2   Stress-strain curves for 15% gels of native corn starch and heat-moisture treated corn starch stored at 5℃ for one, three, five and seven days Measurement temperature 25℃. Compression speed:0.5mm/s.

and Young's modulus of NCS and HMCS gels increased whilst breaking strain decreased with increasing storage time at 5℃.   This has been observed for many starch gels [4,5].

Creep curves for 15% NCS and HMCS gels which were kept at 5℃ for 3days.were analysed by a four element model consisting of a Voigt element and a Maxwell element combined in series, and the results are shown in Table1.

	NCS	HMCS-120	HMCS-130
Hooke's elasticity (Pa)	2.3×10^2	3.9×10^4	7.7×10^4
Voigt's elasticity (Pa)	2.6×10^3	7.9×10^5	1.7×10^6
Voigt's viscosity (Pa · s)	3.3×10^6	9.6×10^6	4.8×10^7
Newton's viscosity (Pa · s)	1.6×10^8	3.1×10^8	5.4×10^8

Table1 Mechanical parameters determined by analysing creep curves for 15% starch gels stored at 5℃ for 3 days

As is clearly shown in this Table, the elastic coefficients of both Hooke's spring element and of Voigt's element increased far more remarkably with comparison to the increase in the viscosities of both Newton's dashpot element and of Voigt's element by the heat-moisture treatment.

It is concluded that the retrogradation of 15% gels of HMCS is promoted by the incrrease in the temperature of the heat-moisture treatment, and this is consistent with the conclusion based on the differential scanning calorimetry [6]. The retrogradation ratio, which was defined as the endothermic enthalpy in the second run heating DSC curve divided by the endothermic enthalpy in the first run heating DSC curve[7], increased much faster for the heat moisture treated starch treated at a higher temperature [6].

Although Donovan and his coworkers [1] reported that the endothermic peak in the heating DSC curve split into two peaks by heat-moisture treatment for wheat and potato starch, we have not observed such a splitting for corn starch [6]. It is well known that the slower scan rate is more suitable to detect the different microstructures from morphological point of view [8]. Since the scan rate of our experiment was much slower than that of their experiment, the different findings should be attributed to the difference in the structure of starch granules of wheat, potato and corn. This should be explored in the future.

Generally speaking, the retrogradation causes difficult problems in the application of starch in the food processing. However, the retrogradation is sometimes useful, for example, for the production of harusame noodle.

REFERENCES
1. J.W.Donovan, K.Lorenz and K.Kulp, Cereal Chem., 60 (1983) 381.
2. K.Lorenz and K. Kulp, Starch, 35 (1983) 123.
3. K.Kudo, Proc. Australia/Japan Symp.Food Sci. Technol. 1993 (1993) 205.
4. C.J.A.M.Keetels, T.Van Vliet and P.Walstra, Food Hydrocoll, 10 (1996) 355.
5. M.Yoshimura, T.Takaya and K.Nishinari, J.Agric.Food Chem., 44 (1996) 2970.
6. T.Takaya, C.Sano and K. Nishinari, Carbohydr. Polym., Submitted
7. K.Kohyama and K.Nishinari, J.Agric.Food Chem., 39 (1991) 1406.
8. B.Wunderlich, Macromolecular Physics. Crystal melting 3, Academic Press, New York, 1980.

HYDROCOLLOIDS – PART 1
Edited by K. Nishinari
2000 Elsevier Science B.V.

347

Starch as a filler, matrix enhancer and a carbon source in freeze-dried denitrifying alginate beads

Y. Tal[a], J. van Rijn[a] and A. Nussinovitch[b]*

Faculty of Agricultural, Food and Environmental Quality Sciences, The Hebrew University of Jerusalem, P.O. Box 12, Rehovot 76100, Israel.

[a]Department of Animal Sciences
[b]Institute of Biochemistry, Food Science and Human Nutrition

*Corresponding author

Freeze-dried, alginate-based beads used for the immobilization of a denitrifying bacterium (*Pseudomonas* sp.) were filled with different concentrations (10, 20, 30 and 40%, w/w) of granular starch. The beads were incubated under denitrifying conditions in laboratory-scale, flow-through columns and monitored for changes in their physical and denitrifying properties. Freeze-dried beads containing high concentrations of starch were found to have better mechanical and denitrifying properties than beads containing low concentrations of this filler. Nitrate removal by the beads was found to be correlated with the starch content of the beads. Nitrite accumulation as a result of incomplete denitrification, increased with the decrease in starch content of the beads. Nitrite in the outlet of the columns was measured in all types of beads during the initial phase of incubation but was undetectable, with exception of beads with the lowest starch content, at later stages of incubation.

1. INTRODUCTION

In recent years, elevated levels of nitrate in ground and surface water have led to increased research efforts toward the development of methods for nitrate removal [1,2]. The use of bacterial denitrification for nitrate removal from contaminated waters appears to be a promising way to compete this problem since as compared to physical and chemical methods this biological method is cheap and reliable [3].

Entrapment of denitrifying bacteria in a proper support matrix is a method that can be employed for nitrate removal. Several natural materials such as agar, κ-carrageenan, chitosan and alginate are favored for this purpose because of their low toxicity to the immobilized bacteria [4]. By preinocculation of bacteria within these entrapment systems an initial lag phase in nitrate removal can be avoided while, in addition, wash-out of bacteria in the effluent water is low due to the favorable bacterial entrapping properties of many of the immobilization agents used for these purposes. Calcium alginate, a nontoxic polyanion, is widely used as an entrapping agent because, in addition to its favorable entrapping properties, it is readily available [4-6]. Problems encountered with the use of calcium alginate are mainly related to mechanical weakening of the immobilization complex with time [7]. Various methods have been employed to increase the stability of alginate-based immobilization complexes. One of these methods involves the replacement of calcium with divalent or trivalent cations with better binding properties to alginate than calcium [6] such as Sr^{2+}, Ba^{2+} and Al^{3+}. However, the potential toxicity of these di- and trivalent cations to bacteria and to the environment limits their full-scale use. The use of solid-body additives (fillers) can also avoid weakening of the alginate beads. This method is well established and is widely used for strengthening composite material made of synthetic polymers such as PVC-polyvinylchloride, PP- polypropylene and PS-polystyrene. Fillers can be divided into two main categories: (1) inert or extender, (2) active or reinforcing. Minerals such as calcium carbonate, bentonite, silica and talc belong to the first whereas carbon components such as graphite and carbon fibers belong to the second category. The addition of filler to a synthetic polymer can alter many of its physical characteristics. Some changes attributed to the addition of fillers are: 1) an increase in density and modulus of elasticity as well as in compressive and flexural strength, 2) an increase of hardness, and 3) an increase of the tensile and shear strength as well as the maximum strength [8]. To the best of our knowledge the application of fillers for improving the mechanical properties of beads made of natural polymers has not been examined. The problem of weakening of entrapment complexes consisting of denitrifying bacteria and natural polymers has been addressed in recent studies by our group [9,10]. One of the conclusions resulting from these studies was that insoluble starch provides an excellent carbon source for denitrifying bacteria within the entrapment complex. The effect of starch as a filler on the mechanical properties of these complexes was not examined and forms the basis of the present study.

2. MATERIALS AND METHODS

2.1 Preparation of the immobilization complex

Alginate (2.0 %, w/w) (Keltone LV, San Diego, CA; mol.wt 70,000-80,000, 61% mannuronic acid and 39% guluronic acid content) was dissolved in double-distilled water. The solution, mixed with a magnetic stirrer for 24 h at room temperature, was strained through a 100-mesh (US) nylon sieve to remove remaining undissolved particles. The final pH of the alginate solution was 6.8. A potato-starch suspension (Sigma, St. Louis, MO) was added to the alginate solutions at final concentrations of 10%, 20%, 30% and 40% (w/w). These solutions were mixed with an axenic culture of the *Pseudomonas* sp., containing 10^7 colony-forming units (CFU)/ml, in a proportion of one part bacterial suspension to nine parts of alginate-starch solution (w/w), giving a final concentration of *Pseudomonas* sp. in the solution of about 10^6 CFU/g. This final solution composed of alginate, starch and suspended bacteria was dropped into a 0.5% (w/w) solution of calcium chloride (Sigma, St. Louis, MO). A spontaneous cross-linking reaction occurred, resulting in spherical beads with an average diameter of about 3 mm. The beads were stored at -80°C for about 24 h before freeze-drying (Repp Sublimator Model 15RSRC-X, Repp Industries, Instr., Gardiner, NY). Freeze-drying was carried out at -50°C and at a pressure of 11×10^{-6} bar.

2.2 Incubation of the entrapped Pseudomonas sp.

The freeze-dried beads were transferred to glass columns (height: 23 cm; diameter: 2 cm; working volume: 25 ml) and kept in suspension using a constant flow of tap water, spiked with 50-100 mg NO_3-N/l, entering the column base. Columns containing beads without *Pseudomonas* sp. served as controls. Each column contained 400 beads and was operated at a retention time of 3.78 h. Incubation temperature was 30°C. Medium reservoir and columns were flushed continuously with purified N_2 gas (99.999% purity). Periodically, water samples collected from the inlet and outlet of the columns were examined for nitrite and nitrate. Beads were collected at selected time intervals and examined for protein content and mechanical properties. Each treatment was conducted in triplicate.

2.3 Mechanical properties and physical characterization of the freeze-dried beads

Mechanical properties of the beads were determined periodically by uniaxial compression with a Universal Testing Machine (model 1100, Instron Ltd., Canton, MA). A special program developed by the manufacturer enabled on-line conversion of voltage into digitized force-deformation, force-time, stress-strain or stress-time files using a 486, IBM-compatible personal computer. The force versus time data were converted to a 'pseudo-stress' versus engineering strain relationship using the following substitutions:

$$\sigma = F/A_0 \tag{1}$$
$$\varepsilon_E = (\Delta D/D_0) \tag{2}$$

where σ is the pseudo-stress, F, the force needed to break or burst the bead, A_0, the cross-sectional area of the original bead, ε_E, the dimensionless engineering strain, ΔD, the total deformation and D_0, the original diameter of the bead. The individual relationships for freeze-dried beads could be fitted to a compressibility model previously developed for the sigmoid stress-strain relationships of cellular solids[11]:

$$\sigma = C_1\varepsilon/[(1+C_2\varepsilon)(C_3-\varepsilon)] \tag{3}$$

where σ and ε are the stress and strain, respectively, and C_1, C_2 and C_3 are the constants calculated by nonlinear regression. The constant C_1 is primarily a scale factor and has stress units. The constant C_2, dimensionless, is a measure of the shoulder's prominence in the stress-strain curve, that is, when $C_2 = 0$ the curve has no shoulder and its slope increases monotonically. The constant C_3, also dimensionless, is a rough measure of steepness of the stress-strain curve in the high-strain region[11].

2.4 Electron microscopy

The structural characteristics of the freeze-dried beads at different time intervals were determined by scanning electron microscopy. The beads were subjected to dehydration by freeze-drying and spattered with gold for 195 sec at 5°C in an Argon atmosphere, using the SEM Coating System (Polaron 515, UK). The beads were examined in a Jeol (model JSM-356, Japan) scanning electron microscope.

2.5 Chemical and physical analyses

Nitrite was measured according to Strickland and Parsons (1968) and nitrate was measured with Schezochrome NAS reagent (Ben Gurion University, Be'er Sheva, Israel) or with a specific nitrate electrode (Radiometer, Denmark), amplified by a pH meter (Radiometer, Denmark, model PHM92). Oxygen and temperature were measured with a YSI (model 58) temperature/oxygen probe (Yellow Springs Instruments, USA). Protein was measured according to Lowry [12] after dissolution of the beads in citrate (five beads in 5 ml of 20g/l citrate solution). Starch determent as glucose by using Anthrone reagent (Sigma, St. Louis, MO)

3. RESULTS AND DISCUSSION

3.1 Physical characterization of the beads.

Typical stress-strain relationships for alginate beads with different starch contents are presented in Fig. 1.

Fig.1.

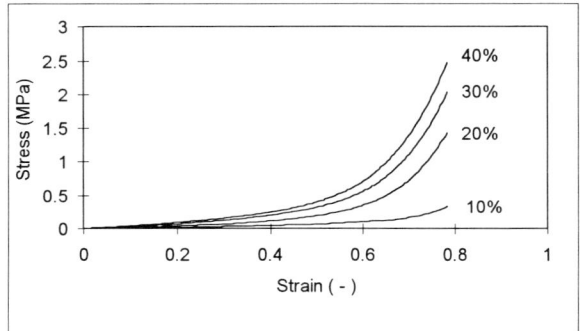

Deviation of compression characteristics of beads from the same treatment did not differ by more than 5%. The compressive stress-strain relationship is different from the regular sigmoid shape of most cellular solids. The absence of a shoulder indicates that increases in stress values cause a parallel increase in strain values. Such a response is similar to cellular solids that have passed through one cycle of compression [11]. From these results it can be concluded that the strength of the bead is directly proportional to its starch content i.e. beads with a higher starch content are stronger than those with a lower starch content. It should be noted that these results are for an initial situation, after preparation of the beads and before incubation under denitrifying conditions. However, similar patterns of stress-strain relationship were obtained when beads were collected at different incubation times (not shown). At deformations above 10%, stress values were significantly different between treatments (not seen directly from Fig. 1). We compared stress values at 75% deformation when the beads were near to their compaction and the stresses tend to reach their asymptotical value. By comparing the stress values of the different beads at ~75% deformation it was found that stress was nearly eight times higher in beads containing 40% starch than in beads with 10% starch (2 Mpa versus 0.25 Mpa). In fact, the bead in such a case resembles a composite material where inclusion of particles increases its mechanical strength[13].
The inclusion of 40% particles (w/w) is equivalent to two thirds of the maximal filling capacity of a matrix when it is randomly filled with spherical identical particles[14]. Incubation of the various beads under denitrifying conditions resulted in decrease in strength with time

(Table 1). Also during incubation, strength could be correlated to the starch content of the beads: the higher the starch content the stronger the beads. Table 1 presents the calculated values of the constants C_1 and C_3 of the compressibility equation (#3) achieved by non-linear regression procedure. For all calculation the regression coefficients R^2 was in the range of 0.95-0.99 and the constant C_2 value was zero (i.e. no indication of a shoulder). C_3 indicates the asymptotical value of the strain at the end of compaction and as such has more or less the same value ~0.9 for all filler inclusions. C_1 has a physical dimension; high values indicate a strong matrix.

Table 1. C_1 and C_3 calculated values[1] of Eq. # 3 for beads including different percentage of starch at selected incubation days[2].

Day of incubation	% Starch in beads	C_1 (MPa)	C_3 (-)
0	10	$0.240^a \pm 0.044$	$0.897^e \pm 0.038$
	20	$0.316^a \pm 0.106$	$0.875^e \pm 0.080$
	30	$0.537^b \pm 0.058$	$0.880^e \pm 0.091$
	40	$0.493^b \pm 0.057$	$0.920^e \pm 0.061$
3	10	$0.234^a \pm 0.042$	$0.934^e \pm 0.049$
	20	$0.318^b \pm 0.043$	$0.990^e \pm 0.060$
	30	$0.521^c \pm 0.054$	$0.969^e \pm 0.020$
	40	$0.512^c \pm 0.021$	$0.954^e \pm 0.044$
6	10	$0.241^a \pm 0.042$	$0.919^e \pm 0.057$
	20	$0.328^b \pm 0.110$	$1.020^e \pm 0.036$
	30	$0.338^b \pm 0.048$	$1.010^e \pm 0.055$
	40	$0.539^c \pm 0.048$	$0.990^e \pm 0.039$
9	10	$0.138^a \pm 0.025$	$0.951^e \pm 0.037$
	20	$0.200^b \pm 0.078$	$1.010^e \pm 0.014$
	30	$0.364^c \pm 0.017$	$1.010^e \pm 0.015$
	40	$0.507^d \pm 0.034$	$1.000^e \pm 0.030$
12	10	$0.107^a \pm 0.006$	$0.994^e \pm 0.017$
	20	$0.178^b \pm 0.036$	$1.039^e \pm 0.008$
	30	$0.205^c \pm 0.019$	$0.994^e \pm 0.032$
	40	$0.458^d \pm 0.026$	$1.012^e \pm 0.014$
15	10	$0.047^a \pm 0.002$	$1.010^e \pm 0.012$
	20	$0.207^b \pm 0.021$	$1.000^e \pm 0.017$
	30	$0.216^b \pm 0.013$	$0.990^e \pm 0.022$
	40	$0.448^c \pm 0.007$	$0.986^e \pm 0.032$

1) Each value represents the average of 5 measurements $\pm$ SD.
2) values with different letters within the same day of incubation are significantly different (p<0.05).

As can be seen from Table 1, beads with more embedded starch exhibited higher C_1 values and these values decreased slower during incubation than values found for beads with less starch. For example C_1 value of beads with 10% starch decreased from day 6 to day 9 by 43% compared to a decrease of only 6% for beads containing 40% starch.

3.2 Denitrification properties of the immobilized bacteria.
Nitrate removal by the beads increased with time in all treatments (Fig. 2). The proliferation of denitrifying bacteria within the beads as observed by the increase in protein content of the

beads (data not shown), most probably underlies this increase in nitrate removal. Beads with 10% starch demonstrated lower rates of nitrate removal than beads with higher starch contents during all days of incubation. Nitrite accumulation resulting from incomplete denitrification, increased with the decrease of starch content of the beads. In beads with 10% starch, nitrite accumulated during the entire examined period (21 days). Nitrite accumulation ceased from day 15 in beads with 20% starch and ceased after six days of incubation in beads containing 30% and 40% starch (not shown).

Fig. 2.

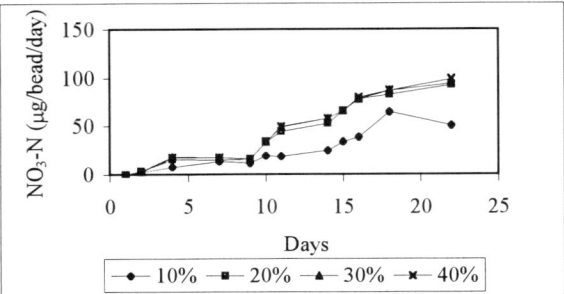

The accumulation of intermediate nitrogen forms (NO_2^-, N_2O and NO) during denitrification is closely related to the availability and type of carbon source [15]. Short-chain volatile fatty acids, sugars and alcohols are preferred carbon sources for denitrifying organisms and not all denitrifiers are capable of using starch as a carbon and electron donor. For the latter reason, use of granular starch as a carbon source in denitrifying immobilization systems is not common. Starch-degrading denitrifiers, like the *Pseudomonas* sp. (JR12) used in this study, provide the possibility to operate the system without the addition of soluble carbon sources to the influent water. Degradation of the granular starch within the complex requires direct contact between the starch and the amylase enzyme of the denitrifiers[16]. Therefore, by providing more starch, more surface area is provided for bacterial adhesion (see Fig. 3.) and consequently nitrate removal capacities are better in beads with a higher starch content.

Fig. 3.

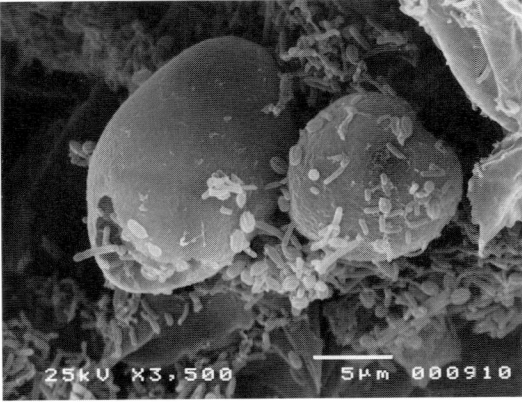

Batch studies with this denitrifier using starch as a carbon source, revealed a molar C/N ratio of 1: 4±1 (not shown). This ratio, together with the total amount of starch in the bead and the average daily nitrate removal per bead, allowed us to calculate the number of days required for depletion of the starch from the different beads (Table 2).

Table 2. Denitrification potential of beads with different starch content.

Starch (%/bead)	Starch (mg/bead)	Glucose[1] (mM/bead)	Maximum NO3 removal[2] (mM/bead)	Days of denitrification[3]
10	5	0.027	0.108	46.9
20	10	0.055	0.220	95.6
30	15	0.083	0.332	144.3
40	20	0.110	0.440	191.3

1) Assuming that all the starch is degraded to glucose
2) Maximum nitrate removal possible with the available glucose. Calculated from the molar ratio of 1:4 between starch (as glucose) and nitrate as determined experimentally.
3) Calculated by dividing the maximum nitrate removal capacity by the average daily nitrate removal rate (0.0023 mM NO_3-N/bead/day, average of 60 days incubation under denitrification condition).

It appears that the starch content in beads containing 40% starch enables denitrification for more than six months as compared to only six weeks for the 10% starch treatment. In our experiments, with exception of the 10% starch treatment, beads remained intact and removed nitrate for more than three months (data not shown). Therefore, based on the above calculation, beads should be filled with at least 20-30% starch in order to enable complete denitrification at prolonged incubation times.

4. CONCLUSIONS

Physical weakening of denitrifying alginate beads is a main limiting factor in the continuous operation of bioreactors of this kind. In this study we confirmed that it is possible to improve the mechanical properties of beads made of natural polymers by using similar methods to those applied in the field of synthetic polymers. Not only the mechanical properties but also the biological properties of the freeze-dried alginate beads were improved by using starch as both a filler and carbon source. Beads containing the highest concentration of starch (40%) revealed a superior performance with respect to both longevity and denitrification as compared to beads with less starch.

REFERENCES

1. D. Lemoine, T. Jouenne, G. A. Junter, 1991, pp. 437-440. In: M. Russ, H. Chmiel, E.D. Gilles, H. J. Knackmuss (eds.), Biochemical Engineering. Gustav Fisher, Stuttgart.
2. M. Volokita, B. Shimshon, A. Abeliovich, M. I. Soares, Wat. Res., 30 (1996) 965.
3. Metcalf and Eddy, Inc. 1991. Wastewater Engineering, Treatment, Disposal, Reuse (3rd ed.). McGraw-Hill International Editions, N.Y., 1334 pp.
4. A. Nussinovitch. 1997. Hydrocolloid Applications: Gum Technology in the Food and other Industries. Blackie Academic and Professional, London. pp. 247-264.

5. A. Nussinovitch, M. Nussinovitch, R. Shapira, Z. Gershon, *Food Hydrocolloids,* 8 (1994) 361.
6. G. Skjak-Braek, Martinsen, A. 1991, In: M. D. Guiry, G. Blunden (eds.), Seaweed Resources in Europe: Uses and Potential. John Wiley and Scns Ltd., N.Y. pp. 221-257.
7. M. B. Cassidy, H. Lee, J. T. Trevors, Journal of Industrial Microbiology, 16 (1996) 79.
8. L. A. Utracki, T. V. U. Khanh. 1992, pp. 207-268. In: I.S. Miles, S. Rostami (eds.), Multicomponent Polymer Systems. Longman Group, U.K.
9. A. Nussinovitch, Y. Aboutboul, Z. Gershon, J. van Rijn, Biotechnol. Prog, 12 (1996) 26
10. Y. Tal, J. van Rijn, A. Nussinovitch, Biotechnol. Prog., 13 (1997) 788.
11. M. Peleg, I. Roy, O. H. Campanella, M. D. Normand, J. Focd Sc., 54 (1989) 947.
12. O. H. Lowry, N. J. Rosebrough, A. L. Farr, R. J. Randall, J. Biol. Chem., 193 (1951) 265.
13. L. E. Nielsen, 1974. Mechanical properties of polymers and composites. Marcel Dekker Inc. New York. Vol. 2, p. 382.
14. A. Nussinovitch, Biotechnol. Prog. 10 (1995) 551.
15. J. van Rijn, Y. Tal, Barak Y, Appl. Environm. Microbiol., 62 (1996) 2615.
16. A. Kimura, J. F. Robyt, Carbohydrate Research, 287 (1996) 255.

6. PROTEINS

HYDROCOLLOIDS – PART 1
Edited by K. Nishinari
2000 Elsevier Science B.V.

357

A comparison of the gelling and foaming properties of whey and egg proteins

E. A. Foegeding[a], L.H. Li[b], C.W. Pernell[a] and S. Mleko[a]

[a]Department of Food Science, North Carolina State University, Raleigh, NC 27695, USA

[b]Kraft Foods, Glenview, IL, 60025, USA

Whey and egg white protein ingredients are often used in similar functional applications which involve gelation or foam formation. We compared the large-strain (fracture) rheological properties of heat-induced (80°C for 30 min) whey protein isolate (WPI) and egg white (EW) gels made under solution conditions which were optimized for gel strength (fracture stress). Gels formed under these conditions (pH 7.0 and 50 mM NaCl for WPI, and pH 9.0 for EW) had similar values for fracture stress over the range of 6 - 18% w/v protein. Gel deformability (fracture strain), had protein-specific trends. Fracture strain for WPI gels decreased as protein concentration increased, whereas fracture strain for EW gels remained constant. The stress - strain relationships for EW and WPI gels were characterized by calculating a ratio between the rigidity (stress/strain) at fracture and the rigidity at 30% of the fracture strain, called the rigidity ratio. Both EW and WPI gels had higher rigidity ratios (1.4 - 1.8) at low protein concentrations then shifted to lower values, of around 1, as protein concentration increased. The various gels formed at different protein concentrations and heating times and temperatures could be fit to a master curve of rigidity ratio vs. fracture rigidity. The yield stress of EW and WPI foams was determined by a vane method. Egg white foams exhibited relatively high yield stresses, even at low protein concentration and short whipping time. Maximum yield stress occurred at 8-10 min in egg white foams, while whey protein isolate foams required ≥ 15 min. The rheological properties of egg white and whey protein isolate foams and gels show protein-specific and protein-independent properties. These observations will be discussed relative to the chemical properties of these proteins.

1. INTRODUCTION

Food protein ingredients are used in a variety of applications, such as processed meats, nutritional beverages, baked goods and dairy products [1]. Whey protein concentrate (35-80% protein), isolate (≥ 90% protein) and dried egg white (albumen) are similar in that they contain mainly globular proteins that are water-soluble and easily dispersed. Whey and egg white protein ingredients are composed of a mixture of proteins and other compounds such as lipids, carbohydrates and minerals. It is often that whey and egg protein ingredients do not have equivalent functional properties when used in a specific food application.

That is to say, substitution of one for the other, usually done on an equal mass basis, does not produce the same results in the finished product. The reason for a lack of equivalent functionality could be due to: different amounts of protein added because substitution was based on equal mass; the effects of non-protein components; differences in protein structure; differences related to the relative proportion of individual proteins; solution conditions required for functionality; or they could be due to how proteins were altered during processing and drying.

The complexity of the problem makes it difficult to determine molecular mechanisms. The situation can be greatly simplified by investigating properties of β–lactoglobulin and ovalbumin, the major protein found in whey and egg white ingredients, respectively. However, mechanisms determined in simple monodispersed protein solutions have to be tested in complex food applications for validation. This is often not done for various reasons including the tremendous time and expense required to produce enough pure protein for making a food product.

The comparison between protein functionality of whey and egg white protein ingredients will be addressed by first comparing compositional differences then addressing the functional differences in heat-induced gelation and foam formation.

2. COMPOSITION

Whey protein concentrates and isolates are refined from whey generated from processes used to make casein ingredients (caseinates and caseins) or cheese. In contrast, egg white ingredients are produced by drying egg white. A comparison of starting materials and the composition of dehydrated starting materials is seen in Table 1.

Table 1
Composition of Cheese Whey and Egg White (%)

	Water	Protein	Lipid	Carbohydrate	Ash
Egg White					
Total mass	88.9	10	0.03	0.5	0.5
Solids		90.7	0.3	4.5	4.5
Cheese Whey					
Total mass	93	0.8	0.5	4.9	0.5
Solids		11.9	7.5	73.1	7.5

Data for cheddar cheese whey and egg white are from [2] and [3], respectively.

Drying produces a high protein ingredient (> 90% protein) from egg white, while dried whey is only 12% protein, with lactose being the major component. Therefore, making whey protein concentrates (≥ 35% protein) and isolates (≥ 90% protein) requires extensive removal of non-protein solids, whereas the focus in producing dried egg white is to remove water without damaging protein functionality.

The major protein found in egg white is ovalbumin (54% of total protein), and β–lactoglobulin is the most abundant whey protein (55% of total protein) [4,5]. The five most abundant proteins in whey and egg white account for 84.5% and 98% of the total protein, respectively. The proteins range in molecular weight from ~12,000 to several million. In general, whey proteins are of lower molecular weight and egg white proteins have the distinction of being glycosylated.

Whey and egg white dispersions are clear to opaque, depending on the degree of protein aggregation and non-protein components (e.g, phospholipids can cause opacity in whey protein dispersions). They are often used in similar functional applications that involve gelation or foam formation. In many cases, differences between whey and egg white proteins are seen in gel rheological properties and foaming ability. These differences could be due to protein sequence and structure, thereby being true protein-specific differences. Alternatively, they could be due to conditions being less optimal for one protein compared to the other. We have attempted to determine why one observes different functional properties between egg and whey protein ingredients.

3. HEAT-INDUCED GELATION

One of the major difficulties in comparing gelation properties among proteins is determining which solution conditions are appropriate.

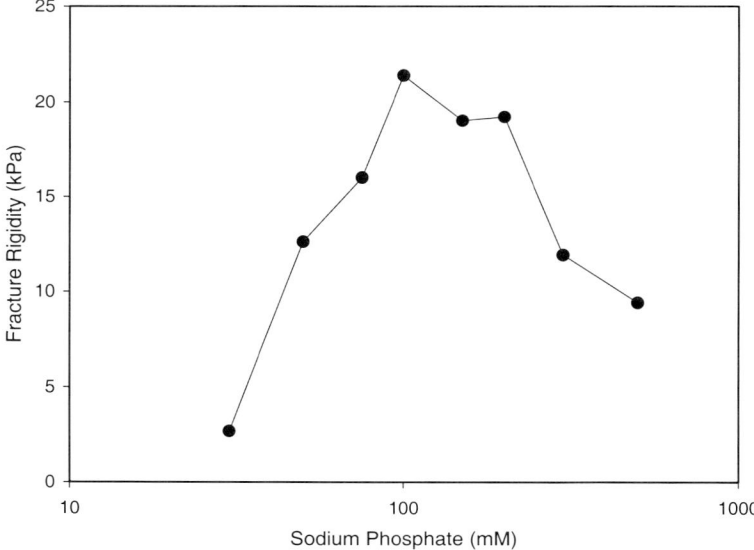

Figure 1. Changes in fracture rigidity with sodium phosphate concentration. Data from [6].

This is illustrated in Figure 1, which shows the change in fracture rigidity (G_f, fracture stress/fracture strain) with sodium phosphate concentration. As phosphate concentration increases, the fracture rigidity increases, plateaus, and then decreases. A parabolic relationship between ion concentration (Sodium phosphate, NaCl, Na_2SO_4) and gel rigidity (G' or G_f) or hardness, has been shown for gelation of whey protein isolate [6] and ovalbumin [7]. This coincides with the transformation from fine-stranded to particulate gel network strutures, with an intermediate "mixed" structure [6,8,9]. Fine-stranded and particulate networks can also be formed under similar salt conditions by changing pH [8,10]. Transformations among network types and rigidity values can be interpreted as the result of changing the solvent quality or chemical potential. This would affect the denaturation temperature in addition to the degree of aggregation for native and denatured proteins. Alternatively, specific protein-ion interactions may be altering the denaturation/gelation process. Therefore, if one observes rheological differences among protein gels, is it due to changes in solution chemical potential or an inherent difference in protein structure?

We approached this problem by comparing egg white (EW) and whey protein isolate (WPI) gels with similar gel strength (true shear stress at fracture or "fracture stress"). Gels were formed under different solution conditions (pH 7.0 and 50 mM NaCl for WPI, and pH 9.0 for EW) by heating at 80°C for 30 min. Egg white and WPI gels had similar values for fracture stress over the range of 6 - 18% w/v protein (Figure 2). The effect of protein concentration on fracture stress for pooled data from EW and WPI gels fit a power law model of:

$$\text{Fracture Stress, kPa} = 0.048 \, (\% \text{ protein})^{2.7}, \text{ with } r^2 = 0.97 \tag{1}$$

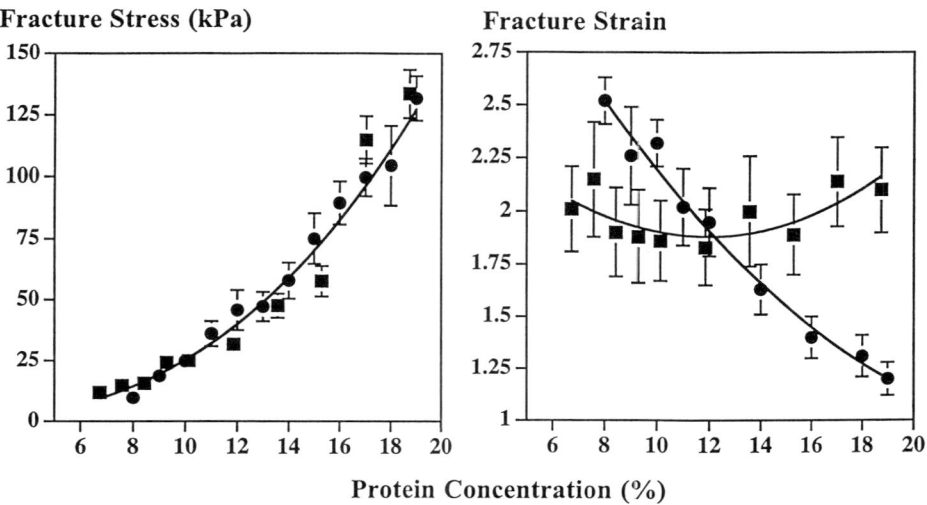

Figure 2. Fracture stress and fracture strain for whey protein isolate (circles) and egg white (squares) gels formed by heating at 80°C for 30 min. Fracture properties, at 22 ± 2°C, were determined by twisting to fracture at 0.05 s^{-1} [6]. Data from [11].

Gel deformability (fracture strain) had protein-specific trends (Figure 2). Fracture strain for WPI gels decreased as protein concentration increased, whereas fracture strain for EW gels remained constant. Deformability of gel networks appears to be protein-specific in gels that show no protein-related differences in fracture stress.

The different fracture strain properties for EW and WPI gels made the fracture rigidity (G_f) relationship protein concentration dependent (Figure 3). At low protein concentrations, EW gels were more rigid than WPI gels because the fracture strain was less than WPI gels. As protein concentration increased, and WPI fracture strain decreased, WPI gels had greater G_f values than EW gels.

Stress vs. strain plots for WPI and EW gels showed general trends of strain hardening. As strain increased, stress increased non-linearly. This behavior was characterized by calculating a ratio between G_f and the rigidity at 30% of the fracture strain. This property was called the *rigidity ratio* and it provided a measure of strain hardening (rigidity ratio > 1) or weakening (rigidity ratio < 1). Both EW and WPI gels had higher rigidity ratios (1.4 - 1.8) at low protein concentrations then shifted to lower values, of around 1, as protein concentration increased (Figure 3). This indicated that the fracture process became more "elastic-like" as protein concentration increased because a Hookean elastic solid would have a rigidity ratio of 1. The trends for EW and WPI gels were similar, with the WPI trend being systematically shifted to higher values at each protein concentration (Figure 3).

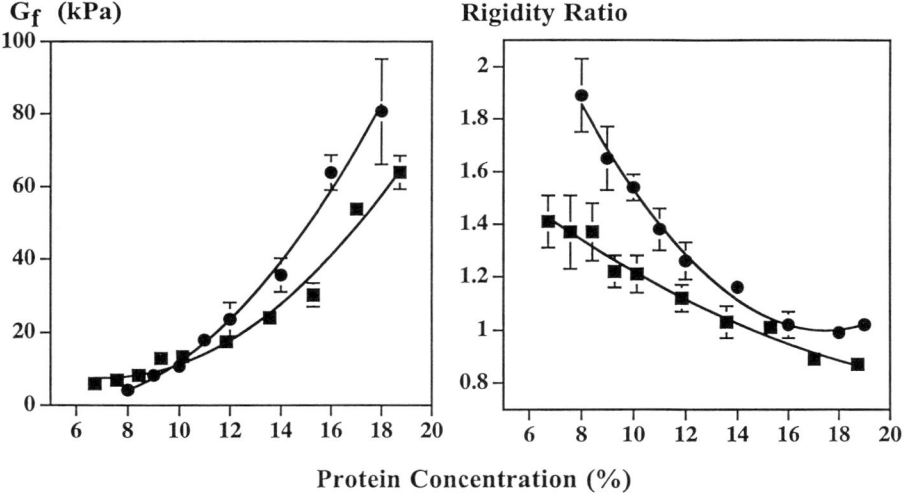

Figure 3. Rigidity modulus and rigidity ratio for whey protein isolate (circles) and egg white (squares) gels formed by heating at 80°C for 30 min. Data from [11].

One possible reason for the differences in rheological properties is a kinetic limitation. This was investigated by looking at the development of rheological properties as heating time at 80°C progressed from 0 – 30 min. For both proteins, self-supporting gels were formed by 5 min and fracture stress was constant by 20 min (data not shown).

Trends were different for fracture strain. Whey protein isolate gels showed a progressive decrease in strain as heating time progresses, while EW gels show little change in strain with heating time (Figure 4).

Gels formed at different protein concentrations (7-19%), heating times (5 – 30 min) and temperatures (65 – 80°C) could be fit to a master curve of rigidity ratio vs. fracture rigidity (Figure 5). The master curves for EW and WPI gels were similar, with the WPI trend being systematically shifted to higher values of rigidity ratio for a given fracture rigidity. These data suggest an inherent difference between EW and WPI gels.

It was possible that the master curve relationship between rigidity ratio and fracture rigidity was only valid because of limitations in the solution conditions evaluated (pH 7.0 and 50 mM NaCl for WPI, and pH 9.0 for EW). This comparison was broadened by altering pH (7-9), and sodium chloride concentration (0 - 1 M). The overall results showed that large-strain rheological properties of EW gels were much less sensitive to changes in pH and sodium chloride concentration than WPI gels [12]. However, the construction of a master curve, regardless of pH, sodium chloride concentration or heating conditions, was possible. This indicated a general relationship between fracture rigidity and the rigidity ratio for the gels made from one type of protein ingredient.

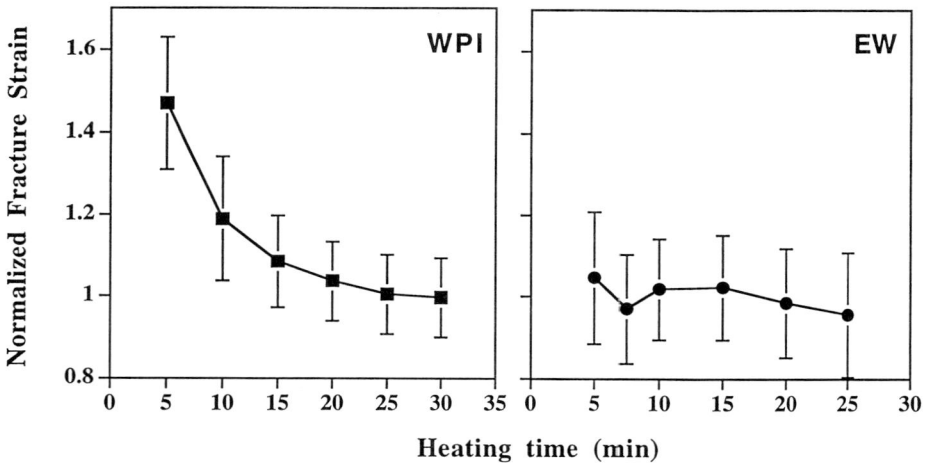

Figure 4. Normalized fracture strain for whey protein isolate (WPI) and egg white (EW) gels formed by heating at 80°C for 5-30 min. Data from [11].

4. HEAT-INDUCED AGGREGATION

Changes in turbidity (optical density at 660 nm) as a function of heating temperature and NaCl concentration are seen in Figure 6.

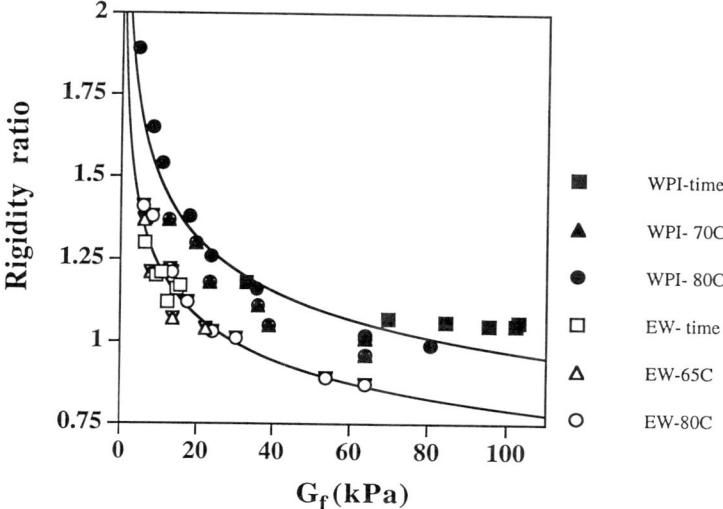

Figure 5. Master curves relating the rigidity ratio to the fracture modulus for whey protein isolate (closed symbols) and egg white (open symbols) gels formed by heating at 80°C for 30 min (circles) or various times (squares). Gels were also formed by 30 min heating of egg white at 65 °C (triangles) or whey protein isolate at 70 °C (triangles). Data from [11].

Whey protein isolate dispersions are at pH 7 and EW dispersions are at pH 9 (these pH values were compared because they were optimal for gelation). The turbidity transitions for WPI show a slight salting out, i.e., increase in turbidity, at temperatures ≤ 65°C, followed by a major denaturation/aggregation transition at 70°C.

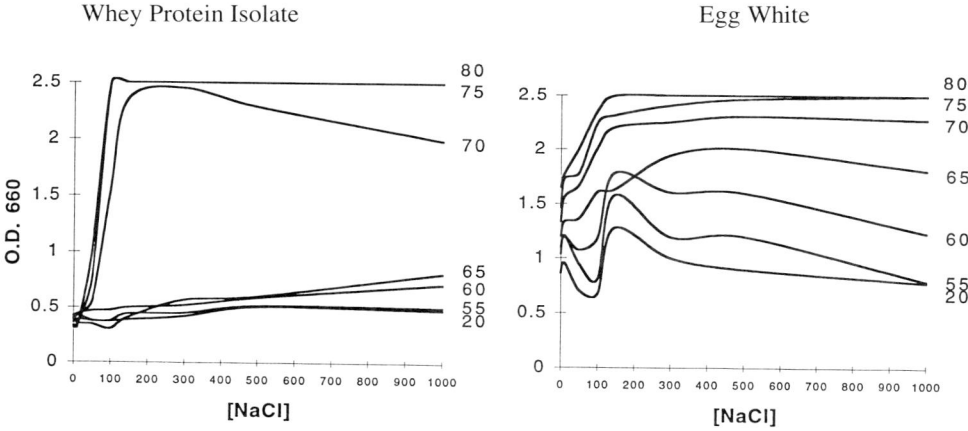

Figure 6. Turbidity of egg white and whey protein isolate solutions (10% w/v) heated at 80°C for 15 min (unpublished data).

In contrast, the turbidity of EW dispersions was very sensitive to NaCl concentration at temperatures ≤ 65°C. Turbidity at NaCl concentrations ≥200 mM showed a more sequential increase in turbidity as temperature increased. While simple turbidity measurements do not produce quantitative information on the aggregates, it does show that the aggregation processes for the two protein types are quite different.

5. FOAM FORMATION

One of the food applications for EW is Angel food cakes, where an EW foam is mixed with flour and sugar to form a cake batter. Whey protein foams, with overrun equal to EW foams, do not produce similar cakes [13]. Cakes made with EW foams increase in volume during baking to a maximum then decrease slightly. In contrast, cakes made with WPI show similar trends in batter expansion but collapse during the latter stages of baking (unpublished data). The reason for the different functional properties is not known, therefore, we have investigated the rheological properties of protein foams to determine if physical properties are important.

The yield stress of EW and WPI foams was determined by a vane method [14]. Yield stress was calculated as a function of vane dimensions and the torque required to initiate movement of the foam [15]. Egg white foams exhibited relatively high yield stresses, even at low protein concentration and short whipping time (Figure 7).

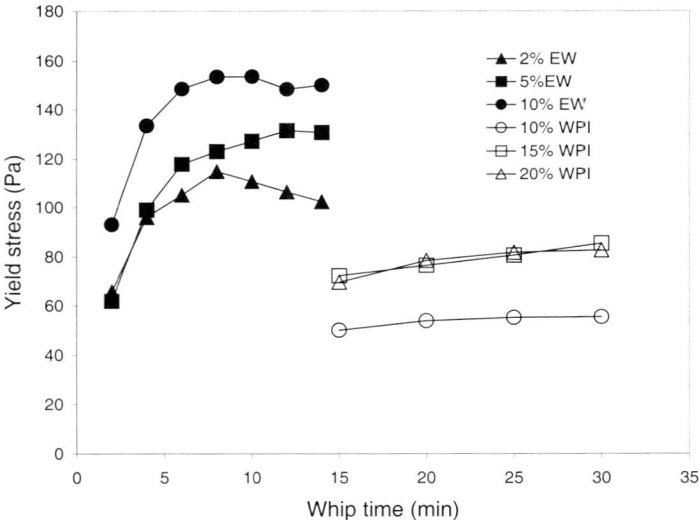

Figure 7. Yield stress of EW and WPI foams formed by whipping pH 7.0 solutions for various times. Data from [14].

Foams formed at three protein concentrations showed similar rates of increase and decrease in yield stress; however, the maximum values of yield stress increased as protein concentration increased, as did the whip time at which the maximum yield stress occurred. Egg white foams consisting of 2%, 5% and 10% protein (w/w) had respective yield stress values of 105.0 ± 4.0 Pa, 125.0 ± 3.0 Pa and 150.0 ± 5.0 Pa . Maximum yield stress occurred at approximately 8 minutes whipping time in the 2% protein foams and at about 10 minutes in the 5% and 10% foams.

Whey protein isolate foams, as compared to EW foams, were far less stiff and required longer whipping times to achieve maximum yield stress values. Maximum yield stress values of 10%, 15% and 20% whey protein foams were 55.0 ± 2.0 Pa, 94.0 ± 1.0 Pa and 87.0 ± 1.0 Pa respectively. The rates of increase in yield stress with whipping time were statistically insignificant over whipping times ranging from 15 –30 minutes in the 10% WPI foams. Whip times of less than 15 minutes produced foams that registered below the 90% confidence limit of the test equipment range. WPI foams containing 15% and 20% protein showed only a modest increase in yield stress with whipping time.

The relationship between yield stress and %overrun for EW and WPI foams is seen in Figure 8.

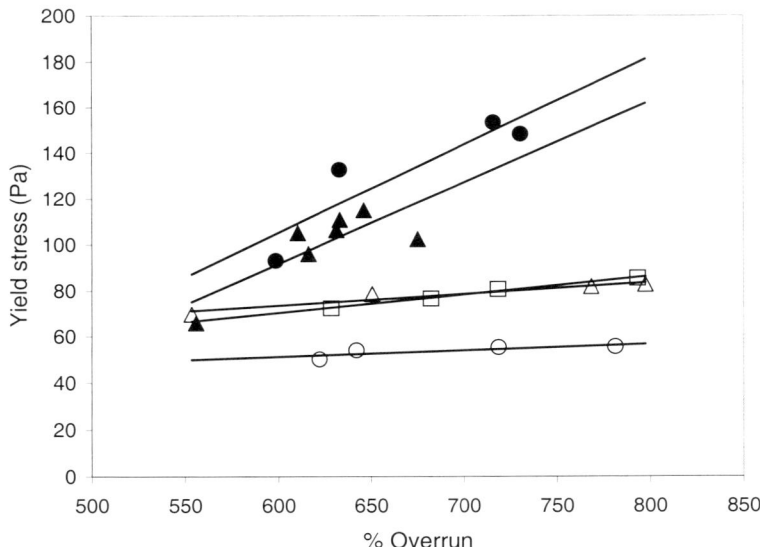

Figure 8. Yield stress and overrun of EW and WPI foams formed by whipping pH 7.0 solutions for various times. Symbols are the same as in Figure 7 and data are from [14].

Overrun increased with whipping time for EW and WPI foams, whereas only EW foams had a major increase in yield stress with whipping time (Figures 7 and 8). It is clear that a given level of overrun, say 600 – 650%, can be achieved by various combination of protein concentration and whipping time for both proteins. In contrast, combinations of yield stress and overrun are protein-specific (with the exception of EW (2% protein) and WPI (20%

protein) foams having a yield stress of ~ 65 Pa and an overrun of 550%). The trends show that the factors responsible for overrun and yield stress are quite different. Moreover, it suggests that different mechanisms, or a different combination of factors describing one general mechanism determines structure and physical properties of WP and EW foams. The true cause remains to be established.

6. CONCLUSIONS

Egg white and whey proteins can form gels and foams that have similar values for gel strength and overrun, respectively. However, these should not be considered similar materials or overall structures. When several properties describing gels and foams are taken into account, protein-specific trends are seen. A reasonable goal is to work towards models that explain the array of physical properties describing protein gels and foams.

7. ACKNOWLEDGEMENTS

This research was funded by grants from the United States Department of Agriculture National Research Initiative Competitive Grants Program, The Southeast Dairy Foods Research Center, and Dairy Management Incorporated.

REFERENCES

1. J.N. deWit, The use of whey protein products, in Developments in Dairy Chem., 1989, pp 323.
2. F.W. Kosikowski, Cheese and Fermented Milk Foods, F.V. Kosikowski and Associates, NY, 1977.
3. W.D. Powrie and S. Nakai. Characteristics of edible fluids of animal origin: Eggs. In Food Chemistry, O.R. Fennema (ed), Marcel Dekker, Inc, NY, 1985.
4. E. Li-Chan and S. Nakai, Crit. Rev. Poultry Biol., 2 (1989) 21.
5. J.N. deWit, J. Dairy Sci., 81 (1998) 597.
6. E.L. Bowland and E.A. Foegeding, Food Hydrocoll, 9 (1995) 47.
7. E. Doi, Trends in Food Sci. Technol., 4 (1993) 1.
8. M. Langton and A-M. Hermansson, Food Hydrocoll., 5 (1992) 523.
9. M. Verheul and S.P.F.M. Roefs, Food Hydrocoll., 12 (1998) 17.
10. S. Mleko, Pol. J. Food Nutr. Sci., 1 (1996) 63.
11. L.H. Li, A.D. Errington, E.A. Foegeding, J. Food Sci., submitted.
12. L.H. Li, S.R. Bottcher, E.A. Foegeding, J. Food Sci., submitted.
13. B. Arunepanlop, C.V. Morr, D. Karleskind, I. Laye, J. Food Sci., 61 (1996) 1085.
14. C.W. Pernell, C.R. Daubert, E.A. Foegeding, J. Food Sci., submitted
15. N.Q. Dzuy and D.V. Boger, (1983), J. Rheology, 27, 321–349.

HYDROCOLLOIDS – PART 1
Edited by K. Nishinari
2000 Elsevier Science B.V.

Structure and rheology of gels formed by aggregated protein particles

T. van Vliet

Wageningen Centre for Food Sciences P.O. Box 557, 6700 AN Wageningen / Food Science Group, Wageningen Agricultural University, P.O.Box 8129, 6700 EV Wageningen, The Netherlands.

Many protein gels are formed by aggregation of protein particles of colloidal size. If particles suspended in a liquid aggregate to form clusters, these clusters often have a fractal structure. Such an aggregation process may ultimately lead to a gel built of fractal clusters. Relatively simple scaling relations can be derived, relating rheological properties such as the shear modulus of the protein network to the volume fraction of primary particles in the network with the fractal dimensionality as part of the scaling exponent. However, for a full description of these relations more detailed information on the structure of the protein gel considered is required. Factors of importance are: (1) the effective size of the building blocks of the fractal clusters, (2) the amount of protein incorporated in the fractal clusters at the moment the gel is formed and (3) the way in which the fractal clusters are linked together. Factors 1 and 3 are strongly affected by the occurrence of rearrangements in the structure of the fractal cluster before and after gel formation. Extensive rearrangements may cause the loss of the fractal character of the clusters.

1. INTRODUCTION

The geometry of many objects in nature is not Euclidian but fractal [1]. Fractal objects are characterized by two characteristics. As a fractal object is magnified, it will look similar independent of magnification. A fractal is thus self similar. The quantitative measure on how a fractal scales with size is expressed by the fractal dimensionality, which is independent of many details of how the object is formed. The fractal dimensionality is related to behaviour at 'large' scales where fluctuations are averaged out; its value is smaller than the dimension of the Euclidian space considered (2 for a plane and 3 for a body). Fractal models have been found to be very useful for describing aggregation, gelation and gel properties of a dispersion of particles [2,3].

Many protein gels are formed by aggregation of protein particles of colloidal size. These particles can be just single globular protein molecules, small aggregates of several of these molecules or large aggregates consisting of a hundred to thousands of protein molecules. If particles in a liquid aggregate undisturbed, clusters are formed with a very tenuous (open, rarefied) structure, these clusters often have a fractal structure. Such an aggregation process may ultimately lead to a gel built of fractal clusters. Examples of protein gels build of fractal clusters are acid-induced and rennet-induced skim milk (casein) gels and whey protein gels [3-6]. In general clusters formed during aggregation of protein particles such as casein micelles are not completely self similar, but on average they exhibit self similarity and their scaling with size can be described by a fractal dimensionality. The clusters formed are so-called stochastic fractals [7].

A description of the aggregation process leading to the formation of clusters with a fractal structure and ultimately a gel enables a better understanding of experimentally observed aggregation and gel formation times than a description that does not take the structure of the formed clusters explicitly into account [7,8]. Moreover it allows deduction of scaling relations for various important properties of the gels formed, such as between concentration and permeability or between concentration and several rheological properties [3,4]. The latter relations have been shown to depend on rather subtle details of the structure of the fractal clusters and how they are linked together [9,10].

The topic of this paper is the rheology of gels formed by aggregation of protein particles leading to the formation of clusters with a fractal structure and ultimately a gel. First a general description will be given on how fractal structures are defined.

1.1. 'Fractal' description of gel formation and structure

The number of primary particles N_p in a cluster with a fractal structure, that just can be contained in a sphere with radius R, scales with R as

$$N_p = \left(\frac{R}{a}\right)^D \tag{1}$$

where D is the fractal dimensionality ($D < 3$) and a the radius of the primary particles. During aggregation a great number of clusters will be formed with at any moment a statistical variation in radius due to the randomness of the aggregation process. The scatter will be smaller for on average larger clusters. The volume of the clusters scales with R^3, so the volume fraction of particles in a cluster is then:

$$\phi_c = \left(\frac{R}{a}\right)^{D-3} \tag{2}$$

Since $D < 3$, ϕ_c decreases with increasing R. The clusters formed are stochastic fractals at the concentrations general studied; they are on average self-similar at length scales between a and R [7]. At a certain radius R_g the average ϕ_c will equal the volume fraction ϕ of the primary particles in the system and the fractal cluster will jointly occupy the total volume; hence a gel will be formed, at which point:

$$<\phi_c> = \left(\frac{<R_g>}{a}\right)^{D-3} = \phi \tag{3a}$$

or:

$$<R_g> = a\,\phi^{1/(D-3)} \tag{3b}$$

where $<R_g>$ is a measure of the average cluster radius at the moment the gel is formed. The gel formed is homogeneous over length scales larger than $<R_g>$ with a dimensionality of 3, it is build up of so-called fractal clusters. In fact $<R_g>$ gives an upper cut-off length of the fractal regime and a a lower cut-off length. The structure of the gel is fractal over length scales between a and $<R_g>$.

Because fractal structures are scale invariant, the gel will also be scale invariant, *i.e.* a gel formed from a system with a high ϕ will resemble that formed from a system with a low ϕ, only the scale will be different. In a gel built of fractal clusters, the size of the largest holes will have a radius of about R_g. Because the flux through a hole (pore) will scale with R^4 (Poiseuille's law) and the number of pores with radius R_g per cross section of a gel with R_g^{-2}, scales the permeability B of a gel with R_g^4 times R_g^{-2} is R_g^2. Combination of this relation with equation 3b gives [3]:

$$B = \left(\frac{a^2}{K}\right) \phi^{2/(D-3)}$$

(4)

where K is a proportionality constant of order 10^2.

So a fractal description of the clusters building a gel allows derivation of scaling relations between a macroscopic property of the gel and parameters characterizing its structure. Equation 4 describes the relation between the permeability of a gel and its structure (expressed in an effective fractal dimensionality D), the volume fraction of particles forming the gel and the size of these particles. It predicts a linear relation between log B and log ϕ, which has been observed for various protein gels as acid- and rennet-induced casein gel and whey protein gels [6,11,12]. From the slope of these curves an effective D can be calculated.

1.2. General expression for the modulus of a gel

If an external force is applied to a gel in the direction x it will deform. The external force will be counteracted by the strands forming the network. The following expression can be derived relating the modulus to the number of stress-carrying strands [10,13]:

$$G = NC\frac{d^2F}{dx^2}$$

(5)

where G is the shear modulus of the network, N the number of stress-carrying strands per unit area in a cross-section perpendicular to x, dF is the change in Gibbs energy when the particles in a strand are moved apart over a distance dx and C is a characteristic length, determined by the geometric structure of the network, which relates the local deformation Δx to the macroscopic shear strain γ. The magnitude of C can only be established with some accuracy for simplified models of the network structure.

At constant temperature $dF = dH - TdS$, where H is enthalpy and S is entropy. This implies that both enthalpic and entropic factors may contribute to the modulus. The entropic contribution is dominant for gels from flexible macromolecules, like gelatin and heat set ovalbumin gels in 6 M urea [14,15]. For gels of stiff macromolecules and of (protein) particles the enthalpic contribution is dominant [10,16].

1.3. Scaling laws relating rheological properties to the structure of 'fractal' gels

The fractal description of the clusters building a gel allows us the derivation of expressions for the parameters N and C in equation 5. Two fractal clusters that aggregate with each other will do so mainly *via* their longest projections. Because on average the structure of a fractal cluster is scale invariant, the number of projections is independent of the size of the fractal cluster, and therefore -as a first approximation- the number of contact points (bonds) between two aggregated fractal clusters will be independent of their size [3]. The contact area will scale with $<R_g>^2$, so after gel formation the number of stress carrying strands per unit cross section will scale as:

$$N \propto <R_g>^{-2} \propto a^{-2} \phi^{2/(3-D)} \tag{6}$$

Equation 6 gives a simple scaling relation between N and the structure of a gel as expressed in an effective D, ϕ and a. There is no similar expression for C, that holds for all gels built of fractal clusters. The dependence of C on ϕ depends on the way the fractal clusters are linked to each other [4,9,10]. For the most simple case that the clusters are connected by straight strands, C is independent of $<R_g>$. Then, by combination of equations 5 and 6, the following scaling expression for G is obtained:

$$G \propto <R_g>^{-2} \frac{d^2F}{dx^2} \propto a^{-2} \phi^{2/(3-D)} \frac{d^2F}{dx^2} \tag{7}$$

In theory, undisturbed cluster-cluster aggregation will lead to tortuous, flexible strands connecting the fractal clusters in a gel. Then C is proportional to $<R_g>^{-(1+y)}$, where y is the "chemical" exponent of the elastic effective strand connecting the fractal clusters with $1<y<1.3$ [17]. However, due to rearrangements after gel formation (see below) the shape of these strands may change. This leads to the next general scaling expression for G with ϕ:

$$G \propto <R_g>^{-(2+z)} \frac{d^2F}{dx^2} \propto a^{-(2+z)} \phi^{(2+z)/(3-D)} \frac{d^2F}{dx^2} \tag{8}$$

where $0<z<2.3$. So z (= $y+1$) has to be known before the fractal dimensionality can be calculated from an experimentally determined relation between G and ϕ. For a hinged connection between the fractal clusters C is proportional to $<R_g>^{-1}$, thus z is 1 [4,9].

During large deformations of a gel flexible connections between fractal clusters will become straightened resulting in C becoming less or even independent of ϕ. For the latter case the following expression is obtained for the dependence of the yield or fracture stress on particle volume fraction in the gel [4]:

$$\sigma_y \propto \phi^{2/(3-D)} \tag{9}$$

2. PROTEIN GELS DESCRIBED AS 'FRACTAL' SYSTEMS

For various protein gels a linear relation has been obtained between the logarithm of the storage modulus, G' and the logarithm of the volume fraction particles, φ. A few examples of results obtained are given in Fig. 1; it is assumed that the volume fraction of particles in the gel is proportional to the protein concentration.

As can be seen in Fig. 1, the slope may depend on gelling conditions and on the type of protein. HCL-induced Na-caseinate gels were prepared by acidification at 2 °C to pH 4.6 followed by quiescent heating to 30 °C. A gel is formed at temperatures above 10 °C. GDL-induced gels were formed by slow acidification at 30 °C due to hydrolysis of added glucono-δ-lactone [4]. Soy protein gels were formed by heating the protein dispersion at 1 K min^{-1} to 95 °C, keeping it for 1 h at that temperature, followed by cooling to 20 °C at 1 K min^{-1} [18].

The difference between the HCL- and the GDL-induced acid casein gels is related to the way the fractal clusters are linked to each other. In the HCL-induced gels they are connected by straight strands while in GDL-induced gels the connection has a hinged character [4]. It causes z in equation 8 to be different for both types of gels, 0 for HCL-induced casein gels and 1 for the GDL-induced casein gels. The straightening in the HCl-induced gels is due to shrinkage of the casein particles caused by the increasing temperature after a gel is formed [19]. It causes, among others things, that the fracture strain of HCL-induced casein gels is about half of that of GDL-induced casein gels.

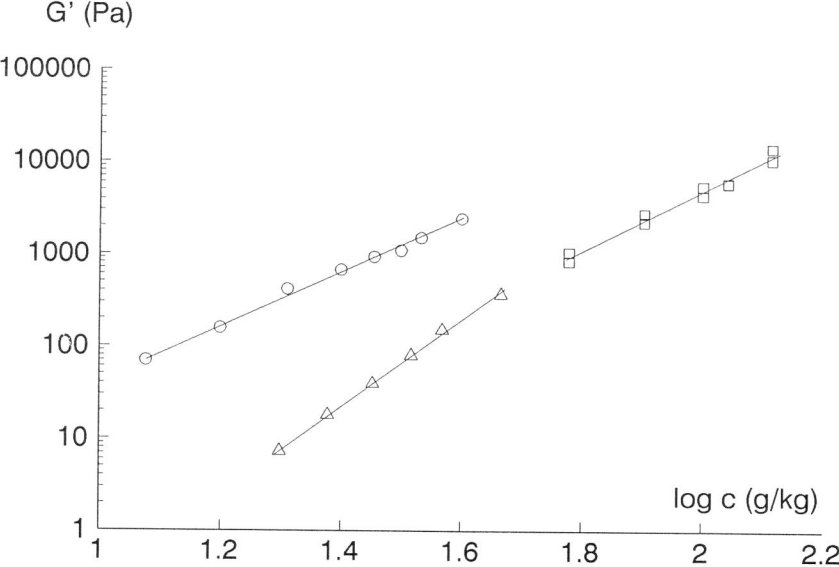

Figure 1. The logarithm of the storage modulus G' as a function of the logarithm of the protein concentration for HCL-induced Na-caseinate gels, pH 4.6, (o); GDL-induced acid Na-caseinate gels, pH 4.6, (Δ); heat-set gels of soy protein isolate, pH 5.2, added NaCl 0.2M (□)

Various literature values for the slope of a log-log plot of G' versus φ or protein concentration are tabulated in Table 1, together with data for the slope of a log-log plot of σ_y versus ϕ. If known, the value of z and the calculated D are also given. For the acid-induced casein gels and the whey protein isolate gels, tabulated values for D are in agreement with D calculated by equation 4 from measurements of the permeability versus protein concentration. For the β-lactoglobulin gel the tabulated D has been obtained form light scattering experiments on dilute whey protein dispersions at pH 5.4 [5]. For these gels, assuming z to be one would result in $D = 2.3$, while a plot of $\log \sigma_y$ versus log protein concentration gave D of 2.4-2.5 [5]. For the acid casein gels plots of $\log \sigma_y$ versus log. protein concentration also gave deviating values for D compared to values obtained by several other methods [4]. This may not be surprising, because equation 9 is only a first approximation. The large deformations involved in fracture experiments may easily involve changes which are not independent of volume fraction

Table 1. Slope of log-log plot of storage modulus, G' or of yield stress, σ_y versus protein concentration for various protein gels. Parameter z characterizes the way fractal clusters are linked to each other. Fractal dimensionality D calculated from slope log-log plots

protein gel and parameter involved	slope	z	D
HCl-induced acid casein gel, G' [4]	2.6	0	2.24
, σ_y [4]	2.5		2.19
GDL-induced acid casein gel, G' [4]	4.6	1	2.35
, σ_y [4]	2.5		2.2
Rennet-induced casein gel, G' [3,20]	2.4	0	2.17
Soy-protein gel, pH 5.2, 0.2 M NaCl, G' [18]	3	-	-
β-Lactoglobulin gel, pH 5.4, 0.16 M NaCl, G' [5]	4.1	~1	~2.2*
, σ_y [5]	3.6		~2.45
, pH 5.3, no salt , G' [21]	3.7		
Whey protein isolate, pH ~ 7, 0.4 - 3 M NaCl, G' [22]	5	-	2.4

*determined from light scattering experiments

2.1. Effect of rearrangements

As indicated already in the discussion of the relation between the modulus and the volume fraction of particles in the gel, more detailed information on the structure of the gel considered is required for a full description of these relations. A factor mentioned already is the way in which the fractal clusters are linked together. Its effect on the rheology has been discussed above. A difference in the way fractal clusters are linked to each other may be caused by rearrangements during, and especially after, gel formation. Such rearrangements may have several consequences such as an increase in the effective size of the building blocks of the fractal clusters, straightening and even yielding / fracture of the strands between the clusters and transport of particles from the outside of the fractal cluster to the inner part.

The effective building blocks of the fractal clusters formed during aggregation may be the primary protein particles, but often they are larger. The latter may be the case if the primary protein particles change their position with respect to each other after bond formation: this may occur, e.g. by rolling around each other until (most) particles has acquired bonds with three or more particles. This process may lead to the formation of dense aggregates, which are clearly larger than the initial primary particles [23]. Because the rate of rearrangement, relative to the aggregation rate will normally decrease with increasing size of these aggregates, at a certain stage ongoing aggregation will dominate structure formation and fractal clusters can still be formed. However, the effective size of the building blocks (i.e., the dense aggregates mentioned) forming the fractal cluster will be substantially larger than the size of the primary protein particles. This gives for equation 1 and 3b

$$\frac{N_p}{N_0} = \left(\frac{R}{a_{eff}}\right)^D \tag{10}$$

$$< R_g > = a_{eff}\, \phi^{1/(D-3)} \tag{11}$$

respectively, where a_{eff} is the radius of the effective building blocks forming the fractal cluster and N_0 the number of primary particles forming such a building block.

A change in effective size of the building blocks occur during the formation of acid-induced casein gels due to the addition of glucono-δ-lactone at temperatures of 30 and 40 °C [23,24]. Some experimental data reported by Lucey et al. [24] are summarized in table 2. The calculations show that at 20 ° C the effective building blocks of the fractal cluster are the original casein particles, but during gel formation at 30 and 40 ° C strong rearrangements has occur, leading to much larger a_{eff}.

Table 2. Calculated size of the building blocks (a_{eff}) (equation 11) of the fractal aggregates in acid-induced Na-caseinate gels made at various temperatures. $<R_g>$ was obtained from confocal micrographs. Protein concentration was constant at 2.51 w/w %. D was taken to be 2.35. Radius primary casein particles 50 nm at 30 °C. Data from Lucey et al [24].

	Gelation temperature (° C)		
	20	30	40
Radius aggregates, R_g (μm)	2.5	10	15
Volume fraction, φ	0.078	0.068	0.060
Radius building blocks fractal cluster, a_{eff} (nm)	50	160	195
Particles in building block, N_0	1	25	40

An increase in the size of the effective building blocks will imply a change from a (more) fine stranded gel to a (more) so-called aggregate (or thick stranded) gel, which will have several consequences.

A larger a_{eff} gives a larger $<R_g>$ (equation 11) and with that larger holes in the gel network, implying a gel network that is inhomogeneous over a longer scale. The larger holes leads to an increased permeability and thereby to an increased propensity to show syneresis [23,24]. For the Na-caseinate gels dealt with in table 2 the larger a_{eff} at 30 and 40 ° C as compared with a_{eff} at 20° C, would lead to an increase of the permeability B by a factor of 10 or 15 (equation 4). The smaller voluminosity (volume per weight) of the casein particles at higher temperature implies that an even stronger increase of B is expected. The measured increase of B by a factor of 6 and 29, respectively, was in reasonable agreement with the calculated one.

Another consequence of an increase in the size of the effective building blocks is that the bending resistance of the resulting thicker strands will in general be higher and so the resistance against deformation of the gel will be affected. For a circular strand with a radius a, which is supported at both ends, the bending force is proportional to a^4 [25]. However, because the number of strands N will scale with a^{-2} and also the magnitude of C may be affected, the effect on G is not straightforward. Due to the larger holes in the network, the fracture strain will decrease in general.

Rearrangements will not stop at the moment a gel is formed. They will continue e.g. due to ongoing formation of physical cross-links between the protein strands in the clusters, fusion of protein particles, etc. [10,23,26]. These processes lead to tensile stresses on the protein strands elsewhere in the network which may occasionally result in breaking of bonds and formation of new bonds at another place. The latter may cause transport of particles from more empty regions to more dense ones, where on average more bonds per particle can be formed. It causes that certain regions (the inner part of the fractal clusters) become more dense and other parts (the contact region between the fractal clusters) less dense. It will results in dense regions connected by thin straight strands, which as a consequence of ongoing rearrangements even may fracture after some time. Ultimately it results in the formation of larger holes and so in a strong increase of B. This process strongly promotes (micro-)syneresis [26]. The size of the holes is not anymore directly related to the size of the fractal clusters. Moreover, the fractal character of the originally fractal clusters will be lost. Such a process occurs for instance in rennet-induced milk gels at higher temperatures (> 20 ° C) and/or a lower pH (range 5.2-6.7) [26-28]. For casein gels the separation in more dense and less dense region may continue, so far that it leads to a decrease in the storage and loss modulus with ageing time (figure 2). The effective number of stress-carrying strands N in equation (5) decreases and this decrease is not compensated enough by an increase in the effective thickness and stiffness of the strands due to fusion of the casein particles.

The extent of rearrangement after gel formation will also affect the way in which fractal clusters are linked together. In general, undisturbed aggregation of fractal clusters will lead to tortuous connections between them. As discussed above, ongoing rearrangements may lead to straightening of strands. Straightening will also be the case if the particles/protein molecules forming the connecting strands shrink after gel formation, e.g. due to a change in temperature, or if they fuse more extensively (e.g. casein particles in milk gels). The consequences for the rheological behaviour of gels are already discussed above.

2.2 Other factors

In the derivation of equations 3 and 11 it was implicitly assumed that all the protein present in the system contributes to gel formation from the start of the aggregation process. Protein gel formation is normally induced by a change in conditions (e.g. temperature or pH) or by enzyme action making the protein particles unstable against aggregation. If the aggregation process proceeds fast compared with the formation of reactive particles, part of the protein will still be in its native state when the gel network is formed. This implies that the volume fraction of protein to be used in equations 3 and 11 is smaller than the total volume

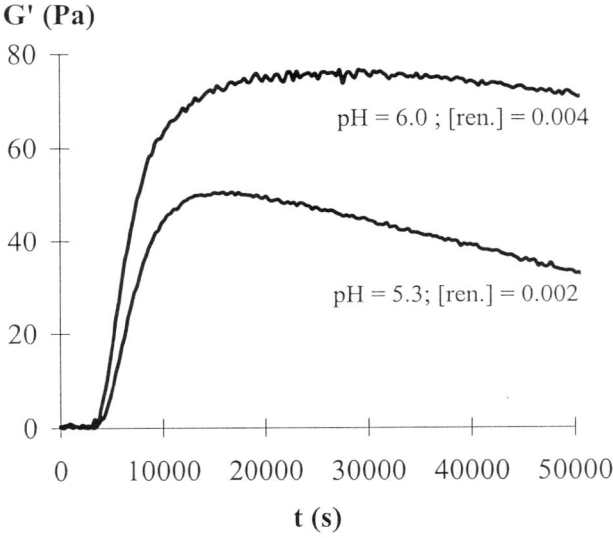

G' (Pa)

pH = 6.0 ; [ren.] = 0.004

pH = 5.3; [ren.] = 0.002

t (s)

Figure 2. Storage modulus (G') as a function of time for casein gels made by acidification and rennet action; rennet concentration and pH indicated. Temperature 30 °C [28].

fraction of protein present. This lower φ determines the size of these clusters in the final network [6,22]. Incorporation of protein in the gel network later on only leads to a more dense network and so to a higher D [28]. The protein strands will become thicker.

For gels built from protein particles those fuse together after some ageing, such as casein particles in milk gels, the building blocks of the fractal clusters cannot be considered anymore as being undeformable under the forces acting on the strands. This will strongly complicate the analysis of the energy term in equations 7 and 8. An alternative is to assign a certain modulus to the strands of aggregated protein particles which is determined by the stiffness of the particles and of their mutual links. For gels consisting of cylindrical strands with an average radius a and an effective Young's modulus, E the following expression for the shear modulus of the gel has been derived [4, 10, 13]:

$$G = NC\pi\, a E \tag{12}$$

3.CONCLUSIONS

The concept that on aggregation of protein particles clusters are formed with a fractal structure and ultimately leading to a gel network built of fractal clusters, has led to a much better understanding of the structure and rheology of gels formed by aggregated protein particles. For instance, several relatively simple scaling relations have been derived relating rheological properties to the volume fraction of protein. However, the exact relations between rheological properties and gel structure depend on various, rather subtle effects. An important factor, that is often not enough appreciated, is that the structure of many gels is not a static but a dynamic property. Rearrangements of the structure during and after gel formation may have large effects on the resulting gel properties.

4. ACKNOWLEDGEMENT

The author gratefully acknowledge the many very valuable discussions with professor P. Walstra on the topic of protein gel formation

REFERENCES

1. B.B. Mandelbrot, The Fractal Geometry of Nature, Freeman, New York, 1983
2. F.Family and D.P. Landau (eds.), Kinetics of Aggregation and Gelation, North-Holland, Amsterdam, 1984.
3. L.G.B. Bremer, T. van Vliet and P. Walstra, J. Chem. Soc., Faraday Trans., 85 (1989) 3359.
4. L.G.B. Bremer, B.H. Bijsterbosch, R. Schrijvers, T. van Vliet and P. Walstra, Colloids and Surfaces, 51 (1990) 159.
5. R.R. Vreeker, L.L. Hoekstra, D.C. den Boer and W.G.M. Agterof, Food Hydrocolloids, 6 (1992) 423.
6. M. Verheul and S.P.F.M Roefs, Food Hydrocolloids, 22 (1998), 17.
7. P. Walstra, in 'Food Rheology and Structure' , E.J. Windhab and B. Wolf (eds.), Proc. 1st Int. Symp. on Food Rheology and Structure, Vincentz Verlag, Hannover, 1997, 107.
8. L.G.B. Bremer, P. Walstra and T. van Vliet, Colloid Surf. A, 99 (1995), 121.
9. T. van Vliet and W. Kloek, in 'Food Rheology and Structure' , E.J. Windhab and B. Wolf (eds.), Proc. 1st Int. Symp. on Food Rheology and Structure, Vincentz Verlag, Hannover, 1997, 101.
10. T. van Vliet, in 'Food Emulsions and Foams, Interfaces, Interactions and Stability' E. Dickinson and J.M.Rodriquez Patino (eds.) Royal Society of Chemistry, Cambridge, in press.
11. S.P.F.M. Roefs, A.E.A. de Groot-Mostert and T. van Vliet, Colloids and Surfaces, 50 (1990) 141.
12. H.J.M. van Dijk and P. Walstra, J. Colloid Interface Sci., 40 (1986) 3.
13. L.G.B. Bremer and T. van Vliet, Rheologica acta, 30 (1991) 98.
14. J.R. Mitchel, J. Texture Studies, 7 (1976) 313.
15. F. van Kleef, J. Boskamp and M. van den Tempel, Biopolymers, 17 (1978) 225.
16. J.R. Mitchel, J. Texture Studies, 11 (1980) 315.
17. W.D. Brown and R.C. Ball, J. Phys. A., 18 (1985) L517.
18. J.M.S. Renkema and T. van Vliet to be published
19. L.G.B. Bremer, Fractal aggregation in relation to formation and properties of particle gels, PhD thesis, Wageningen Agricultural University, Wageningen, The Netherlands, 1992.
20. P. Zoon, T. van Vliet and P. Walstra, Neth. Milk Dairy J., 42 (1988) 249.
21. M. Stading, M. Langton and A-M Hermansson, Food Hydrocolloids, 7 (1993) 195.
22. M. Verheul, S.P.F.M Roefs, J. Mellema and C.G. de Kruif, Langmuir in press.

23. T. van Vliet, J.A. Lucey, K. Grolle and P. Walstra, in 'Food Colloids; Proteins, Lipids and Polysaccharides' E. Dickinson and B. Bergenståhl eds. Royal society of Chemistry, spec. publ. no. 192, Cambridge,1997, 335.

24. J.A. Lucey, T. van Vliet, K. Grolle, T. Geurts and P. Walstra, Int. Dairy J., 7 (1997) 389.

25. W.C. Young, Roark's formulas for stress & strain, 6th Ed. Mc. Graw-Hill, New York,1989.

26. T. van Vliet, H.J.M. van Dijk, P. Zoon and P. Walstra, Colloid and Polymer Sci., 269 (1991) 620.

27. S.P.F.M. Roefs, T. van Vliet, H.C.J.M. van den Bijgaart, A.E.A. de Groot-Mostert and P. Walstra, Neth. Milk Dairy J., 54 (1990) 159.

28. M. Mellema and T. van Vliet, to be published.

29. W.Kloek, Mechanical properties of fats in relation to their crystallization, PhD thesis, Wageningen Agricultural University, Wageningen, The Netherlands, 1998.

HYDROCOLLOIDS – PART 1
Edited by K. Nishinari
2000 Elsevier Science B.V.

379

Heat set proteins – Models for the concentration and temperature dependence of the gelation time

Simon B Ross-Murphy[a] and Atsumi Tobitani[b]

[a]Biopolymers Group, Division of Life Sciences, King's College London, Campden Hill Road, London W8 7AH, UK

[b]Snow Brand Milk Products Co. Ltd. Technical Research Institute, 1-1-2, Minamidai, Kawagoe, Saitama 350, JAPAN

Globular protein gels have been extensively investigated over many years, although the great majority of studies have used rather crude and/or mixed samples of protein. Moreover, most previous work has concentrated on examining structural and rheological properties of fully cured gels. In the present paper we discuss heat-induced gelation of the globular protein bovine serum albumin (BSA) not far from critical gel conditions together with aspects of kinetic gelation theory. We will concentrate on a semi-empirical description of the parameters that alter the gelation or gel time, t_c and employ a model which describes both the effect of temperature and polymer concentration on t_c.

In the experimental work gel formation was achieved by isothermal heating, and the changes in rheological properties during the gelation process monitored by dynamic shear rheometry as a function of time. The measurements are carried out at different temperatures and protein concentrations to clarify their effects on the gelation. The results will be displayed in various diagrams (including a new form of sol/gel state diagram, which one of us (SBRM) has termed a Tobitani diagram) and discussed in terms of both phase behaviour and gelation kinetics. The resemblance between the new kinetic state diagram and the equilibrium phase diagram of a two phase upper critical solution temperature (UCST) system will be discussed.

1. BIOPOLYMER GELS

A fundamental distinction can be made between systems which, when form gels from essentially "disordered" random coil biopolymers and those systems which form, and are maintained in an essentially ordered state, such as gels formed from globular proteins.[1,2]

1.1 Gel Networks Formed From Initially Disordered Polymers

The term gel (attributed to Thomas Graham) originates from gelatin(e), a whole class of materials derived by hydrolytic degradation of collagen, the major structural component of many animal forms of life. The fundamental structural unit of intact collagen is the tropocollagen rod. This is a triple helical structure composed of three separate polypeptide chains. Gelatins normally dissolve in water ($>\sim40^{\circ}$C). On recooling, transparent gels are formed (provided the concentration is greater than some critical concentration, C_0, typically 0.4 to 1.0%). The gels contain extended physical cross-links or "junction zones" formed by a partial reversion to "ordered" triple helical collagen-like sequences, separated along the chain contour by peptide residues in the "disordered" conformation.

Marine polysaccharide forming gels include (ι- and κ-) carrageenan, agar(ose), and the alginates. Much evidence suggests that the first two of these form thermoreversible gels by an extension of the gelatin mechanism. Pectin and starch gels are the most important members of the gelling plant polysaccharides. Pectins of low degree of esterification gel with divalent ions. The more esterified materials gel under conditions of low pH and decreased water activity, when intermolecular electrostatic repulsions are reduced; in this case the junction zones are thermoreversible at say 40°C.

Starch consists of two different polysaccharides, one, predominantly linear, being amylose and the other branched, but otherwise structurally analogous, amylopectin. We preclude discussion, since starch gels are described in detail elsewhere in this volume. However, both amylose and amylopectin solutions can gel, and much of the subtlety of starch behaviour is undoubtedly related to the limited compatibility, and mutual gelation of the two polymeric components.

A number of polysaccharides of interest occur outside the cells of microbes, either covalently attached or secreted into the growth media. These are the microbial exopolysaccharides, and a number of these have been described. At the moment, on a volume production basis, the two major members of this group are two anionic polysaccharides, gellan and xanthan. Gellan gels in the presence of multivalent cations like the gelling carrageenans.

1.2 Gel Networks From Ordered Globular And Rod-Like Biopolymers

Almost all of this group of materials are formed from animal and vegetable proteins. In some cases the resultant gels involve, at least partial, denaturation which does not occur in vivo (heat set proteins) whilst in others the biological function of the protein is to gel under certain physiological conditions (blood clotting = fibrin network formation). In the present article we will consider only the heat-set gels formed from globular proteins just by heating above say 5% concentration.[3] Perhaps the most familiar example of this is the boiling of an egg (essentially gelation of ovalbumin), but heating e.g. serum albumins (SA), chymotrypsin, globins, whey and vegetable (soy) protein solutions will form similar gels.

Until the last 10 years or so, much of the published work on such heat set gels has been restricted to bovine serum albumin, BSA, since pure samples of this can be obtained relatively cheaply. Even now, a distinction has to be made between detailed studies on pure systems, and the comparatively large number of published papers on rather crude mixed samples, such as "whey protein isolate".

Nevertheless, some generalities can be established, which apply both to single and mixed component heat set protein systems. For systems heated to not much higher than the protein unfolding temperature (~70°C), many studies, using electron microscopy, X-ray scattering and spectroscopic techniques, have confirmed that the globular conformation of the native protein is only slightly perturbed. It appears that denaturation partially disrupts the protein without modifying the overall shape very significantly, but exposes some intra-globular hydrophobic residues. At low enough concentrations these can refold all but reversibly, but above a certain concentration there is competition between intra- and intermolecular β-sheet formation. If the latter predominates, gels are formed which are fibrillar, and whose fibrils are approximately 1-2 times the width of the original globule, as illustrated in Fig.1

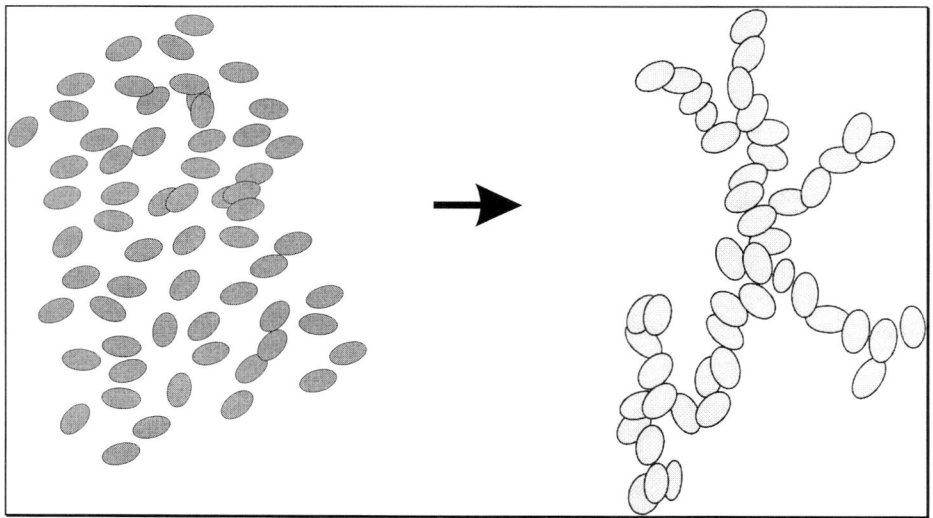

Figure 1 Corpuscular gel formation illustrated for heat-set protein gels – the "hot" or "molten globule" mechanism

The balance between linear and branched chain growth depends upon both pH and ionic strength I, and a range of "gels" can be prepared ranging from transparent through translucent to biphasic gels and turbid coagulates without macroscopic strength. Experimental "phase" diagrams can be constructed showing the boundary between sols, clear gels and turbid gels as a function of protein concentration, pH and I. In fact boiled "egg white" is only white because of the concentration of salt in the ovalbumin solution, suitable dialysis can produce a transparent "white". Prolonged heating of protein gels to >85°C produces a more drastic change, and some intermolecular covalent disulphide bonds are formed; these gels can no longer be regarded as merely physical networks.

Protein gels can also be produced by other means of denaturation than simply heat, including treatment with non-solvents (alcohols), or "hydrogen bond breakers" (urea). Subjecting ovalbumin solutions to both urea and heat (>80°C) apparently produces very substantial peptide unfolding. Finally, gels can be produced enzymatically, as in the antibody-antigen reaction and in the, effectively spontaneous, transformation of fibrinogen into fibrin gels (blood clot formation).

2. THE GEL TIME

The terms "gel time" and/or "gelation time" are familiar to most works in the area of polymer gels and networks. However, defining these terms more precisely, and making rigorous rather than purely empirical rules for the dependence of gel time on, for example, polymer concentration are not aspects that have been examined in sufficient detail. Here we define the gel time to be the time at which the gel point conversion, p_c is reached during a kinetic gelation process. For chemical gels, when critical conversion can be estimated reasonably well from the stoichiometry and branching functionality of the precursor species, then gel time can be estimated reasonably well, by investigating the kinetic order of the cross-linking process. For physical gels such as those from globular proteins, where the gel time has, arguably, more significance, the situation is more complex.

From the viewpoint of theory we can make some general remarks. For example, below the critical gel(ation) concentration, C_0, then our gel time t_c will be infinite. As the initial concentration of reactant species (here assumed to be proportional to polymer concentration) C, increases we would expect t_c to decrease. For heat set gels, the gel time decreases with increasing temperature.

The experimental determination of gel time is not always easy. In ideal circumstances, the Winter-Chambon, common frequency dependent power law behaviour for G' and G", can be applied.[4] However, for some systems this approach is not applicable. For example with certain biopolymer gel systems, G' appears to be > G" well before gelation, and no cross-over or power law region is seen.[5] For this reason, other more pragmatic approaches have been applied. These include identifying the time when G' has increased to a value greater than the experimental noise level, or even the time when a maximum in G" appears.

Here we represent, albeit briefly, measurements on the temperature and concentration dependence of gel time for the heat-set gelation of the globular protein BSA.[6,7] These clearly show that previous attempts to factorise the gelation time into separate temperature and concentration dependencies are flawed. Indeed there appears to be a strong interaction between these factors, this observation has considerable significance, and helps us to develop an empirical model, which can lead to a useful generalisation of behaviours.

2.1. The Gel Time-Ideal Theory

As has been know since the early 1940s, following Flory and Stockmayer, any gelation process must have a critical conversion value, see for example refs. [1,2 and 4]. If we accept, as mentioned above that t_c is the critical gelation time during a kinetic gelation process when the critical or percolation threshold conversion, p_c, is reached, it is easy to derive the concentration dependence of t_c. It may also allows us to examine the temperature dependence for a single Arrhenius process, in the following manner.

For the simplest second order kinetic (irreversible) gelation process we can write that the time required to reach a given degree of chemical conversion is proportional to $(1/C)$.[8] More generally, for the (albeit unlikely) case of an mth order kinetic process, the corresponding time would be proportional to $(C)^{1-m}$.[9] Consequently if, as in the ideal random branching case, the gel conversion p_c is independent of the initial concentration of reactants, the original Flory assumption, this suggests that:

$$t_c \propto \frac{1}{C^{m-1}} \propto \frac{1}{C^n} \tag{1}$$

Since the value of the exponent m is to some extent arbitrary, in succeeding discussion, and in order to maintain consistency, we assume n=m-1, to give the form on the right hand side of Eqn.1. Although the derivation is different, the same result follows from the theory of Oakenfull and Scott [10] and also that of te Nijenhuis [11] some years earlier. All such approaches rely only on a simple kinetic law, with the assumption that the kinetic order is concentration independent and will not be altered provided $0 < p_c < 1$.

2.2. Complications

In practice the exact value of the percolation threshold conversion, p_c, is obviously very important. Most significantly p_c is not independent of C, as assumed above, because of "wastage" reactions such as cyclisation. Consequently p_c will tend to increase as C decreases, as the balance of intramolecular and intermolecular cross-linking reactions shifts towards ring formation. This was well known to Flory himself and is the basis of, for example, the Jacobson-Stockmayer theory for ring-chain competition (see, for example [12]). Indeed, at high dilutions no gel will be formed, even when the extent of reaction is driven towards unity.

For physical gels the situation is still more complicated. Here the relationship between concentration and conversion has to be derived on the basis of an apparent equilibrium between p and C. On the basis of such an equilibrium model, calculations by Clark, [13] have shown that the power law dependence of $1/t_c$ is much more pronounced close to C_0, while before this $1/t_c \sim C$. He concludes that the improved theory does not help significantly in improving the quality of fit to real data.

2.3. The Temperature Dependence Of Gel Time

Now we can ask how t_c, the gelation time, will depend upon temperature, T? Te Nijenhuis [14] fitted gelation time, (in his terminology the induction time) vs. temperature data for gelatin to a simple model, by plotting $t_c^{-1/2}$ against T. This plot, which corresponds to Eqn. 2 below, with q = -2, was quite linear. The generalised form with exponent q was introduced by one of the present authors [15] on the basis of a simple scaling argument. It relies upon the existence of a critical temperature T_c below which (at least for heat set gels) no gel can be formed.

$$t_c \propto \left(1 - \left[\frac{T}{T_c} \right] \right)^q \tag{2}$$

2.4. Concentration and Temperature Dependence

We now describe work by the present authors on the concentration and temperature dependence of the gelation time for heat set gels of the globular protein BSA. For experimental details we refer the reader to these papers [6,7].
By assuming an Arrhenius form for the dependence of temperature and the simple kinetic behaviour for concentration we have already discussed, the following equation for t_c was derived:

$$t_c \propto \frac{1}{C^n} \exp\left(\frac{E_a}{RT} \right) \tag{3}$$

Here E_a is an activation energy for gelation. We note that there is no critical temperature T_c here, so we are assuming a gel can form at any temperature, presumably albeit slowly unless $T >\sim T_c$. It was also established, that for the samples investigated, in particular heat set bovine serum albumin (BSA) gels, experimental errors of t_c did not follow a Gaussian distribution, while those of $\ln (t_c)$ did so. Therefore, by taking natural logarithms and introducing coefficients a_1, a_2, a_3 and a_4, an equation could be written where:

$$\ln(t_c) = a_1 + a_2 \ln(C) + a_3 T + a_4 \ln(C) * T \tag{4}$$

The term in ln(C)*T with coefficient a_4 is necessary because it is clear from examination of the data alone and Figure 2, that there is an interaction between temperature and concentration terms. In other words, the lines in Figure 2 are not parallel. Thus, the interaction between C and T is described by an additional term, although we have not, so far, found a clear explanation of physical meaning for this. One explanation, certainly pertinent to cold-set gels such as gelatin, is that the minimum length of a junction zone will tend to be temperature dependent. Although this model was derived by a rather simple procedure, it can explain experimental data very satisfactorily, as is demonstrated below. It should be emphasised that the model can successfully estimate gelation time in terms of both the polymer concentration C and temperature T; no previous model could achieve this.

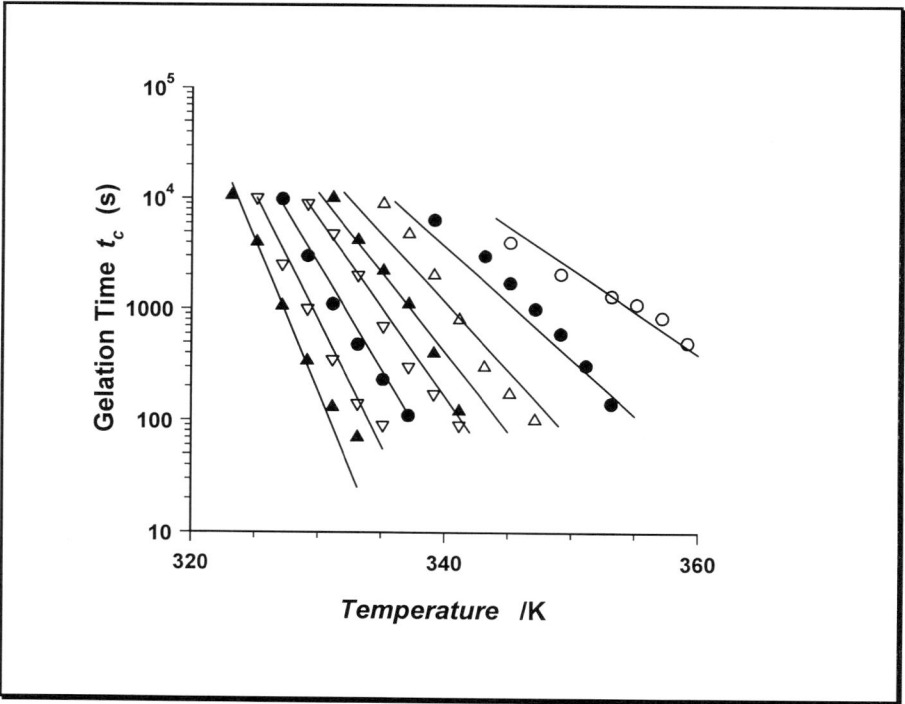

Figure 2 Gelation time versus absolute temperature for different protein concentrations. Solid lines calculated from Eqn. 4. The protein concentrations are, from right to left, respectively, 6% , 7% , 8% , 9% , 10% , 12% , 15% and 20% w/w BSA

2.5. Gelation State (Tobitani) Diagram

By interpolating the geltime determined at a range of concentrations it is possible to construct a phase diagram (strictly a state diagram) of gelation temperature versus concentration (Fig.3). In recent times, one of the present authors (SBRM) has begun referring to this novel presentation as a "Tobitani diagram". This clearly illustrates the qualitative likeness between thermal gelation of globular proteins and the upper critical solution temperature (LCST) type phase diagram seen for certain fluid (and polymer

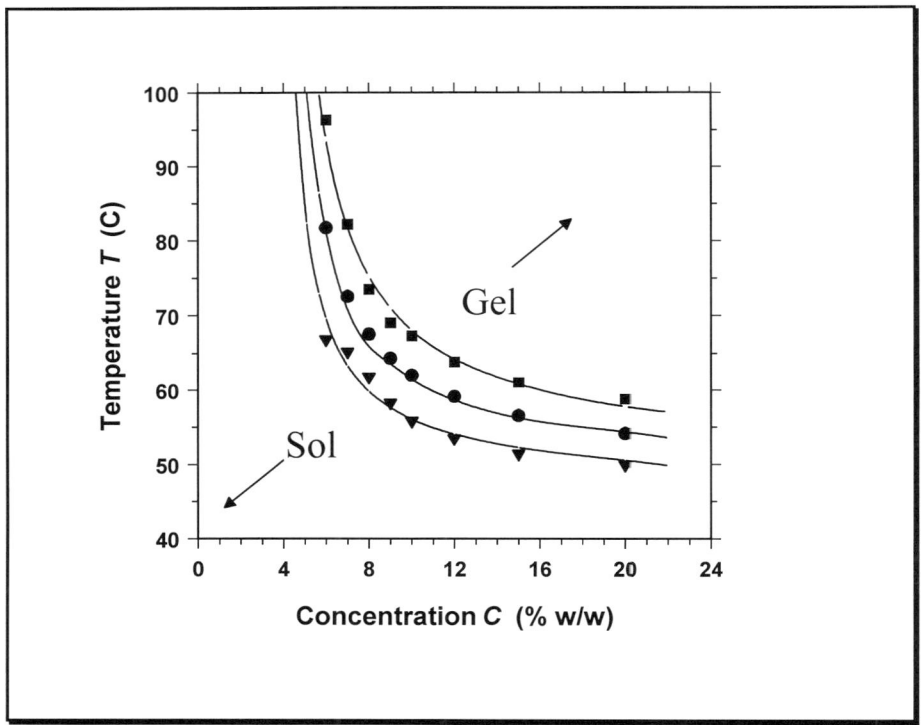

Figure 3 Temperature versus concentration ("Tobitani") state diagram for BSA illustrating the sol /gel boundary. Symbols represent threshold times for gelation (squares, 100s; circles, 1000s; triangle down, 10000s); curves calculated from Eq. 4

solution) systems. We note that, in this figure, and in Figure 2 the temperature range goes above 85^{0}C, so there is a possibility of extra processes, such as disulphide cross-linking occurring. The evidence of this from Figure 2 is, however, equivocal. We await theoretical developments to confirm whether or not the overall apparent behavioural similarity can be made formal. The approaches introduced by F. Tanaka over the last 5 years or so, may prove very useful here.[16,17]

3. CONCLUSIONS

Further work still needs to be performed in this area. Although a considerable number of papers have appeared on the characterisation of globular protein gels, many of these have considered only rather crude samples. We hope the present article will stimulate further interest amongst workers on other hydrocolloid systems to become interested in kinetic gelation measurements.

REFERENCES

1. A.H. Clark and S.B. Ross-Murphy, *Adv. Polym. Sci.* **83** (1987) 57
2. .K. te Nijenhuis, *Adv. Polym. Sci.* **130** (1997) 1
3. A.H. Clark and C.D. Lee-Tuffnell, in *Functional Properties of Food Macromolecules* (Eds. J.R. Mitchell & D.A. Ledward) Elsevier Applied Science Publishers, London 1986 p.203
4. H. H. Winter and F. Chambon, *Journal of Rheology*, **30** (1986)367
5. S.B. Ross-Murphy, *Rheologica Acta*, **30**, (1991) 401
6. A.Tobitani and S.B. Ross-Murphy, *Macromolecules* **30** (1997) 4845
7. A.Tobitani and S.B. Ross-Murphy, *Macromolecules* **30** (1997) 4855
8. P.W. Atkins, *Physical Chemistry 5th Ed.*, Oxford University Press, Oxford, UK 1984
9. A. Tobitani and S.B. Ross-Murphy in *Wiley Polymer Networks Review Vol. 1* Editors K. te Nijenhuis and W. Mijs, John Wiley, Chichester, Sussex, UK 1997 p.39
10. D. Oakenfull and A. Scott, *J. Food Sci.* **49** (1984) 1093
11. K. te Nijenhuis, *Colloid and Polym. Sci.*, **259** (1981) 522
12. S.B. Ross-Murphy and R.F.T. Stepto in *Cyclic Polymers*, J. A. Semlyen, Ed. Elsevier Applied Science, Barking, 1986, p. 320
13. A.H. Clark, *Polymer Gels and Networks.* **1** (1993)139
14. K. te Nijenhuis, Doctoral Thesis, University of Delft, Netherlands, 1979
15. S.B. Ross-Murphy, *Carbohydrate. Polymers* **14** (1991) 281
16. F. Tanaka and W.H. Stockmayer, *Macromolecules* **27** (1994) 3943
17. F. Tanaka and K. Nishinari, *Macromolecules* **29** (1996) 3625

HYDROCOLLOIDS – PART 1
Edited by K. Nishinari
2000 Elsevier Science B.V.

389

Dynamic light scattering study of supramolecular structure of annelid extracellular multi-subunit hemoglobin

K. Kubota[a], M. Yamaki[b]*, S. Ebina[b] and T. Gotoh[c]

[a]Faculty of Engineering, Gunma University, Kiryu, Gunma 376-8515, Japan,

[b]Nagayama Protein Array Project, ERATO, JRDC, 5-9-1 Tokodai, Tsukuba 300-2635, Japan,

[c]Faculty of Integrated Arts and Science, University of Tokushima, Tokushima 770-0814, Japan

*Present Address: Yeast Laboratory, National Food Research Institute, 2-1-2 Kan-nondai, Tsukuba, 305-8642, Japan

Hierarchic organization of annelid extracellular multi-subunit hemoglobin (because of its gigantic size, it is called giant Hb) which is composed of about 200 polypeptide chains (144 globin chains and about 36 non-heme chains called linkers) has long been mysterious why such a gigantic structure is formed orderly. The most important point in understanding the mysterious hierarchic organization of giant Hb is to know the role of linkers in holding together the entire two-tiered hexagonal form. The effects of addition of monosaccharides on the dissociation of giant Hb from the marine-worm *Perinereis aibuhitensis* were monitored using dynamic light scattering (DLS), transmission electron microscopy (TEM), and circular dichroism (CD) measurements. Changes in Stokes radius and more clearly the size distribution analysis of the Hb based on the DLS measurements showed that Hb preferentially dissociates into hexagonal units (called submultiples), which was consistent with the results of TEM and CD measurements. The results thus show that linkers specifically "clamp" submultiples together to organize the two-tiered form through carbohydrate gluing. Thus, a submultiple behaves like an ordinary protein, whereas the intact Hb behaves like a miniature supramolecular system. This clamp model is plausible because it inherently involves catastrophe of the molecular stoichiometry at the two-tiered hexagonal formation level because carbohydrates are under the post-translational regulation and therefore contain structural ambiguity and flexibility.

1. INTRODUCTION

Catastrophe in molecular stoichiometry is natural in the higher levels of the biological hierarchy: proteins - supramolecular systems - organelles - cells - tissues/organs - individuals. Such catastrophe (break-down of stoichiometry) eventually introduces structural ambiguity. The question is how and at which level the stoichiometry breaks (becomes catastrophic). Lectin-carbohydrate type binding in the evolution from proteins to supramolecular systems was first indicated recently in annelid extracellular multi-subunit hemoglobin (giant Hb) from a marine-worm and called carbohydrate-gluing [1,2]. This indication inherently implies an appearance of catastrophe in this gigantic protein because carbohydrates are under "post-translational" regulation and contain structural ambiguity. Moreover, ambiguity in the number of linkers exists, whereas stoichiometry strictly stands for the composition of globin chains in giant Hb of an earthworm [3-5].

Giant Hb is composed of numerous polypeptide chains, including globin chains (four types) and non-heme chains (two types, called linkers) [6]. The long history of research on giant Hb originated with the first discovery in 1933 by Svedberg of the enormous size of a red-pigmented protein in the blood of a worm [7]. Subsequent studies demonstrated another hierarchic structuring of these chains within Hb. Here, the nomenclature we use for these chains of a marine-worm is a, A, b, and B for the four globins and L1 and L2 for the two linkers [8,9]. The hierarchic structuring of giant Hb proceeds in steps: (i) a globin monomer (a) and a disulfide-bonded globin trimer (A-b-B) form a globin tetramer (a · A-b-B); (ii) three globin tetramers form a submultiple, $i.e.$, $3(a · A-b-B)$; and (iii) twelve submultiples and linkers form intact two-tiered hexagons, $i.e.$, 12 submultiples · nL_2 [6,8-11], where L_2 represents a linker dimer and n is around 18 for a marine-worm [2,11]. All three steps involve non-covalent structuring [12]. The final two-tiered hexagonal arrangement is called the "bracelet model" [6], which is based on the critical role of linker chains for the structuring. Suzuki and Riggs suggested that a repeated cysteine motif in the linkers plays a significant role in the structuring via electrostatic binding [13].

A surprising effect of carbohydrate gluing as a major contributor to the structuring mechanism of linker chains was discovered using transmission electron microscopy (TEM): a reversible dissociation of the Hb (from the marine-worm, $Perinereis\ aibuhitensis)$ into smaller units by the addition of monosaccharides, such as mannose or N-acetyl galactosamine (GalNAc) [1]. This effect was reinforced by other evidence that we have collected since then, including the presence of carbohydrates in linkers (L1 and L2 chains) and two types of globins (a and A chains), irreversible dissociation of the Hb due to deglycosylation, and sequential homology of linkers with wheat germ agglutinin, which is a plant lectin [1]. We also confirmed the specific binding of lectins to L1, L2, a, and A [2].

These findings suggest that carbohydrate gluing is the consequence of lectin-carbohydrate type binding in the Hb. More recently, we have localized L2 predominantly at the outer and inner boundaries between neighboring submultiples in the hexagonal form [14].

In the present study, we determined the structuring level at which carbohydrate gluing governs Hb. The TEM technique alone was inadequate for determining the structure of the components after the dissociation [1]. Due to its low sensitivity and more crucially to the strong tendency of the dissociated units to adhere to the chromatography matrix, molecular-sieve chromatography also is inadequate for detecting the expected changes in Hb. We then used dynamic light scattering (DLS) measurements in conjunction with TEM and circular dichroism (CD) measurements. Decay rate distribution of the correlation functions was examined relating to the particle size (Stokes radius) using Laplace inversion method, and the dissociation behavior of the giant Hb was traced. As a result, we could clarify carbohydrate-gluing as a clamping strategy for organizing submultiples together to arrange the intact two-tiered hexagonal form and thereby clarify that catastrophe occurs at this step.

2. EXPERIMENTAL

2.1 Materials

Giant Hb from the marine worm *Perinereis aibuhitensis* was used after purifying it by the sequential use of centrifugation, ammonium sulfate fractionation, gel-filtration, and ultracentrifugation as described previously [15]. Chemicals and monosaccharides used were all commercial materials [1,2,14].

2.2 DLS measurements

The DLS measurements of the giant Hb in 50 mM phosphate buffer (pH 7.3 and 9.0) in an 8-mm-path sample tube were done at 25 °C (±0.05) using a spectrometer (constructed in-house), the detail of which has been described elsewhere [16]. The typical concentration of the Hb was about 0.5 mg/ml. Because the Hb is red, the DLS measurements were done at a 632.8 nm-line using a He/Ne laser as the incident light source in vacuum. At the sample cell position, the power of the laser was about 10 mW. The sample solutions were centrifuged at 12,000 rpm for 2 h, filtered through a membrane filter (0.2-μm pore size) in a dust-free clean box, and immediately used in the DLS measurements. The autocorrelation function of the scattered light intensity, $G^{(2)}(\tau)$ with τ being the delay time, yields the initial decay rate $\bar{\Gamma}$, which corresponds to the transitional diffusion coefficient D as $D = \bar{\Gamma}/K^2$, where $K= (4\pi/\lambda)\sin(\theta/2)$, λ is the wavelength in the medium, and θ is the scattering angle. The autocorrelation functions of scattered light intensity, $G^{(2)}(\tau)$, obtained by the homodyne mode were analyzed by the cumulant expansion and CONTIN methods. $G^{(2)}(\tau)$ has the following form related to the normalized electric field correlation function, $g^{(1)}(\tau)$,

$$G^{(2)}(\tau) = A [1 + \beta \mid g^{(1)}(\tau) \mid ^2] \tag{1}$$

where A is a baseline and β is a machine constant relating to the coherence of detection. Generally $g^{(1)}(\tau)$ is expressed by the distribution function $G(\Gamma)$ of the decay rate Γ as

$$g^{(1)}(\tau) = \int G(\Gamma) \exp(-\Gamma\tau) \, d\Gamma \tag{2}$$

where $\int G(\Gamma)d\Gamma = 1$. That is, $g^{(1)}(\tau)$ is the Laplace transform of $G(\Gamma)$. $g^{(1)}(\tau)$ is expanded by the cumulant expansion,

$$g^{(1)}(\tau) = \exp(-\bar{\Gamma}\tau) [1 + (\mu_2/2!) \tau^2 - (\mu_3/3!) \tau^3 + \cdots] \tag{3}$$

$\bar{\Gamma}$ (average decay rate) and $\mu_2/\bar{\Gamma}^2$ (normalized variance) are related to $G(\Gamma)$ by

$$\bar{\Gamma} = \int \Gamma G(\Gamma)d\Gamma \tag{4}$$

$$\mu_2/\bar{\Gamma}^2 = \int [(\Gamma - \bar{\Gamma})^2/\bar{\Gamma}^2]G(\Gamma)d\Gamma \tag{5}$$

The third cumulant method was used to retrieve a reliable average decay rate $\bar{\Gamma}$ in this work. We assumed Hb to be globular, and thus obtained apparent values of Stokes radii R_s of the Hb according to the Einstein-Stokes equation, $R_s = k_BT/6\pi\eta \bar{D} = k_BT \, K^2/6\pi\eta \bar{\Gamma}$. k_B and η are the Boltzmann constant and the solvent viscosity, respectively. These Stokes radii were corrected for viscosity and refractive index of the monosaccharide solution in 50 mM phosphate buffer at 25 °C [16]. As shown by eq. 2, $g^{(1)}(\tau)$ is the Laplace transform of the decay rate distribution $G(\Gamma)$. Because respective decay rate corresponds to Stokes radius, the size distribution of the sample can be evaluated by the Laplace inversion of $g^{(1)}(\tau)$. Laplace inversion is known to be an ill-posed problem in nature mathematically, and various methods have been proposed to overcome such an ill-posed problem. CONTIN 2DP (ALV-CONTIN) method was used in this work to examine the size distribution.

2.3 Hb observation using TEM

We used a previously described TEM technique in which a drop of the Hb solution was first placed on a carbon film and then stained with uranyl acetate [14].

2.4 CD measurements

The CD measurements of the Hb in 50 mM phosphate buffer, pH 7.2 (0.65 mg/ml) were done at 25 °C in the absence and presence of monosaccharide using a 2-mm light path cell with a CD spectrophotometer (J-720W model, JASCO).

3. RESULTS

Although the TEM results show the dissociating process of Hb as a function of the monosaccharide concentration, the TEM alone was inadequate for determining the structure of the components after the dissociation [1]. We took DLS measurements. After confirming that there was no dependence of the Stokes radius on scattering angle (from 30 to 90 degrees), we used a scattering angle of 30 degrees in the DLS measurements. The results for the Γ-distribution show that the original Hb was monodisperse; the Stokes radius was 13.9 nm, which coincides well with the diameter of the intact Hb (29.0 $\pm$ 0.2 nm) [1]. In preliminary experiments using 50 mM of either GalNAc or mannose, we discovered an approximate 1.5% increase in the averaged Stokes radius obtained from $\overline{D}$. Although the change of the averaged Stokes radius with the addition of monosaccharides is not large enough, such a change was observed to be very systematic against the concentration of monosaccharide and was reproducible enough. Subsequent time-course experiments with various concentrations of either GalNAc or mannose showed that the change in the Hb radius at low monosaccharide concentrations was unstable for the first hours, after which it reached a constant value. A typical profile for the structural transition as a function of the change in Stoke radius shows that at a low concentration range the Hb initially increased in radius, reaching a peak at ~50 and ~100 mM for GalNAc and mannose, respectively, and then decreased. The corresponding TEM micrographs show that as the monosaccharide concentration increased, the Hb first swelled by loosening the hexagonal texture and then partially dissociated into smaller units. The stronger effect of GalNAc than mannose seen in the profile confirmed our previous result by TEM. The profile was predominantly large-sized components, thereby making it difficult to detect smaller-sized components of the Hb from the results of the averaged radius alone because the scattered intensity is proportional to (number) x (radius)^6 for spherical particles [17].

Therefore we determined the size distribution of the components from Hb based on the electric-field correlation function $g^1(\tau)$ in order to clarify the behavior of each component in the presence of monosaccharides. Using the Laplace inversion with a non-negative constraint, i.e., $g^{(1)}(\tau) = \int \exp(-\Gamma\tau) \, G(\Gamma) \, d\Gamma$, we calculated $G(\Gamma)$, which is the distribution function of the decay rate Γ of the components. In the DLS measurements, translational diffusion of the Hb was the only factor governing the fluctuation of the scattered intensity of the samples studied, because no other mode (e.g., rotational or intramolecular motions) exists in the present case that can cause a fluctuation of scattered light. Therefore, the Γ distribution corresponds to the size distribution of the Hb according to the Einstein-Stokes equation. Unlike the experimental data for the control samples (i.e., with no monosaccharides) and for the samples with GalNAc, the data for the samples with mannose were difficult to analyze due to relatively large concentration of mannose. The empirical Γ was corrected for solvent viscosity and refractive index of the respective monosaccharide in 50 mM phosphate buffer (pH 7.3), and called Γ_{corr}. The Γ_{corr} distribution is shown in Fig. 1, where $\Gamma G(\Gamma)$ instead of $G(\Gamma)$ is used because a logarithmic scale is more convenient for showing the intensity of the respective mode. Note that $\Gamma G(\Gamma) \, d(\log \Gamma_{corr}) = G(\Gamma) \, d\Gamma_{corr}$ was proportional to the scattered light intensity of the decay mode between Γ_{corr} and

$\Gamma_{corr} + d\Gamma_{corr}$.

Without GalNAc at pH 7.3, the Hb exhibited a sharp distribution of its size centered at $\Gamma_{corr} \sim 800 \text{ s}^{-1}$, which corresponded to the size of the intact Hb ($R_s \sim 13.9$ nm). In the presence of 50 mM GalNAc, the Hb had a broader distribution. In the presence of 150 mM GalNAc, the Hb had two distinct distributions, one corresponding to the swollen form (namely, the central value of log Γ_{corr} was $\sim 700 \text{ s}^{-1}$) and therefore a slightly broader distribution than that for the intact form, and the other consisting of smaller units (the central value log Γ_{corr} was $\sim 1700 \text{ s}^{-1}$) corresponding to submultiples (Stokes radius is ~ 6.5 nm). Thus, exposure of the Hb to GalNAc at pH 7.3 caused partial dissociation of the Hb

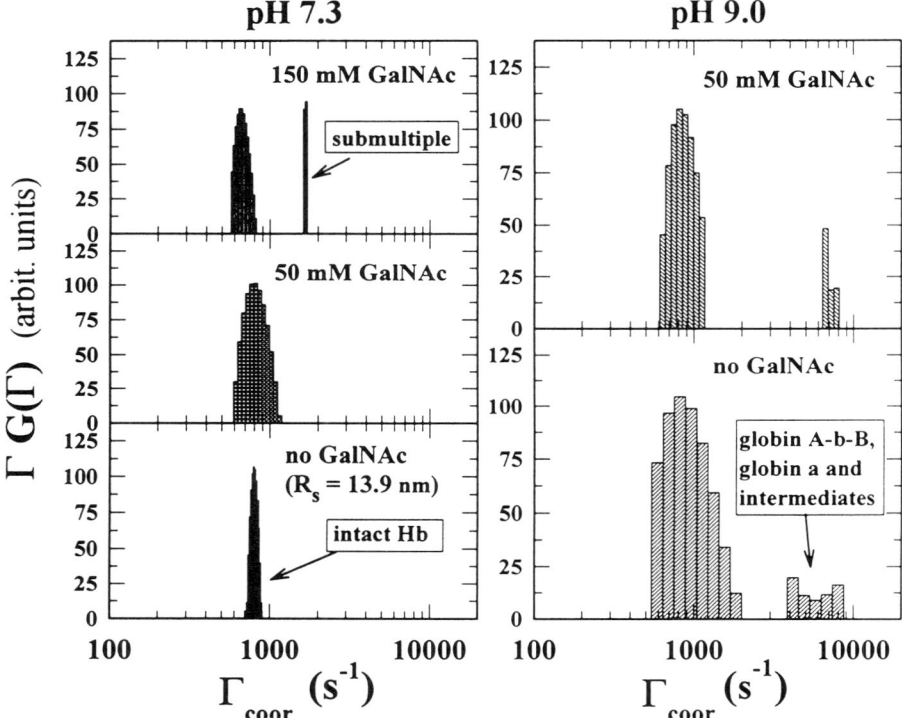

Figure 1. Decay rate distribution of *Perinereis* Hb at various GalNAc concentrations and pH 7.3 and 9.0 obtained by DLS measurements. Distributions are represented by $\Gamma G(\Gamma)$ vs. Γ_{corr}. The Γ distribution was determined by the least-square-fit of the correlation function using a non-negatively constrained Laplace inversion method (ALV-CONTIN). Because there are no decay modes of the internal motions other than the translational diffusion detectable by DLS, Γ simply relates to the Stokes radius, and therefore corresponds to the size distribution of Hb.

into submultiples. Importantly, there was no sign for further dissociation of submultiples into smaller units. The extensive dissociation due only to the exposure to pH 9.0 in the absence of monosaccharide was clearly seen in the appearance of two broad distributions: one corresponded to a mixture of the swollen intact form and submultiples and another might be a mixture of globin A-b-B, globin a, and intermediates between submultiples and these globins. With 50 mM GalNAc at pH 9.0, the Hb components in both distributions displayed significant narrowing in the distribution. It showed GalNAc facilitated the dissociation of the submultiples and the intermediates into globin A-b-B and a at pH 9.0. In both profiles of pH 7.3 and 9.0, the swollen intact form of Hb seems to remain in some extent. However, the weight fraction of it is extensively reduced because the scattered intensity is proportional to (weight) x (radius)^3 and (number) x (radius)^6 as well [17]. For example, at 150 mM GalNAc and pH 7.3, only about 20 % weight (2 % number) of the total sample remains as the swollen intact form.

In order to make clear whether carbohydrate gluing also governed the helical conformation of globin chains, we also took far UV CD measurements for Hb in the presence of either GalNAc or mannose. The spectrum for the intact Hb showed two minima (at ~210 and ~222 nm), indicating a large amount of α-helices in the globin chains. Note that there is no report identifying the conformation of linkers. The adsorption by the monosaccharides made it difficult to detect the first minimum (namely, at ~210 nm), whereas the second minimum (namely, at ~222 nm) was clearly seen even in the presence of a high concentration of monosaccharides, such as 150 mM of GalNAc or mannose. Although there might be a slight decrease in the ~222 nm minimum for both GalNAc and mannose, these results clearly show that neither GalNAc nor mannose caused significant unfolding of the helices. There was a slight change in the aromatic region of CD spectra in the presence of the monosaccharides, which suggests a conformational perturbation of aromatic residues.

4. DISCUSSION

The present results of DLS measurements, in agreement with the results of TEM micrographs and CD spectra, suggest the following points. The helical conformation of the globin chains is maintained, although the added monosaccharide causes the compact shape of the intact form Hb to swell and to dissociate largely into submultiples by perturbing the carbohydrate gluing. Furthermore, when either 150 mM GalNAc is added, only ~20 % weight and ~2 % number of the Hb remains as a swollen form, while the majority is dissociated into smaller units. GalNAc does not affect the ordinary inter-protein interactions, such as electrostatic binding [13] and hydrophobic interactions, but does affect the carbohydrate gluing selectively by the competitive inhibition. The GalNAc alone does not dissociate submultiples into smaller units. The GalNAc does not completely dissociate the Hb into submultiples. This simply implies that linkers bind submultiples together not only by carbohydrate gluing but also by the ordinary inter-protein interactions.

All of our results (DLS, CD, TEM) show that carbohydrate gluing clamps submultiples together. Note that this clamping is reversible [1]. Evidence of reassembly to submultiples in the absence of linker chains was first indicated by repeated procedures of assembly and reassembly by using a denaturant, where linker chains were gradually disappeared but the reassembly to submultiples steadily occurred [18]. We confirmed this reassembly by combining isolated globin A-b-B with isolated globin a (unpublished result). This reassembly also coincides with our recent localization of L2 chain predominantly at boundaries between neighboring submultiples [14]. We therefore propose a "clamp model" in which carbohydrate gluing of linker chains clamps submultiples together to form the two-tiered hexagonal arrangement. Such a model is critical because the arrangement is regulated by the "post-translational" level. For another type of giant Hb (i.e., from the earthworm *Lumbricus terrestris*), Vinogradov *et al.* described an apparent paradox of molecular stoichiometry: a noticeable ambiguity in the number of linker chains, but uniformity in the number of globin chains [3,4]. This paradox, however, coincides with our clamp model because the model inherently introduces such a paradox into the structure. Other evidence of this paradox is the recent discovery by Zal *et al.* of a 47:53 ratio of the linker chains to globin chains for giant Hb from the deep marine-worm *Alvinella pompejana*, and a ~30:70 ratio for giant Hb from the ordinary marine-worm and earthworm [19]. Such an evidence suggests that no comprehensive rule exists for the stoichiometry between linker chains and globin chains. Again, this is consistent to the clamp model. Catastrophe is natural in the molecular stoichiometry in the higher levels of the biological hierarchy: proteins - supramolecular systems - organelles - cells - tissues/organs – individuals. Our results show that giant Hb may provide an answer to the question of how the stoichiometry becomes catastrophic and at which level: at supramolecular system by carbohydrate gluing. This suggests that carbohydrate gluing may be a competent strategy for introducing an ambiguity in biological structures that is essential for many sophisticated functions of biological systems.

ACKNOWLEDGMENTS

We thank Mr. T. Takakuwa (JASCO) for the CD measurements, and Ms. K. Matsubara (ERATO) for helping in the preparation of giant Hb samples and Y. Iizuka (Gunma

University) for her assistance in the DLS measurements. We also thank Dr. K. Imai (Osaka University) for his critical discussion on this study. Part of this paper has been published in the journal of Arch. Biochem. Biophys [20].

REFERENCES

1. S. Ebina, K. Matsubara, K. Nagayama, M. Yamaki and T. Gotoh, Proc. Natl. Acad. Sci. USA, 92 (1995) 7367.
2. K. Matsubara, M. Yamaki, K. Nagayama, K. Imai, H. Ishii, T. Gotoh and S. Ebina, Biochim. Biophys. Acta 1290 (1996) 215.
3. P.D. Martin, A. R. Kuchumov, B. N. Green, R.W.A. Oliver, E.H. Braswell, J.S. Wall and S.N. Vinogradov, J. Mol. Biol. 255 (1996) 154.
4. J.N. Lamy, B.N. Green, A. Toulmond, J. S. Wall, R. E. Weber and S.N. Vinogradov, Chem. Rev. 96 (1996) 3113.
5. T. Suzuki and T. Gotoh, J. Mol. Biol. 190 (1986) 119.
6. S.N. Vinogradov, S.D. Lugo, M.G. Mainwarning, O.H. Kapp and A.V. Crewe, Proc. Natl. Acad. Sci. USA 83 (1986) 8034.
7. T. Svedberg, J. Biol. Chem. 103 (1933) 311.
8. T. Gotoh and T. Suzuki, Zool. Sci. 7 (1990) 1.
9. T. Suzuki, T. Takagi and T. Gotoh, J. Biol. Chem. 256 (1990) 12168.
10. T. Suzuki, O.H. Kapp and T. Gotoh, J. Biol. Chem. 263 (1988) 18524 .
11. B.N. Green, T. Suzuki, T. Gotoh, A.R. Kuchumov and S.N. Vinogradov, J. Biol. Chem. 270 (1995) 18209.
12. P.K. Sharma, A.R. Kuchumov, G. Chottard, P.D. Martin, J.S. Wall and S.N. Vinogradov, J. Biol. Chem. 271 (1996) 8754.
13. T. Suzuki and A.F. Riggs, J. Biol. Chem. 268 (1993) 13548.
14. M. Yamaki, K. Matsubara, A. Shibuya, T. Gotoh and S. Ebina, Archiv. Biochem. Biophys. 335 (1996) 23.
15. A. Tsuneshige, K. Imai, H. Hori, I. Tyuma and T. Gotoh, J. Biochem. 106 (1989) 406.
16. K. Kubota, H. Urabe, Y. Tominaga and S. Fujime, Macromolecules 17 (1984) 2096.
17. P. Stepanek, (1993) in Dynamic Light Scattering (Brown, W., ed.) pp. 177-241, Oxford Scientific Press, New York.
18. M.G. Mainwaring, S.D. Lugo, R.A. Fingal, O.H. Kapp, A,V. Crewe and S.N. Vinogradov, J. Biol. Chem. 261 (1986) 10899.
19. F. Zal, B.C. Green, F.H. Lallier and A. Toulmond   Biochemistry 36 (1997) 11777.
20. M. Yamaki, K. Kubota, K. Matsubara, S. Ebina and T. Gotoh, Arch. Biochem. Biophys. 355 (1998) 119.

HYDROCOLLOIDS – PART 1
Edited by K. Nishinari
2000 Elsevier Science B.V.

Gelation properties and interactions of fish proteins

Nazlin K. Howell

School of Biological Sciences, University of Surrey, Guildford, Surrey, GU2 5XH, U.K.

The rheological properties of fresh and frozen lean fish species namely cod and a fatty species mackerel were compared by small deformation viscoelastic measurements. The increase in G' values and protein aggregation observed during frozen storage were related to changes in the physical-chemical properties of the proteins as assessed by amino acid analysis, atomic force microscopy, differential scanning calorimetry and Raman spectroscopy. Biochemical mechanisms responsible for the changes have been discussed in relation to the role of formaldehyde, lipid oxidation products and ice crystal growth.

1. INTRODUCTION

Fish proteins exhibit strong gelling properties which are important in the manufacture of many products including surimi, comminuted products, ready prepared meals, canned and pickled products, extruded snack foods and petfoods. The rheological properties of fish are affected by processing including chilling, freezing temperature and time of storage. Frozen fish may be used as fillets or in the above products; thus optimum handling and processing of fish is vital for preserving texture. The objectives of this paper are:

- to compare the rheological properties of fresh and frozen lean and fatty fish;
- to relate the rheological changes to physico-chemical properties and
- to identify biochemical mechanisms responsible for texture changes.

1.1. Background

Many fish species such as cod and hake are reported to undergo toughening on frozen storage. Because of the commercial importance of these lean species, a number of studies have been undertaken in order to elucidate the mechanism of interactions leading to protein denaturation and aggregation. The mechanisms proposed are:

Protein denaturation arising from the ice crystal formation

Protein denaturation may be caused by ice crystals as a result of freezing. It is well established that the extent and size of ice crystals is affected by the rate of freezing and the temperature during frozen storage; small ice crystals are reported to form intracellularly at low temperatures of freezing whereas large ice crystals may be formed extracellularly at higher temperatures of freezing. The formation of ice crystals may lead to disruption of cells and release of enzymes and may be accompanied by dehydration which increases the solute concentration and divalent cation levels. The overall result is a decrease in protein extractability and myosin ATPase activity; increase in hydrophobicity, aggregation and loss of muscle functionality which we have identified in a number of species in our laboratory including cod, haddock, whiting, hake, mackerel and horse mackerel.

Protein-lipid interactions

Proteins can also complex with lipids and lipid oxidation products which may contribute to protein denaturation and aggregation. In a series of studies undertaken in our laboratory we have identified the transfer of free radicals from oxidised lipids and fatty fish oil to a number of amino acids e.g. lysine, arginine, leucine and proteins including ovalbumin, lysozyme and fish myosin by Electron Spin Resonance (ESR) spectroscopy. Aggregation of the amino acids and proteins followed radical formation in the proteins and was monitored by fluorescence increase [1-3]. Further studies in this field are being undertaken but a discussion of this aspect is beyond the scope of this paper.

Protein-formaldehyde interaction

Many bony fish (teleosts) particularly the gadoid family including cod, pollack, haddock, whiting, hake and cuss contain an osmoregulator trimethylamine oxide (TMAO). There are two main pathways for the breakdown of TMAO [4]. First, in the presence of psychrotrophic microrganisms on poor storage, TMAO can break down to trimethylamine (TMA). Thus the presence of TMA may be used as a spoilage indicator [5]. Second, on frozen storage, maximally between -5 and $-10^{\circ}C$, TMAO can break down into dimethylamine (DMA) and formaldehyde in the presence of an enzyme TMAOase and cofactor NADH. Alternatively, the decomposition of TMAO to TMA, DMA and formaldehyde can occur chemically by catabolites of cysteine initiated by Fe^{2+}. Formaldehyde vapour is considered to be toxic, although it has not been proven that the low level of formaldehyde, up to 200 ppm, produced in frozen fish is toxic. DMA can react with dietary nitrite to produce toxic nitroso-DMA compounds. Therefore the presence of these compounds is undesirable in fish products [6]. In addition, formaldehyde has been considered to form complexes with proteins and result in a tough rubbery texture which makes the product unpalatable. To date the interaction of formaldehyde has been tested in model systems [7,8] where complexes with proteins are confirmed. However, investigation on whole fish fillets has been lacking. We have recently examined differences between a gadoid fish producing formaldehyde e.g. cod and a gadoid fish producing negligible formaldehyde namely haddock [5,9]. Some of the findings from these studies are reported below.

2. MATERIALS AND METHODS

2.1. Materials

Lean fish species cod (*Gadus morhua*) and haddock (*Melanogrammus aeglefinus*) as well as a fatty fish Atlantic mackerel (*Scomber scombrus*) were supplied by Marine Laboratories,

Aberdeen as matched pairs of fillets with each half wrapped in polythene and stored at -10 or -30°C for about a year.

2.2. Methods

Formation of aggregates. Fish fillets were periodically extracted consecutively with sodium chloride; SDS and finally a mixture of SDS and β-mercaptoethanol to yield a soluble fraction and a precipitate from each extraction [9].

Rheology. Small deformation rheology was undertaken on a Rheometrics 100 constant stress rheometer using parralel plate geometry with 3mm gap; stress 1Pa and frequency 1 rad per second. A temperature sweep was made from 20 to 90°C and back to 20°C on homogenised fish samples [10].

Differential scanning calorimetry studies were undertaken on a Setaram DSC7 using about 0.8 mg of homogenised fish samples from 20 to 80°C [11,12].

Amino acids in soluble and insoluble extracts were measured on an Waters HPLC Pico Tag Analyser [9].

Atomic force microscopy was undertaken according to Badii *et al.*[13].

2.3. Results

a)

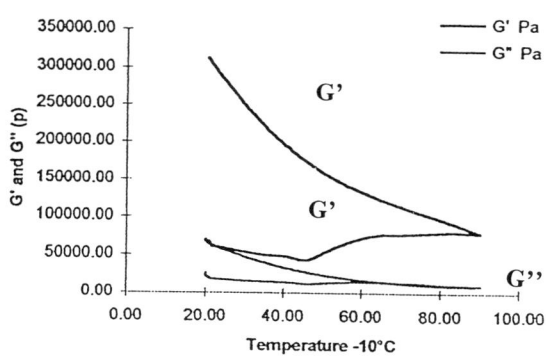

b)

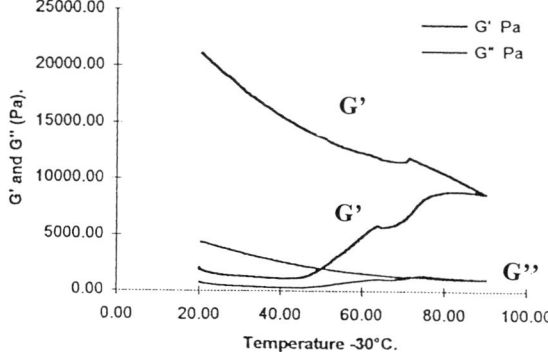

Fig. 1. Temperature sweep (20-90 and 90-20°C) for homogenised Atlantic mackerel (*Scomber scombrus*) fillets stored at a) -10°C and b) -30°C for five months [12].

Rheological studies indicated that there was an increase in G' values after heating and cooling of homogenised cod or haddock previously stored at -10°C for about 30 weeks compared with fillets stored at -30°C. Similar increases were also noted for a fatty species mackerel (Fig.1) indicating that the problem of toughening on frozen storage is common to most types of fish and is not only due to differences in formaldehyde level. Toughening may be due to ice crystal formation, dehydration during storage and the effect of lipid oxidation products on protein denaturation due to the presence of free radicals and interaction with secondary products. These aspects are currently being studied by the author.

Texture changes in fish were accompanied by protein aggregation. The aggregates formed, particularly those insoluble in SDS or β-mercaptoethanol, were greater in cod and haddock stored at -10°C compared with fish stored at -30°C. The nature and amount of aggregate produced did not differ for cod or haddock after one year [9]. Analysis of amino acids from the supernatant or aggregates indicated differences in the ratio of amino acids present in each fraction thus reflecting the differences in the proteins precipitated with time (Fig.2). These differences in amino acids in the supernatant and aggregate fractions were also confirmed by SDS-polyacrylamide gel electrophoresis indicating that myosin heavy chain denatured first followed by actin and myosin light chains. Both cod and haddock indicated similar results.

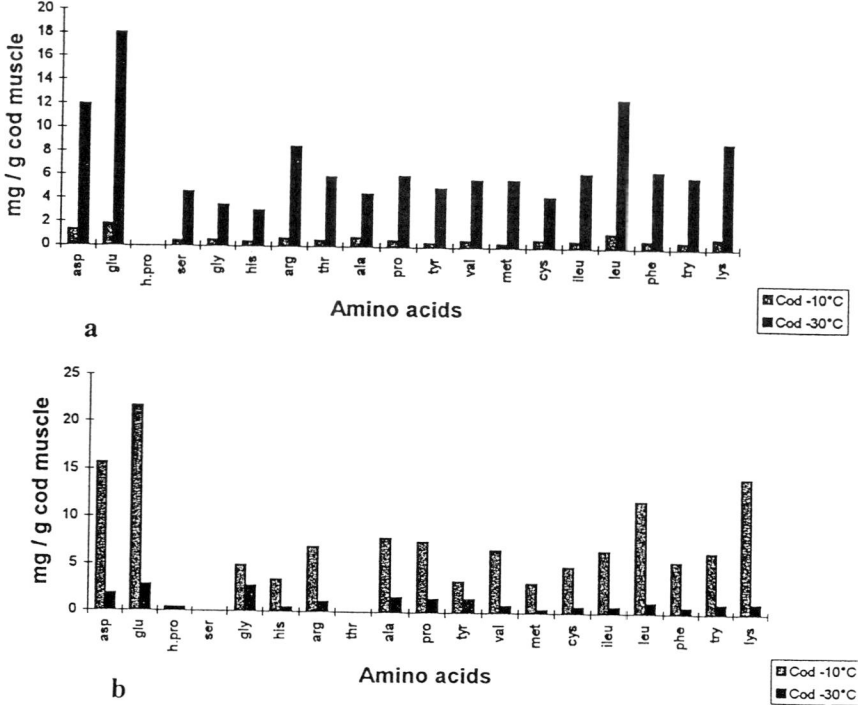

Fig.2. Concentration of amino acids in a) salt soluble extracts (S1) and b) insoluble protein (P1) obtained from cod stored at -10 and -30°C for one year [9].

Differential scanning calorimetry can provide a means of monitoring denaturation of proteins during frozen storage in whole fish and confirmed that the three transitions (range 25-40°C) observed for myosin in cod or haddock stored at -30°C were not observed for fish stored at -10°C. There was a reduction in the enthalpy and in the transition temperature. The transitions for actin (67.57 °C) and sarcoplasmic proteins (48.97°C) were affected to a lesser degree (Fig. 3).

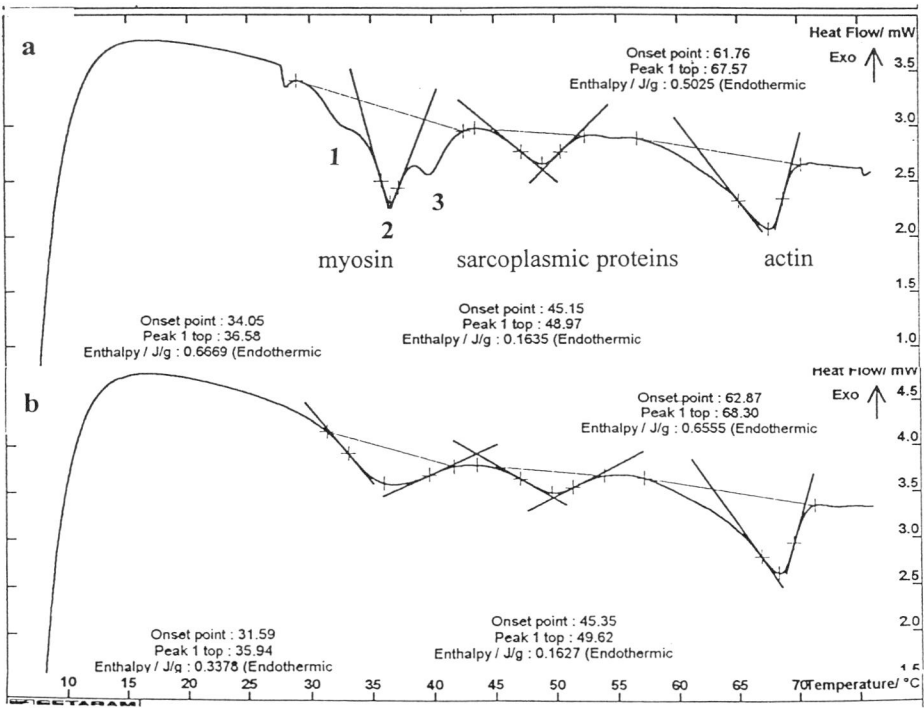

Fig.3. Differential scanning calorimetry (DSC) scans for cod stored for 65 weeks at a) -30°C and b) -10°C

A closer examination of cod and haddock fillets undertaken by atomic force microscopy revealed the presence of aggregates (white elevations) in the fish stored at -10°C compared with that stored at -30°C for eleven months where fewer aggregates were observed (Fig.4). The micrographs also indicated the spaces (dark patches) in between the muscle fibres in fish stored at -10°C which may be attributed to shrinkage of fibres during freezing and storage. Alternative methods of investigating aggregation have been undertaken by transmission electron microscopy in conjunction with immunocytochemistry using antibodies produced against the aggregates; this shows that fish tissue stored at -10°C indicated more binding of the gold-labelled antibodies compared with fish stored -30°C.

404

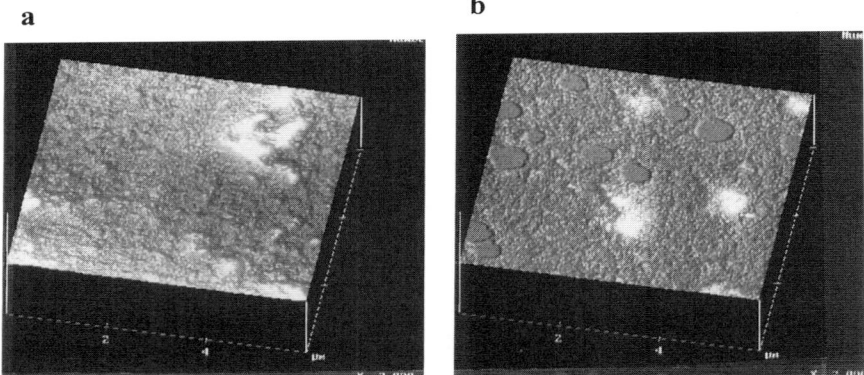

Fig. 4. Atomic force micrographs of cod flesh stored at a) -30°C and b)-10°C for one year [13]

Finally, aggregates which were insoluble in SDS and β-mercaptoethanol were investigated by FTIR-Raman spectroscopy to identify the nature of cross-links formed. Preliminary evidence suggests the involvement of various groups including changes in the helix (amide 111 region; aliphatic, aromatic, and carboxyl groups [14].

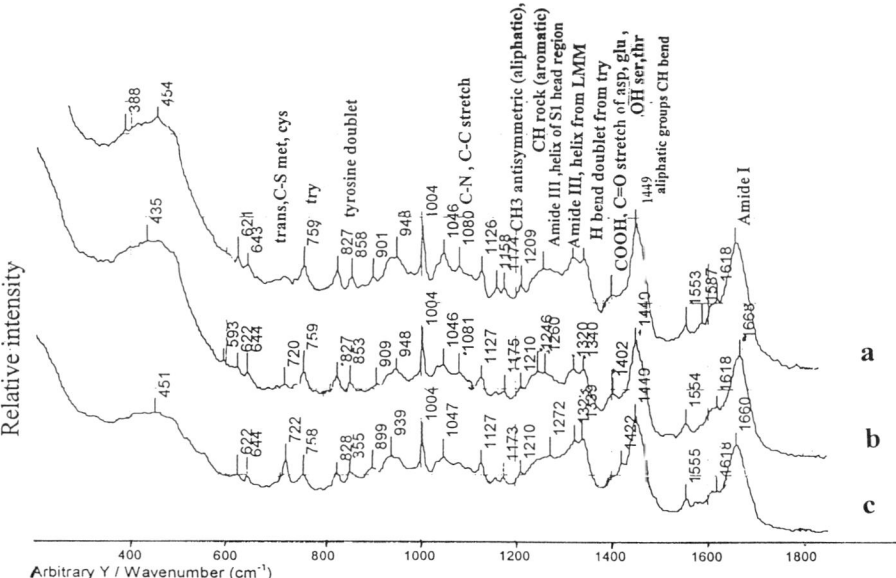

Fig. 5. FTIR-Raman spectra in the 400-2000 cm^{-1} region for (a) fresh cod, (b) cod stored at -30°C for 15 months and (c) cod stored at -10°C for 15 months. Bands indicating differences in the three samples are assigned [14].

4. CONCLUSIONS

Major differences in texture were observed by small deformation rheology for lean fish cod and haddock stored at -10°C compared with fish stored at -30°C. The toughening observed was evident to even a greater extent in fatty fish species mackerel. The texture changes were attributed to the denaturation of the myofibrillar proteins, particularly myosin, as assessed by differential scanning calorimetry. Protein denaturation led to a progressive increase in the formation of salt-insoluble, SDS-insoluble and SDS-β-mercaptoethanol insoluble aggregates with increased time and temperature of storage, indicating the presence of strong covalent linkages on prolonged storage of fish. These linkages as well as the secondary structure changes and hydrophobic groups were further identified by Raman spectroscopy. Differences in the amino acid composition of the aggregates were observed, confirming the presence of myosin in the aggregates. In addition to the isolation of aggregates in model systems, covalently-linked aggregates in whole fillets were also visualised by atomic force microscopy and electron microscopy/immunocytochemistry. However, after one year of storage, differences were not exhibited between cod, a fomaldehyde producing lean fish and haddock which is also a gadoid fish which produces negligible formaldehyde. Although formaldehyde may be important in the early stages of storage in the formation of the non-covalently bound aggregates other factors play a major role in the toughening of fish during frozen storage. The results strongly indicate that texture changes in frozen fish may be attributed primarily to changes in fish due to freezing conditions and also to protein-lipid interactions; both aspects are currently being investigated by the author.

REFERENCES

1. N.K. Howell and S. Saeed. In: Antioxidants in Human Health and Disease. (Ed. T.K. Basu, N.J. Temple and M.L.Garg). CAB International, Oxford, UK. (1999) 43-54.

2. S. Saeed and N.K. Howell In: Advances in Magnetic Resonance in Food Science. (Ed. P.S.Belton, B.P. Hill and G.Webb). The Royal Society of Chemistry, Cambridge, UK (1999) 135-143.

3. S. Saeed, S.A. Fawthrop and N.K. Howell, Electron spin resonance (ESR) study on free-radical transfer in fish lipid-protein interactions. (1999). Submitted.

4. I. Mackie Food Reviews International, **9** (1993) 575-610.

5. N.K. Howell, Y. Shavila,. M. Grootveld, and S. Williams, J. Sci Food and Agric. 72, (1996) 49-56.

6. N.K. Howell. In: Ingredient Interactions-Effects on Food Quality. (Ed. A. Gaonkar). Marcel Dekker, N.Y. (1995) 269-289.

7. A. Iyambo. The structure and physicochemical properties of frozen Atlantic cod (*Gadus morhua*) and Namibian hake (*Merluccius capensis*).PhD Thesis.University of Surrey. (1994).

8. M. Tejada, M. Careche, P. Torrejon, M.L. Del Mazo, M.T. Solas, M.L. Garcia and C Barba, J.Agric. Food Chem., 44 (1996) 3308-3314.

9. F. Badii, and N.K. Howell, Effect of frozen storage on the physico-chemical properties of cod and haddock proteins. (1999). Submitted.

10. G.L Friedli and N.K. Howell, Gelation properties of deamidated soluble wheat protein. Food Hydrocolloids, **10** (1996) 255-261

11. F. Badii, and N.K. Howell, Effect of frozen storage on the rheological and structural

properties of cod and haddock proteins, (1999). Submitted.

12. S. Saeed, Lipid oxidation mechanisms and lipid-protein interactions in frozen mackerel (*Scomber scombrus*). PhD Thesis, University of Surrey. U.K. (1998).

13. F. Badii, P. Zhdan, and N.K. Howell, Atomic force microscopy of frozen cod and haddock. (1999). Submitted.

14. F. Badii, H. Herman and N.K. Howell, Raman spectroscopy studies on frozen cod and haddock. (1999). Submitted.

5. ACKNOWLEDGEMENTS

The author would like to thank postdoctoral fellows Dr. Farah Badii and Dr. Suhur Saeed as well as colleagues at the University of Surrey including Dr. P. Zhdan; Dawn Chescoe, Dr. S. Hampton, Dr. H. Herman and N. Walker. This research project was financed by The Commission of the European Communities within the STD Framework Contract No. TS3*-CT94-0340 and FAIR Contract No CT-95.1111 both awarded to and co-ordinated by Dr. Nazlin K. Howell.

HYDROCOLLOIDS – PART 1
Edited by K. Nishinari
2000 Elsevier Science B.V.

Enhanced gel formation of whey proteins and egg white proteins by α-lactalbumin

S. Hayakawa[a] and M-L. Anang[b]

[a] Department of Biochemistry and Food Science, Faculty of Agriculture, Kagawa University, Miki-cho, Kagawa 761-0795, Japan.

[b] Faculty of Animal Science, University of Diponegoro, Semarang, Indonesia.

The interaction of α-lactalbumin (α-La) and β-lactoglobulin (β-Lg) in heat-induced gelation of protein mixture was examined at different concentrations of glutathione (GSH) and NaCl. Multiple regression analysis indicated that both GSH and NaCl significantly contributed to gelation. The promotion of gelation with GSH was assumed to result from GSH reacting with intramolecular disulfide bonds in α-La and β-Lg. It was postulated that α-La and β-Lg were interacting in the gelation of the mixed proteins, mainly through SH-disulfide bond interactions.

Gel strength of the mixed proteins of whey and egg white proteins was determined after heating, and the interaction of α-La and ovalbumin upon heating resulted in highly polymerized proteins. Gel strength of the mixture of 4% α-La and 4% ovalbumin was twice those of 8% ovalbumin, although modified α-La at Cys6-Cys120 (3SS α-La) had low enhancement effects on ovalbumin gelation. Competitive ELISA using monoclonal antibody to α-La showed decreased binding reactivities after heating α-La with ovalbumin at high protein concentrations. The decrease of total SH groups in the gel mixed with α-La was much larger than when mixed with 3SS α-La.

It is suggested that a specific disulfide bond Cys6-Cys120 in α-La contributed to interactions between ovalbumin and α-La in heat-induced gels.

1. INTRODUCTION

Gelation is an important function of proteins in food systems. Different proteins produce gels which vary in textural characteristics [1]. Ovalbumin is the most abundant protein in egg albumen that contributes gel properties [2,3]. Whey proteins are also widely used as ingredients in food products, primarily to form heat-induced gels [4,5]. β-Lg and BSA are the principal gelling proteins in whey [2,4,6], whereas α-La can not form gel alone [7,8]. Results showed α-La improved the gelation properties of other proteins in whey [7-10].

The monomeric α-La is stabilized by intramolecular 4 disulfide bonds [11] and these are probably important in heat-induced gelation, especially when α-La is heated with a protein containing sulfhydryl (SH) groups [7-9,12,13]. However, the contributions of disulfide bonds in α-La to the mechanisms of gelation are unknown.

Our current objective was to elucidate the contribution of disulfide bonds in α-La in enhancing the effects of α-La on heat-induced gelation of β-lactoglobulin or ovalbumin.

2. MATERIALS AND METHODS

2.1. Preparation of proteins.

α-La and β-Lg were fractionated from whey protein isolates by the ammonium sulfate precipitation. Ovalbumin was prepared from hen egg white by ammonium sulfate precipitation. 3SS α-La reduced at a specific disulfide bond (Cys6-Cys120) was prepared by the treatment of dithiothreitol in the presence of $CaCl_2$ followed by blocking free SH with iodoacetamide [14].

2.2 Gel preparation and gel strength determination.

α-La and β-Lg solutions at pH6.3 were mixed ratios of 2:8, 5:5 and 8:2. NaCl concentrations were adjusted to 25, 50, 100, 125 and 150 mM, and glutathione (GSH) concentrations were varied to 0, 25, 50 and 75 mM. The final protein content of each mixed solution was adjusted to 8%. α-La and 3SS α-La were mixed with ovalbumin at different ratios. The pH of the mixed solutions was adjusted to 7.0. Aliquots of the various mixed solutions were placed in a small petri dish covered with a glass plate, and then heated in a water bath at 80°C for 15 min.

Gel strength was determined using a rheometer. The sample was tested with a probe made of a razor measuring 0.36 x 0.03 cm at the edge [9]. Gel strength was expressed as the force in kPa applied to the probe edge when the surface yield point was reached. Determination of gel strength was repeated 6 times for each sample.

2.3. Total SH groups determination.

Total SH groups was determined using DTNB reagent containing 6 M urea and 2% SDS. The absorbance was read at 412 nm. The amounts of SH groups were calculated using a molar extinction coefficient of 1.36×10^4 M^{-1} cm^{-1}.

2.4. Production of monoclonal antibodies.

Mice (BALB/c) were immunized with α-La. Mouse spleen cells were fused with myeloma (Sp2/O) cells with PEG, and the hybridoma cells were selected in HAT medium. Each hybridoma was cloned twice by limiting dilution. Three monoclonal antibodies designated MAb2, MAb3 and MAb4 were obtained and characterized by enzyme-linked immunoassay (ELISA).

2.5. Competitive ELISA.

α-La mixed with ovalbumin was heated at low concentration of NaCl, and then 20 μg α-La/ mL of mixed solutions serially diluted was examined by competitive ELISA. The binding reactivity of protein with antibody was expressed as % inhibition.

3. RESULTS AND DISCUSSION

3.1. Gelation of the mixture of α-La and β-Lg.

The interaction of α-La and β-Lg in heat-induced gelation of protein mixture was estimated by multiple regression analyses. The coefficient of determination (R²) for the mixtures of α-La and β-Lg in the ration of 2:8, 5:5 and 8:2 were 0.91, 0.83 and 0.74, respectively and both GSH and NaCl significantly (p < 0.01) contributed to gelation. Maximum gel strengths for mixtures of α-La and β-Lg in the ratio of 2:8, 5:5, and 8:2 were obtained at GSH and NaCl concentrations of 0 mM and 86 mM, 43 mM and 116 mM, and 45 mM and 83 mM, respectively.

Figure 1 shows the three- dimensional surface plots for the gel strength of the mixed proteins as a function of GSH and NaCl concentrations. These plots indicated an increasing trend in the maximum gel strength as the proportion of α-La in the mixed proteins increased. Maximum gel strength of the mixtures of α-La and β-Lg in ratios of 2:8, 5:5 and 8:2 were 120, 294 and 342 kPa. The presence of larger amount of α-La in the mixed proteins would contribute to the formation of rigid gel, probably through SH-disulfide bond interactions.

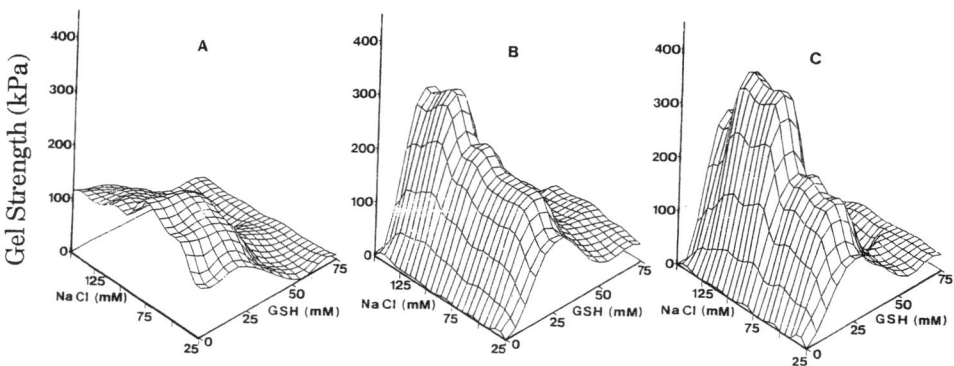

Figure 1. Three-dimensional surface plots of gel strength of the protein mixtures as functions of sodium chloride (NaCl) and glutathione (GSH) concentrations.
(A) α-La : β-Lg = 2 : 8, (B) α-La : β-Lg = 5 : 5, (C) α-La : β-Lg = 8 : 2.

3.2. Gelation of mixture of α-La/ovalbumin and 3SS α-La/ovalbumin.

Gel strength of the mixed proteins of whey and egg white proteins was determined after heating at 80℃ for 15 min at pH 7.0 in the presence of 86 mM NaCl. The mixture of α-La and ovalbumin resulted in high gel strengths. Mixing ratio of α-

La and ovalbumin affected gel strengths. A gel strength of 206 kPa was obtained for 8% ovalbumin. With α-La alone, up to 8% protein did not form a gel, but addition of α-La to ovalbumin enhanced markedly the gel strength of the mixed proteins. Gel strength Increased as proportions of α-La increased. The gel strength of the mixture was 403 kPa, almost double that of 8% ovalbumin. When 3SS α-La was added to 4% ovalbumin, quite soft gel was observed (Fig. 2).

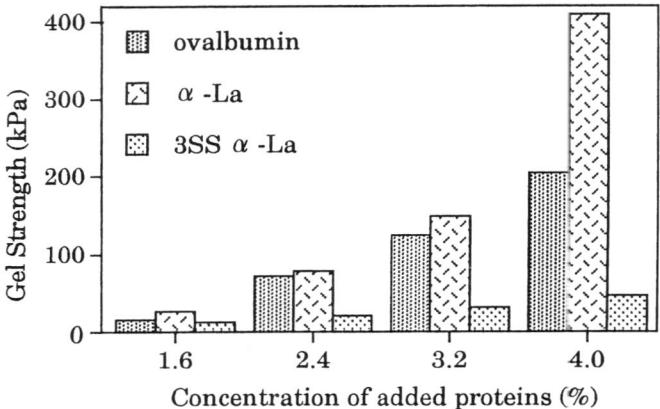

Figure 2. Effect of addition of either 3SS α-La or α-La to 4% ovalbumin on gel formation.

3.3. Binding reactivity of α-La with antibody

To determine the binding reactivity of α-La with monoclonal antibody in heat-induced gelation, competitive ELISA was performed. Monoclonal antibody, MAb2 reacted with unheated and heated α-La, whereas unheated and heated 3SS α-La had no reactivity with MAb2. The conformation of holo type 3SS α-La reduced at the disulfide bond between Cys6-Cys120 is stable and close to that of the native α-La [14]. Therefore, it can be considered that MAb2 recognizes the region around the disulfide bond Cys6-Cys120. ELISA patterns of the mixture of α-La/ ovalbumin at high concentrations in the absence of additional NaCl showed binding reactivity of proteins. Unheated and heated α-La showed a high inhibition, reflecting its high binding reactivity. However, the mixtures of α-La/ovalbumin exhibited decreasing reactivity upon heating. As the proportion of α-La increased, the reactivity decreased. These results supported that interaction of α-La with ovalbumin inhibited antibody from reacting with epitope in α-La.

3.4. Changes in total SH groups in heated protein mixture

The contribution of a specific disulfide bond Cys6-Cys120 in α-La was also confirmed by determination of total SH groups (Fig. 3) which indicated the effects of adding either α-La or 3SS α-La to 4% ovalbumin. Total SH groups of unheated

4% ovalbumin was 86 μ mole/g (corresponding to about 4 sulfhydryl groups in ovalbumin molecule). Heat treatment of 4% ovalbumin caused only a slight decrease in total SH. The decrease of total SH groups was due to formation of disulfide bonds through either oxidation of SH groups or SH-disulfide interchange reactions.

By the addition of α-La to ovalbumin, total SH groups decreased as concentration of α-La increased upon heating. Total SH groups of the mixture of 4% α-La and 4% ovalbumin was 35 μ mole/g ovalbumin. However, addition of 3SS α-La to ovalbumin caused a slight decrease in total SH. Decreasing total SH groups was related to the disulfide bond formations which occurred not only in the ovalbumin, but also in the mixed protein system. This supported observations of SH groups in a mixture of α-La with either β-Lg or BSA [7,8]. Disulfide bonds of α-La participated in intermolecular disulfide bond formation during heating of the mixed proteins, especially through SH-disulfide interchange reactions. The relationship between gel strength and total SH groups was confirmed by the results (Fig. 2 and 3).

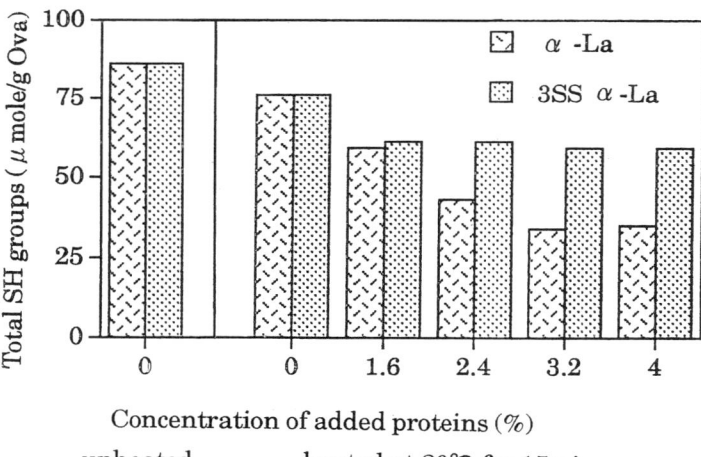

Figure 3. Effect of addition of either 3SS α-La or α-La to 4% ovalbumin on total SH groups.

3.5. Contribution of specific disulfide bond in α-La to the polymerization.

The presence of 4 disulfide bonds in α-La seemed to be important to enhance gelation. Reduction of a specific disulfide bond Cys6-Cys120 caused a sharp decrease in α-La enhancement of gelation. The disulfide bond is important to stabilize the structure of proteins and reduction of disulfide bonds may affect the structural and functional properties of proteins. The disulfide bond Cys6-Cys120 in α-La contributes to the stabilities of native and intermediate folded state [15].

The disulfide bond Cys6-Cys120 is located on the surface of α-La molecule, and is more reactive than other three disulfide bonds in α-La [14]. Accordingly, it was reasonable that reduction of a disulfide bond Cys6-Cys120 in α-La would reduce the capability of α-La to participate in gelation with ovalbumin. The contribution of a specific disulfide bond Cys6-Cys120 in α-La was implied by its participation in the enhancement of gel strength of the mixed proteins.

REFERENCES

1. S.Barbut and E.A.Foegeding, J.Food Sci. 58 (1993) 867.
2. G.R.Ziegler and E.A.Foegeding, Advances in Food and Nutrition Res., J.E. Kinsella (Ed.), pp. 204-297. Academic Press Inc., San Diego, 1990.
3. T.J.Herald and D.M.Smith, J. Agric. Food Chem. 40 (1992) 1737.
4. D.M.Mulvihill and J.EKinsella, Food Tech. 41(9) (1987) 102.
5. J.E.Matsuura and C.Manning, J. Agric. Food Chem. 42 (1994) 1650.
6. Y.A.Kim, G.W.III.Chism and M.E.Mangino, J. Food Sci. 52 (1987) 124.
7. N.Matsudomi, T.Oshita, E.Sasaki and K.Kobayashi, Biosci. Biotech. Biochem. 56 (1992) 1697.
8. N.Matsudomi, T.Oshita, K.Kobayashi and J.E.Kinsella,J. Agric. Food Chem. 41 (1993) 1053.
9. A.M.Legowo, T.Imade and S.Hayakawa, Food Res. Int. 26 (1993) 103.
10. M.E.Hines and E.A.Foegeding, J. Agric. Food Chem. 41 (1993) 341.
11. J.R.Brunner, Food Proteins, J.R. Whitaker and S.R. Tannenbaum (Eds.), pp. 175-208. Avi Publishing Co. Westport, CT, 1977.
12. J.N.deWit and G.Klarenbeek, J. Dairy Sci. 67 (1983) 2701.
13. S.Utsumi, S.Damodaran and J.E.Kinsella, J. Agric. Food Chem. 32 (1984) 1406.
14. K.Kuwajima, M.Ikeguchi, T.Sugawara, Y.Hiraoka and S.Sugai, Biochemistry 29 (1990) 8240.
15. M.Ikeguchi and S.Sugai, Biochemistry 31 (1992) 12695.

HYDROCOLLOIDS – PART 1
Edited by K. Nishinari
2000 Elsevier Science B.V.

413

Viscoelastic behavior of dehydrated egg white gel

A. Nakamura, K. Hara[a], A. Matsumoto[b] and N. Hiramatsu

Department of Applied Physics, Faculty of Science, Fukuoka University, Nanakuma, Jonan-ku, Fukuoka 814-0180, Japan

[a]Institute of Environmental Systems, Faculty of Engineering, Kyushu University, Hakozaki, Higashi-ku, Fukuoka 812-8581, Japan

[b]Osaka Municipal Technical Research Institute, Morinomiya, Joto-ku, Osaka 536-8553, Japan

The opaque heat-treated egg white gel turns into a transparent substance by dehydration, which looks like a glass. We carried out the measurement of the viscoelastic stiffness of the dehydrated egg white gel with increasing temperature in order to examine the mechanical properties, which is one of the most popular examinations of the glass transition of polymers. In the measurement, the storage modulus decreased remarkably and the loss modulus showed a broad peak around the temperature T_g where a thermal anomaly had been observed. As these features are usually observed in the noncrystalline polymers around the glass transition temperature, the occurrence of the glass transition at T_g was confirmed in the dehydrated egg white gel and also the dehydrated egg white gel could be regarded as a glass.

1. INTRODUCTION

It is known that soft and turbid heat-treated egg white gel (HTEWG) becomes fragile and transparent by dehydration, like the glass state of noncrystalline polymers [1]. During the dehydration process of HTEWG, the log weight showed a decrease proportional to the measurement time and the slope altered at a certain time (t'_g) [1]. The linear decrease with a steep slope in the early period of the dehydration process is due to the loss of free water, while that after t'_g with a gentle slope results from the loss of bound water [1].

The interaction between the network and the solvent in the gel also influenced the elastic property of the gel during the dehydration process, as well as the weight and the volume. Koshoubu *et al.* measured the complex elastic stiffness of HTEWG during the dehydration process, and found that the amplitude of the complex elastic stiffness increased up to $\sim 10^3$ times the initial value around t'_g, and that the elastic loss tangent (tan δ) showed a peak just before t'_g [2]. Such features resembled the evolution of the elastic properties during the glass transition in noncrystalline polymers with decreasing temperature. Their experimental results can be explained as follows. Because free water can move almost freely in the gel network, it acts as a buffer of mechanical motion of the networks, namely, free water keeps the gel soft. The loss of free water during the dehydration process reduces the distance between the network polymer chains due to the capillary force of the residual solvent; the interaction among approaching network molecules increases and this hinders their thermal motion. Such a situation is similar to the freezing of micro-Brownian motion with decreasing temperature in the case of glass transition; therefore, an elastic anomaly similar to glass transition with decreasing temperature will take place in the dehydration process of HTEWG. From the elasticity measurements, the important role of free water in the maintenance of the network polymer chain fluctuation was confirmed, as well as the similarity in the mechanisms of the dehydration process of HTEWG and the glass transition of the noncrystalline polymer.

On the basis of the similarity in the formation mechanism of the dehydrated HTEWG (DHTEWG) and glass, a comparison of the properties of DHTEWG with those of common glass attracted much interest. One of the most interesting subjects was whether the "glass transition" would take place in DHTEWG with increasing temperature. Therefore, Kanaya *et al.* performed the differential thermal analysis (DTA) measurements of DHTEWG with elevating temperature [3], and found an endothermic peak of which the intensity and the position increased with elevating the rate of temperature increase. Moreover, DHTEWG shows no sharp X-ray diffraction peak except for the broad amorphous halo [3-5].

In addition to these property changes, a broad excitation in the low frequency region of the Raman spectrum has been observed in both DHTEWG and dehydrated polyacrylamide (PAAm) gel [6-9]. (The PAAm gel, one of the most popular synthesized gels with a composition simpler than that of HTEWG, also becomes hard and transparent upon dehydration, and the weight and elastic stiffness show a similar behavior to those of HTEWG during dehydration [10].) The feature of a broad excitation in DHTEWG and dehydrated PAAm gel resembled the boson peak commonly observed in glass or amorphous materials, though its origin has not been elucidated yet. These above-mentioned facts resembled the features usually observed in many glasses [11].

However, we also noted some differences between the dehydrated gels and usual noncrystalline polymers. The glass transition temperatures of the dehydrated gels were much lower than those of the usual noncrystalline polymers [9], and the low-lying Raman band in the dehydrated PAAm gel was still observed far above the glass transition temperature determined by DTA measurement [9]. These results seemed somewhat different from the usual glass transition in the noncrystalline polymers. Therefore, the more basic investigations on the property change in the transition of the dehydrated gel has been indispensable in order to make clear the nature of this transition. In these circumstances, in the present study, we carried out the measurement of elastic stiffness of DHTEWG with increasing temperature, which is one of the most popular examinations of the glass transition of organic polymers [12].

2. EXPERIMENTAL

The sample was prepared as shown in Figure 1. The heat-treated egg white gel was prepared by boiling egg white for 15min at 98℃. Then, we obtained DHTEWG by dehydrating the gel in a refrigerator (temperature; 5℃, humidity;60%) for 2 weeks. The sample for measuring the complex elastic stiffness was prepared by cutting DHTEWG into a sheet (length 30.0~ 40.0mm, width 7.0mm, thickness 1.5mm). The water content (the weight ratio of water to the network in the gel [10]) of these samples were around 0.14, namely, almost all of the water in the samples was evaporated. At that time, they became hard and looked like plastics. In order to measure the complex elastic stiffness, we utilized a commercial apparatus (Seiko Denshi Model DMS110) in the three-point bending mode at a constant frequency (1 Hz). The measurements were performed in a temperature range from -30 to +170℃ with increasing temperature at a rate of 8℃/min. In order to prevent from cracking by the loss

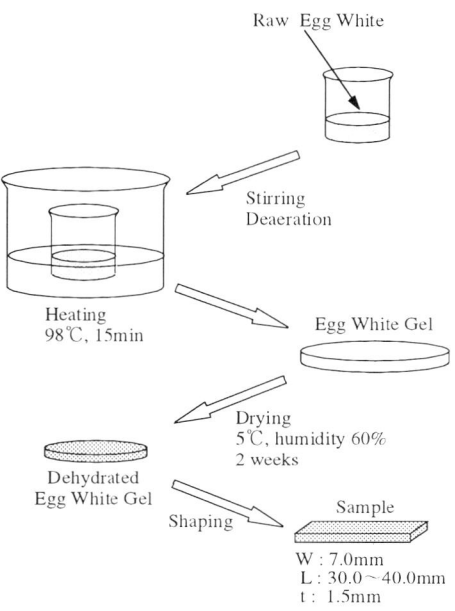

Figure 1. Schematic diagram of the sample preparation in the present study.

of the moisture in the specimen, it was covered with very thin silicone grease layer.

3. RESULTS AND DISCUSSION

Figure 2 shows evolution of the real, imaginary part and loss tangent of the complex elastic stiffness with increasing temperature. Both the marked decrease of the real part and a broad peak of the imaginary part were observed around 50℃ (T_g) where the endothermic peak had been observed in the previous temperature-increasing DTA measurement [3]. Then, a sharp peak of the elastic loss tangent was also observed at the higher temperature.

Because the imaginary part of the complex elastic stiffness corresponds to the dissipation of the mechanical energy [12], the results in the present study has revealed the occurrence of the mechanical transition around the temperature where the thermal anomaly occurred [3]. It is well known that the glass transition of the noncrystalline polymers is characterized by the softening of the system in addition to the thermal anomalies which we observed in the previous study [3]. Therefore, the occurrence of the glass transition in DHTEWG has become beyond doubt from the present results [12]. Besides, the order of magnitude of the anomalies of the real, imaginary part of

416

the complex elastic stiffness and the loss tangent corresponded to that of the noncrystalline polymers [9].

However, supposing that the similar transition occurs in the dehydrated PAAm gel around T_g, the present result will indicate the complicated structure of the dehydrated gel because the low-lying Raman mode is observed far above T_g [9]. More detailed investigations, including the elastic measurement of the dehydrated PAAm gel, are important and in progress.

Acknowledgments

This work was partly supported by the Japanese Private School Promotion Foundation, Grant-in Aid for Scientific Research (C) from the Ministry of Education, Science, Sports and Culture.

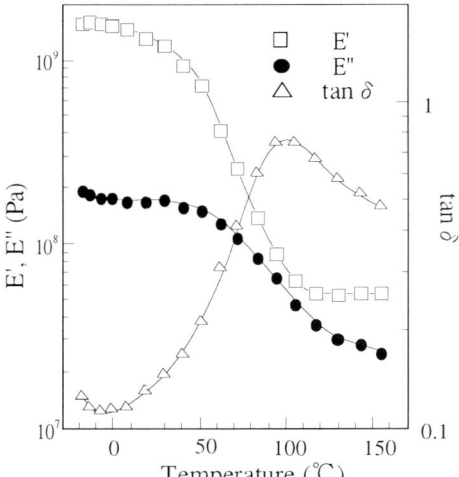

Figure 2. The change in the complex elastic stiffness of dehydrated heat-treated egg white gel with increasing temperature.

REFERENCES

1. E. Takushi, L. Asato and T. Nakada: Nature, **345** (1990) 298.
2. N. Koshoubu, H. Kanaya, K. Hara, S. Taki, E. Takush_ and K. Matsushige: Jpn. J. Appl. Phys. **32** (1993) 4038.
3. H. Kanaya, T. Nishida, M. Ohara, K. Hara, K. Matsushige, E. Takushi and Y. Matsumoto: Jpn. J. Appl. Phys. **33** (1994) 226.
4. H. Kanaya, K. Ishida, K. Hara, H. Okabe, S. Taki, K. Matsushige and E. Takushi: Jpn. J. Appl. Phys. **31** (1992) 3754.
5. H. Kanaya, K. Hara, E. Takushi and K. Matsushige: Jrn. J. Appl. Phys. **32** (1993) 2905.
6. K. Hara, T. Masuike, A. Nakamura, H. Okabe and N. Hiramatsu: Jpn. J. Appl. Phys. **34** (1995) 5700.
7. A. Nakamura, H. Okabe, K. Hara and N. Hiramatsu: Jpn. J. Appl. Phys. **35** (1996) L43.
8. A. Nakamura, K. Hara and N. Hiramatsu: Jpn. J. Appl. Phys. **37** (1998) L143.
9. K. Hara, A. Nakamura and N. Hiramatsu: Jpn. J. Appl. Phys. **36** (1997) L1184.
10. T. Masuike, S. Taki, K. Hara and S. Kai: Jpn. J. Appl. Phys. **34**(1995) 4997.
11. for example; R. A. Ramos: Phys. Rev. **B49** (1994) 702.
12. L. E. Nielsen: *Mechanical Properties of Polymers* (Reinhold Publishing Co., London, 1962).

HYDROCOLLOIDS – PART 1
Edited by K. Nishinari
2000 Elsevier Science B.V.

Evaluation of the fractal dimension for aggregates in heat induced BSA gels

H. Kumagai[a], T. Hagiwara[b], O. Miyawaki[a], and K. Nakamura[a]

[a]Department of Applied Biological Chemistry, The University of Tokyo,
1-1-1 Yayoi, Bunkyo-ku, Tokyo 113, Japan.

[b]Department of Food Science and Technology, Tokyo University of Fisheries, 4-5-7 Konan,
Minato-ku, Tokyo 108-8477, Japan

The fractal structure of aggregates in heat-induced BSA gels prepared with and without salt ($CaCl_2$ or NaCl) was examined. From the concentration dependence of the gel elasticity determined from a uniaxial compression test of the gel, the fractal dimensions D_f of the aggregates in the gels were evaluated, using the theory of Shih *et al.*. It was confirmed that the gels with and without salt showed weak- and strong-link behavior in the theory of Shih *et al.*, respectively. The obtained values of D_f were about 2 for strong-link gels and about 2.7 for weal-link gels. In addition, as for the gels containing salt (weak-link type), from the analysis of the gel image obtained with confocal scanning laser microscopy of the gels, the fractal dimensions were also evaluated, the value being close to that evaluated from the gel elasticity measurements. These results indicate that the elastic behavior of the aggregate gels is a reflection of fractal structure of the aggregates in the gels.

1. INTRODUCTION

The aggregate structure in heat-induced protein gels would influence the macroscopic physical properties for gel. Many studies have been reported in which the macroscopic properties of aggregated protein gels vary with the aggregation conditions; *e.g.*, pH and ionic strength; however, only the correlation between the conditions and the macroscopic physical properties has been repeatedly discussed so far. To systematically understand the behavior of the macroscopic physical properties of protein gels, the relationship between the structure of the aggregates and the macroscopic physical properties should be investigated.

Recently, fractal analysis has attracted attention as a quantitative analytical method that can characterize many kinds of disordered shapes [1]. A fractal is a self-similar structure which can be characterized by a non-integer dimension; the fractal dimension D_f[1,2].

In this study, the fractal structure of aggregates in heat-induced bovine serum albumin (BSA) gels was examined.

2. MATERIALS AND METHODS

2.1. Theory of Shih *et al.* [3]

The structure of a colloidal gel (aggregated gel) is approximated as closely packed fractal flocs, and the elastic property of the gel is dominated by that of the flocs [3]. Depending on the strength of the links between the neighboring flocs in comparison with that in the flocs, the links are classified into two types; strong-link and weak-link. In the strong-link regime, the links between the neighboring flocs have a higher elasticity than those in the flocs. For the gel with a strong-link (hereafter referred to as a strong-link gel), the dependence of the elasticity E and the limit of linearity γ_0 (γ_0 is the upper limit value of strain (γ) where the stress σ is proportional to γ) of the gels on the particle (in this study, protein) concentration ϕ can be described as follows:

$$E \propto \phi^{(3+x)/(3-D_f)} \tag{1}$$

$$\gamma_0 \propto \phi^{-(1+x)/(3-D_f)} \tag{2}$$

where D_f is the fractal dimension of the flocs ($D_f \leq 3$), and x is the backbone fractal dimension of the flocs, which varies between 1.0 and 1.3 [3]. On the other hand, in the weak-link regime, the links in the flocs have a higher elasticity than those between the neighboring flocs: for the gel with a weak-link (hereafter referred to as a weak-link gel), the dependence of the E and γ_0 on particle concentration ϕ can be expressed as follows:

$$E \propto \phi^{1/(3-D_f)} \tag{3}$$

$$\gamma_0 \propto \phi^{1/(3-D_f)} \tag{4}$$

2.2. Preparation of Gels for Elasticity Measurement

BSA (Boehringer Mannheim GmbH, Mannheim, Germany; ref.238040) was dissolved in the four kinds of buffer; (a) 50 mM HEPES (*N*-2-hydroxyethylpiperazine-*N'*-2-ethanes ulfonic acid) buffer (pH 7.0, no salt was added), (b) 50 mM acetate buffer (pH 5.1, NaCl was added to make the ionic strength of the buffer 0.1 M), (c) 50 mM HEPES buffer (pH 7.0, $CaCl_2$ was added to be 30 mM), and (d) 50 mM HEPES buffer (pH 7.0, $CaCl_2$ was added to be 5 mM). The solutions were degassed under vacuum for 3 min to remove air. The pH of the solutions was then adjusted to the pH of the buffers, using NaOH and HCl solutions.

2.3. Elasticity Measurements

The elasticity of the gels was determined from an uniaxial compression test with Rheoner RE-3305 (Yamaden, Co., Tokyo, Japan). A cylindrical gel was vertically compressed with a flat plunger (30 mm diameter) at a compression rate 1.0 mm/s. The strain γ of the gel was determined as the ratio of the deformation to the initial height of the gel. The elasticity E was calculated from the linear part of the stress-strain curve at $\gamma < 0.01$.

2.4. Evaluation of the Fractal Dimension from Elasticity Measurements

First, γ_0 was plotted against ϕ. Because γ_0 decreases with increasing ϕ for a strong gel and increases for a weak-link gel as shown in Eqs (2) and (4), the link type for the gel can be identified from the sign of the slope of the plot for $\log\gamma_0$ *vs.* $\log\phi$. As the limit of linearity γ_0, the strain value where the deviation was 5 % between the ordinate value of the stress-strain curve σ and γE was taken.

For the weak-link gels, the value of the fractal dimension D_f is evaluated from the slope of the $\log E$ *vs.* $\log\phi$ plot, using Eq. (3) for a weak-link gel. For the strong-link gel, to determine the value of D_f from Eq. (1), the value of the fractal dimension for the effective backbone of the aggregate x is necessary. However, because x took a value from 1.0 to 1.3 [2], the obtained value of D_f varies little in the range of the value of x=1.0~1.3. Therefore, in this study, the minimum and maximum values of D_f were estimated from the slope of the $\log E$ *vs.* $\log\phi$ plot, using the value of x as 1.0 to 1.3, for a strong-link gel.

2.5. Evaluation of fractal dimension D_f from the images of protein aggregates in a gel

Samples were heated between two glass plates (a gap is 0.18mm) at 95°C for 10 min. Thereafter, the sample was cooled to 25°C and stored for 24h. The gel strips were removed and cut into 5 mm squares. The obtained gels were immersed in the buffer containing 0.001 wt% fluorescein isothiocyanate (a fluorescent labeling agent for proteins) for 1h with gentle shaking and subsequently washed in fresh buffer for 1h. The stained gels were then mounted on a slide glass with a spacer around the gels. Then, a cover glass was placed on top of the spacers and fixed with nail polish. A confocal laser scanning microscope model MRC600 (Bio-Rad Laboratories, Inc., California, USA) was used for observing aggregates in the gel.

The obtained confocal microscopy images were digitized with the public domain NIH Image program ver.1.59 [4] on the Macintosh platform. From the digitized image, the fractal dimension D_f was calculated by the box counting method [5,6] as follows: A square mesh of a certain size L is laid over the object on the digitized image. The fractal dimension of the protein aggregates on the image, D, is determined using Eq. (5) from the slope of the double logarithmic plot for N(L) *vs.* L; computer software for fractal analysis [5] was used.

$$N(L) \propto L^{-D} \tag{5}$$

where N(L) is the number of mesh boxes that contain part of the image. The fractal dimension of protein aggregates of three dimensions D_f can be calculated from the following equation [2]:

$$D_f = D + 1 \tag{6}$$

3. RESULTS [7,8]

Figure 1(A) shows the double logarithmic plot of γ_0 *vs.* ϕ for the BSA gels prepared with the BSA solution of pH 7.0 (containing 30 mM $CaCl_2$). Because γ_0 tended to increase with increasing ϕ, this gel is confirmed to be a weak-link gel, as explained before. Figure 1 (B) shows the double logarithmic plot of E *vs.* ϕ for the same samples as those in Figure 1(A). From the slope of the plot, using Equation (3) for weak-link gels, the fractal dimension D_f was evaluated to be 2.82. These gels had a turbid appearance.

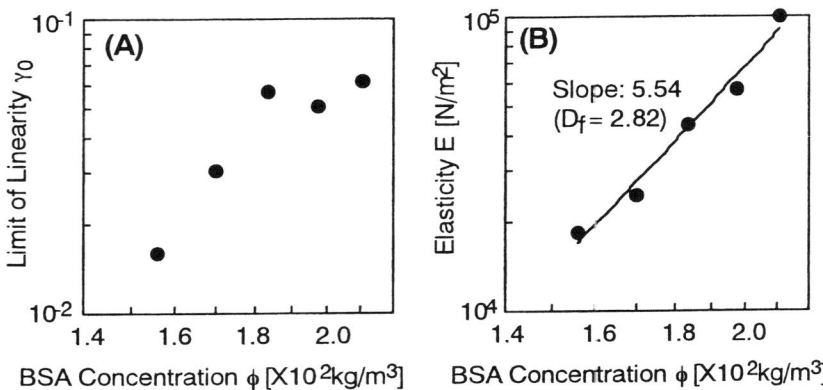

Figure1 The double-logarithmic plots of γ_0 vs. ϕ (A) and E vs. ϕ (B).
Solvent: 50mM HEPES buffer (pH 7.0, 30mM $CaCl_2$)

Figures 2 (A) and 2 (B) show the double logarithmic plots of γ_0 *vs.* ϕ and E *vs.* ϕ, respectively, for the BSA gels prepared with 50 mM HEPES buffer (no salt was added). From the slope of the $\log\gamma_0$ *vs.* $\log\phi$ plot, this gel is confirmed to show a strong-link behavior, and the value of the fractal dimension D_f was estimated to be $D_f = 2.00 \sim 2.07$, using Eq.(1). The gel had a transparent appearance. Moreover, aggregates of micrometer order were not observed in the gels using the confocal laser scanning microscope, suggesting that the order of the aggregate size was smaller than a micrometer.

The Table 1 summarizes the results of the fractal analysis of the BSA gels from the elasticity measurements. The value of D_f for the strong-link gel was about 2, while those for the weak-link gels were 2.6-2.8.

Figure 3 shows the double logarithmic plots of the box number $N(L)$ *vs.* box size L for a gel prepared by the same way as that in Figure 1 (BSA concentration was 197 kg/m³). The plot shows a power law dependence of the box number $N(L)$ on box size L, as predicted by Equation (5). From the slope of the plot, the value of D of the gels was 1.81, the values of D_f being 2.81. This value of D_f was very similar to that obtained from the dependence of the elasticity of the gel on the concentration (Figure 1), indicating that the elastic behavior of this gel is a reflection of the fractal structure of the aggregates in the gels. The obtained value of D_f for this gel was larger than that of the aggregates in dilute BSA solution reported in a previous study [9]. In addition, the value of D_f obtained from the image of the aggregates was almost constant, irrespective of the BSA concentration in the concentration range examined, though the data are not shown.

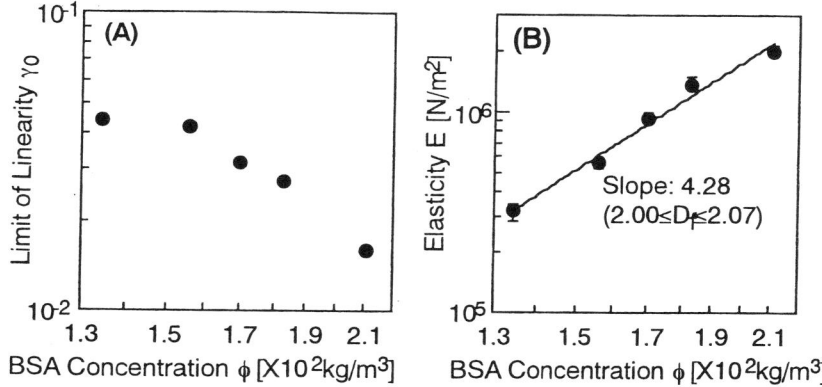

Figure 2 The double-logarithmic plots of γ_0 vs. ϕ (A) and E vs. ϕ (B).
Solvent: 50mM HEPES buffer (pH 7.0, no salt was added)

Table 1 Summary of the fractal analysis of BSA gels prepared with various buffers.

buffer conditions	link type	fractal dimensions D_f
pH 7.0, no salt was added	strong	2.00-2.07
pH 7.0, 30mM $CaCl_2$	weak	2.82
pH 7.0, 5mM $CaCl_2$	weak	2.82
pH 5.1	weak	2.61

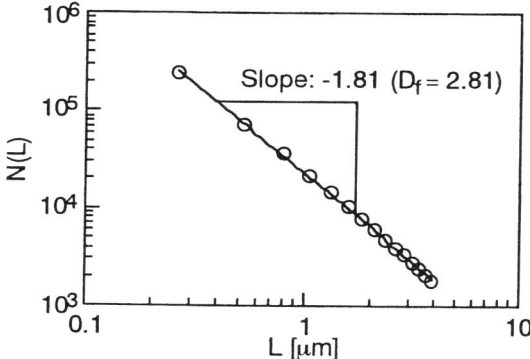

Figure 3 Estimation of fractal dimension D_f by the box counting method.
Buffer: 50mM HEPES buffer (pH 7.0, 30mM $CaCl_2$)
BSA concentration: 197 kg/m^3

4. DISCUSSION

Shih *et al.* [3] reported that boehmite alumina colloid gels showed a strong-link behavior.

However, a gel with a weak-link behavior has not been reported. In the present study, the BSA gel without salt showed a strong-link behavior (Table 1), while the other BSA gels showed a weak-link behavior. That is, it was confirmed that the protein aggregate gels showed both strong- and weak-link behavior by varying the aggregation conditions.

As can be seen from Table 1, the values of D_f for BSA gels with salt were approximately 2.6-2.8, while that of D_f for the gel prepared without salt were about 2. We have already analyzed the fractal structure of aggregates formed by heating dilute BSA solution (BSA concentration, 0.1wt%; ionic strength, 0.1M) with the static light scattering method [9]; the fractal dimension of the aggregates at pH 7.0 (apart from the isoelectric point of BSA (pH, 4.9)) was about 2.1, the value of which agreed with that predicted by the reaction-limited cluster-cluster aggregation model [2,10], while at pH 5.1, the fractal dimension was about 1.8, which agreed with that predicted by the diffusion-limited cluster-cluster aggregation model [2,11]. The results suggested that the amount of average charge of BSA molecules influence the fractal structure of aggregates. Meakin et al. [12] stated that the fractal dimension of aggregates increased by "restructuring" during aggregation. The larger values of the fractal dimension for the weal-link gels would also be caused by the restructuring. Salt addition would restructure protein aggregates and make the value of D_f increase.

As for the BSA gel prepared with 50 mM HEPES buffer containing no salt, from confocal scanning laser microscopy, a clear image of the aggregates was not obtained, suggesting that the size of the aggregates was smaller than the smallest size which the confocal scanning microscope used could observe. To analyze the structure of the aggregates for these gels, another experimental method rather than confocal scanning microscopy should be developed.

REFERENCES

1. B.B. Mandelbrot, The Fractal Geometry of Nature; Freeman, San Francisco, 1982.
2. T. Viscek, Fractal Growth Phenomena, World Scientific, Singapore, New Jersey, London, Hong Kong, 1989.
3. W.-H. Shih, W.Y. Shih, S.-I. Kim, J. Liu, and I.A. Aksay, Phys. Rev. A, 15 (1990) 4772.
4. W. Rasband, NIH Image Manual,National Institutes of Health, USA., 1996.
5. B.H. Kaye, Image Analysis Techniques for Characterizing Fractal Structure. In Fractal Approach to Heterogeneous Chemistry; Avnir, D., Ed.; John Wiley & Sons: Chichester, New York, Brisbane, Toronto, and Singapore, 1989.
6. P. Bourke, Fractal Dimension Calculator User Manual Version 1.5; Auckland University, New Zealand, 1993.
7. T. Hagiwara, H. Kumagai, and T. Matsunaga, J. Agric. Food Chem., 45 (1997) 3807.
8. T. Hagiwara, H. Kumagai, and K. Nakamura, Food Hydrocolloids, 12 (1998)29.
9. T. Hagiwara, H. Kumagai, and K. Nakamura, Biosci. Biotech. Biochem., 60 (1996) 1757.
10. W.D. Brown and R.C. Ball, J. Phys. A: Math. Gen., 18 (1985) L517.
11. W.D. Weitz and M. Oliveria, Phys. Rev. Lett., 52 (1984) 1433.
12. P. Meakin and R. Jullien, J. Chem. Phys., 89 (1988) 246.

HYDROCOLLOIDS – PART 1
Edited by K. Nishinari
2000 Elsevier Science B.V.

423

Analysis of heat-induced BSA aggregates by scattering methods

T. Hagiwara[a], H. Kumagai[b], T. Suzuki[a] and R. Takai[a]

[a]Department of Food Science and Technology, Tokyo University of Fisheries,
4-5-7 Konan, Minato-ku, Tokyo 108-8477, Japan

[b]Department of Applied Biological Chemistry, The University of Tokyo,
1-1-1 Yayoi, Bunkyo-ku, Tokyo 113-8657, Japan

The structure of aggregates formed by heating dilute BSA (bovine serum albumin) solution was analyzed using a laser light scattering method. The fractal structure was observed on the scale length from decades to hundreds nanometers. With increase in salt concentration of the solution, the value of fractal dimension D_f for the aggregates tended to increase, suggesting that the aggregates were reconstructed by salt addition.

The aggregate structure in BSA gels prepared from concentrated solution was also investigated using a small angle X-ray scattering method on the scale length from several to decades nanometers. It was suggested that the aggregates in the gels were formed through the process different from the conventional cluster-cluster aggregation model mentioned above.

1. INTRODUCTION

The protein aggregates in a gel would considerably influence the physical properties of the gel. We have shown that the elastic behavior of the heat-induced BSA (bovine serum albumin) gels was a reflection of the fractal structure of the protein aggregates on the scale length of micrometers [1]. In addition, we have observed that the value of the fractal dimension D_f for heat-induced BSA gels varied according to the condition of the solvent used in preparation of gels. However, the method for controlling the fractal dimension D_f is not satisfactorily known. To control the fractal dimension of aggregates, it is necessary to elucidate the mechanism for the development of fractal structure of aggregates.

Scattering methods are very effective to analyze colloidal structure below hundred nanometers. The analysis of aggregate structure on such scale using the scattering methods would help us to guess the mechanism of development of the fractal structure in protein aggregate gels. Light scattering gives information on dilute colloidal structure on the scale length from decades to hundreds nanometers. The investigation on the dilute system is very important because dilute colloidal system has been theoretically examined on the basis of the cluster-cluster aggregation model. In a previous study, we have already analyzed the fractal structure of aggregates formed by heating dilute bovine serum albumin (BSA) solution with the static light scattering (SLS) method [2]. The results showed that controlling the pH of the solution, the value of D_f had almost the same value as predicted by the cluster-cluster aggregation model. However, the effect of salt on D_f in the dilute system has not been examined. On the other hand, small angle X-ray scattering (SAXS) gives information on the scale length from several to decades nanometers.

In this study, by using laser light scattering, the structure of aggregates formed by heating BSA dilute solutions at various salt concentrations was investigated. The aggregate structure in BSA gels prepared from concentrated solution was also analyzed using SAXS method

2. MATERIALS AND METHODS

2.1. SLS

BSA was obtained from Boehringer Mannheim GmbH (Mannheim, Germany; ref.238040). All other chemicals were of reagent grade.

BSA aggregate suspensions (BSA concentration; 0.1wt%) were prepared by almost the same method as that described in the previous work [2]. BSA was dissolved in 50 mM HEPES buffer (pH 7.0). The ionic strength of the buffer I_s was varied from 0.2 to 1.0M by the addition of NaCl to the buffers. The solutions were kept at 25℃ for about 20min and then heated at 95℃. The samples were used for SLS measurement. The aggregates prepared by heating BSA solutions with $CaCl_2$ were also prepared. BSA was dissolved in 50 mM HEPES buffer (pH 7.0) containing 5mM $CaCl_2$; the solution was heated in the same way as mentioned before and used for SLS measurement.

The fractal dimension D_f of the aggregates was evaluated using the following equation [2]:

$$I(q) \propto q^{-D_f} \qquad (1 < D_f < 3) \tag{1}$$

where $I(q)$ is the total scattered intensity and q is the length of scattering vector defined by

$$q = (4\pi n_s/\lambda)\sin(\theta/2) \tag{2}$$

where n_s, λ, and θ are the refractive index of the solvent, the wavelength of the light source in vacuum, and the scattering angle, respectively. The scale length d can be calculated using q as follows [6]:

$$d = 2\pi/q \tag{3}$$

SLS measurements were done similarly to that described in the previous work [2] on a System 4700 (Malvern);the light source was a 30mW He-Ne laser(wavelength $\lambda = 632.8$nm; NEC Co., Ltd.). The value of q was varied from 5×10^6 to 3×10^7 m^{-1} (scale length d; 2.1×10^{-7} to 1.3×10^{-6} m). From Eq. (2), D_f was evaluated from the slope of the double logarithmic plot of $I(q)$ vs. q.

2.2. SAXS

BSA were dissolved in three kinds of buffer: (A) 50 mM acetate buffer (pH 5.1, $I_s = 0.1$ M);

(B) 50mM HEPES buffer (pH 7.0, I_s=0.2 M); (C) 50mM HEPES buffer (pH 7.0, I_s=0.012 M). The ionic strength of the buffer I_s was adjusted by the addition of NaCl to the buffers The solutions were degassed under vacuum for 3 min to remove air.   The pH of the solutions was then adjusted to the initial pH of the buffers, using NaOH and HCl solution.   The final BSA concentrations were (A) 16.6%; (B) 17.4%; (C) 17.5%, respectively.   After preheated at 50℃ for 60min, the solution was gelled by heating at 95℃ for 10 min in 0.7 mm diameter sample capillary tube (Mark-Rhörchen, Berlin, Germany).

The SAXS measurements were done with a Kratky Compact Camera System (Anton Parr GmbH, Austria).   The X-ray source was CuK_α (wavelength λ = 0.154 nm, Phillips Co., the Netherlands).   The scattered intensities I were detected with a proportional counter as a function of q defined by Eq.(2); value of n_s can be taken as unity in case of SAXS [6].   The value of q was varied from 8.2×10^7 to 5.6×10^9 m^{-1} ($1.1 \times 10^{-9} < d < 7.7 \times 10^{-8}$ m)

The effect of the slit collimation system from a Kratky camera was corrected by using the method of Guinier and Fournet [7] after subtraction of the blank scattering from the capillary tube.

3. RESULTS

3.1. SLS

Figure 1 shows the double logarithmic plots of I(q) *vs.* q for the sample prepared from a BSA solution containing NaCl.   Both plots were linear, indicating that aggregates prepared from BSA solutions containing NaCl were fractal.   The values of D_f for (A) and (B) were evaluated to be 2.19 and 2.84, respectively.

Table 1 summarizes the values of D_f of aggregates prepared from BSA solutions containing NaCl at various I_s, including the result of the previous work (I_s=0.1 M) [2]. The values of D_f were larger for longer heating time at an identical value of I_s and tended to increase with increasing I_s. The values of D_f were larger than the maximum of D_f, 2.1 predicted by the conventional "cluster-cluster aggregation model" mentioned before [3-5].

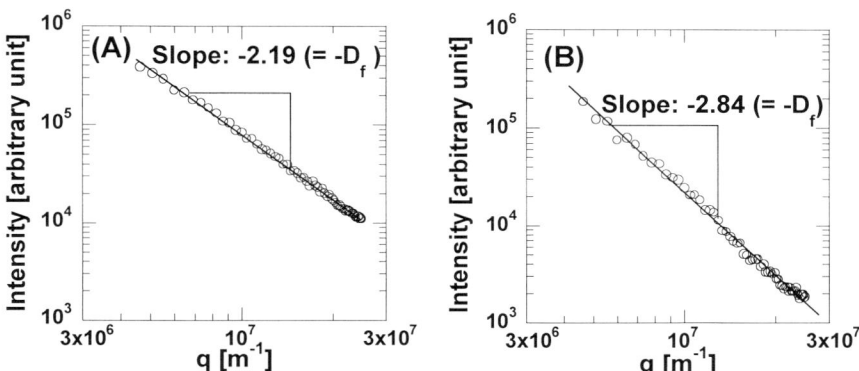

Figure 1 Dependence of scattered light intensity I for heated BSA solutions on q. (A) I_s=0.2 M; heating time was 90 min.(B) I_s=0.6; heating time was 13 min.

Table 1   Fractal dimension of BSA aggregates containing NaCl.

Ionic strength I_s [M]	NaCl concentration [M]	Heating time [mim]	Fractal dimension D_f
0.1	0.0868	90	2.11 [2]
0.2	0.186	40	2.08
0.2	0.186	90	2.19
0.2	0.186	120	2.27
0.3	0.286	17	2.08
0.3	0.286	25	2.39
0.4	0.386	20	2.53
0.5	0.486	16	2.50
0.6	0.586	13	2.84
1.0	0.986	10	2.12
1.0	0.986	11	2.71

In Figure 2, I $vs.$ q plots for aggregates containing $CaCl_2$ are shown. The linear relationship was not observed, indicating that the aggregates in these samples were not fractal. However, the aggregates in these samples may be fractal, considering that aggregates in BSA gels containing $CaCl_2$ were fractal over the range of about 0.2-10 micrometers from image analysis [1].   Another experimental methods that can clarify the aggregates structure at the scale length larger than that by SLS used in this study, should be developed to confirm that the aggregates formed with $CaCl_2$ are fractal.

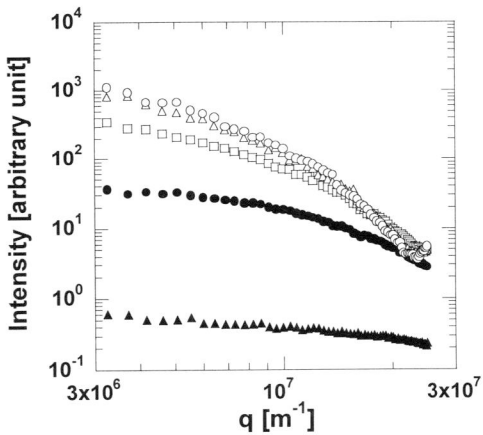

Figure 2 Dependence of scattered light intensity I on q for heated BSA solutions containing $CaCl_2$ at various heating times.
Heating Tome: ○, 45 min; △, 30 min; □, 10 min; ●, 5 min: ▲, 2 min.

3.2. SAXS
Figure 3 shows the double logarithmic plots of I(q) vs. q for the sample prepared using the

buffer (A) 50 mM acetate buffer (I_s = 0.1 M); (B) 50mM HEPES buffer (I_s=0.2 M); (C) 50mM HEPES buffer (I_s=0.012 M). All of the three plots were not single straight lines, indicating that these gels were not fractal predicted by Eq.(1).

The scattering pattern for the gels prepared at pH5.1, shows a peak at $q \doteqdot 2.5 \times 10^8 m^{-1}$. From the value of q at the peak top, the corresponding scale length d were evaluated to be about 25 nm, using Eq.(3). At $q>2.5 \times 10^8$ m^{-1} (scale length d < 25 nm), the plot had a linear region, whose slope value was close to -4; $I \propto q^{-4}$, which corresponds to the well-known Porod's law for the scattering profile characteristic to a smooth surface object [7,8]. The scattering profile of the gels prepared with 50 mM HEPES buffer had no peak.

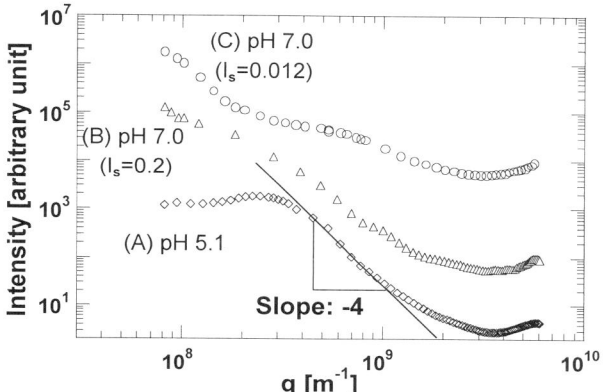

Figure 3   Small angle X-ray Scattering curves for the BSA gels.
Buffer: (A) 50 mM acetate buffer (pH5.1);
(B) 50mM HEPES buffer (pH 7.0, I_s=0.2 M);
(C) 50mM HEPES buffer (pH 7.0, I_s=0.012 M).

3.4. DISCUSSION

Scattering methods (light, X-ray) have no need for processing samples before observation such as in electron microscopy, and are useful techniques for the characterization of the native gel microstructure, however, there have been only few studies investigating the structure of BSA aggregates by scattering methods.

The fractal structure of aggregates in the dilute system has been examined by the cluster-cluster aggregation model [3,4]; diffusing particles or aggregates in certain medium stick to each other at a probability P on contact. According to the computer simulation of the cluster-cluster aggregation model, the values of the fractal dimension D_f for the aggregates are 1.8 at P=1 (the diffusion-limited cluster-cluster aggregation; DLCCA) and 2.1 at P<<1 (the reaction-limited cluster-cluster aggregation; RLCCA)[3]. As stated before, we have already analyzed the fractal structure of aggregates formed by heating dilute BSA solution at the condition of I_s=0.1 M (NaCl concentration was 0.0868 M)[2]; the value of D_f varied from 1.8 to 2.1 by controlling solution pH. On the other hand, increasing the salt concentration as shown in Table, the values of D_f could be larger than the maximum of D_f, 2.1 predicted by the

428

conventional "cluster-cluster aggregation model"[3-5]. According to Meakin *et al.*[9], the value of D_f can be larger than 2.1 by "restructuring" of aggregates during aggregation. The restructuring of aggregate structure by NaCl addition would cause the larger values of D_f as shown in Table. In the preceding study, the values of D_f for BSA gels with salt addition were approximately 2.6-2.8[1], while those of D_f for the gels prepared without salt addition were about 2. Salt addition would also cause restructuring protein aggregates in the gels and result in a higher value of D_f.

The peak of the SAXS profile for the gels at pH 5.1 suggested that the gels had a periodic structure[6,7] whose size was about 25 nm.; it is larger than that of one BSA molecule (about 3 nm). In addition, the existence of region of Porod's law at the scale length below 25 nm, indicated that the periodic structure had a smooth surface. From these findings, it is suggested that the aggregates in this gel were composed of the subunit structure (size; about 25nm) packed tightly with the protein molecules as shown in Figure 4.

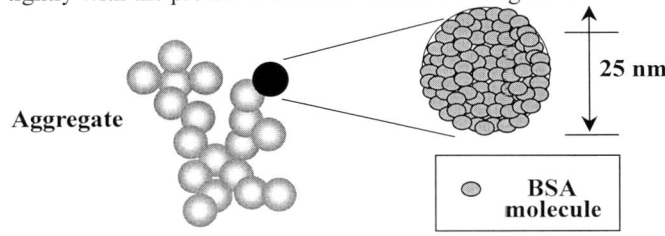

Figure 4 Schematic diagram of aggregates structure in BSA gels
prepared with 50 mM acetate buffer.

For all of three gels examined in this study, the fractal structure was not observed on the scale length accessible by the SAXS measurement, indicating that the aggregates in these gels were formed through the process different from the conventional cluster-cluster aggregation model mentioned above.

REFERENCES

1. H. Kumagai, T. Hagiwara, O. Miyawaki and K. Nakamura, Evaluation of the Fractal Dimension for Aggregates in Heat Induced BSA Gels, Proc. of 4th International Conference on Hydrocolloids, Osaka, Japan, 1998.
2. T. Hagiwara, H. Kumagai and K. Nakamura, Biosci. Biotech. Biochem., 60 (1996), 1757.
3. T. Vicsek, "Fractal Growth Phenomena", World Scientific, Singapore, New Jersey, London, and Hong Kong, 1989.
4. W. D. Brown and R. C. Ball, J. Phys. A: Math. Gen., 18 (1985) L517.
5. D. A. Weitz and M. Oliveria, Phys. Rev. Lett., 52 (1984) 1433.
6. O. Glatter and O. Kratky, "Small Angle X-ray Scattering", Academic Press, London, 1982.
7. A. Guinier and G. Fournet, "Small Angle Scattering of X-rays.", Wiley, New York, 1955.
8. H. D. Bale and P. W. Schmidt, Phys. Rev. Lett., 53 (1984) 596.
9. P. Meakin, P. and R. Jullien, J. Chem. Phys., 89 (1988) 246.

HYDROCOLLOIDS – PART 1
Edited by K. Nishinari
2000 Elsevier Science B.V.

429

Rheological studies on preheated β-conglycinin and glycinin

Takao Nagano[a], Takeshi Akasaka[b], and Yoichi Fukuda[b]

[a]Faculty of Education, Ehime University, 3, Bunkyo-cho, Matsuyama-shi,
  Ehime, 790-8577, Japan

[b]Fuji Oil Co., Ltd., Kinunodai, Yawara, Tsukuba-gun, Ibaraki, 300-2497, Japan

Preheated soybean protein isolates (SPIs), which have different compositions of β-conglycinin and glycinin, were prepared by a typical method for manufacturing commercial SPI. Rheological properties were studied by dynamic viscoelastic measurements and penetration testing. The gelation properties of preheated β-conglycinin-rich and glycinin-rich SPIs were investigated using dynamic viscoelastic measurements. In the absence of NaCl, an increase in the storage modulus G' was observed for preheated β-conglycinin-rich SPI, but not for preheated glycinin-rich SPI. In the presence of 2.5% NaCl, an increase in G' was observed for both preheated SPIs in the order β-conglycinin-rich SPI > glycinin-rich SPI. The gel properties of preheated SPIs of β-conglycinin and glycinin fractions were investigated by a penetration test. The initial slope in force-penetration curves increased and the breaking length decreased for preheated β-conglycinin gel by adding NaCl. Preheated glycinin fraction formed only a weak gel in the absence of NaCl, however, gel forming ability was drastically improved by adding NaCl. These results indicate that the gelation and gel properties of preheated β-conglycinin and glycinin are quite different and changed drastically by adding NaCl.

1. INTRODUCTION

The object of this study is to clarify gel properties of preheated β-conglycinin and glycinin. β-Conglycinin and glycinin are major storage proteins of soybeans and are believed to reflect the functional properties of soybean proteins. It is well known that β-conglycinin and glycinin have different structure and functional properties [1-3].

Preheat treatment such as pasteurizing is an important and useful method in food systems. It is reported that the gel forming ability of soybean proteins is improved by preheat treatments [4,5]. The gel properties of each preheated β-conglycinin and glycinin are not well understood. In this study, we prepared preheated soybean protein isolates (SPIs) which different compositions of β-conglycinin and glycinin by a typical method for manufacturing commercial SPI. Rheological properties were studied by dynamic viscoelastic measurements and penetration testing.

2. MATERIALS AND METHODS

β-Conglycinin-rich and glycinin-rich SPIs were prepared from two new soybean lines that have extremely different compositions of β-conglycinin and glycinin [6]. Proportion of β-conglycinin and glycinin was determined by sodium dodecyl sulfate-polyacrylamide gel electrophoresis (SDS-PAGE) with densitometry. The composition of β-conglycinin and glycinin were 56% and 20% for β-conglycinin-rich SPI, and 16% and 62% for glycinin-rich SPI, respectively. The total of β-conglycinin and glycinin was similar in β-conglycinin-rich and glycinin-rich SPI. β-Conglycinin and glycinin fractions were obtained from an ordinary variety of soybeans, Enrei, as described previously [7]. The purity of β-conglycinin and glycinin fractions was also determined by SDS-PAGE with densitometry. The purity of β-conglycinin and glycinin fractions was 85% and 91%, respectively. Each solution of SPI was pasteurized at 140 ℃ for 20s using steam and spray-dried at 75-80℃ to yield preheated SPI.

Dynamic viscoelastic measurements were performed using a CLS-500 rheometer (Carri-Med Ltd., Surrey, U.K.). Samples of 12% SPI was subjected to shear oscillations at 1 Hz frequency and 0.025 strain. The storage modulus G' was determined at pH 7 while temperature was increased from 30 to 80 ℃ and next lowered to 20 ℃ at 2 ℃/min. Penetration measurements were performed using a Rheoner RE-33005 (Yamaden Co. Ltd., Tokyo) attached a 2 kgf load cell. The paste of 18% protein was heated at 80 ℃ for 30 min in oil bath and cooled with running water, and then stored at 4 ℃ for more than 18 h. SPI test pieces for the penetration test had a cylindrical shape with a diameter of 20 mm and a height of 20 mm. A cylindrical plunger (diameter 1.5 mm) with a ball (diameter 5 mm) at the end penetrated the sample at a speed of 300 mm/min.

3. RESULTS

3.1. The gelation properties of β-conglycinin-rich and glycinin-rich SPIs

Preheated β-conglycinin-rich and glycinin-rich SPIs were prepared from new soybean lines with extremely different β-conglycinin and glycinin content. The gelation properties of the β-conglycinin-rich and glycinin-rich SPIs were investigated using dynamic viscoelastic measurements. Figure 1 shows the results in the absence of NaCl. An increase of the storage modulus G' was observed for preheated β-conglycinin-rich SPI but not for preheated glycinin-rich SPI at 12% protein concentration. Figure 2 shows the results in the presence of 2.5% NaCl. The value of G' was increased for preheated β-conglycinin-rich SPI with adding 2.5% NaCl. An increase of G' was observed for preheated glycinin-rich SPI in the presence of 2.5% NaCl (Fig. 2) although no increase of G' was observed in the absence of NaCl (Fig. 1). The gel forming ability improved by adding 2.5% NaCl for both β-conglycinin-rich and glycinin-rich SPIs.

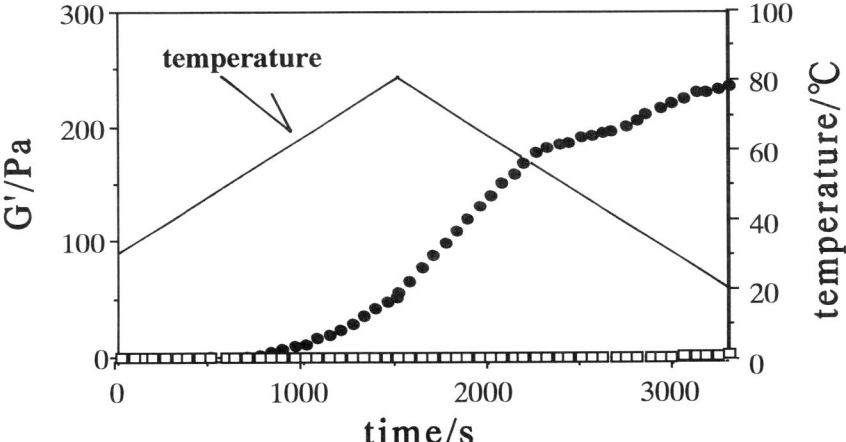

Figure 1. Gelation of 12% (w/w) preheated SPI at pH 7 in the absence of NaCl. The temperature was increased from 30 to 80 ℃ and lowered to 20 ℃ at 2 ℃/min. ●, β-conglycinin-rich; □, glycinin-rich.

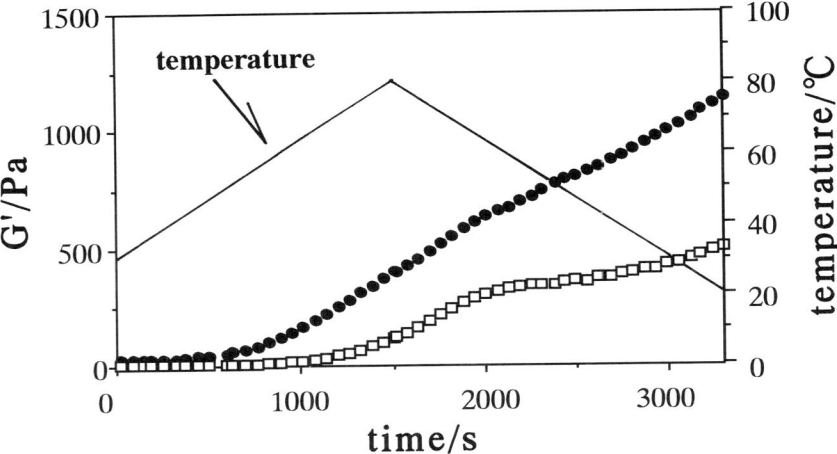

Figure 2. Gelation of 12% (w/w) preheated SPI at pH 7 in the presence of 2.5%
NaCl. The temperature was increased from 30 to 80 °C and lowered to
20 °C at 2 °C/min. ●, β-conglycinin-rich; □, glycinin-rich.

3.2. The gel properties of preheated SPIs of β-conglycinin and glycinin fractions

Preheated SPIs of β-conglycinin and glycinin fractions were prepared by a typical method for manufacturing commercial soybean protein isolate (SPI). The gel properties of preheated β-conglycinin and glycinin fractions were studied by penetration testing at 18% protein concentration in the absence and presence of 2.5% NaCl. Table 1 indicates the maximum force, breaking length, and initial slope at small strain range.

The preheated β-conglycinin fraction formed a heat-induced gel in the absence of NaCl. By adding 2.5% NaCl, the initial slope increased, and the breaking force and the breaking length decreased. Preheated glycinin fraction formed only a weak gel in the absence of NaCl, however, the breaking force and breaking length of preheated glycinin gel increased by adding 2.5% NaCl. In the absence of NaCl, the values of breaking force and breaking length of preheated β-conglycinin fraction were larger than these of preheated glycinin fraction. On the contrary, in the presence of 2.5% NaCl, the values of breaking force and breaking length of preheated β-conglycinin fraction were smaller than these of preheated glycinin fraction. Both in the absence and presence of 2.5% NaCl, the initial slope was larger for the gel of the preheated β-conglycinin fraction than for the gel of the preheated glycinin fraction.

Table 1
Gel properties of preheated SPIs at 18% protein concentration in the absence and in the presence of 2.5% NaCl.

		β-conglycinin fraction,	glycinin fraction,
−NaCl	Breaking Force (g)	576±26	127±16
	Breaking Length (mm)	18.0±0.3	14.9±0.7
	Initial Slope (g/mm)	13	3
+NaCl	Breaking Force (g)	387±9	603±35
	Breaking Length (mm)	10.4±0.1	16.0±0.3
	Initial Slope (g/mm)	48	12

a) SPI test pieces for the penetration test had a cylindrical shape with a diameter of 20 mm and a height of 20 mm.
b) A cylindrical plunger (diameter 1.5 mm) with a ball (diameter 5 mm) at the end penetrated the sample at a speed of 300 mm/min.
c) Mean value ± standard deviation of five replications.

4. DISCUSSIONS

We prepared preheated SPIs with different content of β-conglycinin and glycinin by a typical method for manufacturing commercial SPI. Rheological properties were studied by dynamic viscoelastic measurements and penetration testing.

Dynamic viscoelastic measurements indicated that gelation properties of preheated β-conglycinin-rich and glycinin-rich SPIs were different. An increase of G' was observed for preheated β-conglycinin-rich SPI but not for preheated glycinin-rich SPI at 12% protein concentration in the absence of NaCl. By adding 2.5% NaCl, the values of G' were increased for preheated β-conglycinin-rich SPI and the increase of G' was also observed for preheated glycinin-rich SPI.

Penetration test showed that preheated β-conglycinin and glycinin fractions had different behavior in force-penetration curves. The initial slope of preheated β-conglycinin fraction increased by adding NaCl. Preheated glycinin fraction formed a very weak gel at 18% protein concentration in the absence of NaCl, however, the gel forming ability was improved by adding NaCl (Table 1). Two different rheological methods showed that gelation and gel properties were changed by adding NaCl for both preheated β-conglycinin and glycinin.

Moreover, the value of G' of preheated SPI gels also increased at 2.5% NaCl in the order glycinin-rich SPI < β-conglycinin-rich SPI (Fig. 2). The initial slope

434

of preheated SPI gels increased in the same order if 2.5% NaCl is added. The initial slope, where the deformation is only very small, gives the rigidity or shear modulus [8]. The results of dynamic viscoelastic measurements are in good agreement with the results of penetration testing.

It was reported that preheat-treatment improved gel forming ability for proteins such as SPI [4], glycinin [5], egg white [9], and milk whey protein [10]. In the previous study [3], the value of G' of unheated β-conglycinin-rich SPI decreased by adding NaCl. In the present study, the value of G' of preheated β-conglycinin-rich SPI increased by adding NaCl (Figs. 1 and 2). Preheat-treatment must induce conformational changes of protein molecules, which may be molten globule state. It is likely that globular proteins in the molten globule state mediate food protein functionality, more so than the native or denatured states [11, 12]. Investigations on protein conformational states of preheated β-conglycinin and glycinin are necessary to explain the difference of their gel properties.

REFERENCES

1. N. S. Nielsen, Structure of Soy Proteins. In New Protein Foods, A. M. Altschul, H. L. Wilcke (eds.), Academic Press Inc., New York, 1985.
2. T. Nagano, T. Akasaka and K. Nishinari, Biopolymers 42 (1994) 1303.
3. T. Nagano, Y. Fukuda, T. Akasaka, J. Agric. Food Chem. 44 (1996) 3484.
4. T. Furukawa and S. Ohta, Nippon Shokuhin Kogyo Gakkaishi (in Japanese) 28 (1981) 451.
5. T. Nakamura, S. Utsumi, T. Mori, J. Agric. Food Chem., 33, (1985) 1201.
6. K. Kitamura, Trends Food Sci. Technol. 4 (1993) 64.
7. T. Nagano, M. Hirotsuka, H. Mori, K. Kohyama and K. Nishinari, J. Agric. Food Chem. 40 (1992) 941.
8. D.G. Oakenfull, Foods Food Ingredients J. Jpn., 167 (1996) 48.
9. A. Kato, H.R. Ibrahim, T. Takagi and K. Kobayashi, J. Agric. Food Chem. 38 (1990) 1868.
10. Y. Kinekawa and N. Kitabatake, Biosci. Biotechnol. Biochem. 59 (1995) 834.
11. E. Doi, Trends Food Sci. Technol., 4 (1993) 1.
12. M. Hirose, Trends Food Sci. Technol., 4 (1993) 48.

HYDROCOLLOIDS – PART 1
Edited by K. Nishinari
2000 Elsevier Science B.V.

Effects of NaCl and temperature on the gelation of soybean glycinin

T. Wongprecha[a], T. Takaya[a], T. Kawase[a], T. Nagano[b], K. Nishinari[a]

[a]Faculty of Human Life Science, Osaka City University
Sumiyoshi, Osaka 558-8585, Japan

[b]Faculty of Education, Ehime University, Matsuyama-shi, Ehime 790-77

Effects of NaCl and heat treatment on the gelation of glycinin dispersions were studied by dynamic viscoelastic measurement, Fourier transform infrared spectroscopy (FTIR) and differential scanning calorimetry (DSC). Storage shear modulus G' and the absorbance at 1618 cm^{-1} in FTIR spectrum for glycinin dispersions as a function of the added NaCl concentration became maximum at 0.5% NaCl. The endothermic peak in DSC heating curves shifted to higher temperatures with increasing concentration of the added NaCl. The shift of the endothermic peak to higher temperatures with increasing concentration of the added NaCl was attributed to the stabilization of glycinin molecules by the addition of salt.

1. INTRODUCTION

11S is composed of 6 acidic subunits and 6 basic subunits packed in two identical hexagons and stacked on the other [1]. Since glycinin has a gelling ability, it plays an important role in food industry. The mechanism of heat induced gelation involves the unfolding or dissociation of the protein and follows by the association reactions which under appropiate conditions results in the formation of the gel network structure [2]. The gelation process comprises progel state and gel state [3].

Gelation process is a complicated process in which the solvent conditions affect the network characteristics. The effects of ionic strength, pH, and heat treatment in the soy protein molecules have been studied on both molecular level and on rheological properties [4-11].

In the present paper, the effects of NaCl and heat treatment on the gelation process were studied by dynamic viscoelastic measurement, Fourier transform infrared spectroscopy (FTIR) and differential scanning calorimetry (DSC). The dynamic shear measurement was performed within linear viscoelastic regime.

2. MATERIALS AND METHODS

Glycinin was kindly supplied by Fuji Oil Co.,Ltd. The dispersions with final concentration of 5 wt% were prepared by dissolving the dry matter in distilled water or 35 mM potassium phosphate buffer by a magnetic stirrer at room temperature. Solid NaCl on a weight percent basis was added to the dispersions until the desired concentrations were reached. The pH of the dispersions was adjusted to the desired pH by adding a small amount of 1 N NaOH or 1 N HCl. The glycinin dispersions were de-aerated by an aspirator before heat treatment.

2.1 Rheological measurement

The storage and the loss shear modulus, G' and G", were determined by a fluid spectrometer RFS II (Rheometrics, Far East, Tokyo). The dispersions were loaded between cone and plate (0.1 rad, 0.05 mm gap, 25.0 mm diameter) which had been kept at 90°C beforehand. The dispersions were heated for 40 min and then cooled down to 20°C at 1°C/min. Silicone oil was used to cover the surface of the upper geometry to prevent the evaporation of water.

2.2 Fourier transform infrared spectroscopy (FTIR) measurement

Infrared spectra were recorded on FT-IR spectrometer 1720X (Perkin Elmer). Spectral measurement was carried out at a resolution of 2 cm^{-1} with 100 scans. Glycinin dispersions were prepared by dissolving glycinin in D_2O solution of 35 mM phosphate, pD 7.6. Glycinin dispersions were set between two CaF_2 disks with 15 μm pathlength and sealed with a Teflon tape. The samples were heated in drying oven (DS-42, Yamato, Tokyo, Japan) at various temperatures for 40 min and cooled down to 25˚C for 1 hour. To obtain the difference spectra, the spectra of unheated dispersions were subtracted from the spectra of heated dispersions.

2.3 Differential scanning calorimetry (DSC) measurement

The DSC heating curves were obtained by a Micro DSC III (Setaram, Caluire, France). The sample pan was filled with approximately 0.85g of glycinin dispersions and hermetically sealed. Distilled water was used as a reference.

3. RESULTS AND DISCUSSION

3.1 The effect of NaCl

The DSC heating curves of 5 wt% glycinin dispersions with different NaCl concentrations in distilled water and in phosphate buffer, pH 7.5 were shown in Figs. 1a and 1b, respectively. The endothermic peak became sharper with increasing concentration of the added NaCl for the dispersions in distilled water. In contrast, the dispersions in phosphate buffer showed little difference of the peak shape with varied NaCl concentrations. The denaturation temperature of the dispersions in phosphate buffer was higher than that in distilled water at the same NaCl concentration. NaCl shifted the endothermic peak to higher temperature both in distilled water and in phosphate buffer. The effect of salt on stabilizing quaternary structure of glycinin [12,13] would result in the shift of the denaturation temperature to higher temperatures in the dispersions with increasing concentration of the added NaCl.

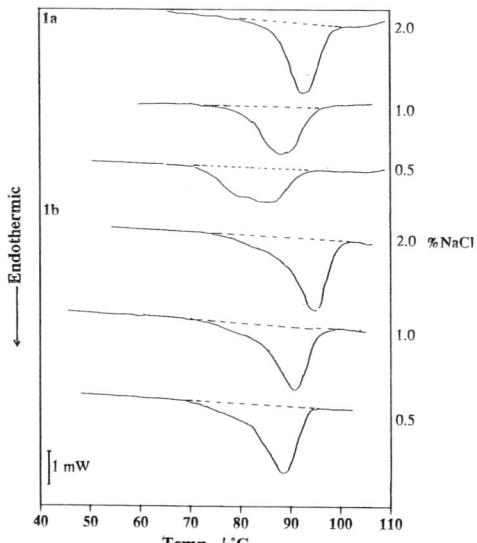

Figure 1. DSC heating curves of 5 wt% glycinin dispersions pH 7.5 with different NaCl concentrations, pH 7.5; a). in distilled water, b). in 35 mM phosphate buffer. Heating rate : 1°C/min.

The storage modulus of the glycinin dispersions in phosphate buffer at various NaCl concentrations is shown in Fig. 2. G' of the dispersions at a low NaCl concentration (0.5%) showed a sharp increase during cooling, whilst G' of the dispersions with higher NaCl concentrations (1-2%), showed a slight increase in cooling period. At low ionic strength, the dissociation into subunits was accelerated. Further increase in ionic strength results in preventing the destruction of quarternary structure into subunits [6]. The other thing that should be considered is that the temperature of the experiment was 90°C. This temperature may not cause the denaturation of the protein molecule in the presence of higher NaCl concentration as effectively as protein molecule at lower NaCl concentration since the DSC heating curves of the dispersions at 0.5%, 1% and 2% showed the endothermic peak temperature at 87.8°C, 90.4°C and 94°C, respectively.

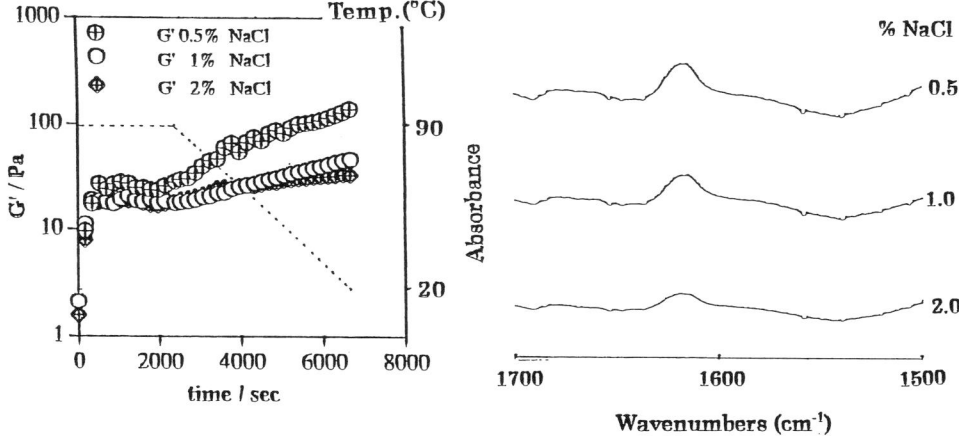

Figure 2. The storage modulus of 5 wt% glycinin dispersion in phosphate buffer, pH 7.6 with diffferent NaCl concentrations. The dispersion was kept at 90°C for 40 min and then cooled down to 20°C at 1°C/min.

Figure 3. Difference spectra of 5 wt% glycinin dispersions in D₂O solution of 35 mM phosphate buffer, pD 7.6 with various NaCl concentrations. The spectra were obtained by subtracting the spectrum of an unheated dispersion from that of a gel which was heated at 90°C for 40 min and cooled down to 20°C.

Fig. 3. shows the FTIR spectra of the glycinin dispersions in D_2O solution of 35 mM phosphate at pD 7.6. The band at 1618 cm^{-1} decreased with. increasing of the added NaCl. The band at around 1620 cm^{-1} indicates the characteristics of the β-sheet structure [14] which is the support structure of the gel network. The highest absorbance of the band at 1618 cm^{-1} of the dispersions with 0.5% NaCl concentration and the reduction of the absorbance as the NaCl concentration increased agreed well with the data obtained from rheological measurement in Fig. 2.

438

3.2 The effect of temperature

The effect of temperature on the storage modulus (G') is shown in Fig. 4. Dispersion heated at 90°C showed the highest G', and G' decreased as the heating temperature decreased. The similar result was also observed in the FTIR measurement, which is shown in Fig. 5. The absorbance of the band at 1618 cm^{-1} decreased as the heating temperature decreased. The higher of the heating temperature, the more exposure of the functional groups, which facilitates the network formation.

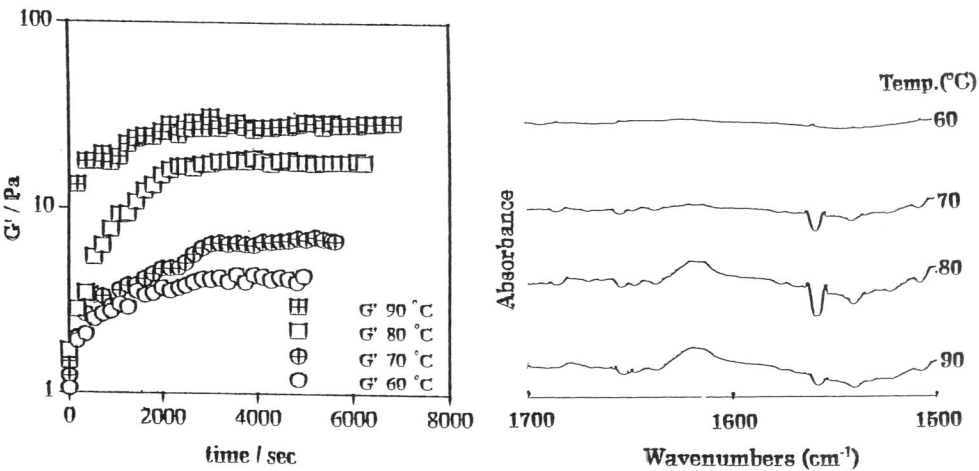

Figure 4. The storage modulus of 5 wt% glycinin dispersion in phosphate buffer, pH 7.6. The
dispersions were heated at 60, 70, 80 and 90°C for 40 min and cooled down to 20°C at 1°C/min.

Figure 5. Difference spectra of 5 wt% glycinin dispersions in D$_2$O solution of 35 mM phosphate buffer, pD 7.6. The dispersions were heated at different temperatures for 40 min.
The spectra were obtained from subtracting the spectrum of an unheated dispersion from that of
gel which was heated at 90°C for 40 min and cooled down to 20°C.

REFERENCES

1. R.A. Badley, D. Atkinson, H. Houser, G. Oldani, J.P. Green, J.M. Stubbs, Biochim. Biophys. Acta, 412 (1975) 214.
2. R.H. Schmidt, Gelation and Coagulation in "Protein Functionality in Foods", ACS symposium series 147, Ed.J. P. Cherry, Am. Chem. Soc., Washingon, DC, (1981) 131
3. N. Catsimpoolas, E.W. Meyer, Cereal Chem., 47 (1970) 559.
4. N. Catsimpoolas, S.K. Funk, E.W. Meyer, Cereal Chem., 47 (1970) 331.
5. W.J. Wolf and T. Tamura, Cereal Chem., 46 (1969) 331.
6. S. Iwabuchi and K. Shibasaki, Agric. Biol. Chem., 45 (1981) 285.
7. S.J. Circles, E.W. Meyer, R.W. Whitney, Cereal Chem., 41 (1964) 157.
8. T. Beveridge, L. Jones, M.A. Tung, J. Agric. Food Chem., 32 (1984) 307.

9.F.S.M. Van Kleef, Biopolymers 25 (1986) 31.
10.T. Nagano, H. Mori, K. Nishinari, Biopolymers 34 (1994) 293.
11.I.S. Chronakis, Food Research International 29 (1996) 123.
12.A.M. Hermansson, J. Texture Stud., 9 (1978) 33.
13.M. Babajimopoulos, S. Damodaran, S.S.H. Rizvi, J.E. Kinsella, J. Agric. Food Chem. 31 (1983) 270.
14.W. K. Surewicz, H.K. Mantsch, D. Chapman, Biochemistry 32 (1993) 389

HYDROCOLLOIDS – PART 1
Edited by K. Nishinari
2000 Elsevier Science B.V.

Rheological study on pH-induced gelation of reconstituted skim milk at different temperatures and concentrations

A. Tobitani, N. Ueda, Y. Shiinoki, K. Joho and T. Yamamoto

Technology and Research Institute, Snow Brand Milk Products Co., Ltd.
1-1-2 Minamidai, Kawagoe, Saitama 350-1165, Japan

pH-Induced gelation of reconstituted skim milk was studied using a dynamic oscillatory rheometer. The acidification was achieved by hydrolysis reaction of glucono-δ-lactone, and the effects of reaction temperature and protein concentration on gelation were examined. The temperature dependence of gelation pH was found significant, while almost no effect of milk concentration was shown. By converting the data of gelation pH vs temperature and concentration, we obtained a sol-gel diagram, in which a gel region exists at higher temperatures.

1. INTRODUCTION

Milk gels are among the very popular materials in our food. One such product is yoghurt, which is produced from bovine milk by bacterial acidification. In manufacturing yoghurt, there are several factors which affect the quality of the final products, such as processing temperature and protein content. These factors are considered to determine the gel formation and therefore need to be investigated in connection with the gelation kinetics, which in the case of yoghurt can be described as a function of time and pH.

In this study, we used reconstituted skim milk acidified with glucono-δ-lactone as a model system for yoghurt. The effects of temperature and protein concentration on gelation were evaluated by the use of a strain-controlled oscillatory rheometer. The results were discussed in connection with gelation thresholds and sol–gel diagrams; this kind of approach has not been attempted before for milk systems. Prior to the measurement, we also tested two different measuring geometries of the instrument in order to check for an occurrence of 'syneresis', which very often appears for milk gels and causes a slip between the sample and the attachment.

2. EXPERIMENTAL

2.1. Materials

Commercial skim milk powder (Snow Brand Milk Products Co Ltd, Japan) was dissolved into deionised water to prepare 5-30 % w/w reconstituted skim milk (abbreviated as RSM). The main components in the powder were as follows (% w/w): protein 35.4, fat 0.8, saccharide 51.7, ash 8.1, water 4.0. The RSM solution was stirred for 24 h at 10°C and pasteurised at 90°C for 10 min, followed by cooling to 10 °C.

In this experiment, acidification of milk was achieved by using glucono-δ-lactone (GDL). This chemical is hydrolysed in water, producing gluconic acid, and, as a result, lowers the pH. The acid production rate depends mainly on the GDL concentration and reaction temperature, although it is also largely affected by the buffer action of RSM in this case.

An appropriate amount of GDL powder (G-4750, Sigma chemical Co, USA) was added to the RSM solution and stirred for 1 min for dissolution. The sample was kept at 10 °C because the subsequent hydrolysis needed to be suppressed. The amount of the added GDL was determined in a range of 1-3 % w/w depending on the RSM concentration and reaction temperature so as to keep the rate of pH decrease as constant as possible.

2.2. Measurements

Gelation processes were monitored by using a strain-controlled oscillatory rheometer (RFS II, Rheometric Scientific FE, USA). The measuring geometries selected were a cone/plate (50 mm in diameter, 0.02 rad in angle) and a Couette (bob: 32 mm in outer diameter, 33 mm in length; cup: 34 mm in inner diameter). After GDL addition, the sample solution was transferred to the rheometer and liquid paraffin was put around the sample for preventing evaporation. Since the temperature of the attachment, chosen from a range of 10-60°C, was kept constant beforehand, the sample temperature was seen to be raised within 30 s, which was thought negligibly small in the present experiment. The strain and frequency were respectively set to 1% and 1 rad/s. The data were obtained at appropriate time intervals of 30 or 60 s.

Prior to the measurements of RSM in various conditions, we tested two different measuring geometries. This was intended to check how the data would be influenced by 'syneresis', which has been frequently observed for milk gels and produces failure due to slip between the sample and the attachment. By this preliminary test, we found that the effects of syneresis could not be eliminated by either of them. Nevertheless the Couette was thought to be more sensitive, and therefore was used for remaining experiments.

In parallel with the rheological measurements, the sample pH was monitored by using a pH-meter (D-14, Horiba Ltd., Japan) equipped with a glass electrode. An approximately 20 ml sample solution was put into a 50 ml centrifugal glass tube immersed in a temperature controlled bath. The pH data were taken from the voltage output of the meter and stored in a computer.

3. RESULTS AND DISCUSSION

3.1. Effects of attachment geometry and syneresis

The complex modulus G* for RSM with the two different attachments are plotted against pH in Fig. 1. It is found that G* using both geometries showed a rapid increase around pH5.4. The measurement with cone/plate then showed a maximum at 5.2 while with the Couette geometry this seemed at pH5.1. The sharp drop in the G* was due to a slip between the sample and the attachment, and this was originally caused by syneresis of the sample gel. Once the slip occurred, the sample could not be properly evaluated any more. This means that further analysis should be limited to the data taken before the drop.

Although the Couette geometry was found to be able to follow the gelation process further than the cone/plate, the difference was not thought significant. This suggested that we could select either of the geometries. In the remaining experiments, however, we used the Couette geometry because it gave higher torque signals and hence better quality of data.

These results suggest that the application of dynamic oscillatory rheometers to measurement of gelation is rather restricted if the sample tends to develop syneresis easily. Although a few attempts have been made recently [1], more improved attachments are necessary for accurate measurements of milk gelation.

3.2. Determination of gelation points

Typical profiles of the storage (G') and loss moduli (G") together with pH are shown in Fig. 2. The sample was 10% w/w RSM measured at 40°C. Initially the G' and G" were scattered owing to the low torque signals, and at approximately 17 min they sharply increased.

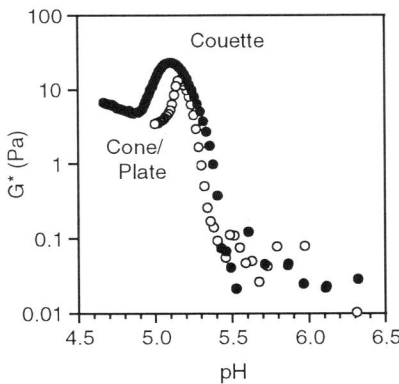

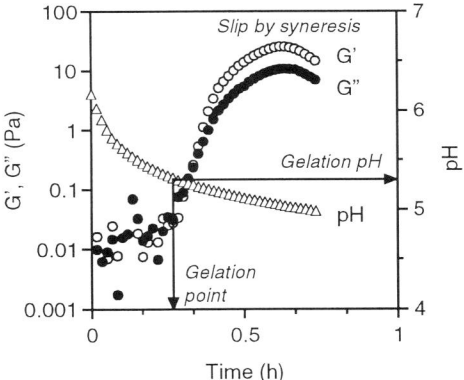

Fig. 1 Effects of attachment geometry on RSM gelation (10% RSM at 40°C)

Fig. 2 Typical profiles of dynamic moduli and pH at a gelation process of 10% RSM (40°C)

They both continued increasing but started decreasing over 30 min. As this drop was caused by slip, the measurement was stopped there.

A number of methods have been proposed for determining the gelation point [2-4]. Although Winter and Chambon's criterion can be said to be the most theoretically rigid of all, its application is often impossible in food systems because of experimental difficulty and pre-gel behaviour such as structuring of the solution [5, 6]. Considering the quality of the present data and the measurement conditions, we chose the following method: the gelation point is defined as the point where G' showed a pronounced increase above the noise level. In this case, the gelation point may be described as a function of time, and also of pH. Therefore we call the gelation point determined against pH as gelation pH.

3.3. Temperature dependence of gelation pH and a suggested gelation mechanism

In Fig. 3, the changes in G' for 10% w/w RSM at various temperatures are displayed as a function of pH. On the basis of the above criterion, the gelation pH were determined at each temperature and summarised as in Fig. 4. It seems that the gelation pH increased as the reaction temperature was raised. In the figure, the results for 20% w/w RSM are also plotted, although the trend does not seem very different.

This strong temperature dependence of gelation pH might be rationalised as follows. At native pH of milk ($\approx$ 6.5), the casein micelles are stabilised by electrostatic repulsion. As the pH is decreased towards the isoelectric point of the main casein components ($\approx$ 4.6), the micelles become more unstable and start to aggregate. The pH when the aggregation begins is thought to be determined by the balance of the electrostatic repulsion and other attractive forces. The present data suggest that this attractive force would be due mostly to the hydrophobic interaction.

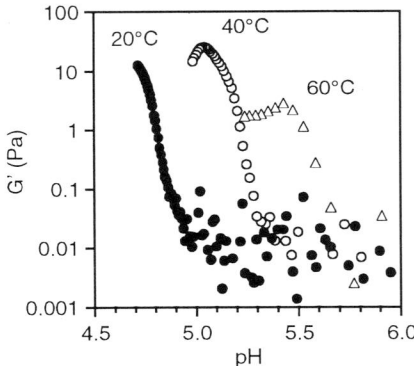

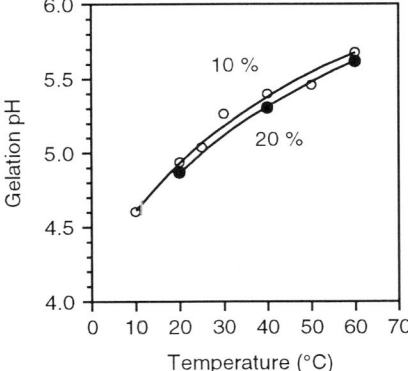

Fig. 3 pH dependence of G' at different temperatures (10% RSM)

Fig. 4 Temperature dependence of gelation pH at different concentrations

Several driving forces are said to be involved in milk gelation, such as hydrogen bonding, electrostatic interaction and hydrophobic interaction. Among them, only the hydrophobic interaction becomes significantly stronger with the increase in temperature. Therefore, if the temperature is raised, the contribution of the hydrophobic interaction becomes more dominant and consequently the gelation is expected to occur more easily. The relationship between gelation pH and the temperature as shown in Fig. 4 appears reasonable from this viewpoint. This should be confirmed by further experiments which examine the effects of hydrophobicity in more direct ways.

3.4. Concentration dependence of gelation pH and conversion to a sol–gel diagram

The gelation pH values at various RSM concentrations are displayed in Fig. 5. It was found that the gelation pH was not much affected by the RSM concentration at any temperatures. By converting these data and plotting in a different set of axes, the sol–gel boundary, or gelation curve, can be determined as shown in Fig. 6. Since the number of data points were not sufficient for the conversion and also because the reproducibility was not satisfactory, the lines are not well enough defined to discuss the shape of the gelation boundaries. Nevertheless, it is at least possible to note that the gel region exists above the lines and the sol region below. This means that gelation occurs at higher temperatures, and even suggests that the RSM may well have a lower critical solution temperature (LCST) diagram.

However, there is one thing which is not easy to be explained: the gelation boundary of RSM does not show any marked concentration dependence. It is theoretically predicted that the gelation threshold, which in many cases is presented as a function of temperature, should depend strongly on the polymer concentration [7-9]. This has been reported for both synthetic polymers [10, 11] and biopolymers [7]. Moreover, some proteins obtained from bovine milk

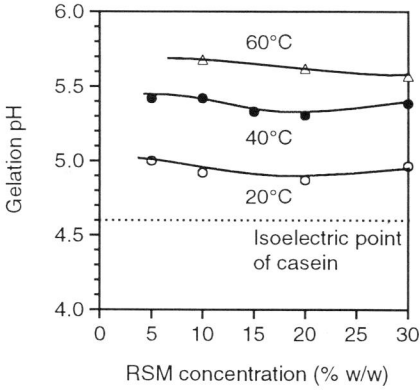

Fig. 5 Concentration dependence of gelation pH at different temperatures

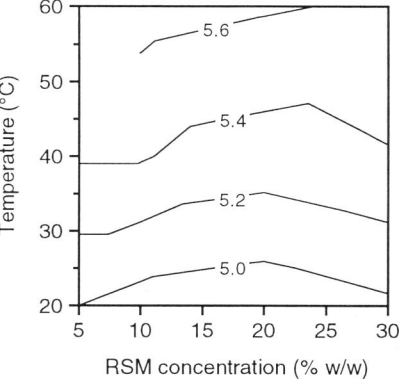

Fig. 6 Sol – gel diagram of RSM at various pH (Sol region: below the lines, gel region: above)

have been shown to demonstrate a strong concentration dependence of their gelation temperature [6, 12].

If the polymer possesses an LCST type diagram, the gelation temperature becomes higher as the polymer concentration decreases, and the gelation does not occur below a certain concentration; that is the critical concentration for gelation (more detailed discussion, see e.g. [13]). For the upper critical solution temperature (UCST) type polymer, the inverse of this gelation behaviour should occur. Since the RSM is found to have a gel region at the higher temperature side, it would be expected to show the LCST type concentration dependence. Even taking account of the effective protein content in the RSM samples, however, the present data did not display such concentration dependence.

The absence of concentration dependence would suggest that it may not be possible to explain the RSM gelation by acidification in the ordinary gelation scheme. Alternatively, another cooperative reaction such as phase separation could be involved in the gelation [9, 14, 15]. It is not yet possible to fully explain the acid-induced gelation of RSM, and therefore further experimental studies are required.

REFERENCES

1. R.K. Richardson and F.M. Goycoolea, Carbohydr. Polym., 24 (1994) 223.

2. H.H. Winter and F. Chambon, J. Rheol., 30 (1986) 367.

3. M. Stading and A.-M. Hermansson, Food Hydrocoll., 4 (1990) 121.

4. S.B. Ross-Murphy, Rheol. Acta, 30 (1991) 401.

5. T. Matsumoto and H. Inoue, Chem. Phys., 178 (1993) 591.

6. A. Tobitani and S.B. Ross-Murphy, Macromolecules, 30 (1997) 4845, 4855.

7. T. Tanaka, G. Swislow and I. Ohmine, Phys. Rev. Lett., 42 (1979) 1556.

8. A. Coniglio, H.E. Stanley and W. Klein, Phys. Rev. B, 25 (1982) 6805.

9. F. Tanaka, Macromolecules, 23 (1990) 3784, 3790.

10. J. Eliassaf and A. Silberberg, Polymer, 3 (1962) 555.

11. H.-M. Tan, A. Moet, A. Hiltner and E. Baer, Macromolecules, 16 (1983) 28.

12. P.L. San Biagio, D. Bulone, A. Emanuele and M.U. Palma, Biophys. J., 70 (1996) 494.

13. S. B. Ross-Murphy and A. Tobitani, in *The Wiley Polymer Networks Group Review: Proceedings 13th Polymer Networks Group Conference - Polymer Networks 96*, Vol. 1, edited by K. te Nijenhuis and W. J. Mijs (John Wiley, UK, 1998), pp. 39-49.

14. J. Arnauts and H. Berghmans, Polym. Commun., 28 (1987) 66.

15. S. Callister, A. Keller and R.M. Hikmet, Makromol. Chem. Macromol. Symp., 39 (1990) 19.

HYDROCOLLOIDS – PART 1
Edited by K. Nishinari
2000 Elsevier Science B.V.

Viscoelastic properties of acid-induced milk gel at different gelation temperatures

R. Niki, H. Motoshima and F. Tsukasaki

Research Center, Yotsuba Milk Products Co., Ltd.
465-1, Wattsu, Kitahiroshima, Hokkaido, 061-1264 JAPAN

Effects of the gelation temperature on the viscoelastic properties of acid-induced skim milk gel were investigated using a dynamic rheological measurement apparatus. The maximum of storage modulus of gel and the rate of gelation increased with increasing gelation temperatures, whereas the gelation time decreased with gelation temperatures. The apparent activation energy of the gelation was calculated from Arrhenius plot of the gelation rate, and its value was approximately 17 kcal/mol.

1. INTRODUCTION

When milk is acidified slowly and its pH approaches the isoelectric point of casein (pH=4.6), aggregation and subsequent coagulation of casein micelles in milk occur and, furthermore, a gel network is formed . This is the principle of production of fermented milk products (yogurt, cheese, quarg, etc.). The objective of this study was to investigate the gelation of milk induced by the acidogen glucono-δ-lactone (GDL). In particular, the effects of gelation temperature on the viscoelastic properties of the milk gel were examined.

2. EXPERIMENTAL

2.1 Materials

Reconstituted skim milk: 12 g of skim milk powder (a product of Yotsuba Milk Products Co., Ltd.) was dissolved in 100 ml of deionized water.

2.2 Measurements

Viscoelastic measurements: 5 ml of sample solution was held in a water bath adjusted thermostatically to the measuring temperature. A weighed portion of GDL powder (0.2g/5ml) was added to the reconstituted skim milk, then the mixture was vigorously stirred for 30 sec and 1.65 ml was transferred to the cell of a rheological measurement apparatus

(Rheolograph-Sol, Toyoseiki Seisakusho Ltd., Tokyo, Japan). The storage modulus (G')
and loss modulus (G") of the samples were measured at 2 Hz, with the amplitude of 0.125
mm, at a constant temperature (in the range of 3 to 30°C) as a function of time after the
addition of GDL. The shear strain employed in the measurements was 0.125.

3. RESULTS AND DISCUSSION

When milk is acidified by an acidogen, the aggregation and gelation of casein micelles
occur as the pH of the milk approaches the isoelectric point of casein (pH=4.6). Principally, the acidification of milk by an acid brings about interactions between casein micelles

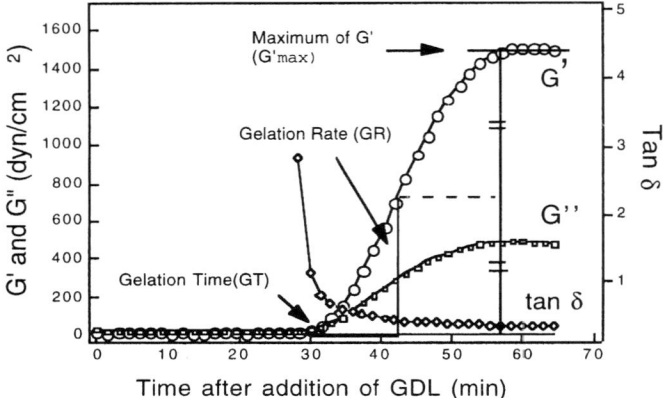

Figure 1. A typical time course of G', G" and loss tangent of
reconstituted skim milk after addition of GDL

by reducing their net negative charges and then the micelles begin to coagulate and form a
gel network.

The behavior of casein micelles in milk upon acidification is markedly dependent on the
method of acidification. The casein micelles in milk coagulate and precipitate when the
pH falls rapidly below 5.0 at temperatures above room temperature, while the micelles
aggregate and form a gel if the pH falls slowly as in the case of lactic fermentation[1].
Glucono-δ-lactone (GDL) has been used as an acidogen to investigate the properties of
acid-induced milk gel, it is slowly hydrolyzed to gluconic acid after its addition to milk.
Consequently, milk can form a smooth gel as in the case of yogurt. For these reasons, the
acidogen GDL was used in this study.

The storage modulus (G') and loss modulus (G") of the samples were determined using a rheometer as a function of time after the addition of GDL. Figure 1 shows the typical time course of changes in G', G" and the mechanical loss tangent (tan δ =G"/G') of reconstituted skim milk gel. G' and G" increased with time after a certain latent time and approached a maximum value (G'max). As shown in Fig.1, we obtained the values of three parameters from the curves of G', that is, the maximum of storage modulus (G' max), the geltime (GT) and the rate of gelation (GR). The gelation time was estimated as the time at which the shear modulus began to deviate from baseline. The rate of gelation (GR) was calculated as follows: the value of one-half G' max was divided by the time elapsed from GT to the time where G' reached the value of one-half G' max.

The time course of changes in G' and G" of reconstituted skim milk after the addition of GDL at different temperatures was examined (Fig. 2). The time course curves (the so-called gelation curve) of G' and G" basically showed the same trends, for example, the GT

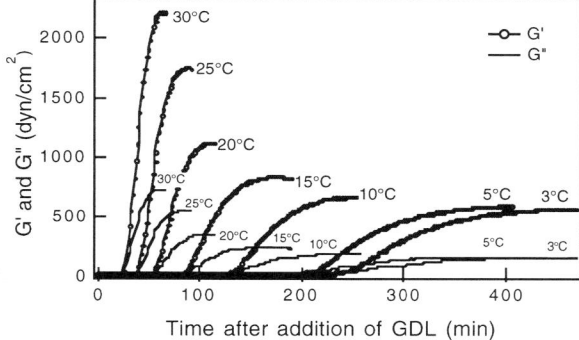

Figure 2. Time course of G' after addition of GDL at different temperatures

values in the case of G' and G" were approximately the same, and also G' and G" reached a maximum value at the same elapsed time after addition of GDL.

For the three parameters, the values obtained from the gelation curves in Fig.2 are tabulated in Table 1, where the results obtained at other gelation temperatures not shown in Fig.2 are also included.

G'max and GR increased with increasing gelation temperature. On the othr hand, GT decreased with increasing gelation temperature. In this connection, Bringe and Kinsella[2] investigated the effects of temperature and pH on the rate of casein coagulation induced by acid and observed that the rate of casein micelle coagulation at any pH decreases upon lowering the temperature from 25°C to 5°C. They assumed that hydrophobic interactions are the major driving force for coagulation and gelation of casein upon acidification, since these interactions reduced as the temperature decreases [2]. Our results in Table 1 support

450

Gelation temp. (°C)	Gelation time (min.)	G'max (dyn/cm²)	Gelation Rate (dyn/cm² · min)
3	228	582	3.9
5	199	578	4.85
8	148	678	6.98
10	121	664	7.64
13	99	762	11.9
15	82	821	14.9
18	64	995	20.3
20	52	1110	28.5
23	42	1440	40
25	36	1740	48.3
28	27	2210	61.4
30	22	2210	66.9

Table 1. Gelation time, maximum of storage modulus and rate of gelation at different temperatures

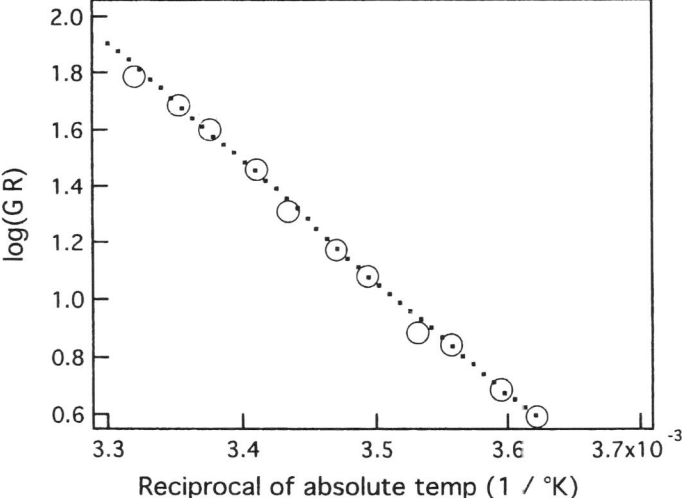

Figure 3. Plots of log (GR) vs. reciprocal of absolute temperature

the assumption of Bringe and Kinsella.

The temperature dependence of gelation rate was investigated in the present study. Arrhenius plots (log (GR) vs. the reciprocal of absolute temperature) are shown in Fig. 3. As described above, the gelation rate of the samples increased with increasing temperature. In the Arrhenius plots, the log (GR) values increased with decreasing temperature below 32°C. The plots each gave a straight line at temperatures between 5°C and 23°C. From the slope of the straight line, the apparent activation energy was calculated and found to be approximately 17 Kcal/mol. Tokita et al.[3] and Tuzynski [4] reported apparent activation energy values of 10 and 21 Kcal/mol, respectively, for rennet gel of skim milk.

REFERENCES

1. J. Heertje, J. Visser and P. Smits, Food Microstructure. 4 (1985) 267.
2. N. A. Bringe and J. E. Kinsella, Forces involved in the enzymic and acidic coagulation of casein micelles, *in* Development in food Proteins-5, B. J. F. Hudson (ed.), Elsevier Applied Sci., (1987) 159.
3. M.Tokita, K. Hikichi, R. Niki and S. Arima, Biorheology, 19 (1982) 209.
4. W. B. Tuzynski, J.Dairy Res., 38 (1971) 115.

HYDROCOLLOIDS – PART 1
Edited by K. Nishinari
2000 Elsevier Science B.V.

453

Characterization of Globin Hydrolyzates by Light Scattering

Xin Qi Liu , Mitsutoshi Nakajima and Yoh Sano

National Food Research Institute, Tsukuba City, Ibaraki 305, Japan

ABSTRACT

Globin hydrolyzates have successfully been prepared with citric acid, in order to improve a better solubility and more excellent ability on gel formation induced by heating than intact globin, and to control the texture of gel. In this study, we studied detailed gel network formation of globin hydrolyzate by light scattering methods. The globin hydrolyzates containing mainly 8 kinds of peptides are obtained by hydrolyzing the globin with 8 M citric acid. In the process of gel formation, a rod-shaped aggregate with a length of 130-140 nm and a molecular weight of about 870,000 Da is formed by a noncovalent bond between the globin β-chain and the peptide β-1. The peptide α-1 originated from globin α-chain has a high hydrophilicity, and shows the properties of association and dissociation depending on concentration and temperature. The hypothetical gel formation mechanism of the globin hydrolyzates is presented as follows: initially the randomly coiled polymers are produced by about 8 molecules of peptide α-1, next the crosslinked structures are constructed between the length of 130-140 nm rod-shaped aggregates, and then the network is gradually formed and the gel is finally formed.

INTRODUCTION

The blood of animal contains about 18% protein and hemoglobin accounting for more than half of the blood protein. Most of the hemoglobin is not utilized in the food industry because of its unattractive color and odor. The preparation methods of decolorized globin, nutritional studies and the functional properties of globin such as solubility, emulsifying properties and foaming properties were studied by several researchers[1,2]. Attempts have been made to incorporate globin into sausage-meat, and a cheese-like emulsion was prepared by using globin as an ingredient. Although the abilities of proteins to form a gel and to provide a structural matrix for holding water applications are useful in food industry, very few studies have been done on the gelling properties of globin. Heat-induced globin gel prepared by each of the above-mentioned methods had a low hardness, and was affected easily by many factors such as pH and ions, and lacked adequate functional properties[3,4].

In this study, globin hydrolyzates have successfully been prepared using citric acid. It was found that the globin hydrolyzates showed a better solubility and more excellent ability on gel formation induced by heating than intact globin, and formed another type of gel which was different from that of intact globin. Furthermore, we studied detailed gel network structures of globin hydrolyzate by scattering methods.

MATERIALS

Globin was prepared from porcine blood cells using Tybor's acidified acetone method with a little modification[1,2,5,6].

The decolorized powder globin was hydrolyzed with 0.8 M citric acid at 95 °C for 15 min, then cooled to room temperature and dialyzed against flowing water with Seamless Cellulose Tubing (Molecular cutoff size 10,000 Da). The dialyzate was air-dried with a spray-dryer (Pulvis minispray-GA-32, Yamato Science, Tuskuba, Japan), then the resulting sample was designated globin hydrolyzate sample[3,4,5,6,7].

RESULTS AND DISCUSSION

Characterization of Acid Hydrolyzed Globin Aggregates

The hydrolysis of globin was performed with 0.8M citric acid. The globin hydrolyzates showed 8 bands by tricine-SDS-PAGE, whose molecular weight ranged from 5,000Da to 15,000Da [5,6,7].

The self-assembling properties of globin hydrolyzate was checked by gel filtration with TSKgel TOYOPEARL HW-60S[5]. In the case of intact globin, a sharp monomer peak was detected at fraction number 18. When gel filtration of the hydrolyzates was done immediately after globin was hydrolyzed, the position of the main peak was the same as intact globin, except for the peaks of the lower molecular weights. The chromatography of the samples standing for 10hr and 24hr at 30°C showed a new peak formed by self-assembly and grew with increasing standing time [5]. On the contrary, the other peaks assigned to the lower molecular weight peptides became smaller with time. These results indicated that the large aggregates were formed easily in the case of the globin hydrolyzate. Comparing with the position of standard protein (Thyroglobulin, MW 669,000Da), the molecular weight of the aggregates was estimated as above 700,000Da. On the other hand, when globin hydrolyzates were treated with 8M urea gel filtration showed that the aggregates became samller in the presence of 8M urea. This result suggested that hydrophobic interaction among polypeptides was essential for the formation of aggregates. In order to identify the constituent of the aggregates, the fraction of aggregates was analyzed with Tricine-SDS-PAGE and protein sequencer. It is clear that the aggregates were composed of two kinds of polypeptides: one of which was β-chain originated from intact globin and the other was peptide β-1 originated from β-Chain of globin by cleavage between 99(Asp) and 100(Pro), which suggested that the aggregates were formed from β-chain and β-1 peptide combined specifically through noncovalent bonding[5].

Tertiary strucure of the aggregates in solution was measured with light scattering method[8,9,10]. According to light scattering theoretical, plot of $KC/R_{\theta \to 0}$ values extrapolated to $\theta \to 0$ against each C, as shown in Figure 1, gives moleuclar weight from the intercept of the ordinate and the second virial coefficient from the inclination of the straight line, respectively. The values obtained were M_w= 872,200Da and A_2=$-$6.075$\times$10^{-5} ml · mol/g. The Mw obtained from gel filtration was more than 700,000Da by comparison with the standand protein (Thyroglobulin 669,000Da). The Mw obtained by light scattering method was in accordance with which obtained from gel filtration. Therefore, it was estimated that the aggregate was formed by 33-34 units, each unit was in the ratio of 1:1 conplex of β-chain and β-1 of molecular weight of 16,000Da and 10,900Da, respectively [5].

The particle scattering factor was theoretically calculated for various kinds of models such as sphere, rod, random coil. etc. We tried many models in order to explain the agreement between scattering data and the theoretical values. The rod-shaped model was much more suitable in this case. Theoretical values were shown in Figure 2 by changing the length of the rods from 100nm to 250nm. The experimental values were in good agreement with the theoretical values of Lw=130-140 nm. It suggested that the aggregates of globin hydrolyzates in solution wass approximated with thin rod like model of the length 130-140nm. The electron micrograph result also showed that the aggregates consisted of the thin rod aggregates, which is in accordance with the light scattering data.

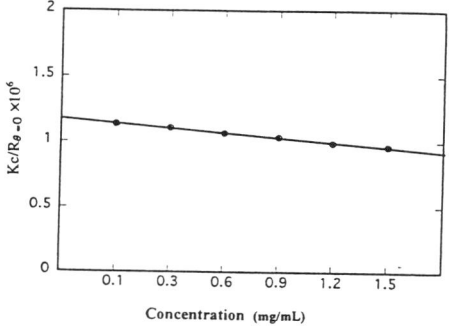

Figure 1. Plot of inverse scattering intensity extrapolated to θ =0 against concentration of the aggregates of globin hydrolyzates in 0.1M citric acid buffer, pH 3.3.

Figure 2. Particle scattering factor P(θ) versus Q(nm-1). Closed circles show the expermental data. The theoretical curve as thin rod shaped model is shown with open circles, whose length ranges from 100nm to 250nm, interval is 30nm.

Self-Assembling Properties of Peptide α-1

We have found that the aggregates of globin hydrolyzates existed in a monomolecular state. It is necessary that the cross-combination should take place between aggregates of globin hydrolyzates to form a network of the gel [6]. Table 1 showed that the gel did not form without adding the peptide α-1, and the gel was formed whenever the peptide α-1 was added. Furthermore, the gel hardness increased with an increase in peptide α-1 concentration, indicating that the peptide α-1 plays a role in transformation of aggregates of globin hydrolyzates to gel. In other words, peptide α-1 can be seen as crosslinker between aggregates of globin hydrolyzates [6,7].

Table 1. Gel formation and breaking stress of rod-shapedaggregates by the addition of peptide α-1.

Aggregates (%) [1]	5.0	4.5	4.0	3.5	3.0
Peptide α-1 (%) [2]	0.0	0.5	1.0	1.5	2.0
Gel formation	-	+	+	+	+
Breaking stress (g)		0.22	0.35	0.40	0.57

The gel was formed by heating at 90 ℃ for 15 min and allowed to stand for 10min at room temperature. Breaking stress of gel was measured with a rheometer. (+): gel was formed. (-): gel was not formed. [1] The rod-shaped aggregates. [2]Peptide α-1.

The peptide α-1 was obtained by cleavage of the peptide bond between 94 (Asp) and 95 (Pro) in the α-chain of globin with 0.8 M citric acid, it were isolated by hydrophobic interaction chromatography [5,6]. The hydrophobic chromatography results also showed that globin hydrolyzates were mixtures consisting of at least three types of polypeptides having different hydrophobicities. The peptide α-1 was a highest hydrophilic

456

polypeptide, which had a molecular weight of 9,839 Da. Furthermore, with using peptide α-1 treated with urea and/or heated at 50 ℃, the molecular weights in solution were measured by a light scattering method as shown in Figure 3. Peptide α-1 was a monomer in the presence of urea, and showed no concentration dependence. The estimated molecular weight was 10,000 Da, which was in agreement with the amino acid sequence data. However, the molecular weight in the absence of urea was 7-9 fold more than that of the monomer, which can be considered that peptide α-1 can self associate and form aggregates. The association depends on concentration of peptide α -1. Based on the observation that the aggregates of peptide α-1 dissociated to monomer when urea was present, we regarded that aggregates of peptide α-1 could be formed by noncovalent bonds. Moreover, when the molecular weight was measured at 50 ℃, the aggregates of peptide α-1 formed were smaller than those formed in untreated peptide α-1, although the molecular weight depends on the concentration of peptide α-1. From these results, it can be concluded that the association of peptide α-1 depends on the concentration of peptide α-1 and temperature [6]. The same results aslo were obtained by the small-angle solution X-ray scattering method [11,12,13]. (Figure 4).

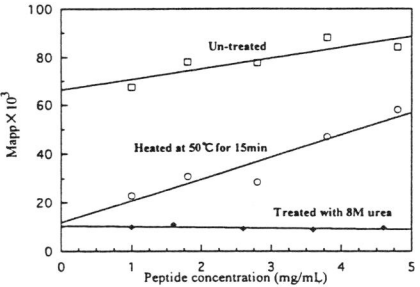

Figure 3. Concentration dependence of apparent molecular weight of peptide α-1. The apparent weight-average molecular weight was measured by the intensity of light scattered at 90° angle with the instrument of a modified Ellipsometer, an Automatic Light Scattering Analyzer AEP-100. The temperature was kept constant by circulating thermostatically controlled water.

Figure 4. Concentration dependence of radius of gyration obtained from Guinier plots of peptide α-1 . The SAXS intensity was measured for 600s for all the solutions and buffers, and the net scattering intensities were calculated by subtracting the scattering intensities of a blank buffer solution from those of the sample solutions.

The evidence that the aggregates of peptide α-1 began to dissociate on heating suggested that the force maintaining the stability of tertiary structure was not the increase of hydrophobic interaction but mainly the breakage of the hydrogen bonds between the aggregates of peptide α-1, and that the surface properties of peptide α-1 may be important in its stability. The distribution of hydrophilic and hydrophobic residues in the α-chain was counted with the Kyte and Doolittle method. The hydrophobic domain is located in residues 100-140 of the α-chain and its hydrophilic domain is located in peptide α-1 (residues 1-94). We concluded that the functional properties of proteins had a close relationship with the primary structure from the function of peptide α-1.

Complex Formation of Peptide α-1 and Globin Hydrolyzates Aggregates

The interactions between the thin rod aggregates and the peptide α-1 were studied by determining the variations in diffusion coefficient D with reaction time using quasi-elastic light scattering [7,10]. The results (Figure 5) showed that in the case of the thin rod aggregates alone there are no changes in the diffusion coefficient D. This results also suggested that the thin rod aggregates exist in monodispersive state and could not form a gel by self-polymerization between the aggregates in solution. But in the case of mixture of thin rod aggregates and peptide α-1 (the mixing ratio 7:3, the final concentration 20 mg/mL) the value dropped more sharply than whole globin hydrolyzates mixture (see Fig. 5 : ●line) indicating that because the rod-shaped aggregates have been already formed, it is not necessary to take time for this formation process, thus the network of gel constructed by the interaction of the thin rod aggregates and the peptide α-1 can be formed within one hour. From these results, we concluded that the gel formation of whole globin hydrolyzates mixture can be divided into two steps. The first step was mainly the formation of the rod-shaped aggregates which originated

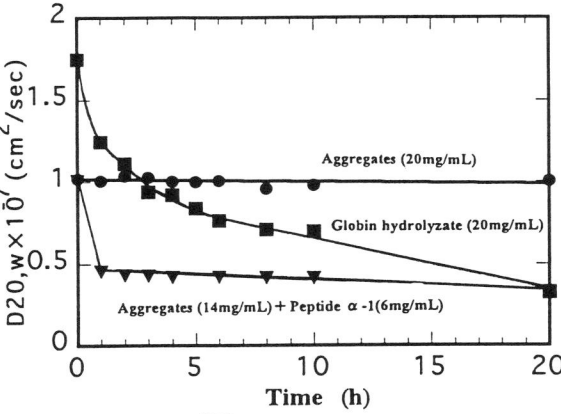

Figure 5. Apparent diffusion coefficients of mixture of globin hydrolyzates (rod-shaped aggregates) after addition of peptide α-1. The samples were pre-heated at 95 ℃ for 15 min.

fromlow molecular weight peptides (globin β-chain and the peptide β-1). In this step the oligomers formation of peptide α-1 was also involved. The second step was the formation process of early network structure based on the interaction between rod-shaped aggregates produced at the first step and the peptide α-1. The transformation from the thin rod aggregates to network may be affected by the concentration of the thin rod aggregates. The formation of network structure probably began in the first step when a certain amount of the thin rod aggregates were accumulated, and this process became gradually dominant in the second step.

Schematic Model of gel formation of Globin Hydrolyzates

Based on the results described above, the schematic model of gel formation of globin hydrolyzates was presented in Figure 6. At first, the globin hydrolyzates containing mainly 8 kinds of peptides are obtained by hydrolyzing the globin with 8 M citric acid. The globin hydrolyzates are composed of undigested globin α-chain, β-chain, the peptide α-1 originated from globin α-chain with a molecular weight of 9,839 Da and the peptide β-1 from globin β-chain with a molecular weight of 10,887 Da [4,5,6].

When the globin hydrolyzates are cooled to room temperature after heating, a rod-shaped aggregate with a length of 130-140 nm and a molecular weight of about 870,000 Da is formed by a noncovalent bond between the globin β-Chain and the peptide β-1

458

originated from the globin β-chain. The peptide α-1 originated from globin α-chain has a high hydrophilicity, and shows the properties of association and dissociation depending on concentration and temperature. The hypothetical gel formation mechanism of the globin hydrolyzates is presented as follows: initially the randomly coiled polymers are produced by about 8 molecules of peptide α-1, next the crosslinked structures are constructed between the length of 130-140 nm rod-shaped aggregates, and then the network is gradually formed and the gel is finally formed [7,14].

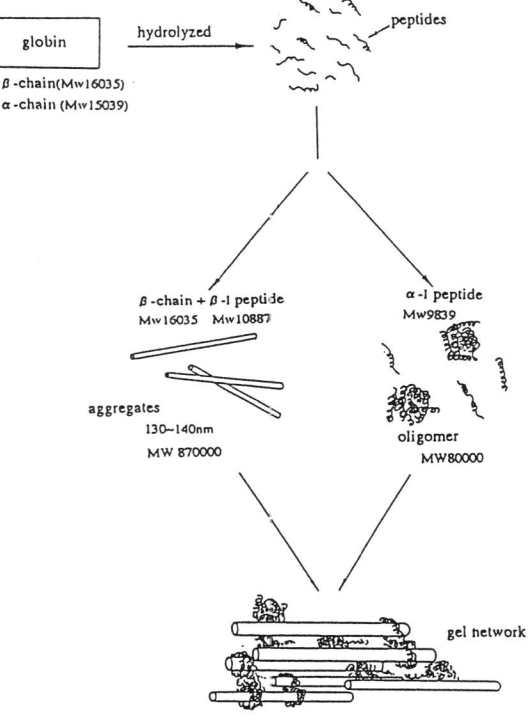

Figure 6. Schematic model of gel formation of whole globin hydrolyzates.

ACKNOWLEDGMENT

This work was supported in part by a grant from PROBRAIN, Japan.

REFERENCES

1) Liu, X.Q.; Yonekura, M.; Tsutsumi, M.1994, *Nippon Shokuhin Kogyo Gakkaishi.* 41(3), 178-183.

2) Liu, X.Q.; Yonekura, M.; Tsutsumi, M.1994, *Nippon Shokuhin Kogyo Gakkaishi.* 41(3), 196-201.

3) Liu, X.Q.; Yonekura, M.; Tsutsumi, M.1995, *Nippon Shokuhin Kogyo Gakkaishi* 42(8), 562-568.

4) Liu, X.Q.; Yonekura, M.; Tsutsumi, M. 1995, *Nippon Shokuhin Kogyo Gakkaishi.* 43(2), 141-145.

5) Liu, X.Q.; Sano, Y. 1996, *J.Agric. Food Chem.* 44(10), 2957-2961.

6) Liu, X. Q.; Tsutsumi, M.; Sano, Y. 1997.*J.Agric. Food Chem.* 45(2), 328-333.

7) Liu, X. Q.; Sano, Y. 1997, *J. Agric. Food Chem.,* 45(5), 1574-1578.

8) Sano, Y. 1988, *J. Colloid and Interface Sci.,* 124, 403-406.

9) Sano, Y. 1990, *J. Colloid and Interface Sci.,* 139, 14-19.

10) Sano, Y. 1993, *Biopolymers.,* 33, 69-74.

11) Sano, Y.; Inoue, H.; Hiragi, Y.; Urakawa, H.; Kajiwara, K. 1995, *Biophysical Chemistry ,* 55, 239-245.

12) Liu, X. Q.; Sano, Y. 1997,*J. Agric. Food Chem.* 45(12), 4535-4539.

13) Sano, Y.; Inoue, H.; Kajiwara, K.; Hiragi, Y.; Isoda, S. 1997,*J. Protein chem.,* 16, 151-159.

14) Liu, X. Q.; Sano, Y. 1998, *Recent Res. Devel. in Agricultural & Food Chem.,* 2, 183-225.

HYDROCOLLOIDS – PART 1
Edited by K. Nishinari
2000 Elsevier Science B.V.

Poly(γ-glutamic acid) from *Bacillus subtilis* as an optically heterogeneous peptide in which D- and L-glutamic acid isomers are copolymerized into a single chain

T. Tanaka and M. Taniguchi

Department of Bio- and Geoscience, Graduate School of Science, Osaka City University, 3-3-138 Sugimoto, Sumiyoshi-ku, Osaka 558-8585, Japan

Poly(γ-glutamic acid) (PGA) produced by *Bacillus subtilis* F-2-01 on an industrial scale was ultimately depolymerized into a peptide fraction (F-I) and a γ-glutamyl oligopeptide fraction (F-II) by the action of fungal PGA-hydrolase (γ-L-glutamyl hydrolase). F-I consisted of both D- and L-glutamic acid (76:24), whereas L-glutamic acid alone was detected with F-II. Incomplete depolymerization of PGA depended on either the complex formation between γ-D-glutamyl and γ-L-glutamyl peptide unit or the presence of covalent binding between D- and L-glutamic acid residues in F-I but not on an end product inhibition or denaturation of the enzyme. Analyses by ultracentrifugation supported the loss of intermolecular interaction between γ-glutamyl peptides that may interfere with the enzymatic hydrolysis of F-I by accelerating aggregate formation. F-I is proposed to exist as an optically heterogeneous peptide unit in PGA, in which D- and L-glutamic acid isomers are copolymerized into a single chain.

1. INTRODUCTION

Poly(γ-glutamic acid) (PGA) is a bacterial polypeptide in which glutamic acid residues are extensively polymerized *via* a linkage between the γ-carboxyl and amino groups [1-5]. PGA has been attracting industrial attension because of its extremely viscous property as well as high water-retaining ability. It is generally accepted that PGA exists as a mixture of optically homogeneous peptides such as poly(γ-D-glutamic acid) (D-PGA) and poly(γ-L-glutamic acid) (L-PGA), as illustrated in Fig. 1. D-PGA was predominantly isolated as the capsular component of *Bacillus anthracis* [6] and *B. polymyxa* [7], whereas L-PGA was isolated from the cell wall component of an alkalophilic strain of *B. subtilis* [8]. L-PGA was also found in nematocysts from *Hydra* as an example of PGA distribution in eucaryotic cells [9]. In these PGA preparations, however, either the L- or D-glutamic acid residue was detected as a minor component in addition to the other optical isomer as the major constituent of the molecule.

A specific endo-type PGA-hydrolase was isolated and purified from the culture supernatant of *Myrothecium* sp. TM-4222 (ATCC 201200) [10] in order to clarify the relationship between the physicochemical property of PGA and its molecular structure. The enzyme, which has an optical specificity to preferentially hydrolyze the γ-glutamyl linkage between L-glutamic acid residues, could depolymerize only a portion of PGA to γ-L-glutamyl oligopeptides, thereby allowing the other portion to remain with the original molecular size [11]. This observation agrees with the homo model (Fig. 1), in which D-PGA is expected to exist as the fraction resistant to enzyme action. Recently, the industrial-scale manufacture of PGA has been achieved with the aid of the high PGA productivity of strain F-2-01 [3]. The commercially available preparation showed an altered susceptibility to the action of PGA-hydrolase, being converted to the smaller peptide fraction (F-I) and γ-L-glutamyl oligopeptide fraction (F-II). The molecular structure of PGA could not be elucidated simply by the conventional homo model. In this study, we characterized the mode of arrangement of D- and L-glutamic acid residues in PGA as a cause of its characteristic physicochemical properties.

2. MATERIALS AND METHODS

PGA was manufactured by Meiji Seika Kaisha Co. (Odawara, Japan) by modifying the conditions in the culture employed for the laboratory-scale preparation [3] as follows. For an enlarged-scale production of PGA, strain F-2-01 was grown in $5 \times 10^3 \mathrm{~m}^3$ medium containing 10% glucose, 8% L-glutamic acid, 0.7% peptone, 0.68% urea, 0.5% $NaNO_3$ and 0.24% KH_2PO_4 at 37°C and pH 7.5 for 6 d with vigorous agitation. PGA-hydrolase was purified from the culture supernatant of a filamentous fungus, *Myrothecium* sp. ATCC 201200, as described previously [11]. Analytical procedures of PGA and its enzymatic and acid hydrolysis products were also described previously [10,11]

3. RESULTS AND DISCUSSION

The reaction mixture containing 100 mg of PGA and 6.4 units of PGA-hydrolase in 100 ml of 50 mM acetate buffer (pH 5.0) was incubated at 37°C. At various time intervals, aliquots were taken to analyze the hydrolysis pattern of PGA by HPLC using a gel filtration column in which poly(α-glutamic acid)s (Sigma) were used as the molecular weight standards [11]. As shown in Fig. 2, PGA was depolymerized during the initial 40-min incubation to give fractions F-I and F-II, apparently distinguishable by their molecular size distributions. Although F-II was gradually depolymerized to the level of oligopeptide, F-I was ultimately converted to a peptide fragment resistant to enzyme action.

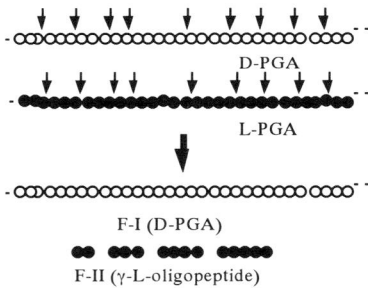

D-PGA

L-PGA

F-I (D-PGA)

F-II (γ-L-oligopeptide)

Fig. 1. Diagrammatic illustration of the proposed structure of PGA and its hydrolysis products by PGA-hydrolase in a homo model, in which D- and L-glutamic acid residues are expressed by (○) and (●), respectively. Each γ-L-glutamyl oligopeptide is represented by the number of L-glutamic acid residues (●) in F-II. Arrows indicate the site of the hydrolytic action of PGA-hydrolase.

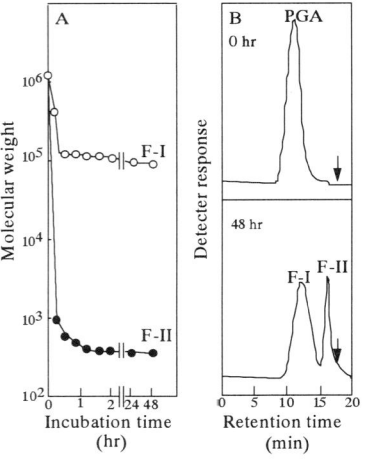

Fig. 2. HPLC-analysis of hydrolysis patterns of PGA by PGA-hydrolase (A) and the elution profile of PGA (B) before and after 48-h incubation with the enzyme. In (A), each plot is made with the elution peak of F-I (○) or F-II (●). Arrows indicate the retention time of the glutamic acid monomer.

F-I and F-II were separated by ultrafiltration using a polysulfone membrane (UP-20; Advantec, Osaka) after the elution profile of each fraction gave no visible change following 48 h incubation. An HPLC-dependent amino acid analytical system [10] detected γ-glutamyl di-, tri- and tetrapeptide in F-II at a molar ratio of 1:5:2 together with the pentapeptide as a minor component. As summarized in Table 1, both D- and L-glutamic acid were detected with PGA, as is often the case with PGA from *B. subtilis*. L-Glutamic acid was still detected with F-I although the fraction had been depolymerized by PGA-hydrolase to the maximum level, while

461

the L-isomer alone was detectable with F-II. These findings were inconsistent with the homo model in which the fraction containing D-glutamic acid was expected to appear as D-PGA without alteration in its original molecular size.

Table 1. Analytical data on PGA and its hydrolysis products by PGA-hydrolase.

Fraction	Proportion in total PGA (%)	Composition of glutamic acid isomer[a] (%)	
		D-Glu	L-Glu
PGA	100	53	47
F-I	73	76	24
F-II	27	0	100

[a]D-Glu and L-Glu indicate D-glutamic acid and L-glutamic acid, respectively.

PGA and F-I were subjected to the ultracentrifugal analysis, as directed by Troy [12], to examine whether or not their molecular weight distributions were affected by intermolecular interactions. These samples had been extensively dialyzed against 50 mM phosphate buffer (pH 7.0) to achieve analysis under conditions of low ionic strength that may accelerate aggregation of γ-glutamyl peptide. Dialysis had been also done against 2.5 M sodium chloride, which was employed to minimize intermolecular interactions in the capsular PGA from *B. licheniformis* [12]. Figure 3 shows a typical plot of the log of concentration versus radial displacement (r^2) for PGA and F-I, respectively, in 2.5 M sodium chloride. Various distinguishable slopes were visualized with PGA and F-I due to deviations from linearity. The weight-average molecular weight was calculated from each slope value obtained with 0.2 and 0.3% solutions of PGA and F-I.

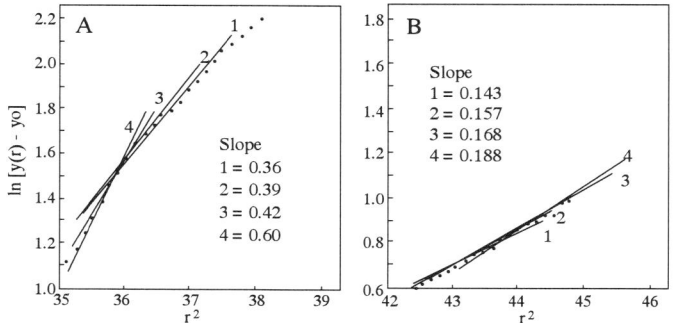

Fig. 3. Typical plot of ln [$y(r) - y0$] versus r^2 from the sedimentation equilibrium method using a 0.3% solution of PGA (A) or F-I (B) in 2.5 M NaCl.

As summarized in Table 2, the Svedberg constant was kept almost at the same level with or without the presence of 2.5 M sodium chloride, indicating the negligible effect of environmental ionic conditions on the molecular size or shape of PGA and F-I. PGA showed a weight-average molecular weight distribution from 160 to 390 k, a range comparable to the 208 and 222 k determined by ultracentrifugation [2] and the 275 k by SDS-polyacrylamide gel electrophoresis [5] for PGAs from other strains of *B. subtilis*. F-I had a significantly lower and narrower molecular weight distribution ranging from 20 to 40 k regardless of whether it was exposed to a low or high ionic strength environment. F-I was thus evaluated as a mixture of γ-glutamyl peptides having the above molecular weight distribution in fully dissociated form. It should be noted that the molecular weight values of PGA and F-I were highly exaggerated in the HPLC-analysis, possibly due to a difference in the 3-dimensional structure between PGA and poly(α-glutamic acid) used as a molecular weight standard.

Table 2. Analysis by ultracentrifugation of Svedberg constants and weight-average molecular weights of PGA and F-I

Sample	Solvent	Svedberg[a] constant	Average molecular weight (x 10^{-3})[b]
PGA	50 mM phosphate buffer	3.34	168, **162**
	2.5 M NaCl	3.40	**234**, **254**, **275**, **393**
F-I	50 mM phosphate buffer	2.06	**23.6**, 28.8, **29.5**, 33.6, 40.7
	2.5 M NaCl	2.04	19.6, **30.5**, 31.3, **33.4**, **35.8**, 39.8, **40.0**

[a] Svedberg constants were given by the sedimentation coefficients determined from the schlieren patterns obtained at 5,000 rpm and at 25°C
[b] Average molecular weight values were calculated from each slope obtained with 0.2% (normal letters) and 0.3% (boldface letters) solutions of the samples.

γ-L-Glutamyl tetrapeptide was not an effective substrate for PGA-hydrolase because it was detectable as one of the major hydrolysis products of PGA (see previous section). F-I was characterized as a repeating unit of γ-glutamyl polypeptides constituting PGA so that it could be excised from the region where L-glutamic acid formed a cluster larger than the tetrapeptide unit. F-I (12 mg) was incubated with 1.0 unit of carboxypeptidase G (γ-L-glutamyl hydrolase; Sigma) in 1.2 ml of 10 mM phosphate buffer (pH 7.2) at 30°C to estimate the number of L-glutamic acid residues at the carboxyl terminal. At various time intervals, portions were withdrawn and analyzed to determine L-glutamic acid content as described previously [3]. The enzymatic hydrolysis of F-I occurred to the maximum extent at a 4 h-incubation time when 73 nmol of L-glutamic acid was liberated from 1 mg of F-I. The yield of L-glutamic acid was calculated to be 2.2 nmol per 1 nmol of F-I on the assumption that F-I is a 30 k (232-mer) peptide, indicating that F-I contained at least 2 or 3 L-glutamic acid residues at the carboxyl terminal and possibly at the amino terminal. The number of L-glutamic acid residues was much lower than expected from the ratio of D- and L-glutamic acid (3:1, see Table 1) originally detected in F-I, even if the above carboxypeptidase reaction had not been fully accomplished. This means that L-glutamic acid residues were mostly distributed along with a linear chain of F-I together with D-glutamic acid residues or clusters of γ-D-glutamyl peptide units. L-Glutamic acid residues could be present in F-I as a cluster shorter than the pentapeptide unit because such a short cluster should be free from the action of PGA-hydrolase. We propose F-I as an optically heterogeneous peptide unit in PGA, in which D- and L-glutamic acid residues are copolymerized into a single chain. Figure 4 illustrates one of the possible structures of F-I assuming that PGA-hydrolase can split a γ-L-glutamyl pentapeptide unit inserted in a larger peptide unit. In Fig. 4, L-glutamic acid residues are also postulated to form a cluster larger than the pentapeptide unit in PGA so that the cluster can yield each γ-L-glutamyl di-, tri- and tetrapeptide as a major constituent of F-II upon enzymatic hydrolysis. The production of these oligopeptides makes the whole structure of PGA more complex than expected from the structure of F-I itself. Such an optically heterogenous strucuture should give it a high viscosity and some other relating properties of industrial interest.

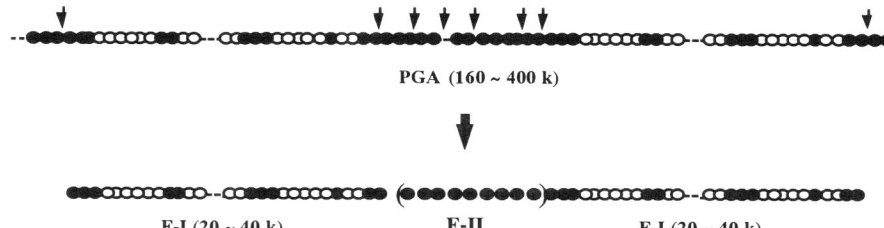

Fig.4. Diagrammatic illustrations of the proposed structure of F-I and the corresponding structure of PGA, which yields both F-I and F-II by the action of PGA-hydrolase. D- and L-glutamic acid residues are expressed by (○) and (●), respectively. Arrows indicate the site of hydrolytic action of PGA-hydrolase.

REFERENCES

1. M. Bovarnick, J. Biol. Chem., 145 (1942) 415.
2. A. Goto and M. Kunioka, Biosci. Biotech. Biochem., 56 (1992) 1031.
3. H. Kubota, T. Matsunobu, K. Uotani, H. Takebe, A. Satoh, A. T. Tanaka and M. Taniguchi, Biosci. Biotech. Biochem., 57 (1993) 1212.
4. C.B.Thorne, C.G. Gomez, H.E. Noyes and R.D. Housewright,J. Bacteriol., 68 (1954) 307.
5. F. Yamaguchi, Y. Ogawa, M.Kikuchi, K. Yuasa and H. Motai, Biosci. Biotech. Biochem. 60 (1996) 255.
6. V. Bruckner, J. Kovaca and G. Denes, Nature, 172 (1953) 508.
7. F.A. Troy, J. Biol. Chem., 248 (1973) 305.
8. R. Aono, Biochem. J., 245 (1987) 467.
9. J. Weber, J. Biol. Chem., 265 (1990) 9664.
10. T. Tanaka, T. Yaguchi, O. Hiruta, T. Futamura, K. Uotani, A. Satoh, M. Taniguchi, M. and S. Oi, Biosci. Biotech. Biochem., 57 (1993) 1809.
11. T. Tanaka, O. Hiruta, T. Futamura, K. Uotani, A. Satoh, M. Taniguchi and S. Oi, Biosci. Biotech. Biochem.,57 (1993) 2148.
12. F.A.Troy, J. Biol. Chem., 248 (1973) 316.

INDEX

466

Hydrocolloids

Part 2

Fundamentals and Applications in Food, Biology, and Medicine

Hydrocolloids

Part 2

Fundamentals and Applications in Food, Biology, and Medicine

Edited by

Katsuyoshi Nishinari

Osaka City University

Osaka, Japan

2000

ELSEVIER

Amsterdam – Lausanne – New York – Oxford – Shannon – Singapore – Tokyo

ELSEVIER SCIENCE B.V.
Sara Burgerhartstraat 25
P.O. Box 211, 1000 AE Amsterdam, The Netherlands

First edition 2000

Library of Congress Cataloging in Publication Data
A catalog record from the Library of Congress has been applied for.

ISBN 0 444 50178 9

∞ The paper used in this publication meets the requirements of ANSI/NISO Z39.48-1992 (Permanence of Paper).

Printed in The Netherlands.

Preface to Part 2

This volume is based on the presentations given at Osaka City University International Symposium 98 - Joint meeting with the 4th International Conference on Hydrocolloids - held on 4-10 October 1998 in Osaka.

The first article in Section 1, a masterly review by Professor Phillips shows the wonderful rich world of hydrocolloids, how they are useful in various fields with many potential future developments especially in processing of foods and in biomedical fields. This contribution is based on a lecture given to the general public.

Section 2 includes the articles treating the fundamental aspects and industrial applications of dispersions, emulsions, suspensions, and surfaces. Although it is frequently said that this is the world of mystery and art rather than science, the articles in Section 2 show the steady advance in the understanding of this world.

Section 3 covers the mixtures of biopolymers which have been the subjects of hot debate these ten years. Our understanding of the interaction of different biopolymers is certain to be interesting and important from the view point of not only science but also industry.

Section 4 consists of articles concerning processing. The effects of shear on the gelation is an important problem, and the recent marvellous achievements are described by Professor Djabourov and Professor Norton. Many interesting problems in food processing are discussed in this section.

Section 5 gathers the articles in biomedical fields. Although only two articles closely related with pharmaceuticals, these articles show the important relation between hydrocolloids and pharmaceuticals. Since hyaluronan plays an important role in the field of orthopaedics, ophthalmology, and cosmetics, six articles together with some other articles related with biorheology are included in this section.

Although most articles in this volume treat the hydrocolloids as functional materials which modify the texture of foods, control the rheological properties of foods, biofluids and pharmaceuticals, most hydrocolloids are at the same time dietary fibres. Section 6 includes contributions to this aspect.

Section 7 gathers the articles describing the problems of sensory evaluation, texture measurements and mastication which are very important to improve the quality of life.

I am sure that this volume provides valuable information and stimulating problems based on the enthusiastic discussions, questions, comments and answers during the conference. All the articles included in this volume have been reviewed and rewritten carefully according to comments and criticisms. I hope that the readers will share the pleasure to get the experience on many exciting aspects and infinite possibility of hydrocolloids.

I would like to thank especially Drs. G.O. Phillips, P.A. Williams, R.A. Williams, K. Furusawa, H. Ohshima, T. Imae, T. Shikata, E. Dickinson, D.S. Horne, B. Launay, E.R. Morris, J.R. Mitchell, S.B. Ross-Murphy, D. Oakenfull, .M. Djabourov, T. Uryu, G. Franz, M. Yonese, T. Norisuye, H. Watanabe, S. Matsumoto, Y. Matsumura, O. Miyawaki, H. Kumagai, Y. Sano, S. Innami, T. Hayakawa, M.A. Bourne, AL. Halmos, K. Kubota, A. Okamoto, T. Morimoto, and F. Nakazawa for their valuable comments.

Katsuyoshi Nishinari
Department of Food and Nutrition
Faculty of Human Life Science
Osaka City University
3-3-138 Sugimoto, Sumiyoshi-ku,
Osaka 558-8585, Japan
Tel : +81-6-6605-2818
Fax : +81-6-6605-3086
e-mail : nisinari@life.osaka-cu.ac.jp

CONTENTS

3. MIXED SYSTEMS

5. BIOMEDICALS

6. NUTRITION

7. SENSORY EVALUATION AND MASTICATION

1. INTRODUCTORY LECTURE

HYDROCOLLOIDS – PART 2
Edited by K. Nishinari
2000 Elsevier Science B.V.

3

Colloids: a partnership with nature

Glyn O. Phillips
Research Transfer Ltd, 2 Plymouth Drive, Radyr,
Cardiff, UK   CF 4   8BL

1. NATURAL FIBRE FOR HEALTH AND LIVING

Throughout the world there is a growing belief that natural foods are an integral part of a healthy life style. It is inevitable, therefore that food producers source an increasing proportion of their raw materials from nature itself. The most significant growth demonstrated in the last European Food Ingredient Exhibition was in preparing healthier foods, particularly to replace animal fat and to introduce more nutritional fibre. A well researched market analysis " Prospect for Food and Drink Ingredients in the European Union (EU)" predicts that the use of fat replacers in the EU will increase by 113.8% from 202,504 tons to 433,038 tons from now until the year 2000. The sectors showing the largest fat substitution potential are yellow fats, biscuits and soups, sauces and dressings. The greatest percentage changes will occur in savoury snacks, cheese and drinking milk cream and condensed milk. However, there will certainly be an increased demand by an increasingly health-conscious consumer for reduced fat and enhanced fibre foods of all types. If this can be achieved using materials which have low calorific value, further health benefits will result. Foods containing such ingredients will need to match the quality of the original product and without adverse dietary effects. This target, cannot be achieved without the scientific use of thickeners, stabilisers and emulsifiers, particularly of the "natural type". This calls for colloids, which can interact with water to form new textures and perform specific functions, and as such can be classed as "hydrocolloids". In 1997 the world market for such hydrocolloids was 2,621 million US dollars and is set to grow significantly to meet the health aspirations of the consumer in the next millennium.

Most of the main hydrocolloids used in the food industry are carbohydrate. There is a growing interest in carbohydrate and its components throughout the world and recommendations have been made to encourage increased carbohydrate consumption. The main sources are cereals, sweeteners, root crops, pulses, vegetables, fruit and milk products. These are the main source of food energy, and as a per cent of energy, total carbohydrate ranges from about 40% to over 80%, with the developed countries such as those in North America, Western Europe and Australia at the low end of the range and developing countries in Asia and Africa at the high end. Starch accounts for 20 - 50% or more of energy where the total carbohydrate is in the high range.

The McGovern Report " Dietary goals for the United States " in 1977 first introduced the term " complex carbohydrates " to distinguish sugars from other carbohydrates. It was used to encourage consumption of what were considered to be healthy foods, such as whole grain cereals etc. Later came the use of the terms " available and unavailable carbohydrates " in order to draw attention that some carbohydrates are not digested and absorbed in the small intestine but rather reaches the large bowel where it is fermented. Subsequently was added the concept of " resistant starch " to describe starch or starch degradation products not absorbed in the small intestine of healthy humans. Now we prefer to refer to these materials in terms of "dietary fibre", the main components of which were first thought to come from the cell walls of plant material in the diet and comprise cellulose, hemicellulose and pectin (the non-starch polysaccharides). We now also draw a distinction between soluble and insoluble nutritional fibre.

The physical properties of fibre allow it to perform both in a physical role to assist laxation, by increasing bowel bulk and speeding up of intestinal transit time. There is also a metabolic function by fermenting through colonic microflora to give short-chain fatty acids (SCFA), mainly acetate, propionate and butyrate. These have a very beneficial effect on colon health through stimulating blood flow, enhancing electrolyte and fluid absorption, enhancing muscular activity and reducing cholesterol levels.

Now that these effects are known, it is the task of the food scientist to provide the hydrocolloids in the most appropriate form for inclusion in the food product. This requires an understanding of their structure and the way in which they act to produce the desired function in the food. Many of the contributions in this book are devoted to achieve this objective.

2. THE WORLD MARKET FOR FOOD HYDROCOLLOIDS

An indication of the size of the market for natural hydrocolloids is shown in Table 1. In 1997 the value was 2,621 million US $.

Table 1 **WORLD MARKET OF HYDROCOLLOIDS 1997**

HYDROCOLLOID	$ MILLION	%	COMMENT
Starches	646	25	Cost effective and label friendly
Gelatin	538	21	achieved despite BSE
Carrageenan	263	10	politics almost over
Pectin	260	10	user recognition and label friendly
Xanthan	205	8	first one tried
Arabic	147	6	still surviving + +
Agar	130	5	dominated by Japan
LBG	97	4	an old favourite
Alginate	87	3	speciality role
Guar	77	3	Most cost effective
CMC	67	3	too chemical sounding others now
MMC	46	2	opened process technology
Konjac	11	0	suprising advance
Tara	5	0	
Gellan	6	0	on the way, but too costly now
Cassia Tora	5	0	
TOTAL	2,621		

Source : Third International Business Conference on Food Hydrocolloids, Nice France 1997

The starches from a variety of natural sources, such as corn and potato remain the leading hydrocolloid, followed closely by gelatin. Now concern about diseases which may be transferred from animal sources has raised some query about this useful material.

Carrageenan, a product from seaweed has been a centre of controversy because of a presumed adverse effect on health, and because of a commercial battle with a new cheaper form of carrageenan from the Philippines, now called "Processed Eucheuma Seaweed" for regulatory purposes. Pectin from apples is still a friendly and useful product. Xanthan is the first of the biosynthetic hydrocolloids to break through into the top five, and leads the way for other bacterial polysaccharides such as gellan to come into regular use. At present it is their price that holds them back. Gum arabic, an exudate from trees in Africa, continues to be used for a multitude of food uses and is favoured as an emulsifier particularly in beverages and as a source of natural fibre in health drinks. It is also a valuable encapsulator of flavours and it remains a food approved adhesive as it has been for more than 2000 years. The traditional thickeners, locust bean gum and guar gum remain competitive and find their way into most sauces, salad dressings and food fillings. The alginates from seaweed can perform a variety of speciality functions and have great flexibility within their structure, and as such are used in a range of applications varying from toothpaste to ice cream and wound dressings. Carboxymethyl cellulose and microcrystalline cellulose represent a different category since they are produced from wood pulp by a variety of physical and chemical processing methods. Since these are a good source of fibre and have a variety of functionalities in food, they seek to competed with the traditional natural hydrocolloids by offering security of supply, quality and price.

Essentially the function of such hydrcolloids is to improve the texture of foods and to assist in presenting various processed food products in a healthy and attractive ready to use form. Table 2 shows how the hydrocolloids form the dominant sector of the texturising agents now in use An important new development is to be able now to form foods which have the mouthfeel of a fat without the presence of fat itself. The fat replacers usually form small gel particles which simulate the behaviour of natural fats in food and so assist to reduce harmful fats in the diet.

Table 2

HYDROCOLLOIDS ARE THE DOMINANT SECTOR
AMONG TEXTURISING AGENTS

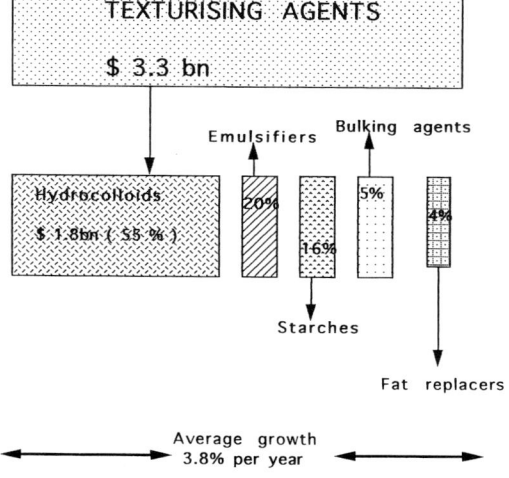

Thus the drive to achieve healthy living has been assisted by the careful use of natural hydrocolloids.

3. HYDROCOLLOIDS FROM TREES AND PLANTS

3.1. Insoluble cellulosic fibre

Few would think of eating wood ! Yet that is exactly what we are now doing. Trees represent the largest source of usable biomass and have the merit of being renewable. If they can be harnessed as a supplement to a healthy diet, it is therefore important. That is precisely what some companies have now done. Wood is made up of three ingredients fused together into one hard matrix. These are cellulose, lignin and hemicellulose. For generations wood has been converted into cellulose for use as paper by separating these three parts of wood. The resulting wood pulp (cellulose) is now a valuable raw material for a variety of food additives.

The FMC Company produce such cellulose products used in food : Avicel and Novagel which are used here as examples of soluble fibre.

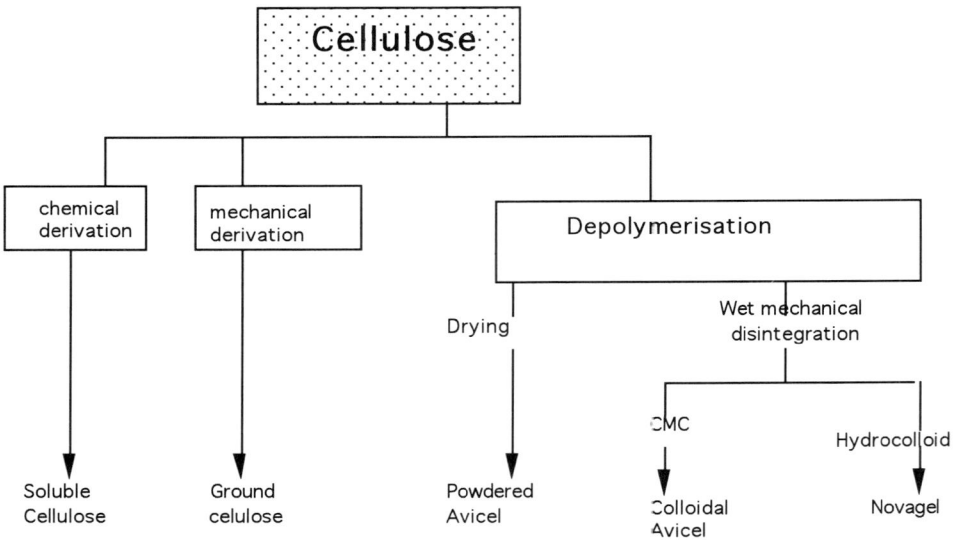

Both these forms of cellulose fibre have uses in food. When properly disperses Avicel cellulose sets up a network with particles less than 0.2 micron in size. These particles are held together by hydrogen bonds which continue to form over a period of 24 hours, resulting in an increase in viscosity. Unlike soluble hydrocolloids the Avicel cellulose binds water to a much lesser extent. It is the physical network that affects the moisture in food systems. The application are varied. The gels made with Avicel cellulose readily break down with shear, but when the shear is removed the gel will reform(thixotropy). Temperature has very little effect on the functionality and viscosity of these dispersions and so can be used during baking, processing and microwave heating.   Additionally the fibre is non-calorific and so can be used as a source of dietary fibre.   Because of these properties Avicel cellulose can be used to stabilize foams and emulsions.

Novogel differs from Avicel in that it capitalises on the unique interaction between guar gum and the microcrystalline cellulose.   It is co-processing of cellulose gel and guar gum that imparts the special fat-like properties.   To illustrate the wide range of application for these two examples of cellulose food additives the information given by FMC for their products is given in Table 3, but is not meant to endorse these products exclusively.   There are other emerging cellulose products.   For example Primacel is a family of products containing fermentation derived cellulose in combination with various co-agents.   Primacel can be used as a highly efficient thickener, stabiliser and film former.   It can be used as a starch extender and in addition can be used to modify texture to give a more creamy mouthfeel in low-fat food systems.

Table 3   Cellulose Food Gels

	Novagel	Avicel	Avicel
Particle Size (when dispersed)	$0.5 - 15.0\mu$	$6.0 - 100\mu$	$0.1 - 2.0\mu$
Funtionality	* Permits fat replacement * Provides creamy mouthfeel * improves body * adds opacity	* Tableting aid * Adds dietary fibre * Anti-caking agent * Extrusion aid * Non-nutritve filler * Adds opacity * Adds solids	* Controls flow * Adds structure * Stabilises o/w emulsions * Stabilises foam * Imparts body * Provides creamy mouthfeel * Improves cling * Imparts freeze/thaw stability * Controls ice crystal formation * Suspends solids * Adds opacity
Primary Applications	* Frozen desserts * Processes cheese * Sauces * Soups * Cheesecake (RCN -10 Sphere -shaped) * Pourable dressings * Spoonable dressings * Frostings * Sauces (RCN - 15 Irregular shaped particles)	* Tableting aid * Spice carrier * Shredded cheese flow aid * Opacifier * Non-caloric fat extender	* Whipped toppings * Heat stable emulsions * Frozen desserts * General food stablise * Spoonable dressings/sauces * Pourable dressings sauces * Dry blended foods * Low moisture foods

3.2. Food fibre from Japan

New hydrocolloid formulations for food use are appearing regularly.   Japan is now contributing its own characteristic materials to this field.   One is konjac glucomannan flour, which is a main component of the tuber of *Amorphophallus konjac*   K. Kock and produces a thermally stable gel (Konnyaku) by addition of alkaline coagulant.   This has been a popular traditional food in Japan since the 1600's. The original home of the Konjac tuber is considered to by South East Asia, and commercial sources in China are now being exploited internationally. Konjac tuber grows in size year by year and konjac flour is produced from 2-year old tubers. The konjac tuber, unprocessed has a harsh taste, but this can be removed from the flour by processing. In the same way in which elastic gelatin-like gels can be produced by the combined use of kappa-carrageenan and galactomannans, such as locust bean gum, konjac flour can produce thermoreversible and very elastic gels with kappa-carrageenan or xanthan gum. It thus possesses considerable potential as a texture modifier. The breaking strength of mixed gels of konjac flour and kappa-carrageenan is about twice that of mixed gels of locust bean gum and kappa-carrageenan.

A soluble soybean polysaccharide (SBP) is being produced in Japan from bean-curd refuse, a by-product of bean curd and soya protein. This new polysaccharide has properties which could make it eventually a powerful competitor of Gum Arabic as an emulsifier and stabiliser, particularly in acidified milk systems. In addition it functions as an antioxidant, so that the drink is able to keep its flavour for a very long time. Since there is an abundant supply of the soya bean raw material it is possible to look forward to period of stability in price and quality with this new material, which has eluded Gum Arabic over the past 20 years. The great strength of SBP is its ability to stabilise protein particle in acidic conditions and so has great potential for stabilising yoghurt drinks.

3.3. Soluble gum arabic fibre

It may, however, be too early yet to write off Gum Arabic, for it remains the most versatile food additive now available because of its range of functionalities. It has been an important commercial product for thousands of years.

Gum Arabic is the acidic polysaccharide exudate currently derived from two Acacia tree sources : *Acacia senegal* and *Acacia seyal*. The former (commercially referred to as **hashab**) is a pale to orange-brown solid which breaks with a glassy fracture, and latter (called **talha** in the industry) is darker, more
friable and is rarely found as lumps in export consignments. Hashab is undoubtedly the premier product, but the lower priced talha has found recently new specialised uses which has boosted its value and price. It is not possible to identify precisely the exact balance between these two products in the market place since it is continually changing.

The Gum Arabic - yielding *Acacias* grow in semi-arid areas and the vast majority of the product which enters international trade originates in the so-called gum belt of Sub-Saharan Africa. The belt occurs as a broad band from Mauritania, Senegal and Mali in the west through Burkina Faso, Niger, northern parts of Nigeria, Chad to Sudan. Ethiopia and Somalia in the Horn of Africa.

Although Gum Arabic has a very long history, going back some 4000 years for embalming bodies in Egypt, modern trade has been dominated by the Sudan. Thus production in Sudan over the years gives a good indicator of consumption world-wide. Towards the end of the 1960's, total Gum Arabic production in Sudan (hashab and talha) was in excess of 60,000 tons. Events, mainly drought, locusts and political instability in the 1970's and 1980's led to fluctuation in both the supply and the price, and as a consequence led to changes in demand. The severe Sahelian drought of 1973/74 resulted in a world shortage and high prices, which in turn accelerated the search for substitutes such as gelatin, maltodextrins and modified starches. A low point of approximately 20,000 tons of Sudanese exports was reached in 1975, which recovered to around 40,000 tons during 1979. A further drought in 1982-84 saw levels of exports fall to below 20,000 tons in the mid-

1980's and early 1990's. Sudan now faces an embargo on its products in the USA for its alleged terrorist supporting activities,

Europe is the biggest regional market for Gum Arabic, and imports averaged 29,300 tons per year over the 7 year period 1989-95, with peaks of 32,100 tons in 1991 and 34,000 tons in 1994. France and the UK are the biggest markets, although both re-exported large proportions of their imports, which averaged 10,000 and 7,900 tons per year respectively. France shows and upward trend over the 7 years while the UK trend has been downward. Germany and Italy were the nest biggest markets, averaging 4200 tons and 3700 tons respectively. Outside Europe, the USA is the largest market for Gum Arabic. Imports averaged 10,000 tons in 1994. Japanese imports averaged 1900 tons over the 7 year period.

Of the other producers Nigeria is the next most important after Sudan averaging exports of between 4000 -7000 tons. Chad comes next, and has increased its production each year since 1990, reaching 5400 in 1995. The upward trend has continued, often due to gum originating in the Sudan finding its way out through Chad. Currently the best estimate of the overall annual useage of Gum Arabic is 40,000- 50,000 tons.

Technically Gum Arabic is a superb product, unmatched by any other food additive in its range of functionality. In food systems alone it can serve as an emulsifier, flavouring encapsulator, stabiliser, thickener, surface-finishing agent, and in addition retards sugar crystallisation and provides adhesion. These diverse functionalities can be related to the various structural features associated with this globular cross-linked polysaccharide containing several neutral monosacharides and an acid moeity, all woven together with a small amount of protein. An indication of its versatility can be gained by reference to particular sectorial areas.

Industrial non-food :

Gum Arabic serves as a film former, and sensitiser to ensure increased wettability in lithography offset plates, etching solutions, fountain solutions and lacquer development. It is both an emulsifier and suspending agent to provide the continuous phase in inks. In water paints and pastels it is a binder to improve appearance and ensure retention of shape. In the ceramic and porcelain processing it is the glaze binder. Water absorption and improved binding is the function in foundry moulds and in fireworks, explosives and cartridge powder it is both a binder and adhesive to provide controlled release. In pesticides and insecticide sprays it improves the spraying quality in the role of the suspending agent.

Pharmaceuticals:

Stabilisation of oil-in-water emulsions calls for its emulsification function, and in lotions and creams it adds the quality of a film former and viscosity controlling agent. In compressed tablets it is both a binder and adhesive and stable end product in syryps. Its encapsulation of oil-soluble vitamins in powder form increases the resistance to oxidation and it is an excellent coating for pills due to its film-forming, glazing, coating and adhesive qualities. Its low calorie and non-sugar features allow its use in diabetic confectionery and medicated cough drops, where is the main ingredient and supplies good texture and prevents crystallisation.

Confectionery :

Gum arabic is extensively used for high quality confectionery. In gum drops, pastilles and candies it provides fibre, is the main binding ingredient, prevents sugar crystallisation and serves in a similar capacity in reduced sugar, no-sugar confectionery, and compressed sugar confectionery. In caramels, toffees and chewing gum it improves chewiness, ensures even fat distribution, and retains flavour.   It remains the material of choice for coated products such as dragees, chocolate centres, nut centres, coated almonds, peanuts etc by bringing in its compete range of functionalities - coating, glazing, adhesion, engrossing, film forming, integration of the oil-water interface.   These qualities are called to play in a wide range of bakery applications.

Emulsification

The emulsification functionality in a wide range of beverages is now well understood. Gum Arabic is not a uniformly molecular discrete species.   Fractionating methods can identify at least 3 fractions.   The main material is an arabinogalactan (AG) and contains very little protein.   This fraction is essential for stabilising the hydrophilic phase.   For hashab some 10% contains both AG and an attached protein, making up an arabinogalactan protein (AGP). A minority fraction, representing only about 1% of the total gum is a low molecular weight glycoprotein.   It is the AGP, or indeed more than one AGP which are preferentially adsorbed on to the oil droplets and prevents them reaggregating to break down the emulsion. Thus a uniform droplet size can be achieved with stability over long storage times.   The mechanism of emulsification is illustrated schematically in Figure 1.

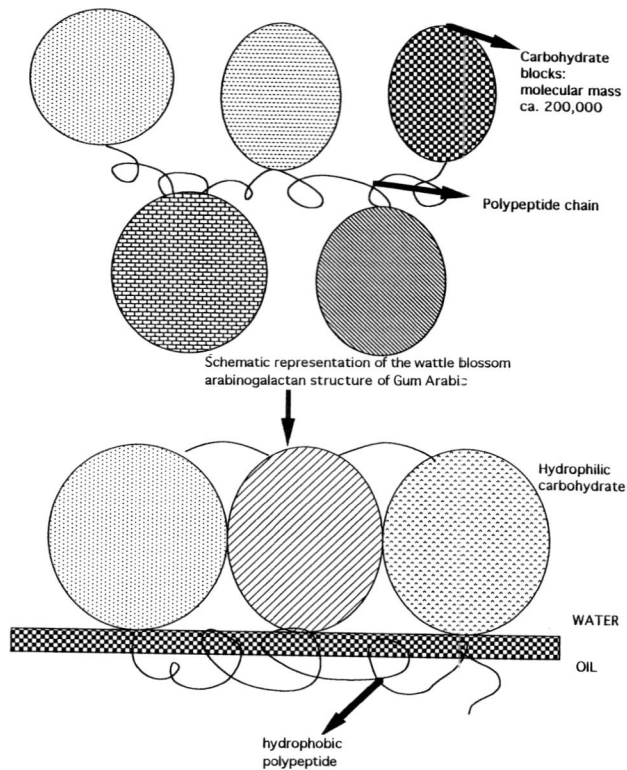

Figure 1. The wattle blossom structure of Gum Arabic and the role
of the protein in oil-water interfaces during emulsification

Gum arabic is therefore widely used in the flavour and beverage industry, for essential oil emulsions (carbonated and non-carbonated soft drinks), instant beverages in powder form, flavoured beverages with added pulp and encapsulation of flavours in powder form. Now there is increase application in the high fibre drinks when its ability to produce beneficial fermentation products in the colon contributes to a healthy nutrition practices.

4. HYDROCOLLOIDS IN WOUND HEALING AND TRANSPLANTATION

Human tissue is a complex hydrocolloid matrix made up of a protein (collagen) and polysaccharides (glycosaminoglycans). When this protective and shock-absorbing part of the body is damaged through trauma or as a result of a burn or disease, infection can enter, with loss of fluid and energy. Immediately there is need to apply a covering which is as close to the natural skin in composition and function as possible. Hydrocolloids, the same or similar, to those used in food can also be harnessed to supply wound dressings and replacements or supplements for human tissue. A natural hydrocolloid which is now finding extensive uses in wound healing and tissue augmentation and repair is hyaluronan and new coss-linked derivatives. It primary structure is based on the disaccharide made up of glucuronic acid and
N-acetyl glucosamine.

D-Glucuronic Acid   N-Acetyl-D-Glucosamine

Matrix engineering , the term for such replacement of tissue functionality was first used by Balazs in 1971 to describe the use of natural and chemically modified biopolymers, derived from hyaluronan, to control, direct and augment tissue regeneration processes. Such biomaterials, therefore, can be regarded as a specialised type of non-inflammatory, biocompatible tissue graft, capable of a wide range of applications for the protection and repair of human tissues. The base material, the polymer hyaluronan is present in virtually every tissue in the human body, and if the natural repeating structure and conformation are preserved and inflammatory fractions removed, the resulting biomaterial is not recognised as immunologically foreign. This technically difficult task was achieved and enabled the routine production of a highly purified non-inflammatory hyaluronan fraction, known by its acronym, NIF-NaHA (Balazs, 1979).

The first clinical studies with NIF-NaHA were started in 1968 using race horses and human patients with osteoarthritis. Here the elastoviscosity of the synovial fluid in the knee joints was impaired, but using NIF-NaHA with a weight-average molecular weight of 2-3 x 10^6, the pathological synovial fluid of low elastoviscosity could be replaced with a fluid that had nearly as high a elastoviscosity as the normal healthy synovial fluid. For this supplementation of the synovial fluid, the term *viscosupplementation* was given (Balazs,

1977).   The medical benefits are now known to depend on the rheological and other physical properties of the material, rather than on the component chemical structure.

The company Pharmacia were given the opportunity to exploit NIF-NaHA world-wide, and the product was given the name *Healon* .   The first clinical use of elastoviscous NIF-NaHA   was   for intraocular lens implantation was first described by Balazs in 1979 and formally reported as a new tool in viscosurgery in 1980.   In a very short time viscosurgery with Healon became routine in opthalmic surgery, and provided elastoviscous protection against mechanical damage due to instruments and implants.   It makes and maintains space for surgical manipulation, prevents tissue adhesion, controls bleeding by its barrier effect, and makes removal of tissue debris, such as pieces of cataractous lens, better controlled.   Now, in addition to Healon, at least 10 different preparations are available to the opthalmic surgeon, but none have the same high viscoelastic properties of Healon, because of their   lower molecular   weight.   The current medical uses of hyaluronan are based on placing hyaluronan matrices directly at sites where either a mechanically functional physical barrier is required (viscosurgery, viscoprotection, viscoseparation) or supplementation of a mechanical dysfuntional tissue.   Consequently, viscosurgery has been internationally accepted as a universal concept, such that the International Standards Organisation now provides standards for "Ophthalmic Viscosurgical Devices for use in Anterior Segment".

It was not until 1987 that viscosupplementation became a safe and effective treatment for osteoarthritis, and two companies Seikagaku in Japan and Fidia in Italy produced a low molecular weight NIF-Na HA and introduced its application in their respective countries. The lower molecular weight meant that these products had to be injected oftener than the more elastoviscous NIF-NaHA tested decades earlier   The need for a better elastoviscous hyaluronan system to produce a viscosity as close as possible to that of synovial fluid of a healthy young adult led to the development by Balazs and Leshchiner working at Biomatrix Inc. USA of a new family of hyaluronan derivatives, which were   called " **hylans** ".

Hylan is a generic term used to refer to a class of hyaluronan derivatives produced by cross-linking of hyaluronan chains via hydroxyl groups of the hyaluronan chains, leaving the carboxylate and the acetamido groups unreacted.   The retention of the carboxylate group is particularly important because these confer the polyanionic character, which is critical to the physicochemical and biological properties.   Now hylan polymers are available in physical forms which range from highly elastoviscous solutions to viscoelastic gels and solids.   This allows the physical properties and residence time of hyaluronan to be controlled without loss of biocompatibility.

The hylan polymers currently approved and used in medical practice are designated hylan A and hylan B.   Hylan A is synthesised *in situ*   by treating hyaluronan-rich tissue sources with aldehydes prior to extraction.   The aldehyde activates the hydroxyl groups of the hyaluronan, forming a protein-mediated cross-link and creating soluble hyaluronan polymers in the subsequent extracts.   Hylan B is synthesised by treating hyaluronan or hylan A with divinyl   sulphone under mild alkaline conditions.   The reaction conditions can be varied to produce materials with properties which range from soft deformable gels to solid membranes and tubes with long retention times.

Based on these materials a range of products have been produced.   Synvisc (hylan G-F 20) is a highly elastoviscous synovial fluid supplement used for the treatment of osteoarthritis, and received clearance by FDA to be marketed in USA and certain   European countries in August 1997.   Hylaform (hylan B) is a solid hydrated gel implanted into soft tissues for viscoaugmentation.   Hylashield (hylan A) is a dilute elastoviscous solution applied to the surface of   the   eye   for   comfort   and   protection   (viscoprotection).   All   are   registered trademarks of Biomatrix Inc USA, the company set up by Dr Endre A. Balazs to find application for hyaluronan and its derivatives.

The clinical benefits of these materials is now well documented. Apart from 7 successful clinical trials. a retrospective analysis of the results of the use of Synvisc in clinical practice in Canada in 1996 with 336 patients who received 1537 injections to treat osteoarthritis of the knee over a two-and-a half-year period showed that 80% of the patients were improved or much improved and the median duration of relief was 8.3 months In addition data were provided that show that a second course was equally effective.   In the clinical trials there were no systemic adverse events, and the local adverse event rate was essentially the same as in   other clinical trials, about 2 to 3%. Now Synvisc is being used world-wide, and positive post-marketing clinical data were reported at the British and Spanish Societies of Rheumatology within the first year of use, when compared with a commonly used non-steroidal anti-inflammatory drug.

Tendon adhesion following injury or after surgical repair is a significant clinical problem. Ever expanding procedures associated with tissue banking are the replacement of the cruciate ligament with human patellar allografts, spinal fusion and revision hip surgery (using bone allografts) where this problem is frequently encountered.   Up to 20% of spinal surgery fails to relieve pain or restore function because of adhesions and arachnoiditis which occurs along the tract of the surgical incision and subsequent disc excision.   Hyaluronan derivatives have been shown to be effective in reducing scar and adhesion formation, and may even enhance tendon healing and have no adverse local or systemic effects.   The use of hylan materials for such viscoseparation and viscoprotection of surface injury during surgery is now being realised.

5. HUMAN HYDROCOLLOIDS

Human hydrocolloids, either in the form of human skin or a membrane which surrounds the placenta (human chorion amnion) is an excellent material for treating burns, leprosy lesions and pressure sores which readily develop in the paraplegic.   In the event of a serious burn it is immediately necessary to cover and seal the wound with a biocompatible covering. This prevents entry of infection and reduces energy and fluid loss.   The field of " Tissue Banking" has now grown up rapidly to provide safe and sterile human tissue dressings for this purpose.   These are termed allografts, and can be obtained from either live of cadaveric donors.   These are truly biological dressings and because of their close relationship to the tissue lost promote granulation of new tissue readily in the case of a first or second degree burn.   If the burn is of the third degree, then excision of the necrotic tissue is necessary and then the wound covered as described.   In such a situation the dressing will last only for a short time before being rejected because it is still a foreign material.   However, during the period of covering, the person's own skin (keratinocyte) cells can be grown to give a sheet of the person's own skin, and this then used as a permanent graft.

Plant hydrocolloids, as have been described, such as alginate, carboxymethyl cellulose have been used as the basis of wound and burns dressings.   However, they are not as effective as material of human or animal origin.   For this reason hyaluronan is under intensive investigation by most of the main wound dressing producers to improve the biocompatibility of their material and make a product more akin to human skin itself. Therefore, collagen if often   used in conjunction with hyaluronan in an attempt to simulate human skin.

SELECTED REFERENCES

1. Gum arabic

Phillips, Glyn O. (1998). Acacia gum(Gum Arabic): a nutritional ibre; metabolism and calorific value. *Food Additives and Contaminants,* 15, 251 - 264

Islam, A.M., Phillips, G.O., Sljivo, Snowden, M.J. and Williams, P.A., (1997). A review of recent developments on the regulatory, structural and functional aspects of gum arabic, *Food hydrocolloids,* 11(4), 493-505.

Phillips, Glyn O. (19980 The classification of natural gums. Par 10. Chemometric characterisation of exudate gums that conform to the revised specification of the guma arabic for food use and the identification of adulterents, **Food Hydrocolloids,** 12, 141-150.

2. Use of the hydrocolloid hyaluronan and its derivatives for tissue repair

Al-Assaf, S., Phillips, G.O., Deeble, D.J., Parsons, B., Starnes, H. & von Sonntag, C. (1995). The enhanced stability of the cross-linked hylan structure to the hydroxyl ($^{\circ}$OH) radicals compared with the uncross-linked hyaluronan. *Radiat. Phys. Chem.* 46, 207-217.

Balazs, E.A. (1968). Viscoelastic Properties of Hyaluronic Acid and biological lubrication. *Univ. Michigan Med. Center J.,* Special Issue, 255-259.

Balazs, E.A. (1979). Ultrapure hyaluronic acid and the use thereof. *US Patent No. 4,141,973,* 27th Febuary 1979.

Balazs, E.A. (1984). Viscosurgical Tools: Principles of Action. *Ocular Surgery News, Nov.* 15, 36-42.

Balazs, E.A. and Leshchiner, E. (1984). Cross-linked gels of hyaluronic acid, *US Patent no.* 4,582,865.

Balazs, E.A. and Denlinger, J.L. (1985). Sodium hyaluronate and joint function. *J. Eq Vet. Sci.* 5, 217-228.

Balazs, E.A., Leshchiner, E., Leshchiner, A. and Band, P (1987). Chemically modified hyaluronic acid and method of recovery thereof from animal tissues, *US Patent no.* 4,713,448.

Balazs, E.A. & Leshchiner, E.A. (1989). Hyaluronan, its cross-linked derivative hylan and their medical application, In "*Cellulosic Utilization*" Research and Rewards in Cellulosic. (Inagaki, H. and Phillips, G.O., Eds.), Elsevier, New York, pp 233-241.

Balazs, E.A., and Denlinger, J.L. (1993). Viscosupplementation: A new concept in the treatment of arthritis. *J. Rheum.* 20, 3-9.

Bragantini, A., Cassini, M., De Bastiani, G. and Perbellini, A. (1987). Controlled single blind-trial of intra-articularly injected hyaluronic acid (HYALGAN®) in osteoarthritis of the knee. *Clin. Trials J. 24,* 333-340.

Gunning, A.P., Morris, V.J., Al-Assaf, S. and Phillips, G.O. (1996). Atomic force microscopic studies of hylan and hyaluronan. *Carbohyd. Poly 30,* 1-8.

Lussier, A., Cividino, A.A., Mc Farlane, C.A., Olszynski, W.P., Potashner, W.J. and De Medicis, A. (1996). Viscosupplementation with hylan for the treatment of osteoarthritis findings from clinical-practice in Canada. *J. Rheumatol. 23*, 1579-1585.

Mensitieri, M., Ambrosio, L., Innace, S. and Nicolais, L. (1995). Viscoelastic evaluation of different knee osteoarthritis therapies. *J. Material Sci.- Materials in Medicine 6*, 130-137.

Miller, D. and Stegmann, E.A. (1980). Use of sodium hyaluronate in anterior segment eye surgery. *Amer. Intraocul. Imp. Soc. J. 6,* 13-15.

Namiki, O., Toyoshima, H. and Morisaki, N. (1982). Therapeutic effect of intra-articular injection of high molecular weight hyaluronic acid on osteoarthritis of the knee. *Int. J. Clin. Pharm. Ther. and Toxicol. 20,* 501-507.

Pape, L.G. and Balazs, E.A. (1980). The use of sodium hyaluronate (Healon®) in human anterior segment surgery. *Opthalmology 87*, 699-705.

Rydell, N. and Balazs, E.A. (1971). Effect of intra-articular injection of hyaluronic acid on the clinical symptoms of osteoarthritis and on granulation tissue formation. *Clin. Orthop. 80,* 25-32.

Scale, D., Wobig, M. & Wolfert, W. (1994). Viscosupplementation of osteoarthritis knees with hylan: A treatment schedule study. *Current Therapeutic Research 55*, 220-232.

Takigami, S., Takigami, M. and Phillips, G.O. (1993). Hydration characteristics of cross-linked derivatives hylan. *Carbohyd. Poly. 22*, 153-160.

Takigami, S., Takigami, M. and Phillips, G.O. (1995). Effect of preparation method on the characteristics of hylan and comparison with another highly cross-linked polysaccharide, gum arabic. *Carbohyd. Poly. 26*, 11-18.

3. Human tissue banking

Phillips Glyn O. (Editor-in- Chief), *Advances in Tissue Banking*   Volume 1, 1997 (ISBN 981-02-3190-3) and Volume 2, 1998 (World Scientific, London and Singapore ISBN 981-02-3534-8)

Acknowledgement :

I wish to thank my colleagues   Professors   Peter A. Williams, Katsuyoshi Nishinari   for their collaboration in the work I have described on food hydrocolloids and for their partnership in organising meetings over the years of this subject. Dr Saphwan Al-Assaf has carried out much of the work on hylan. Dr Endre A. Balazs I thank for introducing me to hyaluronan and for his steadfast cooperation over the years.

2. DISPERSIONS, EMULSIONS AND SURFACES

HYDROCOLLOIDS – PART 2
Edited by K. Nishinari
2000 Elsevier Science B.V.

On-line measurement of aggregation and flocculation

R A Williams and X Jia

Particle and Colloid Engineering Group

Camborne School of Mines, University of Exeter, Redruth, Cornwall TR15 3SE, UK

The need for practical on-line measurement methods for monitoring the state of dispersion of concentration colloidal suspensions is discussed. Developing methods based on laser light reflection and electrical tomography are described and illustrated for emulsion and solid/liquid mixtures. Ultimately the interpretation of measurement signals derived from in-situ sensor may also require the use of an appropriate phenomenological model for the aggregation dynamics - one such approach is described.

1. INTRODUCTION

The ability to measure the size and shape of dispersed and aggregated particulates *in-situ* is required if these properties are sensitive to the hydrodynamic environment. This is often the case for hydrocolloid systems whose properties are shear- and time-sensitive. Nevertheless, many commonly accepted and standard laboratory methods are based on *ex-situ* methods which require sampling and dilution of process stream. This is problematic, since the act of removing samples from their natural flow environment can change the characteristics of the particulates. Dilution and further handling and sampling can induce considerable changes to the observed properties of friable or deformable aggregates. The paper will present recent developments in alternative strategies employing on-line analysis using two types of sensing method that have been used to assess the behaviour of industrial manufacturing systems. These methods are based on optical sensing (using back-scattered scanning laser microscopy or SLM) and electrical sensing (using microelectrical resistance tomography or MERT). The principles, application and future prospects for use of these techniques will be discussed.

2. OVERVIEW OF INSTRUMENTAL METHODS

A number of methods are in use to measure aggregation in-situ. Some of the basic techniques are summarised in Table 1. This list is by no means comprehensive since a variety of analytical methods are in existence. Further, there is substantial variation in the implementation of specific techniques. For instance, the use of rheo-optics and electro-optics are particularly useful and can be deployed in several ways (e.g. through birefringence of transmitted light, diochroism of transmitted light, attenuation of light, intensity of scatter of scattered light, fluorescence, acoustic birefringence, etc). Perturbation of the flow conditions can also be used to differentiate properties (elongational flow, stop-flow, centrifugal flow etc). Latterly, the development of acoustic spectroscopic methods would appear to provide potentially useful means of measuring aggregation effects, but as yet the field is relatively unexplored [1].

Table 1. Some common techniques for measurement of aggregation and dispersion

Optical	Acoustic	Electrical
Turbidity	Acoustic attenuation	Conductivity
Diffraction or angular scattering	Acoustic wave spectroscopy	Permittivity
Dynamic light scattering and FODLS	Electroacoutsic spectroscopy	Electromagnetic induction
Diffuse waves spectroscopy		
Photon migration spectroscopy		
Laser reflection		
Particle imaging		

Two of the above methods based on light reflection and electrical conductivity will now be considered.

3. SCANNING LASER MICROSCOPY (SLM)

Scanning laser microscopy (SLM) provides a means of assessing the approximate size and number density of particulates in a size range of 1-1000 μm every few seconds. The principle of the measurement is shown in Figure 1, whereby a laser beam focused outside or close to sapphire window at the probe tip is rastered. As particulates pass through the vicinity of the focal volume, light may be reflected back into the probe according to the refractive index of the phases and the length of the intersected chord. Thus pulses of back-scattered light can be sensed and the information transformed into size and number count statistics. The deduction of actual particle size (typically over the range 0.8-1000 μm) can only be achieved reliably for model systems or through calibration knowing the shape factor. The device can operate at high volume fraction (> 60 v/v %). In our work, measurements have been made using a probe that was inserted into stirred tanks and also pairs of probes in a pipeline. Studies of both the formation and the breakage of flocs have been undertaken. In the case of stirred tank applications, the measurements are used to measure the apparent rate of particle growth under different shear and chemical conditions. Examples, which have been given elsewhere [2], for polymer flocculated silica and the preparation of an inorganic solid (ferric hydroxide), show the importance of the flocculant addition protocol (addition rate, micro-mixing, etc). The *strength* of aggregates that have been formed can also be inferred by simultaneous use of the number and size data, from which breakage constants can be deduced. This is useful for measuring break-up in a pipeline handling friable flocculated materials. The experimental approach can also assist in optimising reagent and processing conditions to generate flocs conforming to user-specified properties (size, shape, packaging density etc) [3].

In the case of emulsions the method can be used to monitor the state of coalescence and rupture in a convenient manner. In the production of commercial cosmetic and food emulsions using membrane emulsification [4] the effect of pump induced rupture of droplets and other effects can be readily assessed. For instance, Figure 2 shows the form of the droplet degradation with time, and Figure 3 gives the computed breakage rate (K, α) and spread functions (μ, σ), based on a population balance model [3]. The results suggest that significant droplet breakage only starts to occur for size greater than 30 μm (Figure 3a), and the mechanism of size reduction is through droplet splitting (Figure 3b). In this way choices can made on the correct design and gap settings etc of pumps for handling different emulsion formulations. This has proved invaluable in scale-up studies.

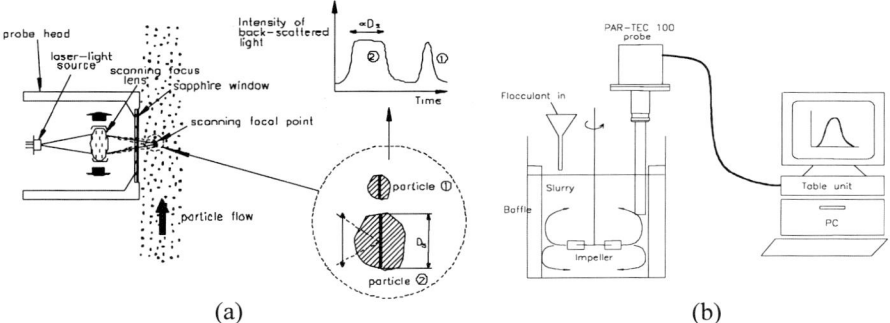

Figure 1. Schematic diagram showing (a) details of the scanning laser microscope probe and (b) an example of a measurement configuration.

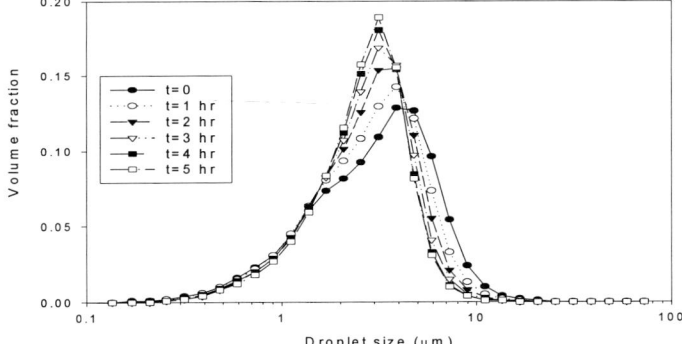

Figure 2. Droplet size distribution in a concentrated oil-in-water emulsion as a function of time during a floc breakage test in a lobe pump rig.

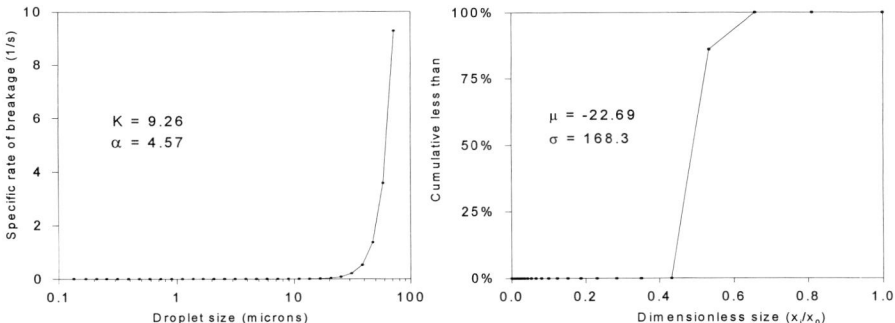

Figure 3. Computed (left) specific rate of breakage and (right) breakage distribution function for a commercial 3-lobe pump.

4. MICROELECTRICAL RESISTANCE TOMOGRAPHY (MERT)

Electrical sensing methods have been used for many years for particle counting but the full potential of using a multiplicity of electrical sensing probes to measure conductivity or permittivity has not been fully exploited. The use of electrode arrays around a pipe and small-scale reaction vessel will be described, from which information on the homogeneity of the dispersions and, sometimes, particle size and shape can be inferred. The techniques are based on tomographic measurement, although the method of analysis may not necessarily involve a reconstruction of an image of the distribution of the electrical properties (which can be ill-posed inverse problem). The principle of electrical resistance tomography will be introduced with an example showing the three dimensional dispersion of gas bubbles in a small reactor. The miniaturisation of the method to allow visualisation of aggregated particulates in dilute and concentrated suspensions flowing in small bore (1 mm) tubes will be considered. The potential of the technique for quantifying the state of coalescence of oil-in-water emulsions will be illustrated based on analysis of three-dimensional 'electrical texture' of the emulsion. This has applications as an on-line sensor for monitoring the quality/state of concentrated colloidal mixtures.

Conventionally, tomographic techniques based on electrical resistance tomography (ERT) have been used to provide images of the material distribution within one or more cross-sections through pipes, agitated tanks, or separation vessels [5]. This is obtained by determining the resistivity profile which can be related to the localised particle concentration. Such data are of value for verification of detailed microscale and fluid dynamic models. ERT has been applied to process-scale systems with a cross-sectional diameter ranging from 25 mm to 3 m. An example is shown in Figure 4, where gas is pumped into a 30 dm^3 baffled mixing tank and measured using ERT. There are 4 sensing planes, each having 16 electrodes (Figure 4a). ERT involves injecting an electrical current through a pair of electrodes and measuring the response signals (i.e. voltages) on the others. Such procedure is repeated for all possible but independent combinations. The collected voltages are then used to solve an inverse problem to find the conductivity distribution across the sensing plane. Details of data collection protocols and reconstruction methods are described elsewhere [5,6]. The resulting tomographic images can be stacked together to provide a pseudo 3D 'image' of the stirred tank, as shown in Figure 4b. The method can be seen to be sensitive enough to discern the change of resistivity distribution which in turn is related to gas distribution, due to different gas injection rate.

If features of less than 1 mm are to be perceived, it is necessary to reduce the electrode projection distance. The ERT electrode sensor assembly is the principal component which differs between a MERT system and a conventional ERT system [7]. Since the vessel under examination is in the order of 1 mm diameter, it is necessary to place a ring of electrodes around this bore, whilst ensuring high tolerances. The electrode design used here incorporates an array of gold electrodes (100 μm x 380 μm) upon a ceramic substrate. Figure 5 shows a base sensor plate on which an array of gold electrodes is deposited on a thin ceramic wafer substrate containing a 1 mm hole (Figure 5a), and the integration of these base sensor plates in a sensor unit ready to be inserted into a pipe (Figure 5b). Figure 5c shows the MERT sensor mounted into a test flow loop which can be used to circulate dispersions or emulsions.

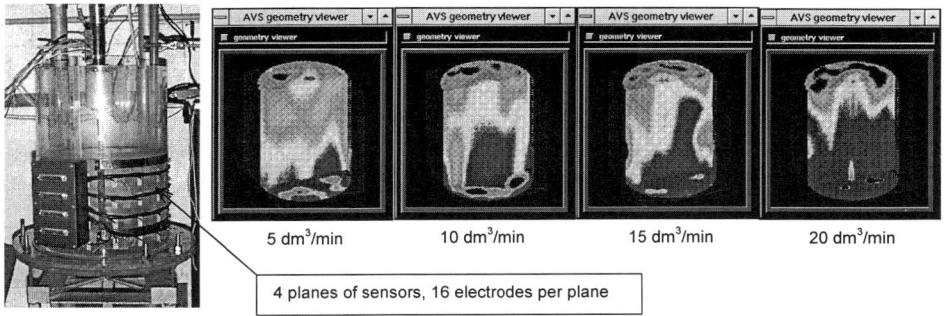

5 dm³/min   10 dm³/min   15 dm³/min   20 dm³/min

4 planes of sensors, 16 electrodes per plane

Figure 4. Effect of sparge rate on resistivity distribution for constant stirring speed of 300 rpm in a baffled tank of 30 dm³, gas flux from the central sparger is 5, 10, 15 and 20 dm³/min respectively.

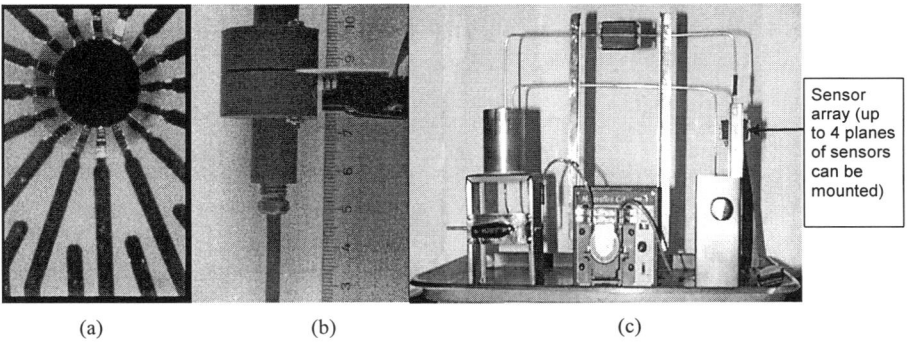

(a)   (b)   (c)

Sensor array (up to 4 planes of sensors can be mounted)

Figure 5. (a) 16 gold electrodes deposited on a ceramic substrate to form a basic electrode plate. (b) The electrode plate(s) are integrated into a sensor unit. (c) The sensor unit has been inserted into a flow pipe containing a flowing emulsion.

One of the advantages of this design is the high degree of reproducibility in the manufacturing process, thus enabling several nearly identical assemblies to be realised. This sensor can be coupled to an ERT instrument (such as the UMIST Mk1b/e [6]) but requires some modification to allow injection of small currents.

In MERT, particle size measurement may be approached by the methods shown in Figure 6 and Table 2 [8]. The first mode aims to recover the boundary of objects in either a true 3D or a time-spaced 3D tomogram, depending on the measurement arrangement. In a true 3D measurement, the corresponding forward and inverse solvers must be 3D models and the electrodes must be mounted along the pipe wall. In a time-spaced measurement, the 3D tomogram is obtained by stacking sequential 2D tomograms, and the speed of flow is assumed or known. Some results of such time-spaced measurements showing the 3D structure of emulsion flow states are demonstrated later (Figure 7).

24

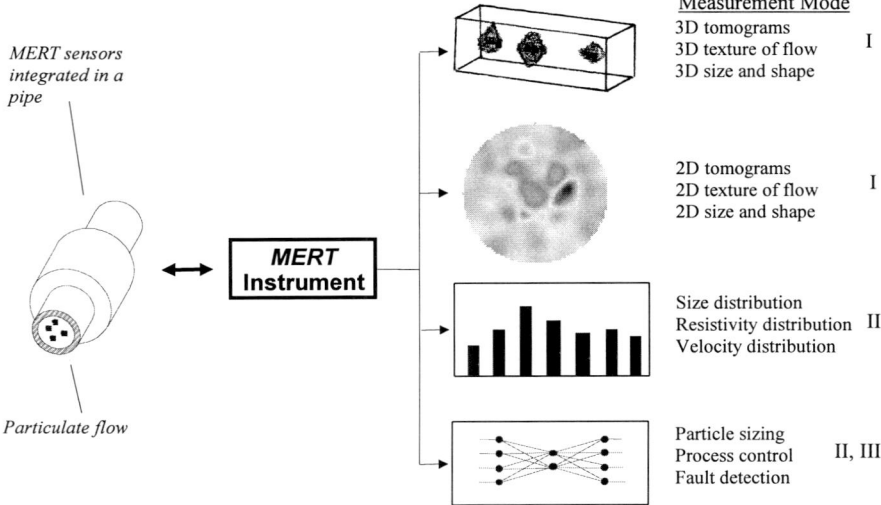

Figure 6. Schematic diagram showing possible applications of MERT

Table 2. Three modes of MERT operation for particle size and shape measurement

Mode	Measurement	Deduced property
I	Full 3D tomogram or stacking of 2D time–spaced tomograms	Volume fraction, shape, size, electrical texture
II	Change in mean conductivity over entire sensing space of the cross–section	Volume fraction, size
III	Particle transit time	Axial chord length

In the second mode of operation, the change in conductivity caused by a non-conducting droplet passing through the sensor orifice will produce a unique change in the boundary potentials at each electrode pair. These changes can be used to infer particle size. A neural network based particle sizer can be used to produce the size information directly from such measurements. Such a neural network sizer is particularly useful in applications where the principal interest is not the detailed visualisation of the flows, but is the particle size distribution in the mixture passing through the pipe. One advantage of adopting this mode of measurement is that it may only require use of a fraction of the total number of electrodes that are required for tomographic image reconstruction.

In the third mode, by recording the *transit time* required for a particle to move through the sensing region, a characteristic length of particles along the axial direction can be obtained, provided that the flow velocity is known.

In comparison, the first mode gives most accurate results, though it is the most time-consuming in yielding a result. The second and third modes complement the first and could ultimately be used to cross check the data interpretation procedure.

When the dispersed particulates are smaller than the reconstructed pixel size, individual particles cannot be uniquely distinguished. Nevertheless, the electrical resistivity profile or texture may be related to some bulk properties of the dispersion. It is proposed that MERT can be used to observe structural features in flowing suspensions, concentrated emulsions, and pastes. Here, the concept is to characterise the formulation properties of complex aqueous-based suspensions and dispersions using a 3D electrical fingerprint [9]. If such fingerprints can be obtained in real-time, they can be used to monitor product quality, for process control purposes and hence to improve process efficiency and to reduce wastage (e.g. due to manufacture of off-specification materials). The electrical fingerprint is the electrical resistivity texture map of the suspension as it flows through a sensor array in a narrow bore tube, as illustrated in Figure 7, based on a physical simulation. It is believed that the texture of the image reflects properties of the suspended particulates (states of aggregation, voidage, particle shape, alignment in flow, etc) and these can be abstracted by analysing the texture and, also, the raw measurement signals. The method offers the potential for operation in opaque and fast moving process systems, but most significantly utilises spatial and temporal information not provided by other methods. For example, Figure 7c shows the spatial distribution of volumes contained within given iso-surfaces, and these features can be analysed (in terms of size, number, shape, co-ordination number etc.) and classified. The voxel characteristics can be correlated to suspension properties, for example, using an on-line modelling method based on a neural network scheme. Such characteristics can then be related to the performance of the manufacturing process producing the suspension or emulsion and, if appropriate, to the perception of the customer or user of the product or material.

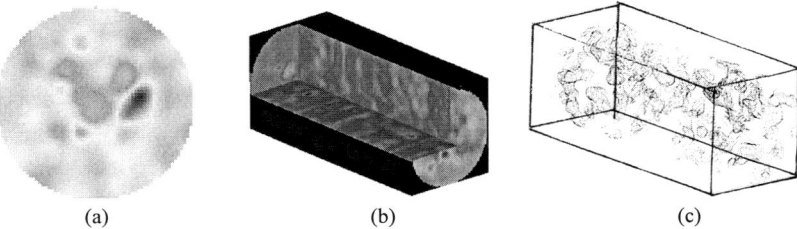

(a) (b) (c)

Figure 7. Concept of fingerprinting the properties of complex suspensions by extracting features from (a) the electrical resistivity texture map, (b) time-spaced pseudo-3D flow texture, and (c) example of texture extraction for further statistical analysis.

One such application of MERT would be to monitor a concentrated emulsion (oil-in-water, or o/w, type) flow. The emphasis here is to observe the transition of state change in the mixture during the flow. The electrical resistivity change in the mixture caused by the coalescence of emulsion droplets is an indication that MERT could provide the information on the change in electrical texture in such a physical (coalescence) process.

5. AGGREGATION MODELLING

The analysis and interpretation of any measurement signal from an instrument requires a model for the *sensor* performance and a model that embodies the essential characteristics of the *process* that is being observed. Sometimes appropriate models relating to the process are not known, so interpretation of the signals can be problematic and subject to error. This is often the case for on line measurement applications. However where *a priori* information is available, the robustness of the interpretation of the measurements can be improved very significantly [10]. Our own recent work has sought to develop practical modelling approaches that described colloidal aggregation processes from which information on the sedimentation and packing behaviour can be obtained.

Traditionally, random-walk aggregation models, such as the diffusion-limited aggregation (DLA) model introduced by Witten and Sander [11], are based on a 'stick at the first contact' scheme and there are no allowances for the possible subsequent breakage and/or deformation of the aggregates. We propose an extension to the basic DLA algorithm to include such allowances through the use of three probabilistic parameters: the probability to attach (ρ_1), to detach (ρ_2), and to deform (ρ_3). Each probability may be related to, or interpreted as the manifestation of, some physical properties of the colloidal system. For example, ρ_1 may be related to the net colloidal interaction (i.e. higher probability indicates a stronger attraction); ρ_2 to hydrodynamic shear or mechanical wear (leading to floc breakage or erosion); and ρ_3 to short-range interactions, thermal fluctuations, collision and ageing effect. It is possible to simulate simultaneous sedimentation by simply restricting particle movements in the vertical-up direction and changing the boundary condition at the bottom of the simulation box to non-penetrating solid-wall.

There can be many different implementations of the extended model. In our cluster-cluster aggregation and sedimentation simulator, the probabilities are incorporated as follows. The probability to attach, ρ_1 applies to each contact made during trial moves of flocs (which include single-particle flocs). The trials are regulated such that in aggregation mode larger flocs move slower than smaller flocs, whereas in sedimentation mode larger flocs settle faster than smaller flocs. Furthermore, in aggregation mode, periodical boundary conditions apply on all the sides of the simulation box. In sedimentation mode, the bottom side uses solid-wall boundary condition instead. The probabilities to detach, ρ_2, and to deform, ρ_3, are in effect the fraction of randomly selected primary particles within each floc, which are subject to detachment and rearrangement respectively. As the name suggests, if a primary particle is selected for detachment, it breaks away from its host floc and becomes either a single-particle floc or attached to another floc. A primary particle, when selected for rearrangement, moves within the floc, i.e. without losing contact with the host floc. There is an option where a trial move leading to more internal contacts is always accepted regardless the value of ρ_3. This tends to generate a more compact structure for the flocs. Some examples, both 2D and 3D, generated under different conditions, are shown in Figure 8. The ability of the model to simulate both aggregation and sedimentation, within a single simulation run, is illustrated in Figure 9. During the sedimentation stage, flocs may still undergo aggregation, depending on the value of ρ_1.

In future we expect the predictions from the models can be run in tandem with measurements on the sedimentation (and other) process. Instrumental measurements will express their outputs in terms of the same parameters used in the process model, thus allowing verification and quantification of the model and instrument. Confidence levels for the fit of the model and measurements will be obtained.

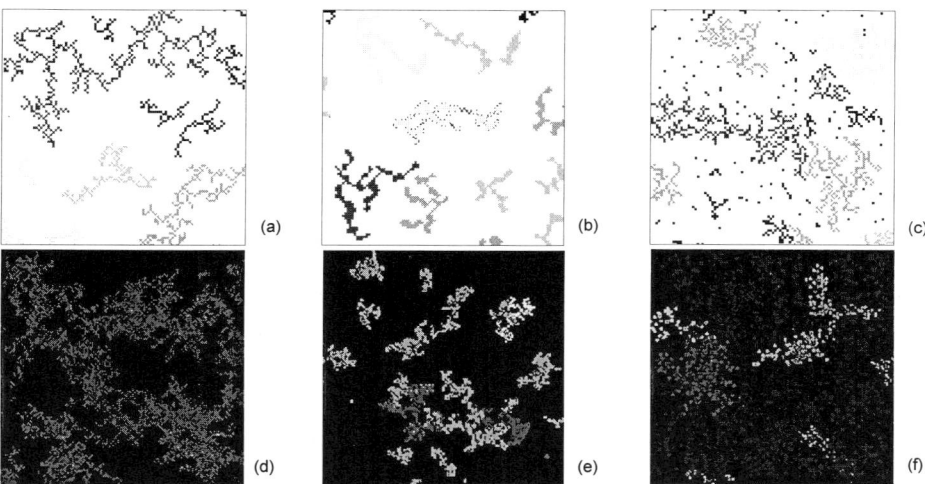

Figure 8. Examples of flocs generated under different conditions. (a) 2D flocs generated with $\rho_1 = 1$ and $\rho_2 = \rho_3 = 0$. (b) 2D flocs generated with $\rho_1 = 1$ and rearrangements that favour a compact structure. (c) 2D flocs generated with $\rho_1 = 0.5$, $\rho_2 = 0.1$ and $\rho_3 = 0.2$. (d) 3D single, 5000-particle floc percolating the simulation box, generated with $\rho_1 = 1$ and $\rho_2 = \rho_3 = 0$. (e) 3D flocs generated with $\rho_1 = 0.5$ and rearrangements that favour a compact structure. (f) 3D flocs generated with $\rho_1 = 0.5$, $\rho_2 = 0.1$ and $\rho_3 = 0.2$.

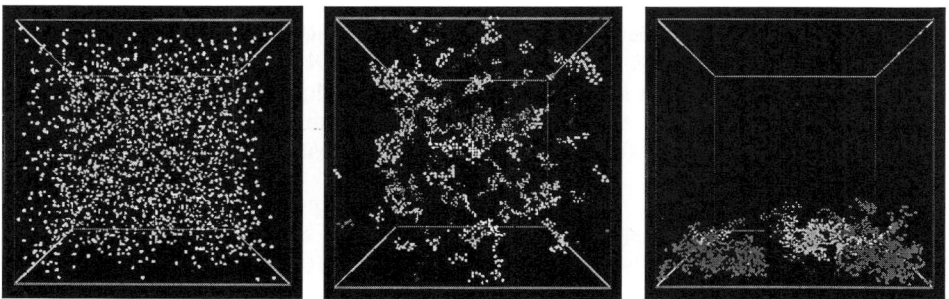

Figure 9. Sample 3D configuration at the start of simulation (left), end of aggregation and onset of sedimentation (middle) and end of sedimentation simulation (right).

6. CONCLUSIONS

Some recent developments in on-line measurement of aggregation and flocculation have been described. SLM is an established optical sensing method, it provides a powerful and semi-quantitative means of following the evolution of aggregating population in sheared suspensions and of abstracting appropriate growth or breakage constants. However, the method is intrusive in that the laser probe must be inserted into the process vessel and have physical contact with the suspension. MERT, on the other hand, is still being actively developed, but it has the potential to offer a low-cost and non-intrusive as well as on-line method to characterise particle size and shape in dilute dispersions. For concentrated colloidal systems, where scanning laser microscopy may no longer be suitable, MERT can still offer useful information in the form of an electrical fingerprint of the suspension, thereby enabling certain electrical-related bulk property to be inferred, even though individual particles cannot be distinguished. Signals can be correlated with the formulation/manufacturing process and with the customer's perception of the product. Method of signal analysis could involve: analysis of texture from a reconstructed image (via statistical methods to obtain characteristic parameters for model-building or to train a neural network); or analysis of the raw measurement signals (as inputs to a neural net, for instance). The low cost of the sensor head itself would allow use of the analysis method at multiple sites in a plant.

REFERENCES

1. M. J. W. Povey, Ultrasonic Techniques for Fluid Characterisation, Academic Press, San Diego, (1997).
2. R. A. Williams and S. J. Peng, Controlled Formation of Particle, Droplets and Bubbles, D J Wedlock (Ed.), Butterworth-Heinemann, Oxford, p327-358, (1994).
3. R. A. Williams, Euro Patent 94907644.2, Nov 97, assigned to British Nuclear Fuels plc, (1997).
4. R. A. Williams, S. J. Peng, D. A. Wheeler, N. C. Morley, D. Taylor, M. Whalley and D. Houldsworth, Chem Eng Res Des, 76 (1998) 902.
5. R. A. Williams and M. S. Beck, Process Tomography-Principles, Techniques and Applications, Butterworth-Heinemann, Oxford, (1995).
6. F. Dickin and M. Wang, Meas. Sci. Technol., 7 (1996) 247-260.
7. P. Gregory, M. Phil. Transfer Report, University of Exeter, (1996).
8. T. M. Shi, R. A. Williams, X. Jia, and F. J. Dickin, Frontiers in Industrial Process Tomography II, Delft, April 1997, (Engineering Foundation, New York), 125-130, (1997).
9. P. Gregory, R. A. Williams, X. Jia, T. M. Shi and S. P. Luke, Frontiers in Industrial Process Tomography II, Delft, April 1997, (Engineering Foundation, New York), 3-8, (1997).
10. R. M. West, X. Jia and R. A. Williams, Proceedings of the First World Congress on Industrial Process Tomography, 14-17 April 1999, Buxton, UK, p444-450, (1999).
11. T. A. Witten and L. M. Sander, Phys. Rev. Lett., 47 (1981) 1400.

HYDROCOLLOIDS – PART 2
Edited by K. Nishinari
2000 Elsevier Science B.V.

Interactions between the oil globules in aqueous media (Case study on the W/O/W emulsions)

S. Matsumoto

Department of Applied Biochemistry,
College of Agriculture,
Osaka Prefecture University,
Gakuen–cho, Sakai, Osaka 599–8531, Japan*

This article reviews fundamental works on characterization of W/O/W–type multiple emulsions including interactions between the dispersed globules of the emulsions in the presence of proteins and saccharides in view of the morphological feature of the dispersed vesicular globules in these types of emulsions. The contents are divided into four sections : formation of W/O/W emulsions, characteristics of oil layer on the surface of the vesicular globules, total interactions between the vesicular globules in the presence of proteins, and effect of saccharides on the zeta potential of the vesicular globules.

1. INTRODUCTION

Emulsion theory is partly an outgrowth of colloid and interface science [1] including study on surfactant solutions [2], and partly a development of the classical art involved in the production of commercial emulsions. For example, food emulsions are delicate and sophisticated systems in which specific properties such as taste, flavour, texture or mouth feel are as much important as the types and stability of the emulsion states [3]. Nevertheless, the historical background of emulsion science may be characterized by the respective study on several stages of the entire destabilization process. That is, the flocculation kinetics on the dispersed globules [4], electrostatic and Van der Waals interactions between the dispersed globules [e.g. 5], depletion effect of polyelectrolytes in the suspending media [6], and hydration interaction between the lipid vesicles in the aqueous media [7].

The oil globules dispersed in W/O/W emulsions can be generalized by a vesicular structure of thin oil layer enclosing a large amount of inner aqueous phase [8], and will be able to immobilize a certain amount of water–soluble ingredients [9, 10]. Therefore, an attempt was made to immobilize a series of proteins in the inner aqueous phase of W/O/W emulsions so as to control surface potential of the vesicular globules without any additive in the aqueous medium [11]. The emulsions were then applied to measure the zeta potential of the globules for estimating total interactions between the vesicular globule at a definite condition of the aqueous medium. The influences of a series of saccharides existing in the two aqueous phases of W/O/W emulsions on the interactions between the vesicular globules have also been investigated [12]. Consequently, it seems that the use of W/O/W emulsions plays as model systems for studying several subjects on the stability of macroemulsions in potential application of multiple emulsion systems.

---

* Present address : 6–902, Hamadera–motomachi, Sakai, Osaka 592–8343, Japan

2. FUNDAMENTAL ASPECTS OF W/O/W EMULSIONS

2. 1. Location of multiple emulsion formation in a phase diagram

It is possible to specify the formation region of multiple emulsions using one of the generalized phase diagrams of the ternary mixture of water, oil and emulsifier [2], as illustrated schematically in Fig. 1. The hydrophilic emulsifiers dissolve in water and form an aqueous micellar solution solubilizing a definite amount of oil, while a large part of oil phase separates from the aqueous micellar solution.   An ordinary O/W emulsion will be prepared in this area, when one makes mechanical disruption on the interface between the two separated phases. On the other hand, a variety of lipophilic (hydrophobic) emulsifiers dissolve in oil and form the reversed micells solubilizing a certain amount of water, while the residual water separates from the reversed micellar solution.   This area will be featured by the occurrence of W/O emulsions, when the interface between the two separated phases is agitated mechanically.

In the phase diagram of the ternary mixture, one can recognize the phase of emulsifiers, denoted as D–phase (detergent phase), as shown in Fig. 1. The solubilizates do not dissolve completely in the D–phase, while the three phases coexist at relatively higher concentration of solubilizates.   The D–phase divides the type of emulsions into O/W and W/O, *i.e.* the phase inversion of emulsions occurs in this area.   In 1925, Seifriz [13] published a paper on multiple emulsions according to his recognition of the existence of W/O/W– and O/W/O–type dispersions during the phase inversion of ordinary macroemulsions.   His finding can be reproduced on the phase diagram, in which the region for obtaining two different types of multiple emulsions will be located closely in either sides of the D–phase in Fig. 1 [14, 15].   That is, W/O/W emulsions form closely under the D–phase, while O/W/O emulsions appear conversely over this phase.

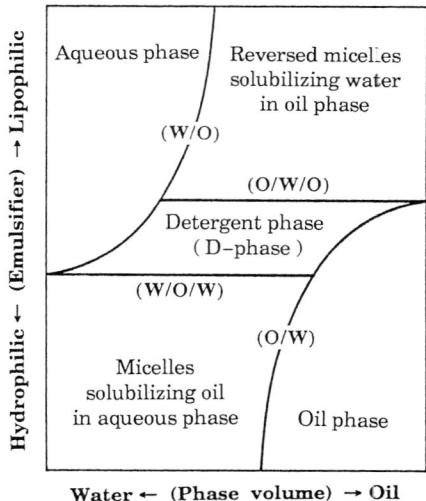

Fig. 1. Changes in solution behaviour of emulsifiers with hydrophile–lipophile balance in a ternary mixture of water, oil, and emulsifier [2].

2. 2. Techniques for providing W/O/W emulsion systems

Although a variety of phase compositions may be employed in different applications of multiple emulsions, the two separated steps of emulsification [16, 17] will be reliable for preparing W/O/W emulsions, especially in the case for encapsulating a certain amount of water–soluble ingredients in the inner aqueous phase. The first step procedure is made initially for providing W/O emulsion by use of hydrophobic emulsifiers. The W/O emulsion prepared is then mixed and stirred with an aqueous solution of hydrophilic emulsifier as the second step procedure, which plays a role in migrating phase location of the mixture from W/O to O/W emulsion region passing through the D–phase. With respect to a series of the experimental results obtained, one of the factors affecting the yield of W/O/W emulsions in this technique is the pronounced dependence on the ratio of hydrophobic to hydrophilic emulsifier existing in the final mixture [8, 16], e.g. more than two times as much Span 80 as Tween 80 is necessary to obtain 90% or higher yields of the W/O/W emulsion, as shown in Fig. 2.

As has been mentioned by Seifriz [13], the occurrence of multiple emulsions is connected with the phase inversion phenomenon of macroemulsions. This suggests another simple technique, i.e. phase inversion method, for preparing these types of emulsions [14, 18]. This technique, however, would be useless to encapsulate any ingredient in the inner phase, which will be partitioned off spontaneously to the suspending fluid during the phase inversion phenomenon. In the case of W/O/W emulsion preparation, the procedure can be completed by stirring the oil phase containing hydrophobic emulsifier, while an aqueous diluted solution of hydrophilic emulsifier is being introduced successively into the oil phase. The mixing system is initiated by forming an ordinary W/O emulsion, and then the continuous oil phase of the emulsion is suddenly substituted by the aqueous phase when the phase location is passing through the D–phase. Consequently, the first W/O emulsion inverts to a mixture of W/O/W– and O/W–type dispersions. It then follows that the W/O/W emulsion disappears gradually with additional increments of the aqueous solution of hydrophilic emulsifier, and inverts completely to an ordinary O/W emulsion.

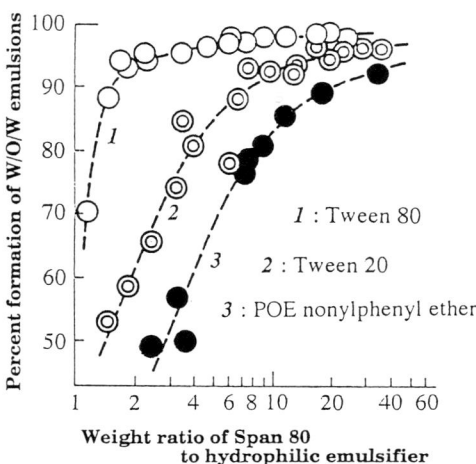

Fig. 2. Plots of the yield of W/O/W emulsions *vs.* weight ratio of Span 80 in the oil phase to hydrophilic emulsifiers in the aqueous phase [16].

2.3. Characterization of dispersed globules in W/O/W emulsions

The morphological feature of W/O/W emulsions was given by Becher [19] in his book, as follows: "A multiple emulsion is one in which both types of emulsions exist simultaneously. What is meant by this is that an oil droplet may be suspended in an aqueous phase, which in turn encloses a water drop, thus giving what might be described as a W/O/W emulsion."

It thus appears that the morphological feature of W/O/W emulsions can be simply generalized as follows : Each dispersed oil globule forms a vesicular structure with one or more compartments of the inner aqueous phase separated from the aqueous suspending medium by a layer of oil phase components.

When a dispersed vesicular sphere of diameter D consists of a spherical inner aqueous phase of diameter d surrounded by a layer of the oil phase, the volume fraction ϕ of the inner aqueous sphere in the dispersed vesicle is calculated by the ratio of the diameters d^3/D^3. Accordingly, the ratio d/D increases markedly with increasing volume fraction ϕ of the inner aqueous sphere, while the thickness $\Delta\,(=[\,D-d\,]/2\,)$ of the layer of oil phase decreases with increasing volume fraction ϕ, as follows [20] :

$$\phi = (\,D - 2\Delta\,)^3 / D^3 = d^3 / (\,d + 2\Delta\,)^3 \tag{1}$$

$$\text{or} \quad \Delta = [\,1 - (\,\phi\,)^{1/3}\,]\,D/2 = [\,(\,1/\phi\,)^{1/3} - 1\,]\,d/2 \tag{2}$$

Equation (2) suggests that the dispersed vesicular spheres in W/O/W emulsions are generally composed of the thin oil layer surrounding the inner aqueous phase, *e.g.* the value for Δ of the oil layer will be estimated as about 0.1 μm for the case of $\phi = 0.5$ and $d = 0.8$ μm.

In fact, a series of microscopic examinations suggest that the oil layer on the surface of the inner aqueous phase in W/O/W emulsions is extremely thin irrespective of the amount of oil phase components and of the method of preparation [20]. This phenomenon may be brought about by the thinning process occurring in the oil layer immediately after the formation of W/O/W emulsions due to the disjoining pressure of oil layer [21]. The surface of the dispersed vesicular globules are consequently covered by the thin oil layer in a steady state, while the surplus of the oil phase components is condensed heterogenously on the surface of the vesicular globules or dispersed in the aqueous suspending fluid as the simple oil droplets.

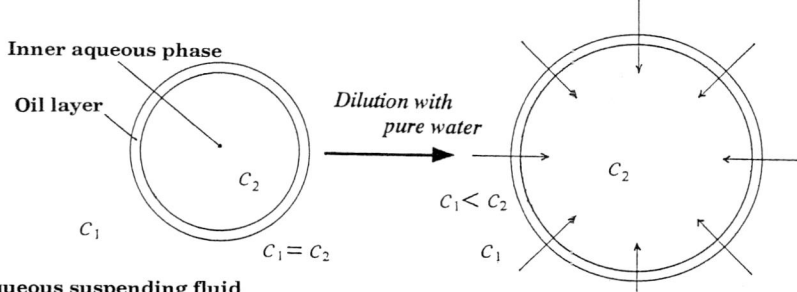

Fig. 3. Schematic illustration of water migration permeating through the oil layer under an osmotic pressure gradient between the two aqueous phases.

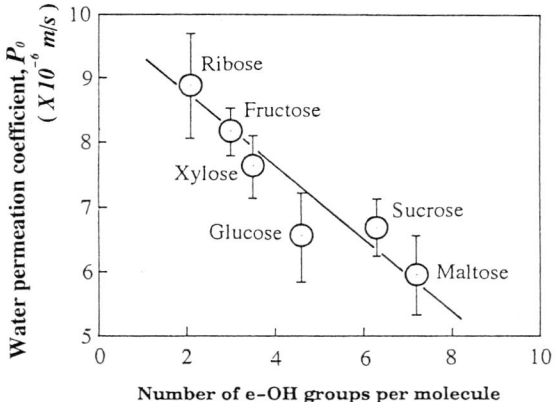

Fig. 4. The number of equatorial hydroxyl groups (e–OH) in a saccharide molecule affecting the water permeation coefficient P_0 of oil layer in W/O/W emulsions [10].

Water molecules could migrate to the inner or outer aqueous phase permeating through the thin oil layer under the osmotic pressure gradients between the two aqueous phases [22]. This causes a swelling or shrinkage of the inner aqueous phase consistent with the migration of water molecules according to the drift of osmotic pressure in the two aqueous phases, as schematically illustrated in Fig. 3. The author and his colleagues [10, 23] tried to estimate the water permeation coefficient P_0 of oil layer in W/O/W emulsions using a series of saccharides, which participate in the osmotic pressure gradient between the two aqueous phases with a concentration of 200 mM. The coefficient P_0 could be calculated from the rate of the volume flux dv/dt of water under the concentration gradient of saccharide, as follows:

$$dv/dt = P_0 A \ (g_2 c_2 - g_1 c_1) V \qquad (3)$$

where A is the surface area of a vesicular globule, V is the partial molar volume of water, g and c are the osmotic coefficient and the concentration of saccharide, and subscripts 1 and 2 refer to the outer side and the inner side of the oil layer. A range of values, from 6×10^{-6} to 9×10^{-6} m/s, could be obtained for the water permeation coefficient P_0, as shown in Fig. 4. It would be worth noting that the values for the coefficient P_0 obtained may be comparable with those for the lipid membrane systems. It is also clear that the coefficient P_0 decreases systematically with increasing number of equatorial hydroxyl groups (e–OH) in a saccharide molecule.

3. INTERACTIONS BETWEEN THE VESICULAR GLOBULES

3. 1. Effect of proteins on zeta potential of the vesicular globules
The results obtained from the preliminary experiments indicated that the dispersion state of the vesicular globules in W/O/W emulsions is much influenced by a small amount of protein immobilized in the inner aqueous phase at a definite pH with a fixed ion strength of the aqueous suspending medium. Proteins, one of the amphoteric polyampholytes and

amphiphilic substances, will adsorb tightly onto the inner surface of the oil layer from the inner aqueous phase. Such behaviour of proteins was confirmed by studies on enzymatic activity in the vesicular globules of W/O/W emulsions [9].

Despite the fact that the vesicular globules are already negatively charged in the queous suspending fluid due to the differences between the affinity of the oil layer and that of the aqueous phase for ions [24], the adsorption of the protein molecules on the inner surface of oil layer may produce changes in the charged surface potential of the vesicular globules. Attempts were, therefore, made to obtain information on the correlation between the zeta potential of the dispersed vesicular globules and the isoelectric point of proteins immobilized in the inner aqueous phase of the vesicular globules [11] with the W/O/W emulsion samples, which were divided into liquid paraffin and olive oil systems, respectively. The former was stabilized with Span 80 (sorbitan monooleate) and Tween 80 (polyoxyethylene sorbitan monooleate), while the latter was prepared by using TGCR (tetraglyceryl condensed ricinolate) and DGMO (decaglyceryl monooleate). Each system was also subdivided into four samples according to the kind of proteins immobilized in the inner aqueous phase of the vesicular globules : (1) control (without protein) ; (2) BSA (bovine serum albumin), pI 4.4–4.9 ; (3) chymotrypsin, pI 8.1–8.6 ; and (4) lysozyme, pI 10.5–11.0.

An assembly with a rectangular cell was provided for evaluating electrophoretic mobility of the dispersed vesicular globules at neutral pH at room temperature (ca. 25°C). Each sample to be tested was diluted 100 times with an aqueous solution of 1 mM KCl immediately before the measurement. Field strength within a range from 3.3 to 10.0V/cm were used to ensure that the electrophoretic mobility was independent of the applied voltage. The zeta potential ζ on the slipping plane of the diffused portion in the electrical double layer around the vesicular globules was calculated by use of the Smoluchowski's equation for electrophoresis on nonconducting spheres, as follows:

$$\mu = \varepsilon_0 D \zeta / \eta \tag{4}$$

where μ is the electrophoretic mobility, ε_0 is the permittivity of free space, and D and η are the relative permittivity and viscosity of the aqueous medium, respectively.

Table 1. Microelectrophoretic data of dispersed vesicular globules in W/O/W emulsions at neutral pH at room temperature [11].

Inner aqueous phase	Mobility ($\mu m \cdot s^{-1} / V \cdot cm^{-1}$)	Zeta potential (mV)	Deviation (mV)
Liquid paraffin system			
Without protein	−2.50	−32.1	±0.9
1% Bovine serum albumin	−2.14	−27.4	±0.5
1% Chymotrypsin	−1.67	−21.4	±0.4
1% Lysozyme	−1.06	−13.7	−−
Olive oil system			
Without protein	−5.56	−71.3	2.8[a]
1% Bovine serum albumin	−4.30	−55.1	3.1[a]
1% Chymotrypsin	−3.53	−45.3	2.2[a]
1% Lysozyme	−2.27	−29.0	2.0[a]

[a] Standard deviation.

Table 1 summarizes the values for the microelectrophoretic mobility μ and zeta potential ζ of the vesicular globules of the samples described above. The effect of proteins on the zeta potential of the vesicular globules appears to follow the different magnitude for the isoelectric point of the proteins immobilized in the inner aqueous phase, although the values obtained with the liquid paraffin system are smaller than those obtained with the olive oil system. This may be explained by the relatively longer distance to the slipping plane from the surface of the globules in the liquid paraffin system, since the globules would be covered by a deep hydration layer due to the existence of polyoxyethylene chains in the adsorbed molecules of Tween 80 [25].

3. 2. Total interactions between the vesicular globules due to the DLVO theory

It should be mentioned that the zeta potential of the vesicular globules can be ranked with the magnitude of the isoelectric points of proteins existing in the inner aqueous phase irrespective of the type of emulsifiers used, as shown in Table 1, while a rapid coagulation occurs irreversibly among the dispersed vesicular globules in the lysozyme samples having the lowest zeta potential of the globules within a series of the samples examined. Therefore, the adsorbed protein molecules on the inside surface of oil layer should play a significant role in the surface potential of the vesicular globules at a definite pH of the two aqueous phases.

An attempt was made to estimate the total potential energy V_T of interactions between the dispersed vesicular globules as a function of the separation H of the slipping planes according to the classical DLVO theory [26], as follows:

$$V_T = 2 \pi \varepsilon_0 D r \zeta^2 \ln [1 + \exp (-\kappa H)] - r A / 12H \qquad (5)$$

where r is the radius of the vesicular globules, A is the Hamaker constant, and κ is the Debye–Hückel parameter given by

$$\kappa = (e^2 \sum n_i z_i^2 / \varepsilon_0 D kT)^{1/2} \qquad (6)$$

Here, n, e, and z are the concentration, elementary charge, and valence of the counter ions, and kT is the thermal kinetic energy.

Figure 5 shows, as an example, the total potential energy of interactions between the dispersed vesicular globules in the liquid paraffin system [11, 27]. It was assumed in fact that there is no energy barrier against the separation between the globules containing lysozyme because of the instability of the dispersion state in each system, as briefly described previously. Consequently, the value of the Hamaker constant A for the liquid paraffin sysytem could be estimated as being about $4 \times 10^{-20} Joules$ for the liquid paraffin system.

3. 3. Effect of saccharides on the stabilization of the vesicular globules

Since the zeta potential of the colloidal particles in aqueous media is generally identical to the electric potential on the slipping plane located over the surface of each particle, the zeta potential seems to be more or less influenced by the condition of hydration on the surface of the particles at a definite surface potential. As reported previously [10], water molecules around the dispersed vesicular globules tend to increase in viscosity in the presence of saccharides in the two aqueous phases with increasing number of equatorial hydroxyl groups (e–OH) in a saccharide molecule in relation to the formation of a tridymite structure with water molecules.

Attempts were, therefore, made to obtain informations on the correlation between the zeta potential of the vesicular globules and the number of e–OH groups of a series of sugars such as ribose (2.1 e–OH), fructose (3.0 e–OH), xylose (3.5 e–OH), glucose (4.6 e–OH), sucrose (6.3 e–OH), and maltose (7.2 e–OH). These saccharides were dissolved in the two aqueous phases with a variety of concentrations up to 1000 mM using the liquid paraffin system of

36

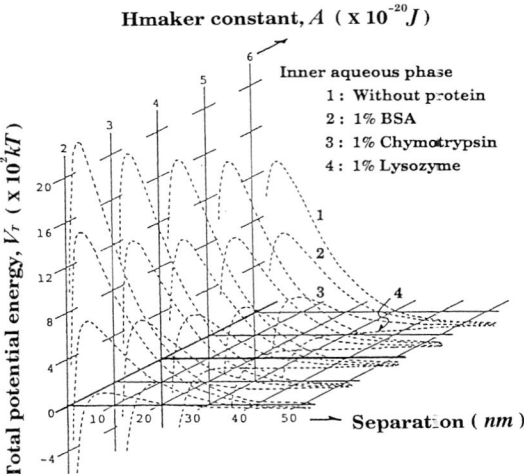

Hmaker constant, A (x $10^{-20}J$)

Inner aqueous phase
1 : Without protein
2 : 1% BSA
3 : 1% Chymotrypsin
4 : 1% Lysozyme

Total potential energy, V_T (x 10^2kT)

Separation (nm)

Fig. 5. Effect of proteins immobilized in the inner aqueous phase of liquid paraffin systems on the total interactions V_T between the vesicular globules at neutral pH at about 25°C [11].

W/O/W emulsions (see Table 1). The assembly and techniques described previously were employed for measuring zeta potential of the dispersed vesicular globules in the sample at neutral pH at room temperature (ca. 25°C), while each sample to be tested was diluted with an aqueous mixed solution of 1 mM KCl and one of the saccharides (from 10 mM to 1000 mM) immediately before the measurement.

Figure 6 shows the effect of saccharide concentrations on the zeta potential of the vesicular globules in comparing the minimal number of e–OH groups (ribose) with the maximal number of those (maltose) within a series of saccharides to be tested [12, 27]. The results obtained indicate that it is not possible to identify the influence of number of e–OH groups on the zeta potential within the experimental error, but the zeta potential of the globules certainly decreases with increasing concentration of saccharides in the two aqueous phases. This phenomenon may suggest that the presence of saccharides in the aqueous medium induces migration of the slipping plane over the surface of each vesicular globule.

The distribution of electrical potential around the surface of the vesicular globules along the vertical axis (x–axis) from the slipping plane could be approximated by use of the Maxwell–Boltzman's probability, while the Debye's length ($1/\kappa$) was already estimated as 9.6 nm for all cases to be examined. Thus, it is possible to calculate the distance of migration of the slipping plane along the x axis from the relationship between the zeta potential and the concentration of saccharides.

Figure 7 shows the movement of the slipping plane calculated from the mean values of the zeta potential at each concentration of saccharides to be examined. The expanding location of the slipping plane during electrophoretic movement of the vesicular globules due to the increasing concentration of saccharides seems to be brought about by the viscous hydration layer surrounding the globules. Each plot in Fig. 7 can be summarized by an expression on the basis of least squares, as follows:

$$x = 7.5 \left[\, 1 - \exp\left(-c\,/\,0.49\right)\,\right] \tag{7}$$

where x (*nm*) is the distance of movement of the slipping plane from the control (without saccharide), and c is the molar concentration of each saccharide in the two aqueous phases.

It should be noted that the lysozyme sample in the liquid paraffin system is stabilized when more than 300 *mM* saccharide are present in the two aqueous phases of the sample. The vesicular globules in the creaming layer of the lysozyme sample containing saccharides are readily redispersed into the aqueous suspending fluid by a gentle agitation of the sample container. This phenomenon may be related to the hydration repulsion [7], which may play a role in stabilization of the vesicular globules in aqueous media.

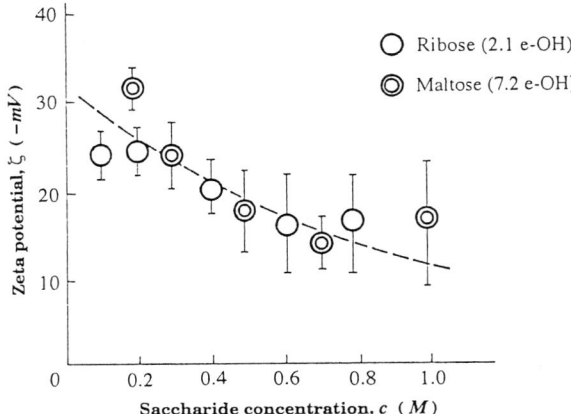

Fig. 6. Plots of zeta potential of vesicular globules *vs.* saccharide concentration in the two aqueous phases of W/O/W emulsions at neutral pH at about 25°C [12, 27].

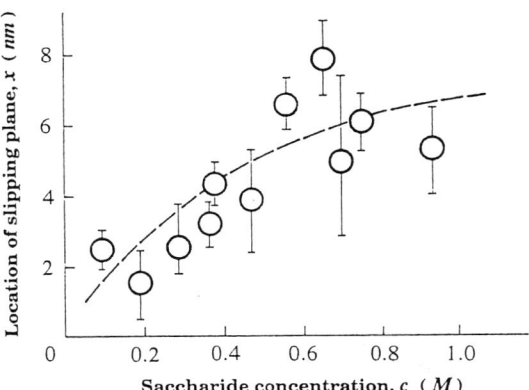

Fig. 7. Movement of slipping plane over the surface of the vesicular globules occurring when increasing concentration of saccharides in the aqueous phase [12, 21, 27].

38

In concluding this article, the author would like to quote from the remarks made by Drs. Gerittsen and Roux at the Second World Congress of Emulsion held in 1997 at Bordeaux, as follows : "We know that in many years, particularly in industrial areas, emulsions are still considered to be an art, sometimes even a black art. From this congress we learn more and more that emulsions also a science. We would like to invite you to continue to work on the development of the science of the art of emulsions. Emulsions of art and science."

REFERENCES

1. P.Sherman (ed.), Emulsion Science, Academic Press, New York, 1968.
2. K.Shinoda and S.E.Friberg, Emulsions and Solubilization, John Wiley, New York, 1986.
3. S.E.Friberg (ed.), Food Emulsions, Marcel Dekker, New York, 1976.
4. A.S.C.Lawrence and O.S.Mills, Discuss. Faraday Soc. No.18 (1954) 98.
5. H.R.Kruyt (ed.), Colloid Science Vol.1, Elsevier, Amsterdam, 1952.
6. P.R.Sperry, H.B.Hopfenberg and N.L.Thomas, J.Colloid Interface Sci., **82** (1981) 62.
7. R.P.Rand and V.A.Parsegian, Biochim. Biophys. Acta, **988** (1989) 351.
8. S.Matsumoto, ACS Symp. Ser. No.272 (1985) 415.
9. S.Matsumoto and W.W.Kang, Agric. Biol. Chem., **52** (1989) 2689.
10. K.Ueda and S.Matsumoto, J.Colloid Interface Sci., **147** (1991) 333.
11. K.Ueda and S.Matsumoto, Bull. Chem. Soc. Japan, **64** (1991) 3163.
12. S.Matsumoto, In K.Nishinari and E.Doi (eds.), Food Hydrocolloids : Structure, Properties and Functions, pp. 399–408, Plenum press, New York, 1994.
13. W.Seifriz, J.Phys. Chem., **29** (1925) 738.
14. S.Matsumoto, J.Colloid Interface Sci., **94** (1983) 362.
15. S.Matsumoto and W.W.Kang, J.Japan Oil Chem. Soc., **38** (1989) 165.
16. S.Matsumoto, Y.Kita and D.Yonezawa, J.Colloid Interface Sci., **57** (1976) 353.
17. S.Matsumoto, M.Kohda and S.Murata, J.Colloid Interface Sci., **62** (1977) 149.
18. S.Matsumoto, Y.Koh and A.Michiura, J.Disper. Sci. Technol., **6** (1985) 507.
19. P.Becher, Emulsions : Theory and Practice (2nd ed.), Reinhold, New York, 1965.
20. S.Matsumoto, In M.J.Schick (ed.), Nonionic Surfactants : Physical Chemistry, pp. 549–600, Marcel Dekker, New York, 1987.
21. S.Matsumoto, In M.Seiller and J.L.Grossiord (eds.), Multiple Emulsions : Structure, Properties and Applications, pp. 19–52, Édition de Santé, Paris, 1998.
22. S.Matsumoto, Proced. 2nd World Congr. on Emulsion Vol.4 (1997) 91.
23. S.Matsumoto, T.Inoue, M.Kohda and K.Ikura, J.Colloid Interface Sci., **77** (1980) 555.
24. R.J.Hunter, Zeta Potential in Colloid Science : Principles and Applications, Academic Press, London, 1981.
25. J.V.Boyd, C.Parkinson and P.Sherman, J.Colloid Interface Sci., **41** (1972) 359.
26. E.J.V.Verwey and L.Th.G.Overbeek, Theory of the Stability of Lyophobic Colloids, Elsevier, Amsterdam, 1948.
27. S.Matsumoto, In H.Ohshima and K.Furusawa (eds.), Electrical Phenomena at Interfaces, pp. 595–604, Marcel Dekker, New York, 1998.

HYDROCOLLOIDS – PART 2
Edited by K. Nishinari
2000 Elsevier Science B.V.

Caseins : interfacial layer properties and their influence on emulsion stability

D. S. Horne

Hannah Research Institute, Ayr, KA6 5HL, Scotland, United Kingdom

The properties of protein-stabilised colloids are largely determined by the structure of the biopolymer layer adsorbed at the particle surface. The milk proteins, α_{s1}-casein and β-casein, are the biopolymers predominantly present at the oil/water interface in many dairy emulsions. This paper summarises important salient features of recent work on the characterisation of the conformations of casein proteins adsorbed at air/water and oil/water interfaces using a range of physical and biochemical techniques. The experimental results are compared with predictions of adsorbed structures on flat hydrophobic surfaces using a modification of Scheutjens-Fleer self-consistent-field theory. Further calculations show that α_{s1}-casein and β-casein differ in adsorbed molecule conformation with α_{s1}-casein adopting a train-loop-train structure and β-casein a tail-train conformation. Predictions of interaction energies show the β-casein conformation always giving rise to repulsion and stability, whereas attraction and hence instability is a possibility at higher ionic strengths with α_{s1}-casein stabilised systems. Experimental evidence is offered in support of these theoretical calculations.

1. INTRODUCTION

The stability and rheological properties of emulsions are largely determined by the interactions between the droplets [1]. The nature and strength of the interactions are, in turn, dependent on the structure and composition of the adsorbed layer at the oil/water interface. In food emulsions, the stabilising layer around the droplets is mainly composed of proteins, frequently derived from milk, particularly the industrial preparation, sodium caseinate. This paper summarises some of the experimental and theoretical advances made in recent years by studying some simpler model systems containing pure caseins adsorbed at macroscopic interfaces or at the surfaces of oil droplets or solid polystyrene particles.

Milk proteins, particularly the caseins are employed as emulsifiers because of their high surface activity, their strong and rapid adsorption at the oil/water interface [1,2]. The protein film creates a physical steric barrier to droplet coalescence and subsequent separation back to the two original phases of oil and water [3].

Attaining a high degree of stability in emulsion systems, is not, however, the only goal. Research in the area of food acceptability has identified texture as a key attribute. The internal architecture of the product defines consumer appreciation. Creating that macroscopic texture demands controlled manipulation of the interactions between the constituent particles at the molecular level. This can only be achieved through a full and complete understanding of the intermolecular and colloidal forces, including those responsible for steric stabilisation between emulsion droplets. In protein-stabilised emulsions, the conformational structure of the adsorbed protein and its contribution to droplet interaction potential are critical factors in this scenario.

2. CHARACTERISTICS OF α_{S1}-CASEIN AND β-CASEIN

The casein proteins in milk are strongly aggregated into polydisperse protein particles of approximate average diameter 200 nm, known as casein micelles [4]. Lowering the milk pH to 4.6 can isoelectrically precipitate these proteins. This decrease in pH also solubilises the other essential component of the natural casein micelle, mineral colloidal calcium phosphate which, in nanometre-sized clusters, functions as a neutralising and cross-linking agent for the casein macropeptide chains in the formation of casein micelles [5]. Redispersing the casein precipitate and raising the pH with sodium salts leads to the production of sodium caseinate. This exists in solution at neutral pH mainly as a dispersion of individual casein molecules and their mixed aggregates, frequently and erroneously referred to as sub-micelles.

The caseins are a family of phosphoproteins. Four members, α_{S1}-, α_{S2}-, β- and κ-caseins, are found in bovine milk. Seventy-five percent of bovine casein is made up of the two major members, α_{S1}-casein and β-casein. These share many properties in common. Both are single chain proteins, approximately 200 residues in length [4]. Both are calcium-sensitive phosphoproteins, with a high content of proline but no cysteine residues. Each carries a negative charge at neutral pH but the non-uniform distribution of charged and hydrophobic residues along their chains make the molecules distinctly amphiphilic. It is in this distribution that differences between these proteins begin to emerge. The N-terminal sequence of ca 50 residues of β-casein is highly hydrophilic and contains all five phosphoserines, with the rest of the molecule being substantially hydrophobic and neutral [4]. α_{S1}-Casein, on the other hand, has three main hydrophobic regions, one at each end of the molecule, both 40-50 residues in length, and a shorter region of ca 20 residues in the middle. Both proteins have a tendency to self-associate in solution but the polymers formed reflect the hydrophilic/hydrophobic distributions just detailed. β-Casein associates into ellipsoidal surfactant-like micelles at concentrations above ca 0.05 wt% whereas α_{S1}-casein shows a tendency to form a worm-like chain polymer under similar conditions [6].

The preponderance of proline residues and the absence of cysteine residues and, hence, of potential disulphide cross-links means that in solution the caseins adopt flexible disordered configurations with little ordered secondary structure. It is this similarity to random-coil polymer behaviour, which allows us to apply the self-consistent-field approach of Scheutjens and Fleer to calculating the conformational and interaction behaviour of these molecules when adsorbed at a hydrophobic interface.

3. CALCULATED PROTEIN CONFORMATIONS

Pure analytical modelling of the conformations of polymers at an interface is not feasible for complex proteins where different parts of the molecule have a different affinity for surface and solvent. However, numerical modelling of adsorbed polyelectrolytes of arbitrary sequence is now possible using the Scheutjens and Fleer approach. Full details involved in implementing the theory are to be found in the original works of Scheutjens and Fleer [7-9] and only a brief qualitative description follows, highlighting variations introduced to cope with the protein molecular structure.

The theoretical model regards the polymer as a string of segments along a single chain. In the protein these are the amino acid residues and for simplicity they are assumed to be identical in size. The adsorbing surface is assumed to have a sharp, flat and impenetrable boundary. The space above the surface is filled by a lattice that serves as a co-ordinate system across which the protein chains may be distributed, generating the many and various conformations the protein may adopt. In the protein not all of the amino acid residues are identical. There are twenty-one amino acid types and we simplify our representation by allocating these to three different categories, hydrophobic, polar (uncharged but with an affinity for water) and ionic (potentially charged). The protein model is then represented as a sequence of these hydrophobic, polar and ionic residues paralleling the known primary structure of the protein molecule.

The creation of the protein molecule as a single chain with no branching, and the connectivity of its residues, is ensured in the calculation procedure by a first-order Markov approximation. Adjacent segments can be no more than one lattice unit apart and back-folding is allowed. The calculation involves the determination of the density distribution of protein residues in the lattice sites. The segment distribution functions are related through Boltzmann factors to the segment potential, which is itself, fixed by the segment density distributions. Hence the calculation is the determination of the self-consistent condition in which segment potentials and segment density distributions are uniquely defined.

The segment potential is made up from a combination of an excluded volume interaction, a non-bonded nearest neighbour interaction of the Flory-Huggins χ-parameter type and an electrostatic interaction which depends on segment density and the local electrical potential. The values of the χ-parameters are assigned so as to give a strong surface binding affinity for the non-polar segments and a net attraction between small mobile ions (Na^+ and Cl^-) and solvent water molecules. Account is taken of the pK_a values of weakly charged groups by allowing the ionising segments to assume more than one internal state [10].

4. PREDICTED PROFILES FOR β-CASEIN

The first computations using the self-consistent-field theory applied to protein were carried out on a β-casein model, varying the protein concentration, the solution pH and the ionic strength [10,11]. The scf theory predicts that the adsorption of the model β-casein polymer produces a monolayer at bulk protein concentrations below a certain condensation limit. The condensation threshold is reduced to lower protein concentrations with increasing ionic strength or decreasing pH. Two different representations of the volume fraction profiles calculated at different solution pH values are shown in Figure 1. The linear plot of Figure 1a shows that the layers close to the surface are densely occupied with volume fractions close to unity. Occupation density falls of rapidly as you move away from the surface, dropping of to around 1% at 10 nm from the surface. Reducing the solution pH from 7.0 leads to an increase in protein density around the mid-region ($z \sim 2 - 3$ nm) with a distinct shoulder appearing in the profile at pH 5.5. This shoulder is indicative of proximity to the condensation limit.

A picture is thus emerging of the majority of the β-casein being entrained close to the interface with a smaller part of it feeling more comfortable in the solvent far from the surface, and it is the contribution of this hydrophilic portion which governs the steric stabilising ability of the system. The second representation (Figure 1b), replotting the same data as the logarithm of the segment density as a function of distance from the surface, shows that this low density region continues out to around 15 nm before dropping abruptly by several orders of magnitude to the bulk solution protein concentration. This 'knee' in this semi-log plot we take as defining the hydrodynamic thickness of the adsorbed layer. This nebulous low-density region entraps solvent water, which would move with the particle were the layer surrounding an emulsion droplet. Inspection of this tail region calculated for the adsorbed layer shows that the hydrodynamic layer thickness is not a strong function of pH, with the calculated plots all lying very close to one another.

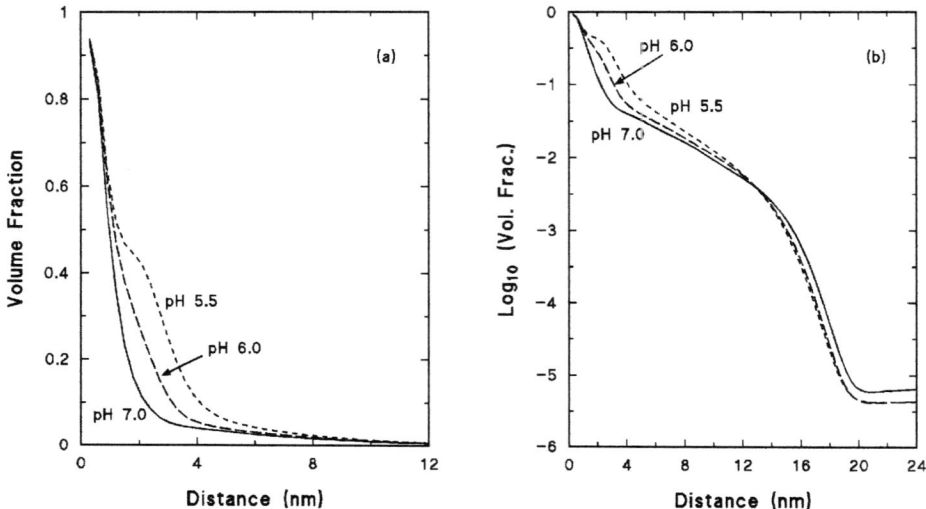

Figure 1. Segment density distributions calculated for adsorbed β-casein at solution pH values indicated. Plot (a) shows a linear representation of density, plot (b) a logarithmic representation as a function of distance from the hydrophobic surface.

The behaviour of adsorbed β-casein experimentally observed in neutron reflectivity on the CRISP instrument at the Rutherford-Appleton Laboratory parallels these trends with pH predicted by the self-consistent-field theory [12-14]. Neutron reflectivity measures the neutron intensity reflected from the film of protein adsorbed at the air/water interface as a function of the angle of incidence. In principle, it should be possible to calculate from this reflectivity data, the density profile of material normal to the interface, the very function predicted by theory. In practice, this inversion demands data of far higher quality than is presently accessible instrumentally. Hence data analysis reverts to either a model-independent Guinier treatment which calculates adsorbed amount with high precision and estimates the thickness of the adsorbed layer, or to fitting a multi-layer approximation to the reflectivity data, minimising the number of parameters and varying these until a 'best fit' is obtained. The model independent analysis showed that for β-casein adsorbed at the air/water interface both the adsorbed amount and layer thickness (δ) increase as pH is decreased (Table 1). Operating to the minimum number of variable parameters, 'best-fits' to the data were achieved with a two-layer model, with a dense thin inner layer next to the interface and a more nebulous outer layer extending in total to a distance of about 5 - 8 nm into the aqueous phase (Table 1). Lowering the solution pH, the 'best-fit' behaviour with the two-layer model suggests that the increasing amounts of adsorbed protein are initially distributed in the diffuse outer layer at pH 6.0 with both volume fraction and layer thickness increasing (Table

1). By pH 5.5, however, the ever-increasing surface coverage can only be accommodated by a doubling of the thickness of the inner layer (Table 1).

Table 1. Surface coverage (Γ), layer thickness (δ), layer parameters derived by fitting a two-layer model to neutron reflectivity data for β-casein adsorbed at the air/water interface at three different pH values. Error estimates for all parameters are also given.

pH	Γ mg m^{-2}	δ nm	φ_{inner}	d_{inner} nm	φ_{outer}	d_{outer} nm
7.0	2.05±0.1	1.65±0.07	0.96±0.12	1.00±0.1	0.15±0.01	4.4±0.2
6.0	2.85±0.1	2.39±0.08	0.95±0.10	1.03±0.1	0.20±0.02	6.1±0.5
5.5	3.90±0.15	2.57±0.08	0.94±0.16	1.85±0.15	0.19±0.02	6.9±0.7

Experimental estimates of hydrodynamic adsorbed layer thickness are available from dynamic light scattering measurements made on β-casein coated polystyrene latex spheres [15]. The uncoated monodisperse spheres are stable in their own right enabling accurate measurements of their size to be made. Following complete coating with the protein and re-measurement of particle size, the adsorbed layer thickness is readily obtained by difference. In practice the formation of the β-casein film is followed by measuring the increase in particle size until the layer thickness becomes constant and complete coverage is achieved. Measurements made at different pH values [11] showed, somewhat unexpectedly, that there was no major collapse of the β-casein hydrodynamic layer as the pH was reduced and the negatively charged residues of the protein molecule were neutralised. Unexpected because previous measurements at pH 7.0 showed such a collapse when the same charges were neutralised by the binding of divalent cations [15] but wholly in line with the predictions of the self-consistent-field theory (Figure 1b). It seems likely that the increasing density of the protein segments within the layer also plays a part via a simple overcrowding mechanism in maintaining hydrodynamic layer thickness as pH is lowered. Equally convincing verification of the theoretical predictions is available for the behaviour of dephosphorylated β-casein where measurements of hydrodynamic layer thickness [15] showed a decrease on enzymatic removal of the phosphate groups, completely in accord with theory [10].

The experimental observations and theoretical predictions of the structure of the adsorbed β-casein layers share some major features and common trends. Both theory and experiment suggest a dense inner layer close to the interface with a volume fraction close to unity. It is also a noteworthy feature of the theory that, whilst producing the dense inner layer which parallels that implied by the neutron reflectivity measurements [16], it also predicts the observed extended hydrodynamically active layer. The behaviour of this calculated layer also reproduces quantitatively the collapse observed on dephosphorylating the protein and the lack of change when pH is reduced.

5. COMPARISON OF STRUCTURES PREDICTED FOR ADSORBED CASEINS

So far in this review we have only considered predictions of the volume fraction profile of the whole molecule. To assist us further in understanding the typical structure adopted by the protein on adsorption, the modelling package also allows estimation of the preferred location of the various parts of the molecule with respect to the surface. This is done by calculating the mean distance from the surface of each of the residues in the molecule [17]. The plots of mean distance as a function of residue number for the two caseins, α_{S1} and β, are depicted in Figure 2 for adsorption at pH 7.0 and ionic strength 10 mM. It is important to remember that each plot does not represent any one conformation but is an average of all conformations adopted by the molecule on adsorption. It should also be noted that the same inter-segment interaction parameters were used for both species, and these were the same as those used in the preliminary calculations for β-casein.

Clear differences in average conformation between β-casein and α_{S1}-casein can be seen in Figure 2, though similarities are also present. In both cases the molecule is strongly bound to the surface at the C-terminus. At the N-terminus, the adsorbed structure is quite different. For β-casein the 50 or so residues from the N-terminus form a long tail which extends away from the surface. The preferred location of the N-terminal amino group is well away from the surface in β-casein, unlike that of α_{S1}-casein where the mean distance profile indicates a region around residues 20 - 30 which is anchored to the surface with the intermediate region between here and the C-terminus train looping out into the solvent.

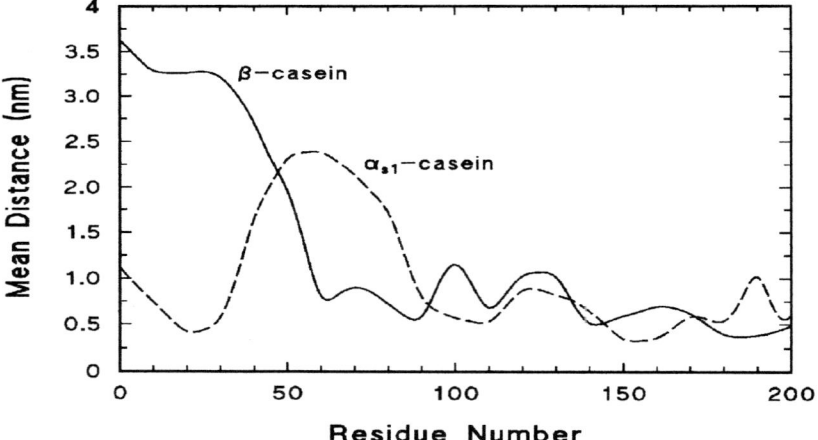

Figure 2. Preferred location of amino acid residues with respect to the hydrophobic surface, expressed as computed mean distance from the surface and plotted as a function of residue number. Theoretical calculations were performed for both proteins for the same solution conditions (pH 7.0, ionic strength 0.010 M).

Extended calculations as a function of pH and ionic strength show little response to either variable in the hydrophobic regions of both molecules [17]. All segments prefer to remain close to the surface. The effect of changes in ionic strength is more complicated; the mean locations of segments in the hydrophilic regions of both molecules unexpectedly increasing as the ionic strength is decreased [17]. However, these distributions represent average behaviour and closer analysis indicates that many conformations of each protein coexist and contribute to this average, and even that there is a relatively high proportion of these with the hydrophilic residues close to the surface. Individual residue locations can be represented by a segment density profile identical in nature to the total protein profiles of Figure 1. Closer examination of the effect of ionic strength on these profiles shows shifts in accord with intuitive expectations, and in agreement with the changes in adsorbed layer thickness observed by dynamic light scattering [15] and sedimentation techniques [18].

These pictures of the adsorbed caseins are consistent with their accessibility or lack of accessibility to proteolytic enzymes in emulsion systems. Thus Leaver and Dalgleish [19] found that with β-casein stabilised emulsions the peptides released showed the lysine residues at positions 25 and 28 to be readily accessible to trypsin, whereas all other possible attack points for the enzyme were less so. Dynamic light scattering measurements of the decrease in hydrodynamic radius of emulsion droplets following exposure to trypsin [20] lend further support to this train-and-tail model predicted by theoretical calculations for β-casein. The structure predicted for α_{S1}-casein is also consistent with the inaccessibility of peptide linkages around residues 21 - 25 to proteolytic enzymes observed by Shimizu *et al* [21] but is somewhat different from the structure speculated by Dalgleish [22] on the basis of the same proteolysis results. Mobility of the phosphoserine-rich tail region in adsorbed β-casein has been confirmed by ^{31}P NMR [23].

The average conformation of the adsorbed protein molecules in α_{S1}-casein and β-casein layers appears to be somewhat different. In general terms, the β-casein layers appear to be characterised by a long tail at the N-terminus, which extends far into solution (Figure 3). The adsorbed α_{S1}-casein molecule has its N-terminus much closer to the surface, forming a loop in the hydrophilic region (Figure 3). However, it should be emphasised once more that these simple pictures are depictions of average behaviour, composites of many coexisting molecular conformations. In β-casein, the predominant conformation is of extended tails but it has been found that a number of conformations exist in which the N-terminus is rather close to the surface [17], reminiscent of the looped structure originally favoured by Leaver and Dalgleish [19,20]. Hence it is likely that looped structures also exist in β-casein layers, though the proportion of each type means that the average conformation favours tails. In α_{S1}-casein, the presence of a greater number of hydrophobic residues near the N-terminus causes a larger proportion of conformations to be attached in this region, promoting the existence of loops rather than tails. Nevertheless, a significant number of coexisting conformations with extended tails present themselves even in the case of α_{S1}-casein.

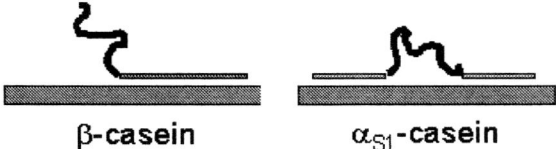

β-casein α_{S1}-casein

Figure 3. Schematic diagrams, based on predictions of scf calculations, depicting typical conformations of β-casein and α_{S1}-casein adsorbed onto planar hydrophobic interface. Bars are simply used to denote hydrophobic regions and do not imply rigidity.

These simple pictures of adsorbed casein molecules also correlate well with the behaviour of the molecules in solution [5]. Though depicted here as adsorbing to a planar hydrophobic surface, taking this surface as the hydrophobic region of another casein molecule allows us to explain the self-association behaviour of the two caseins, the surfactant-like micelle structure of β-casein and the worm-like chain of α_{s1}-casein [6]. Further growth of the polymers is limited by electrostatic interactions. These can be reduced or eliminated by increasing ionic strength, specific binding of calcium or reduction of the solution pH leading to further aggregation and precipitation. A further extension along this theme produces a plausible model for the internal structure of the casein micelle [5].

6. INTERACTIONS BETWEEN PROTEIN-COATED SURFACES

Ultimately the stability of protein-coated emulsion droplets is determined by the interaction potential between the droplets. If the net potential energy is positive at all surface separations, the emulsions will be stable. But, if the interaction energy is attractive over some range of separation, then the droplets will tend to cluster and flocculate [24]. Whether flocculation occurs in practice depends on the overall balance of attractive Van der Waals forces, steric forces and electrostatic forces [3]. It is now recognised that the latter division is somewhat simplistic and that the last two contributions are difficult to separate. They arise mainly from the interactions of the adsorbed proteins and because cross-coupling effects occur, particularly in determining the extension of the steric barrier, they can ultimately only be considered in combination.

We have used the self-consistent-field theory to estimate the interaction energy between a pair of parallel hydrophobic surfaces in the presence of adsorbing α_{s1}-casein or β-casein [25]. Again the protein, solvent and small ions are introduced into the gap between parallel hydrophobic surfaces and the system allowed to reach full thermodynamic equilibrium. No modifications to the theory are required to carry out these calculations as the earlier studies on isolated surfaces simply equate to the situation where the surfaces are separated by an infinite gap. Closing the gap and permitting equilibrium adsorption to be attained allows the interaction energy between the coated surfaces to be calculated as a function of the surface separation.

It was found that the calculated interaction energy remained positive and therefore repulsive for all separations for β-casein irrespective of pH or ionic strength. Direct experimental verification of the existence of long-range repulsive forces between hydrophobized mica surfaces coated with β-casein has recently been obtained using the interferometric surface force apparatus [26]. On the other hand, for α_{s1}-casein layers, the theory predicts major qualitative differences in behaviour between the two types of coating, particularly at moderate ionic strengths. This has clear implications for the relative colloidal stability of systems involving these two proteins.

With both proteins we find a large positive energy at close separations, corresponding to strong surface-surface repulsion. For the same solution conditions, the range and strength of repulsion are distinctly greater for the β-casein layers. This is illustrated in Figure 4 for the interactions calculated at pH 7.0 and an ionic strength of 10 mM. The effect of reducing the pH towards the protein isoelectric point is to reduce the strength and range of the interlayer repulsion for both proteins. This implies a substantial electrostatic contribution to the calculated interaction energy. We have seen previously that the charged N-terminal tail of β-casein extends far into the aqueous phase and it seems reasonable to infer that the interaction between such protruding tails would provide the combination of steric and electrostatic contributions responsible for the long range repulsive potential of β-casein at neutral pH. For adsorbed α_{s1}-casein, for which there is no extended tail but only a charged loop lying preferentially closer to the surface than the β-casein tail, the pH-dependent effect of the repulsive electrostatic contributions occurs at closer separations, presumably through loop-loop interactions.

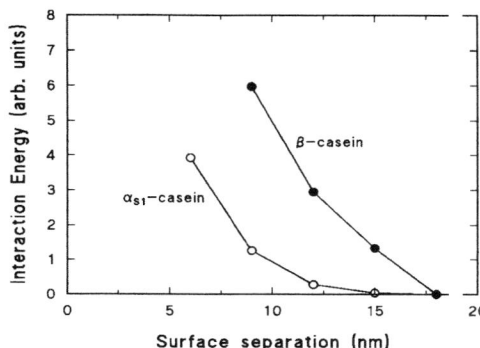

Figure 4. Potential energy calculated for interaction of two protein-coated surfaces as a function of surface-surface separation. Proteins were allowed to reach equilibrium adsorption under solution conditions of pH 7.0 and ionic strength 0.010 M.

The main difference in predicted interaction potential for the two adsorbed caseins occurs at higher ionic strength. Whilst the potential energy remains repulsive for β-casein at all separations, the theory predicts for α_{s1}-casein layers a region of negative interaction potential corresponding to surface-surface attraction under conditions of moderately high ionic strength (50 - 200 mM). The strength of this attraction increases linearly with ionic strength and would be significant at solution conditions equivalent to milk serum (pH 6.8, I = 80 mM). The underlying physical mechanism thought to be the origin of this attraction is the possible bridging of the gap by the α_{s1}-casein molecules with their two sticky patches. If the gap is small enough, then with one patch stuck on a surface, the patch at the opposite end of the molecule has the option of sticking onto the same surface forming a loop, or onto the opposing surface giving a bridge. This bridging effect computes into the interaction energy as an attractive contribution, entropic in origin, in thermodynamic terms. Though some α_{s1}-casein bridging probably occurs under all solution conditions, at low ionic strength the attractive effect of bridging is presumably overwhelmed by steric/electrostatic repulsive contributions. At high salt concentrations, however, when electrostatic interactions are screened, the bridging effect presumably becomes of sufficient importance to give a substantial net attraction at intermediate surface-surface separations. This predicted attraction is in broad qualitative agreement with what is actually observed in model oil/water emulsion systems [27-29]. Those prepared with β-casein as sole emulsifier are stable towards NaCl addition (> 2 M) whereas those prepared with α_{s1}-casein become extensively flocculated in 0.1 - 0.2 M NaCl. Casanova and Dickinson [28] have also shown that β-casein in model mixed films of α_{s1}-casein and β-casein, has a strongly protective effect in overcoming the tendency of α_{s1}-casein emulsions to flocculate at these low ionic strengths.

Whilst this agreement in outcome between theory and experiment is gratifying, it should be remembered that the physico-chemical origin of the two effects is not identical. The calculations report the result of an equilibrium being achieved whereas the stability experiments describe the outcome of kinetic phenomena. We should also remember that the computed interaction does not include any contribution from the ubiquitous Van der Waals attractive forces. It only includes that part of the overall interaction that is due to the casein chains. For oil droplets in an emulsion, a direct comparison with theory is therefore only possible if the correct Van der Waals contribution is added. It should also be emphasised that no attempt has been made to adjust any segment parameters to optimise agreement with experiment. The same segment parameter values have been used throughout and were selected only as reasonable in chemical interaction terms. The significance of predictions based upon the theoretical model therefore lies not in absolute values of the adsorption properties or interaction energies, but rather in the trends in behaviour with pH and ionic strength, and in the contrasts between the different caseins which must originate in their different primary sequence structure. Such an approach will give useful insight into how to design new protein emulsifiers or how to explain the emulsion-stabilising properties of caseinate systems, ultimately allowing us to manipulate stability and network formation to achieve desired structure and texture.

REFERENCES

1. E. Dickinson, J. Chem. Soc. Faraday Trans., 94 (1998) 1657
2. E. Dickinson, A. Mauffret, S. Rolfe and C.M. Woskett, J. Soc. Dairy Tech., 42 (1989) 18.
3. E. Dickinson and G. Stainsby, Colloids in Food, Applied Science, London, 1982.
4. H.E. Swaisgood, In Developments in Dairy Chemistry 1. Proteins (P.F. Fox, ed.) Applied Science, London, p. 1, 1982.
5. D.S. Horne, Int. Dairy J., 8 (1998) 171.
6. D.G. Schmidt, In Developments in Dairy Chemistry 1. Proteins (P.F. Fox, ed.) Applied Science, London, p. 61, 1982
7. J.M.H.M. Scheutjens and G.J. Fleer, J. Phys. Chem., 83 (1979) 1619.
8. J.M.H.M. Scheutjens and G.J. Fleer, J. Phys. Chem., 84 (1980) 178.
9. J.M.H.M. Scheutjens and G.J. Fleer, Macromolecules, 18 (1985) 1882.
10. F.A.M. Leermakers, P.J. Atkinson, E. Dickinson and D.S. Horne, J. Colloid Interface Sci., 178 (1996) 681.
11. P.J. Atkinson, E. Dickinson, D.S. Horne, J. Leaver, F.A.M. Leermakers and R.M. Richardson, in Food Colloids: Proteins, Lipids and Polysaccharides (E. Dickinson and B. Bergenståhl, eds.) Royal Soc. Chem., Cambridge, p. 217, 1997.
12. E. Dickinson, D.S. Horne, J.S. Phipps and R.M. Richardson, Langmuir, 9 (1993) 242.
13. P.J. Atkinson, E. Dickinson, D.S. Horne and R.M. Richardson, in Proteins at Interfaces II. Fundamentals and Applications (T.A. Horbett and J.L. Brash, eds.) ACS Symposium Series Vol. 602, American Chemical Society, Washington, p. 311, 1995.
14. P.J. Atkinson, E. Dickinson, D.S. Horne and R.M. Richardson, J. Chem. Soc. Faraday Trans. 91 (1995) 2847.
15. D.V. Brooksbank, C.M. Davidson, D.S. Horne and J. Leaver, J. Chem. Soc. Faraday Trans. 89 (1993) 3419.
16. P.J. Atkinson, E. Dickinson, D.S. Horne, F.A.M. Leermakers and R.M. Richardson, Ber. Bunsenges. Phys. Chem. 100 (1996) 994.
17. E. Dickinson, D.S. Horne, V.J. Pinfield and F.A.M. Leermakers, J. Chem. Soc. Faraday Trans. 93 (1997) 425.
18. Y. Hemar and D.S. Horne, J. Colloid Interface Sci. in Press.
19. J. Leaver and D.G. Dalgleish, Biochim. Biophys. Acta, 1041 (1990) 217.
20. D.G. Dalgleish and J. Leaver, J. Colloid Interface Sci., 141 (1991) 288.
21. M. Shimizu, A. Ametani, S. Kaminogawa and K. Yamauchi, Biochim. Biophys. Acta 869 (1986) 259.
22. D.G. Dalgleish, Food Res. Int., 29 (1996) 541.
23. L.C. ter Beek, M. Ketelaars, D.C. McCain, P.E.A. Smulders, P. Walstra and M.A. Hemminga, Biophys. J., 70 (1996) 2396.

24. E. Dickinson, in Food Structure - Its Creation and Evaluation (J.M.V. Blanshard and J.R. Mitchell, eds.) Butterworths, London, p. 41, 1988.
25. E. Dickinson, V.J. Pinfield, D.S. Horne and F.A.M. Leermakers, J. Chem. Soc. Faraday Trans., 93 (1997) 1785.
26. T. Nylander and M. Wahlgren, Langmuir, 13 (1997) 6219.
27. E. Dickinson, R.H. Whyman and D.G. Dalgleish, in Food Emulsions and Foams (E Dickinson, ed.) Royal Society of Chemistry, London, p. 40, 1987.
28. H. Casanova and E. Dickinson, J. Agric. Fd. Chem., 46 (1998) 72.
29. E. Dickinson, M.G. Semenova and A.S. Antipova, Food Hydrocolloids, 12 (1998) 227.

HYDROCOLLOIDS – PART 2
Edited by K. Nishinari
2000 Elsevier Science B.V.

Adhesion process of egg PC vesicles on mica surface; studies by atomic force microscopy

K. Furusawa, H. Egawa and H. Terashima*

Departments of Chemistry and Applied Physics* , University of Tsukuba, Ten-noudai 1-1-1, Tsukuba, Ibaraki 305-0006, Japan

The dynamic topography change of PC and PE liposomes from a vesicle form to a flat bilayer through their direct adhesion on a mica surface have been recorded by Atomic Force Microscope *in situ*.   In the PC liposome, the second bilayer was hardly formed and only a saturated first bilayer was extended over the all mica surface.   On the other hand, in PE liposome system, a vertical growth of lipid molecules was recognized.   This different behavior has been interpreted by the existence of hydration layer around the PC surface.

1. INTRODUCTION

Biological membranes are very complex.   However, there are many aspects that can be understood in terms of the fundamental physical and colloid chemistry.   The study on geometical deformation process from the vesicle form to a flat bilayer through adhesion of the vesicles on a solid surface is a typical subject in such a field and a lot of studies using the different techniqes, e.g., spectroscopy[1], NMR[2], and the surface chemical methods[3-5], have been reported. Further, as a new technique, it has been demonstrated that the atomic force microscopy (AFM) can be used also to study the dynamic deformation in surface topography during adhesion of vesicle particles.   This technique allowes us to visualize the adhesion process directly with nanometer lateral and vertical resolution in an aqueous environment[6-9].

In this study, the adhesion process of PC and PE liposomes on a mica surface has been observed *in situ*   by the AFM to understand the deformation mechanism from the vesicle form to the flat bilayer.

2. EXPERIMENTAL

As a solid substrate, a moscovite mica was chosen and used immediately after cleavage in a clean atmospher. Two kinds of phospholipid, phosphatidylcholine (PC) from egg yolk and phosphatidylethanolamine (PE) from bovine brain , were used without further purification.

Unilamellar PC vesicles were prepared by the usual extrusion method. The particle size of resulted PC vesicles was measured by the dynamic light scattering and resulted as 200 nm. On the other hand, PE vesicles were prepared by the sonication method with filtration to exclude the large sizes and dialysis to exclude the small ones, because PE molecules adsorbed strongly into the polycarbonate filter. The size of the resulted PE vesicles was 100 nm.

The ζ-potentials of PC and PE vesicles were measured under the different salt conditions by means of electrophoretic velocity technique . On the other hand, the ζ-potential of mica surface was determined by means of the plane interface technique[10,11] .

The AFM observation was performed with Nano Scope II . All the images were obtained by the contact mode using triangular cantilevers (Si_3N_4-oxide type). The initial surface (cleaved mica surface) characterization was performed in buffer solution (10^{-2} M $MgCl_2$ aqueous solution) by using a fluid-cell. To observe the adhesion process of liposome on mica, the buffer was replaced by a liposome suspension . Here, the starting point of adhesion (t = 0) was defined by injection time of the liposome into the cell to exchange the contents .

3. RESULTS AND DISCUSSION

The electrostatic interaction between PC liposome and mica surface is an important factor in determining the adhesion rate. So, ζ-potentials of the liposome and mica surface have been measured under the different $MgCl_2$ conditions. As can be seen from Fig.1, all the vesicle samples have negative ζ-potentials under a low ionic condition. This indicates that even amphoteric PC liposome has a net negative

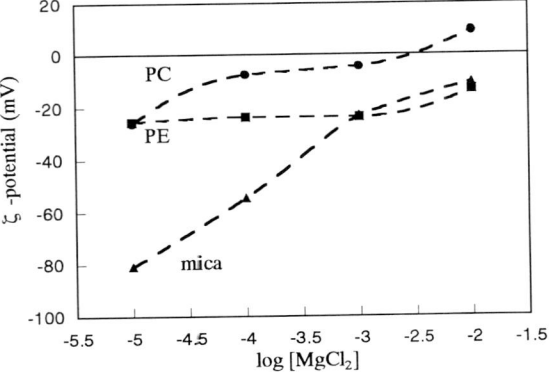

Fig. 1, ζ-potentials of liposomes(PC and PE) and mica surface bathed in various $MgCl_2$ solutions.

charge on their surface, which will come from some acidic impurities included in PC molecules. However, some addition of metal cation induces the negative value to positive side. Especially, Mg^{2+} ion considerably reduces the ζ-potential and finally reversed the sign only for PC liposome.

3.1. Deformation Process of PC Vesicle on Mica Surface

Fig.2 shows a series of AFM image of PC vesicles adhered on mica at different elapsed times after injecting the diluted PC liposome . It is appeared that the adhesion process is continued slowly and scarcely at 240 min after injection (at 120 min after 2nd injection), the mica surface was occupied completly by PC flat bilayer. At 5 min after injection, the aggregates on mica are shaped partly a spherical form (d=200 nm) which will be remained the original vesicle shape. With increasing the occupied area, the deposited aggregates became to an asymmetrical form and finally after 200-250 min, the aggregates were modified to a saturated flat bilayer. The film thickness was about 4 nm, showing a single PC bilayer is developed on mica surface, i.e.,a conformational deformation from the spherical vesicle form to a flat bilayer membrane has occurred on mica surface. However, 2nd PC bilayer has never be formed.

To clarify the mechanism of deformation process, the adhesion experiments under the different salt conditions have been carried out. In Fig.3, the results of time dependent bilayer coverage following to the successive injection, are also indicated under the three different $MgCl_2$

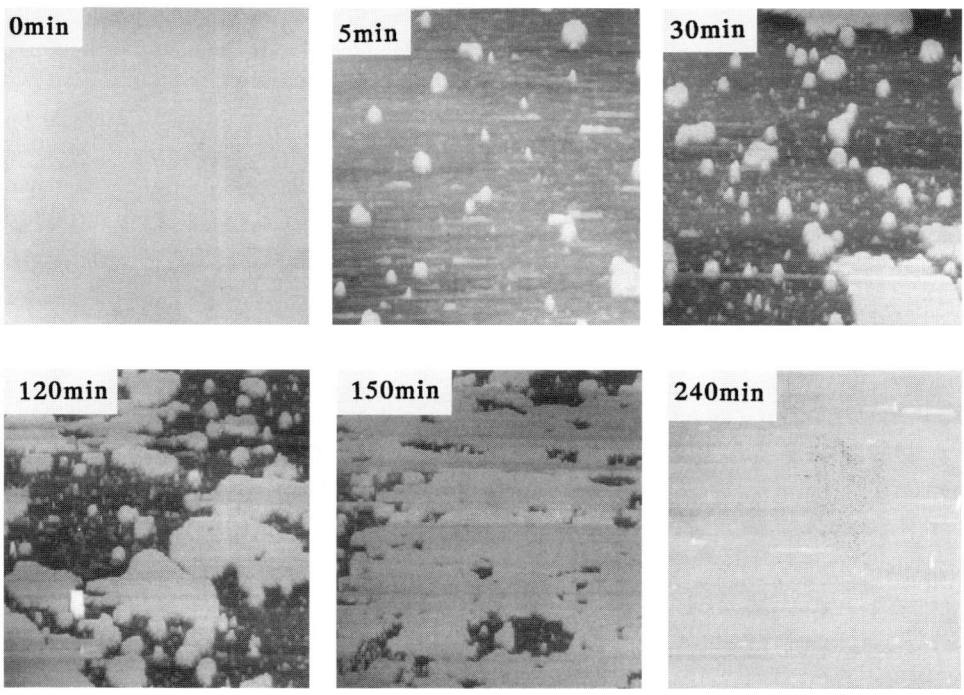

1 μ m

Fig.2, A series of AFM images taken during the adhesion of PC liposome diluted by $10^{-2}M$
$MgCl_2$ solution. The time indicates for exposure period of mica in PC vesicle suspension.

concentrations.

It is found that the time dependent coverages in 10^{-3} ~ 10^{-4} M MgCl$_2$ solutions increased linearly with increasing the elapsed time regardless of their succesive injection. However, the coverage in 10^{-2} M MgCl$_2$ solution increased strongly

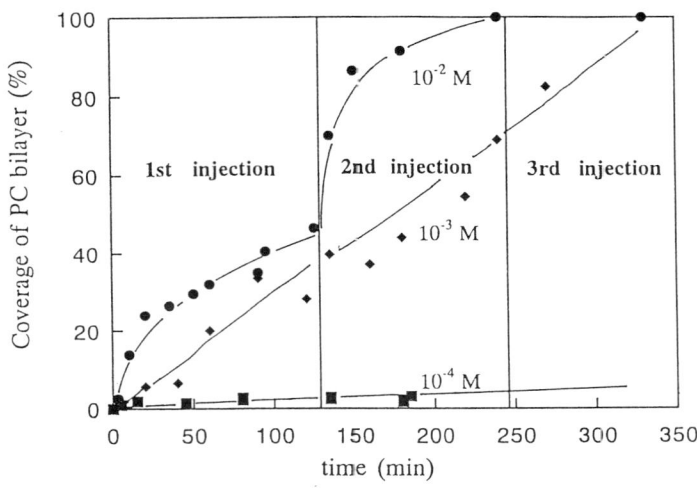

Fig. 3, Time dependentnbilayer coverage following to the succesive injection in the three different MgCl$_2$ ccncentration solutions.

just after the second time injection i.e., under the 1st injection, the coverage saturated after 120 min and scarcelly arrived to 50 % coverage. However, by the 2nd time injection the coverage increased steeply again and arrived to 100 % coverage after another 120 min. These results indicate that the deformation rate of liposomes will be influenced by the two factors, one is the electrostatic repulsion between the liposome particle and mica plate because the ζ-potentials of vesicle and mica decrease with increasing MgCl$_2$ concentration, and second is the vesicle numbers near the mica surface which influences on flequency of collision of adsorbable vesilcles. However, it is a very characteristic that we never observed 2nd bilayer formation on mica from any adhesion conditions.

3.2. Deformation Process of PE Vesicle on Mica Surface

It has been realized that the membrane surfaces of PC and PE molecules experience a different repulsive force at close contact[12]. The force is called "hydration force" .Fig.4 shows a typical AFM image of PE vesicles where the ζ-potentials of PE liposomes and mica are nearly zero (see Fig.1) and a very weak electrostatic repulsion will be opperating . It is appeared that in early deformation stage of PE vesicle, the same flat bilayer has developed as in the case of PC vesicle system. With the elapse of

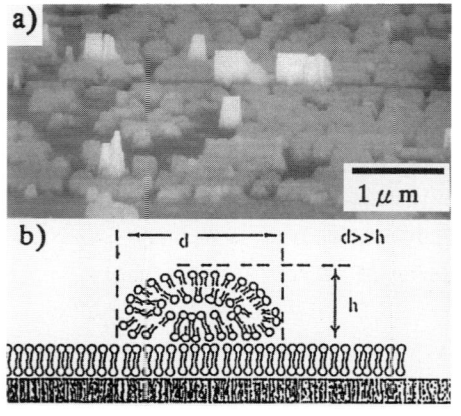

Fig. 4, A typical AFM image of PE vesicles adhered on mica at 30 min after injecting PE vesicle suspension diluted by 10^{-2} M MgCl$_2$ solution(a) and schematic picture of unruptured PE liposome(b).

time , however, a specific 2nd bilayer of PE molecules which looked like as a sharped point ,was partly appeared on the 1st PE bilayer. It is recognized that the lateral fusion between these sharped points proceed slowly and a clear outline between two neighbors remained for a long time. The other specific in PE vesicle adhesion,is to exist a conical relif as shown in Fig.4 . This indicates that the conical relief in the PE system would be regard as an unruptured and crushed PE liposome adhered on PE 1st layer. Therfore,

Fig. 5, A special AFM image which was scanned repeatedly by the AFM probe at a high force(30 nN) and a high speed(120 Hn) within a definite area.

it is assumed that the adhesion affinity between the 1st and the 2nd PE bilayers will be weak than that of PE 1st bilayer on mica surface. This fact can be clarified by the following special scanning experiment by using the AFM, i.e., mica surface covered by PE 1st and 2nd bilayers was scanned repeatedly by using the AFM probe at high force (-30 nN) and at a high speed (120 Hz) within a definite area. Fig. 5 shows a typical AFM image obtained by this special scanning experiment. As can be seen, only the 2nd bilayer pieces were swept away from the 1st bilayer surface, because the thickness measured from the defect image is corresponded to that of a single bilayer membrane (about 4 nm). Furthermore, it is realized that even though succesive scanning for the exposed area by using the same high force and the same high speed, a further deep defect could not be created on any defect surfaces. All these results indicate that the adhesion of PE head groups on mica is more strong than the adhesion between PE head groups themselves and a some interaction force between the phospholipid membranes, probably a hydration force (layer) on their surfaces, will influence on the deformation process of PE vesicles on mica surface[12].

3.3. Deformation Process of Mixed(PC +PE) Vesicles on Mica Surface

Fig.6 shows a series of AFM images of mixed (PE + PC) bilayers adhered on mica surface using the various mixed suspensions with different PE/(PC+PE) ratios. As can be seen, with increasing PE content in the sample, a pointwise second bilayer was found, i.e., from the system of PE/(PE+PC) = 0.3, a sharped pointwise second bilayers was appeared and their coverage increased with increasing PE contents in the mixture.

According to the SFA[7] studies using phospholipid membranes, it is defined that there is a hard hydration layer only on PC membrane surface and this hydration layer will resist their real contact and prevent their layer adhesion into primary minimum of their interaction potential. On the other hand, such a hydration layer effect between PE membranes is relatively small and is expected that a close contact between PE membranes will occur. So, it is well understood that the different adhesion topograph observed in PC and PE vesicle systems will be based on existence or no existence of hydration layer on their membrane surfaces.

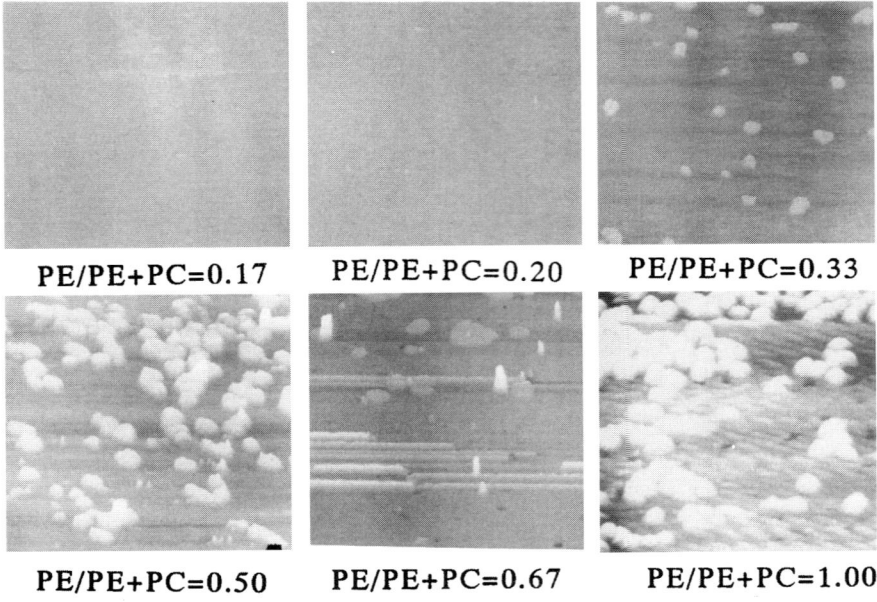

| PE/PE+PC=0.17 | PE/PE+PC=0.20 | PE/PE+PC=0.33 |

| PE/PE+PC=0.50 | PE/PE+PC=0.67 | PE/PE+PC=1.00 |

1 μ m

Fig. 6, A series of AFM images on mica adhered from mixed liposomes with different PE/(PE+PC) contents.

REFERENCES

1. E.Kalb, S.Frey anf L.K.Tamm, Biochim. Biophys. Acta, 1103(1992) 307.
2. T.M.Bayer and M.Bloom,Biophys. J., 58(1990) 357.
3. S.Jackson, M.D.Reboiras, IG.Lyle and M.N.Jones, Faraday Discuss. Chem. Soc., 85(1986) 291.
4. J.Radler, H.Kiefer and F.Jahrig, Biophys. J., 69(1995)1447.
5. P.Nollert, H.Strey and E.Sackmann, Langmuire, 11(1995)4539.
6. G.Binning, C.F.Qante and C.Gerber, Phys. Rev. Lett.,56(1986)930.
7. S.Manne, J.P.Cleveland, H.E.Caub, G.D.Stucky and P.K.Hansma, Langmuire, 10(1994) 4409.
8. J.Mou, J.Yang, C.Huang and Z.Shao, Biochemistry, 33(1994) 9981.
9. S.L.S.Stipp, Langmuire, 12(1996)1884.
10. R.W.Haddleson, A.L.Smith, in Proceedings of a Symposium on Form, Academic Press, New York, 163(1975).
11. H.Sasaki, A.Muramatsu, A.Arakatsu and S.Usui, J.Colloid Interface Sci., 142(1991)266.
12. J.Marra and J. Israelachvili, Biochemistry, 24(1985)4608.

HYDROCOLLOIDS – PART 2
Edited by K. Nishinari
2000 Elsevier Science B.V.

Reactions of lipoxygenase from cucumber cotyledon in oil-in-water emulsions

Y. Matsumura[a], N. Matsuo[a], J. Kimata[a], K. Matsui[b] and T. Mori[a]

[a]Research Institute for Food Science, Kyoto University, Gokasho, Uji, Kyoto 611-0011, Japan.

[b]Department of Biological Chemistry, Faculty of Agriculture, Yamaguchi University, Yamaguchi 753-0841, Japan.

The reaction of lipoxygenase from cucumber cotyledon was studied in the oil-in-water emulsion system containing methyl linoleate as the substrate lipid at pH 7.0. The high oxygenation rates were observed in the emulsions stabilized by β-casein and sodium deoxycholate, whereas the oxygenation rates were low in the emulsions stabilized by Tween 20 and sucrose esters of high HLB number. The addition of Tween 20 and sucrose esters caused the decrease of lipoxygenase activity in the β-casein stabilized emulsion and the decrease of interfacial tension of the β-casein-adsorbed plane interface. There was a strong correlation between the lipoxygenase activity and the interfacial tension level. On the other hand, the positional specificity of lipoxygenase in the production of hydroperoxides was not affected by the addition of surfactants. Based on these results, we discussed the relationship of the lipoxygenase activity and the physical states of adsorbed layer on the substrate oil droplet surface.

1. INTRODUCTION

Lipoxygenases(EC. 1. 13. 11. 12) catalyze the oxygenation of polyunsaturated fatty acids of their esters containing one or more (1Z, 4Z)-pentadiene systems. Lipoxygenases are thought to play important roles in plant systems(for instance, fruit ripening, senescence, and the defense against microbial invasion, etc.)[1] and food systems(the generation of objectionable as well as pleasant flavor, enhancement of texture of bread dough and bleaching of white breads, etc.)[2].

The assay of lipoxygenase activity is carried out normally using free fatty acids as substrates. However, some of lipoxygenases, for instance soybean lipoxygenase-2 and -3 prefer esterified substrates[3]. Since majority of fatty acids in biological and food systems are esterified to form triacylglycerols, phopholipids, and glycolipids, it is possible that lipoxygenases act on directly such esterified lipids. Of esterified lipids, non-ionic types such as triacylglycerol and fatty acid methyl ester normally form oil-in-water emulsions, when they are dispersed in water using emulsifiers. Lipoxygenase reactions in emulsion systems have been scarcely studied so far. In the present paper, we studied the reaction of lipoxygenase from cucumber cotyledon in oil-in-water emulsions stabilized by proteins or low molecular weight surfactants. This enzyme is thought to directly oxygenate triacylglycerols stored in oil body[4].

2. MATERIALS & METHODS

2.1. Materials

Cucumber cotyledon lipoxygenase(LOX) was prepared from six-day-old light grown cucumber cotyledons according to the method of Matsui et al[5]. β-casein was purchased from Sigma Chemical Company(St. Louis. MO). Tween 20 was a high purity sample(Surfact-Amp 20) obtained from Pierce Chemicals(U.K.). Six sucrose esters having varied HLB number from 7 to 18 were supplied by Mitsubishi Kagaku Foods Co.Ltd(Tokyo Japan). Methyl linoleate(>99 %) and other reagents of analytical reagent grade were purchased from Wako Pure Chemicals Co. Ltd(Osaka, Japan).

2.2. Preparation of emulsions

Aqueous solutions of proteins(0.5 wt%) or low molecular weight surfactants(2 wt%) such as Tween 20, sodium deoxycholate and sucrose esters were prepared in 10 mM sodium phosphate buffer(pH 7.0). An oil-in-water emulsion was prepared from 5 wt% methyl linoleate and 95 wt% aqueous solution of emulsifier as follows. The oil and aqueous phase were mixed and homogenized for 3 min in a high-speed blender(Nihon Seiki Kaisha Ltd) operating at 1.3×10^4 r.p.m. followed by the ultrasonication using ultrasonicator(Nihon Seiki Kaisha Ltd) operating at maximum power for 2 min. A laser diffraction particle analyzer(Horiba Seisakusho Ltd, Model LA-500) was used to determine the oil droplet-size distributions of emulsions. The emulsions thus obtained were diluted before the following experiments to adjust the methyl linoleate content to 0.294 wt%(10 mM). The concentration of β-casein in the diluted emulsion was 0.0294 wt%(12.25 μM).

2.3. Assay of LOX reaction

The assay of LOX reaction in emulsions was performed at 25℃ with a oxygen electrode(Hansatech D.W. Oxygen Electrode Unit equipped with a recorder). The reaction was initialized by adding 3 μl LOX solution (0.254 μg enzyme) to 1 ml emulsion prepared in the previous section.

2.4. Effects of addition of surfactants on LOX reaction in emulsions stabilized by β-casein.

The emulsion prepared in the section of 2.2. was divided into several aliquots, and to each was added the various amounts of Tween 20 or 0.2 wt% of sucrose esters. It was confirmed that there was no change in droplet-size distribution by the addition of surfactants. After the storage at 25℃ for 2 h, each emulsion was used for the LOX assay.

2.5. Interfacial tension measurements

Interfacial tension at oil-water interface was measured using a Wilhelmy-plate-type surface tensiometer(Kyowa Model CBVP). In this experiment, the concentration of β-casein in aqueous phase(10 mM sodium phosphate buffer, pH 7.0) was 0.001 wt%(0.42 μM). The various amounts of Tween 20 or constant amount(0.0067 wt%) of sucrose esters were added to the aqueous phase to investigate the effects of surfactants on interfacial tensions.

2.6. High performance liquid chromatography(HPLC) analysis

HPLC analyses of hydroperoxides of methyl linoleate produced by LOX were performed with a Zorbax SIL column(Shimazu Seisakusho Ltd, 4.6 mm x 250 mm) after the reduction of hydroperoxides with triphenylphosphine. Elution was carried out with *n*-hexane / *iso*-propanol(99:1, v/v) at a flow rate of 1 ml/min.

3. RESULTS

3.1. LOX reaction in emulsions stabilized by proteins and surfactants

Figure 1 shows the oxygenation induced by LOX in emulsions stabilized by β-casein and surfactants. For the β-casein stabilized emulsion, rapid oxygen consumption occurred after the addition of LOX. The rate of oxygenation which was calculated from the slope of reaction curve was 52.3 μM/min. The high oxygenation rate(51.8 μm/min) was similarly observed for the β-lactoglobulin-stabilized emulsion(the reaction curve is not shown). Oxygenation rate slightly decreased in the emulsion stabilized by sodium deoxycholate, but it was still high(47.3 μm/min) .

When the sucrose ester S-1670 (HLB number:16) and Tween 20 were used as emulsifiers, the resultant emulsions were not good substrates for LOX(Figure 1). The oxygenation rates were 6.1 μm/min and 0.9 μm/min for the emulsions stabilized by S-1670 and Tween 20, respectively. The low reaction rate is not attributable to the small interfacial area of substrate emulsions, because S-1670 and Tween 20 produced finer emulsions than proteins and sodium deoxycholate(data not shown).

3.2. Effects of Tween 20 on LOX activity

We investigated the inhibition of LOX reaction in the β-casein stabilized emulsion by Tween 20, because there was a large difference in reaction rate between the β-casein and Tween 20-stabilized emulsions in the previous section. Figure 2 shows the results on the change of oxygenation rate by the addition of varying amounts of Tween 20 to the emulsion stabilized by β-casein. The amount of Tween 20 is expressed as the molar ratio of Tween 20 and β-casein(R). As described in the section 2.2., the concentration of β-casein in the emulsion was 12.25 μM.   The reaction rate was expressed as the percent of the rate at R=0.

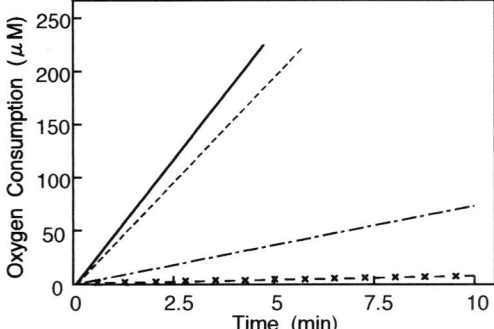

Figure 1. Oxygen consumption induced by LOX in emulsions stabilized by β-casein(▬), sodium deoxycholate(- - -), sucrose ester S-1670(─ - ─), and Tween 20(─ x ─).

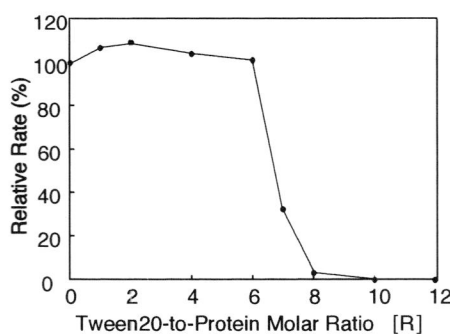

Figure 2. Effects of Tween 20 addition on LOX activity in the emulsion stabilized by β-casein. The oxygenation rate was expressed as the percent of the rate without Tween 20(R=0).

The rate of oxygenation was not affected by the addition of Tween 20 in the range of low molar ratios(R≦6), but the rate remarkably decreased at R=7 and became less than 1 % of the original level in the region of R≧8(the concentration of Tween 20 at R=8 was 98 μM, i.e., 0.012 wt%). The effects of Tween 20 addition on interfacial tension of β-casein solution(0.001 wt%, 0.42 μM) was also investigated. Without Tween 20, the interfacial tension at the oil-water interface was 22.0 mN/m, but the tension decreased to 2.1 mN/m by the addition of Tween 20 at R=8. These results suggest that the decrease of interfacial tension by Tween 20 was closely related to the inhibition of LOX reaction.

3.3. Effects of sucrose esters on LOX activity

In order to shed more insights into the relationship between LOX activity and interfacial tension of substrate oil droplet surface, we used sucrose esters with various HLB number from 7 to 18. As described in the METHODS, 0.2 wt% surfactants were added to β-casein stabilized emulsions and oxygenation rate induced by LOX was measured after 2h. In the case of interfacial tension measurements, 0.0067 wt% sucrose esters were added to β-casein(0.001 wt%) solutions. The results are shown in Figure 3. The higher HLB sucrose esters caused more decrease of interfacial tension and LOX activity. There was a strong correlation(R²=0.99) between LOX activity and interfacial tension. This result suggests the possibility that lipoxygenase reaction is controlled by the interfacial tension level of substrate oil droplet surfaces.

In order to investigate the effects of sucrose esters on the positional

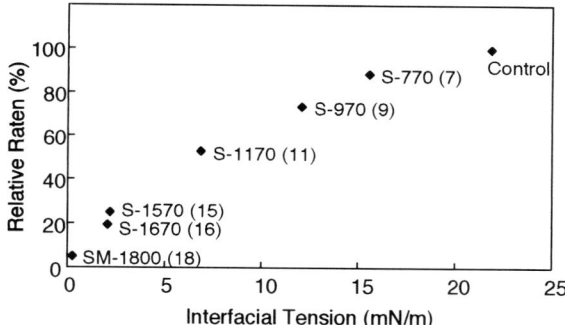

Figure 3. Correlation between LOX activity and interfacial tension in the presence of 6 sucrose esters. The numbers in parentheses are HLB numbers. 'Control' indicates the result without sucrose esters.

Table 1   Composition of hydroperoxide isomers by LOX reaction in emulsions

Emulsifiers	Hydroperoxide isomers			
	13(9Z,11E)*	13(9E, 11E)	9(10E. 12Z)	9(10E, 12E)
β-casein	65.5	9.7	14.8	10.1
+ S-770	66.0	9.5	14.2	10.2
+ S-1170	67.1	8.4	15.1	9.4
+ S-1670	67.7	8.4	14.9	9.0

*Abbreviation;13(9Z,11E) indicates 13-hydroperoxy-(9-cis,11-trans)-octadecadienoic acid methyl ester.

specificity of LOX in the production of hydroperoxides, the hydroperoxides in the emulsions of Figure 3 were analyzed. As shown in Table 1, the composition of hydroperoxide isomers were almost constant irrespective of the presence or absence of sucrose esters. This result means that the chemical reaction mechanism of LOX, that is, the process of the site specific removing of hydrogen and addition of oxygen is not influenced by sucrose esters.

4. DISCUSSION

We have shown that LOX can act on substrate lipids emulsified by proteins very well(Figure 1). It is thought that adsorbed proteins form two-dimensional network structure at the interfaces in which the lateral diffusion of protein molecule is highly restricted[6]. There should be small spaces within the network which are large enough to allow the passing of lipid molecules. LOX maybe draws substrates lipids from oil droplets through these spaces. The data of three dimensional structure of lipoxygenases have shown the presence of the 'cavity' functioning as a path for a substrate lipid from the exterior of the enzyme molecule to the catalytic site[7]. It is thought that LOX uses this cavity for the uptake of substrate lipid into the interior catalytic site.

As well known, most of lipases have no such cavity, but they have the 'lid domain' covering catalytic site[8]. For lipases, therefore, the conformational change induced by the adsorption at the interface is essential for the open of the lid and the following exposure of catalytic site to substrate lipids. Such 'interfacial activation' is impossible when substrate lipids are covered by surface active proteins such as caseins and β-lactoglobulin[9]. It is worth emphasizing that interfacial behavior of LOX is different from that of lipases with respect to 'interfacial activation', although both enzymes act on emulsified lipids

The addition of low molecular weight surfactants such as Tween 20 and sucrose esters with high HLB(S-1670, S-1800) to the protein-stabilized emulsion, however, modifies the adsorbed layers of oil droplet surfaces, thereby influencing LOX activity. LOX can not draw the substrate lipid from oil droplet surfaces densely covered by the surfactants. This may be the reason for the inhibition of LOX by Tween 20 and the sucrose esters (Figures 2 and 3). Alternative explanation is that direct interaction of these surfactants and LOX affects LOX activity. However, this mechanism is unlikely because kinetic analyses in micellar systems have suggested that the effects of surfactants on LOX activity can be ascribed to physicochemical interaction of surfactants with substrates [10].

The positional specificity of oxygenation products did not change by the addition of sucrose esters such as S-1170 and S-1670(Table 1), although these surfactants decreased the reaction rate of LOX(Figure 3). These results suggest that the surfactants prevent the access of LOX to the oil droplet surfaces and/or the uptake of substrate lipids by the enzyme, but do not influence the chemical process in relation to the regiospecificity of hydroperoxidation.

ACKNOWLEDGMENT

This research was partially supported by Grant-in-Aid from Program for Promotion of Basic Research Activities for Innovative Biosciences.

REFERENCES

1. Y.-L. Peng, Y. Shirano, H. Ohta, T. Hibino, K. Tanaka and D. Shibata, J. Biol. Chem., 269(1994)3755.
2. A.M. Spanier, H. Okai and M. Tamura(eds.), Food Flavor and Safety, ACS Symposium series, 528(1993) 192.
3. B. Axelrod, T.M. Cheesbrough and S. Laakso, Methods Enzymol., 71(1981)441.
4. K. Matsui and T. Kajiwara, Lipids, 30(1995)733.
5. K. Matsui, E. Tsuru, T. Kajiwara and T. Hase, Plant Physiol., 109(1995)337.
6. E. Dickinson, J. Chem. Soc. Farady Trans., 94(1998)1657.
7. J.C. Boyington, B.J. Gaffney and L.M. Amzel, Science, 260(1993)1482.
8. H. van Tilbeurgh, M.-P. Egloff, C. Martinez, N. Rugani, R. Verger and C. Cambillau, Nature, 362(1993)814.
9. Y. Gargouri, G. Pieroni, K. Sugihara, C. Reviere, L. Sarda and R. Verger, J. Biol. Chem., 260(1985)2268.
10. M.J. Schilstra, G,A, Veldink and J.F.G. Vliegenthart, Lipids, 29(1994)225.

HYDROCOLLOIDS – PART 2
Edited by K. Nishinari
2000 Elsevier Science B.V.

Interaction between egg PC vesicles and emulsion droplets

B. Yang[a], H. Matsumura[b], H. Kise[a], and K. Furusawa[c]

[a] Institute of Material Science, University of Tsukuba, Tsukuba, Ibaraki 305, Japan

[b] Electrotechnical Laboratory, AIST, MITI, Tsukuba, Ibaraki 305, Japan

[c] Department of Chemistry, University of Tsukuba, Tsukuba, Ibaraki 305, Japan

The aggregation process of emulsion droplets after adding phosphatidylcholine(PC) vesicles was investigated by determining the ζ-potential and the enlarging size ratio (P) of emulsion droplets at respective step in 10^{-4} M $LaCl_3$ aqueous solution. The aggregation behavior could be explained well by using the DLVO theory.

1.INTRODUCTION

The interaction between phospholipid vesicles and emulsion droplets is an important subject in the fields of food science, biology, and medicine. Especially, the emulsions are used prosperously in the field of food technology. For examples, mayonnaise and cream are O/W emulsions and margarine is a typical W/O emulsion. To obtain the stable emulsion, some emulsifier must be added into the system. But, the ionic surfactant is known to have toxicity to living body. So, some bio-surfactants are desirable. It has two characters; the harmless to body and the high conformity with body. Lipids from living body are bio-surfactants and they may play an important role in this field. As the lipid vesicles have the same structure with bilayers of cell membranes, the lipid vesicles have been extensively applied in various fields of cell biology as a model for cell membranes [1-3]. Especially, in recent years, the studies of interaction between vesicle and oil-water interface are an important topic [4,5], which indeed enhanced the studies of interaction between vesicles and emulsion droplets .

In this work, we have studied the interactions between PC vesicle and hexadecane emulsion by utilizing some colloid chemical techniques, such as dynamic light scattering, electrophoresis and fluorescence measurements.

2. EXPERIMENTAL

2.1. Materials

Egg yolk phosphatidylcholine (PC) was purchased from Sigma Chemical Co. Ltd., (USA). Inorganic chemical (LaCl$_3$) and oil (hexadecane) were analytical reagent grade and supplied by Wako Pure Chemical Industry (Japan).

2.2. Vesicles and emulsion preparation

The PC vesicles with different sizes (d=100, 200, and 300nm) were prepared by the extrusion method[6] using two stacked polycarbonate filters having respective pore size. The concentration of lipid was analyzed by the Bartlett method[7]. The O/W emulsion of hexadecane was prepared by the sonication method and then stored at 4~10℃ overnight.

2.3. Aggregation behavior

Aggregation of emulsion droplets was detected by measuring the enlarging size rate (P) using a dynamic light scattering apparatus (Otsuka Elect. ELS-800) after adding PC vesicles to the emulsion dispersion. Here, the P is given by $P = (D_t - D_0) / D_0$, where, D_t is the mean diameter of emulsion droplet at time t after adding PC vesicles into the emulsion, and D_0 is the mean diameter of the emulsion droplet at the starting time (t = 0).

2.4. Electrophoresis

The electrophoretic mobility measurements of vesicles and emulsion droplets were carried out by a microelectrophoretic apparatus (Zeecom; Microtech Nichion. Co. Japan). ζ-potentials were calculated by O`Brien-White equation [8].

2.5. Leakage test of fluorescence dye through the PC vesicle

To know the relationship between vesicle destabilization and leakage, the change of fluorescence intensity was measured by the fluorophotometer (F-3010; Hitachi. Co. Japan) after adding PC vesicles encapsulated fluorophore/quencher molecules into the emulsions. Details of this method are described in elsewhere [9,10].

3. RESULTS AND DISCUSSION

Generally, the electrostatic repulsive force between the particles plays an important role on the colloid stability, like latices, emulsion and lipid vesicles. So, we have investigated the ζ-potential of vesicles and of emulsions in the aqueous LaCl$_3$ solutions with various concentrations (see Figure.1).

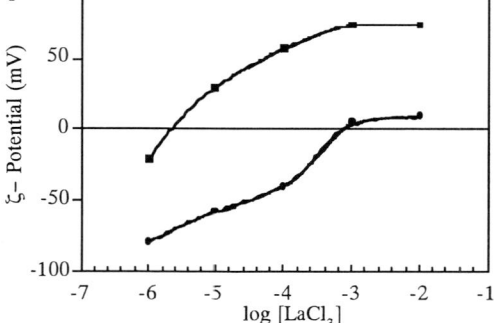

Figure 1. ζ–potential of PC vesicles (■) and emulsion droplets (●) bathed in various LaCl$_3$ concentration solutions.

The PC vesicle and emulsion have opposite charges at $5 \times 10^{-5} \sim 10^{-3}$ M concentration range of $LaCl_3$, so, the aggregation behavior of (vesicle +emulsion) system was investigated at 10^{-4} M $LaCl_3$ concentration. In this situation, the ζ-potentials of emulsion and vesicle were -43.7mV and 57mV respectively. Hence, it is expected that a strong electrostatic attraction will be operated between them.

Next, a set of aggregation experiment were carried out by using three different PC vesicles (d=100, 200, and 300 nm) under the various PC concentration solutions at 10^{-4} M $LaCl_3$. In Figure 2, the enlarging size ratios (P) of emulsion after adding PC vesicle (200 nm) with various concentrations are shown as a function of elapsed time. As seen from the figure, the P values depend on the amounts of PC vesicle and there is an optimum amount of PC vesicle for the aggregation of emulsion. To clarify this reason, we plotted the P value at 900 second elapsed times against the PC concentration (Figure 3). The P value shows a maximum at the respective PC concentration of each size vesicle. Here, we call this maximum as a "maximum aggregation concentration (MAC)" of the PC vesicle. Interestingly, for the three sizes of PC vesicle, the changing behavior of P value to the PC concentration showed a similar tendency, only the MAC shifted toward higher PC concentration

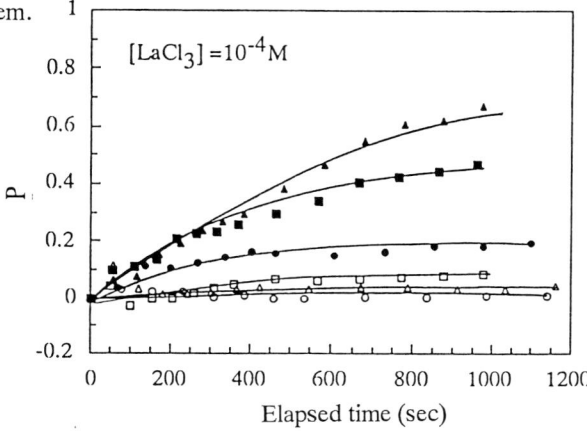

Figure 2. The enlarging size rate (P) of emulsion vs. the elapsed time curves under different PC concentrations at 10^{-4} M $LaCl_3$.
(△) 0 mM, (●) 0.0016 mM, (▲) 0.008 mM, (■) 0.016 mM, (□) 0.08 mM, (○) 0.8 mM.

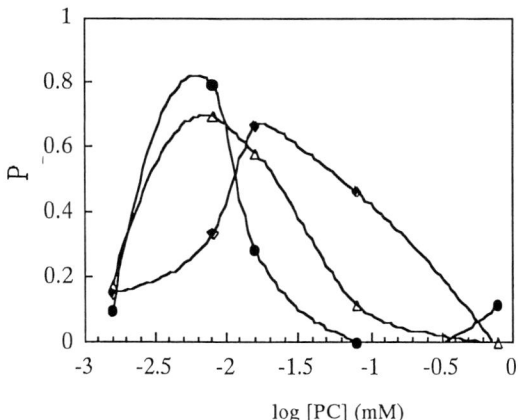

Figure 3. Enlarging size rate (P) vs. PC concentration curves for three typed PC vesicles (d = 100 nm (●), d = 200 nm (△), d =300 nm (◆)).

with increasing the vesicle size.

In order to make clear the situation and the structure of PC vesicles encountered with the emulsion droplets, we carried out the electrophoresis and the fluorescence measurements.

Figure 4 shows the ζ-potential of emulsion droplets after adding PC vesicle (d=200 nm) with various concentrations. The ζ-potential of emulsion droplet decreased quickly with increasing PC concentration and reached zero value at a very low PC concentration. It indicates that the positively charged PC vesicles by La³⁺ adsorption adhered strongly on the negatively charged emulsion surface. Here, the arrow in Figure 2 indicates the point corresponding to the value of ζ-potential at the MAC.

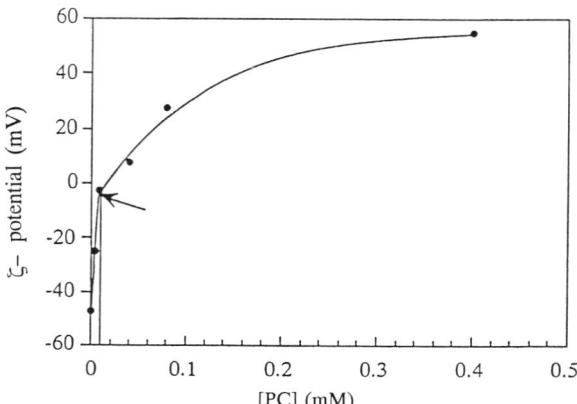

Figure 4. ζ-potential of emulsion as a function of PC concentration

Furthermore, it is appeared that the ζ-potential of emulsion at the saturated state of PC adhesion (55mV) is nearly equal to the value of PC vesicles themselves (57mV). This implies that the emulsion surface after saturated adhesion of PC vesicles shows the same electrostatic nature of PC vesicles.

The vesicle aggregation process can be analyzed also from the leakage test of fluorescence dye encapsulated in the vesicle. Figure 5 shows the time dependence of the fluorescence intensity (I) of fluorophore-quencher encapsulated in the vesicles. As can be seen, the fluorescence intensity (I) increased gradually with the elapsed time in the system with emulsion droplets. On the other hand, no change in the intensity (I) was observed from the system without hexadecane emulsion. This implies

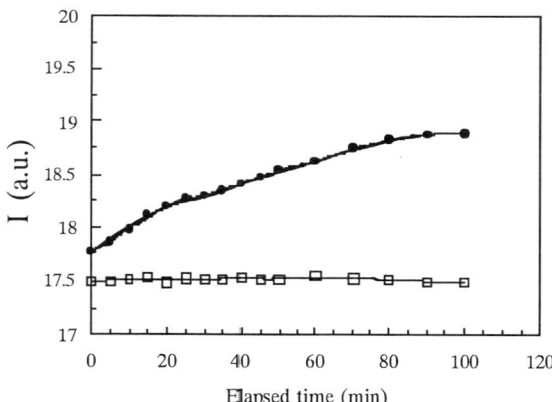

Figure 5. Variation of fluorescence intensity vs. elapsed time. (□)without emulsion (●) with emulsion

that the fluorescence dye leaked from the inside of vesicles to the bulk solution, which will be due to the destruction of vesicle form when the vesicles were encountered with emulsion droplets. From our surface tension experiments [11], it is confirmed that the decrease in interfacial tensions of hexadecane/water ($\Delta\gamma$ = -52 mJ/m^2) induced by addition of PC vesicles, is close to the value in a monolayer state of PC molecules on hexadacane/water interface ($\Delta\gamma$ = -51.8 mJ/m^2). All these results indicat that the PC vesicles will be destroied from the vesicle form to a monolayer-like structure on the oil/water interface after encountered with the emulsion droplets.

According to the DLVO theory [12-13], the total interaction potential energy (V_T) between two emulsion droplets after addition of PC vesicles is given by the following equation,

$$V_T = -\frac{Aa}{12H} + \frac{\varepsilon a\Phi^2}{2}\ln\left[1 + \exp\left(-\kappa H\right)\right] \qquad (1)$$

where the first term is the van der Waals potential energy and the second term is the electrostatic potential energy. ε is the dielectric constant of the suspending medium, Φ the ζ-potential of emulsion after adsorbed by the PC vesicles with various concentration, which are taken from Figure. 4. a is the radius of emulsion droplet, 1/κ the Debye-Hückel reciprocal length parameter, H the surface-surface distance between two emulsion particles and A is the Hamaker constant.

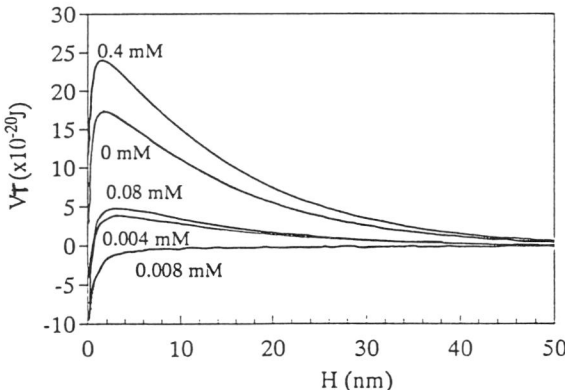

Figure 6. Plots of V_T versus H, for two spheres of equal radius (390 nm). Curves are drawn for different PC concentration system by assuming A = 1.12 $\times 10^{-21}$ J and κ = 0.08 nm^{-1}.

As seen from Figure 6, when the PC concentration is increased, the potential barrier decreases at first, meaning that the repulsive potential between the emulsion particles becomes weak and aggregation occur easily. At 0.008 mM of PC concentration, the potential barrier disappears completely. Further increase of the PC concentration, however, the potential barrier appears again and has a high barrier at a higher PC concentration than 0.4 M. This means that the repulsion between emulsion particles becomes strong, hence the aggregation never occurs. As this, it is realized that the aggregation process of emulsion droplets after adding PC vesicles with various concentration can be explained well by using the classical DLVO theory.

In conclusion, the mechanism of aggregation between emulsion droplets by PC vesicles is as following. The PC vesicles approach to the emulsion surface by electrostatic attraction, the vesicles adsorb and collapse on the emulsion surface into a monolayer-like structure. The negative charge on the emulsion surface is neutralized by positive charge on the head group of PC + La^{3+}, hence the electrostatic repulsion between emulsion droplets decreases and the aggregation of emulsion droplets easily occurs. When the ζ-potential reaches to zero, the aggregation behavior shows a maximum. At higher PC concentration than this, the emulsion surface is covered with large amounts of PC + La^{3+}, and the aggregation is prevented by the electrostatic repulsion between the PC adsorbed layer.

REFERENCES

1. H. Hauser, M. C. Philips and R. M. Marchbanks, Biochem. J, 120 (1970) 329.
2. L. Weiss, S. Nir, J. P.Harlos and J. R. Subjeck, J. Theor. Biol, 51 (1975) 439.
3. M.S.Perin, V.A.Fried, G.A.Mignery, R.Jahn, T.C.Sudhof. Nature, 17, 345 (6272) (1990) 260.
4. R. J. Davies, M. N. Jones, Biochim. Biophys. Acta, 858 (1) (1986) 135.
5. I. Ueda, J. S. Chiou, P. R. Krishna, H. Kamaya, Biochim. Biophys. Acta, 1190 (2) (1994) 421 .
6. M. J. Hope, M. B. Bally, G. Webb, P. K. Cullis., Biochim. Biophys. Acta, 812 (1985) 55.
7. G. R. Bartlett, J. Biol. chem, 234 (1959) 466.
8. R. W. O`Brien, L. R. White, J. Chem. Soc. Fraday Trans, 2, 74 (1978) 1607.
9. M. N. Dimitrova and H. Matsumura , Colloids and Surf. B: Biointerfaces, 8 (1997) 287.
10. H. Elens, T. Bentz and F. C. Szoka, Biochemistry, 24 (1985) 3099.
11. B. Yang, H.Matsumura, K. Furusawa, J. Colloids and Surfaces B: Biointerfaces. (in press).
12. B.V.Derjaguin, L.Landau, Acta Physicochim. URSS , 14 (1941) 663.
13. E.J.W.Verwey, J.Th.G.Overbeek, "Theory of the Stability of Lyophobic Colloids", Elsevier, Amsterdam (1948).

HYDROCOLLOIDS – PART 2
Edited by K. Nishinari
2000 Elsevier Science B.V.

69

Dynamic electrophoresis of colloidal particles in concentrated suspensions

Hiroyuki Ohshima

Faculty of Pharmaceutical Sciences and
Institute of Colloid and Interface Science,
Science University of Tokyo,
12 Ichigaya Funagawara-machi, Shinjuku-ku,
Tokyo 162-0826, Japan

The particle volume fraction (ϕ) dependence of the dynamic electrophoretic mobility μ of spherical colloidal particles in a concentrated suspension as well as the colloid vibration potential (CVP) and electrokinetic sonic amplitude (ESA) in the suspension are discussed on the basis of Kuwabara's cell model. From an Onsager relation for concentrated suspensions it is suggested that the CVP or ESA depends much stronger on ϕ than predicted from the ϕ dependence of μ.

1. INTRODUCTION

When a suspension of charged colloidal particles is irradiated with a sound wave, a macroscopic electric field is generated in the suspension. This field is called the colloid vibration potential (CVP). Inversely, when an oscillating electric field is applied to the suspension, a macroscopic sound wave, whose amplitude is called the electrokinetic sonic amplitude (ESA), is generated in the suspension. The measurement of these electroacoustic quantities is a novel technique for the electrokinetic analysis of concatenated suspensions of colloidal particles. It is known that CVP and ESA are both proportional to the dynamic electrophoretic mobility of the particles in an oscillating electric field. In recent papers [1, 2] we have derived the general expression for the dynamic electrophoretic mobility of spherical colloidal particles in concentrated suspensions on the basis of Kuwabara's cell model [3]. The obtained dynamic mobility μ depends on the frequency ω of the applied electric field and the particle volume fraction ϕ as well as on the reduced particle radius κa (where κ is the Debye-Hückel parameter and a is the particle radius) and the zeta potential ζ. The purpose of the present article is to review our work on the dynamic electrophoretic mobility in concentrated suspensions with the particular emphasis on its dependence on the particle volume fraction ϕ.

2. DYNAMIC ELECTROPHORETIC MOBILITY

Consider a swarm of identical spherical colloidal particles of radius a in a liquid containing a general electrolyte. All the particles move with the same velocity $U\exp(-i\omega t)$ in an applied oscillating electric field $E\exp(-i\omega t)$, where ω is the frequency and t is time. The dynamic electrophoretic mobility m of the particle is defined by

$$U = \mu E \tag{1}$$

where $U = |U|$ and $E = |E|$. We employ a cell model [3] in which each sphere is surrounded by a concentric spherical shell of an electrolyte solution, having an outer radius of b such that the particle/cell volume ratio in this unit cell is equal to the particle volume fraction ϕ throughout the entire suspension, viz.,

$$\phi = (a/b)^3. \tag{2}$$

The dynamic electrophoretic mobility μ depends not only on the zeta potential ζ of the particle and κa but also on the frequency ω and the particle volume fraction ϕ.

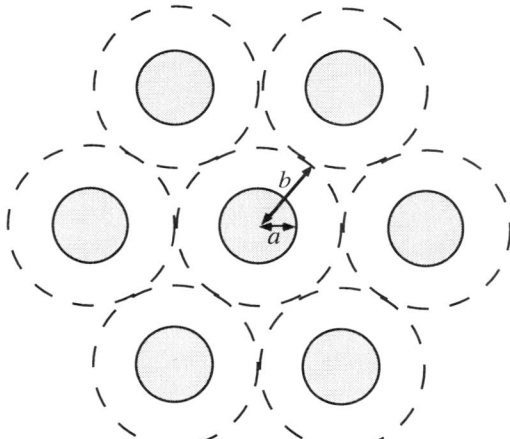

Figure 1. Spherical particles of radius a in concentrated suspensions in the cell model [3]. Each sphere is surrounded by a virtual shell of outer radius b. The particle volume fraction ϕ is given by $\phi = (a/b)^3$.

The main assumptions in our analysis are as follows. (i) The applied field E is weak so that U is proportional to E and terms of higher order in E may be neglected. (ii) The slipping plane (at which the liquid velocity relative to the particle becomes zero) is located on the particle surface (at $r = a$). (iii) No electrolyte ions can penetrate the particle surface. (iv) The relative permittivity ε_p of the particle is much smaller than that of the liquid ε_r so that ε_p is practically equal to zero. (v) In the absence of the applied electric field the particle surface is uniformly charged with a surface charge density σ and has a corresponding zeta potential ζ. (vi) The fluid vorticity is zero at the outer surface of the unit cell [3].

On the basis of these assumptions we have solved the electrokinetic equations for this system, that is, the Navier-Stokes equation and Poisson-Boltzmann equation and finally arrived at the following expression for the dynamic mobility of spherical particles in concentrated suspensions:

$$\mu = \frac{\varepsilon_r \varepsilon_o \zeta}{\eta[M(a) - \Gamma]} \left\{ \frac{2}{3} \left[1 + \frac{1}{2(1 + \delta/\kappa a)^3} \right] S_1 + S_2 \right\}, \tag{3}$$

with

$$\Gamma = \frac{2(\gamma a)^2 (\rho_p - \rho_o)}{9\rho_o}, \tag{4}$$

$$M(a) = \frac{H(a) + i\gamma a(1 - R)}{1 - \phi + [3\phi/(\gamma a)^2](1 - i\gamma a R)}, \tag{5}$$

$$H(a) = 1 - i\gamma a - \frac{(\gamma a)^2}{3}, \tag{6}$$

$$S_1 = \frac{1}{1 - \phi} \left[-\frac{(\kappa a)^2}{3\phi^{2/3} P} + \frac{\gamma^2(1 + \kappa a Q) + \kappa^2(1 - i\gamma a R)}{(\gamma^2 + \kappa^2)\{1 - \phi + \frac{3\phi}{(\gamma a)^2}(1 - i\gamma a R)\}} \right], \tag{7}$$

$$S_2 = \frac{2(1+\phi/2)(\kappa a)^2}{9\phi^{2/3}(1-\phi)P}, \tag{8}$$

$$P = \cosh[\kappa(b-a)] - \frac{1}{\kappa b}\sinh[\kappa(b-a)], \tag{9}$$

$$Q = \frac{1 - \kappa b \cdot \tanh[\kappa(b-a)]}{\tanh[\kappa(b-a)] - \kappa b}, \tag{10}$$

$$R = \frac{(1+i\gamma b)e^{-i\gamma(b-a)} + (1-i\gamma b)e^{i\gamma(b-a)}}{(1+i\gamma b)e^{-i\gamma(b-a)} - (1-i\gamma b)e^{i\gamma(b-a)}}, \tag{11}$$

$$\gamma = \sqrt{\frac{i\omega\rho_o}{\eta}} = (i+1)\sqrt{\frac{\omega\rho_o}{2\eta}}, \tag{12}$$

$$\delta = \frac{2.5}{1 + 2\exp(-\kappa a)}, \tag{13}$$

where ρ_p is the mass density of the particle, ρ_o is the mass density of the liquid, η is the viscosity, e is the elementary electric charge, k is Boltzmann's constant, and T is the absolute temperature. Equation (3) is applicable when the zeta potential is low and the particle permittivity is very small.

In the limit of $\kappa a \to \infty$, in particular, Eq. (3) reduces to

$$\mu_D(\infty,\phi) = \frac{\varepsilon_r\varepsilon_o\zeta}{\eta(1-\phi)} \frac{1 - i\gamma aR}{H(a) + i\gamma a(1-R) - \Gamma[1-\phi+\frac{3\phi}{(\gamma a)^2}(1-i\gamma aR)]}. \tag{14}$$

We plot the magnitude of m for $\kappa a = 10$ and ∞ in Figs. 2 and 3, where calculation was made with the help of Eqs. (3) ($\kappa a = 10$) and (14) ($\kappa a = \infty$) for several values of $\rho_o a^2 \omega/\eta$ ($= |\gamma^2|a^2$) at $(\rho_p-\rho_o)/\rho_o = 0.1$.

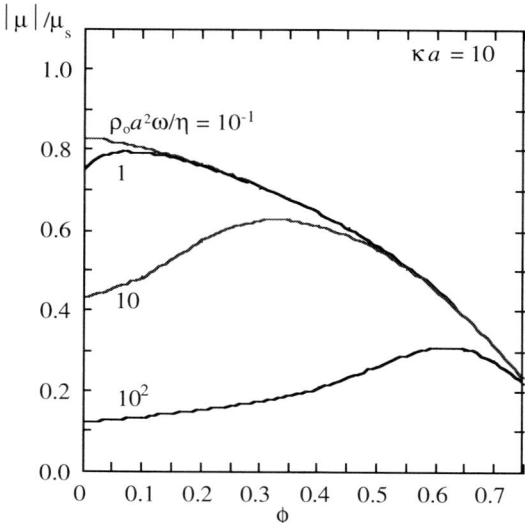

Fig. 2. Magnitude of the dynamic electrophoretic mobility μ, scaled by μ_s (where $\mu_s = \varepsilon_r\varepsilon_o\zeta/\eta$ is Smoluchowski's mobility) as a function of the particle volume fraction ϕ for several values of $\rho_o a^2\omega/\eta$ ($=|\gamma^2|a^2$) at $(\rho_p-\rho_o)/\rho_o =0.1$ and $\kappa a =10$ [1].

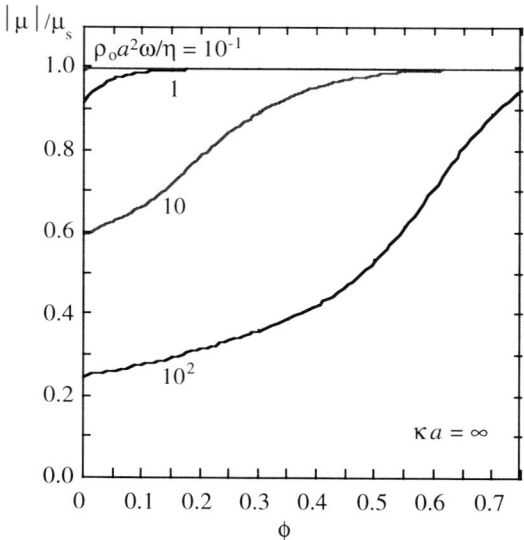

Fig. 3. Same as Fig. 2 but for $\kappa a =\infty$ [1].

3. DISCUSSION

In electroacoustic measurements what is actually measured is not the dynamic electrophoretic mobility itself but the colloid vibration potential (CVP) or electrokinetic sonic amplitude (ESA). Ohshima [4] has recently derived the following relation between sedimentation potential E_{SED} and static electrophoretic mobility μ:

$$E_{SED} = -\frac{\phi(1-\phi)}{(1+\phi/2)}\frac{(\rho_p - \rho_o)}{K^\infty}\mu(\kappa a,\phi)\boldsymbol{g}, \tag{15}$$

where K^∞ is the electrical conductivity of the electrolyte solution in the absence of the particles and $\boldsymbol{g}$ is the gravity. This relation differs from the corresponding relation for the dilute case, viz.,

$$E_{SED} = -\frac{\phi(\rho_p - \rho_o)}{K^\infty}\mu\boldsymbol{g} \tag{16}$$

by a factor $(1 - \phi)/(1 + \phi/2)$.

From the analogy between E_{SED} and CVP or ESA, the same ϕ dependence as Eq. (15) should hold CVP (or ESA) and dynamic electrophoretic mobility μ. That is, CVP (or ESA) is proportional to $\phi(1 - \phi)/(1 + \phi/2)\mu$ instead of $\phi\mu$, as shown recently by Ohshima and Dukhin [5] (See also Ref. [6]). Because of the presence of the factor $(1 - \phi)/(1 + \phi/2)$, the CVP or ESA depends much stronger on ϕ than predicted from the ϕ dependence of μ.

REFERENCES

1. H. Ohshima, J. Colloid Interface Sci., 195 (1997) 137.
2. H. Ohshima and K. Furusawa (eds.), Electrical Phenomena at Interfaces, 2nd Edition, Chap . 2, Marcel Dekker, New York, 1998.
3. S. Kuwabara, J. Phys. Soc. Japan, 14 (1959) 527.
4. H. Ohshima, J. Colloid Interface Sci., 208 (1998) 295.
5. H. Ohshima and A. S. Dukhin, J. Colloid Interface Sci., in press.
6. A. S. Dukhin, Newsletter, January, 1999, Dispersion Technology, Inc., http://www.dispersion.com

HYDROCOLLOIDS – PART 2
Edited by K. Nishinari
2000 Elsevier Science B.V.

75

Viscoelastic behavior of bimodal hardcore suspensions

T. Shikata and Y. Morishima

Department of Macromolecular Science, Osaka University,
Toyonaka, Osaka 560-0043, Japan

Basic viscoelastic parameters, such as the zero-shear viscosity, the relaxation time and so on for the bimodal suspension with the hardcore potential were discussed in terms of a simple model based on the average radius.

1. INTRODUCTION

The viscoelastic behavior of unimodal spherical suspensions has been investigated extensively in recent years by use of sophisticated samples for which the interparticle potentials are very close to the hardcore one.[1,2]   In these suspensions, the particles do not associate with each other, although such association usually occurs in ordinary non-aqueous suspensions.   In all suspensions, particles undergo Brownian motions.

The question arises as to what happens to suspensions consisting of two kinds of spherical unimodal particles with different particle radii; a *bimodal suspension*. Although a number of rheological studies on bimodal suspensions have been carried out, most of the investigated bimodal suspensions have strong interparticle interactions such as attractive dispersion forces.   Long range repulsive electrostatic interactions between particles are controlling interactions in aqueous suspensions.   Because of the existence of complicated interparticle potentials beyond control, it is difficult to discuss accurately the contribution of the particle radius distribution to the over all rheology of bimodal suspensions.   If one employs a bimodal suspension with the hardcore potential, the changes in viscoelastic features are directly dependent on the particle radius distribution.[3]

1.1. Viscoelastic Features of Unimodal Hardcore Suspensions

Because the viscoelastic features of unimodal spherical suspensions with the hardcore potential are fundamental bases for a understanding of bimodal suspensions, we first summarize them below.   i) The reduced zero-shear viscosity (η_0/η_m) which is defined as the ratio of the zero-shear viscosity to the medium viscosity is a function of the volume fraction (ϕ) of suspended particles as eq 1, but is independent of the particle radius (r). $F_0(\phi)$ is a steep increasing function of f.   ii) The reduced high frequency limiting viscosity (η_∞/η_m) is also a function of only f as in eq 2. $F_\infty(\phi)$ is also a steep function of ϕ.   However, the dependence of $F_\infty(\phi)$ on ϕ is a little weaker than that of $F_0(\phi)$, therefore, the relationship $F_0(\phi) > F_\infty(\phi)$ always holds.

$$\eta_0/\eta_m = F_0(\phi) \tag{1}$$

$$\eta_\infty/\eta_m = F_\infty(\phi) \tag{2}$$

iii) The average relaxation time (τ_w) corresponds well to a time constant named *Peclet time* (τ_p) which is a measure of the time necessary for the suspended particle to migrate for a distance equal to its radius due to the Brownian motion. The translational short range diffusion constant ($D(\phi)$) at finite volume fraction is expressed by eq 3; the *modified Stokes-Einstein low*, where k_BT denotes the product between the Boltzmann constant and the absolute temperature. Then, τ_w is approximated by eq 4. iv) The high frequency limiting modulus (G_∞) and the steady state compliance (J_e^0) of the unimodal hardcore suspension can be reduced to functions of only ϕ (eq 5) by converting them to the reduced high frequency limiting modulus ($G_\infty r^3/(k_BT)$) and to the reduced reciprocal steady state compliance ($r^3/(J_e^0 k_BT)$), respectively.[4]

$$D(\phi) = k_BT/(6\pi\eta_\infty r) \tag{3}$$

$$\tau_w \approx 0.5\tau_p = r^2/(6\,D(\phi)) = \pi\eta_\infty r^3/(k_BT) \tag{4}$$

$$G_\infty r^3/(k_BT) = g_\infty\,(\phi) \qquad\qquad [r^3/(J_e^0\,k_BT) = j_e\,(\phi)\,] \tag{5}$$

2. EXPERIMENTAL

2.1. Materials

Three kinds of silica particles (KE series) used were supplied by Nippon Shokubai Co., Ltd. (Osaka, Japan). We ascertained that images of electron micrographs showed radii very close to those of the company data and quite narrow radius distributions. The radii of these particles are 65 nm ; KE13, 90 nm ; KE20, and 215 nm ; KE40, respectively.

We employed ethylene glycol as a suspending medium. Because it possesses a refractive index close to that of the silica particles used, interparticle potentials between suspended particles due to dispersion forces were effectively reduced.[1] To reduce slightly remaining electrostatic interaction between particles, we added potassium chloride to the suspensions at a concentration of 2.0×10^{-4} g/g.

Three suspensions were prepared at ratios (r_L/r_S) of a large particle radius to small one of 1.4; KE20/KE13, 2.4; KE40/KE20, and 3.3; KE40/KE13. The total concentration of the silica particles in these suspensions was kept at 52 wt%, while the weight composition of the small particle (X_S) in the suspensions was altered from 0 to 1 by a step of 0.2.

2.2. Rheology Measurements

Dynamic viscoelastic measurements were carried out at several temperatures from -40°C to the room temperature with a cone-plate type rheometers. The frequency (ω) range covered was 0.01 to 100 rads^{-1}. We made master curves of storage (G') and loss (G") moduli at -10°C as the standard temperature for all the suspensions examined.

3. RESULTS

Figure 1 shows the dependence of G' and G" on ωa_T for a bimodal suspension with r_L/r_S = 2.4 and X_S = 0.8 as a typical example. The ωa_T dependencies of G' in all the bimodal suspensions examined were very similar to each other and also similar to those of a unimodal suspension, although the G_∞ and the τ_w values depended on both the r_L/r_S ratio and X_S. Actually, all the G' curves for the bimodal suspensions can be well superposed onto that of the unimodal suspension of KE13, by shifting along both the axes by the factors of a_C and b_C

as seen in Figure 2. This fairly good superposition of the G' curves for all the bimodal suspensions examined onto that of the unimodal suspension suggests that a bimodal suspension with $r_L/r_S \leq 3.3$ behaves as a unimodal suspension consisting of particles with the average radius.

The η_0 and η_∞ values for the bimodal suspensions examined are plotted in Figure 3 as functions of the composition X_S. Both the η_0 and η_∞ values at $X_S = 0$ and 1 should be identical to each other. However, the data at $X_S = 0$ and 1 in Figure 3 do not agree exactly, especially in the suspension with $r_L/r_S = 3.3$. A reason for the discrepancies for the unimodal suspensions examined may be the slight difference in the hydrodynamic specific volume of each silica particle. Another reason could be a slight experimental error to determine the total weight fraction of suspended particles.

From the density of the silica particle used, the specific volume can be directly estimated as ~ 0.5 cm^{3}g^{-1}. However, the hydrodynamic specific volume (q), which is effective in rheological investigation, is generally larger than that. By inversely applying the η_0 and η_∞ values for unimodal suspensions to $F_0(\phi)$ and $F_\infty(\phi)$ functions respectively, the effective ϕ value was evaluated as 0.46. From this value, q of the suspended particles was estimated to be ~ 0.6 cm^{3}g^{-1}, which is a normal value for ordinary silica particles.

4. DISCUSSION

4. 1. Simple Model for Bimodal Hardcore Suspensions

Here, we propose a model[3] based on the idea that the viscoelasticity of the bimodal suspension can be rescaled into that of the unimodal suspension consisting of the hypothetical particles. When the ratio r_L/r_S is not too large, the relaxation modes of the Brownian motions for the two particles cannot be distinguished into two separated modes, but the suspension will show relaxation modes which can be attributed to the Brownian motion of a hypothetical particle with an average radius. In this modeling, we assume that all the viscoelastic parameters of the bimodal suspension can be estimated by use of empirical equations summarized in the introduction above.

4. 2. Relaxation Time

To evaluate the average relaxation time, τ_w, of a bimodal suspension, we assume that τ_w is identical to a half of the Peclet time τ_P as in the behavior of the unimodal suspension. In a bimodal suspension, we must modify eq 4 because r^2 can not be determined uniquely. In the case of the bimodal suspension, r^2 must be replaced with the number average of the square particle radius ($<r^2>$), then, eq 6 is obtained. Moreover, according to the modified Stokes-Einstein law the average translational diffusion constant ($<D>$) at finite ϕ can be expressed with the number average of the particle radius ($<r>$) as eq 7. From eqs 6 and 7, τ_w can be expressed by eq 8.

$$\tau_P = <r^2>/(\,6<D>\,) \tag{6}$$

$$<D> = k_BT/(6\pi\eta_\infty<r>) \tag{7}$$

$$\tau_w \approx 0.5\pi\eta_\infty<r>\,<r^2>/(k_BT) \tag{8}$$

$$a_c = <r>\,<r^2>/r_{KE13}{}^3 \tag{9}$$

Now, we consider the shift factor a_C necessary to obtain the superposition of G' curves for the bimodal suspensions onto that of the unimodal suspension of KE13 shown in Figure 2. The meaning of a_C is the ratio of τ_p for a bimodal suspension to that of the unimodal KE13 suspension. Because η_∞ is essentially independent of X_S for all the bimodal suspensions examined in this study as seen in Figure 3, eq 9 is easily obtained. Experimental a_C values for all the bimodal suspensions are also plotted as a function of $<r><r^2>/r_{KE13}^3$ double-logarithmically in Figure 4. Since agreement between eq 9 and the experimental data is quite good, eqs 8 and 9 possess the essential feature in the dependence of τ_w on X_S for the bimodal suspension with $r_L/r_S \leq 3.3$.

4. 3. Relaxation Strength

The reduced reciprocal steady state compliance of a bimodal suspension can be expressed using the number average of the cubic particle radius ($<r^3>$), as in a manner of eq 10, because this value is a measure of the storage mechanical energy in a volume comparable to the number average size of the suspended particles.

$$<r^3>/(J_e^0 k_B T) = j_e(\phi) \tag{10}$$

$$b_C = <r^3>/r_{KE13}^3 \tag{11}$$

$j_e(\phi)$ in eq 10 is the universal function of f in the unimodal hardcore potential suspension (cf. eq 5): $j_e(\phi)$ is estimated at ca. 1.3 for this study because $\phi = 0.46[1]$.

The relationship between eq 10 and b_C necessary to obtained superposition of the G' curves of bimodal suspensions onto that of the unimodal suspension of KE13 shown in Figure 2 can be also realized. The meaning of b_C is that of the ratio of the J_e^0 values for bimodal suspensions to that of the unimodal suspension of KE13. Thus, eq 11 can be derived, since $j_e(\phi)$ is constant in this study. Eq 11 can be tested by plotting the experimental b_C values as a function of $<r^3>/r_{KE13}^3$, as shown in Figure 5. The agreement between eq 11 and the experimental data is good enough to convince one that eq 11 has the essential features in the dependence in the relaxation strength on X_S for a bimodal suspension with $r_L/r_S \leq 3.3$.

4. 4. Zero-Shear Viscosity

The zero-shear viscosity, η_0, is generally written as eq 12 for a system with a non-relaxing part of η_∞ .[5] By substituting eqs 8 and 10 to eq 12, the final expression of η_0 for the bimodal suspension with $r_L/r_S \leq 3.3$ is obtained as eq 13.

$$\eta_0 = \tau_w/J_e^0 + \eta_\infty \tag{12}$$

$$\eta_0 = \eta_\infty \{ 1 + 0.5\pi j_e(\phi)<r><r^2>/<r^3>\} \tag{13}$$

In the derivation of eq 13, we did not refer to the dependence of η_∞ on X_S. As seen in Figure 3, the η_∞ value for the bimodal suspensions examined is a function of only ϕ, not X_S.

Eq 13 implies that the essential reason for the change in η_0 depending on X_S is not only the X_S dependence on η_∞, but also its dependence on $<r><r^2>/<r^3>$. Especially, in the bimodal suspensions examined in this study, the η_∞ values have very weak X_S dependence, therefore, the term $<r><r^2>/<r^3>$ is the decisive parameter. Now, we calculate η_0/η_m by use of eq 13 for the bimodal suspension examined, as a function of X_S. Broken lines in Figure 3 represent values of η_0/η_m calculated from eq 13. To evaluate the η_0/η_m value, we assume $0.5\pi j_e(\phi) = 2$ in eq 13, because $j_e(\phi)$ is essentially independent of X_S, and is estimated to be ca. 1.3 from J_e^0 data reported for a unimodal hardcore suspension around $\phi = 0.46$[1]. Moreover, we use experimental η_∞/η_m values shown in the same figure to calculate η_0/η_m. Eq 13 reproduces reasonably well the dependence of the η_0/η_m values on X_S in a bimodal suspension with $r_L/r_S \leq 3.3$, as seen in Figure 3.

The model proposed here only takes into account a change in the relaxation time due to the Brownian motion of suspended particles, and it can reproduce the experimental η_0/η_m data quite successfully. This strongly suggests that the contribution of the Brownian motion is essential to the viscoelastic features in the hardcore suspension even in multimodal suspensions.

Another possible contribution to the viscoelastic features of a bimodal suspension is the depletion effect[6]. The depletion effect always works as an attractive interaction, and it results in an increase in η_0. In principle, there might be a contribution of the depletion effect to the viscoelastic behavior of our bimodal suspensions. However, in our experiments the contribution of the depletion effect to the viscoelastic behavior is negligible, because we find only minimums in the relationship between η_0/η_m and X_S.

In bimodal suspensions with r_L/r_S much higher than 3.3, relaxation modes for two kinds of suspended particles would be observed separately, because the relaxation times for the two particles would not be averaged into one set of relaxation modes, as observed in this study. In that case, the proposed model would no longer be successful.

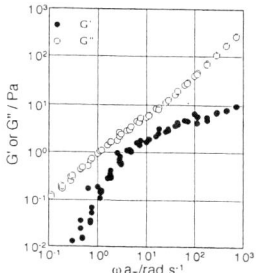

Figure 1. G' and G" versus ωa_T for a bimodal suspension of $r_L/r_S = 2.4$ and $X_S = 0.8$.

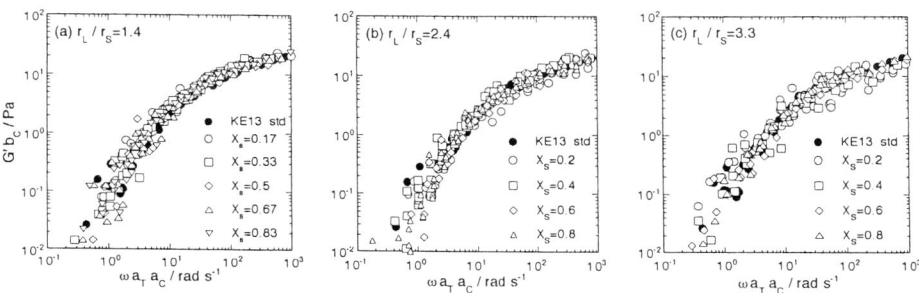

Figure 2. Superposition of G' curves for bimodal suspensions of $r_L/r_S = 1.4$; (a), 2.4 ; (b) and 3.3 ; (c) onto those of KE13.

80

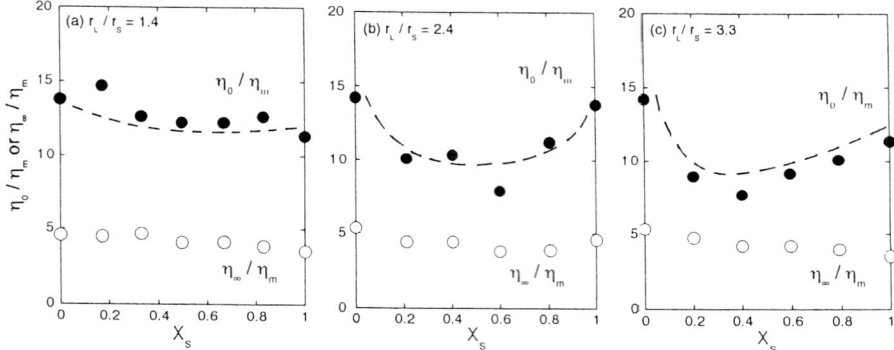

Figure 3. Dependence of η_0/η_m and η_∞/η_m on X_S for the bimodal suspensions of $r_L/r_S = 1.4$; (a), $r_L/r_S = 2.4$; (b) and $r_L/r_S = 3.3$; (c). Broken lines represent calculated η_0/η_m values by use of eq 13.

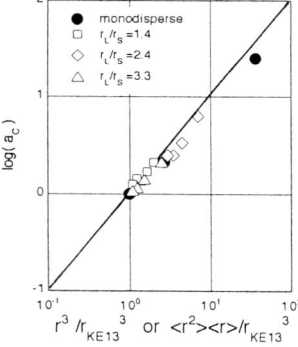

Figure 4. The shift factors, a_C, to obtain Figures 2 versus r^3/r_{KE13}^3 (closed symbols; unimodal suspensions) or $<r><r^2>/r_{KE13}^3$ (open symbols; bimodal suspensions).

Figure 5. The shift factors, b_C, to obtain Figures 2 versus r^3/r_{KE13}^3 (closed symbols; unimodal suspensions) or $<r><r^2>/r_{KE13}^3$ (open symbols; bimodal suspensions).

REFERENCES

1. T. Shikata, and D. S. Pearson, *J. Rheol.*, **38**(1994)601.
2. J. C. van der Werff, C. G. de Kruif, C. Blom and J. Mellema, *Phys. Rev.*, **A39**(1989)795.
3. T. Shikata, H. Niwa and D. S. Pearson, *J. Rheol.*, **42**(1998)765.
4. R. A . Linoberger, and W. B. Russell, *J. Rheol.*, **38**, 1885 (1994).
5. J. D. Ferry, *Viscoelastic Behavior of Polymers*, 3rd ed., Wiley, New York, 1980.
6. S. Asakura, and F. Oosawa, J. Chem. Phys., **22**(1954)1255.

HYDROCOLLOIDS – PART 2
Edited by K. Nishinari
2000 Elsevier Science B.V.

Microdynamics of threadlike micellar systems

T. Shikata, S. Imai and Y. Morishima

Department of Macromolecular Science, Osaka University
Toyonaka, Osaka 560-0043, Japan

Microdynamic behavior of threadlike micelles was examined by using a fluorescence probing method. The rates of molecular motions in the threadlike micelle are about a half of those in the spherical micelle formed with the same surfactant.

1. INTRODUCTION

It is well known that some cationic surfactants from enormously long threadlike micelles in aqueous solution with additives.[1-4]   The long threadlike micelles make concentrated entanglement networks and show pronounced viscoelastic behavior as polymer molecules do in semi-dilute to concentrated conditions.   In the concentrated polymer system, the slowest relaxation mode shows a box type relaxation spectrum which means the broad distribution of relaxation modes.[5]   On the contrary, the slowest relaxation mode of the threadlike micellar system show a Maxwell element like very sharp relaxation spectrum, which means it possesses single relaxation time in the slowest mode.[3]   This difference in the slowest relaxation mechanism reflects essential difference in structural features of these threadlike matters.   In the case of the polymer chain molecule, chain units are tightly bound each other by the covalent chemical bonding.   On the other hand, the threadlike micelle are formed with not so strong intermolecular interaction between surfactant molecules called the *hydrophobic interaction*.

We have proposed a model[4] called *a phantom crossing model* to understand the unique viscoelasticity of the threadlike micellar system.   In the model we assume that every entanglement point has a lifetime equal to the mechanical relaxation time, and two threadlike micelles cross each other at an entanglement point when the lifetime is over.

When one looks at dynamics of the threadlike micelle between entanglement points in which a time scale is much shorter than the slowest relaxation time, one cannot find out any differences in the viscoelastic behavior of the concentrated polymer system and of the entangling threadlike micellar system.[6]   The reason for no differences in the viscoelasticity of the two systems might be that both the polymer chain and the threadlike micelle behave as flexible thread which can bend rather freely.

The last part of consideration of the dynamic behavior of the threadlike micelle is *microdynamics*.[7]   In the threadlike micelle, surfactants and additives should have rapid molecular motions.   They always alter their positions and rotate quickly to reduce the memory of position and orientation in the micelle, since the micelle forming origin is the intermolecular interaction not the chemical bonding . This study addresses such microdynamic behavior of the threadlike micelle. By choosing sophisticated molecules which can probe the molecular motions of the micelle forming substances, the microdynamics of the threadlike micelle can be revealed. A fluorescence anisotropy relaxation measurement was employed to investigated rotational relaxation times of the fluorescent probe molecules working as substitutes for micelle forming surfactant and additive molecules. We have used a combination between cetyltrimethylammonium bromide (CTAB) and sodium salicylate (NaSal)[3] as micelle forming substances so far for investigation of several physicochemical features in threadlike micellar system. However, the fluorescence of NaSal is too strong to examine the fluorescence behavior of molecular probes incorporated in the threadlike micelle.[7] Thus, we employ sodium p-toluenesulfonate (NapTS), which is weakly fluorescent, as an additive to enhance the threadlike micelle formation of CTAB.

Threadlike micelles formed in an aqueous solution of CTAB and NapTS (CTAB:NapTS/W) are known to exhibit similar rheological behavior to that observed in the aqueous CTAB and NaSal system.[8] Thus, the microdynamics of threadlike micelles formed in the CTAB and NapTS system might be essentially the same as that in the CTAB and NaSal system.

2. EXPERIMENTAL

2. 1. Materials

A cationic surfactant, CTAB, was purchased from Wako Chemical Co. Ltd. (Osaka, Japan) and was purified by recrystallization in a mixture of methanol and acetone. Na pTS was also purchased from the same company and was used without further purification. Highly deionized water with a specific resistance $\geq 14 M\Omega cm^{-1}$ obtained by a MilliQ SP system was used as a solvent.

Fluorescent molecules, sodium 2-hydroxy-3-naphtoate (NaHN) and sodium 9-anthracenecarboxylate (NaAC), were purchased from Wako Chemical Co. Ltd, and 1-[4-(trimethylamino)phenyl]-6-phenylhexa-1,3,5-triene bromide (TMA-DPH) was purchased from Molecular Probes, Inc. (Oregon, USA). Another fluorescent molecule, cetylacridiniumorange bromide (CAOB), was synthesized according to a method proposed by Miethke and Zanker[7,9], and was recrystallized in methanol.

The concentration of CTAB (C_D) was kept constant at 10 mM, while the concentration of NapTS (C_S) was altered from 0 to 300 mM. The concentrations of NaHN and NaAC were about 1 mM and were much lower than the C_D value. On the other hand, the concentrations of TMA-DPH and CAOB were lower than 1mM.

2. 2. Methods

Steady state fluorescence anisotropy (r) of the fluorescent molecular probes incorporated in the micelle was determined at 25 °C with a conventional fluorescence photometer equipped with a polarizer in front of the excitation window and an analyzer in front of the emission one. Excitation wave lengths for the probes, NaHN, NaAC, CAOB and TMA-DPH, were 365, 389, 498 and 360 nm, respectively. Furthermore, emission wave lengths for these probes were 512, 415, 512 and 433 nm, respectively. The value of r was estimated with eq 1 below with fluorescence intensities, I_{vv} and I_{vh} : The subscripts (v; vertical and h; horizontal) of the intensities mean the direction of the polarizer and of analyzer, respectively. The factor of g means the correction factor for the polarizing character of the fluorescence photometer used and can be estimated by $g = I_{vh}/I_{hh}$.

$$r = (I_{vv} - gI_{vh})/(I_{vv} + 2gI_{vh}) \tag{1}$$

$$1/r = (1/r_0)(1 + 3\tau_{life}/\tau_\phi) \tag{2}$$

Fluorescence lifetimes (τ_{life}) of the molecular probes in the micelle were measured by use of a conventional time correlation single photon counting apparatus equipped with a high pressure hydrogen flash lamp. The half width of the flash pulse was ca. 2 ns.

Rotational relaxation times (τ_ϕ) of the fluorescent molecular probes were evaluated with eq 2 through the values of r, τ_{life} and instantaneous fluorescence anisotropy (r_0), which was determined in a extraordinary viscous medium like glycerin at low temperature.

3. RESULTS AND DISCUSSION

If one assumes a radius of the spherical micelle of CTAB in aqueous system to be 2.5 nm, the rotational relaxation time for the spherical micelle can be evaluated as ca. 50 ns with the Stokes-Einstein relationship. Moreover, the time constant for over all rotation and/or bending motion of the threadlike micelle should be longer than this value. By the way, the values of τ_ϕ for all the fluorescent molecular probes in this study *via* eq 2 were shorter than 5 ns. Here, we consider a simple case that a probe molecule is incorporated in a spherical micelle and has the rotational relaxation time (τ_ϕ^m) for the microdynamics, and the micelle has the rotational relaxation time (τ_ϕ^r) for over all rotation. In this case, the value of τ_ϕ can be expressed in a manner of $1/\tau_\phi = 1/\tau_\phi^m + 1/\tau_\phi^r$. From this consideration, the τ_ϕ value in the case of this study essentially corresponds to τ_ϕ^m, because the τ_ϕ^r value; ca. 50 ns, is much longer than τ_ϕ^m.

3. 1. Fluorescent Molecular Probes for pTS⁻

Fluorescent probe molecules, NaHN and NaAC, are perfectly incorporated in the micelle in the form of dissociated anions, HN⁻ and AC⁻. This is confirmed by measuring fluorescence lifetimes, τ_{life}, of these probe molecules. The τ_{life} value for NaHN in the pure water was 1.5 ns and 4~6 ns in the micelle, and the τ_{life} for NaAC in the pure water was 1.0 ns and 2~3 ns in the micelle.

Because the chemical structures of NaHN and NaAC are not so different from that of NapTS, they might play a role of substitutes for NapTS in the micelle. NaHN has two hydrophilic groups such as OH and COO⁻ , so that it can be incorporated in the micellar interior sticking out the two hydrophilic groups toward the bulk aqueous phase as schematically shown in Figure 1(a). The direction of the transition moment of NaHN is aligned as seen in the same figure. The value of r_0 for NaHN was determined to be 0.34 in glycerin at -10 °C, and was not so different from the limiting value of 0.4 at the condition for which the transition moment and the axis of fluorescence emission was parallel. Therefore, the rotational motion around an axis shown in Figure 1(a) may be the most effective motion to relax the fluorescence anisotropy.

NaAC can be also a substitute fluorescent molecular probe for NapTS, and it might be incorporated in the micellar interior as shown schematically in Figure 1(b). Because the transition moment of NaAC is placed in the direction connecting 1 and 9-carbon atoms as shown in the same figure, the translational motion along the curvilinear surface of the micelle and/or wobbling motion might be the effective molecular motion to relax the fluorescence anisotropy. Thus, the rotational relaxation times, τ_ϕ, for NaHN and NaAC incorporated in the micellar interior should provide distinct information on the rates of the molecular motion of NapST in the micelle in different directions.

The relationship between τ_ϕ for NaHN incorporated in the micelle and a ratio of C_S/C_D is plotted in Figure 2. From the NMR investigation of a CTAB and NapTS system in D_2O, it was revealed that in the threadlike micelle Br^- of CTAB was replaced by dissociated pTS^- from NapTS. The composition between CTA^+ and pST^- in the threadlike micelle was determined to be unity. Thus, the ratio of C_S/C_D can be a good parameter indicating the shape of micelles. A region with the C_S/C_D value from 0 to unity is occupied by spherical to short rodlike micelles. On the other hand, a region with $C_S/C_D > 1$ is occupied by long threadlike micelles grown enough. Thus, the stepwise increase in τ_ϕ data around $C_S/C_D = 1$ seen in Figure 2 implies a change in the rates of molecular motions of the HN^- anion in the micelle caused by a change in the shape of micelle.

Because the τ_ϕ value for HN^- incorporated in the micelle implies the rotational relaxation time around the axis shown in Figure 1(a) as pointed out above, the data in Figure 2 mean that the rotational rate of HN^- in the threadlike micelle is about a half of that in the spherical one. This fact reflects difference in the inner viscosities of the two kinds of micelles, since the rate of the rotational Brownian motion is inversely proportional to the viscosity of a medium surrounding the probe molecule. Here, we conclude that the inner viscosity of the threadlike micelle is twice higher than that of the spherical one.

Figure 3 shows the relationship between τ_ϕ for AC^- incorporated in the micellar interior and the C_S/C_D ratio. Since the τ_ϕ value for AC^- in the pure water is ca. 0.1 ns, the rate of the rotational motion in the micelle is much slower than in the pure water. This means that the molecular motions of AC^- are highly restricted in the micelle. However, a clear stepwise change in the τ_ϕ data for AC^- can not be found as seen in Figure 2. The τ_ϕ data for AC^- in the micelle might be the average values of the relaxation time of the translational motion along the curvilinear surface of the micelle and that of a quick wobbling motion as discussed before. Because the translational motion along the curvilinear surface of the micelles should have stronger dependence on the inner viscosity of the micelle than the wobbling motion, the τ_ϕ data in Figure 3 with weaker C_S/C_D dependence than the data in Figure 2 suggests that the contribution of the wobbling motion can not be ignored in the τ_ϕ data for AC^-.

3. 2. Fluorescent Molecular Probes for CTA^+

TMA-DPH and CAOB have chemical structure not so different from that of CTAB. TMA-DPH has the fluorescence lifetime, τ_{life}, shorter than 0.3 ns in water, whereas τ_{life} becomes 1~1.5 ns in the micellar interior. On the other hand, CAOB is insoluble in water, however, it can be well incorporated in the micelle formed with CTAB and shows the τ_{life}

value of 1~1.5 ns. Therefore, both TMA-DPH and CAOB do work well as substitute fluorescent probe molecules in the micelle in dissociated forms, TMA^+-DPH and CAO^+.

Because the direction of transition moment of TMA-DPH is aligned in the long molecular axis connecting two phenyl rings and the value of r_0 is 0.39, the effective molecular motion in the micelle for the fluorescence anisotropy relaxation is the translational motion along the curvilinear micellar surface and/or the wobbling motion. On the other hand, the direction of transition moment of CAO^+ is parallel to an axis connecting nitrogen and 9-carbon atom in an acridinium ring. Thus, the translational motion along the micellar surface and/or the wobbling one should be also effective to relax the fluorescence anisotropy for CAO^+ in the micelle.

The τ_ϕ values for TMA^+-DPH and CAO^+ are plotted as functions of the C_S/C_D ratio in Figure 4. The C_S/C_D ratio dependence of the data in this figure roughly resembles that in Figure 2. This means that rates of molecular motions for both TMA^+-DPH and CAO^+ alter with the change of shape of the micelle due to the C_S/C_D ratio as well as HN^-. Therefore, we conclude that in the threadlike micelle the translational motion along the micellar surface and/or the wobbling motion of the CTA^+ cation is twice slower than in the spherical micelle.

In the case of NaAC, in which both the translational motion along the micellar surface and the wobbling motion can possibly be the effective origin for the fluorescence anisotropy relaxation, the C_S/C_D dependence of τ_ϕ shown in Figure 3 does not show obvious stepwise increasing as seen in Figure 4. The reason for this difference would be difference in the size of fluorescence probes. Since the size of NaAC is smaller than both TMA^+-DPH and CAO^+, the quick wobbling motion of AC^- in the micellar interior is more intense than those of TMA^+-DPH and CAO^+. Therefore, the fluorescence anisotropy for AC^- is relaxed by the wobbling motion, in part. On the contrary, translational motion along the curvilinear micellar surface may be more effective to the fluorescence anisotropy relaxation for larger fluorescent probes of TMA^+-DPH and CAO^+ than the wobbling motion.

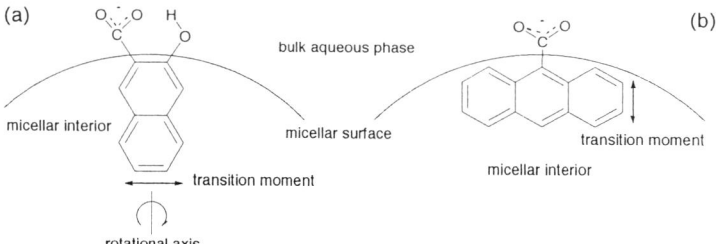

Figure 1. Schematic representation of location and the direction of the transition moments for fluorescent molecular probes, 2-hydroxy-3-naphtoate anion (HN^-); (a) and 9-anthracenecarboxylate anion (AC^-); (b) in the micelle.

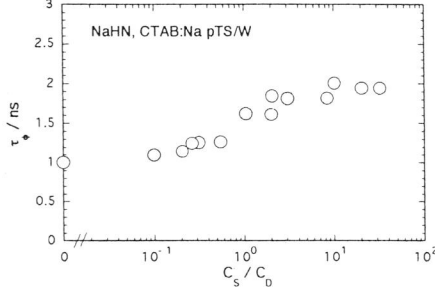

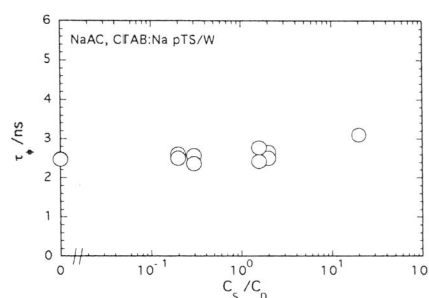

Figure 2.   The dependence of τ_ϕ on a C_S/C_D ratio for HN⁻ in the CTAB:NapTS/W system at 25 °C.

Figure 3.   The dependence of τ_ϕ on the C_S/C_D ratio for AN⁻ in the CTAB:NapTS/W system at 25 °C.

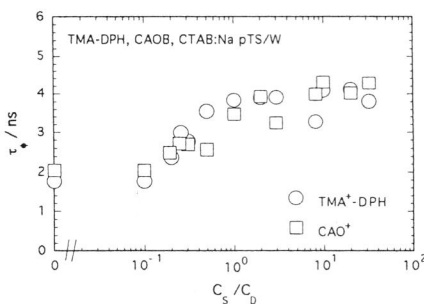

Figure 4.   The dependence of τ_ϕ on the C_S/C_D ratio for TMA⁺-DPH and CAO⁺ in the CTAB:NapTS/W system at 25 °C.

REFERENCES

1. S. Glavsholts, *J. Colloid Interface Sci.*, **57**(1976)575.
2. H. Hoffmann, and H. Rehage, *Mol. Phys.*, **5**(1989)1225.
3. T. Shikata, H. Hirata and T. Kotaka, *Lamgmuir*, **3**(1987)1081.
4. T. Shikata, H. Hirata and T. Kotaka, *Lamgmuir*, **4**(1988)354.
5. J. D. Ferry, Viscoelastic Properties of Polymers, 3rd ed., Wiley, New York, (1980).
6. T. Shikata, in preparation.
7. T. Shikata, and Y. Morishima, *Lamgmuir*, **13**(1997)5229.
8. T. Shikata, unpublised data.
9. E. Miethke, and V. Z. Zanker, *Phys. Chem.* (Frankfurt am Main), **18**(1958)375.

HYDROCOLLOIDS – PART 2
Edited by K. Nishinari
2000 Elsevier Science B.V.

Production of super-monodispersed lipid microspheres with microchannel system

M. Nakajima*, T. Kawakatsu, H. Nabetani, Y. Kikuchi and Y. Sano

National Food Research Institute, Ministry of Agriculture, Forestry and Fisheries
Kannondai 2-1-2, Tsukuba, Ibaraki 305-8642, Japan

A new emulsification technology was developed for producing super-monodispersed lipid microspheres. Oil in water (O/W) microspheres were generated by permeating an oil phase into a continuous water phase through a silicon microchannel which was designed and prepared using semiconductor technology. The microprocessing of microspheres was monitored and controlled using a microscopic video system. Super-monodispersed O/W microspheres were found to be made by the microchannel and the glass plate with use of an appropriate surfactant.

1. INTRODUCTION

Scientific and technological efforts on mechanical emulsification using mixers, colloid mills, homogenizers, and sonicators have focused on making more homogeneous and stable microspheres (emulsions and microparticles) [1]. Membrane filtration can be applied to obtain greater homogeneity. Olson *et al.* (1979) used several polycarbonate membranes to remove large microspheres [2]. Suzuki (1993) reported that the size distribution of microspheres was narrow as the microspheres were filtered repeatedly [3]. Membrane emulsification was first reported by Nakashima and Shimizu (1993) and Nakashima *et al.* (1993), in which microspheres were made by filtrating an internal phase into a continuous phase [4, 5]. The significant point of their method is that a membrane was used for microspheres creation and not separation. Since the size of microspheres could be controlled by membrane pore size, the method is advantageous for making a monodispersed microspheres in comparison with conventional methods. The method is now commercially used to produce low fat margarine [6].

The present address of T. Kawakatsu is Department of Chemical Engineering, Tohoku University, Sendai 980-8579
Contact author: M. Nakajima*, Tel: +81-298-38-7997, Fax: +81-298-38-8122, e-mail: mnaka@nfri.affrc.go.jp

Kikuchi *et al.* developed a system in which a microscope was attached to a video recorder (microscope video system) for viewing optically accessible microchannels formed in a single crystal silicon substrate which was manufactured using semiconductor technology [7]. The microscope video system was used for diagnosing blood cell deformability by observing permeability through the silicon microchannel. The large advantage of this method is real-time optical observation of the behavior of micron-sized materials. We have recently proposed a new emulsification technology using the microchannel system [8].

In this paper, basic characteristics of this new emulsification technology for producing super-monodispersed microspheres was investigated using the silicon microchannel and the microscope video system. Visual images are recorded during the creation of O/W (oil in water) microspheres.

2. MATERIALS & METHODS

2.1 Materials

High oleic sunflower oil (triolein, >90% purity) was obtained from Nippon Lever B.V. (Tokyo, Japan). Food grade sorbitan monolaurate, L1C (HLB: 8.6) and sorbitan monooleate, O10 (HLB: 4.3) were obtained from KAO Chemicals (Wakayama, Japan). Special reagent grade oleic acid, sodium dodecyl sulfate (SDS, HLB: 40), KCl, NaCl and reagent grade soybean lecithin were purchased from Wako Pure Chemical Ind. (Osaka, Japan). Kerosene was purchased from Japan Energy Corporation (Tokyo, Japan).

2.2 Visual microprocessing system

Figure 1 shows the microscope video system, the silicon microchannel module and the silicon microchannel plate for microchannel emulsification. An inverted metallographic microscope (MS-511B, Seiwa Kougaku Seisakusho Ltd, Japan) and a 3CCD camera (HV-C20; Hitachi, Japan) were used to observe the microchannels, and images were recorded by using and a VHS video player (WV-TW2; Sony Corporation, Tokyo, Japan). The total magnification was about x1000. The silicon microchannel plate measures 15 mm x 15 mm. The average channel width, channel wall height and the terrace height are 6 μm, 4.5 μm and 60 μm, respectively. The silicon material was originally hydrophobic, however, after the processes of photolithography and orientation-dependent etching the silicon plate became hydrophilic [7, 8].

2.3 Microchannel emulsification

Figure 2 schematically shows the flow mechanism in the module and through the microchannel. The silicon microchannel plate was tightly covered with a flat glass plate and the channel image was obtained through the glass plate. The silicon microchannel module was initially filled with the continuous phase liquid of the desired emulsion and the internal phase liquid was pressed into the module by lifting the liquid chamber filled with the internal phase liquid. During the microprocessing of emulsion cells, the operating pressure and the flow rate were regulated by changing the height of the chamber with real-time optical observation.

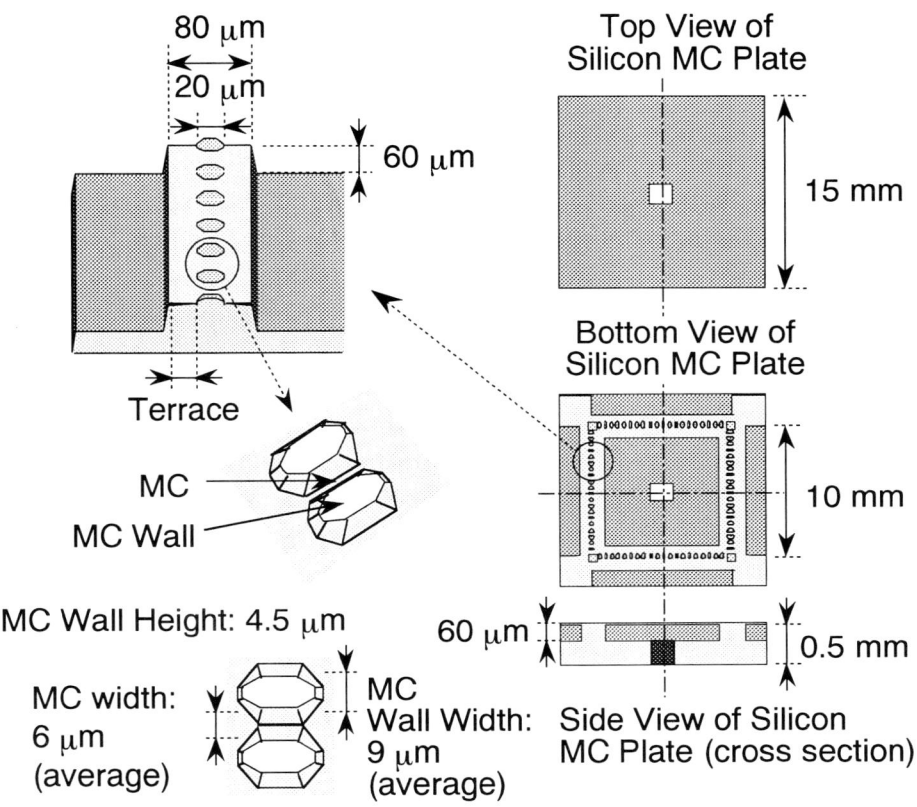

Fig. 1 Silicon Microchannel Plate

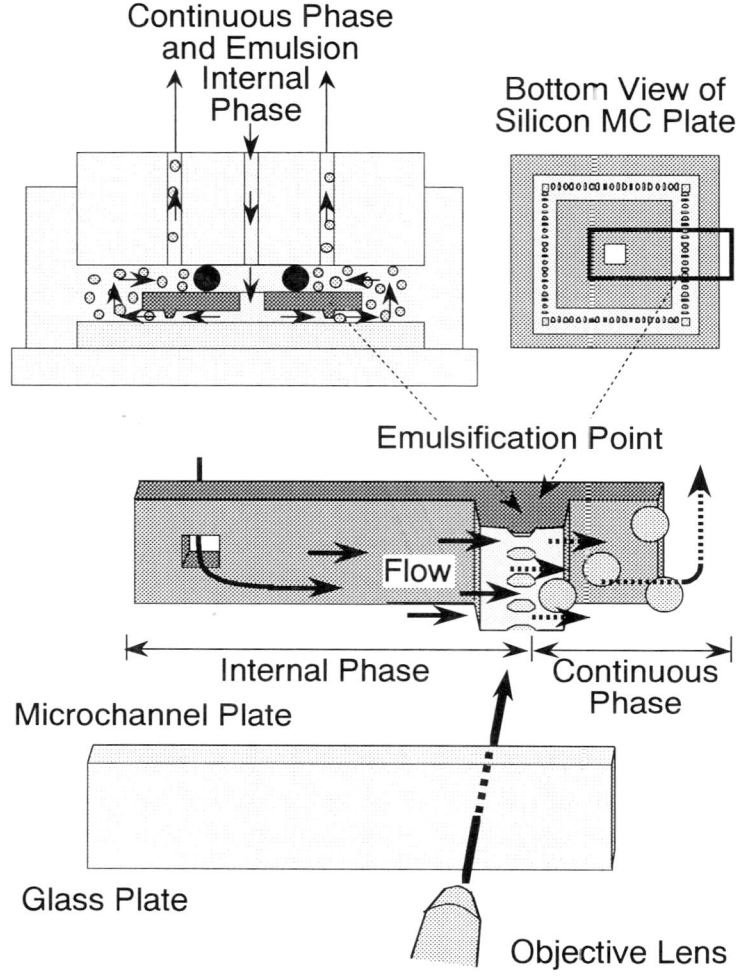

Continuous Phase
and Emulsion
Internal
Phase

Bottom View of
Silicon MC Plate

Emulsification Point

Flow

Internal Phase

Continuous
Phase

Microchannel Plate

Glass Plate

Objective Lens

Fig. 2 Schematic Flow Mechanism
        through the Silicon Microchannel

3. RESULTS & DISCUSSION

Triolein was used as the internal oil phase with 0.3 wt% sorbitan monolaurate. SDS was also used for making O/W microspheres, and was dissolved in the water phase in contrast to sorbitan monolaurate which was dissolved in oil.

As the silicon microchannel is hydrophilic, it is more easily wetted with water than oil. Therefore, for oil (triolein) injection into the microchannel plate module and permeation through the microchannel, it was necessary to apply pressure by increasing the height of the liquid chamber. Figures 3a-3d show the process of microchannel emulsification for making O/W microspheres using 0.3 wt% sorbitan monolaurate in triolein (internal oil phase). At 2.65 kPa the oil phase intruded into the terrace of the microchannel (Fig. 3a). The boundary line between water and oil phases gradually moved to the entrance of the microchannel and at 3.97 kPa the line contacted the entrance (Fig. 3b). At 4.41 kPa, it had fully contacted with the entrance (Fig. 3c). The operating pressure was slowly increased with simultaneous observation of the boundary position, and when the pressure reached 8.38 kPa, the oil phase broke through the channel and the creation of regular sized cells started (Fig. 3d). The breakthrough site appeared to be random. The speed of microspheres creation at 8.38 kPa was very high (about 2 microspheres per second for each creation point) and gradually decreased as the operating pressure was reduced. Microspheres creation continued until the operating pressure reached 4.85 kPa. The microspheres were regular with a size of 22.5 μm (Fig. 3e). Standard deviation of the size was 0.2 μm, which suggests super-monodispersibility of the microspheres. Sotoyama *et al.* (1998) obtained that W/O microspheres having average size of 12.4 μm and standard deviation of 5.3 μm by membrane emulsification [10]. Although each operating condition was quite different, the present microchannel emulsification was found to be so effective to obtain super-monodispersed microspheres. The obtained size of the microspheres was 3.75 times larger than the average microchannel width (6 μm). This in good agreement with the membrane emulsification results obtained by Nakashima and Shimizu [4]. They empirically found that the average size of microspheres was 3.3 times larger than the membrane pore size. The size of microspheres was independent of the operating pressure in the range examined. During the emulsification, generated microspheres were gradually accumulated and amount of water relatively reduced in the permeate side. When the water became insufficient to cover the surface of O/W microspheres, irregular ones began to appear (Fig. 3f).

As a substitute for sorbitan monolaurate, SDS was dissolved in the continuous water phase. With this system, microspheres were also created. Table 1 shows the effect of pressure (head difference) on microchannel processing of O/W microspheres. The pressure difference needed to make O/W microspheres decreased as the SDS concentration increased since the interfacial tension between water and triolein became lower.

The method to produce W/O microsphere was reported elsewhere [8]. It is promising that the microchannel emulsification method allows possibilities to create super-monodispersed lipid microspheres. Recent study has revealed that the size of the microspheres depends greatly on the channel size, which is similar to membrane

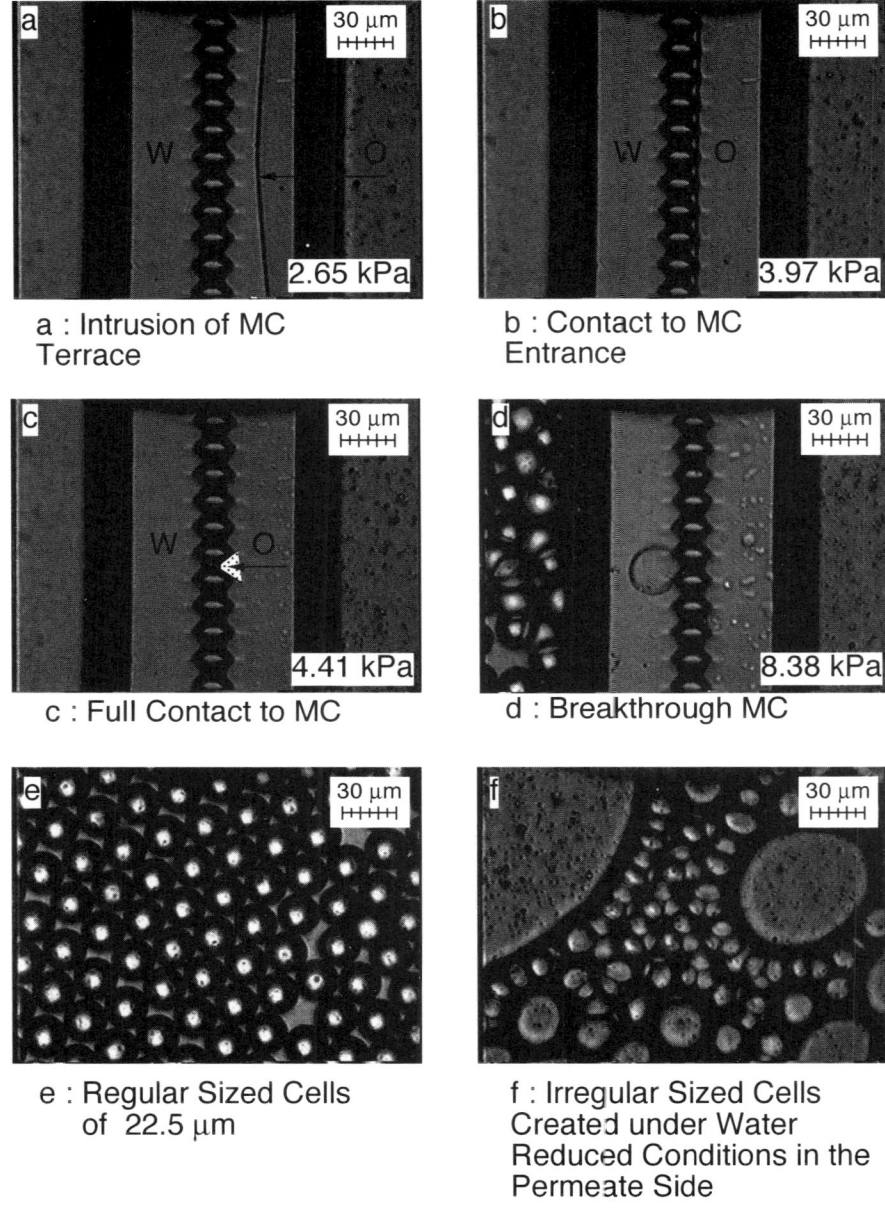

a : Intrusion of MC Terrace

b : Contact to MC Entrance

c : Full Contact to MC

d : Breakthrough MC

e : Regular Sized Cells of 22.5 μm

f : Irregular Sized Cells Created under Water Reduced Conditions in the Permeate Side

Fig. 3   Microchannel Emulsification for O/W Cells in Triolein with 0.3 wt% Sorbitan Monolaurate/Water System emulsification. The effects of operating conditions, water and oil phase components and the structure and surface modification of the silicon microchannel should be investigated further.

Table 1
Effect of pressure on microchannel processing for creation of monodispersed O/W microspheres

Internal phase /Continuous phase (surfactant)	Intrusion into terrace of microchannel: Pressure [kPa]	Full contact to entrance of microchannel: Pressure [kPa]	Breakthrough the microchannel and creation: Pressure [kPa]	Lower limit for creation after breakthrough: Pressure [kPa]
Triolein (0.3 wt% L10) /Water	2.7	4.4	8.4	4.9
Triolein /Water (0.1 wt% SDS)	3.7	4.1	8.5	5.4
Triolein /Water (0.2 wt% SDS)	1.8	1.9	2.4	1.5

ACNOWLEGMENT

This research was funded by the Program for Promotion of Basic Research Activities for Innovative Biosciences.

REFERENCES

1. E.S.R. Gopal, Principles of Emulsion Formation, in Emulsion Science, edited by P. Sherman, Academic Press, London and New York, pp. 1-75. (1968).
2. F. Olson, C.A. Hunt, F.C.Szoka, W.J. Vail, and D. Papahadjopoulos, Biochim. Biophys. Acta 557 (1979) 9.
3. K. Suzuki, (1993) The Food Industry 36: 26-31.
4. T. Nakashima, and M. Shimizu, Kagaku Kogaku Ronbunshu 19 (1993) 984.
5. T. Nakashima, M. Shimizu, and M. Kukizaki, Kagaku Kogaku Ronbunshu 19 (1993) 991.
6. R. Katoh, (1993) Shokuhin to Kaihatsu 28: 9-13.
7. Y. Kikuchi, K. Sato, and T. Kaneko, Microvascular Res. 44 (1992) 226.
8 T. Kawakatsu, Y. Kikuchi, and M. Nakajima, JAOCS. 74（1997）317.
9. S.M. Sze, Lithography and etching, in Semiconductor devices physics and technology, edited by Willy International, John Willy and Sons, Inc., New York, p. 428. (1985)
10. K. Sotoyama, Y. Asano, K. Ihara, K. Takahashi, and K. Doi, Nippon Shokuhin Kagaku Kogaku Kaishi 45（1998）253.

HYDROCOLLOIDS – PART 2
Edited by K. Nishinari
2000 Elsevier Science B.V.

Structure and viscoelastic properties of surfactant/water colloidal systems

T. Mori[*a] and T. Matsumoto[*b]

*a Fundamental Research Laboratories, Noevir Co. Ltd., Okada-cho 112-1, Youkaichi, Shiga 527-8588, Japan

*b Division of Forest and Biomaterials Science, Kyoto University, Kyoto 606-8502, Japan

Rheological properties and structure of the lamellar phase formed in Aerosol OT (Sodium bis - (2 - ethylhexyl) sulfosuccinate; AOT)/water/n-decane colloidal systems have been investigated using dynamic viscoelastic measurements and small angle X-ray scattering (SAXS).

The lamellar repeat distance L is linearly dependent on the function Φ ($= (\phi_W /\rho_W + \phi_D /\rho_D)/\phi_A$), where ϕ and ρ are volume fraction and density, subscript W, D and A mean water, n-decane and AOT respectively, which suggests that the surface area S_A per AOT molecule is independent of the composition, and the thickness l_A of AOT bilayer is independent of AOT/water ratio, but swells proportional to in amount of n-decane. From the slope and intercept of the plots of L against Φ, the values of l_A, density of AOT bilayer ρ_A, and S_A can be obtained respectively as $17.8\,\text{Å}$, 1.28 g/cm³, and 64.5 Å^2.

The values of storage modulus G' and loss modulus G'' are almost constant over a wide range of frequency, and G' always higher than G'' for all the samples. These facts suggest that the lamellar structure is three-dimensionally extended over the systems. The values of G' dose not increase linearly with increasing AOT content. In AOT/water binary system, the relationship between G' and L seems to be tied by two straight lines. In AOT/water/n-decane ternary system, G' has minimum value in low AOT/n-decane ratio. These results would suggest some special structural change which can not be detected by SAXS measurements.

1. INTRODUCTION

Aqueous colloids of numerous surfactants are widely used in various industries and fields. These surfactants form various ordered structures in aqueous solutions, such as spherical micelle, threadlike micelle, hexagonal, or lamellar and more complex structure. Recently, a certain kind of surfactant with two or more hydrophobic chains have been noticed because they formed bilayer structure considering to the primitive model of biomembrane and there are many studies for these systems from the thermodynamical or structural viewpoints [1—4]. On the other hand, there is little investigations for the viscoelastic properties of bilayer systems [5—9].

AOT is an anionic surfactant which has double hydrophobic chains. AOT can dissolve much water to oil without co-surfactant, so it is used for the studies of inverse microemulsions [10]. It has been also known that lamellar liquid crystalline phase is formed in wide concentration range of AOT in AOT/water/oil colloidal systems [11—13].

In this paper, we discuss the structure and dynamic viscoelasticity of the lamellar phase formed in AOT/water/n-decane colloidal systems, in comparison with AOT/water binary systems.

2. EXPERIMENT

2.1. Materials

AOT (99%) and n-decane were purchased from Fluka and Tokyo Kasei, respectively. And AOT used after dried in a vacuum oven at room temperature. AOT and n-decane were dispersed in distilled water at various compositions. In ranging from the AOT content of 0.2 - 0.7 and the n-decane content of 0 - 0.05 in weight fraction, it was confirmed under crossed polarizer that these systems have anisotropic phase.

The disperse systems were provided to SAXS and dynamic viscoelastic measurements after keeping at 25℃ for one week or more, in order to develop fully the lamellar structure.

2.2. Measurements
SAXS

SAXS measurements were performed with the 6 m point-focusing SAXS camera at the High Intensity X-Ray Laboratory, Kyoto University, with CuKa radiation (wave length $\lambda = 1.54$ Å) . The sample was loaded in a cell with mica windows and the sample thickness was 1.5 mm.

Dynamic Viscoelasticity

The viscoelastic properties were measured by means of a cone-plate type rheometer (RDA II , Rheometric Scientific, Tokyo) which was devised to prevent vaporization of water. The diameter and angle of cone used were 50 mm and 0.04 rad . Measurements were carried out at 25℃ under dynamic strain amplitude γ of 0.0075, in which all the samples showed linear viscoelasticity.

3. RESULT AND DISCUSSION

3.1. SAXS

One or two sharp diffraction peaks which seems to be attributed to the Bragg space due to the anisotropic liquid phase, can be apparently observed in all the systems. For sharp peaks, we discuss the lamellar structure using our model.

For AOT/water binary systems, we already discussed in the previous report [14].

Similar analysis can also be applied to the AOT/water/n-decane ternary systems, assuming that all the AOT molecule used to form the bilayer, all of n-decane added exists in the bilayer, the surface area per AOT molecule S_A would be independent of the composition of samples, and the thickness of bilayer l_A increases in proportion with the amount of n-decane.

The lamellar repeat distance L is provided by the next equation (1) with our model.

$$L = K \cdot \Phi + l_A \quad\text{——— (1)}$$

Here, Φ and K are given as follows.

$$\Phi = \frac{1}{\phi_A} \left\{ \frac{(1 - \phi_A - \phi_D)}{\rho_W} + \frac{\phi_D}{\rho_D} \right\}$$

$$K = \frac{2M}{N_A \cdot S_A} = l_A \cdot \rho_A$$

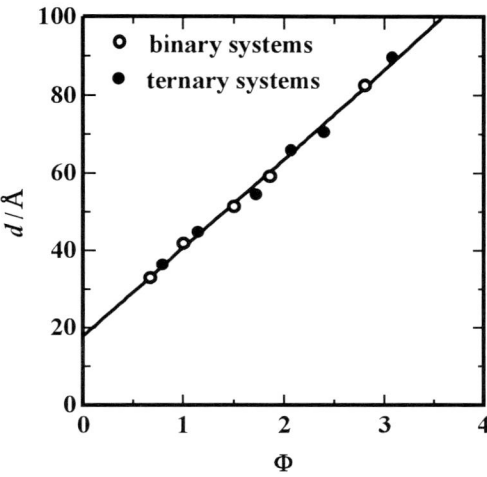

Fig.1 Bragg space d (i.e. L) plotted against Φ

where **M** are weight fraction and molecular weight of AOT, N_A is Avogadro's number, ϕ and ρ are volume fraction and density, and subscript **W, D** and **A** mean water, n-decane and AOT bilayer, respectively.

According to the equation (1), the relationship between L and Φ can be represented by a straight line with slope **K**.

In Fig.1, Bragg space d ascribed to the first diffraction peaks, which would be corresponded to L in equation (1), was plotted against Φ. The plot can be expressed by a straight line, that is, lamellar repeat distance is proportional to Φ, which suggests our model for the lamellar structure is reasonable. Fig.1 shows that the lamellar repeat distance increases continuously with increasing Φ value. From the slope and intercept of the straight line, the values of l_A, ρ_A and S_A can be evaluated as 17.8 Å, 1.28 g/cm³, and 64.5 Å^2, respectively.

It should be noted that the thickness of the AOT bilayer l_A is slightly smaller than twice of the length of one AOT molecule, ca. 12 Å. It seems that the hydrophobic parts of AOT molecules somewhat overlap each other like a fingerjoint, and that the molecules are loaded bending these hydrophobic chains or the molecules are arranged with a slight tilt in the bilayer.

3.2. Dynamic Viscoelasticity

Storage modulus G' and loss modulus G" were measured at various angular frequencies ranging from 0.01 s^{-1} to 100 s^{-1} for the AOT /water/n-decane colloidal systems at 25℃. The values of G' and G" are almost constant over a relatively wide range of frequency, and

98

G' is always higher than G" in all the samples. Considering that all the systems show the lamellar structure, the flatness of dynamic moduli against ω suggests that the well developed lamellar structure is three dimensionally extended over the systems.

Fig.2 shows the logarithms of G' at ω = 1s⁻¹ plotted against the lamellar repeat distance L calculated by equation (1). In the AOT/water binary systems, G' decreases with increasing of the lamellar repeat distance L, and the relationship between G' and L seems to be tied by two straight lines which have a large slope in the low L region (L<40 Å) and a small slope in the high L region (L>40 Å). In the AOT/water/n-decane ernary systems, the relationship between G'

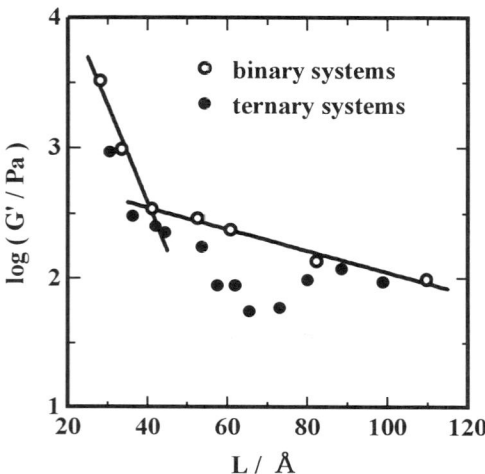

Fig.2 Storage modulus G' at ω = 1 s⁻¹ plotted against the lamellar repeat distance L calculated by equation (1)

and L can be represented by straight lines with a large slope in the low L region and a small slope in the high L region. But it shows a minimum value near $d = 70$ Å in ternary systems. Although the difference of the structure is expected in these regions from the rheological viewpoint, the d value changes smoothly near to 40 and 70 Å. That is, the relation between d and Φ can be smoothly represented by a straight line as shown in Fig.1. The SAXS measurements did not detect any special structural change near to d of 40 and 70 Å.

According to the SAXS measurement and the model analysis, the surface area per AOT molecule is independent of the composition of the systems. And the thickness of AOT bilayer changed with addition of n-decane. In the AOT/water binary systems, the structure of bilayer is not changed with AOT content. It would be thought from these results that the rapid increase of G' near to 40 Å with increasing the AOT content is attributed to a size change of domain or morphology, which can not be detected by SAXS measurement. Similarly, it would be expected that n-decane induced the structural change near to 70 Å in the AOT/water/n-decane ternary system. Though the origin of this discrepancy is not clear at the present time, it can be expected that the rheological measurements are more sensitive to the structural change than the SAXS measurements [15-17].

REFERENCES

1. J. Marignan, J. Appell, P. Bassereau, G. Porte, Prog. Coll. Polym. Sci., 101 (1998) 76
2. T. Shimobouji, H. Matsuoka, N. Ise, H. Oikawa, Phys. Rev. A, 39 (1989) 4125
3. R. Strey, R. Schomacker, D. Roux, F. Nallet, U. Olsson, J. Chem. Soc. Faraday Trans,

86 (1990) 2253

4. N. Nakamura, T. Tagawa, K. Kihara, I. Tobita, H. Kunieda, Langmuir, 13 (1997) 2001
5. T. Matsumoto, Coll. Polym. Sci., 270 (1992) 492
6. T. Matsumoto, J. Soc. Rheol. Jpn., 19 (1991) 158
7. T. Matsumoto, D. Ito, H. Kohno, Coll. Polym. Sci., 267 (1989) 946
8. M. Gemma, M. Valiente, E. Rodenas, Langmuir, 12 (1996) 5202
9. P. Versluis, J.G. van de Pas, J. Mellema, Langmuir, 13 (1997) 5732
10. M. Kotlarchyk, S.H. Chen, J.S. Huang, M.W. Kim, Phys. Rev. Lett., 53 (1984) 941
11. K. Fontell, J. Colloid Interface Sci., 44 (1973) 318
12. E.I. Franses, T.J. Hart, J. Colloid Interface Sci., 94 (1983) 1
13. H. Kunieda, T. Sato, J. Jpn. Oil Chem. Soc., 28 (1979) 627
14. T. Mori, T. Tada, T. Matsumoto, J. Soc. Rheol. Jpn., 125 (1997) 223
15. T. Matsumoto, H. Inoue, Chem. Phys., 178 (1993) 591
16. T. Matsumoto, H. Inoue, Chem. Phys., 207 (1996) 167
17. H. Inoue, T. Matsumoto, Coll. Surfaces, 109 (1996) 89

HYDROCOLLOIDS – PART 2
Edited by K. Nishinari
2000 Elsevier Science B.V.

101

Effect of gravity on the apparent contact angles

H.Sakai and T.Fujii

Faculty of Human Life Science, Osaka City University
Sugimoto 3, Sumiyoshi-ku, Osaka 558-8585 JAPAN

The effect of gravity on the rough solid-liquid interface has been studied theoretically in the sessile drop method. Then, it was found that its interfacial tension is enhanced by gravity when gas is adsorbed at it. As a result, the apparent equilibrium contact angle, which is considered not to be influenced by gravity so far, can be raised by gravity for rough surfaces. The calculated dependence of contact angles on gravity under the ordinary conditions of the sessile drop method is large enough to detect by experiment.

1. INTRODUCTION

Recently, studies on the wetting of a solid surface are very active not only from fundamental aspects but also from practical ones. The settlement of a sign problem of line tension[1] and the development of super liquid-repellent surfaces[2,3] are examples. These advances are largely due to improvements in measurement techniques of contact angles. The surface free energy of a solid, which is not directly measurable, is calculated from the contact angle data with several liquids[4]; here, it is necessary to analyze the measured contact angles adequately.

The equilibrium contact angle θ_{eq} of a liquid placed on a flat solid surface is given by Young's equation[5,6]

$$\cos\theta_{eq} = \frac{\gamma_{sg} - \gamma_{sl}}{\gamma_{gl}} \tag{1}$$

where γ_{sg}, γ_{sl}, and γ_{gl} are the solid-gas, the solid-liquid, and the gas-liquid interfacial tensions (or the interfacial free energies per unit area), respectively.

When a solid surface is rough with the roughness factor r, the ratio of actual to projected area, the solid-gas and solid-liquid interfacial tensions become r-fold. Then, the equilibrium contact angle ϕ_{eq} is given by

$$\cos\phi_{eq} = \frac{r\,\gamma_{sg} - r\,\gamma_{sl}}{\gamma_{gl}} = r\cos\theta_{eq} \tag{2}$$

It is called Wenzel's equation and is valid when the size of the roughness structure is much smaller than the contact radius of a liquid drop[5]. ϕ_{eq} is the angle measured at the geometric level of the rough surface and is called the apparent contact angle(Figure 1). On the other hand, θ_{eq} is called the intrinsic contact angle.

According to Eq.2, under the condition of $\theta_{eq} > 90°$, the apparent contact angle ϕ_{eq} increases with increasing θ_{eq} or r and ϕ_{eq} should be $180°$ at $r\cos\theta_{eq} = -1$. However, when

ϕ_{eq} approaches $180°$, it is difficult for a liquid to penetrate into the troughs of the rough solid surface and the surrounding gas (air, nitrogen gas, etc.) is adsorbed at the solid-liquid interface. This situation is depicted in Figure 1. The adsorbed gas reduces the solid-liquid interfacial tension and ϕ_{eq} approaches $180°$ very slowly[2,3,5], not straight in the cosine form predicted by Eq.2. That is, Eq.2 or Wenzel's equation fails under the gas adsorption. Here, in the setup of Figure 1, the gas has to lift up a drop against gravity to get into troughs under it. Therefore, gravity works against the gas adsorption. Then, the solid-liquid interfacial tension is also affected by gravity through gas adsorption, and finally the apparent contact angle depends on gravity. This scenario is not considered in the previous wetting theories and the contact angle has been considered not to depend on gravity so far[5,7].

In this paper, after explaining the mechanism of the gas adsorption in detail, the dependence of the apparent contact angle on gravity under the ordinary conditions of the sessile drop method[8], which is widely used to measure a contact angle, is discussed.

2. MECHANISM OF GAS ADSORPTION

According to Eq.2, $\phi_{eq}=180°$ when $\theta_{eq}=106°$ and $r=4$. However, the calculations in the following sections show that the gas adsorption does occur at the sinusoidal rough solid surface of Figure 1. The free energy of the system takes its minimum when about 80 percent of the solid-liquid interface is covered by the adsorbed gas and ϕ_{eq} becomes $157°$.

The mechanism of gas adsorption is explained as follows. Under the adsorption, the liquid touches only the gentle slope parts of the solid surface because troughs of the surface are filled with the adsorbed gas. Therefore, the adsorption effectively smooths down the solid surface. The roughness factor, which is originally 4, reduces to about 2 in the above case. This lowers the solid-liquid interfacial tension. However, note that the solid-gas and liquid-gas interfaces are created associated with the gas adsorption because the gas phase intervenes between the solid and the liquid under the drop. Thus, the decrease of the solid-liquid interfacial free energy due to the roughness factor reduction must exceed the increase of the free energy associated with the creation of solid-gas and liquid-gas interfaces in magnitude in order to lower the total free energy of the system and stabilize the state of adsorption. Therefore, a solid surface with steep concave and gentle convex is preferable for gas adsorption. The behavior of gas adsorption is affected by the roughness structure. In an extreme case, the gas cannot be adsorbed at all at a sawtooth solid surface no matter how rough it may be because the surface slope is constant and there is no reduction in the roughness factor by the adsorption. This is one of the main reasons that most of the wetting theories[9,10,11] adopt the sawtooth surface model: this model shows no adsorption in equilibrium, so the calculation becomes simple.

The deviation of the apparent contact angle from Wenzel's is observed in experiment[2,3,5]. However, to compare experimental results and theoretical ones, much attention should be paid to the contact angle hysteresis. Owing to the liquid surface tension or viscosity, it takes time for a liquid to settle down into the energetically preferable state when a drop is put on a rough solid surface. The adsorbed gas makes the equilibrium contact angle smaller than Wenzel's. However, gas can be adsorbed transitionally as a hysteresis. In such a non-equilibrium state, the solid surface may show contact angles larger than Wenzel's. Here, the equilibrium contact angle only is considered throughout this paper.

3. THEORY

Let us consider a sessile drop on a sinusoidal rough surface. Figure 1 is a cross section of the system through the origin. For simplicity, to make the shape of a drop axisymmetric, the solid surface is assumed to be circularly symmetrical about the z axis,

$$z = z_0 \cos(2\pi\rho/\rho_0) \tag{3}$$

in cylindrical coordinates (ρ, Ψ, z) and a drop is placed on the origin. The solid surface is assumed to be rigid, insoluble, nonreactive, and homogeneous. The acceleration of gravity is parallel to the $-z$ direction. Here, the drop shape is assumed to be a spherical cap. Distortion from a spherical cap due to gravity is neglected. The effect of gravity on the drop shape was discussed in reference 7 in detail. Only the effect of gravity on the gas adsorption is considered in this paper. The line tension at the three-phase contact is also neglected. As for adsorption, the adsorbed gas-liquid interface under the drop is assumed to be flat as depicted in Figure 1; its length parallel to the ρ axis is d. Moreover, the gas is assumed to freely go into or get out of troughs under the drop, ignoring adsorption or desorption processes themselves.

By preceding assumptions, the configuration of the drop on the rough solid surface with the intrinsic contact angle θ_{eq}, and the period ρ_0 and the amplitude z_0 of the roughness, is described by two parameters, the apparent contact angle ϕ and the interface length d.

At constant temperature and pressure, the difference of the total free energy of the two configurations (ϕ^j, d^j), (ϕ^k, d^k) is given by[12]

$$\Delta F^{jk} = \gamma_{gl}(A_{gl}^k - A_{gl}^j) + \gamma_{sl}(A_{sl}^k - A_{sl}^j) + \gamma_{sg}(A_{sg}^k - A_{sg}^j) + U^k - U^j \tag{4}$$

where U is the potential energy of the drop. A_{gl}, A_{sl}, and A_{sg} are the total areas of the interfaces between the gas-liquid, solid-liquid, and solid-gas phases, respectively. Now, the area of the solid surface, $A_{sl}+A_{sg}$ is constant irrespective of configuration. Substituting this and Eq.1, Eq.4 is rewritten as

$$\Delta F^{jk} = \{\gamma_{gl}(A_{gl}^k - A_{sl}^k \cos\theta_{eq}) + U^k\} - \{\gamma_{gl}(A_{gl}^j - A_{sl}^j \cos\theta_{eq}) + U^j\} \tag{5}$$

Here, let us define the relative free energy F_{rel} at a configuration:

$$F_{rel} = \gamma_{gl}(A_{gl} - A_{sl}\cos\theta_{eq}) + U \tag{6}$$

The detailed procedure to obtain the equilibrium configuration is as follows. For all the possible configurations (ϕ, d), i.e., $\phi = 90 \sim 180°$ and $d = 0 \sim \rho_0$, F_{rel} is calculated numerically by using given values of γ_{gl} and θ_{eq}. The configuration at which F_{rel} takes its minimum value is the equilibrium configuration (ϕ_{eq}, d_{eq}) for a given γ_{gl} and θ_{eq}. In the equation of F_{rel}, the interfacial areas A_{gl} and A_{sl} are calculated from the assumption that the drop shape is a spherical cap. Note that A_{gl} includes the area of the adsorbed gas-liquid interface under the drop when $d \neq 0$. The value of U is calculated by integrating the product of liquid density and the gravitational acceleration g (=9.8 m/s^2) over the volume of the drop. As for the case without gravity, $U = 0$.

The important point is how large the magnitude of the potential energy term of Eq.6 is in comparison with that of the interfacial free energy term. This determines that how strongly the drop configuration is affected by gravity. Note that the interfacial free energy is roughly proportional to the area on the one hand, but the potential energy is proportional to the volume on the other. Therefore, the potential term can be predominant as much as one like for larger and larger liquid volumes and densities. Moreover, the liquid volume removed from the troughs by the adsorption process will be large and the degree of lift-up will be large for larger roughness factors, or larger roughness structures (i.e., large ρ_0 and z_0). These factors also make the potential term large. Here, the effect of gravity on the apparent contact angle is calculated under the plausible conditions in the sessile drop method to make sense.

In the next section, the equilibrium configuration (ϕ_{eq}, d_{eq}) is calculated numerically under the following conditions. Water (density: 1, γ_{gl}=73.48 mN/m)[13] is considered as a drop liquid which is in equilibrium with water-saturated air. The volume of the drop is $(4\pi/3)R_0{}^3$, where R_0=1~5mm. As for the solid surface, two sets are used: ρ_0=50 μ m, z_0=20.7 μ m (which corresponds to r =2) and ρ_0=50 μ m, z_0=47.6 μ m (r =4).

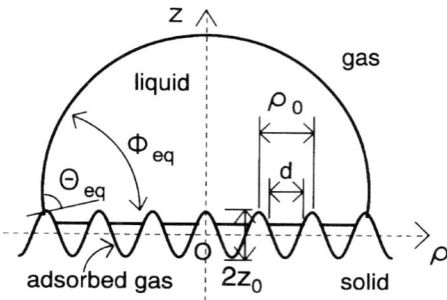

Figure 1 Cross section of a sessile drop on a sinusoidal rough solid surface. The acceleration of gravity is parallel to the - z direction.

4. RESULTS AND DISCUSSION

Figure 2 shows the apparent contact angle ϕ_{eq} of a drop of R_0=5 mm when θ_{eq} is increasing from $90°$ to $180°$. The solid curves indicate the angles with gravity. The dashed curves are those without gravity. On the whole, ϕ_{eq} increases with rising θ_{eq}. At the beginning, ϕ_{eq} is proportional to θ_{eq} in the cosine form. There is no adsorption (d_{eq}=0) in these straight parts and their proportional constants are 2 when r =2, and 4 when r =4. Within this range, the solid lines agree with the dashed lines. Thus, the apparent contact angle without gas adsorption does not depend on gravity.

At a certain value of θ_{eq}, the gas adsorption occurs ($d_{eq}\neq0$) and ϕ_{eq} begins to deviate from Wenzel's. Above that value of θ_{eq}, $\cos\phi_{eq}$ is no more proportional to $\cos\theta_{eq}$. Here, the degree of deviation with gravity is smaller than that without gravity. This is because the gas adsorption is suppressed by gravity and the solid-liquid interfacial tension reduced by adsorbed gas is enhanced. Then, ϕ_{eq} is raised by gravity. The apparent contact angle under the gas adsorption does depend on gravity. Here, when r =2, the difference in ϕ_{eq} between with and without gravity is less than $1°$. Thus, it is negligible in comparison with the uncertainty(about $1°$) in the contact angle measurement. However, the difference for r =4 is not negligible.

For r =4, drop size dependence of the apparent contact angle is calculated as shown in Figure 3. At θ_{eq}=106° (or $\cos\theta_{eq}$=-0.276), gas is adsorbed whether with or without gravity. ϕ_{eq} without gravity (open triangles) is constant ($\fallingdotseq$157°) despite the drop size. On the other hand, ϕ_{eq} with gravity (open circles) increases gradually as R_0 increases. For $R_0 \geqq 2$ mm, it becomes more than $1°$ larger than that without gravity. This is a drop-size effect. It is comparable in magnitude to other drop-size effects like correction of a contact angle by line tension. Note that, however, this gravity effect becomes predominant for larger drops. In contrast, correction by line tension becomes negligible for larger drops[1,5]. Note also that the roughness scale of the solid surface is important. For example, if we use a one-tenth scale, i.e., $\rho_0 = 5 \mu$ m and $z_0 = 4.76 \mu$ m, which remains $r = 4$, the liquid volume removed from the troughs by the adsorption process becomes small and correction by gravity is as small as $0.1°$ (solid circles), which is negligible as shown in Figure 3. The dependence of a contact angle on gravity is affected by the roughness scale as well as the value of r.

In the foregoing considerations, ϕ_{eq} for $\theta_{eq} > 90°$ only is paid attention. Under that condition, the adsorption of gas at the solid-liquid interface is relevant. However, for $\theta_{eq} < 90°$, opposite phenomenon is relevant: liquid adsorbs at the solid-gas interface and the adsorbed liquid reduces the solid-gas interfacial tension. ϕ_{eq} approaches $0°$ very slowly, not straight in the cosine form predicted by Eq.2 when θ_{eq} is decreasing from $90°$ to $0°$. This time the gravity effect is negligible because the mass of gas lifted up is small. The previous adsorption theories predict the deviations from Wenzel's due to gas and liquid adsorption in a symmetrical way (see Figure 1 of reference 2). In contrast, the present theory predicts that the deviations from Wenzel's are asymmetrical in the setup of the sessile drop method because that of the gas adsorption is suppressed by gravity. There has been no report that an equilibrium contact angle is affected by gravity so far. Nevertheless, asymmetrical deviations from Wenzel's were observed[2,3,5]. Until now, these observations have been simply explained in terms of a contact angle hysteresis. However, the present calculations show that the gravity dependence through gas adsorption is one of the causes of those asymmetrical deviations.

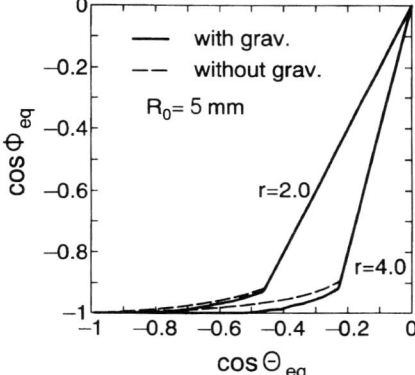

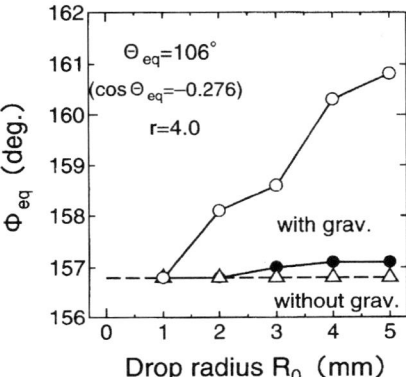

Figure 2 Relations between the apparent contact angle ϕ_{eq} and the intrinsic contact angle θ_{eq}; gravity is considered(solid line) or ignored(dashed line) in calculation.

Figure 3 Drop size dependence of ϕ_{eq}. $\bigcirc$: the calculated result with gravity when $\rho_0 = 50 \mu$ m and $z_0 = 47.6 \mu$ m. $\triangle$: result without gravity. $\bullet$: result with gravity when $\rho_0 = 5 \mu$ m and $z_0 = 4.76 \mu$ m.

5. SUMMARY

We have shown that, by calculating the wetting behavior in the setup of the sessile drop method, the apparent contact angle of a drop on the rough solid surface is raised by gravity under the condition of gas adsorption. Its dependence is determined by the liquid volume, the density and the roughness structure of a solid surface. The asymmetrical deviations of the contact angle from Wenzel's[2,3,5] are explained in terms of this gravity effect.

When measuring a contact angle, liquid drops with diameters of several mm are used preferably in order to reduce undesirable factors, such as the three-phase line tension at the drop perimeter, volume reduction of a liquid due to vaporization, and defects of a solid surface[14]. However, it was found from this study that if a large drop is used under the condition of gas adsorption, the effect of gravity has to be considered when analyzing the measured contact angle. Until now, large contact angles exceeding $150°$, which cause gas adsorption in the equilibrium state, are seldom encountered. Recently, super liquid-repellent solid surfaces, for example, which show more than $170°$ against water, are being developed[2,3]. The gravity effect discussed in this paper is important for a wetting analysis of such surfaces.

REFERENCES

1. J.Gaydos and A.W.Neumann, "Applied Surface Thermodynamics(Surfactant science series 63)" (A.W.Neumann and J.K.Spelt, Eds.),p.169.Marcel Dekker,1996.
2. T.Onda, S.Shibuichi, N.Satoh and K.Tsujii, *Langmuir* **12**(1996)2125.
3. S.Shibuichi, T.Onda, N.Satoh, K.Tsujii, *J.Phys.Chem.* **100**(1996)19512.
4. N.T.Correia, J.J.Moura Ramos, B.J.V.Saramago and C.G.Calado, *J.Colloid Interface Sci.* **189**(1997)361.
5. A.W.Adamson, "Physical Chemistry of Surfaces" 5th ed. John Wiley & Sons, New York, 1990.
6. R.E.Johnson, Jr. and R.H.Dettre, "Wettability (Surfactant science series 49)" (J.C.Berg,Ed.),p.1.Marcel Dekker, New York,1993.
7. H.Fujii and H.Nakae, *Philosophical Magazine* **A72**(1995)1505.
8. F.K.Skinner, Y.Rotenberg and A.W.Neumann, *J.Colloid Interface Sci.* **130**(1989)25.
9. D.Li and A.W.Neumann, "Applied Surface Thermodynamics(Surfactant science series 63)" (A.W.Neumann and J.K.Spelt, Eds.),p.109.Marcel Dekker,1996.
10. R.D.Hazlett, *J.Colloid Interface Sci.* **137**(1990)527.
11. J.D.Eick, R.J.Good and A.W.Neumann,*J.Colloid Interface Sci.* **53**(1975)235.
12. R.E.Johnson, Jr. and R.H.Dettre, "Contact Angle, Wettability, and Adhesion(Advances in Chemistry Series 43)" (F.M.Fowkes, Ed.),p.112. American Chemical Soc., Washington,D.C.,1964.
13. "CRC handbook of Chemistry and Physics 74th ed." (D.R.Lide, Ed.), CRC Press, Boca Raton,1993-94.
14. J.K.Spelt and E.I.Vargha-Butler, "Applied Surface Thermodynamics(Surfactant science series 63)" (A.W.Neumann and J.K.Spelt, Eds.),p.379.Marcel Dekker, New York,1996.

HYDROCOLLOIDS – PART 2
Edited by K. Nishinari
2000 Elsevier Science B.V.

107

Gel-like rheological behavior of mesophases in photoreactive azo-dye/water/KCl   systems

T. Imae,[a*] Y. Ikeda,[a] I. Spring,[b] C. Thunig,[b] and G. Platz[b]

[a]Department of Chemistry, Faculty of Science, Nagoya University, Nagoya 464-8602, Japan

[b]Institute für Physikalische Chemie I, Universität Bayreuth, Bayreuth 95440, Germany

Phase diagram has been drawn for ternary systems of azo-dye Levafix Goldgelb E-G (F9)/water/KCl. The ternary systems of dilute F9 concentrations displayed three isotropic (L_1, sb, and $L_{3m}Bl$) phases, an anisotropic (L_α) phase, and an inhomogeneous (L_α + qlc) phase. The rheological behavior has been investigated by oscillatory measurements and shear stress-strain examinations. In addition of viscoelastic behavior, apparent yield stress was observed for $L_{3m}Bl$ and L_α phases. While the rheological behavior was independent of KCl concentration, it strongly depended on the F9 concentration. The storage modulus, loss modulus, and apparent yield stress were scaled against F9 concentration over $L_{3m}Bl$ and L_α phases, indicating the existence of gel-like structure.

1.   INTRODUCTION

Dyes are useful as dyestuffs, paints, and coloring materials on various industrial applications. Especially, azo-dyes are photoreactive compounds which induce photochromic cis-trans isomerization. Many investigations have been carried out for dispersions in organic solvents or for Langmuir monolayers and Langmuir-Blodgett films of water-insoluble azo-dyes, azo-dye hybrids, and azo-dye polymers [1-3]. On the other hand, amphiphiles including azo-dye as a rigid spacer form bilayers in aqueous dispersions [4] and supramolecular assemblies in aqueous methanol solutions [5-7]. The amphiphilic azo-dye in aqueous methanol solutions forms rodlike swarms at low amphiphilic azo-dye concentrations and liquid crystals at high amphiphilic azo-dye concentrations [7].

Although azo-dye Levafix Goldgelb E-G (F9) is not a typical amphiphile, it has hydrophilic ionic groups and, therefore, is water-soluble. Liquid crystalline mesophases are observed at F9 concentrations above 15 wt% in aqueous solutions and at very low F9

---

[*] To whom correspondence should be addressed.

concentrations in the ternary F9/water/methanol systems. The ternary systems have been investigated by polarized optical microscopic, electon microscopic, and small-angle neutron scattering methods [8]. The formation and structures of supramolecular assemblies have been discussed in relation to methanol content in isotropic and anisotropic phases. Stripe-like assemblies and their piles are constructed in the phase of low methanol content. On the other hand, fibrils having linear molecular arrangement and their bundles are formed at high methanol content. Moreover, liquid crystals in anisotropic gels consist of organization of stripes or fibrils. The photo-induced optical anisotropy has been investigated for aqueous F9 systems which form chromonic mesophases upon addition of methanol or NaCl [9].

In the present work, the phase diagram for ternary systems of F9/water/KCl is presented, and the rheological properties of mesophases in the ternary systems are investigated. The relation of mesophases and rheological behaviors is discussed.

Chart 1

$SO_3^-Na^+$ $SO_3^-Na^+$ HO $N=N$ Cl N C=O N-H COO$^-$Na$^+$

2. EXPERIMENTAL SECTION

A sample of F9 was donated by Hoechst AG, Germany and purified: An aqueous 6 wt% F9 solution was filtered through a glass filter and freeze-dried. The product was refluxed in ethanol for one hour. The methanol solution was stored overnight at 5 °C. The resultant orange precipitates were filtered and dried in vacuo. KCl is a commercial product. MiliQ water was used for the preparation of the ternary systems. The mixtures prepared by an unit of wt (w/w) % were incubated at 25 °C for 2 weeks. The phase diagram was determined visually and from the polarized microscopic observation.

Microscopic observation was carried out on a Nikon OPTIPHOT-2 optical microscope equipped with FUJIX HC-300i digital camera and Mitsubishi CP700DSA color printer. Rheological measurements were performed with a Bohlin CS 10 stress-controlled rheometer and an OCRD oscillating capillarity and densitometer from Anton Paar K. G., Graz, Austria. The viscoelastic properties were determined by oscillatory measurements from 0.001 Hz to 10 Hz. For shear stress-strain measurements, the Bohlin rheometer with a cone-plate system was used. Yield stress values were obtained by plotting the strain as a function of the stress. The sample solutions were stored after preparation for one day at constant temperature. The measurements were started one hour after filling the sample solutions in the measuring devices.

3. RESULTS AND DISCUSSION

Figure 1 shows the phase diagram at 25 ℃ for ternary F9/water/KCl system of F9 concentrations (C_{F9}) less than 10 wt% and KCl concentrations (C_{KCl}) less than 20 wt%. At low KCl concentrations below 3 - 4 wt%, the solutions of lower F9 concentrations were fluid and isotropic (L_1), as expected. Those of higher F9 concentrations were also fluid and isotropic at stationary state but induced birefringence under shear. This phase is called sb. The sb phase was also observed at high KCl concentrations above 3 - 4 wt% and low F9 concentrations. The same phase has been reported even for the F9/water/methanol and F9/water/NaCl systems [9]. When the F9 concentration was increased for solutions above 3 - 4 wt% KCl concentration, the viscous isotropic solutions exhibited shear-induced birefringent waves, between crossed polarizers, which looked rainbow-like, as given in Fig. 2(left). This phase, which is noted as

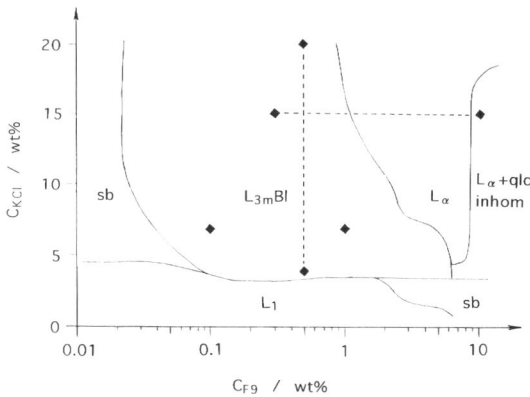

Figure 1. Phase diagram of the ternary F9/water/KCl systems at 25 ℃. ◆ and ◆---◆ represents the points and regions measured rheology, respectively.

L_1 = isotropic,
sb = isotropic with shear induced birefringence without shear waves,
$L_{3m}Bl$ = isotropic with shear induced birefringence and shear waves,
L_α = stationary birefringent,
L_α + qlc inhom = L_α + quasi-liquid crystalline inhomogeneous.

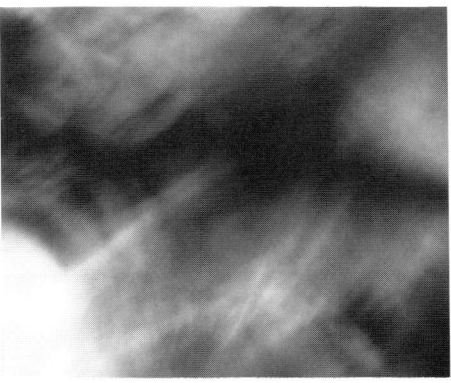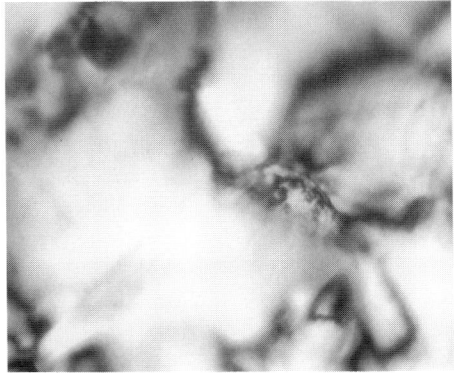

Figure 2. Polarized optical microscopic photographs of the ternary F9/water/15 wt% KCl systems. F9 concentration: left, 0.5 wt% (after shearing); right, 6 wt%.

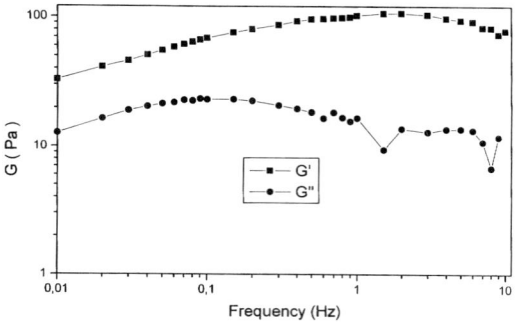

Figure 3. Storage modulus G' and loss modulus G" as a function of the frequency
for a ternary F9(1 wt%)/water/KCl(7 wt%) system at 25 ℃.

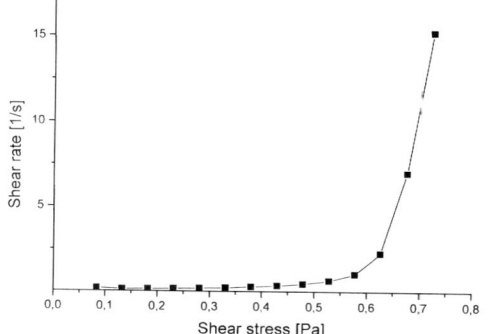

Figure 4. Shear rate as a function of the shear stress
for a ternary F9(0.1 wt%)/water/KCl(7 wt%) system at 25 ℃.

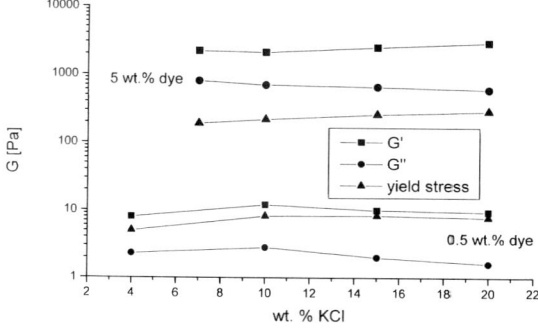

Figure 5. Storage modulus G' and loss modulus G" at 0.1 Hz and yield stress as a function of
the KCl concentration for the ternary F9(0.5 and 5 wt%)/water/KCl systems at 25 ℃.

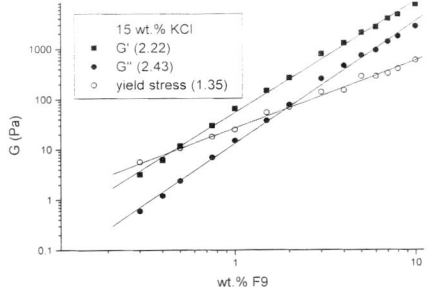

Figure 6. Storage modulus G' and loss modulus G" at 0.1 Hz and yield stress as a function of the F9 concentration for the ternary F9(0.5 and 5 wt%)/water/KCl systems at 25 °C.

L$_{3m}$Bl, has been reported for the F9/water/methanol systems but not for the F9/water/NaCl systems. The jelly-like L$_\alpha$ gel phase was observed for the systems of 1 - 9 wt% F9 concentrations at KCl concentrations above 3 - 4 wt% and displayed stationary birefringence between crossed polarizers. As seen in Fig. 2(right), the L$_\alpha$ phase showed no Schlieren (herring bone and mosaic) texture in contrast to the corresponding phase in the ternary F9/water/methanol system [8]. The L$_\alpha$ phase displayed uniform birefringent color over wide range of vision on polarized optical microscope. At high F9 region above 9 wt%, the phase was inhomogeneous due to the coexistence of L$_\alpha$ and quasi-liquid crystalline (qlc) phases. The qlc phase exhibited properties between pure crystal and liquid crystal. The phase appeared to be crystal but behaved as liquid crystal under high shear forces, as observed in the F9/water/NaCl systems [9].

The rheological properties were estimated by oscillatory measurements and shear stress-strain investigations. Figure 3 elucidates the characteristic rheological behavior of an aqueous F9 solution containing considerable amounts of KCl. The storage modulus G' was significantly higher than the loss modulus G" for all measured frequency > 0.01 s^{-1}. There were weak dependence of moduli on the frequency: The modulus G' increaseed with increasing frequencies from 0.01 to 1 Hz, and the modulus G" displayed maximum around 0.1 Hz.

The results of the oscillatory measurements indicated that the systems should display considerable yield stress. For the direct estimation of the yield stress values, stress-strain measurements were carried out. Typical curve is shown in Fig. 4. An abrupt transition from hardly flowing to low viscous flow behavior was found. The shear stress value at this transition can be defined as apparent yield stress. The value was 0.65 Pa for the ternary F9(0.1 wt%)/water/KCl(7 wt%) system at 25 °C.

The flow behavior below the apparent yield stress has been investigated for the dye system of pseudoisocyaninchloride in water [10], where structural relaxation times were found in the order of minutes. It can be expected that the system in the present work displays similar long-time behavior. From this expectation, it follows that the yield stress of the F9/water/KCl system may depend on the measured time scale. In all cases examined here, there can be extremely long structural relaxation time, which was not detected because the measuring time was too short. In spite of these restrictions, it can be concluded without any doubt that the rheological behavior of the F9/water/KCl system is determined by gel-like F9 aggregates, which are present even at extremely low F9 concentrations.

Figure 5 shows moduli G' and G" at 0.1 Hz and apparent yield stress as a function of the KCl concentration for the ternary F9(0.5 wt%)/water/KCl systems at 25 °C. Above 4 wt% KCl, there remains only a weak dependence of the rheological data on further increasing KCl concentration. This result indicates that the gel structure is not essentially influenced by increasing KCl concentration above concentrations where the gel was just formed, that is, the transition from L_1 phase to $L_{3m}Bl$ phase occurred. Similar behavior was observed even for the L_α region in the ternary F9(5 wt%)/water/KCl systems at 25 °C, as given in Fig. 5.

The moduli G' and G", and apparent yield stress of the F9 network were strongly dependent on the F9 concentration, as seen in Fig. 6. A single scaling law represents the rheological behavior of G', G", and apparent yield stress as a function of the F9 concentration over the $L_{3m}Bl$ and L_α phases. The storage modulus scales with the F9 concentration to 2.22. This is exactly in the same value which was found for the dye system of pseudoisocyaninchloride in water [10]. The loss modulus and the apparent yield stress also display single scaling laws with exponent of 2.43 and 1.35, respectively. The rheological scaling behavior corresponds well to that of a network of entangled macromolecules [11,12] and to viscoelastic surfactant systems [13]. In solutions of macromolecules, one expects scaling exponents for G' between 1.9 to 2.4. In viscoelastic surfactant solutions, the values around 2 are frequently found.

The rheological behavior indicated that the same gel structure should be conteined in the different phase regions of the F9/water/KCl systems. It is concluded that flow birefringence and permanent birefringence are dominated by the concentraticn-dependent dynamic behavior of this gel state.

REFERENCES

1. D. Mobius, *Ber. Bunsenges. Phys. Chem., 82* (1978) 848.
2. J. Heesemann, *J. Am. Chem. Soc., 102* (1980) 2167, 2176.
3. N. Katayama, Y. Ozaki, T. Seki, T. Tamaki, and K. Iriyama, *Langmuir, 10* (1994) 1898.
4. T. Kunitake, N. Nakashima, M. Shimomura, Y. Okahata, M. Kano, and T. Ogawa, *J. Am. Chem. Soc., 102* (1980) 6644.
5. T. Imae, C. Mori, and S. Ikeda, *J. Chem. Soc., Faraday Trans. 1, 78* (1982) 1359, 1369.
6. T. Imae, S. Ikeda, and K. Itoh, *J. Chem. Soc., Faraday Trans. 1, 79* (1983) 2843.
7. T. Imae and S. Ikeda, *Mol. Cryst. Liq. Cryst., 101* (1983) 155.
8. T. Imae, I. Gagel, C. Thunig, G. Platz, T. Iwamoto, and K. Funayama, *Langmuir, 14* (1998) 2197.
9. C. Hahn, I. Gagel, C. Thunig, G. Platz, and A. Wokaun, *Langmuir, 14* (1998) 6871.
10. H. Rehage, G. Platz, B. Struller, and C. Thunig, *Tenside, Surf. Det. 33* (1996) 242.
11. M. S. Turner, M. E. Cates, *Langmuir, 7* (1991) 1590.
12. R. Granek, M. J. Cates, *J. Chem. Phys., 96* (1992) 4758.
13. H. Hoffmann, in Structure and Flow in Surfactant Solutions, ed. by Herb, A. C., Prud'homme, R. K., ACS Symposium Series 578, American Chemical Society, Washington; D. C. 1994.

HYDROCOLLOIDS – PART 2
Edited by K. Nishinari
2000 Elsevier Science B.V.

Functional interaction between membranous surface and colloidal inside of yeast cells as reflected by an isopreniod-promoted mitochondrial generation of reactive oxygen species

T. Tanaka, H. Nakamura K. Machida and M. Taniguchi

Department of Bio- and Geoscience, Graduate School of Science, Osaka City University , 3-3-138 Sugimoto, Sumiyoshi-ku, Osaka 558-8585, Japan

The mechanism of isopreniod farnesol (FOH)-promoted generation of reactive oxygen species (ROS) was studied in terms of signal transduction between membranous surface and colloidal inside of the yeast cells. FOH-treated cells were characterized with 5-8 fold increase in the level of ROS generation at the initial 30-min incubation while none of ROS generating response was observed against other isoprenoid compounds like geraniol, geranylgeraniol and squalene. Dependence of FOH-induced growth inhibition on such an oxidative stress was confirmed by its protection with the coexistence of antioxidant such as α-tocopherol (α-TOH), probucol and N-acetylcysteine (NAC). To assess the role of mitochondrial function in FOH-induced ROS generation, the FOH-sensitivity of a $[rho^0]$ petite mutant was compared with that of the wild-type grande strain. FOH could accelerate ROS generation only in cells of wild-type strain but not in those of the respiration-deficient petite mutant, representing the role of mitochondrial electron transport chain as its origin. Among the respiratory inhibitors, ROS generation could be effectively canceled with myxothiazol which inhibits oxidation of ubiquinol to ubisemiquinone radical by the Rieske iron-sulfur center of complex III but not with antimyicin A, an inhibitor of electron transport functional for further oxidation of ubisemiquinone radical to ubiquinone in Q-cycle of complex III. Cellular oxygen consumption was inhibited immediately upon extracellular addition of FOH whereas FOH and its possible metabolites failed to directly inhibit any of oxidase activities detected with isolated mitochondrial preparation. These findings strongly suggested that FOH did not directly interact but indirectly interacted with mitochondrial electron transport chain via a mechanism of signal transduction.

1. INTRODUCTION

Farnesol (FOH) is an isoprenoid alcohol which may be endogenouly generated within the cells by enzymatic dephosphorylation of farnesyl pyrophosphate (FPP), an intermediate of the metabolic pathway to yield sterols and other isoprenoid compounds from mevalonate [1, 2]. FPP plays another important role as a precursor of protein prenylation such as post translational modification of oncogenic RAS proteins and other GTP-binding proteins [3]. Recently, FOH is attracting much attention since it causes apoptotic cell death of human acute leukemia CEM-C1 cells [4] and HL-60 cells [5]. Interference with a phosphatidylinositol-type signaling has been proposed to be a cause of apoptosis in FOH-treated mammalian cells. In our previous study, FOH was found to exhibit a static growth inhibitory effect on the yeast $Saccharomyces cerevisiae$ in which intracellular diacylglycerol (DAG) level was significantly decreased with accompanying down regulation of cell cycle gene expression [6]. Mammalian and yeast cells may share a common mechanism in their responses to FOH regardless of whether these organisms are subjected to a fatal or static damage with it.

Reactive oxygen species (ROS) are highly toxic oxidants including hydrogen peroxide, superoxide anion and hydroxyl radical which are inevitably produced to a certain extent under aerobic condition. ROS generation is remarkably enhanced in K-ras-transformed murine cells with the inhibitor of protein farnesylation such as (α-hydroxyfarnesyl) phosphonic acid and with lovastatin which should also affect this reaction by inhibiting 3-hydroxy 3-methylglutaryl

CoA reductase [7]. However, no evidence has been obtained to elucidate the mechanism of ROS generation under the condition where the protein farnesylation or the corresponding mevalonate biosynthetic reaction is inhibited. It is highly probable that FOH-induced events depend on oxidative stress in both yeast and mammalian cells since FOH can ultimately participate in the reaction of protein farnesylation. In this study, the mechanism of FOH-induced growth inhibition of *S. cerevisiae* was studied in terms of its promotive effect on ROS generation. Mitochondrial electron transport chain was considered as a target to be inhibited for ROS generation by FOH. We hereby consider the possibility that FOH can indirectly inhibit the mitochondrial function *via* interference with a phosphatidylinositol-type signaling.

2. MATERIALS AND METHODS

Yeast strains and media. S. cerevisiae strains used in this study were X2180-1A(*MATa*) (grande) and its isogenic [*rho*°] petite mutants which had been generated by the treatment with ethidium bromide [8]. Unless stated otherwise, the yeast cells were grown overnight in YPD medium with vigorous shaking and were inoculated into freshly prepared medium to give an initial cell density of approximately 10^7 cells/ml. Cells were then grown with or without each effector or inhibitor with vigorous shaking and portions were withdrawn at various time intervals to measure the cell density at A_{610}. The cell suspension (10^7 cells/ml) gave an A_{610} value of approximately 1.0.

Measurement of ROS production. Cellular ROS production was examined by the method depending on intracellular deacylation and oxidation of 2', 7'-dichlorodihydrofluorescein diacetate (DCFH-DA) to the fluorescent compound 2', 7'-dichlorofluorescein (DCF). After preincubation of the yeast cells (10^7 cells/ml) in YPD medium with 40 µM DCFH-DA at 30°C for 60 min, the cell suspensions (1.0 ml) were withdrawn and further treated with each chemical for the indicated time, being washed and resuspended with 100 µl of PBS. Fluorescence intensity of the cell suspension (100 µl) containing 10^7 cells was read using a Cytoflow 2300 fluorescence spectrophotometer (Milipore Co.) with excitation at 480 nm and emission at 530 nm. Arbitrary unit was directly given by the fluorescence intensity.

Preparation of mitochondria and cytosol fraction. Cells of strain X2180-1A were aerobically grown in 10 l of SSM medium at 30°C for 15 hr. Mitochondria were isolated from the cell lysate which had been prepared by the enzymatic digestion with Zymolyase 20 T (Seikagaku Kogyo Co.) according to the method of Glick and Pon [9]. Cells from overnight culture were further incubated in 100 ml of YPD medium (10^7 cells/ml) containing 200 µM FOH for 15 min, being collected, suspended in 1.0 ml of cold lysis buffer (0.3 M D-sorbitol, 0.1M NaCl, 5 mM MgCl$_2$, and 10 mM Tris-HCl, 1mM PMSF pH7.4) at the cell density of 10^9/ml and disrupted by repeated vortex with glass beads [10].

Assay of cellular and mitochondrial respiratory activity. Cells of strain X2180-1A were grown in SSM medium with vigorous shaking at 30°C for 15 hr, being collected and suspended in HEPES buffer (pH 7.4) containing 50 mM glucose at the cell density of 10^7 cells/ml. After preincubating the cell suspension with shaking at 30°C for 10 min, the respiratory activity of yeast cells was measured polarographically with an oxygen electrode (Model 100, Rank Brothers, Ltd., Cambridge, UK). The rate of oxygen consumption by isolated mitochondoria was also measured polarographically in 2 mM HEPES buffer (pH 7.4) containing 0.6 M mannitol, 1 mM KCl, 2 mM MgCl$_2$, 1 mM EDTA and each respiratory substrate. The final mitochondrial protein concentration was adjusted to 250 µg/ml and DNP was added at 40 µM whenever needed.

Chemicals. The following chemicals were purchased from Sigma: geraniol (GOH), FOH, farnesylacetate, FPP, geranylgeraniol (GGOH), antimycin A, myxothiazol, thenoyltrifluoroacetone (TTFA), rotenone, α-tocopherolacetate, probucol, N-acetylcystein (NAC), TMPD and 1-oleoyl, 2-acetyl-*sn*-glycerol (OAG) as a membrane permeable DAG analog. DCFH-DA was a product of Molecular Probe. A stock solution of each chemical was routinely prepared in either phosphate-buffered saline (PBS), ethanol, acetone or dimethylformamide owing to its solubility. Farnesoic acid and farnesal were kindly provided by Dr. G. Asanuma (Kurare, Co., Japan). The other chemicals were of analytical reagent grade.

3. RESULTS AND DISCUSSION

3.1.Correlation between FOH-induced growth inhibition and ROS generation.

Table 1. Dose-depenednt growth inhibition and induction of ROS production by FOH and the relating isoprenoid compounds in *S. cerevisiae.*.

Addition	Concentration (μM)	Relative cell growth (%)	ROS production (arbitrary units)
None	-	100	824 ± 122
Geraniol	200	91	667 ± 84
Farnesol	6.25	104	860 ± 132
	12.5	31	1,146 ± 156
	25	5	5,280 ± 345
	50	0	5,703 ± 192
	100	0	7,440 ± 310
Geranylgeraniol	200	98	626 ± 98

Table. 2. Protective effects of various anti-oxidants on FOH-induced growth inhibition and ROS production in *S. cerevisiae.*.

Addition	Conc. of anti-oxidant (μM)	Relative cell growth (%)	ROS production (arbitrary units)
None	-	100	860 ± 124
FOH	-	1	5686 ± 355
FOH + α-TOH	6.25	10	2663 ± 133
	12.5	37	1367 ± 122
	25	85	769 ± 89
FOH + Probucol	6.25	6	2556 ± 244
	12.5	25	1567 ± 136
	25	90	669 ± 78
FOH + NAC	5×10^3	3	3512 ± 233
	1×10^4	41	1194 ± 144
FOH + L-Ascorbate	2×10^2	1	5488 ± 346
	1×10^4	3	3090 ± 322

As summarized in Table 1, FOH significantly promoted cellular ROS generation in a dose-dependent manner, representing its clear relation to the growth inhibitory effect. ROS generation was kept almost at the control level in the medium with other isoprenoid compound showing none of growth inhibitory effect even at a quite high concentration (200 μM). We next examined the protective effects of various types of anti-oxidants against FOH-induced growth inhibition as well as ROS generation, and found a clear correlation between protection of these events. As summarized in Table 2, the FOH-induced events could be mostly canceled with 25 μM α-tocopherol (α-TOH), a naturally occurring lipophilic anti-oxidant which can easily penetrate the plasma membrane and protect free and membranous lipids against oxidative damage [11]. Among the synthetic anti-oxidants, probucol was as effective as α-TOH for protection of FOH-induced growth inhibition as is the case with its protective effect on generation of hydrogen peroxide in macrophages [12]. NAC was only partly effective even at the concentration of 10 mM. These findings revealed oxidative stress due to ROS generation inside the cytoplasmic membrane as to be a primary cause of FOH-induced growth inhibition in the yeast cells.

3.2. Resistance of respiration-deficient petite mutant against FOH-induced events.

To assess the role of mitochondrial function in FOH-induced ROS generation, FOH-sensitivity of a $[rho^0]$ petite mutant was compared with that of the wild-type grande strain. As shown in Fig. 1 B, the mutant cells could grow in YPD medium depending on fermentation equally with or without 200 μM FOH. In accordance with the fact, the cellular ROS generation was kept at the control level even after 30-min incubation of the mutant cells with FOH (Fig. 1 A). Electron transport chain could be a target of FOH since $[rho^0]$ petite mutant generally lacks cytochrome b of complex III and various subunits of cytochrome c oxidase in addition to F_1F_0-ATPase.

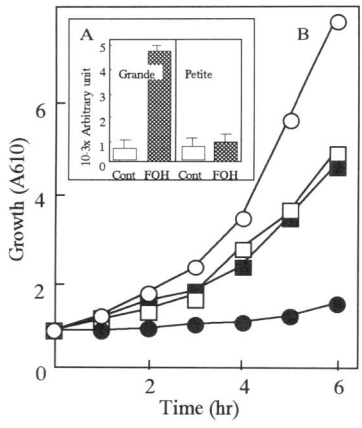

Fig.1. Effects of respiratory competence on ROS production (A) and FOH-induced growth inhibition (B). Cells of parental grande strain (○, ●) and [rho0] petite mutant (□, ■) were grown in YPD medium with (●, ■) or without (○, □) FOH at 30°C for 6 hr. Values are means ± S.D. (n = 4).

3.3. Effect of blocking electron flow at complex I, II, III and IV of respiratory chain on the FOH-induced ROS generation.

Wild-type cells had been pretreated with respiratory inhibitor(s) for 10 min to identify the critical site for ROS generation during the following incubation with 25 μM FOH. Each respiratory inhibitor was used at the concentration entirely inhibiting the cellular oxygen consumption but not influencing the cell growth on fermentation in YPD medium. As shown in Fig. 2, ROS generation was reduced by almost 50% when yeast cells had been pretreated with rotenone which specifically inhibits complex I and TTFA, a specific inhibitor of complex II [13]. ROS generation and growth inhibition could be protected with neither antimycin A which inhibits cytochrome reductase of complex III nor KCN, a typical inhibitor of complex IV [14]. Among various respiratory inhibitors tested, the cellular ROS generation was most effectively canceled with myxothiazol which inhibits oxidation of ubiquinol to ubisemiquinone radical via the Rieske iron-sulfur center of complex III [15]. Under the condition with myxothiazol, the yeast cells treated with FOH (25 μM) could normally grow on fermentation in YPD medium in which the relative cell growth (A_{610} = 3.9) was around 50% of the control level at 6 hr. This strongly supported that the protection of ROS generation by myxothiazol depended only on its role as a respiratory inhibitor.

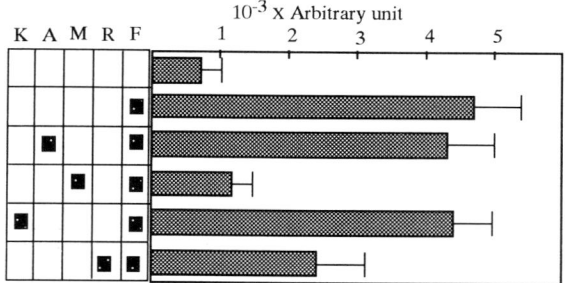

Fig. 2. Protective effects of blocking electron transport chain in FOH-induced ROS generation. Prior to the addition of 25 μM FOH (F), cells had been pretreated with a mixture of 50 μM rotenone and 1 mM TTFA (R), 20 μM antimycin A (A), 30 μM myxothiazole (M), 2.5 mM KCN (K), respectively, for 10 min. Bars are means ± S.D. (n = 4).

3.4. Effects of FOH on cellular oxygen consumption and mitochondrial oxidase activities.

The inhibitory effects of FOH on mitochondrial oxidase activities were examined by measuring the rate of oxygen consumption by isolated mitochondrial preparation. Rotenone, antimycin A and KCN inhibited the corresponding oxidase activities as expected. It was surprising that none of significant inhibition was observed even with 200 μM FOH against the oxygen-consuming reactions catalyzed by NADH oxidase, succinate oxidase or cytochrome c oxidase. These oxidase activities were not inhibited with any of the most probable metabolites of FOH such as FPP, farnesal and farnesoic acid. These results revealed that FOH-induced ROS generation or inhibition of oxygen consumption was not due to direct inhibition of the mitochondrial electron transport at least by FOH or its metabolites.

3.5. Involvement of a phosphatidylinositol-type signaling in FOH-induced ROS generation.

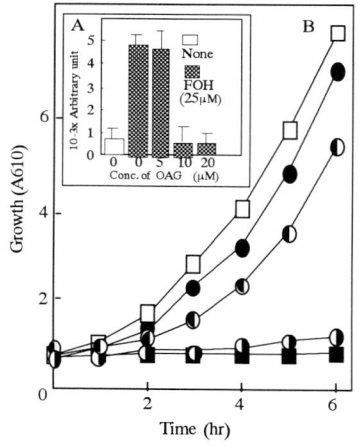

Fig.3. Protective effects of OAG on ROS production (A) and FOH-induced growth inhibition (B). In (B), Cells (107cells/ml) were grown in YPD medium with (■) or without (□) 25 μM FOH. FOH was further added at 25 μM to YPD medium containing OAG either at 5 μM (◯), 10 μM (◯) or 20 μM (●). Control assay was run without any other ingredient in YPD medium. Values are means ± S.D. (n = 4).

As shown in Fig. 3, FOH-induced ROS generation and growth inhibition was apparently diminished to the control level with the coexistence of 10 μM OAG. This agreed with our previous finding that the PKC activator restored FOH-induced growth inhibition of *S. cerevisiae* cells in which endogenous DAG level had been drastically decreased immediately upon incubation with FOH [6]. These findings supported the idea that FOH inhibited the mitochondrial electron transport *via* interference with a phosphatidylinositol-type signaling without causing direct inhibitory effect on mitochondria itself.

REFERENCES

1. D.C. Crick, D.A. Andres and C.J.Waechter, Biochem. Biophys. Res. Commun., 237 (1997) 483.
2. J.L. Goldstein and M.S.Brown, Nature 343 (1990) 425.
3. J.A. Glomset, M.H. Gelb and C.C.Farnsworth, Trends Biochem. Sci., 15 (1990) 139.
4. J.S. Hang, C.M. Goldner, E.M. Yazlovitskaya, P.A. Voziyan and G. Melnykovych, Biochim. Biophys. Acta., 1223 (1994) 133.
5. H. Ohizumi, Y.Masuda, S.Nakajo, I.Sakai, S.Ohsawa and K.Nakaya, J. Biohem., 117 (1995) 11.
6. K. Machida, T.Tanaka, Y.Yano, S.Otani, and M.Taniguchi, Submitted for publication (1998).
7. M. Santillo, P.Mondola, A.Gioielli, R.Seru, S.Iossa, T.Annella, M.Vitale and M.Bifulco, Biochem. Biophys. Res. Commun. 229 (1996) 739.
8. T.D.Fox, L.S.Folley, J.J.Mulero, T.W.McMullin, P.E.Thorsness, L.O.Hedin and M.C.Costanzo, Methods Enzymol., 194 (1991) 149-165.
9. B.S.Glick and L.A.Pon, Methods Enzymol., 260 (1995) 213.
10. J.D.Fishbein, R.T.Dobrowsky, A.Bielawska, S.Garrett, and Y.A.Hannun, J. Biol. Chem. 268 (1993) 9255.
11. M.P.Barroso, C.Gomez-Diaz, G.Lopez-Lluch, M.M.Malagon, F.L.Crane and P.Navas, Arch. Biochem. Biophys. 3432 (1997) 243.
12. M.Fukuda, H.Ikegami., Y.Kawaguchi., T.Sano, and T.Ogihara, Biochem. Biophys. Res. Commun., 299 (1995) 953.
13. R.R.Ramsay, B.A. Ackrell, C.J.Coles, T.P.Singer, G.A.White and G.D.Thorn. Proc. Natl. Acad. Sci. USA 78 (1981) 825.
14. E.C. Slater, Methods Enzymol. 10 (1967) 48.
15. J.A.Thompson, K.M.Schullek, S.B.Turnipseed and D.Ross, Arch. Biochem. Biophys. , 323 (1995) 463.

3. MIXED SYSTEMS

HYDROCOLLOIDS – PART 2
Edited by K. Nishinari
2000 Elsevier Science B.V.

Rheological study of some mixed hydrocolloid systems displaying associative interactions

B. Launay, G. Cuvelier, C. Michon and V. Langendorff

ENSIA, Département Science de l'Aliment
1, Avenue des Olympiades -91744- MASSY, France

The rheological properties of four mixed systems are reviewed : pig skin gelatin/iota-carrrageenan, in comparison with lime hide gelatin/iota-carrageenan, xanthan/carob gum and milk/iota-carrageenan. Contrary to lime hide gelatin, it is shown that pig skin gelatin forms electrostatic complexes with iota-carrageenan, and that the gel is a coupled network. Xanthan and carob gum chains associate, even at very low concentrations, in subunits which aggregate spontaneously and constitute the "building blocks" of the network. Two types of interactions likely contribute to this network : xanthan/carob gum and xanthan/xanthan. In milk, it is found that iota-carrageenan chains are adsorbed on the casein micelles only when the chains are in helices. Some specific properties of the gels, compared to iota-carrageenan/milk permeate gels are explained, in particular their increased melting temperature and, at very low iota-carrageenan concentration, their lack of thixotropic recovery.

1. INTRODUCTION

When two polymers in solution are mixed, phenomena due to their thermodynamic incompatibility are frequently observed, in particular with uncharged polymers [1]. Except in very dilute solutions, incompatibility may lead to phase separation, at the macroscopic or microscopic level. In the latter case, separation of the two coexisting phases may be difficult. When phase diagrams may be obtained, two types of incompatibility are distinguished. The incompatibility is called segregative when each phase is enriched in one of the two components, as for some mixtures of two non-ionic polymers, of two similarly charged polyelectrolytes or of a polyelectrolyte plus a non-ionic polymer [2, 3]. Mixtures of anionic polysaccharides with gelatin at pHs above its pI [4], gelatin and amylopectin [5], lime ossein gelatin and maltodextrin [6, 7] are some examples where this type of incompatibility is assumed. However, when both polymers form thermoreversible gels, clear mixtures may be obtained beyond the gelation temperature in a large range of composition but, at low temperatures, macroscopic phase separation is prevented by gelation, as it has been claimed for mixtures of lime hide gelatin and iota-carrageenan [8]. In the other type of incompatibility, called associative, one of the separated phases is rich in both polymers, the other one being composed of almost pure solvent. Attraction between polymers leading to complex formation are observed in mixtures of opposite charged polyelectrolytes, by example anionic polysaccharides and gelatin at pHs below its pI [6] : mixtures of pig skin gelatin and iota-carrageenan have been shown to behave in this way [8-12].

From a theoretical and practical point of view, the understanding of interactions between colloidal particles and a polymer in solution is essential to study the behavior of mixtures of casein micelles with a polymer. Generally speaking, if there is adsorption of the polymer on the surface of the particle, sterical stabilization [13] or, on the contrary, flocculation due to bridging [14], may be observed. If there is no affinity between the two components, adding the polymer at a sufficient concentration brings about particles flocculation by a depletion mechanism, inducing phase separation [15]. This mechanism is assumed to explain the flocculation of casein micelles in the presence of xanthan (unpublished results) or of guar gum [16], by example.

The aim of this work is to present some properties of gels formed mainly or partly through associative interactions between two polymers, gelatin/iota-carrageenan and xanthan/carob gum, and between casein micelles and iota-carrageenan. In the first and third cases, there are two or one, respectively, polymers which are able to form thermoreversible gels but, for xanthan/carob gum mixtures, gelation is due to a synergy between the two polymers. Even if it is admitted that associative interactions are implied in every case, these systems have very different structures and their rheological behavior will depend, among others, on the properties of each isolated component, on the precise nature of interactions and on gelation conditions.

2. GELATIN/IOTA-CARRAGEENAN

Two gelatin samples, pig skin gelatin (PS, acid-processed, pI $= 8.8$, $\overline{M}w = 1.8 \ 10^5$) and lime hide gelatin (LH, pI $= 4.5$, $\overline{M}w = 10^5$) and a iota-carrageenan sample (iC, 7% kappa - form, $\overline{M}w = 7.10^5$), supplied by SKW (formerly SBI) were used. Phase diagrams at 70°C were obtained and, even if bulk phase separation was rarely observed, from visual examination of the mixtures and from their behavior following dilution, it may be concluded that, in most conditions, PS-iC interactions are attractive and LH-iC interactions segregative [8, 17]. This conclusion has been confirmed by using the method proposed by Snoeren [18] for demonstrating the adsorption of kappa-carrageenan on kappa-caseins. This method relies on the change in the visible light absorption spectrum of methylene blue following its interaction with kappa-carrageenan. If kappa-casein is added to a methylene blue/kappa-carrageenan mixture, it enters into competition with kappa-carrageenan and it releases free methylene blue that can be evaluated from the change in the spectrum. The same type of interactions are formed at 25°C between iC and methylene blue but, whereas the addition of PS gelatin released also methylene blue, only minor changes were observed by adding LH gelatin [8, 17].

Following mixing of the two polymer solutions in 0.2M NaCl at 70°C, the mixed solution is cooled in the rheometer (Rheometrics Fluid Rheometer RFR 7800) unto 40°C (2°C min.$^{-1}$), held a given time (15 min.) at this temperature, cooled from 40°C to 25°C (0.6°C min.$^{-1}$) and aged 3 hours at this temperature before heating (1°C every 30 min.).

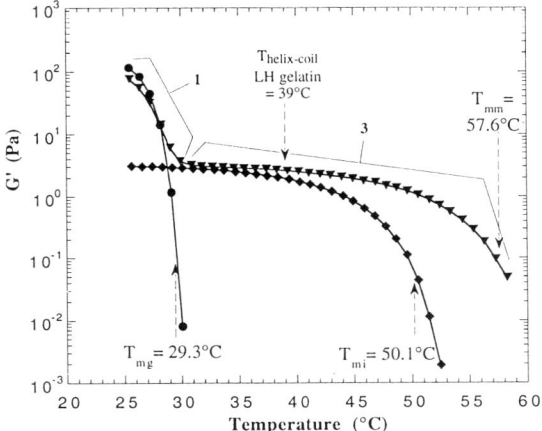

Figure 1.   Melting of three gels = LH (8%, •), iC (0.2%, ♦) and their mixture (▼)
aged 3h at 25.2°C. For meaning of regions 1 and 3, see text [8].

Figure 1 shows the melting of a mixed LH/iC gel (pH7, 0.2 M NaCl, clear gel), in comparison with pure LH and pure iC gels. A two steps melting curve is observed (domains 1 and 3 on fig.1), the first one ending at about the gel-sol transition temperature of a LH gel (29.3°C), without displaying any inflexion at the helix-coil transconformation temperature of the gelatin (40°C). However, the gel-sol transition temperature of the mixed system is 7.5°C higher than the one of the pure iC gel. These critical temperatures were determined from the scaling laws found for chemically crosslinking polymers by Winter and Chambon [19] at the sol-gel transition and applied by Cuvelier and Launay [20] to physical gels (see fig.3) :

$$G'(\omega) \sim G''(\omega) \sim \omega^{\Delta} \tag{1}$$

Therefore, it has been assumed that the gel is structured by two independent and interpenetrated networks, one between LH chains, and the other between iC chains. The shift in the gel-sol transition temperature, from 50.1°C to 57.6°C, is attributed to a stabilization of the iC network due to the polyelectrolyte effect of disordered gelatin chains on the stability of iC helices [17]. If the same experiment is made with a PS-iC gel (fig.2), at intermediate temperatures a third domain (2 on fig.2) may be defined, ending at the helix-coil transition temperature of gelatin (40°C, determined by polarimetry).

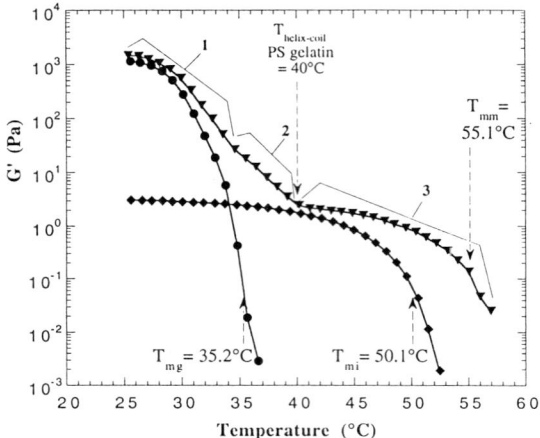

Figure 2.   As figure 1, gelatin PS. For meaning of regions 1, 2, and 3, see text [8].

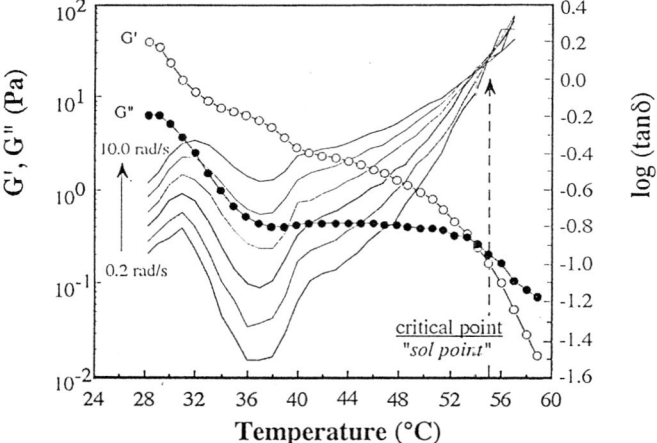

Figure 3.   Melting of a PS (8%)/iC (0.2%) mixed gel aged 1h30 at 28.2°C.
The critical gel-sol transition temperature (55°C) is determined, as for a single
component gel, using the criterion given by eq.1 [11].

On figure 3, the values of tgδ (= G''/G') for a similar gel during melting are shown : a
marked decrease in tgδ starts at about 33°C [11]. Taking into account the previous results, it
is attributed to the association of free gelatin chains, still in helices, with the iC network,

inducing a strengthening of the network. Therefore, it is assumed that the mixed PS-iC network is of the coupled type.

3. XANTHAN/CAROB GUM

From the initial work of Dea and Morrison [21], it is generally assumed that thermoreversible gels of xanthan (X) /carob gum (C) mixtures are formed through synergistic interactions between xanthan chains and unsubstituted regions of galactomannan chains. Although time-temperature treatment of the mixtures may have strong effects on the properties of the gels [22], it is well admitted that it is not necessary to heat above the transition temperature of xanthan to get a gel [23, 24].

To demonstrate that there is an association phenomenon between the two polymers, we have studied their viscous properties, before and after mixing, in the diluted domain [25-27]. We have used a commercial food grade sample of xanthan (Na form, 30% pyruvylated, 100% acetylated) and a purified carob gum sample (mannose/galactose ratio = 2.8), kindly supplied by Rhone-Poulenc (France) and Meyhall Chemical (Switzerland), respectively. Solutions were prepared separately in 0.1 M NaCl by dispersion at laboratory temperature, heating at 80°C, then they were mixed at this temperature, cooled and diluted with 0.1 M NaCl. They were heated again at 75°C, poured in the coaxial cylinders of the viscometer (Low Shear 30, Contraves, Zurich) and the measurements started when the temperature attained 25° C. Taking into account the ionic strength of the solvent, the xanthan chains were probably always remaining in the ordered state [28]. At low (1/9) or high (9/1) X/C ratios, the intrinsic viscosities are not significantly different from the ones calculated by a simple mixing rule, but the Huggin's coefficients are a little larger (fig.4) :

$$(\eta_{sp}/C)_{calc.} = [\eta]_{calc.} + \lambda_{calc.} [\eta]^2_{calc.} C \qquad (2)$$

with $[\eta]_{calc.} = A [\eta]_x + (1-A) [\eta]_c$, $\lambda_{calc.} = A \lambda_x + (1-A) \lambda_c$ and $A/(1-A) = X/C$

At intermediate X/C ratios, a thixotropic behavior may appear, even at very low concentrations (by example at about 7.10^{-3} g dl^{-1} for X/C = 5/5). However, after a shearing step at 70 s^{-1} this behavior disappears and, for a short period of time, a Newtonian behavior is found again. As the corresponding viscosities were consistent with those determined in the absence of thixotropy (see fig.4), the Huggin's relationships were obtained in the full concentration range with a good accuracy.

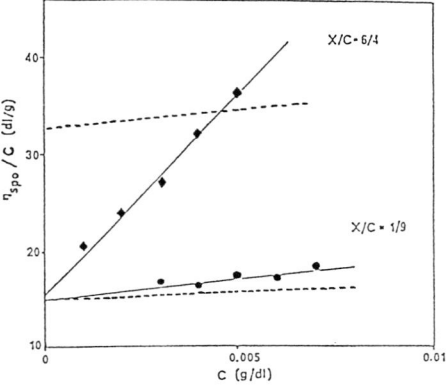

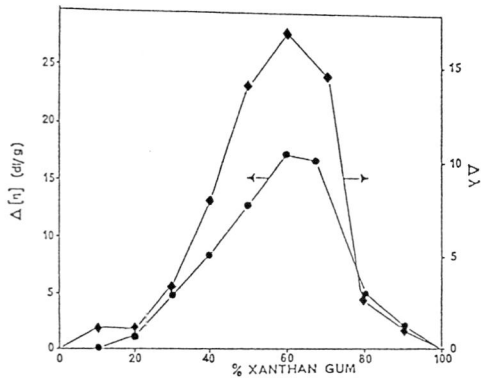

Figure 4.   Reduced viscosity versus concentration at two X/C ratios. Dotted lines calculated from eq.2 , [27].

Figure 5.   Differences between experimental and calculated (eq.2) values of [η] and λ versus X/C ratio [27].

As shown in figure 4 for X/C = 5/5, the experimental intrinsic viscosities and Huggin's coefficients are much smaller and much larger, respectively, than the values deduced from equation 2 : X and C chains associate in compact subunits, prone to organize themselves in aggregates which are stable under shear (until at least 70 s^{-1}). When the X/C ratio is changed, the differences Δ [η] and Δλ between the experimental and calculated values [25, 27] appear to be tightly correlated and pass through a maximum for X/C ~ 6/4 (fig.5).

The aggregated subunits may stick together at rest or under slow shearing, forming shear sensitive "superaggregates" and inducing thixotropic properties, even at very low concentrations. Depending on the characteristics used to define the appearance of a continuous network, the limit curves of the state diagram change [25, 26] but their shapes remain the same and, in all instances, the critical concentrations are much lower than those calculated on the basis of a full occupancy criterion (C[η]~ 1) using [η]$_{calc.}$ or, moreover, [η]$_{exp.}$ (fig.6).

When [η] and λ are determined at increasing temperatures (from 15°C to 65°C), the values of Δ[η] and Δλ decrease progressively and tend to zero between 60°C and 65°C. It has also been shown that, at this temperature, and maybe already from 50°C - 55°C, it is no more possible to form a continuous network, whatever the concentration [27].Therefore, the X-C chains in the subunits are probably associated through hydrogen bonds and these "elementary associated units" are the building blocks of the continuous network, according to the following schema [26, 27] :

$$[(X)_n - (C)_p] \underset{C\uparrow}{\overset{C\downarrow}{\rightleftarrows}} \text{aggregates} \underset{\substack{\text{low shear rate} \\ C\uparrow \\ \text{rest time}}}{\overset{\text{high shear rate}}{\rightleftarrows}} \begin{array}{c}\text{"superaggregates"} \\ \text{continuous network} \\ \text{weak gel}\end{array}$$

Where n and p are the number of chains of Xanthan and Carob, respectively per elementary subunit.

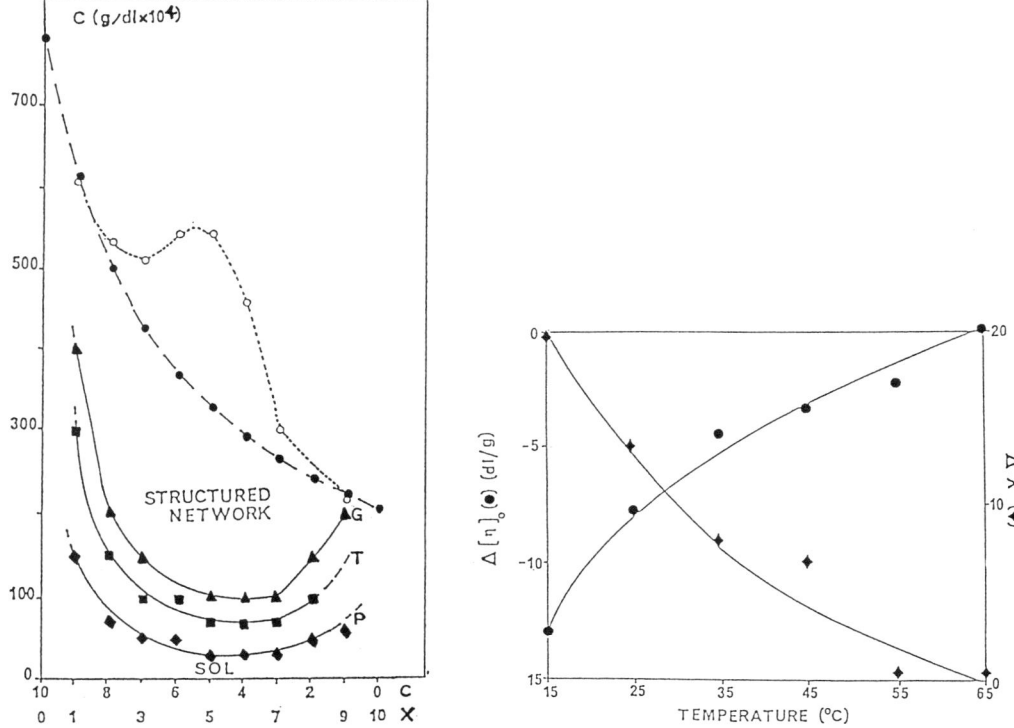

Figure 6. State diagrams versus X/C ratio, G, T : well marked gel properties or weak thixotropic behavior after 15 min. aging time, respectively, P : stress peak at the begining of shearing at 0.9 5 s⁻¹ after 2h aging time. Limits corresponding to C[η] = 1 with [η]calc. (●, eq.2) or [η]exp. (O) [26].

Figure 7. Differences between experimental and calculated (eq.2) values of [η] and λ versus temperature of measurement for X/C : 5/5 [27].

At higher concentrations, strong, thermoreversible gels are formed on cooling and their dynamic viscoelastic properties have been studied [26, 27]. At 25°C, the "elastic" modulus G' depends also strongly on X/C, with a maximum at the same ratio (~ 6/4) as the one corresponding to maximum synergy in the diluted domain. However, when total polymer concentration increases, the maximum progressively flattens between X/C = 4/6 and X/C = 7/3 (fig.8). The same type of viscoelastic behavior has been observed with X/guar gum mixtures [29, 30].

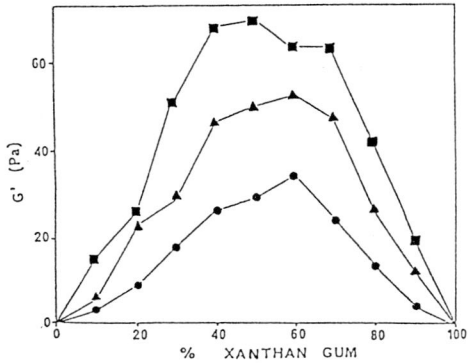

Figure 8. Relationship between G' and X/C ratio. Total polymer concentration : 0.5g/dl (■), 0.4g/dl (s), 0.3g/dl (●) . Carrimed CS 100 (cone-plate, 6.3 rad s^{-1}, 25°C) [27].

Figure 9. Effect of standing time at 25°C on the frequency spectrum of a xanthan/carob gum gel (X/C = 7/3, total polymer concentration 0.5g/dl, Rheometrics Fluid Rheometer RFR 7800, with coaxial cylinders, from 65°C to 25° C in 30 min.) [31] .

By using ethanolic precipitation of carob gum to prepare fractions having various degrees of branching, an increase in G' is also observed when the proportion of unsubstituted mannose increases (unpublished results). When stored at 25°C, the viscoelastic properties of the gels may slowly change, before reaching a stable state after a few hours [31, 32], as shown in figure 9, but this effect is much reduced when the gel is formed at a high cooling rate. It is admitted that, in the semi-diluted domain, xanthan chains display parallel packing association, particularly at high ionic strengths [33, 34]. By electron microscopic studies, Lundin and Hermansson [22] have observed superstrands of xanthan helices, with an increased tendency to form bundles in the presence of carob gum. We tentatively assume that, at high cooling rates, the xanthan chains are rapidly trapped by the formation of stable X-C junction zones, while, at low cooling rates, there is more time to get larger but unstable X-X bundles, resulting in a less dense network which might then slowly reorganize. Schorch [29] has shown that the formation of xanthan mesophases is increased in the presence of guar gum. In conclusion, it is likely that the properties of xanthan/carob gum gels rely on two type of associative interactions, between X and C chains and between X chains [31, 32] and that there is some competition between kinetic trapping by formation of X/C junction zones and microphase separation.

4. CASEIN MICELLES/IOTA-CARRAGEENAN

The unique interaction between casein micelles and carrageenans, principally iota-carrageenans, is widely used in the food industry to make milk gels and to stabilize milk fat emulsions, ice creams and chocolate milks. Dalgleish and Morris [35] have discovered by electrophoretic mobility and light scattering measurements that lambda, kappa and iota-carrageenans may be adsorbed on casein micelles and that their charge densities are a key parameter. In 1976, Snoeren [18], using the method of displacement of the methylene blue/kappa-carrageenan complex previously mentioned, demonstrated that kappa-carrageenan interacts with kappa-casein, mostly located at the periphery of the micelles, but not with αS_1- or β - caseins. However, the exact mechanisms implied in the structuration of milk gels by iota-carrageenans are still a subject of debate.

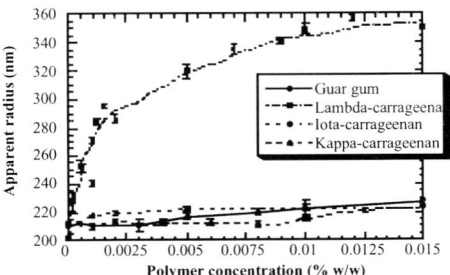

Figure 10. Apparent Stokes-Einstein radius of casein micelles (with standard deviation on 10 measurements) versus polymer concentration (milk diluted at 1%, 60°C) [36].

Figure 11. As on figure 10, for iota-carrageenan at 25° C [36]

Langendorff [36] has recently confirmed by electrophoretic mobility determinations at 25°C that the casein micelles may be progressively covered by iC chains : their mobility is unaffected until a very low critical concentration is reached, then it decreases progressively to a plateau value. Photon correlation spectroscopy of casein micelles in the presence of lambda-, kappa-, iota-carrageenan, or guar gum also indicates (fig.10) that, at 60°C, only the lambda-carrageenan is adsorbed. However, the same method shows that, at 25°C, addition of iC chains increases the Stokes-Einstein radius of the casein micelles (fig.11).

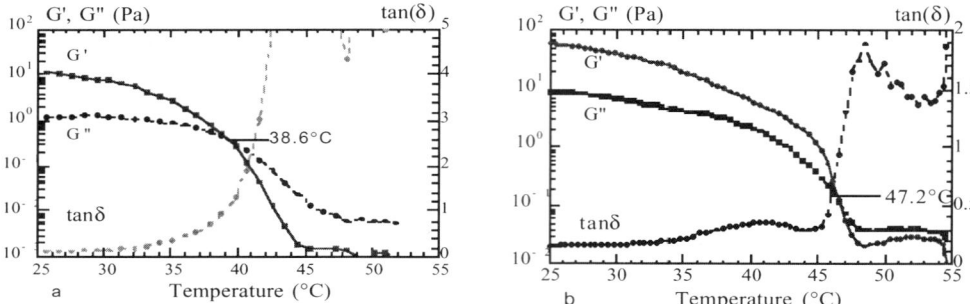

Figure 12. Gelation curve during cooling at 0.5°C min.$^{-1}$ of 0.5% iota-carrageenan in milk permeate (a) or in milk (b) (Rheometrics Fluid Spectrometer RFS II, coaxial cylinders, 1 rad s^{-1}) [36].

Therefore, it is assumed that the coil-helix transition of iC chains is essential to their electrostatic adsorption on casein micelles, this transition resulting in an increased charge density of the polymer chain [36]. These results may explain some peculiarities of iC/casein micelles gels [37, 38]. Figure 12 shows the differences between the gelation curves of iC/milk pemeate and iC/milk mixtures. In milk permeate, there is a progressive increase in G' and G" with a sol-gel point at about 38.6°C, a temperature close to the one of a pure iC gel in 0.2 M NaCl. In milk, the sol-gel transition is attained from a much higher temperature (47°C) and, in addition, a flattening of the G' versus temperature curve may be noted between 35°C and 40°C, tgδ passing through a local maximum at 40°C. Examination of tgδ versus temperature at several frequencies allows a more accurate determination of the sol-gel transition temperatures (fig.13). In iC/milk systems, two critical temperatures, 47°C and 37°C, may be defined : the first one corresponds to the transconformation temperature and the second one is close to the sol-gel transition temperature of iC, in milk permeate in both cases. By opposition, there is only a single critical temperature (39°C) for iC/milk permeate. Therefore, two modes of structuration might take place when iC is added to milk : a network of iC chains reinforced by the presence of casein micelles and an other network, much weaker, implying iC chains in helical conformation adsorbed on casein micelles.

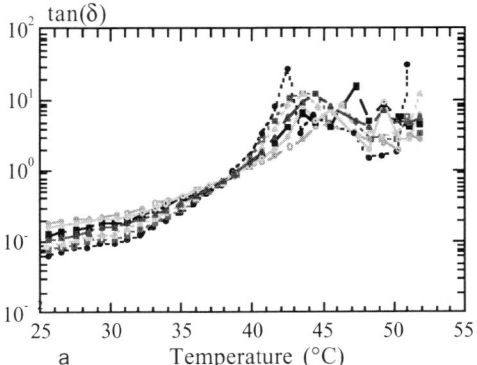
a

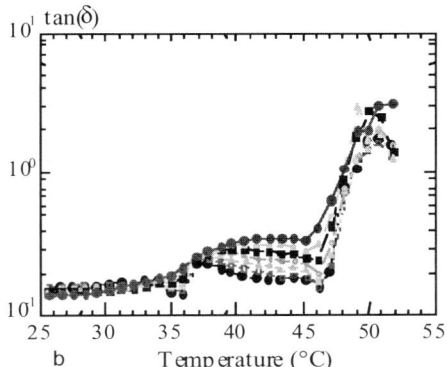
b

Figure 13. As in figure 12, cooling at 0.1°C min.$^{-1}$,
frequency sweep from 0.1 to 100 rad s^{-1} [36].

This model explains why, at a sufficiently high iC concentration, iC/milk gels display thixotropic recovery following shearing, while there is no recovery at low iC concentrations (fig. 14) : in the latter conditions, an iC continuous network cannot be formed, the iC chains being preferentially adsorbed on the casein micelles. These covered micelles are stuck together in a weak network that cannot reorganize following shearing and, in addition, this network is remarkably thermostable [36, 39].

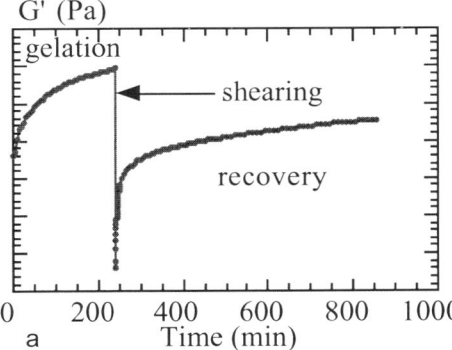
a

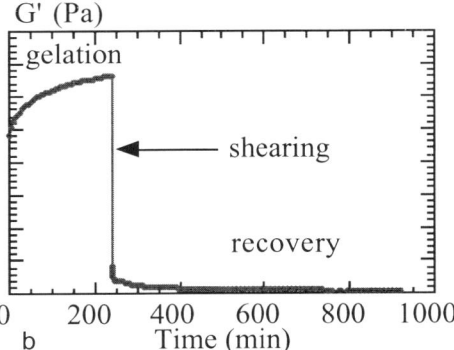
b

Figure 14. Thixotropic recovery of iota-carrageenan/milk gels after 2 min. shearing at 100 s^{-1} .
Gels aged 4h at 25°C and containing more (a) or less (b) than 0.2% iota-carrageenan [39].

5. CONCLUSION

In addition to other mechanisms, in particular excluded volume effects leading to phase separation or to depletion flocculation, associative interactions between hydrocolloids frequently play a crucial role in the structuration of many food systems. These physical associations are mainly dependent on electrostatic interactions or on the formation of hydrogen bonds. By controlling the balance between these mechanisms and the order of appearance of the associated transitions depending, among other factors, on the time-temperature-shearing histories and on the solvent properties, various structural organizations of the food systems could be tailored.

ACKNOWLEDGEMENTS

Some of the results have been obtained thanks to research grants from Rhone-Poulenc and SKW-Biosystems (formerly SBI). The scientific involvement in several aspects of this work of A. Parker (SKW-Biosystems) and C.G. de Kruif (NIZO, Holland) is gratefully acknowledged.

REFERENCES

1. P.J. Flory, In Principles of Polymer Chemistry, pp. 541-593, Cornell University Press, Ithaca, 1953.
2. E. Dickinson, In G.O. Phillips, P.A. Williams and D.J. Wedlock (eds.) Gums and Stabilisers for the Food Industry -4, pp. 249-263, IRL Press, Oxford, 1987.
3. L. Piculell and B. Lindman, Advances in Colloid and Interface Science, 41, 1992, 149.
4. V.B. Tolstoguzov, In J.R. Mitchell and D.A. Ledward (eds.) Functional Properties of Food Macromolecules, pp. 385-416, Elsevier Appl. Sci. Pub., London, 1986.
5. C. Durrani, D.A. Prystupa, M. Donald and A.H. Clark, Macromolecules, 26, 1993, 981.
6. S. Kasapis, E.R. Morris, I.T. Norton and M.J. Gidley, Carbohydrate Polymers, 21, 1993, 249.
7. S. Kasapis, E.R. Morris, I.T. Norton and R.T. Brown, Carbohydrate Polymers, 21, 1993, 261.
8. C. Michon, G. Cuvelier, B. Launay and A. Parker, In E. Dickinson and B. Bergenstahl (eds.) Food Colloids Proteins, Lipids and Polysaccharides, pp. 316-325, The Royal Soc. Chem., Cambridge, 1997.
9. C. Michon, G. Cuvelier, B. Launay, A. Parker and G. Takerkart, Carbohydrate Polymers, 28, 1995, 333.
10. C. Michon, G. Cuvelier, B. Launay and A. Parker, Journal de Chimie Physique, 93, 1996, 828.
11. C. Michon, G. Cuvelier, B. Launay, A. Parker and G. Takerkart, In G.O. Phillips, P.A. Williams and D.J. Wedlock (eds.) Gums and Stabilisers for the Food Industry -8, pp. 247-255, IRL Press, Oxford, 1996.

12. C. Michon, G. Cuvelier, B. Launay and A. Parker, Carbohydrate Polymers, 31, 1996, 169.
13. E. Dickinson and P. Walstra, Food Colloids and Polymers : Stability and Mechanical Properties, 427 p., The Royal Soc. Chem., Cambridge, 1993.
14. R.J. Hunter, Foundations of Colloid Science, vol.1, 673 p., Clarendon Press, Oxford, 1993.
15. V.B. Tolstoguzov, In E. Dickinson and P. Walstra (eds.) Food Colloids and Polymers, pp. 94-102, The Royal Soc. Chem., Cambridge, 1993.
16. S. Bourriot, C. Garnier and J.L. Doublier, Les Cahiers de Rhéologie 15, 1997, 284.
17. C. Michon, Doctorate thesis in "Sciences Alimentaires", ENSIA, Massy, 1995.
18. T.H.M. Snoeren, PhD thesis, University of Wageningen, 1976.
19. H.H. Winter and F. Chambon, Journal of Rheology, 30, 1986, 367.
20. G. Cuvelier and B. Launay, Makromolecular Chemie Macromolecules Symposium, 40, 1990, 23.
21. I.C.M. Dea and A. Morrison, Advances in Carbohydrate Chemistry and Biochemistry, 31, 1975, 241.
22. L. Lundin and A.-M. Hermansson, Carbohydrate Polymers, 26, 1995, 129.
23. P.A. Williams, D.H. Day, M.J. Langdon, G.O. Philipps and K. Nishinari, Food Hydrocolloids, 4, 1991, 489.
24. R.O. Mannion, C.D. Melia, B. Launay, G. Cuvelier, S.E. Hill, S.E. Harding and J.R. Mitchell, Carbohydrate Polymers, 19, 1992, 91.
25. G. Cuvelier, C. Tonon and B. Launay, Food Hydrocolloids, 2, 1987, 311.
26. G. Cuvelier and B. Launay, Carbohydrate Polymers, 8, 1988, 271.
27. C. Tonon, G. Cuvelier and B. Launay, In W. Burchard and S. Ross-Murphy (eds.) Physical Networks, Polymer and Gels, pp. 335-343, Elsevier, London, 1990.
28. B. Launay, G. Cuvelier and S. Martinez-Reyes, Carbohydrate Polymers, 34, 1997, 385.
29. C. Schorsch, Doctorate thesis in "Sciences Alimentaires", ENSIA, Massy, 1995.
30. C. Schorsch, C. Garnier and J.L. Doublier, Carbohydrate polymers, 34, 1997, 165.
31. G. Cuvelier and B. Launay, In G.O. Phillips, D.J. Wedlock and P.A. Williams (eds.) Gums and Stabilisers for the Food Industry -3, pp. 147-158, Elsevier Appl. Sci. Pub., London, 1986.
32. G. Cuvelier, Doctorate thesis in "Sciences Alimentaires", ENSIA, Massy, 1988.
33. I.H. Smith, K.C. Symes, C.J. Lawson and E.R. Morris, International Journal of Biological Macromolecules, 3, 1981, 129.
34. G. Cuvelier and B. Launay, Carbohydrate Polymers, 6, 1986, 321.
35. D.G. Dalgleish and E.R. Morris, Food Hydrocolloids, 2, 1988, 311.
36. V. Langendorff, Doctorate thesis in "Sciences Alimentaires", ENSIA, Massy, 1998.
37. V. Langendorff, G. Cuvelier, B. Launay and A. Parker, Food Hydrocolloids, 11, 1997, 35.
38. V. Langendorff, G. Cuvelier, B. Launay, A. Parker and C.G. De Kruif, Les Cahiers de Rhéologie, 15, 1997, 584.
39. V. Langendorff, G. Cuvelier, B. Launay, A. Parker and C.G. De Kruif, In Structuration (Aliments, Ingrédients et Procédés (AGORAL 98), pp. 352-358, Lavoisier Tec. et Doc., Par 1998.

HYDROCOLLOIDS – PART 2
Edited by K. Nishinari
2000 Elsevier Science B.V.

Rheology of biopolymer co-gels

E.R. Morris

Cranfield University, Silsoe College, Silsoe, Bedford MK45 4DT, United Kingdom

The possible consequences of favourable and unfavourable enthalpic interactions in gelling mixtures of two different biopolymers are illustrated by examples from recent research.

1. INTRODUCTION

In systems containing two different biopolymers, enthalpic interactions between unlike chains will normally be either more favourable or less favourable than interactions between like chains of each type. When the heterotypic interactions are enthalpically unfavourable, there will be a tendency for the system to segregate into regions where the individual chains are surrounded by others of the same type, whereas enthalpically-favourable heterotypic interactions will promote association between the two polymers [1]. The aim of this paper is to illustrate the effect of associative and segregative interactions on the structure and rheology of gelling biopolymer mixtures, using examples drawn mainly from recent work in Silsoe.

2. ASSOCIATIVE INTERACTIONS

In a few systems, association appears to occur by formation of specific heterotypic junctions analogous to the homotypic junctions in single-component polysaccharide gels. "Synergistic" interactions of this type have been discussed in detail in a recent review by the present author [2], and will not be considered further here. A more general mechanism of association, however, is by electrostatic attraction [3] between polyanions (e.g. negatively charged polysaccharides) and polycations (e.g. proteins below their isoelectric point).

Figure 1 shows the effect of progressive incorporation of low-methoxy pectin (DE 31.1) on the DSC heating curves obtained [4] for a gelling concentration (2.0 wt %) of type B (alkali-extracted) gelatin at a pH well below its isoelectric point of ~4.9 (pH 3.0). In the absence of pectin, gel melting is accompanied by a single endothermic transition which is complete by ~35°C. Addition of pectin causes a large increase in the magnitude of the main peak, and the development of a second endotherm at higher temperature (~50°C). Both effects could be eliminated by addition of salt (0.1 M NaCl), consistent with screening of electrostatic attraction between (positively charged) gelatin and (negatively charged) pectin.

Titration curves for the individual polymers showed that the concentration of pectin required for exact charge balance with 2.0 wt % gelatin at pH 3.0 is ~1.35 wt %. As shown in Figure 2, the DSC transitions reached maximum intensity at around this point, and mixtures prepared at compositions close to charge balance formed dense flocs, consistent with precipitation of an electrically-neutral complex. We attribute the massive enhancement in transition enthalpy in this region of composition to opportunistic interchain association within the flocculated particles.

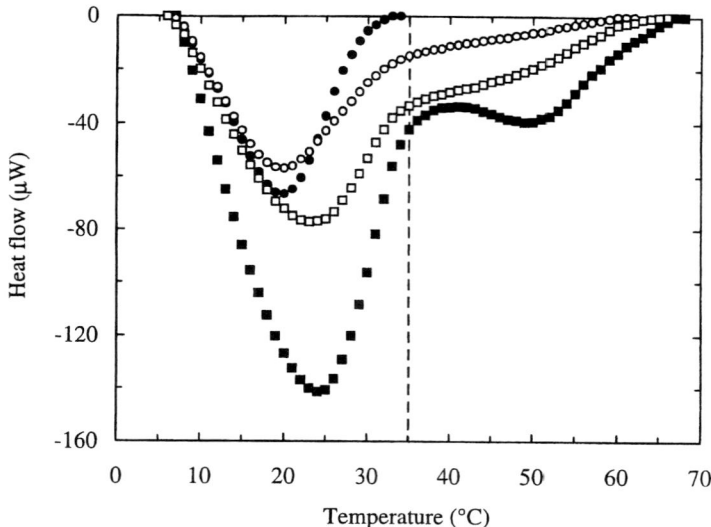

Figure 1. DSC heating scans (0.1°C/min, after 100 min at 5°C) for 2.0 wt % type B gelatin (pI ≈ 4.9) at pH 3.0, in the presence of low-methoxy pectin (DE 31.1) at concentrations (wt %) of 0 (●), 0.5 (O), 1.0 (■) and 2.5 (□).

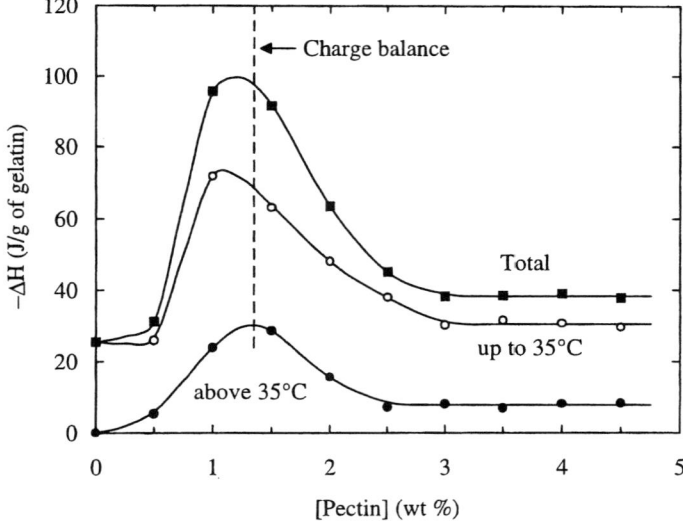

Figure 2. Effect of pectin concentration in mixtures with 2.0 wt % gelatin (type B; pH 3.0) on the transition enthalpy (ΔH) values obtained by integration of DSC peak area up to 35°C (O), above 35°C (●), and total (■).

Mixtures prepared with either component in substantial (electrical) excess, however, remained homogeneous [4], suggesting solubilisation of the electrostatic complex by surplus charge. For these homogeneous preparations, the magnitude of the first endotherm (below 35°C) drops to a value close to that observed for gelatin alone (~30 J/g) and the second endotherm (at ~50°C) also becomes roughly constant (Figure 2) at $\Delta H \approx 8$ J/g of gelatin when pectin is present in large excess, indicating that the two transitions correspond, respectively, to normal disordering of gelatin triple helices and subsequent dissociation of electrostatic junctions between the two polymers.

Rheological measurements of network strength (storage modulus, G') for the homogeneous samples (at pectin concentrations above ~2.0 wt %) were consistent with conversion from a gelatin network augmented by additional crosslinking through pectin, to a (much weaker) pectic acid network [5] augmented by additional crosslinking through gelatin as the relative proportion of pectin was increased.

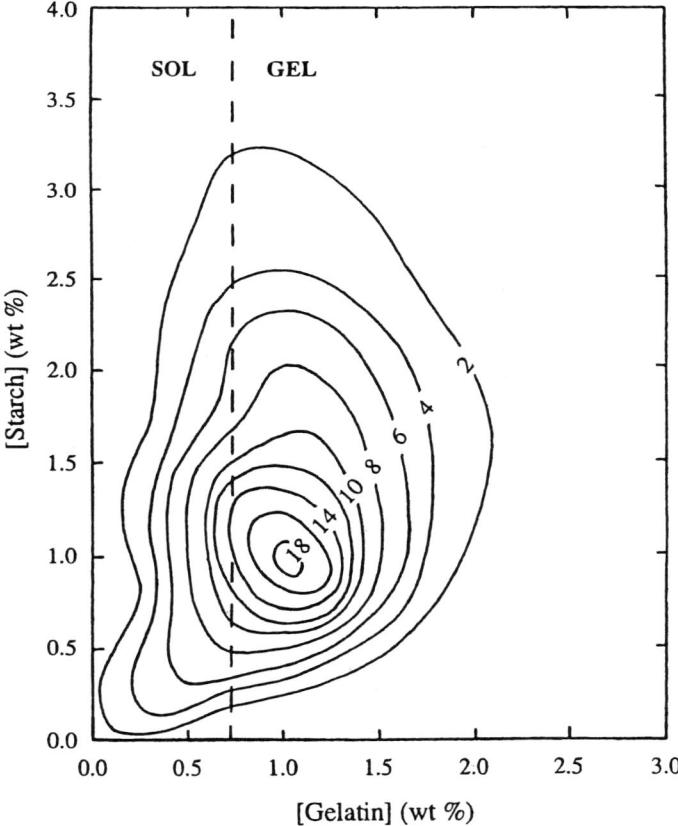

Figure 3. Contour plot showing the variation of turbidity (τ/cm^{-1} ; 400 nm) with concentration of gelatin and starch, and approximate boundary between solution and gel states ($- - -$), for mixtures (pH ≈ 6.2) of type A gelatin (pI ≈ 8.4) and oxidised starch after 24 h at 18°C.

As a further illustration of the effects of electrostatic association, Figure 3 shows the composition-dependence of turbidity for mixtures of oxidised starch (partially-depolymerised amylopectin, with ~1 carboxyl group per 40 residues) and type A (acid-extracted) gelatin (pI ≈ 8.4) at its natural pH of ~6.2. Maximum turbidity occurs when both polymers are present at a concentration of ~1.0 wt %, but decreases steeply with further increase in concentration of either component [6]. As found for the enthalpic interactions in the gelatin–pectin system, development of turbidity could be eliminated by incorporation of salt, again consistent with screeening of electrostatic attraction. We suggest that the reduction in turbidity at high polymer concentration arises in the same way, with the increase in ionic strength from the charged groups on the polyelectrolye chains causing "autoinhibition" of electrostatic association.

Reduction in turbidity on heating occurred in two discrete steps, below and above 35°C, again indicating melting of gelatin helices followed by dissociation of electrostatic junctions. As indicated in Figure 3, however, gelation occurred at essentially the same concentration of gelatin (~0.75 wt %) irrespective of the amount of starch present, and the co-gels gave G' values similar to those obtained for gelatin alone, indicating that the electrostatic associations, although causing massive turbidity, make only a minor contribution to overall crosslinking.

In summary, associative electrostatic interactions cause precipitation of a neutral complex for compositions close to charge balance. Away from charge balance, the complex is solubilised by the surplus charge on the polymer present in electrical excess, and the strength of the dominant network may be slightly enhanced by additional crosslinking through electrostatic junctions. Association is inhibited by increase in ionic strength, including that conferred by the charged groups of the polymer chains.

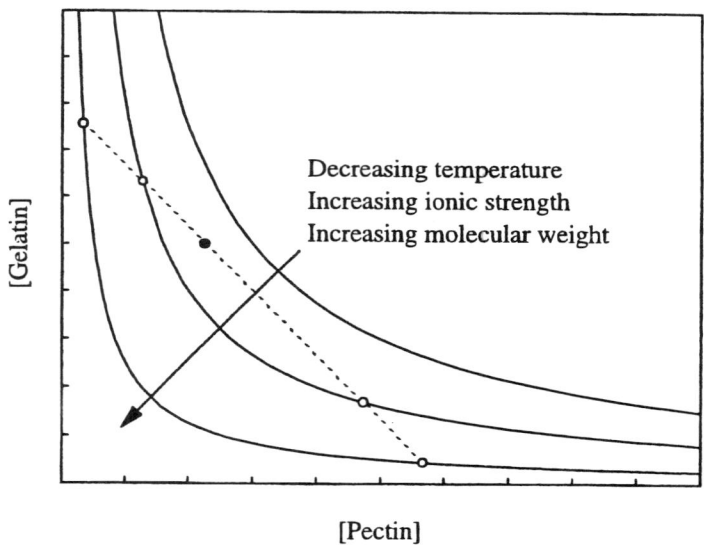

[Pectin]

Figure 4.   Schematic phase diagram, illustrating the effect of temperature, ionic strength (salt concentration) and effective molecular weight on the position of the binodal (———) and hence on the equilibrium composition (O) of co-existing phases formed from the same starting composition (●) along a common tie line (········).

3. SEGREGATIVE INTERACTIONS

In the absence of electrostatic attraction, or of the specific heterotypic junctions mentioned above, the behaviour of biopolymer mixtures is usually dominated by segregative interactions, which may induce separation into two phases, each enriched in one polymer and depleted in the other. Phase separation in the solution state can often be detected by immediate development of turbidity on mixing, due to formation of a "water-in water" emulsion in which one phase exists as a continuous matrix with the other dispersed through it as liquid droplets. The composition of the co-existing phases obtained from different starting compositions can be defined [7] by a "binodal" or "cloud-point curve", which represents the boundary between monophasic and biphasic states of the system (where the enthalpic advantage of segregation and entropic advantage of mixing are in exact balance).

Lowering temperature, which reduces the relative significance of entropy ($\Delta G = \Delta H - T\Delta S$), promotes more complete segregation of the two polymers, shifting the binodal towards the concentration axes (Figure 4). Increase in molecular weight, giving a smaller number of species free to move independently and therefore again decreasing the entropy weighting, has the same effect. In mixtures where one polymer is neutral and the other is charged, preservation of electrical neutrality requires the charged polymer to segregate with its associated counterions, and since the number of counterions is, of course, very much greater than the number of charged chains, the entropic disadvantage of segregation is correspondingly greater than in mixtures of two neutral polymers. There is an analogous (although less severe) entropic barrier to segregation in mixtures where both polymers have charge of the same sign. Phase separation can, however, be induced by addition of extraneous salt [8], since the imbalance in counterion concentration becomes progressively less significant as the overall ionic strength is increased.

3.1. Polymer blending laws

After initial phase separation, mixtures of non-gelling polymers normally show gradual resolution into two clear layers (in response to differences in density between the phases), but for gelling biopolymers the "water-in-water emulsion" structure can be trapped by network formation, giving a biphasic co-gel with one phase continuous and the other dispersed.

Considerable progress has been made in rationalising and predicting the overall physical properties of such co-gels by using the Takayanagi isostrain and isostress blending laws [9] (Equations 1 and 2) to relate the overall modulus of the composite gel (G_C) to the individual moduli (G_X and G_Y) and volume fractions (ϕ_X and ϕ_Y) of the constituent phases. These models were developed initially for sandwich structures with thin layers of two different condensed polymers arranged either in parallel (isostrain conditions) or in series (isostress conditions) relative to the direction of deformation, but they are expected [10] to apply reasonably well to composites with "filler" particles (phase Y) dispersed through a continuous matrix (phase X).

ISOSTRAIN: $\qquad G_C = \phi_X G_X + \phi_Y G_Y$ $\qquad$ if $G_Y < G_X$ $\qquad\qquad$ (1)

ISOSTRESS: $\qquad 1/G_C = \phi_X/G_X + \phi_Y/G_Y$ $\qquad$ if $G_Y > G_X$ $\qquad\qquad$ (2)

In a recent series of experiments, we assessed the error introduced by this approximation by using gelatinisation of starch in solutions of gelling biopolymers (gelatin [11] or agarose [12]) to generate dispersed particles of known phase volume in a continuous gel matrix. To avoid complications from disintegration of the swollen granules or release of amylose into the polymer matrix, a crosslinked waxy (maize) starch was used.

After gelatinisation, the samples were split; part was loaded onto an oscillatory rheometer and cooled to gel the polymer matrix, giving the overall modulus (G'_C); part was centrifuged at a temperature above the gel point, to sediment the starch granules, and the polymer supernatant was loaded onto the rheometer and gelled under the same conditions as were used for the mixtures, thus yielding the modulus of the continuous phase (G'_X). The increase in polymer concentration by transfer of water into the starch granules during gelatinisation was determined by comparing the observed values of G'_X with a standard calibration curve of modulus versus concentration for the polymer alone, giving the relative phase volumes $(\phi_X$ and $\phi_Y)$. The only unknown value in Equations 1 and 2 is then the modulus of the swollen granules (G'_Y). This was determined [12] by finding the relative concentrations of polymer and starch required to give identical moduli for the polymer matrix and composite gel (where $G'_X = G'_Y = G'_C$).

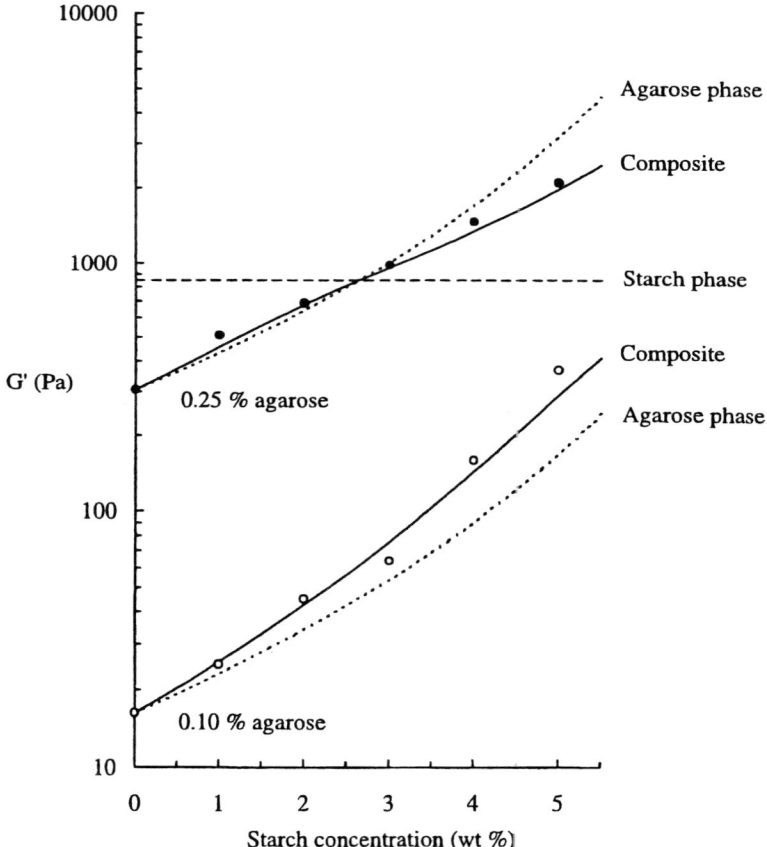

Figure 5. Observed values of composite modulus (G'_C) for gelatinised starch in combination with 0.10 (O) and 0.25 (●) wt % agarose, in relation to the modulus (G'_Y) of the starch phase (– – –), the moduli (G'_X) of the agarose phase (-----), and the calculated values of G'_C (——) from the isostrain model (for $G'_X > G'_Y$) or isostress model (for $G'_X < G'_Y$).

As illustrated in Figure 5, the observed values of G'_C were found to agree closely with values calculated from G'_X, G'_Y, ϕ_X and ϕ_Y by application of the appropriate blending law. Thus the Takayanagi models appear to provide a reliable way of calculating the overall modulus of biphasic biopolymer systems in which one phase is continuous and the other dispersed, if the isostrain model is used when the continuous phase is the stronger component ($G'_X > G'_Y$) and the isostress model is applied to the converse situation ($G'_X < G'_Y$).

For systems in which both phases remain continuous (as might occur, for example, if a biopolymer solution was allowed to gel within the pores of a sponge), the theoretical relationship shown in Equation 3 has been proposed [13] for calculation of the overall modulus from the individual moduli of the constituent phases.

BICONTINUOUS: $$(G_C)^{1/5} = \phi_X(G_X)^{1/5} + \phi_Y(G_Y)^{1/5} \qquad (3)$$

3.2. Phase composition in biopolymer co-gels

For mixtures of gelling biopolymers where phase separation occurs in solution, it might seem reasonable to expect the composition and relative volumes of the individual phases in the final co-gels to be the same as in the pre-gel solution state, particularly if gelation is induced rapidly (e.g. by quench cooling). We have recently tested this expectation [14], using gelling mixtures of (type B) gelatin and oxidised starch (with incorporation of 0.1 M NaCl to eliminate any possibility of complications from the associative electrostatic interactions discussed previously). The work was carried out as part of a collaborative LINK project, and the solution-state phase diagram was obtained by two other participating groups (Department of Chemistry, University of York, and Unilever Research, Colworth Laboratory, Bedford, UK).

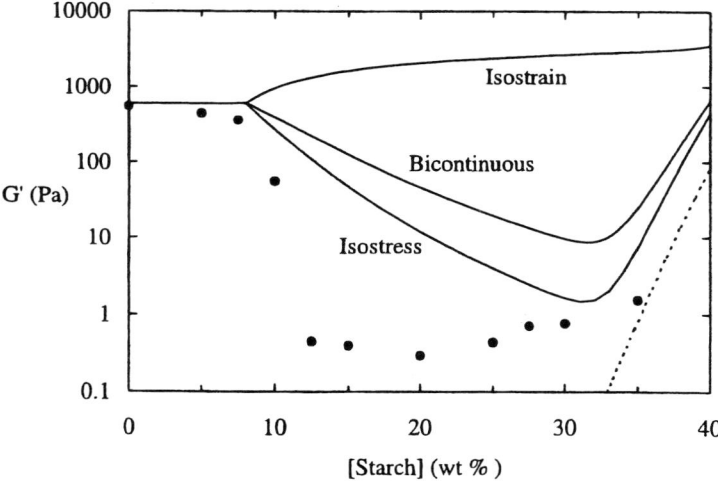

Figure 6. Observed values of G' (●) for co-gels of oxidised starch with 2.0 wt % gelatin, in comparison with calculated values (———) derived from phase composition in solution by application of the isostrain, isostress and bicontinuous network models (Equations 1 - 3). The concentration-dependence of G' for oxidised starch alone is also shown (-----).

Samples were loaded onto an oscillatory rheometer at 60°C, quenched to 5°C, and measured after holding for 130 min to allow network structure to develop. Modulus–concentration curves for the individual polymers were obtained under the same time–temperature regime, and were used in conjunction with the gelatin and starch concentrations from the phase diagram to derive the values of G'_X and G'_Y expected from phase composition in solution. Phase volumes (ϕ_X and ϕ_Y) were obtained directly from the relative lengths of the tie-line segments on either side of the starting composition [7]. Overall moduli were then calculated using the polymer blending laws given in Equations 1 - 3. Figure 6 shows the results obtained for 2.0 wt % gelatin in combination with 0 - 35 wt % oxidised starch.

For compositions in the monophasic region of the phase diagram (up to ~8 wt % starch) the moduli remain close to that of 2.0 wt % gelatin alone, but at higher starch concentrations they drop sharply to values far below those calculated by the isostress relationship (Equation 2), which sets a lower bound [9,10] on composite moduli for any topological arrangement of the constituent phases. It seems clear, therefore, that the starch-rich phase forms the continuous matrix (as would be anticipated from the relative proportions of the two polymers), but that its gelatin content is much lower than expected from the solution-state phase diagram.

As indicated in Figure 4, the reduction in temperature required to induce gelation, and the increase in effective molecular weight of gelatin in the early stages of gel formation, before development of a continuous network, are both likely to contribute to enhanced segregation between the solution and gel states. We suggest, however, that a more important, general mechanism may be migration of sub-gelling concentrations of the minor component in each phase to join the network formed by the same material in the other phase.

Changes in phase structure between the solution and gel states can go beyond simple enhancement of polymer segregation. In studies of the co-gelation of gelatin with potato maltodextrin, we have found some regions of composition in which the samples undergo phase inversion during cooling (from maltodextrin-continuous to gelatin-continuous [15]), and others in which the starting solutions are monophasic, but phase-separate as the temperature is decreased, giving a dispersed maltodextrin phase in a continuous gelatin matrix [16].

3.3. Solvent partition

It is evident from Figure 6 that the phase composition of biopolymer co-gels cannot be determined reliably by studies of the pre-gel solution state. An alternative approach is to assume complete segregation of the two polymers into their respective phases (which, as discussed above, may be a realistic approximation), and to characterise the distribution of solvent between them by a "solvent avidity parameter" (p), defined [17] as the ratio of solvent/polymer in one phase divided by the corresponding ratio for the other phase.

Partition of solvent is the central issue in analysis of the structure and rheology of biphasic gels [18], since it dictates not only the relative phase volumes (ϕ_X and ϕ_Y) but also the polymer concentrations in each phase (which determine G'_X and G'_Y). In co-gels with a dispersed phase of "filler" particles in a continuous matrix, the increase in concentration, and strength, of the continuous phase due to loss of solvent to the dispersed phase is particularly crucial, because the strength of the continuous matrix dominates the overall rheology. For composites where the filler particles are much softer or much harder than the surrounding matrix (by more than about a factor of 10 [11]), the isostrain and isostress blending laws (Equations 1 and 2) reduce to:

ISOSTRAIN: $\qquad G_C = \phi_X G_X \qquad$ i.e. $G_C/G_X = \phi_X \qquad$ if $G_Y \ll G_X \qquad$ (4)

ISOSTRESS: $\qquad 1/G_C = \phi_X/G_X \qquad$ i.e. $G_C/G_X = 1/\phi_X \qquad$ if $G_Y \gg G_X \qquad$ (5)

The onset of close-packing, beyond which the dispersed phase can no longer be regarded as discontinuous, will normally occur when the dispersed particles occupy about two-thirds of the total volume (i.e. at $\phi_Y \approx 0.67$; $\phi_X \approx 0.33$). Thus the overall modulus (G_C) is unlikely to differ by more than a factor of ~3 from that of the continuous phase, irrespective of the strength of the filler particles. In particular, the reduction in G_C from melting of a dispersed phase of gel particles in a continuous matrix would not be expected to exceed a factor of ~9 (from, at most, $G_C \approx 3G_X$ to, at least, $G_C \approx G_X/3$).

Application of the "p factor" approach, and the use of changes in modulus during heating to explore phase continuity, are illustrated here for co-gels formed by calcium-induced gelation of low-methoxy pectin (1.0 wt %) in mixtures with type B gelatin at pH 3.9, which is high enough [4] to eliminate the associative electrostatic interactions demonstrated in Figures 1 and 2.

Mixtures incorporating $CaCl_2$ at stoichiometric equivalence to the carboxyl groups of the pectin were prepared at 85°C and gelled by cooling to 5°C. Loss of network structure on heating (after 100 min at 5°C) occurred in two discrete steps, the first coincident with the gel–sol transition of gelatin and the second with melting of calcium pectinate [4]. As illustrated in Figure 7 for the highest concentration of gelatin studied (10 wt %), two-step melting persists on incorporation of salt (NaCl) at concentrations up to ~0.25 M, and (as discussed above) the reduction in modulus over the temperature-range of the gelatin transition is far larger than the maximum anticipated for melting of dispersed particles. The obvious interpretation is that gelation of calcium pectinate, which is the first process to occur on cooling, gives a continuous network spanning the whole system, and that subsequent gelation of gelatin at lower temperature creates a bicontinuous network structure, with consequent large reductions in modulus over the melting-range of both constituents.

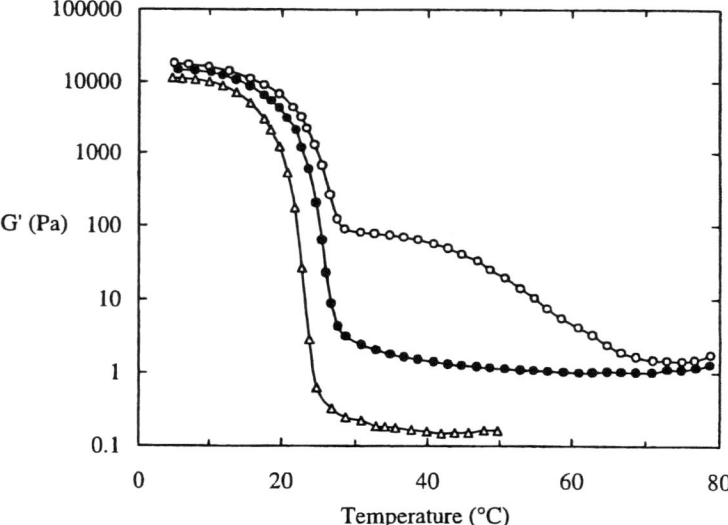

Figure 7. Variation of G' during heating (1°C/min after 100 min at 5°C) for mixtures (pH 3.9) of 10 wt % gelatin and 1.0 wt % low methoxy pectin with stoichiometric Ca^{2+} in the presence of NaCl at concentrations of 0.25 M (O) and 0.30 M (●), and for 10 wt % gelatin alone (Δ).

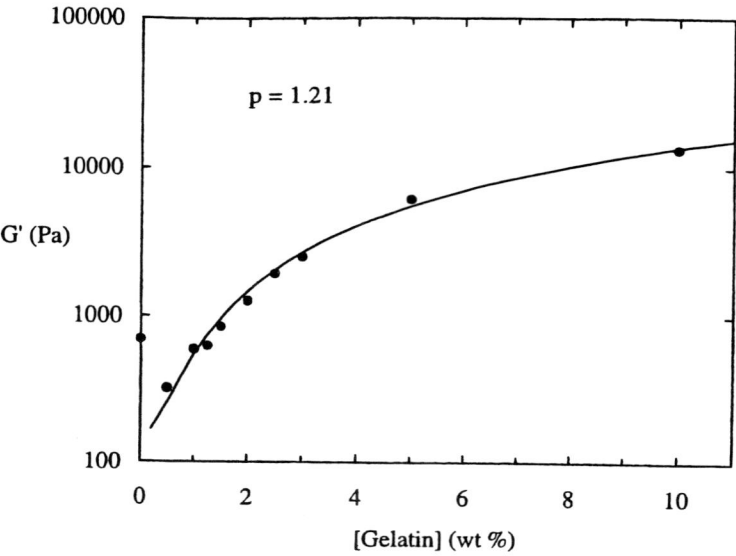

Figure 8. Observed (●) and calculated (———) moduli for co-gels of 1.0 wt % calcium pectinate (stoichiometric Ca^{2+}) with varying concentrations of gelatin in 1 M NaCl.

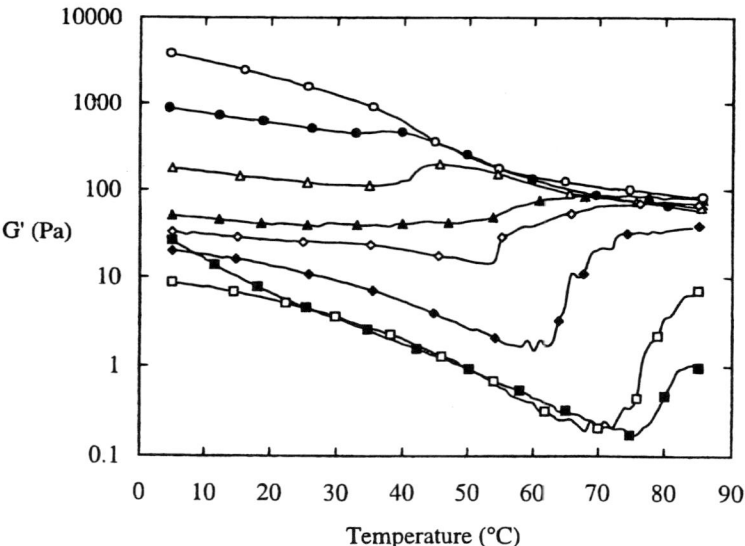

Figure 9. Changes in G' for 2.0 wt % low-methoxy pectin with 100 % stoichiometric Ca^{2+} during cooling (1°C/min) in the presence of oxidised starch at concentrations (wt %) of 0 (O), 3 (●), 5 (Δ), 10 (▲), 12 (◇), 15 (◆), 20 (□) and 35 (■).

On further increase in salt concentration (to 0.3 M; Figure 7), however, reduction in modulus on heating was confined entirely to the temperature-range of the gelatin transition, indicating salt-induced phase separation (Figure 4) to a continuous gelatin phase with dispersed particles of calcium pectinate. This interpretation was tested [4] by quantitative analysis of observed moduli (after 100 min at 5°C) for mixtures incorporating a fixed, high concentration of NaCl (1 M). As discussed above, the analysis assumed complete segregation of the two polymers, and the partition of solvent between them was characterised by a single adjustable parameter, p, defined as the ratio of water/polymer in the gelatin phase divided by the corresponding ratio for the pectin (calcium pectinate) phase. Relative phase volumes at each trial value of p were used to determine the polymer concentration in each phase, and the corresponding moduli were obtained from standard calibration curves. For solvent distributions where the calculated modulus of the continuous gelatin phase was higher than that of the dispersed calcium pectinate phase, co-gel moduli were derived using the Takayanagi isostrain model (Equation 1), and the isostress model (Equation 2) was used for the converse situtation.

As shown in Figure 8, good agreement between observed and calculated values of G' for mixtures spanning a 20-fold range of gelatin concentration (from 0.5 to 10 wt %) was obtained using a single value of p (1.21). A similar standard of agreement was obtained, with the same value of p, by use of the bicontinuous network model (Equation 3) for mixtures prepared without addition of NaCl.

3.4. Precipitation and network collapse

In the gelatin–calcium pectinate system discussed above, phase structure in solution could not be examined directly to obtain an equilibrium phase diagram because the calcium pectinate component forms a "weak gel" network at high temperature, which then consolidates into a "true" gel on cooling. For mixtures with oxidised starch [19], however, we have observed a sharp reduction in modulus during cooling (Figure 9), which increases in magnitude and moves to progressively higher temperature with increasing concentration of starch.

One obvious interpretation is that the system is undergoing phase separation, with the starch-rich phase forming the continuous matrix and the pectin component confined largely to dispersed particles. An alternative possibility, however, is that the collapse of gel structure is, in effect, a precipitation process, with segregative interactions between the two polymers driving the pectin chains into large aggregated bundles which make little contribution to network crosslinking. One line of evidence favouring the precipitation model is that reduction in ester content, which would be expected to facilitate Ca^{2+}-mediated aggregation but inhibit salt-induced phase separation (by increasing the counterion requirement of the pectin), reduces the Ca^{2+} concentration required to trigger network collapse [19].

Partial precipitation of one component from a single-phase mixture has been observed in two previous investigations in this laboratory. Concentrated mixtures of gelatin and potato maltodextrin, at a temperature above the onset of gelation of either component, were found to give classic phase separation into clear solutions with compositions lying along a well-defined binodal. Mixtures prepared at concentrations in the monophasic region below the binodal, however, gave dense precipitates of maltodextrin [20], demonstrating that phase separation and precipitation can occur as separate, distinct responses to segregative interactions between two different polymers. In a later study [21] of mixtures of maltodextrin with whey proteins, phase separation and precipitation (of protein) were again seen as separate processes. In both cases, the extent of precipitation was directly proportional to the concentration of the other polymer and, for the gelatin–maltodextrin system, the mass of maltodextrin precipitated was also found to vary with the square of maltodextrin concentration.

The final moduli of the calcium pectinate–oxidised starch mixtures after cooling to 5°C (Figure 9) could be fitted, to within experimental error, by a somewhat analogous relationship:

$$[starch][free]^2 = K \tag{6}$$

where K can be regarded as a quasi "solubility product" relating the concentration of "free" (unprecipitated) pectin to the concentration of the other polymer. A relationship of the same form was also found to give close agreement between observed and calculated moduli for mixtures of calcium pectinate with maltodextrin [22], and we are continuing to explore the generality of such behaviour in our current research.

REFERENCES

1. L. Piculell, I. Iliopoulos, P. Linse, S. Nilsson, T. Turquois, C. Viebke and W. Zhang, in G.O. Phillips, P.A. Williams and D.J. Wedlock (eds.), Gums and Stabilisers for the Food Industry 7, IRL Press, Oxford, UK, 1994, pp. 309-322.
2. E.R. Morris, in S.E. Harding, S.E. Hill and J.R. Mitchell (eds.), Biopolymer Mixtures, Nottingham University Press, Nottingham, UK, 1995, pp. 247-288.
3. V.B. Tolstoguzov, Food Hydrocolloids, 4 (1991) 429-468.
4. P. M. Gilsenan, R. K. Richardson and E. R. Morris, Biopolymers, submitted.
5. P. M. Gilsenan, R. K. Richardson and E. R. Morris, Carbohydr. Polym., submitted.
6. M.W.N. Hember, L.I. Khomutov, N.A. Lashek, E.R. Morris, N.I. Panina, N.M. Ptitchkina, and S.A. Roberts, Carbohydr. Polym., submitted.
7. V.B. Tolstoguzov, Food Hydrocolloids, 9 (1995) 317-332.
8. L. Piculell, S. Nilsson, L. Falck and F. Tjerneld, Polym. Commun., 32 (1991) 158-160.
9. M. Takayanagi, H. Harima and Y. Iwata, Mem. Fac. Eng. Kyushu Univ., 23 (1963) 1-13.
10. J.A. Manson and L.H. Sperling, Polymer Blends and Composites, Plenum Press, New York, 1976.
11. N.A. Abdulmola, M.W.N. Hember, R.K. Richardson and E.R. Morris, Carbohydr. Polym., 31 (1996) 53-63.
12. Z.H. Mohammed, M.W.N. Hember, R.K. Richardson and E.R. Morris, Carbohydr. Polym., 36 (1998) 27-36.
13. W.E.A. Davies, J. Phys. D. Appl. Phys., 4 (1971) 1325-1339.
14. S.A. Roberts, R. K. Richardson and E. R. Morris, Food Hydrocolloids, submitted.
15. S. Alevisopoulos, S. Kasapis and R.M. Abeysekera, Carbohydr. Res., 293 (1996) 79-99.
16. S. Kasapis, E.R. Morris, I.T. Norton and C.R.T. Brown, Carbohydr. Polym., 21 (1993) 261-268.
17. A.H. Clark, in J.M.V. Blanshard and P. Lillford (eds.), Food Structure and Behaviour, Academic Press, London, 1987, pp. 13-34.
18. E.R. Morris, Carbohydr. Polym., 17 (1992) 65-70.
19. D.R. Picout, R.K. Richardson, C. Rolin, R.M. Abeysekera and E.R. Morris, Carbohydr. Polym., submitted.
20. S. Kasapis, E.R. Morris, I.T. Norton and M.J. Gidley, Carbohydr. Polym., 21 (1993) 249-259.
21. P. Manoj, S. Kasapis, M.W.N. Hember, R.K. Richardson and E.R. Morris, Food Hydrocolloids, submitted.
22. D.R. Picout, R.K. Richardson, and E.R. Morris, Carbohydr. Polym., submitted.

HYDROCOLLOIDS – PART 2
Edited by K. Nishinari
2000 Elsevier Science B.V.

Polysaccharide gelation in the presence of high concentrations of competing polymer: evidence for counter ion entropy effects

M. Puaud, S.E. Hill and J.R. Mitchell*

Division of Food Sciences, School of Biological Sciences, University of Nottingham, Sutton Bonington Campus, Nr Loughborough, Leics, LE12 5RD, England
*To whom correspondence should be addressed

Solutions of the gelling polysaccharides agarose, κ-carrageenan and gellan gum were prepared with gum arabic, gelatin, maltodextrin and low molecular weight grades of sodium alginate, carboxymethyl cellulose (CMC) and hydroxypropylmethyl cellulose (HPMC). The concentration where the cosolute prevented gelation of the polysaccharide was measured. It was found that carrageenan and gellan gum could form gels (continuous networks throughout the solution) at the highest level (14%) of HPMC whereas this cosolute inhibited agarose gelation. In contrast agarose gelled in the presence of 14% alginate or CMC, whereas they inhibited carrageenan and gellan gum gelation. These results are consistent with the hypothesis that phase separation in a mixture of a polyelectrolyte and a non-polyelectrolyte is less likely because of the unfavourable entropy change due to an uneven distribution of counter ions. Agarose and carrageenan gelled in the presence of 14 % gum arabic whereas gellan gum did not. The latter result observed can hardly be explained using the counter ions entropy concept. All three polysaccharides gelled in the presence of 14 % maltodextrin.

1. INTRODUCTION

When two biopolymers are mixed in solution phase separation frequently occurs [1-3]. This is because, for polymers, the entropy gain on mixing is small and is often insufficient to balance the enthalpic disadvantage. Two main types of phase separation can be identified associative, where the two biopolymers can be found in one phase the other phase being solvent rich, and segregative where the two biopolymers are predominantly found in separate phases. Although generally associative phase separation is interpreted in terms of attraction between polymers and segregative phase separation in terms of repulsion between the two polymers, in an excellent review of this subject Piculell et al [3] have made the point that this is not always the case. Applications of aggregative separation to systems of importance to food include systems containing carrageenan and gelatin where the pH is below the isoelectric point of the latter [4]. This is discussed elsewhere in this volume.

148

The concentration at which biopolymers show segregative phase separation depends on a number of factors. In general this concentration decreases with increasing molecular weight, ionic strength, hydrodynamic volume and decreasing temperature [2,3]. The reasons for the importance of some of these factors can be understood through the use of Flory-Huggins theory to predict the phase diagram [3,5]. This and related arguments show that a mixture of a polyelectrolyte and a non-polyelectrolyte are less likely to phase separate than a mixture of two charged or two uncharged polymers [3,6,7]. This is because in the former case, phase separation results in an uneven distribution of counter ions which is entropically unfavourable. In low salt environments the counter ion entropy dominates other influences such as molecular weight but at high added salt levels the counter ion influence loses its importance and the size of the single phase region is reduced. These ideas are illustrated in Figure 1. As the molecular weight decreases and the ionic strength increases (added salts), the area of miscibility (above the curve) is reduced. The entropy of the counter ions and of the polymer itself accounts for this behaviour.

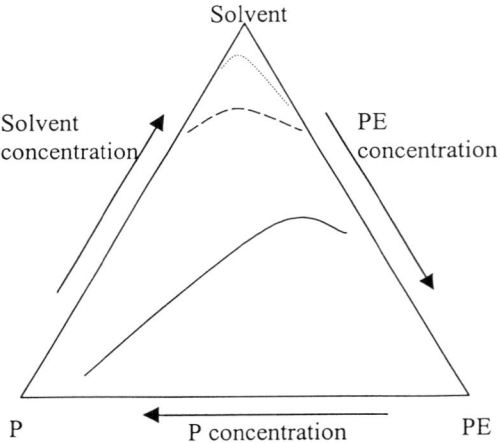

Figure 1. Illustration of the effects of polyelectrolyte molecular weight and salt concentration on the phase diagram of a mixture of a polyelectrolyte (PE) and non-charged polymer (P). Salt free systems, with low or high molecular PE (—), system with low-molecular PE(- -), and system with high-molecular PE in 1M NaCl (····). From Piculell et al, 1991 and Iliopoulos et al, 1989.

This phase separation phenomenon is of particular interest when one or both of the polymers can gel. When only one of the polymers is a gelling agent gelation will occur when the concentration of this polymer in the continuous phase is above the minimum gelling concentration (C_2^*). This will occur if there is no phase separation and the concentration of the gelling polymer is above C_2^* or if there is phase separation and a continuous phase is formed with a concentration above C_2^* [7]. This is illustrated in Figure 2 [8]. Segregative phase separation will prevent gelation of the mixture if the non-gelling polymer forms the continuous phase and the concentration of the gelling agent in this phase is lower than C_2^*. This is likely to occur if the non-gelling polymer is present at high concentrations compared with the gelling polymer.

Zone 1 and 2 represent area of miscibility. Areas 5 and 6 have a continuous P2 rich phase whereas areas 4 and 7 present a continuous P1 rich phase. The tie line drawn separates areas 4 and 5 (where the gelling concentration C_2^* in both phases is not reached) from areas 6 and 7 (where the concentration of P1 in the P1 rich phase is high enough for gelation to occur). The gelation of the overall mixture is therefore only possible in areas 6 and 2.

However one must remember that the respective positioning of the C_2^* and the phase inversion line (dotted line) could be different leading to a complete new phase diagram. Furthermore, a bicontinuous area where none of the polymer can clearly be described as non-continuous might in some cases replace the inversion line.

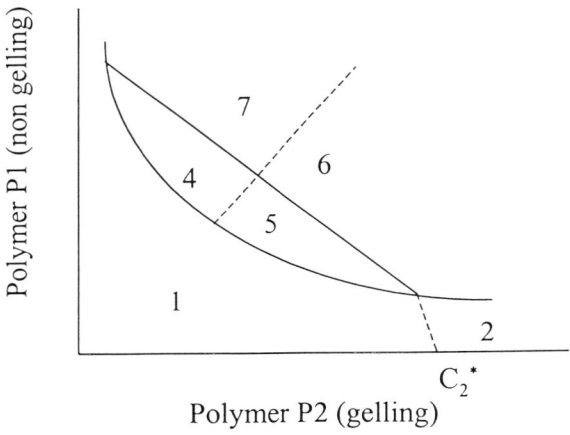

Figure 2. Phase diagram for a mixture of a gelling polymer (P2) and a non-gelling polymer (P1). Gelation occurs in the zones 2 and 6. Zone 3 would be the image of zone 2 for a mixture of two gelling polymers (see text for further explanation). From Zasypkin et al. *Food Hydrocolloids* (1997) 11, 159-170

It might therefore be expected that charged polymeric cosolutes inhibit the gelation of charged gelling agents and uncharged cosolutes inhibit the gelation of uncharged gelling agents. If however one of the components in the mixture is charged and the other uncharged then at low salt levels gelation will occur at high levels of the cosolute because as is illustrated in Figure 1, single-phase behaviour is expected except at very high concentrations. This hypothesis is discussed in the light of measurements on the gelation behaviour of agarose (a non-polyelectrolyte), and κ-carrageenan and gellan gum (polyelectrolytes with different charge densities) in the presence of a range of polymeric cosolutes from which solutions of high concentration (14% (w/v)) can be prepared in water.

2. MATERIALS AND METHODS

κ -Carrageenan (Satiagel MEO5), agarose (Sigma Type I-A A0169) and high acyl gellan gum (Kelcogel LT100) were used as gelling agents. The cosolutes used were gum arabic (Sigma G9752), hydroxypropylmethyl cellulose (HPMC) (Shimatsu grade 606), low viscosity carboxymethyl cellulose (CMC) (Finnfix 2, Metsa), low viscosity sodium alginate (Manucol LB, Kelco, 8.4% w/w sodium, 0.4% w/w calcium) and maltodextrin (Cerestar MD20). The solvent was water when agarose was the gelling agent, 0.05M KCl when κ-carrageenan was used, and 0.02M $CaCl_2$ plus 0.005M sodium citrate when gellan gum was present. At this level of $CaCl_2$ the high mannuronate very low viscosity alginate remained soluble. For the carrageenan and agarose systems the polysaccharides were dissolved in the appropriate solvent and heated with stirring and maintained at 80°C for one hour (carrageenan) and 95°C for 30 minutes (agarose). For gellan gum the temperature of the polysaccharide in water was first raised to 80°C and then the $CaCl_2$ was added. The concentration of gelling agent employed was 0.5g/100ml for κ-carrageenan and agarose, and 0.3 g/100ml for gellan gum. The cosolute was added during the second half of the heating period. The concentration of cosolute in the final solution was 0, 2%, 4%, 6%, 8%, 10%, 12% and 14%. The hot solution was poured into a glass tube (diameter 1.4 cm) covered and allowed to set overnight. The pH of all samples were controlled to pH 6 ±1.0. Gelation was assessed by visual examination. The tube was gently inverted and the flow properties of the solution examined.

3. RESULTS AND DISCUSSION

The data in Figure 3 where a low viscosity CMC and alginate are used show that agarose can gel these systems up to the highest concentration of alginate or CMC used whereas carrageenan and gellan fail to form a gel at this concentration. In this case agarose behaves as the "dominating" gelling agent

Carrageenan and gellan can gel in the presence of the high levels of HPMC whereas gelation of agarose is inhibited. The first two gelling agents are forming continuous networks in the presence of HPMC. Maltodextrin did not inhibit gelation of any of the three gelling polysaccharides at levels up to and including 14%. For gum arabic different behaviour is observed. Both agarose and carrageenan gel this polysaccharide at the highest levels of gum arabic inclusion whereas the gelation of gellan is inhibited at low levels of gum arabic.

Overall this data gives some support to the hypothesis outlined in the introduction. The similar behaviour observed when carboxymethyl cellulose and alginate are used as the cosolute would support the counter ion entropy idea. Since these polymers have different structures most other interpretations would predict a different response to the three gelling agents for the two "filler" materials. The greater inhibition by gellan would be explained in terms of the higher molecular weight of this polysaccharide compared with carrageenan.

With the exception of a system containing a polyelectrolyte and a non-charged polymer at low salt concentrations (Figure 1) phase separation becomes more likely with increasing molecular weight of one or both of the components. The results for the non-charged HPMC are also consistent with the hypothesis. Although the prevention of agarose gelation by low levels of HPMC is in our view almost certainly due to phase separation, another interpretation

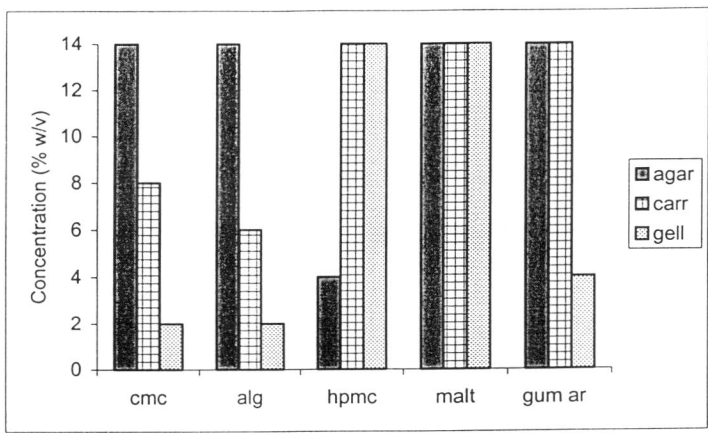

Figure 3. Maximum levels of cosolute in the range investigated (0-14%) allowing gelation of agarose(agar), kappa carrageenan (carr) and low acyl gellan gum (gell) at the concentrations given in the text. Cosolutes: low viscosity CMC (cmc), low viscosity alginate (alg), low viscosity HPMC (hpmc), maltodextrin (DE 20) (malt), gum arabic (gum ar).

could be used to explain gelation of carrageenan and gellan at high HPMC levels. The gelling polysaccharides have setting temperatures (~70°C) whereas HPMC will not be fully soluble until a temperature of 40°C is achieved. It is therefore possible that gellan and carrageenan gel because at the setting point there is no competing polymer in solution. The maltodextrin used would have a number average DP of only 5 and at this low molecular weight compatibility might be expected whatever the electrostatic environment. Kasapis et al [9], with a higher molecular weight maltodextrin reported a phase diagram for a gelatin maltodextrin system showing a maximum maltodextrin concentration for the two phase region of 15% although some interesting precipitation phenomena were observed below the binodal at low maltodextrin concentrations.

The results for gum arabic are interesting. A possible interpretation in terms of charge density could be attempted e.g. gellan gum has an intermediate charge density between carrageenan and agarose. If gum arabic has a similar charge density to gellan gum, then using the counter ion entropy idea, phase separation in carrageenan and agarose systems would not be expected, whereas phase separation in the presence of gellan gum would. We consider it more likely however that the inhibition of gellan gum gelation by gum arabic would be explained in terms of specific interactions between the two components. Whether this is due to gum arabic chelating calcium ions, interfering sterically with the association of the gellan helices, precipitating the gellan by driving gellan interchain association too far as found in some cases for the gelatin maltodextrin system [9] or preventing the formation of the ordered gellan form, is not clear. Although the latter possibility is the most interesting we consider it to be the least likely of the possible interpretations.

4. CONCLUSION

We may conclude that the different phase behaviour of mixtures of polyelectrolytes and non-polyelectrolytes described by Piculell and co-workers [3,6] may provide a basis for predicting the different inhibitory effects on gelation of high concentrations of polymer cosolutes, although in some cases it is necessary to invoke specific interactions between the polymers to interpret the results. It should of course be appreciated that a more extensive study involving an investigation of salt, molecular weight and concentration effects would be required to confirm these preliminary conclusions.

ACKNOWLEDGEMENTS

We would like to acknowledge Pr. E.R.Morris, Dr. G.Sworn and Dr. N.Morrison for helpful discussions.

REFERENCES

1. V.Ya. Grindberg and V.B. Tolstoguzov *Food Hydrocolloids*. 11 (1997) 145
2. V.B. Tolstoguzov *Functional Properties of Food Macromolecules* ed J.R. Mitchell and D.A. Ledward. Elsevier Applied Science Publishers, London, p 385, 1986
3. L. Piculell, K. Bergfeldt and S. Nilsson *Biopolymer Mixtures* edited S.E. Harding, S.E. Hill and J.R. Mitchell, Nottingham University Press, p13, 1995
4. C. Michon, G. Cuvelier, B. Launay, A. Parker and G. Takerkat *Carbohydrate Polymers* 28 (1995) 333
5. C.C. Hsu and J.M. Prausnitz *Macromolecules* 7 (1974) 320
6. L. Piculell, S. Nilsson, L. Falck and F. Tjerneld *Polymer Communications* 32 (1991) 158
7. I. Iliopoulos, D. Frugier, R. Audebert *Polym. Prep.* 30 (1989) 371
8. O.V. Zasypkin, E.E. Braudo and V.B. Tolstoguzov *Food Hydrocolloids* 11(1997) 159
9. S. Kasapis, E.R. Morris, I.T. Norton and M.J. Gidley *Carbohydrate Polymers* 21 (1993) 249

HYDROCOLLOIDS – PART 2
Edited by K. Nishinari
2000 Elsevier Science B.V.

A comparative study of milk gels formed with κ-carrageenan or low-methoxy pectin

D. Oakenfull[a], K. Nishinari[b] and E. Miyoshi[b]

[a]Food Science Australia, North Ryde, NSW 2120, Australia
[b]Department of Food and Nutrition, Faculty of Human Life Sciences, Osaka City University, Osaka 559, Japan

Gelation and thickening of dairy products is one of the most important food applications of the carrageenans, particularly κ-carrageenan. Low-methoxy (LM) pectins also form gels with milk and these too have food applications - mainly in dairy desserts. We have used rheometry and differential scanning calorimetry (DSC) to study the mixed gels formed with skimmed milk powder (SMP) and κ-carrageenan and LM pectin. Our results indicate very different gelation mechanisms for the two polysaccharides.

κ-carrageenan forms a complex with casein micelles and it appears to act as a molecular 'Velcro' - interaction between free ends of bound κ-carrageenan molecules linking casein micelles to form a gel network. At high ratios of κ-carrageenan to SMP, a purely κ-carrageenan network also appears to form, presumably consisting of more extended cross-linked κ-carrageenan structures within which the casein micelles are enmeshed. In contrast, there is no direct interaction between casein micelles and LM pectin. The casein appears simply to act as a source of calcium-ions which promote gelation of the pectin.

1. INTRODUCTION

Gelation and thickening of dairy products is one of the more important food applications of the carrageenans and low-methoxyl (LM) pectins [1]. It has long been known that ι- and κ-carrageenan can form complexes with casein micelles [2-10] - believed to be primarily a consequence of highly specific interaction between the sulphate groups of the carrageenan and a positively charged region of κ-casein, possibly located [7] between residues 97 and 112. Xu and her colleagues [11] have proposed that such interactions are the first step in the gelation of milk by κ-carrageenan. The κ-carrageenan interacts with the casein micelle, but only part of the carrageenan chain, they suggested, is adsorbed, most of it remaining free in solution in the form of loops or tails. As the solution is cooled to below the carrageenan's helix-coil transition temperature, the free κ-carrageenan chains link via double helices forming a gel network incorporating casein micelles (see Fig. 1).

LM pectins form gels in the presence of calcium ions [12]. Whole milk typically contains about 13 mM total calcium, much of which, as calcium phosphate, is intimately involved with the structure and stabilisation of casein micelles [13]. However, some of this calcium is also present in free solution and it is therefore not unreasonable to suppose that, in milk, gelation of LM pectins occurs through the calcium ions naturally present.

But the possibility remains that caseins, and other proteins, may also be active in the gelation process.

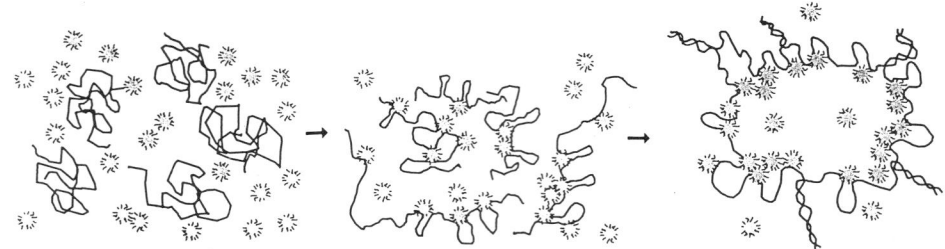

Figure 1. Schematic diagram showing binding of κ-carrageenan to casein micelles and their role in the formation of the mixed gel (as proposed by Xu and her colleagues [11]).

We report here parallel investigations of mixed gels of κ-carrageenan or a LM pectin with skimmed milk powder (SMP). Our results confirm that with κ-carrageenan, casein micelles appear to be linked to form a gel network by association of free loops and tails of κ-carrageenan molecules adsorbed to the micelle surface. However, with LM pectin, there is no direct interaction between pectin and casein micelles. The mixed gel appears to consist of a LM pectin matrix in which the milk protein is embedded.

2. MATERIALS AND METHODS

2.1. Materials

Low heat skimmed milk powder (SMP), κ-carrageenan (Sigma) and the LM pectin (Sanofi 'pectin red 3g'; 18-30% DE) were analysed for calcium, sodium and potassium by ICP-AES (carried out by the CSIRO Division of Exploration and Mining) with the results given in Table 1.

Table 1
Metal ions present in skim milk powder and LM pectin (mg/g) as determined by ICP-AES

	Ca	Na	K
Skim milk powder	13.9	4.4	17.5
κ-carrageenan	1.96	1.16	10.9
LM pectin	1.3	29	1.4

2.2. Sample Preparation

SMP was dissolved in deionised water and heated for 30 min. in a water bath at 95°C. LM pectin and κ-carrageenan were also dissolved in deionised water and heated for 30 min. in a water bath at 95°C to ensure complete dissolution. Appropriate weights of the solutions were mixed hot with deionised water for the preparation of gels. The pH of the mixtures was 6.45 at 25°C.

2.3. Calcium-Ion Activity

Calcium-ion activities were measured with a Radiometer ion-selective electrode standardised against calcium chloride (10 and 0.1 mM). Activity coefficients were calculated from the Debye-Hückel equation [14].

2.4. Differential Scanning Calorimetry

DSC measurements were carried out with a Setaram micro DSC-III calorimeter, Caluire, France. Approximately 900 mg of the sample solution was sealed hermetically into the DSC pan and the pan then accurately weighed. For each sample a reference pan was filled with distilled water to within ±30 micrograms of the weight of the sample pan. The two pans were then placed inside the calorimeter, heated to 95°C and held at this temperature for 10 min to annihilate the thermal history. The temperature was then lowered to 5°C at 1.0 K/min. and raised again at the same rate to 95°C. Transition temperatures (heating and cooling) were estimated as the mid-peak temperatures (T_h and T_c). Transition enthalpies (ΔH) were estimated from peak areas.

2.5. Rheometry

Oscillatory rheological measurements were made with a Dynamic Stress Rheometer DSR from Rheometrics Co, Ltd, NJ, USA. Parallel plate geometry was used, of diameter 50 mm. The hot sample was poured directly onto the plate of the instrument. The temperature dependence of G' and G'' was observed at a frequency of 1 rad/s. The temperature was then lowered to 5°C at 1.0 K/min. and raised again at the same rate to 80°C. All measurements were made within the linear viscoelastic regime.

Difficulty was encountered with syneresis which was very pronounced in the mixed gels. We therefore also made complementary measurements of absolute shear modulus by the method of Oakenfull, Parker and Tanner [15]. This method is not affected by syneresis because it relies on insertion of a probe into a gel formed in a cylindrical container. Gels (10 g samples) were formed in 'scintillation vials' (radius 12.5 mm). The probe (of radius 1.5 mm) was inserted step-wise into the gel with a series of 5 s pulses at a speed of 0.0847 mm/s. The equilibrium force exerted on the probe was measured at each step and the apparent Young's modulus (Y = stress/strain) calculated from the slope of force *vs* penetration. The time required between pulses to achieve equilibrium depended on the composition of the gel. The absolute shear modulus (G) was calculated from the formula $G = 0.0208.Y$ [15]. This method is equivalent to measurement of G' at zero frequency. In these experiments the gels were equilibrated at 15°C for 18 hours and measured at the same temperature (in a constant temperature room).

3. RESULTS AND DISCUSSION

3.1. Differential Scanning Calorimetry

Fig. 2 shows heating and cooling DSC curves for κ-carrageenan (5 g/kg) with different additions of SMP. The cooling curves showed distinct single exothermic peaks; the heating curves also showed single peaks, but these were broader and not well defined. Values for the enthalpy of the cooling transition (ΔH_c) were estimated from cooling curve peaks. Fig. 3 shows how ΔH_c varies with concentration of κ-carrageenan for two concentrations of SMP (25 and 50 g/kg) compared with equivalent data for κ-carrageenan alone. Within experimental error, the three sets of data lie on the same straight line, suggesting that the addition of SMP has very little (if any) effect on the enthalpy of gelation. From the slope of ΔH_c vs concentration, the enthalpy of gelation (ΔH_g) can be calculated. Regression analysis indicated that ΔH_g might be about 10% lower in the presence of SMP, but the differences were not statistically significant. For κ-carrageenan alone, our value (36 ± 1 J/g) is within the range of values reported by Rochas and Rinaudo [16] for melting of potassium-set gels of κ-carrageenan (35-41 J/g), at the lower end of the range reported by Watase and Nishinari [17] (33-53 J/g) but a little larger than reported by Gekko and colleagues [18] (30-33 J/g).

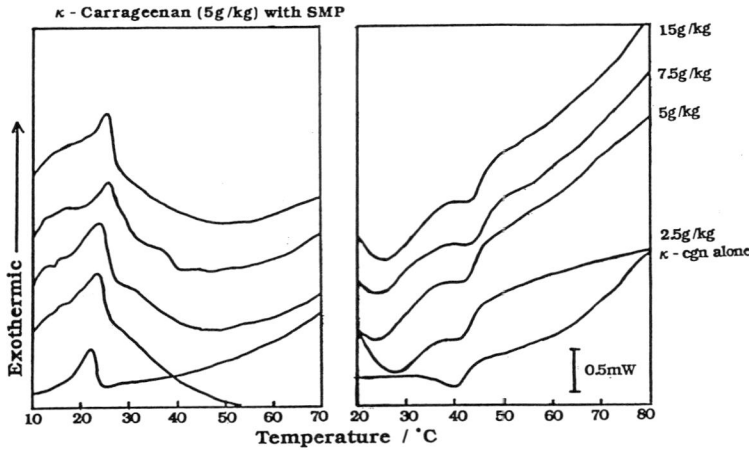

Figure 2. DSC heating and cooling curves for κ-carrageenan (5 g/kg) with different additions of SMP (0-15 g/kg). The left-hand curves follow the same sequence of concentrations of SMP as those on the right.

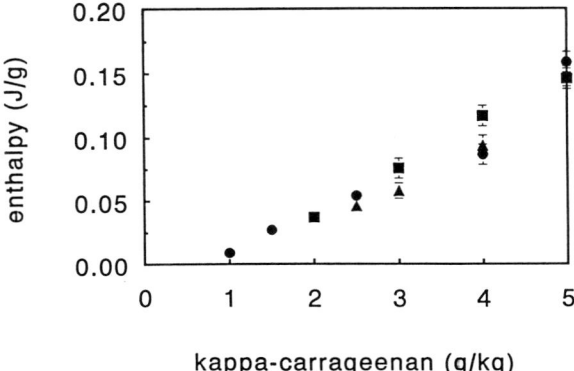

kappa-carrageenan (g/kg)

Figure 3. Enthalpy of the DSC cooling transition (ΔH_c; J per gram of solution) estimated from cooling curve peaks for mixed gels of SMP and κ-carrageenan. ΔH_c plotted against concentration of κ-carrageenan for SMP added at 0 ($\blacktriangle$), 25 ($\blacksquare$) and 50 g/kg ($\bullet$).

The observation that addition of SMP has no experimentally detectable effect on the calorimetric enthalpy of gelation (ΔH_g) can be explained in two ways. A gel network might be formed either (a) from κ-carrageenan alone, with the casein micelles having no significant interaction with the polysaccharide and merely acting as a filler and a source of calcium-ions which promote gelation of the κ-carrageenan or (b) by a mechanism analogous to that proposed by Xu and her colleagues (Fig. 1) [11] in which κ-carrageenan molecules are bound to casein micelles which they link to form a mixed gel network by association of carrageenan double helices - but with only small segments of κ-carrageenan chain adsorbed, leaving the bulk of the polysaccharide as loops and tails, free to form double helices such that the magnitude of ΔH_g appears experimentally to be uninfluenced by proximity to the casein. The more obvious effects of SMP on thermal behaviour during gel melting argue against the first explanation. The melting peak becomes progressively broader with increasing addition of SMP (Fig. 2). That κ-carrageenan is known to interact with casein micelles [2-10] also makes it seem very unlikely that casein takes no part in the gel network.

3.2. Rheometry

3.2.1. κ-carrageenan: The variation of storage modulus (G') and loss modulus (G") with temperature, within a cooling-heating cycle, is shown in Fig. 4 for κ-carrageenan alone (5 g/kg) and its mixtures with skimmed milk powder added at between 2 and 50 g/kg (not all the data are shown). On cooling, G' and G" both increased sharply at about 25-30°C (T_c) and G' became greater than G", indicating that gelation had occurred. On reheating, there was a sharp drop in G' and G" at around 45-55°C (T_h) as the gel melted. At low to intermediate concentrations of SMP (2-15 g/kg), the heating curves showed broadening of the transition, as observed with DSC. However, at higher concentrations of SMP (25-40 g/kg), this effect disappeared and the curves reverted to the smooth, sharp transition observed with κ-carrageenan alone. T_c and T_h measured rheometrically coincided, within experimental error, with the corresponding values found calorimetrically, as shown in Fig. 5, thus confirming that the DSC cooling and heating peaks coincide with onset and disappearance of gelation, respectively.

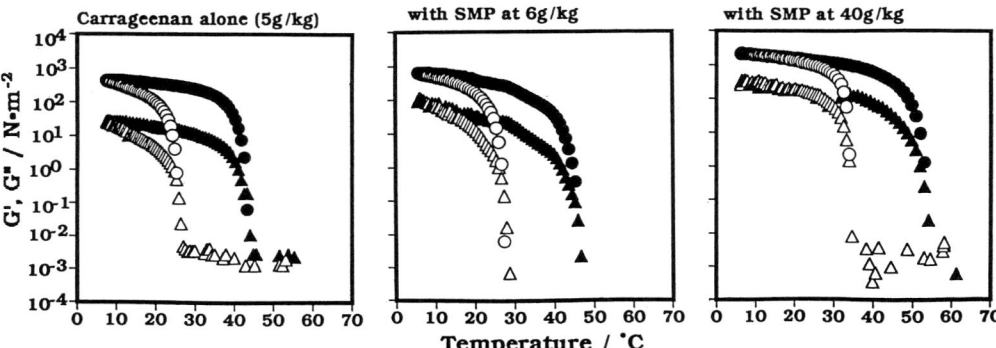

Figure. 4. Storage modulus (G', ● ○) and loss modulus (G", ▲ △) at 1 rad/s as a function of temperature for κ-carrageenan (5 g/kg) with different additions of SMP. Open symbols are for cooling, closed symbols for heating.

In Fig. 6 we show the effects on G' of adding increasing concentrations of SMP to a fixed concentration (5 g/kg) of κ-carrageenan. G' increases with increasing concentration of SMP, but not linearly. In contrast, with a fixed concentration of LM pectin, G' increases linearly with concentration of SMP as shown in Fig. 7. Here, the SMP appears simply to be acting as a source of calcium-ions which promote gelation [19]. The curvature of the equivalent plot for κ-carrageenan (Fig. 6) therefore suggests a more complex interaction with the milk protein.

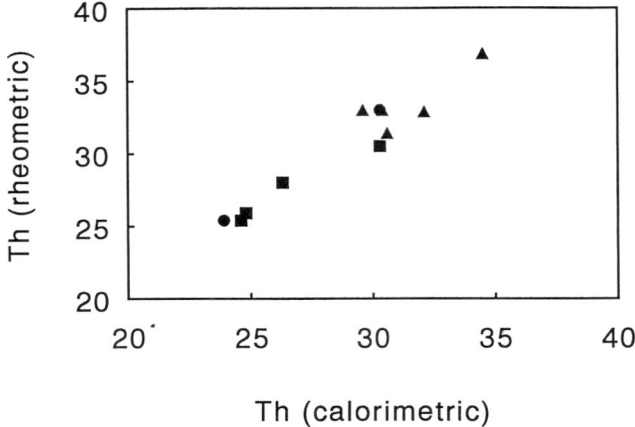

Figure 5. Plot of DSC heating transition temperature (mid-peak temperatures, T_h) against the corresponding rheometrically determined transition temperature (temperature at which G'=G") for different combinations of κ-carrageenan and SMP (▲ 0, ■ 25 and ● 50 g/kg SMP).

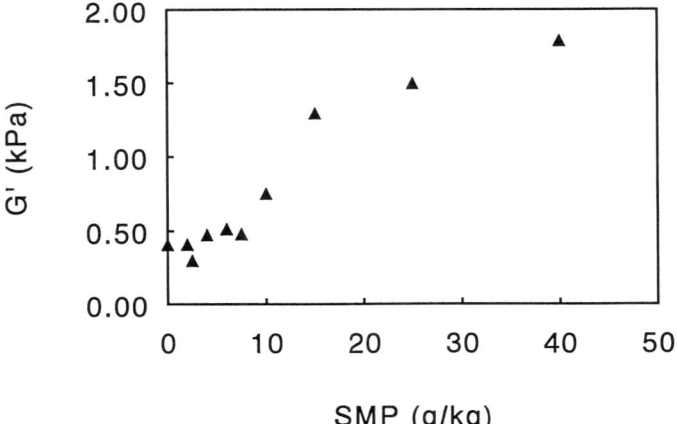

Figure 6. Storage modulus (G') at 15°C and 1 rad/s for κ-carrageenan (5 g/kg) as a function of concentration of the added SMP.

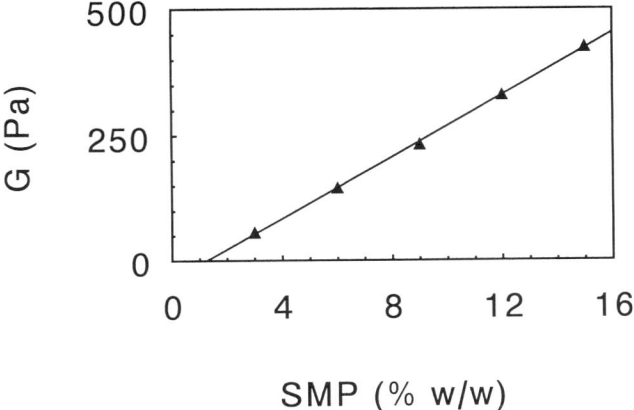

Figure 7. Static shear modulus (G) at 15°C for LM pectin (10 g/kg) as a function of concentration of the added SMP.

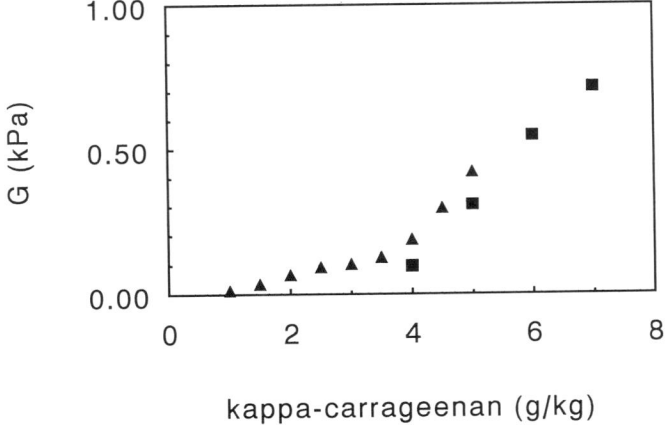

Figure 8. Static shear modulus (G) as a function of concentration of κ-carrageenan for a fixed concentration of SMP (25 g/kg) at 15°C (▲). Equivalent data for κ-carrageenan alone have been added as ■ (see text).

In Fig. 8 we show the effect on the static shear modulus (G) of adding increasing concentrations of κ-carrageenan to a fixed concentration of SMP (25 g/kg). In this case, the curve is biphasic with an abrupt change in slope at approximately 3.5 g/kg κ-carrageenan. For the steep part of the curve, with relatively high concentrations of κ-carrageenan, modulus data for κ-carrageenan alone can be superimposed on the curve for the mixed gel (also shown in Fig. 8). This result can again be interpreted in terms of the gelation mechanism suggested previously. At low ratios of κ-carrageenan to SMP, the gel network consists of concatenations of casein micelles associated *via* bound κ-carrageenan molecules. Once the adsorptive capacity of the casein for κ-carrageenan has been reached, the solution contains free κ-carrageenan molecules which can form an additional network of their own, the modulus then increases more steeply with increasing concentration of κ-

carrageenan. Drohan and colleagues [20] have also noted that at high concentrations of κ-carrageenan, G' is not strongly influenced by concentration of milk protein and that gelation appears to be predominantly the result of association of κ-carrageenan helices. The point of inflexion (*ca* 3.5 g/kg κ-carrageenan) gives an estimate of the adsorptive capacity of casein for κ-carrageenan. SMP contains approximately 26% (w/w) casein [21] which gives an adsorptive capacity of *ca* 0.5 g/g. This is qualitatively in agreement with Dalgleish and Morris' electrophoretic studies [5] which indicate an adsorptive capacity of *ca* 0.28-0.4 g/g.

In contrast, for LM pectin, with fixed concentrations of SMP (100 and 150 g/kg), the shear modulus increased monotonically with the concentration of pectin (c), as shown in Fig. 9. (The approximate dependence of G on c^2 is typical of polysaccharide gels at relatively high concentrations [22].)

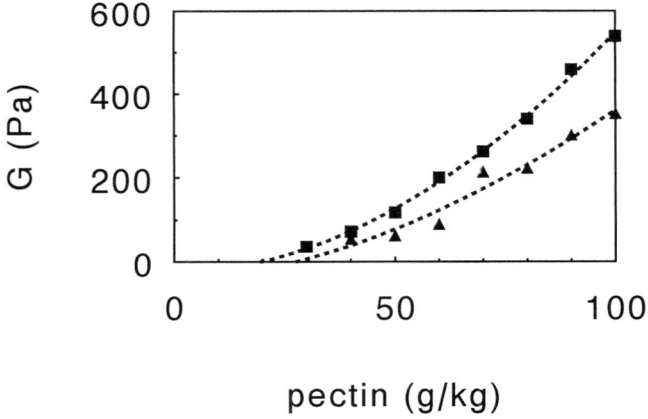

Figure. 9. Static shear modulus (G) at 15°C as a function of concentration of LM pectin with SMP added at 100 (▲) and 150 (■) (g/kg).

3.3. Gelation of κ-carrageenan or LM pectin with SMP Separated by a Dialysis Membrane

κ-carrageenan (2-5 g/kg) was sealed in dialysis tubing and equilibrated with SMP (50 g/kg). The κ-carrageenan formed gels within the dialysis membrane. As only small molecules and ions can pass through the membrane, gelation presumably occurs through 'capture' of calcium-ions from the SMP. Solutions of SMP have a small equilibrium concentration of free calcium ions [19,23]. These ions are free to diffuse through the membrane and bind to κ-carrageenan; equilibrium is restored by loss of calcium from the casein micelles. Fig. 10 shows G' for gels prepared by dialysis compared with gels prepared under equivalent conditions, but with the components intimately mixed. The gel strength was greater when the κ-carrageenan and SMP were intimately mixed than when they were separated by a dialysis membrane.

In contrast, with LM pectin, the gel strength was always greater when the pectin and SMP were separated by a dialysis membrane (Fig. 11). This result suggests that there is interaction between the protein and polysaccharide in the mixed system - but one that strongly disrupts the gel network structure.

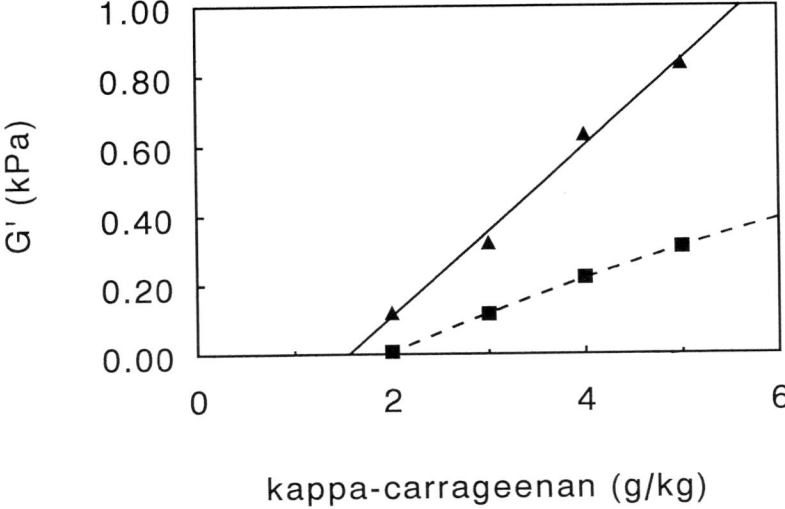

Figure 10. Storage modulus (G′) at 15°C and 1 rad/s for κ-carrageenan and SMP (50 g/kg). Results are compared for the two components intimately mixed (▲) or separated by a dialysis membrane (■) (see text).

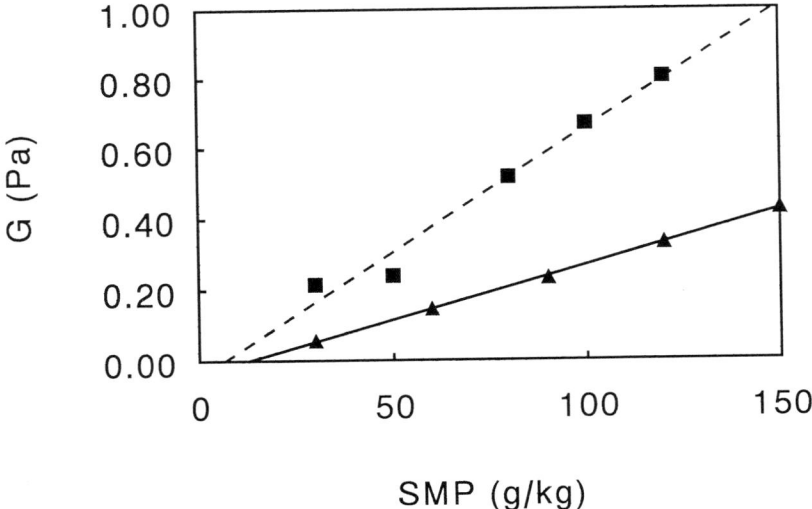

Figure 11. Static shear modulus (G) at 15°C for LM pectin (10 g/kg) and SMP. Results are compared for the two components intimately mixed (▲) or separated by a dialysis membrane (■) (see text).

3.4 Gelation Mechanisms

For κ-carrageenan, in broad outline our results support the mechanism proposed by Xu and her colleagues [11] - i.e. that only part of the κ-carrageenan chain is bound to the casein micelle, with the rest free in solution as loops or tails, able to form a gel network by forming double helices. We are suggesting, however, that perhaps more of the κ-carrageenan remains free to form helices than is indicated by Xu and her colleagues' model. Also, at high ratios of κ-carrageenan to SMP, a purely κ-carrageenan network also appears to form, presumably consisting of more extensive cross-linked κ-carrageenan structures within which the casein micelles are enmeshed.

LM pectin behaves very differently. For equal concentrations of SMP, higher concentrations are required to form gels than for κ-carrageenan [19] - and the κ-carrageenan gels are vastly stronger than the gels formed by LM pectin for the same concentration of polysaccharide [19]. This strongly suggests that there is no direct interaction between the milk proteins and LM pectin, and that the gel network is formed by LM pectin molecules only. The casein appears primarily to be a source of the calcium-ions that enable LM pectin molecules to cross-link and form a gel network. The mixed gel would then consist of a LM pectin matrix in which the milk proteins are embedded. Nonetheless, casein is more than simply a source of calcium ions. The milk proteins have a strong influence on the gelation process and the mixed gel is weaker than would be expected from the shear modulus of the LM pectin alone. This result can be explained by thermodynamic incompatibility between the protein and the polysaccharide causing some degree of separation into protein-rich and polysaccharide-rich phases [24].

REFERENCES

1. M. Glicksman, in *Food Hydrocolloids*, Vol. II, M. Glicksman (ed.), CRC Press, Boca Raton, pp. 73-113, 1983.
2. L.F. Hood and J.E. Allen, *J. Food Sci.*, **42** (1977) 1062.
3. P.M.T. Hansen, *J. Dairy Sci.*, **51** (1968) 192.
4. J. Grinsrod and T.A. Nickerson, *J. Dairy Sci.*, **51** (1968) 834.
5. D.G. Dalgleish and E.R. Morris, *Food Hydrocoll.*, **2** (1988) 311.
6. V. Langendorff, G. Cuvelier, B. Launay and A. Parker, *Food Hydrocoll.* **11** (1997) 35.
7. T.H.M. Snoeren, T.A.J. Payens, J. Jennince and P. Both, *Milchwissenschaft*, **30** (1975) 383.
8. C.F. Lin and P.M.T. Hansen, *Macromolecules*, **3** (1970) 269.
9. B.J. Skura and S. Nakai, *Can. Inst. Food Sci. Technol. J.*, **14** (1981) 59.
10. K. Ozaka, R. Niki and S. Arima, *Agric. Biol. Chem.*, **48** (1984) 627.
11. S.Y. Xu, D.W. Stanley, H.D. Goff, V.J. Davidson and M. Le Maguer, *J. Food Sci.*, **57** (1992) 96.
12. C. Rolin, in *Industrial Gums*, R.L. Whistler and J.N. BeMiller (eds.), Academic Press, San Diego, Chapter 10, p. 257, 1993.
13. R.E. Hargrove and J.A. Alford, in *Fundamentals of Dairy Chemistry*, B.H. Webb, A.H. Johnson and J.A. Alford (eds.), AVI Publishing Co., Westport, Chaper 2, p. 58, 1974.
14. R.A. Robinson and R.H. Stokes, *Electrolyte Solutions*, Butterworths, London, p.229, 1959.

15. D.G. Oakenfull, N.S. Parker and R.I. Tanner, in *Gums and Stabilisers for the Food Industry, 4*, G.O. Phillips, P.A. Williams and D.J. Wedlock (eds.), IRL Press, Oxford, pp 231-239, 1988.
16. C. Rochas and M. Rinaudo, *Carbohydrate Res.*, **105** (1982) 227.
17. M. Watase and K. Nishinari, *Makromol. Chem.*, **188** (1987) 2213.
18. K. Gekko, H. Mugishima and S. Koga, *Int. J. Biol. Macromol.*, **9** (1987) 146.
19. D. Oakenfull and A. Scott, in *Gums and Stabilisers for the Food Industry, 9*, G.O. Phillips, P.A. Williams and D.J. Wedlock (eds.), IRL Press, Oxford, pp. 212, 1998.
20. D.D. Drohan, A. Tziboula, D. McNulty and D.S. Horne, *Food Hydrocoll.*, **11** (1997) 101.
21. B.H. Webb and A.H. Johnson, *Fundamentals of Dairy Chemistry*, AVI Publishing Co., Westport, 1965.
22. D. Oakenfull, *CRC Crit. Rev. Food Sci. Nutr.*, **26** (1987) 1.
23. M.-A. Augustin and P.T. Clarke, *J. Dairy Res.*, **58** (1991) 219.
24. V.B. Tolstoguzov, in *Food Proteins and their Applications*, S. Damodaran and A. Paraf (eds.), Marcel Dekker, New York, pp. 171-198, 1997.

HYDROCOLLOIDS – PART 2
Edited by K. Nishinari
2000 Elsevier Science B.V.

165

Effect of dextran on the thermal and rheological properties of sago starch

P. A. Williams and F. B. Ahmad

Centre for Water Soluble Polymers, North East Wales Institute, Plas Coch, Mold Road, Wrexham LL11 2AW, U.K.

The effect of the presence of dextran on the thermal and rheological properties of sago starch has been studied. Differential scanning calorimetry showed that the gelatinisation process shifted to slightly higher temperatures and became much broader when the dextran concentration exceeded 10%. On standing at 45°C for 48h phase separation was observed above a critical total polysaccharide concentration of ~7%. Starch gelation did not occur at dextran concentrations >1% and this was attributed to inhibition of amylose leaching and/or phase separation.

1. INTRODUCTION

The gelation properties of starch have been reported to be enhanced by the presence of polysaccharides with higher hot paste viscosities and storage moduli having been reported compared to starch alone [1-9]. The increase in the storage moduli for corn starch in the presence of guar, LBG and xanthan has been attributed to phase separation due to incompatibility of the starch and the added polysaccharides [5]. Phase separation was also reported for wheat starch systems in the presence of gelatin, iota-carrageenan, xanthan, kappa-carrageenan and low methoxy pectin [10]. The presence of polysaccharides has been shown to have little effect on the gelatinisation temperature or the gelatinisation enthalpy of starch [8, 11, 12].

This paper is concerned with gaining a fundamental understanding of the effect of the presence of dextran on the gelatinisation and rheological properties of sago starch.

2. MATERIALS

The sago starch was obtained from Mukah, Sarawak, Malaysia and was found to have a moisture content of 13.9%. The amylose content was 30% and the amylose and amylopectin molecular masses were determined by light scattering using 1M KOH as solvent and found to be 1.24×10^6 and 8.64×10^6 respectively.

Dextran was obtained from Sigma Chemical Company and had a reported molecular mass of 2×10^6.

3. METHODS

3.1. Differential scanning calorimetry (DSC)
DSC measurements were performed using a micro DSC (Setaram, Lyon, France). Dextran solutions of various concentrations were prepared and sago starch was added with stirring to give a concentration of 10%w/v. An aliquot of the dispersion (~0.95g) was transferred to the DSC sample cell, the top sealed and the mass recorded. An exactly equal mass of water was added to the reference cell. The cells were heated in the DSC apparatus from 5° - 99°C at a rate of 0.5°C / minute and the heat flow was recorded.

3.2. Rheological properties
The storage and loss moduli (G' and G") were determined using a Carri-Med CSL500 controlled stress rheometer (TA Instruments Leatherhead, U.K.). Starch samples (6%) were prepared on a dry weight basis by adding the appropriate amount to distilled water in a beaker, adjusting the pH to 5.5 and heating for 30 minutes at 95°C with constant stirring (400rpm). The samples were corrected for evaporation loss and then an aliquot was placed into the rheometer measuring system (cone and plate, 4 cm diameter, 2° angle) which was equilibrated at 25°C. Measurements were performed at a frequency of 1Hz with an amplitude of 1 mrad as a function of time over a 6 hour period. At the end of the run a frequency sweep was performed between 0.1 and 10 Hz.
For gel strength (GS) measurements the samples were prepared as above and transferred to a 30ml jar and kept at 25°C in a water bath for 6 h. GS was determined using the Stevens Texture Analyser (Leatherhead U.K.) using a 1 cm diameter probe.

3.3. Phase behaviour
Varying amounts of starch were suspended in dextran solutions of varying concentration and the pH was adjusted to 5.5. The dispersions were heated for 30 minutes at 95°C with constant stirring (400rpm) and tranferred into glass tubes with a screw cap. The samples were kept in an oven at 45°C and inspected after 48h.

4. RESULTS

4.1 Differential scanning calorimetry
Figure 1 shows DSC endotherms for starch/dextran mixtures. In the presence of dextran, the midpoint temperature for gelatinisation, T_{gel}, increased from 70.1°C for starch alone to 72.8°C in the presence of 20% dextran. The onset gelatinisation temperature, T_0 , was not significantly affected but the conclusion temperature, T_c , increased with increasing dextran concentration, i.e. the endothermic peaks became broader. The gelatinisation enthalpy remained almost constant at 16.5 J/g.

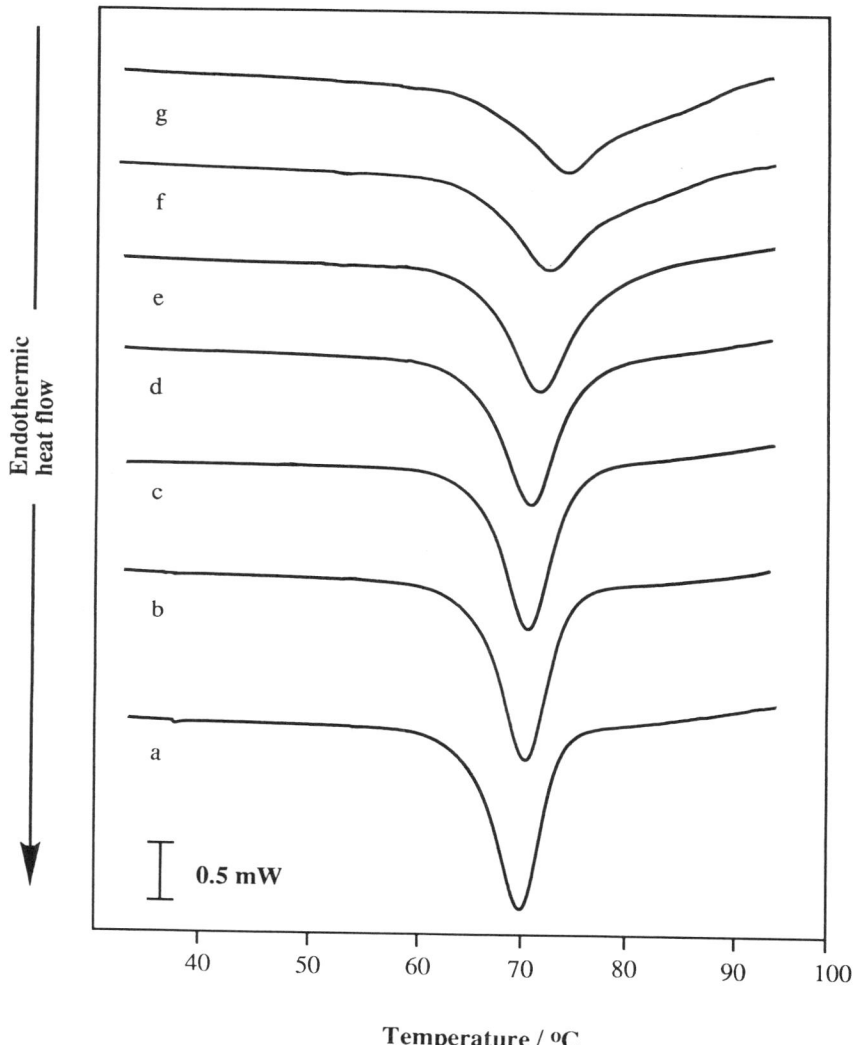

Figure 1: Effect of dextran on the DSC endothermic peaks for 10% sago starch. (a). no dextran (b). 0.5% dextran; (c). 1% dextran; (d). 2.5% dextran; (e). 5% dextran; (f.) 10% dextran; (g.) 20% dextran.

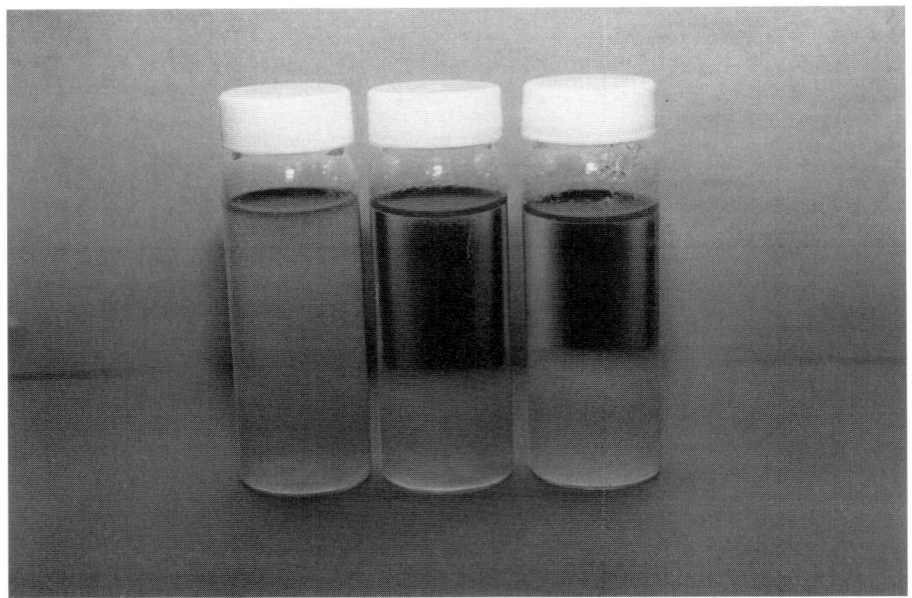

Figure 2. Starch / dextran mixtures after leaving for 48h at 45ºC (a) 7.5% starch, (b) 6% starch / 5%dextran, (c) 6% starch / 10% dextran.

4.2. Phase behaviour

Figure 2 shows the phase behaviour for aqueous starch / dextran mixtures kept at 45º C. In tube (a) which contains starch alone no phase separation was evident, but in tubes (b) and (c), containing 6% starch / 5%dextran and 6% starch / 10% dextran respectively, phase separation occurred. The starch formed a turbid liquid-like layer at the bottom of the tubes as confirmed by iodine staining.

Figure 3 shows the phase diagram of starch / dextran mixtures. The filled circles represent the experimental points where phase separation was first observed and the line is the binodal separating the 1 - phase and 2 - phase regions. For all mixtures a single phase was observed for concentrations below the binodal line and two phases above the binodal line. The critical total polysaccharide concentration for phase separation to occur was ~7%.

Light microscope studies on starch / dextran gels showed that for starch alone a homogeneous gel was obtained (data not shown) and the gel stained blue with iodine. In the presence of dextran, phase separation was observed especially at higher total polymer concentrations and the starch formed the dispersed phase while the dextran formed the continuous phase.

4.3. Rheology

Figure 4 shows the effect of dextran on G' as a function of time for sago starch. The symbols represent the experimental points while the lines have been obtained using the first order rate equation [13]

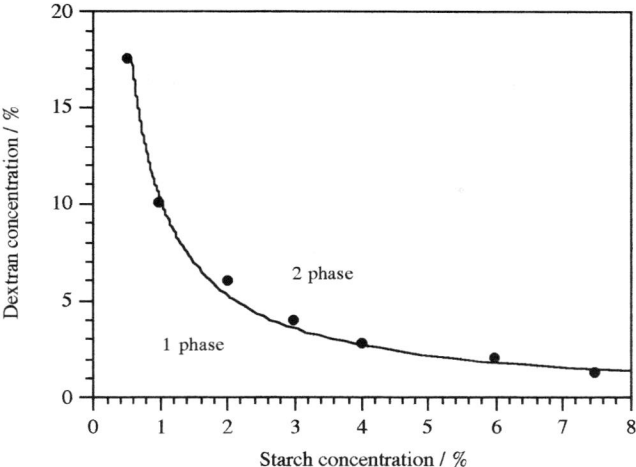

Figure 3. Phase diagram for sago starch/dextran mixtures. Samples were kept at 45º C for 48 h.

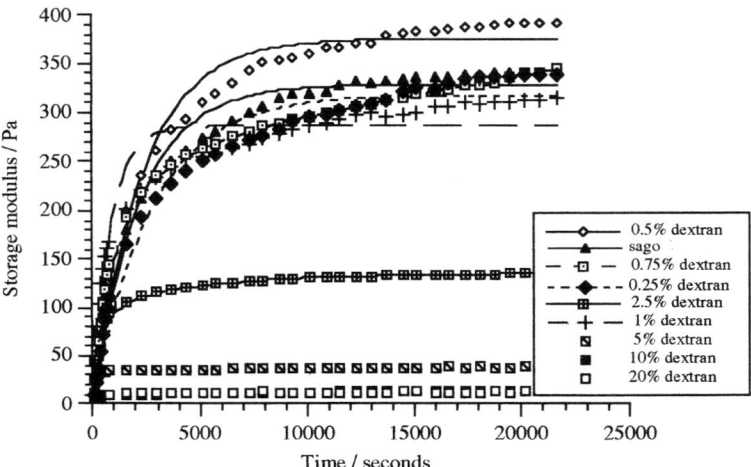

Figure 4. Storage modulus as a function of time for 6% sago starch with varying amounts of dextran.

$$G_t = G_{max} (1 - e^{-kt})$$

where G_{max} is the plateau value of G', k is the rate constant and t is time.

At low dextran concentrations (< 1%) G' was found to rise sharply initially and attain a pseudoplateau value after ~ 4h. The value of G'_{max} was quite similar to starch alone. As the dextran concentration was increased to 2.5% and above, G'_{max} decreased significantly and gelation was inhibited. This is illustrated more clearly in figure 5 which shows G'_{max} as a function of dextran concentration.

Figure 6a shows the mechanical spectra for 6% sago starch in the presence of up to 1% dextran. For these mixtures G' was always higher than G" and while G' was almost independent of frequency G" increased slightly with frequency. The spectra are typical of strong gels. Figures 6b and 6c show the mechanical spectra for 6% starch in the presence of 5% and 20% dextran respectively. For the mixture with 5% dextran at frequencies < 1 Hz, G' was slightly higher than G" while for frequencies > 1 Hz, G" was slightly higher than G'. For starch containing 20% dextran G" >> G' over the entire frequency range studied confirming that gelation had not occurred.

Figure 7 shows the rate constant of gelation for starch as a function of dextran concentration calculated as discussed above. The figure shows that the rate constant increased at dextran concentrations of 1% and above.

Figure 8 shows the gel strength of starch/dextran mixtures. For 6% starch with < 1% dextran the gel strength was close to starch alone but for mixtures with 2.5% dextran the gel strength was reduced significantly. The gel strength results agree well with the small deformation oscillation measurements.

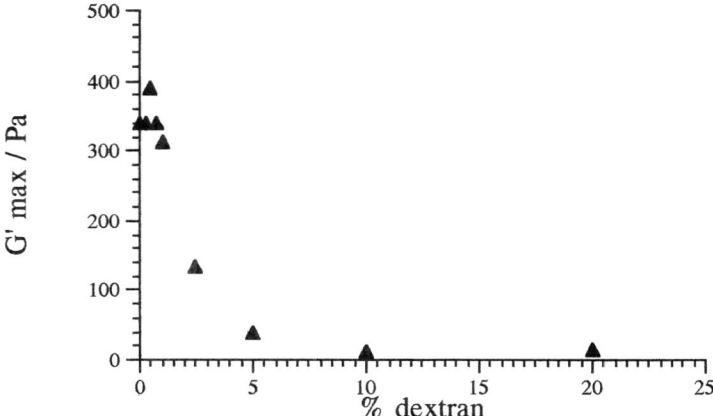

Figure 5. G'_{max} for 6% starch dispersions in the presence of varying dextran concentrations.

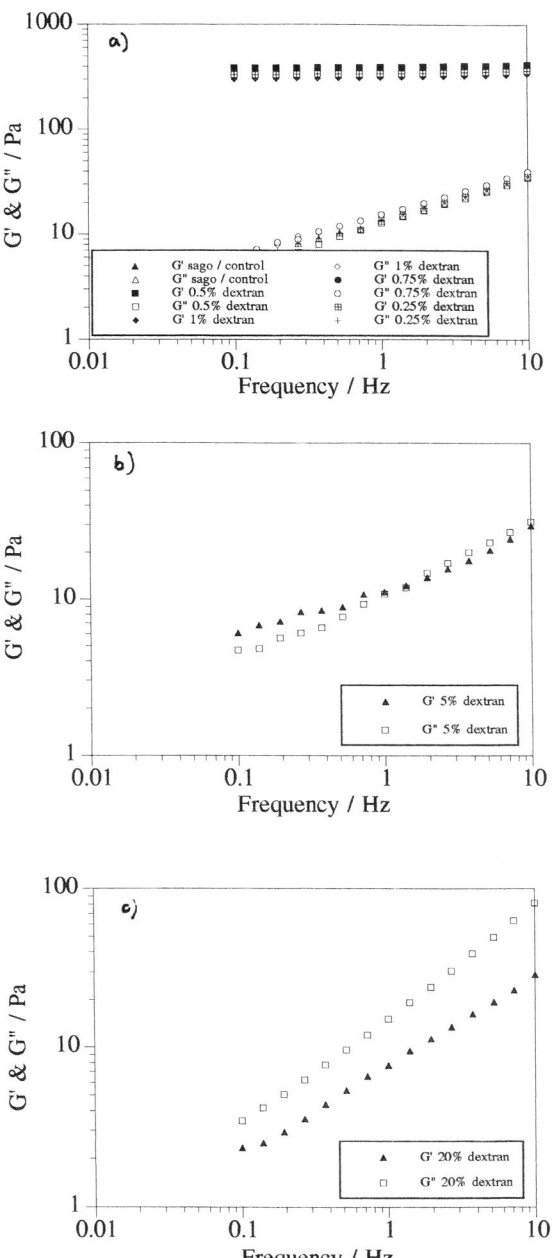

Figure 6: G' and G" for 6%starch dispersions in the presence of (a) 0-1% dextran, (b) 5% dextran (c) 20% dextran

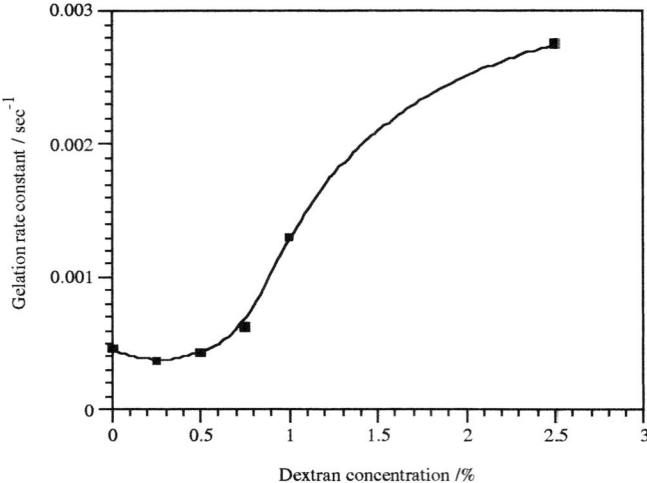

Figure 7: Gelation rate constant for 6% sago starch as a function of dextran concentration.

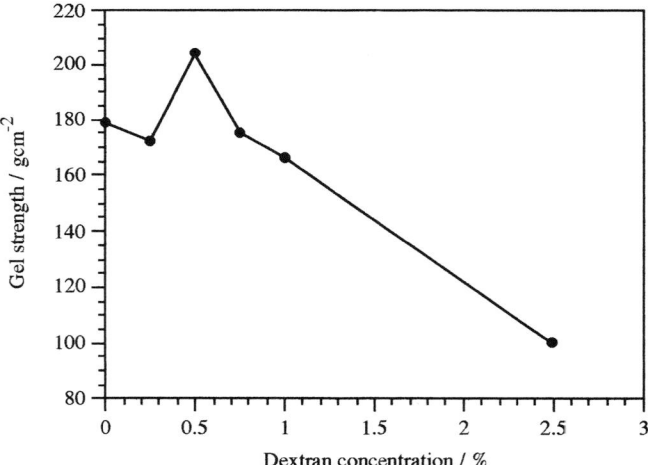

Figure 8. Gel strength of 6% sago starch as a function of dextran concentration.

5. DISCUSSION

5.1. Differential scanning calorimetry

For all the starch/dextran combinations studied the temperature at which the gelatinisation process started, T_0, was almost unaffected by the presence of the dextran while the midpoint temperature of the gelatinisation process, T_{gel}, and the conclusion temperature, T_c, increased particularly at high dextran concentrations.

The increase in ΔT in the presence of added dextran is analogous to the effect observed on increasing the starch concentration and has been attributed by Ferrero *et al* to the reduced availability of water required for the disruption of the crystalline regions within the granule [12]. In studies on corn starch in the presence of xanthan, guar, LBG and sodium alginate these workers showed that the effect was more pronounced with ionic polysaccharides. They found that for 1:10 polysaccharide:starch, T_{gel} was almost unaffected while for 1:1 starch: polysaccharide, T_0, T_{gel}, and T_c increased. Similar studies on corn starch/konjac glucomannan [11] and sweet potato starch in the presence of various cellulose derivatives [7] showed that T_{gel} and T_c increased with increasing polysaccharide concentration.

In our studies ΔH was found to remain almost constant suggesting that the presence of dextran did not significantly influence the swelling and gelatinisation properties. The results are in good agreement with studies on other starch/polysaccharide mixtures [7, 11, 12]

5.2 . Phase behaviour

Phase separation in mixtures of polymers is a general phenomena and depends on various factors such as concentration, molecular mass, temperature, pH etc. For gelatinised starch dispersions the miscibility of the linear amylose and branched amylopectin is generally limited. This was shown by Kalichevsky & Ring [14] in studies on amylose-amylopectin mixtures in aqueous solution. On prolonged heating the amylose-amylopectin solution and standing at 80° C for 48 h two liquid phases were formed. Studies by Svegmark & Hermansson [15] showed that phase separation of amylose-amylopectin in wheat starch occurred more easily compared to potato starch. In this study, 5 - 7.5% sago starch suspensions heated at 95° C for 30 min and kept at 45^oC or 80^oC for 48 h did not result in any observable phase separation. However, samples left for about a week at 80^oC did result in some precipitation.

For 6% starch in the presence of dextran, phase separation was observed on standing the mixture at 45^oC at dextran concentrations above 2.5%. The gels formed at 25^oC were shown to have a starch-rich included dispersed phase and dextran-rich continuous phase. Kalichevsky *et al*. [16] showed similar effects for a mixture of dextran and amylose held at 80^oC. They found that the mixture became translucent after mixing and with time separated into two phases where the upper layer was dextran-rich and the most dilute system showing phase separation contained 2.5% dextran and 2 % amylose. They also found that after mixing a sufficiently concentrated solution of amylose and dextran at 75^oC the mixtures existed as 'water-in-water' emulsions with droplets, either dextran-rich or amylose-rich, dispersed in a continuous phase enriched in the other polymer, consistent with our observations.

5.3. Rheology

For 6% starch in the presence of up to 1% dextran gelation occurred. G'_{max} and GS depended on the dextran concentration. For 6% starch/dextran mixtures, when the dextran concentration was > 2.5% gelation was inhibited. This was also the concentration above which phase separation was observed for samples stored in an oven at 45°C for 48h. It is possible that the dextran acts by inhibiting the amylose from leaching from the granule. Photomicrographs of starch/dextran gels showed that at high dextran concentrations the starch existed as droplets dispersed in a dextran continuous phase and hence formation of a three dimensional amylose network was prevented. Similar observations were reported for dextran/gelatin mixtures by Tolstoguzov et al., [17] and dextran/amylose mixtures (Kalichevsky et al., [16]. The former workers found that in the presence of 0.2% dextran (molecular mass 5×10^5) the compliance for 10% gelatin gels decreased but with higher dextran concentrations the compliance increased and depended on the dextran molecular mass. Kalichevsky et al. [16] showed that for mixtures of amylose and dextran, the rigidity of amylose gels decreased at high dextran concentrations but a higher rigidity was observed at low dextran concentrations.

6. CONCLUSIONS

Dextran has a significant influence on the rheological properties of sago starch. When present at concentrations >1%, gelation is inhibited and this is likely to be a consequence of the fact that phase separation is occurring and / or amylose leaching is inhibited. The effect(s) become more dominant as the dextran concentration increases and at high dextran concentrations (20%) gelation is prevented completely.
Dextran has little influence on the gelatinisation process at low concentrations but when present at concentrations >10% occurs over a broader temperature range. It is likely that this effect is due to the reduced availability of water molecules which act as plasticisers disrupting the crystalline structure within the starch granule.

7. ACKNOWLEDGEMENTS

FBA would like to thank the Commonwealth Scholarship Commission / British Council for financial support and to Universiti Malaysia Sarawak for study leave. The authors are also indebted to Professor John Mitchell for some helpful discussions.

REFERENCES

1. D.D. Christianson, J.E. Hodge, D. Osborne, & R.W. Detroy, . Cereal Chem. **58**, (1981) 513.
2. O. Descamps, P. Langevin, & D.H. Combs, Food Technology , April, (1986) pp 81.
3. S.U. Sajjan, & M.R.R. Rao, Carbohydr. Polym. **7**, (1987) 395.
4. R.J. Tye, Food Hydrocolloids **2**, (1988) 259.
5. M. Alloncle, J. Lefebvre, G. Llamas, & J.L. Doublier,Cereal Chem. **66** (1989) 90.
6. M. Alloncle, & J.L. Doublier, Food Hydrocolloids **5** (1991) 455.
7. K. Kohyama, & K. Nishinari, Journal of Food Sci. **57**, (1992) 128.

8.   M. Annable, M.G. Fitton, B. Harris, G.O. Phillips,& P.A. Williams, P.A. Food Hydrocolloids **8** (1994) 351.
9.   Y.A. Bahnassey, & W.M. Breene,Starch **46** (1994) 134.
10.   I.A.M. Appleqvist, C.R.T. Brown, T.C. Goff, S.J. Lane, & I.T. Norton, In: Gums and Stabilisers for the Food Industry 8, G.O. Phillips, P.A. Williams, and D.J. Wedlock, (eds) IRL Press Oxford, (1996) pp99.
11.   M. Yoshimura,T. Takaya & K. Nishinari,J. Agric. Food Chem. **44,** (1996) 2970.
12.   C. Ferrero, M.N. Martino, & N.E. Zaritzky, J. of Thermal Analysis **47,** (1996) 1247.
13.   M. Yoshida, K. Kohyama, & K. Nishinari, In 'Gums and Stabilisers for the Food ndustry 5', G.O. Phillips, D.J. Wedlock, and P.A. Williams, eds IRL Publishers Oxford (1990) pp 193.
14.   M.T. Kalichesvky & S.G. Ring, Carbohydr. Res. **162,** (1987) 323.
15.   K. Svegmark, & A.M. Hermansson, A.M. Food Structure **10** (1991) 117.
16.   M.T. Kalichesvky, P.D. Orford, & S.G. Ring,Carbohydr. Polym. **6,** (1986) 145.
17.   V.B. Tolstoguzov, V.P. Belkina, V.Ja Gulov, E.F. Titovo, & E.M. Belavzeva, Starch **26,** (1974) 130.

HYDROCOLLOIDS – PART 2
Edited by K. Nishinari
2000 Elsevier Science B.V.

Phase separation and structure of films formed from the gelatin-starch-water system

N.Ptitchkina, N.Panina and L.Khomutov

Department of Polymers, Chemical Faculty, Saratov State University, Astrakhanskaya str. 83, Saratov 410026, Russia

The T-W state diagram of the gelatin-starch-water system, obtained at a fixed total polymer concentration (5 wt %) with variation of temperature (T) and weight-ratio of gelatin/starch (W), consists of three regions: a single-phase solution state (region A), a geterogeneous region of two co-existing liquid phases (B), and the gel state (C). In the present work we have used two methods (DTA and X-ray diffraction) to study some properties of gelatin-starch films formed by isothermal drying at 80^0C, 50^0C, and 20^0C (i.e. with the initial state of the system before drying corresponding to, respectively, regions A, B, and C of the phase diagram).It was found that films formed from region A had an amorphous structure except for those with a dominant starch component where some indications of crystalline ordering could be observed. This crystalline structure of starch (size 3.6-5.1 nm; melting temperature 280 - 285^0C) was also observed for films formed from regions B and C of the phase diagram; when gelatin was the dominant component, these films were characterized by the presence of crystalline structures which melted at 228^0C (for films formed from region B) and 236^0C (for films formed from region C); the gelatin melting temperature did not depend on the composition of the films.

1. INTRODUCTION

Polymer mixtures, which are used in many technological processes,   experience diverse influences - thermal, chemical, mechanical and so on - and give products with characteristics that depend on the initial phase state of the system. Therefore the study of phase states and their limits (i.e. construction of phase diagram) is a problem not solely of theoretical interest, but also of practical importance [1].

On the other   hand, if the diagram of phase states is known, research on the response of the system to external   influences takes on particular significance. Such investigations ensure finally the high standard of predictions in the sequence "phase state - processing - product having predetermined properties".

In an open system at T=$Const$, evaporation of solvent is one of the external influences on solution. In this study, we consider the gelatin-starch-water system. Previous work [2] has shown that the phase diagram of this system consists of three regions: a single-phase state,a heterogeneous region of two co-existing liquid phases, and the gel state. Evaporation of water leads, under certain conditions, to formation of the gelatin-starch films. The objective of the research was to examine the relationship between some properties of these films and the initial phase state of the mixed solutions from which they were obtained.

2. EXPERIMENTAL

2.1. Materials

Commercial samples of food-grade gelatin and soluble potato starch (from, respectively, the Kazansky gelatin plant and Zaspensky starch plant, Russia) were used directly as supplied with no further purification.

Individual stock solutions were prepared by dissolving gelatin and starch at, respectively, 80^0C and 100^0C in distilled water, mixed at 80^0C, and stirred for 15 min, to obtain gelatin-starch-water systems of the required composition. The total polymer concentration (C_{tot}) was 5.0 wt%, the value used previously [2] in determination of the T-W phase diagram. The values of W (weight ratio of gelatin/starch) used were 9:1, 8:2, ..., 2:8, 1:9.

Films were formed on a glass plate under conditions of isothermal drying. Three types of film (A, B, and C) were studied. They differed in the temperature of drying: 80^0C, 50^0C, and 20^0C for films of type A, B, and C, respectively. The initial state of the system before drying was: (i) single-phase solution (films of type A), (ii) two-phase liquid state (type B), and (iii) the gel state (type C). The film thickness was 50 μm.

2.2. Measurements

Film structure was studied by three methods: differential thermal analysis (DTA) of classic type, thermal gravimetric analysis (TGA) and X-ray diffraction. The DTA curves were recorded with a DSC-D calorimeter (production of the company "Universal", Russia) over the temperature range 40^0C-340^0C at a heating rate of 8 ^{0}/min using the air as an atmosphere; sample mass was 14 mg. The TGA measurements were made with a Paulik-Paulik-Erdey deri-vatograph (Hungary) at a heating rate of 10^0/min in an air-atmosphere; sample mass was 100 mg.

Diffractograms of films were obtained with a DRON-3 device (from the former Czechoslovakia) using FeK_{α}–radiation.

3. RESULTS AND DISCUSSION

Figures 1a, 1b, and 1c show DTA curves obtained for films of types A, B, and C, respectively, at different values of gelatin/starch weight-ratio (W). It can be seen that a broad endothermic region with its minimum at 100-110^0C is observed for all films. This endotherm could be eliminated by futher drying of the films at 190^0C, indicating that it arises from evaporation of residual moisture .

For type A films formed from gelatin alone, the heating trace shows a second endothermic region in the temperature range above 250^0C (Figure 1a; curve 1). We attribute this endotherm to thermal degradation of gelatin. This assumption is in agreement with the data of thermal gravimetric analyses. The TGA data showed that the loss of sample mass was 3 % at 250^0C, 7% at 270^0C and continued to increase with increase in temperature; the highest value of the gelatin degradation rate (4.5 mg/min) was observed at 310^0C.

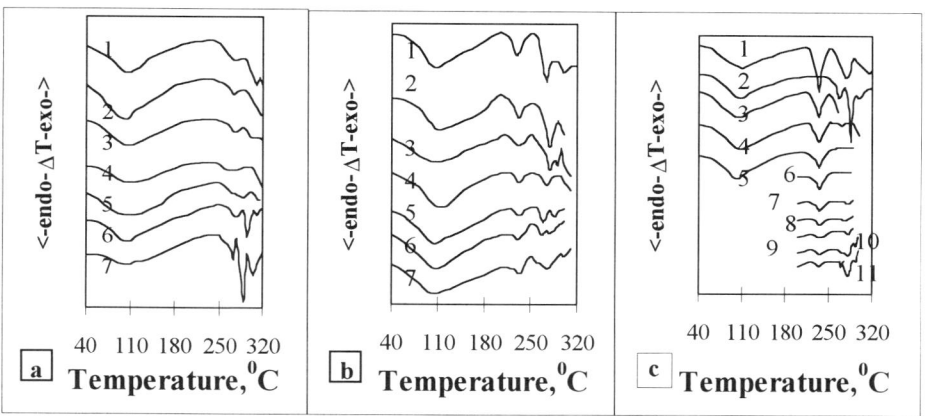

Figure 1. DTA curves for films formed at (a) 80⁰C, (b) 50⁰C and (c) 20⁰C
with gelatin/starch weight-ratio of $W=$ (a) 10:0 (1), 7:3 (2), 5:5 (3), 4:6 (4),
3:7 (4), 2:8 (6), 1:9 (7); (b) 10:0 (1), 9:1 (2), 8:2 (3), 7:3 (4), 6:4 (5),
5:5 (6), 4:6 (7); (c) 10:0 (1), 0:10 (2), 9:1 (3), 8:2 (4), 7:3 (5), 6:4 (6),
5:5 (7), 4:6 (8), 3:7 (9), 2:8 (10), and 1:9 (11)

As the proportion of starch in the films is increased (i.e. as the gelatin content is lowered),
the magnitude of the endothermic region at $T>250^{0}C$ decreases. When starch becomes the
dominant component, an endothermic triplet appears in the DTA curves. This triplet is also
present in DTA curves for films of types B and C at high content of the starch, indicating that

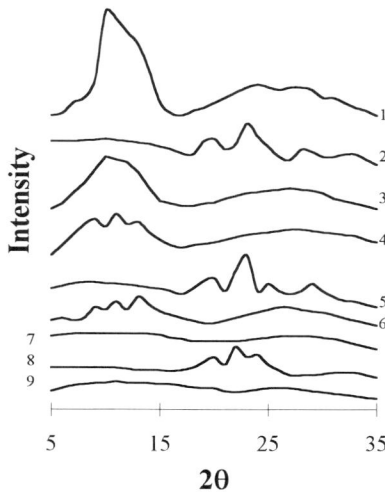

Figure 2. Diffractograms of films formed from the gelatin-starch-water system ($C_{tot}=$
5 wt%) at 20⁰C (1-3), 50⁰C (4-6) and 80⁰C for $W=$ 10:0 (1,4,7); 8:2 (3,6,9);0:10 (2,5,8)

it arises from melting of ordered structures of starch that develop during film formation. The presence of such structures is confirmed by X-ray diffraction data (Figure 2; curves 2, 5 and 8), which indicate a crystalline size of 3.6 - 5.1 nm. According to the DTA data, the average melting temperature of these structures is 280-285°C. This is appreciably higher than the value reported by Whittam and co-workers ($T_m^0 = 257^0 C$) [3]. The difference can obviously be explained by differences in the method of preparation (and, hence, in the quality) of the crystalline structures.

The X-ray patterns show no crystalline ordering for "hot" (type A) films from gelatin alone or from gelatin-starch mixtures of high gelatin content (Figure 2; curves 7 and 9). For "cold" (type C) gelatin films , a broad peak centred at $2\theta = 10.1^0$, corresponding to a distance of 1.1 nm, is observed (Figure 2; curve 1). This peak becomes broader (Figure 2; curve 3) with decrease in gelatin content (increase in starch content at constant C_{tot}). The broadening indicates a higher proportion of amorphous regions (i.e. incorporation of starch does not compensate for the associated reduction in the content of crystalline regions attributable to gelatin.

For "warm" (type B) gelatin films (cast at 50°C), two new peaks are observed (Figure 2; curve 4) on either side of the peak at $2\theta = 10.1^0$. They correspond to a new type of structural organization with distances 1.2 and 0.94 nm. Their intensity decreases with decreasing of gelatin content (Figure 2; curve 6).

The DTA data (Figures 1b and 1c) are in agreement with the X-ray diffraction picture. Indeed, the endothermic peaks at 228°C (for type B films) and 236°C (for type C) can be attributed to the melting of crystalline structures formed by the gelatin component of the films [4]. These structures differ not only in the melting temperature, but also in the form of the peaks, which are sharper for type C, and in the area enclosed by the DTA trace and the baseline. As shown in Figure 3, the peak areas are larger for films of type C than for those of type B (i.e. the crystalline structures in films cast from the gel state have greater thermodynamic stability than those obtained from solutions). The melting temperatures of the gelatin crystalline structures do not, however, depend on the composition of the films, indicating that these structures are formed with no contribution from the starch component. Similar behaviour has been observed [2] in gelatin-starch gels: the gelation temperature and gel strength of the mixed systems were dominated by the gelatin component, with no indication of network formation by starch.

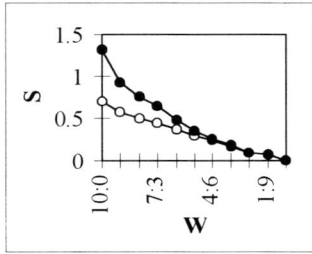

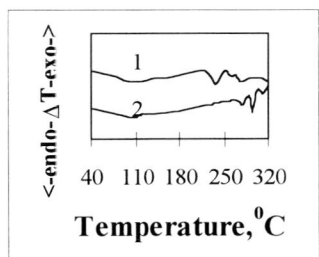

Figure 3. Variation of peak area (S ,in arbitrary units) for the gelatin endotherm in DTA curves formed at 20°C (o) and 50°C (o) with weight-ratio of gelatin/starch (W).

Figure 4. DTA heating curves of films formed from the upper (1) and lower (2) layers of the separated gelatin-starch-water system ($C_{tot} = 5$ wt% ; $W = 7:3$).

One more experimental finding is worthy of mention. In region B of the phase diagram, the gelatin-starch-water system consysts of two liquid phases which , over time, resolve into two layers [2]. We separated them and formed films from each. DTA curces of these films, at $W =$ 7:3 , are shown in Figure 4. It can be seen that curve 1 (recorded for the film cast from the upper layer) contains the endothermic peak   corresponding to the melting of gelatin crystalline structure ($T_m = 228^0C$), while curve 2 (from the lower layer) is characterized by the presence of the endothermic triplet typical of starch. It seems clear, therefore, that the upper layer is rich in gelatin and the lower layer is rich in starch.

REFERENCES

1. V.I. Klenin. "Thermodynamics of Systems with Flexible Polymers", Publish. House of Saratov State University, Saratov, Russia, 1995 (in Russian).
2. L.I. Khomutov, N.A. Lashek, N.M.Ptitchkina, and E.R. Morris, Carbohydr. Polym., 28 (1995) 341.
3. M.A. Whittam, T.R. Noel, and S.G. Ring, in E.Dickenson (ed.) Food Polymers, Gels and Colloids, pp. 277-288, Royal Society of Chemistry, Cambridge, 1991.
4. N.A. Lashek and L.I. Khomutov, Visocomolec. Soed. 31 (1989) 680 (in Russian).

HYDROCOLLOIDS – PART 2
Edited by K. Nishinari
2000 Elsevier Science B.V.

183

Interaction of kappa-carrageenan and β-casein

Y. Sano and T. Hiyoshi

National Food Research Institute, Tsukuba City, Ibaraki 305, Japan

Complex formation between κ-carrageenan and β-casein micelle, in the presence of calcium ions under non-gelling conditions, was investigated with laser light scattering and small-angle X-ray scattering measurements. The molar binding coefficient of β-casein micelle to κ-carageenan, radius of gyration, Kratky plots and the distance distribution function of the complex suggest that several β-casein micelles combine here and there to the core molecules of the κ-Carrageenan.

1. INTRODUCTION

Especially, in the field of food, it is important to develop new functional materials which control physical texture, such as hardness and softness. In particular, polysaccharides are widely used in the dairy industry as stabilizing, thickening and gelling agents. κ-carrageenan, a red seaweed extract composed of alternating α-(1-3)-D-galactose 4-sulfate and β-(1-4)-3,6 anhydro-D-galactose, has gel forming and milk protein stabilizing ability. The mechanisms of gelation and stabilization are considered a consequence of the presence of the 3,6-anhydride which enhances the gel forming ability by increasing the capability of forming a double helix. Moreover, the loose ends of these double helices serve as junction zones to connect other strands of carrageenan to form the three-dimensional network [1]. Using electron microscopy, Chakraborty and Hansen [2] observed a relationship between casein stabilization and double helix junction zones formed by carrageenan. The stabilization of casein against calcium ion precipitation is achieved by entrapping small casein aggregates and keeping them apart so as to prevent further aggregation. Chakraborty and Randolph [3] proposed that the specificity of the interaction of κ-carrageenan with calcium-sensitive proteins lay in the specific conformational features unique to κ-carrageenan. Ozawa et al. [4] investigated the relationship between gelling and casein stabilizing ability of κ-carrageenan using viscosity and turbidity measurements. They showed that the viscometric behavior of κ-carrageenan-β-casein mixture in the presence of calcium ions was governed by the properties of β-casein rather than that of κ-Carrageenan and that the mixture began to gel at lower concentration of calcium than 5mM, suggesting the participation of β-casein in the formation of the κ-carrageenan structure. Ozawa et al. [5] also showed that the existence of β-casein in the system promoted the gelation of κ-carrageenan in the presence of calcium ions and that β-casein increased the strength of calcium gels of κ-carrageenan with increasing sodium ions and strengthed the κ-carrageenan-calcium gel at neutral pH and suggested that β-casein may participate in the gelation of κ-

carrageenan through the mediation of calcium ions. Snoeren et al. [6] showed that an electrostatic attraction occurs between carrageenan and κ-casein, but not β-casein. He assumed that this was due to the positive charges between residues 97 and 112 of κ-casein, which are absent from β-casein.

β-casein exhibits the characteristics of an amphiphile, since the N-terminal portion of β-casein is rich in polar and the C-terminus consists of dominant non-polar units(hydrophobic residues). Because of an amphiphilic character, it forms a stable micelle depending on factors such as casein concentration, the ionic strength of the buffer and temperature.

In the presence of calcium ions the formation of β-casein micelle is induced and its size increases with increasing concentration of calcium as indicated by small-angle X-ray scattering (SAXS) measurements [7]. However, the structural characterization of the complex formation between κ-carrageenan and β-casein micelle is now open. We investigated the mixture of them in the presence of calcium under non-gelling conditions with SAXS. The results suggested the complex models which several β-casein micelles combined here and there to the core molecules of the κ-carageenan with the data of molar binding coefficient of β-casein micelle to κ-carageenan, radius of gyration, Kratky plot and the distance distribution function of the mixture.

2. MATERIAL AND METHODS

2.1. Materials

Crude β-casein was separated from acid whole casein by urea fractionation and purified chromatographically on a DEAE-cellulose column where the column was equilibrated in 0.22M phosphate buffer (pH 7.0) in advance. The protein was eluted using a 0-0.3 M linear gradient of NaCl in the same phosphate buffer and freeze-dried after sufficient dialysis against deionized water at 2oC. Dried β-casein was dissolved in 20mM imidazole/HCl buffer at pH 7.0 by stirring in the cold (4oC) overnight.

κ-carrageenan was purchased from Sigma Chemical Co. It was dissolved in deionized water. After passing through the ionic exchange resin column, the counter ion was fixed with sodium ion to the original pH. This sample was fractionated with ethanol to five fractions. The concentration of κ-carrageenan was determined by weight and checked with CTAB method [8].

2.2. Turbidity measurements

Turbidimetric measurements of β-casein and the mixture with κ-carrageenan in the presence of calcium ions were carried out with a Shimadzu UV-1600 spectrophotometer [9,10].

2.3. Laser light s cattering measurements

Measurements of the intensity of the scattered light were carried out with a modified Ellipsometer, an Automatic Light Scattering Analyzer AEP-100 [11]. The linearly polarized monochromatic incident light passes through the cylindrical scattering cell, and the light scattered at any scattering angle is detected throgh the linear analyzer by a photomultiplier in a telescope arm that can rotate from 45o to 135o by a stepping motor. The reduced scattering intensity as a function of the scattering angle was checked with a polystyrene sample (prepared by the Pressure Chemical Co. Batch 50124) with a molecular weight of 254,000.

2.4. Small-angle X-ray scattering (SAXS) measurements

SAXS experiments were carried out with the optics and detector system of SAXES in the Photon Factory, KEK, Tsukuba, Japan [12,13]. A wavelength of 0.149 nm was used. The counting time was 600 sec for each measurements. The net scattering intensities were calculated by subtracting the intensities of a buffer solution from those of sample solutions.

3. RESULTS AND DISCUSSION

3.1. Reversible β-casein micelle formation in the presence of calcium ions

Milk contains about 32mM calcium of which 10 mM calcium exists in the form of free ions or calcium phosphate in milk serum. So, it is important to investigate the association of β-casein in the presence of calcium ions. β-casein is sensitive to calcium because of its five phospho-serine residues. It is believed that calcium ions form a bridge between β-casein micelles and induce the precipitation of those complexes.

However, the structural characterization of β-casein micelles in the presence of calcium ions is now open. In this study, we carefully investigated β-casein micellar structure in the presence of calcium ions by means of SAXS measurements. According to turbidity measurements, calcium-precipitated of β-casein micelles can be redissolve at low temperatures, suggesting these reactions to be reversible and not be irreversible, as is shown in Fig. 1. In Fig. 2, the scattering intensity I(Q) is plotted against the magnitude of the scattering vector Q.

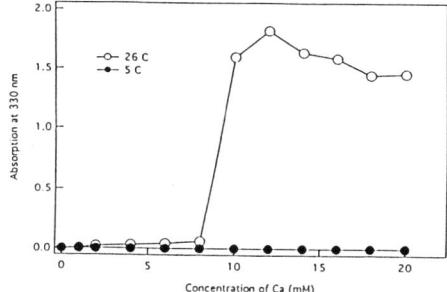

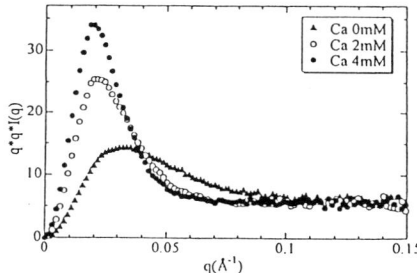

Fig. 1. Reversibility of temperature dependent turbidity of β-casein in the presence of calcium ions.

Fig. 2. Kratky plots for β-casein (10mg/ml) in the presence of calcium ions at 20°C.

A rapid decrease of observed scattering intensities I(Q) at higher Q values would indicate a rather compact structure of β-casein micelles in the presence of calcium ions. The radius of gyration of compact molecules is determined from the initial slope of the so-called Guinier plot where ln I(Q) is plotted against Q^2. Extrapolation of I(Q) to Q=0, which is proportional to the molecular weight of the scatterer. The radius of gyration Rg and the weight average molecular weight of β-casein micelles increased with calcium concentration, as shown in Table 1.

The radius of gyration indicates an approximate size of a solute molecule, but its shape can not be specified by this single parameter. In synthetic polymer chemistry the Kratky plot is usually employed to evaluate the shape and the order of extension of a linear or branched-chain polymer. Even if polymers have the same Rg value in solution, Kratky plots reveal differences in shape and chain extension (e. g., a spherical shape

Table 1. Characterization of beta-casein in the presence of calcium ions

Ca (mM)	Rg (nm)	Mw×10⁻⁵	n-mers	D_{max}(nm)
0	6.05	4.65	19	25.5
2	7.79	14.0	58	25.4
4	8.22	20.0	83	27.3

yields a sharp peak, a Gaussian chain a horizontal line and a rigid rod a straight line with a slope one). As is shown in Fig. 2, the Kratky plot indicated that in the presence of higher concentrations of calcium ions the Kratky plot showed a more distinct single peak and a higher peak, indicating greater compactness of β-casein micelles in the presence of calcium ions. Only a small peak was observed under no calcium conditions.

Another index of polymer chain conformation is a distance distribution function, P(r). This function is obtained by Fourier-transformation of the X-ray scattering intensity function, I(Q), multiplied with Q.r, and represents a statistical distribution function of a pair of points being separated by a distance of r within a polymer molecule. The distance distribution function which is shown in Fig. 3 showed symmetric shape in each case. The size of β-casein micelles seem at almost independent of calcium ion concentration present and was about 26 nm.

The detailed analysis of SAXS data showed that the size of β-casein micelles increased through the enhancement of the hydrophobicity of the C-terminal portions because of shielding of the electric repulsion between the negative charge of micelle due to N-terminal portions in the presence of calcium, and that the internal structure of β-casein micelles could change more compact spherical shapes reversibly even in the presence of calcium ions. The P(r) function in the present work exhibited the characteristic of sphere and not oblate ellipsoid suggested as a previous model of the micelles [14].

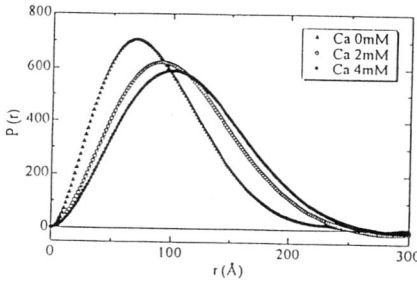

Fig. 3. Distance distribution function P(r) for β-casein.

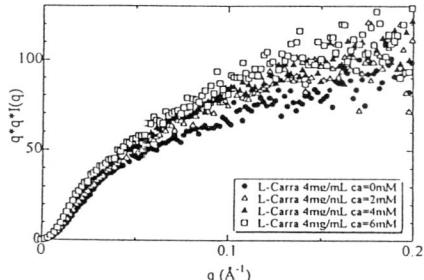

Fig. 4. Kratky plots for κ-carrageenan in the presence of cacium ions.

3.2. Structural characterization of κ-carrageenan in the presence of calcium ions

SAXS data showed that Rg of fractionated κ-carrageenan (Na-type) was almost independent of concentration. Rg increased only 1.4 times with an increase of calcium ion concentration to over 6 mM. The radius of the cross section was obtained from SAXS data as 0.38 nm in the absence and 0.46 nm in the presence of 6 mM calcium ion. The Kratky plots of κ-carrageenan in the presence of calcium ions are shown in Fig. 4, indicating that carrageenan has a fairly flexible structure slowly decreased in flexibility with the increase in calcium ion concentration.

3.3. Complex formation between β-casein and κ-carrageenan mixture in the presence of calcium ions

The complex formation between β-casein micelle and κ-carrageenan mixture was firstly examined by static laser light scattering. In the studies of interaction of nonidentical polymers such casein and carrageenan, either of two methological scheme are generally used; one is the measurement of the concentration dependence of the scattered intensity while keeping constant the ratio of the concentrations of both polymer components. In this case, the scattering intensity for the mixed polymer solution is measured by assuming them to be solutions of one solute whose weight concentration is equal to the sum of the two components. The second procedure is the measurement of the dependence of the scattered light on the concentration of one polymer at a constant concentration of the other polymer, i.e., the latter polymer is regarded as part of the mixed solvent. We have used the first method. The results were shown in Fig.5 with open rectangles for κ-carrageenan and open circles for β-casein, respectively, and closed rectangles for mixed solution of β-casein and κ-carrageenan at weight mixing ratio (g casein/g carrageenan) of 20.

In the electromagnetic scattering from a solution containing L-macromolecular components, the weight-average molecular weight M_w and the z-average radius of gyration obtained are given by the weighted sum of those of L-th components. The following equation can be obtained for a mixed solution of two macromolecular solutes described by the subscripts 1 and 2 [15]:

$$M_w \, \phi^2 = M_1 \, w_1 \, \phi_1^2 + M_2 \, w_2 \, \phi_2^2$$

where M_i, w_i and ϕ_i are molecular weight shown in Fig. 5, weight fraction (=weight concentration of i component/total weight concentration for mixed solution) and refractive index increment of component i, respectively and ϕ is refractive index increment of mixed solution given by weight averages from two components, i.e., $\phi^2 = w_1 \, \phi_1^2 + w_2 \, \phi_2^2$. The results calculated according to these equations, assuming no interaction between β-casein micelle and κ-carrageenan, that is, by setting 1=β-casein and 2=κ-carrageenan,

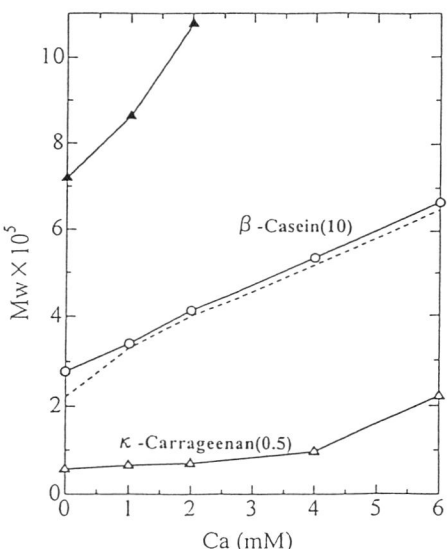

Fig. 5. Weight-average molecular weight M_w obtained by static light scattering for κ-carrageenan ($\triangle$), β-Casein ($\bigcirc$) and mixed solution of them ($\blacktriangle$), respectively, at mixing ratio (g casein/g carrageenan) of 20.

are shown in Fig. 5 by dotted curves and showed a large deviation from the experimental valiues. This means that β-casein molecules interacts with κ-carrageenan molecules strongly and that a complex is formed between them.

If the affinity between β-casein micelle and κ-carrageenan is strong, the complex is formed to as great an extent as one of these components is mostly used up. In this instance, the concentration of κ-carrageenan component is small, so the system can be considered that a part of casein micelle is left free in the solution and that many casein micelles are bound to one carrageenan molecule. Therefore, the weight-binding coefficient can be obtained by using the experimental value of M_{app} and all other quantities with using above-mentioned equation. [15]. The molar binding coefficient of β-casein micelles to κ-carrageenan molecule obtained was 2.87 for 0mM, 3.15 for 1 mM and 3.57 for 2 mM calcium ions, respectively.

On the other hand, SAXS data for the mixture in the presence of calcium ions under non-gelling conditions was almost the same as β-casein micelles only, because the contribution of κ-carrageenan to the scattering length density was negligible compared with the contribution of β-casein micelles.

These results suggest the complex of β-casein and κ-carrageenan has several β-casein micelles combined here and there to the core molecules of the κ-carrageenan.

REFERENCES

1. D.A.Rees, Adv. Carbohyd. Chem. Biochem., 24(1967)267.
2. B.K.Chakraborty and P.T.Hansen, J. Dairy Sci., 54(1971)754.
3. B.K.Chakraborty and H.E.Randolph, J. Food Sci., 37(1972) 719.
4. K.Ozawa, R.Niki and S.Arima, Agric. Biol. Chem., 48(1984)627.
5. K.Ozawa, R.Niki and S.Arima, Agric. Biol. Chem., 49(1985)3123.
6. T.H.M.Snoeren,T.A.J.Payens,J.JeurnikandP.Both,Milchwissenschaft,30(1975)393.
7. T.Hiyoshi, Y.Sano and R.Niki, The 4th Workshop on Principle of Protein Architecture, Abstract 196 (1997).
8. Y.Sano, Bull. Natl. Inst. Agrobiol. Resour. 1(1985)53.
9. Y.Sano, Carbohydrate Polym., 33(1997)125.
10. X.Q.Liu and Y.Sano, J. Protein Chem., 17(1998)478.
11. Y.Sano, J. Colloid Interface Sci., 139(1990)14.
12. Y.Sano, H.Inoue, Y.Hiragi, H.Urakawa and K.Kajiwara, Biophy. Chem., 55(1995)239.
13. Y.Sano, H.Inoue, K.Kajiwara, Y.Hiragi and S.Isoda, J. Protein Chem., 16(1997)151.
14. K.Kajiwara, R.Niki, H.Urakawa, Y.Hiragi, N.Donkai and M.Nagura, Biochim. Biophy. Acta, 955(1988)128.
15.Y.Sano, Bull. Natl. Inst. Agrobiol. Resour., 2(1986)1; Y.Sano et al., J. Biochem., 115(1994)1058; Bull. Chem. Soc. Jpn., 45(1972)1011.

HYDROCOLLOIDS – PART 2
Edited by K. Nishinari
2000 Elsevier Science B.V.

189

Thermal properties of alginic acid-polylysine molecular composites

K. Nakamura[a], E. Kinoshita[a], T. Hatakeyama[a] and H. Hatakeyama[b]

[a] Otsuma Women's University, 12, Sanban-cho, Chiyoda-ku, Tokyo 102-8357, Japan

[b] Fukui University of Technology, 3-6-1, Gakuen, Fukui , 910-0028, Japan

Alginic acid (Alg) - polylisine (PLys) molecular composites (Alg-PLys) were prepared by ionic crosslinking under the acidic condition.   In order to investigate the effect of molecular conformation on the above composites formation, several kinds of Alg's with different guluronic (G) and mannuronic acid (M) ratios were used.

Thermal properties of Alg-PLys   in the dry and wet states were measured by differential scanning calorimetry (DSC).   The glass transition temperature of Alg-PLys composites was observed at a temperature higher than the Tg's which was calculated by assuming a simple mixing law of two components.   Amorphousity of the composites was evaluated via restrained amount of non-freezing water, which was calculated from the enthalpies of melting and crystallization of water.   The obtained results indicate that the molecular motion of Alg-PLys composites composed by G rich Alg is easily enhanced due to the bulky molecular conformation.

1. INTRODUCTION

Alginic acid (Alg) is a copolysaccharide extracted from *brown masse algae* consisting of D-mannuronic acid (M component) and L-guluronic acid (G component) [1].   Sodium alginate   (NaAlg) is a representative polyelectrolyte used widely in various fields, such as food and textile industries.   Although NaAlg is readily dissolved in water, water insoluble alginates can be prepared when sodium ions of NaAlg are replaced by di- and trivalent cations. Among various ions, such as Fe^{2+}, Cu^{2+}, Ca^{2+} and Al^{3+}, Ca^{2+} has received particular attention, since it is thought that a unique "egg-box structure" is formed when G components enclose Ca ions [2].

We have studied the thermal properties of water-NaAlg and water-CaAlg systems and found that the functional properties of alginates are affected by the presence of bound water [3~6].   Polylysine (PLys) is a polycation and is used as an antiseptic agent in food. In this study, alginic acid-polylysine molecular composites (Alg-PLys) with various PLys/Alg ratios were prepared. Thermal properties of the composites were investigated by differential scanning calorimetry (DSC).

2. EXPERIMENTAL

2.1. Sample Preparation

Sodium alginate (NaAlg) samples for bioreactor use having 0.5 and 1.0 of M/G ratio were commercially obtained from Kibun Food Chemipha Co Ltd. The degree of substitution (DS)

was 1.0. ε-Polylisine (PLys) was obtained from Tisso Co Ltd. The degree of polymerization was 25 to 35, and molecular weight was about 4700. Acetic acid was also commercially obtained.

Figure 1 shows the preparation scheme of Alg-PLys molecular composites. The method was as follows. (1) 1 weight % of NaAlg and 2 weight % of PLys aqueous solutions were prepared. (2) PLys solution was slowly added to NaAlg solution using a homogenizer. (3) Mixing ratio (PLys/Alg) of both solutions was varied from 0 to 1.0 mol/mol. (4) Mixed solution of Alg and PLys were cast on glass plates and dried at room temperature. Films, thus prepared form no crosslinking. (5) The above Alg-PLys films were immersed in 3% acetic acid solutions for 0.5 to 24 hours. Crosslinking reaction proceeded by the reaction of acetic acid to Na cation and water insoluble films were obtained. (6) The films were dried in air at room temperature. The thickness of the films ranged from ca. 0.04 to 0.09 mm.

2.2. Measurement

Thermal properties of the films in the dry and wet states were measured using a differential scanning calorimeter (DSC, Seiko Instruments Inc., DSC-220C). The sample films of 4 mm diameter were packed in an aluminum open vessel. Sample weight was about 10 mg. Samples were annealed at 120 °C in a DSC sample holder for 10 min. After cooling to room temperature, the sample was immediately heated at 10 °C/min until 140°C. The second heating was carried out after quenching from 140°C to room temperature in order to avoid the effect of residual moisture in the sample. The second run was used for analysis. Glass transition temperature (Tg) was defined as the cross point of baseline and transition line of glassy and rubbery regions, according to the method reported previously [7]. Heat capacity difference at Tg (ΔCp)

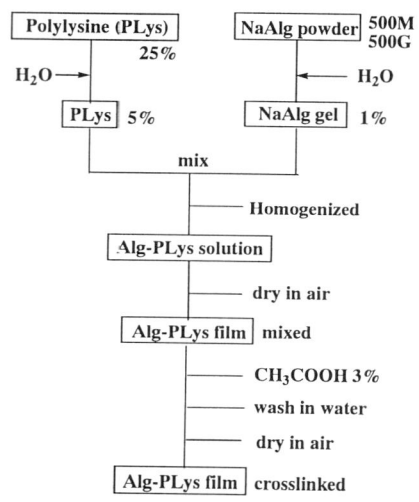

Figure 1. Preparation scheme of Alg-PLys molecular composites.

was defined as the difference between Cp's of glassy and rubbery states of the sample [7]. Tg of the samples in the wet state was measured using a sealed type Al sample vessel. The sample having a certain water content (Wc = weight of sorbed water / weight of the dry sample (g/g)) was cooled from room temperature to -150 °C and then heated to about 70°C. Crystallization and melting enthalpies of water sorbed on the samples were calculated and Tg was also evaluated. Non-freezing water content (Wnf) was calculated by the method reported previously [5,8]. Wnf was calculated by the following equation.

$$Wnf = Wc - Wf \qquad (1)$$

Where, Wf is the amount of free water calculated from the enthalpies of crystallization and melting of sorbed water on the samples.

3. RESULTS AND DISCUSSION

3.1. Glass Transition

The degree of cross linking of the samples was controlled by varying reaction time in the presence of acetic acid, as mentioned in the experimental section. DSC curves of the samples dried after being immersed in the acetic acid, showed a baseline gap due to glass transition at a temperature ranging from 100 to 120 °C. Glass transition temperature (Tg) values change as a function of immersion time in acetic acid. Figure 2 shows the relationships between Tg of Alg-PLys (PLys/Alg=1.0), ΔCp and immersion time in 3% acetic acid

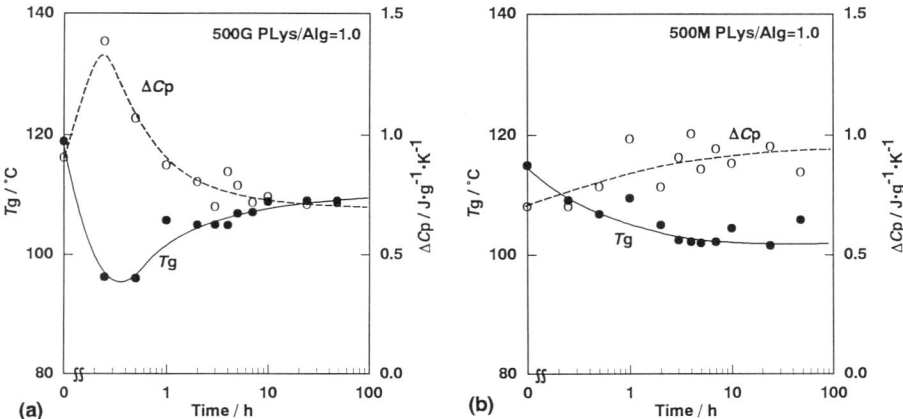

Figure 2. The relationship between Tg of Alg-PLys and immersion time in 3% acetic acid.
(a) in the case of G rich Alg-PLys   (b) in the case of M rich Alg-PLys

solution. In the G rich Alg-PLys (Fig.2a), Tg markedly decreased with increasing immersion time and reached the minimum value at about 0.5 h , increased at a time longer than 0.5 h and then leveled off at around 20 h. However, in the case of M rich Alg-PLys (Fig. 2b), Tg decreased gradually with increasing immersion time. It is also shown in Figure 2 that ΔCp values markedly depend on immersion time. The variation of Tg and ΔCp shown in Figure 2 indicates that the higher order structure of Alg-PLys is changed by both the conformational structure of Alg chains, bulky guluronic or linear mannuronic structure, and the degree of crosslinking.

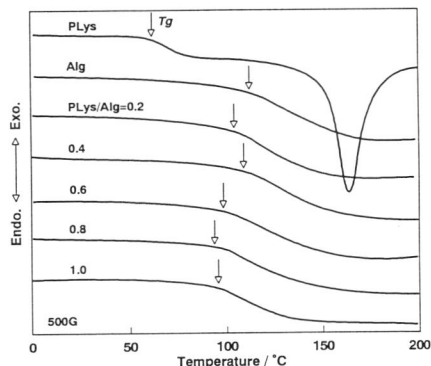

Figure 3. DSC curves of PLys and Alg-PLys near Tg.

In order to investigate the effect of mixing ratio on phase transition behavior, both M rich and G rich Alg's were mixed with PLys with various mixing ratios at constant immersion time (24 h). Figure 3 shows the representative DSC curves of Alg-PLys with various mixing ratios. DSC curves of Alg (DS=1.0) and PLys(DS=1.0) are also shown for the references. As shown in the DSC curve of PLys, the melting peak is observed at 161.6°C, indicating that the crystalline region exists in original PLys. X-ray crystallographic structure of PLys has been reported [9]. Spectroscopic analysis indicated that there exists three kinds of structure, α, β and random forms [10]. When PLys is mixed with Alg, the crystalline region completely disappears as clearly seen in DSC curves of mixtures in which only Tg can be observed at about 110°C. At a higher temperature region, no other phase transition was detected. The exothermic due to the decomposition was observed at a temperature higher than 200°C.

Figure 4 shows the relationships between Tg and PLys/Alg ratio. In the G component rich Alg-PLys, Tg decreased with increasing PLys/Alg ratio. In the M component rich Alg-PLys, Tg increased until 0.2 of PLys/Alg and then decreased with increasing PLys/Alg ratio. Tg's of all M rich samples are higher than those of G rich samples in the whole range of mixing ratio. The calculated values postulating additive rule of Tg's of Alg and PLys were lower than those of observed as shown in the figure. The results indicate that the Tg increase was induced by crosslinking, at the same time, conformational structure of Alg chains markedly affect the polymer complexation. It is reasonable that M rich Alg form well extended structure showing high Tg values.

ΔCp's of G rich Alg-PLys increased from 0.5 to 1.4 J/gK with increasing PLys/Alg. In contrast, ΔCp's of M rich Alg-PLys were maintained almost constant value at about 0.85 J/gK. Figure 5 shows relationship between Tg and ΔCp. ΔCp decreases with increasing Tg, although data are scattered in some extent. As already reported, ΔCp is low for the polymers having two dimensional chemical structure. Tg's of such a polymer are observed at a high temperature. It is thought that complete random structure is difficult to be arranged in the rubbery state. The molecular mobility of Alg-PLys, especially M rich Alg-PLys, is also thought to be restricted via ionic crosslinking even in the rubbery state, and free random motion is difficult to be enhanced.

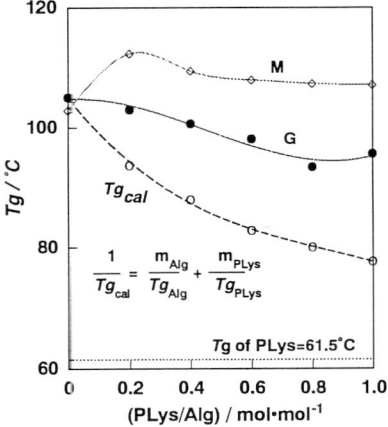

Figure 4. The relationship between Tg of Alg-PLys and PLys/Alg ratio. Dotted line shows Tg value (Tg_{cal}) by mixture rule.

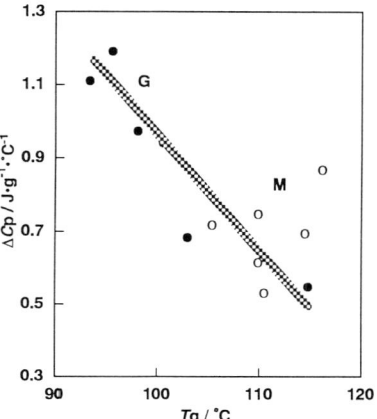

Figure 5. The relationship between ΔCp and Tg of Alg-PLys.

3.2. Non-freezing Water

Water molecules diffuse in the amorphous region of polymers and a fraction of water is tightly bound by the hydrophilic group or ions in the polymer. On this account, when the amount of bound water is quantitatively measured, number of molecular chains capable to be accessed by water molecules are evaluated. In this study, a small amount of water was added to the sample and amount of sorbed water was analyzed by DSC.

Figure 6 shows DSC curves of water sorbed on G rich Alg-PLys containing various water contents. Mixing ratio was 1.0 and immersing time was 24 hours. As shown in the figure, no phase transition was observed in DSC heating and cooling curves in a Wc less than 0.5 (g/g). This fact suggests that 0.5 (g/g) water is strongly restricted by the hydrophilic group in the amorphous region of the sample, where water molecules diffuse freely. Crystallization and melting peaks of sorbed water were observed in the Wc more than 0.5 (g/g). Melting peak started at a temperature lower than 0 °C.

Figure 7 shows the relationships between enthalpies of crystallization (ΔHc) and melting (ΔHm) of sorbed water of G-rich Alg/PLys (mixing ratio=0.4, immersion time 24 hours) and Wc. As can be seen, ΔH (J/g) linearly increased with increasing Wc. The point extrapolated to X axis (Wc) did not accord with 0, suggesting that non-freezing water exists in the system. This Wc is defined as the limited amount of non-freezing water content (Wnf_{lim}). It is also clear that ΔHc values are smaller than ΔHm values in the whole range of Wc, indicating that the systems are super cooled in the cooling process. At the same time, recrystallization of a fraction of non-freezing water may occurs during heating [11]. The

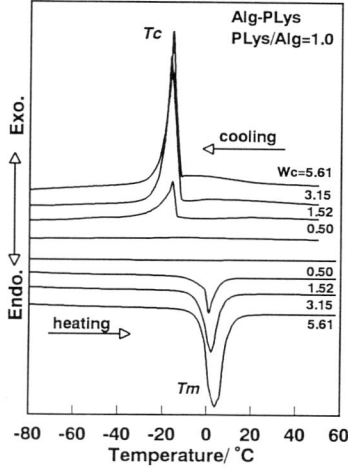

Figure 6. DSC curves of water sorbed on Alg-PLys containing various Wc.

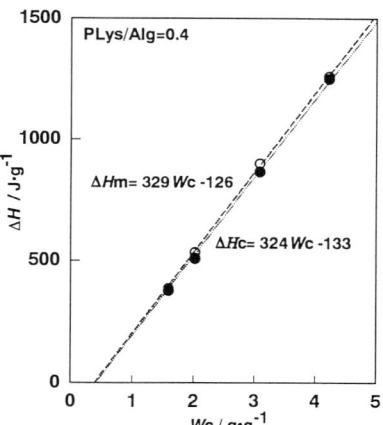

Figure 7. The relationship between ΔH calculated from crystallization and melting peaks of DSC curves.

non-freezing water content (Wnf) of all samples with various mixing ratios was estimated by the above method. Wnf increased linearly with increasing Wc until about the limited Wnf and then leveled off when Wc exceeds the limited value. Wnf calculated from ΔHm (Wnf_m) was always smaller that from ΔHc (Wnf_c) regardless of the sample variation. Wnf values were obtained from both ΔHm and ΔHc. In Figure 8, Wnf values calculated from

ΔHm is representatively shown as a function of PLys/Alg mixing ratio. The maximum Wnf values of each component were 0.525 g/g for G rich Alg, 0.674 g/g for M rich Alg and 0.764 g/g for PLys, respectively. If simple additivity was established, Wnf values of Alg-PLys systems should be found between Wnf of PLys and Alg. As seen in Figure 8, Wnf's of Alg-PLys systems in the both G are 5~6 mol/mol and M rich Alg-PLys systems are 4~5. The values are smaller than that of Alg. Effect of PLys on Wnf values seems to be completely neglected. This strongly suggests the number of free hydroxyl groups is reduced by complex formation. Wnf values of M rich Alg-PLys were higher than those of G rich Alg-PLys. Moreover, both curves are conceived and showed the minimum at about 0.6 of PLys/Alg.

Wnf values were also derived from ΔHc. ΔHc were always smaller than ΔHm, as already mentioned in the former section. In order to compare the Wnf$_c$ and Wnf$_m$, the difference (ΔWnf = Wnf$_c$ − Wnf$_m$) was calculated. Figure 9 shows the relationship between ΔWnf and PLys/Alg mixing ratio presented by mol/mol. ΔWnf increased with increasing PLys/Alg mixing ratio in both G rich and M rich Alg-PLys systems. At the same time, it is noted that ΔWnf values are almost identical in G rich and M rich Alg-PLys until 0.6 g/g and then split into two lines. The ΔWnf increase of G rich Alg-PLys is larger than that of M rich Alg-PLys systems in the high mixing ratio. This suggests that the amount of super cooled water is restrained much greater in G rich Alg-PLys systems when PLys component increases. It is reasonable to consider that G rich Alg takes a bulky and irregular arrangement when PLys content exceeds 0.6 mol/mol and that the higher order structure of Alg-PLys is varied as a function of mixing ratios.

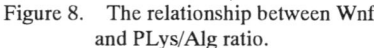

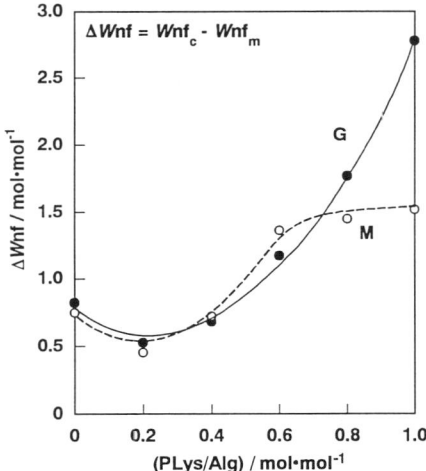

Figure 8. The relationship between Wnf and PLys/Alg ratio.

Figure 9. The relationship between ΔWnf and PLys/Alg ratio.

4. CONCLUSIONS

The above results suggest that Alg molecules form polymer complexes through introduction of polylisine as a polycation. The higher order structure of Alg-PLys varies in a complex manner according to the molecular conformation of Alg, crosslinking density and mixing ratio.   DSC results shows that the crystalline structure of polylisine disappears by the formation of complex with Alg and that at the same time the conformation of Alg takes bulky molecular arrangements   when the mixing ratio is high.   Glass transition behavior in the dry state and the nonfreezing water content in the wet state of the samples indicate that M rich Alg-PLys takes a regular arrangement, in contrast G component rich Alg forms bulky irregular molecular complexes.

REFERENCES

1.   E.D.T. Atkins, I. A. Nieduszynski, W. Mackie, K. D. Parker, and E. E. Smolko, Biopolymer, 12 (1973) 1879.
2.   D. A. Rees,   and E. J. Welsh, Angew. Chem. Int. Ed. Engl., 16 (1977) 214.
3.   K.Nakamura, T. Hatakeyama, and H. Hatakeyama, SEN-I GAKKAISHI   (J. Soc. Fiber Science and Technology, Japan), 47 (1991) 421.
4.   K.Nakamura, T. Hatakeyama, and H. Hatakeyama, in J.F.Kennedy et al. (Eds.), Wood and Cellulosics, Ellis Horwood Ltd.,   Chichester, 1987,   Ch.10, p97.
5.   T. Hatakeyama, H. Hatakeyama and K.Nakamura, Thermochim. Acta,   253 (1995) 137.
6.   K. Nakamura, Y. Nishimura, T. Hatakeyama, H. Hatakeyama,   Thermochim. Acta, 267 (1995) 1.
7.   S.Nakamura, M. Todoki, K. Nakamura and H. Kanetsuka, Thermochim. Acta, 136 (1988) 163.
8.   T. Hatakeyama, K.Nakamura and H.Hatakeyama, Thermochim. Acta, 123 (1988) 153.
9.   J. Hiraki, J. Antibact. Antifung. Agents, 23 (1995) 349.
10.   S. Shima, Dr. Thesis (1985).
11.   H. Yoshida, T. Hatakeyama, H. Hatakeyama. In Viscoelasticity of biomaterials, W Glasser and H. Hatakeyama Eds, Am. Chem. Soc., Washington, DC (1992).

HYDROCOLLOIDS – PART 2
Edited by K. Nishinari
2000 Elsevier Science B.V.

197

Rheological study and a phase diagram on mixture of corn starch and konjac glucomannan

M.Yoshimura[a], T.J.Foster[b], I.Norton[b], K.Nishinari[c]

[a] School of Humanity for Environment Policy and Technology, Himeji Institute of Technology, Himeji, Hyogo, Japan

[b] Unilever Research, Bedford, MK44 1LQ, UK.

[c] Department of Food&Nutrition, Faculty of Human Life Science, Osaka City University, Sumiyoshi, Osaka, Japan

Abstract

Storage (G') and loss (G'') shear moduli have been measured for mixtures of corn starch (CS) and konjac glucomannan (KM) (total polysaccharide concentration:3.50 wt%, CS / KM ratios: 10/0, 9/1, 8/2, 7/3, 6/4). The viscoelastic behaviour of aqueous dispersions of CS (2.10-3.50 wt%, 3.50 wt%=10/0) was an intermediate between a weak gel and an elastic gel, and aqueous dispersions of KM (0.35-1.40 wt%) showed a behaviour similar to a concentrated polymer solution. The behaviour of dispersions of CS and KM mixtures was an intermediate between a concentrated polymer solution and a weak gel. Storage modulus of a CS/KM mixture (constant KM content) as a function of the added CS showed a maximum at a certain CS content. These results suggest that phase separation of the two polymers occur. This was confirmed by confocal laser scanning microscopy, and a phase diagram has been constructed. The microstructure of these mixtures was found to be CS continuous under most conditions. We explain this as a consequence of the starch forming a weak gel even when the KM phase volume appears to dominate the microstructure. Under the conditions of this study the KM behaves as a concentrated polymer solution which is trapped by a weak gels of CS.

1. INTRODUCTION

Starch has found widespread application in the food industry. The addition of hydrocolloid to starch is known to increase the viscosity of starch and influence the gelatinization and retrogradation of starch [1-10]. The increase in viscosity of starch-hydrocolloid mixtures has been attributed to thermodynamic incompatibility which leads to mutual exclusion of the polymers. Incompatibilities between amylose and dextran [11], amylose and amylopectin [12,13], amylose and galactomannan [4], potato maltodextrin and locust bean gum, arabic gum [7], potato starch and xanthan [14] have been reported. It is reported that most hydrocolloids interact with amylose and accelerate their gelation [4,7,14]. Recently, the interaction of KM with other hydrocolloids (e.g.,carrageenan, xanthan gum, gellan gum) has been studied.   It was reported that KM interacted synergistically with carrageenan [15-22], xanthan gum [15,23-26] and gellan gum [27]. In the present study, we established a phase diagram for CS and KM mixtures and confirmed to know the interaction of to be one of thermodynamic incompatibility.   The structure of CS and KM mixtures was observed by using confocal laser scanning microscopy.

2. EXPERIMENTAL

2.1 Material

Corn starch (CS) is a gift from Sanwa Cornstarch Co.Ltd. (Nara,Japan). Konjac glucomannan (KM) is obtained from Shimizu Chemical Co. (Hiroshima, Japan).   Moisture content of KM was 7.9%.   Moisture content and protein content of CS were 12.5 % and 0.28 % respectively.   Amylose content of CS was 25 %.

2.2. Dynamic viscoelasticity

Powders of CS and KM mixture (total polysaccharide content; 3.50 wt%) were dispersed in distilled water and heated in an oil bath to 95 ℃ for 15 min and then held at 95- 98 ℃ for 30 min.   The mixture was stirred continuously.   The mixing ratios of CS and KM were CS/KM=10/0, 9/1, 8/2, 7/3 and 6/4.   Aqueous dispersions of KM (0.35-1.40 wt%) and dispersions of CS (2.10-3.50 wt%, 3.50 wt%=10/0) were also prepared.   Dynamic viscoelastic measurements were carried out using a Fluids Spectrometer RFS2 (Rheometrics Co. Ltd., NJ, USA) with a plate geometry (25 mm diameter) and strain of 4 %.   The hot sample mixture was poured directly onto the plate of the instrument, which had been kept at 5 ℃.   Immediately the storage shear modulus G' and the loss shear modulus G" were measured as a function of time.   After the accomplishment of the plateau values (after 24h), frequency sweep experiments were performed at 5 ℃ .

2.3.Phase diagram and confocal laser scanning microscopy

The phase diagram was obtained by mixing dispersions of CS and KM of various concentrations.   Powders of CS and KM were dispersed in distilled water and stirred at room temperature separately.   Then the dispersions were heated to 95 ℃ and then held at 95-98 ℃ for 30 min being stirred continuously.   The dispersions were held above 70 ℃ to prevent the gelation.   Dipersions with various ratios were poured to centrifuge tubes.   Centrifugation was carried out at 70 ℃ at 10000 rpm for 2 hours.   After centrifugation, the dispersion of mixture was separated into three phases.   The height of the separated phases and overall height of the tube contents were measured, and the phase volumes were calculated.   Biorad MRC 600 Confocal laser scanning microscopy was used to confirm the occurrence of the phase separation of mixture.   Sample dispersions were stained with rhodamine.

3. RESULTS AND DISCUSSION

Figures 1,2 and 3 show the mechanical spectra for aqueous dispersions of CS, KM, and CS and KM mixture after the accomplishment of plateau values of storage shear modulus.   For aqueous dispersions of CS (2.10-3.50 %) (Fig.1) G' remained almost constant within the frequency range observed, while G" increased slightly with increasing frequency.   Moreover, tan δ (= G"/G') for aqueous dispersions of CS are between 0.02 and 0.2.   This behaviour may be classified rheologically as the intermediate between a weak gel and an elastic gel [28,29]. The behaviour of aqueous dispersions of KM (Fig.2) are typical of concentrated polymer solutions.   In concentrated polymer solutions, at lower frequencies G" is larger than G' and both moduli increase with increasing frequency, whilst at higher frequencies G' is larger than G" and both moduli become almost frequency independent.   G" for dispersions of CS and KM mixtures (Fig.3) decreased monotonically with decreasing frequency at lower frequencies than 10^{-1} rad/s, however, G' did not decrease so much as G".   This behaviour may be classified rheologically as an intermediate between a concentrated polymer solution and a weak gel [28,29].   Although G' for dispersions of CS and KM mixtures increased with increasing content of KM at higher frequencies, the plateau of G' observed for a dispersion of CS at lower frequencies tended to disappear with increasing content of KM.   For example at lower frequencies around 10^{-2} rad/s, G' for 3.15 % dispersion of CS (filled squares in Fig.1) is larger than that for a dispersion of 3.15%CS+0.35%KM (CS/KM=9/1, G', filled squares in Fig.3), indicating that KM inhibits the formation of three dimensional network structure of CS.   At higher frequencies this situation is reversed, while at lower frequencies G" for a dispersion of mixture is larger than G' for CS dispersion indicating that the contribution of KM to the mixture is rather viscous than elastic.   This is interpreted as follows; since an aqueous dispersion of CS behaves as intermediate between a weak gel and an elastic gel (i.e. weak network structure), while a dispersion of KM behaves only as a concentrated polymer solution and does not form a network i.e., it is only an entanglement of KM chains.   Since the molecular weight of KM seems to be very high, G' and G" of CS and KM mixtures increase with increasing content of KM, but it was not possible to make a mixture with higher KM content than CS/KM=6/4.   From these observations, it is suggested that KM does not interact synergistically with CS to promote the formation of ordered structure and that the phase arrangement of the two components in the system determines the composite properties.   These are determined predominantly by KM, but as the CS content increases it contributes more to the system in the low frequency range.

Fig.4 shows the storage and loss shear moduli of a 0.6wt% KM dispersion as a function of concentration of the added CS. G' and G" increased with increasing concentration of CS up to 0.5wt%, but at higher concentration G'and G" decreased and then increased.

Fig.5 shows a phase diagram for dispersions of CS and KM mixtures.   After centrifugation above 70 ℃, CS and KM mixture separated into three phases, two phases were liquid and the third phase was debris. The upper phase was KM, while lower phase and debris were CS. A phase diagram is defined by a binodal and tie-lines. The binodal determines the region of miscibility and immiscibility, while tie-lines give the volume fraction of the phases of a de-mixed system as well as their composition. The boundary of the two-phase region is not symmetrical, but is shifted towards the axis representing CS concentration. Kalichevsky[11] reported that when the two polymers have different molecular weights, phase diagrams don't exhibit symmetry, and the binodal shifts towards the lower molecular weight polymer. It was reported that the molecular weight of KM was very high [15]. This is indicating that KM has a higher affinity for the solvent, i.e.,water. Immiscibility becomes greater with increasing molecular weight, so the high molecular weights of these polysaccharides tend to lead to the phase separation.

Fig.6 shows confocal scanning micrographs of 4.8 % CS and 0.13 % KM mixture (A), and of 1.8 % CS and 0.355 % KM mixture (B). The point of A is in CS rich phase, while the point of B is in KM rich phase as shown in Fig.5. A white strand network represents CS, while a black interspersed phase stands for KM. Micrographs of these mixtures were found to be the starch continuous (A) or bicontinuous (B). This mixture (B) would be expected to be KM continuous, which provides an irregularity to conventional thinking, presumably as a consequence of the weak gelling / percolating nature of the CS network. This again provides evidence of a CS influence on the composite properties, as discussed above (Fig.3 and 4)

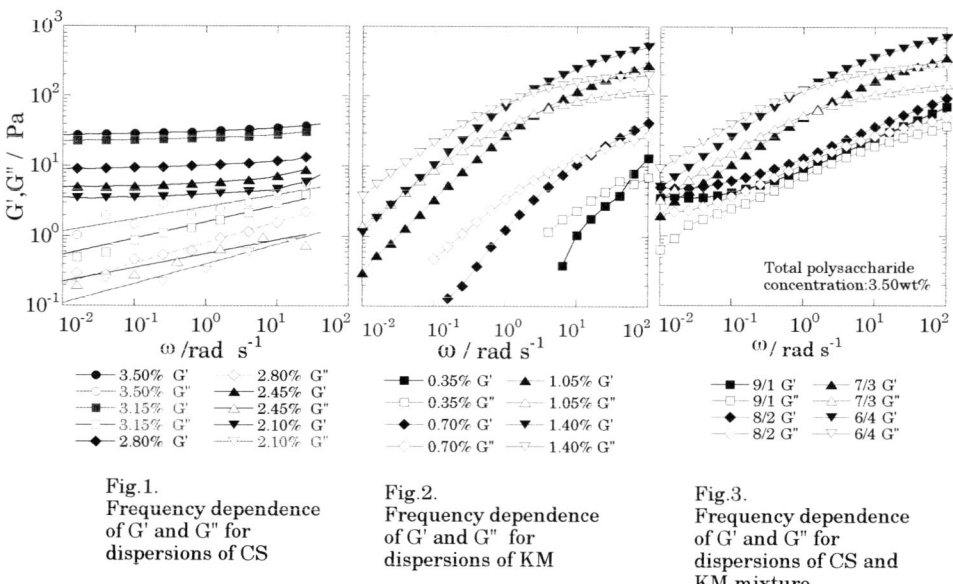

Fig.1.
Frequency dependence
of G' and G" for
dispersions of CS

Fig.2.
Frequency dependence
of G' and G" for
dispersions of KM

Fig.3.
Frequency dependence
of G' and G" for
dispersions of CS and
KM mixture

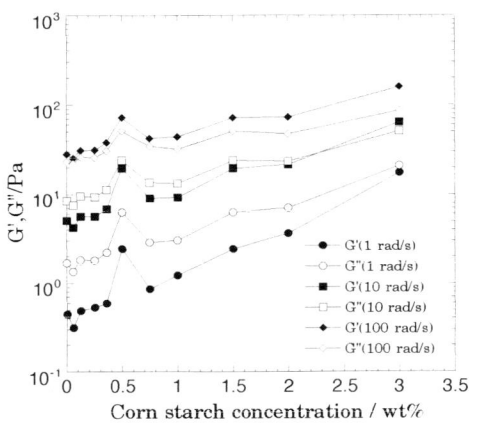

Fig.4. Storage and loss shear moduli of
0.6 wt% konjac glucomannan
dispersions containing corn starch
as a function of corn starch
concentration

Fig.5. Phase diagram of CS and KM mixture

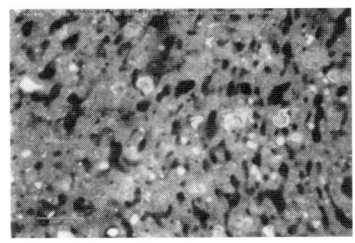

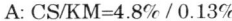

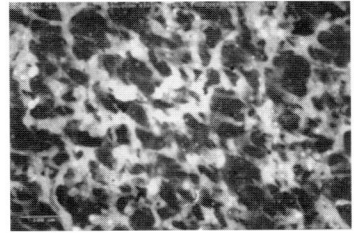

A: CS/KM=4.8% / 0.13%   B: CS/KM=1.8% / 0.355%

Fig.6. Confocal laser scanning microscopy of CS and KM mixture

REFERENCES

1. D.D.Christianson, J.E.Hodge, D.Osborne & R.W.Detroy   Cereal
   Chem., 58 (1981) 513.
2. S.U.Sajjan & M.R.R.Rao   Carbohydr.Polym., 7 (1987) 395.
3. R.J.Tye   Food Hydrocoll., 2 (1988) 259.

202

4. M.Alloncle, J.Lefebvre, G.Llamas & J.L.Doublier Cereal Chem.,
    66 (1989) 90.
5. M.Alloncle & J.L.Doublier Food Hydrocoll., 5 (1991) 455.
6. K.Kohyama & K.Nishinari J.Food.Sci., 57 (1992) 128.
7. P.Annable, M.G.Fitton, B.Harris, G.O.Phillips & P.A.Williams
    Food Hydrocoll., 8 (1994) 351.
8. A.B.Yousria & M.B.William Starch, 46 (1994) 134.
9. M.Yoshimura, T.Takaya & K.Nishinari J.Agric.Food Chem., 44 (1996)
    2970.
10. M.Yoshimura, T.Takaya & K.Nishinari Carbohydr. Polym., 35 (1998) 71.
11. M.T.Kalichevsky, P.D.Orford & S.G.Ring Carbohydr. Polym., 6 (1986) 145.
12. M.T.Kalichevsky & S.G.Ring Carbohydr. Res., 162 (1987) 323.
13. V.M.Leloup, P.Colonna & A.Buleon J.Cereal.Sci., 13 (1991) 1.
14. B.Conde-Petit, A.Pfirter & F.Escher Food Hydrocoll., 11 (1997) 393.
15. K.Nishinari, P.A.Williams & G.O.Phillips Food Hydrocoll., 6 (1992) 199.
16. P.Cairns, M.J.Miles & V.J.Morris Carbohydr.Polym., 8 (1988) 99.
17. P.Cairns, E.D.T.Atkins, M.J.Miles & V.J.Morris
    Int.J.Biol.Macromol., 13 (1991) 65.
18. P.A.Williams, S.M.Clegg, M.J.Langdon, K.Nishinari &
    G.O.Phillips In "Gums & Stabilisers for the Food
    Industry 6", eds.G.O.Phillips, D.J.Wedlock & P.A.Williams, Oxford
    University Press, Oxford, New York, Tokyo, pp.209 (1992).
19. P.A.Williams, S.M.Clegg, M.J.Langdon, K.Nishinari &
    L.Piculell Macromolecules, 26 (1993) 5441.
20. K.Kohyama, H.Iida, T.Ochi, S.Ohashi & K.Nishinari In" Food
    Hydrocolloids; Structure, Properties, and Functions", eds. K.Nishinari &
    E.Doi, Plenum Press, New York, pp.457 (1993).
21. K.Kohyama, H.Iida & K.Nishinari Food Hydrocoll., 7 (1993) 213.
22. K.Kohyama, Y.Sano & K.Nishinari Food Hydrocoll., 10 (1996) 229.
23. G.J.Brownsey, P.Cairns, M.J.Miles & V.J.Morris
    Carbohydr.Res., 176 (1988) 329.
24. P.A.Williams, S.M.Clegg, D.H.Day, G.O.Phillips & K.Nishinari
    In" Food Polymers, Gels and Colloids", ed. E.Dickinson, RSC Spec.
    Publication, London, pp.339 (1991).
25. P.A.Williams, D.H.Day, M.J.Langdon, G.O.Phillips &
    K.Nishinari Food Hydrocoll., 4 (1991) 489.
26. K.P.Shatwell, I.W.Sutherland, S.B.Ross-Murphy & I.C.M.Dea
    Carbohydro.Polym., 14 (1991) 131.
27. K.Nishinari, E.Miyoshi, T.Takaya & P.A.Williams
    Carbohydr.Polym., 30 (1996) 193.
28. E.R.Morris In"Gums & Stabilisers for the Food Industry 2" ,eds.
    G.O.Phillips, D.J.Wedlock & P.A.Williams, Pergamon Press, Oxford, New
    York, pp.57 (1983).
29. A.H.Clark & S.B.Ross-Murphy Adv.Polym.Sci., 83 (1985) 57.

HYDROCOLLOIDS – PART 2
Edited by K. Nishinari
2000 Elsevier Science B.V.

Physical and chemical characteristics of alginate-starch sponges

D. K. Rassis, I. S. Saguy and A. Nussinovitch

Institute of Biochemistry, Food Science and Nutrition
The Hebrew University of Jerusalem
Faulty of Agricultural, Food and Environmental Quality Sciences
P.O. Box 12, Rehovot, 76100
Israel

The structural and mechanical properties of alginate-starch gel after immersion in sucrose solution was studied. Sucrose concentration in the center of the specimen increased exponentially with time. Immersion resulted in decreased volume and weight of the wet gels, followed by an increase in relative density of the dried samples and decrease in their porosity. Pore population within the freeze-dried gels revealed the presence of open- and closed pores. The latter were present mainly at the intermediate time after the commencement of sucrose diffusion and before its termination. Compression tests of the dry gels showed dependency of the strength and brittleness on sucrose diffusion time.

1. INTRODUCTION

Hydrocolloid-based cellular solids can be produced either by freeze-dehydration of gels [1,2] or after their immersion in different carbohydrate solutions, changing their physical and chemical properties [3]. The resultant dried cellular solids are an interconnected network of solid struts or plates that form the edges and faces of sponge cells [4]. When freeze-drying is used to produce sponges from gels, water loss contributes to the creation of pores in the resultant cellular solid. The preparation procedures could modify the mechanical properties of these sponges. For instance, internal gas bubbles embedded in wet agar gels significantly reduced the mechanical integrity of the dried sponges and affected their porosity. However, in alginate sponges the same process caused minor mechanical changes [2]. Oil included in alginate gels weakened the mechanical strength of the dried sponge, lowered its stress and stiffness at failure as reflected by the deformability modulus and changed the size distribution and structure of pores of the dried sponge [5]. Water plasticization of sponges changes their stress-strain behavior [6].

The main objectives of this study were: to demonstrate the feasibility to govern alginate-starch sponge physical properties (e.g., strength, brittleness and porosity); to study changes in the dry sponges after immersion and sucrose diffusion into the wet gel before drying; to study changes in gel dimensions during diffusion and to evaluate sugar-concentration changes with immersion time. This knowledge could ultimately lead to the construction of crunchy products according to specific requirements. The capability to be able to modify porosity and structure of dried gels by physical and chemical procedures should furnish a valuable tool for the simulation of a variety of numerous other cellular foods differing in their properties.

2. MATERIALS AND METHODS

2.1. Preparation of cellular solids

Sodium alginate powder LV, 2.5% (w/w) was mixed with double-distilled water for 7 h at ambient temperature (25 ± 2°C). Then, 1.5% (w/w) sodium hexa metaphosphate (SHMP, BDH, Poole Dorset, UK) was added, and the mixture was further stirred for 30 min and heated to ca. 40°C, prior to the addition of 1.5% (w/w) $CaHPO_4$ (Riedel-de Haen, Seelze, Germany) that was incorporated for 60 min. The mixture was cooled to 20 ± 1°C and 3.0% (w/w) fresh glucono-δ-lactone solution, GDL, (Sigma) was mixed in. The mixture was poured into glass cylinders and left overnight (15 h) at 4°C for gelation. The gel samples were immersed in 60°Bx sucrose solution at ambient temperature (22 ± 2°C). Immersion periods ranged from 0 to 96 h. The samples were freeze-dried in a pilot plant unit (Model 15 RSRC-X, Repp Industries Inc., Gardiner, NY), operating at 33.3 Pa and -45°C.

2.2. Particle and bulk density determination

The particle density of the cellular solid was determined with a multi-pycnometer (Quanta Chrome Corp., Syosset, NY). The bulk density was determined by volumetric displacement using glass beads (425 to 600 μm, Sigma), and porosity was calculated [7]:
$$\varepsilon = 1 - (\rho_b/\rho_p) = 1 - \rho_r \quad (1)$$
Where: ε – porosity; ρ_b– bulk density; ρ_p– particle density; ρ_r is defined as the relative density derived from the ratio ρ_b/ρ_p [4].

2.3. Electron microscopy micrographs (SEM)

Dry gels after sucrose diffusion were studied on SEM. Micrographs were obtained by cutting through the dry cellular solid with a double-edged razor blade and exposing the internal surface features. Samples were examined a SEM (JEOL JSM 25S SEM, Tokyo, Japan) at an accelerating voltage of 15 kV and a working distance of 48 mm. Micrographs were magnified at 1000.

2.4. Mechanical tests

Samples were compressed to 80% deformation between lubricated parallel plates, at a constant deformation (displacement) rate of 10 mm/min, on an Instron Universal Testing Machine (Model 1011, Instron Corporation, Canton, MA). A program developed at the Instron Corporation (Canton, MA) and modified in our laboratory performed the data acquisition and deformation conversion of the continuous voltage of the Instron vs. time output into digitized stress vs. engineering strain relationships:
$$\sigma = F/A_0 \quad (2)$$
$$\varepsilon_E = \Delta H/H_0 \quad (3)$$
Where: σ - stress; ε_E - engineering strain; F - the momentary force; ΔH - momentary deformation; $(H_0-H(t))$; A_0 and H_0 - the cross sectional area and height of the original specimen, respectively.

2.5. Diffusion measurements and statistical analyses

Diffusion of sucrose concentration into the gels was measured with a table refractometer (Carl Zeiss, Berlin Germany) by removing the samples from the sugar solution and wiping carefully with a paper towel. Refractometric values within the gel and the surrounding sucrose solution were also determined vs. time of immersion.

Triplicate samples for each treatment were used and the experiments were repeated. Nonlinear regression and one-way ANOVA (Systat Ver. 5.03 Systat Inc. Evanson, IL) were utilized to calculate the effective diffusion and significant differences.

3. RESULTS AND DISCUSSION

3.1. Weight and Volume Changes in Wet Gels

After gelation, the samples were immersed for 0, 1, 2.5, 4, 6, 8, 15, 23, 48 and 96 h in 60°Bx sucrose. Weight loss of the gels during immersion were expressed as a weight ratio, W_t/W_0 (weight at immersion time, t, and initial value prior to immersion, respectively) is listed in Table 1.

Table 1: Effect of immersion time in 60°Bx sucrose solution on weight and volume of alginate gels (average ± standard deviation; triplicates)

Immersion time (h)	Weight ratio (W_t/W_0)	Weight ratio change (%)	Volume ratio (V_t/V_0)	Volume ratio change (%)
1.0	0.86 (±0.00)	14	0.83 (±0.02)	17
2.5	0.79 (±0.01)	21	0.77 (±0.03)	23
4.0	0.75 (±0.01)	25	0.73 (±0.01)	27
6.0	0.70 (±0.01)	30	0.67 (±0.02)	33
8.0	0.67 (±0.00)	33	0.62 (±0.01)	38
15.0	0.60 (±0.00)	40	0.54 (±0.01)	46
23.0	0.57 (±0.00)	43	0.51 (±0.01)	49
48.0	0.56 (±0.00)	44	0.50 (±0.02)	50
96.0	0.57 (±0.00)	43	0.51 (±0.01)	49

The average initial weight of the gels was ca. 7.2 g and their volume ca. 6.2 cm [3]. A 14% weight loss was noted after 1 h of immersion. The gel-weight ratio decreased to 0.70, 0.60 and 0.57 of the initial value after immersion for 6, 15 and 96 h, respectively. The weight changes in gels were the overall result of water diffusion out of the gel to the surrounding sucrose solution and the penetration of sucrose into the network of the immersed gels. The volume ratio decreased to 0.83, 0.67, 0.54 and 0.51 of its original value after immersion for 1, 6, 15 and 96 h, respectively. The volume changes were more prominent (6%) than the weight changes. The changes in volume and weight were caused by a dynamic process as a result of the osmotic-pressure differences between the gel and the surrounding medium. This pressure is composed of three forces - rubber elasticity, polymer-polymer affinity and ion pressure [8]. The osmotic-pressure probably had a higher influence on the volume of samples than on their weight. Volume and weight changes could explain the increase in bulk density and decrease of porosity of the cellular solids after sample freeze-dehydration, as will be demonstrated later.

3.2. Sucrose Diffusion During Immersion

After the gel specimens were removed from the sucrose solution, the total concentration of soluble solids in the sample core was evaluated. Prior to immersion, the average refractometric value of the 60°Bx sucrose solution of

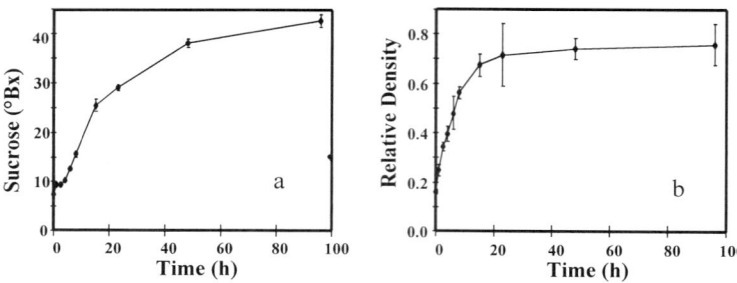

Figure 1. Effect of immersion in 60°Bx sucrose solution. (a) Sucrose diffusion into gels during immersion in 60°Bx sucrose solution: (b) Relative density of dried cellular solids.

alginate-starch gels was 7.5°Bx. Due to sucrose diffusion the concentration increased to 9.3 15.6, 25.5, 29.1, 38.1 and 42.6°Bx after immersion for 1, 8, 15, 23, 48 and 96 h, respectively (Fig. 1a).

The refractometric values after immersion for 48 and 96 h were almost identical, reaching an asymptotic value. Sucrose penetration into the gels was described according a diffusion equation for a semi infinite solid in one direction [9]:

$$(C - C_o) / (C_f - C_o) = 2/\pi \sum \{ [(-1)^n / (n+\tfrac{1}{2})] \cdot \exp [-(n+\tfrac{1}{2})^2 \pi^2 D_{eff} t / L^2]$$
$$\cdot \cos [(n+\tfrac{1}{2}) \pi x/L]\} \quad (4)$$

Where: C, C_o and C_f are sucrose concentration at a given time, t, initial and final time, respectively; D_{eff} - effective sucrose diffusivity; L is the length of the semi infinite solid in the x direction.

Applying nonlinear regression, the values of the effective diffusivity at the center of the gel was derived as $4.6 \cdot 10^{-10}$ m^2/s with an asymptotic standard error of $5.6 \cdot 10^{-11}$ and R^2 of 0.997. Similar values, $7.8 \cdot 10^{-10}$ and $3.3 \cdot 10^{-10}$ m^2/s were reported for diffusion of glucose (25°C) into a carrageenan (2%) gel [10], and sugar (61%) in an agar gel/milk bilayer system [11], respectively.

3.3. Relative Density

The average relative density of the samples prior to immersion was 0.16 (±0.01), increasing parallel to sucrose diffusion into the gels followed by their shrinkage (Fig. 1b). Immersion for 6 h tripled the initial relative density value, reaching a maximum value of 0.76 (±0.08) after immersion for 96 h. Changes in relative density were faster during the first 15 h of immersion, and leveled off as immersion time increased. The changes in relative density and sucrose diffusion had the same trend: a fast increase during the first 15 h of immersion, followed by a slower change. The changes in relative density were caused by the diffusion of sucrose into the gels, which explains the exponential behavior of these changes. Nonlinear regression yielded D_{eff} of $1.51 \cdot 10^{-9}$ m^2/s, an asymptotic standard deviation of $5.6 \cdot 10^{-11}$ m^2/s and R^2 =1.00.

3.4. Open and Closed Pores

Diffusion of sucrose into the gels affected the distribution of the open and closed pores in the dried cellular solids. Changes in the porosity of freeze-dehydrated cellular solids after different immersion times show that porosity decreased from an initial value of 0.84 (±0.01) to 0.38 (±0.07) after immersion in sucrose solution for 96 h (Fig. 2).

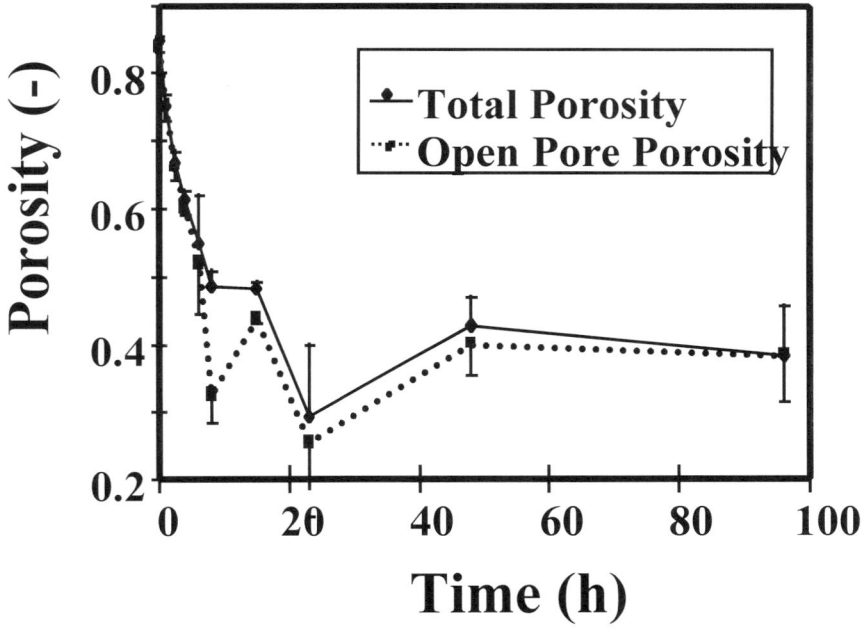

Figure 2. Porosity of open and total pores

The difference between the total porosity and the open-pore porosity is the sum of the closed pores within the structure of the dried specimen. Up to 4 h of immersion, no difference between the total and open porosity was observed. A higher total porosity was observed between 6 and 48 h of immersion. However, only at 8 and 15 h was this difference significant (P<0.002). During immersion of 8 to 15 h, the specimens' appearance changed from cylindrical to shrunken. The shrinking at mid-sample height and the center of the bases occurred during a few hours, in which a greater number of closed pores were detected. When shrinking was completed additional sugar diffusion decreased the number of closed pores.

3.5. Pore Size Distribution

The pores detected in the dry gels (sponges) were of different size and shapes (Figure 3a). Many

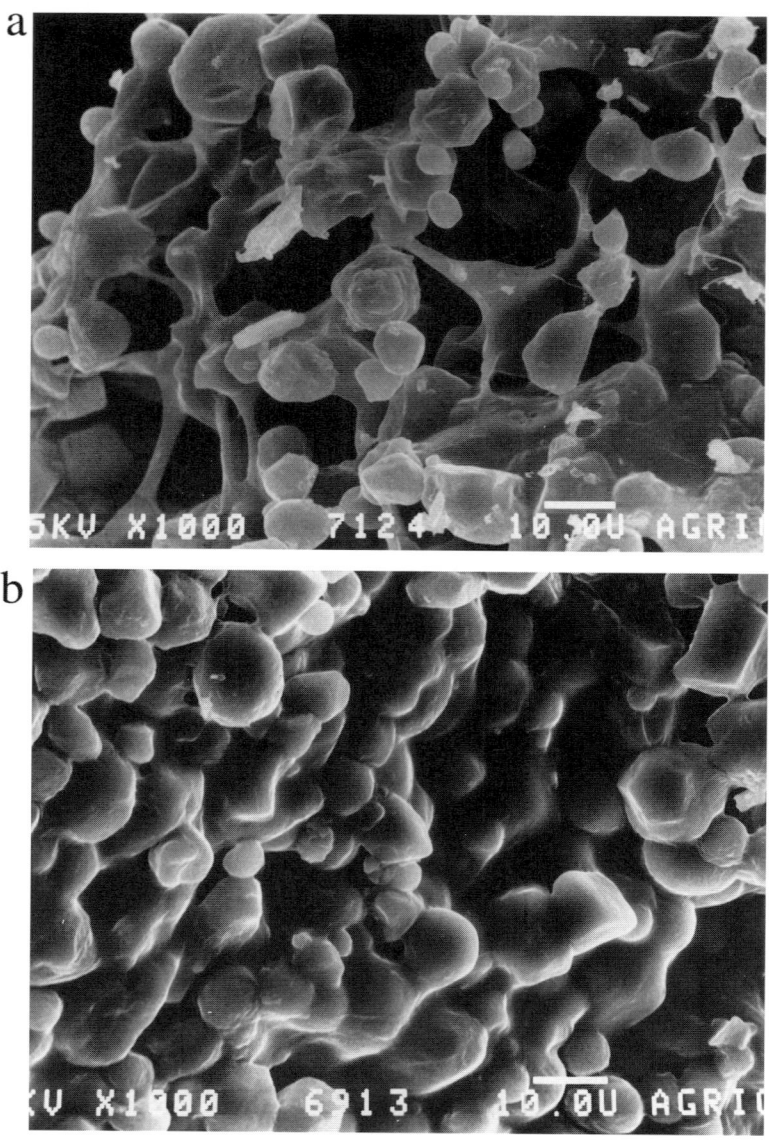

Figure 3: Typical SEM of a cellular solid composed of alginate (2.5%) and starch (15%), magnification x1000.(a) Pre-immersion. (b) Immersion for 48 h in 60°Bx sucrose solution. void spaces were detected between the starch (sphere like) granules (25 to 50 $\cdot$ 10^{-6} m), embedded in the matrix created by the alginate network. Changes in the resultant cellular solid caused a decrease in void spaces after immersion for 48 h in 60°Bx sucrose solution are

shown in (Fig.3b). In addition, a "cover" coating of sucrose wrapping the primary matrix can be seen. The sucrose deposits mostly in an amorphous state [12]. Some of the spherical granules changed to more angular shapes by the sucrose deposition, probably as a result of the osmotic pressure. The changes in the cellular solid matrices, as revealed by volume and weight decrease, are also reflected in the SEM micrographs.

3.6. Stress-Strain Relationship

Dry hydrocolloid cellular samples ("sponges") produced from gels immersed in sucrose solution of 60°Bx for 0, 4, 48 and 96 h, were freeze-dried and compressed to ca. 80% deformation. A typical stress-strain relationship of such cellular solids includes three easily observed parts. Similar curves can be obtained for samples not previously immersed in sucrose solution (Fig.4). The first stage of the curve was linear, indicating that up to a small deformation of ca. 12%, the solid behaved as an elastic solid. In the second part of the curve, only relatively small changes in stress (0.6 - 1.0 MPa) were detected, followed by larger changes in strain (0.1 - 0.7), possibly a result of cell collapse. The third part of the curve showed a prominent increase in stress with a little or no change in strain. The steep increase in stress was result of densification of the dry matter of which the cellular solid was composed.

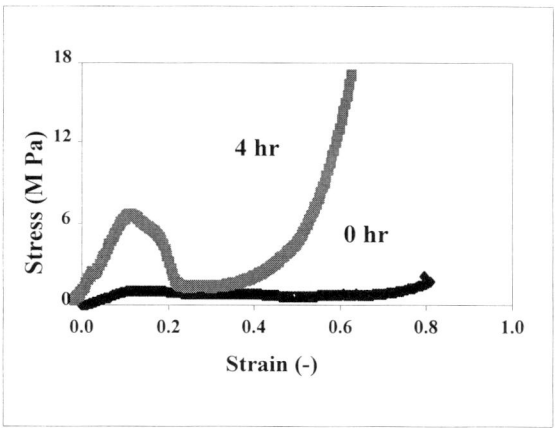

Figure 4: Compressive stress vs. engineering strain: pre- and 4 h immersion in 60°Bx sucrose solution.

The stress-strain behavior of such cellular solids after embedding the alginate-starch gel in sucrose solution for 4 h followed by drying (Fig. 4). After the diffusion ended, the gel had a harder layer of about 2 to 3 mm that contained sucrose was detected. The gel, and later the sponge, comprised of an inner core of untouched cellular solid coated by a layer absorbed with sucrose. When this solid was compressed, a different sigmoid stress-strain relationship was observed. Up to a level of ca. 0.15 engineering strain, stress increased as expected from such solids due to the sucrose casing. The later reduction in stress may have been caused by the softer texture of the inner solid. It should be emphasized that the cellular solid was much stronger after sucrose diffusion in comparison to the non-treated gel. The maximal stress levels observed for the shoulder in the stress-strain curves was approximately 1.0 and 6.0 MPa for 0 and 4 h of immersion, respectively.

4. REFERENCES

1. D.K. Rassis, I.S. Saguy and A. Nussinovitch. J. Agric. Food Chem., 46 (1998) 2981.
2. A.Nussinovitch R. Velez-Silvestre and M. Peleg. Biotechnol. Prog., 9 (1993) 101.
3. D. Rassis, A. Nussinovitch and I.S. Saguy. Inter. J. of Food Sci. Technol., 32 (1997) 271.
4. L.J. Gibson and M.F. Ashby. In: Cellular Solids, Structure and Properties. Pergamon Presss, New York, NY. (1988) pp. 120.
5. A. Nussinovitch and Z. Gershon. Food Hydrocol., 11 (1997) 281.
6. G.E. Attenburrow, R.M.Goodband; L.J. Taylor, and P.J. Lillford. J. Cereal Sci., 9 (1989) 61.
7. S.N. Marousis and G.D. Saravacos. J. Food Sci., 55 (1990) 1367.
8. T. Tanaka, Scientific American., 244 (1981) 110.
9. J.B. Gross and M. Ruegg. In: Physical Properties of Foods-2. Jowitt, R., Escher, F., Kent, M., McKenna, B., Roques, M. (Eds.), Elsevier, London, UK, (1987) pp, 71.
10. M. Hendrickx, C. Vanden Abeele; C. Engels and P. Tobback J. Food Sci., 51 (1986) 1544.
11. F. Warin, V. Gekas, A. Voirin and P. Dejmek. J. Food Sci., 62 (1997) 454.
12. Y.H. Roos. Phase Transitions In Foods. Academic Press, New York, NY. (1995) pp,110.

HYDROCOLLOIDS – PART 2
Edited by K. Nishinari
2000 Elsevier Science B.V.

Effect of chitosan on the gelation of κ-carrageenan under various salt conditions

F.M. Goycoolea[a], W. Argüelles-Monal[b], C. Peniche[c] and I. Higuera-Ciapara[a]

[a] Center of Research for Food and Development, P.O. Box 1735 Hermosillo, Sonora 83000 Mexico. Tel +52-62-800-057; Fax +52-62-800-055. E-mail: fgoyco@cascabel.ciad.mx

[b] IMRE, Universidad de La Habana, 10400 Cuba

[c] Centro de Biomateriales, Universidad de La Habana, La Habana 10400, Cuba

The behaviour of non-stoichiometric polyelectrolyte complexes of Na^+-κ-carrageenan with short chitosan segments in presence of NaCl is rationalised in terms of a reduction of the charge density of κ-carrageenan chains with a subsequent stabilisation of their ordered conformation leading to a reinforcement of the gel network. Hydrophobic interactions between complexed segments may contribute to a more extensively connected network. By contrast, in KCl, selective cation-driven aggregation of κ-carrageenan seems to compete with chitosan for complex formation, thus effectively weakening the carrageenan gel network.

1. INTRODUCTION

Polyelectrolyte complexes (**PEC**) obtained by mixing two oppositely charged polyelectrolytes, normally lead to an insoluble precipitate due to strong coulombic interactions. However, when the **PEC** involves two polyelectrolytes of different molecular size, soluble non-stoichiometric heterotypic structures can be formed under controlled conditions, which are regarded as non-stoichiometric polyelectrolyte complexes (**NPEC**). In such complexes, the larger polymer chain behaves as a 'host' macromolecule to the shorter one (or 'guest') which attaches into stretches of the host chain at random. Furthermore, it has been recognised that the host polymer should be a strong polyelectrolyte in order to retain the interpolymer complex in solution [1]. Recently **PEC**s have received a great deal of attention, due to their practical relevance in chemistry, biotechnology (*e.g.* microencapsulation, membrane technology, etc.), pharmacy (*e.g.* controlled drug release devices) and biomaterials engineering (*e.g.* immobilised pancreas islet cells).

The **PEC** between chitosan (bearing amine groups) and κ–carrageenan (carrying one sulphate group per disaccharide unit) seems to be one of the most interesting because of the behaviour of κ–carrageenan as a strong polyelectrolyte and its known gelling capacity. A series of insoluble chitosan/carrageenan **PEC** systems have recently been described [2], with emphasis on the charge density and the conformation of the carrageenan macromolecule. However, the behaviour of chitosan/κ–carrageenan in a soluble **NPEC** during gelation, in the presence of added counterions, to our knowledge, has not yet been addressed.

The aim of this study was to investigate the effect of complexing hydrolysed chains of chitosan of 20 or 75 β-D-glucosamine hydrochloride residues with κ–carrageenan in the disordered conformation, to form a soluble **NPEC**, with respect to the behaviour of κ–carrageenan during the sol-gel transition in the presence of sodium or potassium counterions.

2. EXPERIMENTAL

2.1. Materials
Chitosan:
- ✓ Source: lobster cephalotorax (*Panulirus argus*)
- ✓ N-acetyl content (molar fraction of N-acetyl groups, F_A): 0.201, ^{1}H-NMR.
- ✓ Mv: 2.3×10^5, measured at 25°C in 0.3 mol·L^{-1} acetic acid/0.2 mol·L^{-1} sodium acetate.
- ✓ Controlled hydrolysis with HNO: 20 and 75 residues of D-glucosamine (respectively CHI-20 and CHI-75). The final chain length was checked by capillary viscometry.

κ–Carrageenan:
- ✓ Supplier: Hercules (X6960)
- ✓ Sodium salt was prepared by ionic exchange with Amberlite 200 (Sigma)
- ✓ Content of ι-sequences: 8 %.

2.2. Methods
The chitosan/κ–carrageenan **NPEC**s were prepared by careful dropwise addition of a small aliquot of 0.039 equiv.·L^{-1} chitosan hydrochloride into 0.0103 equiv.·L^{-1} carrageenan solution under vigorous mixing at ca. 90°C, in either NaCl or KCl solutions. The hot mixture was directly transferred to the rheometer. The composition of the **NPEC**s, expressed as the molar ratio (**Z**) of NH_3^+ to SO_3^- functional groups, varied in the range between 0.028 and 0.084.

The **NPEC**s formed as above were allowed to gel in a Rheometrics RFSII Fluids Spectrometer fitted with a truncated cone-plate tool (cone angle: 0.0397 rad, diameter: 50 mm, truncation gap: 53 μm) and a circulating environmental system for temperature control. The viscoelastic properties were monitored by small deformation oscillatory testing at varying frequency ($\omega = 1$ to 100 rad·s^{-1}) during cooling and heating programs. Low-amplitude oscillatory measurements were made within the linear viscoelastic region ($\gamma = 0.15$), as verified by strain sweeps of the gels at 4°C.

3. RESULTS

3.1. Gelation in 0.25 mol·L^{-1} NaCl
The evolution of the storage modulus at $\omega = 1$ rad·s^{-1} during cooling of a κ–carrageenan solution alone and in combination with CHI-20 in **NPEC**s of varying composition (**Z**=0.028 to 0.084), at identical carrageenan concentration in 0.25 mol·L^{-1} NaCl is presented in Figure 1. Inspection of the individual traces reveals a general elevation of the final G' values of the **NPEC** at low temperatures as the proportion of complexed chitosan increases. Moreover, the onset temperature of gelation (Tg) of the **NPEC**s is clearly shifted to higher temperature, regardless of the **NPEC** composition. It should be noted that in two of the complexes (**Z** = 0.042 and 0.084) the G' moduli registered at temperatures above Tg are greater than the rest of the complexes and κ-carrageenan alone.

The criterion adopted to define the critical *gel point* (*i.e.* the percolation threshold at the formation of an incipient continuous network of infinite molecular weight), was to mark where a power-law variation of the dynamic mechanical moduli is obeyed. In other words, it is considered as the point where $G'(\omega) \propto G''(\omega) \propto \omega^\Delta$, hence *tan* $\delta = G''/G' = $ const. [3, 4].

At this point, individual traces of *tan* δ registered at varying frequency versus temperature intersect at a critical value, from which the gelling temperature was calculated. Figure 2 shows that for κ–carrageenan and for the **NPEC** of **Z**=0.084, at the critical gelation temperature (*i.e.* the gel point), there is indeed little dependence of the *tan* δ values on ω, and such critical temperature closely corresponds to that where the *tan* δ curves cross over. Similar analysis were conducted for the other **NPEC** series of CHI-20. Overall

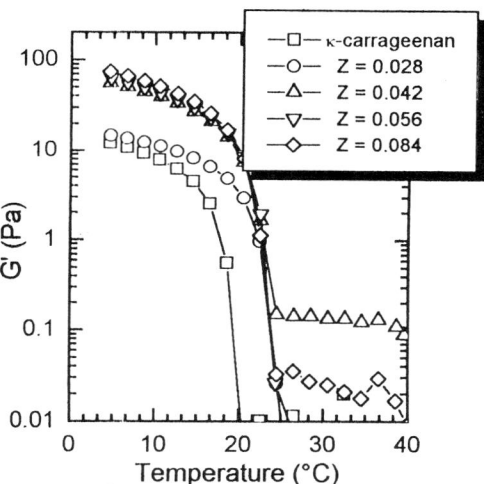

Figure 1. Temperature-dependence of G' (1 rad·s⁻¹; γ = 0.15; 1°C·min⁻¹) for κ-carrageenan in NaCl alone and in complexes with chitosan (CHI-20) of varying **Z** (as indicated in label).

similar results to those described above, were observed for **NPEC**s containing CHI-75.

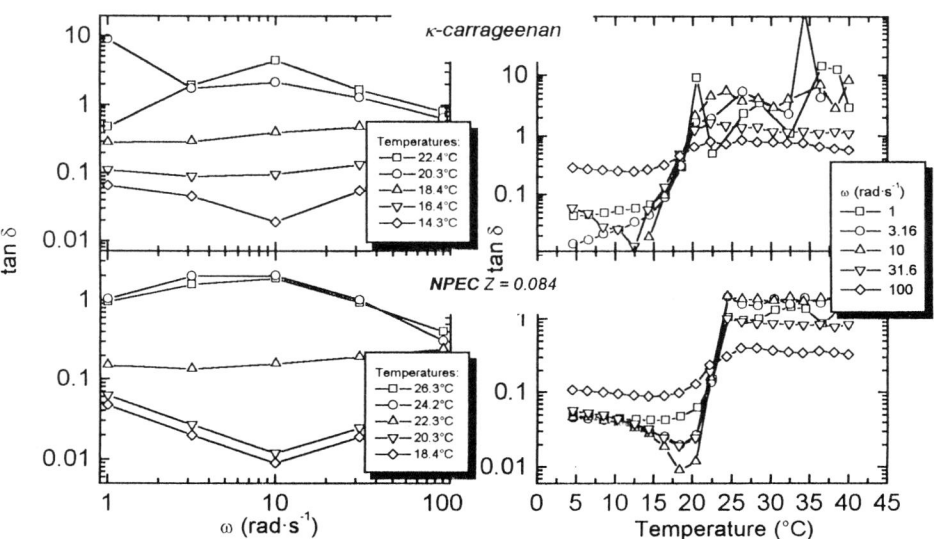

Figure 2. Frequency-dependence of tan δ (γ = 0.15) at varying temperature in the vicinity of the critical gel point (left frames) and variation of tan δ with temperature on cooling (1°C·min⁻¹) at varying frequency of oscillation (right frames) for κ-carrageenan alone and for a complex (**Z** = 0.084) from Figure 1.

In Figure 3 is shown the variation in the storage modulus at $\omega = 1$ rad·s⁻¹ corresponding to gel formation and melting during respectively cooling and heating scans of κ–carrageenan and chitosan/carrageenan **NPEC** of **Z** = 0.084. The difference in the cooling and heating traces reflects the thermal hysteresis expected for helix-helix aggregation of κ–carrageenan in 0.25 mol·L⁻¹ NaCl [5]. This thermal hysteresis has a greater magnitude in the presence of chitosan in the **NPEC** of **Z**=0.084, and similar results were obtained for the other **NPEC**s.

Figure 4 shows data for G' and G" of the gels formed at 4°C for κ–carrageenan alone and **NPEC**s with CHI-20 and CHI-75 as a function of the amount of complexed chitosan. Notice that in both cases, the **NPEC** with CHI-20 and CHI-75, there is a plateau level in G' values the gels tend to at 4°C as the **Z** value in the **NPEC** increases. **NPEC**s with CHI-20 however, seem to level off at **Z**>0.042, whereas **NPEC**s with CHI-75 reach their plateau at **Z**>0.028. The mechanical spectra (variation of G', G" and η* with ω) recorded on completion of the cooling scans (results not shown) were typical for a polysaccharide gel network. The presence of chitosan resulted in an overall increase in the viscoelastic moduli, but no significant change in the spectral shape of the gels was observed (*i.e.* no dependence of G' and G" and linear decrease of η* on frequency with a slope close to -1).

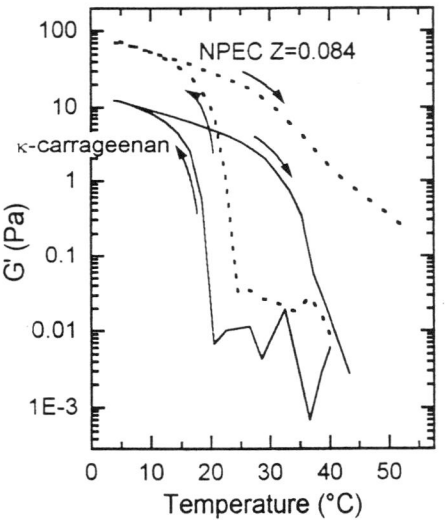

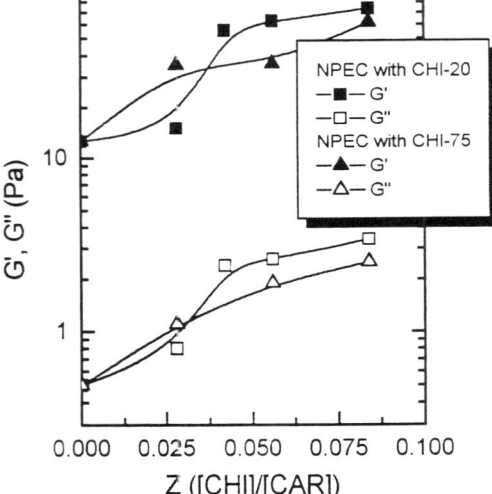

Figure 3. Variation of G' on cooling and heating (as indicated by arrows; 1°C·min⁻¹) for κ–carrageenan alone and for a complex (**Z**=0.084) from Figure 1.

Figure 4. Variation of G' (4°C; 1 rad·s⁻¹; γ = 0.15) with **Z** for complexes of κ–carrageenan with chitosan (CHI-20 and CHI-75) in NaCl. The value of G' at **Z** = 0 corresponds to κ–carrageenan alone.

3.2. Gelation in 0.03 mol·L⁻¹ KCl

The temperature course of gel formation for κ–carrageenan complexed with chitosan (CHI-20) and alone in 0.03 mol·L⁻¹ KCl, was also monitored by small-deformation oscillatory measurements. The traces of G' at ω=1 rad·s⁻¹ are drawn in Figure 5. Under these conditions, the final G' moduli values are almost the same within the expected error for the **NPEC**s and for κ–carrageenan. Only a very small elevation in the final G' values is observed for the

NPEC of Z=0.028, as compared to the rest of the gels. It is evident that there is no detectable change in gelation temperature (*i.e.* the onset of the steep rise in moduli). The sol-gel transition for these systems lied within the range of 32 to 34°C.

In Figure 6 is shown the variation in the storage modulus corresponding to gel formation and melting during respectively, cooling and heating scans of κ–carrageenan and the chitosan/carrageenan **NPEC** of Z=0.028. The difference in the cooling and heating evolution of G' values reflects again thermal hysteresis, expected for helix-helix aggregation of κ–carrageenan in 0.03 mol·L^{-1} KCl [5]. In this case, the ion-driven transition is promoted at significantly lower ionic strength than in the systems prepared in the presence of NaCl. This is due to the well known specific binding of K$^-$ cations to the double-helical form of κ–carrageenan, thus effectively promoting conformational ordering and aggregation at significantly lower concentrations than do non-specific ions, acting solely by charge screening [6]. In contrast with NaCl gels, this thermal hysteresis is unaffected by the presence of chitosan in the **NPEC** of Z=0.028, and similar results were obtained for the other **NPEC**s.

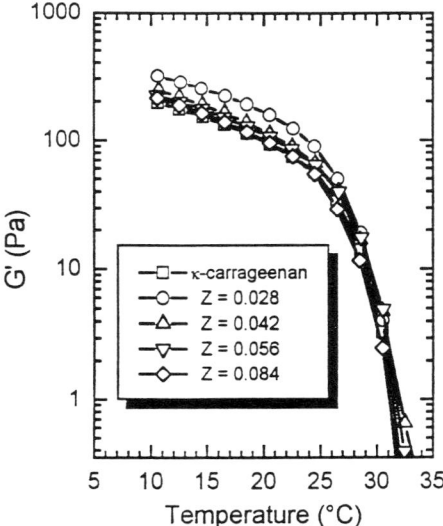

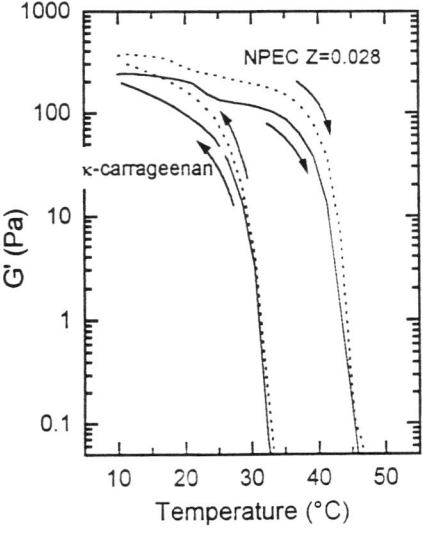

Figure 5. Temperature-dependence of G' (1 rad·s^{-1}; γ = 0.15; 1°C·min^{-1}) for κ-carrageenan in KCl alone and in complexes with chitosan (CHI-20) of varying Z (as indicated in label).

Figure 6. Variation of G' on cooling and heating (as indicated by arrows; 1°C·min^{-1}) for κ-carrageenan alone and for a complex (Z = 0.028) from Figure 5.

4. DISCUSSION

Chitosan, one of the few cationic industrial polysaccharides, has deserved special attention in the formation of **PEC**s. Carrageenans in turn, are strong polyanions, whose composition and charge density, vary with botanical source and method of isolation. One of the key features of carrageenans, namely of κ– or ι–carrageenan, is their ability to undergo a coil-to-double helix conformational transition leading to helix-helix aggregation and the development of a gel network. This process is sensitive to the type and amount of external counterions. In this study we have investigated the effect of complexing small stoichiometric amounts of short chitosan sequences of β-D-glucosamine into κ–carrageenan chains in the disordered coil state. The complexation was carried out under ionic conditions known to promote conformational ordering and gel formation of the carrageenan macromolecules. We have gathered evidence to indicate that chitosan binds to κ–carrageenan even in the presence of a high concentration of non-specific counterions (Na^+ and Cl^-). Under such ionic conditions charge screening is known to stabilise the coil-to-helix transition and hence to induce aggregation of κ–carrageenan, leading to gel formation, provided the concentration of carrageenan is above the critical gelling concentration (c_o). The complexation of chitosan with κ–carrageenan coils leads to a reduction of charge density and hence to the stabilisation of the ordered form of this gelling biopolymer. However, the obtained gel network is stronger than that of κ–carrageenan at the same equivalent concentration. This effect is rationalised in terms of the formation of smaller helical junction zones, therefore greater in number, thus effectively leading to a denser gel network. It seems reasonable to argue that the complexation of small amounts of chitosan onto κ-carrageenan in the coil state, leads to the formation of hydrophobic regions along the carrageenan chain, which self-associate [7] at temperatures well above Tg (Figure1). The formation of such hydrophobic **PEC** junctions seems to reinforce the carrageenan network. This would also be in agreement with wider thermal hysteresis between setting and melting processes (Figure 3), as well as with greater G' values recorded for the CHI-20 than for the CHI-75 **NPEC**s as the **Z** value increases as illustrated in Figure 4 (*i.e.* since the number of CHI-20 chain species introduced into the complex is larger than that of CHI-75 at identical concentration). These differences are a logical consequence of the cooperative nature of the complex formation. By contrast, in the presence of K^+ counterions –firmly established to bind specifically to the carrageenan helix– the addition of chitosan leads to the development of a slightly weaker gel network as the amount of chitosan in the complex increases. This is rationalised as a consequence of a competition for charged sites, thus effectively interfering with the overall degree of K^+–driven aggregation in the κ-carrageenan network [8].

ACKNOWLEDGMENTS
W.A.M. wishes to recognise the financial support from CONACYT, Mexico ("Programa de Colaboración México-Cuba"), and his colleagues from CIAD for their kind invitation to collaboration. We also thank Prof. M. Rinaudo and Prof. E.R. Morris for helpful discussions.

REFERENCES
1. V.A. Kabanov and A.B. Zezin, *Soviet Sci. Rev., Ser. B, Chem. Rev.*, **4** (1982) 207.
2. A.M. Hugerth, N. Caram-Lelham and L.O. Sundelof, *Carbohydr. Polym.*, **34** (1997) 149.
3. H.H. Winter and F. Chambon, *J. Rheol. (N.Y.)*, **30** (1986) 367.
4. K. Nijenhuis and H.H. Winter, *Macromolecules*, **22** (1989) 411.
5. C. Rochas and M. Rinaudo, *Biopolymers*, **19** (1980) 1675.
6. L. Piculell, S. Nilsson and P. Strom, *Carbohydr. Res.*, **188** (1989) 121.
7. V.A. Kabanov and A.B. Zezin, *Pure Appl. Chem.*, **56** (1984) 343.
8. E.R. Morris, D.A. Rees and G. Robinson, *J. Mol. Biol.*, **138** (1980) 349.

4. PROCESSING

HYDROCOLLOIDS – PART 2
Edited by K. Nishinari
2000 Elsevier Science B.V.

The production, properties and utilisation of fluid gels

I.T. Norton, C.G. Smith, W.J. Frith and T.J. Foster.

Unilever Research Colworth, Colworth House, Sharnbrook, Bedford MK44 1LQ, UK.

An understanding of the production/formation and properties of fluid gels of single and mixed biopolymer systems is discussed.

Considering model microgel suspensions has enabled a better understanding of the processes occuring during fluid gel formation.

The production of fluid gels is a nucleation and growth process. It seems reasonable to predict that the particles grow on a very fast time scale from the original nuclei, via aggregation of helices in the initial stages of gel network formation, and reach a size that is restricted by the flow field imposed on the system. The volume occupancy, moduli and charge of particles also influence the properties of these particulate systems.

A model is presented showing the important experimental parameters that need to be considered if the microstructure of a single or mixed biopolymer system is the be understood or even controlled.

1. INTRODUCTION

Biopolymers are of growing interest as they can be used in industry to provide structure stability and texture to a wide range of products (ranging from drilling muds for enhanced oil recovery, through shampoos and shower gels to structuring of emulsion based food products etc.). Many of the fabricated products have traditionally been stabilised by controlling emulsion properties using ingredients or the production process to provide colloidal interactions. One specific example of how the market is changing is in emulsion based foods, as there is a tendency to reduce the amount of fat in the diet, where the understanding of how to structure the aqueous phase has taken on added significance. In some cases, eg. low fat dressings, the viscosifying action of a water soluble polysaccharide like xanthan gum is used, however, the texture that such a polysaccharide imparts is often described as being 'thick' and 'slimy' (which is generally undesirable). Therefore there is a need to understand product microstructures such that the contribution of the individual structural elements within a composite material is known, and can therefore be enhanced or even replaced [1]. This can be achieved if methods are developed such that we can manipulate the physical properties of allowed structuring agents. As a point of comparison the microstructure of fluid gels [2] and phase separating biopolymer water-in-water emulsions [3,4,5,6] can be made to look like an oil-in-water (or water-in-oil) emulsion (figure 1). The most straightforward way to make particulate fluid gels is to pre-form the gel and then shear or chop it to the required particles size. One of the major problems faced using this methodology however is the degree of syneresed water which is produced, which make the products unstable. A way to prevent this is to form particles starting with the same nominal polymer concentration, but now perturbing

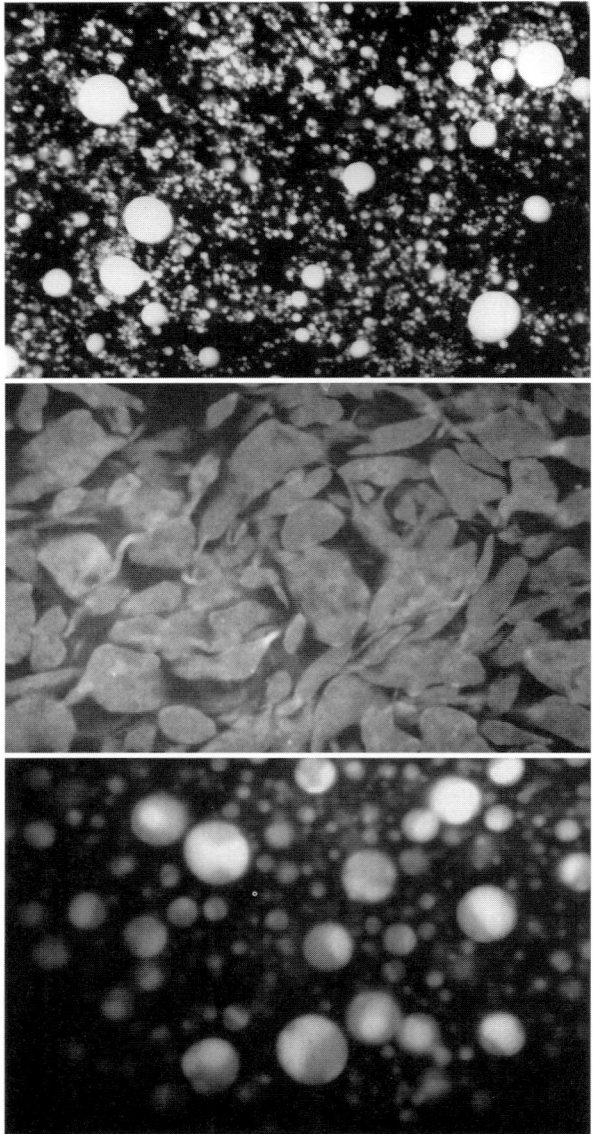

Figure 1. From top to bottom the microstructures of :
- a 40% oil-in-water emulsion,
- a 3% agar fluid gel shear cooled at 40s^{-1};
- a 3% gelatin water-in-water emulsion.

(25 μm _____)

the sample during gelation. For control of resultant properties it is important to understand i) how such microstructures of biopolymer gels are formed, ii) the derived properties and iii) how to utilise such composites. This work will show how general rules for operation have been developed.

2. EXPERIMENTAL

2.1. Materials
The materials used in this study are agar (Luxara 1253, Arthur Branwell & Co. Ltd), gelatin (acid pre-treated pig skin, 250 Bloom, Croda Colloids), maltodextrin (Paselli SA2, Avebe).

2.2. Methods and Measurements
The biopolymers were dissolved with stirring (Silverson Lab. mixer) in water or salt solutions and heated to 95°C (60°C for gelatin). Viscosity measurements were made in a concentric cylinder (couette flow) device (modified with an insert to eliminate the low shear region at the base of the geometry) was used to make fluid gels by quench cooling from 95°C to 5°C at various shear rates.

Model microgel particles were produced by emulsifying a 5% w/w solution of agar/agarose in sunflower oil at 90°C using Hypermer B246 (ICI Surfactants) as an emulsifier. The w/o emulsion was then cooled to gel the biopolymer and the resulting microgel was purified and harvested by repeated dilution and centrifugation in deionised water. Flow curves were determined using a controlled stress rheometer (SR200, Rheometrics Scientific Inc., USA). A parallel plate geometry was used, with roughened surfaces to prevent wall slip. The sample was subjected to a conditioning stress of 10Pa between each measured stress, in order to provide a uniform stress history for each measurement and to counteract sedimentation of the particles.

Single and mixed fluid gel dispersions (100ml) were placed in the well of a microscope slide, treated with aqueous Rhodamine solution (2ml, 0.05%w/w)) and an optical slice (0.7mm thick, 50mm under the surface) examined with a Confocal Scanning Light Microscope (BioRad MRC 600) using a x 60 objective and video recorder (Video Mag. x2000) and laser excitation wavelength of 488nm.

3. RESULTS AND DISCUSSION

3.1. Production
Previous work has addressed the concept of fluid gels and the molecular events taking place during their formation. Such systems are produced by gelling a biopolymer (polysaccharide [2,7,8,9] or protein[10]) in a shear field. It was shown [2] that the shear regime used in the formation of the fluid gel particles does not break the polysaccharide chains as the enthalpy of transition from DSC is the same for a fluid gel and a quiescently cooled gel. NMR T_2 measurements indicate that slightly smaller aggregates of ordered helices are produced in the shear field.

It is believed that the process of formation of particles is one of nucleation and growth where the viscosity of the evolving structure increases (figure 2) as the number and volume fraction of particles increases. The size (and number) of particles are both shear rate [5,11] and concentration dependant [9]. As concentration and/or shear rate are increased the local shear stresses experienced by the particle nuclei increases. It is believed that such shear stresses exceed the failure stress of the early stages of gelation, where the nucleation step of molecular ordering takes place on the millisecond time scale [12] and the aggregation of the ordered species then occurs in the following seconds/minutes. It therefore seems reasonable to predict that the particles grow on a very fast time scale from the original nuclei, via aggregation of helices in the initial stages of gel network formation, and reach a size that is restricted by the flow field imposed on the system. The rate at which this particle formation process occurs is strongly dependent on the depth of quench, as it has been shown that at higher temperatures the time for complete ordering [12] and through space connection for gelation [13,14] can take hours or days.

We propose that the temperature at which we see a peak in viscosity is that at which

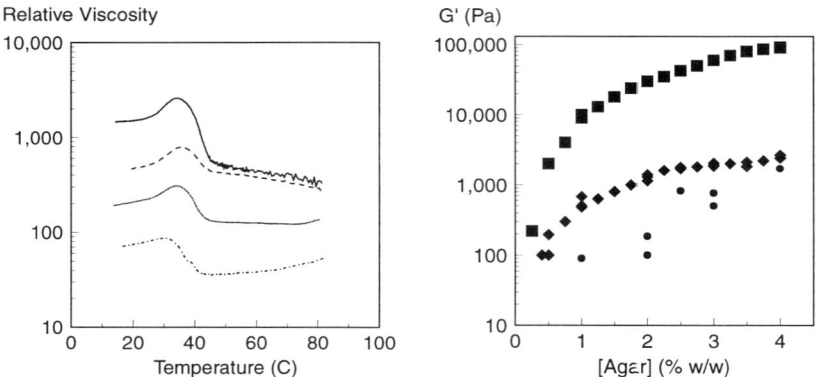

Figure 2 (LEFT). Different concentrations of agar cooled at $1.5°Cmin^{-1}$ while shearing. 3% at $100s^{-1}$ (____), 3% at $500s^{-1}$ (_ _ _), 2% at $500s^{-1}$ (.....), 1% at $500s^{-1}$ (_._._).
Figure 3 (RIGHT). Increase in modulus as a function of concentration for agar gels formed by cooling quiescently (■) or in a shear field at $40s^{-1}$ (◆) and $75Cs^{-1}$ (●).

the total number of particles is determined, and that the subsequent decrease in viscosity is a 'shrinkage' of those particles as intra-particle molecular ordering progresses. If the applied flow field is removed prior to the completion of the molecular ordering/aggregation processes then the particles will associate to form a gel network as the intra-particle aggregation is allowed to become inter-particle aggregation. Indeed if the cross-linking sites are labile (as for instance in the case of gelatin) then the fluid gel is not stable and the particles will aggregate. For lower concentrations of agar the peak is not seen due to a lower particle volume fraction (discussed below).

3.2. Properties

The properties of fluid gels are very polymer concentration and process (shear rate) dependant (figure 3). It has been shown in the past [2] that fluid gel suspensions exhibit yield stress behaviour and flow when such a value of applied stress is exceeded. To obtain a better understanding of the physical properties of these suspensions model spherical microgel particles of agar and agarose were studied [15, 16,17]. It was assumed that the gel properties inside such microgel particles are the same as those of a quiescently cooled gel of the same concentration on a larger, more commonly studied, distance scale. On this basis the elastic modulus of the microgel particles used was 400 kPa and 150 kPa for agarose and agar respectively. From figure 4 (a&b) it can be seen that the measured viscosity is a function of both particle volume fraction and shear rate. In both cases, as the volume fraction is increased there is evidence of shear thinning at low shear rates. This appears to arise from weakly attractive dispersion forces between the microgel particles [16]. As the shear rate is increased the suspensions display a shear thickening region. This is more pronounced when the volume fraction is increased and when the notional particle gel moduli are higher (agarose *versus* agar).

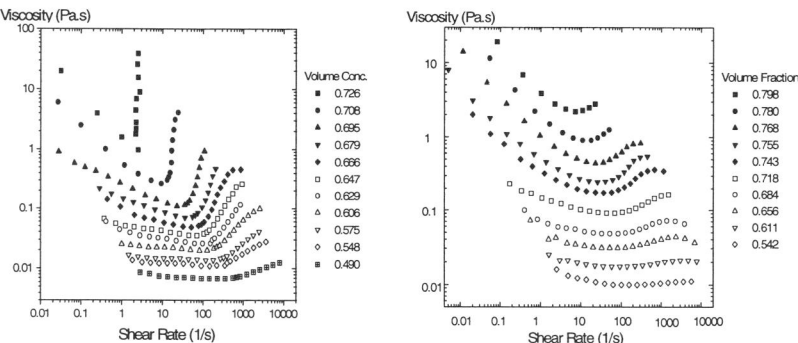

Figure 4a&b. Volume concentration dependence of flow curves for agarose (LEFT) and agar (RIGHT) microgel suspensions.

Many suspensions are known to display such shear-thickening (or dilatant) behaviour [17,18,19,20] which is beleived to arise from shear induced formation of particle clusters [19,21]. At higher shear rates a second shear thinning region appears, which is more evident in the softer agar suspensions, leading us to conclude that this behaviour is a result of particle deformability.

If the high shear relative viscosity (η_{R_∞}) is replaced in the Krieger equation [22]

$$\eta_{R_\infty} = (1 - (\phi/\phi_M))^{-k}$$

by the minimum viscosity observed as a function of shear rate (η_{RM}) then it can be seen that, for both systems, the maximum packing fraction is higher than expected for mono-sized spherical particles (figure 5). Static light scattering reveals that the particles have a broad distribution in particle size which explains this. The difference in agar and agarose is again thought to arise from the differing deformability of the agar and agarose particles.

When considering fluid gels we can now start to understand the properties as a function of volume fraction of particles. This is both concentration and shear rate dependant as seen in a reduction in viscosity (figure 2) and modulus (figure 3). At constant shear rate or concentration the volume fraction is lower as the concentration decreases or shear rate increases, respectively. The effect of shear rate is less evident at high polymer concentration (figure 3 - 4% Agar) presumably because at this concentration even the highest shear rate employed produces a sufficient number of particles such that they are very close to the maximum packing fraction.

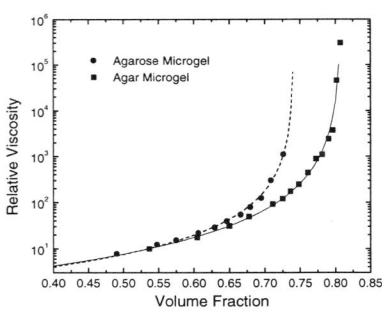
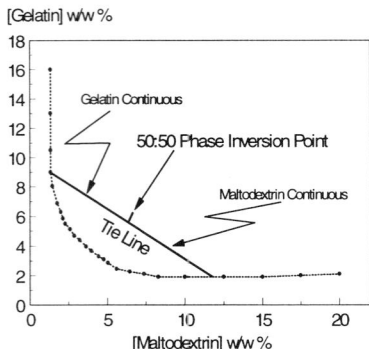

Figure 5 (LEFT). Concentration dependence of η_{RM} for suspensions of agar and agarose microgels.

Figure 6 (RIGHT). Phase diagram for gelatin and maltodextrin measured at pH 6.7 and 60°C.

3.3. Utilisation

The utilisation of fluid gels is a result of their colloidal rheological properties described above. This enables them to impart emulsion-like properties without the presence of oil (eg. skin creams, water continuous vaseline-like products and water continuous spreads [7]) or controlled yield stress behaviour (eg. herb suspension in dressings [1]).

3.3.1. Shear Effects in Mixed Biopolymers Systems

It is often found that the required product properties are not sufficiently fulfilled by using only one biopolymer. Additional information on how biopolymers mix and demix and the influence of gelation of either of the biopolymers, in the presence or absence of a shear field, is essential to enable construction of the desired microstructures.

It is now well established that incompatible biopolymer mixtures behave as water-in-water emulsions when in the ungelled, liquid-liquid state [3,4,5,6]. They seem to follow the rules of droplet break up [23] where the ratio of the internal droplet phase and continuous phase viscosities, and interfacial tension determines the amount of droplet disruption [24]. Stokes' law is also followed, with the density difference between the two phases, the droplet size and continuous phase viscosity influential in the rate of bulk separation [24]. Bulk separation is achieved because the droplets coalesce, but this too is phase sense dependant, which indicates that droplet surface effects are also important [24].

Micrographs have shown phase separation and even phase inversion of quiescently cooled biopolymer mixtures [3,25] where one polymer was held at constant concentration while the other was varied. Although phase inversion can be measured in such systems by microscopy or gel melting points (this is dependant on the continuous phase being above the critical concentration for gelation at the time of measurement [5,11]), both phase compositions and phase volumes are being varied. This work has been developed by using phase diagrams (constructed under given experimental conditions) so that the phase compositions and phase volumes can be varied independently [5]. As one travels along a tie-line phase inversion is seen at ~ 50:50 phase volume (figure 6) [11]. Phase diagrams are a measure of biopolymer A : biopolymer B thermodynamic compatibility under certain experimental conditions (temperature, pH, ionic content). Theoretical models such as the Flory-Huggins model for a two polymer - one solvent system can be used to describe the system if the molecular weight and second virial coefficient of one of the two polymers is known [26].

When incompatible mixtures of gelling biopolymers are cooled under shear we have found phase inversion to occur if the first gelling biopolymer forms the continuous phase in the ungelled mixture [5]. The point of such an inversion appears to be coincident with the onset of gelation of the first gelling biopolymer. Indeed if the viscosity is followed while cooling and shearing an agar continuous agar:gelatin mixture, a similar peak to that seen for the single agar system is observed (figure 2). Particles begin to form as the continuous agar phase starts to gel, at the molecular level, seemingly oblivious to the fact that a second biopolymer is present. By shear cooling an agar:gelatin mixture down to different temperatures (ie. different degrees of agar ordering) and then cooling to 5°C quiescently we have seen that particles are visible after the initial stages of agar ordering and, as for the single agar systems, as the amount of ordering, while shearing, increases the phase volume of particles visibly increases. At the peak in viscosity the argument above would indicate that all of the particles are formed and therefore complete separation is achieved with hairy/swollen agar particles dispersed in a gelatin continuous phase.

The same effect is seen when gelatin is the first gelling biopolymer [5,11], however a gelatin dispersed phase has a tendency to percolate throughout the whole structure if the second gelling biopolymer does not have a significant amount of structure on the time scale of the shearing experiment [5]. This is due to the lower temperature of gelatin gelation and the dynamic nature of the gelatin network, as the gels often are held at temperatures close to the gelatin equilibrium melting temperature. Indeed, we have shown [2] that this re-healing effect can be slowed down or even stopped if a gelatin fluid gel is held at low temperatures (~2°C). Such an understanding has enabled specific product design in low fat food systems [27]. Recent work has shown similar effects with colloidal heat set whey protein gels when mixed with gelatin. Low shear promotes the measured gel modulus, which is ascribed to shear induced secondary aggregation of the protein structure, however at high shear a weak system is produced due to aggregate break up in shear [28].

4. MODEL FOR MICROSTRUCTURE FORMATION

Single and mixed fluid gel composites can be produced as in figure 7, where the microstructures can be controlled by the following conditions:

(A) and (B) - Properties dependant upon i) concentration, ii) cooling rate and iii) shear rate (for fluid gels only),

(C) - Phase inversion takes place. The particles size is determined by the shear rate.

(D) - i) Quiescent cooling: Thermodynamic incompatibility (phase separation) occurs upon cooling or is induced by gelation. If, upon cooling, further phase separation takes place and the original composition gave phase ratios close to critical (50:50) then, depending on how the subsequent separation evolves, it can be envisaged that the phase ratios could flip to the other side of critical, followed by phase inversion. This would not be an equilibrium position however, and the system would be dependant upon rearrangement (via coalescence) which could be time limiting due to interfacial properties (discussed above) and would probably be arrested away from equilibrium if one (or both) of the biopolymers gels during cooling, becoming kinetically trapped.

      ii) Shear cooling: Particle size is determined by shear rate.

(E) - Fast rates of cooling or gelation traps the system, by gelation of the first gelling biopolymer, as a homogeneous mixture. Slower rates of cooling or gelation will lead to (D) - i). The competition between phase separation has invoked a lot of current research activity in the area of food polymer [29,30,31] and synthetic polymer [32] science, and the influence of shear on such systems is the topic of a current collaborative European research initiative[33].

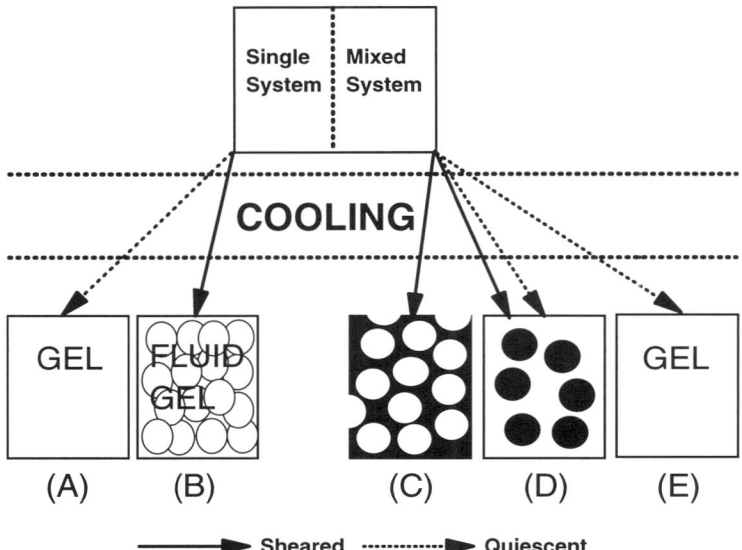

Figure 7. A model for the control of microstructure of single and mixed fluid gels.

Acknowledgements

The authors would like to thank P. Knight, D.A Jarvis, A. Aldred and D.P Ferdinando for their experimental contributions and C.R.T Brown, A.H. Clark and J. Underdown for helpful discussions and suggestions.

References

1. C.R.T.Brown, S.Daniels, M.G.Jones and I.T.Norton (1995) Patent No: WO 96/02151.
2. I.T.Norton, T.J.Foster and C.R.T.Brown in *Gums and Stabilisers for the Food Industry 9*, eds P.A.Williams and G.O.Phillips, RSC, (1998) 259-268.
3. A.H.Clark, R.K.Richardson, S.B.Ross-Murphy and J.M.Stubbs, *Macromolecules*, **16** (1983) 1367-1374.
4. V.B.Tolstoguzov in *Functional Properties of Food Macromolecules*, eds J.R.Mitchell and D.A.Ledward, Elsevier, London, (1986) 385-415.
5. C.R.T.Brown, T.J.Foster, I.T.Norton and J.Underdown in *Biopolymer Mixtures*, eds S.E.Harding, S.A.Hill and J.R.Mitchell, Nottingham University Press, (1995) 65-85.

6. A.Syrbe, P.B.Fernandes, F.Dannenberg, W.Bauer and H.Klostermeyer in *Food Maocromolecules and Colloids*, eds E.Dickinson and D.Lorient, RSC, Cambridge, (1995) 328-340.

7. C.R.T.Brown, N.Cutler and I.T.Norton, (1989) European Patent No: EP 0355 908 B1.

8. N.D.Hedges and I.T.Norton, (1990) European Patent No: EP 0432 835 A1.

9. I.T.Norton, D.A.Jarvis and T.J.Foster, *International Journal of Biological Macromolecules*, (1999) in preparation.

10. W.de Carvalho and M.Djabourov, *Rheol Acta*, **36** (1997) 591-609.

11. T.J.Foster, C.R.T.Brown and I.T.Norton in *Gums and Stabilisers for the Food Industry 8*, eds G.O.Phillips, P.A.Williams and D.J.Wedlock, Oxford University Press, (1996) 297-309.

12. I.T.Norton, D.M.Goodall, K.R.J.Austen, E.R.Morris and D.A.Rees, *Carbo Polym*, **25** (1986) 1009-1029.

13. Z.H.Mohammed, M.W.N.Hember, R.K.Richardson and E.R.Morris, *Carbo Polym*, **36** (1998) 15-26.

14. P.Aymard, T.J.Foster and A.H.Clark, Poster Presentation at Gums and Stabilisers for the Food Industry 9, 7-11 July 1997.

15. W.J.Frith and A.Lips, *Proceedings of the XIIth International Congress on Rheology*, Quebec City, (1996) 558.

16. W.J.Frith, A.Lips and I.T.Norton, *Proc. 4th Royal Society-Unilever Indo-UK Forum 'Structure and Dynamics of Materials in the Mesoscopic Domain.'* Imperial College Press, in Press.

17. W.J.Frith, A.Lips, J.Melrose and R.C.Ball, *'Modern aspects of colloidal dispersions'*, eds R.H.Ottewill and A.R.Rennie, Kluwer (1998) 123.

18. H.A.Barnes, *J. Rheol.*, **33** (1989) 329.

19. W.J.Frith, P.D'Haene, R.Buscall and J.Mewis, *J. Rheol.*, **40** (4) (1996) 531.

20. J.Bender and N.J.Wagner, *J. Rheol.*, **40** (1996) 899.

21. J.R.Melrose, J.H.vanVliet and R.C.Ball, *Phys. Rev. Lett.*, **77** (22) (1996) 4660-4663.

22. I.M.Krieger and T.J.Dougherty, *Trans Soc Rheol*, **3** (1959) 137-152.

23. H.P.Grace, *Chem Eng Commun*, **14** (1982) 225.

24. T.J.Foster, J.Underdown, C.R.T.Brown, D.P.Ferdinando and I.T.Norton in *Food Colloids: Proteins, Lipids and Polysaccharides*, eds E.Dickinson and B.Berdenstahl, RSC, (1997) 346-356.

25. S.Kasapis, E.R.Morris, I.T.Norton and C.R.T.Brown, *Carbo Polym*, **21** (1993) 261.

26. C.M.Durrani, D.A.Prysyupa, A.M.Donald and A.H.Clark, *Macromol*, **26** (1993) 981-987.

27. C.R.T.Brown, R.A.Madsen, I.T.Norton and L.H.Wesdorp (1993) European Patent No: EP 574973.

28. P.Walkenstrom and A.-M.Hermansson, *Food Hydrocoll*, **12** (1998) 77-87.

29. A.M.Donald, C.M.Durrani, R.A.L.Jones, A.R.Rennie and R.H.Tromp in *Biopolymer Mixtures*, eds S.E.Harding, S.A.Hill and J.R.Mitchell, Nottingham University Press, (1995) 99-116.

30. R.H.Tromp, R.A.L.Jones and A.R.Rennie, *Macromol*, **28**(12) (1995) 4129-4138.

31. S.Alevisopoulos, S.Kasapis and R.Abeysekera, *Carbo Res*, **293**(1) (1996) 79-99.

32. R.G.Larson, *Rheologica Acta,* **31**(6) (1992) 497-520.

33. European Commission FAIR project CT96 1015 (1996) *Mixed Biopolymers: Mechanism and Application of Phase Separation.*

HYDROCOLLOIDS – PART 2
Edited by K. Nishinari
2000 Elsevier Science B.V.

Shearing effects on physical network formation

M. Djabourov, I. Capron, S. Costeux and M. Kané

Laboratoire de Physique et Mécanique des Milieux Hétérogènes (CNRS UMR 7636) ESPCI, 10 Rue Vauquelin 75231 Paris Cedex 5 France

Fluids under flow undergoing a phase transformation are expected to suffer a profound modification of their natural way of structuring. A brief survey of the literature reveals that the influence of shear on the structure of complex fluids has received an increased attention during the last decade and a variety of systems have been considered, in relation with their phase behaviour: polymer blends, colloidal suspensions, surfactant solutions, emulsions... These problems are mostly encountered in industry during the processing steps. A review is presented and illustrated by recent experimental results on the effect of shear on physical gelation (for gelatin gels), on crystallisation (for paraffinic crude oils and model solutions), on orientation of flexible or rigid (xanthan) polymers in semi-dilute solutions and finally on the microstructure of two phase fluids (phase separated mixtures of alginate and caseinate in aqueous solutions). The mechanisms controlling the size of inclusions (polymer clusters, droplets, dispersions...) or the orientation of molecules, during these processes are underlined and the consequences upon the rheological properties are suggested.

1. INTRODUCTION

Shearing plays a major role in industrial processes, where fluids are mixed, pumped, transported along pipes, while they are submitted to specific thermal treatments. In many cases, the thermal treatments induce physical phase transitions in the media which then unavoidably take place under shear. The thermodynamic phase transitions or the spontaneous evolution of non equilibrium states occur either in a well controlled process, and thus can be considered as a necessary step for elaborating special textures (polymer materials, paints, food...) or may a be an undesired phenomenon which perturbs the process. In laboratory work, one can use conventional rheometers and imagine different protocols in order to simulate the situations encountered in industry, with the advantage of imposing well controlled conditions, for the flow, the thermal treatment or any other parameter relevant to the process. Shear flows are the most common ones, but elongational flows should also be considered. Consequently, rheological experiments can serve both to induce disturbances to the phase transition and to explore the properties of the medium after perturbation. In order to develop reliable models for the mechanisms of structural modifications induced by shear, it is important to couple various methods, rheology being one of them and molecular spectroscopy, microscopy, scattering techniques... being the complementary ones.

The aim of this paper is to briefly recall the main mechanisms which are at the origin of the shearing effects in complex fluids, trying to classify these complex fluids into suspensions of rigid spherical particles, solutions of elongated rigid or flexible molecules (polymers), aggregating colloids (flocculation) and emulsions (non miscible fluids). In most of the examples quoted above, shear induces a decrease of the viscosity (shear-thinning behaviour). There are however examples where shear thickening occurs (increase of the viscosity under shear), but the case will not be considered here. In the second part of the paper, our recent experimental work is presented, which illustrates some of the ideas presented above for various systems : physical gelation under shear for gelatin gels, crystallisation in solution, alignment of flexible or rigid polymers in semi-dilute solutions (rheo-optical experiments) and finally phase separated solutions of binary polymer mixtures under shear. We conclude with an outlook to future work.

2. ORIGIN OF SHEARING EFFECTS

The different cases which are presented belong either to the colloidal or to the polymer field, but the basic concepts are the same.

2.1. Crowding effects in suspensions

It has been established that suspensions of colloidal rigid particles which exhibit Brownian motions, behave under flow as shear thinning fluids. The explanation for this effect was first given by Krieger and Dougherty [1]. In their model, shear thinning arises from concentration fluctuations and crowding effects in suspensions of solid volume fractions above 20 %. Indeed, due to Brownian movements, instant pairs of particles form and break spontaneously. In presence of a low shear rate, the pairs or doublets of particles tend to rotate as dumbbells. At high shear rates the dumbbells are progressively dissociated and the particles rotate independently as singlets. Krieger and Dougherty assumed a linear relation between the viscosity of the suspension and the fraction of doublets and singlets and derived, with additional simplifications, the equation relating the viscosity of the suspension to the shear stress. The viscosity in the Krieger-Dougherty model writes :

$$\frac{\eta - \eta_{\infty}}{\eta_0 - \eta_{\infty}} = \frac{1}{1 + \tau / \tau_c} \tag{1}$$

τ is the macroscopic stress and τ_c the characteristic stress. The latter depends on the size of the individual particles a and on temperature T :

$$\tau_c = \frac{kT}{\beta a^3}$$

where β is a numerical factor. The characteristic shear stress τ_c controls the appearance of shear thinning of the medium. The two limits η_0 and η_∞ correspond respectively to zero stress and infinite stress.

An equivalent equation can be derived for the shear thinning effects versus the shear rate :

$$\frac{\eta - \eta_\infty}{\eta_o - \eta_\infty} = \frac{1}{1 + \dot{\gamma}/\dot{\gamma}_c} \tag{2}$$

where the characteristic shear rate $\dot{\gamma}_c$ is given by:

$$\dot{\gamma}_c = \frac{kT}{6\pi a^3 \eta_s} \tag{3}$$

The characteristic shear rate $\dot{\gamma}_c$ is the inverse of the time for Brownian motion of the particle of size a diffusing on a mean square distance of the order of its squared radius. The viscosity is then related to the ratio of the actual shear rate of the flow to the characteristic shear rate of the suspension $\dot{\gamma}/\dot{\gamma}_c$. In some cases $\dot{\gamma}_c$ is expressed by the Péclet number which is precisely:

$$P_e = \frac{\frac{6\pi\eta_s a^3}{kT}}{(\dot{\gamma})^{-1}} = \frac{\text{characteristic time of Brownian motion}}{\text{characteristic time of the flow}} \tag{4}$$

The characteristic shear rate $\dot{\gamma}_c$ corresponds to the Péclet number $P_e \approx 1$. For $P_e \ll 1$, the suspension has a viscosity η_O, at large $P_e \gg 1$, the viscosity reaches the lowest limit η_∞. An alternative derivation of this equation is due to Cross [2] which established the well known equation (Cross equation) which is widely used in polymer flows:

$$\frac{\eta - \eta_\infty}{\eta_O - \eta_\infty} = \frac{1}{1 + (\dot{\gamma}/\dot{\gamma}_c)^m} \tag{5}$$

The exponent m ($0 < m < 1$) appears as a phenomenological parameter.

2.2. Orientation of rod-like molecules

Rigid anisotropic molecules are easily oriented by external fields, such as electric, magnetic or hydrodynamic fields. At rest, anisotropic rigid molecules are in a random orientational state due to Brownian motions. Under a shear flow, the velocity gradient tends to orient them in the direction of the stream lines, whereas Brownian motion acts against orientation. Under shear, molecules spend a greater part of time at a preferred orientation with respect to the stream line. The degree of orientation is related to the ratio of the shear rate $\dot{\gamma}$ to the rotational diffusion coefficient of the rods, D_{rot}. The latter has the dimension of the inverse of a time (s⁻¹) and thus is comparable to the characteristic shear rate defined in the preceding section for rigid spherical particles. The orientation of the molecules in the flow creates an anisotropy of the index of refraction due to the difference of electronic polarizabilities of the molecules between their axial and transversal directions. Flow birefringence appears, which can be detected experimentally. This effect can be used as a direct mean for measuring the rotational diffusion coefficients and for deriving molecular parameters. The orientation angle of the molecules with respect to the stream lines, called the extinction angle χ, varies with the shear rate and is very sensitive to the length of the rods and thus to any polydispersity effects. The subsequent decrease of the viscosity due to alignment of the rigid molecules has been predicted theoretically by Doi and Edwards [3] for semi-dilute solutions. The viscosity, like the extinction angle, are functions of $\dot{\gamma}/D_{rot}$.

2.3. Deformation and orientation of flexible polymers

Shear thinning is also observed in semi-dilute solutions of flexible polymers. The theoretical modelling of this effect is delicate. In dilute solutions, the shear rate dependence of the viscosity is ascribed to molecular orientation and deformation. Various molecular theories of Kuhn, Rouse and Zimm yield to an expression of the rotational diffusion coefficients as the reciprocal of the longest molecular relaxation time $(t_{relax})^{-1}$ of the chain :

$$t_{relax} = \frac{[\eta]\eta_s M}{kN_A T} \qquad (6)$$

where M is the molar mass, $[\eta]$ the intrinsic viscosity, η_s the solvent viscosity, N_A the Avogadro number. Orientation of the molecules is accompanied by deformation of the coils whose shape changes from sphere to ellipsoid. The orientation angle between the longest axis of the ellipsoid and the flow velocity direction χ varies with the reduced shear rate :

$$\dot{\gamma}_r = \dot{\gamma} \ t_{relax} \qquad (7)$$

according to the relation

$$\chi = \frac{1}{2} \arctan \frac{C}{\dot{\gamma}_r} \qquad (8)$$

with the numerical factor C being: $C= 1$, Kuhn model, $C= 2.5$ Rouse model, $C= 4.88$ Zimm model [4]. Concerning viscosity, deformation and orientation operate in opposite ways: deformation of the coils increases flow resistance, whereas orientation decreases the friction and thus the viscosity. If these two effects compensate, viscosity should be independent on the shear rate. Whenever orientation effects predominate, shear thinning is observed.

2.4. Growth of fractal clusters

Initially dispersed colloidal suspensions flocculate when the Brownian particles undergo attractive interactions. Tenuous networks or clusters are formed to which the shear may produce additional effects : i) the elastic clusters are deformed and break and ii) the shear modifies the rate on encounter of the clusters (Wessel and Ball [5]). Due to attractive interactions between the particles (electrostatic, Van der Waals, bridging between individual particles...) there is a maximum value for the torque, Γ_{max} that the cluster can support without breaking. This torque limits the maximum size R_{max} that a cluster can grow in presence of a shear rate $\dot{\gamma}$:

$$R_{max} \sim (\Gamma_{max})^{1/3} (\eta_s \dot{\gamma})^{-1/3} \qquad (9)$$

This equation predicts a power law behaviour with an exponent -1/3 for the maximum size of the clusters versus the shear rate. If one assumes that the clusters have a fractal structure with a fractal exponent D_f, in the limit of dilute suspensions, the viscosity varies linearly with the volume fraction of the clusters (a being the size of the primary particles) :

$$\eta - \eta_s \propto \left[\frac{\eta_s a^3 \dot{\gamma}}{\Gamma_{max}} \right]^{\frac{3-D_f}{3}} \tag{10}$$

The model predicts a shear thinning behaviour due to the size change of the clusters (change of their volume fraction) with the shear rate, for a given volume fraction of primary particles.

2.5. Break-up of droplets under shear in emulsion-type systems

In phase separated systems, emulsions, non-miscible blends of polymers... the microstructure is controlled by hydrodynamic forces. The microstructure is composed of droplets of a dispersed phase into a continuous phase. The typical size of the droplets is between 1 to 100 μm. The microstructure may appear spontaneously during phase separation or result from an emulsification process, when two immiscible liquids are mixed to obtain a dispersion of one liquid in the other. Droplet break-up and coalescence occur during processing. If the emulsions or phase separated systems remain stable for some time, they can be characterised by their microstructure (droplet size distribution) and their linear dynamic properties.

A liquid droplet dispersed in another immiscible liquid has a spherical shape at rest. Under a steady flow, the droplet will deform until it reaches a steady shape. Depending on the so-called capillary number, which is given by C_a which is given by :

$$C_a = \frac{\eta_{cont} \dot{\gamma} R}{\sigma}$$

where σ is the interfacial tension between the two liquids, R the radius of the droplet and η_{cont} the viscosity of the continuous phase. Break up occurs when a critical value of the capillary number C_a^c is reached. Let p be the ratio of viscosities of the inclusions (dispersed phase), η_{incl}, versus the continuous phase, η_{cont}

$$p = \frac{\eta_{incl}}{\eta_{cont}}$$

C_a^c corresponds to a dimensionless shear rate at which the droplet can no longer assume a steady shape and is broken by shear. It has been established [6] from experimental, theoretical and numerical approaches on single droplets, that there is a relation between the critical capillary number C_a^c and the ratio of viscosities p. For Newtonian liquids, if the interfacial tension σ and the viscosity of the continuous phase η_{cont} are known, the critical capillary number is known from this relation and the product $\dot{\gamma} R$ is fixed. The maximum size of the droplets thus decreases inversely to the shear rate. Visco-elastic measurements can be performed on emulsions in the linear regime and they reflect the microstructure of the emulsion. The theoretical modelling which has been recently developed by Palierne [7] allows to interpret the data, even for non Newtonian liquids, and to derive, from the dynamic measurements, the distribution of radii of the droplets, when the interfacial tension is known independently.

3. EXPERIMENTAL DETAILS

The rheological measurements were performed with a Carrimed CSL^2 100 or an AR 1000 rheometer working with a cone and plate geometry (diameter 4 cm, angle 2°). The rheo-optical experiments were performed with a home built instrument adapted on a Haake RS 100 constant stress rheometer, with a coaxial cylindrical geometry with a gap of 1mm, and a height of 57 mm (optical path). The He-Ne laser beam crosses the gap between cylinders parallel to the axis. The top and the bottom lids of the cylinders are quartz windows which have no stay birefringence. The amplitude and phase of the detected signal give access to the macroscopic birefringence and to the average orientation angle.

The samples used in these various experiments are gelatin from SKW Biosystems, whose molecular characteristics have been reported elsewhere [8], poly(styrene sulphonate) (PSS) from National Starch and Chemicals ($M_w= 10^6$ g/mole), xanthan from SKW Biosystems. The samples (PSS, xanthan) used for the rheo-optical experiments have been filtered under Millipore 1μm before used. All the samples were dissolved into distilled water. Other aqueous phase separated solutions were prepared alginate from Kelco and Na-caseinate from DMV. Finally, a non aqueous suspension is reported: a crude oil sample containing a small percent of paraffinic long chains (between C_{20} and C_{30}) which was provided by Elf EP.

4. RESULTS

The concepts recalled in Section I are illustrated with examples which can be compared to phase transitions under shear: gelation under shear for a physical polymeric gel, crystallisation for paraffinic crude oils, appearance of birefringence for semi-dilute polymer solutions and finally phase separated binary polymer solutions submitted to shear. Most of the results presented are rheological experiments, some were also investigated by the rheo-optical technique. It is important, as it has been underlined in the introduction, that different techniques are used to corroborate the interpretation of the rheological data. Flow birefringence experiments, like those presented in this paper, require fully transparent fluids and thus cannot be used for all types of solutions.

4.1. Gelation under shear
The problem that we considered in this experiment is the phenomenon of gelation for a physical polymeric gel, which is well known for its applications in the food or the photographic industries, this is the gelatin gel. Gelation is obtained by cooling solutions below 30°C, which induces the coil to helix transition of the gelatin chains. As triple helices are formed (collagen type structure), stabilised by hydrogen bonds, the conformational transition induces gelation by physically crosslinking the chains. The question that we addressed was how application of a

well controlled shear would disturb the formation of the gel network. Under flow, a competition takes place between formation of the clusters by chain to chain crosslinking and their disruption by shear forces. Rheological experiments [8] were performed for the same temperature history (cooling and keeping the sample at a fixed temperature) by imposing different protocols: either a permanent shear stress or a permanent shear rate. In order to characterise the structure of the fluid at any stage of the process, we added dynamic measurements (linear regime) and instant flow curves (non linear regime) during very brief interruptions of the process. In order to avoid memory effects provoked by the non-linear measurements, a new sample was taken for successive instant measurements. The time evolution of the flow curves (given for instance by the viscosity versus shear rate plots in double logarithmic scale) was very much significant of the state of the fluid.

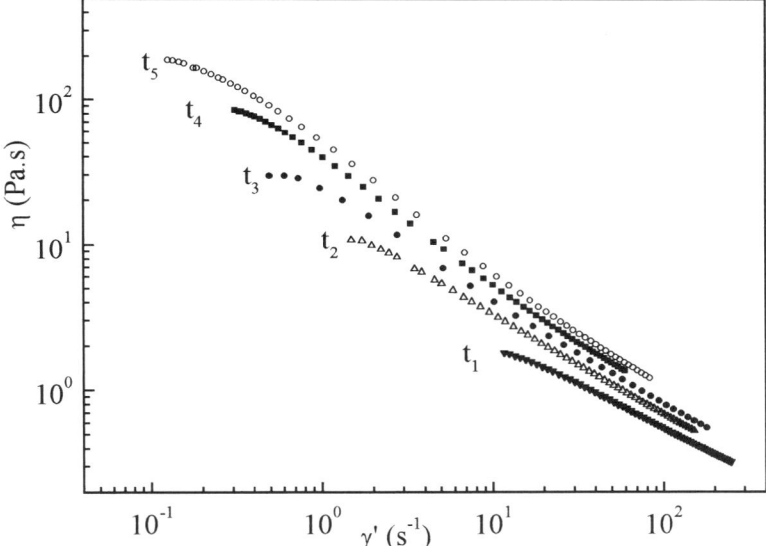

Figure 1.   Flow curves during gelation of a gelatin solution of concentration 6.5% at a temperature of 26°C. The different measurements are made at times t_1, t_2 ... during the kinetics of gelation under a permanent shear stress of 40 Pa.

    In Figure 1 one can see the beginning of the process where the instant rheograms are shown successively at times t_1, t_2, t_3...for a solutions evolving under a constant shear stress. One clearly notices that the flow curves are shifted towards low shear rates. Recalling the Cross equation (5) presented in section 2, one can fit theses curves and derive the characteristic shear rate $\dot{\gamma}_c$ of the solution. $\dot{\gamma}_c$ decreases on several decades in the course of gelation. This parameter can be related to the size of the clusters. After some time the sol-gel transition takes place at a particular moment which is detected by an abrupt increase of the viscosity. The whole rheological investigation [8] allowed us to define a schematic evolution for the cluster formation under shear. This is shown in Figure 2: i) under a fixed shear stress, the size of the microgels increased with time, until percolation is observed. A particulate gel was built, in contrast with a homogeneous network; ii) under an imposed shear rate, the size remained constant and did not allow the formation of a gel. No gelation (liquid like behaviour) is observed after 15 hours for a gelatin solution of 6.5% concentration under a shear of 100 s^{-1} which normally gels in 15 min ! This schematic representation of the process will be analysed by using independent techniques, such as rheo-optical techniques as described before.

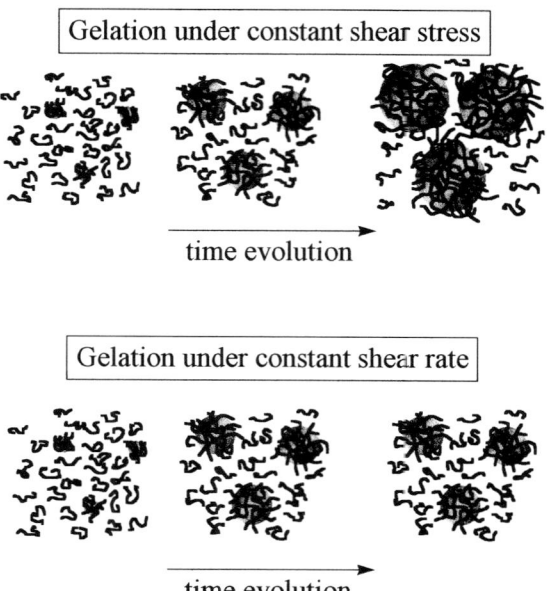

Figure 2.   Schematic representation of the time evolution of the gelatin microgels during gelation at constant temperature, under imposed shear stress or imposed shear rate.

4.2. Crystallisation under shear

We deal here with a practical problem encountered in oil production. Crude oils which contain around 20% of n-paraffins $>C_{10}$ exhibit complex flow behaviours which strongly depend on the thermal and flow conditions. When the oil is cooled below a certain temperature (around 30°C for instance) the paraffinic longest chains become insoluble and start to crystallise. These are called waxy crude oils. The flow characteristics for these oils are time and history dependent. When the oil is cooled in quiescent conditions, a strong gel is formed. Such a transition is shown in Figure 3. Within a few degrees the elastic modulus of the gel increases over six decades (up to 10^6 Pa). However, when the oil is cooled under flow, as it happens usually in oil production, only a moderate increase of the viscosity is seen, such as in Figure 4. The oil has been cooled under a fixed shear rate, as indicated on the figure. The apparent viscosity is plotted versus temperature. One can see that it varies strongly with the shear rate, for a given range of temperatures and a fixed cooling rate. When the process of cooling takes place under a fixed shear stress, one observes a gelation transition which can be compared to gelatin gels gelation under shear.

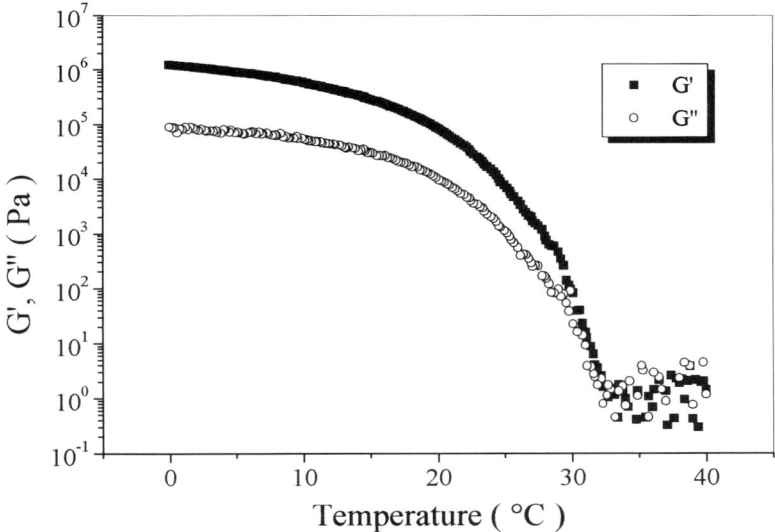

Figure 3. Gelation of a crude oil containing a small amount of paraffins. The shear moduli G' and G" were measured at the frequency of 1 Hz during cooling at a rate of 0.5°C/min.

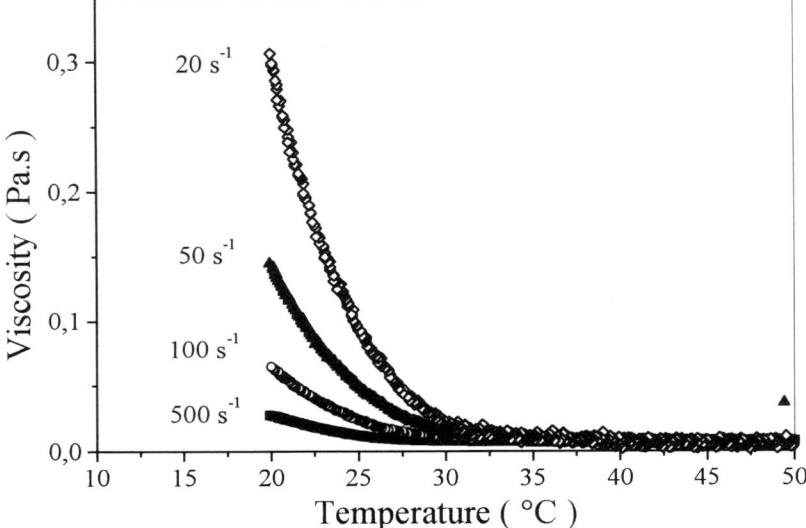

Figure 4. The apparent viscosity of the crude oil measured as a function of temperature during cooling under fixed shear rates as indicated.

Although these are preliminary results the process of crystallisation under shear can be interpreted by following the ideas of formation of clusters with a fractal structure. The shear rate controls the size of the clusters, the shear thinning effect is due to variation of the volume fraction of the clusters. The volume fraction of primary particles may be assimilated to the crystalline content for each temperature. The clusters which were broken by shear do not re-aggregate. In parallel to the rheological data microscopic investigations of the structure are under way in order to characterise the architecture of the network and the size and structure of the clusters under shear. The crystallisation of model paraffinic solutions exhibit a similar behaviour and are also currently investigated.

4.3. Flow birefringence of semidilute polymer solutions

The first results [9] obtained with our rheo-optical instrument concern semi-dilute solutions of flexible and rigid macromolecules under shear. The main parameters that we derived are the macroscopic birefringence and the average orientation angle χ for these solutions. For PSS solutions at two different concentrations, 5% and 10 %, Figure 5 shows that the extinction angles decrease from 45° to 36° when the shear rate increases from 0 to 10^3 s^{-1}.

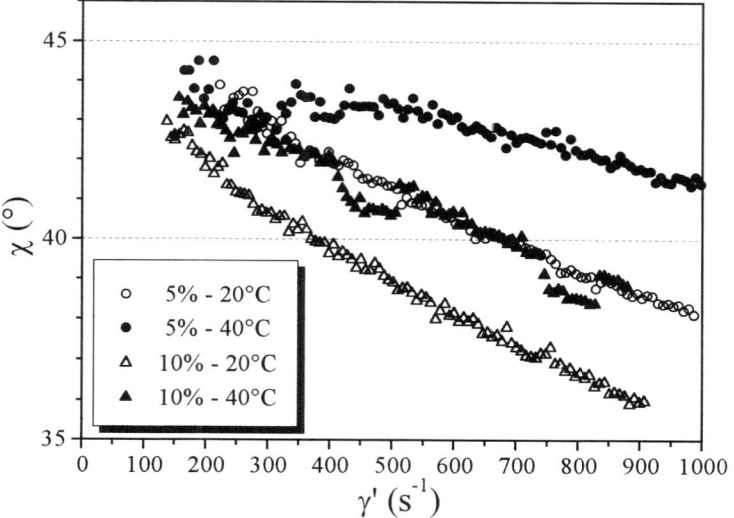

Figure 5.   The extinction angle versus the shear rate for PSS solutions at two concentrations (5% and 10 %) and two temperatures (20°C and 40°C).

This orientation is relatively weak, while the shear rate covers the whole accessible range in currently used instruments. The other example is xanthan: the extinction angle and corresponding viscosities are plotted versus shear rate in Figure 6a and b. The decrease of χ with the shear rate is much more important than in PSS and accordingly an large shear thinning appears for the viscosity. These preliminary results demonstrate a strong influence of the chemical nature of the polymers (neutral or polyelectrolyte) and of the local rigidity of the chains on the orientation of the molecules. Turing back to the problem of gel formation under shear these effects should be taken into account. Gelatin gelation under shear is one of the systems that we intend to investigate with this device.

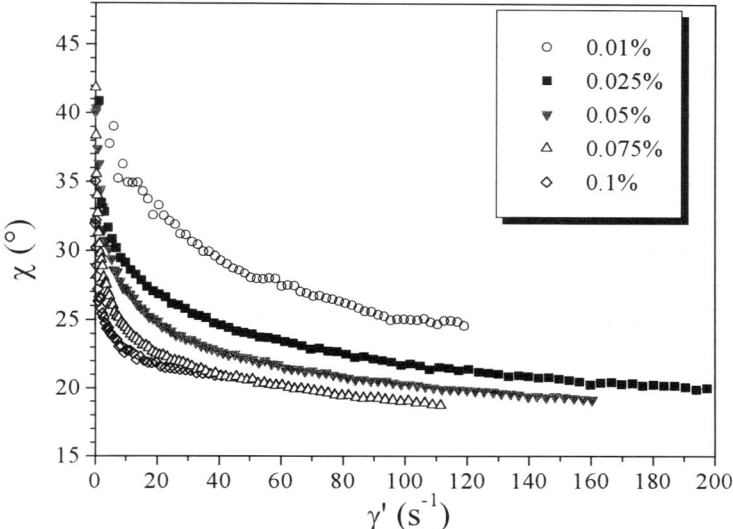

Figure 6a. Extinction angle versus shear rates for xanthan solutions of various concentrations as indicated, at room temperature.

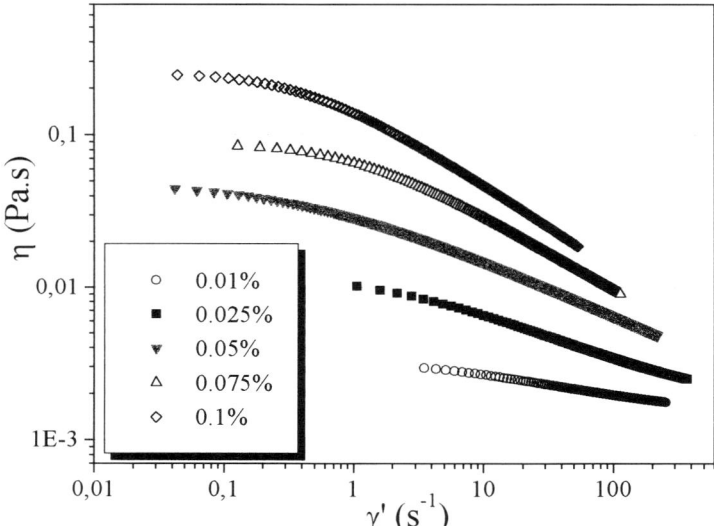

Figure 6b. Viscosity versus shear rate for the xanthan solutions at the same concentrations, at room temperature.

4.4. Phase separated solutions under shear

The last case that we examine is a phase separated aqueous solution containing alginate and caseinate. An initial mixture in the two phase region of the solution was prepared. Then the solution was centrifuged and the two phases were collected, one being a caseinate rich phase and the other an alginate rich phase, both phases containing the two biopolymers. Various amounts of each phase were then mixed together varying the volume fraction of one phase from 0 to 100%. Phase separated solutions of alginate and caseinate [10] show micro-structures which are close to emulsions, with droplets of various sizes scaling in the range 1 to 100 μm. By applying different shear rates it is possible to change the microstructure in agreement with the ideas presented in section 2. Then, the spectrum of the emulsion can be determined. Knowing the volume fraction of each phase and its dynamic spectrum, the interpretation of the dynamic spectrum of the emulsion can be performed according to the Palierne theory [7]. In Figure 7 we illustrate this effect. A phase separated solution containing a volume fraction of 10% of the caseinate rich phase which is the less viscous one, was submitted to intensive shearing at either 0.03 s^{-1} or at 500 s^{-1} then the dynamic spectrum was performed. It appears in the figure that at the low frequency limit of the spectrum, the storage modulus G' exhibits a visible change, while the loss modulus G" remains insensitive to the pre-shear. The linear viscoelastic behaviour [11] of two phase polymer blends in the melt has attributed to the relaxation of shape of the droplets. Changes in size of the droplets induce a visible shift of the characteristic frequencies. The full interpretation of the data requires the knowledge of the interfacial tension between the phases.

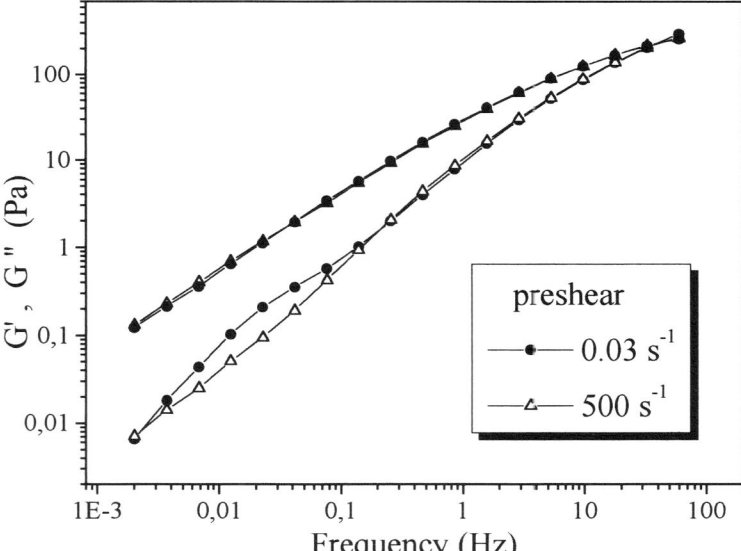

Figure 7. Dynamic spectra for an alginate-caseinate solution in the two phase region. The solution has been submitted to two different shear rates (0.03 and 500 s^{-1}) before measurement of the dynamic spectra. The continuous lines are guides for the eye.

5. CONCLUSION

This review underlines the effects of shear on the molecular conformation or the microstructure of complex fluids containing polymers and colloids. Although the basic mechanisms have been identified and classified, real processes involving aggregation, physical gelation, phase separation...involve a juxtaposition of these mechanisms and thus are more difficult to analyse. Combining rheological and structural techniques one should be able to better understand such systems. Models relating the local structure and the flow properties of these solutions can then be tested and improved. This area of research seems very promising for the future.

REFERENCES

1. I.M. Krieger and T.J. Dougherty, *Trans. Soc. Rheol.*, **III** (1959) 137.
2. M.M. Cross, *J. Colloid. Sci.* **20** (1965) 417.
3. M. Doi and S.F. Edwards, *J. Chem. Soc. Farad. Trans. II,* **74** (1978) 560; M. Doi and S.F. Edwards, *J. Chem. Soc. Farad.* Trans. *II* **74** (1978) 918.
4. A. Link, M. Zisenis, B. Prötzl and J. Springer, *Makrom. Chem. Makrom. Symp.,* **61** (1992) 358.
5. R. Wessel and R.C. Ball, *Phys. Rev. A*, **46** (1992) R3008.
6. B.J. Bentley and L.G. Leal, *J. Fluid Mech.* **167** (1986) 241 ; H.P. Grace, *Chem. Eng. Commun.*, **14** (1982) 225.
7. J.F. Palierne, *Rheol. Acta*, **29** (1990) 204
8. W. de Carvalho and M. Djabourov, *Rheol. Acta*, **36** (1997) 591.
9. S. Costeux, M. Djabourov and J.C. Charmet, Proceedings of the 5th European Rheology Conference, Progress and Trends in Rheology, I. Emri Ed, Darmstadt, Steinkopff, (1998) 459.
10. J.C.G. Blonk, J. van Eendenburg, M.M. G. Koning, P.C.M. Weisenborn and C. Winkel, *Carbohyd. Polym.*, **28** (1995) 287.
11. P. Scholtz, D. Froelich and R. Muller, *J. Rheol.*, **33** (1989) 481.

HYDROCOLLOIDS – PART 2
Edited by K. Nishinari
2000 Elsevier Science B.V.

243

Kinetic analysis of freeze - induced coagulation of soyprotein

M. Urai and O. Miyawaki

Department of Applied Biological Chemistry, The University of Tokyo, 1-1-1 Yayoi, Bunkyo-ku, Tokyo 113-8657, Japan

Freeze-induced concentration of protein triggered freeze-denaturation of soyprotein, which was a rate process strongly dependent on time and temperature. A kinetic model was proposed to describe the freeze-denaturation in consideration of freeze-induced concentration of solute, which accelerated the reaction rate substantially. The theoretical model was effective to describe the influence of temperature and the protective effect of glucose against denaturation. From the kinetic analysis, the reaction cascade of freeze-induced coagulation of soyprotein seemed to be limited by the initial interaction between the two protein molecules.

1. INTRODUCTION

Freezing of solutions causes denaturation of proteins, which causes various changes of protein properties. Shikama and Yamazaki [1] investigated the freeze-denaturation of catalase and showed the existence of critical temperature range for the enzyme inactivation. Chilton et. al. [2] studied on the effect of freezing on enzyme activity of lactate dehydrogenase and pointed out subunit dissociation as a mechanism for the enzyme inactivation. Curti et. al. [3] reported the existence of the optimal temperature for freeze-denaturation of L-amino acid oxidase and proposed a reaction mechanism of denaturation. Tamiya et. al. [4] analyzed freeze-denaturation of lactate dehydrogenase, malate dehydrogenase, alcohol dehydrogenase, glucose-6-phosphate dehydrogenase, and pyruvate kinase and suggested aggregation, unfolding, and transconformation as mechanisms of inactivation. Carpenter et. al. [5] tried to separate the freezing- and drying-induced denaturation of lyophilized lactate dehydrogenase and pyruvate kinase and applied their theory of preferential exclusion of water from protein molecules [6] as a mechanism of the cryoprotection by polyethylene glycol.

Apart from enzymes, various other proteins have been studied on their freeze-denaturation. Hashizume *et. al.* [7] investigated coagulation of soyprotein by freezing and proposed a mechanism, in which intermolecular reactions occur through S-S bonds as a result of concentration by freezing. Yamazaki *et. al.* [8] reported freeze-denaturation of ricin D as a consequence of acid concentration and change in water structure by freezing. Watanabe *et. al.* [9] studied on protection effect of non-ionic surfactant against freeze-denaturation of rabbit skeletal myosin showing that the protective effect of surfactant cannot be explained by the theory of preferential exclusion of water from protein [6]. Sotelo and Mackie [10] reported the effect of formaldehyde, which is formed as a result of enzymatic break down of trimethylamine during frozen storage, on the aggregation behavior of bovine serum albumin under freezing. Kitazawa *et. al.* [11] reported on the insolubilization of carp myofibrils during frozen storage as a result of actin-myosin complex formation.

Thus, investigation of the freeze-denaturation of proteins is so diversified and various mechanisms were proposed. In this paper, we quantitatively analyzed freeze-denaturation of soyprotein isolate from the standpoint of freeze-induced concentration, as a common basis for all the mechanisms proposed for freeze-denaturation of proteins. Direct interaction of soyprotein molecules was analyzed by a kinetic model in consideration of this "concentration effect" of proteins by freezing.

2. THEORETICAL

Freezing denaturation of soyprotein is described by a model in which n protein molecules reacts one another to coagulate for precipitation.

$$nC_N \rightarrow C_D \downarrow \tag{1}$$

where C_N and C_D are molar concentrations of native and denatured protein, respectively. The reaction order is assumed to be n-th order so that the reaction rate is:

$$-(dC_N / dt) = kC_N^n \tag{2}$$

where t is time and k is reaction constant. When protein solution is frozen, protein as a solute is microscopically concentrated through ice crystal formation as shown Figure 1. Because of this "concentration effect", protein concentration will be increased by a factor of α and the reaction volume will be decreased by a factor of $1/\alpha$, where α is a concentration factor in freeze-induced concentration. Then, Equation (2) will be rewritten as follows:

$$-(\mathrm{d}C_N / \mathrm{d}t) = (1/\alpha)k(\alpha C_N)^n = k\alpha^{n-1}C_N^n \tag{3}$$

This equation is easily integrated if α is constant:

$$C_N / C_{N0} = \frac{1}{\{1+(n-1)k\alpha^{n-1}tC_{N0}^{n-1}\}^{(1/(n-1))}} = \frac{1}{\{1+Kt\}^{(1/(n-1))}} \tag{4}$$

$$K=(n-1)k\alpha^{n-1}C_{N0}^{n-1} \tag{5}$$

where C_{N0} is the initial concentration of denaturable protein and K is considered to be an apparent rate constant.

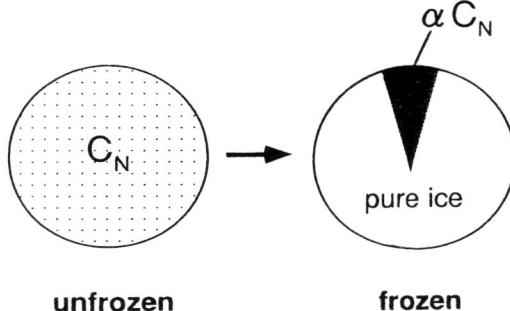

Figure 1 Freeze-induced concentration of solute

3. MATERIALS AND METHODS

3.1 Preparation of soyprotein solution

Soyprotein isolate (SPI: New Fujipro SE, Fuji Oil) was dissolved in distilled water at a concentraton of 8 wt%. The suspension was homogenized for 3 min at 18,000 rpm (Homogenizer, Nihon Seiki) and centrifuged at 10,000 rpm for 15 min at 4 °C. The supernatant was diluted by 11-fold and was used as a sample to be frozen.

3.2 Measurement of freeze-denaturation

The above sample was divided into many small plastic tubes (2ml) and were frozen at −10, -18, and −50°C, respectively, in refrigerators with or without addition of glucose as a cryoprotectant. The frozen sample tube was taken out at an interval, thawed at 25°C and turbidity was measured at 600nm. Differential scanning calorimetry (Perkin Elmer, DSC-7) was applied to analyze the frozen state of water in the sample.

4. RESULTS AND DISCUSSIONS

Figure 2 shows the freeze-induced denaturation process of soyprotein as expressed by the normalized concentration ratio of native protein measured by the optical density at 600 nm. Because of the formation of coagulation, the ratio of native protein decreased with time showing that the denaturation of soyprotein is a time-dependent rate process. The rate was also strongly dependent on temperature. With a decrease in temperature, the denaturation rate decreased. In Figure 2, solid and dotted lines are the optimized theoretical results described by Equation (4), which had a good correlation with experimental data. This shows the effectiveness of the present theoretical model of freeze-denaturation. In the optimization, the reaction order n was also determined to be 2. In this case, Equation (4) becomes very simple as follows:

$$C_N / C_{N0} = \frac{1}{1 + k\,\alpha\,t C_{N0}} \tag{6}$$

This apparent reaction order of 2 suggests that the rate limiting step of the reaction cascade of protein coagulation is the initial interaction between two protein molecules.

Equation (3) shows that reactions are accelerated under freezing when the reaction order is higher than one. In the present case, the estimated reaction order was 2 so that the reaction was much accelerated by freezing. The highest reaction rate was obtained at the highest temperature below freezing point as long as the concentration factor α is unchanged. In the present case, this temperature was $-10°C$. From the optimized reaction rate at different temperatures, the activation energy of the coagulation reaction of soyprotein was obtained to be 68.6 kJ/mol.

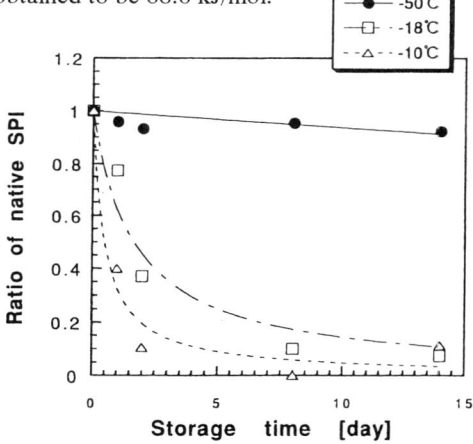

Figure 2   Effect of temperature on freeze-denaturation rate of SPI.

Chemical reaction accelerated by freezing is reported in the literature.   Pincock and Kiovsky [12] reported that the reaction of ethylene chlorohydrin with sodium hydroxide in frozen aqueous solution is 1000 times faster than in supercooled solution at the same temperature.   Takemura *et. al.* [13] reported on the acceleration of the rate of nitrite oxidation by freezing in aqueous solution, which might cause the existence of nitrate in acid rain.

In Equation (4), the effect of temperature on the concentration factor α should be discussed.   To this end, the soyprotein solution used was analyzed by differential scanning calorimetry.   As a result, freezing of water in the soyprotein solution completed above $-5°C$ or so.   Therefore, the concentration factor α was confirmed to be unchanged in the temperature range between -10 and $-50°C$.

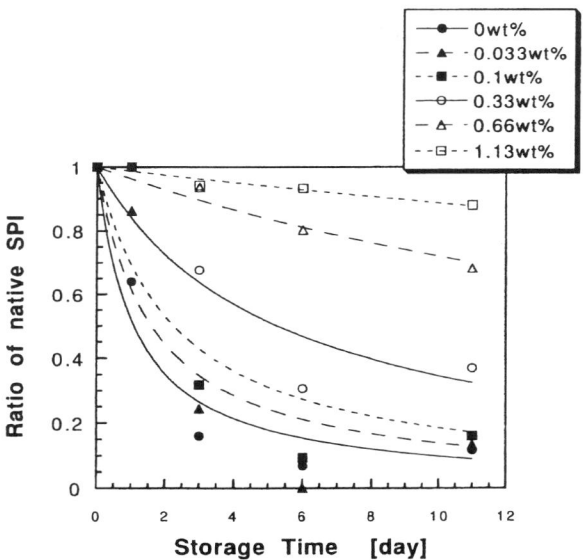

Figure 3   Effect of glucose added on freeze-denaturation rate of SPI at -18 °C.

For the protection from freeze-denaturation, sugars have been frequently used as a cryoprotectant.   Figure 3 shows the effect of glucose on the freeze-denaturation of soyprotein.   It is apparent that the addition of glucose is very effective to prevent the freeze-denaturation of soyprotein.   In this figure, lines are theoretical results described by Equation (5).   The theoretical model was also useful to describe the cryoprotective effect of glucose against freeze-denaturation of soyprotein.   Thus, the mechanism of cryoprotective effect of glucose against freeze-denaturation is proved to be primarily the "dilution effect" because of the coexistence of glucose with protein under freezing. Because of this, the concentration factor α will decrease with an increase in glucose concentration, which caused cryoprotective activity.

In conclusion, freeze-induced coagulation of soyprotein was explained well by the kinetic model in consideration of freeze-induced concentration of protein. Because of this concentration effect, the reaction was accelerated under freezing. Effect of freezing temperature and the protective effect of glucose against protein denaturation was also explained by the present model.

REFERENCES

1. K. Shikama and I. Yamazaki, Nature, 190 (1961) 83.
2. O. P. Chilson, L.A. Costello, and N. O. Kaplan, Federation Proc., 24, suppl.15 (1964) 55.
3. B. Curti, V. Massey, and M. Zmudka, J. Biol. Chem., 243 (1968) 2306.
4. T. Tamiya, N. Okahashi, R. Sakuma, T. Aoyama, T. Akahane, and J. J. Matsumoto, Cryobiology, 22 (1985) 446.
5. J. F. Carpenter, S. J. Prestrelski, and T. Arakawa, Archiv. Biochem. Biophys., 303 (1993) 456.
6. T. Arakawa and S. N. Timasheff, Biochemistry, 21 (1982) 6536.
7. K. Hashizume, K. Kakiuchi, E. Koyama, and T. Watanabe, Agric. Biol. Chem., 35 (1971) 449.
8. N. Yamazaki, T. Hatakeyama, and G. Funatsu, Agric. Biol. Chem., 52 (1988) 2547.
9. T. Watanabe, N. Kitabatake, and E. Doi, Agric. Biol. Chem., 52 (1988) 2517.
10. C. Sotelo and I. M. Mackie, Food Chem., 47 (1993) 263.
11. H. Kitazawa, Y. Kawai, K. Yamazaki, N. Inoue, and H. Shinano, Fisheries Sci., 63 (1997) 635.
12. R. E. Pincock and T. E. Kiovsky, J. Am. Chem. Soc., 88 (1966) 4455.
13. N. Takenaka, A. Ueda, and Y. Maeda, Nature, 358 (1992) 736.

HYDROCOLLOIDS – PART 2
Edited by K. Nishinari
2000 Elsevier Science B.V.

Gum coating of cheeses

N. Kampf and A. Nussinovitch

Institute of Biochemistry, Food Science and Nutrition
The Hebrew University of Jerusalem
Faulty of Agricultural, Food and Environmental Quality Sciences
P.O. Box 12, Rehovot, 76100
Israel

Semi-hard and dry white brined cheeses were coated by hydrocolloid films based on κ-carrageenan, alginate and gellan. The cheeses were immersed in the gum solution followed by cross-linking of the gum solution to produce a coated product. A few coating films were then dried by air-flow to induce better adherence of the coating to the coated object. The coated cheeses were stored at 4°C and relative humidity of 73%. Weight loss, gloss, roughness of surface area, changes in mechanical properties vs. time (i.e. stress and strain at failure, stiffness and elastic properties), peel-bond strength of the coating film from the cheese and sensory evaluation were studied for each coated system. For the semi-hard cheese, all kinds of coating reduced weight loss during 46 days of storage. With regard to weight loss, no significant differences between the various coatings were found. The coating contributed to a better color and gloss. The roughness of the coated cheese decreased after coating, since surface ruggedness was filled by the film. Advantages in the textural properties of the coated cheese were observed. Since the coated cheese lost less water by evaporation, a desirable softer and a less brittle texture was detected. Recoverable work of coated and non-coated cheeses seemed to decrease from ~48% to 34% after 24 days of storage; no advantage of the coated cheese was observed. The peel-bond strength of film based on alginate was ~1g force/cm. In the case of the dry white brined cheese, no deliberate drying of the coated film was performed. All coatings reduced weight loss of the cheese with a significant advantage of the κ-carrageenan-based coating. The coatings contributed to a lower reduction in pH, thus a higher-quality cheese was obtained. In addition, the coated cheeses were softer and less brittle when compared to the non-coated cheese. The percentage of recoverable work of cheeses did not change even after 24 days of storage. The coatings increased the gloss of the cheese fivefold in comparison to the non-coated cheese. In a sensory evaluation, the coated cheese was found to be advantageous over the non-coated system. In general, cheese coating based on hydrocolloid films improved textural and sensorial properties when compared to non-coated cheeses. The coatings did not influence the taste of white brined cheese, and the technology of coating is simple and relatively cheap.

1. INTRODUCTION

Cheese is a complex food product consisting mainly of casein, fat and water. The fat content is influenced by the percentage of fat in the milk used for manufacture. The milk may maintain its original fat content, may be skimmed to varying degrees or enriched by addition of cream. A list of the variety of cheeses manufactured in the various countries of the world would contain several hundreds of different names, although the number of distinct types is a few tens [1]. Coating cheeses by synthetic films for moisture regulation and protection against contamination is a well-known procedure. Cheese is waxed or film-wrapped to create a barrier against mold entry, to reduce the rate of moisture loss, to prevent oiling off and to make the cheese more attractive and easier to handle. Some materials for coating cheeses include cellophane, cellophane-polyethylene, saran, parakote, pliofilm, cryovac, tifoil and aluminum foil. Another example is the *elast* coating, which is based on aqueous dispersions of butyl rubber and a copolymer of vinyl and vinylidene chloride [1,2].

The objectives of this research were to study physical properties such as weight loss, gloss, roughness and mechanical properties and to provide sensory evaluation of semi-hard and white brined cheeses after they have been coated by edible films prepared from κ-carrageenan, alginate and gellan.

2. MATERIALS AND METHODS

2.1 Coating procedures

Semi-hard and dry white brined cheeses were coated with a film of food-grade alginate, Mw 70,000-80,000, 61% mannuronic acid and 39% guluronic acid contents (Kelgin, LV, Kelco Division of Merck & Co. , USA). Cheese cubes, 11 x 11 x 3 cm length x width x height, were immersed in a 2% (w/w) sodium alginate solution. Residual alginate was then allowed to drip off, before immersing the cheese in a 2% (w/w) solution of calcium lactate for about 30 s to induce a spontaneous cross-linking reaction [3]. Small portions of the wet film surface were inspected by a magnifying glass. Drying of the coatings (either alginate, carrageenan or gellan) was performed with warm air (~40°C) until a moisture content of about 15% remained in the film. Later, the coated cheeses were stored at 4°C and relative humidity of 73%. Weight loss was monitored by periodical weighing. Results are the average of 20 weighings, at an accuracy of ±0.1g. Other coatings were based on 0.2% (w/w) food-grade gellan (Kelcogel, Kelco Division of Merck &Co., USA) and 1% (w/w) κ-carrageenan (Sigma Chemical Co., St. Louis, MO, USA).

2.2 pH, peel testing and roughness measurements

The pH of dry white brined cheese specimens was examined by using a digital pH meter (HI 8424, Hanna Instruments, Liemena, Padua, Italy).

Peel testing, i.e. the force necessary to peel the coating, was used to estimate the adhesion of the film to the cheese. The coating was peeled at 90° from the substrate, and the adhesion strength was estimated by the force per unit width necessary to peel the coating [4]. The peeling test can be applied only to dried films, therefore it was performed only for films coating the semi-hard cheeses.

Roughness was measured using a portable surface-roughness tester (Mitutoyo Corporation, Japan). Readings were given as the arithmetic mean deviation of the profile,

Ra in mm for an evaluation length of 12.5 mm at a speed of 0.5 mm/s [5]. This test was applied only to the semi-hard cheese due to its texture's suitability to such examinations, while the dry brined cheese is coated by a gel layer and therefore such measurements are not applicable.

2.3 Uniaxial compression and elasticity measurements

Cylindrical cheese specimens, 1.5 x 1.5 cm diameter × height, taken from the cheeses by cork-borers were compressed to failure in an Instron universal testing machine, model 1100 (Canton, MA). The force-versus-time data was converted to a true stress, σ_{cor} (kPa) versus Hencky's strain ε_H (-) relationship using the following substitutions

$$\sigma_{cor\,(t)} = F(t)\,H(t)/A_oH_o$$

where σ_{cor} is the corrected momentary stress, ε_H is the dimensionless momentary Hencky's or natural strain and is equal to ln (Ho/Ht), F(t) is the momentary force at time t, Ho the initial specimen length, H(t) the height of the deformed specimen and Ao is the cross-sectional area of the original specimen

2.4 Gloss measurements

The gloss of film-coated cheese was examined using a flat surface glossmeter, capable of measuring gloss at coincident angles of 20, 60 and 85° (Triple Angle Novo-gloss, Rhopoint Instrumentation Ltd, Hann, Germany) [6]. All results were recorded in gloss units, which are relative to a highly polished plane surface of black glass, which served as the standard and was arbitrarily assigned a gloss value near to 100, differing according to the angle used. A total of 10 readings were made at randomly selected parts of each cheese-surface, using the most suitable angle of 60°.

For elasticity measurements, cylindrical cheese specimens were subjected to a uniaxial compression-decompression cycle between two lubricated plates (Fig. 1) [7].

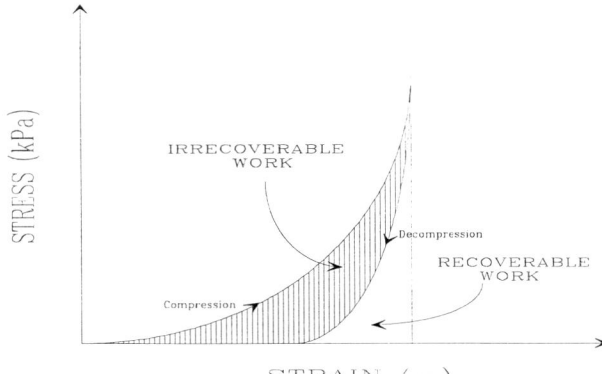

Figure 1: Schematic representation of stress - strain relationships in a single compression-decompression cycle.

3. RESULTS AND DISCUSSION

The composition of the two cheeses to be coated is presented in Table 1. The chemical analysis was supplied by the local manufacturer. The semi-hard yellow cheese contained 30% fat, 43% moisture and 1.7% salt, and the dry white brined cheese had 9% fat, 66.4% moisture and 1.6% salt. Both cheeses were coated by three hydrocolloid-based coatings and changes in their physical characteristics were monitored with time.

	Type of Cheese	
Properties	Semi-hard	White-brined
pH	5.21	6.00
Dry matter (%)	57	33.6
Fat (%)	30	9
Salt (%)	1.7	1.6
% salt in brine	20	16
Additives	sorbate	sorbate

Table 1: Composition of the cheeses used for the coating experiments (Data supplied by the manufacturer)

As a result of the drying, the cheese was coated by a dried transparent edible film adhered to its surface (Fig. 2).

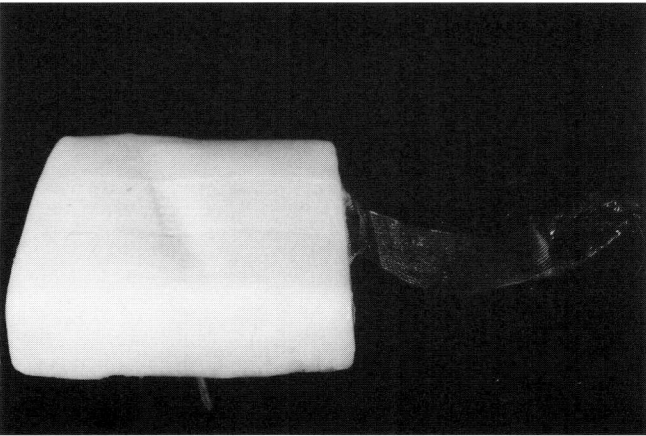

Figure 2: Semi-hard cheese coated by dried alginate-based film.

After 144 h all three coatings seemed to adhere well to the semi-hard cheese, and no separation of the coating from the coated cheese was detected. After 6 days the color of the cheese changed a bit to a darker yellow, indicating changes in its moisture content. As stated, weight loss was monitored by periodical weighing. The weight loss of the semi-hard cheese was about 13.4% after it was stored for 23 days at 4°C and 73.3% relative humidity. All different kinds of coating reduced weight loss (Fig. 3). A maximal 2% difference in weight between the uncoated and the coated cheeses was observed in favor of the coated cheeses. No significant differences between the various coatings were found. In other words, to decide which hydrocolloid coating of the three tested in this study is the best, other characteristics of the coatings should be considered.

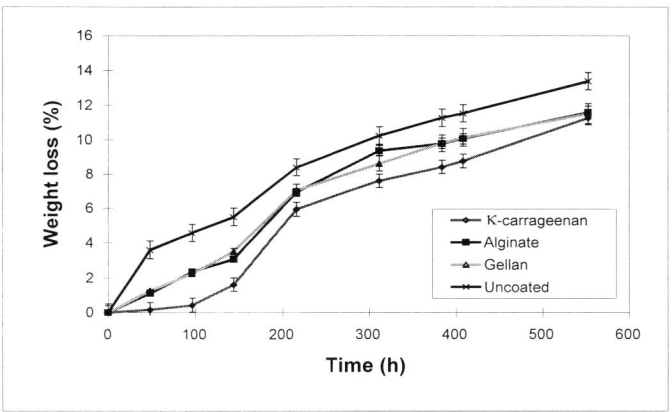

Figure 3: Weight loss of uncoated and coated semi-hard cheeses.

As a result of weight loss and partial drying, changes in the mechanical properties of the cheeses occurred. Immediately after coating, the stress and Hencky's strain at failure for the coated and uncoated cheese were the same (Table 2).

After 18 days of storage, the stress at failure for the uncoated cheese increased from 50.3 to 58.2 kPa, indicating an increase in strength as a result of drying. Hencky's strain at failure decreased from 1.14 to 0.90, thus the cheese became more brittle and changed its texture. After 18 days, the stress at failure for the alginate-coated cheese remained the same, but a decrease in strain to 0.82 was observed.

In the case of the dry white-brined cheese, no deliberate drying of the coating film was performed (Fig. 4). After ~18 days, uncoated cheese lost 6.2% of its initial weight in comparison to 4.7, 5.2 and 2.0% for gellan, alginate and carrageenan coatings, respectively. In other words, dry white-brined cheese coated with the κ-carrageenan-based coating had a significant advantage over the other coatings. Since in this cheese no visible changes occurred, weight loss was examined only at the end of the experiment.

	Mechanical Properties	
	Stress at failure (kPa)	Hencky's strain at failure (-)
Cheese type/time after coating		
Uncoated semi-hard	50.3 ± 3.3^a	1.14 ± 0.15^b
Uncoated semi-hard (18 days after coating)	58.2 ± 2.2^c	0.90 ± 0.13^g
Coated semi-hard (immediately after coating)	50.3 ± 3.3^a	1.14 ± 0.15^b
Coated semi-hard (18 days after coating)	50.9 ± 8.2^a	0.82 ± 0.16^f
Uncoated dry brined	23.4 ± 3.1^d	0.77 ± 0.02^e
Uncoated dry brined (24 days after coating)	16.2 ± 5.9^h	0.71 ± 0.06^e
Coated dry brined (immediately after coating)	22.2 ± 5.9^d	0.71 ± 0.06^e
Coated dry brined (24 days after coating)	15.2 ± 0.7^h	0.72 ± 0.03^e

Table 2: Mechanical properties of alginate-coated and uncoated semi-hard and dry brined cheeses. Each result is the average of five determinations ±SD. Different superscript letters indicate significant differences at P<0.05.

The mechanical properties of the dry white brined cheese showed values of 23.4 kPa and 0.77 strain at failure for the uncoated cheese. These results did not change significantly after the coating. After 24 days the coated cheeses were softer and less brittle when compared to the non-coated cheese (Table 2). In addition, the percentage of recoverable work (elastic properties) of cheeses did not change significantly even after 24 days of storage and was ~46-48%. As was found for the semi-hard cheeses, the coatings increased the gloss of the cheese fivefold in comparison to the non-coated cheeses. Sensory evaluation revealed that the coated cheese were advantageous over the non-coated system with regards to its textural qualities.

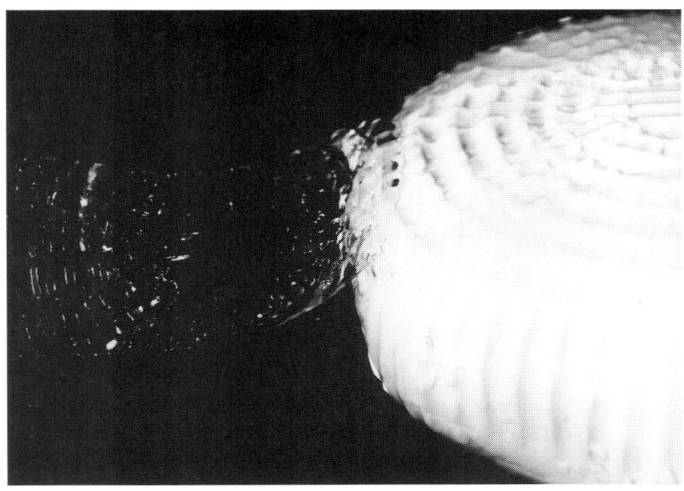

Figure 4: White brined cheese coated by alginate film

In general, cheese coating based on hydrocolloid films improved textural and sensorial properties when compared to non-coated cheeses. The coatings did not influence the taste of white-brined cheese, and the technology of coating is simple and relatively cheap.

Acknowledgement

This project was supported in part by "Tnuva".

REFERENCES

1. L. Tumerman and B.H. Webb, in *Fundamentals of Dairy Chemistry*, (Eds B.H. Webb and A.H. Johnson), The Avi Publishing Company, Inc. NY, (1965) pp. 506.
2. F. Kosikowski, in Cheese and Fermented Milk Foods, published by the author, distributed by Edwards Brothers, Inc. Ann Arbor, Michigan, (1970) pp. 313.
3. A. Nussinovitch and N. Kampf, Leben.-Wiss. und-Technol., 26 (1993) 469.
4. O. Ben-Zion and A. Nussinovitch, Food Hydrocoll., 11 (1997) 429.
5. V. Hershko, E. Klein and A. Nussinovitch, J. Food Sci., 61 (1966) 769.
6. G. Ward and A. Nussinovitch, Food Hydrocoll., 11 (1997) 357.
7. N. Kampf and A. Nussinovitch, Food Hydrocoll.,11 (1997) 261.

HYDROCOLLOIDS – PART 2
Edited by K. Nishinari
2000 Elsevier Science B.V.

Effect of food emulsifiers on stability of O/W emulsion during freeze-thaw treatment

M. Miura[a], Y. Ishikawa[a], H. Kuzui[b] and S. Kokubo[c]

[a] Department of Agricultural Engineering, Faculty of Agriculture, Iwate University,
 3-18-8, Ueda, Morioka 020-8550, Japan
[b] Specialty Chemicals R&D Center, Mitsubishi Chemical Corp., 1, Tohocho, Yokkaichi
 510-0000, Japan
[c] Market Development Group, Mitsubishi-Kagaku Foods Corp., 1-3-9, Ginza, Chuo-ku,
 Tokyo 104-0061, Japan

The effect of sucrose esters of fatty acids (SE) on freeze-thaw stability of oil-in-water (O/W) emulsion has been studied. Emulsion stability was assured by gravitational and centrifugal creaming methods. Droplet size distributions, dielectric relaxation spectrum and unfreezable water contents of fresh O/W emulsions were obtained. Dynamic viscoelasticity as a function of temperature have been used to monitor the demulsification behavior.

1. INTRODUCTION

Emulsions form the basis of many food products including homogenized milk, coffee whitener, ice cream, butter and margarine, mayonnaise, cake batter, and sugar confectioneries, to name a few. Emulsions are inherently unstable. Even if a system that appears to be perfectly stable, with a shelf-life of several years, the total number of droplets, their size distribution, and their arrangement in space, are all changing imperceptibly with time. The primary processes leading to instability, are creaming, flocculation and coalescence [1].

The demulsification occurs by a freeze-thaw processing, and fats and oils are tending to separate from O/W emulsions such as coffee whitener and mayonnaise. It is necessary to improve the freezing-thawing tolerance of emulsification for production of frozen emulsified foods. The theories of freeze-thaw stability are not adequate for making quantitative predictions for food emulsions, and so at the present time the technologist must fall back on empirical knowledge in attempting to reduce or eliminate the problem of destabilization in any real system. SE's are nonionic surfactants manufactured from sucrose and fatty acids. Because of their low toxicity, SE is widely used as emulsifiers in the food, cosmetic and pharmaceutical industry. Thus, the objectives of this work are to clarify the influence that SE exert on the freeze-thaw stability.

2. MATERIALS AND METHODS

2.1. Materials

To understand clearly the influence of the freeze-thaw processing on the emulsification stability of O/W emulsion, the emulsions were prepared from a fat that is not easy to prepare an emulsion with an excellent emulsification stability. Hydrogenated rapeseed oil (mp 34.0℃, solid fat content 36.2) was obtained from Miyoshi Oil & Fat Co., Ltd., Japan. Sucrose esters of stearic acid (Ryoto Sugar Ester S-570, S-1170, S-1570 and S-1670) were commercial samples from Mitsubishi-Kagaku Corp., Japan. They were mixture of the palmitic and stearic acid esters of sucrose; compositions of fatty acids esterified with sucrose were approximately 30% palmitic acid and 70% stearic acid. Monoester contents of S-570 (SES5), S-1170 (SES11), S-1570 (SES15) and S-1670 (SES16) were approximately 30%, 55%, 70% and 75%, respectively. And they had the values of hydrophile-lipophile balance (HLB) of approximately 5, 11, 15 and 16, respectively.

2.2. Preparation of O/W emulsions

SE's having lower HLB values were dissolved in hydrogenated rapeseed oil at 50.0℃. SE's having the higher values were dissolved in deionized water. The oil and water were combined at the constant rate of addition of the water, and mixed together by a mixer (RW20DZM P4 IKA, Janke & Kunkel Co.) operating at 80 rpm for 20 min. O/W emulsions (29.7 wt% hydrogenated rapeseed oil, 69.3 wt% deionized water, 1.0 wt% emulsifier, pH 7) were then prepared using an ULTRA-TURRAX homogenizer (LR-A25, Janke & Kunkel Co.) at 50.0℃. The homogenizer was operated for 6~10 min at the rotational speed of 20,500 rpm (Shaft: S25N-10G-VS). The operation times were 8, 6, 6 and 10 min for the preparation of emulsions with SES5, SES11, SES15 and SES16, respectively.

2.3. Droplet size distribution

Three different preparations of the emulsion containing 1.0 wt% SE and 29.7 wt% oil were studied. When the emulsification was completed, samples were taken for droplet size measurement. The size distribution of oil droplets in each emulsion was determined with a laser scattering & diffraction particle size distribution analyzer (LS230, Coulter Electronics Ltd.) based on Mie scattering theory and Fraunhofer diffraction theory.

2.4. Unfreezable water content

Unfreezable water contents of all emulsions were determined using a differential scanning calorimeter (SSC5200H, Seiko Instruments Co., Japan). Aliquots of the emulsion (21~35 mg) were sealed in silver sample pans. A pan containing distilled water was used as a reference. Differential scanning calorimetry (DSC) curves were obtained by the cooling program of initial temperature of 20.0℃ and final temperature of -20.0℃ at cooling rate of -8.0℃・min^{-1}, the isothermal program of -20.0℃ for 15 min, and the heating program of initial temperature of -20.0℃ and final temperature of 70.0℃ at heating rate of 2.0℃・min^{-1}. Subzero glass-to-rubber transition temperature (T'_g) of a maximally-freeze-concentrated solute/unfrozen water glassy matrix surrounding the ice crystals in a frozen solution was obtained from DSC derivative curve [2]. Unfreezable water contents (W_g') were calculated in the endothermic and exothermic processes.

2.5. Dielectric properties

Dielectric relaxation spectrum of all emulsions were obtained from a dielectric thermal analyzer (DETA-3 SI 1255, Rheometric Scientific. Inc.) using a parallel plate cell with a plate diameter 20 mm, a gap size of 770 μ m and blank electrostatic capacity of 3.612 pF. Complex dielectric constant (ε^*) was measured, and dielectric constant (ε'), dielectric loss constant (ε'') and dielectric loss tangent (tan δ) were calculated.

2.6. Dynamic viscoelasticity

Rheological measurements were carried out in a stress-controlled dynamic stress rheometer (SR-500, Rheometric Scientific. Inc.) using a parallel plate geometry with a plate diameter 40 mm and a gap size of 50 μ m. Temperature of the sample was controlled between -15.0 and 40°C with a Peltier element. All the samples were left for 5 min before running any measurements, in order to allow some stress relaxation and the required temperature of 20°C to be reached. A disolvent trap was used to create a water atmosphere around the sample so as to prevent its drying out during the time taken to complete the measurements. Oscillatory tests was conducted at 1 Hz, and at a maximum stress amplitude of 2 mPa (20°C to freezing point), 38 Pa (freezing point to -15°C) and 170 mPa (20°C to 40°C).

2.7. Freeze-thaw treatment

The emulsion (40 g) in a cylindrical vessel made of aluminum (38 mm in diameter, 68 mm in height) was set up in a thermal hygrostat chamber (PL-1SP, TABAI ESPEC Corp., Japan). The temperature in the chamber was maintained for 4 hours at 20.0°C and an initial temperature of the sample had been adjusted. Then, the chamber was cooled down to -15.0°C at cooling rate of -2.0°C·min^{-1}, and the sample was frozen and stored at the same temperature for 24 hours. The sample was thawed in the chamber continuously up to 20.0°C at heating rate of 2.0°C·min^{-1} and maintained for 5 hours to adjust the final temperature of the sample. The temperature of the sample was measured with a thermocouple temperature logger (U Logger L822, UNIPULSE Co., Ltd., Japan).

2.8. Emulsion stability

Immediately after the freeze-thaw treatment or without treatment, each O/W emulsion was centrifuged at $50 \times g$ for 10 min at 25°C to separate the aqueous phase from the O/W emulsion. After emulsification, aliquots of the emulsion were immediately removed for measuring emulsifying stability. The emulsion stability index (ESI)[3] was measured after centrifugation and expressed as ESI [%] $= 100 \times (F_f - F_s)/F_f$, where F_f was the weight of water in emulsion formulation and F_s was the weight of water separated. The fresh emulsion poured into a test tube (16.5 $\phi \times 125$ mm) in height of 75 mm. A test tube (30.0 $\phi \times 200$ mm) was used for the freeze/thaw-treated emulsion. Emulsion stability was evaluated from the height of separated water after 24 hours at 25°C by the category scale of 4 stages from stable emulsion (A) to unstable emulsion (D).

3. RESULTS AND DISCUSSION

3.1. Droplet size distribution

Oil droplet size distributions of O/W emulsions prepared with sucrose esters of stearic acid were shown in Table 1. Before freeze-thaw treatment, it was observed that emulsions prepared with oil-soluble emulsifier, notably SES5 and SES11, had smaller mean droplet size and narrow particle-size distribution. In general, the particle size of the emulsion increased after freeze-thaw treatment. Especially, the mean diameter of the emulsion using water-soluble sucrose esters, that is, SES15 and SES16, had about twice of the diameter of the emulsion without freeze-thaw treatment. Moreover, the standard deviation of the particle size quadrupled in the emulsion prepared with SES15, and increased to about three times in the SES16 sample. This shows that as well as droplet-size polydispersity, there may also be considerable heterogeneity in the nature of the interfacial layer around the emulsion droplets.

Table 1 Changes in droplet size of O/W emulsions prepared with SE before and after freeze-thaw treatment

Freeze-thaw	Sample	Droplet diameter [μm]			
treatment		Mean	Median	Mode	SD
Before	SES5	18.07	18.97	21.69	8.53
	SES11	19.66	20.40	23.81	8.97
	SES15	28.08	27.08	28.69	14.80
	SES16	27.26	24.83	28.69	18.72
After	SES5	27.35	20.75	21.69	26.43
	SES11	32.44	24.68	26.14	29.32
	SES15	55.77	33.77	23.81	60.64
	SES16	48.16	26.64	23.81	64.05

SD, standard deviation

3.2. Unfreezable water content

It was expected that the demulsification might be happened by destabilization of oil/water interfacial phase according to the ice crystal growth during freeze-thaw processing. Therefore, emulsion with much unfreezable water content may be more stable. The contents of the unfreezable water calculated from the freezing and thawing processes were shown in Table 2. In each sample, no remarkable difference was observed in the content of the unfreezable water. Unfreezable water content of the emulsion was poor predictor of the stability in the O/W emulsion prepared with hydrogenated rapeseed oil.

Table 2 Unfreezable water contents and T_g' of O/W emulsions

Sample	W_{fg}' [-]	W_{tg}' [-]	T_g' [°C]
SES5	0.30	0.15	−11.2
SES11	0.25	0.12	−11.0
SES15	0.27	0.13	−11.6
SES16	0.31	0.12	−11.4

W_g'=(Total water content [mg] − Freezable water content [mg])／Total water content [mg]
W_{fg}', value from freezing process; W_{tg}', value from thawing process
T_g', subzero T_g of a maximally-freeze-concentrated aqueous solution

3.3. Dielectric characteristics

Plots of dielectric constant and dielectric loss tangent vs. frequency for all the emulsions to be studied were shown in Figure 1. The frequency dependency of dielectric constant showed that a dielectric relaxation had occurred in a mechanism in these emulsions. It is suggested that all the emulsions had one oil-water boundary. Significant differences were observed in the peak of the dielectric loss tangent for the 4 emulsions. The peak of the emulsion prepared with SES5, an oil-soluble emulsifier, was smaller than other samples and existed in the lower frequency region. Monoester in SE may migrate to oil-water interface from the oil phase, and be adsorbed. Di-/tri-esters in SE may penetrate into the oil phase and form tightly packed interfacial layers with monoester. Polyesters in SE will have the capability to hinder crystallization and polymorphic transformation of fat. It appears that the content ratio of monoester/di- and tri-esters/polyesters plays an important role in improving of freeze-thaw tolerance for O/W emulsion.

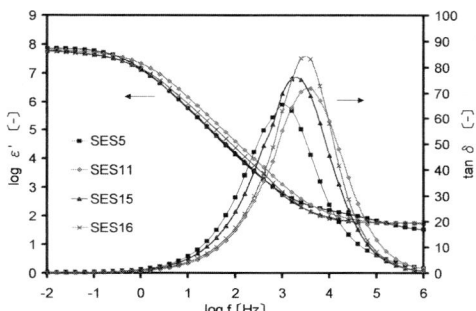

Figure 1. Frequency dependence of the dielectric constant and dielectric loss tangent of O/W emulsions at 25°C.

3.4. Dynamic viscoelasticity

Figure 2, plots of dynamic viscoelasticity (storage modulus, loss modulus and mechanical loss tangent) versus temperature, show a freezing behavior of emulsion prepared with SES11. A remarkable increase in storage modulus (G') and loss modulus (G") was observed in the temperature range $-6.0 \sim -6.4$°C in the cooling process from 20 to -15°C. An increase in G' which seemed that the demulsification occurred and oil droplets showed flocculation and coagulation, or flocculation, coagulation and coalescence was seen in the temperature range of $10 \sim 0$°C in a commercial coffee whitener without freeze-thaw tolerance (data not shown). However, such behavior was not seen for emulsions prepared with SE. Plots of dynamic viscoelasticity versus temperature during thawing process for the emulsion prepared with SES11 are shown in Figure 3. G' and G" exhibited remarkable decreasing in the temperature range $-0.6 \sim 0.8$°C in the heating process from -15°C to 40°C. It is also noted that no decrease in G', which seemed that oil crystal melting, was observed. In the emulsions prepared with SE, whether the demulsification occurred in either of the freezing process or the thawing process could not be deduced by the dynamic viscoelasticity examination.

3.5. Emulsion stability

As indicated in Table 3, lipophilic SE (SES5 and SES11) produced more stable emulsion. The stability index may refer in part to creaming and flocculation, as well as coalescence. The emulsion stability was evaluated by gravitational and centrifugal creaming methods in this study to obtain a suitable index for freeze-thaw tolerance of the emulsion. However, creaming was accelerated by the centrifugation and no accurate values were obtained by the centrifugal creaming method. The emulsions prepared with these SE had smaller mean droplet size, standard deviation of droplet size and peak height of dielectric loss tangent.

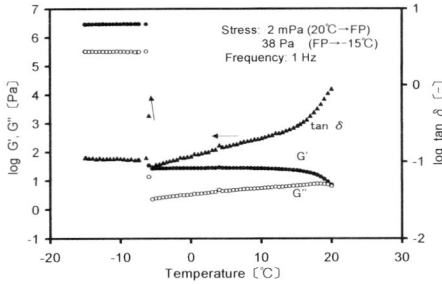

Figure 2. Changes in the dynamic viscoelasticity of O/W emulsion prepared with SES11 during freezing process.

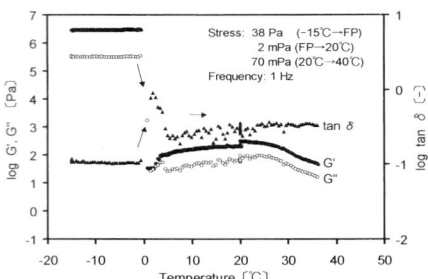

Figure 3. Changes in the dynamic viscoelasticity of O/W emulsion prepared with SES11 during thawing process.

Table 3 Effect of freeze-thaw treatment on emulsion

Freeze-thaw treatment	Sample	Emulsion stability	
		Appearance*	ESI**
Before	SES5	A	0.25±0.02 (n=18)
	SES11	A	0.33±0.03 (n=12)
	SES15	C	0.42±0.07 (n=14)
	SES16	B	0.41±0.05 (n=14)
After	SES5	A	0.84±0.03 (n=10)
	SES11	A	0.96±0.02 (n=10)
	SES15	C	0.58±0.05 (n=10)
	SES16	C	0.62±0.03 (n=10)

* Height of separated water during storage for 24 h at 25℃
Before : A, 0～9 mm; B, 10～19 mm; C, 20～29 mm
After : A, 0～5 mm; B, 6～10 mm; C, 11～15 mm

$$** \text{ Emulsion stability index} = \frac{\text{Weight of water in formulation [mg]} - \text{Weight of water separated [mg]}}{\text{Weight of water in formulation [mg]}} \times 100$$

ACKNOWLEDGMENT

The authors thank Miyoshi Oil & Fat Co., Ltd. for providing us with oils. We also thank Akita Research Institute Food and Brewing, Rheometric Scientific, F.E. Ltd. and Coulter K.K. for technical supports.

REFERENCES

1. E. Dickinson and G. Stainsby, E. Dickinson and G. Stainsby (eds.), Advances in Food Emulsions and Foams, Elsevier Applied Science, London, pp.1-44, 1988.
2. H. Levine and L. Slade, T.M. Hardman (eds.), Water and Food Quality, Elsevier Applied Science, London, pp.71-134, 1989.
3. S.C. Hung and J.F. Zayas, *J. Food Sci.*, **56** (1991) 1216-1223.

HYDROCOLLOIDS – PART 2
Edited by K. Nishinari
2000 Elsevier Science B.V.

Factor affecting the quality of choux paste
- deterioration of puffing property with aging-

N. Imazuya[a], K. Nishimura[b], M. Kubo[c] , T. Ueda[c] and K. Katsuta[d]

[a]School of Education, Seiwa College, Nishinomiya, Hyogo 662-0827, Japan.
[b]Department of Food Science & Nutrition, Doshisha Women's College of Liberal Arts, Kamigyo-ku, Kyoto 602-0893, Japan.
[c]Material Property Research Center, Nippon Paint Co. Ltd., Neyagawa, Osaka 572-8501, Japan.
[d]Department of Food Science & Nutrition, Nara Women's University, Kitauoyanishi-Machi, Nara 630-8506, Japan

When choux paste (CP) was stood at room temperature, the expansion of choux crust was impeded and the deterioration of choux quality progressed with lapse of the aging time.   In this study, the relationship between the rheological properties of CPs with and without aging and the expansion of them during baking was investigated.   The influences of dextrin added to the CP on their properties were also studied.

When the CP was aged at $35^{\circ}\mathrm{C}$ for 3hrs, the dynamic modulus (G') and the loss modulus (G") for the ACP were decreased. The values of both moduli decreased with increasing the concentration of dextrin content of CP (DXCPs).   During baking, the ratio of expansion in diameter for ACP became larger and that in height became smaller rather than those for normal CP.   Increasing the amount of dextrin added, the beginning to spread in width of CPs occurred but less suitable shapes of choux were observed.   From these results, it might be concluded that the degradation of starch with aging introduced the deterioration of qualities of CP corresponding to the decrease of viscoelasticities for the paste.

1. INTRODUCTION

Choux crust is a popular baked confectionery. The word of choux came from "choux" which means cabbages in French, because there are some cracks on the surface of products and the expanding of figures looks like the vegetable.   Generally choux crust prepared as follows; shortening or butter are molten in boiled water, wheat flour is added and   heated for several minutes while stirring.   Then egg solution is added little by little into the heated mixture while stirring.   An aliquot of the mixture (CP) is baked in a oven for the choux crust.   Since two steps on heating process are needed to prepare the choux crusts, choux is an unique confectionery compared to the other baked products such as cookies, breads and cakes.

When the CP is stood at room temperature for a while, the expansion of CP decrease and the quality of crust becomes poor   [1,2] .   We had investigated the ability of expansion of choux crust prepared with wheat starch, instead of wheat flour, and reported that the deterioration of CP with aging might be introduced by the degradation of wheat starch with enzyme (α-amylase) in egg resulting in the increase of lower molecular substances [3-6].

There are some reports on the rheological properties and the texture of CP and crusts [7-9]. However, there are no researches on the changes in the rheological properties of CPs during baking. The quantitative treatment of expanding phenomena of CPs during heating also were not performed. In this study, therefore, the temperature dependence of dynamic viscoelastic properties for CP with and without aging was measured and their expansion behavior was monitored. To confirm our idea which the lower molecular substances in starch may induce the deterioration and result in poor quality of choux, the influences of presence of dextrin in CP on the rheological and expanding properties were investigated.

2. MATERIALS AND METHODS

2.1. Materials
Soft wheat flour was kindly supplied by Torigoe Milling Co.(Fukuoka, Japan), wheat starch and dextrin by Sanwa Starch Co.(Nara, Japan), and shortening by Tsukishima Food (Tokyo, Japan), respectively. Fresh eggs were purchased in a market.

2.2. Preparation of choux paste (CP)
The procedure of making CP employed in the present study was previously reported by Imazuya and Kisaki [3]. The normal (control) CP is defined as CCP in this study. A part of the CCP was wrapped with a polyethylene film and kept at 35℃ for 3 hrs in an incubator (Mini-subzerd MC-810; Tabai, Japan) and this aging CP was defined as ACP.

2.3. Preparation of CP model
The CP model with starch (instead of flour) was prepared by the modified method based on the procedure reported by Abe and Matsumoto [10]. To the starch in the CP model, 10 and 40 % Dextrin were replaced (DXCP).

2.4. Portrait analysis of visual shape
The changes in visual shape of CPs during baking at 180℃ for 20 min were photographed by video camera, input the data into a computer and the lengths of maximum diameter (D) and heights (L) of CP were measured with lapse of the time using a portrait analysis program. The changes in the temperature of CP during heating were also monitored by thermo-couple (CA: Chromel-Alumel).

2.5. Dynamic viscoelastic measurements
A rheometer (Rheosol G-3000, UBM Co., Kyoto, Japan) was used for the dynamic viscoelastic measurements of CP before and during heating. An aliquot of CP was placed in the gap (50 μ m) between the cone (1.946 deg., 3.986 cm ϕ) and plate, and measurement of frequency dependence of dynamic moduli was carried out at room temperature (25℃). For the

temperature dependence studies, parallel plate (1.797 cm ϕ) was used. The exposed sample surface between the test fixture was covered with silicone oil.

3. RESULTS AND DISCUSSION

3.1. Rheological properties of CP at room temperature

Figure 1 shows the frequency dependence of storage shear modulus (G') and loss modulus (G") for CP with and without aging at 35℃ for 3hrs. The measurements were made at 25℃ and with a sinusoidal oscillatory strain, γ =0.016. Since the viscoelastic parameters of CP altered with time, a narrow range of frequency (0.1 through 1 Hz) were applied. The magnitudes of both moduli for ACP showed the lower value than those for control (CCP), suggesting the production of low molecular weight substances. The large scattering, i.e., poor reproducibility, of the data for ACP probably indicate that it is heterogeneous system.

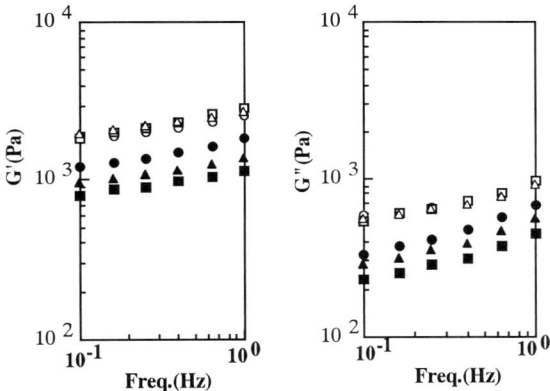

Figure 1. Effect of aging on frequency dependence of G' and G" for choux paste at room temperature (25℃).
Open symbols show the control CP (CCP) and solid symbols the aged CP (ACP) at 35℃ for 3hrs. The differences between symbols show the scattering of three times tests.

3.2 Effects of aging on expanding properties of CP

The changes in diameter (D) and height (L) of CP with times during baking were illustrated in Figure 2. An aliquot (10 g) of ACP just before baking (0 min) had increased in width on the oven plate and decreased in the height compared to the CCP. At the end of baking (20 min), the height was 492 mm for CCP and 445 mm for ACP, and the diameter was 595 mm for CCP and 695 mm for ACP, respectively.

Figure 3 shows the expansion ratios as a function of temperature in the CP observed by a thermo-couple during baking. The expansion behavior of both CPs gave similar profiles; the expanding of CP began to increase rapidly at around 85℃ in the height or 103℃ in the diameter.   The magnitude of expansion ratio for ACP was slightly larger than CCP at 25-60℃, and the maximal value at 111℃ for ACP was greater than CCP.   In the past some workers [1,11,12]   have reported that the vapor pressure or the development of dissolved air in CP caused the film of starch to extend in CP.   The results obtained here indicate that expanding profile of CP could be divided into two stages as follows; in the first (initial) stage, ～ 111℃, CP has mainly increased in height by generation of the vapor and then, in second stage, blown up horizontally by pressure arisen from vapor filled in choux crust.   The amount of air in the sample might effect on the expansion of CP during baking.   The density of sample should be needed to measure.

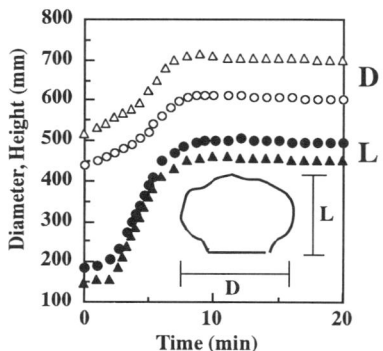

Figure 2. Changes in visual shape of CP during baking at 180℃.

  ○   diameter (D) of CCP △ (D) of ACP
  ●   height (L) of CCP   ▲ (L) of ACP

Figure 3. Changes in expanding ratio of CP as a function of CP temperature during baking at 180℃.
     Symbols are the same as used in Figure 2.

3.3. Effect of dextrin on expanding and rheological property of model CP

At room temperature, i.e., just before baking, the initial diameters of CP rose and the initial heights reduced in the presence of dextrin (Figure 4).   At the end of baking (20 min),   the height was 504 mm for control CP, 539 mm for CP containing 10% dextrin (10%DXCP) and 342 mm for 40%DXCP, whereas the diameter was 678 mm for control CP, 809 mm for 10%DXCP, and 787 mm for 40%DXCP.   For the higher concentration of dextrin in CP, the increase of diameter and the decrease of height in the expanding ratio were enhanced (Figure 5). The expansion ratio of diameter and height for any CPs gradually increased up to around 100℃. The control and 10%DXCP expand remarkably at around 100℃ horizontally and contracted deliberately after passing the maximum.   Whereas, horizontal expansion of 40%DXCP occurred slowly, reached a maximum and diminished.   If the CP is immediately cooked as well as control CP and 10% DXCP, the paste expands to both direction of vertical and horizontal because the starch films were resistant to tensile stress.   Whereas, the CP consisting of low molecular weight substances such as 40% DXCP mainly puffed in the vertical dimension.

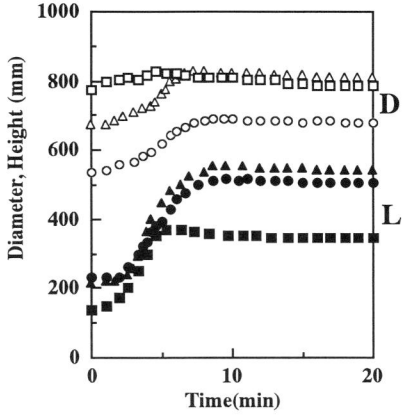

Figure 4. Changes in visual shape of model CP containing dextrin (DXCP) during baking at 180℃.

CP is prepared with wheat starch (instead of flour). Symbols show the amount of dextrin contained to CP.

- o   diameter (D) of CP without dextrin
- •   height (L) of CP without dextrin
- △   (D) of 10%DXCP   ▲   (L) of 10%DXCP
- □   (D) of 40%DXCP   ■   (L) of 40%DXCP

Figure 5. Changes in expanding ratio of DXCP as a function of CP temperature during baking at 180℃.

The temperature was obtained by thermo-couple (CA). Symbols are the same as used in Figure 4.

The changes in storage shear modulus (G') for model CP and DXCP with temperature could be classified into four stages (Figure 6). The value of G' decreased (stage Ⅰ), increased (stage Ⅱ, 50 through 80℃), reduced again (stage Ⅲ) and elevated again (stage Ⅳ). The behavior in stage Ⅰ may be introduced by the melting of solid fat (shortening). The increase of G' in stage Ⅱ is influenced by some factors such as swelling of the starch granules, gelation and / or aggregation of egg protein. In stage Ⅲ, the value of G' decreased probably because of the " filling-in " between starch particles and the fusion of crystallites in starch. Also the most important event in this stage may be the evaporation of water, because a wheat coherent structure had not occurred. After the complection of filling-in and fusion, a wheat coherent structure might be formed and developed, then the value of G' subsequently rise with vapor pressure in the paste.

The maximal values of G' for all samples are indistinguishable, because the magnitude of the value does not necessarily follow the concentration of dextrin. However, there are significant differences between the values of G' at room temperature (25℃) for the control model CP and those for DXCP and the magnitude depend on the concentration of dextrin. In stage Ⅰ and stage Ⅱ, the differences are still maintained. These viscoelastic behaviors are closely related to the expansion properties of CP. The lower magnitude of G' is a mirror of the delicate structure which is introduced by low molecular weight substances.

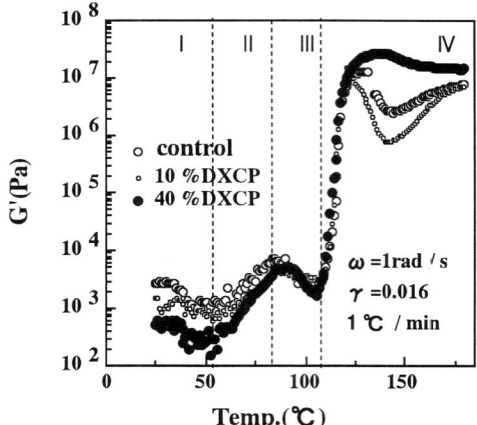

Figure 6. Temperature dependence of G' for model CP containing dextrin (DXCP)

4. CONCLUSION

In this study, quantitative measurement of the expansion behavior for CP was accomplished. The expansion behavior was influenced by the rheological behavior for the paste. The results obtained by both expansion and rheological measurements nearly proved our assumption that deterioration of expansion of CP is introduced by the degradation of starch in CP.

REFERENCES

1. M. Tsuji, Y. Tani, Y. Katayama, and S. Kurosawa, *Doshisha, J. Home Econ. Jpn*, 14 (1980) 31.
2. S. Ohkita and M.Yamada, *J. Cookery Sci. Jpn*, 23 (1990) 73.
3. N. Imazuya and H. Kisaki, *J. Cookery Sci. Jpn*, 29 (1996) 87.
4. N. Imazuya and H. Kisaki, *J. Cookery Sci. Jpn*, 29 (1996) 93.
5. N. Imazuya and H. Kisaki, *J. Cookery Sci. Jpn*, 29 (1996) 115.
6. K. Nishimura, N. Imazuya and S. Nakai: *Food Sci. Technol.* Int. Tokyo, 4 (1998) 18.
7. F. Matsumoto and N. Abe, *J. Home Econ. Jpn*, 13 (1962) 240.
8. E. Mori and K. Endo, *J. Home Econ. Jpn*, 39 (1988) 659.
9. S. Ohkita, M. Yamada and K. Endo, *J. Home Econ. Jpn*, 46 (1995) 549.
10. N. Abe and F. Matsumoto, *J. Home Econ. Jpn*, 15 (1964) 245.
11. K. Shiratori, F.Matsumoto, *J. Home Econ. Jpn*, 30 (1979) 231.
12. Y. Fuchimoto, Y. Shimiya, K. Sakai, K. Hatae and A. Shimada, *J. Home Econ. Jpn*, 41 (1990) 1049.

HYDROCOLLOIDS – PART 2
Edited by K. Nishinari
2000 Elsevier Science B.V.

Effects of mixture ratio of *kudzu* (arrowroot) starch and sesame contents on the physical properties of *gomatofu* (sesame tofu)

E. Sato[a] and R. Ito[b]

[a] Department of Human Life Science, Women's College of Niigata, 471 Ebigase, Niigata -shi, Niigata 950-8680, Japan

[b] Kadoya Oil Co., 6188 Tonosho, Kagawa 761-4100, Japan

Gomatofu (sesame tofu), one of the mixed gel consisting of *kudzu* (arrowroot) starch and sesame, possesses an extremely original textural characteristics which is soft, smooth, but springy, and the textural properties are greatly influenced by starch and sesame milk contents or preparing conditions. Hence, the effects of mixture ratio of *kudzu* starch and sesame contents on the textural properties of *gomatofu* were investigated by rheological measurement, sensory evaluation and SEM observation. The larger the amounts of both *kudzu* starch and sesame contents there were, the more hardness, adhesiveness and gumminess increased.

All of four texture parameters increased with increasing amount of the *kudzu* starch, but cohesiveness decreased with increasing content of sesame. It was considered that sesame contributed to the strength and stability of *kudzu* starch gel.

1. INTRODUCTION

Gomatofu (sesame tofu) is one of the traditional Japanese healthy foods and is representative of all *shojin* (vegetarian) dishes. There are many kinds of sesame which we can use to prepare *gomatofu*, such as, white, black, gold ones and roasted or unroasted ones and so on.

Gomatofu has a characteristic texture which is soft, springy and smooth. Sato and co-workers [1,2] investigated the effect of preparing conditions, the mixing rates and cooking times, on the physical properties of *gomatofu* (the suspension of *kudzu* starch and sesame milk), and reported that the magnitudes of hardness and gumminess of the sample cooked for 25 min increased with a cooking time, and the value of adhesiveness depended greatly on the mixing rate in the penetrating test. The best preparing condition was 250 rpm for mixing rate and 25 min for heating time. It was also reported in our previous papers [1-3] that SEM observation had indicated that the samples prepared by mixing at 250 rpm for 25min had a uniform network structure, but the sesame oil got together in the samples mixed at 350 rpm for 45 min and the globular size became larger with cooking time. The sample prepared by mixing at 250 rpm for 25min showed the best palatability organoleptically   [3].

In this study, therefore, the effects of mixture ratio of *kudzu* starch and sesame contents on the textural properties of *gomatofu* were investigated by rheological measurements,   sensory evaluation,   and a scanning electron microscopy (SEM).

2. EXPERIMENTAL

2.1. Materials and preparation of sample

The highly   pure (99.0%) *kudzu*   starch   was   purchased   from   Inoue-tengyokudou (Nara, Japan).   The shucked   and   unroasted   white   sesame   which   was   produced   in China (1996)   was   used.

2.1.1. *Kudzu* starch content

Forty grams   of   sesame seed   and   450g   of   water   were   constant   for any samples but *kudzu* starch contents changed into 30 through 60g .

2.1.2. Sesame seed content

Forty grams of   *Kudzu* starch   and   450g   of   water   were   constant   for any samples but sesame seed changed contents into 0, 20, 40, 60, 80g respectively.

2.1.3. Preparation

Four hundred thirty five grams   of   sesame milk (40g sesame /450g water)   and   52.1g of sesame residue were obtained from 40g of sesame seed and 450g of water by mixing for 3 min and   filtration.   The   suspension   of   these   materials   (sesame   milk   and   *kudzu* starch)   were prepared using a simmering method at the mixing rate of 250 rpm for 25 min of cooking time.   The suspension of materials was immediately poured into a glass ring case (20×20 mm ϕ).   After samples were cooled at room temperature for one hour,   they were employed to measurement.

2.2. Measurements

A   creep   meter (Rheoner RE-3305, Yamaden Co., Ltd. Tokyo, Japan)   was   used   for the measurement of textural properties of *gomatofu* under uni-axial compression.   The samples were compressed to 60% deformation by a plunger (30 mm ϕ).   Texture parameters of hardness, cohesiveness, adhesiveness and gumminess were obtained from the mean value and standard   deviation   of   10   times   test.   Creep   measuremens   were   also   carried   out using   a   Rheoner set by a thermostatted chamber connected to a water bath to maintain the measuring   temperature(20 ℃).   The   samples   were   compressed   with   20g   of   load by a plunger (40mm ϕ) at a elevating speed of 1.0 mm/s.

2.3. Scanning electron microscopy(SEM)

Scanning   electron   microscopy   (SEM-800, Hitachi,   Co)   observations   were   made.

The samples were fixed by glutaraldehyde and osmium acid using a critical point drying technique.

2.4. Sensory evaluation

Sensory evaluation was performed by scoring methods.   The panel was composed of 20 students of Niigata women's college.

3. RESULTS AND DISCUSSION

3.1. Texture

The texture parameters of *gomatofu* prepared with various contents of *kudzu* starch and sesame are shown in Figure 1.   With increasing content of both *kudzu* starch and sesame, the magnitudes of hardness, adhesiveness and gumminess increased.   The magnitude of cohesiveness increased with increasing amount of *kudzu* starch,   but it decreased with increasing content of sesame.   Namely, it was considered that the larger amount of the *kudzu* starch increased the magnitudes of all of four texture parameters,   but the larger amount of sesame contents decreased that of cohesiveness .

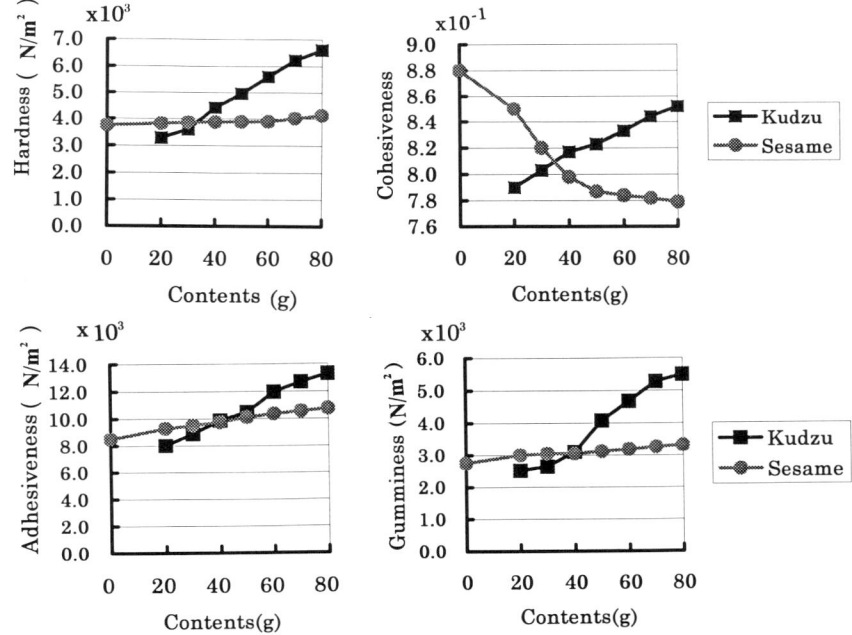

Figure 1. Effect of contents of Kudzu starch and sesame on texture property of *gomatofu*
　━■━ *kudzu* : Alternation of *kudzu* starch contents (40g of sesame were constant)
　━●━ Sesame : Alternation of sesame contents (40g of *kudzu* starch were constant)
　Gomatofu was prepared with a mixing rate of 250 rpm for 25 min.

3.2 .Viscoelastisity

The creep curve for the *gomatofu* could be analyzed approximately by using a 4-element model, which consisted of Elastic modulus of Hookean body (E_0), Elastic modulus of Voigt body (E_1), Viscosity of Voigt body (η_1), and Viscosity of Newtonian body (η_N).   The viscoelastic parameters of *gomatofu* prepared with different sesame contents   are shown in Table 1.   The elastic moduli (E_0, E_1) of *gomatofu* increased,   and its viscosities (η_N, η_1) and retardation time decreased with increasing sesame content.

Table 1.   Data showing the viscoelastic parameters of *gomatofu* prepared with different sesame contents

K (g)	W (g)	S (g)	SM (g)	E_0 10^2N/m^2	E_1 10^4N/m^2	η_1 10^5Pa·S	η_N 10^7Pa·S	τ_1 sec
		0	0	2.21 ± 0.15	2.12 ±0.15	6.16 ±0.25	3.36 ±0.12	29
		20	14	2.53 ± 0.19	2.35 ±0.19	5.03 ±0.24	3.03 ±0.17	21
40	450	40	28	2.57 ± 0.16	2.60 ±0.18	4.82 ±0.30	2.64 ±0.25	19
		60	41	2.60 ± 0.18	2.68 ±0.19	4.30 ±0.28	2.58 ±0.20	16
		80	55	2.65 ± 0.21	2.86 ±0.20	3.78 ±0.30	2.34 ±0.25	13

K: *Kudzu* starch   W: Water   S: Sesame
S M : Solid material in sesame milk, E_0 : Elastic modulus of Hookean body
E_1 : Elastic modulus of Voigt body, η_1 : Viscosity of Voigt body
η_N : Viscosity of Newtonian body, τ_1 : Retardation time

3.3. SEM observations

Figure 2   presents   the SEM photographs   of   *gomatofu*.   It   revealed   that   the *kudzu* starch gel had a thick branched structure,   and   *gomatofu* had a fibrous microstructure which surrounds globular sesame oils.   The larger amounts of sesame involved in the *kudzu* starch gels,   the finer fibrous structure was formed.   It   was found   that syneresis from the *kudzu* starch gel occurred during 2hrs after preparation, while it did not occur during the same period for *gomatofu* (sesame contents 60, 80g).   From these results,   it is considered that the creep behavior reflects the structure of *gomatofu* by SEM observations.   Morris [4]   studied the gels consist of a gel matrix filled with particulate inclusions,   and reported that the types of fillers encountered in foods would include gas bubbles, liquid droplets, or cellular components.   The characteristics of the filler may change during processing : the size,   shape and deformability of granules change during cooking or baking.   It   is assumed   that   *gomatofu* was a phase separated model [4]   which had networks consisting of *kudzu* starch and sesame components (protein and lipid).   Sesame (sesame protein,   lipid and water system) played an important role in a filler reinforcement effect [5] of starch granules for *kudzu* starch gel.

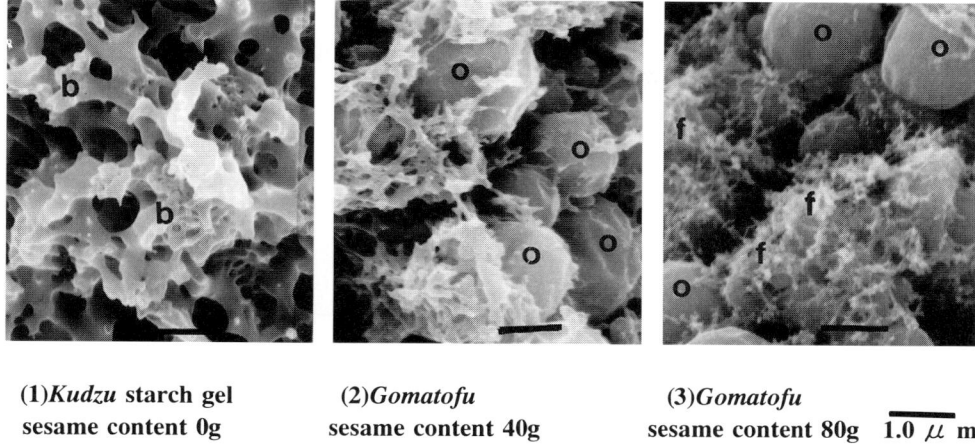

(1)*Kudzu* starch gel	(2)*Gomatofu*	(3)*Gomatofu*	
sesame content 0g	sesame content 40g	sesame content 80g	1.0 μ m

Figure 2.   SEM photographs obtained by fixing with glutaraldehyde and osmium acid
for *kudzu* starch gel and *gomatofu* prepared with mixing rate at 250rpm for
25min.

(1) *Kudzu* starch gel (*kudzu* starch 40g, sesame 0g, water 450g)x10,000
(2) *Gomatofu* (*kudzu* starch 40g, sesame 40g, water 450g)x10,000
(3) *Gomatofu* (*kudzu* starch 40g, sesame 80g, water 450g)x10,000
b   shows branched net work structure,   o   oil droplets,   and   f   microstructure

3.4   Sensory evaluation

The results of the scoring method are shown in Figure 3.   With increasing
content of both *kudzu* starch and sesame,   the higher score for hardness was obtained.   With
regard to mouthfeel,   the highest score was obtained when the sample prepared with 30g of
kudzu starch and 40g of sesame.   Springiness was decreased with increasing sesame content.

Total acceptance of *gomatofu* prepared with   40~60g of sesame   (in the case of 40g of
kudzu) was highly evaluated.   In our previous work [6], with regard to hardness,   a good
positive correlation   (r = 0.92, p $<$ 0.001) was obtained   between   parameters (obtained
by instrumental measurements) and sensory scores with increasing both *kudzu* starch and sesame   content.   Also   good   correlation   (r = 0.79, p $<$ 0.001)   was   obtained   between
adhesiveness and viscosity with a larger amount of sesame,   and its correlation (r = - 0.96,
p $<$ 0.001)   was obtained between hardness and mouthfeel with a smaller amount of *kudzu*
starch.   When the sesame contents were altered,   a high negative correlation (r = - 0.93, p
$<$ 0.001)   was obtained between cohesiveness and hardness by sensory scores.   From the
result of the sensory test, *gomatofu* prepared with 40~50g of *kudzu* starch, 40~60g of sesame
and 450g of water was palatable in softness, mouthfeel, and springiness.

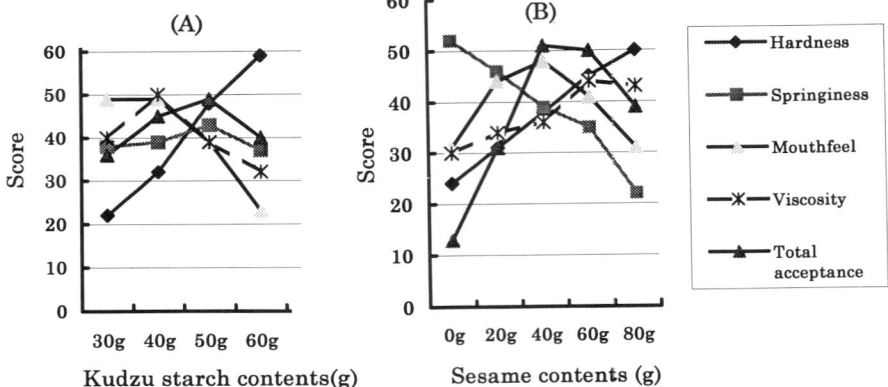

Figure 3. Sensory evaluation for *gomatofu* prepared with various contents of *kudzu* starch and sesame by scoring methods

   (A) Alternation of *kudzu* starch contents (40g of sesame were constant)

   (B) Alternation of sesame contents (40g of *kudzu* starch were constant)

It is considerd that the larger amount of the *kudzu* starch increased the magnitudes of all of four texture parameters, but the larger amount of sesame decreased the cohesiveness. It is suggested that sesame contributes to the strength and stability of the *kudzu* starch gel , and the larger amounts of sesame involved in the *kudzu* starch gels, the finer fibrous structure was observed. From the result of sensory test, good correlation was obtained between texture parameters and sensory score. *Gomatofu* prepared with ingredients having a ratio of *kudzu* starch contents (1), sesame contents (1~1.5), and added water(10~11) were palatable insoftness, mouthfeel and springiness.

REFERRENCES

1. E. Sato, E. Miki, S. Gohtani, Y. Yamano, Journal of the Japanese Society for Food Science and Technology, 42, (1995) 737 .
2. E. Sato, E. Miki, S. Gohtani, Y. Yamano, Journal of the Japanese Society for Food Science and Technology, 42, (1995) 871 .
3. E, Sato, R. Ito, Y. Yamano, Effects of preparing on the texture of *gomatofu*, submitted to Journal of the Japanese Society for Food Science and Technology.
4. V.J.Morris, "Gums and Stabilizers for the Food Industry 3" p87-99 (1985).
5. S, Matsumoto : Food Texture and Rheology, Sherman,P.,ed (Academic Press, London), 291 (1979).
6. E, Sato, R. Ito, Y. Yamano, Effects of ingredient ratio of materials and sensory evaluation on the texture of *gomatofu*, submitted to Journal of the Japanese Society for Food Science and Technology.

HYDROCOLLOIDS – PART 2
Edited by K. Nishinari
2000 Elsevier Science B.V.

275

Effects of sugars on stability of egg foam and their rheological properties

A. Ochi[a], K. Katsuta[a], E. Maruyama[a], M. Kubo[b], and T. Ueda[b]

[a]Department of Food Science & Nutrition,   Nara Women's University
Kitauoyanishi-Machi, Nara 630-8506, Japan
[b]Material Property Research Center, Nippon Paint Co. Ltd.,
Neyagawa, Osaka 572-8501, Japan

The effects of sugars on the stability of egg foams were investigated using a rheological technique, pulsed NMR measurements and some other tests.

The linear viscoelastic region for foam systems under uniaxial compression was very narrow. The creep behavior for foam systems could be divided into three parts; elastic, viscoelastic and steady flow part. The creep compliance and steady flow part of the foam systems decreased with increasing sugar concentration.

The amounts of draining from the foam systems increased and the protein contents in drainage liquid decreased with standing time. The rate of draining was impeded when sucrose was added to albumen. The density of the foam systems became larger on addition of saccharides, because the particle size of the gas cells became smaller and more homogeneous in the presence of sugar. The spin-spin relaxation times of foam systems obtained by the CPMG (Carr-Purcell-Meiboom-Gill) method became shorter on addition of sugar.

It might be considered that saccharides prevent the mobility of water between the films of foam, therefore foam systems might be stabilized.

1. INTRODUCTION

We have studied the relationship between the expansion of cakes during baking and the temperature dependence of rheological properties of batter. The increase of specific volume of cake batter is mainly caused by expansion of gases in the foams during baking. Therefore, the quality of cakes is highly influenced by the foaming properties of the egg, the most important process on making cake batter is the preparation of the foams.

Forming behavior has extensively been studied on food foams; for example, Dickinson has indicated that the effectiveness of egg-white as a foaming agent in meringue or cake batter arose from successful teamwork between the various constituent proteins and glycoproteins in chicken egg albumen [1]. Walstra and Smulders have stated that making bubbles is easy, but making stable small ones might be difficult [2]. It has been reported that sugar stabilized the foam [3-5], but the mechanism is still unclear. Especially, the rheological property of foams and effects of the presence of sugar on it remain to be investigated, although there are a few reports by Fujioka and Matsumoto [6-8]. In this present study, the rheological property of foams and the effects of saccharides on stability of foam systems were investigated.

2. MATERIAL AND METHODS

2.1. Material
Frozen albumen was kindly supplied by QP Co. Ltd. (Tokyo, Japan). The protein contents in albumen was 12.3%. Reagent grade saccharides were used without further purification in this study.

2.2. Preparation of foam systems
Frozen egg white was thawed in a water bath. Saccharides (9 or 18g) were weighed, added to the albumen solution (100g) and held at room temperature for 15 minutes before foaming. Egg white was whipped using a Mixer (KEN MIX KM-230, Aikosha Manufacturing Co. Ltd., Tokyo, Japan) for 90sec.

2.3. Static viscoelastic (creep) measurements
The static viscoelastic (creep) properties under uniaxial compression were determined over 300 sec by a Creep Meter, (Rheoner RE-33005, Yamaden Co. Ltd., Tokyo, Japan) at 25℃ for 300 sec. In order to avoid the effect of drainage and to maintain the homogeneous systems, foams were poured into a vessel (8.5 φ ×6.0 cm) having holes (1.0 mm φ) in the bottom. The measurements were made, after samples were pre-compressed at 5 gf (gram force) for 1 sec and the load was immediately removed to obtain the plane surface.

2.4. Physico-chemical experiments
2.4.1. Density
The foam systems were carefully filled in the vessel (8.6 φ ×8.0 cm) and weighed.

2.4.2. Pulsed NMR measurements
The foam systems were cautiously poured into glass tubes and pulsed NMR measurements were performed using a pulsed NMR spectrometer (JNM-MU25A Model, JEOL, Ltd., Tokyo, Japan). The spin-spin (T_2) relaxation times were measured by the CPMG (Carr-Purcell-Meiboom-Gill) method with pulsed sequence at 2μ s and resonance frequency at 25 MHz.

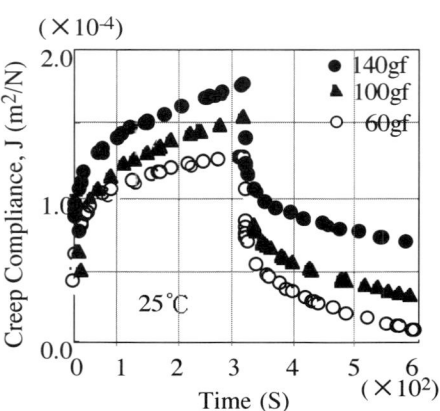

Figure 1. Effect of applied force on the creep compliance for foam systems containing 18% sucrose.

2.4.3. Amounts of drainage and protein contents in drainage

Samples of foams were immediately filled in the glass funnel, and changes in the amounts of drainage were observed at the various periods of standing time. The protein contents in drainage were measured using the micro Biuret method. The measurements of absorption at 310 nm was made using a spectrophotometer (UV-3100 PC, Shimadzu, Ltd., Tokyo, Japan).

3. RESULTS AND DISCUSSION

3.1. Rheological properties

With regard to the dynamic viscoelastic properties of foam systems, it was found that the dynamic moduli (G' or G") obtained by shear (rotatory oscillation) experiments did not present the correct magnitudes (data not shown) because of slippage between samples and test fixture. The foam systems showed syneresis or drainage and this is a reason why slippage occurred.

The force-deformation (strain) curves for foam systems with and without sugar

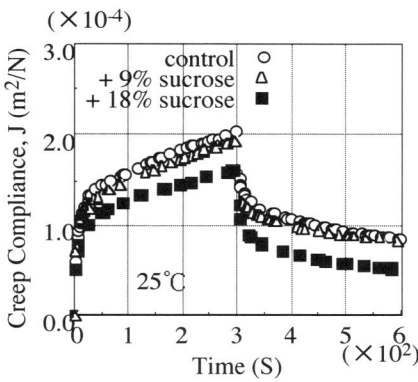

Figure 2. Effect of sucrose on creep compliance for foam system.

under uniaxial compression showed that their viscoelastic behavior was non-linear under the applied forces. Figure 1 shows the creep compliance (and recovery) curves for various compression stress (force). Fujioka and Matsumoto [6-8] studied the static viscoelastic properties of egg albumen foam systems, and stated that the stress relaxation curves of any samples could be approximated by a single Maxwell model and the initial relaxation modulus increased with increasing sugar concentration, while the creep curves could be described by a simple combined model consisting of four elements (Burger body). The result obtained in this study indicates that the rheological behavior is non-linear, hence viscoelastic parameters were not estimated. Under the same force (120 gf), sugar reduced the slope of the steady flow in the creep compliance curve (Figure 2). This means that the viscosity of the foam systems is higher in the presence of sugar.

3.2. Density and pulsed NMR

The density of foam systems containing sugars is remarkably higher compared to the control (Figure 3). The albumen-sucrose foam system might consists of smaller size gas cells and of more homogeneous or close-packed ones. The most effective sugar in increasing the density was glucose. This is probably due to the large number of

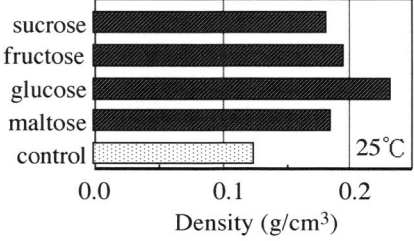

Figure 3. Effect of sugars on density of foam systems. The concentration of sugar added were 18 %.

equatorial-OH groups in the glucose [9].

As seen in Figure 4, spin-spin relaxation time of foam systems became shorter rather than egg albumen (soln). Egg white consists mainly of proteins (10.5%) and water (85%) [3]. When egg white was foamed, the mobility of free (bulk) water in the space between films might be suppressed.

The logarithm of spin-spin relaxation curves for any foam systems showed an almost straight line and the slope became steeper with increasing sugar contents. This also means that the mobility of free (bulk) water in the liquid films might be suppressed by sugars.

3.3. Drainage

Food foam systems like the egg white foam are very unstable, because the collapse of the liquid films progress with aging. And the drainage from foam systems is appeared.

Figure 5 shows the amounts of drainage with standing time from foam systems prepared by 100g of egg white solution. The time dependencies of drainage showed a two-steps kinetics. The reaction can be expressed by the two terms of first-order equation as follows,

$$m(t) = m_{1\infty}[1 - \exp\{-k_1 \cdot (t-t_1)\}] + m_{2\infty}[1 - \exp\{-k_2 \cdot (t - t_2)\}] \quad (1)$$

where m is the amounts of drainage, $m_{1\infty}$ and $m_{2\infty}$ are the infinite values, t the retention time, t_1 and t_2 the onset times for draining and k_1 and k_2 the rate constants, respectively. The calculated data are shown in Table 1.

The onset time for drainage in the initial reaction, t_1, was delayed by sucrose in proportion to its concentration, whereas

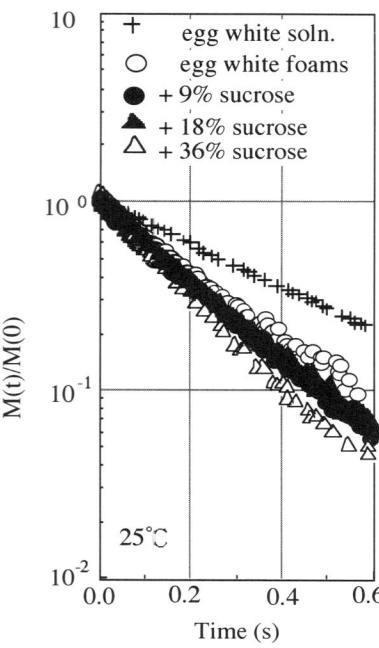

Figure 4. Effects of sugar on spin-spin relaxation behavior at 25 °C.

The measurements were made by the CPMG methods at 25 MHz. M(0) means the initial magnitute when t=0.

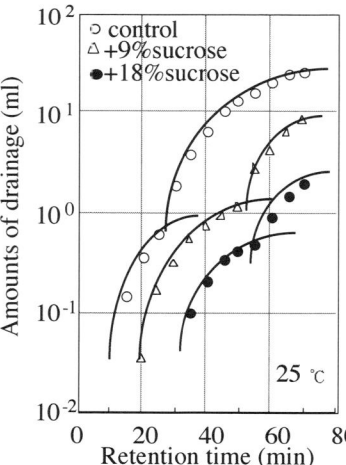

Figure 5. The changes in amounts of drainage from foam systems against standing time.

the onset time of second stage, t_2, might be independent from the concentration of sugar. The magnitudes of k_1 for any systems was smaller than those of k_2. Both rate constants, k_1 and k_2, were reduced in the presence of sugars, and decreased with increasing concentration. These results show that sugars prevent the drainage from foam system because of increase of viscosity [8,10] and of decrease of mobility of bulk water in liquid films, but the existence of a two-step kinetics for the draining reaction can not be explained by those reasons.

Table 1   Data showing the kinetic parameters for the drainage.

	k_1	k_2	$m_{1\infty}$	$m_{2\infty}$	t_1	t_2
	$(\text{min}^{-1} \times 10^{-2})$			(ml)	(min)	
Control	4.25	62.19	41.0	600	11.8	29.0
9%Sucrose	3.57	34.80	35.0	340	20.4	47.8
18%Sucrose	1.87	10.00	18.0	92	34.5	42.0

The parameters were obtained by curve fitting to the Equation (1).

Stainsby [3] reported that the large density difference between the phases ensured that gravitationally-induced drainage occurred, and pointed out that capillary-driven drainage from lamella into plateau borders. The pressure in the liquid near a highly curved border is lower than in the adjacent regions [3]. He also said that the rate of drainage usually felt with time, but even initial rates have little meaning when an unknown amount of drainage has already taken place during foam formation.

The time dependence of the amounts of protein in drainage was also divided into two stages (Figure 6). The changes in the amounts of both protein and drainage with standing time were divided into two steps at around 22 minutes of retention time.

The foams are consisted by the liquid films protein adsorbed to the surface between gas phase [8]. Major protein in albumen is ovalbumin, ∼ 54% of total egg albumen solid [3], and ovalbumin plays an important role as a surfactant in egg white foam systems. Mita et al. [10] stated that the drainage and stability of foams are essentially different phenomena, because the drainage is caused by thinning of the liquid film of foams without rupture in the early stages of the life of foams. The thickness of the liquid films are maintained by the electrostatic interactions (repulsive force) and the surface active force of

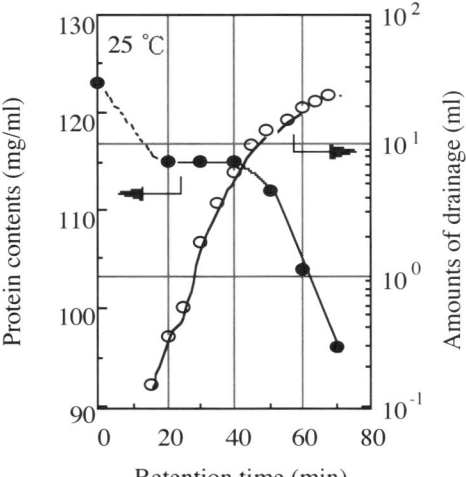

Figure 6. Amounts of drainage and protein in drainage.

ovalbumin [8]. The progress of thinning of liquid film of foams induces the coagulation of protein by hydrophobic interaction [11]. The coagulums might be insoluble, because the protein contents in drainage decreased as seen in Figure 6. It follows to lower of interfacial tension of the foams and break up foams [2]. Therefore, the second stages in the draining reaction had the large rate constants suggesting the rapid reaction.

4. CONCLUSION

Sugars have a large number of equatorial-OH groups, hence they induce a decrease of mobility of water in the aqueous solution. Sugar also increase the viscosity of solvent between foams. Thus sugar inhibits the drainage from bulk phase of foam systems. Retention of water in and between the membrane prevents the hydrophobic interaction of proteins and diminishes the collapse of foams.

In this investigation, it was difficult to obtain the exact parameters of dynamic viscoelastic property for foams when rotatory shear strain was applied. A rotatory shear measurement is unsuitable not only for foam, but also for the other samples (for instance, gels), if the sample shows drainage or syneresis. Our current work is concerned with obtaining dynamic parameters using a longitudinal oscillatory measurements.

REFERENCES

1. E. Dickinson, In *An Introduction to Food Colloids*, pp.135-137, Oxford Sci. Pub., New Yolk , 1992.
2. P. Walstra, I. Smulders, In E. Dickinson and B. Bergenstahl (eds.) *Food Colloids*, pp.367-381, The Royal Soc. Chem., Cambridge, 1996.
3. G. Stainsby, In J.R. Mitchell and D.A. Ledward (eds.) *Functional Properties of Food Macromolecules*, pp. 315-353, Elsevier, London, 1986.
4. A.-M. Hermansson, B. Sivik and C. Skjoldebrand, *Lebensm.-Wiss. u-Technol.*, **4** (1971) 201.
5. P.J. Halling, *CRC Crit. Rev. Food Sci. Nutr.*, **15** (1981) 155.
6. T. Fujioka and S. Matsumoto, *J. Cookery Sci. Jpn.*, **28** (1995) 91.
7. T. Fujioka and S. Matsumoto, *J. Text. Studies*, **26** (1995) 411.
8. T. Fujioka and S. Matsumoto, *J. Cookery Sci. Jpn.*, **27** (1994) 7.
9. H. Uedaira and H. Uedaira, *Bull. Chem. Soc. Jpn*, 61 (1989) 1.
10. T. Mita, K. Nikai, T. Hirooka, S. Matsuo and H. Matsumoto, *J. Colloid Interface Sci.*, **59** (1977) 172.
11. J.F. Zayas, In *Functionality of Proteins in Food*, pp.287-292, Springer-Verlag, New Yolk, 1997.

HYDROCOLLOIDS – PART 2
Edited by K. Nishinari
2000 Elsevier Science B.V.

Application of transglutaminase for food processing

C.Kuraishi, H. Nakagoshi, H. Tanno and H. Tanaka

Food Research and Development Laboratories, AJINOMOTO CO.,INC., 1-1 Suzuki-cho, Kawasaki-shi, 210 JAPAN

Transglutaminase(TG) is an enzyme that catalyzes crosslinking reaction of protein molecules. TG is widely distributed in the nature and has been found in various animal tissues, fish, plants, and microorganisms. We have succeeded in the mass production by fermentation method and in commercialization of TG for the first time in the world. This microbial transglutaminase(MTG) has various effects on the physical properties of food proteins[1]. MTG has been put to practical use for meat and seafood processing, surimi products, noodles and pasta, and tofu. Some of these applications are shown in this paper.

1. GENERAL REACTIONS CATALYZED BY TRANSGLUTAMINASE

Transglutaminase(TG) is an enzyme that catalyzes the acyl-transfer reaction between the γ-carboxyamide group of glutamine residues in peptide-bonds and primary amines[2] (Figure 1-a). When TG acts upon protein molecules, they are cross-linked and polymerized through ε-(γ-glutamyl)lysyl peptide bonds[3](Figure 1-b). In the absence of suitable primary amines or in the case that the ε-amine of lysine is blocked by chemical reagents, it is possible to make water act as the acceptor, and the glutamyl residue changes to a glutaminyl residue by deamidation through transglutaminase reaction(Figure 1-c).

The most effective reaction to be used for food industry is reaction (b), crosslinking of protein molecules. The crosslinking reaction may be found in both intermolecular and intramolecular proteins.

282

(a) $\mathrm{Gln-\overset{|}{\underset{\underset{O}{\|}}{C}}-NH_2}$ + RNH$_2$ → $\mathrm{Gln-\overset{|}{\underset{\underset{O}{\|}}{C}}-NHR}$ + NH$_3$

(b) $\mathrm{Gln-\overset{|}{\underset{\underset{O}{\|}}{C}}-NH_2}$ + $\mathrm{NH_2-\overset{|}{\underset{|}{Lys}}}$ → $\mathrm{Gln-\overset{|}{\underset{\underset{O}{\|}}{C}}-NH-\overset{|}{\underset{|}{Lys}}}$ + NH$_3$

(c) $\mathrm{Gln-\overset{|}{\underset{\underset{O}{\|}}{C}}-NH_2}$ + HOH → $\mathrm{Gln-\overset{|}{\underset{\underset{O}{\|}}{C}}-OH}$ + NH$_3$

Figure 1. General reactions catalyzed by transglutaminase
(a) acyl-transfer reaction
(b) crosslinking reaction
(c) deamidation

2. ε-(γ-GLUTAMYL)LYSYL PEPTIDE BOND

In the crosslinking reaction of MTG, intermolecular and intramolecular ε-(γ-glutamyl)lysyl peptide[ε-(γ-Glu)Lys] bonds are formed. Gel structure containing proteins was strengthened by ε-(γ-Glu)Lys bonds. Firmness of gel increased with increasing the number of ε-(γ-Glu)Lys bonds (Figure 2)[4].

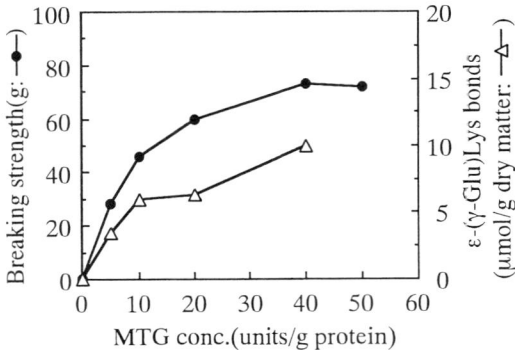

Figure 2. Effect of MTG treatment on gel strength of soy protein and amount of ε-(γ-Glu)Lys bonds of soy protein isolate incubated at pH7.0 at 37°C for 1hr. Protein concentration is 10%(w/w%). ε-(γ-Glu)Lys bonds content was measured by the method previously described[5].

When the excess amount of MTG was added, the firmness of the gel was decreased.   A gel formed with a caseinate solution(10 w/w%) treated with MTG 20units/g protein at 37°C at pH6.5 was more breakable than a gel formed with MTG 15unit/g protein.   It was suggested that there might be a limit to the ability of ε-(γ-Glu)Lys bonds to improve gel strength, and an excess of ε-(γ-Glu)Lys bonds partially break down or weaken the structure of gel (Figure 3)[4].

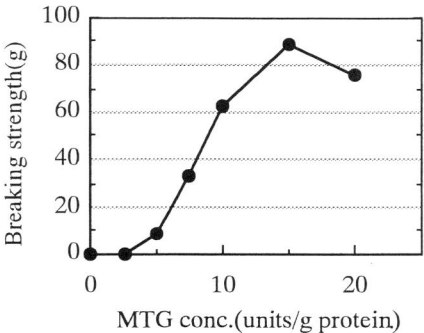

Figure 3. Effect of MTG on gel strength of caseinate solution incubated at pH 6.5 at 37°C for 1hr.   Protein concentration: 10(w/w%).

3. TEXTURE IMPROVEMENT OF RETORTED SAUSAGE

Crosslinks, G-L bonds, which are caused during MTG reaction are covalent bonds that are stable unlike electrostatic and hydrophobic interactions.   Therefore, even after retort treatment, foods treated with MTG can maintain its original texture better than non-treatment ones.   The breaking strength of the retorted sausage decreased by 20% as compared with the non-retorted sausage.   The MTG-treated(2.2unit/g protein) sausage after retort cooking also decreased in firmness, but was still firmer than non-MTG sausage before retort cooking. And the deformation of MTG-treated sausage after retort cooking is larger than that of non-MTG sausage after retort cooking(Figure 4).   MTG treatment recovered the decreased quality of the texture of retorted sausage and improved elasticity.

284

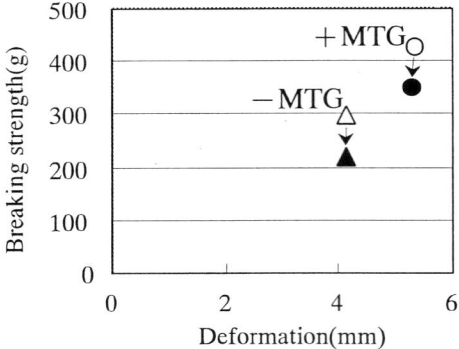

Figure 4. Effect of MTG on breaking strength and deformation of retorted sausage.
△: without MTG before retort cooking. ▲: without MTG after retort cooking. ○: with MTG 2.2 units/g protein before retort cooking. ●: with MTG 2.2units/g protein after retort cooking. Sausages were cooked at following condition: Drying 60°C 25min, Somke 60°C 10min, Steam 75°C 30min. Retort treatment, Fo=3.9

4. TEXTURE IMPROVEMENT OF PASTA

Texture of Noodles and pasta was improved by MTG treatment. Noodles with MTG could keep its texture even after acid- or heat- treatment for longer shelf-life. With the addition of MTG, texture of various noodles, Chinese noodle, Udon(Japanese noodle made from wheat flour), Soba, and pasta were improved. When the pasta was treated with MTG 0.2units/1g flour, the breaking energy, firmness, was increased from $27.6 \times 10^4 erg/cm^2$ to $32.6 \times 10^4 erg/cm^2$ (Figure 5).

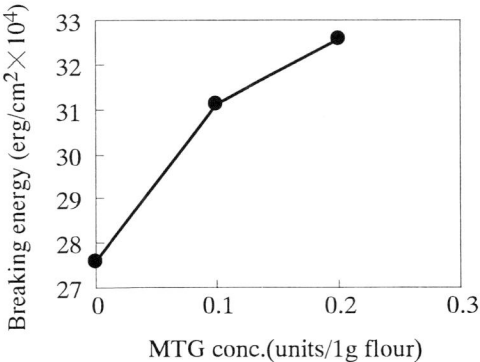

Figure 5. Effect of MTG on firmness of pasta. Pasta were made from durum 20% and wheat flour 80%. The dough was left for 60min at 20°C, for enzyme reaction, after adding MTG.

5. RETORT-RESISTANT TOFU

Soy bean proteins are also good substrates for MTG. Tofu gel treated with MTG had good consistency and improved smooth texture. It was also possible to produce retort-resistant tofu by using MTG.

The tofu gel without MTG shrank and lost >35% of its weight through retort cooking. It was suggested that the structure of tofu was changed because of heat treatment, and that the water entrapped in the microstructure of the tofu was released. The MTG-treated tofus also shrank and decreased in weight, but the weight loss was improved. This results suggests that tofu gel treated with MTG, which forms more stable covalent ε-(γ-Glu)Lys bonds, are able to hold more water in spite of temperature changes (Figure 6) [6].

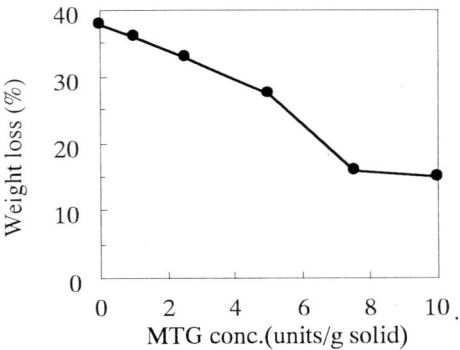

Figure 6. Effect of MTG on retort-induced weight decrease of tofus with different MTG concentration. MTG and GDL were added to soymilk(solid: 9.8%), and the soy milk mixture was packed and incubated for 30min at 55°C. After the incubation, the packed tofu were heated at 90°C for 30min to inactivate the enzyme. Retort cooking was performed until the Fo value reached 5.0.

REFERENCES

1. C. Kuraishi, J. Sakamoto, and T. Soeda, ACS Symposium series 37, (1996) 29.
2. J.E.Folk, Adv.Enzymol.,54(1983) 1.
3. M. Motoki, K. Seguro, N. Nio, and K. Takinami, Agric.Biol.Chem. ,50(1986) 3025.
4. H. Sakamoto, Y. Kumazawa, and M. Motoki, J.Food Sci., 59(1994) 866.
5. Y. Kumazawa, K. Seguro, M. Takamura, and M. Motoki, J.Food Sci., 58(1993) 1062, 1083.
6. M. Nonaka, H. Sakamoto, S. Toiguchi, K. Yamagiwa, T. Soeda, and M. Motoki, Food Hydrocoll., 10(1996) 41.

HYDROCOLLOIDS – PART 2
Edited by K. Nishinari
2000 Elsevier Science B.V.

Analyzing effects of environmental factors on viscosity of xanthan gum solution aided by experimental design

Y. Hayase, T. Aishima, K. Kidzu and T. Nagahori

Kikkoman Corporation, 399 Noda, Noda, Chiba 278-0037, Japan

Effects of xanthan gum, sucrose, salt, vinegar and their interactions on the viscosity of model solutions were quantitatively analyzed by utilizing full factorial designs and the following analysis of variance and multiple linear regression analysis.

1. INTRODUCTION

In designing liquid foods, "good viscosity for consumers" is a key point. For food manufacturing, viscosity of liquid foods should be controlled so as to meet required sensory properties and usability. Today, various polysaccharides are widely used to control the viscosity of foods. Generally, liquid foods are complicated multicomponent systems consisting of various materials. Thus, the viscosity of liquid food is influenced by many factors, such as ions, sugars and pH, besides the polysaccharide itself. For manufacturing liquid foods having desired viscosity, interactions of such factors should be simultaneously considered. This research aims to clarify the relationship between the viscosity of liquid foods and their constituents. Full factorial designs (FFD) were used to simulate liquid foods since the FFD can suggest us the most informative mixing ratios of ingredients [1, 2]. The analysis of variance (ANOVA) and multiple linear regression analysis (MLR) were applied to quantitatively analyzing effects of each factor and their interactions on the viscosity.

2. MATERIALS AND METHODS

2.1. Materials

Ingredients used were xanthan gum (Dainippon Pharmaceutical Co., Ltd, Osaka), sucrose (Dai-Nippon Meiji Sugar Co., Ltd., Tokyo), salt (Sin Nihon Salt Co., Ltd., Tokyo) and vinegar (Marukansu Vinegar Co., Ltd., Kobe). Distilled-deionized water was used.

2.2. Full factorial design

Two-level three-factor full factorial designs (FFD, 2^4) having the center point were used. As the four-dimensional FFD cannot be illustrated, a two-level (+, -) three-factor (A, B and C) FFD with the center point is shown in Figure 1 [1]. In the cube, black circles indicate experimental points. The ranges for the FFD were settled by referring to the formulations of ordinary liquid foods. Twenty model solutions corresponding to the cube points and four replicates at the center point suggested by the FFD were prepared as shown in Table 1. Prior to measuring viscosity, the order of experimental points or model solutions was randomized in order to avoid the time dependent effect.

2.3. Measurement of viscosity

The viscosities of twenty model solutions prepared according to FFD were measured with a rotational viscometer (HAAKE Viscotester VT500). The viscometer was operated in an automatic mode where the speed increased (or decreased) up to a maximum shear rate of 260 s^{-1} in two minutes at 25°C. The apparent viscosity at shear rate 50 s^{-1} was used as the response for the subsequent statistical analyses (Figure 2).

2.4. Analysis of variance and multiple linear regression analysis

Effects of individual factors and their interactions on the viscosity were examined by the analysis of variance (ANOVA) using Unscrambler ver.7.0 (CAMO AS, Trondtheim, Norway). The MLR and illustration of response surfaces were performed by TRIAL RUN 1.0 (SPSS Inc., Chicago, IL).

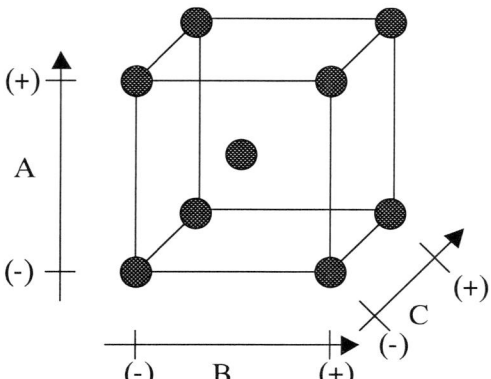

Figure 1. Two-level (+, -) three-factor (A, B, C) FFD with the center point

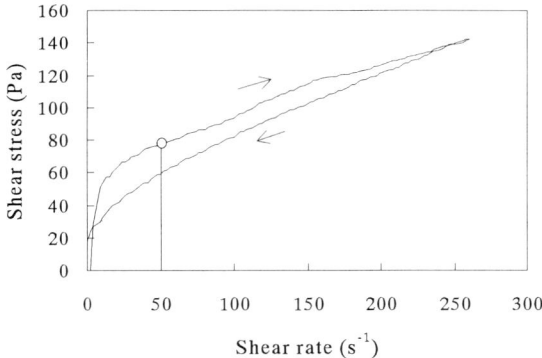

Figure 2. Apparent viscosity at shear rate 50 s^{-1}

Table 1 Experimental conditions based on the FFD and responses for them

	Factor				Response
Exp.	salt	sucrose	vinegar	gum	viscosity (mPas)
1	2.5(%, w/w)	12.5(%, w/w)	0.76(%, v/w)	0.25(%, w/w)	47
2	7.5	12.5	0.76	0.25	51
3	2.5	37.5	0.76	0.25	91
4	7.5	37.5	0.76	0.25	107
5	2.5	12.5	2.25	0.25	42
6	7.5	12.5	2.25	0.25	53
7	2.5	37.5	2.25	0.25	91
8	7.5	37.5	2.25	0.25	114
9	2.5	12.5	0.76	0.75	229
10	7.5	12.5	0.76	0.75	232
11	2.5	37.5	0.76	0.75	376
12	7.5	37.5	0.76	0.75	430
13	2.5	12.5	2.25	0.75	227
14	7.5	12.5	2.25	0.75	258
15	2.5	37.5	2.25	0.75	372
16	7.5	37.5	2.25	0.75	436
17	5.0	25.0	1.51	0.50	167
18	5.0	25.0	1.51	0.50	163
19	5.0	25.0	1.51	0.50	167
20	5.0	25.0	1.51	0.50	160

3. RESULTS

3.1. ANOVA for FFD

The viscosities of twenty model solutions are shown in Table 1. From the results of ANOVA shown in Table 2, xanthan gum was the most important factor for the viscosity followed by sucrose and their mutual interaction. Salt was statistically significant whereas vinegar did not show any significant contribution to the viscosity. Among interactive effects, that of xanthan gum and sucrose was the most significantly contributed but those of others were only negligible. Concerning the viscosity of xanthan gum solution, positive contribution of salt has been widely indicated [3] but a relationship found between salt and viscosity was significant only at 5% level in this analysis. However, if the presence of salt in the solution would be meaningful, the viscosity of solution, which did not contain any salt, should have been compared with that of solution containing more or less salt. Therefore, the ranges of factors for the second FFD were settled as follows; sugar: 0-50 (%, w/w), salt: 0-10 (%, w/w), vinegar: 0-3 (%, v/w) and xanthan gum: 0-1 (%, w/w).

Table 2   ANOVA table for the FFD

	SS	DF	MS	F-ratio	p-value
Salt	0.003	1	0.003	6.003	0.0368
Sugar	0.048	1	0.048	110.094	0.0000
Vinegar	0.000	1	0.000	0.139	0.7179
Gum	0.241	1	0.241	551.682	0.0000
Salt×Sugar	0.001	1	0.001	1.635	0.2331
Salt×Vinegar	0.000	1	0.000	0.409	0.5386
Salt×Gum	0.001	1	0.001	1.372	0.2715
Sugar×Vinegar	0.000	1	0.000	0.026	0.8766
Sugar×Gum	0.013	1	0.013	30.095	0.0004
Vinegar×Gum	0.000	1	0.000	0.071	0.7960
Error	0.004	9	0.000		

3.2. ANOVA for the second FFD

The results of ANOVA for the second FFD are shown in Table 3. The effect of xanthan gum was the largest again. Comparing with the results of ANOVA for the first FFD shown in

Table 3 ANOVA table for the second FFD

	SS	DF	MS	F-ratio	p-value
Salt	0.391	1	0.391	12.495	0.0064
Sugar	0.679	1	0.679	21.663	0.0012
Vinegar	0.000	1	0.000	0.001	0.9756
Gum	1.599	1	1.599	51.052	0.0001
Salt×Sugar	0.217	1	0.217	6.912	0.0274
Salt×Vinegar	0.000	1	0.000	0.000	0.9854
Salt×Gum	0.367	1	0.367	11.708	0.0076
Sugar×Vinegar	0.001	1	0.001	0.038	0.8497
Sugar×Gum	0.600	1	0.600	19.164	0.0018
Vinegar×Gum	0.000	1	0.000	0.014	0.9075
Error	0.282	9	0.031		

Table 2, differences among magnitudes of their effects were much smaller. Further, considerable increase in the effect of salt concentration was indicated. Although the significant effect of salt and xanthan gum interaction on the viscosity was observed, neither vinegar nor other interactions showed any significant relation to the viscosity.

3.3. MLR models and their response surfaces for the second FFD

The two-factor response surface illustrating the relationship between the viscosity and concentrations of xanthan gum and sugar clearly indicated the existence of their interactive effect (Figure 3). Figure 4 shows the similar relationship calculated between the viscosity and

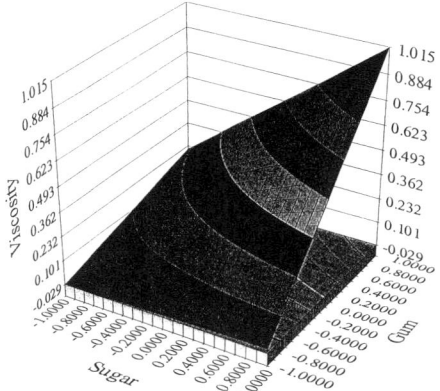

Figure 3. Two-factor response surface: Viscosity = 0.299 + 0.206***×Sugar + 0.316***×Gum + 0.194**×Sugar×Gum, adjusted R^2=0.856. ** P<0.01, *** P<0.001.

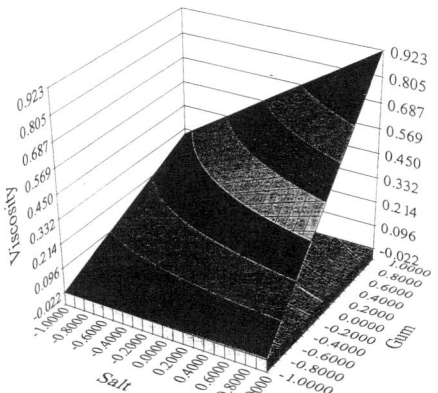

Figure 4. Two-factor response surface: Viscosity = 0.299 + 0.156**×Salt + 0.316***×Gum + 0.151**×Salt×Gum, adjusted R^2=0.856. ** P<0.01, *** P<0.001

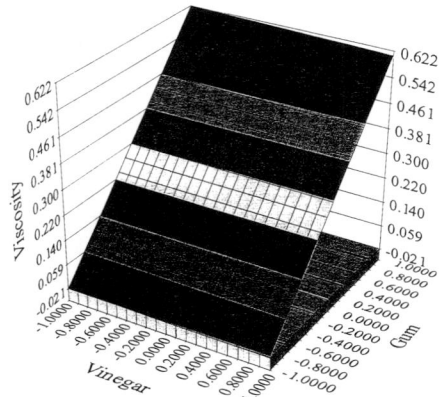

Figure 5. Two-factor response surface: Viscosity = $0.299 - 1.392 \times 10^{-2} \times$ Vinegar + $0.316^{***} \times$ Gum $-5.288 \times 10^{-3} \times$ Vinegar $\times$ Gum, adjusted $R^2 = 0.856$. *** $P < 0.001$.

concentrations of xanthan gum and salt. However, no interactive effect between the concentrations of xanthan gum and vinegar was indicated by the response surface (Figure 5).

4. DISCUSSION

Experimental designs have shown their great versatility for systematically simulating complicated multicomponent systems, *i.e.*, liquid foods. By applying both ANOVA and MLR to analyzing relationships between the measured viscosities and concentrations of individual factors and their interactions, every factor effect can be separately and quantitatively expressed as the form of numerical terms. Thus, experimental designs and the following statistical analysis will be a powerful tool to clarify the complicated relationships existing between the viscosity of liquid foods and their constituents and enable us to manufacture highly preferable and usable products.

REFERENCES

1. E. Morgan, Chemometrics: Experimental Design, John Wiley & Sons Ltd., Chichester, England, 1991.
2. A. Hamza-Chaffai, J. Food Sci., 55 (1990) 1630.
3. K.S. Kang and D.J. Pettitt, Xanthan, gellan, welan and rhamsan, in: Industrial Gums, Third Edition, R.L. Whistler and J.N. BeMiller (eds.), New York, 1993, pp.341-397.

5. BIOMEDICALS

HYDROCOLLOIDS – PART 2
Edited by K. Nishinari
2000 Elsevier Science B.V.

295

Biological and pharmaceutical activities of sulfated poly- and oligo-saccharides

T. Uryu, K. Katsuraya, K.-J. Jeon and Y. Gao

Institute of Industrial Science, University of Tokyo, Roppongi,
Minato-ku, Tokyo 106-8558, Japan

Highly anti-HIV active sulfated polysaccharides and sulfated alkyl oligo-saccharides were synthesized. Action mechanism of the anti-HIV activity for these sulfated compounds was examined by use of polylysine as a model polypeptide of an HIV protein and a synthetic portional protein of the HIV envelope protein by NMR spectroscopy.

1. INTRODUCTION

Since we found out that sulfated 1,3-β-linked glucans such as sulfated lentinan and curdlan have high anti-HIV (human immunodeficiency virus) activity but low anticoagulant activity,[1-3] relationships between the structure and biological activities have been investigated as well as action mechanisms of the sulfated polysaccharides as an AIDS (acquired immunodeficiency syndrome) drug. Sulfated polysaccharides such as 1,5-α-linked ribofuranans and 1,6-α-linked dextrans exhibiting both high anti-HIV activity and high anticoagulant activity were unsuitable for an AIDS drug.[4-7] In addition, curdlan sulfates prepared by sulfating agents other than piperidine N-sulfonic acid showed considerably high anticoagulant activities.[8] In order to examine dependence of the position of the sulfate group on the anti-HIV and anticoagulant activities, regioselectively sulfated curdlans were prepared.

Succeeding to the synthesis of sulfated polysaccharides, sulfated alkyl oligosaccharides were molecular-designed. Sulfated laminari- and malto-oligosaccharides having alkyl groups longer than dodecyl at the reducing end exhibited high anti-HIV activities, when the number of glucose residues was more than 5.[9-11] Influences of other structural factors on the anti-HIV activity were studied in detail.

A natural sulfated polysaccharide heparin binds to antithrombin-III to manifest its anticoagulant activity.[12] In this case, a negatively charged portion of the heparin ionically interacts with a positively charged portion in the protein. Furthermore, it has been revealed that heparin ionically interacts with tryptase to generate a complex by the formation of which the tryptase is activated.[13,14] Since it is assumed that the anti-HIV activity of curdlan sulfate is also caused by such ionic interactions with HIV proteins, NMR spectroscopic studies were for the first time performed to measure the ionic interactions between negatively charged polysaccharides and positively charged polylysines which are regarded as the model compound for a HIV protein.

Interactions of curdlan sulfate with a synthetic oligopeptide corresponding to a part of HIV envelope protein gp120 was also measured by NMR spectroscopy.

2. RESULTS AND DISCUSSION

2.1. Structure of curdlan sulfates with high anti-HIV activity but low anticoagulant activity

Almost all sulfated polysaccharides having both high molecular weights and high degrees of sulfation exhibited high anti-HIV activities.[14] However, another biological activity of sulfated polysaccharides, i.e., anticoagulant activity, was different among kinds of polysaccharides. Unlike 1,5-α-linked ribofuranan and xylofuranan sulfates, and 1,6-α-linked dextran sulfates with high anticoagulant activities,[4-6] 1,3-β-linked lentinan and curdlan sulfates showed considerably low anticoagulant activities, even when they had high molecular weights.[1-3]

For curdlan sulfates differing in molecular weight, the degree of sulfation, and position of sulfate groups, anti-HIV and anticoagulant activities are shown in Table 1. Curdlan sulfate (CS) (no. 1) having weight-average molecular weight of 79 x 10^3 and S content of 15.2% exhibited a very high anti-HIV activity of a drug concentration for complete inhibition (IC_{100}) of 3.3 μg/mL.[2] This curdlan sulfate showed a low anticoagulant activity of 14-16 unit/mg, although it had a high molecular weight. It was reported that curdlan sulfate no. 2 with lower molecular weight had a very low anticoagulant activity of <10 unit/mg.[3] Since these curdlan sulfates were sulfated in dimethyl sulfoxide solution with piperidine N-sulfonic acid, sulfation occurred at 100% of 6-hydroxyls and 40-50% of 2-hydroxyls but not at 4-hydroxyls. On the other hand, curdlan sulfates nos. 3 and 4 which were sulfated with chlorosulfonic acid and sulfur trioxide-pyridine complex, respectively, exhibited considerably high anticoagulant activities (39 and 36 unit/mg) as

Table 1. Anti-HIV Activity of Curdlan Sulfates[a]

no.	proportion of sulfate group[b]			M_n (x10^3)	S (%)	DS[c]	anti-HIV activity		anticoagulant activity (unit/mg)
	at C6	at C4	at C2				EC_{50}[d] (μg/mL)	CC_{50}[e] (μg/mL)	
1	1.0	0	ca.0.5	79[e]	15.2	nd	3.3[g]	>5000[h]	14-16
2	1.0	0.05	0.4	46	14.4	1.6	3.3[g]	>5000[h]	<10
3[i]	nd	nd	nd	50	nd	1.9	<1.0	nd	39
4[j]	nd	nd	nd	31	—	2.6	1-10	—	36
5	1.0	0	0.55	8.8	14.3	1.55	0.04	>1000	14
6	1.0	0	0.52	10.8	14.0	1.52	0.16	>1000	19
7	0	1.0	0.39	8.6	13.2	1.39	0.16	>1000	15
8	0	1.0	0.51	9.3	13.9	1.51	nd	nd	16
9	0.87	k	k	7.1	15.4	1.79	0.19	>1000	10
10	0.63	k	k	7.7	13.1	1.45	0.19	>1000	11

a) Sulfating agent: piperidine N-sulfonic acid. b) Determined by ^{13}C NMR and elemental analysis. c) Degree of sulfation. d) Anti-HIV activity: drug concentration effective for 50% inhibition of virus infection in 5-day HIV-infected MT-4 cell culture. e) Cytotoxicity: drug concentration for 50% cytotoxicity in 5-day MT-4 cell culture. f) Weight-average molecular weight. g) Anti-HIV activity: drug concentration for 100% inhibition of virus infection in 6-day HIV-infected MT-4 cell culture. h) Cytotoxicity: drug concentration for 100% cytotoxicity in 6-day MT-4 cell culture. i) Sulfating agent: chlorosulfonic acid. j) Sulfating agent: sulfur trioxide-pyridine complex. k) Existing but difficult to estimate.

well as high anti-HIV activity.[8] In curdlan sulfates no. 3 and 4, the position of sulfate groups was not regiospecific. As a result, it was necessary to examine whether the anticoagulant activity of curdlan sulfates depends on the position of sulfate groups when they had similar degrees of sulfation.

Making use of higher reactivity of the primary 6-hydroxyl, regioselectively sulfated curdlans having $\overline{M}_n$ 6×10^3 - 10.8×10^3 were prepared.[16] In this experiment, low molecular weight curdlan was used for the sulfation to proceed smoothly under mild conditions. Results are summarized in Table 1.

CS no. 6 having the sulfate group at 6 (100%) and 2 positions (52%) exhibited a high anti-HIV activity of $EC_{50} = 0.16$ µg/mL. Similarly, CS no. 7 sulfated at 4- (100%) and 2-hydroxyls (39%) had $EC_{50} = 0.16$ µg/mL. CS no. 9 having non-regiospecific sulfate groups at 6, 4, and 2 positions also showed a high activity of $EC_{50} = 0.19$ µg/mL. Therefore, it was concluded that the anti-HIV activity to a large extent depends on the degree of sulfation and that curdlan sulfates with the degree of sulfation higher than 1.35 exhibit high anti-HIV activities.

On the other hand, the anticoagulant activity of curdlan sulfates represented a different tendency, i.e., dependence on the molecular weight. As the molecular weight increased from 7×10^3 to 10.8×10^3, the anticoagulant activity linearly raised from 10 unit/mg to 19 unit/mg, which is shown in Figure 1. The anticoagulant activity did not depend on the position of sulfate groups.

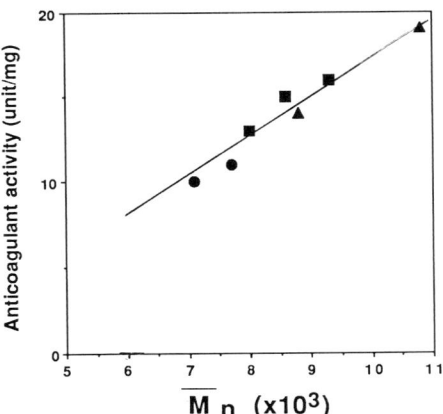

Figure 1. Dependence of anticoagulant activity on the molecular weight of curdlan sulfates. (●) 642S; (■) 62S; (▲) 42S.

We have revealed that all kinds of sulfated polysaccharides having both high anti-HIV and high anticoagulant activities can be synthesized. In addition, it was revealed that low anticoagulant sulfated polysaccharides were limited to curdlan sulfates. Therefore, it was concluded that the low anticoagulant for the curdlan sulfate might originate from potentially helical conformation of the curdlan sulfates which was kept even after sulfation of curdlan with helical backbone.

Curdlan sulfate was examined on the efficacy of an AIDS drug in Phase I/II test in the United States.[17] After it was intravenously injected into each subject for 100, 200, or 300 mg, a large increase in the number of CD4 cells in the blood was observed. However, the increase was a short-time phenomenon and the number became to an original value within a month.

2.2. Synthesis of anti-HIV active sulfated alkyl oligosaccharides

In general, active sites in biologically active polysaccharides consist of oligosaccharide portions. In addition, surface-active agents such as sodium dodecyl sulfate and polyethylene glycol are routinely used to destroy lipid bilayers of cells and bacteria. Combining the above two facts, we thought out the synthesis of sulfated alkyl oligosaccharides possessing a structure similar to the surface-active agent.[9]

As shown in Scheme 1, alkyl oligosaccharide peracetate was obtained by reacting an oligosaccharide peracetate with a long chain alcohol by use of Lewis acid catalyst in a medium yield. Discovery of this reaction lead us to the synthesis of the target compound. After deacetylation of the alkyl oligosaccharide peracetate, an OH-free oligosaccharide was sulfated to afford a sulfated alkyl oligosaccharide.[10]

Scheme 1. Synthetic Route of Sulfated Alkyl Oligosaccharide

For sulfated alkyl maltotetraosides bound by C12-C18 alkyls, their anti-HIV activities were depicted in Figure 2. All compounds exhibited low activities, indicating that four glucose residues were insufficient to produce sulfated alkyl oligosaccharides with high anti-HIV activities.

Using maltopentaose with more glucose residues as an oligosaccharide, the anti-HIV activity of sulfated alkyl maltopentaosides was determined. As the result is depicted in Figure 3, a hexyl derivative showed rather low activity of $EC_{50}=2.0$ µg/ml, while homologs having dodecyl (C12) to octadecyl (C18) groups possessed high activities of EC_{50} less than 1 µg/ml. Of them, the hexadecyl derivative had a high activity of $EC_{50}=0.27$ µg/ml. Therefore, it

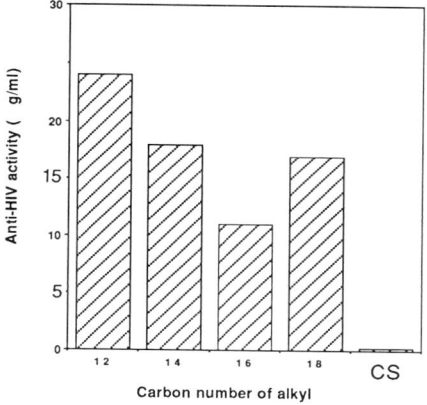

Figure 2. Effects of alkyl length on the anti-HIV activity in sulfated alkyl maltotetraosides. CS designates highly active curdlan sulfate.

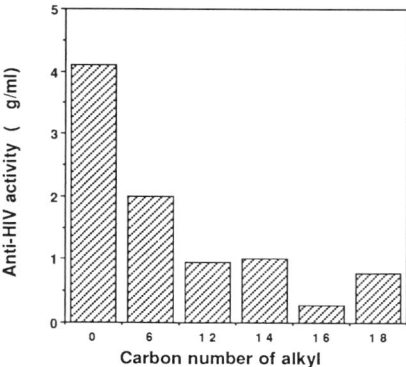

Figure 3. Effects of alkyl length on the anti-HIV activity in sulfated alkyl maltopentaosides.

was suggested that sulfated alkyl oligosaccharides having glucose residues more than 5 and alkyl groups longer than dodecyl possess high anti-HIV activities. However, the compounds containing longer alkyls than hexadecyl exhibited cytotoxicity.

Sulfated alkyl oligosaccharides composed of oligosaccharides of 5 to 9 glucose residues and of dodecyl (C12) or octadecyl (C18) exhibited high anti-HIV activities in the range of EC_{50}=0.10-0.63 μg/ml. The result is summarized in Table 2. For example, sulfated dodecyl laminaripentaoside (L5C12S) had a high activity of EC_{50}=0.10 μg/ml and a low cytotoxicity of CC_{50}>1000 μg/ml. Although an octadecyl homolog showed a high activity, it exhibited a considerably high cytotoxicity of CC_{50}=220 μg/ml. Both heptaoside and nonaoside showed a similar tendency on the activity and cytotoxicity.

Next, effects of the alkyl structure were examined. A branched 3,3-dimethylbutyl derivative had a fairly high activity of EC_{50}=0.78 μg/ml, indicating that this activity might be caused probably because of total carbon number of 6 and bulkiness of the group. Since a hexylphenyl derivative (L5PhC6S) had an alkyl corresponding to the group length of about 10 carbons, it showed a very high activity of EC_{50}=0.20 μg/ml. On the other hand, compounds bound by short alkyls such as (-)-2-methylbutyl and cyclohexylmethyl exhibited slightly lower activities.

Table 2. Anti-HIV Activity of Sulfated Alkyl Oligosaccharides

sample designation	sulfated alkyl oligosaccharide		DS[a]	anti-HIV activity EC_{50} μg/ml	cytotoxicity CC_{50} μg/ml
	number of glucose residues	carbon number or name of alkyl			
Laminarioligosaccharide-linear alkyl					
L5C12S	5	12	3.0	0.10	>1000
L5C18S	5	18	nd	0.63	220
L7C12S	7	12	nd	0.14	>1000
L7C18S	7	18	2.8	0.20	180
L9C12S	9	12	nd	0.18	>1000
L9C18S	9	18	2.4	0.59	240
Laminaripentaoside-branched, cyclo, or aralkyl					
L5C33DM4	5	3,3-dimethylbutyl	nd	0.78	>1000
L5C(-)AmS	5	(-)-2-methylbutyl	nd	1.3	>1000
L5CcHMS	5	cyclohexylmethyl	nd	1.3	>1000
L5PhC6S	5	4-hexylphenyl	nd	0.20	>1000
Laminaripentaoside-oxyethyene- or sulfate-containing alkyl					
L5EG3C2S	5	ω-ethoxytri(ethyleneoxy)	nd	110	>1000
L5C12OS	5	ω-sodium sulfatododecyl	nd	18	>1000

a) Degree of sulfation.

As it has been clear that hydrophobicity of the alkyl group to a large extent affects the anti-HIV activity, compounds composed of alkyls containing hydrophilic triethyleneoxy (L5EG3C2S) and ω-sodium sulfato (L5C12OS) groups were prepared and examined on the activity. As expected, the former showed almost no activity, and the latter did a very low activity of $EC_{50}=18$ μg/ml.

Concerning to another biological activity, i.e., an anticoagulant activity, all sulfated alkyl laminari-oligosaccharides had no or very low anticoagulant activity.

2.3. Studies on action mechanism of sulfated poly- and oligo-saccharides as an anti-HIV agent by NMR spectroscopy

2.3.1. Interactions of curdlan sulfate with polylysine

Sulfated polysaccharides such as dextran sulfate,[18] heparin,[19] and glycosaminoglycan[20] have been found to possess antiviral activities. It is known that heparin interacts with antithrombin-III to produce an anticoagulant complex.[21] In this interaction, an active site of heparin containing sulfamide and sulfate groups binds to a positive charge-accumulating portion of antithrombin-III,[12] triggering the anticoagulation.

Since HIV envelope glycoprotein gp120 includes several positively charged portions, it is assumed that curdlan sulfate (CS) ($\overline{M}$ w 79x10^3) might ionically interact with such portions. In this experiment, polylysine hydrobromide (PL) ($\overline{M}$ w 10.7x10^3) was used as a model compound of the protein. When CS and PL were mixed in an NMR tube, gellike material was formed. ^{1}HNMR spectra are shown in Figure 4. In spectrum of a mixture with molar ratio [CS]/[PL]=0.5, new peaks appeared in addition to original ones.[22] For example, εCH_2 protons were seen as two peaks appearing at 2.84 and 3.03 ppm which corresponded to the gellike material and the original polylysine, respectively. In the molar ratio of 0.8, proportion of the gel was 100%, and it decreased with increasing molar ratio. In ^{13}CNMR spectra, the same tendency was observed.

According to elemental analysis of the gellike complex formed in the molar ratio of 0.8, about 80% of Na^+ and Br^- which were included in the starting materials were liberated from the gel. Therefore, it was suggested that sodium curdaln sulfate reacted with polylysine hydrobromide to from crosslinks by precipitating NaBr, leaving the remaining ions into crosslinked networks. As a result, the interaction of CS with the protein occurred in the form of crosslinking.

2.3.2. Interactions of sulfated dodecyl laminaripentaoside with polylysine

We examined whether action mechanism of sulfated alkyl oligosaccharides as an anti-HIV agent is started by adhering to HIV envelope protein as was the case in the sulfated polysaccharide. When sulfated dodecyl laminaripentaoside (L5C12S) (degree of sulfation 3.0) was mixed with polylysine (PL) ($\overline{M}$ w 4000), gels were formed. In the ^{1}HNMR spectra, an εCH_2 absorption corresponding to the gel appeared at 0.2 ppm upfield to the original one. In the molar ratio [L5C12S]/[PL]=1, proportion of the gel was maximum. The proportion of the gel decreased with increasing molar ratio. However, when polylysine with $\overline{M}$ w 8000 was used, NMR absorptions were very weak probably because of restricted local motions in stiff gels. Role of the long alkyl group for the anti-HIV activity was not clear by NMR spectroscopy.

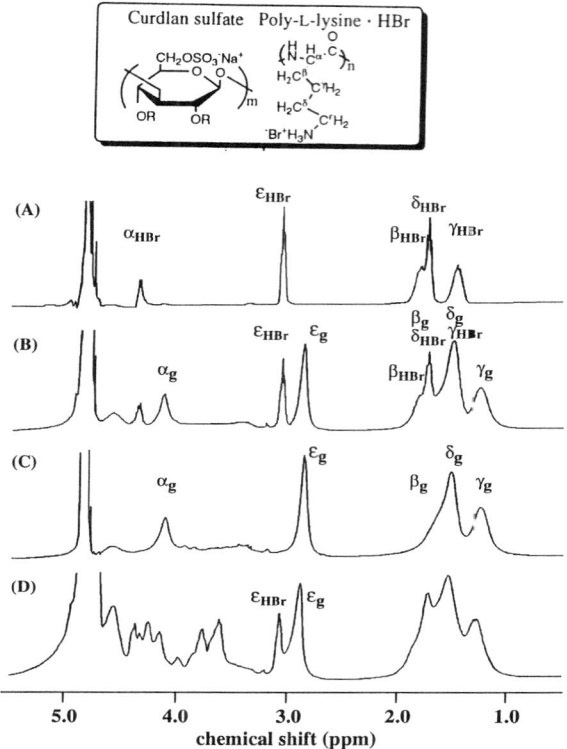

Figure 4. 400 MHz ^{1}HNMR spectra of (A) polylysine hydrobromide and (B) gellike complexes formed by mixing curdlan sulfate with polylysine in the molar ratio of 0.5, (C) 0.8, and (D) 1.0. The concentration of PL is 2.8% (w/v).

2.3.3. Interactions of curdlan sulfate with an HIV portional protein

One of the positively charged portions in HIV envelope glycoprotein gp120 is a portion from the amino acid residue 506 to 518.[23] We examined whether this portion interacts with curdlan sulfate.

A polypeptide sequence doubled by the sequence of 506 to 518 was synthesized for the polypeptide to have an appropriate molecular weight. The following polypeptide (D518) composed of 26 amino acids sequence was used. (Thr)(Lys)(Ala)(Lys)(Lys)(Arg)(Val)(Val)(Gln)(Arg)(Glu)(Lys)(Arg)-(Thr)(Lys)(Ala)(Lys)(Lys)(Arg)(Val)(Val)(Gln)(Arg)(Glu)(Lys)(Arg)

When CS was mixed with the polypeptide, gels were formed. In ^{1}HNMR spectra, both εCH_2 protons of lysine and arginine exhibited appearance of new peaks due to the gel at 2.84 and 3.03 ppm in addition to the original absorptions at 3.03 and 3.24 ppm, respectively. In the molar ratio [CS]/[D518]=0.27, proportion of the gel was 100%. The NMR behaviors were similar to those of the curdlan and polylysine mixture. Therefore, it was revealed that curdlan sulfate strongly interacts with a positively charged portion of HIV envelope glycoprotein gp120.

3. EXPERIMENTAL

Experimental procedures are described in our previous papers except for sections **2.3.2** and **2.3.3**.

REFERNCES

1. K. Hatanaka, T. Yoshida, T. Uryu, O. Yoshida, H. Nakashima, N. Yamamoto, T. Mimura, and Y. Kaneko, Jpn. J. Cancer Res., **80**, 95 (1989).
2. Y. Kaneko, O. Yoshida, R. Nakagawa, T. Yoshida, M. Date, S. Ogihara, S. Shioya, Y. Matsuzawa, N. Nagashima, Y. Irie, T. Mimura, H. Shinkai, N. Yasuda, K. Matsuzaki, T. Uryu, and N. Yamamoto, Biochem. Pharm., **39**, 793 (1990).
3. T. Yoshida, K. Hatanaka, T. Uryu, Y. Kaneko, E. Suzuki, H. Miyano, T. Mimura, O. Yoshida, and N. Yamamoto, Macromolecules, **23**, 3717 (1990).
4. K. Hatanaka, I. Nakajima, T. Yoshida, T. Uryu, O. Yoshida, N. Yamamoto, T. Mimura, and Y. Kaneko, J. Carbohydr. Chem., **10**, 681 (1991).
5. T. Yoshida, Y. Katayama, S. Inoue, and T. Uryu, Macromolecules, **25**, 4051 (1992).
6. T. Yoshida, C. Wu, L. Song, T. Uryu, Y. Kaneko, T. Mimura, H. Nakashima, and N. Yamamoto, Macromolecules, **27**, 4422 (1994).
7. C. Flexner, P. A. Barditch-Crovo, D. M. Kornhauser, H. Farzadegan, L. J. Nerhood, R. E. Chaisson, K. M. Bell, K. J. Lorentsen, C. W. Hendrix, B. G. Petty, and P. S. Lietman, Antimicrob. Agents Chemother., **35**, 2544 (1991).
8. T. Yoshida, Y. Yasuda, Y. Kaneko, T. Mimura, H. Nakashima, N. Yamamoto, and T. Uryu, Carbohydr. Res., **276**, 425 (1995).
9. T. Uryu, N. Ikushima, K. Katsuraya, T. Shoji, N. Takahashi, T. Yoshida, K. Kanno, T. Murakami, H. Nakashima, and N. Yamamoto, Biochem. Pharm., **43**, 2385 (1992).
10. K. Katsuraya, T. Shoji, K. Inazawa, H. Nakashima, N. Yamamoto, and T. Uryu, Macromolecules, **27**, 6695 (1994).
11. K. Katsuraya, T. Shibuya, K. Inazawa, H. Nakashima, N. Yamamoto, and T. Uryu, Macromolecules, **28**, 6697 (1995).
12. G. B. Villanueva, J. Biol. Chem., **259**, 2531 (1984).
13. S. C. Alter, D. D. Metcalfe, T. R. Bradford, and . B. Schwartz, Biochem. J., **248**, 821 (1987).
14. P.J. B. Pereira, A. Bergner, S. Macedo-Ribeiro, R. Huber, G. Matschiner, H. Fritz, C. P. Sommerhoff, and W. Bode, Nature, **392**, 306 (1998).
15. T. Uryu, Y. Kaneko, T. Yoshida, R. Mihara, T. Shoji, K. Katsuraya, H. Nakashima, and N. Yamamoto, Carbohydrates and Carbohydrate Polymers, Ed. M. Yalpani, ATL Press, Mount Prospect, 1993, pp.101-115.
16. Y. Gao, A. Fukuda, K. Katsuraya, Y. Kaneko, T. Mimura, H. Nakashima, and T. Uryu, Macromolecules, **30**, 3224 (1997).
17. M. Gordon, M. Guralnik, Y. Kaneko, T. Mimura, M. Baker, and W. Lang,

J. Med., **25**, 163 (1994).

18. A. J. Nahmias, S. Kibrick, and P. Bernfeld, Proc. Soc. Exp. Bio. Med., **115**, 993 (1964).
19. A. Vaheri and K. Cantell, Virology, **21**, 661 (1963).
20. H. Voss, K. H. Sensch, and P. Panse, ZEntralbl. Baᴣerior. Orig. A, 229, 1 (1974).
21. U. Lindahl, G. Backstrom, and L. Thunberg, J. Bio. Chem., **258**, 9826 (1983).
22. K.-J. Jeon, K. Katsuraya, Y. Kaneko, T. Mimura, and T. Uryu, Macromolecules, **30**, 1997 (1997).
23. L. Ratner et al., Nature, **313**, 277 (1985).

HYDROCOLLOIDS – PART 2
Edited by K. Nishinari
2000 Elsevier Science B.V.

Applications of capillary electrophoresis for analysis of liposome dispersions

K. Kawakami, Y. Nishihara, and K. Hirano

Formulation R & D Laboratories, Shionogi & Co., Ltd., 12-4 Sagisu 5-chome, Fukushima-ku, Osaka 553-0002, Japan.

Capillary electrophoresis (CE) has been demonstrated to be a new powerful tool for investigating the various characteristics of liposomes, such as size distributions, compositional homogeneity of membranes, and membrane rigidity, etc. Our results imply that CE can be a useful technique for the quality control of liposomal products.

1. INTRODUCTION

CE [1] is a newly developed method which enables the separation of solutes by their size and amount of charges in a capillary which has a diameter of ca. 50 - 150 μm (Figure 1). The principle of the separation is based on free zone electrophoresis. Electrophoretic mobility of particles μ is given by [2]

$$\mu = f(\kappa a)\frac{2\varepsilon_r \varepsilon_o \zeta}{3\eta}, \tag{1}$$

where κ, a, $f(\kappa a)$, ε_r, ε_o, ζ and η are the Debye-Hückel parameter, the radius of particles, Henry's coefficient, the relative permittivity of continuum phase, the permittivity of a vaccum, the zeta-potential, and viscosity of the continuum phase, respectively. From this equation, we can speculate that the mobility increases with an increase of the incorporated charged components when the radius is fixed, and also, with the increase of radius when the amount of charges is fixed because Henry's coefficient is the increasing function from 1 to 1.5 with the increase of κa. The migration time t_m is expressed as

$$t_m = \frac{L}{E(\mu_e + \mu)}, \tag{2}$$

Figure 1. Schematic representation of CE separation for colloidal particles.

where L, E, and μ_e is the effective length of the capillary, the electric field, and the mobility of the electroosmotic flow, respectively. From these equations, it is apparent that the observed migration time gives us information about the amount of charges and the size of the liposomes.

We introduce here some possible applications of CE for the analysis of liposome dispersions. First, the size distribution of liposomes was investigated as the simplest application. This method has the possibility of being useful for investigating the bimodal size distributions, which have often been failed to be analyzed by the conventional light scattering method.

Second, we report the application of CE to detect the compositional homogeneity of liposome membranes [2], which has practical importance from the industrial viewpoint. The concept of this study is as follows. In the case of charged guest molecules distributed on monodispersed liposome membranes, the amount of charge on each particle (zeta-potential of each particle) is the only factor which controls the migration time in the analysis. Therefore, all particles would be detected as a single peak if the guest molecules are homogeneously dispersed on the membranes, i.e., the amount of charge per particle is equivalent. However, several distinguishable peaks will be observed if they are heterogeneously dispersed. If the guest molecules are noncharged, taking the signal ratio obtained at two different wavelengths would enable us to investigate the homogeneity if they have UV or visible light absorption. If they are homogeneously dispersed, the signal ratio obtained at two different wavelengths (one of them should be that at which maximal absorption of the guest molecules occurs) will show a constant value. However, this would not be the case with heterogeneous distribution. The difference in the dispersity of loaded guest molecules was successively distinguished.

The membrane rigidity was also observed to have a slight but reproducible effect on the migration time [3]. It is likely to be related to the deformation of the particle shape. Since the experimental observation of the membrane rigidity has not been reported very frequently, this technique can be a powerful tool for such investigations.

2. MATERIALS AND METHODS

2.1. Materials

Didecanoylphosphatidylcholine (DC$_{10}$PC), dimyristoylphosphatidylcholine (DMPC), dipalmitoylphosphatidylcholine (DPPC), dipalmitoylphosphatidylglycerol (DPPG), and distearoylphosphatidylglycerol (DSPG) were purchased from Nippon Fine Chemicals. Cholesterol was supplied from Nacalai Tesque. Vitamin E acetate (VE) was obtained from Shionogi & Co. All chemicals were at least of reagent grade and used as supplied.

2.2. Preparation of liposomes

Liposome dispersions were prepared by the freeze-dry method in which weighed membrane constituents were dissolved in *tert*-butyl alcohol above the phase transition temperature of phospholipids. The alcohol solution was freeze-dried, and the obtained powder was dispersed in 10 mM phosphate buffer (pH 8.0), also above the temperature. The size of the liposome particles was controlled with Extruder (Biomembranes) using polycarbonate membranes, except that small liposomes of 20 nm in diameter were obtained by using Nanomizer (Sayama Trading). For the investigation of the compositional homogeneity, the direct-dispersal method was employed as an alternative one, which was done by dispersing all components directly in buffer solution with heating, followed by the same process of sizing treatment. Size distributions were characterized by laser light scattering method (Coulter N4 plus).

2.3. CE analysis

CE analysis was performed by Hewlett Packard 3D CE system using a fused silica capillary column (50 μm ID×56 cm). The running buffer was the same as the solution buffer. The precondition was performed before each run, which consisted of 20 minutes flushing with buffer at 60°C, 2 minutes with methanol at 35°C, and 2 minutes with 0.1 M NaOH at 35°C. These steps were planned to prevent the column from clogging although this did not often occur. Injection (from cathode side) was done by the constant pressure method (50 mbar) for 10 seconds unless otherwise mentioned, followed by pushing inward with the buffer for 100 seconds at the same pressure. The voltage was set at 30 kV and the temperature was at 25°C. Voltage, current, and temperature were monitored during the measurement to confirm that they all remained at constant values.

3. RESULTS

3.1. Effect of size and charges on the migration time

Theoretically, the migration time will be shorter when the amount of charges is less for the negatively charged particles, or the size of the particle is smaller. In Figure 2 and 3, the effect of charges and diameter of liposomes on the migration time is presented. The diameter was confirmed to have nearly monodisperse distributions for all samples. In these figures, the dependence on the injection volume is also shown. For both conditions of the injection volume, the effect of the charges agreed well with the theory except that there was a dependence on the injection volume, and this also seemed to be the case for the size dependence.

3.2. Size distributions

From the results mentioned above, we can expect that the size distributions of liposomes may be analyzed by CE measurement, if the charged components are homogeneously

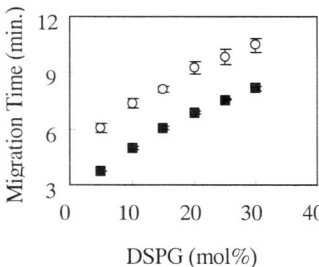

Figure 2. Dependence of the proportion of charged molecules (DSPG) on migration time. Diameter was adjusted to 100 nm. ○ : 30 sec. injection, ■ : 10 sec. injection.

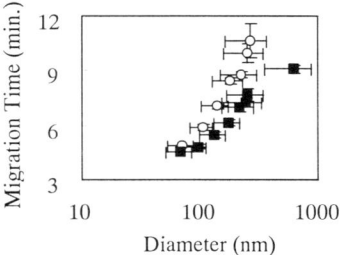

Figure 3. Dependence of the diameter of liposomes on migration time. DPPC/DSPG = 10/1 in molar ratio. ○ : 30 sec. injection, ■ : 10 sec. injection.

308

distributed. In Figure 4, we present the electropherograms of liposome dispersions which have diameters of 100 nm and 200 nm, and that of the equimolar mixture of those dispersions. As can be seen, the electropherogram of the mixture is approximately the summation of peaks of the monodisperse samples, suggesting that CE analysis can be a powerful tool for investigating the size distribution of liposomes as implied by Roberts et al. [4]

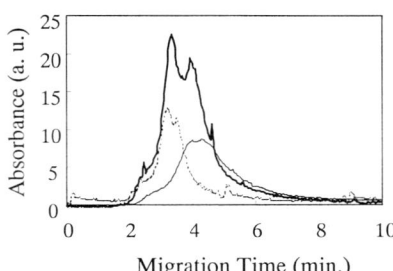

Figure 4. Electropherograms of liposomes (DPPC/DSPG = 10/1), which have diameter of 100 nm (---), 200 nm (—), and the equimolar mixture of these liposomes (—). Temperature was 50 °C.

3.3. Compositional homogeneity of membranes

Figure 5 presents the electropherograms of the liposome dispersions composed of DMPC/DPPG (DPPG is regarded as the guest molecule here) prepared by two different methods. Signal detection was done at 220 nm. The diameter was controlled at 20 nm because smaller liposomes showed sharper peaks. As can be seen, the freeze-dried sample presented a single peak, while the directly-dispersed one showed a collection of several broad peaks. Since the diameter of both samples was confirmed to be monodisperse, this difference must be due to the difference in compositional homogeneity, i.e., differences in the amount of charge on the individual particles.

As the noncharged guest molecule, VE, which is known as a highly oil-soluble drug, was loaded on DMPC/DPPG membranes. To avoid charge heterogeneity, lipids (DMPC and DPPG) were freeze-dried with *tert*-butyl alcohol in advance even for the directly-dispersed sample. The wavelengths selected for detection were 220 nm and 285 nm at which maximal absorption of VE was attained. The diameters of the particles were adjusted to 200 nm because broader peaks make the analysis easier. Figure 6 shows the signal ratio of two wavelengths from the two types of liposome dispersions. Clearly, the directly-dispersed sample gave a more disordered signal ratio than the freeze-dried one, suggesting that VE molecules are heterogeneously dispersed by the direct-dispersal method.

3.4. Dependence on membrane composition

Another finding was that the migration time depended on the membrane composition.

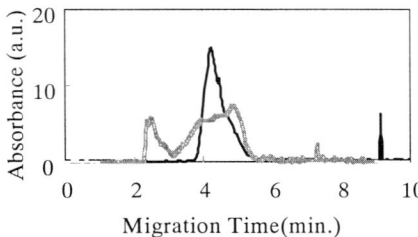

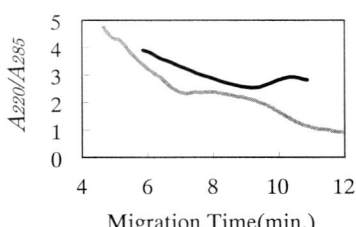

Figure 5. Electropherogram of DMPC/DPPG(16/3) liposome, prepared by two methods: —, freeze-dry. —, direct-dispersal.

Figure 6. Signal ratio of DMPC/DPPG/VE (16/3/1) liposome, prepared by two methods: —, freeze-dry. —, direct-dispersal.

Table 1 shows the observed migration times when various compositions were selected. As can be seen, the addition of cholesterol to DPPC/DSPG liposomes made the migration time longer in gel state, and shorter in liquid crystal state. This result may be elucidated in terms of the well-known altering effect of cholesterol on membrane rigidity. When the lipid membrane was softened, it seemed that the deformation of the spherical shape of the liposomes occurred and therefore the mobility of the liposome particles decreased. Addition of the short chain lipids (DC$_{10}$PC), which is also known to soften membranes [5], made the migration time longer, as expected. This was also observed when the chain length of all the constituent lipids was shortened.

Table 1. Migration time of various liposomes

Composition (molar ratio)	t_m (25°C)	t_m (50°C)
DPPC/DSPG = 10/1	5.09	3.00
DPPC/DSPG/Cholesterol = 8/1/2	6.35	2.77
DPPC/DSPG/DC$_{10}$PC = 10/1/0.5	6.75	3.25
DMPC/DPPG = 10/1	5.92	3.11

Diameter was 100 nm for all liposomes.

4. DISCUSSION

4.1. Behavior of liposomes in the capillary

Although it was shown that the behavior of liposomes in the capillary can be approximately interpreted as that of ideal colloidal particles, some significant differences were also shown to exist. Equation 1 reveals that the difference in the electrophoretic mobility of liposomes should be 1.5 times at most when the diameter is altered, however, it can be estimated from the experimental results that it is at least 2.1 times even in the diameter range presented in Figure 3. Therefore, there exists another reason of reducing the mobility of larger liposomes. Since we observed broadening of the peak widths (data not shown) as the diameter increased, it is very likely that disruption of the plug flow occurs due to occupation of the large cross-sectional area by liposomes. This assumption also can explain why a large amount of injection volume led to long migration times.

4.2. Size distributions

The most common technique for measuring the size distribution of colloidal particles has been the light scattering method. However, interpretation of the obtained data sometimes fails when the distribution is bimodal. On the other hand, since CE analysis is based on a very simple principle, it offers a great possibility of being a very powerful tool for measuring the size distributions. Actually, the bimodal sample used for the CE analysis shown in Figure 4 was analyzed as monodisperse by the light scattering method.

4.3. Compositional homogeneity of membranes

The most frequently used method for determining compositional homogeneity is differential scanning calorimetry. However, there are limitations to its application mainly due to the small transition enthalpy of lipids. Although fluorescent probes can also be used to evaluate several kinds of homogeneity, it is suspect that probes themselves are distributed homogeneously in some cases, and this method cannot be adopted for quality control because there are few cases in which guest molecules in the products can be treated as fluorescent probes.

In the previous section, it was presented that CE analysis could detect the compositional homogeneity for both charged and noncharged guest molecules. In the case of charged molecules distributed, the difference of the electropherograms for homogeneous and heterogeneous distribution was clear. However, in the case of noncharged, molecules distributed, we need a quantitative discussion. Quantitative evaluation of the degree of the disorder can be done by calculating the variance S^2 by

$$S^2 = \frac{\int A(t)(R_{ave} - R(t))^2 dt}{\int A(t)dt},$$

(3)

where $A(t)$, $R(t)$, and R_{ave} were the signal obtained at 220 nm, the signal ratio (A_{220}/A_{285}), and the averaged value of signal ratio, respectively. The calculated values were reproducible for the freeze-dried sample although the shapes of the signal ratio were not always the same. Generally, freeze-dried samples gave smaller values than 0.1, while directly-dispersed samples always gave larger values than 0.3. Therefore, it is quantitatively shown that the directly-dispersed sample has more heterogeneous membrane than the freeze-dried one. Some extent of the disorder in the signal ratio observed for the freeze-dried sample is also attributed to a certain degree of heterogeneity, since the VE-free sample gave completely constant signal ratio (data not shown).

4.4. Other applications

Another possible application is the evaluation of membrane rigidity from the observed migration time as already shown. Experimental results revealed that the "softer" liposomes require a longer migration time, which is likely to be elucidated in terms of the deformation of the shape of liposomes. We also investigated the temperature-dependence on the migration time and found that it was not proportional to the medium viscosity and the discontinuous point was observed near the phase transition temperature (data not shown), suggesting that it may be possible to obtain information about the phase transition behavior. (This was not observed for polystyrene standard particles.) In addition, at higher temperature, the liposome structure seems to have been destroyed. This was likely to be due to the pulling out of the charged lipids, suggesting that it may also be possible to evaluate the self-assembly energy of the bilayer structure. The biggest challenging problem to realize these studies seems to be the improvement in quantitative reproducibility.

REFERENCES

1. L.A. Holland, N.P. Chetwyn, M.D. Perkins, S.M. Lunte, Pharm. Res., 14 (1997) 372.
2. K. Kawakami, Y. Nishihara, K. Hirano, J. Colloid Interface Sci., 206 (1998) 177.
3. K. Kawakami, Y. Nishihara, K. Hirano, Langmuir, in press.
4. M.A. Roberts, L.L. Brown, W.A. MacCrehan, R.A. Durst, Anal. Chem., 68 (1996) 3434.
5. J. Lemmich, K. Mortensen, J.H. Ipsen et al. Eur. Biophys. J., 25 (1997) 293.

HYDROCOLLOIDS – PART 2
Edited by K. Nishinari
2000 Elsevier Science B.V.

Conformation and solution properties of water-soluble polysaccharides: case study of hyaluronic acid

T. Norisuye

Department of Macromolecular Science, Osaka university
Toyonaka, Osaka 560-0043, Japan

Chain-stiffness and excluded-volume effects in solutions of hyaluronic acid (sodium salt) are discussed by analyzing data for the mean-square radius of gyration $<S^2>$, the scattering function, and the intrinsic viscosity $[\eta]$ in aqueous NaCl with different salt concentrations C_s at 25°C on the basis of current theories for the wormlike chain with or without excluded volume. For molecular weights lower than 1.2×10^4 these properties are consistently explained by the unperturbed wormlike chain. Intramolecular excluded-volume effects begin to appear, almost regardless of C_s, when the molecular weight exceeds 1.5×10^4. The quasi-two-parameter theory for volume effects in nonionic chains is shown to be applicable to $<S^2>$ and $[\eta]$ of the charged polysaccharide unless C_s is lower than 0.01 M. The estimated persistence length, which is 4 nm at infinite ionic strength, increases with lowering C_s more remarkably than predicted by existing theories for the electrostatic persistence length, while the excluded-volume strength is in reasonable agreement with the Fixman-Skolnick theory.

1. INTRODUCTION

The global conformation and properties of a polyelectrolyte with intrinsically weak stiffness in solution are influenced greatly by electrostatic stiffening and excluded-volume effects. When the polyelectrolyte is modeled by the Kratky-Porod wormlike chain [1], a special case of Yamakawa's helical wormlike chain [2], the former effect is predicted to appear as an increase in the persistence length with lowering ionic strength [3,4], but its experimental estimation generally requires knowledge about the latter effect. Thus the electrostatic excluded-volume problem, which remains unsettled, is most basic to the characterization of charged poly-saccharides in solution.

This paper discusses excluded-volume effects in aqueous sodium chloride solutions of hyaluronic acid (sodium salt) by summarizing our recent studies [5-8] on the conformation and properties of the charged polysaccharide. It focuses on the applicability of the quasi-two-parameter (QTP) theory for volume effects in wormlike or helical wormlike chains to the mean-square radius of gyration $<S^2>$ and the intrinsic viscosity $[\eta]$ in aqueous NaCl with different salt concentrations C_s at 25°C. According to recent experimental studies [2,9], this theoretical scheme allows an almost quantitative, consistent description of excluded-volume effects on $<S^2>$ and $[\eta]$ of nonionic, linear polymers, both flexible and stiff, so that its applicability to ionic chains deserves to be investigated by experiment. Detailed analyses of these properties and the particle scattering function $P(\theta)$ based on the wormlike chain are presented below. The electrostatic contributions to the persistence length and the excluded-volume strength are also discussed in relation to theoretical predictions [3,4,10-12].

2. MATERIALS AND METHODS

Hyaluronate samples (Shiseido Co., Japan) were degraded to different molecular weights by heating at 110°C for 25 min - 48 h, and some of the degraded samples were hydrolyzed according to the method of Cleland [13]. The resulting samples and the original one with $[\eta]$ (in 0.2 M aqueous NaCl at 25°C) of 10 to 2560 cm^3 g^{-1} were extensively fractionated by repeated fractional precipitation [5,7,8]. From a number of fractions thus prepared, appropriate middle were chosen and converted to the Na salt form.

Light scattering intensities were measured for high molecular weight fractions in 0.02, 0.1, and 0.5 M aqueous NaCl at 25°C on a Fica-50 light-scattering photometer with vertically polarized incident light of 436 or 546-nm wavelength to determine M_w (the weight-average molecular weight), A_2 (the second virial coefficient), and $<S^2>_z$ (the z-average $<S^2>$). Values of M_w for lower molecular weight fractions were determined by sedimentation equilibrium with 0.5 M aqueous NaCl at 25°C as the solvent. Ratios of the z-average to weight-average molecular weight, M_z/M_w, were also estimated from equilibrium concentration profiles to be 1.1 ± 0.1.

Small-angle X-ray scattering (SAXS) experiments were performed for four fractions with M_w of 3.8 x 10^3 - 1.1 x 10^4 in 0.02 and 0.1 M aqueous NaCl at 25°C using a new SAXS apparatus equipped with an imaging plate detector [14]. Intensity data at fixed scattering angles θ were extrapolated to infinite dilution with the aid of the square-root plot to obtain $P(\theta)$.

Intrinsic viscosities at zero shear rate were determined for 17 fractions in aqueous NaCl with 10 different salt concentrations of 0.005 to 2.5 M at 25°C using a four-bulb low-shear capillary viscometer of the Ubbelohde type or conventional capillary viscometers. They were considered free from electroviscous effects for the reason mentioned elsewhere [6].

3. RESULTS

Figure 1 shows $<S^2>_z$ data from light scattering and SAXS for Na hyaluronate samples in 0.02 M aqueous NaCl as a double-logarithmic plot against M_w. The two data sets are fitted by a smooth curve whose slope decreases from about 1.6 (for M_w < 10^4) to 1.26 (for M_w > 2 x 10^5) with increasing M_w. This change in slope indicates that the polysaccharide chain in the aqueous salt has an unmistakable semiflexibilty [15].

Figure 2 shows the C_s dependence of $[\eta]$ for Na hyaluronate samples with different molecular weights (data for some samples are omitted here for clarity). The curves fitting the

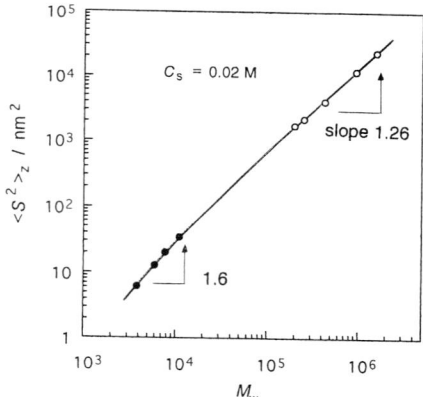

Figure 1. Molecular weight dependence of $<S^2>_z$ from SAXS (filled circles) and light scattering (unfilled circles) for Na hyaluronate in 0.02 M aqueous NaCl at 25°C.

plotted points for high molecular weights rise with decreasing C_s, but they flatten as M_w decreases. The molecular weight dependence of $[\eta]$ at fixed C_s is illustrated in Figure 3. The data points at each C_s follow a curve which is slightly convex upward for $M_w < 10^4$ and $M_w > 2 \times 10^5$ and almost linear in between. The slope in this intermediate M_w range is 0.84 at $C_s = 2.5$ M and 1.14 at $C_s = 0.005$ M.

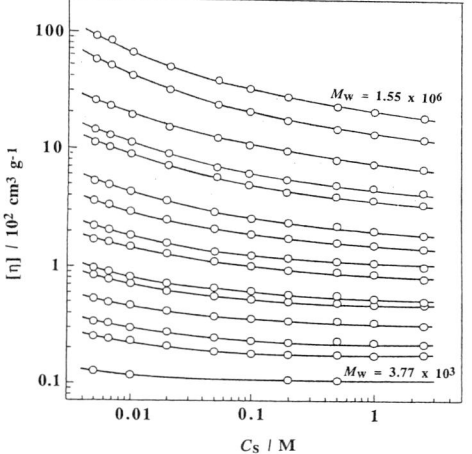
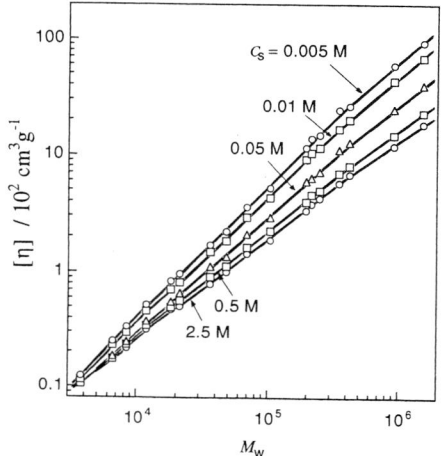

Figure 2. Dependence of $[\eta]$ on C_s for Na hyaluronate samples with different molecular weights in aqueous NaCl at 25°C.

Figure 3. Molecular weight dependence of $[\eta]$ for Na hyaluronate in aqueous NaCl with indicated C_s at 25°C.

4. DISCUSSION

4.1. Analysis of Viscosity Data

We analyze the $[\eta]$ data in Figure 2 on the basis of the Yamakawa-Fujii-Yoshizaki (YFY) theory [16,17] for the intrinsic viscosity $[\eta]_0$ of an unperturbed wormlike chain combined with the QTP theory [2] for excluded-volume effects in nonionic wormlike or helical wormlike (HW) bead chains. The former theory contains three parameters, L (the contour length of the chain), λ^{-1} (the Kuhn segment length or more generally the stiffness parameter in the HW chain [2]), and d (the hydrodynamic diameter). The first parameter is related to the molecular weight M by $L = M/M_L$, with M_L being the molar mass per unit contour length.

The QTP theory predicts that the cube of the viscosity expansion factor, α_η^3 ($\equiv [\eta]/[\eta]_0$), for nonionic linear chains is a universal function of the scaled excluded-volume parameter $\bar{z}$ defined by

$$\bar{z} = (3/4)K(\lambda L)z \tag{1}$$

with

$$z = (3/2\pi)^{3/2}(\lambda B)(\lambda L)^{1/2} \tag{2}$$

Here, z is the conventional excluded-volume parameter, $K(\lambda L)$ is a known function of λL [2], and B is the excluded-volume strength for the interaction between a pair of beads. In the coil limit where λL is infinitely large, $K(\lambda L)$ becomes equal to 4/3 and $\bar{z}$ reduces to z. On the other hand, for $\lambda L < 1$, $K(\lambda L) \approx 0$ and hence $\tilde{z} \approx 0$ even for nonzero B. In the wormlike chain limit of the HW chain we have [2]

$$\lambda^{-1} = 2q \tag{3}$$

$$B = \beta/a^2 \tag{4}$$

where q is the persistence length, β the binary cluster integral, and a the bead spacing. If the Barrett function [18] is adopted, α_η^3 in the QTP scheme reads

$$\alpha_\eta^3 = (1 + 3.8\tilde{z} + 1.9\tilde{z}^2)^{0.3} \tag{5}$$

As outlined above, $[\eta]$ for a given M is determined by M_L, q, d, and B. Since it is insensitive to q and B for very low M, we first determined M_L and d together with a first approximation value of q from our $[\eta]$ data for low molecular weight samples and Cleland's oligomer data [13,19] in 0.2 M aqueous NaCl using the YFY theory. With $M_L = 405$ nm^{-1} and $d = 1.0$ thus obtained, we then searched for a set of q and B that allows the YFY theory with eq 5 to give the closest fit to the $[\eta]$ data in the entire M_w range studied. Excluded-volume effects became appreciable when M_w exceeded 1.5 x 10^4.

The $[\eta]$ data at other salt concentrations were similarly analyzed, but in actuality, the d of 1.0 nm determined above at $C_s = 0.2$ M was assumed to be applicable to any C_s. We found that excluded-volume effects become significant at an M_w of about 1.5 x 10^4, almost regardless of C_s.

4.2 Molecular Parameters and Conformation

The wormlike-chain parameters and the excluded-volume strength obtained from the above analysis are summarized in Table 1. The M_L values are all in the range between 400 and 410 nm^{-1}. With the molar mass (401 g mol^{-1}) of the repeating unit of Na hyaluronate, they yield 1.0 nm for the monomeric length h along the chain contour. This h value is consistent with the chemical structure of the polysaccharide [13], indicating that the wormlike chain is a good model for Na hyaluronate in the C_s range studied. The other parameters, q and B, are seen to increase monotonically with lowering C_s. This implies that both stiffness and volume effects are responsible for the increases in $[\eta]$ with decreasing C_s observed for high molecular weight samples in Figure 3. In the region of C_s higher than 0.2 M, however, q hardly changes with C_s while B increases with decreasing C_s. Thus, in this C_s region, the observed viscosity increases may be ascribed primarily to excluded-volume effects. The q value of about 4 nm at such high ionic strength indicates that the intrinsic stiffness of the hyaluronate chain is relatively low [15].

Table I. Wormlike-chain parameters and excluded-volume strength for Na hyaluronate in aqueous NaCl at 25°C

C_s/M	M_L/nm^{-1}	q/nm	B/nm	C_s/M	M_L/nm^{-1}	q/nm	B/nm
0.005	400	9.7	19	0.1	410	4.8	4.8
0.0067	400	8.5	17	0.2	405	4.2	4.0
0.01	400	7.4	13	0.5	400	4.1	2.8
0.02	405	6.0	9.0	1	400	4.0	2.3
0.05	410	5.0	7.0	2.5	405	4.0	1.6

Another important finding from the analysis of $[\eta]$ is that the hyaluronate chain (at $C_s \leq$ 0.005 M) is essentially unperturbed by excluded-volume effect if M_w is lower than 1.5 x 10^4.

We found that this is consistent with the $<S^2>_z$ data ($3.8 \times 10^3 < M_w < 1.2 \times 10^4$) from SAXS at $C_s = 0.02$ and 0.1 M. In fact, these data at the respective C_s agreed (within experimental error) with the unperturbed mean-square radii of gyration calculated for the wormlike chains (see eq 6) with the q and M_L values in Table 1, though somewhat systematically larger than the calculations probably because of samples' slight polydispersity [8]. The unperturbed wormlike chain behavior of low molar mass hyaluronate chains is further confirmed by $P(\theta)$ as described below.

In panels (a) and (b) of Figure 4, the experimental scattering functions (the circles) for a hyaluronate sample with $M_w = 1.10 \times 10^4$ ($L = 27.2$ nm) at $C_s = 0.02$ M and for a sample with $M_w = 5.92 \times 10^3$ ($L = 14.4$ nm) at $C_s = 0.1$ M, both in the form of the Kratky plot, are compared with the theoretical curves (the solid lines) for the unperturbed wormlike chains [20] with the q and M_L values in Table 1; the effect of chain thickness on $P(\theta)$ was negligible in the range of k (the magnitude of the scattering vector) studied [8]. The solid curves are seen to come close to the data points in the respective panels. Thus, the q and M_L values estimated from $[\eta]$ are consistent with the $P(\theta)$ data.

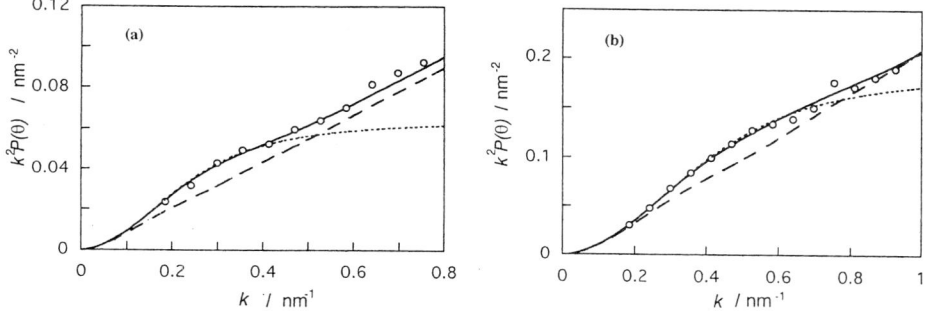

Figure 4. Comparison between the theoretical and experimental scattering functions for Na hyaluronate samples with $M_w = 1.10 \times 10^4$ in 0.02 M aqueous NaCl at 25°C (a) and $M_w = 5.92 \times 10^3$ in 0.1 M aqueous NaCl at 25°C (b). The solid line in each panel represents the theoretical values [20] for the unperturbed wormlike chain with the parameters in Table 1. See the text as for the dotted and dashed lines.

The dotted curve in either panel of Figure 4 represents the Debye function for the Gaussian chain [21] with the same $<S^2>$ as that of the wormlike chain under consideration and the dashed one, the scattering function for the thin rod [21] with the same L as that of the wormlike chain. We see in panel (a) that, as k increases, the experimental scattering function continuously changes from the Gaussian to rodlike behavior. This is a typical feature of the unperturbed wormlike chain. Similar behavior was also found for the same sample ($M_w = 1.10 \times 10^4$) at $C_s = 0.1$ M. As M_w lowered, the approach to the asymptotic rod behavior at both 0.02 and 0.1 M became less definitive in the k range examined. Panel (b) shows an example of this. In conclusion, our data of $<S^2>_z$, $P(\theta)$ and $[\eta]$ for Na hyaluronate samples with $M_w < 1.2 \times 10^4$ in 0.02 and 0.1 M aqueous NaCl are all consistently explained by the unperturbed wormlike chain.

4.3 Comparison between Theory and Experiment

In Figure 5, our $[\eta]$ data at the indicated salt concentrations are compared with the YFY theory with eq 5 for the perturbed wormlike chains with the parameters in Table 1. The

316

theoretical curves appear somewhat above the data points for M_w between 3×10^4 and 1.5×10^5, but except for $C_s = 0.005$ M, the general agreement is fairly good.

At $C_s = 0.005$ M, the theoretical curve has a pronounced downward curvature, whereas the experimental log $[\eta]$ vs. log M_w relation at the same C_s is almost linear with a viscosity exponent of 1.14 for M_w between 1×10^4 and 2×10^5, as already mentioned (see Figure 3). This difference results in an appreciable upward deviation of the theoretical curve (for the chosen parameters) from the data points in the M_w range between 3×10^4 and 1.5×10^5, the deviation amounting to 10 - 20%. The contribution from $[\eta]_c$ to the exponent 1.14 is about 0.79 when estimated from the YFY theory with the parameters in Table 1. The difference of 0.35 in exponent, which comes from excluded-volume effects, exceeds the asymptotic value 0.30 predicted by eq 5. A similar argument applies to $C_s = 0.0067$ M, at which the agreement between the theoretical and experimental $[\eta]$ was not good either. Thus the high experimental exponents for $[\eta]$ in a limited M_w range from 3×10^4 to 1.5×10^5 at $C_s < 0.01$ M seem to be responsible for the poor agreements between the QTP theory and the data (see ref. 7 for further discussion).

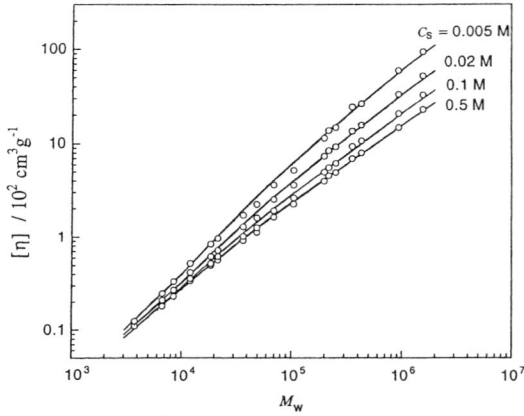

Figure 5. Comparison between the measured $[\eta]$ at the indicated C_s and the theoretical curves computed from the YFY theory [16,17] for unperturbed wormlike chains and eq 5 with $d = 1.0$ nm and the parameters in Table 1.

The radius expansion factor α_s ($= (<S^2>/<S^2>_0)^{1/2}$ is also a function only of $\bar{z}$ in the QTP scheme, where $<S^2>_0$ (the unperturbed mean-square radius of gyration) for the wormlike chain is given by [22]

$$<S^2>_0 = (qL/3) - q^2 + (2q^3/L) - (2q^4/L^2)[1 - \exp(-L/q)] \qquad (6)$$

Adopting the Domb-Barrett function [23] for α_s^2, we have

$$\alpha_s^2 = [1 + 10\bar{z} + (\tfrac{70\pi}{9} + \tfrac{10}{3})\bar{z}^2 + 8\pi^{3/2}\bar{z}^3]^{2/15}[0.933 + 0.067 \exp(-0.85\bar{z} - 1.39\bar{z}^2)] \qquad (7)$$

In Figure 6, the theoretical curves computed from eqs 6 and 7 with the parameters in Table 1 are compared with our light scattering $<S^2>_z$ data at $C_s = 0.02$, 0.1, and 0.5 M, where the theoretical $<S^2>_0$ has been replaced by its z-average calculated by use of the z-average L (= M_z/M_L) with $M_z/M_w = 1.1$ (the average obtained from sedimentation equilibrium); we note that $<S^2>_0$ is determined substantially by the first two leading terms of eq 6 for large L/q with which the present analysis is concerned. All the curves come close to the data points for the respective salt concentrations. Though the slope 1.21 of the line for $C_s = 0.02$ M is appreciably smaller than that (1.26) of the corresponding experimental relation, the differences between the theoretical and experimental $<S^2>_z$ values are very small in the molecular weight range studied here. Hence, we may conclude that with the parameters derived from $[\eta]$, the combination of eqs 6 and 7 describes the $<S^2>_z$ data for Na hyaluronate in the three aqueous salts. In other words, the QTP theory is applicable to $<S^2>_z$ and $[\eta]$ of the charged polysaccharide in aqueous NaCl with $C_s \leq 0.02$ M.

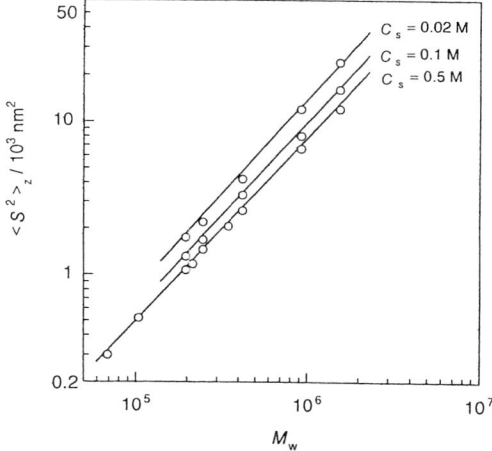

Figure 6. Comparison between the experimental $<S^2>_z$ at the indicated C_s and the theoretical curves calculated from eqs 6 and 7 with the parameters in Table 1.

4.4. Electrostatic Contribution to the Persistence Length

According to Odijk [3] and Skolnick and Fixman [4], q of a charged wormlike chain is the sum of q_0 (the intrinsic persistence length) and q_{el} (the electrostatic persistence length), i.e., $q = q_0 + q_{el}$. These groups independently derived an analytical expression for q_{el} of a thin wormlike chain with line charge distribution on the basis of the Debye-Hückel approximation. More elaborate analyses were made by Le Bret [10] and Fixman [11], who computed q_{el} for a wormlike cylinder with radius r, dielectric constant ε, and smeared charge distribution (on the surface), using the complete Poisson-Boltzmann equation. The results are tabulated in Le Bret's paper as a function of σ (the linear charge density) and κr for $\varepsilon/\varepsilon_0 = 0$ and ∞. Here, ε_0 is the dielectric constant of water and κ is the reciprocal of the Debye screening length defined by $\kappa^2 = 8\pi Q N_A C_s/1000$ for aqueous 1 - 1 electrolyte, with Q and N_A being the Bjerrum length (0.714 nm for water at 25°C) and the Avogadro constant, respectively; C_s is expressed in units of M. All the above theories predict q_{el} to be a decreasing function of C_s and vanish at infinite ionic strength, so that q at this limit equals q_0. From the data of q in Table 1 q_0 of Na hyaluronate (at infinite ionic strength) is estimated to be 4.0 nm. It may be regarded as a constant independent of C_s in first approximation.

In Figure 7, our q data are compared with the Le Bret theory [10] for $q_0 = 4.0$ nm, $\sigma = 1.0$ nm^{-1}, $\varepsilon/\varepsilon_0 = 0$, and $r = 0.5$ nm; note that σ is equal to $1/h$ since the repeating unit of Na

318

hyaluronate carries one charge. Except for the region of $C_s^{-1/2} < 5$ M$^{-1/2}$ in which q is determined substantially by q_0, the theoretical q_{el} values are distinctly larger than the experimental ones, the discrepancy being as large as 70% at $C_s \lesssim 0.02$ M. Interestingly, however, both theoretical and experimental q_{el}'s, when plotted double-logarithmically against C_s (not shown), are essentially parallel with a slope of - 0.6 for $C_s < 0.1$ M. This slope happens to be close to those (about - 0.5) estimated for several intrinsically flexible or weakly stiff polyelectrolytes [24,25] without consideration of excluded-volume effects.

The Odijk-Skolnick-Fixman (OSF) theory [3,4] predicts the C_s^{-1} dependence (or more generally the κ^{-2} dependence) of q_{el} differing from the Le Bret theory, but it (without ion condensation [26]) gives q values close to the Le Bret line, as indicated by a dashed curve in Figure 7; note that the charge parameter σQ for the hyaluronate chain is less than unity (about 0.7). Skolnick and Fixman's calculation [4] for a discrete charge model gives a curve that is indistinguishable from the dashed line (not shown here). A point to note is that the discrete charge distribution significantly lowers the OSF theoretical q_{el} only for C_s above 0.2 M where q of the hyaluronate chain is dominated by q_0. In short, all the above-mentioned theories fail to describe the present q_{el} data and appear to be only applicable to (long) intrinsically stiff chains [27].

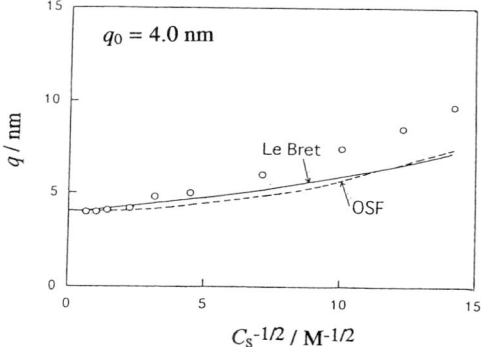

Figure 7. Experimental values (circles) for the total persistence length of Na hyaluronate in aqueous NaCl compared with the Le Bret theory [10] (the solid line) and the Odijk-Skolnick-Fixman theory [3,4] (the dashed line).

4.5. Excluded-Volume Strength

The circles in Figure 8 show that B for Na hyaluronate in aqueous NaCl increases almost linearly with increasing $C_s^{-1/2}$. If extended to infinite ionic strength, this linear relation yields a value of about 1 nm for the nonionic excluded-volume strength, which is one order of magnitude smaller than B at $C_s^{-1/2} \geq 7$ M$^{-1/2}$ (i.e., at $C_s \lesssim 0.02$ M). Thus, β (= a^2B) in this low C_s region is dominated by its electrostatic contribution β_{el}. As is well known, the conventional bead-bead interaction model in the Debye-Hückel approximation gives theoretical β_{el} values that are a few orders of magnitude larger than experimental estimates at such low ionic strength [12,28,29]. This is also the case for Na hyaluronate, for which the discrepancy amounts to one order of magnitude at $C_s = 0.005$ M [6].

Fixman and Skolnick [12], evaluating the excluded volume β' for a pair of rodlike segments with line charge distribution in the Debye-Hückel approximation, explained the early observed large discrepancies [28,29] in β_{el} as due to the direct application of the bead-bead interaction model to long-range charge interactions at low ionic strength. Since their theory was often used for evaluation of z in experimental studies of q_{el} [25,30,31], its comparison with our B values (at $C_s \leq 0.01$ M) is of interest.

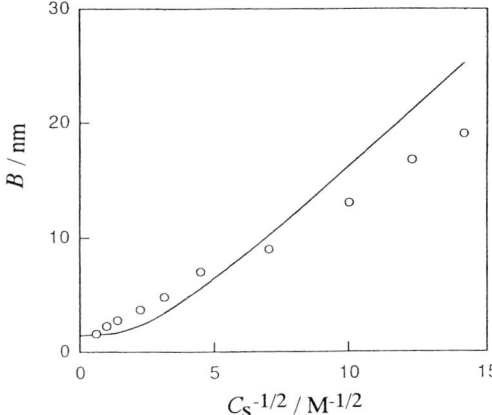

Figure 8. Values of the excluded-volume strength B for Na hyaluronate (circles) compared with the Fixman-Skolnick theory [12] (the curve) for $\sigma = 1.0$ nm^{-1} and $d_c = 1.0$ nm.

If β' for a pair of segments of length l, each consisting of n_0 beads ($l = n_0 a$), is equated to $n_0^2 \beta$ [12], it follows that $\beta/a^2 = \beta'/l^2$. The Fixman-Skolnick theory for β' is then expressed in terms of B as

$$B = (\pi/2)d_c + (2/\kappa)J(y) \tag{8}$$

where d_c is the hard core diameter of each rodlike segment and $J(y)$ is a known function of y defined by

$$y = 2\pi\sigma^2 Q\kappa^{-1} \exp(-\kappa d_c) \tag{9}$$

In eq 8, the first and second terms represent the hard core and electrostatic contributions, respectively, and the attractive contribution is assumed to be negligible. If y is greater than 3, $J(y)$ is approximated by $J(y) = (\pi/4)(\ln y + 0.7704)$ and B does not depend on d_c.

The curve in Figure 8 represents eq 8 with $\sigma = 1.0$ nm^{-1} and $d_c = 1.0$ nm. It comes close to the plotted points for $C_s^{-1/2}$ below 10 M$^{-1/2}$. Thus we find that the Fixman-Skolnick theory allows a fairly satisfactory description of α_η^3 for Na hyaluronate at $C_s \geq 0.01$ M if it is combined with the QTP theory.

5. CONCLUSIONS

Our data for $<S^2>_z$, $P(\theta)$, and $[\eta]$ for Na hyaluronate samples with $M_w < 1.2 \times 10^4$ in 0.02 and 0.1 M aqueous NaCl are all consistently explained by the unperturbed wormlike chain. Above $M_w = 1.5 \times 10^4$, intramolecular excluded-volume effects are significant. The present analysis shows that the quasi-two-parameter theory [2] is a fairly satisfactory approximation to both $<S^2>$ and $[\eta]$ of the charged polysaccharide unless C_s is lower than 0.02 M. Below $C_s = 0.01$ M, this theory combined with the Yamakawa-Fujii-Yoshizaki theory [16,17] fails to explain the experimental viscosity exponents in a limited range of molecular weight from 3 x 10^4 to 1.5 x 10^5. The electrostatic contribution to the persistence length determined from the analysis of the $[\eta]$ data is systematically larger than predicted by current theories [3,4,10,11], while the estimated values for the excluded-volume strength (or the binary cluster integral) are in moderate agreement with the Fixman-Skolnick theory [12] for a pair of charged rodlike segments.

REFERENCES

1. O. Kratky and G. Porod, *Rec. Trav. Chim.*, 68 (1949) 1106.
2. H. Yamakawa, Helical Wormlike Chains in Polymer Solutions, Springer, Berlin, 1997.
3. T. Odijk, *J. Polym. Sci: Polym. Phys. Ed.*, 15 (1977) 477.
4. J. Skolnick and M. Fixman, *Macromolecules*, 10 (1977) 944.
5. K. Hayashi, K. Tsutsumi, F. Nakajima, T. Norisuye, and A. Teramoto, *Macromolecules*, 28 (1995) 3824.
6. K. Hayashi, K. Tsutsumi, T. Norisuye, and A. Teramoto, *Polym. J.*, 28 (1996) 922.
7. K. Tsutsumi and T. Norisuye, *Polym. J.*, 30 (1998) 345.
8. N. Mizukoshi and T. Norisuye, *Polym. Bull.*, 40 (1998) 555.
9. T. Norisuye, A. Tsuboi, and A. Teramoto, *Polym. J.*, 28 (1996) 357.
10. M. Le Bret, *J. Chem. Phys.*, 76 (1982) 6243.
11. M. Fixman, *J. Chem. Phys.*, 76 (1982) 6346.
12. M. Fixman and J. Skolnick, *Macromolecules*, 11 (1978) 363.
13. R. L. Cleland, *Biopolymers*, 23 (1984) 647.
14. Y. Nakamura, K. Akashi, T. Norisuye, A. Teramoto, and M. Sato, *Polym. Bull.*, 38 (1997) 469.
15. T. Norisuye, *Prog. Polym. Sci.*, 18 (1993) 543.
16. H. Yamakawa and M. Fujii, *Macromolecules*, 7 (1974) 128.
17. H. Yamakawa and T. Yoshizaki, *Macromolecules*, 13 (1980) 633.
18. A. J. Barrett, *Macromolecules*, 17 (1984) 1566.
19. R. L. Cleland and J. L. Wang, *Biopolymers*, 9 (1970) 799.
20. T. Yoshizaki and H. Yamakawa, *Macromolecules*, 13 (1980) 1518.
21. See, for example, ref. 20.
22. H. Benoit and P. Doty, *J. Phys. Chem.*, 57 (1953) 958.
23. C. Domb and A. J. Barrett, *Polymer*, 17 (1976) 179.
24. M. Tricot, *Macromolecules*, 17 (1984) 1698.
25. S. Ghosh, X. Li, C. E. Reed, and W. F. Reed, *Biopolymers*, 30 (1990) 1101.
26. G. S. Manning, *J. Chem Phys.*, 51 (1969) 924.
27. See ref. 10.
28. I. Noda, T. Tsuge, and M. Nagasawa, *J. Phys. Chem.*, 74 (1970) 710.
29. M. Nagasawa and A. Takahashi, Light Scattering from Polymer Solutions, M. B. Huglin, Ed., Academic Press, New York, N. Y., 1972.
30. L. Wang and H. Yu, *Macromolecules*, 21 (1988) 3498.
31. E. Fouissac, M. Milas, M. Rinaudo, and R. Borsali, *Macromolecules,* 26 (1992) 5613.

HYDROCOLLOIDS – PART 2
Edited by K. Nishinari
2000 Elsevier Science B.V.

Rheological and related properties of hyaluronate solutions

A.Okamoto

Research Center, Denki Kagaku Kogyo Co., Ltd,
3-5-1 Asahi-cho, Machida-shi, Tokyo 194-8560, Japan

The rheological and related properties of synovial fluids from normal(n=61), osteoarthritis(n=72) and rheumatoid arthritis(n=36) human knee joints were investigated. The viscoelasticity of the synovial fluids was compared with buffer solutions of hyaluronates with various molecular weight. The curves of the storage and loss moduli of the synovial fluids were well fitted to those of buffer solutions and the crossover point of synovial fluids coincides well with the calculated value for the buffer solution with the same concentration of hyaluronate with the same molecular weight. The difference in the rheological and related properties of synovial fluids from young, old and pathological human knees will be discussed.

1. INTRODUCTION

Hyaluronate is a water soluble glycosaminoglycan consisting of repeating disaccharide units of D-glucronic acid and N-acetyl-D-glucosamine. It is a main component of the intercellular matrix of most connective tissues such as umbilical cord, synovial fluid, septal cartilage, etc. Especially for its prominent viscoelastic properties, hyaluronate acts as a lubricant and shock absorber in synovial fluid[1,2].

Until recently, hyaluronate was to be extracted from rooster combs, shark skins, bovine vitreous humor mainly for its use as moisturizing agent in cosmetic industry. Nowadays, it is produced via fermentation with *Streptococcus equi* and

S. zooepidemicus[3]. It was first used in clinical practice around 1970 for the treatment of joints in race horses and today intra-articular hyaluronate is given to patients with osteoarthritis. In 1980s hyaluronate also became a common device in intraocular surgery. Recently it has become apparent that hyaluronate is not at all an inert biological material[4].

The purpose of this work was to provide data of rheological and related properties of synovial fluids from normal, osteoarthritis and rheumatoid arthritis human knee joints being compared with buffer solutions of hyaluronates with various molecular weight.

2. MATERIALS AND METHODS

2.1. Materials

Synovial fluids were collected from normal(n=74), osteoarthritis(n=72) and rheumatoid arthritis(n=36) donors' knee joints at Kitasato University East Hospital.

Hyaluronate samples were purified after extraction from the culture medium of Streptococcus equi in the laboratory of Denki Kagaku Kogyo Co., Ltd. The hyaluronate solutions of different polymer concentrations were prepared by dissolving it in 2mM phosphate buffered saline solution at pH 7 stirring for 20 hours at about 25 ℃.

2.2. Dynamic Viscoelasticity and Intrinsic Viscosity

The dynamic viscoelastic properties of the fluids were measured at 37 ℃ with a CSL500 Controlled Stress Rheometer(Carri-med, England). The geometries used were cones of 4- and 6-cm diameters and cone angle of 2 ° with a solvent trap system from Carri-med. The dependence of the intrinsic viscosities($[\eta]$) of hyaluronates with different molecular weight on shear rates was measured using three types of viscometers: Zimm-Crothers type ultra-low shear rotational viscometer, two-bulb spiral capillary viscometer and two standard capillary viscometers of the Ubbelohde Type.

2.3. Molecular Weight and Concentration

The determination of the molecular weights of hyaluronate in synovial fluids was carried out at 40℃ in a Tosoh HLC-8120GPC(with a specially ordered column from Tosoh) connected in a series to a Wyatt DAWN DSP MALLS (multi-angle laser light scattering) photometer with a flow cell. The concentration of hyaluronate were determined by the use of a calibration curve from hyaluronate solutions with known concentration and corrected by subtracting the concentration of contaminating proteins in the synovial fluids which was estimated from the values (reducing as albumin) of absorbance and refractive index using a spectrophotometer and a differential refractometer connected to the GPC system.

3. RESULTS

3.1. Molecular weight dependence

Figure 1 depicts the shear rate dependence of $[\eta]$ for hyaluronate samples tested in 2mM phosphate buffered saline solution at 25 ℃. With increasing molecular weight, $[\eta]$ depends more strongly upon the shear rate. The values of $[\eta]$ at zero shear ($[\eta]_0$) of the hyaluronate solutions were determined from the Newtonian plateau. Figure 2 shows the relation between $[\eta]_0$ and Mw for the hyaluronate solution. Data points fit well with a straight line expressed by the Mark-Houwink equation reported by Laurent[5].

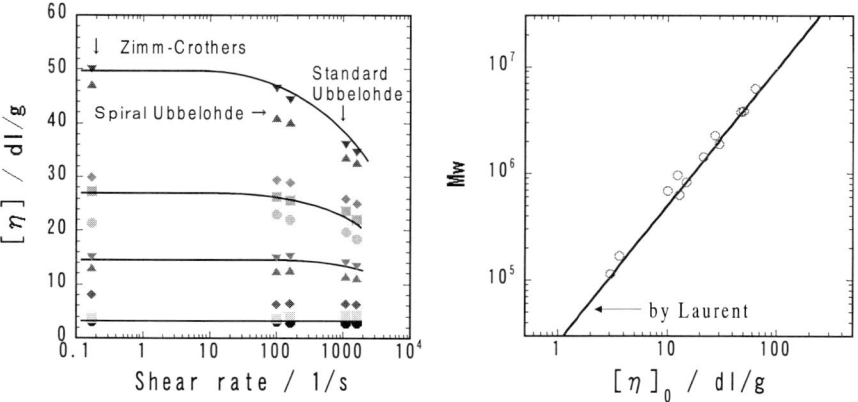

Figure 1 Shear rate dependence of $[\eta]$

Figure 2　Relation between $[\eta]_0$ and Mw

The frequency dependence of storage modulus, G', and loss modulus, G", for 1% phosphate buffered saline solution of hyaluronate with different molecular weight are shown in Figure 3. The shape of the curves was observed to be independent of the molecular weight. Hence, all of the curves at a given molecular weight could be shifted in a way as to superpose following the method of reduced variable(see Figure 4). The value assigned to Mw_0 was 2240000. The frequency shift factor, a_M, may be regarded as empirically evaluated constant[6].

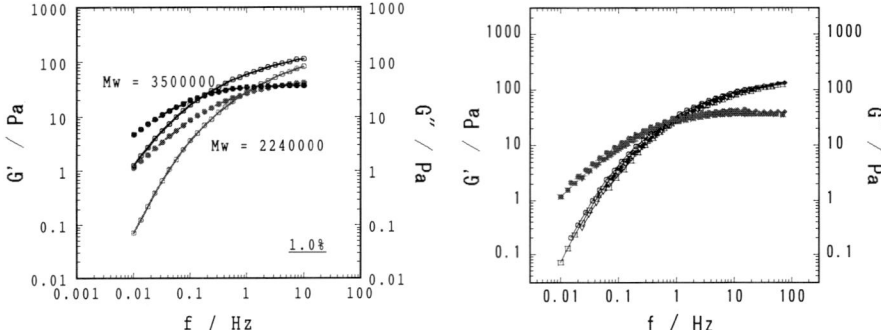

Figure 3   Dynamic viscoelasticity of HA solutions with different molecular weights

Figure 4   Master curves of dynamic viscoelasticity using the shift factor a_M : Mw = 2240000, 3500000, 2490000, 2680000, 3820000, 3860000

3.2. Concentration dependence

The frequency dependence of storage modulus, G', and loss modulus, G", for phosphate buffered saline solutions of different concentration of hyaluronate with molecular weight 2240000 are shown in Figure 5. All of the curves at a given concentration could be superposed upon the curves at $1\%(C_0)$ using the frequency shift factor, a_c, and the modulus shift factor, b_c (see Figure 6).

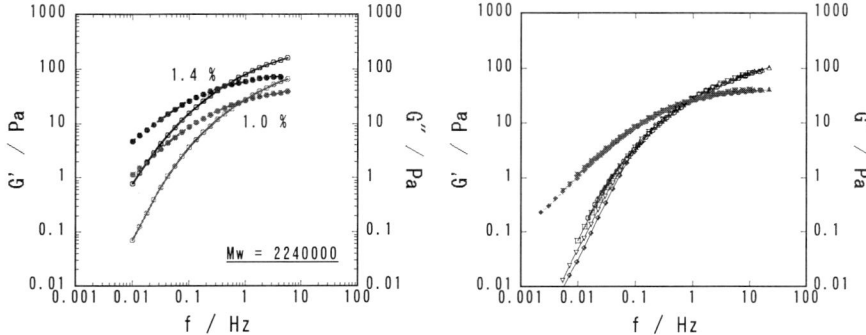

Figure 5   Dynamic viscoelasticity of HA solutions with different concentrations

Figure 6 Master curves of dynamic viscoelasticity using the shift factors **a** c and **b** c : concentration = 0.6, 0.8, 1.0, 1.2, 1.4 %

3.3. Synovial fluids

The frequency dependence of storage modulus, G', and loss modulus, G", for synovial fluids from normal(young), normal(representative) and osteoarthritis(representative) human knee joints is shown in Figure 7.   As the frequency increases, the curves of loss and storage moduli cross each other in the normal fluids but not in the representative pathological fluid.

The crossover points(Gcross=G'=G", $f$cross), as a parameter for the viscoelasticity, for synovial fluids from normal(n=61 in 61) human knee joints are depicted being divided into three different age groups in Figure 8.

The crossover points for normal synovial fluids are compared with those for synovial fluids from osteoarthritis(n=21 in 72) and rheumatoid arthritis(n=5 in 36) human knee joints(see Figure 9).

The molecular weight and the concentration of hyaluronate in synovial fluids from normal(n=61), osteoarthritis(n=72) and rheumatoid arthritis(n=36) knee joints measured by GPC-MALLS are plotted in Figure 10[7].

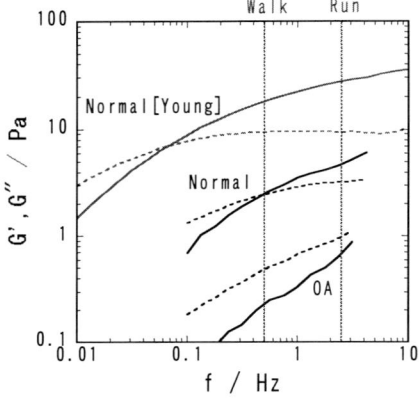

Figure 7   Dynamic viscoelasticity
of synovial fluids

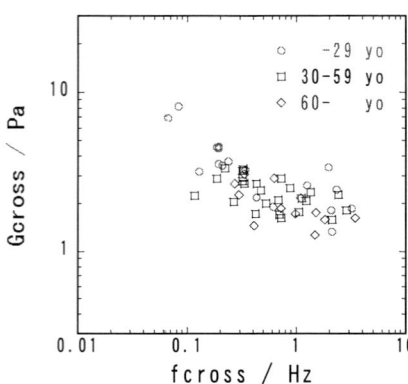

Figure 8   Crossover point of
normal synovial fluid

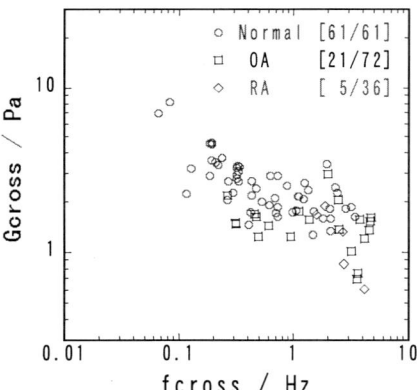

Figure 9   Crossover point of
synovial fluids

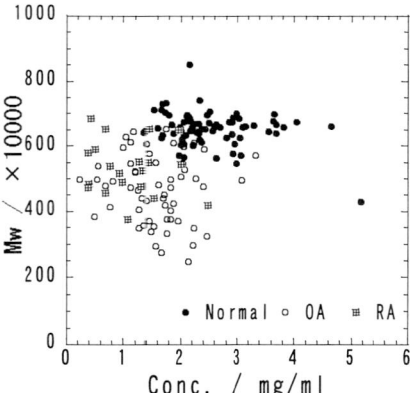

Figure 10   Plot of molecular weight
and concentration of hyaluronate in
synovial fluids

4. DISCUSSION

4.1. Comparison of synovial fluids with hyaluronate solutions

As shown in the Figure 4, curves of G' and G", or the crossover point(Gcross=G'=G", $f$cross), for the solution of hyaluronate with different molecular weight are superposed upon the master curves, or the standard crossover point, using the frequency shift factor, a_M.

The exponent in the molecular weight dependence of a_M was obtained from the slope of the plots as -3.7(see Figure 11). The crossover frequency, $f$cross, goes with molecular weight just as the longest relaxation time, τ_{MAX}:$f$cross $\sim \tau_{MAX}^{-1}$ $\sim Mw^{-3.4}$[8,9]. The theoretical value of the exponent in the molecular weight dependence of the longest relaxation time is 3 and is weaker than the experimental one; the measured exponent is higher than 3, ranging from 3 to 3.7 [10].

The concentration dependence of the crossover point is expressed by two shift factors, the frequency shift factor, a_c, and the modulus shift factor, b_c, as already stated. The exponents in the concentration dependence of a_c and b_c were calculated from slopes of the plots in Figure 12 and 13: $a_c \sim C^{-2.6}$ and b_c $\sim C^{-2}$ or Gcross $\sim C^2$. The concentration dependence of τ_{MAX} is delicate since the friction constant depends on the concentration in a nontrivial way[10]. It seems difficult to discuss on the value of 2.6.

The modulus shift factor b_c or Gcross goes with concentration just as the plateau modulus G^0, $G^0 \sim C^{9/4}$ in semidilute solutions and experimentally $G^0 \sim C^2$ in concentrated solutions[10]. The exponent value of 2 for Gcross is within these values.

Since the crossover point is determined by the concentration and molecular weight of hyaluronate in solutions unequivocally, one can determined the concentration and molecular weight of hyaluronate in solutions from the crossover point.

We could estimate the concentration and molecular weight of hyaluronate in synovial fluids from normal(young) and normal(representative) human knee joints, 0.53%(from 0.47% by GPC),7930000(6600000 from GPC) and 0.31%(0.29% from GPC),6670000(6710000 from GPC), respectively(see Figure 14). These results suggest that in synovial fluids hyaluronate molecules are scarcely affected by substrates like proteins and its viscoelastic properties in synovial fluids are almost the same as in the buffered saline solution. This result

328

is consistent with Balazs' finding that the human knee synovial fluid and the protein-free hyaluronate prepared from other tissues are identical from a rheological point of view[11].

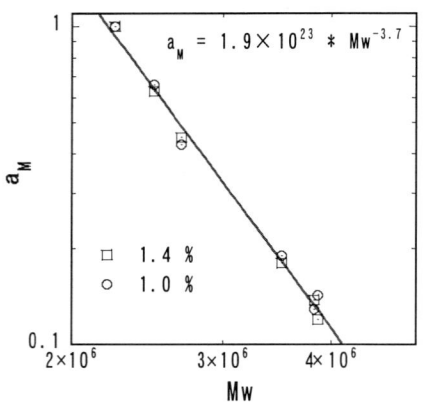

Figure 11　Relation between molecular weight and a_M

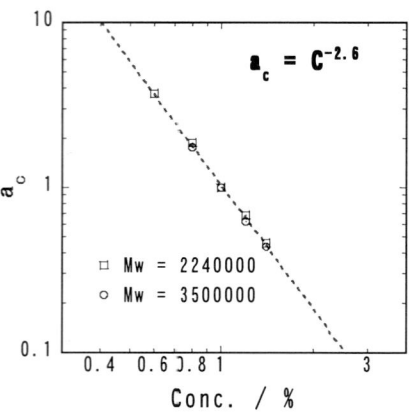

Figure 12　Relation between concentration and a_c

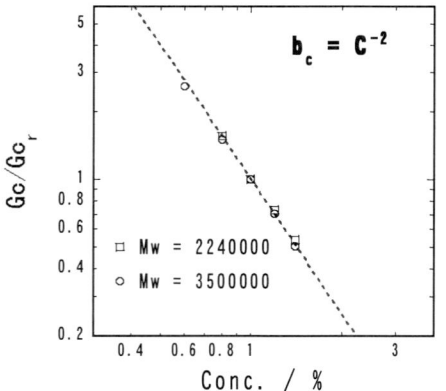

Figure 13　Relation between concentration and Gc/Gc_0

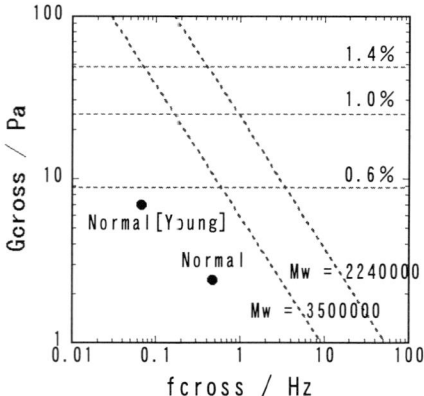

Figure 14 Determination of molecular weight and concentration of HA in synovial fluid from crossover point

4.2. Effect of aging

As shown in Figure 7, the viscoelastic properties of synovial fluids from normal(young) human knee joints has the highest, and as the frequency increases, the curves representing the loss and storage moduli cross each other at frequency lower than the crossover frequency for old one. This means that the normal young fluid is predominantly viscous at very low strain frequency and shifts to be predominantly elastic at lower strain frequency(~ 0.1 s^{-1}).

The crossover points of synovial fluids from normal human of different age are depicted in Figure 8. It is important to note that the crossover points of the synovial fluids from human knee joints under the age of 29 scatter wide range of both frequency and modulus area. While those from the knee joints of human over the age of 60 gather into the narrowest area of lower frequency and lower modulus values.

4.3. Pathological deterioration

The crossover points for normal synovial fluids are compared with those for pathological synovial fluids in Figure 9. Balazs reported that all rheological properties of pathological synovial fluid, such as the dynamic viscous and elastic moduli and the crossover point of the two moduli are much below normal values and further that , in the pathological joint, the synovial fluid does not have those rheological properties for the protection of the synovial tissue and cartilage against mechanical stress[12].

The crossover point are observed for all normal synovial fluids, 21 of 73 samples from OA knees and only 5 of 35 samples from RA knees. On the whole, the crossover points from normal synovial fluid locate both higher frequency and higher modulus region as compared with pathological ones.

The molecular weight and the concentration of hyaluronate in synovial fluids from normal(n=61), osteoarthritis(n=72) and rheumatoid arthritis(n=36) knee joints measured by GPC-MALLS are plotted in Figure 10. As expected from the results of the crossover point, the molecular weight and the concentration of hyaluronate in normal synovial fluids are higher than those in pathological ones. The molecular weight of hyaluronate in the normal synovial fluid is about 6500000, but the concentrations of hyaluronate in the normal one are spread over wide range from 1.5 mg/ml to 5 mg/ml. It is worth noting that the molecular weight of hyaluronate in RA knees appears to be higher and the concentration of hyaluronate in RA knees are generally more dilute as compared to OA knees.

5. CONCLUSION

The molecular weight dependence of the crossover point ,or the shift factor a_M, coincides well with that of τ_{MAX}. The concentration dependence of the crossover point, or the modulus shift factor b_c, is related to that of plateau modulus G°, $G^\circ \sim C^2$.

The molecular weight and the concentration of hyaluronate are deducible from the crossover point obtained from the dynamic viscoelastic measurement. Thus obtained values are coincide well with those from GPC.

The crossover points of synovial fluids from young donors scatter widely both viscoelasticity and frequency region, mainly due to the change in the concentration of hyaluronate. While the crossover points from old or pathological donors gather into lower viscoelasticity and lower frequency region.

6. ACKNOWLEDGEMENT

We would like to thank Professor Y.Tsukamoto and Dr.T.Endo for a valuable collaboration on the synovial fluids from many donors, Professor K.Nishinari and Professor K.Kubota for helpful suggestions which contributed to the improvement of the work, and Mr.M.Kawata for his tremendous effort to carry out this work.

7. REFERENCES

1. R.R.Myers et al., Biorheology,3(1966)197.
2. J.E.Gomez et al., Biorheology,30(1993)409.
3. D.C.Armstrong et al., Appl.Env.Microbiol.,63(1997)2759.
4. J.R.E.Fraser, T.C.Laurent et al., J.Intern.Med.,242(1997)23.
5. T.C.Laurent et al., Biochim.Biophys.Acta,42(1960)476.
6. M.Kawata et al., To be submitted.
7. T.Endo et al., To be submitted.
8. H.H.Winter, Private communication.
9. H.H.Winter et al., Rheologica Acta,31(1992)75.
10. M.Doi and S.F.Edwards, The theory of polymer dynamics,Clarendon Press, Oxford, 1986.
11. E.A.Balazs,Univ.Michigan Med.Ctr.J.[Special Issue](1968)255.
12. A.Heifer (ed.),Disorders of the Knee,J.B.Lippincott Co., Philadelphia,1974.

HYDROCOLLOIDS – PART 2
Edited by K. Nishinari
2000 Elsevier Science B.V.

331

Changes of the properties of colloid in arthritic synovial fluid

T. Kitano[*]

Department of Orthopaedic Surgery, Osaka City University Medical School, 1-4-3 Asahimachi, Abeno-ku, OSAKA 545-8585 JAPAN

The colloids in joint fluid play parts in homeostasis of articular cartilage, protection from microorganism, joint lubrication, and etc.. The study of human joint from the point of view of colloid science will contribute to the clinic al medicine, e.g. orthopaedic surgery, sports medicine, and rehabilitation medicine. Measurements of ζ-potential of particulate biomaterial in inflammatory synovial fluid with changes in pH and protein constituents revealed that the value of ζ-potential particulate biomaterial depend on pH and protein constituent. Bacterial adhesion study revealed that the initial process of bacterial adhesion on natural and artificial joint surfaces depend on the polarity component of surface free energy. Friction test revealed that albumin constituent in protein rich hyaluronan solution reduces the coefficient of friction of UHMWPE to SUS to minimum value at pH 8.0 at the walking speed.

1. SYNOVIAL FLUID

1.1. Bioparticles in synovial fluid

Hyaluronan (HA) is a high molecular weight polysaccharide present in the extracellular matrix of most living tissues. Proteoglycans is major components of articular cartilage *matrix*. Proteoglycan aggregate is broken down and the fragment is released from the matrix into the synoivial fluid. Albumin, globulin, phospholipid, and blood cells are derived from blood vessel in synovium.

1.2. Synovial fluid of normal and inflammatory human joint

In the inflammatory conditions, e.g. osteoarthritis (OA), rheumatoid arthritis (RA), traumatic synovitis, the permeability of the vessels in synovium increase. That is, the

Correspondence to: Toshio Kitano M.D.; e-mail: tkitano@med.osaka-cu.ac.jp concentration of these proteins and blood cells increase in the inflammatory conditions. These bioparticles in synovial fluid play important parts in the functions of natural human joints.

The concentration of hyaluronic acid ranged from 3.0 to 3.5 mg/ml in normal synovial fluid, from 0.28 to 0.56 mg/ml in inflammatory synovial fluid. The pH of normal synovial fluid ranged between 7.3 and 7.43 [1]. The pH of inflammatory synovial fluid ranged between 7.4 and 8.5 [4]. The concentration of all protein constituents increases in *acute* inflammatory phase and the concentration of albumin decreases to near the control level in *chronic* phase [2].

1.3. What do the colloids in joint fluid play a part in?

The first part is *homeostasis* of articular cartilage. There is no blood vessel in articular cartilage. Nutrients and oxygen from synovium reach the chondrocyte by diffusion through the cartilage matrix via synovial fluid.

The second part is *protection* from microorganism. Natural joint is aseptic area. Leukocyte and immunoglobulin take parts in protecting vulnerable articular cartilage from microorganism. It has been suggested that protein rich hyaluronan solution work the barrier from microorganism itself.

The third part is *joint lubrication*. Natural human joint has very low coefficient of friction. The dynamic coefficient of friction (μ_D) of cartilage on cartilage is from 0.002 to 0.005 while the dynamic coefficient of friction of ice on ice at $0°C$ is 0.1. The dynamic coefficient of friction between artificial joint surfaces which is replaced for destroyed human joint due to degenerative joint disease is from 0.07 to 0.12 [3].

2. THE PROPERTIES OF COLLOID IN ARTHRITIC SYNOVIAL FLUID

The author's works relate to colloid in synovial fluid are reviewed,

1) Measurements of ζ-potential of particulate biomaterial in inflammatory synovial fluid with changes in pH and protein constituents

2) What occurs in the initial process of bacterial adhesion to natural and artificial joint surfaces?

3) Participation of constituents of and pH changes in protein rich hyaluronan solution in joint lubrication.

3. ζ-POTENTIAL OF PARTICULATE BIOMATERIAL IN INFLAMMATORY SYNOVIAL FLUID [4]

3.1. ζ-Potential

The ζ-potential of particulate biomaterial was measured in normal and inflammatory pH ranges with three types of protein rich hyaluronan solution. The effects of the ionic fluxes with the adsorbed proteins on the ζ-potentials are thought to act through changes in *protein* adherence and conformational structural changes

3.2. Method of measuring ζ potential of particulate biomaterials

Using a microelectrophoretic method, ζ-potentials of three kinds of particulate biomaterials in three types of protein-rich hyaluronan solution with changes in pH. Electrophoretic mobility was measured using a particle microelectrophoretic apparatus (Model Mark II, Ranks Brothers, Cambridge, UK) with the following media. Medium 1 was tris buffer and sodium hyaluronate, medium 2 was containing bovine albumin, and medium 3 was containing bovine γ-globulin. The ζ-potentials of each material were calculated from electrophoretic mobility in each of four different media using the Henry equation (1):

$$\upsilon = \frac{2\varepsilon\zeta}{3\pi\eta}[f(\kappa a)] \tag{1}$$

where υ is the electrophoretic mobility, ε and η are the dielectric constant and the viscosity coefficient of the solution surrounding the colloid particle respectively, and $f(\kappa a)$ is the correction factor. The pH - ζ potential curve was drawn for the pH ranged from 6.5 to 8.6

3.3. The pH-ζ potential curve

The addition of γ-globulin in medium produces much more difference in ζ-potential among materials than the addition of albumin dose. Ultra high molecular weight polyethylene(UHMWPE) exhibited the increased absolute ζ-potential value in pHs above 7.5 in medium 3. It has been hypothesized that the ζ-potential of biomaterials affects biotribologic phenomena, for example, the process of infection, and lubrication mechanism.

4. BACTERIAL ADHESION TO NATURAL AND ARTIFICIAL JOINT SURFACES [5]

4.1. Bacterial adhesion to biomaterial

The physicochemical aspects of the interaction between the surfaces of biomaterials and bacterial cell membrane was investigated to address the questions: what component of the surface free energy of a biomaterial plays a dominant role in the initial process of bacterial adhesion? ; Do cell adhesion molecules play any roles in bacterial adhesion to biomaterials?

4.2. Relationship between Polarity Component of Surface Free Energy and Cell Counts

Number of adherent bacterial cells (*S. aureus* Cowan I) to biomaterials was counted under a scanning electron microscope. The surface free energy of three components, that is, dispersion, polarity, and hydrogen bond component, was calculated between biomaterial and bacterial cell surface using extended Fowke's equation. A correlation was found between the polarity component of surface free energy and the number of bacterial cell adherent to biomaterials that have no constituents of hydroxyapatite.

4.3. The effect of antibodies to cell adhesion molecules

The effect of antibodies to cell adhesion molecules was studied using antibody pre-incubated bacterial cell suspension. As for the effect of the anti-cell adhesion molecule antibody, the number of adherent cells to Stainless steel was not reduced with any concentration of antibody to Vitronectin receptor. On the other hand, the number of adherent cells to hydroxyapatite was reduced with 0.1 and 0.8mg/ml of antibody to Vitronectin receptor. The result suggested that bacterial adhesion be prevented by specific blockade of cell adhesion molecule receptors.

5. PARTICIPATION OF CONSTITUENTS OF AND pH CHANGES IN PROTEIN RICH HYALURONAN SOLUTION IN JOINT LUBRICATION [6]

5.1 Joint lubrication

Lubrication acts to reduce frictional resistance and to separate surfaces, thereby reducing local stress concentrations. Many modes of lubrication exist to provide the minimal friction of articular joint, for example, fluid film lubrication, boundary lubrication, and combination of these modes. In *hydrodynamic* lubrication, the design objective in many rapidly rotating or sliding machine parts, the viscosity of the lubricant is of primary importance. A viscous lubricant cannot instantaneously be squeezed out from the gap between two surfaces; *squeeze-film* produces a pressure that is capable of supporting high load. Fluid expressed from the porous surface produce *weeping* lubrication. This is believed to be an important lubrication mechanism in normal articular cartilage. In *elastohydrodynamic* lubrication, the elasticity of the bearing surfaces allows adaptation of irregularities without producing plastic deformities. In *boundary* lubrication, a monolayer of glycoprotein is adsorbed on the articular surfaces. This monolayer protects the articular surfaces from abrasion wearing. [3]

5.2 Materials and methods

Coefficient of friction between UHMWPE and Stainless steel was measured with the mechanical spectrometer changing lubricant constituents and pH values. The lubricants we tested are as follow.

Lubricant 1 is tris buffer and sodium hyaluronate

Lubricant 2 is lubricant 1 and bovine albumin

Lubricant 3 is lubricant 1 and bovine γ-globulin

Lubricant 4 is lubricant 1 and Lα-dipalmitoyl phosphatidylcholine

5.3. Relations between coefficient of friction and viscosity of synovial fluid

5.3.1 Angular velocity - coefficient friction curve

Angular velocity - coefficient friction curves show

1) Coefficient of friction increased at whole range of angular velocity with decrease in pH of lubricant 2 with albumin. 2) Coefficient of friction at high angular velocity increased with increase in pH value of lubricant 3 with γ-globulin increased. 3) The difference of pH value of Lα-DPPC dispersed in lubricant affected a little. The albumin constituent in protein rich hyaluronan solution reduced the coefficient of friction between UHMWPE and SUS to minimum value at pH 8.0 at the walking speed.

5.3.2 Shear rate - viscosity curve

Shear rate - viscosity curves show,

1) The difference of pH value in lubricant 2 and 3 affect viscosity at whole range of shear rate.

2) The difference of pH value in lubricant 1 and 4 dose not affect viscosity at any range of shear rate.

5.3.3 Lubrication mode of UHMWPE to SUS

There are two components in lubrication of UHMWPE to SUS, that is, boundary component at middle and low angular velocity and another component of lubrication related with viscosity at high angular velocity. The difference of pH value in hyarulonan solution with albumin and γ-globulin affects coefficient of friction of UHMWPE to SUS

6. CONCLUSION

6.1.Colloids in normal and inflammatory synovial fluid

While colloids in normal joint fluid participate in homeostasis of joint cartilage and biotribological properties, these microparticles in inflammatory joint fluid take *undesirable* parts in joint disorder. For example, wear particle derived from artificial joint surface would cause osteolysis, which result in the failure of artificial joint replacement. This is one of the most important problems in the area of orthopaedics.

REFERENCES

1. D. J. McCarty, ed., Arthritis and applied conditions: a textbook of rheumatology, Lea and Febiger, Philadelphia, 1989, 69-90.
2. J. Delecrin, M. Oka, S. Takahashi, et al., Clin Orthop, 307 (1994) 240.
3. S. R. Simon (ed.), Orthopaedic Basic Science, AAOS, 1994, 434-467
4. T. Kitano, H. Ohashi, Y. Kadoya, et al., J Biomed Mater Res, 42 (1998) 453.
5. T. Kitano, Y. Yutani, A. Shimazu, et al., Int J Artif Organs, 19 (1996) 353.
6. T. Kitano, L. H. Wang, G. A. Ateshian, et al., Trans Orthop Res Soc, 43 (1997) 760.

HYDROCOLLOIDS – PART 2
Edited by K. Nishinari
2000 Elsevier Science B.V.

337

Strong contraction of crosslinked hyaluronate gel by cationic drugs

C. Yomota, S. Okada

Drug Division, National Institute of Health Sciences Osaka Branch
Hoenzaka 1-1-43, Chuo-ku, Osaka 540, Japan

The cooperative interaction of the crosslinked HA gel and cationic drugs was found and also the possibility of the delayed release of cationic substances was investigated.

In the case of promazine, the weight of HA gel decreased to only 4% of the initial weight above the concentration of 0.008M, and the shape of the contracted gel was kept the initial cube. When the contracted HA gel by promazine was transferred in pure water, only 7-10% of absorbed promazine released.

1.INTRODUCTION

Hyaluronate(HA) is a charged, linear polysaccharide with the repeating disaccharide units consisted of N-acetyl-ß-D-glucosamine and ß-D-glucuronate, and found in the connective tissue of vertebrates, the synovial fluid of joints and the vitreous humor of the eye. It is well known that HA plays an important role in several physiological function. Already hyaluronate has bee used for osteoarthritis by intraarticular administration and in ophthalmic surgery such as anterior segment surgery. In these background, hyaluronate is one of the polysaccharides successfully applicable to the biomedical applications. The crosslinked HA gel has been reported that its ability to contain other substances is not strong enough to use a pharmaceutical reservoir.

It has been reported that some anionic polymer gels bind cationic surfactants cooperatively[1-2] and on the other had, some kinds of drugs have the similar properties as cationic surfactants. In this report we studied the interaction of the crosslinked HA gel and cationic drugs, and the release of cationic substances from the HA gel was investigated.

2.EXPERIMENTAL

2.1.Materials

Highly purified hyaluronate from rooster comb(Mw:900,000)(HA-a) was prepared by Seikagaku kogyo Co., Ltd.[3]. All samples contained more than 99% hyaluronate, less than 0.02% protein. And the other hyaluronate(HA-b)is supplied by Kibun Food Chemifa Co., Ltd. as a dry powder. Ethyleneglycol diglycidylether(EGDGE, Nippon Oil&Fats Co., Ltd) was used as a cross-linking agent. Dodecyltrimethylanmonium bromide(DOTMA, Tokyo Kasei Co., Ltd.) and 4 kinds of promazine, chlorpromazine, trifluoromazine, promazine, promethazine(Sigma Co., Ltd.) were used without purification.

2.2.Binding Isotherms of Cationic Surfactant

Surfactant-selective electrodes developed recently[4] were applied to determine the free surfactant concentration in the HA-a solution. The membrane potential was measured with a Metrohm 654 potentiometer at 25°C.

2.3.Preparation of the HA gel

The HA gel was prepared as reported previously[5-6]. 1g of HA-b was dissolved in 5ml of 1N NaOH. To this solution 1.46g EGDGE in 1ml ethanol was mixed and heated at 60 °C for 15 min. The obtained gel was placed into the mixture of water/ethanol (1:1v/v) and neutralized by the addition of 0.2N HCl, then replaced with distilled water and stored.

2.4.Measurements

A cube of HA gel, 3-4g, was immersed in the cationic substances solution at 25°C, and the shrinking behavior was followed by measuring the wet weight of the gels. The binding amount of cationic substances was calculated from their free concentration estimated by HPLC or using cationic surfactant selective electrode prepared as reported[6].

3.RESULTS AND DISCUSSION

3.1.Binding isotherms of DOTMA for hyaluronate

The amounts of DOTMA bound to hyaluronate in the solutions were determined by using the cationic surfactant-selective electrode in the presence of NaCl. Figure 1 shows the binding

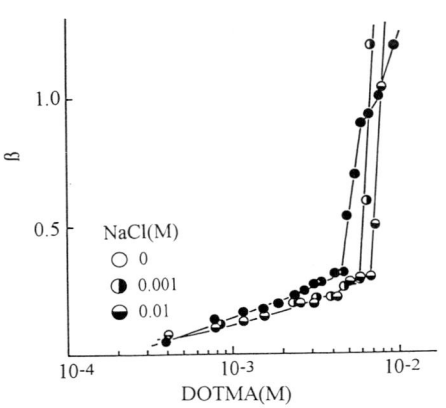

Fig.1 Binding isotherms of DOTMA+ for Hyaluronate in the presence of NaCl.

isotherms of DOTMA for hyaluronate solution(0.001eq./kg). DOTMA abruptly bound to HA in a narrow concentration range, 0.005-0.006M far below the surfactant's critical micelle concentration (0.016M), even in the excess salt. The binding process of the surfactant onto an oppositely charged polyelectrolyte is characterized by two processes[1, 2, 4, 7]. One is an electrostatic salt formation of the surfactant molecules with oppositely charged polyion. The other is a hydrophobic interaction between adjacently bound surfactant molecule. The steepness of binding isotherm is characteristic of a cooperative process that the hydrophobic interaction between alkyl chains of bound surfactants brings about the marked cooperative nature of surfactant-polyion interaction. In case of HA, an addition of the NaCl did not result in the remarkable shift of the minimum surfactant concentration at which the binding starts. In other words, the electrostatic shielding effect of the salts is not so distinct as other investigated anionic polysaccharides[4,6] and as other synthetic polymers[1]. In many cases, an increase in ionic strength makes the slope of the binding curve, the index of the cooperative effect, steeper [1,4,6]. However there can be seen only a slight effect in Fig.1. These particular binding behavior of DOTMA to HA may be due to the low charge density of HA, one carboxyl group per disaccharides.

3.2.Shrinking behavior of HA gel by the addition of DOTMA

When a water swollen HA gel is immersed in a DOTMA solution, the gel was observed to shrink with time.(Fig.2) Above the concentration of 0.006M DOTMA, the weight of the HA gel decreased to only 4% of the initial weight. The shape of the obtained small HA gel was kept to be initial cubic. The minimum concentration of DOTMA to cause the contraction of the HA gel, 0.006M, was almost coincident to the point where binding isotherms start to increase as shown in Fig.1. And the presence of NaCl seems to cause only the slight increase of DOTMA concentration enough to shrink the HA gel similar to Fig.1. It can be seen from these results the same phenomena of binding and cooperative binding as mentioned above in the HA solution also appeared in the crosslinked HA gel. And strong hydrophobic interaction among the bound DOTMA on the HA gel network resulted in the contraction of the gel.

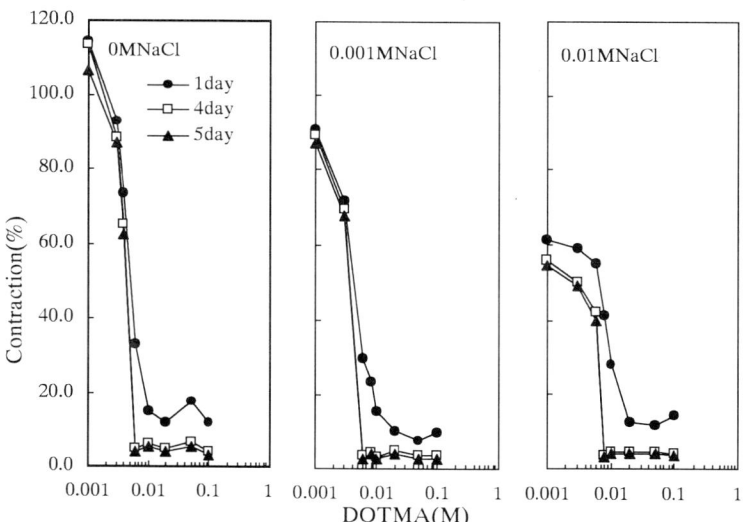

Fig.2 Volume contraction of the HA gel by addition of DOTMA[+].

3.3.Effect of phenothiazines to the HA gel

It is well known that many kinds of drugs have the properties like surfactants, and self association(micelle) in aqueous solution have been investigated[7-9]. Therefore as cationic surfactants3, some drugs were expected to cause the strong shrinking of the HA gel. As shown in Fig.3, when the HA gel was immersed into the promazine solution, the contraction

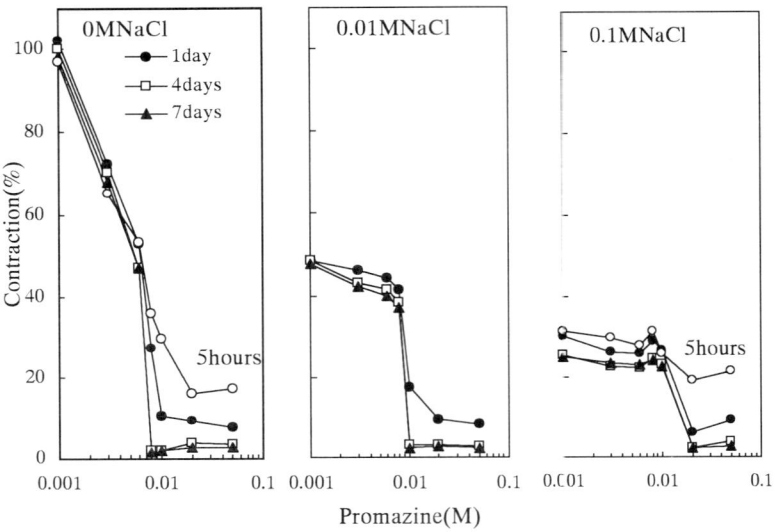

Fig.3 Volume contraction of the HA gel by addition of Promazine.

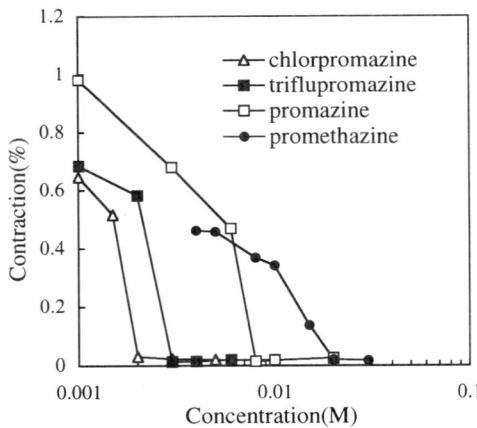

Fig.4 Volume contraction of the HA gel by addition of 4 kinds of phenothiazine.

was also observed above the concentration of 0.008M promazine far below the cmc of promazine(0.036M). The addition of NaCl resulted in an extensive shift of the minimum concentration at which the HA gel started to steeply decrease its volume.

Further other 4 kinds of phenothiazines were also investigated and the results are shown in Fig.4. The concentration to cause the shrinking of the gel were much different in 4 phenothiazines. Their physicochemical properties have been well studied[8-11]and the values of cmc obtained by pH measurement[8] are listed in Table I with the minimum concentration enough to contract

the HA gel(cc). As can be seen in Table I, the values of cmc in phenothiazines are much different and correlate to cc. Osada et al.[1] have reported the binding isotherms of surfantant molecules with a anionic polymer gel and shown that cooperativity and stability of the binding significantly increase with increasing alkyl chain length of surfactants. Similary also in the case of phenothiazines, the shrinking behavior of the HA gel affected by the hydrophobicity of the bound drugs indicated as the values of cmc.

Table I Critical micelle concentration(cmc) of
phenothiazines and minimum phenothiazine
concentration to cause the contraction of the HA gel (cc)

	cmc(M)	cc(M)	cc/cmc
chlorpromazine	0.0175	0.0012	0.069
triflupromazine	0.0185	0.0028	0.151
promazine	0.036	0.0065	0.181
promethazine	0.046	0.0153	0.333

3.4.Binding of promazine to the HA gel and release in salts solutions

By measuring the free concentration of promazine in the immersed solution by HPLC, the binding amount to the HA gel was calculated. The HA gel seemed to shrink as soon as more drugs bound to 5-10% of the carboxyl residue on the gel. When the HA gel was transferred into water, from the highly contracted gel obtained above 0.01M promazine, only 7-10% of absorbed promazine was released in 2days. While 50-60% was released from the more swelling gel obtained in the lower concentration of promazine(Fig.5). And as shown in Fig.6, the contracted gel was immersed in NaCl solutions, more drug released as NaCl concentration increased. And at the point indicated by a arrow, the gels were moved into PBS solutions, then drug was released abruptly by switch of the ionic strength.

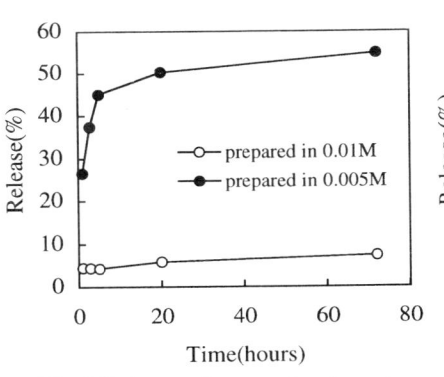

Fig.5 Release of promazine in water from HA gel immersed in 0.005M and/or 0.01M promazine solutions.

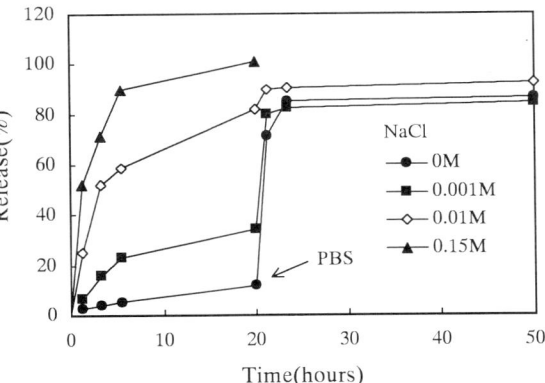

Fig.6 Release of promazine in NaCl solutions from the contracted HA gel prepared in 0.01M promajine.

From these results, it was clarified that the contracted HA gel can retain the pharmaceuticals such as phenothiazines and release as the increase of the ionic strength. The crosslinked HA gel investigated in this paper was degraded by hyaluronidase and the shrink gel needed longer time for enzymatic degradation. The crosslinked HA gel may be a useful biodegradable pharmaceutical reservoir.

REFERENCES

1. H. Okuzaki, Y. Osada, Macromolecule, 27(1994)502.
2. H. Okuzaki, Y. Osada, Macromolecule, 28(1995)4554.
3. A. Hamai, K. Horie, T. Yamaguchi, JPN, Tokkyo Koho, Japan Patent 8621241
4. K. Hayakawa, J. C. T. Kwak, J. Phys. Chem., 86(1982)3866.
5. N. Yui, T. Okano, Y. Sakurai, J. Controlled Release, 22(1992)105.
6. C. Yomota, Y. Ito, M. Nakagaki, Chem. Pharm. Bull., 35(1987)798.
7. K. Shirahama, S. Sato, M. Niino, N. Takisawa, Colloids and Surfaces, 112 (1996)233.
8. M. Nakagaki, S. Okada, Yakugakuzasshi, 98(1978)1311.
9. D. Attwood, O. K. Udeala, J. Pharm. Pharmacol., 27(1974)395.
10. D. Attwood, P. Fletcher, J. Pharm. Pharmacol., 38(1986)67.
11. D.Attwood, P.Fletcher, J. Phys. Chem., 94(1990)6034.

ACKNOWLEDGEMENT

Financial support was provides by the Japan Health Sciences Foundation.

HYDROCOLLOIDS – PART 2
Edited by K. Nishinari
2000 Elsevier Science B.V.

343

Viscoelasticity of synovial fluids and additive effect of hyaluronate

M. Kawata[a], A. Okamoto[a], T. Endo[b], Y. Tsukamoto[b]

[a]Research Center, Denki Kagaku Kogyo Co., Ltd., Tokyo, Japan

[b]Dept. of Orthop. Surg., Kitasato Univ. East Hosp.

The viscoelastic measurement was performed on synovial fluids from normal (61 samples), osteoarthritis (OA; 72 samples), and rheumatoid arthritis (RA; 36 samples) human knee joints. As increasing the frequency, both storage modulus, G' (elasticity), and loss modulus, G" (viscosity) increase similarly to the moduli of aqueous solution of HA. The values of G' and G" are highest for synovial fluid from normal joint and lowest for that from RA over entire frequency range. The effect of addition of hyaluronate of different molecular weight to synovial fluids from OA and RA was investigated. Both elasticity and viscosity of pathological synovial fluid increase on addition of hyaluronate. It is noteworthy that by adding of higher molecular weight hyaluronate the viscoelasticity of pathological synovial fluids approaches to that from normal one.

1. INTRODUCTION

The high viscoelasticity of synovial fluid is known to play an important role in joints. The viscoelasticity of synovial fluid is mainly due to hyaluronate (HA). Sodium hyaluronate is a water-soluble glycosaminoglycan and a linear polyelectrolyte. It consists of repeating disaccharide units of sodium D-glucuronate and N-acetyl-glucosamine with β $(1 \rightarrow 4)$ interglycosidic linkages. HA is a major component of biopolymers found in cartilage, eye vitreous humor, and synovial fluid, and plays a very important role due to its marketed viscoelasticity. Further, HA has many functions, such as inhibiting the modulation of lymphocytes to lymphoblasts and preventing target cells from killing sensitized lymphocytes. In synovial fluid, HA is generally believed to be essential for lubrication of joints and for shock absorption. Viscosity of synovial fluid gives lubrication that contributes to smooth movement of joint. On the other hand, elasticity of synovial fluid fill the role of a shock absorber that protects the synovial tissue and cartilage against permanent deformation. Therefore, the viscoelastic properties of HA have attracted considerable attention. The viscoelasticity of HA solutions has been extensively studied by many research groups [1-8]. It has been shown that synovial fluid acts as a viscous liquid in the lower-frequency region (corresponding to slow joint movement), but shows elastic behavior in the higher-frequency region (corresponding to rapid joint movement). Bothner et al. reported on the formation of a highly entangled network of HA chains at concentration about 1% [5]. HA has been used as a surgical aid in ophthalmic surgery and in the medical treatment of arthritis, due to its viscoelastic properties. The viscoelasticity of HA depends both on its concentration and molecular weight (MW). In synovial fluid, there are various components such as HA, proteins and salts. Tirtaatmadja et al. reported that the rheological properties of hyaluronic acid and recombined hyaluronic acid-protein solutions show an undeniable contribution of protein to the flow behavior of synovial fluid in joints [4]. However, there has been no evidence to date that the presence of proteins in synovial fluid significantly alters the viscoelastic properties of HA. Furthermore, HA solution without protein exhibits a molecular relaxation mechanism similarly to that of synovial fluid when it is exposed to strain of various frequencies.

The synovial fluids aspirated from inflammatory joint disease or rheumatoid arthritis have been consistently shown to have lower viscosity than synovial fluid from normal joints. This difference has been explained by the degradation of high-molecular-weight hyaluronate and/or by a decrease in the concentration of high-molecular-weight HA during inflammatory disease. Thus, pathological synovial fluids seem to contain HA segments with various chain lengths. The influence of HA segments on the viscoelastic properties of HA solutions is an important problem from both rheological and clinical perspectives. In this study, we measured viscoelasticity of 169 synovial fluid samples from normal and pathological human knee joints. Then the effect of addition of hyaluronate of different molecular weight to the pathological synovial fluid was investigated.

2. MATERIALS

Synovial fluids were collected from normal (61 samples), osteoarthritis (OA; 72 samples), and rheumatoid arthritis (RA; 36 samples) knee joints of donors.

HA samples extracted from the culture medium of Streptococcus equi were purified in the laboratory of Denki Kagaku Kogyo Co., Ltd. The amounts of proteins was maintained less than 0.04%. The two samples having different molecular weight MW, 3000000 (high) and 1200000 (low) obtained from multi-angle laser light scattering measurements, are dissolved in a saline containing 2 mM phosphate buffer (pH 7) to 1 % concentration.

3. METHODS

3.1. Viscoelasticity Measurement

The complex modulus of synovial fluid was measured at 37 ℃ as a function of frequency with CSL500 Controlled Stress Rheometer (Carri-med, England) using cone-plate geometries with the diameter of 4 cm or 6 cm, and the angle of 2 ° .

3.2. Additive effect of hyaluronate

In the experiment for OA, equal amount of synovial fluid and HA solution were mixed, then the mixture was kept at 37 ℃ for 9 hours. For RA, the mixing ratios of synovial fluid and HA solution were 3:1 and 1:1, and the measurement was performed after keeping the mixture at 37 ℃ for 15 hours.

4. RESULT AND DISCUSSION

4.1. The Comparison with Data of Balazs

Balazs reported on dynamic viscoelasticity of synovial fluids from normal knee joints of a young (20 yr.) and a old (67 yr.) humans and from an osteoarthritic joint of an old (63 yr.) human [Figure 1] [9]. Balazs focused upon the crossover point of storage modulus (G') and loss modulus (G"). In all three fluids, both G' and G" increase, as the frequency increases. It is important to note that the frequencies at which these measurements were made are within the range to which the fluid is exposed in the course of the normal movement of the knee joint (flexing under no load, walking , running). The synovial fluids from normal joint act as viscous fluids at lower frequency while predominantly elastic at higher frequency. In the pathological knee joint, since crossover is not observed, the synovial fluid behaves as a predominantly viscous in the entire frequency range. The synovial fluid occupies narrow channels between the soft tissues of the joint, and it is sandwiched between the two cartilage surfaces. When the joint is flexed slowly, corresponding to the movement with the lower frequency, the synovial fluid moves like viscous liquid in the channels, and behaves as a lubricant. On the other hand, under movement with higher frequency, say running, the synovial fluid stays in the channels and behaves like an elastic solid which absorbs the shock of mechanical stress between the surfaces of the articular cartilage storing the energy mechanically. In pathological joints, the synovial fluid does not have such elastic properties but is always viscous and the synovial fluid might not stay in the channels nor store the energy mechanically against quick movement.

Spectra of viscoelasticity of synovial fluids from two normal tests and an osteoarthritic test are shown in Figure 2. The synovial fluid from the young human exhibits pronounced viscoelasticity with highest moduli over entire frequency range and the crossover point locates at lowest frequency. In the case of the dynamic viscoelastic measurement by Balazs for synovial fluid from young human, abrupt decrease in the viscosity at higher frequency region was observed. In our case, however, such a marked change was not detected for all tests. The viscoelasticity of the synovial fluid from Old human is lower than from young human, and crossover point shifted to higher frequency. The crossover point is not observed for the osteoarthritic joint, similarly to the result presented by Balazs.

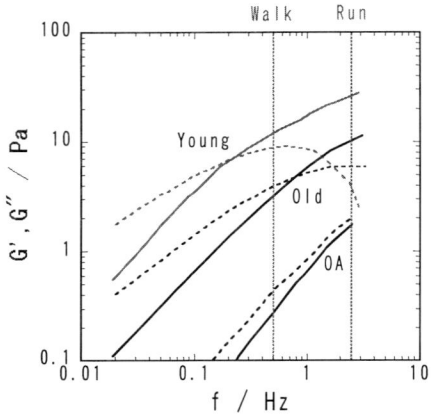

Figure 1   Dynamic viscoelasticity reported by Balazs

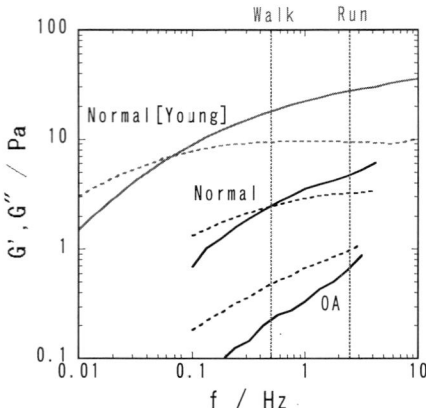

Figure 2   Dynamic viscoelasticity of synovial fluid

4.2. Crossover Point

The crossover points of the G' and G" are observed for all normal synovial fluids, for 21 of 73 samples from OA, and for only 5 of 35 samples from RA [Figure 3]. The crossover points for normal fluids from young human scatter wide range of both frequency and modulus regions. While those from old human gather into lower modulus and higher frequency area. On the whole, normal synovial fluid include higher concentration of HA with higher molecular weight as compared to those of RA and OA.

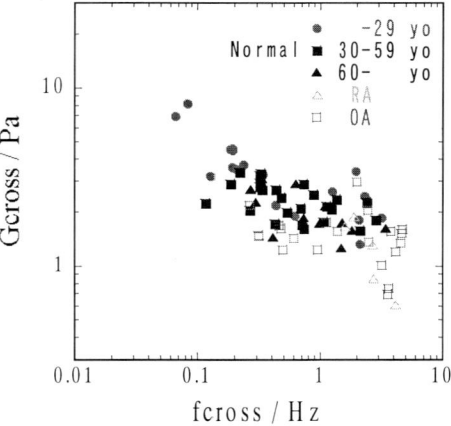

Figure 3   Crossover points of synovial fluid

4.3. Additive Effect of HA upon OA Synovial Fluid

The effect of addition of HA with different MW to OA synovial fluid is shown in figure 4. The G' and G" curves of OA synovial fluid moved to the region of higher modulus by the addition of HA and the crossover point was appeared. The addition of higher MW HA shifts the crossover point of OA to lower frequency and higher modulus region approaching to the crossover point normal synovial fluid. Whereas the addition of lower MW HA, the crossover point of OA shifts to higher modulus but to higher frequency region apart from the normal synovial fluid.

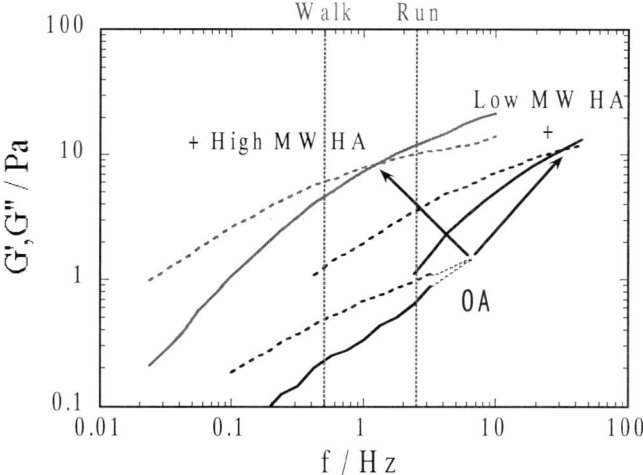

Figure 4   Additive effect of HA to OA synovial fluid

4.4. Additive Effect of HA to RA synovial fluid

The effect of addition of HA with different MW to RA synovial fluid was investigated [Figure 5]. The values of G' and G" at the frequency of 0.5 Hz corresponding to the movement of knee joints at walking were compared with normal synovial fluid, RA, and RA with and without additional HA with different MW. Although the RA synovial fluid shows very low viscoelasticity as compared to the normal synovial fluid and hardly plays the role of lubricant and shock absorber as pointed above, after the addition of HA to the RA synovial fluid, the viscoelasticity increases markedly. The effect of addition of HA to the viscoelasticity of the RA synovial fluid is more remarkable for HA with higher MW. The RA synovial fluid with equal amount of higher MW HA shows almost the same viscoelasticity as normal synovial fluid. On the other hand, addition of lower MW HA increase the viscosity slightly while the elasticity does not change appreciably.

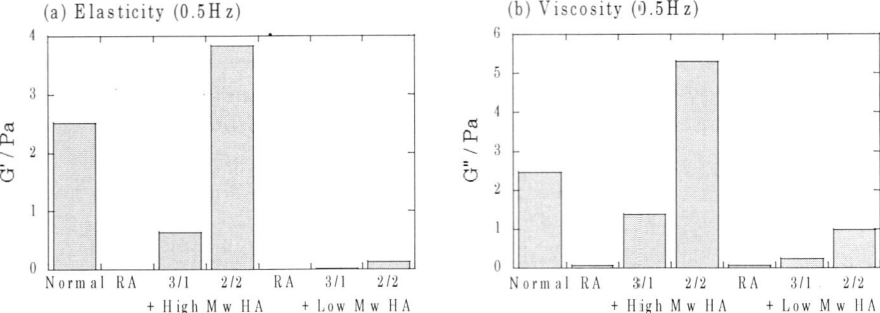

Figure 5   Additive effect of HA to RA synovial fluid

5. CONCLUSION

The viscoelastic measurement was performed on synovial fluids from normal (61 samples), osteoarthritis (OA; 72 samples), and rheumatoid arthritis (RA; 36 samples) knee joints of donors. The synovial fluids from normal joints show the viscoelasticity with higher modulus at entire frequency and the crossover points of G' and G" locate at lower frequency region. When HA is added to the pathological synovial fluid, both elasticity and viscosity increase. Remarkably, the addition of HA with higher molecular weight shifts the viscoelasticity of pathological synovial fluid to that of normal synovial fluid.

REFERENCES

1. J. Schurz, Kolloid-Ztschrft., 155 (1967) 45.
2. D. A. Gibbs, E. W. Merrill, K. A. Smith, and E. A. Balazs, Biopolymers, 6 (1968) 777
3. E. R. Morris, D. A. Rees, and E. J. Welsh, J. Mol. Biol., 138 (1980) 383.
4. V. Tirtaatmadja, D. V. Boger, and J. R. E. Fraser, Rheol. Acta, 23 (1984) 311.
5. H. Bothner and O. Wik, Acta Otolaryngol (Stockholm), 442 (1987) 25.
6. T. Yanaki and T. Yamaguchi, Biopolymers, 30 (1990) 415
7. S. C. De Smedt, P. Dekeyser, V. Ribitsch, A. Lauwers, and J. Demeester, Biorheology, 30 (1993) 31.
8. Y. Kobayashi, A. Okamoto, and K. Nishinari, Biorheology, 31 (1994) 235.
9. E. A. Balazs, Univ. of Michigan Med. Ctr. July [Special Arthritis issue], December (1968) 255.

HYDROCOLLOIDS – PART 2
Edited by K. Nishinari
2000 Elsevier Science B.V.

Effects of sodium chloride and sucrose on the conformational and rheological properties of sodium hyaluronate solutions

Y.Mo[a], T.Takaya[a], K.Nishinari[a], R.Takahashi[b], K.Kubota[b] and A.Okamoto[c]

[a] Department of Food and Nutrition, Faculty of Human Life Science, Osaka City University, Sumiyoshi, Osaka 558-8585, Japan

[b] Department of Biological and Chemical Engineering, Faculty of Engineering, Gumma University, Kiryu, Gunma 376-8515, Japan

[c] Research Center, Denki Kagaku Kogyo Co., Ltd., Asahi-machi, Machida-shi, Tokyo 194-8560, Japan

Effects of sodium chloride and sucrose on the conformational and rheological properties of sodium hyaluronate (NaHA) solutions were investigated by dynamic light scattering and viscoelastic measurements. Results of light scattering suggested that the NaHA molecules were constricted and the interaction between NaHA segments were enhanced by the addition of sucrose while the addition of sodium chloride reduced the coil dimension of NaHA. Viscoelastic measurements also suggested that sodium ions reduced the coil dimension. Sucrose stabilized the entanglement coupling between hyaluronic acid molecules and promoted the formation of temporary network. Shear thinning behaviour of NaHA solutions was discussed.

1. INTRODUCTION

Sodium hyaluronate is a linear polysaccharide consisting of disaccharide repeating sequence. The two saccharide-residues are D-glucuronic acid and N-acetyl-D-glucosamine, which is linked by ß(1-3) and ß(1-4) each other. Sodium hyaluronate is a major macromolecular component of the intercellular matrix of most connective tissues such as cartilage, eye vitreous humor, synovial fluid, and acts as a lubricant and shock absorber in synovial fluid. Balazs [1] suggested that intraarticular application of highly purified (protein content <0.3%) concentrated (10 to 20mg/ml) NaHA that contains fairly large molecules (molecular weight 1 to 3 million) of this biopolymer can influence the healing and regeneration of cartilage and soft tissues of joint. It is used as a medicine for arthritis and a surgical aid in ophthalmic surgery.

The rheological characteristics of HA molecules forming entanglement coupling and a temporary network in solution were widely studied [2,3]. Typical molecular weight of HA has been reported as $2\sim3\times10^6$, and the critical concentration above which the molecules overlap each other is roughly between 0.54 and 1.0mg/cm^3 [1,2]. The molecules either form loops around each other or form knots which act as physical junctions, and result in the fluid behaving as an elastic fluid at very high frequencies.

The flow properties of normal synovial fluid derived from human and cattle joints have been reported [1,4]. It is known that normal synovial fluids are shear-thinning showing a power-law behavior at high shear rates, and a constant zero-shear viscosity (η_0) at lower shear rates. The critical shear rate ($\dot{\gamma}_c$) is the reciprocal of a time constant and marks the onset of shear rate dependent behavior. It could be used as a diagnostic aid for synovial fluids, e.g. in the case of degenerative and chronic arthritis, the reciprocal of $\dot{\gamma}_c$ becomes lower [4,5].

Morris et al. [6] reported that the coil dimension reduces with increasing ionic strength in dilute HA solutions at neutral pH by intermolecular charge screening; however, with increasing ionic strength in high polymer concentrations, transient network structure is built up by the suppression of electrostatic repulsions. Kobayashi et al. [7] found that storage and loss shear moduli were decreased by the addition of NaCl, but increased by adding sucrose. In the present work, we use dynamic and steady viscoelastic measurements to study the effects of sodium chloride or sucrose on the rheological properties of NaHA solutions.

2. EXPERIMENTAL

2.1. Materials

The samples of powdered NaHA, extracted from the culture medium of *streptococcus equi* and were purified in the laboratory of Denki Kagaku Kogyo Co., Ltd.. The molecular weights of three samples used for rheological measurements were determined from the intrinsic viscosity using the Mark-Houwink parameters [8] :1.6, 1.98 and 2.02×10^6. Another fraction of Mw= 2.26×10^6 was used for light scattering measurements. Sodium chloride and sucrose of the reagent grade were purchased from Wako Pure Chemical Industries, Ltd. (Osaka, Japan) and used without further purification. The NaHA-solutions of different polymer concentrations were prepared by dissolving in solution of sodium chloride or sucrose and stirred at room temperature for 20 hours.

2.2. Measurements

Light scattering measurements were carried out using a home made spectrometer. The light source was an Ar ion laser operated at 488.0nm. The temperature of the sample solution was regulated within 0.01°C [9].

Steady viscoelastic measurements were performed by a RFSII (Rheometrics, Inc., NJ, U.S.A.). The diameter and angle of the cone were 2.5cm and 0.1rad, respectively. The shear rates ranged from 0.01 to 1000 s^{-1} in the steady tests. The temperature was fixed at 5°C or 20°C.

The intrinsic viscosity [η] was determined using an Ubbelohde capillary viscometer at 25 °C. The flow time for water in the viscometer used was ~141s.

3. RESULTS AND DISCUSSION

3.1.Conformation of HA molecules

Figure 1 shows the variation of NaHA molecular dimensions with respect to the concentration of added sucrose in the presence of 0.2M NaCl determined by light scattering. Sucrose may interact with NaHA leading to the complexation via sucrose and the NaHA chains are contricted. The ratio of R_g/R_h reflects the segment distribution of NaHA, and the increase of it with the addition of sucrose suggests that some conformational change occurs by sucrose, and NaHA molecules behave apparently so that the flexibility of NaHA chains decreases. Intersegmental interaction should be enhanced by the presence of sucrose.

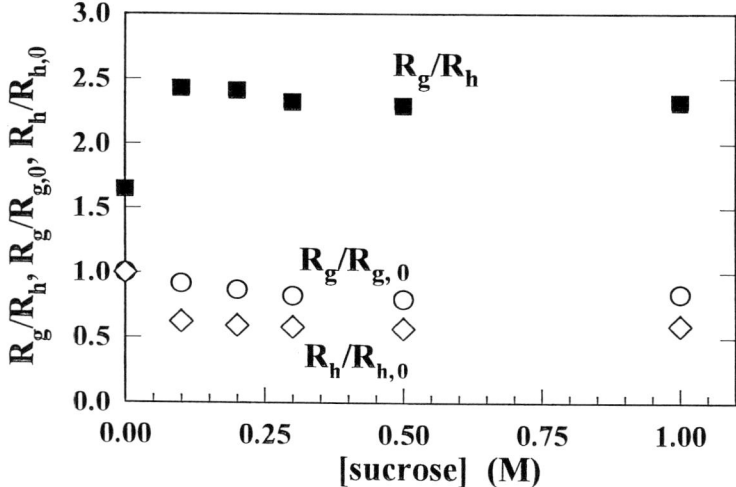

Figure1. Radius of gyration and hydrodynamic radius of NaHA as a function of concentration of added sucrose in the presence of 0.2M NaCl.

The reduced viscosity of NaHA solutions against NaHA concentration without NaCl showed a steep increase with decreasing NaHA concentration at dilute region (data not shown). The intrinsic viscosity could be determined by the usual extrapolation of concentration to zero concentration in the presence of NaCl. Sodium ions shielded the electrostatic repulsion between molecular coils. The intrinsic viscosity decreased with increasing concentration of added NaCl. Balazs and Laurent [10] have reported the similar experimental result. Huggins constant k' for chain polymers at the theta point has been estimated to be 0.5~0.7 [11]. It is strongly related to the formation of molecular aggregates. The experimental findings that k' of NaHA solutions was larger than unity suggest the existence of aggregation.

352

3.2. Flow behaviours

Figure 2 shows the viscosity of NaHA solutions of various concentrations as a function of shear rate. The viscosity of NaHA soluions was independent of shear rate at lower shear rates, i.e. NaHA solutions showed a Newtonian behaviour, but the viscosity began to decrease with increasing shear rate above a certain shear rate, i.e. NaHA solutions showed a shear thinning behaviour.

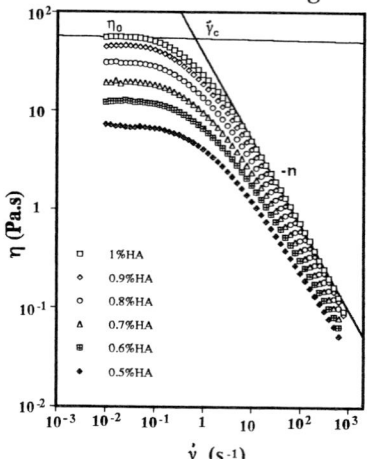

Figure 2. Shear rate dependence of viscosity of NaHA solutions of various concentrations.

At low shear rates, intermolecular entanglements for concentrated solution disrupted by the imposed deformation are replaced by new interactions between different partners; however, with increasing shear rate, the rate of externally imposed movement becomes greater than the rate of formation of new entanglements and the extent of re-entanglement decreases [12]. The viscosity (η) remains constant at a maximum value (the zero-shear viscosity, η_0) in Newtonian region, but decreases in non-Newtonian region. The non-Newtonian flow behaviour of polymer solutions is often fitted to a power law: $\eta \sim \dot{\gamma}^n$, where n was called pseudoplasticity index and corresponds to the slope of the curve in the shear-dependent viscosity domain. The critical shear rate ($\dot{\gamma}_c$) corresponds to the onset of the shear-thinning behaviour.

Double logarithmic plots of concentration (C) and zero-shear specific viscosity value (η_{sp0}) for NaHA solutions showed a pronounced change of slope at a critical concentration (C*) [13]. The critical concentration for NaHA solutions determined in the present work was in good agreement with C*≅1mg/ml(~0.1wt%) [1,2]. At concentrations below C*≅0.1wt%, NaHA molecular chains are free to move individually; however, at concentrations above C*≅0.1wt%, NaHA molecular chains start to overlap with each other and form a transient network structure.

The overlap concentration C* for NaHA solutions with NaCl was higher than that without NaCl, and shifted to higher polymer concentrations with increasing concentration of added NaCl. On adding sodium ions, the electrostatic repulsion between NaHA molecules is shielded and the coil dimension of NaHA molecules is reduced. Therefore, the coil overlap concentration C* shifts to higher concentrations of NaHA solutions. The zero-shear specific viscosity was plotted as a function of the overlap parameter $C[\eta]$, and the onset of coil overlap of NaHA molecules in the presence of NaCl was found to occur at a $C*[\eta]$ value of 4.6, and η_{sp0} was about 15. The addition of sucrose in the presence of 0.2M NaCl shifted $C*[\eta]$ to 3.4, and η_{sp0} was about 25 [13]. The overlap concentration decreased and η_{sp0} at the overlap concentration increased by the addition of sucrose. As is well known, the intrinsic viscosity depends on the shear rate of the used viscometer because of the shear thinning behavior and the values of $[\eta]$ in plotting the zero shear specific viscosity might be affected by it. However, the measurement conditions are the same in both measurements on the addition of NaCl and sucrose, and the chain dimension in the presence of sucrose is smaller than that without sucrose. Therefore, the characteristics that the overlap concentration decreased and η_{sp0} increased by the addition of sucrose can be concluded safely, and are in good agreement with the picture of enhanced interaction within NaHA chain and between NaHA molecules via sucrose. One of the possibilities resulting in such an interaction could be a promotion of hydrogen bonding via sucrose. This point should be further examined in future.

Figure 3(a) shows the critical shear rate value $(\dot{\gamma}_c)$ of the onset of non-Newtonian (shear thinning) behavior as a function of concentration for NaHA solutions with NaCl. When the concentration of NaHA solutions is higher than 0.12wt% (concentrated solution), a sharp decrease of C value was found with increasing of NaHA concentration. With increasing concentration of the added NaCl, C value became larger. NaHA molecules overlap each other in the concentrated solution and with increasing ion concentration the molecular coils begin to entangle only at higher concentrations because of the contraction of molecular coils.

The domains shown in figure 3 represent the semi-dilute regime of the polymer and correspond to about $C*\cong0.12$wt% and above. The concentration dependence of critical shear rate in the presence of sufficient salt was $\dot{\gamma}_c\sim C^{-3}$ which is in good agreement with experimental results by Fouissac et al [3] whilst that in the absence of salt or at a low salt concentration (0.1 MNaCl) was $\dot{\gamma}_c\sim C^{-0.6}$ or C^{-1}. NaHA molecules behave like flexible chains even in the presence of a small amount of salt (0.05M NaCl) although NaHA molecules are known to be stiffened random coils in the absence of salt.

The critical shear rate corresponding to the onset of the shear thinning behavior C as a function of molecular weight was found to decrease with increasing molecular weight of NaHA [3]. The fact that C decreased with increasing concentration of added NaCl is understood as a result of contraction of chain dimension by the addition of NaCl, i.e. NaHA molecules behave like a lower molecular weight NaHA by the addition of NaCl. The difference in C for NaHA solutions with different concentrations of NaCl became smaller at higher NaHA concentrations, because the molecular entanglement effect becomes more important than the electrostatic interaction at higher polymer concentrations.

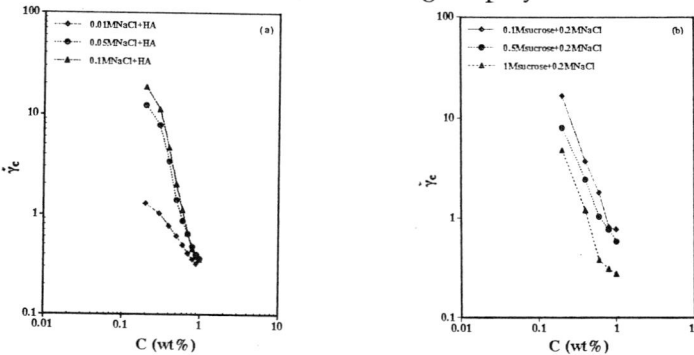

Figure 3 The critical shear rate value ($\dot{\gamma}_c$) of the onset of viscoelastic behaviour as a function of concentration for HA solutions with NaCl.

The critical shear rate value of the onset of viscoelastic behaviour as a function of concentration for NaHA solution with sucrose in the presence of 0.2M NaCl is shown in Figure 3(b). $\dot{\gamma}_c$ decreased with increasing concentration of NaHA solutions and became less concentration dependent in more concentrated ranges. The exponent m in the equation of $\dot{\gamma}_c \sim C^{-m}$ is about 2 for the NaHA solutions in the presence of sucrose with 0.02M NaCl. The dependence of m on the concentration of sucrose was negligible.

The critical shear rate $\dot{\gamma}_c$ is well approximated by the inverse of the longest relaxation time τ_r of Rouse model, which is given by $6(\eta_0 - \eta_s)M/\pi^2 R Tc$. Therefore $\dot{\gamma}_c \sim C^{-3}$, if $\eta_0 \sim C^4$ is substituted for the semi-dilute regime. However, Rouse model is valid only for dilute solutions.   The dynamics of polymer chains in the concentrated solutions and melts has been treated successfully by a tube model [14,15]. The longest relaxation time is a time necessary for a chain to wriggle out from a tube, and is given by a scaling treatment $\tau_r \sim C^{(3-3v)/(3v-1)}$ where v is the exponent in the relation $Rg \sim N^v$ (N is the number of segments). Therefore, $\tau_r \sim C^3$, $\eta_0 \sim C^6$ for $v=0.5$, and $\tau_r \sim C^{1.5}$, $\eta_0 \sim C^{15/4}$ for $v=0.6$.

Experimetal findings in the present work that $\dot{\gamma}_c \sim C^{-3}$ i.e. $\tau_r \sim C^3$ are consistent but $\eta_0 \sim C^4$ is not consistent with the above-mentioned theoretical prediction. According to de Gennes, the diameter of a tube is determined by the correlation length ξ of entangled network chains $\xi \sim C^{-3/4}$ for semi-dilute solutions [15]. Then, $\tau_r \sim C^{3/2}M^3$ and $\eta_0 \sim C^{15/4}M^3$. If the diameter of the tube varies as $\sim C^{-1}$, then $\tau_r \sim C^2$ M^3 and $\eta_0 \sim C^5 M^3$. Similarly, if the diameter varies as $C^{-0.5}$, then $\tau_r \sim C M^3$ and $\eta_0 \sim C^3 M^3$. Again, these are not consistent with the experimental findings. This should be explored in the future.

REFERENCES

1. E. A. Balazs, In Disorders of the Knee. (Ed. Helfet, A.) T. B. Lippincott Company, Philadelphia. (1974) 63.

2.S. C. de Smedt, P. Dekeyser, V. Ribitsch, A. Lauwers, and J. Demeester, Biorheology, 30 (1993) 31.

3. E. Fouissac, M. Milas, and M. Rinaudo, Macromolecules, 26 (1993) 6945.

4. J. Schurz, Prog. Polym. Sci.,16 (1991) 1.

5. J. E. Gomez, and G. B.Thurston, Biorheology, 30 (1993) 409.

6. E. R. Morris, D. A. Rees, and J. E. Welsh, J. Mol. Biol., 138 (1980) 383.

7. Y. Kobayashi, A. Okamoto, and K. Nishinari, Biorheology, 31 (1994) 235.

8. T. C. Laurent, M. Ryan, and A. Pietruszkiewicz, Biochim. Biophys. Acta., 42(1960) 476.

9. D. Ito, and K. Kubota, Macromolecules, 30 (1997)7828.

10. E. A. Balazs, and T. C. Laurent, J. Polym Sci. Lett. Ed., 6 (1951) 66.

11. T. Sakai, J. Polym. Sci., A2,6 (1968) 1535.

12. W.W. Graessley, Adv. Polymer Sci., 16 (1974)1.

13. Y. Mo, T. Takaya, K. Nishinari, K. Kubota, and A. Okamoto, Biopolymers, inpress.

14. M. Doi, and S. F. Edwards, The Theory of Polymer Dynamics, Clarendon Press,Oxford (1986)

15. P-G. de Gennes, Scaling Concepts in Polymer Physics, Cornell Univ. Press, Ithaca and London (1979)

HYDROCOLLOIDS – PART 2
Edited by K. Nishinari
2000 Elsevier Science B.V.

Biotransformation of monoterpenoids in common cutworm larvae (*Spodoptera litura* fabficius)

M. Miyazawa[a]*, S. Kumagae[a], T. Nakamura[b], T. Mineshita[b], T. Wada[a], H. Yanagihara[a] and H. Kameoka[a]

[a]Department of Applied Chemistry, Faculty of Science and Engineering, Kinki University, Kowakae, Higashiosaka-shi, Osaka 577-8502, Japan

[b]Laboratory of Food Science Tezukayama College, 3-1-3, Gakuen-minami, Nara 631-8502, Japan

α-Terpinene was mixed in an artificial diet at a concentration of 10 mg/g of diet, and the diet was fed to the last instar larvae of common cutworm (*Spodoptera litura*). Metabolites were recovered from frass and analyzed spectroscopically. The α-terpinene was transformed mainly to 4-isopropyl-1,3-cyclohexadienoic acid and cumic acid. Similarly, (+)-limonene was transformed mainly to (+)-*p*-menth-1-ene-8,9-diol (uroterpenol) and (+)-*p*-mentha-1,8-dien-7-oic acid (perillic acid), (-)-limonene was transformed mainly to (-)-*p*-menth-1-ene-8,9-diol (uroterpenol) and (-)- *p*-mentha-1,8-dien-7-oic acid (perillic acid). On the other hand, γ-terpinene was transformed mainly to *p*-mentha-1,4-dien-7-oic acid and *p*-cymen-7-oic acid (cumic acid) by same way. The results indicate that the intestinal bacteria probably participated in the metabolism of α-terpinene. The aerobically active intestinal bacteria transformed α-terpinene to 4-isopropyl-1,3-cyclohexadienemethanol, and the anaerobically active intestinal bacteria transformed α-terpinene to *p*-cymene.

1. INTRODUCTION

Studies relating to the search for biologically active substances from natural products have been carried out extensively. Terpenoids constitute a large class of natural products that are

---

*Author to whom correspondenece should be addressed (fax +81-6-6727-4301; e-mail miyazawa@apch.kindai.ac.jp)

isolated to be used as biologically active substances (allelochemicals).

A great majority of these terpenoids are produced as plant secondary metabolites, and these terpenoids have been shown to have biological activity against plants, microorganisms, and insects.

Biotransformation is a useful way to produce biologically active terpenoids. We have investigated the biotransformation of terpenoids in mammals and by microorganisms. In our study, the biotransformation of terpenoids was attempted by the larvae of common cutworm (*Spodoptera litura*). The reasons for using the larvae of *S. litura* as a biological catalyst are as follows: lepidopteran larvae feed on plants containing terpenoids as thier diet and therefore possess a high level of enzymatic activity against terpenoids; the worm consumes a large amount of plants, making it possible to obtain more metabolites; and the worm is easy to rear on a laboratory scale.

2. MATERIALS AND METHODS

2.1. Chemicals

The α-terpinene (1), (+)- and (-)- forms of limonene (7) and γ-terpinene (10) were purchased from Tokyo Kasei Kogyo Co., Ltd. (Tokyo, Japan).

2.2. Gas Chromatography (GC)

A Shimadzu GC-14A gas chromatograph equipped with a flame ionization detector, an OV-1 fused-silica capillary column (25 m length, 0.25 mm i.d.), and a split injection of 50:1 were used. Nitrogen at a flow rate of 1 mL/min was used as a carrier gas. The oven temperature was programmed from 80 to 240 °C at 4 °C/min The injector and detector temperatures were 250 °C. The peak area was integrated with a Shimadzu C-R3A integrator.

2.3. Gas Chromatography/Mass Spectrometry (GC/MS)

A Shimadzu GC-15A gas chromatograph equipped with a split injector was combined by direct coupling to a Shimadzu QP1000A mass spectrometer. The same type of column and the same temperature program as just described for GC were used. Helium at 1 mL/min was used as a carrier gas. The temperature of the ion source was 280 °C, and the electron energy was 70 eV. The electron impact (EI) mode was used.

2.4. Infrared (IR) Spectroscopy

The IR spectra were obtained with a Perkin-Elmer 1760X spectrometer. CHCl$_3$ was used as a solvent.

2.5. Nuclear Magnetic Resonance (NMR) Spectroscopy

The NMR spectra were obtained with a JEOL GSX-270 (270.05 MHz, ^{1}H; 67.80 MHz, ^{13}C) spectrometer.

2.6. Rearing of Larvae

The larvae of *S. litura* were reared in plastic cases (200×300 mm wide, 100 mm high, 100 larvae/case) covered with a nylon mesh screen. The rearing conditions were as follows: 25 ℃, 70% relative humidity, and constant light. A commercial diet (Insecta LF; Nihon Nosan Kogyo Co., Ltd. Japan) was given to the larvae from the first instar. From the fourth instar, the diet was changed to an artificial diet composed of kidney beans (100 g), brewer's dried yeast (40 g), ascorbic acid (4 g), agar (12 g), and water (600 mL) [3].

2.7. Administration of Monoterpenoids

The artificial diet without the agar was mixed with a blender. Then, monoterpenoid was added directly into the blender at 1 mg/g diet. Agar was dissolved in water and boiled and then added into the blender. The diet was then mixed and cooled in a tray (220Å~310 mm wide, 30 mm high). The diet containing **10** was stored in a refrigerator until the time of administration. The last instar larvae (average weight, 0.5 g) were moved into new cases (100 larvae/case), and the diet was fed to the larvae in limited amounts. Groups of 500 larvae were fed for 2 days with 500g of the diet containing **10** (483 mg), then the artificial diet not containing **10** was fed to the larvae for an additional 2 days. The frass were collected daily (total 4 days) and stored in a solution of CH$_2$Cl$_2$ (500 mL). **1**, (+)- and (-)-**7** were administered to the last instar larvae as well as **10**.

2.8. Isolation and Identification of Metabolites from Frass

The frass were extracted three times with CH$_2$Cl$_2$ each time. The extract solution was evaporated under reduced pressure, and the 1681 mg of the extract was obtained. The extract was distributed between 5% NaHCO$_3$ aq. and CH$_2$Cl$_2$, the CH$_2$Cl$_2$ phase was evaporated, and the neutral fraction (864 mg) was obtained. The alkali phase was acidified with 1N HCl and distributed between water and CH$_2$Cl$_2$. The CH$_2$Cl$_2$ phase was evaporated, and the acidic fraction (450 mg) was obtained. The acidic fraction was dissolved in CH$_2$Cl$_2$ (20 mL), and CH$_2$N$_2$ (5 mL) was added to the solution. The solution was evaporated, and the methylated fraction (478 mg) was obtained. The methylated fraction was analyzed by GC-MS, and methylated **11** and methylated **6** were occured in this fraction. The methylated fraction was subjected to silica-gel open-column chromatography (silica gel 60, 230-400 mesh, Merck) with a 9:1 n-hexane:CHCl₃ solvent system, and methylated **11** (28 mg) and methylated **6** (34 mg) were isolated. Methylated **11** and methylated **6** were identified by a comparison of established MS and ^{1}H NMR data. **1**, (+)- and (-)-**7** were treated by same way. Metabolites **2**, **3**, **5** and **6** were obtained from **1**. Similarly, (+)-**8** (66mg) and (-)-**9** (41mg) were obtained from (+)-**7**. On the other hands, (-)-**8** (57mg) and (-)-**9** (33mg) were obtained from (-)-**7**.

2.9. Incubation of Intestinal Bacteria with Monoterpenoids

This experiment was intentionally carried out under sterile condition. Petri dishes, pipets, and solutions were autoclaved. A GAM Broth (Nissui Pharmaceutical Co., Ltd., Japan) was adjusted to pH 9.0 and placed in Petri dishes at 10 mL/Petri dish. The fresh frasses (5 g) of the last instar larvae were suspended in physiological saline (100 mL), and the suspension (1 mL) was pipetted in the medium. The medium without frass was also prepared for a blank experiment. These media were incubated (18℃, darkness, 2 days) under aerobic and anaerobic conditions. After growth of bacteria, monoterpenoid (10 mg/Petri dish) was added to the medium and the incubation was continued. The percentage of metabolites in the medium was determined 12, 24, and 48 h after addition of monoterpenoid. The medium was acidified with 1N HCl and distributed between Et₂O and saturated solution of salt. The Et₂O phase was evaporated, and the extract was obtained. For the quantitative analysis of metabolites, the GC analysis was used as an internal standard with monoterpenoid.

3. RESULTS AND DISCUSSION

3.1. Metabolites from Frass

Metabolic reaction by the larvae of *S. litura* was observed as follows: substrate was administered to the larvae as their diet; then, metabolite was detected and isolated from the frass of larvae. In the previous paper, α-terpinene (**1**) was mixed in the diet of larvae at a high concentration (10 mg/g diet) to increase the production of potential metabolites [1]. However, intermediary metabolites (alcohols and aldehydes) were not isolated, though alcohols were detected by GC analysis. These results suggested that intermediary metabolites were hardly excreted into the frass. In the present study, a concentration of 1 mg/g diet was therefore chosen as optimum for administration. The optimum means the

Scheme 1. Metabolites of α-Terpinene (1) by the Larvae of *S. litura*.[a]

[a]Percentage was calculated from the peak area in the GC spectra of the extract of frass. 100% was defined as total metabolites of **1**.

Scheme 2. Metabolites of (+)- and (-)-Limonene (7) by the Larvae of *S. litura*.[a]

[a]Percentage was calculated from the peak area in the GC spectra of the extract of frass. 100% was defined as total metabolites of (+)- and (-)-**7**.

Scheme 3. Metabolites of γ-Terpinene (10) by the Larvae of *S. litura*.[a]

[a]Percentage was calculated from the peak area in the GC spectra of the extract of frass. 100% was defined as total metabolites of **10**.

concentration consume a substrate completely. In the biotransformation of γ-terpinene (**10**), the two metabolites isolated from the frass were identified as *p*-mentha-1,4-dien-7-oic acid (**11**) and *p*-cymen-7-oic acid (**6**) (see Scheme 3). The majority of metabolites was **11** (46%) and **6** (48%). Percentage was calculated from the peak area in the GC spectra of the extract of frass. 100% was defined as total metabolites of **10**. Substrate **10** and intermediary metabolites (alcohols and aldehydes) were not detected in the frass by GC analysis. Metabolites **11** and **6** were produced by oxidation at the C-7 position of **10**, and allylic oxidation was the only metabolic pathway. In the biotransformation of α-terpinene (**1**), the four metabolites isolated or detected from frass were identified as 4-isopropyl-1,3-cyclohexadienemethanol (**2**), 4-isopropyl-1,3-cyclohexadienoic acid (**3**), cumic alcohol (**5**), and cumic acid (**6**) (see Scheme 1). The majority of metabolites were end metabolites **3** (71.7% in metabolites of **1**) and **6** (7.8%), and the remainder were intermediary metabolites **2** (3.8%) and **5** (0.5%). These metabolites were produced by oxidation at the 7-position of **1**, and allylic oxidation was the main metabolic pathway. In the biotransformation of (+)-**7** the two metabolites isolated from the frass were identified as (+)-(4*R*)-*p*-menth-1-ene-8,9-diol (**8**) and (+)-(4*R*)-*p*-mentha-1,8-dien-7-oic acid (**9**) (see Scheme 2). The majority of metabolites were (+)-**8** (52%) and (+)-**9** (43%). Percentage was calculated from the peak area in the GC spectra of the extract of frass. 100% was defined as total metabolites of (+)-**7**. Substrate (+)-**7** and intermediary metabolites (alcohol, aldehyde, and epoxide) were not detected in the frass by GC analysis. Metabolite (+)-**8** was produced by oxidation at the 8,9-double bond of (+)-**7**, and metabolite (+)-**9** was produced by oxidation at the C-7 position of (+)-**7**. In the biotransformation of (-)-**7**, similarly, the two metabolites isolated from the frass were identified as (-)-(4*S*)-*p*-menth-1-ene-8,9-diol (**8**) and (-)-(4*S*)-*p*-mentha-1,8-dien-7-oic acid (**9**) (see Scheme 2). The majority of metabolites were (-)-**8** (50%) and (-)-**9** (44%). These results were similar to those for (+)-**7**.

3.2. Intestinal Bacteria

We reported the participation of intestinal bacteria in the metabolism of **1** [1]. The aerobically active intestinal bacteria transformed **1** to 4-isopropyl-1,3-cyclohexadiene-methanol (**2**), and the anaerobically active intestinal bacteria transformed **1** to *p*-cymene (**4**). The in vitro metabolism of **10** by intestinal bacteria was also examined in a similar manner of **1**. However, **10** was not metabolized at all (no reaction). These results suggested that the intestinal bacteria was unparticipated in the metabolism of **10**. (+)- and (-)-**7** also resulted in similar **10** [2].

3.3. Metabolic Pathways

In our study of biotransformation of **1**, the larvae mainly transformed **1** to **3** and **6** (Scheme 1) [1]. Similarly, the larvae transformed (+)-**7** to (+)-**8** and (+)-**9**, (-)-**7** to (-)-**8** and (-)-**9** (Scheme 2) [2]. On the other hands, **10** to **11** and **6** (Scheme 3). The two substrates **1** and **10** are isomeric with each other on the difference of position of endo-cyclic double bond (1,3- and 1,5-diene, respectively). These results indicate that there is little difference in the metabolic pathways (the oxidation at the C-7 position and the dehydrogenation) between **1** and **10**. However, the biotransformation of **1** differ from **10** in terms of the enzymatic activity (the concentration of substrate in diet), the proportion of products, and the participation of intestinal bacteria. Substrate **10** was not remained in the frass of larvae fed to **10**, on the other hand, a large quantity of substrates **1** was remained in the frass of larvae fed to **1**, respectively, at a concentration of 10 mg/g diet. The proportion of two metabolites of **10** was about 5:5 of 7-oic acid (**11**) and dehydrogenated 7-oic acid (**6**), on the other hand, the proportion of two metabolites of **1** was about 9:1 of 7-oic acid (**3**) and dehydrogenated 7-oic acid (**6**). Furthermore, only **1** was metabolized by intestinal bacteria.

REFERENCES

1. M. Miyazawa, T. Wada and H. Kameoka, J. Agric. Food Chem. 44 (1996), 2889.
2. M. Miyazawa, T. Wada and H. Kameoka, J. Agric. Food Chem. 46 (1998), 300.
3. K. Yushima, S. Kamano and Y. Tamaki, Rearing Methods of Insects; Japan Plant Protection Association: Tokyo, Japan, (1991); pp 214-21

HYDROCOLLOIDS – PART 2
Edited by K. Nishinari
2000 Elsevier Science B.V.

Physico-chemical property of Silkworm(*Bombyx mori*) blood and Hasumon Yoto(*Spodoptera litura*) blood

T. Nakamura[a], A. Yamamoto[a], H. Nankai[a], M. Miyazawa[b] and T. Mineshita[a]

[a]Department of Food Science, Tezukayama College, Gakuen-minami 3-1-3, Nara 631-8585, Japan

[b]Department of Applied Chemistry, Faculty of Science and Engineering, Kinki University, Kowakae Higashiosaka, Osaka 577-8502, Japan.

To account for the dispersion state of insect blood cell, flow property of blood of *Bombyx mori* and *Spodoptera litura* was studied by viscosity measurement. *Bombyx mori* blood showed a non-Newtonian flow behavior and a high viscosity value was observed at a low shear rate. The viscosity value of blood obtained from the beginning stage of five age of silkworm decreased, and this stage was the same stage as containing the largest amount of N-Acetyl Neuraminic acid(NANA) in blood. Non-Newtonian flow behavior and viscosity value of *Spodoptera litura* blood were shown as smaller than those of *Bombyx mori* blood, and these flow properties depended on the dispersion state of blood cell particle. From the result of an electron micrograph, it was shown that shape and aggregate structure of blood of *Bombyx mori* were more complicated than those of mammal. From this result, schematic model of silkworm blood cell was derived.

1. INTRODUCTION

Physico-chemical property of Silkworm(*Bombyx mori*) blood and Hasumon Yoto (*Spodoptera litura*)blood was studied by using low shear cone-plate viscometer, electron microscope, and GC.Mass Spectrography. There is only a few papers[1~4] concerning this theme about these insects blood. Silkworm has a short life cycle of 40~50 days and it will grow up within a short period. It is a very interesting insect because it has various stages of metamorphosis, as example, stage of spreading of thread, cocoon, and pupa. It is also very interesting fact that such a small insect can spin marvelous silk thread of about 1,000m per one head of silkworm. Silk thread will be reflected by silkworm blood, therefore, some species of silkworm and various stages of age in silkworm obtained as blood sources were used throughout this experiment.

2. EXPERIMENTAL

In this experiment, three different types of sources of silkworm blood were used. One is the source of the original wild type of silkworm(Original type) and it has a yellow color, second type of source is obtained from the regular type of silkworm(Regular type) and it has a white color, and the third one is the same origin as the second type(ATLANTIS) but it was obtained from hatched egg brought back from the Space Shuttle put on the orbit of ATLANTIS. This sample was supplied by Dr.Toshiharu Furusawa, Professor of Kyoto Technological and Textile University, who jointed as a member of the Universal Shuttle Experimental Planning STS-84. A viscometer used throughout this experiment was a cone-plate viscometer(Contraves LS-40) and it is available to measure the viscosity of blood, especially at a lower range of shear rate. Scanning Electron microscope(Hitachi Scanning Electron microscope S-510B type) was used to obtain the image of size, shape and aggregate structure of the silkworm blood cell.

3. RESULTS AND DISCUSSION

Figure 1 shows the viscosity versus shear rate plots of silkworm blood obtained from various stages in the age of regular type of silkworm. Figure 1(a) shows the viscosity of whole blood of the final stage of four age in the silkworm age, Figure 1 (b) shows the same result as above of the beginning stage of five age, Figure 1 (c) shows the same relationship as above of the middle stage of five age, and it is shown the least viscosity value in all stages. Figure 1 (d) shows the same plots as above of the final stage of five age in silkworm blood. All stages of these silkworm blood showed non-Newtonian flow behavior at the entire range of shear rate($\dot{\gamma}$), and the viscosity(η) of silkworm blood changed with the silkworm age.

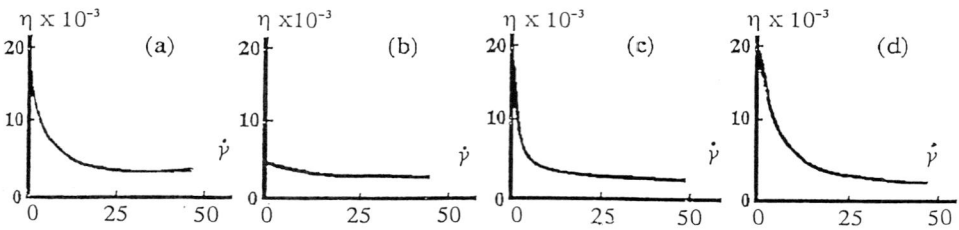

Figure 1   *Bombyx mori* blood viscosity(η) versus shear rate(γ) plots of various stages in *Bombyx mori* age of Regular type   (a) Final stage of four age   (b) Beginning stage of five age   (c) Middle stage of five age   (d) Final stage of five age.

Generally, viscosity is defined as the ratio of shear stress(τ) and shear rate($\dot{\gamma}$). Shear stress versus shear rate plots of various sources are shown in Figure 2. Shear rate dependence of shear stress of silkworm blood obtained from the regular type is shown in Figure 2(a), and the same dependence obtained from the regular type is shown in Figure 2(b), and that obtained from ATLANTIS is shown in Figure 2(c). Both non-Newtonian flow behavior and high viscosity value at the low shear range were observed in these figures. On the other hand, silkworm blood brought back with Space Shuttle from NASA shows the least viscosity value and slight non-Newtonian flow behavior. This fact means that the difference between the flow property of silkworm blood in each source will depend on the dispersion state of blood cell particle in each blood.

Figure 3 shows the shear stress versus shear rate plots of *Spodoptera litura*   blood. This insect does not make cocoon, and the viscosity value of fresh blood was very small as shown in Figure 3(a), but after storage in refrigerator at 5^0C and 24 hours, the viscosity value increased as shown in Figure 3(b), because blood cell formed aggregate structure with hydrophilic binding. After terpen was added in this blood, the viscosity decreased as shown in Figure 3(c), because the disturbance of the aggregate structure of blood cell occurred due to the addition of small molecule of terpen.

Figure 4 shows the relationship between NANA content in silkworm blood and silkworm age. It is shown that NANA content in blood changes with the silkworm age, and it decreases on the stage of three days previous to spinning age.

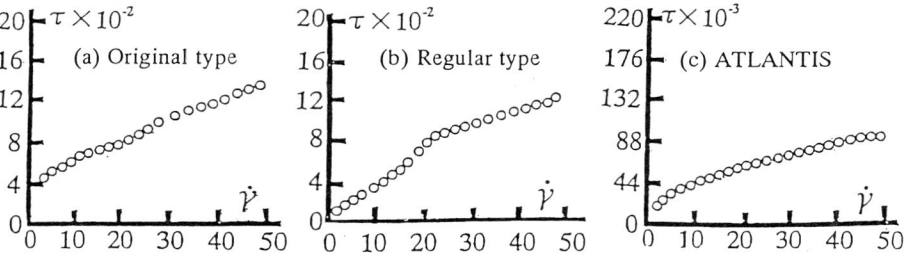

Figure 2 Relationship between shear stress(τ) versus shear rate( ) of various species of *Bombyx mori* blood.

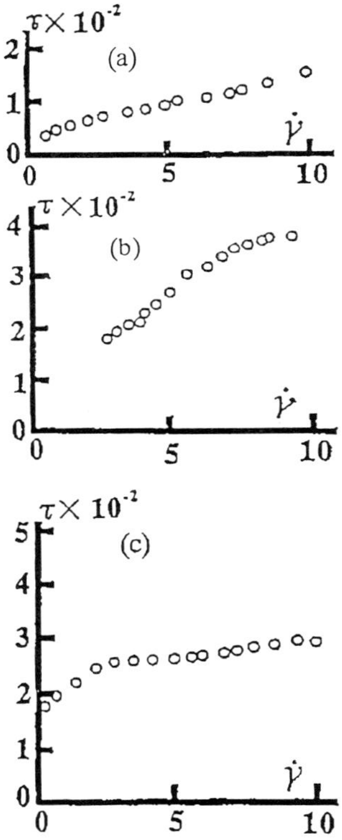

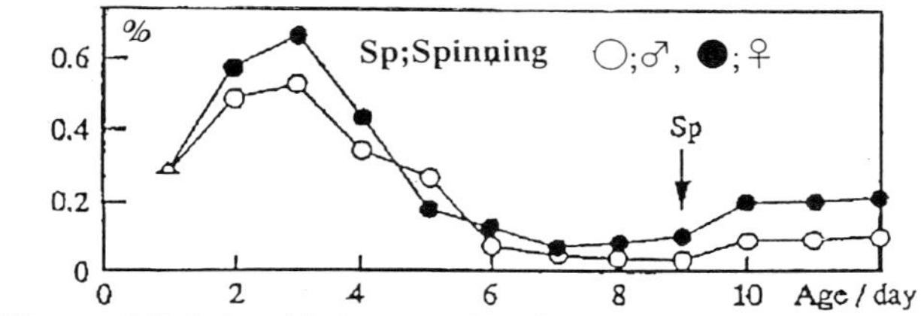

Figure 4 Relationship between NANA content in*Bombyx mori* blood versus various stages of *Bombyx mori* age.

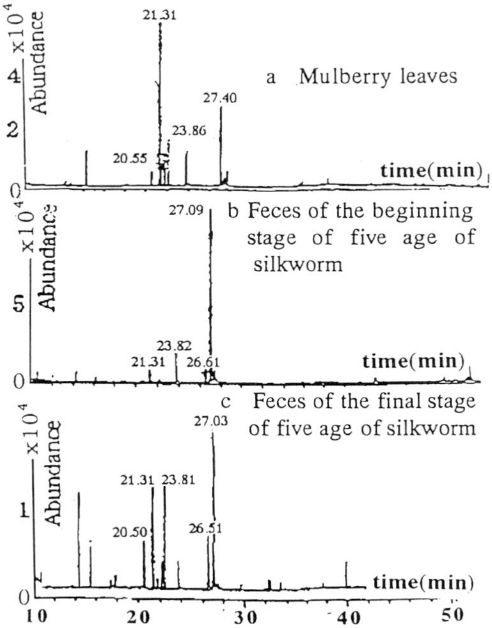

Figure 3 Shear stress versus shear rate plots of *Spodoptera litura* blood.
(a) Fresh blood (b) Blood(after 24hours)
(c) Blood+Terpen (1mg/1ml)

Figure 5 GC.Mass Spectrogram of *Bombyx mori* feces

On this stage of the beginning of five age in silkworm, NANA[5] content in blood decreases and shows minimum value in this silkworm age. This stage cf silkworm age coincides with the stage which showed the minimum viscosity value of silkworm blood. This fact means that the silkworm in this stage has to spend NANA to the physiological requirement for metamorphosis of silkworm, therefore, the blood viscosity value decreases and shows the minimum value in this stage. This change in chemical component will depend on the physiological requirement for the metamorphosis and, it will be necessary for the self-protection of silkworm to prepare to form cocoon.

Figure 5 is the result of GC.Mass spectrogram of silkworm feed and feces. The lipophilic substance in silkworm feces was analyzed by GC.Mass Spectrograph. (a) is the result of mulberry leaves, (b) is the result of feces of a silkworm obtained from the middle stage of five age, (c) is the result of feces of a silkworm obtained from the final stage of five age. It is shown that the largest amount of linoleic acid was obtained in mulberry leaves, and in feces, palmitic acid content is larger than that of linolenic acid. Palmitic acid content in feces obtained from the middle stage of five age of silkworm was much larger than that obtained from the final stage of five age of silkworm. To account for this flow property, size, shape, and numbers of the cell particle of the silkworm blood were measured by scanning electron microscope.

Figure 6 shows the electron micrograph of silkworm blood cell obtained from various sources of silkworm. Essentially, the blood cell particle consists of spherical part which has some curvature on the surface and fibrous part. In this experiment, average diameter of the spherical part is estimated by the electron micrograph as 4.8 μm. However, most of the popular form of silkworm blood particle is round and amorphous sphere shape and short length of fibrous shape.

Figure 7 shows the electron micrograph of aggregate structure formed by silkworm blood cell, and it is composed with spherical round part and fibrous rod part, and they conjugate and entwine each other. These aggregate structures are classified into four types of conjunction pattern of blood particles as shown in Table1.

Table 1   Amount of various aggregate structures of silkworm blood cell obtained from the middle stage of five years of silkworm(Regular type).

Natural SEM	A	B	C	D	E
Numbers	20	13	14	9	13
Amount x10^{-6}/cm^3	363	163	175	113	163

A: Spherical form,
B: Spherical form+Fibrous form(1+1)
C: Spherical form+Fibrous form(2+1)
D: Spherical form+Fibrous form(3+1)
E: more than D

Various types of aggregate structure were counted from this electron microgram. Some aggregates are formed by the conjugation with one spherical part and one fibrous part, and the other type of aggregate structure formation is conjugated with two spherical parts and one fibrous part. Another type of aggregate structure is formed with three spherical parts and two fibrous parts, and the other type of aggregate is formed with combining more than three spherical parts and more than two fibrous parts of particles. All of these aggregate formation is very important and characteristic property for silkworm to spin and make a thread of silk. Although their conjugated shapes are different and size distribution is larger than those of another mammals blood, more complicated aggregate structure formation will occur in a silkworm. Comparing with the viscosity value of silkworm blood obtained from the regular type and the original type of silkworm, viscosity at infinite shear rate is almost same value, and the viscosity at zero shear rate is the larger value in the regular type than that of original one. Consequently, aggregate structure of the silkworm blood obtained from the original type is less than that of the regular one, and ATLANTIS is the least in all samples.

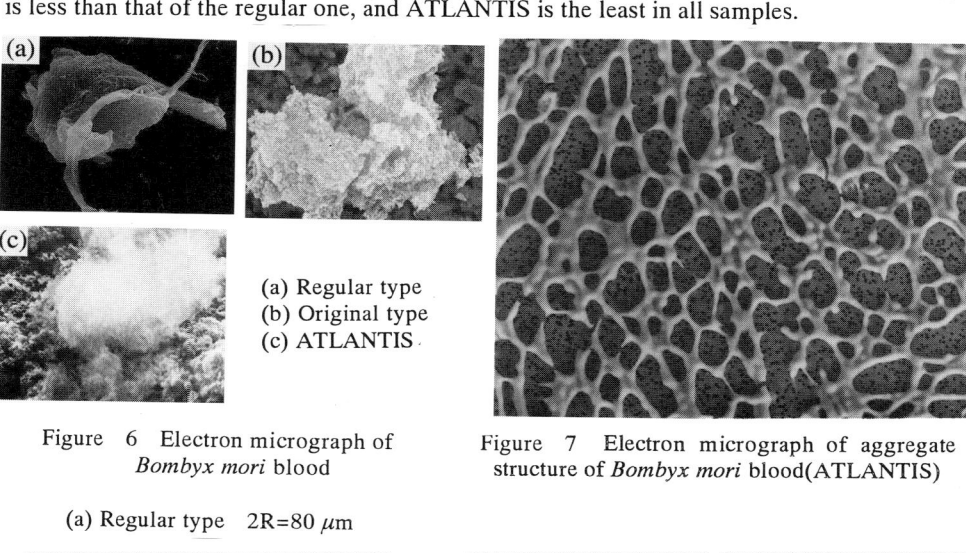

(a) Regular type
(b) Original type
(c) ATLANTIS

Figure 6   Electron micrograph of
Bombyx mori blood

Figure 7   Electron micrograph of aggregate structure of *Bombyx mori* blood(ATLANTIS)

(a) Regular type   2R=80 μm

(b) Original type   2R=80 μm

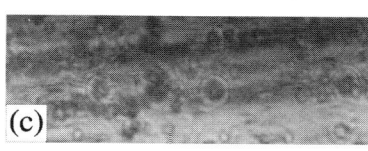

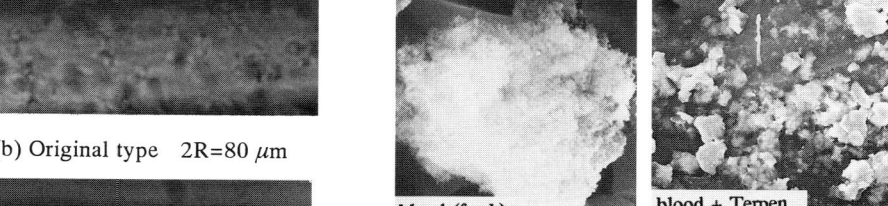

blood (fresh)   blood + Terpen
                        (1mg / 1ml)

Figure 8   Electron micrograph of *Spodoptera litura* blood cell.

(c) ATLANTIS   2R=90 μm

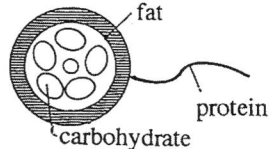

fat

protein

carbohydrate

Figure 10   Schematic model of a blood cell

Figure 9   Photomicrograph of silkworm blood flowing in a capillary

Figure 8 shows an electron micrograph of *Spodoptera litura* blood. This blood is treated by osmium in fresh blood, and another blood was added terpen in fresh blood.

370

Comparing with this blood and silkworm blood, spherical shape was observed also in this case, but, fibrous shape could not be observed in this case. From the electron micrograph of terpen added blood, it is observed that the blood cell was smaller than that treated by osmium in fresh blood, because aggregate structure of blood cell was disturbed by added terpen. The dispersion state of the blood is more remarkably shown with particles flowing in a capillary.

Figure 9 shows the photomicrogram of the image of various sorts of silkworm blood flowing in a capillary tube. As shown in (a), large aggregate structure of blood particle obtained from the regular type of silkworm was flowing in both side of the center and near the capillary wall. (b) shows the same results of the blood obtained from the original type of the silkworm. In this case, blood particle formed the small aggregate structure, and they were flowing at the center of the capillary tube. On the contrary, the blood particles obtained from hatched egg brought back with the Space Shuttle from NASA show the small aggregate structure flowing in the near side of the capillary wall, as shown in (c). In this case, irradiation effect and gravity change effect on size and shape of the silkworm blood particles will be exerted remarkably, and as the result of it, blood particles were cut off and separated each other and the larger aggregate structure could not be observed, but the smaller could be observed. Consequently, these aggregate structure formation will change also with the capillary bore-size of the viscometer.

Figure 10 shows a schematic model of *Bombyx mori* blood cell particle. The shape of a blood cell is composed with spherical part and fibrous part. Spherical part is composed with three main parts. The most inside layer is composed with hydrophilic constituent like as carbohydrate. Outside of the layer is surrounding with lipophilic membrane like as fat and lipid. And, the most outside layer is surrounding with proteins, and some of them are binding with fibrous part, therefore, they will expand and constrict .

ACKNOWLEDGMENTS

The authors gratefully acknowledge to Professor S.Takagi, Kinki University for his helpful discussion. The authors also thank Professor S.Mineshita, Dr.S.Kusakari and Dr.M.Wada for providing the electron microscope. Authors wish to express our appreciation to Professor T.Furusawa, Kyoto Technological and Textile University, for his generous gift of a precious sample. This work was indebted to Tezukayama Gakuen for that fund.

REFERENCE

1. T.Nakamura et al, Proceedings of 6th ASEAN Food Conference p.55(1997).
2. T.Nakamura et al, FOOD HYDROCOLLOIDS Structure, Properties, and Function (Ed.by K.Nishinari and E.Doi) Plenum Press, New York, p.395(1994).
3. T.Nakamura et al, DEVELOPMENTS IN FOOD ENGINEERING Part 1 (Ed.by T.Yano, R.Matsuno, and K.Nakamura) BLACKIE ACADEMIC and PROFESSIONAL, London, p.54(1994).
4. T.Nakamura et al, Biorheology (B&R)(in Japanese), **13**, (1999) 1.
5. Y.Kato et al, J.Sericul.Sci, Japan **67**, No,4.p319(1998).

HYDROCOLLOIDS – PART 2
Edited by K. Nishinari
2000 Elsevier Science B.V.

371

Alginate coating of <u>Xenopus laevis</u> embryos

N. Kampf, C. Zohar and A. Nussinovitch

Institute of Biochemistry, Food Science and Nutrition
The Hebrew University of Jerusalem
Faulty of Agricultural, Food and Environmental Quality Sciences
P.O. Box 12, Rehovot, 76100
Israel

Xenopus laevis eggs were coated, immediately after squeeze-stripping and fertilization, with a thin layer (~50 μm) of film based on one of three different types of alginates which varied in their mannuronic/guluronic acid ratio. The alginate was cross-linked with either Ca or Ba ions at three different concentrations. The developmental, survival and hatching of this embryos and the swelling of their natural jelly coats or hydrocolloid coatings were studied over 8 days, while embryos were maintained in flowing, aerated and circulated water ratio of 85 ml per embryo, or at a very diminished ratio of 1.6 ml sterile or non-sterile Modified Marc's Ringer (MMR) solution per embryo. All experiments were conducted in triplicate at 20±1°C. The coating conferred major advantages when the ratio between the embryos and the surrounding medium was at a minimum under non-sterile conditions, perhaps due to the film's resistance to diffusion. In the studied systems, the coating seemed to postpone embryo hatching to a more developed stage. In addition, the coating served as a barrier to microbial contamination and thus improved survival prospects.

Key words: Alginate, Coating, *Xenopus laevis*, Embryos, Survival.

1. INTRODUCTION

Cells can be entrapped within a gel matrix. A wide range of characteristics are attributed to gels as an entrapment medium. On the one hand, they include macromolecules held together by relatively weak intermolecular forces, such as hydrogen-bonding or ionic cross-bonding by divalent or multivalent cations. On the other hand, strong covalent bonding, where the lattice in which the cells are entrapped is considered as one vast macromolecule, is limited only by the particle size in the immobilized cell preparation. During immobilization, 10^4 to 10^9 microorganisms (bacteria, yeast or fungal spores having a maximal diameter of 5 microns) can be entrapped within 1 ml of gelling agent. In such cases, a maximal 6.5% of the volume is occupied by the microorganisms. In other words, ~93.5% of the volume is not occupied by the cells, or, if the cells are evenly distributed throughout the gel volume, then each individual cell is coated by a very thick layer of gel in comparison to its own natural dimensions. In contrast, the idea of coating a single cell with a thin layer of film, comprising only a fraction of its diameter, could be of interest. The difference between coating and entrapping is the thickness of the coating layer being very thin in the former, thick in the latter. Taking this definition into account, it seems that most, if not all reports on coating are in fact describing cell entrapment within a gel matrix. To obtain a true coating, special micro-coating procedures need to be invented, or, at least as a start, attempts need to be made with very large cells. For this study, we chose one of the biggest cells in nature: the fertilized frog egg. The objective of this research was to study the influence of different alginate coatings (thin films that are glued to the outer surface of the egg) on the survival of <u>Xenopus laevis</u> embryos, in terms of their biological advantages and disadvantages under different conditions of cross-linking and storage.

2. MATERIALS AND METHODS

Frog maintenance, egg ovulation and fertilization procedure was carried out according to Wu and Gerhart (1991), Hedrick and Nishihara,(1991) and Nieuwkoop and Farber, (1994).

2.1 Coating procedure

After fertilization, the embryos were dropped into a 1% alginate solution (made by dissolving Na-alginate in one-third-strength MMR solution): alginate compositions, supplied by the manufacturer, are given in Table 1. Embryos were then sucked into a 1.5-mm diameter tube and dropped into the cross-linking agent. The alginates were cross-linked with either Ca or Ba ions (available as $CaCl_2$ or $BaCl_2$ salts (Sigma Chemical co., St. Louis, MO)) at three different concentrations: 0.25, 0.5 or 1% (w/w) (equal to 25, 50 and 100 mM $CaCl_2$, respectively or 12.5, 25 and 50 mM $BaCl_2$, respectively). The salts were dissolved in one-third-strength MMR solution to maintain the egg's physiological osmotic pressure. After dipping in the cross-linking agent for ~20 s, coated embryos were washed once and then stored in sterile one-third-strength MMR solution.

2.2 Storage conditions

All embryos were kept for 196 h under one of three different storage conditions:
1) Closed (sterile) petri dishes containing 30 embryos in 50 ml of one-third-strength MMR solution at a volume ratio of 1.6 ml per embryo.
2) Open petri dishes containing 30 embryos in 50 ml of one-third-strength MMR solution at a volum ratio of 1.6 ml per embryo.
3) Aerated, circulated , stirred and dechlorinated tap water at a volume ratio of 85 ml per embryo.

All experiments were conducted in triplicate at $20 \pm 1°C$ (maintained by air conditioner), and the embryo's developmental stages were monitored after fertilization. Every 4 to 8 h, larval hatching from the natural jelly coat or artificial alginate coating was determined. Survival of the larvae was determined by observing movement. Dead or non-hatched embryos were not included in the survival calculations. Percent survival after hatching was calculated as the surviving hatched larvae out of the total number of embryos.

Changes in the egg's natural jelly coat's dimensions and the artificial alginate coat's thickness were measured up to ~48 h under a binocular using a grid-measuring lens. Scanning electron microscopy of eggs was performed in a Jeol JSM 35C (Tokyo, Japan).

Immediately after laying, the egg was glued to a polypropylene stub and tested under low-vacuum conditions.

Microbial mass in the embryo's one-third-strength MMR medium was assayed as presence of microbial ATP using a dairy products sterility test kit (LUMAC® B.V. Landgraaf, The Netherlands). Free ATP was degraded by adding 10 µl of ATPase enzyme (SOMASE™) to 50 µl of embryo medium and incubating at room temperature for 15 min. Then, the enhancement of microbial cell wall and membrane permeability to ATP was established by adding L-NRB® reagent for 30 s. Finally, the presence of microbial ATP was assayed for 10 s by coupled reaction of luciferin-luciferase enzymes [4]. Emitted light was measured by luminescence photometer (BIOCOUNTER®,M 2500, Landgraaf, The Netherlands). The correlation between the actual number of microorganisms and the light emitted from the above-mentioned assay was found using a total plate count culture composed of 1% agar (Difco, MI, USA), 0.5% yeast extract (Difco) and 3% tryptic soy broth (Difco).

3. RESULTS AND DISCUSSION

In a first set of experiments, X. laevis fertilized eggs were coated with three different types of alginate. The properties of these alginates are summarized in Table 1: they differed with respect to their molecular weights, viscosities, gel strengths and the content ratios of guluronic (G) to mannuronic (M) acid. The molecular weight, and the proportion and arrangement of M and G are expected to affect a particular alginate's behavior. The percentage of M in the alginates used for coating ranged from 29 to 35 in the alginates extracted from *Laminaria hyperborea,* to 61 in the alginate extracted from *Macrocystic pyrifera*. Each egg was covered with a thin layer of calcium- or barium- alginate gel. Alginate was chosen for this study because its coatings are easy to produce (see M&M), and they have been used successfully for many products [5-9].

Company	Product Name	Origin	Molecular Weight	Viscosity	Gel Strength	% Dry Solids	G:M Ratio
Sigma Chemical Co. St. Louis, USA	Alginic acid sodium salt, Low visc.	Macrocystic pyrifera	60,000-70,000	228 (cP) at a concentration of 2%	Not detected	~ 88	39:61
Pronova Biopolymer. a.s. Drommer, Norway	Alginic acid sodium salt (Protanal LF 10/60)	Laminaria hyperborea	123,000	50 (cP) at a concentration of 1%	59.9 g (water)	87.8	71:29
Pronova Biopolymer. a.s. Drommer, Norway	Alginic acid sodium salt (Protanal LF 20/60)	Laminaria hyperborea	185,000	126 (cP) at a concentration of 1%	56.9 g (water)	86.5	65:35

Table 1: Alginate compositions (supplied by the manufacturer).

374

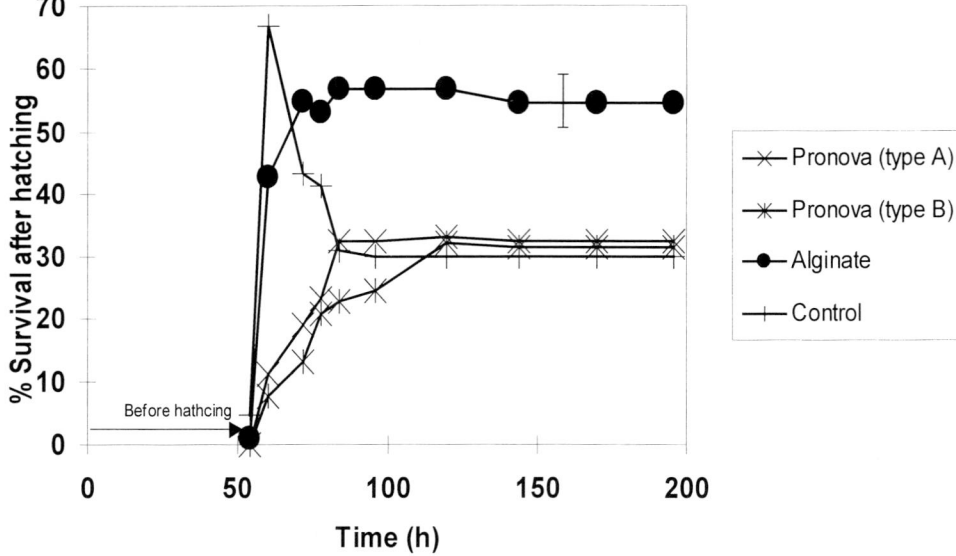

Figure 1: The effect of alginate type on the survival of X. laevis embryos vs. elapsed time. The ± 5% bar indicates the experimental uncertainty.

Moreover, as can be assumed from the vast experience accumulated from cell-entrapment experiments, alginate gels maintain cell viability [10]. We performed our first coating and storage experiments under so-called "harsh" conditions (storage conditions # 1), thereby making it easy to conclude whether a particular coating is beneficial relative to uncoated embryos: the conditions were modified from those recommended by Wu and Gerhart, (1991) and Phillips, (1979). However, we increased the proportion of eggs to medium solution such that instead of including 10 embryos per 50 ml medium, we introduced 30 embryos per 50 ml and allowed only passive natural aeration to take place, thereby increasing the stress on the coated embryos. Embryo's medium was contained within sterile container and conditions. Coated embryos were also introduced into the same medium, except that the sterile medium was exposed to non-sterile conditions (storage conditions # 2). Coated embryos were also maintained under the "ideal" conditions reported by Wu and Gerhart, J. (1991) and Phillips, R.J. (1979) to check their performance in a more favorable environment (storage conditions # 3).

The survival of embryos vs. time under storage conditions #1 is shown in Fig. 1. The survival percentage is equivalent to the accumulated number of hatching embryos to a maximal or asymptotic survival value, and is the number of embryos left after they begin to die. The accumulated survival percentage [1] of non-coated embryos was ~4.6, 54 h after fertilization, increasing to 66 after 60 h (Fig. 1). Percent survival then decreased to 41 after 78 h and reached an asymptotic value of 30 between 84 and 196 h. Reduced survival percentages could be due to the secretion of nitrates or other substances into the medium by the developing embryos. In parallel to the survival-prospects study, embryo development was monitored by comparing their developmental stage (observed through a binocular lens) to that of non-coated embryos [3]. No difference between the two was observed, implying that the coating film does not hamper embryo development.

A large difference between the alginates was observed: the alginate with a high proportion of M held better prospects for embryos hatching. The asymptotic survival value for the high-M coating was ~53-56% vs. 22 to 32% for the high-G coatings. This is due to the fact that the higher the G content, the stronger the gel (i.e. the film coating the embryo). Coated embryos appeared to develop in a normal fashion, similar to non-coated embryos, however the strong coating (high G) prevented hatching embryos from bursting the thin coating film and thus 120 h after fertilization, they perished. No significant differences were found between the two alginates extracted from the *L. hyperborea*. Significant differences in survival rate were observed between the high-M and high-G alginates. It is important to note that regular hatching is a consequence of enzymatic action as well as natural mobility of the developing embryo [12]. Mobility helps the embryo emerge from its jelly coat, but is not enough to break through a high-G coating film. An additional difference was observed between the uncoated and coated embryos. The former reached their maximal survival rate a short time after hatching began. For the coated systems, a maximal value was reached ~25 h later. This means that some delay in hatching was effected by the coating process.

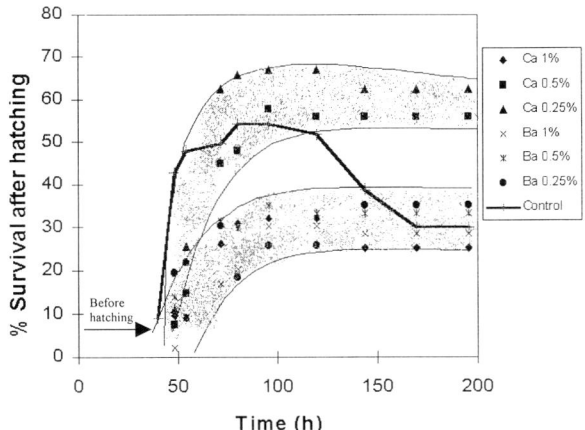

Figure 2: The effect of cross-linking agents on the survival of X. laevis embryos vs. elapsed time in the case of storage condition #1. Stippled areas emphasize coatings with which no significant difference between percent survival was detected.

This delay is important for laboratories interested in performing longer-term experiments with embryos. Another advantage is that the embryo hatches at a much more developed stage in relative to non-coated embryos. Thus the embryo is less prone to mechanical damage or microbial contamination. Bacteria have been reported to stick to the surface of the J_3 outer layer of the jelly coat and that removal greatly reduces their number. Coating embryos could therefore eliminate the need for including neomycin sulfate in the media [12]. Based on these preliminary coating experiments and the conclusion that embryos are not capable of breaking through films with a high G content, further coating experiments were carried out only with the high-M alginate.

Sodium alginate can be cross-linked with several divalent ions. We checked the performance of the high-M alginate coating after cross-linking with different concentrations of Ca or Ba. The embryos were immersed in the same medium (one-third MMR solution) but the conditions were not sterile, and the embryos were prone to microbial contamination. Fig. 2 demonstrates the relative successes of the different coatings.

Coatings produced with alginate cross-linked with 0.25 and 0.5% CaCl$_2$ were most successful, i.e. a higher percentage of hatching and survival was observed relative to the controls (non-coated) or the other variously coated embryos. Lower concentrations of Ba or Ca, i.e. 0.0625-0.125%, were avoided because they did not produce a "nice" coating (as per the definition of a "coating", see Introduction). Ba is known to produce stronger gels with alginate than Ca at the same alginate concentration. In addition, the higher the

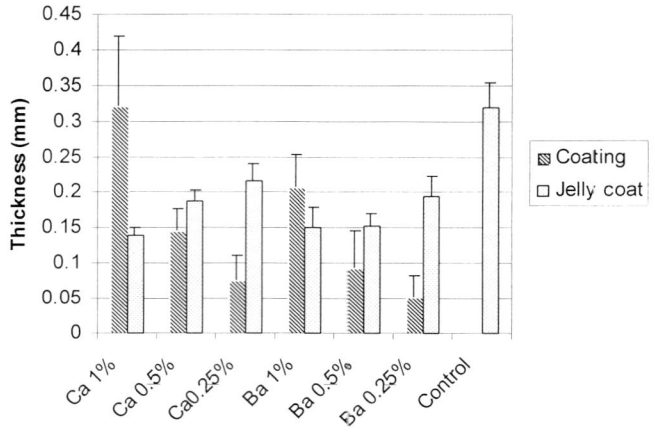

Figure 3: Influence of salt type and concentration on the thickness of the alginate coating and the embryo's jelly coat, 4 h after fertilization.

concentration of the cross-linking agent with the same predetermined alginate concentration, the stronger the gel. As noted earlier, a stronger coating limits the percent of hatched embryos. Another explanation for our findings is that diffusivity decreases with increasing alginate concentration or gel strength. A third, potentially more important explanation is the toxicity of Ba ions to embryos, as reported by Spangenberg and Cherr (1966).

Fig. 3 presents the thickness of the film and jelly coat for coated embryos. Coating thickness was not more than 16% of the embryo's natural Ferret diameter, including the coating (from 0.07 to 0.2 mm), and in general, not thicker than the embryo's natural jelly coats. During the course of natural fertilization, the jelly coat swells when it is immersed in water [14]. In this study, the alginate coating limited the swelling of the jelly coat. After 4 h of observation, we noted that the thinner the coating, the more swollen the natural jelly coat. The amount of cross-linking agent in the system was much higher than the stoichiometric amount necessary to cross-link the alginate [15,16]. After the spontaneous cross-linking, the strength of the coating film increased and its thickness decreased. After 24 h, film thickness was reduced by ~10 to 40% for the different cross-linking agents used, while the film strengthened. The final outcome of this effect was a limitation of the natural jelly coat's swelling, which was either slowed or prevented by the strengthening of the coating. After ~24 h, the film appears to reach its maximal strength [16] and the jelly coat stops swelling. The coating prevents the jelly coat from reaching its optimal thickness, as compared to non-coated embryos.

The medium in our case was prone to microbial contamination because the petri dishes were stored open, under non-sterile conditions. It was interesting to note the effect of the alginate coating on the microorganism's development as recorded in relative light units (RLU) vs. time. RLU can easily be transformed to microbial counts with a conversion factor. Using a such conversion we found that about 20 h after the coating experiments began, total counts were on the order of 10^1 to 10^2, reaching values of 2 to 5×10^3 after 48 h, and average values of 0.7 to 1.5×10^4 after 72 h. One striking observation was that the non-coated embryos were much more contaminated than their coated counterparts. Normally, microorganisms are glued to the jelly coat, causing considerable contamination of the non-coated embryo [17]. The thin film coating the embryo prevented microorganisms from being glued directly to the jelly coat, thereby reducing contamination. In addition, it is important to note that the alginate-based coating is not a good medium for microorganism development. Moreover, the fact that the coated embryos hatched at a more mature stage than their non-coated counterparts made them more resistant to microbial contamination. Finally, it must be remembered that bacterial growth, which naturally results in oxygen inhibition, causes death, particularly in newly emerged young frogs [17]; in this light, the contribution of the coating becomes much more important.

Light and electron microscopy observation indicated that the alginate coating is glued directly to the exterior of the embryos, i.e. the J_3 layer, with no observable gap between the two (Fig.4).

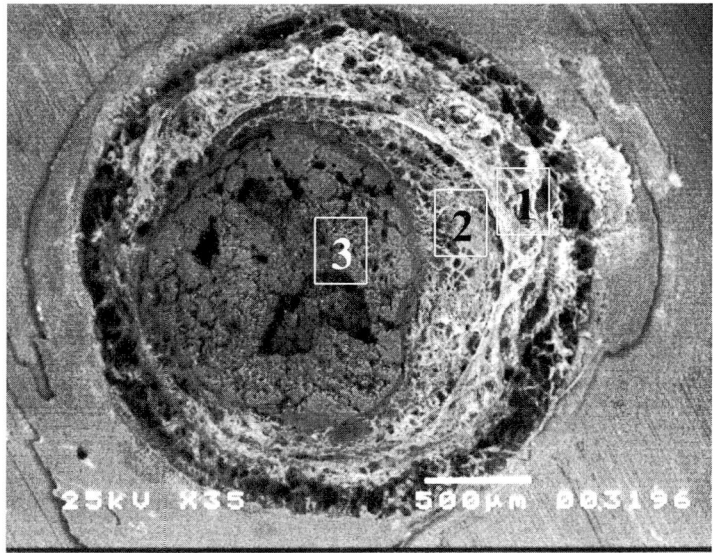

Figure 4: SEM micrograph of X. laevis embryo: 1) Alginate coating. 2) Jelly coat. 3) Embryo.

To study the effect of different conditions on the coated embryos' survival, they were introduced into the same medium, which, this time, was sterile. The results of these experiments are shown in Fig. 5. Two main treatment groups appear to emerge: the first reaching asymptotic survival rates of 64 to 70% from 70 h after fertilization, and the second reaching smaller asymptotic survival values of 34 to 52% at the same time point. This latter group was comprised of coatings cross-linked with 0.5 and 1% BaCl$_2$, again demonstrating barium's toxicity. Since the medium was sterile, the advantages of successful coating were less salient. Although the controls (non-coated) had an initially higher hatching percentage than the coated embryos, the survival prospects of the embryos coated with alginate cross-linked with calcium (0.25, 0.5 or 1%) or barium (0.25%) were better. This can be due to defense against mechanical damage and hatching at a later stage when the embryo is more developed.

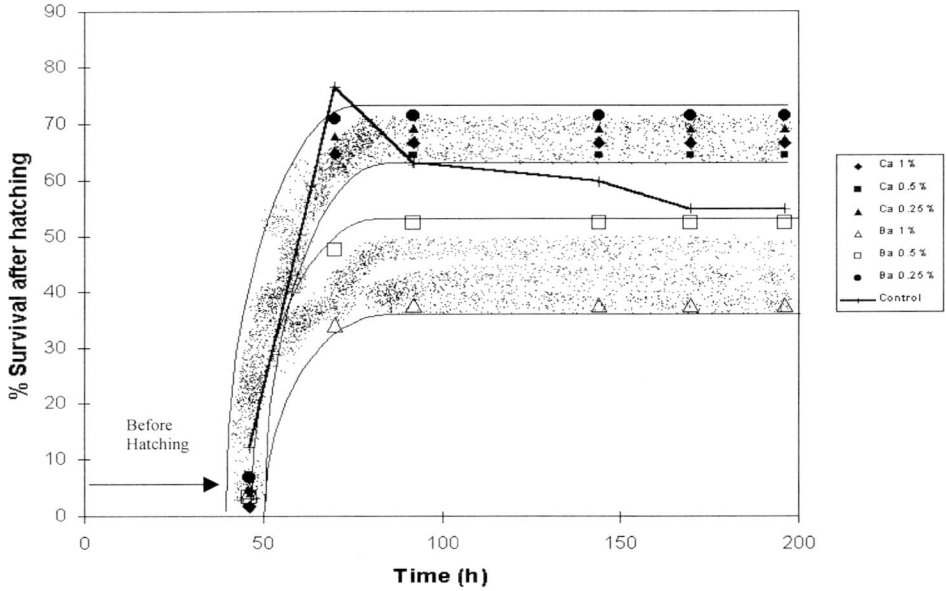

Figure 5: The effect of cross-linking agents on the survival of X. laevis embryos vs. elapsed time in the case of storage condition #2. Stippled areas emphasize coatings with which no significant difference between percent survival was detected.

The final storage experiments consisted of immersing the coated embryos into dechlorinated, aerated, circulating tap water (see M&M). This situation more closely resembles that found in nature. A significant difference between the controls and coated systems was observed. The control exhibited an asymptotic survival percentage of nearly 58, whereas the coated embryos reached no more than 31%. However, in this case the coated embryos held a unique advantage. Survival reached an asymptotic value at least 40 h after the control. Such a system can therefore be used in laboratories for long-term experiments.

4. CONCLUSION

Alginate-coating of embryos is different from entrapment in that the coating around the embryo is thinner, comprising no more than 15% of the embryo's diameter. The thin alginate film is glued to the natural jelly coat directly or via a calcium bridge. It protects the embryo against bacterial contamination and mechanical damage, and under stressful conditions, improves its performance. Under "comfortable" conditions, its main contribution is that it extends the time for which the embryos can be utilized in the laboratory. Coated embryos hatch at a later stage and are therefore more developed and better able to resist natural hazards.

REFERENCES

1. M. Wu and J. Gerhart, Methods Cell Biol. 36 (1991) 3.
2. J.L. Hedrick and T.J. Nishihara, Electron Microscopy Technique. 17 (1991) 319.
3. P.D. Nieuwkoop and J. Farber, In Normal table of <u>Xenopus laevis</u> (Daudin). pp. 162. New York & London: Garland Publishing, Inc. 1994.
4. G. Waes, Milchwissenschaft. 12(39) (1984) 707.
5. A. Nussinovitch and N. Kampf, Lebensmittel-Wissenchaft und-Technologie, 26 (1993) 469.
6. A. Nussinovitch and V. Hershko, Carbohydrate Polymers, 30 (1996) 185.
7. V. Hershko, D. Weisman and A. Nussinovitch, J. of Food Science, 63 (2) (1998) 317.
8. V. Hershko and A. Nussinovitch, J. Agricultural and Food Chemistry, 46 (8) (1998) 2988.
9. N. Kampf and A. Nussinovitch, Polymer Networks, June 28 - July 3, Trondheim, Norway. (1998).
10. F. Lim and A.M. Sun, Science. 210 (1980) 908.
11. R.J. Phillips, J. The Institute of animal Technology. 30(1) (1979) 11.
12. E.J. Carroll (Jr.) and J.L. Hedrick, Developmental biology. 38 (1974) 1.
13. J.V Spangenberg and G.N Cherr, Environmental Toxicology and chemistry. 15(10) (1966) 1769.
14. R.S. Seymour, Israel J. of Zoology. 40 (1994) 493.
15. M. Glicksman, In: Gum Technology in the Food and Other Industries. pp. 152. Chapman and Hall, London, UK. 1969.
16. A. Nussinovitch, In: Gum Technology in the Food and Other Industries. pp.176. Chapman and Hall, London, UK. 1997.
17. J.S. Davys, The breeding of Xenopus laevis on a large scale in the laboratory. Animal Technology, 37(3) (1986) 217.

6. NUTRITION

HYDROCOLLOIDS – PART 2
Edited by K. Nishinari
2000 Elsevier Science B.V.

383

Dietary fiber and gastrointestinal functions

S. Innami, J. Shimizu and K. Kudoh

Department of Nutritional Science, Faculty of Applied Bio-Science, Tokyo University of Agriculture, Setagaya-ku, Tokyo 156-8502, Japan

Dietary fiber exhibits various physiological functions by expressing own physico-chemical characteristics at passage through gastrointestinal tract. This paper dealts with gastrointestinal functions of several dietary fibers with hydrophilic and hydrophobic properties. Curdlan and gellan gum, gell-forming polysaccharides produced by bacteria, are first described in relation to fermentation and lipid metabolism in rats. Next, the relation between dietary fiber and the intestinal immune system which has been hardly so far elucidated was investigated in respect to B lymphocyte responses in the intestinal mucosa of rats. Male rats of the Sprague-Dawley strain were fed diets containing cellulose powder, curdlan and gellan gum at 5% for 4 weeks. Shortened gastrointestinal transit time, increased weight of cecal contents and feces were observed in the gellan gum group compared with curdlan group. The amounts of short-chain fatty acids and lactic acid in cecum were markedly increased in the curdlan group. The hepatic cholesterol concentration in the curdlan group was significantly decreased compared with that in the cellulose and gellan gum groups. These results reveal that curdlan is easily hydrolized and fermented by intestinal bacteria but gellan gum not easily. The reduction of hepatic cholesterol concentration in the curdlan group may be attributable to a remarkable increase of fermentation products by intestinal bacteria. In the next experiments, rats were given diets containing cellulose, arabic gum, corn husk (Celfur), glucomannan, curdlan and indigestible dextrin at 5% for 3 weeks. Some kinds of dietary fiber induced proportions of κ-light chain and IgA-presenting B lymphocytes in the mucosa of small intestine and cecum, and increased IgA secretion to cecal contents and feces, but the degree of responses differs depending on the type of dietary fiber. These results show that some dietary fibers are possible to be involved in the intestinal immune system of rats, suggesting a new development on physiological functions of dietary fiber.

1. INTRODUCTION

Shortly after Trowell [1] proposed a revolutionary concept "Dietary Fiber", Burkitt & Trowell [2] presented an important hypothesis at 1975 in which dietary fiber (DF)-deficient diet or insufficient diet may relate to occurrences of non-infective intestinal diseases such as constipation, diverticular disease, diaphragm hernia, appendicitis, intestinal polyp and colon cancer, and metabolic diseases such as gallstone, coronary heart disease, diabetes and obesity. This hypothesis has been extensively elucidated by experimental, clinical and epidemiological studies, although some remain unclear. It has been known that the role of DF in human diseases depends on its behavior at passage through the gastrointestinal tract. The behavior within gut is regulated by the composition and content of the ingested DF. Various physiological functions of DF are attributed to the following several physico-chemical properties: water holding capacity, swelling, diffusion suppressing ability, binding ability, and ease or resistant to bacterial degradation and fermentation. Thereby, DF is regarded as acting on the modifications of nutrient absorption, entero-hepatic sterol metabolism, intestinal microflora and decomposition and fermentation in intestinal tract, and further on the modulations of motility of digestive tract, gastrointestinal transit time, and bulk and weight of stool. Consequently, it is recognized that DF refers effectively to prevent or improve the non-infective intestinal diseases and metabolic diseases.

In the present paper, curdlan (CL) and gellan gum (GG), which are gel-forming polysaccharides produced by bacteria and their physiological functions are not sufficiently clarified, are first described in relation to production of organic acids such as short-chain fatty acids (SCFA) in the intestinal tract and lipid metabolism in rats. Furthermore, the relation between DF and the intestinal immune system which has been hardly so far elucidated is picked up as the next subject. The latter is newly developing and important field in DF research. Then, the effect of water soluble and insoluble DF on B lymphocyte responses in the intestinal mucosa of rats is investigated.

2. EFFECT OF GELLATING POLYSACCHARIDES ON INTESTINAL FERMENTATION AND LIPID METABOLISM IN RATS

It has been known that many gellating polysaccharides exhibit various important physiological functions in animals and humans. For example, pectin, glucomannan and guar gum have been found to reduce blood cholesterol and triglyceride levels[3], and also to suppress an elevation of postprandial blood glucose level[4]. CL and GG picked up here have been used as food additives for various food processing in food industry. As either having gell forming ability, they are utilized as materials for producing various dessert food or cooled cakes. CL [5-6], a polysaccharide produced by *Alacaligenes faecalis* var. *myxogenes*, is β-1,3-glucan linked to straight chain D-glucose. It is insoluble in water, but it becomes gelled when its aqueous suspension is heated. On the other hand, GG [7] produced by *Psedomonas elodea*, is a water-soluble polymeric polysaccharide consisting of two D-glucose molecules and one each L-rhamnose and D-glucuronic acid molecule. GG is also known to easily form gels under the condition of low concentration of cations.

Male 4-weeks old Sprague-Dawley rats were divided into three groups : control (cellulose powder, CP), CL and GG and were given the experimental diets containing each 5% level of these fibers for 4 weeks [8]. Serum concentrations of total cholesterol, HDL-cholesterol, triacylglycerol and phospholipids were determined with Test Wako Kit (Wako Pure Chemical Industries, Ltd., Osaka). Concentrations of total cholesterol and triacylglycerol in the liver were determined by the Zak method and the Fletcher method, respectively. Bile acids in feces were determined by gas-liquid chromatography (Model GC-12A, Shimadzu Co., Kyoto). As the internal standard, 23-nor-deoxycholic acid was used. The amounts of SCFA in cecal contents were analyzed by programmed-temperature gas chromatography (Shimadzu GC-12A, Shimadzu Co., Kyoto) equipped with glass column (3.2mm × 1.8m) packed with 10% SP-1200/1% H_3PO_4 on 80/100 mesh Chromosorb WAW (Superlco Inc., PA, USA). Lactic acid in cecal contents was measured using an F-Kit (Boehringer Mannheim Co., Tokyo).

Table 1 showed that the cecal weight of the CL group was approximately twice as much as that of the control group. Conversely the GG group showed a lower weight which was not significantly different from the control group. The weight of cecal contents in the CL group increased with statistical significance compared with that in the control and GG groups. Cecal contents of the CL group was highly viscous, but that of the GG group was like gel. As cecal contents in the GG group consisted of gelled particles, it is presumed that GG administered in powder form became gelled in the rat stomach. It is further assumed that physical properties of gels decreased the weight of the cecal tissues and contents and increased the number of fecal lumps. The pH of the cecal contents of the CL group was significantly lower than those of the control and GG groups. Gastrointestinal transit time in the GG group was significantly shortened compared with those in the other two groups. Many of water soluble DF are characterized to be easily degradated and fermented by intestinal bacteria [9] and extend the gastrointestinal transit time. The amounts of SCFA and lactic acid in the cecal contents of the CL group significantly increased compared with those of the control and GG groups (Table 2). Lowerd pH of the cecal contents is recognized to be attributable mainly to the increased concentrations of organic acids such as lactic acid and succinic acid [10]. GG is considered to be hardly degraded by intestinal bacteria.

Table 1 Effect of curdlan and gellan gum on characteristics of digestive tracts, fecal weight and gastrointestinal transit time (GTT) in rats.[1,2]

Items	Cellulose	Curdlan	Gellan gum
Cecum			
Weight (g)	0.50[a]	1.1C[b]	0.45[a]
Contents (g)	2.11[a]	4.37[b]	1.86[a]
pH	7.35[a]	6.12[b]	7.47[a]
Colon plus rectum			
Weight (g)	1.08[a]	1.39[b]	1.36[b]
Length (cm)	19.7	20.9	20.1
Feces			
Lumps (number/day)	16.3[a]	12.4[a]	32.7[b]
Wet weight (g/day)	2.20[a]	1.39[b]	2.42[a]
Dry weight (g/day)	1.87[a]	0.88[?]	1.60[c]
Moisture (%)	15.2[a]	35.5[b]	33.1[b]
pH	7.62[a]	5.69[b]	7.23[c]
GTT (min)	684[a]	673[a]	514[b]

[1] Values are means (n =7).   Values in the same row not sharing a common superscript letter are significantly different at $p < 0.05$.

[2] Reprinted from : Reference number [8] by J. Shimizu et al.

Table 2 Effect of curdlan and gellan gum on amounts of SCFA and lactic acid in cecal contents of rats.[1,2]

Items	Cellulose	Curdlan	Gellan gum
	(μ mol/cecal contents)		
Acetic acid	80[a]	199[b]	62[a]
Propionic acid	35[a]	83[b]	19[a]
Butyric acid(n- + iso-)	18[a]	49[b]	13[a]
Total SCFA	139[a]	337[b]	97[a]
Lactic acid	3.0[a]	25.8[b]	1.2[a]

[1] Values are means (n =8).   Values in the same row not sharing a common superscript letter are significantly different at $p < 0.05$.

[2] Reprinted from : Reference number [8] by J. Shimizu et al.

Table 3 showed that no significant inter-group differences in the total cholesterol and HDL-cholesterol concentrations in serum were observed. Serum triacylglycerol concentration significantly increased in the GG group compared with the control group and tended to rise in the CL group. Total cholesterol concentration in the liver of the CL group decreased significantly compared with those of the control and GG groups (Table 3).

Table 3 Effect of curdlan and gellan gum on serum and liver lipid concentrations and liver weight in rats.[1,2]

Items	Cellulose	Curdlan	Gellan gum
Serum			
Total cholesterol (mg/dl)	129	123	130
HDL-cholesterol (mg/dl)	71.0	72.4	62.0
Triacylglycerol (mg/dl)	180[a]	241[ab]	304[b]
Phospholipids (mg/dl)	256	250	275
Liver			
Weight (g)	17.9	16.3	16.9
Total cholesterol (mg/g)	3.62[a]	2.65[b]	3.56[a]
Triacylglycerol (mg/g)	54.7[a]	31.5[b]	45.8[ab]
Phospholipids (mg/g)	26.4	24.5	27.5

[1] Values are means ($n = 7$): Values in the same row not sharing a common superscript letter are significantly different at $p < 0.05$.

[2] Reprinted from : Reference number [8] by J. Shimizu et al.

The hepatic triacylglycerol concentration reduced significantly in the CL group and tended to decrease in the GG group compared with that of the control group. Fecal neutral sterol excretions were significantly higher in the CL and GG groups than in the control group, especially significant increase in the CL group. There were no significant inter-group differences in the total fecal bile acid excretions. In the control and GG groups, major bile acid was hyodeoxycholic acid, while in the CL group it was β-muricholic acid. Lithocholic acid excretion in the CL group was significantly higher than those in the other two groups. The proportion of secondary bile acids and the CDCA/CA ratio were significantly lower in the CL group than those in the control and GG groups. The theory that effects of DF on cholesterol metabolism are mediated by the increased steroid excretion into feces [11] has been quite influential so far, but there are also reports that no relevance is observed between the two [12-13]. There is a report that SCFA metabolites by intestinal bacteria in the intestinal tract, participate in affecting cholesterol metabolism [14]. This is because propionic acid among SCFA was observed to lower cholesterol synthesis in vitro experiment using cultured hepatocytes. The amount of propionic acid in the cecal contents of the CL group was significantly higher than those of the CP and GG groups. Topping and Pant [15] pointed out that comparison is difficult because the concentration of propionic acid differs in in vitro and in vivo experiments.

3. EFFECT OF DIETARY FIBER ON INTESITINAL IMMUNE RESPONSES IN RATS

Intestinal tract is an important immune organ for animals and human.
In fact, it has gut associated lymphoid tissues (GALT) such as Peyer's patches (PP), which is responsible for local immunity against allergens from diets or invasion of bacteria and viruses.   B lymphocytes which is one of the responsible cells in charge of intestinal immunity, include κ -light chain and IgA-presenting lymphocytes.   The latter produces and secretes IgA, the principal antibody for immunization of the intestinal tract.   However, it has been hardly so far investigated how DF would influence the intestinal immune system.   This problem is newly developing and important field in DF research.   Then, the effect of water soluble and insoluble DF on B lymphocyte responses in the intestinal mucosa of rats is reported [16].

Male 4-weeks-old Sprague-Dawley rats were used.   DF samples used for the experiments were cellulose (CP, Oriental Yeast Co., Osaka), arabic gum (AG, Wako Pure Chemical Industries, Ltd., Osaka), corn husk (Celfur, CF, Nihon Shokuhin Kako Co., Tokyo), purified indigestible dextrin (PIDD, Sanmatsu Industry Co., Tokyo), CL (Takeda Chemical Industries, Co., Osaka) and glucomannan (GM, Shimizu Kagaku Co., Hiroshima).   CP was used as the control.   Rats were given the experimental diets containing each 5% level of DF samples for 3 weeks.   After gating lymphocyte classes of the cell groups obtained, the proportions of κ -light chain (T-039, Cosmo Bio Co., Tokyo) and IgA (MARA-1-F, Experimental Immunology Unit, Brussels)-presenting lymphocytes were analyzed by a flowcytometer (Coulter Epics Elite, Coulter Electronics, Miami, FL) using monoclonal antibody labeled with fluorescent isothiocyanate (FITC). The proportions of cells presenting CD4 (MCA 55F, Serotec, Oxford), CD8(MCA 48P, Serotec) and CD3 (U-MR5301, Caltag Laboratories Co., CA) were similary measured using a flowcytometer.

Table 4 showed that the weights of the cecum and cecal contents were significantly increased in the AG and PIDD groups compared with the CP group.   The weight of the colon was significantly decreased in the AG group as compared with the CP and CF groups.   The pH value of the cecal contents was significantly lower in the other groups except the CF group than that in the CP group.   The amounts of SCFA and lactic acid in the cecal cotents were shown in Table 5.   The amount of propionic acid increased significantly or tended to increase in the AG and PIDD groups as compared with the CP group.   n-Butyric acid amount showed a tendency to increase in the CF and PIDD groups as compared with the CP group.   The amount of lactic acid in the PIDD group or the AG group increased significantly or tended to increase as compared with the CP group, respectively.

Table 4 Body weight gain, food intake, feed efficiency, weight of organs and pH of cecal contents.[1,3]

Item	CP[2]	AG[2]	CF[2]	PIDD[2]
Body weight gain (g)	200	189	201	203
Food intake (g)	462[a]	453[a]	506[b]	471[a]
Feed efficiency	0.43[a]	0.41[ab]	0.39[b]	0.43[a]
Small intestine (g)	9.36	9.77	9.19	8.95
Cecum (g)	0.61[a]	0.94[b]	0.58[ac]	1.15[b]
Cecal contents (g)	2.42[a]	4.12[b]	2.13[a]	3.75[b]
pH of cecal contents	7.64[a]	6.73[b]	7.48[a]	6.23[c]
Colon (g)	1.07[a]	0.87[b]	1.08[a]	1.02[ab]

[1] Values are means (n= 6 or 7).   Values in the same row not sharing a common superscript letter are significantly different at $p<0.05$.

[2] CP, cellulose; AG, arabic gum; CF, celfur; PIDD, purified indigestible dextrin.

[3] Reprinted from : Reference number [16] by K. Kudoh et al.

Table 5 Amounts of SCFA and lactic acid in cecum.[1,3]

Acid	CP[2]	AG[2]	CF[2]	PIDD[2]
	(μ mol/cecum contents)			
Acetic acid	28.2	29.1	15.9	32.3
Propionic acid	2.01[a]	6.14[b]	1.15[a]	5.20[b]
iso-Butyric acid	0.24	0.12	0.20	0.13
n-Butyric acid	2.07	1.57	2.62	3.34
iso-Valeric acid	0.25	0.15	0.20	0.11
n-Valeric acid	0.20	0.16	0.32	0.13
Total SCFA	32.8	37.2	19.6	40.6
Lactic acid	25.0[a]	43.0[ab]	22.5[a]	69.4[b]
Total acids	67.3	80.2	53.5	138

[1] Values are means (n= 6 or 7).   Values in the same row not sharing a common superscript letter are significantly different at $p<0.05$.

[2] See footnote to Table 4.

[3] Reprinted from : Reference number [16] by K. Kudoh et al.

The proportion of κ -light chain-presenting lymphocytes in the mucosa of intestinal tract was not different in the small intestine, but the CF group showed a tendency to increase as compared with the CP group.   In the cecum, those of the three DF groups significantly increased as compared with the CP group. The proportion of

IgA-presenting lymphocytes in the mucosa of small intestine and cecum showed a tendency to increase in the CF and PIDD groups as compared with the CP group. In colonic mucosa, the proportions of κ-light chain and IgA-presenting lymphocytes were not different among the DF groups.

Table 6   Proportion of κ-light chain and IgA-presenting lymphocytes in intestinal mucosa.[1,3]

Mucosa	CP[2]	AG[2]	CF[2]	PIDD[2]
		(κ-light chain%)		
Small intestine	55.2	61.4	76.9	57.0
Cecum	46.7[a]	59.4[b]	63.3[b]	66.1[b]
Colon	51.3	63.4	55.0	63.8
		(IgA%)		
Small intestine	47.9	56.8	69.2	67.8
Cecum	56.5	52.9	64.3	63.4
Colon	63.2	63.1	60.1	68.5

[1] Values are means (n= 6 or 7).   Values in the same row not sharing a common superscript letter are significantly different at $p<0.05$.

[2] See footnote to Table 4.

[3] Reprinted from : Reference number [16] by K. Kudoh et al.

Most orally administered DF undergoes decomposition and fermentation by intestinal microflora to form organic acids, acidify intestinal contents and improve the environment.   SCFA is known to affect the host in various ways.   There are many reports [17-18] on the effects of SCFA, particularly of butyric acid, on cell proliferation.   SCFA is believed to promote the proliferation and differentiation of normal cells and to suppress pathological cells such as cancer cells.   It is known that some lactic acid-producing bacteria have an immunopotentiating function. *Bifidobacterium* [19-20] is reported to enhance immunopotentiation by increasing IgA antibody production in vitro as well as in vivo by oral administration to germ-free mice, and Rao [21] assumed that changes in intestinal microflora by DF affect the immunological functions.   Lim et al [22] recently reported that the IgA concentrations in serum and the culture medium of mesenteric lymph nodes (MLN) of rats fed pectin were much higher than those fed cellulose.   They observed that the proportions of CD4[+] T lymphocytes in MLN and spleen and the CD4[+]/CD8[+] ratio increased in the pectin group respectively as compared with the control group, and suggested that DF may be responsible for modulating the intestinal immune function.

We [23] observed in our another experiment that proportion of IgA-presenting lymphocytes significantly increased in cecal mucosa of rats fed GM and CF. We [23] further found a significant increase of IgA secretion into cecum in the groups fed fermentable GM and CL, but not in the group fed CP. But amounts of IgA in feces of the GM and CL groups were significantly decreased or tended to decrease compared with that of the CP group. It was assumed that enhancement of IgA secretion from cecal mucosa of the GM and CL groups might be attributed to the increase of intestinal flora or their metabolites resulted from decomposition and fermentation of DF in the cecum. Since it is known that secretary IgA inhibits invasion of allergens, virus and bacteria, increase of IgA secretion to intestinal tract may strengthen the gut immune system. These results may endow a new important role on physiological functions of DF in biodefensive mechanism.

REFERENCES

1. H.C. Trowell, Atheroscleosis.16 (1972) 138; Am.J.Clin.Nutr.25 (1972) 926.
2. D.P. Burkitt and H.C. Trowell, Refined Carbohydrate Food and Disease,Some Implications of Dietary Fibre, Academic Press, London,1975.
3. J.W. Anderson, Dietary Fiber in Health & Disease, p.126, D. Kritchevsky and C. Bonfield (eds.), St. Paul, USA, 1995.
4. D.A. Jenkins, A.L. Jenkins, T.M.S. Wolever, V. Vuksan, A.V. Rao, L.U.Thompson and R.G. Josse, Dietary Fiber in Health & Disease, p.137, D. Kritchevsky and C. Bonfield (eds.), St. Paul, USA, 1995.
5. T. Harada, M. Masuda, K. Fujimori and I. Maeda, Agric. Biol. Chem., 30 (1968) 196.
6. T. Harada, A. Misaki and H. Saito, Arch. Biochem. Biophys., 124 (1968) 292.
7. D.M. Maurice and S.P. Srinivas, J. Pharm. Sci., 81 (1992) 615.
8. J. Shimizu, M. Wada, T. Takita and S. Innami, J. Nutr. Sci. Vitaminol., 45 (1999) 251.
9. J.R. Jacobs, Dietary Fiber; Chemistry, Physiology and Health Effects, p.389, D. Kritchevsky, C. Bonfield and J. W. Anderson (eds.), New York and London, 1990.
10. S. Hoshi, T. Sakata, K. Mikuni, H. Hashimoto and S. Kimura, J. Nutr., 124 (1994) 52.
11. B.H. Arjmandi, J. Ahn, S. Nathani and R.D. Reeves, J. Nutr., 122 (1992) 246.
12. J.W. Anderson, D.A. Deakins and S.R. Bridges, Dietary Fiber; Chemistry, Physiology and Health Effects, p.339, D. Kritchevsky, C. Bonfield and J. W. Anderson (eds.), New York and London, 1990.
13. N. Nishimura, H. Nishikawa and S. Kiriyama, J. Nutr. 123 (1993) 1260.
14. R. S. Wright, J.W. Anderson and S. R. Bridges, Proc. Soc. Exp. Biol. Med., 195 (1990) 26.

15. D.L. Topping and I. Pant, Physiological and Clinical Aspects of Short Chain Fatty Acids, p.459, J. H. Cummings, J. L. Rombeau and T. Sakata (eds.), Cambridge Univ. Press, London, 1995.
16. K. Kudoh, J.Shimizu, M. Wada, T. Takita, K. Kanke and S.Innami, J. Nutr. Sci. Vitaminol., 44 (1998) 103.
17. R. L. Joanne and P. K. Pamela, J. Nutr., 123 (1993) 1522.
18. R. H. Whitehead, G. P. Young and P. S. Bhathal, Gut, 27 (1986) 1457.
19. H. Yasui and M. Ohwaki, J. Dairy Sci., 74 (1991) 1187.
20. S. Yamazaki, K. Machii, S. Tsuyuki, H. Momose, T. Kawashima and K. Ueda, Immunology, 56 (1985) 43.
21. A.V. Rao, Dietary Fiber in Health & Disease, p.257, D. Kritchevsky and C. Bonfield (eds.), St. Paul, USA, 1995.
22. B. O. Lim, K. Yamada, M. Nonaka, Y. Kumamoto, P. Hung and M. Sugano, J. Nutr., 127 (1997) 663.
23. K. Kudoh, J. Shimizu, A. Ishiyama, M. Wada, T. Takita. Y. Kanke and S. Innami, J. Nutr. Sci. Vitaminol. , 45 (1999) 173.

HYDROCOLLOIDS – PART 2
Edited by K. Nishinari
2000 Elsevier Science B.V.

393

Effects of partially hydrolyzed guar gum on the morphological surface structure of intestinal mucosa in the rat

M. Tetsuguchi[a], S. Mamiya[b], T. Inden[b], M. Katayama[c] and Y. Sugawa-Katayama[a]

[a] Faculty of Human Life Science, Osaka City University, 3-3-138, Sugimoto, Sumiyoshi-ku, Osaka 558-8585, Japan
[b] Nutritional Foods Division, Central Research Laboratories, Taiyo Kagaku Co., Ltd., 1-3, Takara-cho, Yokkaichi 510, Japan
[c] Department of Applied Biological Chemistry, Osaka Prefecture University, 1-1, Gakuen-cho, Sakai 599-8531, Japan

We intended to know how the partially hydrolyzed guar gum affects the morphological fine structure of intestinal mucosa in rat. The *Sprague-Dawley* male rats were fed a diet containing 5% partially hydrolyzed guar gum (experimental diet) for 4 weeks; a significant increment in the weight of the cecum and in its contents, and a significant decrement in the fecal weight were observed in comparing with those fed a 5% cellulose diet (control diet). The transit time of gastrointestinal tract was extended by the partially hydrolyzed guar gum supplementation. The surface structures of ileal and cecal mucosa changed markedly to the abnormal in the rats fed the experimental diet; their microvilli of experimental diet group were more tightly packed than those of the control, and more intestinal contents adhered to the tops of the microvilli in the experimental diet group. When rats were fed a diet containing 1% partially hydrolyzed guar gum together with 4% cellulose, the effects of the partially hydrolyzed guar gum were reduced, significantly.

In conclusion, excess of a partially hydrolyzed guar gum exerted effects on the ileal and cecal mucosal surface structure.

1. INTRODUCTION

Some undigestable polysaccharides have been used as food additives. In the digestive tracts, they are expected to play some roles as dietary fibers [1, 2]. Out of the various dietary fibers, partially hydrolyzed guar gum, prepared by digesting guar gum with β-1,4-endomannanase, shows less viscosity (approx. 1/1000) than plain guar gum [3]. It has been reported that physiological activity decreases when polysaccharides are hydrolyzed to lower molecular weight substances [4, 5]. The partially hydrolyzed guar gum shows various physiological effects such as improvement of glucose tolerance [6], enhancement of mineral absorption [7], hypo-cholesterolemic and hypolipidemic effects [8, 9] and improvement of the intestinal microflora balance [10], but there is no report on whether partially hydrolyzed guar gum having an effect on the intestinal mucosal surface structure. We have observed that the degree of the morphological changes of gastrointestinal mucosa in rat fed pectin depended upon the concentration of pectin in the diet [11]. The changes of morphological structure were also dependent on the type of dietary fiber [1, 2], and were likely to contribute to the effect of fiber on the digestive and absorptive functions.

We intended to examine the effects of partially hydrolyzed guar gum on the morphological fine structure of intestinal mucosa in rats, and obtained results suggesting those effects owing to its viscosity in the digestive tract.

2. MATERIALS AND METHODS

2.1. Experimental animals
Three-week-old *Sprague-Dawley* male rats (Japan SLC, Inc.) were fed laboratory chow (CE2, Clea Japan, Inc.) for four days before experimental feeding.

Experiment I [12]: Twelve rats were equally divided into two groups, 5% *cellulose* diet and 5% *partially hydrolyzed guar gum* diet groups. The rats were fed the respective diets for four weeks *ad libitum*. They were housed individually in a stainless cage placed in an air-conditioned room maintained at 22-24°C with a 12-hour light-dark cycle. Throughout the experimental period, the body weight and food intake were monitored every 2 days. The transit time of food through the gastrointestinal tract was measured five days before the end of the feeding period.

Experiment II [13]: Six rats were fed a diet containing 1% partially hydrolyzed guar gum together with 4% cellulose for 4 weeks. Other conditions were the same to the Experiment I.

2.2. Sampling of tissue specimens
After four weeks on the diets, the animals were anesthetized by intraperitoneal injection of Nembutal (pentobarbital sodium, 5mg/100g body weight) and dissected from the abdomen. Tissue samples for scanning electron microscopy were excised from the ileal, cecum and colon.

2.3. Transit time of intestinal contents
Five days before the end of the feeding period, the rats were fasted for nine hours for measurement of the transit time. Each rat was allowed to eat 2g of the respective diet containing 3% carmine for 30 minutes, after which the respective carmine-free diets were given *ad libitum*. The transit time was calculated as follows;
Transit time (h) = (F+L)/2, where F and L were the first and last appearance times(h), respectively, of carmine in the feces as detected visually after the start of ingestion of the carmine diet.

2.4. Scanning electron microscopy
The excised samples of the intestinal tracts were cut into approximately 1 cm² pieces and rinsed in cold saline, carefully. The samples for scanning electron microscopy were taken at 2-3 cm proximal to the cecum (ileum), center of the cecum (cecum) and 2-3 cm distal to the cecum (colon), respectively. They were pre-fixed by infiltration with 2.5% glutaraldehyde in 0.1M phosphate buffer (pH 7.4) for one hour and post-fixed with 1% osmic acid in the same buffer overnight. After fixation, they were dehydrated with ethanol of graded increasing concentrations, infiltrated with 3-methylbutyl acetate and dried with a critical point dryer (Hitachi HCP-2) with liquid carbon dioxide. The samples were mounted on aluminium stubs with electro-conductive silver paste (Dotite) and coated with gold or platinum-palladium with an ion-sputter (Hitachi E-101). The morphological structure of the surface of intestinal mucosa was observed under a scanning electron microscope (Hitachi S-800) at the accelerating voltage of 20kV.

2.5. Reagent
The partially hydrolyzed guar gum was prepared according to the procedure of Taiyo Kagaku Co., Ltd. Other reagents were JIS Guaranteed Grade and Analytical Grade, or their equivalent.

2.6. Statistic analysis
All results were performed by one-way analysis of variance (ANOVA) and were subjected as ±SE (standard error of means) from six rats. Significant differences between groups were tested by using Duncan's multiple-range test and were assessed at $p<0.05$.

3. RESULTS

3.1. Transit time and weight of the feces   (Table. 1)[12, 13]

The transit time of intestinal contents through the gastrointestinal tract was significantly longer in the rats fed the partially hydrolyzed guar gum diet than that of the *cellulose* group.   The feces of the *partially hydrolyzed guar gum* group were more soft and longer than those of the *cellulose* group.   Their mean wet weight of the *partially hydrolyzed guar gum* group was significantly lower than that of the *cellulose* group.

3.2. The weights of cecum and colon, and of their contents   (Table. 2) [12, 13]

The cecum of rats fed the partially hydrolyzed guar gum diet was markedly enlarged.   The weight of the cecum and cecal contents were significantly greater in the rats fed the partially hydrolyzed guar gum diet than those fed the cellulose diet.   However, the weights of the colon and colonic contents were significantly lower in the rats fed the partially hydrolyzed guar gum diet than those in the *cellulose* group (Experiment I).   When rats were fed a diet containing 1% partially hydrolyzed guar gum together with 4% cellulose, the effects of the partially hydrolyzed guar gum were reduced (Experiment II).

Table 1   Transit time and wet weight of the feces.

		Experimental I	Experimental II
	5% Cellulose	5% Partially hydrolyzed guar gum	1% Partially hydrolyzed guar gum + 4% Cellulose
Transit time (h) [1]	30.2 ±2.3 [a]	72.8 ±0.7 [b]	36.5 ±0.3 [c]
Fecal wet weight (g/day)	1.78±0.14 [a]	0.90±0.11 [b]	1.64±0.11 [a]

[1] : Transit time of the contents through the gastrointestinal tract.
Values with different letter are significantly different (Duncan's multiple range test, $p<0.05$).

Table 2   Wet weights of the cecum and colon, and of their contents (g).

	Wet weights of			
	Cecum	Cecal contents	Colon	Colonic contents
5% Cellulose	0.56 ± 0.02 [a]	3.38 ±0.11 [a]	0.90 ±0.02 [a]	0.84 ± 0.12 [a]
5% Partially hydrolyzed guar gum	1.06 ± 0.02 [b]	7.52 ±0.50 [b]	0.74 ±0.05 [b]	0.52 ± 0.12 [b]
1% Partially hydrolyzed guar gum+ 4% Cellulose	0.73 ± 0.06 [c]	4.25 ± 0.36 [a]	0.93 ±0.05 [a]	0.92 ± 0.16 [a]

Values with different letter are significantly different (Duncan's multiple range test, $p<0.05$).

3.3 Morphological changes   [12]

Ileum (Fig. 1):   Scanning electron micrographs of the ileum of the rats fed the cellulose diet showed typical leaf-shaped villi of the small intestine, and individual villi appeared smooth and round in the configuration.   Their microvilli showed regular arrangement.   On the other hand, the villi of the rats fed the partially hydrolyzed guar gum diet were thinner compared to those of the *cellulose* group, and the microvilli were carrying something (probably intestinal contents) on the tip and more tightly packed.

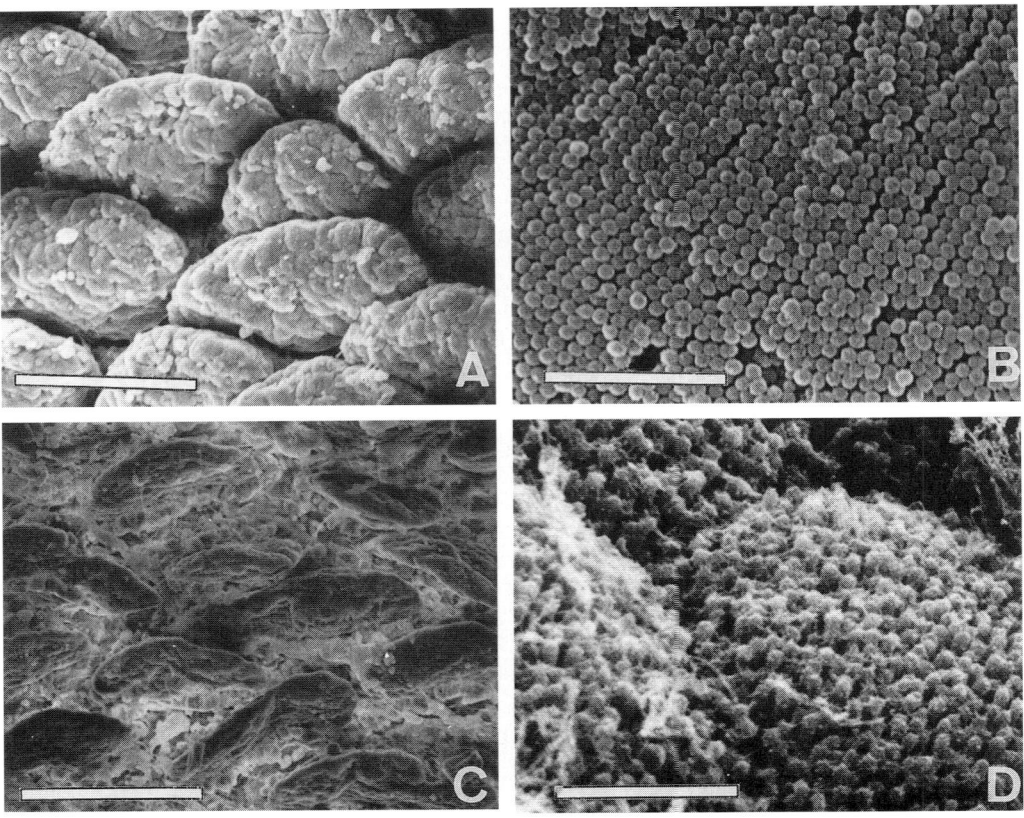

Fig. 1   Scanning electron micrographs of the ileal mucosal surface.

From the top, the photographs are the cellulose group (A, B) and
partially hydrolyzed guar gum group (C, D).   The light bars on the
left in A and C, and right in B and D represent 100 μ m and 1.0 μ m,
respectively.

Cecum (Fig. 2):   Scanning electron micrographs of the cecum in the rats fed the cellulose
diet showed normal histological feature.   In the rats of the *partially hydrolyzed guar gum* group,
the mucosal surface of the cecum was markedly abnormal.   Moreover, the cecal mucosal
surface was covered with the contents and the mucosal surface showed to be wavy under the
contents.   The microvilli of the rat of the *partially hydrolyzed guar gum* group were tightly
packed and covered with the contents.

Colon: There were no significant differences in the surface structure of the colonic
mucosa between the *partially hydrolyzed guar gum* and *cellulose* groups.

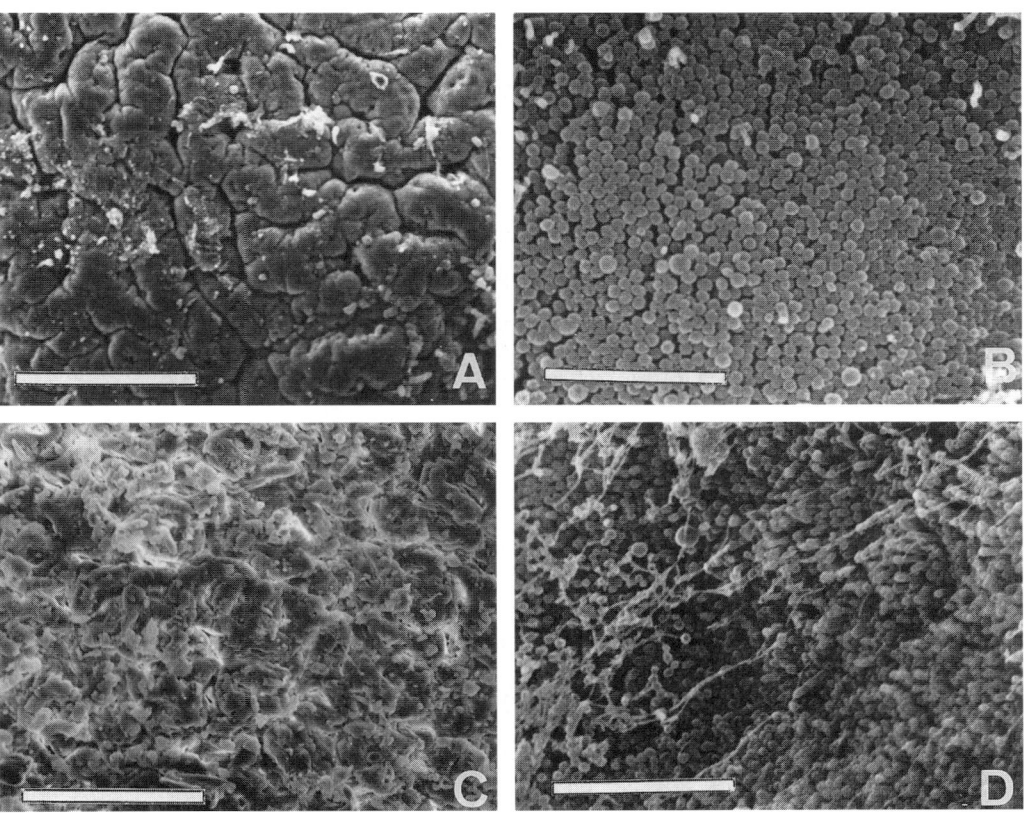

Fig. 2   Scanning electron micrographs of the cecal mucosal surface.

From the top, the photographs are the cellulose group (A, B) and
partially hydrolyzed guar gum group (C, D).   The light bars on the
left in A and C, and right in B and D represent 100 μ m and 1.0 μ m,
respectively.

4. DISCUSSION

The lower digestive tract, especially the cecum, was markedly influenced by the partially
hydrolyzed guar gum addition in the diet.   The weights of the cecum and cecal contents were
significantly greater in the rats fed the partially hydrolyzed guar gum diet than those fed the
cellulose diet.   Fig. 1 and 2 show that the ileal and cecal mucosa in the rats fed the partially
hydrolyzed guar gum diet were irregular in shape, and their microvilli were packed tightly and
covered by the contents on the surface in comparison with those of the rats fed the cellulose diet.
These morphological changes may be caused by alteration of the properties of intestinal contents:
The cecal contents of the *cellulose* group appeared solid, whereas those of the *partially
hydrolyzed guar gum* group were fluid and flowed out of the cecum on dissection.

In the cecum of the rats fed the partially hydrolyzed guar gum, the cecal contents may be fermented by the bacterial flora to produce a greater amount of short-chain fatty acids. The increase in number of the epithelial cells may supposedly be cased by short-chain fatty acids produced in the cecum, resulting in the morphological changes of the mucosal surface structure. In fact, when we measured the short-chain fatty acids in the cecal contents, they were greater in the rats fed the partially hydrolyzed guar gum diet than those of the *cellulose* group (data is no publication).

The transit time of the rats fed the partially hydrolyzed guar gum diet was significantly longer than that of the *cellulose* group (Table. 2). It is considered that the cecal contents of the rats fed the partially hydrolyzed guar gum diet remained for the long time in the cecum and transit to the colon became slower. This may explain the finding that the fecal weight of the *partially hydrolyzed guar gum* group was significantly lower than that of the *cellulose* group (Table. 1).

Our morphological observation showed that the partially hydrolyzed guar gum in Experimental I as well as Experimental II brought distinct effects on the surface of the large intestinal mucosa, suggesting the roles of their viscosity. The similar phenomena had been found by Katayama and Izuta [11], Elsenhans et. al. [14] and Ikegami et. al. [15]; these were explained that the gel-forming properties or viscous properties of dietary fiber made the diffusion of substrates to the enzymes restricted in the digestive tract.

So the conclusion, the partially hydrolyzed guar gum feeding exerted effects on the morphological fine structure of the ileal and cecal mucosal surface fine structure, suggesting these effects being related to the viscosity of the intestinal contents.

REFERENCES

1. M. Tetsuguchi, S. Nomura, M. Katayama and Y. Sugawa-Katayama, *J. Nutr. Sci. Vitaminol.*, **43** (1997) 515.
2. M. Tetsuguchi, Y. Yamashita, M. Katayama and Y. Sugawa-Katayama, *J. Nutr. Sci. Vitaminol.*, **44** (1998) 601.
3. Y. Yamamoto, S. Yamamoto, I. Miyahara, Y. Matsumura, A. Hirata and M. Kim, *Denpun Kagaku* (in Japanese), **37** (1990) 99.
4. S. Kiriyama, H. Morisaki and A. Yoshida, *Agric. Biol. Chem.*, **34** (1972) 641.
5. S. Kiriyama, A. Enishi, A. Yoshida, N. Sugiyama and H. Shimahara, *Nutr. Rep. Int.*, **6** (1972) 231.
6. A. Golay, H. Schneider, D. Bloise, L. Vadas and J. Ph. Assal, *Nutr. Metab. Cardiovasc. Dis.*, **5** (1995) 141.
7. H. Takahashi, Ik. Yang. Sung, Y. Ueda, M. Kim and T. Yamamoto, *Comp. Biochem. Physiol.*, **109A** (1994) 75.
8. H. Takahashi, Ik. Yang. Sung, C. Hayashi, M. Kim, J. Yamanaka and T.Yamamoto, *Nutr. Res.*, **13** (1993) 649.
9. T. Ide, H. Moriuchi and K. Nihimoto, *Ann. Nutr. Metab.*, **35** (1991) 34.
10. T. Okubo, N. Ishihara, H. Takahashi, T. Fujisawa, M. Kim, T. Yamamoto and T. Mitsuoka, *Biosci. Biotech. Biochem.*, **58** (1994) 1364.
11. Y. Sugawa-Katayama and A. Izuta, *Oyo Toshitsu Kagaku* **41** (1994) 335.
12. M. Tetsuguchi, M. Katayama and Y. Sugawa-Katayama, *J. Jpn. Assoc. Dietary Fiber Res.*, **43** (1997) 515.
13. M. Tetsuguchi, M. Katayama and Y. Sugawa-Katayama, in prep.
14. B. Elsenhans, R. Blume and W.F. Caspary, *Am. J. Clin. Nutr.* **34** (1981) 1837.
15. S. Ikegami, F. Tsuchihashi, H. Harada, N. Tsuchihashi, E. Nishide and S. Innami, *J. Nutr.* **120** (1990) 353.

HYDROCOLLOIDS – PART 2
Edited by K. Nishinari
2000 Elsevier Science B.V.

Hypocholesterolemic effects of levan in rats

Y. Yamamoto[a], Y. Takahashi[a], M. Kawano[a], M. Iizuka[b], T. Matsumoto[b], S. Saeki[a] and H. Yamaguchi [a]

[a]Department of Food and Nutrition, Faculty of Human Life Science and
[b]Department of Biology, Faculty of Science, Osaka City University,
 Sugimoto 3-3-138,   Sumiyoshi, Osaka 558 8585, Japan

A significant hypocholesterolemic effect was observed in rats fed cholesterol-free diets containing 1 or 5% of high molecular weight (ca. 2,000,000) levan for 4 weeks.   The hypocholesterolemic effect was accompanied by a significant increase in fecal excretion of sterols and lipids.   High molecular weight levan was hydrolyzed by gastric juice and was changed to a low molecular weight (ca. 4,000) levan without producing any fructo-oligosaccharides.   Low molecular weight (ca. 6,000) levan was not hydrolyzed by either pancreatic juice or small intestinal enzymes.   This suggests that, in vivo, low molecular weight levan derived from the high molecular weight material is not further digested, and reaches the colon intact.   The fermentation of low molecular weight levan (ca. 6,000) by several strains of bifidobacteria was not observed.   These results suggested that the hypo-cholesterolemic effect of levan may result from the prevention of intestinal sterol absorption, and not from the action of the fermentation products.

1. INTRODUCTION

Non-digestible oligomers of fructose and other saccharides are known to have physiological and biochemical effects.   In contrast to the extensive studies on the effects of β-(2-1)fructan like   fructooligosaccharides and inulin [1-3], few investigations of levan, which is composed of β-(2-6)linkaged fructose, have been carried out.   We have successfully synthesized levan having high molecular weight (ca. 2,000,000) from sucrose by using bacterial levansucrase immobilized on a honeycomb-shaped ceramic support [4].   In the present study, the in vitro digestibility and fermentability of high molecular weight (ca. 2,000,000) levan and its effect on the metabolism of lipids in growing   rats   which   were fed cholesterol-free diets were investigated.

2. MATERIALS AND METHODS

2.1. Preparation of high and low molecular weight levan

Levan having high molecular weight (ca. 2,000,000) was prepared by bacterial levansucrase immobilized on a honeycomb-shaped ceramic support (SM-10) as reported previously [4]. Judging from H- and C-nuclear magnetic resonance spectroscopy, the resulting levan consists of β-(2-6) fructosyl units with small amounts of branch β-(2-1) linkages. The molecular weight of the levan measured by HPLC (Asahipak GS-710H) was about 2,000,000, which is equivalent to more than 10,000 monosaccharide moieties. From methylation analysis of the levan, it was concluded that the number average of fructose residues in the linear part was approximately 9. Low molecular weight levan (ca. 6,000) was synthesized by incubating a mixture of sucrose and *Bacillus natto* levan sucrase in the presence of 6M NaCl. Isolation of levan from the reaction mixture was done by gel permeation chromatography (Bio Gel P-2) after heating the reaction mixture at 100 °C for 10 min to inactivate the levansucrase.

2.2. Animals, diets, and lipids analysis

Male 3-weeks-old Sprague-Dawley rats (Clea Japan, Tokyo) were given *ad libitum* access to food for 4 weeks in a room kept at 24 ± 1 °C and with a 12:12 h light : dark cycle. The composition of the experimental diets was as follows (wt%): casein, 20.0; corn oil, 5.0; mineral mixture, 5.0; vitamin mixture, 2.0; and corn α-starch to make 100. The control diet contained 5.0 % cellulose, and the levan diets contained 1.0 or 5.0 % high-molecular weight levan (ca. 2,000,000) at the expense of corn starch. The mineral mixture contained the following (g/100 g mixture): $CaHPO_4$ $2H_2O$, 14.56; KH_2PO_4, 25.72; NaH_2PO_4, 9.35; NaCl, 4.66; Ca-lactate, 35.09; Fe-citrate, 3.18; $MgSO_4$, 7.17; $ZnCO_3$, 0.11; $MnSO_4$ $4H_2O$, 0.12; $CuSO_4$ $5H_2O$, 0.03; KI, 0.01. The vitamin mixture contained the following (mg/100 g mixture): vitamin A acetate, 100; vitamin D_3, 0.25; vitamin B_1 HCl, 120; vitamin B_2, 400; vitamin B_6 HCl, 80, vitamin B_{12}, 0.05; vitamin C, 3,000, vitamin E, 500; vitamin K_3, 520; biotin, 2; folic acid, 20; Ca pantothenate, 500; p-aminobenzoic acid, 500; nicotinic acid, 600, inositol, 600; choline chloride, 20. The animals were anesthetized with pentobarbital, and blood was collected from trunk blood into a centrifuge tube. Concentrations of total cholesterol, triacylglycerol and glucose in serum were determined using commercial diagnostic kits from Wako Pure Chemical Industries, Ltd. Total lipids in feces of the final 2 days were extracted and were measured gravimetrically. Sterols in feces were determined using diagnostic kit for cholesterol determination.

2.3. In vitro digestibility of levan

Artificial digestion of levan was carried out at 37 °C by incubating a levan solution with digestives. For digestion by the saliva or gastric juice, high molecular weight levan (ca. 2,000,000) was incubated with saliva containing α-amylase from human saliva or with gastric juice containing pepsin, respectively. For the digestion by pancreatic juice or small intestinal juice, low molecular weight levan (ca. 6,000) was incubated with pancreatic juice or small intestinal juice, respectively. Pancreatic juice contained pancreatin from porcine, and small intestinal juice contained a powder material extracted with acetone from rat small intestine. All of the digestive enzymes used in this experiment were purchased from Wako Pure Chemicals.

The hydrolysis products of levan, oligolevan and fructose were monitored by HPLC (Shimpack SCR 101N), and the retention times of these products were determined. Results were expressed in terms of the retention times of the main products of hydrolysis. The qualitative analysis of hydrolyzed compounds of levan was carried out using thin layer chromatography (TLC). The changes in molecular weight distribution of the hydrolyzed levan were measured using Tohso TSKgel G6000PW and TSKgel G3000PW columns.

2.4. Fermentation of levan by bifidobacteria and some other bacteria

In vitro fermentation of low molecular weight levan (ca. 6,000) and levanbiose was investigated using different bacterial species including 8 strains of bifidobacteria in anaerobic batch culture fermenters. The decrease in pH of the culture medium was taken as a measure of utilization of these substrates by the bacteria.

2.5. Statistical analysis

Data were expressed as mean $\pm$ SEM for six rats. All statistical analyses were performed by one-way analysis of variance, and the differences between means were tested using Duncan's multiple range test when the F value was significant. A P-value of 0.05 was considered significant.

3. RESULTS AND DISCUSSION

3.1. Growth and hypolipidemic effect

No significant differences in the body weight gain were found between rats fed levan and the control rats over the experimental period (data were not shown). As shown in Figure 1, the serum cholesterol level of rats fed 1 %

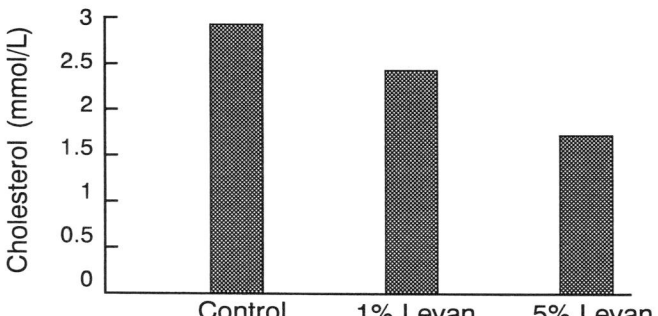

Figure 1. Effects of Levan on the Serum Cholesterol
Values are means±SEM (n=6). These values are
significantly different from each other (p<0.05).

and 5% levan diets was significantly lower than that of the control rats ($p<0.05$). The cholesterol level decreased with increase in the dietary levan content, and the differences between 1% and 5% levan were significant ($p<0.05$). On the other hand, neither serum triacylglycerol nor glucose level were affected by the dietary levan (data were not shown).

The effects of levan on the fecal excretions of total sterol and total lipids are shown in Table 1. The total sterol excretion in 1% and 5% levan was significantly higher than the control ($p<0.05$). Total lipids excretion was also significantly higher in levan groups than the control ($p<0.05$).

Table 1. Effects of Levan[1] on Excreted Total Sterol and Lipids in Feces[2]

Diets	Total sterol	Total lipid
	(mg/g dry feces)	
Control	10.09 ± 0.83^a	59.21 ± 3.88^a
1%Levan	21.19 ± 1.42^b	104.25 ± 7.96^b
5%Levan	19.64 ± 1.52^b	105.15 ± 3.52^b

1. High molecular weight levan (ca. 2,000,000) was used.
2. Values are means ± SEM (n=6). Values within a column with different superscript letters are significantly different from each other (P<0.05).

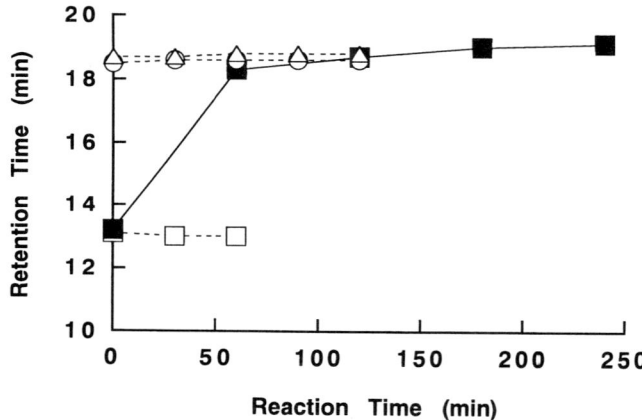

Figure 2. Hydrolysis products of lavan by digestive enzymes. The hydrolysis products of lavan were monitored by HPLC, and are shown in terms of their retention times.
    □ High MW levan (ca. 2,000,000) was hydrolyzed by salivary juice.
    ■ High MW levan (ca. 2,000,000) was hydrolyzed by gastric juice.
    ○ Low MW levan (ca. 6,000) was hydrolyzed by pancreatic juice.
    △ Low MW levan (ca. 6,000) was hydrolyzed by small intestinal juiceal

Table 2. Utilization Test[1] of Sugars by Bifidobacteria and Some Other Bacteria

Bacteria	Value of pH[2]				
	Glucose	Fructose	Levan[3]	Levanbiose	Control[4]
Mitsuokella multiacidus	4.64±0.04 a	4.61±0.06 a	6.71±0.05 b	4.59±0.05 a	6.46±0.04 b
Clostridium sordellii	5.95±0.03 a	6.10±0.02 a	6.69±0.06 b	6.62±0.02 b	6.50±0.02 b
Eubacterium aerofaciens	4.65±0.05 a	4.65±0.05 a	6.47±0.04 b	4.67±0.05 a	6.45±0.02 b
Bacterioides vulgatus	5.47±0.04 a	5.84±0.03 b	6.61±0.02 c	5.90±0.03 b	6.51±0.03 c
Bifidobacterium infantis	4.65±0.03 a	4.67±0.04 a	6.43±0.04 b	4.77±0.03 a	6.50±0.02 b
Bifidobacterium liberorum	4.63±0.06 a	4.67±0.03 a	6.47±0.03 b	4.67±0.04 a	6.56±0.01 b
Bifidobacterium lactentis	4.63±0.05 a	4.65±0.05 a	6.49±0.03 b	4.69±0.04 a	6.53±0.02 b
Bifidobacterium animalis	4.63±0.03 a	6.51±0.03 b	6.54±0.04 b	4.76±0.03 a	6.51±0.03 b
Bifidobacterium breve	4.55±0.04 a	4.60±0.05 a	6.40±0.02 b	4.66±0.04 a	6.54±0.02 b
Bifidobacterium longum	4.66±0.07 a	4.66±0.04 a	6.46±0.03 b	4.66±0.05 a	6.53±0.02 b
Bifidobacterium bifidum	4.90±0.04 a	4.66±0.03 b	6.51±0.05 c	6.52±0.03 c	6.56±0.03 c
Bifidobacterium adolescentis	4.65±0.05 a	4.65±0.04 a	6.45±0.04 b	4.97±0.06 c	6.55±0.02 b
Lactobacillus acidophilus	4.70±0.04 a	5.10±0.04 b	6.42±0.03 c	6.54±0.02 c	6.55±0.02 c

1) The initial pH of all cultures was 7.0.

2) Values are means±SEM (n=3). Values within a row with different superscript letters are significantly different from each other (P<0.05).

3) Low molecular weight levan (ca. 6,000) was used.

4) Control was fermented without test carbohydrate.

404

3.2. Digestibility of levan

The main products of hydrolysis were expressed as a function of the reaction time (Figure 2). High molecular weight (ca. 2,000,000) levan was not hydrolyzed after 60 min exposure to amylase-containing saliva. It was hydrolyzed by gastric juice after 60 min of reaction time, and led to a low molecular weight levan (ca. 4,000). Pancreatic juice and small-intestinal enzymes did not hydrolyze the low molecular weight (ca. 6,000) levan. These results suggest that, in vivo, the low molecular weight levan is derived from the high molecular weight one by the action of gastric juice, but is not further hydrolyzed by digestive enzymes, and reaches the colon intact.

3.3. Fermentation of levan by bifidobacteria and some other bacteria

It is possible that levan may be a substrate for potentially beneficial bacteria, e. g., bifidobacteria and lactobacilli. We used the changes in pH of the culture medium to judge the utilization of levan by the bifidobacteria. The low molecular weight (ca. 6,000) levan was used as substrate. As shown in Table 2, the levan was not fermented by these bacteria, and the pH of the culture medium at the end of fermentation was similar to that of the control. In contrast, levanbiose and fructose were used by bifidobacteria, and the pH of the culture medium dropped as drastically as when glucose was used.

In this experiment, we showed a cholesterol-, but not triacylglycerol- or glucose-lowering effect of levan in rats fed a cholesterol-free diet. We have never seen any other papers reporting this effect of levan. In this experiment, the fecal excretions of total sterols and lipids were significantly higher in levan-fed rats than in controls. These results suggest that levan may bind or entrap the steroids in the intestine, and disturb their reabsorption. The finding that levan binds taurocholate in vitro [5] supports this hypothesis. Then, the hypocholesterolemic effects of levan may result from the disturbed enterohepatic circulation of steroids. However, further studies should be carried out to elucidate the exact mechanisms of the hypocholesterolemic effects of levan.

ACKNOWLEDGEMENT

This work was partly supported by the Sugar Industry Association (Japan).

REFERENCES

1. M. Levrat, C. Remesy, and C. Demigne, J. Nutr., 121 (1991) 1730.
2. K. Yamashita, K. Kawai, and M. Itakura, Nutr. Res., 4 (1984) 961.
3. H. Hidaka, Y. Tashiro, and T. Eida, Bifidobacteria Microflora, 10 (1991) 65.
4. M. Iizuka, H. Yamaguchi, S. Ono, and N. Minamiura, Biosci. Biotech. Biochem., 57 (1993) 322.
5. S. Cho and M. Kadota, Fukuoka Joshi Daigaku Kaseigakubu Kiyo,15 (1984) 33.

HYDROCOLLOIDS – PART 2
Edited by K. Nishinari
2000 Elsevier Science B.V.

405

Effects of xyloglucan on lipid metabolism

K. Yamatoya,[a] M. Shirakawa[b] and O. Baba[c]

[a] Food, Food Additives and Chemicals Division, Dainippon
  Pharmaceutical Co., Ltd, 25-6, Yanaka, 3-Chome, Taito-ku, Tokyo
  110-0001, Japan
[b] Food, Food Additives and Chemicals Division, Dainippon
  Pharmaceutical Co., Ltd, 33-94, Enoki-Cho, Suita 564-0053, Japan
[c] Tokyo Kasei Gakuin Junior College, 22 Sanbancho, Chiyoda-ku,
  Tokyo 102-0075, Japan

Xyloglucan is found in the primary cell walls of higher plants, where it
plays important biological roles. It also exists in certain seeds, for example
tamarind, as a reserve polysaccharide. Tamarind seed xyloglucan has been
used as a thickener or stabilizer in the food industry in   Japan.
Since it has a β -1,4-glucan backbone, xyloglucan is not digested by
human digestive enzymes and acts as a dietary fiber. We report here the
effects of tamarind xyloglucan on lipid metabolism in rats. Both intact
xyloglucan and hydrolyzed xyloglucan reduced plasma and liver lipids
in rats fed   a high-cholesterol diet. Both also significantly reduced
adipose tissue weight. Hydrolyzed xyloglucan reduced plasma and liver
lipids in rats that were fed a high-fat diet. These results indicate that the
intake of both intact and hydrolyzed xyloglucan improves lipid metabolism
in rats. These hypolipidemic effects may be important for the treatment and
prevention of geriatric diseases, including diabetes and cardiac disorders.

1. INTRODUCTION
    Xyloglucan is a major structural polysaccharide in the primary cell walls
of higher plants [1]. Cell growth and enlargement are controlled by the
looseness of a thin net of microfibrils made of cellulose. Xyloglucan
cross-links these cellulose microfibrils and provides the flexibility necessary
for the microfibrils to slide. It has been suggested that the cleavage of
cross-linking xyloglucan by endolytic enzymes is necessary   for   cell
enlargement during growth [1]. Some of the xyloglucan oligosaccharides
released by β -glucanase have biological functions and are capable of
promoting plant cell growth [2].
    Seeds of the tamarind tree (Tamarindus indica) contain   a
xyloglucan. Tamarind seed xyloglucan is used as a food thickener, stabilizer
and gelling agent in Japan. In solution, it exhibits typical Newtonian flow
properties similar to starch and is very stable against heat, pH and   shear.
Xyloglucan polysaccharide has a β -$(1 \rightarrow 4)$-linked D-glucan backbone that is
partially substituted at the O-6 position of its glucopyranosyl residues with
α -D-xylopyranose. Some of the xylose residues are β -D- galactosylated at
O-2.   The   unit   structures   of   tamarind   xyloglucan, xyloglucan

heptasaccharide   (Glu ₄ Xyl ₃),   octasaccharide   (Glu ₄ Xyl ₃ Gal)   and
nonasaccharide (Glu ₄ Xyl ₃ Gal ₂) are shown in Fig.1.

Xyloglucan is an indigestible polysaccharide that acts as a dietary
fiber. Dietary fiber has been shown to control blood glucose and reduce
cholesterol [3-6]. These effects are important for the treatment and
prevention of chronic diseases, including diabetes and cardiac disorders.
Some water-soluble dietary fibers are so viscous that it is difficult to add
them to food in amounts sufficient to have physiological effects
without changing the texture of the food. To solve this problem, a
strategy has been developed which involves the partial hydrolysis of these
polysaccharides, which reduces their viscosity to allow for easy intake while
retaining their physiological activities [7].

In the present study, the effects of tamarind xyloglucan and its
enzyme hydrolysate (which has a biological function in plants) on lipid
metabolism were compared in rats fed a high-cholesterol diet. The effects of
hydrolyzed xyloglucan on lipid metabolism were also examined in rats that
were fed a high-fat diet.

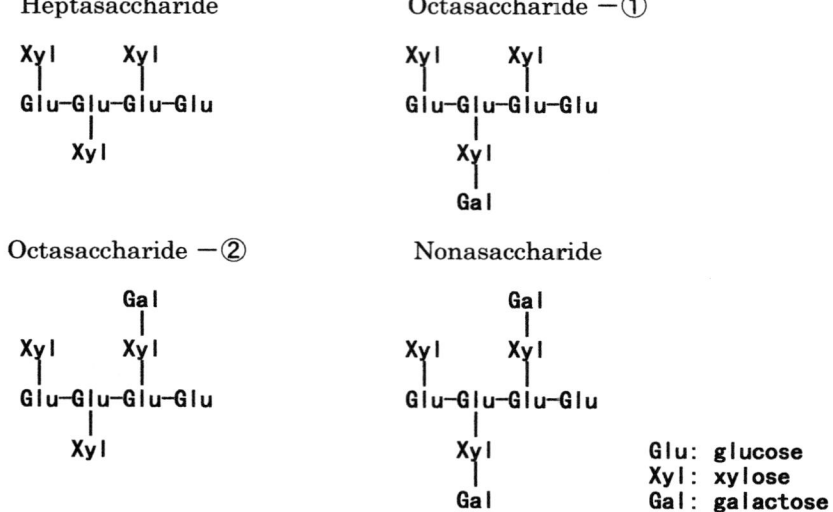

Fig. 1   The unit structures of tamarind xyloglucan

2. MATERIALS AND METHODS

Xyloglucan consisted of tamarind seed xyloglucan (Dainippon
Pharmaceutical Co., Ltd., Japan), and hydrolyzed xyloglucan was
prepared using endo- β -1,4-glucanase [8].

In experiment 1, male Wistar rats (4 weeks old, body weight
approximately 100 g) were used. One group (n= 7) was fed a
high-cholesterol diet for 28 days (control) and other groups (each n= 7)
were given a high-cholesterol diet containing xyloglucan (5%) or hydrolyzed
xyloglucan (5%) instead of cellulose for 28 days (test). The composition of
the high-cholesterol diet was: casein, 20.0; DL-methionine, 0.3; cornstarch,

40.75; sucrose, 15.0; corn oil, 13.0; mineral mix. (AIN-76), 3.5; vitamin mix. (AIN-76), 1.0; choline bitartrate, 0.2; cholesterol, 1.0; sodium cholate, 0.25 and cellulose, 5.0.

Male Wistar rats (6 weeks old, body weight approximately 130 g) were used in experiment 2. One group (n= 6) was fed a high-fat diet for 35 days (control) and another group (n= 6) was given a diet containing hydrolyzed xyloglucan(5%) instead of cellulose for 35 days (test). The composition of the high-fat diet was: casein, 25.0; DL-methionine, 0.3; cornstarch, 29.8; sucrose, 18.0; corn oil, 2.0; lard, 15.0; mineral mix. (AIN-76), 3.5; vitamin mix. (AIN-76), 1.0; choline bitartrate, 0.2; cholesterol, 0.2, and cellulose, 5.0.

In both experiments, all animals were allowed free access to food and drinking water. Feed efficiency was defined as body weight gain divided by food intake. Rats were maintained and handled according to the recommendations of the Institutional Ethics Committee (Tokyo Kasei Gakuin). After the rearing period, rats were fasted overnight. Blood was collected from each rat through the abdominal aorta under pentobarbital anesthesia. The blood was centrifuged at 1,100 g for 15 min, and the plasma components were analyzed using an automatic analyzer(Hitachi-736, Hitachi Ltd, Japan). Animals were dissected, and adipose tissues were removed from both the retroperitoneal region and gonadal region and weighed. Hepatic lipid was extracted from homogenized liver tissue (about 100 mg) with 5 ml of a chloroform-methanol (2:1) mixture. Total lipid was quantitatively determined gravimetrically. Cholesterol was determined according to Zak's method [9]. Triglyceride was determined by the method of Van Handel and Zilversmit [10]. Phospholipid was determined using a test kit (Test kit Wako, Wako Junyaku, Japan).

The results are expressed as the mean ± standard deviation. The significance of differences between the control and test groups was assessed by the t-test.

3. RESULTS

After hydrolytic reaction, a mixture of hydrolyzed xyloglucan heptasaccharide, octasaccharide and nonasaccharide was obtained. The constituent ratio of these three saccharides was 13:39:48.

In experiment 1, the addition of xyloglucan and hydrolyzed xyloglucan did not affect body weight gain, food intake, feed efficiency or liver weight (Table 1). Both significantly increased cecum weight and reduced adipose tissue. Table 2 shows plasma and hepatic lipid levels, respectively.

Table 1　Body weight gain, food intake, feed efficiency, liver weight, cecum weight and adipose tissue weight in rats

	Control	Xyloglucan	Hydrolyzed xyloglucan
Body weight gain(g)	134.8±8.0	128.7±6.2	126.9±9.4
Food intake(g)	416.7±11.2	409.6±16.2	405.9±19.8
Feed efficiency	0.321±0.017	0.314±0.011	0.312±0.011
Liver weight(g)	12.8±0.6	11.2±0.8	12.3±1.1
Cecum weight(g)	2.6±0.4	7.2±0.8(**)	6.8±1.2(**)
Adipose tissue weight(g)	9.3±1.4	7.7±1.18(*)	6.2±1.4(**)

Values are means±SD for 7 rats. * p<0.05, ** p< 0.01: significantly different from the corresponding value in rats fed a control diet.

Table 2 Plasma and liver lipids in rats fed a high-cholesterol diet

Lipids	Control	Xyloglucan	Hydrolyzed xyloglucan
Plasma (mg/dl)			
Total cholesterol	131±27	86±11(**)	123±22
Triglyceride	19.1±4.8	16.3±9.7	10.0±3.0(**)
β-lipoprotein	277±147	113±31(**)	281±110
Free fatty acid (mEq/l)	1.10±0.22	0.95±0.15	0.99±0.11
Liver (mg/g)			
Total lipid	248±19	165±17(**)	235±17(*)
Total cholesterol	133±21	70±11(**)	122±14
Triglyceride	49.5±6.1	41.5±5.84	40.7±5.7(*)

Values are means±SD for 7 rats. * $p<0.05$, ** $p< 0.01$: significantly different from the corresponding value in rats fed a control diet.

Xyloglucan (intact) significantly reduced plasma total cholesterol, plasma β-lipoprotein, liver total lipid, liver cholesterol and liver triglyceride. Hydrolyzed xyloglucan significantly reduced plasma triglyceride and liver triglyceride.

In experiment 2, the addition of hydrolyzed xyloglucan did not affect body weight gain, food intake, feed efficiency or liver weight (Table 3). Hydrolyzed xyloglucan significantly increased cecum weight and reduced adipose tissue weight. Tables 4 shows plasma and hepatic lipid levels, respectively. Among the plasma lipids, total lipid, cholesterol, triglyceride and β - lipoprotein were reduced 14 - 17% by hydrolyzed xyloglucan. Among the hepatic lipids, total lipid, cholesterol, triglyceride and phospholipid were significantly reduced by hydrolyzed xyloglucan.

Table 3 Body weight gain, food intake, feed efficiency, liver weight, cecum weight and adipose tissue weight in rats

Plasma lipids	Control	Hydrolyzed xyloglucan
Body weight gain(g)	173.2±9.8	164.6±15.0
Food intake(g)	529.7±30.9	497.2±25.5
Feed efficiency	0.328±0.017	0.331±0.026
Liver weight(g)	15.7±1.5	15.1±1.1
Cecum weight(g)	4.2±0.3	6.2±1.9(*)
Adipose tissue weight(g)	21.0±2.5	15.0±1.4(**)

Values are means±SD for 6 rats. * $p< 0.05$, ** $p< 0.01$: significantly different from the corresponding value in rats fed a control diet.

Table 4   Plasma and liver lipids in rats fed a high-fat diet

Plasma lipids	Control	Hydrolyzed xyloglucan
Plasma (mg/dl)		
Total lipid	1,646±105	1,367±258(*)
Total cholesterol	168±17	144±25
Triglyceride	774±65	656±189
β -lipoprotein	1,020±111	848±219
Liver (mg/g)		
Total lipid	107.7±7.0	87.6±3.2(**)
Total cholesterol	17.1±2.3	11.6±2.4(**)
Triglyceride	49.4±3.6	41.9±4.4(*)
Phospholipid	20.6±3.9	15.3±1.3(*)

Values are means±SD for 6 rats.   * $p<0.05$, ** $p< 0.01$: significantly different from the corresponding value in rats fed a control diet.

4. DISCUSSION

These results indicate that the intake of both intact and hydrolyzed xyloglucan improve lipid metabolism in rats.   These hypolipidemic effects may be important for the   treatment   and   the   prevention   of   geriatric diseases, including diabetes and cardiac disorders. Due to its low viscosity (5.9cps for 30% aqueous solution of hydrolyzed xyloglucan; cf. 150cps   for 1% xyloglucan), hydrolyzed xyloglucan is easy to intake in a daily life.

The mechanism by which water-soluble dietary fibers lower cholesterol levels remains to be established.   This hypocholesterolemic effect of fibers may   be   due   to   two   common   attributes: they   are   viscous   and / or fermentable by intestinal bacteria [11].

In a high-cholesterol diet (experiment 1), the absorption of   cholesterol may dominate cholesterol metabolism. Under these   conditions,   viscous intact xyloglucan rather than hydrolysate has a greater effect in lowering cholesterol. The hypolipidemic effects of   lower-molecular-weight   dietary fibers may not be solely due to   the   inhibition   of   lipid   absorption.   In studies of hydrolyzed guar gum (galactomannan), which also   has   a   low viscosity, hypolipidemic effects have   been   observed   in   both   rats   and humans [7, 12,13]. Hydrolyzed guar gum has a   larger   M.W.   than   the hydrolyzed xyloglucan used in this study.   Gallaher et al. suggested   that increasing the viscosity of intestinal contents can be effective in   reducing cholesterol, but that only moderate increases are necessary to achieve this effect [11].

Another possible mechanism related to low-viscosity dietary fibers   is that they may affect the endogenous metabolism of lipids. The production of short-chain fatty   acids   upon   the   fermentation   of   fibers   has   been proposed to explain the link between fiber intake and cholesterol lowering [6].   Reducing the solubility of   bile acids by short-chain fatty   acids   and increasing sterol excretion may reduce   cholesterol   reabsorption.   It   has been postulated that propionate has an   inhibitory   effect   on   cholesterol synthesis [14]. Our preliminary study   indicated   that   cecal   short-chain fatty acids were increased by both intact and hydrolyze xyloglucan intake in rats: acetic acid: cellulose 2960, xyloglucan 4683, hydrolyzed   xyloglucan

410

3962. Propionic acid: cellulose 702, xyloglucan 1104, hydrolyzed xyloglucan 1351 (μ g/cecum, unpublished data). Hartemink et al. proposed a model for xyloglucan degradation by intestinal microflora [15]. Xyloglucan is hydrolyzed by endo-β-1,4-glucanase to oligosaccharides (Fig.1). These oligosaccharides are further hydrolyzed by some strains of Bifidobacterium and Lactobacillus species which might produce short-chain fatty acids. Since the weight of adipose tissue was significantly lower with hydrolyzed xyloglucan than with intact xyloglucan, some factor other than viscosity also affects lipid metabolism.

The solution properties of dietary fibers such as viscosity or emulsion stability are important for their physiological effects. These are closely related to the conformation of the polysaccharide molecule. The SAXS (Small-Angle X-ray Scattering) profiles of hydrolyzed xyloglucan fit those calculated for tri-axial bodies of ellipsoid. For example, xyloglucan heptamer has a flat structure characterized by an ellipsoid with semi-axes of 0.22 nm x 0.62nm x 1.80nm [16]. An octasaccharide or nonasaccharide had a shorter long axis than a heptasaccharide, which indicates that the lack of galactose units promoted aggregation. In fact, removing galactose from xyloglucan by an enzyme generates a gelling property [17]; i.e., galactose side chains are key factors in controlling the conformation or water-holding properties. The next step in this research is to study the relationship between the conformation and solution properties. Our final goal is to tailor the polysaccharide molecule to achieve the optimum function.

REFERENCES

1. T. Hayashi, Ann. Rev. Plant Physiol. Plant Mol. Biol., 40 (1989)139.
2. G. J. McDougall and S. C. Fry, Plant Physiol., 93(1990)1042.
3. D. J. A. Jenkins, T. M. S. Wolever, A. R. Leeds, M. A. Gassull, P. Haisman, J. Dilawari,D. V. Goff, G. L. Metz and K. G. M. M. Alberti, Br. Med. J., 1 (1978)1392.
4. E. Tsuji, K. Tsuji, and S. Suzuki, Jpn. J. Nutr., 33 (1975)273.
5. K. L. Roehring, Food Hydrocolloids., 2 (1988)1.
6. D. Lairon, Eur. J. Clin. Nutr., 50 (1996)125.
7. K. Yamatoya, International Food Ingredients, 4 (1994)15.
8. K. Yamatoya, M. Shirakawa, K. Kuwano, J. Suzuki and T. Mitamura. Food Hydrocolloids, 10 (1996)369.
9. B. Zak, Am. J. Clin. Pathol., 27 (1957)583.
10. E. Van Handel and D. B. Zilversmit, J. Lab. Clin. Med., 50 (1957)152.
11. D. D. Gallaher, C. A. Hassel, K-J. Lee and C. M. Gallaher, J. Nutr., 123 (1993)244.
12. K. Yamatoya, K. Sekiya, H. Yamada, and T. Ichikawa, J. Jpn. Soc. Nutr. Food Sci., 46 (1993)199.
13. K. Yamatoya, K. Kuwano and J. Suzuki, Food Hydrocolloids, 11 (1997)239.
14. W. J. L. Chen, J. W. Anderson and D. Jennings, Proc. Soc. Exp. Biol. Med., 175 (1984)215.
15. R. Hartemink, K.M.J. Van Laere, A.K.C. Mertens, and F.M. Rombouts, Anaerobe, 2 (1996)223.
16. Y. Yuguchi, M. Mimura, H. Urakawa, K. Kajiwara, M. Shirakawa, K. Yamatoya and S. Kitamura, "Green Polymers", Indonesian Polymer Association, (1997)306.
17. M. Shirakawa, K. Yamatoya and K. Nishinari, Food Hydrocolloids, 12 (1998) 25.

HYDROCOLLOIDS – PART 2
Edited by K. Nishinari
2000 Elsevier Science B.V.

Intake effect of resistant starch on degradation and fermentation in gastrointestinal tract: high-amylose cornstarch versus prepared resistant starch

T. Hayakawa, T. Okumura and H. Tsuge

Department of Food Science, Faculty of Agriculture, Gifu University
1-1 Yanagido, Gifu, 501-1193 JAPAN

Intake effects of prepared resistant starch (RS) on degradation and fermentation were compared with those of high-amylose cornstarch (HAS). HAS and RS showed similar effects on fecal weight (increased), the length of colon + rectum (increased), pH of cecal content (decreased) and cecal tissue and content weight (increased). Total contents of short-chain fatty acids in cecal contents and feces of the HAS and RS groups increased as compared to the group fed a fiber-free diet or a cellulose powder diet. Propionate content was significantly lower in the HAS group than in the RS groups. It was speculated from Toyopearl HW-50 column chromatograms on cecal and fecal starch fraction and from fecal glucose output that rate of degradation will affect cecal fermentation.

1. INTRODUCTION

A dietary starch component has been recognized as an energy source and no particular effects had been expected. Although certain carbohydrates such as oligosaccharides have functional properties, these substances exert their effects in gastrointestinal tracts. Recently it was pointed out that some portion of starch in food resists digestion and reaches large bowel[1]. This portion includes physically inaccessible starch molecule due to residual husk or relatively intact grain structure, enzyme-resistant starch granules like raw potato starch and retrograded starch during processing and storage[2]. It has been called resistant starch[3]. At that time resistant starch was found as a fraction of starch that could be solubilized from dietary fiber (DF) residue by $2N$ KOH or dimethylsulfoxide (DMSO). Although no critical method has been available, contents of resistant starch in starchy food have been reported on several starchy food (ex. , 1% in breads and 3% in cornflakes). Intake effects of resistant starch on gastrointestinal functions also have been reported[4, 5]. Most evident effects were fecal bulking effects and fermentation in the large bowel[4]. In addition to these, effects on lipid metabolism have also been reported[5]. From these results, DF-like behaviors are expected to resistant starch.

We have been studying the physiological effects of prepared resistant starch (RS) which had been obtained by enzymatic hydrolysis of retrograded high-amylose cornstarch (HAS). We showed that this RS preparation had a fermentable property like water-soluble dietary fibers and also a fecal bulking effect like water-insoluble dietary fibers[6]. HAS has been used in many studies as a resistant starch source because of easiness of use. Although constituent of resistant starch is glucose only, there might be a difference in its intake effects on gastrointestinal functions depending on the source or method to prepare. At present there are no data about physiological difference between original HAS and prepared RS. So we investigated the intake effects of prepared RS mainly in terms of degradation and fermentation as compared to HAS.

2. MATERIALS AND METHODS

2.1 Preparation of resistant starch from high-amylose cornstarch

HAS was obtained from Shikishima Starch Co. (Suzuka, Japan). RS was prepared as follows: HAS suspension was preliminary gelatinized in water bath for 10 min followed by autoclaving for 15 min and kept in cold room (4°C) for 3 days, this procedure was repeated 3 times. Alpha-amylase (from porcine pancreas) and glucoamylase (from *Aspergillus niger*) were added 3 times during digestion period of 24 h at 0 h, 6 h and 12 h at 40°C. Resulting residue was collected, dried and passed through 75 mesh sieve. DF content of this prepared RS was evaluated using AOAC method[7]. This prepared RS contained 39%DF.

2.2 Animal, diets and experimental design

Male Wistar rats (4 weeks old) were obtained from SLC Japan (Hamamatsu, Japan). They were preliminary fed an AIN-76-based diet which was prepared by substituting starch with sucrose of AIN-76 diets[8] for ca. one week before start of the experiment. Rats were fed the experimental diets for 3 weeks in a meal-feeding schedule (feeding time; 18:00-09:00). Cellulose powder (group CP), HAS (group HAS) or prepared RS (group RS) were added to the diet to give 6% DF level at the expense of sucrose in the AIN-76-based diet. Fiber-free diet was also prepared (group FF). On the 14th day, gastrointestinal transit time was measured using carmine in each diet (0.5%, w/w) as a marker with fecal color checked every 30 min after 22:00. Fresh fecal sample was collected on the 14th day, and weighed.

A portion was used for measurement of short-chain fatty acids (SCFA) and another portion was dried overnight at 105°C. Feces were also collected for 5 days before sacrifice. On the final day, rats were anaesthetized by diethylether and blood was drawn from abdominal aorta. Liver, cecum and colon + rectum were excised. Liver and colon + rectum were rinsed with physiological saline before weighing. Cecum were weighed before and after collection of cecal contents. SCFA contents in fresh feces and cecal contents were analyzed by GLC according to the method by Remesy and Demigne[9]. Fecal glucose was extracted and assayed using Glucose Test-Wako (Wako Pure Chemical Industries, Ltd, Osaka, Japan). Cecal contents and feces of the groups HAS and RS were solubilized with DMSO and diluted with 0.2 %NaCl-0.02N NaOH and centrifuged at 5000xg for 10min at 20°C. Each supernatant was applied on a Toyopearl HW-50 (Tosoh corporation, Tokyo) column which was eluted with 0.2 %NaCl-0.02N NaOH. Eluent was analyzed by phenol-sulphate method (total sugar) and modified Park-Johnson Method (reducing sugar). Degree of polymerization (DP) was calculated from the both measurements.

2.3 Statistical analysis

Statistical significance of variance was evaluated by Duncan's Multiple Range Test after one-way ANOVA. Data not sharing the same superscript letter have statistically different at $p<0.05$.

3. RESULTS

There were no significant differences in final body weight and body weight gain among experimental groups. Effects of diets, however, were evident on the weight and size of organs examined. Weight and length of colon + rectum increased in the HAS and RS groups (Figs. 1 and 2). On the contrary weight of colon + rectum decreased in the FF group. Weight of cecal tissue and its content in the HAS and RS groups were significantly higher than the other groups (Fig. 3). Cecal pH in the HAS and RS groups was significantly lower than that in the CP and FF groups (Fig. 4). Cecal SCFA contents were also higher in the HAS and RS groups although propionate was significantly lower in the HAS group than in the RS group (Fig. 5). Fecal propionate concentration was lowest in the HAS group. Concentration of fecal n-butyrate in the HAS group was significantly higher than that of the CP group, however, it was significantly lower than that of the RS group (Fig. 6).

414

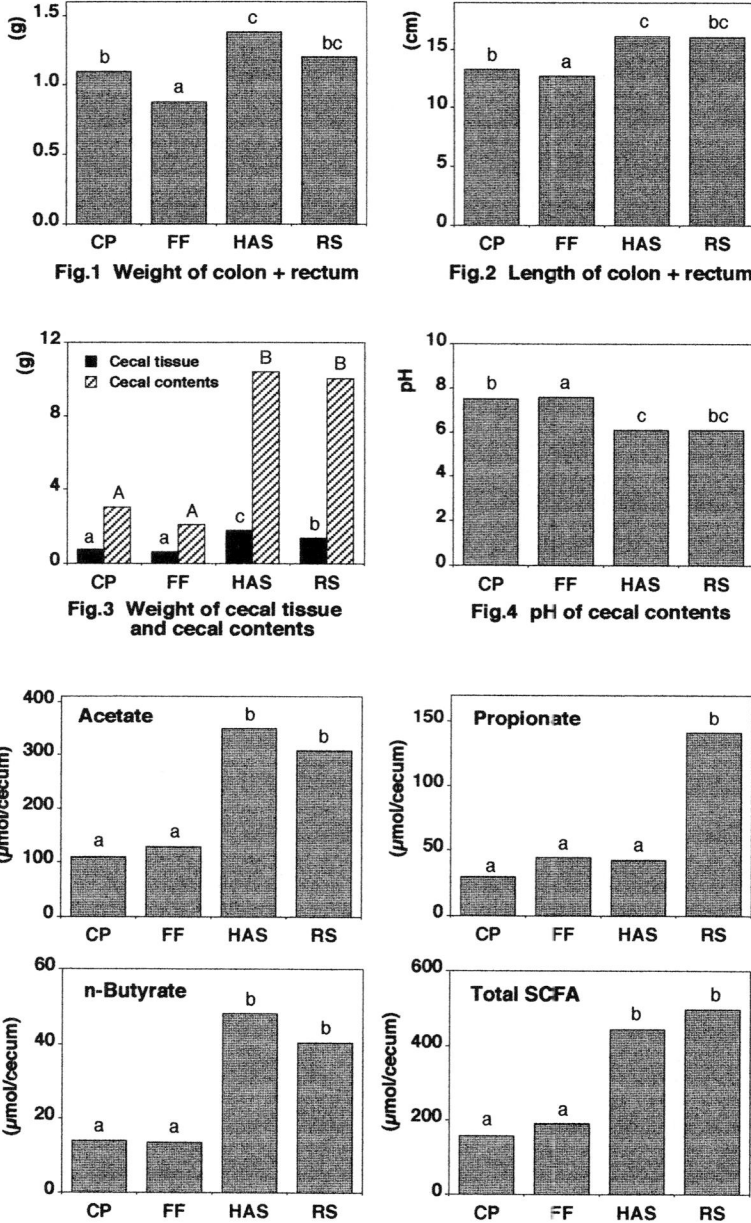

Fig.1 Weight of colon + rectum

Fig.2 Length of colon + rectum

Fig.3 Weight of cecal tissue
and cecal contents

Fig.4 pH of cecal contents

Fig.5 Content of short-chain fatty acids in cecum

Fecal water-holding capacity was lowest in the FF group, low in the CP group, but it was higher in the HAS and RS groups than in the former two groups (Table 1). Fecal dry weights of the latter two groups were higher than that of the FF group, but they were lower than that of the CP group. Fecal wet weights of the latter two groups, however, were almost the same as that of the CP group owing to their high water-holding capacity. Free glucose in feces was obviously found in the HAS (10.3mg/g feces) and RS (3.0mg/g feces) groups. From chromatographic data, degradation products of HAS were much more than those of RS in cecal contents and feces (Figs 7-8).

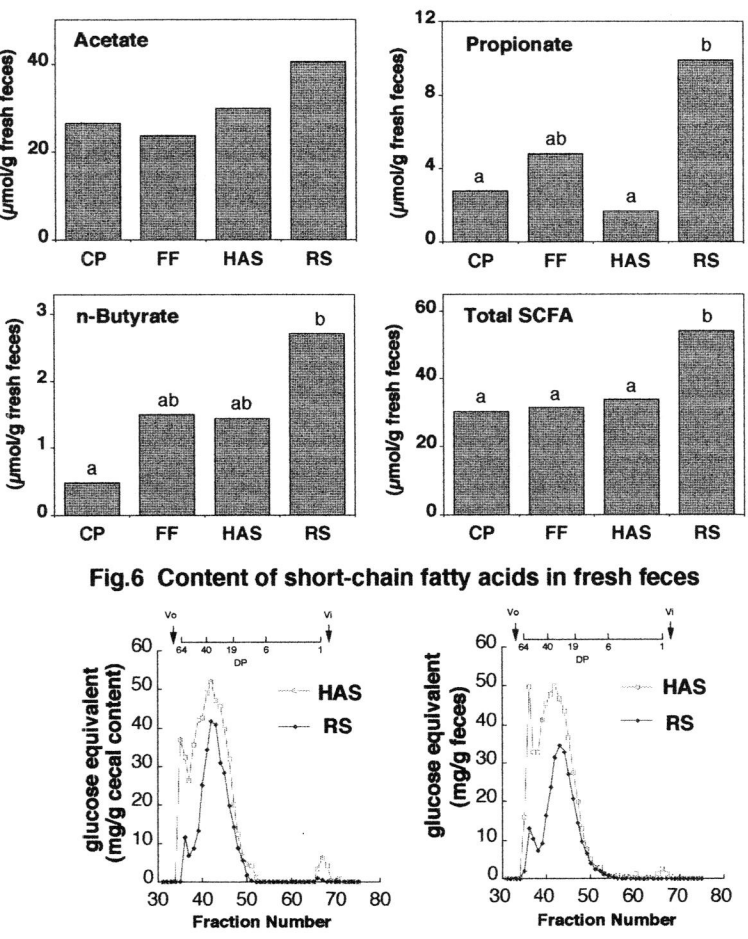

Fig.6 Content of short-chain fatty acids in fresh feces

Fig.7 Elution profiles of cecal starch fraction

Fig.8 Elution profiles of fecal starch fraction

Table 1. Fecal wet weight and water-holding capacity

Group	Wet weight (g)	Dry weight (g)	Water-holding capacity
CP	2.12±0.17[b]	1.32±0.05[c]	1.61±0.60[a]
FF	0.56±0.09[a]	0.35±0.02[a]	1.57±0.16[a]
HAS	2.91±0.69[b]	1.00±0.18[bc]	2.85±0.18[b]
RS	2.46±0.43[b]	0.83±0.13[b]	2.95±0.20[b]

4. DISCUSSION

In this experiment dietary level of DF was adjusted at 6.0%. Physiological effects of HAS and RS were similar in several parameters; e.g. fecal bulking effect, fermentation in the large bowel, cecal pH lowering effect and fecal glucose output. However, there exists obvious difference in some effects between HAS and RS. These differences were evident in fecal and cecal SCFA patterns and in Toyopearl HW-50 chromatogram of respective cecal and fecal extracts. In the HAS group, production of propionate was as low as that in the CP group and was much lower than in the RS group. Fecal glucose output, however, was significantly higher in the HAS group than in the RS group. From these results it was speculated that rate of degradation would affect cecal fermentation pattern. In this report we indicated that there exists a difference in cecal fermentation depending on resistant starch source. This difference might depend on the balance of starch as substrate and rate of fermentation in the large bowel. How rate of fermentation and fermentation pattern correlates each other is a matter of concern but remains uncertain.

REFERENCES
1. H.N. Englyst and J.H. Cummings, Am. J. Clin. Nutr., 42 (1985) 778.
2. D. Sievert and Y. Pomeranz, Cereal Chem., 66 (1989) 342.
3. H.N. Englyst, H.S. Wiggins and J.H. Cummings, Analyst 107 (1982) 307.
4. J.M. Gee, R.M. Faulks and I.T. Johnson, J.Nutr., 121 (1991) 44.
5. E.A.M. DE Deckere, W.J. Kloots and J.M.N. VAN Amelysvoort, J.Nutr., 123 (1993) 2142.
6. T. Hayakawa, Y. Sato, S. Hayashi and H. Tsuge, Abstracts of 16th International Congress of Nutrition (1997) p72.
7. L. Prosky, N.G. Asp, I. Furda, J.W. DeVries, T.F. Schwezer and B.F. Harland, J. Assoc. Off. Anal. Chem., 68 (1985) 677.
8. Ad Hoc Committee, J Nutr., 107 (1977) 1340.
9. C. Remesy and C. Demigne, Biochem. J., 141(1974) 85.

HYDROCOLLOIDS – PART 2
Edited by K. Nishinari
2000 Elsevier Science B.V.

417

Cholesterol lowering effect of the methanol insoluble materials from the quinoa seed pericarp

Y. Konishi[a]*, N. Arai[a], J. Umeda[a], N. Gunji[a], S. Saeki[a], T. Takao[b], R. Minoguchi[b], and G. Kensho[b]

[a] Faculty of Human Life Science, Osaka City University, 3-3-138, Sugimoto, Sumiyoshi-ku, Osaka 558-8585, Japan
[b] Dai-Nippon Meiji Sugar Co. Ltd., 2-269-4, Kaijin-cho, Funabashi, Chiba 273-0022, Japan

We studied the effect of the pericarp of quinoa grain (*Chenopodium quinoa* Willd.) on blood and liver cholesterol levels in mice. When mice were fed on the pericarp (1.5% and 3%) supplemented diet containing 1% cholesterol for 5 weeks, the hypercholesterolemia induced by dietary cholesterol was strongly alleviated. Liver cholesterol levels were also decreased. The methanol insoluble fraction of the pericarp exhibited plasma cholesterol lowering activity, but the methanol soluble fraction did not, suggesting that the activity is due to substance other than the methanol soluble saponins which are massively contained in quinoa seed pericarp.

1. INTRODUCTION

Quinoa (*Chenopodium quinoa* Willd.) is a lesser known food crop of Andean origin, but has recently attracted attention around the world, because the seeds have high protein content (about 16%) with a well-balanced amino acid composition [1, 2]. As quinoa is used for food, the pericarp is usually removed by abrasion, because it contains saponins [2]. The major aglycones of saponins in quinoa seed pericarp fraction (QPF) has been identified to be oleanoic acid, hederagenin, and phytolaccagenic acid [3-5], which was also confirmed by us by TLC and HPLC (Takao *et al.*, unpublished data). QPF is also rich in cell wall materials. Crude fiber content of QPF (5%) is higher than in whole seeds (1.7%) (Takao *et al.*, unpublished data). However, there is no information on chemical nature and composition of quinoa fiber including dietary fiber.

We are interested in QPF in aspects of physiological functions, because QPF contains saponins and fiber. It is generally known that various source of

---

* To whom correspondence should be addressed.
  Tel & Fax: +81-6-6605-2813; E-mail: konishi@life.osaka-cu.ac.jp

saponins [6] and cell wall materials (especially in nonstarch polysaccharides and lignin) [7] alleviate hypercholesterolemia.

In this study, we examined the effects of QPF on plasma and liver cholesterol concentrations in mice. We found that the QPF strongly prevented hypercholesterolemia and the accumulation of cholesterol in the liver of mice fed on a cholesterol-supplemented diet. We discuss on substance with the activity.

2. MATERIALS AND METHODS

2.1. Quinoa pericarp fraction (QPF)

QPF was obtained by abrasion of quinoa seeds (7.5% yield in weight) using a Toshiba CRM 500 rice pearling machine. The chemical composition of QPF is as follows: moisture (7.5%), crude proteins (3.6%), lipids (3.1%), ash (12.1%), carbohydrates (68.7%), and crude fiber (5.0%). Crude fiber content was determined by the official method of AOAC [8].

2.2. Fractionation of QPF

QPF (30 g) was defatted with ethylether, and then mixed with methanol (500 ml), stirring continuously at room temperature overnight. The methanol soluble and insoluble fractions were separated by centrifugation (3000 rpm, 10 min). The former was dried using a rotary evaporator, and the latter was washed with acetone, then ethylether, followed by air-drying. The yields of the methanol soluble and insoluble fractions were 11% and 89%, respectively. The methanol soluble fraction exhibited about 90% of total haemolytic activity, possibly due to saponins, which was determined by Reichart *et al.* [9].

2.3. Animals and diets.

Male ICR mice (7 weeks old, about 35 g body weight, Clea Japan) were housed individually in cages under controlled lighting (light on from 8:00 to 20:00). The control diet consisted of 1% cholesterol, 25% casein, 5% corn oil, 4% salt mixture, 1% vitamin mixture, 5% cellulose, 28.8% sucrose, and 30.2% potato α-starch. The 1.5% and 3% QPF supplemented diets were composed of the control diet with the addition of 1.5% and 3% QPF, respectively. The methanol soluble and insoluble fraction supplemented diets were prepared by adding a portion (equivalent to 3% QPF) of the fractions in the control diet.

In experiment 1, 23 mice were fed *ad libitum* for 1 week on control diet. Then they were separated into 3 groups and fed for 5 weeks on the control diet, and 1.5% and 3% QPF supplemented diets. In experiment 2, mice were fed on the control diet for 1 week, after which they were fed on the control diet supplemented with 3% QPF, methanol soluble and insoluble fractions for 8 days.

2.4. Measurements of total and HDL-cholesterols, triglycerides, and phospholipids

Blood (about 200 μl) was withdrawn every week at 10:30-12:00 with a heparinized capillary tube from the tail vein of mouse subjected to light anesthesia with ethylether. Plasma was obtained by centrifugation (3000 rpm, 10 min). At the end of feeding, blood was withdrawn from the inferior vena cava to determine plasma HDL-cholesterol. Livers were also excised to measure the total cholesterol after KOH digestion at 100 ℃. Total cholesterol, HDL-cholesterol, triglycerides, and phospholipids were measured using enzymatical assay kits, listed as follows: Cholesterol C-Test Wako, HDL-cholesterol Test Wako, Triglyceride G-Test Wako, and Phospholipid B-Test Wako (Wako Pure Chemicals, Japan).

2.5. Statistical analysis

All data were expressed as means ± SE. Statistical analysis of the data was conducted by using StatView (Version 4.0). Data were treated by one-way ANOVA. When ANOVA indicated significant differrence ($P<0.05$), the mean values of the treatment were compared by Fisher's least significant difference test at $P<0.05$.

3. RESULTS AND DISCUSSION

3.1. Effects of QPF on blood cholesterol levels of mice fed on cholesterol supplemented diet (Experiment 1)

No significant effect of QPF on body weight gains was observed after the 5 week-experimental period (Table 1). However, food intakes of the QPF groups were significantly higher than that of the control group (Table 1).

Table 1. Effects of QPF diet on body weight gains and food intakes of mice

	Body weight gains (g)	Food intakes (g)
Control (8)	0.9 ± 0.79	146.4 ± 3.5
1.5% QPF diet (7)	1.6 ± 0.91	172.9 ± 7.3*
3% QPF diet (8)	0.5 ± 0.84	156.0 ± 6.7

Mice (initial body weight of 39-40 g) were fed on control and the QPF supplemented diets for 5 weeks. Values are mean ± SE of the number of mice shown in parentheses.
* $P<0.01$ vs control.

Figure 1 shows the changes in plasma total cholesterol levels. When mice were fed on control diet for 1 week, plasma cholesterol levels increased at about 1.4-fold, showing hypercholesterolemia. The hypercholesterolemia was lasted thereafter in the control group, while it was diminished by 1.5% and 3% QPF supplemented diets. The effect continued for 5 weeks, showing that the cholesterol levels were comparable to those of mice fed cholesterol-free diet.

Initial levels of plasma phospholipids and triglycerides were 82.6 ± 4.8 (mg/dl) and 314.7 ± 7.9 (mg/dl), respectively. After 5 weeks, plasma phospholipids levels of the control, 1.5% QPF, and 3% QPF groups were 64.3 ± 6.2, 83.4 ± 11.8, and 65.7 ± 6.7 (mg/dl), respectively, and triglycerides levels were 234.1 ± 17.4, 242.0 ± 12.7, and 207.0 ± 14.2 (mg/dl), respectively. There were no significancant differences among the groups.

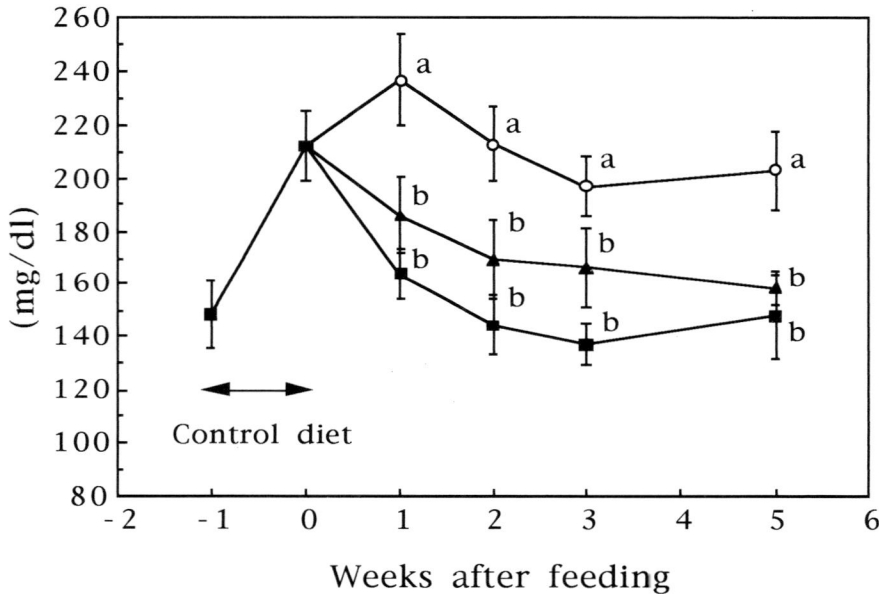

Fig. 1 Changes in plasma total cholesterol levels of mice fed on QPF containing diets.
Mice were fed on a 1% cholesterol containing diet for 1 week, then fed on 0% (○), 1.5% (▲), and 3% QPF (■) supplemented diets for 5 weeks. Different letters (a and b) in this figure indicate significant differences (P<0.05).

Five weeks later, no significant differences in HDL-cholesterol levels were observed among the three groups, although the HDL-cholesterol/total cholesterol ratios in the 3% QPF group were significantly higher than that of control (Table 2).

As shown in Table 3, the liver total cholesterol levels of QPF groups were significantly lower than that of the control group. In the control group, hepatomegaly by cholesterol accumulation was observed.

Thus, QPF specifically prevented plasma hypercholesterolemia and hepatic cholesterol accumulation induced by ingestion of dietary cholesterol. The mechanism of QPF' action is unclear, although QPF is suggested to interact with cholesterol in the intestinal tract, where cholesterol absorption is inhibited.

Table 2 Effects of QPF diets on plasma HDL-cholesterol levels

	Total Cholesterol (A)	HDL-Cholesterol (B)	(B/A)
	(mg/dl plasma)		x 100
Control (8)	203.0 ± 14.9	94.8 ± 7.1	47.3 ± 3.4
1.5% QPF diet (7)	158.5 ± 6.9*	88.7 ± 7.3	56.7 ± 5.4
3% QPF diet (8)	148.0 ± 15.7*	89.1 ± 8.7	62.6 ± 6.0*

Mice were fed on control and the QPF supplemented diets for 5 weeks. Values are mean ± SE of number of mice shown in parentheses. * $P<0.05$ vs control.

Table 3 Effects of QPF diets on liver weight and liver cholesterol level

	Liver weight (g)	Liver cholesterol (mg/g liver)
Control (8)	3.01 ± 0.18	79.8 ± 7.6
1.5% QPF diet (7)	2.39 ± 0.10*	23.5 ± 4.0**
3% QPF diet (8)	2.31 ± 0.13**	13.9 ± 2.1**

Mice were fed on control and the QPF supplemented diets for 5 weeks. Values are mean ± SE of number of mice shown in parentheses. * $P<0.05$ vs control; ** $P<0.01$ vs control.

3.2. Effects of the methanol subfractions of QPF (Experiment 2)

QPF was fractionated into the methanol soluble and insoluble fractions to characterize the QPF's hypocholesterolemic activity. About 90% of total haemolytic activity, possibly derived from saponins, occurred in the methanol soluble fraction.

As shown in Table 4, cholesterol lowering activity was exclusively exhibited by the methanol insoluble fraction of QPF. In addition, we have observed that the hypocholesterolemic activity was recovered in the ethanol insoluble fraction of the water extract of QPF (manuscript in preparation). These suggest that the activity is not ascribed to saponins, the compounds that had been shown to be hypocholesterolemic action in various studies [6]. One can assume that the cholesterol lowering activity of QPF is due to non-starch and water soluble polysaccharides like pectins. We have recently extracted about 3% pectins from QPF, of which the physicochemical properties *in vivo* are under investigation.

Table 4   Effects of the QPF methanol soluble and insoluble fractions on plasma cholesterol levels in mice

| | BW gain | Food intakes | Plasma total cholesterol (mg/dl) | |
		(g in 8 days)	0 day	8 th day
Control	1.3 ± 0.4^a	39.9 ± 1.0	212.1 ± 17.8	201.4 ± 16.2^a
3% QPF	-1.2 ± 1.1^{bc}	39.5 ± 4.5	212.6 ± 14.6	157.0 ± 14.9^{bc}
MeOH sol	1.4 ± 0.5^a	42.4 ± 2.1	212.5 ± 12.7	197.9 ± 12.8^{ac}
MeOH insol	0.2 ± 0.3^{ac}	42.7 ± 2.7	211.9 ± 9.7	152.4 ± 11.6^b

Mice (six each per group) were fed a 1% cholesterol containing diet (control diet) for 1 week and then fed on 3% QPF, the MeOH-soluble, and -insoluble diets for 8 days. Different superscript letters indicate significant differences ($P<0.05$).

REFERENCES

1.   National Research Council, "Lost Crops of the Incas," National Academy Press, Washington, D.C. (1989).
2.   M. J. Koziol, J. Food Comp. Anal., **5** (1992) 35.
3.   F. Mizui, R. Kasai, K. Ohtani and O. Tanaka, Chem. Pharm. Bull., **36** (1988) 1415.
4.   K. G. Ng, K. R. Price and G. R. Fenwick, Food Chem., **49** (1994) 311.
5.   C. Cuadrado, G. Ayet, C. Burbano, M. Muzquiz, L. Camacho, E. Cavieres, M. Lovon, A. Osagie and K. R. Price, J. Sci. Food Agric., **67** (1995) 169.
6.   D. Oakenfull and G. S. Sidhu, Eur. J. Clin. Nutr., **44** (1990) 79.
7.   A. Stark and Z. Madar, "Dietary fiber" in "Functional Foods," ed. I. Goldberg, Chapman & Hall, pp.183-201 (1994).
8.   AOAC, "Official Methods of Analysis," 12th ed., Association of Official Analytical Chemists, Washington, D.C. (1975).
9.   R. D. Reichert, J. T. Tatarynovich and R. T. Tyler, Cereal Chem., **63** (1986) 471.

7. SENSORY EVALUATION AND MASTICATION

HYDROCOLLOIDS – PART 2
Edited by K. Nishinari
2000 Elsevier Science B.V.

Why so many tests to measure texture?

M.C.Bourne[a]

[a]Institute of Food Science, Cornell University, Geneva, New York 14456, USA

Many different instruments using different principles are used to measure food texture and they are all needed because: 1) people use different parts of the body and different chewing patterns for different foods. Fingers, lips, tongue, incisor teeth, cuspid teeth and molar teeth use different test principles. 2) There are many different texture notes in foods and different principles are needed to detect different notes.

Single test instruments measure only one texture note while universal testing machines can be programmed to measure from one to several texture notes. For both classes of instrument it is necessary to select a test principle that correlates highly with the one used by people for that food. A scheme is outlined that helps the food scientist quickly find the test principle that matches the food to be tested.

1. INTRODUCTION

Novices faced with the problem of measuring texture of foods or other consumer products can easily become bewildered by the voluminous literature on texture and the multiplicity of instruments and procedures used to measure textural quality. They often ask the questions, "Why so many different tests? Why does texture have to be so complicated?"

Another frequent question is "What instrument should I use to measure the texture of my product?" This paper will explain why this question should be one of the last questions asked instead of one of the first.

2. WHAT IS TEXTURE?

There are numerous different definitions of texture. The International Organization for Standardization defines texture as "all the rheological and structural (geometrical and surface) attributes of a food product perceptible by means of mechanical, tactile, and where appropriate, visual and auditory receptors" [1].

The definition that I have formulated to cover all aspects of texture is, "The textural properties of a food are that group of physical characteristics that arise from the structural elements at the food, are sensed by the feeling of touch, are related to the deformation, disintegration and flow of the food under force, and are measured objectively by functions of mass, time, and length" [2]. Because "texture" is a group of properties, not a single entity, it is necessary to have a number of different configurations using different test principles in order to specify the texture.

Most active researchers in food texture agree that "texture" is a sensory attribute measured by human senses, primarily the tactile sense and that it is a group of properties, not a single property. While many different kinds of instruments are used to measure "texture," the final calibration of these instruments has to be against people because only people can measure texture. Instruments only measure physical properties and these must correlate highly with people's measurement of texture to be of use.

A distinction needs to be made between "physical properties" and "textural properties." There are many physical properties, most of which can be measured precisely and reproducibly, often by official standard methods. However, not all physical properties are textural properties. If a physical property is not detected by the human senses it is not a textural property and it is a waste of time to measure it and claim that it is "texture." This point is explained schematically in Figure 1.

A useful analogy is found in the electromagnetic spectrum which ranges from the long wavelength radio waves through the visible light spectrum to the very short gamma waves. Most food laboratories have infrared and ultra-violet spectrophotometers and for good reasons. However, they would never use them to measure the color of food because they operate outside the visible spectrum. To measure color, one must measure wavelengths that people see with their eyes; they are blind to all other wavelengths. I believe that some physical and rheological properties are not "seen" by the human tactile sense or other senses and therefore should not be equated to textural properties, even though engineering design requirements may make it necessary to measure them.

3. HOW PEOPLE MEASURE TEXTURE

3.1. Non-destructive Procedures

These measurements are usually made by the hand and because they are non-destructive are frequently made in the food store. They generally consist of gently compressing the food between the thumb and one or more fingers. Since it is a deformation test, it falls squarely within the field of rheology.

3.2. Destructive Procedures

These measurements are usually made in the mouth, although they can be performed in other ways, for example, by using utensils such as knives and spoons. The process of mastication is highly destructive and is designed to smash the food into small particles, to make a thin slurry with saliva that is easy to swallow. Therefore, the majority of texture measurements are highly destructive.

There are three kinds of teeth, each of which has a different shape and function. The incisors located in the front-center are wedge-shaped and are used to cut or incise foods. The cuspids which are located at the corners of the mouth have a single crown with a pointed cusp and are used to tear foods. The molars are located at the back of the mouth and each one has several cusps whose broad surfaces are used to crush and grind the food. The degree to which each kind of tooth is used for mastication depends largely on the nature of the food and the next step needed for destruction. For soft foods such as yogurt, the teeth may not be used at all; the food is manipulated by the tongue against the hard palate until the bolus is thin enough to be swallowed comfortably.

The temperomandibular joint around which the mandible articulates has the unique capability of moving in three planes. People automatically adopt a chewing cycle appropriate for the food in the mouth, and they transfer the food between those teeth that will accomplish the next step in the mastication process.

The structure of the teeth and other mouth parts and the musculature that powers the complex patterns of articulation of the mandible are described in detail in every textbook of human anatomy.

The answer to the question "why so many tests to measure texture?" is that 1) there are numerous texture notes in foods and 2) people manipulate different foods in different ways that depend on the nature of the food. There is no ISO procedure for masticating food.

An example of how people treat food in different ways is reported in a study in which 131 people were asked to test the firmness of nine pairs of different foods without eating them [3]. Depending on the food, the respondents used a viscosity test, a deformation test, a puncture test with the thumb or a flexure test using two hands. Four totally different test principles were used and all were called "firmness."

Because people use many different techniques to masticate their food, many kinds of instrumental tests are needed that will match the different test principles people use.

4. PRINCIPLES OF TEXTURE MEASUREMENT

The major principles are:

1. Puncture	6. Tensile
2. Cutting – shear	7. Snapping – Bending
3. Crushing	8. Deformation
4. Extrusion	9. Flow
5. Torque	10. Texture Profile Analysis

Other principles that indirectly measure textural properties such as density, chemical analysis, optical density, etc. are sometimes used but will not be discussed here.

All of the above test principles are used successfully on some foods; none of them are used for all foods. Therefore, it is essential to select a test principle that matches what people do to their food. One cannot expect to satisfactorily measure "texture" of a food by using a different test principle than what people use.

5. SELECTION OF A SUITABLE INSTRUMENT

A procedure has been described that leads the novice through the numerous options for measuring texture to that option that is most likely to correlate highly with sensory assessment of the textural quality of that food [2]. A series of questions needs to be answered to find a suitable instrument.

1. What is the nature of the product? Liquid and semi-liquid foods need some kind of flow test while solid foods will need a destructive test.

2. Where will the test be performed? In-Store tests must be non-destructive but most laboratory tests need to be destructive to match the process of mastication. Any instrument can be used in the laboratory but instruments located on the factory floor may have to endure heat, water, steam, dust, vibration and other hazards.

3. What is the purpose of the test? A rapid test is needed for quality control on work in process, but a more relaxed time schedule can be used for product development and other purposes.

4. Cost. This includes the cost of purchasing and maintaining the instrument and also the labor costs of operating it.

Answering the above set of questions should have narrowed the field to a small number of instruments. Universal Testing Machines (UTM's) are now widely used because they can be configured to use a number of different test principles. When a UTM is being considered the types of test principles should by now be narrowed to a small number.

The next series of actions should identify the most promising test procedure fairly rapidly.

1. Eliminate unsuitable test principles. For example, snappy foods such as crackers are obviously unsuited to an extrusion test, while flexible foods such as meat are unsuited to a snapping test.

2. Identify two or three test principles that seem to have promise. Use each of these principles on samples of your product over the full range of textures that can be encountered with that food from excellent texture to poor texture. A scatter plot of instrumental measurement against sensory measurement is useful in identifying the best principle which is the one that gives a steep slope of the instrument versus sensory graph with a close fit of the data points to the trend line.

3. Having identified the most effective test principle, it is now time to decide which instrument will be suitable, or for the UTM's which attachment is most suitable.

4. Finally, refine the test conditions to optimize and standardize the procedure. Factors such as preparation of the sample, size of test sample, temperature at which it is tested, compression speed, whether the food is isotropic or anisotropic, and number of replicate tests needed should be studied to find optimum conditions.

Each of the factors in item 4 above range from critical to unimportant depending on the type of food and the test principle. For non-Newtonian semi-solids it is important not to disturb the weak structure while for Newtonian liquids this is not a problem. For the puncture test the size of the specimen is unimportant so long as semi-infinite geometry is maintained, while for texture profile analysis the size and shape of the sample is critical. The texture of some foods is markedly changed with temperature while for other foods the temperature effect is minor. Likewise, the compression speed has a major effect for some foods and almost no effect for other foods. Isotropic foods can be tested in any orientation while ansiotropic foods should always be aligned in the same direction. Some foods have a fairly uniform texture and require only a small number of replicate tests while other foods have a high inherent variability from unit to unit, or even within the same unit and require a large number of replicates to ensure statistically useful results.

There are no general rules to cover what needs to be done to standardize and optimize the test procedure. Each food has its own unique characteristics and therefore all of the items described above need to be examined to see whether they are critical, somewhat important, or not important.

REFERENCES

1. International Organization for Standardization. I.S.O. 5492, 1992
2. M.C. Bourne. 1982 Food Texture and Viscosity. Concept and Measurement. 325pp. Academic Press, New York. Reprinted 1994
3. A.S. Szczesniak and M.C. Bourne 1969. Sensory evaluation of food firmness. J. Texture Studies 1 (1969) 52.

HYDROCOLLOIDS – PART 2
Edited by K. Nishinari
2000 Elsevier Science B.V.

431

Relationship between instrumental texture measurements and sensory attributes

A.L. Halmos
RMIT University
Department of Food Science
Melbourne, Australia

The ultimate physical property of foodstuffs is the sensory attribute of texture as encountered in the mouth. This property can determine whether it is acceptable or otherwise to the consumer. In some culinary cultures, the Japanese in particular, the texture is a primary attribute. To quantify this property is difficult as the flow behaviour in the mouth is very complex. The shear rates or deformations encountered is three dimensional and not constant. Therefore, instrumentation has attempted to simulate mastication rather than a holistic behaviour in the mouth. This has lead to empirical relationships that are very narrow in scope, and attempt to address several characteristics simultaneously. Each foodstuff has its own set of correlations and their relationship cannot be cross correlated. Worse still, even within a particular type of food, hard yellow cheeses for example, the correlation of some of the textural properties are good while others fail. An example of the hardness, cohesiveness, adhesiveness of full fat yellow cheeses will be discussed.

On the liquid side, the relationship is still being defined. Viscosity in all its complexities exist in food stuffs. Correlation of the flow of materials over the tongue and its textural mouth feel is very difficult to define, and attempt to correlate with rheological model parameters has not yet been successful. This is within the realms where rheological models can be often correlated with components such as fat concentrations. However, it is the mouth feel that is of importance to consumers, and with the health conscious modern world, fat substitutes to mimic mouth feel will become highly sought, but instrumentation to provide a means of measuring the characteristics will be needed for design and quality control of the product. The sensory attributes of milks as a function of fat content will be used to illustrate this point.

The conclusion that can be made is that the behaviour of the foodstuffs in the mouth is not yet understood, nor is the method by which the mouth senses the different textures. Therefore attempts to simulate and quantify the sensations with instruments is unrealised as yet and therefore, the empirical approach must dominate at this stage. However, work should continue to understand the fundamentals of processes in the mouth, food behaviour under those circumstances and the stimuli created. Only then will we really be able to use instruments to predict sensory evaluation of the consumer.

1.   INTRODUCTION

The application of food rheology goes beyond the understanding of flow behaviour or phenomena to explain and control outcomes of a flow situations. While these are important, there is a more obvious relationship and complexity of the material behaviour. This is referring to the consumer expectation as encountered by the sensory attributes of food. The usual sensory attributes one considers are colour, flavour and odour. These are obvious as they relate to the senses of the eyes, tongue and nose. However, one very important sensory attribute is related to the sense of touch, and in this case usually in the mouth. That is texture of the foodstuff. Texture is not a characteristic that stands on its own in western culture, though in some parts of the world, such as Japan, it is a primary characteristic. The characteristics of texture contributes to the sense of flavour as it enhances, or retards, flavour release. Therefore, it is the most complex and least understood of the food attributes. Texture is considered to be one of the four principal quality factors in food (Bourne [2]), the other three being appearance, flavour and nutrition. However, texture has no standard definition despite numerous attempts by food scientists to define it. This is probably because texture means different things to different people and cultures and in different foods. Moreover, studies have shown that it is an attribute often taken for granted, it is only when texture falls completely out of range that people start to complain strongly about the food quality, even more strongly than other attributes.

The contribution of rheology to defining textural attributes of food is enormous. First consider the act of drinking. The flow of liquids over the tongue provides a mouth feel which, aside from the heat exchange between the walls of the mouth, the sensation of the flow is would be expected to be influenced by the flow characteristics of the fluid, under hitherto undefined flow regime. The shear rates encountered would be difficult to define, and to my knowledge no-one has tried. The flow is three dimensional with various gaps and motions of teeth, tongue and throat superimposing on top of the bulk pressure flow. Associated with this is the flavour exchange from the components of the fluid. Most flavour components are fat, rather than water soluble. Therefore, the fat content and structure of the fluid would contribute, if not control the flavour release of the liquid. This can be characterised by the fat distribution, eg. the fat globule size and distribution in the stream. The flavour release would be controlled by the surface area of contact and the rate of flavour component transfer.

Next consider eating a piece of solid food. The textural properties range from soft, to hard through all the forms of hardness, elasticity, brittleness etc. these are perceived in the mouth, again loosely termed mouth feel. However, the need to quantify these properties are part of rheology where instruments such as the Texture Analyser is used. However, the meaning of these results is still not fully understood, and while the instrument attempt to simulate mastication, one has to ask the question, under what conditions. The action of chewing is carried out by the temperomanditular joint, the most complex joint in the human body. It is capable of compression, as well as shear in two directions simultaneously. So the motion creates stress and strain simultaneously. Can someone simulate this motion, and hence design an instrument that will simulate mastication more accurately. Nakazawa has carried out some novel experiments in trying to measure the movement of teeth during mastication by actually inserting a false tooth and plotting its movement. At Leatherhead, a device was created that measures the motion of the jaw, from the outside, to understand the mastication process. It is experiments like this that will characteristics of the food under the conditions of deformation, resulting in the true ability of rheology to product the sensory (textural) attributes of solid food.

The complexity in foodstuffs is exacerbated by the fact that many foodstuffs not only exhibit viscous properties, but also possess yield stresses and elastic properties. These specific characteristics that are so important for the textural features of a food item are not characterised instrumentally. A recent design of a new instrument (Field et al)[15] is the Micro Fourier Rheometer (MFR) which could be a new and novel way to quantify these element. The instrument uses a random squeezing of a sample between two parallel plates. Using Fast Fourier Transformer and a calculated transfer function, complex modulus, loss modulus and dynamic viscosity is determined for any given excitation. The frequency of operation may be restricted and the elastic nature, as well as yield stress and viscosity may be obtained for the sample. The results are still to be validated although the characteristic obtained will be valid only at low shear rates.

Once it is understood how to characterise foodstuffs and predict behaviour during mastication, texture modifiers or controllers can be created and used to design food to behave both during processing and consumption as required. Protein structures are already being used to attempt to simulate fat and hence provide texture required while maintaining nutritional requirements. Rheology has never had so many challenges and complexities, and now that fundamental understanding of simple polymer systems has been tackled, the more difficult complex area of food rheology deserves the attention of the top minds.

2.   RESULTS AND DISCUSSION

DAIRY PRODUCTS
Liquid milks

Liquid products contain every rheological complexity possible. It traverses the properties including yield stresses, time dependence and viscoelasticity. Each property has its benefits and full understanding is still out of reach. There are four areas where rheology is important for manufacturers. These are the quality control of ingredients and finished products, the design and evaluation of processes; the consumer acceptance of product in terms of sensory behaviour and functionality, and the design of new products. The use of rheology to understand the structure and behaviour of products is also important.

To illustrate the rheological complexity of products, bovine milk will be used. Milk is a very complex liquid which has a structure still to be fully understood. The components of milk are several hundred in number, which can be classified into seven groups: proteins, lipids, carbohydrates, minerals, pigments and vitamins, enzymes and others. From a physicochemical perspective milk is a dispersion of whey proteins, casein corpuscular micelles and fat globules in a solution of salts, lactose and other minor organic components. The physical state of the dispersed phase has a double colloidal form; a suspension and an emulsion. This provides the most divergent product structure of any food ingredient.

Prawiro[10] investigated the rheology of milk is such that at low fat levels, (0.1 to 4.7%) the milk is characterised as Newtonian or near-Newtonian. This is consistent with Wayne and Shoemaker[19] results. These liquid milks have viscosities at 25°C of 2.1 to 2.6 mPa.s. the corresponding total solids levels range from 10 to 13%. (See Table 1.) These values were obtained using a U-tube viscometer. There appears to be no trend in these figures and so one cannot draw any conclusions at this stage. However, the variations could be due to the structure of the milk, ie. the fat globule size distribution. This hypothesis is currently being examined.

Table 1. Properties of liquid milks at the low end of the fat scale.

Fat Content (% w/w)	Solids Content (% w/w)	Density (g/mL)	Viscosity (mPa.s)
0.1	10.1	1.033	2.1
1.2	11.4	1.046	2.2
1.4	12.1	1.048	2.6
3.8	12.4	1.016	2.3
4.7	13.2	1.029	2.5

In attempting to measure the sensory attributes of these milks, a simple 9 point test was set up for "creaminess". The panellists were asked to grade the milks as "Absence of Creaminess" to "Extremely Creamy". Fifteen panellists were enlisted. The results showed that there is a mouth feel gradation consistent and statistically significant trend with fat content. However, rheological measurements are not sensitive enough to distinguish at that level. At higher fat level the story is different. The different fat level were obtained by simple mixing of milks and creams and then accurately measuring the total fat content. At this concentration (above 10%), the instrumental method now clearly identifies different characteristics.

Table 2. Ostwald de Waele parameters of varying fat levels of liquid milks

Fat Content (% w/w)	Fluid Consistency Index K (Pa.s^n)	Flow behaviour Index n
3.79	0.0035	0.95
5.01	0.0036	0.95
10.03	0.0086	0.89
11.98	0.0091	0.89
14.01	0.0149	0.88
20.10	0.8089	0.31
30.02	3.869	0.22
40.05	21.32	0.21
44.67	147.5	0.08

Figure (1) illustrates the rheogram of the low fat content milks. There is little or no difference between the milks. The gradient, or flow behaviour index is a constant 0.96 ± 0.02, which can be considered near Newtonian behaviour. Figure (2) shows the higher fat level rheograms where the pseudoplastic nature of these materials are clearly indicated. All these results were obtained on a Carri-Med CSL 100 rheometer, with a double concentric cylinder for the low fat levels with shear rate range of 10 to 2000 sec^{-1}, and a cone and plate for the higher fat level milks with shear rate range of 0.01 to 2000 sec^{-1}. The results were fitted with an Ostwald de Waele model and are presented in Table 2. On examining the relationship of flow behaviour index as a function of fat content, Figure (3), the sudden change around 20% fat content in the behaviour of the fluid implies a change of structure. This structural change is probably due to the suspension of fat globules and its size distribution. Work is going on to elucidate this phenomenon, but it is salutary to contemplate that the human senses are more sensitive than the instruments.

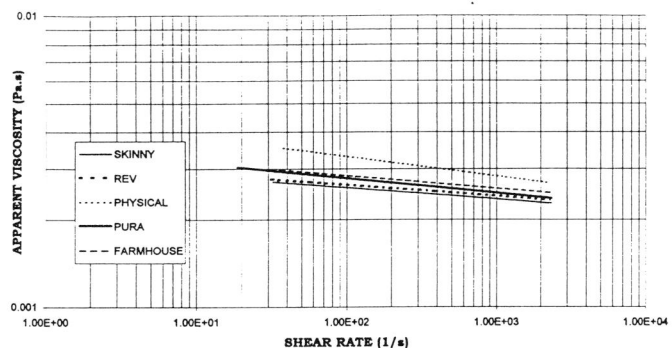

Figure 1. Viscosity of commercial milks (T=25°C)

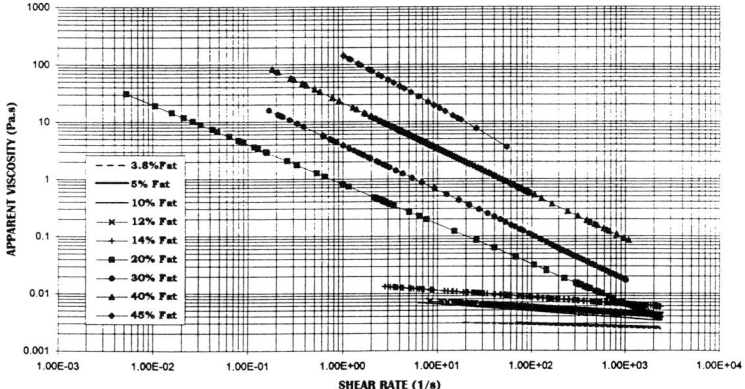

Figure 2. Flow curves of milk and cream at varying fat concentration (T=25°C)

436

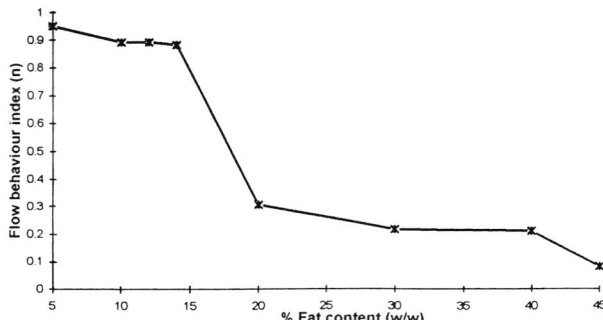

Figure 3. Serial of flow behaviour index for milk and cream at varying fat content

Cheeses

Prentice et al[11] stressed the importance of measuring rheological properties of cheese for the cheese-maker in seeking physical measurements which may assist him with a means of quality control, or to assist him ultimately with a fully automated system that works independently of the human factor. Another long term aim could be to study the structure of cheese itself. Szczesniak [16] originally reviewed the definitions of texture used by workers in the field of food rheology. She identified from those definitions two important elements of texture; the physical structure of the material (its geometry) and the way the material handles and feels in the mouth (its mechanical and surface properties). She also developed the system of texture nomenclature with the aim of bridging the gap between fundamental rheological principles and popular nomenclature. The system is illustrated in Table 3. This proposed classification has been used by many researchers later with objective and subjective methods of texture characterization.

More recently, Bourne[2] also reviewed other researchers' definitions of texture and he defined texture as a group of physical characteristics that arise from the structural elements of food, are sensed by the feeling of touch, are related to deformation, disintegration, and the flow of the food under a force, and are measured objectively as functions of mass, time, and distance. This set of definition is closer aligned to rheological concepts. However, there is still room for clarification as these definitions are still subject to location and perceptions.

Objective instrumental measurement have been divided into fundamental, empirical, and imitative tests. Most of the instruments used by the food industry are empirical or imitative, while fundamental tests are rarely used because they generally give poor correlation with sensory evaluations of texture quality. Most of the early rheological investigations have been empirical tests. Empirical tests measure parameters, often poorly defined, that practical experience indicates to be related to textural quality. Penetrometers, tensile/compressor testers, consistometers, shear measures, and other devices (Szczesniak[17]) are empirical in nature and were the basis of many studies. The texturometer was described by Friedman et al[8]. This device cyclically compressed a bite size sample to 25% of its original height, thereby imitating jaw movement. Strain gauges and a strip-chart recorder produced a force time curve from which a Texture Profile Analysis (TPA) could be derived. The texturometer was applied to measurement of the mechanical textural parameters: hardness, cohesiveness, viscosity, elasticity, adhesiveness, brittleness, chewiness and gumminess. The other instrument is the Instron Universal Testing Machine which is designed for cheese testing. The fixture is comprised of compression anvils where the samples are compressed to simulate chewing. Bourne[3] adapted the Instron Universal Testing Machine for TPA studies, and Shama and Sherman[12] published the first studies of this instrument with cheese. Both the texturometer and the Instron gave good correlation between instrumental values and subjective evaluation.

Table 3. Relationship between textural parameters and popular nomenclature

Mechanical characteristics / Primary parameters	Secondary parameters	Popular terms
Hardness		Soft → firm → hard
Cohesiveness	Brittleness	Crumbly → crunchy → brittle
	Chewiness	Tender → chewy → tough
	Gumminess	Short → mealy → pasty → gummy
Viscosity		Thin → viscous
Elasticity		Plastic → elastic
Adhesiveness		Sticky → tacky → gooey
Geometrical Characteristics / Class		Examples
Particle size and shape		Gritty, grainy, coarse, etc.
Particle shape and orientation		Fibrous, cellular, crystalline, etc.
Other characteristics / Primary parameters	Secondary parameters	Popular terms
Moisture content		Dry → moist → wet → watery
Fat content	Oiliness	Oily
	Greasiness	Greasy

Imitative tests are tests whereby the measurement imitate the conditions to which the material is subjected in practice (Szczesniak[18]). The two most important instruments using imitative tests are General Foods Texturometer and Instron Universal Testing Machine. These two instrument are quite similar in concept though their mechanical features are different. Some fundamental tests such as force-compression test led to significant increase in the application of rheological theory to food analysis (Bagley and Christianson[1]). The fundamental tests also provide bases for development of more meaningful empirical tests.

Instrumental methods of texture analysis are more suitable for routine applications (Eves et al[6]). Objective measurements allow for better analysis of the data and are reproducible over a long period, providing the major advantage of objective measurement over subjective measurement. Szczesniak[18] established standard rating scales of hardness, brittleness, chewiness, gumminess, viscosity, and adhesiveness for quantitative evaluation of food texture.

Correlation was good between sensory and instrumental (texturometer and viscosimeter) evaluation of texture.

Brandt et al [4]developed the texture profile method that uses the A. D. Little flavour profile method as a model. The texture profile is defined as the organoleptic analysis of the texture complex of a food in terms of its mechanical, geometrical, fat, and moisture characteristics, the degree of each present, and the order in which they appear from t he first bite through complete mastication. Civille et al[5] published the guidelines to training a texture profile panel that includes panel selection, panel training, expansion of the basic texture profile method for specific products, panel performance, maintaining a good panel, and reporting texture profile results. The texture profile method will be used as basis for designing the sensory test for texture measurement in this project.

The most complete system of sensory texture measurement is the General Foods Sensory Texture Profiling technique (Bourne[2]). He commented that this technique is an extremely powerful tool and is highly recommended. Another extensively used method of sensory analysis of food is Quantitative Descriptive Analysis (QDA) (Stone et al,[14]). Skinner[13] mentioned that both strategies (TPA and QDA) require extensive training and maintenance of laboratory panels and this is the major drawback of subjective measurement of texture. However, since the final judge of food texture is the consumer, sensory evaluation of texture cannot and should not be discarded.

Bourne[2] mentioned that the most thorough piece of work correlating instrumental measurements with sensory assessment was performed by the texture group at the General Foods Corporation approximately 25 years ago. He showed the correlation between hardness as measured by the General Foods Texturometer and by a trained panel. Similar correlations were obtained for the textural parameters of fracturability, chewiness, gumminess, and adhesiveness. He also outlined a system that enables one to use the type of instrument that has a high probability of yielding good correlations with sensory evaluations.

The relationship between rheology and composition of Cheddar cheeses and texture as perceived by consumers was studied (Jack et al[9]). Procrustes analysis of compositional and instrumental data on a group of 19 Cheddars showed that certain important Instron variables and compositional variables discriminated between samples in terms of textural characteristics. But neither analyses discriminated between the samples in the same way as the consumer perceived texture. Sensory data showed discrimination between samples in terms of mouth feel or tertiary textural maturity and textural maturity. The results indicated the need for advances in this area to develop techniques which meet the need of both industry and food research while providing an precise representation of texture perceptions since there is little correlation between both instrumental and compositional analyses and perceptions of texture by untrained consumers.

Cheese is a common and popular food in Australia but the measurement of its texture still rely heavily on subjective measurements. There was not much work published on work similar to this project which was designed to study texture of cheese by both instrumental and sensory methods followed by attempting correlation of these two methods. The ultimate aim of this series of work was to suggest instrumental measurement that could be used to predict texture of cheese, this will assist the cheese manufacturer in quality control of cheese so that texture measurement of cheese become less dependent on human factor in subjective methods. Projects with this objective was initiated by Foo[7].

Cheese samples for both instrumental and sensory analyses of texture were commercial sample obtained from Kraft Foods Ltd. (Australia). They include the following six varieties :

    1. Processed Cheddar block
    2. Coon* (Mild Cheddar)
    3. Cracker Barrel* - (Extra Tasty Cheddar)
    4. Swiss
    5. Romano
    6. Havarti

* Registered Trade name of Kraft Foods Ltd (Australia). Natural cheddar variety. Typically, the mild cheddar is aged for 3 months, while the extra tasty variety is aged for 9 to 12 months. Samples submitted to both instrumental and sensory test were of size of 1 cm^3, cut by hand, and were equilibrated to room temperature (20°C) before start of test.

    With exception of the cheese samples, which were obtained from Kraft, the rest of the foods used in project, as the basis for ranking, were bought form local supermarkets. A total of 3 sessions were held, one for each of the texture parameter measured using instrument. Prior to doing any sensory tests , the panels were requested to study the classification of texture characteristics developed by Szczesniak[18] and to be clear of the definitions of the mechanical characteristics in particular those which were under study.

    Tables (4), (5) and (6) summarise the instrumental reading on the TA-XT2 Texture analyser. The results were obtained from a double compression test on the cheese samples and results were extracted from resulting graphical output, as illustrated on Figure (4). The results quoted are the mean of eight replicas. The reading obtained for range showed the difference between the largest and smallest reading for each cheese variety. The range for hardness was largest for Romano, indicating that the sample might not be homogeneous thus the poor reproducibility of results. The range for cohesiveness was small for all the cheese samples, showing a better reproducibility for this parameter.

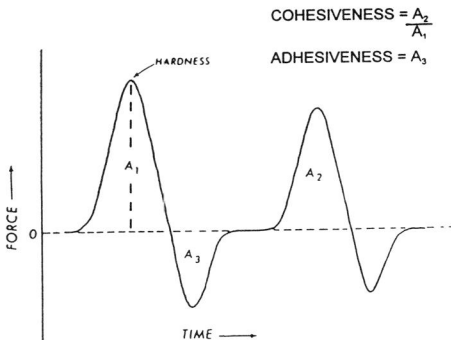

COHESIVENESS = $\frac{A_2}{A_1}$

ADHESIVENESS = A_3

Figure 4. A typical plot of the Force-Time relationship from a Texture Analyser

    The readings obtained for mean showed that the instrument is capable of measuring the difference in the texture characteristics of the six types of yellow cheese. The readings obtained for hardness ranged from 65.23 kPa for Havarti to 482.25 kPa for Romano, indicating that Havarti was the softest while Romano was the hardest among the six cheeses measured. Generally, there were obvious differences in hardness between the samples except for Coon and Extra Tasty which gave very close hardness readings. The readings obtained for cohesiveness ranged from 0.21 for Extra Tasty to 0.57 for Havarti, suggesting that Extra Tasty was the least cohesive and Havarti was the most cohesive. Likewise for adhesiveness, the readings obtained ranged from 0.83 kPa.s for Romano to 7.39 kPa.s for Extra Tasty thus Romano was would be the least adhesive and Extra Tasty would be the most adhesive among the cheeses measured.

    The results obtained from instrument measurements indicate that the Texture Analyser is able to discriminate between the cheese samples for their texture characteristics of hardness, cohesiveness and adhesiveness. However, there is some question as to the adhesiveness obtained from the graph due to the area being very small and as well as the variation on sample to sample. The sensitivity of the instrument in measuring of adhesiveness should be further examined using food items of varying adhesiveness and observed the response compared to cheese.

It was known at this stage that no instrument test could be compared to the process of mastication because of the complexity of the latter. Thus the objective of the instrument test should be to be as close as practicable in simulating the action in the mouth to improve its accuracy in predicting the sensory quality under study. It is suggested that future work in this aspect could use similar procedures used by Shama and Sherman[12]. Their studies with Gouda and White Stilton cheeses illustrated that when an objective comparison is being made between foods with different textural properties, or even between different grades of the same food, different instrumental test conditions may be necessary in each case if the relevant masticatory patterns are to be simulated. The tests they carried out with an Instron enable one to establish a three dimensional judgment zone which defines the force-compression-crosshead speed conditions to be employed if the mechanical conditions for instrumental evaluations are to simulate those pertaining to sensory evaluation.

This study has also illustrated that instrument measurement is objective, rapid, requires less labour and several texture characteristics could be measured in one test. Recommendation to replace subjective measurements with instrument measurement could only be made if the instrument readings correlate well with sensory readings.

The standard scales for the three texture parameters should have been modified so that they can represent the whole range of texture in the cheese samples used in the study. Instead, the standard scales made up of food items identical to those developed by General Foods were used not as reference scale but rather for training the panellists so that they become familiar with the type of texture before they evaluated the actual test samples. In this way, the local "experts" defined their own scales for the sensory attributes.

In general, the scale for hardness was to represent the first bite of mastication. This is simulated on the force peak of the first compression cycle of the force - time curve. Cohesiveness is the concept of how well the foodstuff will maintain its macro characteristic without breaking up. Hence it is represented by the ratio of the amount of work absorbed in the first bite relative to the second bite, represented by the areas under the curve. Adhesiveness is the ability to adhere to the mouth or teeth, and hence the work needed to separate the mouth or the plunger form the sample.

A panel of seven members was recruited with requisition that they do not have dentures. The panel consisted of 3 male and 4 female from age of 20 to 35 years. The results from the training using standard scales show similar panel rating as the General Foods for hardness and adhesiveness. But for cohesiveness, the rank of Melba toast and cracker switched position and this was most probably due to materials being made by different manufacturers. It was observed during the sessions that the panellist has least problem assessing hardness, followed by cohesiveness and they had most difficulty with the adhesiveness scale. The two problems faced when evaluating the cheese samples for adhesiveness include :

(1) The usual technique used for standard adhesiveness scale could not be applied because the cheese sample could not stick to the palate for evaluating the force used by the tongue to remove it. This problem was overcome by experimenting with a few more techniques before deciding on one, which the panellists felt was most useful in helping them to judge adhesiveness. The

techniques tried involve placing the sample between the molar teeth, bite down evenly and evaluate the force require to remove it from the teeth. The number of bites was varied from 1 to 5 times and 3 bites was selected because after 3 bites, the sample was broken to an amount sufficient to stick onto the teeth. (2) Even after having a standard technique, the panellist spent most time on this scale before they could come to consensus because scores varied greatly when they assessed the samples independently.

The values obtained for instrument and sensory measurement of hardness, cohesiveness and adhesiveness are summarized in Table 4, 5 and 6 respectively. Plots of instrument readings against sensory score for each of the texture parameter are shown in Figures 5, 6 and 7 respectively.

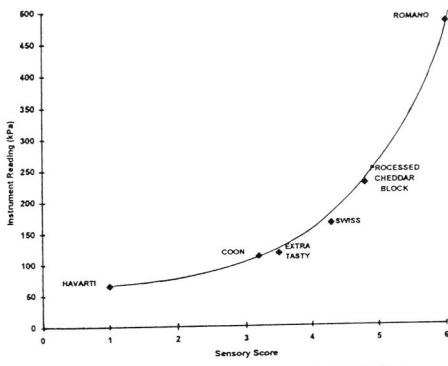

Figure 5. Correlation Between Instrument And Panel On The Hardness Scale

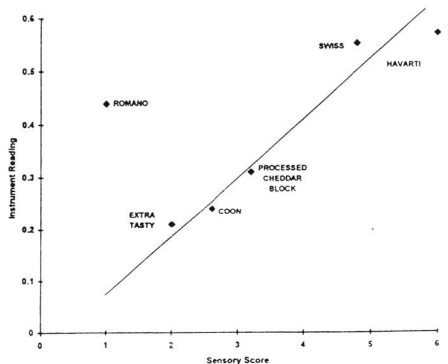

Figure 6. Correlation Between Instrument And Panel On The Cohesiveness Scale

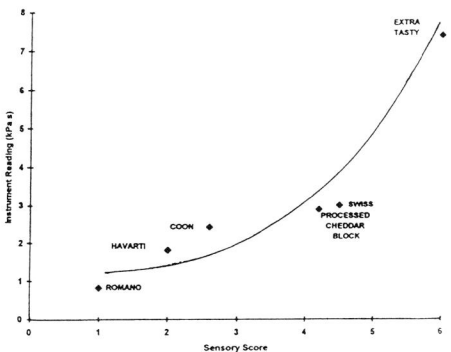

Figure 7. Correlation Between Instrument And Panel On The Adhesiveness Scale

Table 4 Instrument Reading And Sensory Score Of Hardness Of Cheese

Sample	Instrument Reading (kPa)	Sensory Score
Havarti	65.23	1.0
Coon	112.63	3.2
Extra Tasty	117.39	3.5
Swiss	165.06	4.3
Processed Cheddar Block	228.53	4.8
Romano	482.25	6

Table 5 Instrument Reading And Sensory Score Of Cohesiveness Of Cheese

Sample	Instrument Reading	Sensory Score
Extra Tasty	0.21	2.0
Coon	0.24	2.6
Processed Cheddar Block	0.31	3.2
Romano	0.44	1.0
Swiss	0.55	4.8
Havarti	0.57	6.0

Table 6 Instrument Reading And Sensory Score Of Adhesiveness Of Cheese

Sample	Instrument Reading (kPa).	Sensory Score
Romano	0.83	1.0
Havarti	1.83	2.0
Coon	2.43	2.6
Processed Cheddar Block	2.89	4.2
Swiss	3.00	4.5
Extra Tasty	7.39	6.0

The relationship of all three pairs of measurements are monotonic in nature, with the exclusion of the cohesion of the Romano cheese. It can be concluded that there exists a correspondence between instrumental and sensory measurement for hardness and adhesiveness. For cohesiveness, correspondence was similarly good for five samples excluding Romano. The instrument reading for Romano is on the higher end of the scale while the panellist rated this sample as the least cohesive among the six samples. Possible causes for this discrepancy include : I) Difficulty in comprehension of the concept of cohesiveness II) Romano is the only cheese that has a rough surface among all the cheese samples used in this study, so perhaps its appearance would have influenced the panel judgment of its texture in terms of cohesiveness III) Faster breakdown of structure by saliva in the mouth compared to the other cheeses. IV) Slower breakdown of structure when compressed between flat surface of cylinder and platform (in instrument measurement) compared to faster breakdown of structure when compressed between uneven surfaces of molar teeth (in sensory measurement), this resulted in a higher reading for cohesiveness by instrument measurement even though both measurements indicated that Romano was the hardest among all the cheese samples. In short, it is conceivable that instrumental methods do not simulate, or do not interpret sensory measurement the same way for different products.

The good relationships obtained for most of the samples of the texture parameter studied indicated that the instrument is measuring the texture in some similar way to human. This result is encouraging because it means that the instrument used in this study has the potential to replace existing subjective methods used for texture measurement of cheese. However, it is also clear that sensory approach places a different emphasis or relativities than instrumental methods. In future work in correlation studies, it is worthwhile to consider the key factors that affect sensory/instrument correlations besides quality of execution of both sensory and instrument tests. These factors suggested by Szczesniak[18] are : (I) similarity of the physical aspects of the sets of measurements; (II) nature of the test material, including its heterogeneity and rheological character; (III) selection of sensory terms and sensory scales.

Instrumental measurement used in the study could discriminate between the six types of yellow full fat cheese for their hardness, cohesiveness and adhesiveness. The instrument readings could also be useful in studying the structure of cheese by looking at the shape of the curves obtained and the relationship between different texture parameters for each cheese type. Sensory measurement showed that adhesiveness of yellow cheese was more difficult to evaluate than hardness and cohesiveness. Correlation between instrumental measurement and sensory measurement was good for hardness and adhesiveness for all the six types of cheese. For cohesiveness, correlation was good for all the cheeses except for Romano.

Since it was demonstrated in this study that the Texture Analyser has the ability to measure some of the texture characteristics in cheese, the next step is to refine the test conditions for each test parameter. Variables such as sample size, degree of compression, rate of compression and sample temperature should be studied to find which samples gives the best resolution between the samples. Those conditions that give the best resolution and reproducibility should then be standardised and recorded for future use. Another investigation in terms of instrumental measurement would be the relationship between the different texture characteristics measured to help understand the structure of cheese better.

In order to improve correlation with instrument measurement, it is recommended that standard scales be established specifically for each product, for instance cheese. The panellists would then be able to use the standard scales as reference when they rate the cheese samples for the parameter under studied. It should be noted that the standard scales developed for cheese would be specific and not used when assessing foods of considerable different texture from cheese.

3. CONCLUSIONS

In conclusion, the examples of everyday products present a gamut of complex, intriguing opportunities for rheological research. The complication of the consumer perspective will make it more difficult to solve than to contend with physical systems alone. However, with the co-operation of psycho-rheology and physical rheology, cross correlations and understandings can be established for the enhancement of science and knowledge.

REFERENCES

1. E.B.Bagley and D.D. Christianson; *Food Technology*; 41(1987): 96.
2. M.C.Bourne; Food Texture and Viscosity-Concept and Measurement; Academic Press; (1982) 2-11.
3. M.C.Bourne; *Journal of Food Science.*, 33 (1968),223-226
4. M.A. Brandt, H.H. Friedman and A.S.Szczesniak; *Journal of Food Science*; 28(1963) ; 397-403.
5. G.V. Civille and A.S. Szczesniak; *Journal of Texture Studies*; 4(1973), 204-233.
6. A. Eves, M.Boyer and D. Kilcast; Food Acceptability; Elsevier Applied Science;(1988) 459-472.
7. A.I. Foo, Final year thesis, B.App. Sc. (Food Science & Tech.) RMIT University (1996).
8. H.H. Friedman, A.S. Szczesniak and J.E. Whitney; *Journal of Food Science*; 28(1963); 390-396.
9. F.R. Jack,A. Paterson and J.R. Piggott; Journal of Food Science and Technology; 28 (1993) 293-302.
10. S. Prawiro; Final year thesis, B.App.Sc. (Food Science & Tech.) RMIT University (1997).
11. J.H. Prentice,K.D. Langley and R.J. Marshall; Cheese-Chemistry, Physics and Microbiology Vol.1; Chapman and Hall; (1993) 307-313.
12. F. Shama and P. Sherman; 1973; *Journal of Texture Studies*; 4 (1973), 44-353.
13. E.Z. Skinner; Applied Sensory Analysis Of Foods; Vol.1(1988); C.R.C. Press Inc.
14. H. Stone,J.J. Sidel,S. Oliver, A. Woolsey and R.C. Sinleton;*Food Technology*; 28 (1974), 24-34.
15. J.S. Field, M.V. Swain and N. Phan-Thien, *J. Non-Newtonian Fluid Mechanics,* 65 (1996), 177-194
16. A.S Szczesniak; *Journal of Food Science*; 28(1963), 410-420.
17. A.S. Szczesniak; *Journal of Food Science*; 28(1963),335-389.
18. A.S. Szczesniak; *Journal of Texture Studies*; 18(1987), 1-15.
19. J.B. Wayne and C.F. Shoemaker; *Journal of Texture Studies*; 19 (1988), 143 - 152

HYDROCOLLOIDS – PART 2
Edited by K. Nishinari
2000 Elsevier Science B.V.

445

Response surface analysis using the AIC statistic for a constrained region inside of the Scheffé simplex lattice

S. Naito[a], H. Moritaka[b], and K. Nishinari[c]

[a]National Food Research Institute, Tsukuba 305-8642, Japan
[b]Faculty of Living Science, Showa Women's University, Setagaya-ku 154-8533, Japan
[c]Faculty of Human Life Science, Osaka City University, Osaka 558-8585, Japan

Scheffé simplex lattice designs have been often used in research on mixture experiments and some restrictions on the mixture component proportions are necessary in many cases. However, multicollinearity frequently arises when highly constrained region is set in the lattice. A new data analysis method for a constrained region inside of the lattice, based on using the AIC statistic (Akaike Information Criterion) in regression analysis is proposed.

1. INTRODUCTION

Scheffé simplex lattice designs [1] have often been used in research on mixture experiments. Many practical cases need some restrictions on the mixture component proportions, and some methods dealing with these restrictions in the Scheffé simplex lattice designs have been developed [2]. However, these methods frequently restrict how to set a constrained region. Therefore, the new data analysis method setting a free experimental region inside of the lattice is proposed and usefulness of our method in the three-components-jelly experiments is considered.

2. DATA ANALYSIS METHOD

A three-components mixture experiment is considered in this study. An experimental region is freely set in the Scheffé simplex lattice according to the purpose. SAS Release 6.09 was used for data analysis and a Basic program on a personal computer was used for drawing contours.

2.1. Searching the best response surface model

A response surface model consists of some variables in the Scheffé third polynomial model:

$y = a_1x_1 + a_2x_2 + a_3x_3 + a_{12}x_1x_2 + a_{13}x_1x_3 + a_{23}x_2x_3 + a_{123}x_1x_2x_3$ (1)

where y is the response variable, a_i is the model coefficient and x_i is the ith component proportion having the following relation:

$x_1 + x_2 + x_3 = 1$ $(x_i \geq 0, \ i = 1,2,3)$ (2)

It is important for setting a free experimental region to deal with high multicollinearity [2]. Variable selection in regression analysis is one of the strategies dealing with this problem [2]. Therefore, the useful variables for the response surface are searched using the AIC (Akaike Information Criterion) statistic [3], one of the statistical variable selection methods. For analyzing Eq. (1) as the regression model, the following model without the constant term is defined:

$y = b_1z_1 + b_2z_2 + b_3z_3 + b_4z_4 + b_5z_5 + b_6z_6 + b_7z_7 + \epsilon$ (3)

where $z_1 = x_1$, $z_2 = x_2$, $z_3 = x_3$, $z_4 = x_1x_2$, $z_5 = x_1x_3$, $z_6 = x_2x_3$, $z_7 = x_1x_2x_3$, b_i is the regression coefficient of the explanatory variable z_i and ϵ is the error term. The AIC statistic is defined as

AIC = -2×log(maximum likelihood)+2×(number of free parameters)

$$= n(\log 2\pi + 1) + n \log \frac{1}{n} \{ \sum_{i=1}^{n} y_i^2 - \sum_{j=1}^{m} \hat{b}_j \sum_{i=1}^{n} z_{ji} y_i \} + 2(m + 1)$$

$$\to n \log \frac{1}{n} \{ \sum_{i=1}^{n} y_i^2 - \sum_{j=1}^{m} \hat{b}_j \sum_{i=1}^{n} z_{ji} y_i \} + 2m = n \ \log(SSE/n) + 2m \ \ (\text{AIC of SAS}) \quad (4)$$

where m = number of explanatory variables, n = number of data sets, $\hat{b}_j$ = estimate of regression coefficient and SSE is the error sum of squares.

According to Eq. (4), a model with a smaller AIC is a better model because the AIC statistic of a model with a smaller error and fewer terms becomes smaller. The constants in the AIC equation are removed in the SAS software as Eq. (4) because the AIC statistic is a relative value, not an absolute value, namely a difference of the AIC statistics is a meaningful indicator. If the difference of AIC between some models is less than 1, these models can be regarded as having the same precision [4]. Consequently, the minimum AIC model is adopted as the best response surface model for a constrained region. When the difference between some models and the minimum AIC model is less than 1, the averaged model is calculated as the best model.

2.2. Evaluating the contribution of each variable

1) The AIC change quantity when the variable is removed from the model is an indicator of the variable contribution in the case of no multicollinearity.

2) The approximate model is derived from substituting high correlations (r>0.8) of the explanatory

variables into the best model with multicollinearity.

3. EXAMPLE 1

Proportions of each ingredient in milk jellies prepared from gellan gum, milk powder and an artificial sweetener, Palsweet are shown in Table 1 [5]. The gellan gum concentration suitable for jelly is restricted in the narrow range. However, wider concentration ranges of ingredients are useful for studying their effects on milk jelly properties. Therefore, the rectangular region was set in the Scheffé simplex lattice for expanding the milk jelly concentration range and an artificial sweetener was used because a small amount of sweetener can increase sweetness (Fig. 1).

Table 1 Proportion x_i and sensory hardness of the mixture of milk jellies

Sample	$\alpha_1(\%)$ Gellan gum	$\alpha_2(\%)$ Milk powder	$\alpha_3(\%)$ Palsweet	Sensory hardness[1]
1	0.8	8.0	2.2	2.0
2	0.8	6.9	3.3	2.8
3	0.8	5.7	4.5	2.1
4	0.5	8.1	2.4	0.6
5	0.5	7.0	3.5	0.0
6	0.5	5.9	4.6	-0.4
7	0.3	8.2	2.5	-1.6
8	0.3	7.1	3.6	-2.1
9	0.3	6.0	4.7	-2.2

1) Average value of 8 panelists. Sample No.5 was the standard. +3: very hard, -3: very soft.

Total concentration is fixed 11% (w/w), i.e. $\sum_{i=1}^{3} \alpha_i = 11$

Model variable $x_i = \alpha_i/(\alpha_1+\alpha_2+\alpha_3) = \alpha_i/11$

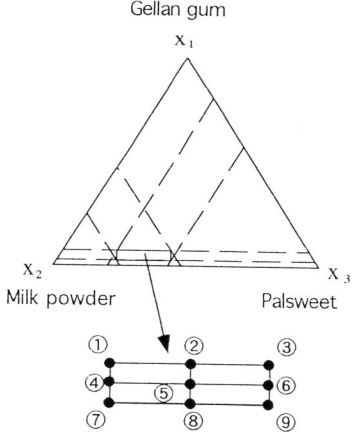

Fig.1 Experimental points in the lattice

This rectangular region has high multicollinearity as shown in Table 2. Regression models were calculated using 8 experimental points except for No.5 and this point was used to examine model significance. Regression models searched by the AIC statistic are shown in Table 3. Since the difference of AIC between the minimum AIC model and the second model is less than 1, the following averaged model was calculated for the best model.

$$y = -136x_1+126x_1x_2-64x_1x_3-25x_2x_3+946x_1x_2x_3 \qquad (5)$$

The regression coefficient of gellan gum in Eq.(5) is -136. This means that hardness decreases with the increasing concentration of gellan gum, but it does not agree with actual phenomena. Since multicollinearity affects regression coefficients, the approximate model was derived from substituting relations shown in Table 2 into Eq.(5). This model consists of the same variables in the Scheffé first polynomial. Then, the approximate model and the Scheffé first polynomial for the sensory hardness are shown below:

$$y \approx 79.9x_1 - 1.1x_2 - 9.3x_3$$
$$= 0.899(88.9x_1 - 1.2x_2 - 10.3x_3) \tag{6}$$

and

$$y = 88.9x_1 - 3.6x_2 - 6.1x_3 \tag{7}$$

Table 2 Correlations between explanatory variables in the Scheffé third polynomial

1) x_1x_2	$= 0.000703 + 0.613x_1$	$r =$	0.93
2) x_1x_3	$= 0.00150 + 0.286x_1$	$r =$	0.77
3) $x_1x_2x_3$	$= 0.00127 + 0.167x_1$	$r =$	0.91
4) x_3	$= 0.939 - 0.979x_2$	$r = -$	0.98
5) x_2x_3	$= 0.390 - 0.307x_2$	$r = -$	0.89
6) x_2	$= 0.943 - 0.972x_3$	$r = -$	0.98
7) x_2x_3	$= 0.0903 + 0.330x_3$	$r =$	0.96
8) $x_1x_2x_3$	$= 0.00205 + 0.475x_1x_3$	$r =$	0.96

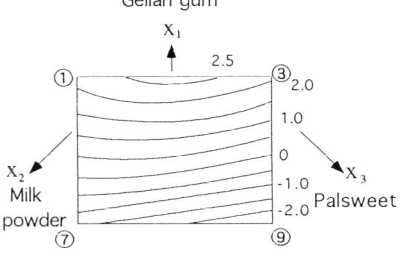

Fig.2 The best model contours of sensory hardness

Table 3 Regression models searched by the AIC statistic

Explanatory variable						AIC	R^2	
x_1		x_{12}		x_{23} x_{123}		-54.13	0.904	} Averaged model
			x_{13} x_{23} x_{123}			-53.29	0.900	
x_1	x_3 x_{12}			x_{23} x_{123}		-52.71	0.905	
x_1		x_{12} x_{13} x_{23} x_{123}				-52.58	0.905	
x_1 x_2 x_3 x_{12} x_{13} x_{23} x_{123}						-49.58	0.906	3rd degree model
x_1 x_2 x_3 x_{12} x_{13} x_{23}						-47.33	0.900	2nd degree model
x_1 x_2 x_3						-43.93	0.884	1st degree model

Eqs. (6) and (7) indicate almost the same result for contribution of each variable in the model. However, the AIC statistic of the best model is significantly smaller than the Scheffé first polynomial. Therefore, the best model, Eq.(5) was used for drawing contours of the response surface (Fig.2).

4. EXAMPLE 2

Proportions of each ingredient in lemon jellies prepared from gellan gum, citric acid and an artificial sweetener, Palsweet are shown in Table 4. A small difference of citric acid concentration affects lemon jelly properties. Then, the converse triangle region differing from the milk jelly region was considered (Fig.3).

Table 4 Proportion x_i and sensory sweetness of the mixture of lemon jellies

Sample	$\alpha_1(\%)$ Gellan gum	$\alpha_2(\%)$ Citric acid	$\alpha_3(\%)$ Palsweet	Sensory sweetness[1]
1	0.9	1.5	4.4	-0.8
2	0.9	1.2	4.7	-0.3
3	0.9	0.9	5.0	0.8
4	0.7	1.3	4.8	0.0
5	0.6	1.5	4.7	-1.2
6	0.6	1.2	5.0	0.9
7	0.3	1.5	5.0	-0.6

1) Average value of 20 panelists. Sample No.4 was the standard. +3: very strong, -3: very weak. Total concentration is fixed as 6.8 % (w/w), i.e. $\sum_{i=1}^{3} \alpha_i = 6.8$ Model variable $x_i = \alpha_i/(\alpha_1+\alpha_2+\alpha_3) = \alpha_i/6.8$

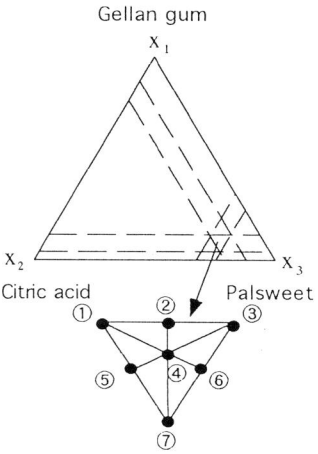

Fig.3 Experimental points in the lattice

The converse triangle region has high multicollinearity as shown in Table 5. Regression models were calculated using 6 experimental points except for No.4 and this point was used to examine model significance. The following minimum AIC model was the best model for the sensory sweetness (Table 6):

$$y = 28.6x_3+417.9x_1x_2-156.0x_1x_3-127.2x_2x_3 \qquad (8)$$

The approximate model and the Scheffé first polynomial are shown below:

$$y \approx -46.2x_1-21.5x_2+73.5x_3$$
$$= 24.32(-1.9x_1-0.88x_2+3.02x_3) \qquad (9)$$

and

$$y = -1.9x_1-18.9x_2+5.1x_3 \qquad (10)$$

According to Eqs. (9) and (10), sweetness of the lemon jelly decreases with the increasing concentration of gellan gum. However, contributions of citric acid and Palsweet are different between Eqs. (9) and (10). It is not clear which model is better. The best model, Eq. (8) was used for drawing contours of the response surface (Fig.4).

Table 5 Correlations between explanatory variables
in the Scheffé third polynomial

1) x_1x_2	$= 0.150x_1$	$r =$	0.81
2) x_1x_3	$= 0.665x_1$	$r =$	0.99
3) $x_1x_2x_3$	$= 0.0959x_1$	$r =$	0.86
4) x_2x_3	$= 0.621x_2$	$r =$	0.96
5) x_1x_2	$= 0.138 - 0.168x_3$	$r = -$	0.90
6) $x_1x_2x_3$	$= 0.0809 - 0.0957x_3$	$r = -$	0.86
7) $x_1x_2x_3$	$= 0.00189 + 0.599x_1x_2$	$r =$	0.99

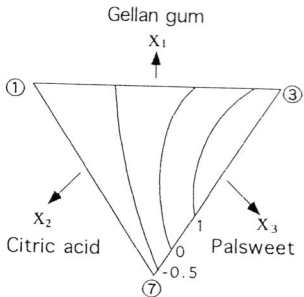

Fig.4 The best model
contours of sensory sweetness

Table 6 Regression models searched by the AIC statistic

Explanatory variable	AIC	R^2	
X_3 X_{12} X_{13} X_{23}	-26.15	0.674	Best model
X_2 X_3 X_{12} X_{13}	-24.21	0.652	
X_1 X_2 X_3 X_{12} X_{13} X_{23}	-22.18	0.675	2nd degree model
X_1 X_2 X_3	-17.20	0.531	1st degree model

5. CONCLUSION

Variable selection using the AIC statistic in regression analysis was effective for searching a response surface model of a constrained region inside of the Scheffé simplex lattice. The best model determined by the AIC statistic was used for drawing contours of the response surface. When a constrained region caused multicollinearity problem, contributions of each variable in the best model were evaluated by the approximate model derived from substituting correlations between explanatory variables into the best model.

REFERENCES

1. H. Scheffé, J.Roy. Stat. Soc., Ser. B, 20 (1958) 344.
2. J.A. Cornell, Experiments with Mixtures 2nd ed., John Wiley and Sons, New York, 1990.
3. H. Akaike, 2nd Inter. Symp. on Information Theory (B.N. Petrov and F. Csaki eds.), Akademiai Kiado, Budapest, (1973) 267.
4. Y. Sakamoto, M. Ishiguro and G. Kitagawa, Information Criterion Statistics, Kyoritsu Shuppan, Tokyo, 1983 (in Japanese).
5. H. Moritaka, S. Naito, K. Nishinari, M. Ishihara and H. Fukuba, J. Texture Studies, 29 (1998) 387.

HYDROCOLLOIDS – PART 2
Edited by K. Nishinari
2000 Elsevier Science B.V.

Effects of an oil phase on the salt taste of oil/water emulsions

Y. Yamamoto and M. Nakabayashi

Department of Food and Nutrition, Faculty of Human Life Science, Osaka City University, Sugimoto 3-3-138, Sumiyoshiku, Osaka 558-8585, Japan

The effect of an oil phase on the sensory intensity of the salt taste of NaCl in an oil/water emulsion was studied using the magnitude estimation technique.   The perceived intensity of the salt taste in emulsions was much higher than that of the solution without oil droplets at the same NaCl concentration.   The salt taste of low-fat emulsion (35 wt % oil) was less intense than that of the high-fat emulsion (70 wt % oil), suggesting that the sensory score depends on the oil content. The NaCl concentration measured in the water phase of the high-fat emulsion, containing 0.9 wt % NaCl, was 2.66 wt %.   The sensory score of the high-fat emulsion, containing 0.9 wt % NaCl, was much lower than that of a 2.66 wt % NaCl solution without oil droplets.   These results suggest that the oil phase enhancement of the perceived intensity of the salt taste in an oil/water emulsion might be due to the increased concentration of NaCl in the water phase and the suppressed contact of NaCl to gustatory cells.

1. INTRODUCTION

Understanding of the dependency of flavor (the combined perception of mouthfeel, taste, and aroma) on composition of food products has increased recently.   Most notably, the effect of polysaccharide thickeners on organoleptic attributes like thickness, sweetness and flavor intensity has been studied during the past few years [1-4].   In foods containing both water and lipids, the lipid phase may also take part in the organoleptic attribute by influencing the distribution of food components in oil, water, and the oil-water interface.   Although there are some reports concerning the dependency of aroma release and aroma perception on the food lipids [5-8], the effects of the lipid phase on the perceived taste intensity were not studied.

There have been many efforts to study the relation between stimulus and sensation, and to make a scale for the numerical estimation of   the subjective   sensory magnitude.   A reliable technique for magnitude estimation was devised by S. Stevens, and has been applied in many areas of sensory analysis, including taste intensity [9-11].   In this technique, panelists are asked to give appropriate scores to the samples by comparing their sensation with that for a standard sample which has been given a standard score.

In this experiment, we have applied the magnitude estimation technique to the effect of the oil phase in an oil/water emulsion on the sensory intensity of the salt taste of NaCl.

2. MATERIALS AND METHODS

2.1.   Emulsion preparation

To prepare an oil/water emulsion, commercial vegetable oil was slowly added to a solution of egg yolk in water while continuously mixed with a homogenizer.   An emulsion with 35 wt % oil was made with 10 g of egg yolk, 55 g of water, and 35 g of oil (low-fat emulsion).   An emulsion with 70 wt % oil was made with 10 g of egg yolk, 20 g of water, and 70 g of oil (high-fat emulsion).   Sodium chloride was dissolved in water before emulsification at the concentrations required to make 0.9 or 2.0 wt % in the emulsions.   Cornstarch (3% by weight) was added to the low-fat emulsion, because it was unstable without any additional thickeners and phase-separated during the experiment.   A solution of cornstarch in water was heated at 80 °C for 5 min, cooled to room temperature, mixed with NaCl and egg yolk, and homogenized with oil.   All emulsions used in this experiment were visually stable during the experiments.

2.2. Viscosity measurements

The steady flow behavior of emulsions was studied using a Haake Viscometer RV-20 (Germany).   As most fluid foods show non-Newtonian behavior,   the finding of the shear rate, to evaluate the objective viscosity of these foods, is a significant problem.   The apparent viscosity of non-Newtonian fluid foods, measured at a shear rate of 50 s^{-1}, was in good agreement with the viscosity derived when a Newtonian fluid was used [3].   So we used the apparent viscosity at shear rate 50 s^{-1} as the objective viscosity for non-Newtonian fluids.

2.3.   Subjective sensory estimation

The magnitude estimation technique [9-11] was used to measure the intensity of perceived salt taste.   About 20 trained panelists were engaged.   Samples were transferred to the mouth with plastic teaspoons.   Panelists were then asked to give appropriate scores to the intensity of salt taste of each sample.   The standard was a NaCl solution (0.9 or 2.0 wt %) in water, without oil droplets, and was scored at 100.   Then, for example, if one perceived twice the intensity of the standard, one should record a score of 200.   The experiment was conducted in a room maintained at 25 ± 1 °C with samples equilibrated to the room temperature.

3. RESULTS AND DISCUSSION

3.1. Objective viscosity and perceived salt intensity of cornstarch solutions

The effect of viscosity on the perceived intensity of salt taste was studied with solutions containing 0.9 wt % NaCl and various concentrations of cornstarch without oil (Figure 1). The apparent viscosity at shear rate 50 s^{-1} was used as the objective viscosity of the cornstarch solutions. The perceived taste intensity of NaCl was not affected by the presence of cornstarch within a limited concentration range. At higher concentrations of cornstarch, a concentration-dependent decrease in sensory score of salt taste was observed. A similar suppression of taste intensity in the thickened or structured products is well known in the food industry, and the mechanisms of taste intensity lowering by hydrocolloid chains have been discussed in relation to the rheological properties of polysaccharide thickeners [1-2].

3.2. Viscosity of the emulsions

The shear-thinning property of low- and high-fat emulsions is shown in Figure 2. The viscosity of these emulsions at 50 s^{-1} was about 400~500 mPa s. A comparison of the viscosity of these emulsions with cornstarch solutions (Figure 1) suggested that the viscosity of emulsions used in this experiment was not so high as to decrease the salt taste sensory score.

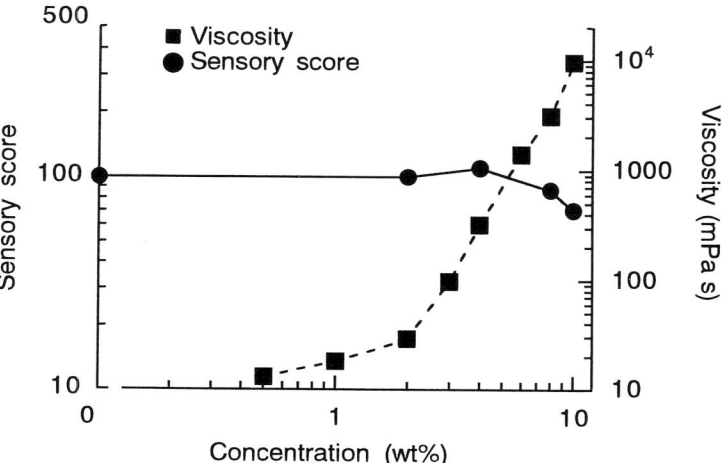

Figure 1. Correlation of objective viscosity and perceived salt taste intensity with cornstarch concentration

Viscosity at 50 s^{-1} and relative salt taste intensity of cornstarch solution containing 0.9 wt% NaCl was expressed as a function of cornstarch concentration.

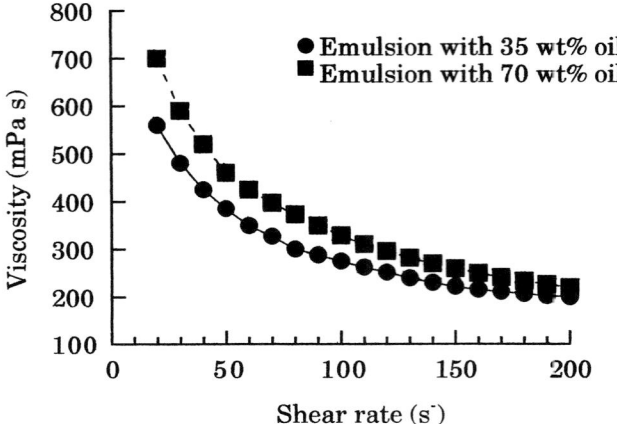

Figure 2. Shear rate dependency of viscosity in an emulsion made with 35 or 70 wt% oil

Emulsion was made with 10 wt% egg yolk and 35 or 70 wt% vegetable oil, and was added 0.9 wt% NaCl. Emulsion with 35 wt% oil was added 3 wt% corn starch.

Table 1   Effects of oil phase on the sensory intensity of salt taste

	Sensory score*
0.9 wt% NaCl Solution	100
Emulsion (35 wt% oil)	135
Emulsion (70 wt% oil)	158
2.0 wt% NaCl Solution	100
Emulsion (70 wt% oil)	150

*Sensory score was expressed as the relative sensory intensity of salt taste on an assumption that 0.9 or 2.0 wt % NaCl solution without oil droplets was scored 100.

3.3. Sensory intensity of emulsions

The results of sensory score estimations are shown in Table 1. The salt taste scores in high-fat emulsions containing 0.9 and 2.0 wt % NaCl were 158 and 150, respectively. These results clearly showed that the perceived intensity of the salt taste in emulsions was much higher than that of the solution without oil droplets at the same NaCl concentration. Furthermore, the salt taste in low-fat emulsions was less intense than that in the high-fat emulsion at the same NaCl concentration (0.9 wt %). These results suggest that the increase in sensory intensity is dependent on the oil content of the oil/water emulsion.

We have never seen any report suggesting the effects of lipids on taste intensity, except for one by Katsuragi and Kurihara, which showed a masking effect of lipoprotein against bitter tastes, with the taste nerve responses in frogs, and taste sensation in humans [12]. Our experiments suggest that an oil phase enhance the sensory intensity of NaCl.

3.4. Effects of NaCl concentration in water phase on taste intensity of emulsions

To study how the oil phase enhances the sensory intensity of NaCl in oil/water emulsions, the distribution of NaCl in water and oil phases of high-fat emulsion was measured after centrifugal separation of oil and water phases. The concentration of NaCl in the water phase was 2.66 wt %, which means that about 80% of the total NaCl in the emulsion is distributed in the water phase. So, the sensory intensity of salt taste of high-fat emulsion containing 0.9 wt % NaCl was compared with that of a 2.66 wt % NaCl solution without an oil phase. As shown in Figure 3, the sensory score of the solution containing 2.66 wt % NaCl was much higher than that of the emulsion containing 0.9 wt % NaCl.

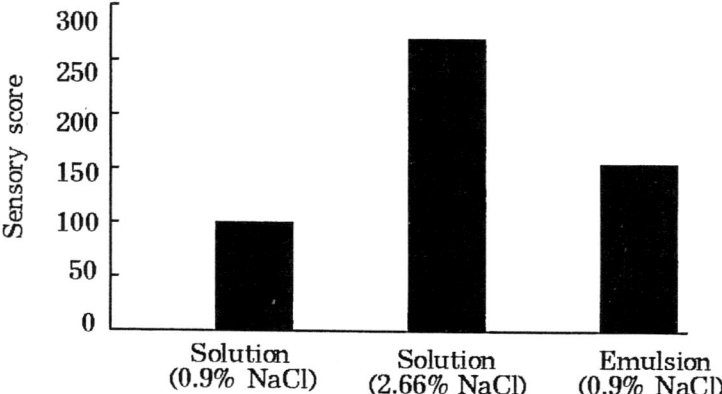

Figure 3. Perceived intensity of salt taste in a solution of NaCl (0.9 or 2.66 wt%) or in an emulsion containing 0.9 wt% of NaCl.

This suggests that an oil phase enhances the perceived intensity of the salt taste in oil/water emulsions, and that the effect of the oil phase is positively dependent on the oil content. To be perceptible, taste compounds must be released into the saliva so that they can contact gustatory cells of the buccal surface. So, lipids in food are expected to influence taste intensity by increasing the concentration of water soluble taste materials in the water phase and suppressing the contact of taste materials with gustatory cells.

ACKNOWLEDGMENTS

The authors are grateful to Prof. Katsuyoshi Nishinari of the Department of Food and Nutrition, Osaka City University, for his helpful suggestions.

REFERENCES

1. R.M. Pangborn, Z.M. Gibbs and C. Tassan, J. Texture Studies, 9 (1978) 415.
2. A.N.Cutler, E.R.Morris and J.J. Taylor, J. Texture Studies, 14 (1983) 377.
3. G.O. Phillips, D.J. Wedlock, and P.A. Williams (eds.), Gums and Stabilizers for the Food Industry, IRL Press, Oxford, 1988.
4. K. Nishinari and E. Doi (eds.), Food hydrocolloids: Structures, Properties, and Functions, Plenum Press, New York, 1994.
5. J.P. Schirle-Keller, G.A.Reineccius and L.C. Hatchwell, J. Food Sci., 59 (1994) 813.
6. P. Landy, J-L Courthaudon, C. Dubois and A. Voilley, J. Agric. Food Chem., 44 (1996) 526.
7. K. Wendin, K. Aaby, A. Edris, M.R. Ellekjaer, R. Albin, B. Bergenstahl, L. Johansson, E.P. Willers and R. Solheim, Food Hydrocolloids, 11 (1997) 87.
8. K.B. de Roos, Food Technology, 51 (1997) 60.
9. S.S. Stevens, Amer. J. Psychol., 69 (1956) 1.
10. S.S. Stevens, Psychol. Rev., 64 (1957) 153.
11. S.S. Stevens and J.R. Harris, J. Exp. Psycol., 64 (1962) 489.
12. Y. Katsuragi and K. Kurihara, Nature, 365 (1993) 213.

HYDROCOLLOIDS – PART 2
Edited by K. Nishinari
2000 Elsevier Science B.V.

Influence of vigorous mastication on the phylo- and ontogenic development of humans

K. Kubota *

Professor Emeritus, Tokyo Medical and Dental University
Guest Professor, Meikai University Faculty of Dentistry

The common ancester of *Pan* and *Homo*, which inhabited tropical forests and had a fruit-eating habit, was divided into western and eastern populations when the Rift-Valley was formed about eight million years ago. The eastern population evolved on the savanna to become humans by shifting from a fruit-eating to an omnivorous habit via a herbivorous and carnivorous habit [7]. However, little is known about the diets of hominids that predate the genus *Homo* [10]. Our data obtained from a PET mastication study shows that vigorous gum-chewing increases human cerebral blood flow in the primary sensorimotor areas by 25-28%, the supplementary motor area and insulae by 9-17%, and the cerebellum and striatum by 8-11%. This suggests that encephalization of the cerebral cortex may have been caused by cerebral blood flow activation.

1. INTRODUCTION

The background considerations of this study are: 1) Does vigorous chewing activate widespread regions of the brain? 2) What was responsible for the encephalization of the brain cortex during human phylogeny? Figure 1 illustrates my new concept of the masticatory system, which is composed of three units: the peripheral effector organs, sensory input and central nervous control [1-2]. Mastication is a rhythmic function involving the coordinated actions of these three units. The human masticatory apparatus, as shown in Figure 2 below, is involved in various body activities, such as chewing, swallowing, digestion, respiration, speech and nonverbal communication, and is probably interrelated with other systemic functions, including locomotion, blood circulation, excretion, endocrine function and reproductive function（Fig. 2）[1].

*To whom all correspondence should be addressed at 1-2-15 Sangenjaya, Setagaya-ku, Tokyo 154-0024, Japan (Fax:+81·3·3412·9487)
Abbreviations : CNS, central nervous control; F, facial nerve nucleus; Lab, labial; Ling,lingual; Mas, mastication; PE, peripheral effector organs; S, system; SI, sensory input; T, trigeminal nerve nucleus

458

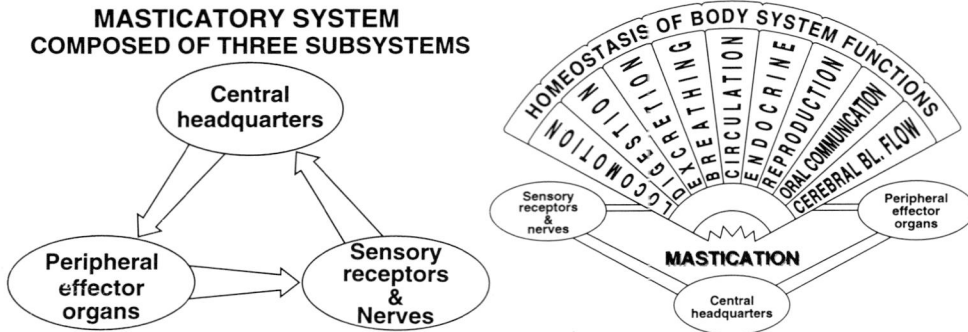

Figure 1. Diagram to show masticatory system comprising three units; peripheral effector organs, sensory input and central nervous control.

Figure 2. Diagram showing possible close interactions between masticatory function and other systemic functions

The masticatory system has evolved to adapt to new food environments, as examplified by the pinnipeds: their rudimentary deciduous teeth are shed into the amniotic fluid during pregnancy and the pups already possess erupted permanent teeth at birth, enabling the young seals to feed on fish in the sea virtually immediately after birth, thus helping to preserve the species. The teeth, one of the effector organs of their masticatory system, have evolved from a diphyodont to a monophyodont form, as shown in Figure 3 [3-6].

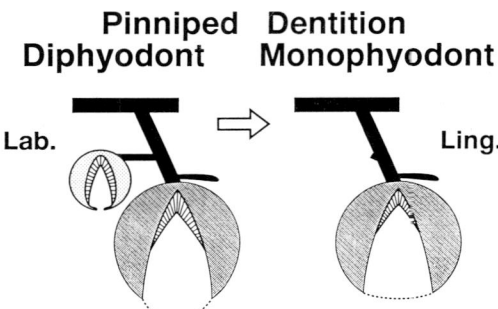

Figure 3. Diagram to explain histologically the adaptation of the masticatory system to aquatic life in the northern fur seal. Their rudimentary deciduous teeth are shed into the amniotic fluid during late pregnancy, and the pups already possess erupted permanent teeth at birth, enabling the young seals to feed on fish in the sea shortly after birth. The teeth have evolved from a diphyodont to a monophyodont form.

Yves Coppens (1994) [7] has described that about eight million years ago, the common ancestor of *Pan* and *Homo*, living in tropical forests, was divided into two groups, western and eastern, by the Rift Valley that crosses the equator from north to south in eastern Africa. The former group thrived and became chimpanzees, whereas the later evolved on the savanna to become humans.

In order to understand the genesis of the marked encephalization of the human cortex, it is necessary to clarify the influences of mastication on the phylogeny of humans.

2. METHOD OF PET MASTICATION STUDY

The interaction between the mastication of chewing-gum and cerebral blood flow was studied in 12 healthy volunteers aged 18 to 40 years. PET-autoradiography was carried out after bolus injection of 1.5 GBq ^{15}O-labelled water with a half life of 2 min [8]. As shown in Figure 4, the PET apparatus, a Headtome IV, was used to examine the regional cerebral blood flow (rCBF) after injection of water labelled with Oxygen-15 as a tracer into the median cubital vein.

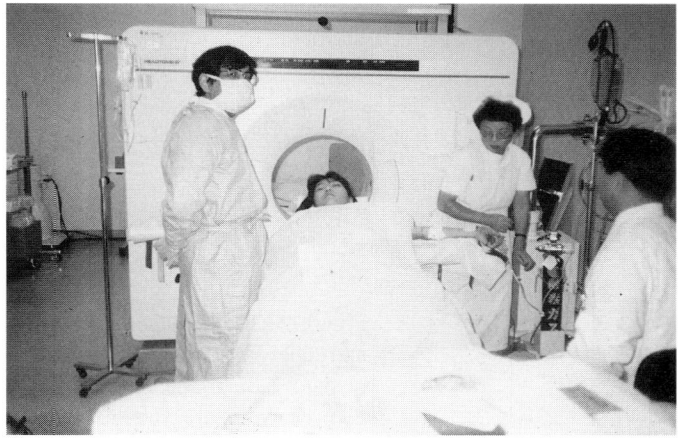

Figure 4. The PET (HEADTOME IV) apparatus used to examine the rCBF after water labelled with Oxygen-15 (H$_2$^{15}O) as a tracer was injected into the median cubital vein.

3. RESULTS AND DISCUSSION

Figure 5 shows PET images obtained during resting and chewing in a 20-year-old woman. It shows images taken at intervals of 13 mm from 20 mm above the orbitomeatal basic plane (OM20) to OM98. Explanatory diagrams for the PET images during gum-chewing are shown on the right.

Chewing-gum mastication significantly increased (P ≦ 0.01) the rCBF by 25-28% in primary sensorimotor areas (Rolandic areas), by 9-17% in the supplementary motor areas and operculum-insulae, and by 8-11% in the cerebellum and striatum. Statistically significant interactions between mastication and cerebral blood flow were demonstrated (P ≦ 0.01).

These increases demonstrated that chewing activates widespread regions of the brain (Fig. 5). The results obtained from this study confirm that mastication of chewing gum activates several different brain areas.

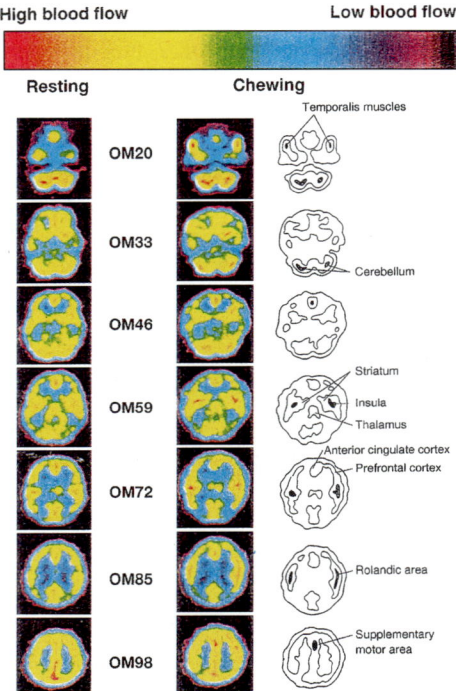

Figure 5. PET images obtained during resting and chewing in a 20-year-old woman. Explanatory diagrams for the PET images during gum-chewing are shown on the right to indicate the anatomical sites investigated.

Figure 6 is a diagrammatic illustration of how chewing activates widespread regions of the brain. The left panel insert is a lateral view of the brain cortex showing activated cerebral blood flow in the Rolandic area during gum-chewing. The right panel shows the activated Rolandic areas and operculum-insulae in a horizontal cross-section through the dotted horizontal line.

A PET mastication study is impossible in children, whose brains are undergoing development. However, vigorous mastication in infancy will activate widespread regions of the developing brain. On the basis of the above and below findings, I considered that the influence of mastication is very important for understanding the marked encephalization of the human brain cortex, since vigorous mastication stimulates the cortex strongly in related areas to accelerate cerebral blood flow. Thus, vigorous mastication may provide the means of developing robust brain and body growth [9].

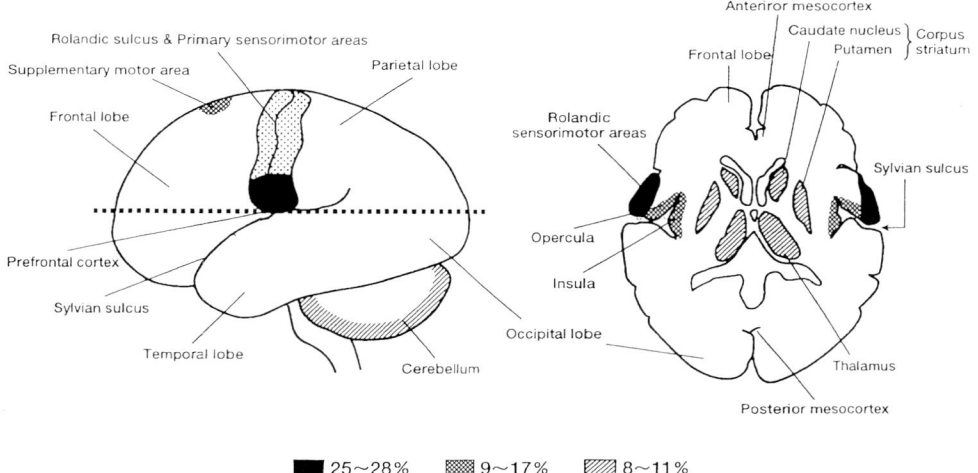

Figure 6. Diagrammatic illustration of how chewing activates widespread regions of the brain.

The western population thrived in forests and became our closest cousins, the chimpanzees, by adopting a vigorous fruit-eating habit

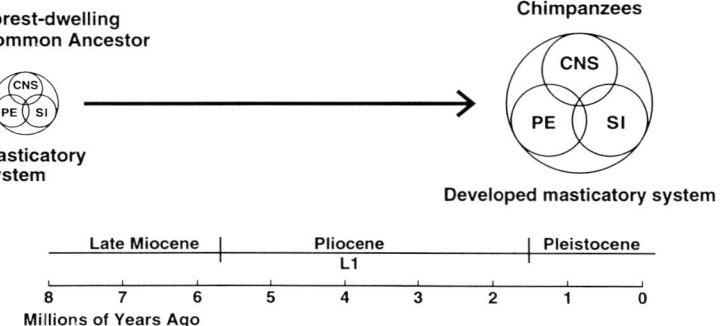

Figure 7. Diagram to explain the forecasted chronological development of the masticatory system during chimpanzee evolution under a vigorous fruit-eating habit, assuming that the western common ancester of *Pan* and *Homo* thrived in forests to become our closest cousins, the chimpanzees, after formation of the Rift-Valley in eastern Africa.

Figure 7 is a diagram to explain the forecasted chronological development of the masticatory system during chimpanzee evolution under their vigorous fruit-eating habit, assuming that the western common ancestor of *Pan* and *Homo* thrived in forests to become our closest cousins, the chimpanzees, after formation of the Rift Valley in eastern Africa.

Figure 8 shows a diagram to explain the chronological development of the masticatory system during human evolution under the influence of vigorous omnivorous mastication of the common ancestor that evolved on the savanna and to become humans after the formation of the Rift Valley. In other words, *Homo sapiens* evolved on the savanna by shifting from a herbivorous to an omnivorous diet via the carnivorous habit [9].

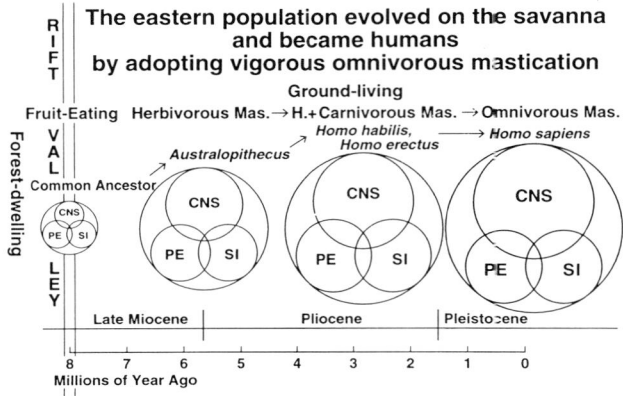

Figure 8. Diagram to explain the forecasted chronological development of the masticatory system during human evolution under the influence of vigorous omnivorous mastication of the eastern common ancestor evolved on the savanna to become humans after the formation of the Rift Valley.

Most recently, Sponheimer and Lee-Thorp [10] have reported the data from carbon isotope analysis of the molar tooth enamel of a 3-million-year-old *Australopithrecus africanus* specimen unearthed in South Africa. They stressed that this early hominid ate C4- and C3-enriched plants and animals, and speculated that the encephalization of early *Homo* was made possible by the consumption of energy- and nutrient-rich animal food material.

The phylogenetic process of human evolution will be reflected in the ontogenetic process of the masticatory system. If this is the case, then from an ontogenetic viewpoint vigorous mastication may also provide considerable generative power for robust brain and body growth, and increase the human lifespan [9].

In contrast, Figure 10 is a diagram to explain the chronological change in the masticatory system under the influence of a modern soft diet. The present diet consumed by humans may result in atrophy of the masticatory system and gradual tooth loss in elderly adulthood. Masticatory dysfunction will then develop, and finally very aged persons who are bedridden will become unable to take food orally.

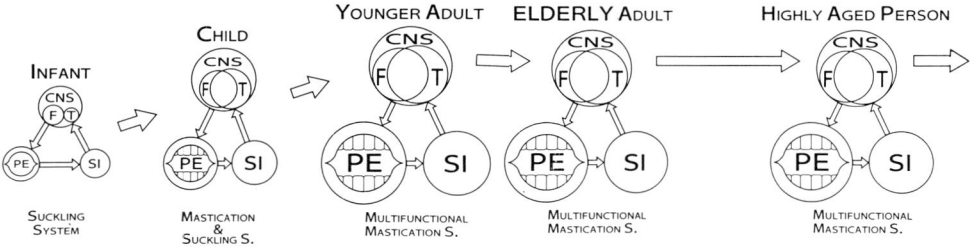

Figure 9. Diagram to explain the forecasted chronological development of the masticatory system in modern humans under the influence of vigorous omnivorous mastication starting from infancy.

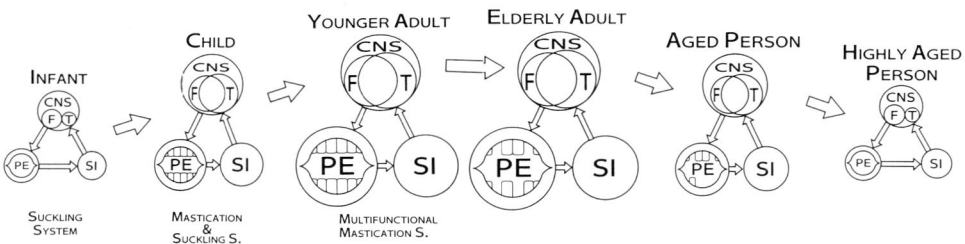

Figure 10. Diagram to explain the chronological change in the masticatory system under the influence of a modern soft diet.

CONCLUSIONS

This phylogenetic concept forms the basis of my argument that if modern humans masticate vigorously from infancy onwards, then they should be able to enjoy a long lifespan, including a healthy, robust old age.

Acknowledgements

A part of this study was supported by a Grant-in-Aid from Lotte Co., Ltd., Tokyo.

REFERENCES

1. K. Kubota, M.-S. Lee, C.-M. Chang, Y. Sonoda, S. Shibanai, and K. Nagae, In: Mechanobiological Research on the Masticatory System (ed. K. Kubota) pp. 69-74, VEB Verlag für Medizin und Biologie, Berlin (1989) GDR.

2. K. Kubota, Mastication and systemic functions. J. Oromaxillofacial Biomechanics, 1 (1995) 5. (In Japanese)

3. K. Kubota, Bull. Tkyo Med. Dent. Univ. 10 (1963) 61.

4. K. Kubota, and K. Matsumoto, Bull. Tokyo Med. Dent. Univ. 10 (1) (1963) 89-93.

5. K. Kubota, Nova Acta Leopoldina NF 58, Nr. 262 (1986) 223.

6. K. Kubota, S. Shibanai and J. Kubota, Tendency to monophyodonty as an adaptation to marine life in the northern fur seal. J. Historical Biology, Special issue (1999). (In press)

7. Y. Coppens, East Side Story: The origin of humankind. The rift valley in Africa holds the secret to the divergence of hominids from the great apes and to the emergence of human beings. Scientific American May (1994) 88-96.

8. T. Momose, J. Nishikawa, T. Watanabe, Y. Sasaki, M. Senda, K. Kubota, Y. Sato, M. Funakoshi, and S. Minakuchi, Archives Oral Biology, 42 (1) (1997): 57-61.

9. K. Kubota, Role of vigorous mastication in the phylo- and ontogenetic development of men. J. Masticat. & Health Soc., 8 (1) (1998): 53-59. (In Japanese with English abstract)

10. M. Sponheimer and J.A. Lee-Thorp, Isotopic evidence for the diet of an early Hominid, *Australopithecus africanus*. SCIENCE, 283 January 15 (1999) 368-370

HYDROCOLLOIDS – PART 2
Edited by K. Nishinari
2000 Elsevier Science B.V.

Sensory control of masticatory jaw movements according to food consistencies in the rabbit

T. Morimoto, T. Inoue, O. Hidaka, Y. Masuda and A. Komuro

Department of Oral Physiology, Osaka University Faculty of Dentistry,
1-8 Yamadaoka, Suita, Osaka 565-0871, Japan

Roles of periodontal and muscle spindle afferents on regulating mastication were examined by using the awake and anesthetized rabbits. After bilateral combined section of the maxillary and inferior alveolar nerves in the awake rabbits, the jaw movements became smaller both vertically and horizontally, their trajectories became irregular, and electromyograms (EMGs) activities of jaw-closing muscle were greatly reduced during mastication. During rhythmic jaw movement induced by repetitive electrical stimulation of cerebral cortex in the anesthetized rabbits, one of polyurethane foam strips of different hardness were inserted between the upper molar and the force transducer. The peak and buildup speed of masticatory force and EMG of masseteric muscle increased in proportion to the hardness of the test strip by strip application. After sectioning of the maxillary and inferior alveolar nerves, the buildup speed of masticatory force was significantly slowed down and the peak force tended to decrease during strip application. Lesioning trigeminal mesencephalic nucleus (MesV) where the cell bodies of the muscle spindle afferents are located, also reduced enhancement of masseteric EMG during strip application. When the MesV was lesioned in combination with sectioning the maxillary and inferior alveolar nerves, the facilitatory responses of masseter nucleus was almost completely abolished. We conclude that both periodontal and muscle spindle afferents are mainly responsible for the regulation of masticatory muscle activities and jaw movements.

1. INTRODUCTION

Chewing movements of the jaw and tongue are influenced by orofacial sensations arising during the chewing of foods of various properties [1, 2]. It has been reported that in various animals including humans, EMGs of the jaw-closing muscles are greater when hard food is chewed than when soft food is chewed [3-6]. These findings suggest that the masticatory force is regulated automatically according to the hardness of food. Sensation arising from sensory receptors in the orofacial area like periodontal membrane and muscle spindle must underlie the food-dependent regulation of masticatory muscle activities and jaw movements [3, 5, 7]. The present study aims to identify what classes of sensory receptors in the orofacial area contribute to this regulation. For this purpose, we performed the following three experiments by using awake and anesthetized rabbits, and demonstrated that both periodontal and muscle spindle afferents are mainly responsible for the regulation of masticatory muscle activities and jaw movements.

2. METHODS

2.1. Awake animal studies

Male rabbits (2.3–3.0 kg) were used. The animals were anesthetized with α-chloralose (60 mg/kg, i.v.) and urethane (0.5 g/kg, i.v.) and a phototransister array which functioned as a laser-sensors of a jaw tracking system during experiment, was attached to the mentum of the mandible. Pairs of Teflon coated stainless steel wires (200 μm in diameter) were inserted bilaterally in the masseter and digastric muscles which are jaw-closing and opening muscles, respectively, for recording EMG activity of those muscles. Two metallic tubes (3 cm long, inner hole diameter 6 mm) were fixed horizontally to the skull with dental cements and three skull screws. At least 2 days after the operation, the head of the animal was painlessly held in a stereotaxic apparatus by means of the two metallic tubes. Jaw movements were monitored by a He-Ne laser jaw tracking system which we have developed. By use of this apparatus, two-dimensional jaw movements on the frontal plane were shown on the oscilloscope, and were further dissolved into the vertical and horizontal movements which were recorded as the separate output currents. EMGs of masticatory muscles were simultaneously recorded with jaw movements using a data recorder and later analyzed using a computer system.

Two kinds of foodstuffs with different physical consistencies were prepared; chow pellets in a cylindrical form (RC-4, Oriental Co., diameter 3.5 mm, length 12 mm) and carrots prepared in quadrangular prism form (side length 3.5 mm, length 12 mm). Four pieces of each kind of food were inserted at one time into the animal's mouth by an experimenter. Before starting the experiments, we prepared pellets containing sucrose or NH_4Cl in order to test whether the difference in food taste affected the jaw-movements and EMG activities. No difference in jaw movement pattern and EMG activities were observed between chewing of these two kinds of food.

2.1. Anesthetized animal studies

The surgical procedure were mostly identical to the awake animal studies. The animals were initially anesthetized by ketamine (16 mg/kg, i.v.) and thiamylal sodium (20 mg/kg, i.v.) and then anesthesia was maintained by a mixture of halothane and oxygen, or sodium thiamylal. A train of square pulses (30 Hz, duration 0.2 ms, < 80 μA) were applied through a glass-coated metal electrode (1–2 M at 1 kHz) inserted into the masticatory area of the cerebral cortex (CMA), and rhythmic jaw movements (CRJMs) were induced. The axially directed masticatory force was measured with a small S-shaped transducer (length 8.5 mm, width 4 mm, height 4.1 mm) which was fixed on the ground surface of the lower posterior teeth. Jaw movements were monitored by optoelectric apparatus (C2399, Hamamatsu Photonics, Hamamatsu, Japan) instead of the He-Ne laser jaw tracking system. Figure 1 shows schematic illustration of the experimental setup used in the anesthetized rabbit.

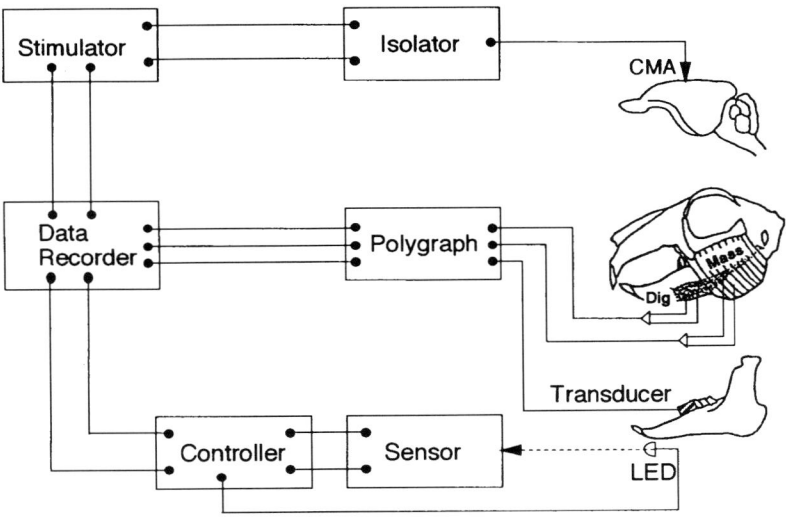

Fig. 1. Schematic illustration of the experimental setup used in the anesthetized rabbit.

3. RESULTS AND DISCUSSION

3.1. Effects of trigeminal deafferentation on mastication in awake rabbits

In the first experiment, we examined the effects of bilateral trigeminal deafferentation on jaw movements and masticatory muscle activities during mastication in awake rabbits.

When pellets or carrots were inserted into the animal's mouth, they were first manipulated at the anterior part of the mouth. Patterns of the jaw movements of this mastication stage were either simple open-close movements or small circular movements. When food was transferred into the posterior part of the mouth, animal began to crush and grind it between the molars on either the right or left side. The basic pattern of the jaw movements at this stage was crescent-shaped loops. They are composed of three phases of the jaw-opening, jaw-closing and power phases. The food was crushed and ground at the power phase. The jaw movements were more rounded and regular for carrot-chewing than those for pellet-chewing.

In order to examine the role of intraoral sensation in regulation of jaw movements and masticatory muscle activities, we examined the effects of sectioning both maxillary and inferior alveolar nerves which innervate periodontal sensations of the upper and lower teeth. After the denervation, the jaw movements became smaller both vertically and horizontally, and trajectories of jaw movements became irregular since the timing of the transition from the jaw-closing phase to the power phase fluctuated. In addition to those changes, the lower jaw did not attain the tooth-contact position or the tooth occluded position shown by dotted lines,

468

although the jaw was raised to the tooth-contact position before nerve section. Those findings strongly suggest that the animal could not chew food strongly after loss of intraoral sensations and that the masticatory efficiency decreased. The number of chewing cycles needed for swallowing increased after the nerve section. Such increase in chewing cycles may compensate for insufficient chewing ability after nerve section. The changes after sectioning both maxillary and inferior alveolar nerves were most serious within one week after nerve section but they gradually recovered thereafter, and the vertical and horizontal excursions returned to about 75% of their initial values within two postoperative weeks. The effects persisted longer for pellet-chewing than for carrot-chewing.

However, sectioning both maxillary and inferior alveolar nerves causes loss of sensation from the lower-half of the face as well as loss of intraoral sensations. Thus, the serious change of jaw movements might be due to loss of facial sensation. Then, we sectioned only the peripheral parts of the both nerves which innervate the face, and left the intraoral sensation intact. In this case, no significant change was observed in jaw movement during mastication. Those observations suggest that loss of intraoral sensations, most likely periodontal sensations, causes severe the change of jaw movements.

Changes in EMG activities of masticatory muscles could also indicate the insufficient chewing ability after nerve section. Before nerve section EMG activities of masseteric muscle were relatively large, however, those were greatly reduced after nerve section (Fig. 2). This reduction of the EMG activities was severe 1 week after nerve section and gradually recovered. However, these effects were not completely recovered even 1 month after the nerve section. In contrast with jaw-closing muscle, jaw-opening muscle was not seriously affected by nerve section.

The effects of trigeminal deafferentation on the masticatory are summarized as 1) decrease in both the vertical and horizontal jaw movements, 2) insufficient occlusion at the power phase, 3) irregular pattern of jaw movements, 4) increase in the number of the chewing cycles before swallowing, and 5) decrease in EMG activities cf jaw-closing muscles. Weijs and Dantuma [6] found that the peak of the masseter muscle activities was attained during the power stroke of a masticatory cycle in which the maximum force was executed and also that enlargement of the medial excursion at the power stroke was associated with increase in the masseteric activities. Insufficient occlusion at the power phase and the decrease in the horizontal excursions observed in deafferent rabbits are thus accounted for by the decrease in jaw-closing muscle activities.

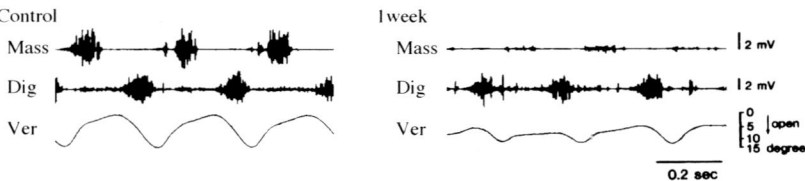

Fig. 2. EMG activities of masticatory muscles and vertical components of associated jaw movements during chewing pellets.

3.2. Effects of intraoral simulation on masticatory force in the anesthetized rabbit

The above study suggests that the periodontal sensation enhances the masticatory force by increasing activities of the jaw-closing muscles and facilitates masticatory efficiency. However, because the EMG activity is affected by various factors such as recording sites, degree of jaw opening [8, 9] and fatigue [10], it is not known whether the EMG activity is always a reliable indicator of the masticatory force. Therefore, we recorded the chewing force directly in the mouth and examined whether stimulation of periodontal sensation enhances the masticatory force. Furthermore, we also evaluated relationships between masticatory force and hardness of the chewing substances. For this purpose, we used anesthetized animals and stimulated the CMA electrically to induce rhythmic jaw movements, instead of natural chewing. This is because it is easier in the anesthetized animal than in the awake animal to induce the same type of rhythmic jaw movements repeatedly and to apply test materials of various hardness between upper and lower molars.

When the cerebral cortex is electrically stimulated, various patterns of rhythmic jaw movements were induced depending on the site of stimulation in the CMA. Among those patterns, only crescent-shaped movements were employed here because they are composed of three phases of the jaw-opening, jaw-closing and occluded phases and resemble most to the normal molar chewing movements [1, 11]. An example of the effects of strip application shown in Fig. 3. One of five test strip of different hardness (Hs: 27, 47, 67, 84 and 91) was inserted between upper 1st molar and force transducer in the period indicated by the thick bar. During strip application the masticatory force increased simultaneously with an increase in the masseteric EMG activity and its duration during strip application compared with those values before strip application. In contrast, the effects on the digastric EMG bursts were relatively minor. The peak and buildup speed of masticatory force, EMG of masseteric muscle and the pattern of jaw movements were affected by hardness of test strips. The peak and buildup speed of masticatory force increased in proportion to the hardness, reaching > 80 N when the hardest strip. The masseteric EMGs also increased in a hardness-dependent manner, whereas the digastric EMGs did not change appreciably. A characteristic change in

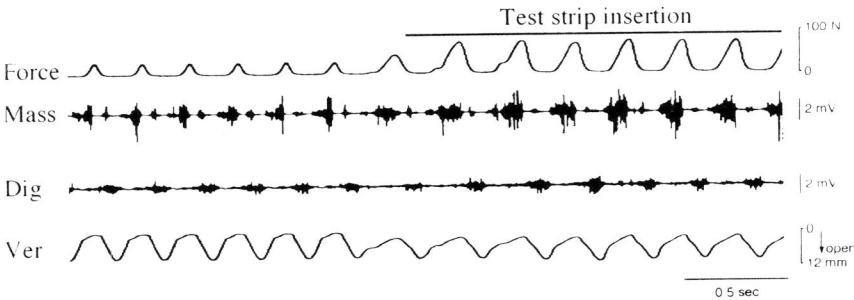

Fig. 3. Effects of strip application on masticatory force, EMGs and vertical components of cortically induced rhythmic jaw movements.

the jaw movement traces was an increase in the minimum gape with the increase in strip hardness.

To examine whether periodontal sensations are involved in those hardness-dependent increases in peak and buildup speed of masticatory force, and masseteric EMGs, the effects of sectioning the maxillary and inferior alveolar nerves were investigated. After denervation, peak masticatory force and masseteric EMGs still significantly increased in a hardness-dependent manner. In contrast, the buildup speed of masticatory force was significantly slowed down and marginal significance was detected in the decrease in peak masticatory force.

Such slowing down of build-up speed of masticatory force after denervation may well be accounted for by the loss of periodontal sensation. The threshold force required to elicit a response from periodontal mechanoreceptors to stimulation of tooth crown is as low as 0.01–0.02 N (~1–2g) in various animals [12, 13]. The low-threshold periodontal receptors are excited immediately when the teeth touch a chewing substance, and most of the molar periodontal receptors in the rabbit are fast adaptive [12]. These properties of periodontal afferents could contribute to a quick buildup of the masticatory force in the beginning of tooth loading, probably via the periodontal-masseteric reflex path [14-16]. On the other hand, masticatory force could be regulated in a hardness-dependent manner even after deprivation of periodontal sensation. Therefore, there must be sensation other than the intraoral sensation which functions to regulate the muscle activities according to food consistencies.

3.3. Blocking of muscle sensation on mastication in the anesthetized rabbit

Muscle spindles in jaw-closing muscle are one of the possible receptors responsible for the hardness-dependent regulation of masticatory force. Muscle spindles are located within the muscles, in parallel with the skeletal muscle fibers and they are sensitive to muscle length and velocity of changes in length. It is known that the sensory nerve innervating muscle spindles in jaw-closing muscles make monosynaptic excitatory connections to motoneurons innervates jaw-closing muscle. When a jaw-closing muscle is stretched, the spindle afferents increase their firing rate, leading to contraction of the same muscle and synergists. This is called the stretch reflex.

The possibility of involvement of the muscle spindles in the hardness-dependent enhancement of jaw-closing activities during strip application, was examined by making a electric lesion in the mesencephalic trigeminal nucleus (MesV) where the cell bodies of the muscle spindle afferents are located. Before making the lesion of MesV, both the stretch reflex response which was elicited by depressing the mandible by 1 mm and also the facilitative response of the masseter muscle to application of a test strip were recorded (Fig. 4A). After lesioning the left MesV, the ipsilateral stretch reflex decreased greatly and the facilitative responses of the ipsilateral masseter muscle to application of the strip were significantly reduced to 80% of the control value after the lesion (Fig. 4B). When the MesV was lesioned in combination with sectioning the maxillary and inferior alveolar nerves, the facilitatory responses of masseter muscles was almost completely abolished. These observations suggest that muscle spindle afferents from jaw-closing muscles are likely to

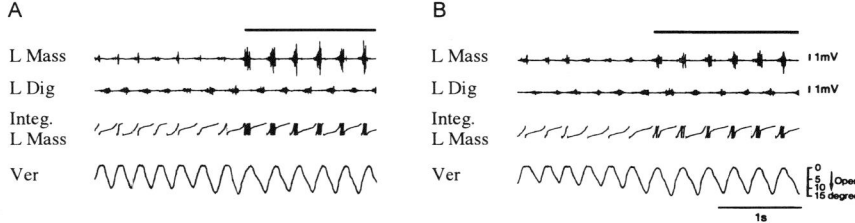

Fig. 4. Facilitative response of masseteric EMGs to strip application before (*A*) and after (*B*) MesV lesion. Thick bars indicate the period of strip application.

contribute to the facilitation of jaw-closing muscle activities during application of test strip. On the other hand, it has been reported that the lesion of the trigeminal mesencephalic nucleus tract in monkeys does not change the chewing pattern of the normal response to variations in food hardness [17]. Taking these findings into consideration, it could be that both periodontal afferents and muscle spindles are responsible for this hardness-dependent change, and that removing only one of them does not have much effect alone.

ACKNOWLEDGEMENTS

This study was supported by Grants-in-Aid for Scientific Research (Nos. 10307045, 09877350, 09832008 and 10557164) from the Japanese Ministry of Education, Science and Culture.

REFERENCES

1. Z.J.Liu, Y.Masuda, T.Inoue, H.Fuchihata, A.Sumida, K.Takada, T.Morimoto, J.Neurophysiol., 69(1993)569.
2. A.Thexton, K.Hiiemae, A.Crompton, J.Neurophysiol., 44(1980)456.
3. O.Hidaka, T.Morimoto, Y.Masuda, T.Kato, R.Matsuo, T.Inoue, M.Kobayashi, K.Takada, J.Neurophysiol., 77(1997)3168.
4. T.Horio, Y.Kawamura, J.Oral Rehabil., 16(1989)177.
5. T.Inoue, T.Kato, Y.Masuda, T.Nakamura, Y.Kawamura, T.Morimoto, Exp.Brain Res., 74(1989)579.
6. W.Weijs, R.Dantuma, Neth.J.Zool., 31(1981)99.
7. T.Morimoto, T.Inoue, Y.Masuda, T.Nagashima, Exp.Brain Res., 76(1989)424.
8. S.J.Lindauer, T.Gay, J.Rendell, J.Dent.Res., 72(1993)51.
9. A.Manns, R.Miralles, C.Palazzi, J.Prosthet.Dent., 42(1979)674.
10. Y.Kawazoe, H.Kotani, T.Maetani, H.Yatani, T.Hamada, Archs Oral Biol., 26(1981)795.
11. T.Morimoto, T.Inoue, T.Nakamura, Y.Kawamura, Archs Oral Biol., 30(1985)673.

12. K.Appenteng, J.P.Lund, J.J.Seguin, J.Neurophysiol., 48(1982)27.
13. A.Taylor, (ed.) Neurophysiology of the Jaws and Teeth, Macmillan, London, 1990.
14. M.Funakoshi, N.Amano, J.Dent.Res., 53(1974)598.
15. L.J.Goldberg, Brain Res., 32(1971)369.
16. J.P.Lund, R.S.McLachlan, P.G.Dellow, Exp.Neurol., 31(1971)189.
17. G.Goodwin, E.Luschei, J.Neurophysiol., 37(1974)967.

HYDROCOLLOIDS – PART 2
Edited by K. Nishinari
2000 Elsevier Science B.V.

473

Evaluation of food texture by mastication and palatal pressure, jaw movement and electromyography

F. Nakazawa and M. Togashi

Physics Lab., Department of Home Economics, Kyoritsu Women's University
1-710 Motohachioji, Hachioji, Tokyo 193-8501, Japan

Food texture was studied by the analyses of human responses to the mastication of food during eating: 1) the three-dimensional movement of a first molar tooth, 2) mastication pressure on the first molar, 3) pressure on three parts of the palate, and 4) electromyography (EMG) of the left and right temporal and masseter muscles.

The mastication course was divided into the forward rhythmic chewing duration and the later irregular movement duration preparatory to swallowing. The rhythmic pulses of mastication pressure were recorded synchronously in each related phase with the tooth movement and with the large EMG signals during the rhythmic chewing duration. In the later irregular movement duration irregular pulses of the tooth movement were observed with a small EMG without mastication pressure pulses. Both durations depended on food texture, the mass of food eaten, and also on the participant.

The maximum velocity of the up and down movement of the molar during eating was obtained as fast as 30 - 100 mm/s depending on the food eaten and on the participant. The maximum force on the first molar was at most around 30 kg .

Simultaneous measurements of palatal pressure and the EMG of mastication muscles made clear whether the food was being eaten by pressing it by the tongue, or biting it with the teeth, depending upon the texture.

1. INTRODUCTION

Food texture was studied objectively by mechanical force-displacement methods and/or subjectively by sensory tests. Although results of the mechanical methods are highly precise, reproducible and quantitative, there are several demerits; namely, (1) there is no saliva, (2) deformation velocity of a food by a machine is slower than chewing velocity in the mouth, (3)

474

the limit of the plunger displacement is arbitrarily fixed by the worker in contrast to chewing the food depending on its texture in the mouth. Therefore, the results of the mechanical methods are sometimes inconsistent with sensory evaluation. The sensory test is useful because food texture is evaluated in the mouth. Efforts to select suitable subjects, to exclude food ordering effects, and to prepare for better physical and mental test conditions were done, however, sensory evaluation often depends on the subjects, and the results are not quantitative but qualitative.

Recently food texture has been studied by various analyses of human responses to chewing foods [1 – 7]. Of these methods, the study on food texture by simultaneous measurements such as pressure on a molar or on palatal, three-dimensional movement of the jaw, and electromyography of mastication muscles during chewing is reported.

2. EXPERIMENTAL METHOD

Because the experimental technique was based on the participants' response to chewing food, the apparatus was planned so that the participants could eat foods naturally. They were not bound to a special machine, but sat down on comfortable chairs keeping their faces in a constant position during eating. Though this plan sometimes lowered experimental accuracy, natural eating by the participants was maintained.

The system of simultaneous measurements of mastication pressure on a first molar or on three points of the palate, three-dimensional movement of the jaw, and electromyography of the left and right masseter and temporal muscles is shown in Fig. 1. The master waveform memory connected to the pressure amplifier and to the Gaussmeter triggered off the slave waveform memory connected to the EMG measurement in less than 1 μs. A delay time of 1 μs is short enough to be considered as zero on the time scale of this measurement.

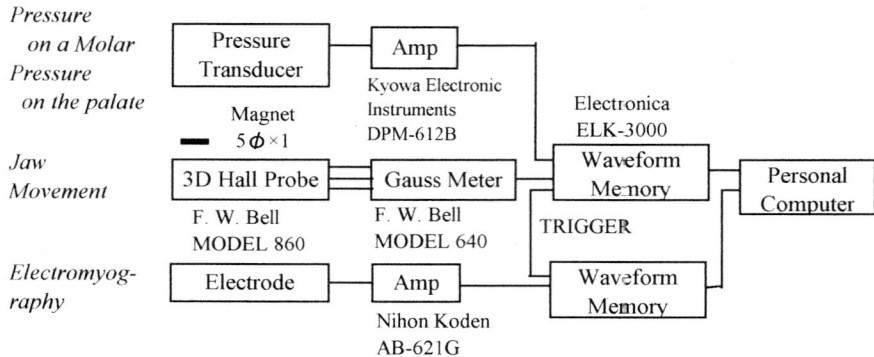

Fig. 1. System for simultaneous measurement of jaw movement, presssure and electromyography.

2.1. Mastication pressure on a molar and on three points of the hard palate

Measurement of pressure on a molar tooth needs a special participant who has lost a lower molar [7]. Someone who had lost two lower molar teeth and usually uses the right side for chewing was located. The first molar tooth plays a most important role during chewing. A pressure transducer of 50 kg/cm^2, 6 $\phi \times 0.5$mm in size, was embedded in an artificial tooth (Fig. 2).

Figure 3 shows three pressure transducers embedded in an artificial resin palate [5]. The maximum pressure on them is 2 kg/cm^2. One transducer is located in the center between the upper incisor teeth, the second in the center between the bicuspid teeth, and the third on the side near the molar usually used. The three transducers were calibrated with the other transducer after being embedded in the resin plate.

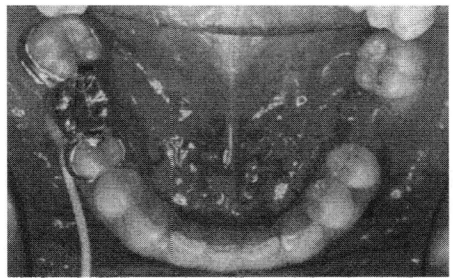

Fig. 2.   Artificial molar for use with a pressure sensor.

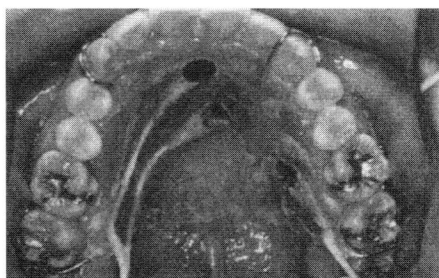

Fig. 3.   Three pressure transducers embedded in a resin plate.

2.2.   Jaw movement [8]

A small strong magnet, a 5 $\phi \times 1$ mm disk type, was pasted to the gum of one of the lower teeth. A participant wasn't bound to a special machine for the measurement but rather was asked to sit keeping his face in a constant position while eating a piece of food. Thus he could chew foods naturally. A Hall probe, which includes three Hall elements perpendicular to each other, was set in front of his mouth separated by about 3 cm from the magnet. Since magnetic permeability of materials such as components of the human body, air, or metals (except ferromagnetic) is almost equal to that of a vacuum, the magnetic field is not disturbed by body elements or air.

Right-handed rectangular coordinates were adopted as shown in Fig. 4. Axes x and y are perpendicular and paralell, respectively, to the row of lower teeth. X is directed to the center of the mouth, coinciding with the axis of the disk magnet. The z axis is perpendicular to the row of the teeth and directed upward. Magnitude of magnetic components was indicated on a 3-channel Gaussmeter and led to a waveform memory. Since the magnetic field of any magnet is unique, measurement of a magnetic field vector gives a unique position vector. Reduction of the rest position vector of the tooth from the change-of-position vector during eating gives the three-dimensional movement of the tooth (gum).

Movement of any tooth can be measured during eating in the same way. Data of 8 ms/point were adopted in the multi-channel waveform memory.

2.3. Electromyography (EMG)

EMG of the masseter and temporal muscles was done using surface electrodes in a shielded room without a band path filter. The other multi-channel waveform memory described above was used for A/D reduction and the record

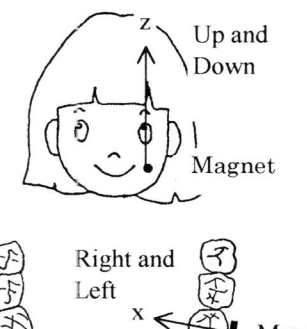
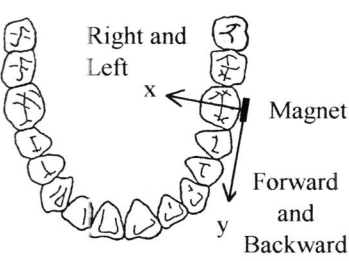

Fig. 4. Right-handed rectangular coordinate for position of a magnet.

of digital data. Data points of 0.5 ms were adopted, since the signal form produced little distortion to affect the results when compared with that of 10μs data points. A recording duration of 20 s (from start of chewing to swallowing) gives 20 s/0.5 ms = 4 × 10^4 points.

3. RESULTS AND DISCUSSION

Since jaw movement is calibrated from the change of magnetic field due to movement of a magnet, an experiment to check whether correct movement of the magnet was obtained was carried out. A disk type magnet was manually traced in a 23 mm diameter circle twice at each position, as shown in Fig. 5. Changes of magnetic field vector components are shown in Fig. 6. From these magnetic vector components, the

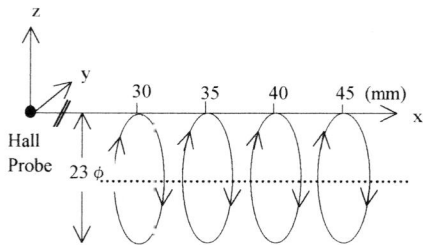

Fig. 5. Movement of a magnet for test experiment. A small magnet turns twice at each position.

position of the magnet could be calibrated, and movement of the magnet on the (y, z) plane at each x position is shown in Fig. 7. Two overlapped circles with a 23 mm diameter are shown, though the SN ratio of the traces is lower at the bigger x.

Approximation that the axis of the disk magnet does not change during chewing gave an

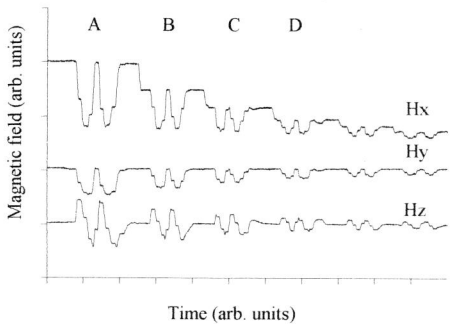

Fig. 6. Change of x, y, z components of magnetic field vectors of the test experiment in Fig. 5. Position of the magnet x is A 30mm, B 35mm, C 40mm, D 45mm from the Hall probe.

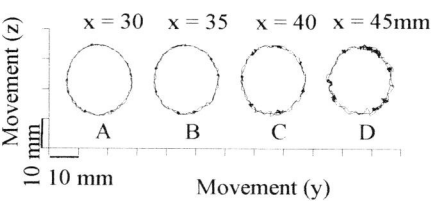

Fig. 7. Calculated trace of the magnet on the (y, z) plane from the magnetic field components shown in Fig. 6. A, B, C and D correspond to the same letters in Fig. 6.

absolute error in the results. Now improvement in the apparatus has been carried out using two Hall probes.

The result of simultaneous measurement of mastication force, three-dimensional movement of the jaw, and electromyogrphy (EMG) of mastication muscles during eating of a dry prune is shown in Fig. 8. The lines are numbered from 1 - 8. Line 1 shows mastication force on the first molar tooth. The tooth worked most during eating. Lines 2, 3, and 4 show three-dimensional movement of the first molar (jaw movement), right and left, forward and backward, and up and down. The participant was asked to chew a prune at the beginning on the molar that is artificial and embedded with a pressure sensor. Therefore, the first pulse of the molar in line 4 precedes the first pressure pulse in line 1; then the rhythmical movement pulses of molar synchronize with the pressure pulses. The second pulse of molar movement corresponds to the movement of the first bite; i.e., to break up the prune. The same measurements for

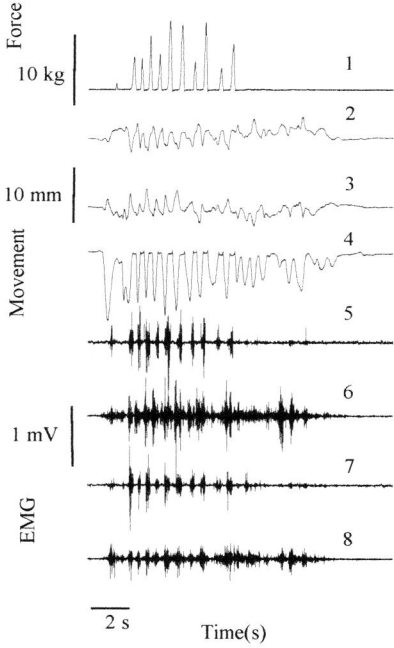

Fig. 8. Mastication force, jaw movement and EMG patterns simultaneously measured during eating a dry prune 1, mastication force; 2, right and left jaw movement(JM); 3, forward and backward JM; 4, up and down JM; 5, left temporal (LT) EMG; 6, left masseter (LM) EMG; 7, RT EMG; 8, RM EMG.

478

various foods showed that texture of the food eaten is most reflected in the first bite, as expected.

After the series of rhythmical force pulses in line 1, irregular pulses of molar movement are observed in lines 2, 3, and 4. This series of irregular pulses likely results from movement of the jaw to make the pieces of dry prune into a bolus for swallowing.

Bigger EMG pulses of the temporal and masseter muscles are rhythmic and synchronized to the force pulses. Smaller EMG pulses are also observed corresponding to irregular molar movement pulses in latter series. Bigger EMG pulses of the masseter muscles observed in lines 6 and 8 around the last series are not always but sometimes generated by the same participant even during chewing of the same food. These are synchronized to swallowing action. It is considered that people sometimes press the teeth tightly upon swallowing, independent of food texture.

Mastication force patterns for processed cheese, boiled potato, raw carrot, mochi (patties of pounded glutinous rice), and a rice cracker are shown in Fig. 9. The force pulse pattern for the first bite shown in Fig. 9 are more different from each other than pulse patterns after the second bite. Since breaking up of the food happens during the first mastication pulse, response to the texture of the food is reflected in the first bite.

According to the pattern of the first bite, namely to food texture, foods were grouped

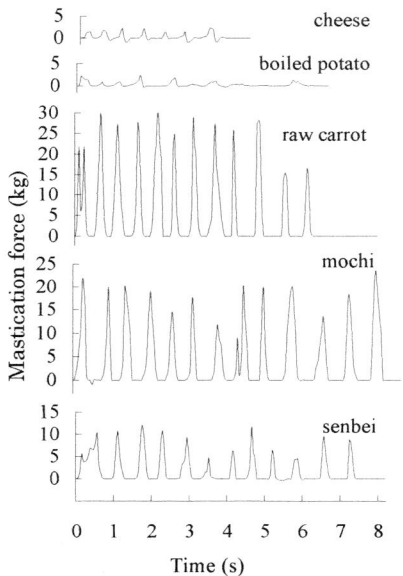

Fig. 9. Mastication force pattern during eating.

Pattern of the first bite		Food
A		bread pancake mochi
B		boiled potato boiled carrot boiled radish cheese , hanpen
C		raw radish, raw carrot cucumber, apple takuan, kamaboko pork sausage
D		senbei almond peanut

Fig. 10. Group of foods according to the mastication force pattern of the first bite.

into four types A, B, C, and D, as shown in Fig. 10. Group A includes foods with small Young's modulus, no breaking point, and high maximum mastication force that are chewy hard to bite off. Spongy breads and cakes and mochi belong to group A. Foods that are soft and easy to eat and to bite off with a small Young's modulus and with a breaking point are included in group B, such as boiled tubers and processed cheese. Raw tubers, fruits, and sausages with relatively high moisture content are grouped in C. They are hard but not too tough to bite off, having a big Young's modulus and clear breaking point. Foods of group D are the hardest of all the groups, having the biggest Young's modulus and a sharp breaking point. The water content of the foods in D is low, and they are pulverized during chewing and scatter in the mouth. The order of the groups according to easiness of eating is B, C, D, A or B, C, A, D depending on the participant.

Fig. 11 shows mastication force patterns for carrot pieces of the same size, raw, boiled 10 min, and boiled 20 min. The first bite pattern, number of pulses, and height of pulses are entirely different. Fig. 12 shows mastication force patterns for apple pieces of different thickness(H), 5, 10 and 15 mm. Break strengths of the first bite of the three samples are almost the same and heights of

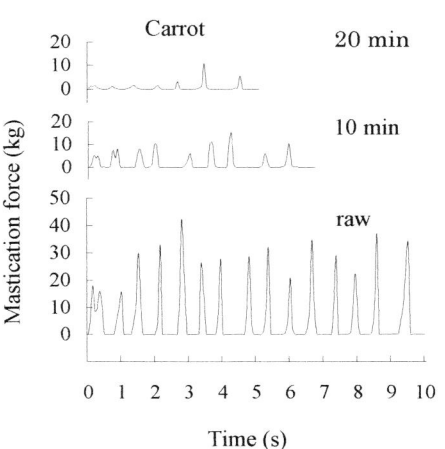

Fig. 11. Mastication force pattern for raw and boiled carrot.

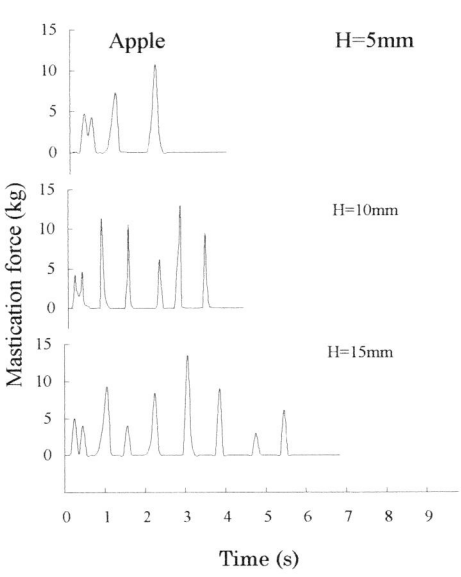

Fig. 12. Mastication force pattern for apple piece of different size.

force pulses are also nearly the same, but the number of pulses in the series is greater for the thick sample than for the thinner one.

480

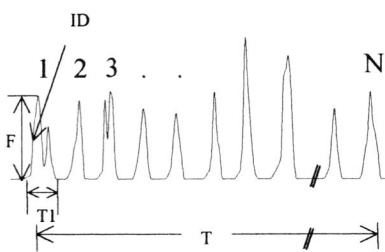

Fig. 13. Definition of quantities
characterizing a mastication force pattern.

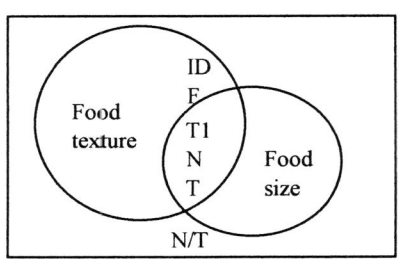

Fig. 14. Grouping of pressure pattern
characterization quantities according to food
texture and food size.

Several quantities are defined for a mastication force pattern in Fig. 13.   ID is the maximum gradient of the first mastication pulse, F is the maximum force of the first pulse, T1 is the time duration of the first pulse, N is the number of mastication pulses, and T is the time duration from the peak of the first pulse to that of the last pulse.   N/T gives the chewing cycle of a food.   These quantities were tested for many foods [9] and the results are in Fig. 14.   ID, F, T1, N, and T depend on food texture and T1, N and T depend on both food texture and food size.   Chewing cycle N/T depends on neither food texture nor size of food, but can be considered the chewing rhythm of human beings. In conclusion, ID and F depend only on food texture.   ID is a quantity related to Young's modulus.   A participant perceives foods with bigger ID as resistant to mastication pressing at the first bite (sensory test).   F is exclusively obtained by human response to food. A participant perceives high-F food as tough and hard to bite off at the first bite (sensory test).   The maximum force at the first bite for various foods is shown in Fig. 15.

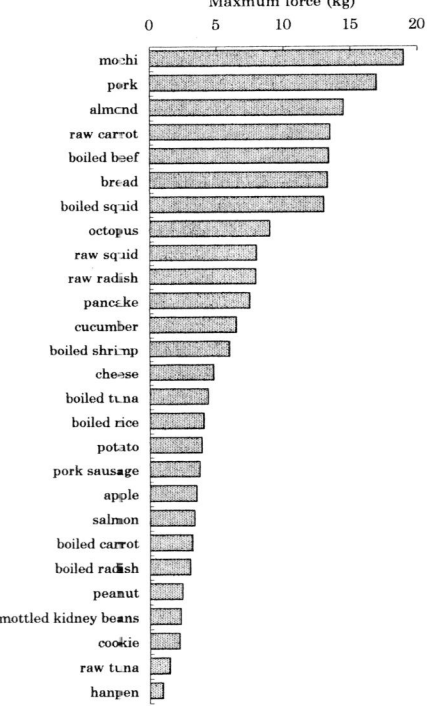

Fig. 15. Maximum mastication force at the first bite.

Three-dimensional movement of the jaw, in this report represented by movement of the first molar on the usually used side in the lower row of teeth, during eating of a piece of

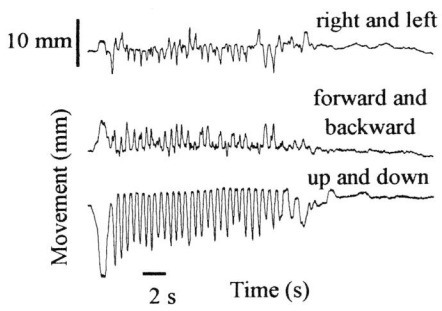

Fig. 16. Three-dimensional movement of a
first molar while eating cheese.

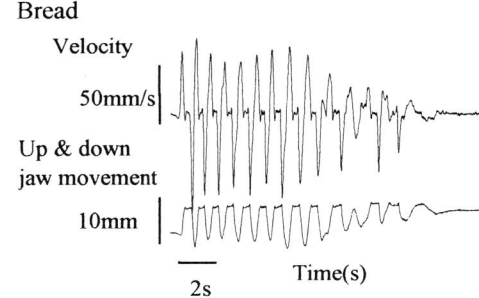

Fig. 17. Up and down jaw(first molar)
movement and velocity.

cheese $1.5 \times 1.5 \times 1.5$ cm is shown in Fig. 16. The molar movement depends not only on food texture and size of the food eaten but also strongly on the participant. One participant moves her molar on the(x, y) plane almost in one direction, another moves it at random, and others change the direction of movement rhythmically from one direction to the reverse.

Differential of movement in z direction gives the velocity of up and down movement of the molar during eating. Velocity and movement of the first molar during eating a piece of bread are shown in Fig. 17. In this case the participant was asked to place a piece of bread on the artificial first molar at eating. Therefore the first upward pulse of velocity gives the pressing velocity of the molar on the bread. The next zero velocity duration corresponds to biting off the bread, and the first downward pulse for opening the mouth follows. Upward pulses are a series showing of the pressing velocity of the molar, and downward pulses are for opening. The maximum pressing velocity of

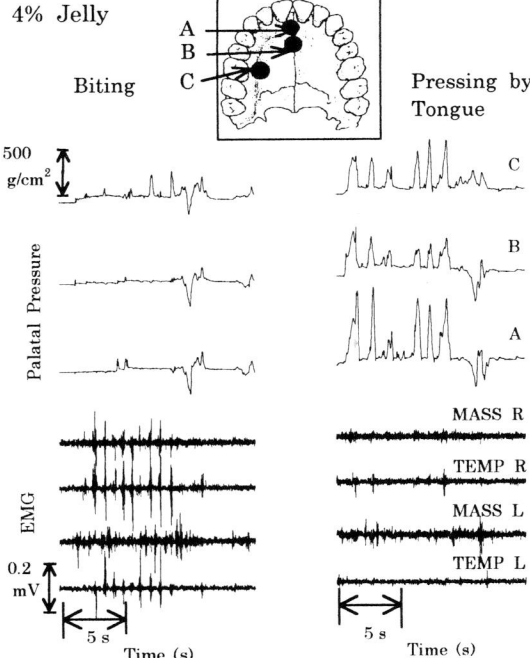

Fig. 18. Palatal pressure pattern and EMG
during biting or pressing by tongue.

the molar at the first pulse is distributed from 30 to 100 mm/s depending on the food and strongly on the participants.   Foods belonging to groups B and C in Fig. 11, namely those that are easy to eat, tend to be eaten with high velocity in contrast to the slow velocity foods such as almonds in group C.

Pressing velocity of a molar tends to slow down in the course of chewing and becomes fairly slow during the preparation period for swallowing.   Discrimination between the chewing process and the preparation period for swallowing can be done by simultaneous measurements of jaw movement and EMG (and mastication force).

Palatal pressures and EMG during the eating of milk gelatin jelly made with 1, 2, 3, and 4% concentrations of gelatin were simultaneously measured.   The results shown in Fig. 18 are for a participant who had to eat a 4% jelly either by pressing with the tongue or biting by the teeth.   Palatal pressure pulses were big, especially on (A); however, EMG pulses of the masseter and temporal muscles were small in the case of pressing by the tongue.   In the case of biting, a few small palatal pressure pulses appeared on (C) pressure sensor near the first molar usually used, and bigger EMG pulses compared with those by the tongue appeared.

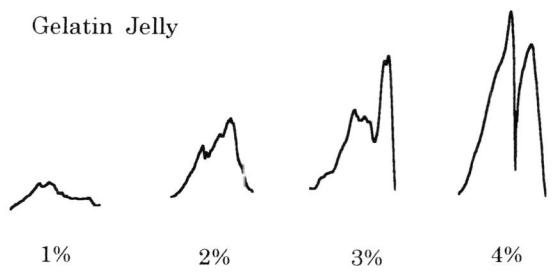

Gelatin Jelly

1%   2%   3%   4%

Fig. 19. Palatal pressure pattern of the first pressing by tongue.

Simultaneous measurements of palatal pressure and EMG of mastication muscles made clear whether the food was being eaten by pressing it by the tongue or biting it with teeth.

When participants were asked to eat the jelly naturally, they ate 1 and 2% jelly by pressing with the tongue and 4% jelly by biting with the teeth.   3% jelly was eaten by either pressing or biting depending on the participant.   Palatal pressure patterns of the first pressing on (A) for 1 to 4% jellies are shown in Fig. 19.   These patterns correspond to pressing on a 20 (L) ×20 (D)×15 (H) mm jelly with the tongue.   4% jelly showing a sharp break seems to be rather more like a solid food than a gel.   Eating of 4% jelly by biting is consistent with the pattern shown in Fig. 19.

Acknowledgements

This work was partially supported by the Special Coordination Fund for Promoting Science and Technology Agency of the Japanese Government.   The author wishes to thank Ms. A. Morita for their experimental work and preparation of figures in this report.   The author gratefully acknowledges Dr. J. Takahashi's permission to cite a part of her doctoral thesis (1993) in this report.

REFERENCES

1. J. B. Palmer, K. M. Hiiemae and J. Liu, Archs. oral Biol. 42 (1997) 429.
2. K. R. Agrawal, P. W. Lucas, J. F. Prinz and I. C. Bruce, Archs. oral Biol. 42 (1997) 1.
3. K. Kohyama and M. Nishi, J. Texture Studies 28 (1997) 605.
4. J. Takahashi and F. Nakazawa, J. Texture Studies 23 (1992) 139.
5. J. Takahashi and F. Nakazawa, J. Texture Studies 22 (1991) 1.
6. J. Takahashi and F. Nakazawa, J. Texture Studies 22 (1991) 13.
7. J. Takahashi and F. Nakazawa, J. Home Econ. 2 (1987) 107.
8. F. Nakazawa, M. Iashito, Y. Iimura, J. Takahashi and M. Masako, J. Home Econ. 48 (1997) 323.
9. J. Takahashi and F. Nakazawa, J. Home Econ. 40 (1989) 489.

INDEX